Lothagam

View of Lothagam from the west.

Lothagam:
The Dawn of Humanity in Eastern Africa

Edited by

Meave G. Leakey and John M. Harris

Columbia University Press
New York

Columbia University Press
Publishers Since 1893
New York Chichester, West Sussex

Library of Congress Cataloging-in-Publication Data

Lothagam: the dawn of humanity in eastern Africa /
[edited by] Meave G. Leakey and John M. Harris
p. cm.
Includes bibliographical references and index.
ISBN 0-231-11870-8 (cloth : acid-free paper)
ISBN 0-231-11871-6 (pbk. : acid-free paper)
1. Vertebrates, Fossil—Kenya—Lothagam Site
2. Paleontology—Miocene.
3. Animals, Fossil—Kenya—Lothagam Site
I. Leakey, Meave G. II. Harris, John Michael.
QE841.L68 2001
566′.096762′7—dc21 2001042433
∞

Columbia University Press books are printed on permanent and durable acid-free paper.

Printed in the United States of America
c 10 9 8 7 6 5 4 3 2 1
p 10 9 8 7 6 5 4 3 2 1

Contents

Lothagam

1

INTRODUCTION

Meave G. Leakey

An island of sediments surrounded by the sandy, windswept plains of the Turkana desert, Lothagam in northern Kenya is one of Africa's most important Late Miocene sites. Its rich red sedimentary rocks, which range in age from 8 to a little less than 4 Ma, preserve an exceptional record of events at a time of dramatic change in the African biota. Expansion of the modern C_4 savanna grassland flora in the Late Miocene coincided with the emergence of faunal elements that would dominate the later Cenozoic—elephants, hippos, giant pigs, grazing antelopes, true giraffes, and humans. Synchronous shrinkage of the equatorial forests led to the loss of many taxa characteristic of the earlier Miocene faunas—including hyrax species and primitive rhinos, giraffids, tragulids, and apes. Regrettably, only a few sites in Africa are representative of the time interval in which this ecological transition took place. Only Lothagam combines a lengthy stratigraphic sequence with diverse and evolving vertebrate assemblages and the presence of early human ancestors. Indeed, the importance of Lothagam lies in its age—a span of prehistory that chronicles a major turnover in the East African biota and documents the emergence of its modern ecosystems.

Lothagam is an uplifted fault block, about 10 km long and 6 km wide, located to the west of Lake Turkana (2° 54′N 36° 03′E) (figure 1.1). Here, two roughly parallel north–south oriented hills are separated by low areas of exposures, with further exposures to the west. The larger eastern hill is a horst that rises more than 200 m above the surrounding plains. The parallel hills protect the exposures from the tons of desert sand that are continually blown across the landscape by the strong easterly Turkana winds. The climate is semiarid. Temperatures at the nearby town of Lodwar, 60 km northwest of Lothagam, range between 23° and 37°C, with a mean temperature over an eight-year period of 35.1°C. The mean annual rainfall, measured at Lodwar between 1947 and 1954, is 150.6 mm (Hopson 1982). With the exception of the Grant's gazelle (*Gazella granti*), the golden jackal (*Canis aureus*), and the Cape hare (*Lepus capensis*), wild mammals are rarely encountered, although an extensive cave system running through the Lothagam deposits provides shelter for the striped hyena (*Hyaena hyaena*) and two species of bats—the tomb bat (*Taphozous mauritianus*) and a pipistrelle (*Pipistrellus* sp.) (L. Leakey et al. 1999).

Figure 1.1 Composite view of the Lothagam sediments taken from the horst.

The area is inhabited by the nomadic Turkana people whose flocks of sheep and herds of goats graze the sparse vegetation. Our fieldwork at Lothagam was enriched by our daily encounters with these tough, resilient people whose beautiful smiles, evocative singing, lively dancing, and friendly outlook belie the hardships of their daily lives.

Lothagam is a unique site preserved by a unique set of circumstances. The initial accumulation of sediments from a large, meandering river was ideal for the preservation of fossils. But had it not been for subsequent massive faulting, which led to the emergence of the horst, the sediments would be buried, like many others, under kilometers of overburden and hidden by an impenetrable carpet of sand. The resistant, fine-grained matrix in which most of the fossils are embedded has contributed to an extraordinary detail of preservation.

Lothagam, with its immense scenic beauty, is perhaps one of the most spectacular sites in the African Rift. Its rich red rocks—carved into dramatic jagged ridges, deep gorges, and winding gullies by thousands of years of weathering and erosion—are a constant source of wonder. The five years that I had the privilege to work at Lothagam were undoubtedly some of the most rewarding of my career.

The Name

"Logatham" is the local Turkana name for the horst that forms the eastern boundary of the site. It is pronounced "Lothsegam." In the Turkana language, Lothagam describes something that is rough, varied, and heterogeneous; it is a reference to the many different rocks that make up the horst—extensive and varied conglomerates, some with enormous boulders, and the several basalt horizons and outcrops of columnar basalt. Early reports named the site Lothagam Hill (Robbins 1967, 1972; Patterson et al. 1970; Smart 1976) but, because Lothagam Hill is the name of the unfossiliferous horst that forms the eastern boundary, the site is now referred to simply as Lothagam.

History

The earliest reports of sediments at Lothagam are those of Champion (1937) and Fuchs (1939), both of whom described exposures consisting of tilted volcanics that were structurally related to the Lothidok range to the north and the Kamutilia Hills to the southwest. Robbins (1967) was the first to note that Lothagam might be an important fossil locality, and it was his reports, resulting from his studies of the Holocene archaeology, that led to the first paleontological expedition in 1967 undertaken by Professor Bryan Patterson of Harvard University. This initial expedition encountered a rich vertebrate fauna, including a mandibular fragment of an early human ancestor. On biostratigraphic evidence the site was estimated to be 6 Ma (Patterson et al. 1970). Patterson led a second expedition to Lothagam in 1968, and the site was worked again several years later, in 1972 and 1973, by Princeton University personnel including Vince Maglio, Dennis Powers, and Charles Smart.

Scientists from the Kenya National Museum's Turkana Basin Palaeontology Project visited the site briefly in 1980 when the project first moved its activities from the eastern to the western shores of the lake. On August 4 of that year, my husband, Richard, who was then director of the National Museums of Kenya and coordinator of the Turkana Basin field expeditions, visited Lothagam briefly with me and a team from the BBC who were filming for the series "Making of Mankind."

Several of the field crew, including Kamoya Kimeu and Peter Nzube, had spent the preceding days at Lothagam in an attempt to locate fossil primates, and they had reported the discovery of three specimens of fossil cercopithecids as well as several other vertebrates. The three monkey specimens were collected along with the partial mandible and skeleton of a squirrel that became the type specimen of *Kubwaxerus pattersoni* (Cifelli et al. 1986).

It was not until ten years later, however, that the expedition was in a position to return to Lothagam to resurvey the area in detail. Early in 1989, Richard was given the responsibility of running Kenya's national parks, which at the time were in serious trouble due to rampant poaching and lack of financial resources. I thus took over from him the coordination of the paleontological field expeditions in the Turkana Basin. During the previous 20 years, these expeditions had concentrated on the Late Pliocene to Early Pleistocene time interval represented by the Omo Group deposits. Those strata had proved to be a uniquely rich source of vertebrate remains, and detailed studies have led to an unusually fine resolution of evolutionary events during this time (Harris 1983, 1991; Harris et al. 1988a, 1988b; Coppens and Howell 1985, 1987a, 1987b). Rich assemblages had also been recovered from the smaller, more tightly time constrained Oligocene and Miocene sites at Losidok (Madden 1972), Buluk (Leakey and Walker 1985), Kalodirr (Leakey and Leakey 1986a, 1986b, 1987), Muororot (Boschetto et al. 1992), and Locherangan (Anyonge 1991).

In my new role as coordinator of the field expeditions, it seemed appropriate to reformulate the expedition's activities and to focus on specific problems and time intervals. The field research over the preceding twenty years had given us a good understanding of the basinal geology and the evolution of the faunal assemblages through the interval from 4 Ma to 1.3 Ma (Brown 1995; Brown and Feibel 1986; Brown et al. 1995; Feibel 1988; Feibel et al. 1989; McDougall 1985; Harris 1983, 1991; Harris et al. 1988a, 1988b).

With the exploration of the northeastern and southwestern shores completed, and with this sound foundation for future studies, I decided that a survey of fossiliferous localities to the south of the Turkwel River was necessary to assess the potential for future fieldwork. The localities included the Miocene sites at Loperot, Aweriweri, and North Napudet; the Pliocene sites at South Turkwel, Longarakak, and Eshoa Kakuongori; and the Late Miocene–Early Pliocene site at Lothagam. At the time there was no good aerial photographic coverage, so that the majority of the fossils we found were left in the field for subsequent retrieval.

The 1989 surveys showed that there was indeed a wealth of fossils remaining to be collected from many of the sites visited and that a considerable amount of work remained to be done. At Lothagam, I was fortunate to discover the skeleton of a large carnivore eroding from the bank of the River Nawata. We left it and a number of cercopithecids, suids, and other vertebrates in the field to collect the following year. In December, we arranged for the Kenya Rangeland Ecological Monitoring Unit (KREMU) to provide aerial photographic coverage of the extensive area between the Kerio River to the east, the Kamutilia Hills to the west, the Turkwel River to the north, and the Kakurio River to the south. This included Lothagam.

The following year, 1990, we began detailed work at South Turkwel (Ward et al. 1999), North Napudet, and Lothagam, spending a little over a month at Lothagam. Subsequent expeditions to Lothagam followed in 1991, 1992, and 1993 and resulted in the recovery of over 1,700 new tetrapod fossils, a good understanding of the geology, and a secure sequence of dates. Preliminary analyses of the geological and faunal studies are summarized by Leakey et al. (1996). The subsequent, more detailed studies provide the substance of this volume.

The Geology and Dating

The first detailed geological survey was undertaken by Bryan Patterson and Bill Sill in 1967. Further detailed investigation was undertaken by Kay Behrensmeyer in 1968. A preliminary report (Patterson et al. 1970) was followed by Behrensmeyer's (1976) summary of the geology, fauna, and dating. Behrensmeyer divided the succession into six major lithostratigraphic units, four of which had previously been designated the Lothagam Group (Patterson et al. 1970) and are of Late Miocene to Early Pliocene age. The Lothagam Group was divided into three members in ascending stratigraphic order; Lothagam 1, Lothagam 2, and Lothagam 3. Lothagam 1 was further subdivided into Lothagam 1A, 1B, and 1C. An olivine basalt, which was interpreted as the Lothagam sill, capped Lothagam 1C, separating Lothagam 1 and Lothagam 2. Dennis Powers completed his Ph.D. dissertation on the geology and magnetostratigraphy of Lothagam and neighboring deposits in 1980. Craig Feibel participated in the Lothagam field expeditions of 1991, 1992, and 1993. Based on their studies, the original designations were replaced with an informal lithostratigraphic framework (Leakey et al. 1996; Feibel this volume: section 2.1, figure 2.5). Wherever possible, local Turkana names have been used for geological units. Thus the lowest portion of the exposed sequence, which is restricted to the horst and consists of interbedded proximal volcaniclastic sediments and lavas (formerly Lothagam 1A), is termed the Nabwal Arangan beds. "Nabwal Arangan" is the water hole in a gorge that bisects the horst, and it means the red water hole, the red color being given by the deep red clays that are washed down the gorge. Stratigraphically above the Nabwal Arangan beds lies the Nawata Formation (previously Lothagam 1B and lower 1C), which includes the earliest fossiliferous strata. "Nawata" is the Turkana name for the long grass that grows in the river draining the northern exposures, and the Turkana use this name for this river. The Nawata Formation is subdivided into lower and upper members (previously lower 1B and upper 1B plus lower 1C, respectively), that are informally referred to as the Lower Nawata and Upper Nawata. The Marker Tuff marks the lower boundary of the Upper Nawata. The superjacent strata (previously upper 1C), are designated the Apak Member of the Nachukui Formation. "Apak" is Turkana for a pass, and at Lothagam it refers to the sandy depression that bisects the western basalt hill, providing people and vehicles with access to the exposures. The Nabwal Arangan beds, the Nawata Formation, and the Apak Member thus replace the earlier Lothagam 1.

Stratigraphically above the Lothagam basalt (the former Lothagam sill) is the Muruongori Member of the Nachukui Formation (replacing Lothagam 2), which is almost certainly a lateral equivalent of the Lonyumun lake sediments exposed to the north and east (Feibel 1988). "Muruongori" is the local name for the western lava ridge, "moru" meaning large hill and "oungori" meaning dark gray. Replacing Lothagam 3 and overlying the Muruongori Member is the Kaiyumung Member of the Nachukui Formation. "Kaiyumung" is a small stream to the west of the site; it is named after a historically significant bull that died there. The uppermost strata of the Kaiyumung Member are truncated by the present-day erosion surface. Small exposures of younger portions of the Nachukui Formation are represented (Feibel this volume: section 2.1) but have yielded few if any fossils. The youngest strata cropping out at Lothagam form a discontinuous veneer over the older units and are attributed to the Holocene Galana Boi Formation, which is geographically widespread over much of the lake basin. Lothagam's vertebrate fossils largely derive from the lower and upper members of the Nawata Formation and from the Apak and Kaiyumung Members of the Nachukui Formation.

Until recently, the age of Lothagam was poorly constrained by questionable radiometric dates (Patterson et al. 1970), paleomagnetic stratigraphy (Powers 1980; Leakey et al. 1996), and biostratigraphic correlations. An estimated age of 5 to 5.5 Ma for Lothagam 1 (Patterson et al. 1970) was based largely on the evolutionary stages of the Proboscidea, with a minimum age of 6 Ma (Hooijer and Maglio 1974). It was noted, however, that the Lothagam 1C fauna was likely to be younger than this (Smart 1976). The Lothagam 3 fauna was recognized as correlative with the Mursi Formation and the lower Shungura Formation of the Omo Group to the north (Maglio 1973). Radiometric dates have recently been reported from the Nabwal Arangan beds, the lower member of the Nawata Formation, the upper Apak Member of the Nachukui Formation, and the Lothagam basalt (McDougall and Feibel 1999). Unfortunately, the Upper Nawata, and the lower Apak and the Kaiyumung Members—the time intervals from which the hominid specimens were recovered—remain poorly constrained.

The Fauna

Over 500 specimens were collected during the course of the earlier expeditions in the late 1960s and early 1970s, including a remarkable diversity of vertebrate fossils. These collections were shipped to international experts in a number of different countries for study, and many of the resultant publications on proboscideans (Maglio 1970, 1973), equids (Hooijer and Maglio 1973, 1974), rhinos (Hooijer and Patterson 1972), hippos (Coryndon 1977), suids (Cooke and Ewer 1972), giraffids (Churcher 1979), crocodiles (Tchernov 1986), a giant squirrel (Cifelli et al. 1986), and an aardvark (Patterson 1975) have proved pivotal to our understanding of the Late Miocene evolution of these lineages. But, because these publications were widely scattered among different international journals and appeared over a protracted period of time, the significance of the Lothagam fauna as a biota has gone largely unappreciated. This volume, with the inclusion of the geology, geochronology, and faunal studies in a single publication, will provide a more comprehensive study of a variety of aspects of this important site.

During the course of the recent expeditions, the faunal collections were quadrupled, so that the total number of specimens now exceeds 2,150 (excluding the fish). The collection of fish (more than 7,000 elements of fish) far exceeds that of all other vertebrates combined due to the efforts of Kathlyn Stewart, who participated in the expeditions of 1991–1993 (see section 3.2 of this volume). The recent collections have precise stratigraphic control in contrast to most of the 500 specimens from the earlier collections. Nearly 400 specimens were collected in 1967; many were referred to one of three units: Lothagam 1, Lothagam 2, and Lothagam 3. Most of the 1967 specimens were from Lothagam 1, which at the beginning of the 1967 season was divided into 1A and 1B, with 1B being further subdivided into lower B1, upper and middle B1, and upper B2 (Behrensmeyer, unpublished note).

Partway through the 1967 season, the stratigraphic divisions were revised; Lothagam 1A became 1B, the lower part of Lothagam 1B became lower 1C, the upper and middle 1B became upper and middle 1C, and the uppermost part of the section became 1D. The fossils collected later in the 1967 season were sometimes referred to the revised stratigraphic units but more often by the original designations. This added confusion to a stratigraphy that was already ill defined and, as a result, few of the 1967 specimens can be accorded secure stratigraphic placement. At this juncture, it is in some cases impossible to assess how the 1967 collections relate to laterally extensive and stratigraphically significant markers, such as the Marker Tuff and the Purple Marker. A few fossils were located on sketch maps, the positions of others are related to geographic features that can be recognized and identified, and some are indicated on stratigraphic diagrams in publications; from these their relative ages may be assessed. Of particular help has been a chart compiled by Kay Behrensmeyer in which all the 1967 specimens are placed in their relative stratigraphic and geographic positions. Occasionally this is at variance with the published positions but, when there is a difference, Kay's chart has been taken as the source for the position of a specimen. In 1968 the terminology of the strata changed again to that published by Behrensmeyer (1976). However, none of the fossils collected in 1968 currently have stratigraphic information; regrettably, these data have been lost together with all of Patterson's field notes. The 1968 fossils include the majority of the elephantids, and, consequently, some of the best elephantid specimens collected at Lothagam lack stratigraphic information. Inquiries to Dr. Vince Maglio (now Dr. Jonathan Dutton), who collected the specimens for his doctoral dissertation, were unsuccessful in solving this problem and only confirmed that this information is lost. The 1972 and 1973 collections are little better in terms of provenance. Although these fossils were documented by detailed grid coordinates for an enlargement of an RAF aerial photograph of Lothagam, no details of the scale of the enlargement, or even of the identification of the photograph that was used, have been recorded.

The extensive new collections have added 65 new mammalian taxa (from the Nawata Formation and Apak Member) to the Lothagam faunal list published by Smart (1976) for Lothagam 1, and 22 new mammalian taxa (from the Kaiyumung) to the faunal list published by Behrensmeyer (1976) for Lothagam 3. These collections also have considerably augmented elements of the fauna previously only known from a handful of specimens. This is particularly true for the carnivores, monkeys, rodents, and birds; 120 cercopithecids have been added to the nine previously accessioned, 111 carnivores to the original nine specimens, 46 rodents to the one previously published (Cifelli et al. 1986), and 31 bird skeletal elements to the one known previously. In addition, many specimens of fossilized eggshell of a large flightless bird were collected, along with numerous fragments (claws and carapaces) of fossilized crabs. Twenty-one new vertebrate species and seven new genera are described in this volume; they include four new species of carnivores and three new bovids, a family which was previously unpublished. Lothagam is the type site for ten vertebrate genera, including seven mammals, and 28 vertebrate species, of which 21 are mammals.

Although additional hominoid and hominin specimens were recovered, these groups remain sparsely represented by two hominoids from the Upper Nawata and four hominins from the Kaiyumung Member. The original hominoid mandible discovered by Bryan Patterson in the lower Apak has been frequently discussed in the literature, with varying opinions as to its taxonomic status (Patterson et al. 1970; Kramer 1986; White 1986; Hill and Ward 1988; Hill et al. 1992; Hill 1993; Leakey and Walker this volume: section 6.2, table 6.16). We had hoped that, with the molecular estimates for the divergence of the ape and human lineages somewhere between 5 and 6 Ma (Caccone and Powell 1989; Hasagawa et al. 1989), Lothagam would be an ideal site to provide evidence of the earliest hominins or perhaps even our last common ancestor with African apes. But the two isolated teeth we found in the Upper Nawata did little to enlighten us in this respect. The specimens recovered from the Kaiyumung Member comprise isolated teeth and tooth fragments but are nevertheless important because few hominin specimens of this age (~3.5 Ma) are known from the Turkana Basin.

The enlarged collections allow a more detailed assessment of those taxa previously recognized in the fauna. And the excellent fossil record in the Nawata Formation and the Apak Member of the Nachukui Formation provide an unusually comprehensive assemblage with which faunas from other Late Miocene–earliest Pliocene sites may be compared. Of particular relevance are Sahabi, Libya, in North Africa (Boaz et al. 1987), the Baynunah fauna of Abu Dhabi (Whybrow and Hill 1999), the Middle Awash Valley in Ethiopia (Kalb and Mebrate 1993; Renne et al. 1999), the Tugen Hills in Kenya (Deino et al. 1990; Hill et al. 1985, 1990; Hill 1999), Kakesio in Tanzania (Leakey and Harris 1987), and Langebaanweg in South Africa (Hendey 1970a, 1970b, 1974, 1981).

The Field Seasons

Fieldwork was conducted during five seasons between 1989 and 1993. The initial survey in 1989 lasted less than a week but served to demonstrate the potential of Lothagam for additional fieldwork in the following years. In spite of its small area, Lothagam is perhaps one of the most physically demanding sites. It experiences exceptionally high temperatures due to the lack of wind and the reflected heat from the rich red rocks, and its deep gullies and steep slippery slopes have to be constantly negotiated in the search for fossils. It is also one of the most rewarding sites on account of its exceptional record of beautifully preserved specimens from a little known but highly significant time interval. Few days passed without the excitement of finding a new species or new details of a species already known. Following are highlights from the various field seasons.

1989

A short field survey was undertaken at Lothagam in mid-August 1989 to assess the potential for future field seasons. We located a number of fossils but only collected a handful—those that were very fragile and unlikely to survive if left in the field. Unfortunately, the following year we found that several specimens left hidden under stones and marked with a discrete stone cairn were missing—having been removed either by local people or by visitors from elsewhere.

1990

After completing fieldwork at South Turkwel and North Napudet, a little over one month was spent at Lothagam in 1990. With a good set of aerial photographs available from the coverage obtained by KREMU the previous December, we were able to accurately record the position of the more than 200 specimens collected. The two photographs that provided the most extensive coverage of Lothagam were enlarged to twice their original size for greater accuracy. Except for one day spent in the

Kaiyumung sediments, only the northern exposures, those to the north of the divide, were explored.

We spent much time excavating the carnivore skeleton that I had discovered in 1989 eroding from the hard clays of a steep cliff on the eastern bank of the river Nawata. It proved to be an exceptionally well preserved skeleton of a new species of mustelid. The excavation of a cave about 12 feet high, 6 feet long, and 4 feet deep led to the recovery *in situ* of the cranium, the mandible, most of the vertebrae, and the fore and hind limbs. A second, almost complete carnivore skeleton, this one a cursorial hyena, was excavated from the bank of a small drainage to the north of the Holocene ridge. The search for hominoids was disappointing. In spite of intensive survey, only a partial M^3 was found. The locality of this specimen was extensively screened but no further pieces were recovered.

1991

The 1991 camp was established at the end of May beside the Koriong River, a small sand river just to the west of the Lothagam exposures. Because the prolonged drought over the previous three years had led to a severe shortage of water in the area, we had to transport our water from Lodwar, which was some 80 km away. During this field season, we surveyed the exposures in both the northern and southern areas and also spent some time in the Kaiyumung Member. Work continued until the end of August.

Kay Behrensmeyer joined the expedition for ten days at the end of July and took time to show us features relevant to the earlier geological interpretation. Together with Patrick N'gang'a, a geologist from the National Museums of Kenya, she drew up a geological map that enabled us to precisely locate the stratigraphic provenance of all the 1990 and 1991 fossils, giving us good provenance data for each specimen. Later, Craig Feibel joined the expedition for several weeks and was able to formulate a more detailed stratigraphy. Dennis Powers had generously given Craig all of his field notes and data to facilitate this study. Craig also found several silty clay lenses with small pumices that he collected in the hope that they might be suitable for radiometric dating. Kathlyn Stewart, a specialist in East African fossil fish, joined the expedition for six weeks, and screened several localities rich in fish, enabling her to make a comparative study of the fish fauna through time. Numerous excellent fossil mammals and reptiles were collected, but we were unable to collect several of those found in situ due to a shortage of time, and we left them for collection the following year. Several specimens of birds and rodents, orders that were very rare in the earlier collections, were also recovered. Once more, the field crew concentrated on its search for fossil hominoids but was again disappointed: none were discovered in the Nawata Formation, and only three isolated teeth and tooth fragments were collected from the Kaiyumung.

1992

The 1992 camp was set on May 8 and fieldwork began in the southern exposures that had been less intensively worked during the previous field seasons. In June the fieldwork moved north to the central area, and in July the northern section was resurveyed. We spent considerable time working in the Kaiyumung Member.

In general, the fossils in this member are rather fragmentary, but there are exceptions and the specimens recovered included an in situ articulated skull and mandible of the large fish-eating crocodile, *Euthecodon brumpti.* In July we discovered a third carnivore skeleton that was eroding from the hard clays in the banks of one of the sand rivers close to the "gateway" where we generally took lunch. Many fragments had fallen into a pit beneath the cliff, which had fortunately trapped the bones. The locality was carefully sieved, and we recovered fragments of the skull, ribs, vertebrae, femora, humeri, and foot bones. We began an excavation in an attempt to retrieve the bones of one of the paws that were visible protruding from the cliff face. The site was difficult to work because the specimen was high in the cliff, the upper surface was very slippery, and the fine silty clay matrix was extremely hard and capped by a thick consolidated sandstone. The majority of the field team left the expedition at the beginning of August, but four remained to continue the excavation. However, when extracting the visible bones, we discovered others continuing into the cliff face. Due to limited time, we were unable to complete the excavation. It was clear that a major excavation would be needed to extract this specimen, which later proved to represent a machairodont, the most common carnivore species in the Nawata Formation.

The total number of new specimens collected in 1992 was over 700, doubling the collection accumulated over the previous two years. Many fragments of fossil eggshell of a large flightless bird were added to the collections, thereby documenting a change in pore basin size between those specimens found above and those found below the Marker Tuff. A single specimen of a diminutive suid, *Cainochoerus* cf. *C. africanus*, was found; *C. africanus* is a species that is well represented at Langebaanweg but was hitherto not recorded elsewhere. Several additional birds and rodents were collected. Kathlyn Stewart continued her study of the Lothagam fish fauna. Thure Cerling collected fragmentary teeth for an

Figure 1.2 Large giraffid footprints discovered by Craig Feibel on the lower surface of the Gateway Sandstone.

analysis of the carbon isotopes in tooth enamel in order to document the diet of the various herbivores; at the same time he collected paleosol carbonate nodules for a similar analysis to detect the photosynthetic pathway of the dominant vegetation. A dramatic change from C_3 to C_4 biomass had been observed in the Late Miocene sediments in North America and the Siwaliks deposits in Pakistan (Cerling et al. 1993). Thure, in collaboration with John Harris, hoped to establish whether a similar change could be detected at Lothagam. A single lower incisor of a hominoid was found by Sila Dominic from the uppermost Upper Nawata, and a hominin half molar was discovered by Samuel Ngui in the Kaiyumung Member, bringing the total number of hominins from the Kaiyumung to four.

In September, after the main expedition had closed, Craig Feibel continued his geological studies, measuring sections and drawing a detailed geological map. He also found additional pumice samples to send to Ian McDougall for dating. McDougall had recently installed new equipment with the capability for single crystal dating. Without this technique it would not be possible to date the several occurrences of tiny pumices that Craig discovered in discrete lenses. During this time Craig also noticed the cloven footprints of a large giraffid in an overhanging ledge beneath the Gateway Sandstone (figure 1.2).

1993

Camp was established on May 25. Not long after the season began, on June 2, Richard's light aircraft crashed shortly after takeoff, necessitating a long sojourn in hospital, first in Nairobi and then in the United Kingdom. Therefore I had to leave the expedition, to be with him, but our daughter, Louise, unhesitatingly took over the leadership, planning, and logistics of the expedition, enabling it to continue in my absence.

Craig Feibel continued his geological studies in June and July, and Ian McDougall joined him in July to locate further pumices for dating and to study the geological context of the samples that Craig had collected previously. Kathlyn Stewart again joined the expedition for three weeks in mid-July and completed her sampling of the fossil fish. Joseph Mworia, the palynologist from the National Museums of Kenya, joined Craig and attempted to locate suitable samples for pollen analysis. Robert Mathenge, an M.S. student at the University of Utah, collected samples for paleomagnetic analysis, and Nassir Malit, a Nairobi University student, worked with the field crew. Emma Mbua, also from the National Museums of Kenya, spent four weeks excavating seven Holocene human skeletons, which she later studied as part of her dissertation for an M. Phil. at the University of Liverpool.

The paleontological prospecting focused on the Apak Member from which relatively few fossils had been collected in previous years. It was hoped that additional hominin specimens would be found in this member, but again we were disappointed. Inquiries carried out by Kay Behrensmeyer and Craig Feibel from members of the 1967 American expedition that had found the Lothagam mandible provided a more precise placement for this enigmatic specimen. We had always assumed that a large sieving area in the uppermost Nawata Formation represented the spot where the mandible had been found. Instead, it was confirmed that this site had been sieved for a specimen that turned out to be a Holocene lag deposit specimen of *Homo sapiens*. The Lothagam mandible had actually come from a spot just above this in the lowermost Apak Member.

In mid-June, Alan Walker organized and supervised the excavation of the saber-toothed cat that we had begun the previous year (figure 1.3). This took considerable time and ingenuity, and the excavation was made more difficult by the thick consolidated sandstone that capped the upper surface and that first had to be removed. Substantial scaffolding was built to gain access to the pieces of the specimen, which were exposed high in the cliff face. The venture was successful and resulted in an almost entire skeleton of the most common carnivore at Lothagam, a species of the sabre-toothed cat *Machairodus*. This is certainly the most complete African specimen of this genus. Of particular interest was the articulated forepaw that had an enlarged claw on the first digit but reduced claws on the remaining digits (figure 1.4). This was the third almost complete carnivore skeleton from Lothagam. These skeletons are described by Lars Werdelin in Chapter 7 of this volume.

The Volume

This volume presents the results of five season's fieldwork, between 1989 and 1993, and the subsequent

Figure 1.3 The 1993 excavation of the skeleton of the Lothagam machairodont. This new species is the most common carnivore in the Nawata Formation.

laboratory studies. The volume has been long in production due to the extensive collection of beautifully preserved fossils and the large number of researchers involved in the analyses. The project has been a truly collaborative, interdisciplinary undertaking, and as such it has proved exceptionally rewarding. The appreciation of the value of such interdisciplinary studies was first realized with the International Expedition to the Omo Valley in 1967, involving French, American, and Kenyan contingents. The practice was continued at East Turkana in the late 1960s and 1970s, and many similar multidisciplinary expeditions have followed. With our increased knowledge and use of advanced analytical techniques, it is essential for field and laboratory studies to involve scientists from many different backgrounds. Techniques that were previously undreamed of—for example, the isotopic analysis of tooth enamel and paleosols, the SEM examination of enamel microwear, and the CT scanning of fossils in order to study the inner recesses of a bone—are now accepted as crucial to a full interpretation of the available evidence. Results from these types of analyses are all reported here. As a result, the research is more sophisticated and the length of time to complete the studies is prolonged. But the information gained is more detailed and the developing picture is more comprehensive.

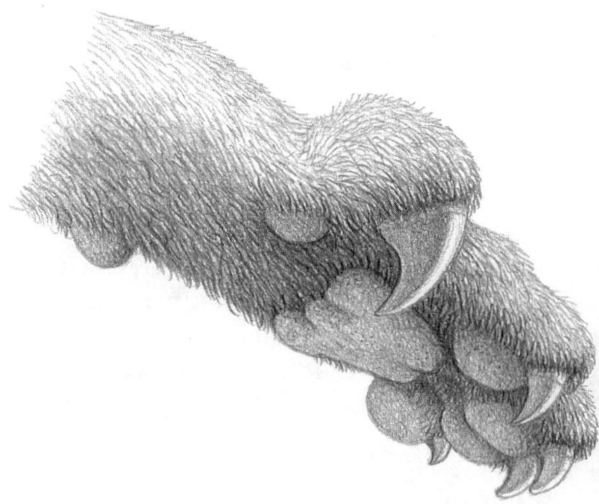

Figure 1.4 Restoration of the paw of the Lothagam machairodont by Mauricio Antón. Contrast the large claw of the first digit with the reduced claws of the remaining digits.

This monograph has been modeled on the excellent volume on Laetoli edited by my mother-in-law, Mary Leakey, who sadly died in 1997, and by John Harris, who is the co-editor of this volume. John has edited two of the series of monographs on East Turkana (Harris 1983, 1991), and this volume has benefited enormously from his expertise. Similar monographic treatments of important Late Miocene and Plio-Pleistocene sites are given on Manonga Valley (Harrison 1997), Semliki Valley (Boaz 1990), Sahabi (Boaz et al. 1987), and Abu Dhabi (Whybrow and Hill 1999). These have proved a useful source of comparison for Lothagam.

The volume provides a compilation of the data currently available on the Late Miocene and Pliocene sediments at Lothagam. The Holocene sediments have not been included, although these are recognized as important for future studies. Chapter 2 describes the geology, and dating, with contributions from Craig Feibel, Jonathan Wynn, and Ian McDougall. Chapters 3 through 11 give descriptions of the fauna, with each chapter and section authored by an expert on the taxa discussed. Discussions of the ecology of both present and past habitats based on the isotopic analyses of Thure Cerling and John Harris follow in Chapter 12. The final chapter, Chapter 13, discusses the significance of the fauna from the biogeographical and paleoenvironmental perspectives.

The monograph includes reconstructions of some of the more common or more unusual species described in each chapter. These reconstructions, drawn by Mauricio Antón, are based on the original Lothagam fossils; Mauricio worked in close collaboration with the respective authors to ensure that the reconstructions would be as accurate as possible. Mauricio has also depicted the prevailing habitats and some of the fauna in three of the time intervals at Lothagam, the Nawata Formation, the Apak, and the Kaiyumung (see figures 13.1, 13.14, 13.15).

I hope that this volume will do justice to the wealth of information preserved in the long sedimentary record at Lothagam and will be of interest to all those who share a common curiosity about our past. The significant faunal and environmental changes that are documented at Lothagam are relevant to our earliest origins and to those of all mammals inhabiting Africa today.

Acknowledgments

A wide-reaching research endeavor such as this, taking place over more than a decade, inevitably involves support and assistance from many different individuals, including donors, colleagues, friends, and family. Space limitations preclude my listing the name of everyone who has contributed to the success of the Lothagam project but to all I record my sincere appreciation.

The field research at Lothagam could not have happened without the sanction and support of the National Museums of Kenya Board of Trustees and the museum director, Dr. Mohamed Isahakia. Dr. Isahakia's enthusiasm for and interest in this project were clearly demonstrated when he personally visited Lothagam in July 1991. Financial support, an essential ingredient of every field expedition, was provided over the 5 years that we worked at Lothagam by Shell Exploration (Kenya) and by the National Outdoor Leadership School in Lander, Wyoming. In particular, I thank Felix Malloy, the managing director of Shell Exploration (Kenya), for his personal interest in this project. The Defender Land Rover donated by the Rover Group in 1991 and the Mercedes-Benz four-wheel drive provided by the National Geographic Society were essential to the project and greatly appreciated.

I am grateful to the local Turkana people who made us so welcome, allowing us to move freely through their area, and particularly to the late Mr. Ekuwom, the head of the family on whose land we camped and who was subsequently buried at our camp site.

The success of the field project was due in large part to the exceptional dedication and expertise of the field crew, who discovered and recovered the remarkable collection of fossil specimens. Their uncomplaining commitment throughout the long, hot days, together with their sharp and experienced eyes, led to the discovery of even minute specimens. Their incorrigible humor and cheerful acceptance of the long hours, the excessive number of flies, and the daily dust storms made each field expedition a special and memorable experience. The members of each of the expeditions are listed elsewhere but I particularly need to thank Kamoya Kimeu, whose many years of experience, leadership, and legendary talents at discovering fossils were indispensable. He set up the camp at the beginning of each season, nestling the tents in the shade of the few available thorn trees, and during my absences in Nairobi he kept the camp and fieldwork functioning smoothly. Benson Kyongo also deserves mention for his skills at nursing the expedition lorry to and from Nairobi at the start and conclusion of the expeditions, and driving the 70 km to and from Lodwar every ten days throughout the field seasons in order to replenish our vital water supply. The camp staff, too, are thanked for their role in keeping the camp running smoothly and for providing substantial and nourishing meals. Peter Nzube is recognized for his exceptional skill in locating elusive fossil monkeys, my own particular interest. In the evenings, he and Kamoya regaled us with entertaining tales of their experiences in earlier years at Olduvai, Lake Natron, Lake Baringo, the Omo Valley, Koobi Fora, and West Turkana. Sila Dominic, Kamoya Kimeu, Mwongela Muoka, Joseph Mutaba, Samuel Ngui, and Kathy Stewart each discovered fossil hominoids (figure 1.5).

Alan Walker directed the excavation of the machairodont skeleton in 1993, a particularly challenging task due to the inaccessible location of the specimen and the very hard rock in which it was entombed. Kathy Stewart

Figure 1.5 The field crew in 1991, taking a rest from carrying a large, articulated carapace of a giant tortoise (*Geochelonia* sp.) to the Land Rover. The fossil is encased in plaster of paris.

spent three seasons in the field providing stimulating companionship, in addition to her talents for recovering thousands of fossil fish elements. During many afternoons in camp sorting the field collections, her good humor and tolerance were severely tested by an army of persistent and irritating flies and by exceptionally strong winds that carried off anything left untethered on the table. Over the five field seasons, several students joined the expedition, carrying out their own projects and assisting with routine work. These included Malou Hanson Hoeck, Catherine Kenyatta, Nasser Malit, Steven Masai, Robert Mathenge, Shanaz Nagri, and Eleanor Weston. Eleanor Weston subsequently gave me considerable help in the lab, drafting detailed overlays for our aerial photographs, which enabled me to accurately locate the position of each fossil specimen.

No paleontological expedition can succeed without a sound geological framework. Kay Behrensmeyer freely shared her geological knowledge gained through the early expeditions and provided us with copies of her notes and sections. Her help in this respect and her stratigraphic plan, on which she recorded the position of many of the 1967 collections, gave us provenance information for many of the early specimens. In 1991, Kay took time from her fieldwork at Amboseli to visit us at Lothagam—to share her initial interpretation of the geology with museum geologist Patrick N'gang'a and me, and to draw up a geological map that enabled us to identify stratigraphic provenance for the 1989 and 1990 collections. Particularly important was her help in locating the exact spot of the 1967 hominoid mandible, which we had erroneously believed to have come from a sieved area that was actually the site of a Holocene hominin. Frank Brown, as always, provided encouragement and support, both in and away from the field. Whenever he came to our camp he never failed to give us advice and help based on his intimate knowledge of the Turkana Basin geology. Major recognition for our current understanding of the Lothagam geology, however, is undoubtedly due to Craig Feibel, who spent considerable time in the field at Lothagam and whose many years of experience in the Turkana Basin were invaluable for his interpretation of the Lothagam geology.

Bob Campbell deserves special mention. Throughout the Lothagam project he repaired, checked, and serviced our old Land Rovers; their continued availability through these expeditions was entirely due to his efforts.

In addition, over the course of several visits to the field he compiled for us a photographic record of the work and the site (figures 1.1, 1.2, and 1.3). Both Fiona Alexander, a personal friend, and Phil Matthews, the chief pilot at the Kenya Wildlife Service, flew members of the expedition and visitors to and from our camp on various occasions. Phil Matthews also helped us safely transport specimens to Nairobi by air, thus avoiding any possible damage from transport on the rough local tracks. I am also particularly indebted to Phil for reacting so quickly to the news of Richard's plane crash and flying me back to Nairobi that same evening.

The essential and indispensable body of people in the Nairobi National Museum also deserves mention. The preparators who removed the matrix and reconstructed the fossils; the curators who accessioned, sorted, and ordered the specimens and who cheerfully assisted the researchers studying specific aspects of the Lothagam collection; and the casting staff, particularly John Ndunda, who rapidly responded to researchers' urgent requests for casts—the efforts of all are acknowledged. Justus Edung, Alfreda Ibui, Ngalla Jillani, Christopher Kiarie, Frederick Kyalo, Benson Kyongo, Wambua Mangao, Joseph Mutaba, Samuel Muteti, and Mary Muungu have been of particular help.

Finally, I must thank my family for their understanding and tolerance of my long absences from home during the field seasons. As always, Richard gave his full support to our endeavors and provided an indispensable backup in Nairobi for urgent requirements. Our evening attempts to contact him on the radio-telephone gave us an essential lifeline to Nairobi, and on many occasions he provided assistance when we urgently needed spare parts or messages passed on to others. Louise, too, gave me indispensable help in 1993 when Richard's light aircraft crashed and it was impossible for me to remain in the field. Louise unhesitatingly took over the leadership of the expedition, enabling it to continue the full season. With characteristic energy, she guided the expedition through the inevitable problems that go with any field project of this nature and accomplished all of the expedition's original goals.

The compilation of this volume has involved the assistance of many. First I must acknowledge the very significant contribution of my co-editor, John Harris. After completing the Laetoli volume in 1987, John vowed that he would never again take part in a similar endeavor. I am enormously grateful that he changed his mind. John's careful and thorough search for a suitable publisher, and his considerable experience and editorial skills, combined with his professional expertise in paleontology, have been a major asset to the volume. Not only has he provided editorial input, but also he has authored and co-authored many of the chapters. Our daily communications by email have made the compilation of the manuscript a particularly rewarding and often amusing experience. John wishes to acknowledge logistical support from the Natural History Museum of Los Angeles County. We both wish to thank all those who have made contributions to the volume by providing expertise in their own particular fields: Craig Feibel (geology), Jonathan Wynn (paleosols), Ian McDougall (dating), Kathy Stewart (Pisces), Roger Wood (Chelonia), Glen Storrs (Crocodylidae), Alisa Winkler (Lagomorpha and Rodentia), Carol Ward (Cercopithecidae postcrania), Mark Teaford (Cercopithecidae microwear), Alan Walker (Hominoidea), Lars Werdelin (Carnivora), and Thure Cerling (isotopes). Mauricio Antón, the artist responsible for all the reconstructions, deserves special recognition. He has shown exceptional patience and tolerance, working closely with the authors and never complaining at repeated requests for changes in his detailed illustrations in our efforts to make the restorations as accurate as possible. His talents have given the volume an additional dimension by vividly bringing the past to the present.

The Leakey Foundation generously awarded us a grant to enable us to engage Mauricio Antón and for him to fly to Nairobi to see the original specimens. The Geological Society of London kindly allowed us to reproduce McDougall and Feibel 1999 as chapter 2.3. Apple Macintosh donated a G3 laptop computer, enhancing my ability to work more closely with John and providing considerable versatility in the compilation of the volume. Judy Harris provided me with a home in Los Angeles while I worked with John on the manuscript, and Bob Campbell helped with many of the photographs and allowed me the use of his SprintScan 35. Finally, both John and I thank Columbia University Press for publishing the volume—in particular, Holly Hodder, who showed great patience in repeatedly extending our deadline for submission.

Field Personnel

1990

Research

Frank Brown
Meave Leakey
Patrick N'gang'a

Field crew

Christopher Epat
Ngeneo Kanyenze
Catherine Kenyatta
Christopher Kiarie

Kamoya Kimeu
Benson Kyongo
Wambua Mangao
Mwongel Muoka
Joseph Mutabe
Kavai Ndunda
Peter Nzube

1991

Research

Kay Behrensmeyer
Frank Brown
Craig Feibel
Meave Leakey
Steven Masai
Robert Mathenge
Patrick N'gang'a
Kathlyn Stewart

Field crew

Christopher Epat
Christopher Kiarie
Kamoya Kimeu
Peter Kiptalam
Frederick Kyalo
Benson Kyongo
Boniface Malika
Wambua Mangao
Sila Mawa
Mwongela Muoka
Joseph Mutaba
Ngui Muteti
Kavai Ndunda
Peter Nzube

1992

Research

Craig Feibel
Meave Leakey
Shanaz Nagri
Kathlyn Stewart
Eleanor Weston

Field crew

Christopher Epat
Paul Joseph Ewor
Kamoya Kimeu
Peter Kiptalam
Benson Kyongo
Boniface Malika
Wambua Mangao
Sila Mawa
Mwongela Muoka
Joseph Mutaba
Ngui Muteti
Peter Nzube

1993

Research

Craig Feibel
Ian McDougall
Meave Leakey (first two weeks only)
Louise Leakey
Robert Mathenge
Emma Mbua
Richard Nassir
Kathlyn Stewart
Alan Walker

Field crew

Justus Edung
Christopher Epat
Christopher Kiarie
Kamoya Kimeu
Benson Kyongo
Boniface Malika
Wambua Mangao
Sila Mawa
Mwongela Muoka
Joseph Mutaba
Ngui Muteti
Peter Nzube

References Cited

Anyonge, W. 1991. Fauna from a new lower Miocene locality west of Lake Turkana, Kenya. *Journal of Vertebrate Paleontology* 11:378–390.

Behrensmeyer, A. K. 1976. Lothagam Hill, Kanapoi, and Ekora: A general summary of stratigraphy and faunas. In Y. Coppens, F. C. Howell, G. L. Isaac, and R. E. Leakey, eds., *Earliest Man and Environments in the Lake Rudolf Basin: Stratigraphy, Paleoecology, and Evolution*, pp. 163–172. Chicago: University of Chicago Press.

Boaz, N. T., ed. 1990. *Evolution of Environments and Hominidae in the African Western Rift Valley.* Memoir No. 1. Martinsville: Virginia Museum of Natural History.

Boaz, N. T., A. El-Arnauti, A. W. Gaziry, J. de Heinzelin, and D. D. Boaz, eds. 1987. *Neogene Paleontology and Geology of Sahabi.* New York: Liss.

Boschetto, H. B., F. H. Brown, and I. McDougall. 1992. Stratigraphy of the Lothidok Range, northern Kenya, and K/Ar ages of its Miocene primates. *Journal of Human Evolution* 22:47–71.

Brown, F. 1995. The potential of the Turkana Basin for palaeoclimatic reconstruction in East Africa. In E. S. Vrba, G. H. Denton, T. C. Partridge, and L. H. Burkle, eds., *Palaeoclimate and Evolution, with Emphasis on Human Origins,* pp. 319–330. New Haven: Yale University Press.

Brown, F. H., and C. S. Feibel. 1986. Revision of stratigraphic nomenclature in the Koobi Fora region, Kenya. *Journal of the Geological Society* (London) 143:297–310.

Brown, F. H., I. McDougall, I. Davies, and R. Maier. 1985. An integrated Plio-Pleistocene chronology for the Turkana Basin. In E. Delson, ed., *Ancestors: The Hard Evidence,* pp. 83–90. New York: Liss.

Caccone, A., and J. R. Powell. 1989. DNA divergence among hominoids. *Evolution* 43:925–942.

Cerling, T. E., Y. Wang, and J. Quade. 1993. Expansion of C_4 ecosystems as an indicator of global ecological change in the Late Miocene. *Nature* 361:344–345.

Champion, A. M. 1937. Physiography of the region to the west and southwest of Lake Rudolf. *Geographical Journal* 89:97–118.

Churcher, C. S. 1979. The large palaeotragine giraffid, *Palaeotragus gemaini,* from Late Miocene deposits of Lothagam Hill, Kenya. *Breviora* 453:1–8.

Cifelli, R. L., A. K. Ibui, L. L. Jacobs, and R. W. Thorington. 1986. A giant tree squirrel from the Late Miocene of Kenya. *Journal of Mammalogy* 67:274–283.

Cooke, H. B. S., and R. F. Ewer. 1972. Fossil Suidae from Kanapoi and Lothagam, northwestern Kenya. *Bulletin of the Museum of Comparative Zoology* 43:149–296.

Coppens, Y., and F. C. Howell, eds. 1985. *Les faunes Plio-Pléistocène de la Basse Vallée de l'Omo (Ethiopie).* Vol. 1. *Perissodactyles, Artiodactyles (Bovidae),* pp. 1–191. Paris: Centre National de la Recherche Scientifique.

Coppens, Y., and F. C. Howell, eds. 1987a. *Les faunes Plio-Pléistocène de la Basse Vallée de l'Omo (Ethiopie).* Vol. 2. *Les Eléphantidés, Proboscidea (Mammalia),* pp. 1–162. Paris: Centre National de la Recherche Scientifique.

Coppens, Y., and F. C. Howell, eds. 1987b. *Les faunes Plio-Pléistocène de la Basse Vallée de l'Omo (Ethiopie).* Vol. 3. *Cercopithecidae de la Formations de Shungura,* pp. 1–169. Paris: Centre National de la Recherche Scientifique.

Coryndon, S. C. 1977. The taxonomy and nomenclature of the Hippopotamidae (Mammalia, Artiodactyla) and a description of two new fossil species. *Proceedings of the Koninklijke Nederlandse Akademie van Wetenschappen,* ser. B, 80:61–88.

Deino, A., L. Tauxe, M. Monaghan, and R. Drake. 1990. Single crystal $^{40}Ar/^{39}Ar$ ages and the litho and paleomagnetic stratigraphies of the Ngorora Formation, Kenya. *Journal of Geology* 98:567–587.

Feibel, C. S. 1988. Paleoenvironments from the Koobi Fora Formation, Turkana Basin, northern Kenya. Ph.D. diss., University of Utah.

Feibel, C. S., F. H. Brown, and I. McDougall. 1989. Stratigraphic context of fossil hominids from the Omo Group deposits: Northern Turkana Basin, Kenya and Ethiopia. *American Journal of Physical Anthropology* 78:595–622.

Fuchs, V. E. 1939. The geological history of the Lake Rudolf Basin, Kenya Colony. *Philosophical Transactions of the Royal Society of London,* ser. B, 229:219–274.

Harris, J. M., ed. 1983. *Koobi Fora Research Project.* Vol. 2. *The Fossil Ungulates: Proboscidea, Perissodactyla, and Suidae.* Oxford: Clarendon Press.

Harris, J. M., ed. 1991. *Koobi Fora Research Project.* Vol. 3. *The Fossil Ungulates: Geology, Fossil Artiodactyls, and Palaeoenvironments.* Oxford: Clarendon Press.

Harris, J. M., F. H. Brown, and M. G. Leakey. 1988a. Stratigraphy and paleontology of Pliocene and Pleistocene localities west of Lake Turkana, Kenya. *Contributions in Science* 399:1–128.

Harris, J. M., F. H. Brown, M. G. Leakey, A. C. Walker, and R. E. Leakey. 1988b. Pliocene and Pleistocene hominid-bearing sites from west of Lake Turkana, Kenya. *Science* 239:27–33.

Harrison, T., ed. 1997. *Neogene Paleontology of the Manonga Valley, Tanzania: A Window into the Evolutionary History of East Africa.* New York: Plenum Press.

Hasagawa, M., H. Kishino, and T. Yano. 1989. Estimation of branching dates among primates by molecular clocks of nuclear DNA which slowed down in Hominoidea. *Journal of Human Evolution* 18:461–476.

Hendey, Q. B. 1970a. A review of the geology and palaeontology of the Plio-Pleistocene deposits at Langebaanweg, Cape Province. *Annals of the South African Museum* 56:75–117.

Hendey, Q. B. 1970b. The age of the fossiliferous deposits at Langebaanweg, Cape Province. *Annals of the South African Museum* 56:119–131.

Hendey, Q. B. 1974. The late Cenozoic Carnivora of the South-Western Cape Province. *Annals of the South African Museum* 63:1–369.

Hendey, Q. B. 1981. Palaeoecology of the Late Tertiary fossil occurrences in "E" Quarry, Langebaanweg, South Africa, and a reinterpretation of their geological context. *Annals of the South African Museum* 84:1–104.

Hill, A. 1993. Late Miocene and Early Pliocene hominids from Africa. In R. S. Corrucini and R. L. Ciochon, eds., *Integrative Paths to the Past,* pp. 123–145. Englewood Cliffs, N.J.: Prentice Hall.

Hill, A. 1999. The Baringo Basin, Kenya: From Bill Bishop to BPRP. In P. Andrews and P. Banham, eds., *Late Cenozoic Environments and Hominid Evolution: A Tribute to Bill Bishop,* pp. 85–97. London: Geological Society.

Hill, A., and S. Ward. 1988. Origin of the Hominidae: The record of African large hominoid evolution between 14 My and 4 My. *Yearbook of Physical Anthropology* 31:49–83.

Hill, A., R. Drake, L. Tauxe, M. Monaghan, J. C. Barry, A. K. Behrensmeyer, G. Curtis, B. F. Jacobs, N. Johnson, and D. Pilbeam. 1985. Neogene palaeontology and geochronology of the Baringo Basin, Kenya. *Journal of Human Evolution* 14:749–773.

Hill, A., S. Ward, and B. Brown. 1992. Anatomy and age of the Lothagam mandible. *Journal of Human Evolution* 22:439–451

Hill, A., P. Whybrow, and W. Yasin al-Tiktiti. 1990. Late Miocene fauna from the Arabian Peninsula: Abu Dhabi, United Arab Emirates. *American Journal of Physical Anthropology* 81:240–241.

Hooijer, D. A., and V. J. Maglio. 1973. The earliest *Hipparion* south of the Sahara in the Late Miocene of Kenya. *Proceedings of the Koninklijke Nederlandse Akademie van Wetenschappen,* ser. B, 76:311–315.

Hooijer, D. A., and V. J. Maglio. 1974. Hipparions from the

Late Miocene and Pliocene of northwestern Kenya. *Zoologische Verhandelingen* 134:3–34.

Hooijer, D. A., and B. Patterson. 1972. Rhinoceroses from the Pliocene of northwestern Kenya. *Bulletin of the Museum of Comparative Zoology* 144:1–26.

Hopson, A. J., ed. 1982. *Lake Turkana: A Report on the Findings of the Lake Turkana Project, 1972–1975.* Vol. 1. London: Overseas Development Administration.

Kalb, J. E., and A. Mabrate. 1993. Fossil elephantoids from the hominid-bearing Awash Group, Middle Awash Valley, Afar Depression, Ethiopia. *Transactions of the American Philosophical Society* 83:1–114.

Kramer, A. 1986. Hominid-pongid distinctiveness in the Miocene-Pliocene fossil record: The Lothagam mandible. *American Journal of Physical Anthropology* 70:457–473.

Leakey, L. N., S. A. H. Milledge, S. M. Leakey, J. Edung, P. Haynes, D. K. Kiptoo, and A. McGeorge. 1999. Diet of striped hyaena in northern Kenya. *African Journal of Ecology* 37:314–326.

Leakey, M. D., and J. M. Harris, eds. 1987. *Laetoli: A Pliocene Site in Northern Tanzania.* Oxford: Clarendon Press.

Leakey, M. G., C. S. Feibel, R. L. Bernor, T. E. Cerling, J. M. Harris, K. M. Stewart, G. W. Storrs, A. Walker, L. Werdelin, and A. Winkler. 1996. Lothagam: A record of faunal change in the Late Miocene of East Africa. *Journal of Vertebrate Paleontology* 16:556–570.

Leakey, R. E., and M. G. Leakey. 1986a. A new Miocene hominoid from Kenya. *Nature* 324:143–146.

Leakey, R. E., and M. G. Leakey. 1986b. A second new Miocene hominoid from Kenya. *Nature* 324:146–148.

Leakey, R. E., and M. G. Leakey. 1987. A new small-bodied ape from Kenya. *Journal of Human Evolution* 16:369–387.

Leakey, R. E., and A. Walker. 1985. New higher primates from the Early Miocene of Buluk, Kenya. *Nature* 318:173–175.

Madden, C. T. 1972. Miocene mammals, stratigraphy and environments of Muruarot Hill, Kenya. *PaleoBios* 14:1–12.

Maglio, V. J. 1970. Four new species of Elephantidae from the Plio-Pleistocene of northwestern Kenya. *Breviora* 341:1–43.

Maglio, V. J. 1973. Origin and evolution of the Elephantidae. *Transactions of the American Philosophical Society,* n.s., 63:1–149.

McDougall, I. 1985. K-Ar and $^{40}Ar/^{39}Ar$ dating of the hominid-bearing Pliocene-Pleistocene sequence at Koobi Fora, Lake Turkana, northern Kenya. *Geological Society of America Bulletin* 96:159–175.

McDougall, I., and C. S. Feibel. 1999. Numerical age control for the Miocene-Pliocene succession at Lothagam, a hominoid-bearing sequence in the northern Kenya Rift. *Journal of the Geological Society* (London) 156:731–745

Patterson, B. 1975. New fossil Orycteropodidae (Mammalia, Tubulidentata) from East Africa. *Orycteropus minutus* sp. nov. and *Orycteropus chemeldoi* sp. nov. *Netherlands Journal of Zoology* 25:57–88.

Patterson, B., A. K. Behrensmeyer, and W. D. Sill. 1970. Geology of a new Pliocene locality in northwestern Kenya. *Nature* 256:279–284.

Powers, D. W. 1980. Geology of Mio-Pliocene sediments of the lower Kerio River Valley, Kenya. Ph.D. diss., Princeton University.

Renne, P. R., G. WoldeGabriel, W. K. Hart, G. Heiken, and T. D. White. 1999. Chronostratigraphy of the Miocene-Pliocene Sagantole Formation, Middle Awash Valley, Afar Rift, Ethiopia. *Geological Society of America Bulletin* 111: 869–885.

Robbins, L. H. 1967. A recent archaeological discovery in the Turkana District of northern Kenya. *Azania* 2:1–5.

Robbins, L. H. 1972. Archeology in the Turkana District, Kenya. *Science* 176:359–366.

Smart, C. 1976. The Lothagam 1 fauna: Its phylogenetic, ecological and biogeographic significance. In Y. Coppens, F. C. Howell, G. L. Isaac, and R. E. Leakey, eds., *Earliest Man and Environments in the Lake Rudolf Basin: Stratigraphy, Paleoecology, and Evolution,* pp. 361–369. Chicago: University of Chicago Press.

Tchernov, E. 1986. *Evolution of the Crocodiles in East and North Africa.* Cahiers de Paléontologie. Paris: Centre National de la Recherche Scientifique.

Ward, C. V., M. G. Leakey, B. Brown, F. Brown, J. Harris, and A. Walker. 1999. South Turkwel: A new Pliocene hominid site in Kenya. *Journal of Human Evolution* 36:69–95.

White, T. D. 1986. *Australopithecus afarensis* and the Lothagam mandible. *Anthropos* (Brno) 23:79–90.

Whybrow, P. A., and A. Hill, eds. 1999. *Fossil Vertebrates of Arabia.* New Haven: Yale University Press.

2

GEOLOGY, PALEOSOLS, AND DATING

2.1

Stratigraphy and Depositional History of the Lothagam Sequence

Craig S. Feibel

The stratigraphic succession exposed at Lothagam comprises some 900 m of conglomerates, sandstones, mudstones and altered tephra, with intercalated lavas. The four major lithostratigraphic units recognized within the sequence document stages in the large-scale tectonic and climatic evolution of the region. Variations in the character of the fluvial strata record a succession of river systems. The Nabwal Arangan beds of Middle to Late Miocene age relate to a high-relief volcanic source terrane nearby and consist largely of conglomerates and lavas. A major fluvial system is documented by strata of the Nawata Formation, with variations in fluvial facies reflecting changes in subsidence rate and water budget through the Late Miocene. Strata of the Apak Member of the Nachukui Formation (Early Pliocene) show a change in source terrane and fluvial style, and are likely related to the ancestral Kerio River system. Upper Apak and Muruongori Member strata are lacustrine in character and correlate with the Early Pliocene Lonyumun Lake phase of the Turkana Basin. The subsequent fluvial deposits of the Kaiyumung Member record yet another fluvial system in the Early to Late Pliocene, which appears to be the ancestral Turkwel River. Early Pleistocene strata attributed to the Kalochoro and Kaitio Members of the Nachukui Formation are primarily lacustrine in character, and reflect conditions in the Lorenyang Lake. The uppermost strata exposed at Lothagam are attributed to the Galana Boi Formation, deposited during a Holocene highstand of Lake Turkana.

The sedimentary strata exposed at Lothagam have received considerable attention from both geologists and paleontologists since they were first recognized by L. H. Robbins in 1965 (Robbins 1967). The significance of these strata lies in the evidence they preserve of rift evolution, patterns of biotic change through the fossil record, and associated clues to the history of environmental change for this part of the African continent. Systematic investigation of the Lothagam strata began with the work of the Harvard and Princeton expeditions (Patterson et al. 1970; Behrensmeyer 1976; Powers 1980) and was extended with work by Meave Leakey's team from the National Museums of Kenya. The geological investigations undertaken as part of the latter project included eight visits to Lothagam by the author between 1991 and 1995. The primary goals of this work were (1) to expand, update, and formally establish a lithostratigraphic terminology for the Lothagam deposits; (2) to locate materials within the sedimentary succession suitable for isotopic dating (McDougall and Feibel 1999); and (3) to collect additional data on the depositional characteristics of the sedimentary sequence in order to improve the paleo-environmental reconstructions for the rich Mio-Pliocene fossil assemblages that have been recovered from Lothagam. This contribution presents a preliminary assessment of depositional environments as they relate to the fossil record.

Physiography and Structure

Lothagam is formed by two prominent north–south ridges and the exposures that occur between and on the flanks of these features. Rising from the low-lying plains between the Kerio and Turkwel Rivers, Lothagam stands as an island in the sand seas of this region (figure 2.1).

The eastern, and most prominent, ridge at Lothagam is a horst, bounded by several steeply dipping faults (figure 2.2). It includes the highest topographic point, Lothagam Peak, as well as subsidiary high points at Awachele and Central Hill. The horst is cut through by a major ephemeral stream, which forms the Nabwal Arangan water hole where it crosses the eastern boundary fault. The horst consists of a thick succession of conglomerates and lavas.

The western ridge at Lothagam, called Muruongori at its highest point in the north, is a much lower relief feature formed as a cuesta of resistant basalt dipping gently to the west. The topographically low saddle near the middle of this ridge, which allows access to the central valley, is known as Apak. The western ridge exposes basalt along its entire length, with steep exposures of the underlying sediments on its eastern side and scattered exposures of the overlying sediments at its western foot.

Between the two ridges is a complex terrain of badlands exposures and cuestas of westward-dipping strata that form the heart of Lothagam. Most of the Miocene strata here strike roughly north–south and dip up to 35° west, while horizontal Late Pleistocene–Holocene deposits cap them unconformably in some areas. Several ephemeral streams drain this central valley. These include the Nawata, which flows north, collecting runoff from the northern half of the central valley; the Nabwal Arangan, which cuts eastward from the central part of the valley; and several smaller drainages in the southern part of the valley.

On either side of the Lothagam ridges, sedimentary strata have been largely planed down to horizontal or gently sloping surfaces that are locally dissected by ephemeral stream drainages. As the prominent winds are from the east, and an ample supply of sand is available from the nearby seasonal Kerio River, much of the eastern plain has been covered with sand dunes. The base of the horst on the east presents a rampart to the migrating sands, and the steep fault-bounded face causes winds to eddy back, leaving a narrow gap between the horst and the encroaching dunes (figure 2.3). However, deflation and migration of dunes on the eastern plain reveals that extensive Plio-Pleistocene sedimentary deposits underlie the Recent sands in this region. The western plain largely comprises beveled Plio-Pleistocene strata, with a local veneer of Late Pleistocene–Holocene lake beds or Recent sands.

The basic structural configuration of Lothagam consists of the eastern horst and a tilted sedimentary succession to the west, with the intercalated basalt providing the resistant cuesta of the western ridge (figure 2.4).

Figure 2.1 Lothagam viewed from the northeast.

The horst itself has at least one major fault that cuts diagonally across it. The central valley succession is cut by a large number of minor faults (typically parallel to strike and difficult to recognize), and one major fault in the western part of the valley offsets both sedimentary strata and the basalt. A series of arcuate faults, dipping northward, occurs in the northern part of the central valley and appears to continue where the sedimentary strata disappear beneath Recent sands. In the northernmost exposures, molluscan sands characteristic of the Plio-Pleistocene succession occur in one small locality. In a broader regional context, the Lothagam succession accumulated within the Kerio half-graben (Morley et al. 1992). This structural unit was active from Middle Miocene through Early Pleistocene times (based on the ages of tilted strata at Lothagam). Footwall uplift along the major boundary fault on the eastern flank of the horst caused elevation of that block and the associated tilting of the sedimentary succession to the west.

Lithostratigraphy

Four major sedimentary intervals have been recognized within the strata exposed at Lothagam. These have been apparent from the earliest investigations, and the various schemes of lithostratigraphic terminology used to discuss these deposits differ mainly in the names chosen, the choice of boundary markers, and the interpretation of temporal and sedimentary affinities used to rank and relate the stratigraphic units. The terminology used here was established by Powers and Feibel (in Leakey et al. 1996) and will be formalized elsewhere. It reflects both a clearer understanding of the characteristics of the deposits and a broader knowledge of depositional history within the Turkana Basin as a whole. As currently understood, the succession consists of one informally designated unit (probably of formational stature, but as yet unstudied) and three formations. Stratigraphic characteristics, subunits, and boundaries of each will be discussed in this chapter, with emphasis on the fossiliferous units (figure 2.5).

Nabwal Arangan Beds

Much of the horst is formed by volcaniclastic cobble- to boulder-conglomerates with minor intercalated lavas, informally designated the Nabwal Arangan beds (Leakey et al. 1996). The unit is named for the water hole near the middle of these exposures. This unit is estimated to be greater than 200 m in thickness (Powers 1980). Recent isotopic age determinations (McDougall and Feibel 1999) demonstrate that much of the Nabwal

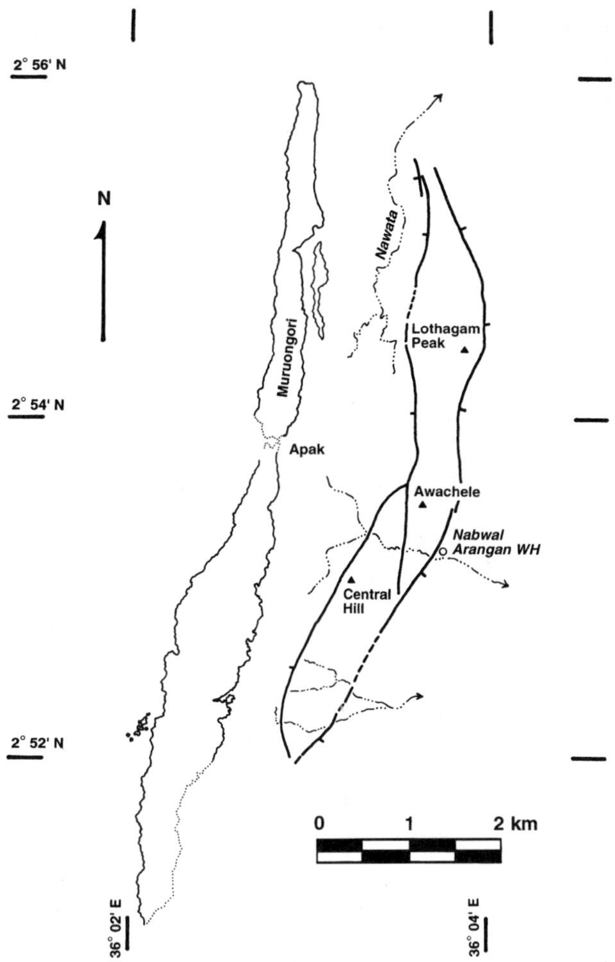

Figure 2.2 Map of the major physiographic features at Lothagam.

Arangan sequence is of Middle Miocene age, while the uppermost basalt flow of the unit yielded an age of 9.1 Ma. This provides important age control on the base of the overlying sequence. The only fossil material recovered to date from the Nabwal Arangan beds is fossilized wood.

Nawata Formation

The sedimentary strata resting on the uppermost basalts of the Nabwal Arangan beds, and extending upward through the top of the prominent analcimolitic Purple Marker, were designated the Nawata Formation by Powers and Feibel (in Leakey et al. 1996). The Nawata Formation is exposed over the eastern two-thirds of the central valley at Lothagam. The base of the formation can be seen on the western flanks of Central Hill, but elsewhere the formation is in fault contact with the underlying strata. In the type section, the formation attains a thickness of 262 m. The formation is named for the prominent ephemeral stream that drains the northern part of the central valley at Lothagam, where the

Figure 2.3 Sand dunes at the foot of the horst on the eastern side of Lothagam.

formation is best exposed. The Nawata Formation has been informally subdivided into two members, using the base of the prominent volcaniclastic Marker Tuff as a boundary. The lower member (or Lower Nawata) is 137 m thick in the type section, while the upper member (or Upper Nawata) attains a thickness of 125 m in the type section (figures 2.6 and 2.7).

The Nawata Formation is a heterogeneous mix of sedimentary rocks, made up mainly of upward-fining sandstone to mudstone intervals, multistoried sandstones, conglomerates, and altered distal tephra. Many of the lithologies, particularly the tephra and mudstones, have been heavily modified by diagenesis. As the focus of interest here is in the original depositional character of these sediments, where possible they will be referred to by their primary character. For example, the Marker Tuff will be discussed as a tephra unit, even though today it retains none of the primary volcanic glass, which has been wholly replaced by clay and zeolite minerals.

Lower Nawata

The lower member of the Nawata Formation is referred to elsewhere in this volume as the Lower Nawata (lower Nawata in Leakey et al. 1996). This member is well exposed in three areas. The first is in the southeast of Lothagam, where it conformably overlies the Nabwal Arangan beds in the southern block of the horst. The second and third areas of exposure are both within the central valley of Lothagam, adjacent to the faults that bound the horst on the west. The northernmost area includes the excellent exposures associated with the Nawata drainage. It is separated from the central area of Lower Nawata exposures by a narrow belt of Upper Nawata strata, which extends across to the horst. This central area includes good exposures east of the Galana Boi beach, but to the south it is heavily mantled by Recent gravels shed off the horst. The type section of the Lower Nawata is a composite section. Exposures in the northern area best reflect the characteristic lithologies of the member, but these are faulted off against the horst. Thus a mappable lithologic couplet is used to correlate this section of middle and upper Lower Nawata strata with a basal section from the southern area of exposures that overlies the uppermost basalt of the Nabwal Arangan beds. In this composite type section (figures 2.6 and 2.7), the member attains 137 m in thickness.

Lower Nawata strata are characterized by thick- to thin-bedded conglomerates, sandstones, and mudstones. They are dominated by detritus from a volcanic source, and they have abundant intercalated altered distal tephra. In the type section, roughly 20 percent of the sequence consists of conglomerates, 34 percent

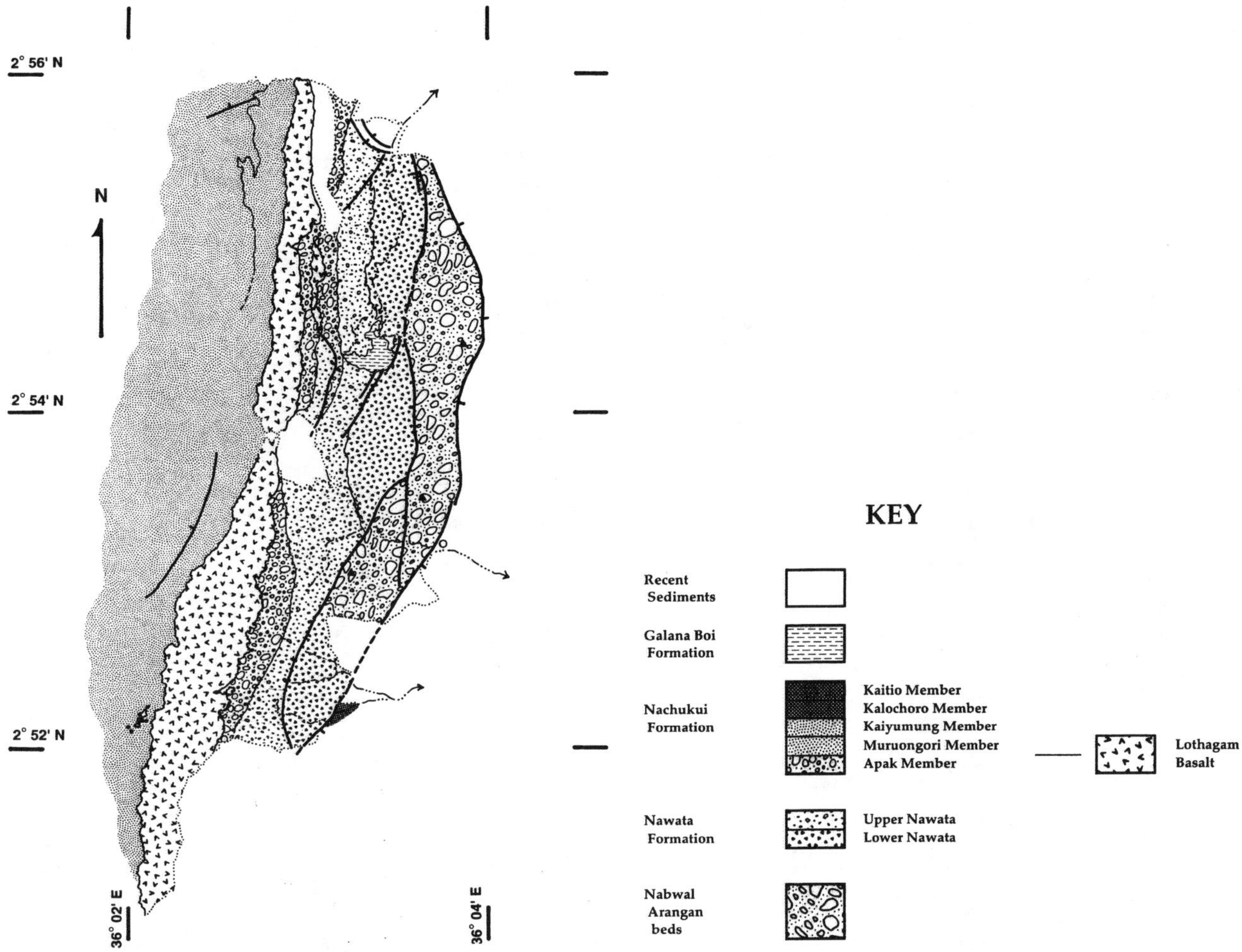

Figure 2.4 Geologic map of Lothagam, field mapped in 1991–1993 based on aerial photograph coverage flown in December 1989. Only major faults are indicated.

sandstones, 36 percent mudstones, and 10 percent altered tephra. Sandstones of the member typically display well-developed low-angle planar (epsilon) cross stratification, along with a variety of medium- to large-scale trough and planar cross bedding. Fossils of the Nile oyster, *Etheria elliptica*, occur commonly as massive reefs within channel sandstone bodies throughout the member. Mudstones of the member typically display the wedge-shaped polygons and large-scale, slickensided dish fractures of vertisols. The basal strata of the Lower Nawata include numerous volcanic-cobble conglomerates, but such coarse lithologies are not seen higher in the member. Prominent in the middle and upper parts of the member are thin, ostracod-bearing

Early Terminology		Powers (1980)		Current Terminology (Powers and Feibel)	
220 Beds				Galana Boi Formation	
				Nachukui Formation	Kaitio Member
					Kalochoro Member
Lothagam 3		Kaiyumung Formation			Kaiyumung Member
Lothagam 2		Muruongori Formation			Muruongori Member
sill					Lothagam Basalt
Lothagam 1	1C	Naiota Formation	feldspathic arenite member		Apak Member
				Nawata Formation	upper member
	1B		volcaniclastic member		lower member
	1A				
Miocene-Pliocene volcanics		Abwaria-alangan beds		Nabwal Arangan beds	

Figure 2.5 Different stratigraphic terminology used for the Lothagam deposits.

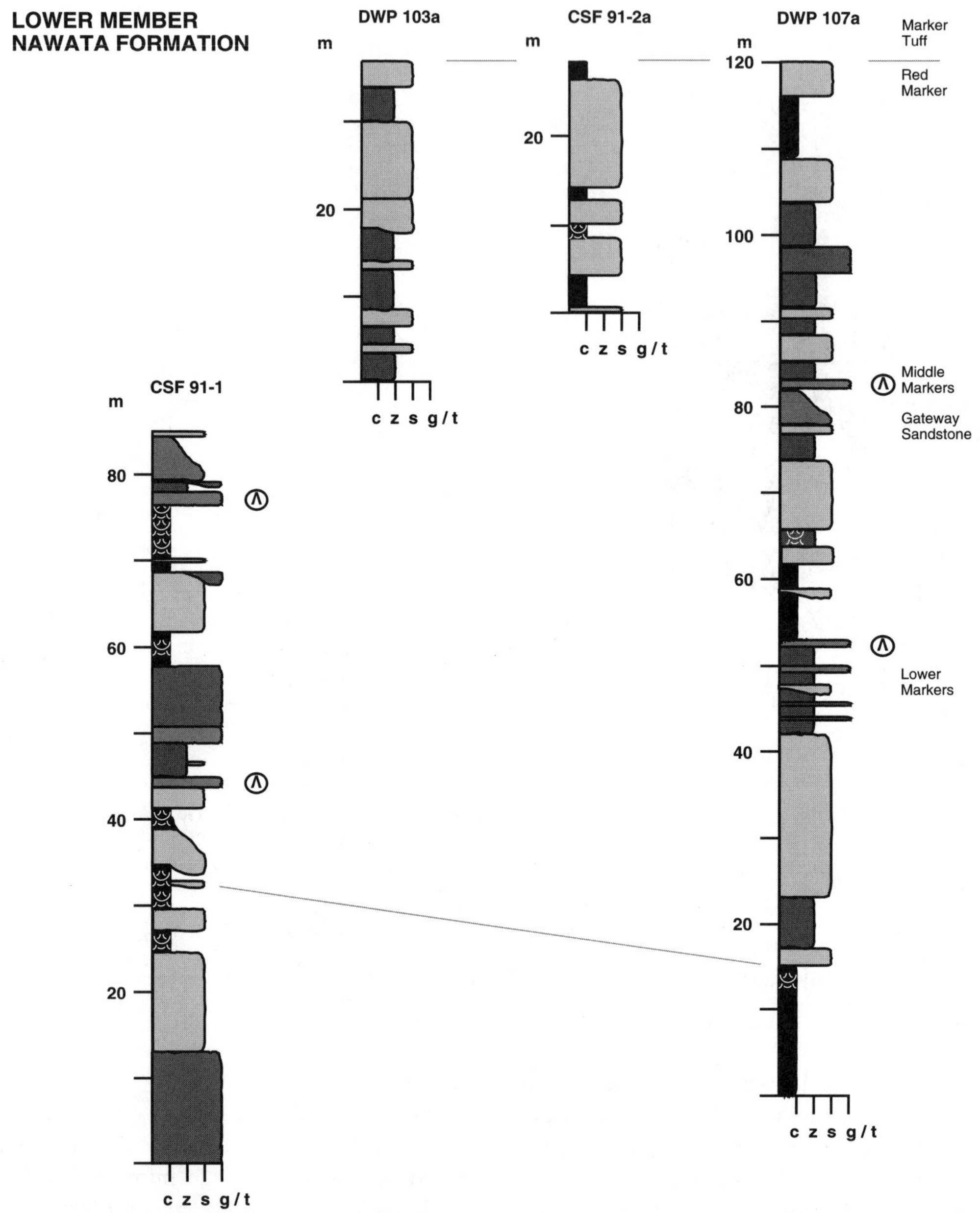

Figure 2.6 Type sections (CSF 91-1, DWP 107a) and reference sections (DWP 103a, CSF 91-2a) of the lower member of the Nawata Formation. For the key to the symbols, see figure 2.10.

limestones. The uppermost unit of the Lower Nawata is a prominent marker complex termed the Red Marker. This is a heterogeneous mix of sandstones and mudstones, which have a strong secondary component of analcime and iron oxides, resulting in its striking color. Other noteworthy marker units within the Lower Nawata include the Lower Markers (an altered tephra complex), the Gateway Sandstone, and the Middle Markers (another tephra sequence). The interval from the Lower Markers through the Marker Tuff has provided the best isotopic age control for the Nawata Formation (McDougall and Feibel 1999) and is also richly fossiliferous.

Upper Nawata

The upper member of the Nawata Formation is commonly referred to as the Upper Nawata (Leakey at al. 1996). The member is exposed as an essentially continuous north–south belt in the central valley, between exposures of the underlying Lower Nawata and the overlying Apak Member, or in fault contact with the horst. The type section of the member, in the northern part of the exposures, measures 125 m in thickness (figure 2.7). The Upper Nawata is characterized by thick, multistoried sandstone bodies, with subsidiary mudstones, and by a paucity of altered distal tephra. The type sec-

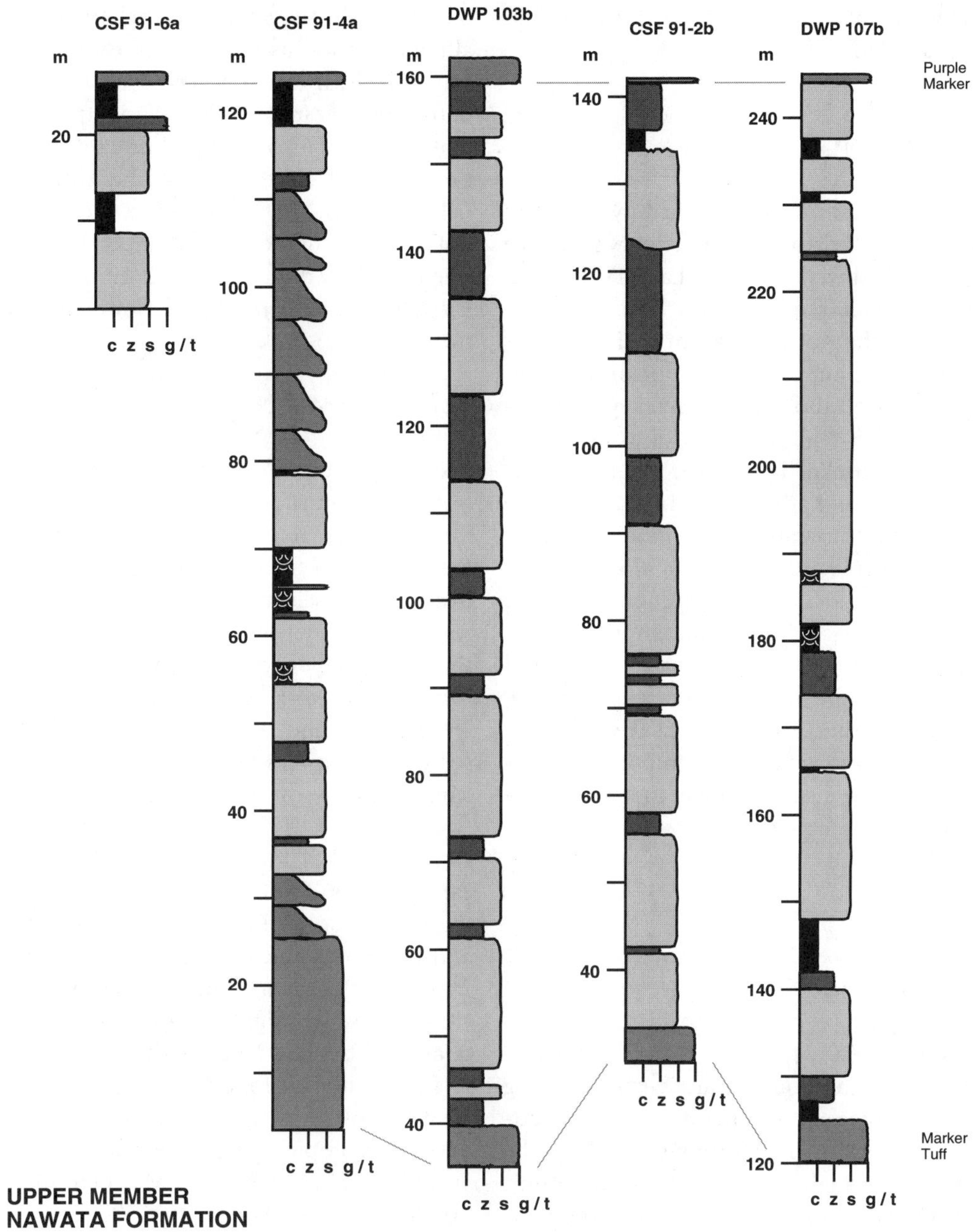

Figure 2.7 Type section (DWP 107b) and reference sections (CSF 91-6a, CSF 91-4a, DWP 103b, CSF 91-2b) of the upper member of the Nawata Formation. For the key to the symbols, see figure 2.10.

tion consists of some 75 percent sandstones, 21 percent mudstones, and 4 percent altered tephra. The Upper Nawata preserves relatively few of the *Etheria* reefs that are characteristic of the Lower Nawata, and few of the thin limestone beds. The basal unit of the member is a tephra complex termed the Marker Tuff. This complex consists of a thin-bedded, relatively homogeneous lower interval (interpreted as airfall tuff), overlain by a massive, poorly sorted sandy tuff (interpreted as a lahar). The latter unit includes rip-up blocks of the basal unit, as well as other cobble- to boulder-sized clasts floating in the tephra. Sandstones of the Upper Nawata are typically characterized by well-developed epsilon cross-stratification and trough cross-beds in the coarser units, grading into small-scale troughs and ripple marks in the overlying finer intervals, with abrupt transitions back to the coarser sands. Mudstones are subordinate, but where present they show characteristics of paleosol development. Analcime and iron oxide cementation of the mudstones is prevalent, implying a component of altered volcanic ash, but discrete tephra beds are rare. There is a change in character near the top of the member, where *Etheria* reefs and tephra deposits become more common, culminating in the prominent Purple Marker, an analcimolitized tephra unit. Vertebrate fossils are moderately abundant in the Upper Nawata, but no isotopically datable units have been recognized above the Marker Tuff.

Nachukui Formation

The Nachukui Formation was established by Harris et al. (1988) for the Plio-Pleistocene sedimentary sequence exposed near the northwest shores of Lake Turkana. A part of the Omo Group sequence, this formation correlates closely with the Koobi Fora Formation east of Lake Turkana and the Shungura Formation of the lower Omo Valley based on lithologic character, geochemically identified marker tephra, and fossil assemblages. Extension of the Nachukui Formation from its originally defined extent to Lothagam is based on three criteria: similarity in stratigraphic and lithological characteristics, tephra correlation, and chronological overlap based on faunal and isotopic age controls. Three new members of the Nachukui Formation were established by Powers and Feibel (in Leakey et al. 1996), and two previously defined members were identified in the Lothagam deposits during this study. The newly defined members of the formation extend its temporal range to greater than 4.22 Ma, while the younger members present reflect Plio-Pleistocene deposition continuing past 1.88 Ma at Lothagam.

Apak Member

The Apak Member, established by Powers and Feibel (and reported in Leakey et al. 1996) to encompass sediments overlying the Purple Marker of the Upper Nawata and beneath the Lothagam Basalt, is exposed along the western side of the central valley and as steep badlands exposures beneath the Muruongori ridge. The Apak was recognized as a discrete lithostratigraphic unit by Powers (1980) based on the predominantly quartzofeldspathic mineralogy of its sandstones. In the course of recent work it was demonstrated that the unit has an important volcanic-derived sediment component in the southern exposures and that its overall stratigraphic organization is more closely allied with that of Nachukui Formation strata. Its designation as a new member of that formation has been further corroborated by a young date of 4.22 Ma on a tephra unit within the member (McDougall and Feibel 1999).

The Apak Member is characterized by thick upward-fining cycles in the 99 m thick type section (figure 2.8).

Here the lithologies are 71 percent sandstone and 29 percent mudstone. In contrast to the underlying Upper Nawata, sand bodies are typically single-storied, dominated by a coarse quartzofeldspathic sand at the base, and fining upward to a pedogenically modified mudstone. The single isotopic age determination from the member (McDougall and Feibel 1999) comes from an altered pumiceous sand in the overbank portion of a fluvial cycle. In the southern exposures, the upper part of the member includes lacustrine strata, with algal stromatolites, coquinas, and abundant fish fossils. This probably reflects the development of the Lonyumun Lake at ca. 4.1 Ma (Feibel 1988). Mammalian fossils are not abundant in the Apak Member, but important specimens such as the Lothagam hominid mandible KNM-LT 329 have been recovered from near the base of the member.

Muruongori Member

Lacustrine strata that overlie the Lothagam Basalt are termed the Muruongori Member of the Nachukui Formation. They are exposed in a north–south belt immediately west of the basalt dipslope. These sediments overlie the basalt in a complex relationship. The basalt was initially interpreted as a sill (Patterson et al. 1970) that postdated deposition of the Muruongori sequence. More recent work suggests that it was a flow that entered a lake and disrupted unconsolidated lacustrine sediments (Powers and Feibel unpublished data). The Muruongori Member is 59 m thick in the type section (figure 2.9) and is composed primarily of claystones and recrystallized diatomites, with some sands and molluscan coquinas. The best exposures of the member are in the northwest, where the characteristic drab claystones and diatomites occur in low badlands exposures. The upper limit of the member is marked by a distinct transition to more brightly colored strata, dominated by quartzofeldspathic sandstones.

Few vertebrate fossils other than fish have been recovered from the Muruongori Member. These strata are dated by isotopic ages on the Apak Member (4.22 Ma) and Lothagam Basalt (4.20 Ma) beneath, and they are lithologically correlated with the Lonyumun Lake phase of the Turkana Basin (ca. 4.1–3.95 Ma; Feibel 1988).

Kaiyumung Member

The Kaiyumung Member of the Nachukui Formation includes the fluvial strata that conformably overlie the Muruongori. They occur in the extreme northwest of the Lothagam exposures, above the Muruongori Member, and continue to the southwest where they appear to be in fault contact with the Muruongori. This fault has brought strata higher in the member into close proximity with the underlying member. To the west, Kaiyumung strata underlie the beveled plain, but lack of exposure has precluded definition of an upper boundary for the member. In the type section, the member attains 94 m in thickness (figure 2.9). The Kaiyumung is characterized by pebble-rich quartzofeld-

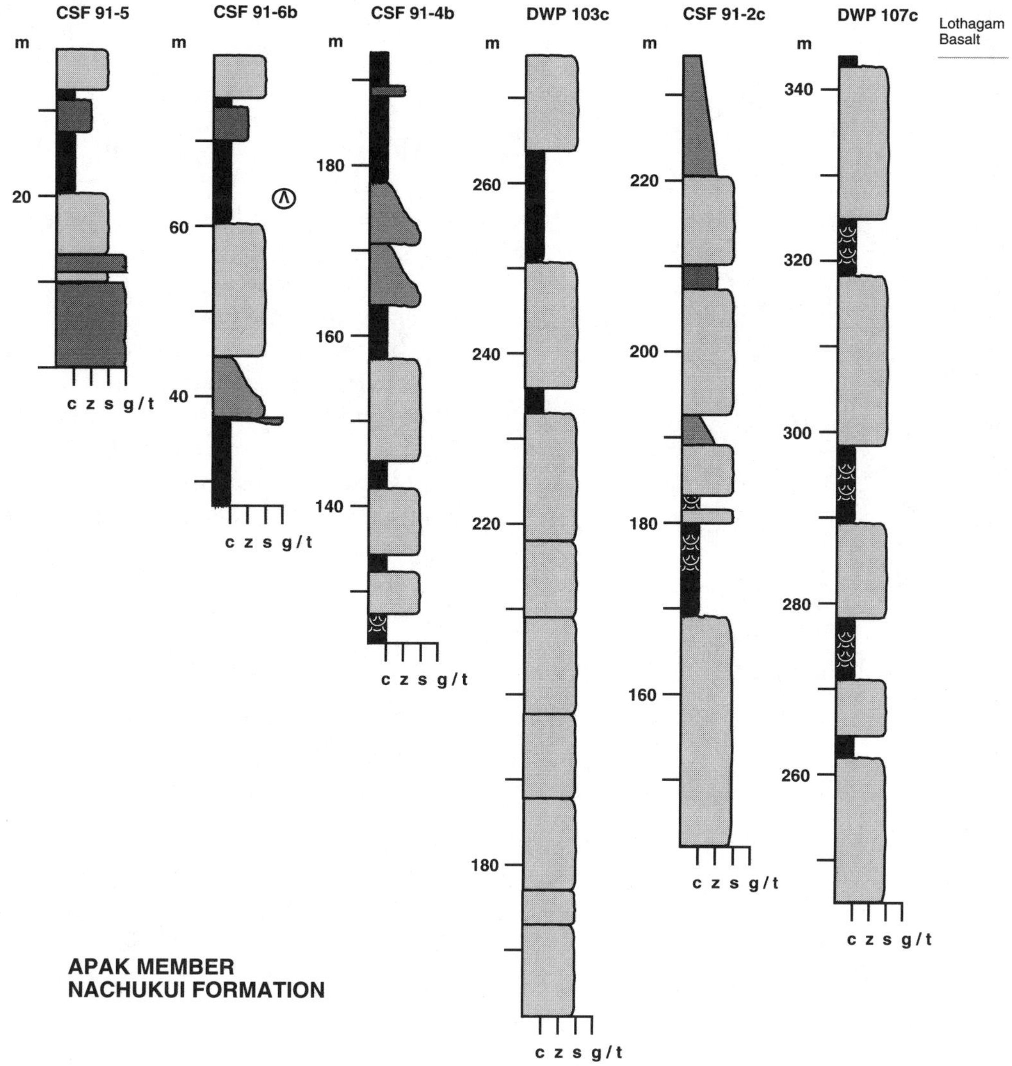

Figure 2.8 Type section (DWP 107c) and reference sections (CSF 91-5, CSF 91-6b, CSF 91-4b, DWP 103c, CSF 91-2c) of the Apak Member of the Nachukui Formation. For the key to the symbols, see figure 2.10.

spathic sandstones and mudstones with well-developed vertic features.

Kaiyumung strata are very rich in fossil vertebrates. At present, however, no isotopically datable materials have been recovered from this member, nor have tephra units been documented to allow tephrocorrelation.

Kalochoro Member

A small exposure in the southeast corner of Lothagam, on the eastern side of the horst, has been identified as belonging to the upper Kalochoro Member of the Nachukui Formation. Some 26 m of claystones, with a resistant ostracod sandstone, underlie the KBS Tuff here (figure 2.10). Kalochoro Member deposits presumably continue down section, but are presently covered by Recent dune sands. The KBS Tuff that caps the member here provides isotopic age control at 1.88 Ma (McDougall 1985). The Kalochoro Member strata at Lothagam correlate to the Lorenyang Lake phase of the Turkana Basin (Feibel 1988) and reflect deposition shortly before KBS time. No vertebrate fossils have been recovered from these exposures.

Kaitio Member

The locality in which Kalochoro Member deposits are exposed also records some 39 m of the Kaitio Member (figure 2.10). The base of this member is defined by the presence of the KBS Tuff (1.88 Ma; McDougall 1985), which occurs here as thin airfall ash in lacustrine clays. Overlying strata include prominent beds of algal stromatolites and molluscan coquinas, as well as gravel-rich quartzofeldspathic sands. The Kaitio exposures at Lothagam reflect lake-level oscillations in the later stages of the Lorenyang Lake (Feibel 1988) and probably were deposited in a short interval after the KBS Tuff was

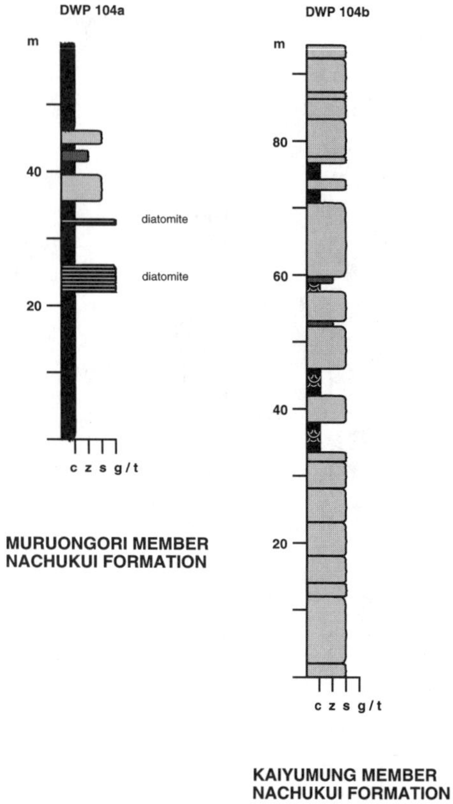

Figure 2.9 Type sections of the Muruongori (DWP 104a) and Kaiyumung (DWP 104b) Members of the Nachukui Formation. For the key to the symbols, see figure 2.10.

erupted. No vertebrate fossils have yet been recovered from this sequence.

Galana Boi Formation

A prominent landmark in the central part of the Lothagam exposures is a tombolo or beach ridge extending unconformably over the Miocene strata. It is capped with coarse gravels and includes sands and coquinas that grade laterally into diatomites. Similar deposits can be found both to the north and south, and a thin veneer of these sediments, often characterized by abundant white shells of the gastropod *Melanoides,* occurs over much of the surrounding area. These deposits are attributed to the Galana Boi Formation, of Late Pleistocene to Holocene age. The aforementioned beach ridge preserves a rich Holocene mammalian fauna and abundant fossil fish, as well as numerous Neolithic burials.

The Galana Boi Formation was defined by Owen and Renaut (1986) at Koobi Fora, where it exhibits lithologies remarkably similar to those seen at Lothagam. A section through the Galana Boi exposures at the tombolo (figure 2.11) illustrates their characteristic features. In some exposures of the Galana Boi, particularly in the southern part of Lothagam, a prominent basaltic ash is intercalated within the lacustrine sequence.

Depositional History

In reconstructing the depositional history of the Lothagam sequence, it is necessary to consider the effects of the dominant controlling factors, including tectonics (affecting basin configurations and subsidence rates), climate (primarily as a water source), sediment supply, and, ultimately, the interactions between these factors. For Lothagam, reconstruction is made more difficult by the limited spatial perspective the locality provides. Covering an area of less than 60 square kilometers, and without any solid regional correlatives until the early Pliocene, the locality provides only a limited glimpse of a much larger landscape. Still, there is much evidence in the sedimentary record that can give clues to the larger picture, as well as much detail to the smaller one.

The coarse-grained, volcanogenic alluvial material and intercalated lavas of the lower Nabwal Arangan beds indicate that an eroding but still active volcanic terrane, with significant relief, existed quite near Lothagam in the Middle Miocene. By ca. 9.1 Ma the character of the conglomerates had changed to predominantly pebble-sized material, and their close association with vertisols and lenticular sands indicates much more subdued local relief. Overall, though, the Nabwal Arangan sediments reflect a local depositional system, and the rather poor sorting of the clastics suggests only seasonal flow in these drainages.

Throughout the Lothagam area, Nawata Formation sediments reflect fluvial deposition, with evidence of channel characteristics along with variations on floodplain sedimentation such as shallow ponds. Although it persisted for nearly 4 million years, the Nawata system did not develop any lakes (at least not at Lothagam), and alluvial fans, if present, were outside the geographic locus of Lothagam itself. The fluvial system that dominated the Nawata was large and, for the most part, perennial. Unfortunately, the very limited geographic perspective of Lothagam does not reveal from where this river was coming, nor to where it was flowing. It may relate to other large fluvial systems in the region, such as the Miocene Loperot River (Mead 1975) and the Plio-Pleistocene Turkana River (Feibel 1994) that flowed through the region en route to the Indian Ocean. The dominant clastic components are volcanic throughout the formation, and further study of their petrology may ultimately lead to a better understanding of their source area.

The basal Lower Nawata strata are alluvial plain sediments. The conglomerates reflect braided channel deposition, and the poor sorting of these gravels implies

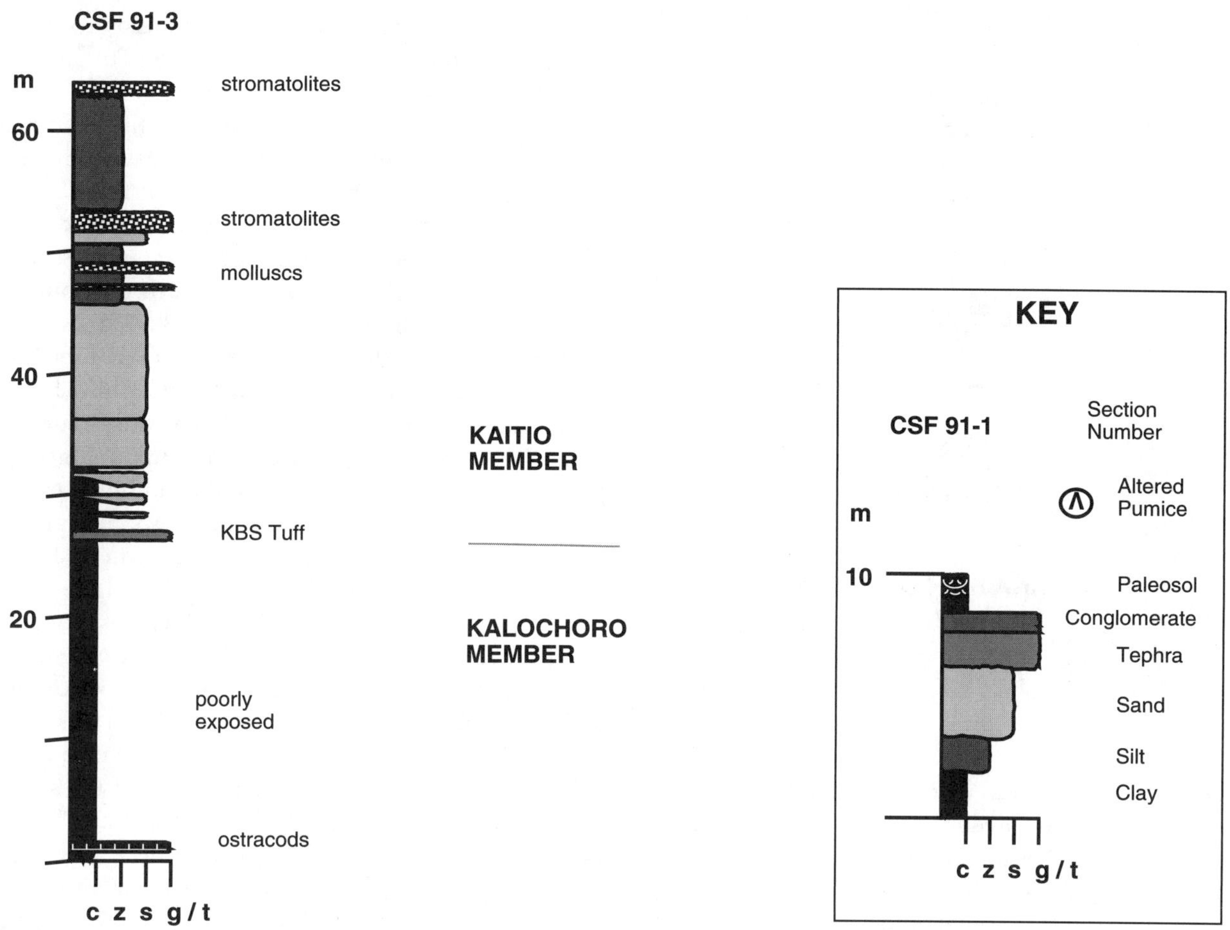

Figure 2.10 Reference sections of the Kaitio and Kalochoro Members (CSF 91-3) of the Nachukui Formation.

they were transported by an ephemeral stream. These strata interfinger, from very early Nawata times, with the characteristic Lower Nawata fluvial deposits, which reflect a very different type of river. The dominant low-angle epsilon cross-stratification indicates a meandering river, the scale of the upward-fining cycles suggests broad but shallow channels, and the abundance of *Etheria* reefs dictates a perennial flow regime. The floodplains of this river system frequently supported shallow ponds, with a characteristic ostracod–*Lanistes–Pila* community. Volcanic ash was regularly introduced into both channel and floodplain environments, either as primary airfall or as secondarily reworked material coming down the river. The essentially indistinguishable isotopic ages on three tephra layers spread through some 25 m of section in this interval (McDougall and Feibel 1999) demonstrate that sedimentation rates, particularly in the upper part of the Lower Nawata, were relatively rapid. The vertic character of the paleosols and the presence of soil carbonate nodules imply at least a pronounced dry season, while most other indicators record relatively wet conditions.

The Marker Tuff records a short-term environmental disruption as the landscape was first mantled by a thick airfall ash and then subsequently buried beneath a thick debris flow or lahar. The Upper Pond complex demonstrates that the depositional system characteristic of Lower Nawata times reestablished itself again briefly in the Upper Nawata, but conditions soon changed. The character of succeeding Upper Nawata strata indicates that subsidence rates decreased dramatically, allowing the fluvial system to regularly rework overbank deposits—recycling mud into clay pebble conglomerates or flushing them downstream to produce the characteristic multistoried sandstones. The excellent preservation of channel facies in the Upper Nawata implies that the dramatic decrease in *Etheria* reefs through the middle part of this unit is real—and probably an indicator of decreased flow, perhaps to or near a seasonal state, at this time. The reappearance of both distal tephra and *Etheria* reefs in uppermost Nawata strata suggest a return to wetter conditions and increased tectonic/volcanic activity. The Purple Marker records two significant events. This unit is the thickest of the fluvially reworked tephra, and pervasive climbing-ripple cross-lamination attests to rapid sedimentation. Thus this represented a major volcanic eruption and tephra fallout. The intense zeolitic alteration of this tuff,

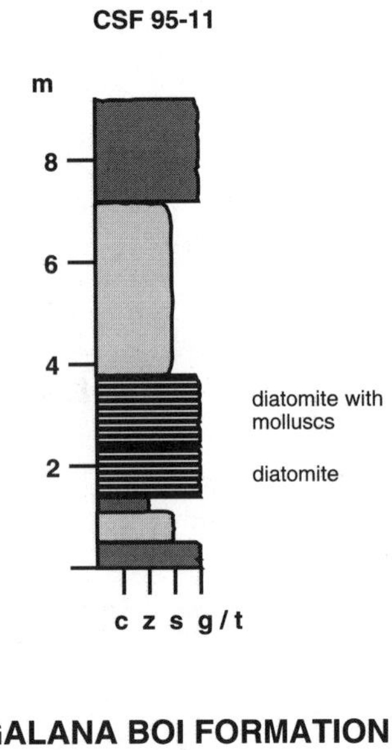

Figure 2.11 Reference section of the Galana Boi Formation at Lothagam (CSF 95-11). For the key to the symbols, see figure 2.10.

coupled with the near ubiquitous association with a dramatic shift to predominantly quartzofeldspathic sands above, implies that this tuff may mark an unconformity within the sequence. Shortly after deposition of the Purple Marker, sedimentation may have ceased for a while, and when accumulation resumed it took on a very different character.

Apak Member sediments reflect a much more complex landscape. Thick upward-fining cycles indicate fairly rapid accumulation rates on a meandering floodplain. The change in the dominant sediment type to quartzofeldspathic material may relate to a change in primary source area. *Etheria* are not recorded in Apak Member channels, which perhaps reflects a more seasonal flow regime in this river, but the well-sorted sands do not suggest strongly ephemeral conditions. The overall characteristics of the Apak Member fluvial system, in conjunction with the isotopic age data, are not incompatible with a single fluvial system being recorded at both Lothagam and Kanapoi at this time.

The shift to lacustrine sedimentation within the upper Apak Member signals the primary difference between the depositional systems in the Nawata Formation and those of the succeeding Nachukui Formation. The lacustrine transgression marks the establishment of the Lonyumun Lake (Feibel 1988), a prominent marker in early Omo Group times and part of an integrated Turkana Basin depositional system. Whereas Nawata depositional patterns may have been restricted to the Kerio half-graben, subsequent patterns relate to a much larger system. A second aspect of the early Nachukui landscape, which strays from the Nawata pattern, is recorded in the Lothagam Basalt. This flow is one of many erupted at about this time throughout the Turkana Basin (Watkins 1986; Harris et al. 1988). It flowed into the lake and interacted complexly with the lacustrine sediments already deposited there, behaving in some ways as a sill (Powers and Feibel, unpublished data).

That the Lonyumun Lake persisted is recorded by Muruongori Member deposits, which are in most ways very similar to Lonyumun Lake strata throughout the Turkana Basin (Brown and Feibel 1986; Harris et al. 1988; Leakey et al. 1995). Subsequent strata of the Kaiyumung Member record what may be a third fluvial system in the Lothagam story—the ancestral Turkwel River. Two aspects of these deposits support this hypothesis. First, the composition of Kaiyumung Member sands is comparable to that seen at South Turkwel (Ward et al. 1999), Napadet Hills, and other parts of the Turkwel system, but differs from that of the lower Kerio Valley (Kanapoi, Nakoret, Eshoa Kakurongori, Longarakak). Second, there is a distressing absence of tephra markers in the Kaiyumung, in stark contrast with the Kerio River sites of Nakoret, Eshoa Kakurongori, and Longarakak. This fact implies a shift in prevailing sediment distribution patterns, with the ancestral Turkwel pushing farther south at this time. Today, Lothagam rests on the divide between the Turkwel and Kerio River systems, so the shift need not have been dramatic, but it had a significant effect on the resulting sedimentary record.

The upper limits of the Kaiyumung Member are poorly constrained because of both poor exposure and limited investigation. Small exposures east and southwest of Lothagam, along with the Kalochoro and Kaitio Member exposures in the fault block of southeastern Lothagam, indicate that Omo Group sedimentation continued, at least episodically, through the early Pleistocene. In common with other parts of the Turkana Basin, Lothagam shows a gap in its sedimentary record from Early/Middle Pleistocene through Late Pleistocene times, when transgression of the Galana Boi lake left an extensive record. By this time, the older strata at Lothagam had been faulted, uplifted, and extensively eroded, and thus much of the movement of the footwall block along the main boundary fault at the east margin of the horst dates to this time.

Acknowledgments

This research was supported by grants from the National Science Foundation (BNS 90-07662) and the L.S.B. Leakey Foundation to the author. Fieldwork at Lothagam was made possible by extensive logistical (and moral) support from Meave Leakey. Special thanks to Harry Merrick and his Koobi Fora Field School for making field vehicles and equipment available. M. M. Smith, J. G. Wynn, T. Muthoka, Nganga Chui, and Nashon Mukongo assisted with the fieldwork and laboratory analyses. D. W. Powers supplied copies of his excellent field notes. A. K. Behrensmeyer provided useful comments and copies of her unpublished work.

References Cited

Behrensmeyer, A. K. 1976. Lothagam Hill, Kanapoi, and Ekora: A general summary of stratigraphy and faunas. In Y. Coppens, F. C. Howell, G. L. Isaac, and R. E. Leakey, eds., *Earliest Man and Environments in the Lake Rudolf Basin: Stratigraphy, Paleoecology, and Evolution*, pp. 163–170. Chicago: University of Chicago Press.

Brown, F. H., and C. S. Feibel. 1986. Revision of lithostratigraphic nomenclature in the Koobi Fora region, Kenya. *Journal of the Geological Society* (London) 143:297–310.

Feibel, C. S. 1988. Paleoenvironments from the Koobi Fora Formation, Turkana Basin, northern Kenya. Ph.D. diss., University of Utah.

Feibel, C. S. 1994. Freshwater stingrays from the Plio-Pleistocene of the Turkana Basin, Kenya and Ethiopia. *Lethaia* 26:359–366.

Harris, J. M., F. H. Brown, and M. G. Leakey. 1988. Geology and paleontology of Pliocene and Pleistocene localities west of Lake Turkana, Kenya. *Contributions in Science* 399:1–128.

Leakey, M. G., C. S. Feibel, R. L. Bernor, J. M. Harris, T. E. Cerling, K. M. Stewart, G. W. Storrs, A. Walker, L. Werdelin, and A. J. Winkler. 1996. Lothagam: A record of faunal change in the Late Miocene of East Africa. *Journal of Vertebrate Paleontology* 16:556–570.

Leakey, M. G., C. S. Feibel, I. McDougall, and A. Walker. 1995. New four-million-year-old hominid species from Kanapoi and Allia Bay, Kenya. *Nature* 376:565–571.

McDougall, I. 1985. K-Ar and $^{40}Ar/^{39}Ar$ dating of the hominid-bearing Pliocene-Pleistocene sequence at Koobi Fora, Lake Turkana, northern Kenya. *Geological Society of America Bulletin* 96:159–175.

McDougall, I., and C. S. Feibel. 1999. Numerical age control for the Miocene-Pliocene succession at Lothagam, a hominoid-bearing sequence in the northern Kenya Rift. *Journal of the Geological Society* (London) 156:731–745.

Mead, J. G. 1975. A fossil beaked whale (Cetacea: Ziphiidae) from the Miocene of Kenya. *Journal of Paleontology* 49:745–751.

Morley, C. K., W. A. Wescott, D. M. Stone, R. M. Harper, S. T. Wigger, and F. M. Karanja. 1992. Tectonic evolution of the northern Kenya Rift. *Journal of the Geological Society* (London) 149:333–348.

Owen, R. B., and R. W. Renaut. 1986. Sedimentology, stratigraphy and paleoenvironments of the Holocene Galana Boi Formation, NE Lake Turkana, Kenya. In L. E. Frostick, R. W. Renaut, I. Reid, and J. J. Tiercelin, eds., *Sedimentation in the African Rifts*, pp. 311–322. Geological Society Special Publication No. 25. Oxford: Blackwell.

Patterson, B., A. K. Behrensmeyer, and W. D. Sill. 1970. Geology and fauna of a new Pliocene locality in northwestern Kenya. *Nature* 226:918–921.

Powers, D. W. 1980. Geology of Mio-Pliocene sediments of the lower Kerio River Valley. Ph.D. diss., Princeton University.

Robbins, L. H. 1967. A recent archaeological discovery in the Turkana District of northern Kenya. *Azania* 2:1–5.

Ward, C. V., M. G. Leakey, B. Brown, F. Brown, J. Harris, and A. Walker. 1999. South Turkwel: A new Pliocene hominid site in Kenya. *Journal of Human Evolution* 36:69–95.

Watkins, R. T. 1986. Volcano-tectonic control on sedimentation in the Koobi Fora sedimentary basin, Lake Turkana. In L. E. Frostick, R. W. Renaut, I. Reid, and J. J. Tiercelin, eds., *Sedimentation in the African Rifts*, pp. 85–95. Geological Society Special Publication No. 25. Oxford: Blackwell.

2.2

Miocene and Pliocene Paleosols of Lothagam

Jonathan G. Wynn

Paleosols preserved in the Lothagam Group sediments preserve a record of the ancient environments in which they formed. The paleosol record described here spans the Miocene-Pliocene boundary and documents several types of paleosols formed in well to poorly drained alluvial settings. Ancient Vertisols throughout the sequence indicate the consistency of regular annual or semiannual dry seasons during the entire interval studied (about 9 to 4.2 Ma). Vegetation throughout the interval appears to have been a mosaic of floodplain savannas dissected by gallery woodlands. Evidence from changes in floodplain paleosol types document a period of increased aridity between about 6.7 and 5 Ma. Two very well developed Luvisols indicate extended periods of depositional stasis at about 6.5 and 5.2 Ma.

This report describes the field and laboratory characterization of paleosols collected during the 1996 field season at Lothagam. The potential for paleoenvironmental reconstruction of Lothagam paleosols is alluring because these soils provide a rare account of the landscapes in which the earliest hominids evolved. Furthermore, recent advances in the stratigraphy and dating of the Lothagam sequence have enabled a more precise evaluation of the local response to global climate change events such as the Late Miocene (Messinian) salinity crises now dated between about 6.7 and 5.5 Ma (Zhang and Scott 1996; Van Couvering et al. 1976).

Through much of the early work on the geology of Lothagam, paleosols received limited attention. Early geological accounts concentrated on the dating and stratigraphical context of fossil collections (Patterson et al. 1970; Behrensmeyer 1976). Powers (1980) provided formal stratigraphic terminology, interpreted depositional and diagenetic settings, and began to recognize pedogenic features such as carbonate nodules, ped structure, and illuviation channels. Recent work has further constrained the stratigraphy and chronology and has contributed a number of precisely dated marker units (Leakey et al. 1996; Feibel, this volume: section 2.1; McDougall and Feibel 1999). This groundwork has provided an excellent foundation for paleopedological interpretation. Research presented in this contribution offers a preliminary view of the range of paleosol types encountered in the Lothagam sequence, and interpretations of their paleoenvironments. Further, more detailed, work is in progress.

Materials and Methods

Paleosols from the Nawata Formation and Apak Member of the Nachukui Formation (figure 2.12) are reported in reference to previously measured stratigraphic sections of Powers (1980) and Feibel (unpublished sections). Superjacent strata of the Muruongori through Kaitio Members are not reported here because paleosols of these time intervals have already been described from deposits elsewhere in the Turkana Basin (Wynn 1998).

Point counts of grain size and mineralogy were made with a Swift automatic point counter using 500 points per specimen. Chemical analyses of major, minor, and trace elements were performed by x-ray fluorescence (XRF). Ferrous iron was analyzed by titrimetric methods. Clay minerals were examined by x-ray diffraction (XRD) and identified using methods outlined by Moore and Reynolds (1989). Crystallinity of clay minerals was assessed using the methods and Weaver "crystallinity index" described by Frey (1987). Profile descriptions, horizon designations, and soil terminology follow the methods of the United States Department of Agriculture (USDA) *Soil Taxonomy* (Soil Survey Staff 1975,

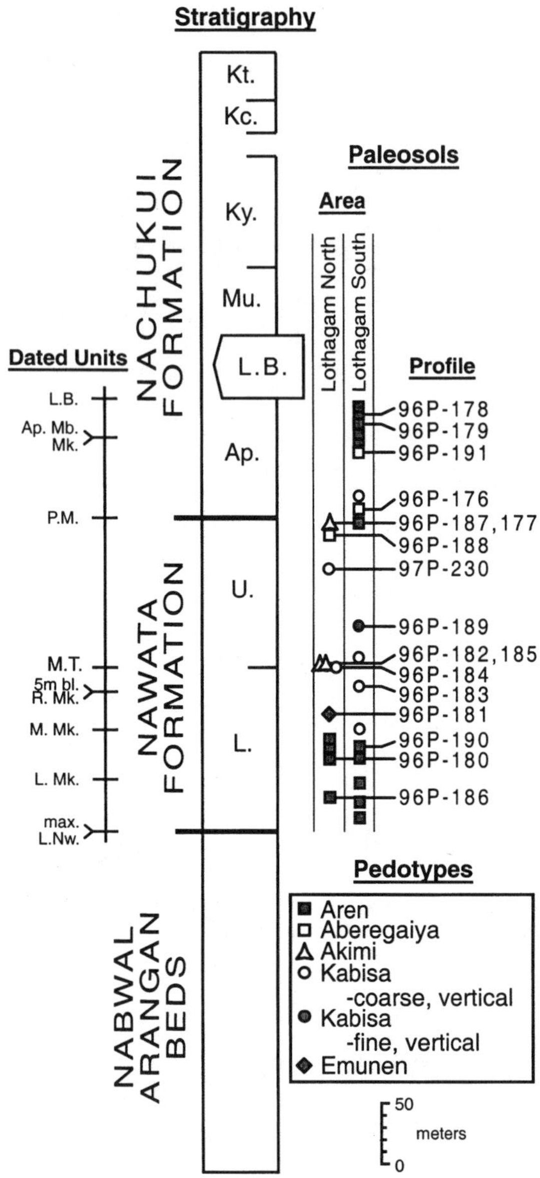

Figure 2.12 Stratigraphic location of paleosols within the Lothagam sequence. Terminology follows that of Powers and Feibel (unpublished); stratigraphic column after Leakey et al. (1996). Formation Member abbreviations are as follows: Nawata Formation, L. = Lower, U. = Upper; Nachukui Formation, Ap. = Apak, L.B. = Lothagam Basalt, Mu. = Muruongori, Ky. = Kaiyumung, Kc. = Kalochoro, Kt. = Kaitio. Abbreviations for dated units from Feibel and McDougall (1999) are: L.B. = Lothagam Basalt, Ap.Mb. Mk. = marker in the Apak Member of the Nachukui Formation, P.M. = Purple Marker, M.T. = Marker Tuff, 5m bl. R. Mk. = horizon 5 meters below the Red Marker, M. Mk. = Middle Markers, L. Mk. = Lower Markers, max. L. Nw. = maximum age of Lower Nawata (from Nabwal Arangan beds). Absolute ages of these units are discussed in Feibel and McDougall (1999). Paleosols of the Lothagam North and South areas are separated by the dune sand near Apak. Paleosols in the Lothagam North area are referenced to sections 103 and 107 of Powers (1980) and to sections CSF 92-1, 2, and 3 of Feibel (unpublished). Those of Lothagam South are referenced to section 105 of Powers (1980).

1992). Petrographic features are described according to the terminology of Brewer (1964), while those of carbonate minerals are according to Wright (1990). A pedotype approach as defined by Retallack (1994) is used to interpret the paleoenvironmental context of paleosol types (pedotypes) with regard to their preserved features. Using this nongenetic approach, distinct pedotypes can be recognized and grouped based on their physical characteristics and followed by interpretations such as classification. Each pedotype described here is classified according to two widely known soil taxonomic systems: USDA's *Soil Taxonomy* (Soil Survey Staff 1975, 1992) and the United Nations Food and Agriculture Organization's (FAO) *Soil Map of the World* (FAO 1977; Fitzpatrick 1980).

Paleosol Descriptions

Many of the Lothagam paleosols conform to previous pedotype designations from the Plio-Pleistocene of the basin (Aberegaiya and Kabisa pedotypes; Wynn 1998). Diagnostic features of these pedotypes are only briefly reviewed here for reference. Three previously unrecognized pedotypes are found in the Lothagam sequence and are presented here in full detail (Aren, Akimi, and Emunen pedotypes). Table 2.1 provides detailed descriptions of the new type profiles. These are illustrated schematically with textural, compositional, and chemical data in figure 2.13. Table 2.2 shows diagnostic features of all previously recognized and new pedotypes found at Lothagam.

Aren Clay Paleosol

The type profile of the Aren clay consists of a 3 m thick homogenous A horizon with distinct, very coarse, angular, blocky ped structure (figure 2.13, cf. figure 2.14C). Slickensided fracture planes mark the surfaces of the wedge-shaped peds. The arrangement and intersection of the fractures indicate mukkara subsurface structure as described by Paton (1974). The Aren type profile is uniformly reddish brown (2.5 YR 4/4), although other profiles vary to light brown (7.5 YR 6/4). Texture of the Aren type profile is extremely clayey and is dominated by smectite with a crystallinity index of 1.07 (Weaver Index, ratio 10 Å to 10.5 Å XRD peaks above background of potassium saturated clays) and minor illite. The clayey matrix is slightly calcareous but lacks nodular or rhizoform carbonate.

Aberegaiya Clay Paleosol

The Aberegaiya type profile consists of a series of A horizons with mukkara structure similar to that of the

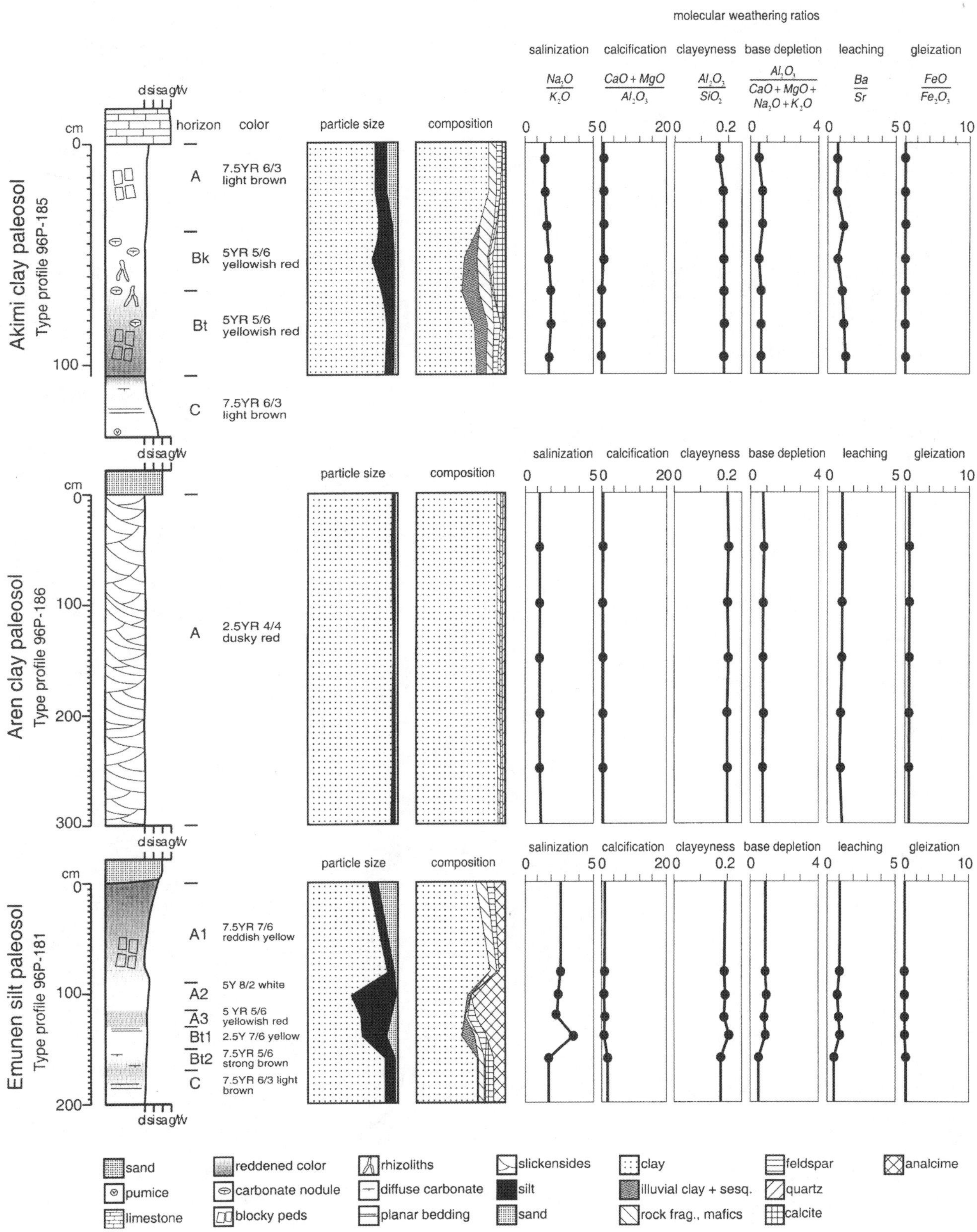

Figure 2.13 Detailed sections of pedotype type profiles. Molecular weathering ratios are plotted according to standard scales that encompass the overall variation of a wide variety of soil and paleosol types. Note that the scale of the Akimi profile differs from the scales of the Emunen and Aren profiles.

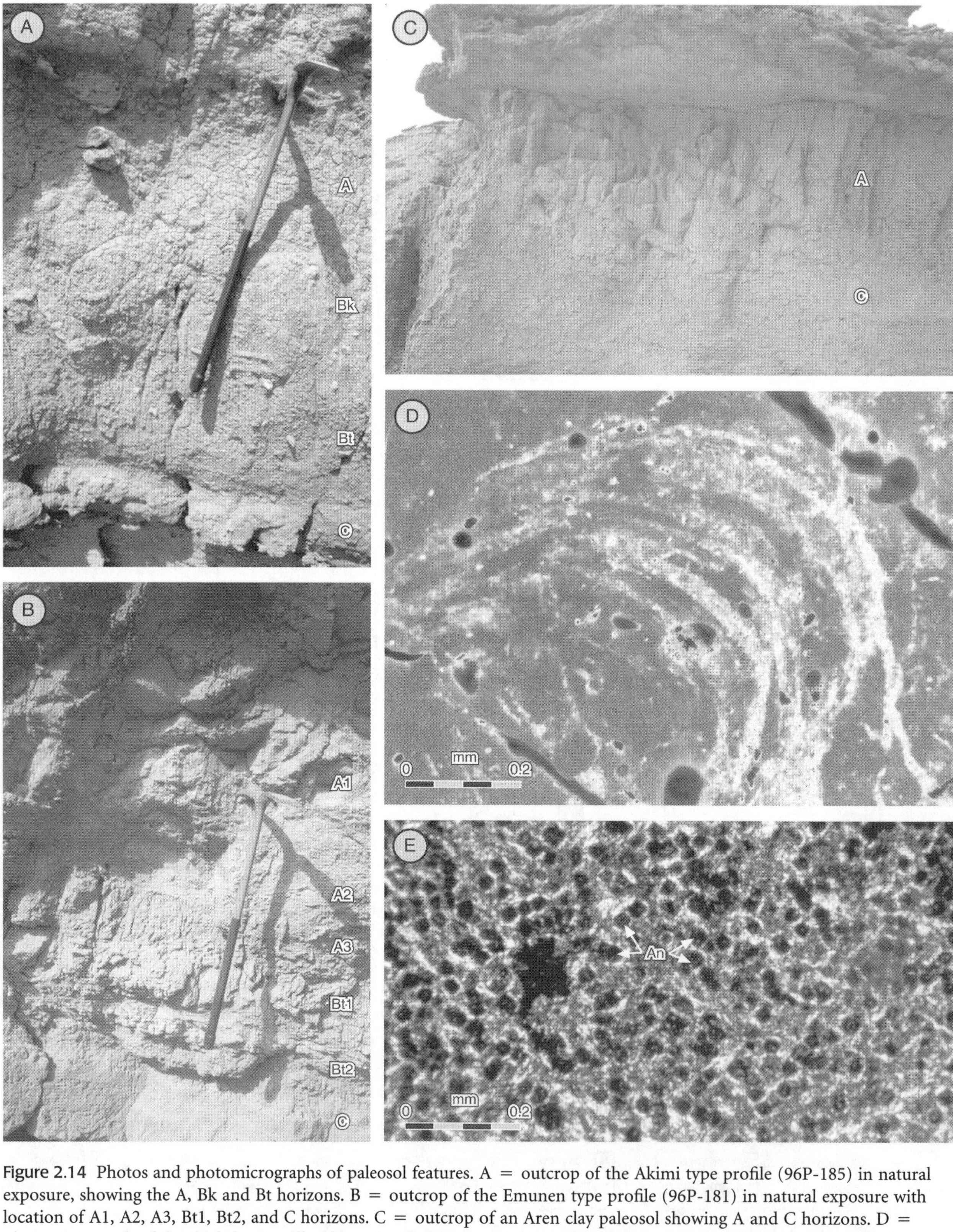

Figure 2.14 Photos and photomicrographs of paleosol features. A = outcrop of the Akimi type profile (96P-185) in natural exposure, showing the A, Bk and Bt horizons. B = outcrop of the Emunen type profile (96P-181) in natural exposure with location of A1, A2, A3, Bt1, Bt2, and C horizons. C = outcrop of an Aren clay paleosol showing A and C horizons. D = photomicrograph of the thick ferri-argillan from the Akimi paleosol that is directly below the Purple Marker (96P-187). E = photomicrograph from the horizon of the Emunen type profile (96P-181) showing the skel-lattisepic microfabric that surrounds euhedral pseudoisotropic (dark) intercalary analcime crystallaria (An). The pick handle in profile photos A and B is 65 cm long. Photomicrographs were taken under crossed nicols.

Aren described above. These paleosols are also very clayey and are dominated by smectite with minor illite. Aberegaiya paleosols differ from the Aren pedotype in having a horizon of nodular carbonate at some depth, generally less than 100 cm. Carbonate generally occurs as diffuse micritic nodules with dense microfabric, floating sand-sized grains, and circumgranular cracks (alpha fabrics of Wright 1990).

Akimi Clay Paleosol

The type profile of the Akimi pedotype consists of A, Bk, Bt, and C horizons (figures 2.13 and 2.14A). The A horizon is light brown (7.5 YR 6/3) with distinct coarse to medium blocky structure and a globular weathering surface in natural exposures. The B horizon is yellowish red (5 YR 5/6) and has a massive to coarse blocky structure. The upper B horizon is marked by calcareous rhizoliths and nodules. Thick, highly birefringent ferri-argillans are also abundant throughout the B horizon (figure 2.14D). The textural composition is very clayey, being composed of very poorly crystalline illite throughout (Weaver indices of 1.25–1.5), probably with an abundant amorphous component.

Kabisa Sand and Kabisa Silt Paleosols

The type profile of the Kabisa pedotype is dominated by sandy parent material that shows little sign of pedogenic alteration except for calcareous rhizoliths. Ped structure is massive, with relict bedding features and no distinct horizonation. Two variants of the Kabisa pedotype are recognized at Lothagam. Sandy Kabisa pedotypes with vertical rhizoliths are common paleosols that recurred throughout the Miocene to Pleistocene of the Turkana Basin (referred to as Kabisa coarse, vertical). A new variant with silty texture is recognized from a single profile at Lothagam (96P-189; Kabisa fine, vertical). Rhizoliths of both variants are vertical to subvertical forms, ranging from 2 mm to 30 cm in cross-sectional diameter. Micromorphological features of Kabisa rhizoliths include *Microcodium* grains, calcified tubules, alveolar-septal fabric, pisolitic features, and septarian cracks (beta fabrics of Wright 1990). Drab haloes (described by Retallack 1990) are common, as are outer rims of sparry calcite cemented sand.

Emunen Clay Paleosol

The type profile of the Emunen pedotype is divided into five horizons based on dramatic color variation that ranges from yellowish red (5 YR 5/6) to white (5Y 8/2; figure 2.13, 3B). Several A horizons are recognized; they have massive to blocky ped structure and variable texture from clay to sandy clay. The microfabric of the A2 horizon is dominated by silt-sized intercalary analcime crystallaria with surrounding clay minerals oriented preferentially (skel-lattisepic fabric; figure 2.14E). The Bt horizons have massive to blocky structure and clayey texture with few illuviation cutans (ferri-argillans). The Bt1 horizon contains analcime and microfabric similar to that of the A2 horizon described above.

Burial Diagenesis of the Paleosols

The most significant and problematic effect of burial diagenesis in the Lothagam paleosols is the authigenic formation of analcime seen in some of the profiles, particularly Emunen and Akimi paleosols. Post-burial authigenesis of analcime, rather than formation within the soil, is clear from its euhedral form, uniform crystalline size, and crosscutting relationships with other soil features. Analcime is thought to have formed by the reaction of saline brines with volcanic glass or phyllosilicates or both (Powers 1980), as is known to have occurred in other similar diagenetic settings (Remy and Ferrell 1989). In the case of the Emunen paleosol, the crystallization of analcime has dramatically altered the mineralogy, chemical composition, and micromorphological features throughout the soil, making it very difficult to interpret the original features of this paleosol. Curiously, those paleosols with analcime appear to contain a unique clay mineral assemblage that is dominated by very poorly crystalline illite and an amorphous component. The nature of this problem will need to be addressed in further research involving more detailed chemical, mineralogical, and textural characterization of the analcime and associated minerals.

The Lothagam paleosols have undoubtedly been affected by other postburial alterations commonly observed in pre-Quaternary paleosols, including decomposition of soil organic matter, burial reddening, compaction, and cementation (Retallack 1991). Although organic carbon contents have not been determined analytically, relatively high color values and chroma and petrographic features attest to the loss of most organic carbon originally present. Drab-haloed root traces in many of the Kabisa paleosols indicate burial gleization that accompanied decomposition of original root material. Burial reddening of fine-grained material accompanies the decomposition of organic matter and is due to the oxidation, dehydration, and recrystallization of pedogenic iron hydroxides and oxyhydrates under increased pressure and temperature. Aren paleosols are likely the most affected by burial reddening. Many similar modern noncalcareous

Vertisols are much darker than the Aren due to the complexing of organic matter and mineral components (Fitzpatrick 1980; Duchaufour 1977). Calculations based on standard compaction equations intended for clayey paleosols indicate that the Lothagam paleosols have been compacted by less than 2 percent (based on 70 percent initial solidity and burial depths between 350 and 700 m; compaction equation of Caudill et al. 1997).

Petrographic evidence also shows that much of the original void space has been filled with cementing materials, including calcite and analcime. Burial illitization of smectitic clays does not appear to have significantly affected any of the paleosols, as evidenced by the low indices of illite "crystallinity" (Kubler indices much less than 2.3; Retallack 1991).

Interpreted Classification and Paleoenvironments

Table 2.2 shows the classification of the Lothagam pedotypes according to two well-known taxonomic systems, the USDA *Soil Taxonomy* (Soil Survey Staff 1975, 1992) and the FAO *Soil Map of the World* (FAO 1977). Comparison of these classifications with local soil maps (Sombroek et al. 1982; FAO 1977) provides regional analogs to the paleosol pedotypes. Taxonomic interpretations are used in conjunction with understanding of modern soil-forming processes (Fitzpatrick 1980; Duchaufour 1977; Buol et al. 1989) in a "factor-function approach" to reconstruct the relative effects of soil-forming factors on each of the paleosol pedotypes (table 2.3).

Classification and Modern Soil Analogs

Vertic properties distinguish the Aberegaiya and Aren pedotypes as Vertisols in both the USDA and FAO systems. Modern analogs to these soils are found on floodplains of large rivers throughout semiarid to subhumid regions of East Africa where there are pronounced dry seasons and dry savanna or thornbush savanna (FAO 1977). Most appropriate analogs to the Aberegaiya paleosols are the Chromic Vertisols of the lower Omo Valley of southwest Ethiopia, which support forb-rich grasslands (Butzer 1971; Carr 1976; FAO map unit Vc26-3a). Other possible analogs include pellic Vertisols of the Tana River floodplain of Kenya with mixed grassland and bushland vegetation (Andrews et al. 1975; FAO map unit Vp49-3a).

Modern Vertisols similar to the Aren pedotype are found in subhumid lowlands of the Kano plains and the upper Athi River basin of Kenya, both of which support semideciduous tree savanna and thorntree savanna (Sombroek et al. 1982; FAO map unit Vp45-2/3a). These soils generally lack shallow calcic horizons and are formed in more humid climates (1000–1500 mm annual precipitation) than are the calcareous Vertisols of the Omo and Tana Rivers (250–1000 mm annual precipitation).

Kabisa pedotypes—which lack diagnostic horizons, have carbonate accumulations, and exist in association with channel sandstones—are distinguished as Calcaric Fluvisols and Calcaric Regosols of the FAO system (Psamments and Fluvents of the USDA system). Similar modern soils include Fluvisols and Regosols developed on young alluvial deposits throughout the lower Turkana Basin which support gallery woodland vegetation dominated by *Acacia* spp. (Hemming 1972; Lind and Morrison 1974; FAO map unit Rc22-2b).

The Akimi Bt horizon with illuvial accumulation of clay and reddened color distinguishes these paleosols as Chromic Luvisols of the FAO system, corresponding to Typic Paleustalfs or Typic Palexeralfs of the USDA system. Analogous modern soils include Chromic Luvisols developed on Precambrian basement and basic igneous rocks in the Zambezi and Luangwa River valleys of the Zambia-Rhodesian highlands. These soils support dry woodland (locally known as Miombo woodland) and dry savanna vegetation in semiarid to subhumid climates with a long annual dry season (FAO 1977; map units Lc3, 52, 53, and 54).

The weakly developed Bt horizon of the Emunen paleosol suggests classification as Chromic Cambisols of the FAO system, corresponding to Ochrepts of the USDA system. Modern analogs to these soils include Chromic Cambisols developed on Cenozoic volcanics in the upper Tana River valley (FAO map unit Bc14-2bc). These relatively young soils support semideciduous tree savanna and thornbush tree savanna in semiarid seasonal climates.

Paleoclimate

One of the most explicit paleoclimatic interpretations of the Lothagam paleosols indicates that the entire sequence of soils must have experienced pronounced annual or semiannual dry seasons. Vertic features of the Aren and Aberegaiya paleosols are known to be direct products of strongly seasonal climates with a dry season of at least 4 months (Young 1976). Formation of illuviation cutans in the Akimi Luvisols and Emunen Cambisols also requires a distinct dry season to allow fine particles to dehydrate and attach to ped surfaces (Fitzpatrick 1980). Likewise, Kabisa paleosols attest to a pronounced dry season during which vertical roots tap into deep groundwater (Cohen 1982).

There are also indications for a relatively dry interval of soil formation from about 6.7 Ma to sometime after 5 Ma (from the Nawata Middle Markers through the lower Apak Member of the Nachukui Formation). During this interval, Aren paleosols are absent or are replaced by Aberegaiya paleosols in similar paleotopographic settings and parent material. This change coincides with a decrease in annual moisture below a threshold of about 1000 mm/yr. This interval is also marked by other paleosols that are indicative of drier conditions, including Kabisa, Akimi, and Emunen paleosols, all of which represent conditions of less than 1000 mm annual precipitation.

Ancient Vegetation

Indications of ancient vegetation are obtained by comparison of the Lothagam paleosols to their modern counterparts. The early sequence (prior to about 6.7 Ma) consists of recurring Aren paleosols on floodplain clays; these sites represent relatively lush semideciduous and thorntree savanna. The existence of a perennial channel during this interval is known from the sedimentology (Leakey et al. 1996), but unfortunately this channel does not appear to be represented in the paleosol record, probably due to the erosional dynamics of large, meandering river systems. The soils of this river system would likely have supported lush gallery woodland, as is evidenced by the interpretations of Leakey et al. (1996).

The dry episode beginning at about 6.7 Ma, as described above, would have been marked by a change in floodplain vegetation to dry savanna and thornbush savanna on Aberegaiya paleosols. These savannas were dissected by gallery woodlands developed on Kabisa paleosols near what were most likely ephemeral channels. Within this interval, Akimi and Emunen soils supported dry woodland and wooded savanna in moderate- to well-drained settings. Sometime after about 5 Ma, the return to tree savannas similar to those of the Lower Nawata was marked by the reestablishment of Aren paleosols in the upper Apak Member.

Paleotopographical Setting

The Lothagam paleosols formed in an alluvial valley with an overall character dominated by a large river with broad, low-gradient floodplains. Poorly drained conditions on the floodplains are indicated by Vertisols of the Aren and Aberegaiya pedotypes. Kabisa sand paleosols formed in well-drained conditions with a permanent but seasonally fluctuating water table, such as in and around ephemeral channels or on levees of perennial channels. Akimi and Emunen paleosols occupied moderate- to well-drained conditions that may have been provided by alluvial fans.

Parent Material

Paleosols of the Lower Nawata Member are formed in predominantly alkaline volcaniclastics, while those of the Upper Nawata and Apak Members are progressively derived from more metamorphic terranes. Abundant analcime in the Emunen paleosol indicates the presence of fresh volcanic material before secondary diagenesis. Most of the paleosols described here formed in mature clayey alluvial deposits, with the exception of the Kabisa sand paleosols, which formed in coarse-grained juvenile alluvium. The predominantly smectitic clays in nearly all fine-grained paleosols must have been weathered from alkaline igneous rocks in upland paleosols which are not preserved in the local depositional setting.

Time for Formation

The features of the Akimi clay paleosols are indicative of remarkably well developed Paleustalfs or Palexeralfs in USDA *Soil Taxonomy*, and these soils probably represent several hundred thousand years of formation. Thick illuviation cutans of profile 96P-187 show alternating bands of iron-oxide rich and depleted layers (figure 2.14D) that may indicate several small-scale climatic fluctuations during its time of formation, perhaps confirming an extremely long period of pedogenesis that encompasses a number of Milankovitch scale climatic cycles. Kabisa and Emunen paleosols are weakly to moderately developed, which indicates pedogenic intervals on the order of hundreds of years.

Conclusions

Preliminary interpretations of the Lothagam paleosols have begun to unveil paleoclimatic and paleovegetational trends in the Late Miocene of East Africa. Changes in the types of floodplain paleosols found in the Lower Nawata Formation through the Apak Member of the Nachukui Formation indicate that a drying episode began about 6.7 Ma and ended sometime after 5 Ma, coincident with recent dates of the Messinian global cooling and salinity crises (Zhang and Scott 1996). This dry interval was accompanied by changes in floodplain vegetation from tree savanna to dry or thornbush savanna. Two well-developed paleosols with illuviation features are indications that extended periods of depositional stasis occurred shortly after the deposition of the Marker Tuff (6.54 ± 0.04 Ma) and at the boundary between the Nawata and Nachukui Formations (possibly near 5.23 Ma).

Acknowledgments

I would like to thank Meave Leakey and the National Museums of Kenya Division of Palaeontology for guidance and logistical support for work at Lothagam. Craig Feibel provided preliminary samples and helpful discussion of the geological context of the paleosols. Justus Edung and Simon Milledge provided valued companionship and aid in the field. Ian Betteridge aided in the preparation of thin sections. Clifford Ambers provided direction in XRD analysis.

References Cited

Andrews, P., C. P. Groves, and J. F. M. Horne. 1975. Ecology of the lower Tana River floodplain. *Journal of the East Africa Natural History Society and National Museum* 151:1–31.

Behrensmeyer, A. K. 1976. Lothagam Hill, Kanapoi, and Ekora: A general summary of stratigraphy and faunas. In Y. Coppens, F. C. Howell, G. L. Isaac, and R. E. Leakey, eds., *Earliest Man and Environments in the Lake Rudolf Basin: Stratigraphy, Paleoecology, and Evolution,* pp. 163–170. Chicago: University of Chicago Press.

Brewer, R. 1964. *Fabric and Mineral Analysis of Soils.* New York: Wiley.

Buol, S. W., F. D. Hole, and R. J. McCracken. 1989. *Soil Genesis and Classification.* Ames: Iowa State University Press.

Butzer, K. W. 1971. *Recent History of an Ethiopian Delta.* Department of Geography Research Paper No. 136. Chicago: University of Chicago Press.

Carr, C. J. 1976. Plant ecological variation and patterns in the lower Omo Basin. In Y. Coppens, F. C. Howell, G. L. Isaac, and R. E. Leakey, eds., *Earliest Man and Environments in the Lake Rudolf Basin: Stratigraphy, Paleoecology, and Evolution,* pp. 432–467. Chicago: University of Chicago Press.

Caudill, M. R., S. G. Driese, and C. I. Mora. 1997. Physical compaction of Vertic Paleosols: Implications for burial diagenesis and paleoprecipitation estimates. *Sedimentology* 44:673–685.

Cohen, A. S. 1982. Paleoenvironments of root casts from the Koobi Fora Formation, Kenya. *Journal of Sedimentary Petrology* 52:401–414.

Duchaufour, P. 1977. *Pedology: Pedogenesis and Classification.* Translated by T. R. Paton. London: George Allen and Unwin.

FAO. 1977. *Soil Map of the World.* Vol. 6. *Africa.* Paris: Unesco.

Fitzpatrick, E. A. 1980. *Soils: Their Formation, Classification, and Distribution.* London: Longman.

Frey, M. 1987. Very low-grade metamorphism of clastic sedimentary rocks. In M. Frey, ed., *Low Temperature Metamorphism,* pp. 9–58. Glasgow: Blackie.

Hemming, C. F. 1972. The South Turkana expedition, 8. The ecology of South Turkana: A reconnaissance classification. *Geographical Journal* 138:15–40.

Leakey, M. G., C. S. Feibel, R. L. Bernor, J. M. Harris, T. E. Cerling, K. M. Stewart, G. W. Storrs, A. Walker, L. Werdelin, and A. J. Winkler. 1996. Lothagam: A record of faunal change in the Late Miocene of East Africa. *Journal of Vertebrate Paleontology* 16:556–570.

Lind, E. M., and M. E. S. Morrison. 1974. *East African Vegetation.* London: Longman.

McDougall, I., and C. S. Feibel. 1999. Numerical age control for the Miocene-Pliocene succession at Lothagam, northern Kenya. *Journal of the Geological Society* (London) 156: 731–745.

Moore, D. M., and R. C. Reynolds. 1989. *X-ray Diffraction and the Identification and Analysis of Clay Minerals.* Oxford: Oxford University Press.

Paton, T. R. 1974. Origin and terminology for gilgai in Australia. *Geoderma* 11:221–242.

Patterson, B., A. K. Behrensmeyer, and W. D. Sill. 1970. Geology and fauna of a new Pliocene locality in northwestern Kenya. *Nature* 226:918–921.

Powers, D. W. 1980. Geology of Mio-Pliocene sediments of the lower Kerio River Valley, Kenya. Ph.D. diss., Princeton University.

Remy, R. R., and R. E. Ferrell. 1989. Distribution and origin of analcime in marginal lacustrine mudstones of the Green River Formation, south-central Uinta Basin, Utah. *Clays and Clay Minerals* 37:419–432.

Retallack, G. J. 1990. *Soils of the Past.* London: Unwin-Hyman.

Retallack, G. J. 1991. Untangling the effects of burial alteration and ancient soil formation. *Annual Reviews of Earth and Planetary Science* 19:183–206.

Retallack, G. J. 1994. A pedotype approach to Latest Cretaceous and Earliest Tertiary paleosols in eastern Montana. *Geological Society of America Bulletin* 106:1377–1397.

Soil Survey Staff. 1975. *Soil Taxonomy.* Washington, D.C.: U.S. Department of Agriculture, Government Printing Office.

Soil Survey Staff. 1992. *Keys to Soil Taxonomy.* SMSS Technical Monograph No. 19. Blacksburg, Va.: Pocahontas Press.

Sombroek, W. G., H. M. H. Braun, and B. J. A. van der Pouw. 1982. *Exploratory Soil Map and Agro-climatic Zone Map of Kenya.* Nairobi: Kenya Soil Survey.

Van Couvering, J. A., W. A. Berggren, R. E. Drake, E. Aguirre, and G. H. Curtis. 1976. The terminal Miocene events. *Marine Micropaleontology* 1:263–286.

Wright, V. P. 1990. A micromorphological classification of fossil and recent calcic and petrocalcic microstructures. In L. A. Douglass, ed., *Soil Micromorphology: A Basic and Applied Science,* pp. 401–407. Amsterdam: Elsevier.

Wynn, J. G. 1998. Paleopedological characteristics associated with intervals of environmental change from the Neogene Turkana Basin, northern Kenya. M.S. thesis, University of Utah.

Young, A. 1976. *Tropical Soils and Soil Survey.* Cambridge: Cambridge University Press.

Zhang, J., and D. B. Scott. 1996. Messinian deep-water turbidites and glacioeustatic sea-level changes in the North Atlantic: Linkage to the Mediterranean salinity crisis. *Paleoceanography* 11:277–297.

TABLE 2.1 Pedological Description for Type Profiles of New Lothagam Pedotypes

Horizon[a]	Depth (cm)	Texture	Color[b]	Other Features	Micromorphology[c]
Type Aren clay paleosol					
A	0–300+	Clay	2.5 YR 4/4 reddish brown	Abrupt smooth contact to overlying sediment; large very angular blocky structure defined by slickensided fracture surfaces; mildly calcareous matrix, lacking nodules or rhizoliths; poorly crystalline smectitic clay	Porphyryskelic argillasepic, to weakly vosepic
Type Akimi clay paleosol					
A	0–40	Clay	7.5 YR 6/3 light brown	Abrupt smooth contact to overlying horizon; large subangular blocky structure; non-calcareous, poorly crystalline illitic clay; diffuse contact to underlying horizon	Porphyskelic argillasepic; few discrete, round to irregular micritic and microsparry carbonate nodules (0.2–1 mm dia.) with sharp boundaries and dense microfabric (alpha-fabric); nodules increase with depth; few sesqui-argillans up to 0.4 mm thick, notably near lower contact; numerous argillic papules (0.2–0.5 mm dia.)
Bk	40–67	Clay	5 YR 5/6 yellowish red	Massive to blocky structure; numerous carbonate nodules and rhizoliths; reddened argillaceous illuviation cutans common; very gradual smooth contact to underlying horizon	Porphyryskelic vosepic and mosepic; common carbonate nodules as above; few argillic papules as above, especially near upper contact; common sesqui-argillans as above
Bt	67–105	Clay	5YR 5/6 yellowish red	Massive to blocky structure; reddened argillaceous illuviation cutans common; abrupt irregular contact to underlying horizon	Porphyryskelic masepic to omnisepic, few carbonate nodules as above; abundant sesqui-argillans as above; common irregular sesquioxidic nodules with distinct boundaries
C	105+	Sandy clay	7.5YR 6/3 light brown	Massive structure, sparry calcareous cement	—
Type Emunen clay paleosol					
A1	0–90	Clay to sandy clay	7.5 YR 7/6 reddish yellow	Abrupt smooth contact to overlying sediment; massive in upper 60 cm, becoming coarse blocky in lower 30 cm; coarsening upward to sandy clay, clear smooth contact to underlying horizon	Porphyskelic masepic, vosepic and skelsepic

continued

TABLE 2.1 Pedological Description for Type Profiles of New Lothagam Pedotypes (Continued)

Horizon[a]	Depth (cm)	Texture	Color[b]	Other Features	Micromorphology[c]
Type Emunen clay paleosol					
A2	90–115	Silty clay	5 YR 8/2 white	Massive to coarse blocky structure, clear smooth contact to underlying horizon; sharp smooth contact to underlying horizon	Skelsepic to clinobimasepic; plasma separations surround numerous analcimitic crystallaria of uniform size (0.02–0.03 mm) and euhedral, equant shape
A3	115–130	Clay	5 YR 5/6 yellowish red	Massive to coarse blocky structure; with globular surface exposure, similar to overlying horizon with reddened color, gradual smooth contact to underlying horizon	Masepic to skelsepic; crystallaria as above
Bt1	130–150	Clay	2.5 YR 7/6 yellow	Massive to blocky structure; globular natural surface exposure; relict bedding in upper 10–15 cm; thin lenses of brown clay	Porhyroskelic clinobimasepic; analcimolitic crystallaria similar to above, but smaller (0.005–0.01 mm); common sesqui-argillans up to 0.5 mm thick
Bt2	150–170	Clay	7.5 YR 5/6 strong brown	Fine to medium blocky structure; globular natural exposure; sharp smooth contact to underlying material	Porphyroskelic insepic; few thin (0.03 mm) neoferrans

[a] Horizon designations and textural classification follow those of the Soil Survey Staff (1992).

[b] Color terminology follows that of Munsell (1990).

[c] Micromorphologic terminology follows those of Brewer (1964) and Wright (1990).

TABLE 2.2 Diagnostic Features and Classification of Paleosol Pedotypes at Lothagam

Pedotype and Type Profile	Name Derivation	Air Photo Coordinates[a]	Diagnosis	Interpreted USDA Classification	Interpreted FAO Classification
Aren 96P-186	Turkana word for "red"	0775/168-055	Physically homogenous horizon of red clay (redder than hue 5YR) with angular blocky structure and wedge-shaped peds defined by numerous intersecting slickensided fracture planes; lacking horizon of nodular carbonate	Typic Haplusterts (Typic Haplotorrerts) (Typic Haploxererts)	Chromic Vertisols (Pellic Vertisols)
Aberegaiya 93P-112	Previously defined for Plio-Pleistocene paleosols, after the location of the type profile	6911/105-125 East Turkana	Thick horizon or sequence of horizons with angular blocky structure, wedge-shaped peds, and abundant intersecting slickensided fracture planes; horizon of nodular carbonate at some depth	Typic Calciusterts (Typic Calcitorrerts) (Typic Calcixererts)	Chromic Vertisols (Pellic Vertisols)
Kabisa (coarse, vertical) 93P-116B	Previously defined for Plio-Pleistocene paleosols, after the location of the type profile	6740/131-160 East Turkana	Horizon of abundant vertical calcareous rhizoliths within coarse grained parent material	Psamments	Calcaric Fluvisols
Kabisa (fine, vertical) 96P-189	Previously defined for Plio-Pleistocene paleosols, after the location of the type profile	0773/115-081	Horizon of abundant vertical calcareous rhizoliths within fine grained parent material	Fluvents (Orthents)	Calcaric Regosols
Akimi 96P-185	Turkana word for "fire"	0775/160-107	Reddened subsurface horizon of illuvial clay (argillic horizon) and sesquioxides	Typic Paleustalfs (Typic Palexeralfs)	Chromic Luvisols
Emunen 96P-181	Turkana word for "color"	0775/161-108	Pale A horizon underlain by B horizon with little illuvual clay (cambic horizon)	Ustochrepts (Xerochrepts)	Chromic Cambisols

[a] Locations are given with respect to a standard reference system for air photos housed at the National Museums of Kenya. Format for these air photo coordinates is: air photo number / *x-y*, where the *x* and *y* coordinates are measured in mm from the photo margins.

TABLE 2.3 Interpretation of Paleoenvironmental Factors Represented by the Pedotypes

Pedotype	Paleoclimate	Organisms	Topography	Parent Matter	Time
Aren	Subhumid, seasonal moisture regime (1,000–1,500 mm/yr); at least one annual dry season greater than 4 months	Grassland to sparsely wooded grassland; units 4a and 4b of F.A.O. map (semi-deciduous tree savanna and open rain-green thorntree savanna)	Poor drainage (slope less than 3 degrees)	Smectitic clay with minor sand and silt	Weakly to moderately developed; greater than 500 yr
Aberegaiya	Semi-arid, seasonal moisture regime (250–1,000 mm/yr); at least one annual dry season greater than 4 months	Grassland to sparsely wooded grassland; units 4e and 4g of F.A.O. map (dry savanna and thornbrush savanna)	Poor drainage (slope less than 3 degrees)	Smectitic clay with minor sand and silt	Weakly to moderately developed; greater than 500 yr
Kabisa (vertical, both fine and coarse)	Arid to semi-arid, seasonal moisture; nearby source of groundwater	Large trees with vertical tap roots; generally gallery woodland	Well-drained, near channel environment with permanent water table	Well-sorted channel or beach sand	Very weakly to weakly developed; less than 1,000 yrs
Akimi	Semi-arid, seasonal moisture (500–1,000 mm/yr); pronounced dry season	Rain-green dry forests including Myombo dry woodland or dry wooded savanna; units 2c, 2d, and 4e of F.A.O. map (large leaved rain-green dry forest, small leaved rain-green dry forest and dry savanna)	Well-drained to moderately well-drained, gentle slopes	Alluvium	Strongly developed; greater than 100,000 yrs
Emunen	Semi-arid, seasonal moisture (500–1,000 mm/yr); likely pronounced dry season	Thornbush savanna; unit 4g of F.A.O. map	Well-drained to moderately well-drained, gentle slopes	Alluvium likely rich in volcanic glass	Weakly developed; less than 1,000 yrs

2.3

Numerical Age Control for the Miocene-Pliocene Succession at Lothagam, a Hominoid-bearing Sequence in the Northern Kenya Rift

Ian McDougall and Craig S. Feibel

Lothagam, located west of Lake Turkana in northern Kenya, is an uplifted fault block comprising a gently westward dipping sequence of volcanic and sedimentary rocks. The lower part of the sequence, lavas and coarse volcaniclastic sediments of the Nabwal Arangan beds, was deposited mainly between 14 Ma and 12 Ma (Middle Miocene), although the uppermost basalt has a K-Ar age of 9.1 Ma. The overlying fluvial sediments of the lower Nawata Formation have yielded ages on five tuffaceous horizons ranging from 7.4 ± 0.1 to 6.5 ± 0.1 Ma, Late Miocene. A tuffaceous horizon in the overlying Apak Member of the Nachukui Formation, yields an age of 4.22 ± 0.03 Ma; ^{40}Ar-^{39}Ar age spectra on the succeeding Lothagam Basalt indicate an age of 4.20 ± 0.03 Ma for its eruption. Much of the rich faunal assemblage from the Nawata Formation derives from the tightly dated lower intervals. Two hominoid teeth from higher in the formation can only be constrained to lie between 6.5 and 5 Ma old. The hominoid mandible, KNM-LT 329, from the lower Apak Member, is older than 4.2 Ma and younger than 5.0 Ma.

Lothagam, located at about 2° 54′ N and 36° 03′ E, is a westward tilted (~10°), uplifted fault block, about 11 by 4 km, rising out of the sandy plains of northern Kenya, just to the west of Lake Turkana (figure 2.15). Lothagam lies within the major region of extension in the northern part of the Kenya Rift (Morley et al. 1992). Volcanism began in this region about 30 Ma ago in the Oligocene (Zanettin et al. 1983; McDougall and Watkins 1988; Boschetto et al. 1992), heralding the active rifting which continues today. From seismic and other evidence, Morley et al. (1992) estimated that extension across the Kenya Rift in the Turkana region has been between 25 and 40 km since rifting began, and that most of this extension has taken place along northerly-trending normal faults, often arranged en echelon. As a result of this extension, quite large half-graben basins formed in which there are substantial accumulations of sediments and volcanics. Seismic reflection data from Lake Turkana show that this kind of extension and faulting remains active, producing half-grabens, where sedimentation currently is focused (Dunkelman et al. 1988, 1989). Movement along the larger normal faults commonly leads to uplifting of their footwalls in flexural isostatic response (Buck 1988; Morley 1989; Morley et al. 1990, 1992). Thus the Lapurr Range, to the west of Lake Turkana, is interpreted to be an uplifted footwall block bounded to the east by a major northerly-trending extensional fault (Morley et al. 1992). On a very much smaller scale, Lothagam is believed to be a similar uplifted tilted block, exposing a sequence previously deposited in the Kerio Basin, a westerly thickening half-graben basin (Morley et al. 1992). Lothagam is bounded on its eastern margin by a major extensional fault dipping to the east, with Lothagam itself cut by a number of westerly dipping normal faults. A somewhat similar uplifted block occurs at Lothidok, about 30 km north of Lothagam, also exposing a sequence that in part is like that at Lothagam (Boschetto et al. 1992).

In Lothagam a succession about 900 m thick is exposed, comprising a lowermost sequence of volcanic and coarse volcaniclastic rocks, followed by a predominantly sedimentary sequence consisting of sandstones, siltstones, and mudstones.

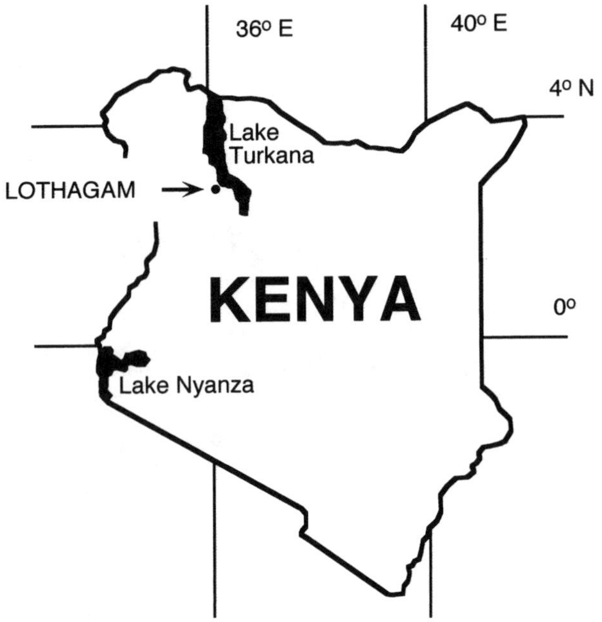

Figure 2.15 Map showing location of Lothagam, west of Lake Turkana in northern Kenya.

A very rich and diverse fauna has been recovered from the Lothagam succession; over 2,000 vertebrate fossils have been accessioned into the collections of the National Museums of Kenya (Leakey et al. 1996; Smart 1976), including a fragmentary hominoid mandible (Patterson et al. 1970). This fauna is of considerable significance as it documents a turnover from Late Miocene forest communities to the early inhabitants of the Plio-Pleistocene in more open bush and woodland. Although there is a general consensus that the Lothagam sequence is Miocene to Pliocene, and that much of the fossil material is Late Miocene, there is limited information as to the age of the succession or its time span.

In this paper we present K-Ar and ^{40}Ar-^{39}Ar data obtained on rocks from Lothagam, providing numerical age control for parts of the succession, which represents about 10 Ma of geological history, commencing about 14 Ma ago in mid-Miocene times. This work has been facilitated by the detailed stratigraphic studies and systematic fossil collecting undertaken over the last decade under the auspices of the National Museums of Kenya. In addition, the present study provides information relating to the interpretation of the units identified in seismic profiles from throughout the area, and it documents part of the extensional history of the northern Kenya Rift.

Geology of Lothagam

Following the initial description of the geology of Lothagam by Patterson et al. (1970), and further documentation by Behrensmeyer (1976) and Powers (1980), Leakey et al. (1996) proposed a more formalized stratigraphic nomenclature, based upon the work of Powers and Feibel (unpublished). This new nomenclature will be utilized throughout the present paper. A generalized geological map of Lothagam is shown in figure 2.16 with a schematic cross section; sample localities are indicated in figure 2.17, and a schematic stratigraphic column is presented in figure 2.18, together with a summary of the age data and magnetostratigraphy.

The lowermost unit in Lothagam, comprising the informally named Nabwal Arangan beds, is mainly fault bounded (figures 2.17 and 2.18). It crops out in a high ridge forming the eastern part of the uplifted Lothagam block. This ridge rises about 200 m above the surrounding sand plains. The Nabwal Arangan beds are at least 280 m thick and consist of massive, dark, phonolitic lava flows intercalated with proximal volcaniclastic cobble to boulder conglomerates, the whole sequence dipping $\sim 10°$ to the west on average. In the fault block located in the southeastern part of Lothagam (figure 2.16), rocks regarded as the youngest representatives of the Nabwal Arangan beds consist of conglomerates interbedded with three basaltic lava flows. Stratigraphically above these deposits on the southern flanks of Central Hill lies the Nawata Formation, the base of which is defined by Powers and Feibel (unpublished) as immediately above the stratigraphically highest basalt flow of the Nabwal Arangan beds; it is not clear whether the contact is conformable or disconformable.

The Nawata Formation, about 240 m thick (figures 2.18 and 2.19), consists mainly of coarse-to-fine sandstones, siltstones, and mudstones, together with significant altered distal volcaniclastic sediments. It is informally subdivided into lower and upper members, each respectively about 120 m thick. Conglomerates dominate the lowermost 30 m of the lower member, followed upward by alternating sandstone and claystone beds with a number of interbedded analcimolitic claystones, altered tuffaceous beds. The clastic component in the lower member is dominated by volcanic material. Locally within some of the analcimolites, flattened, altered pumice clasts are found in lenticular accumulations. Several of these volcanic units have been named; these include the Lower Markers, Middle Markers, Red Marker, and Marker Tuff (figures 2.18 and 2.19). The Marker Tuff is defined as the basal unit of the upper member of the Nawata Formation, the top of which is the distinctive analcimolitic Purple Marker (figures 2.18 and 2.19). The upper member comprises mainly alternating sandstone and claystone units with minor pyroclastic components. The clastic component in the upper member shows increasing proportions of material derived from metamorphic rocks. The Nawata

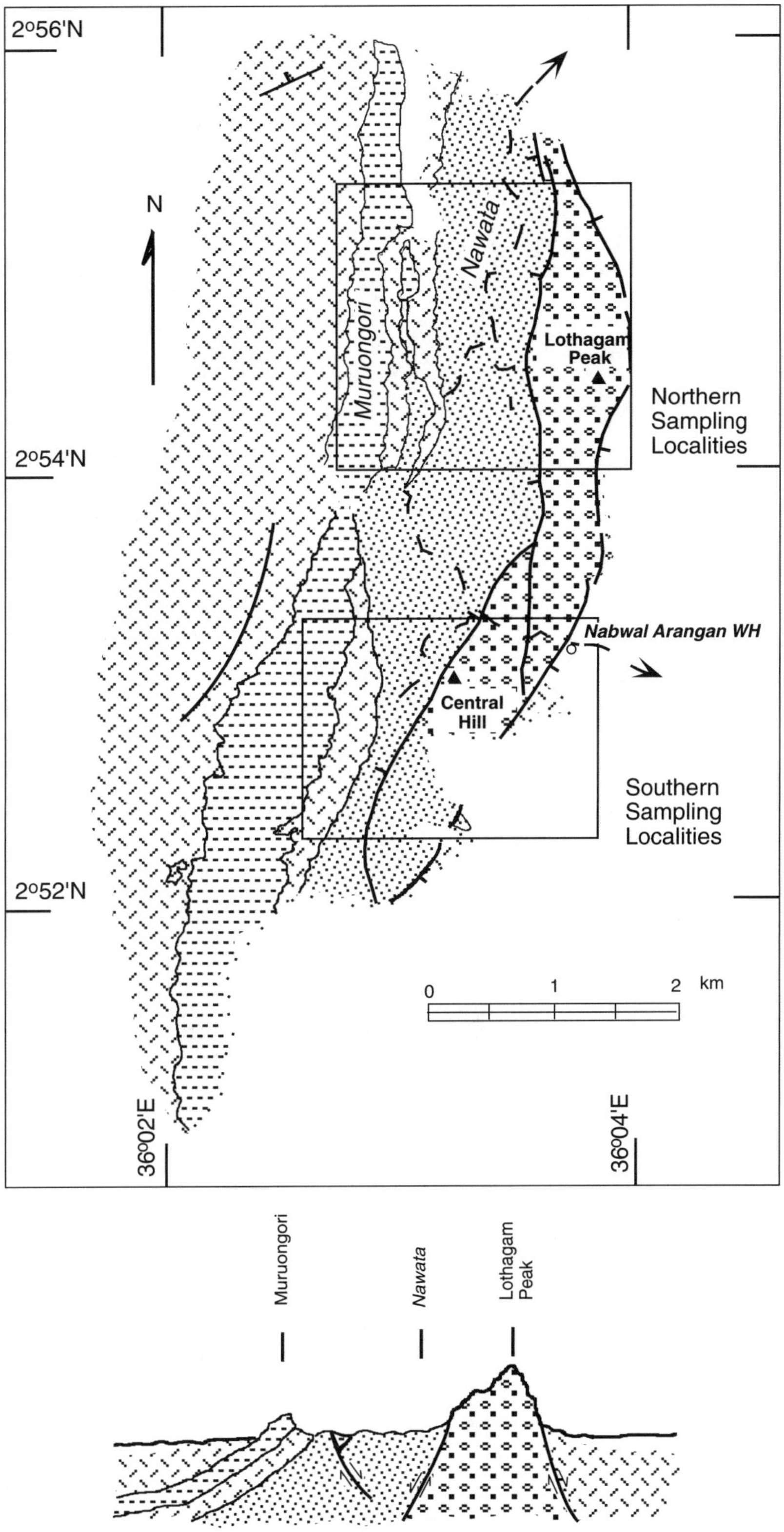

Figure 2.16 Geological map of Lothagam from Feibel (unpublished), modified from Patterson et al. (1970), showing major stratigraphic units in relation to geographic features, and areas covered by detailed locality maps (figure 2.17), together with a schematic cross section in an east–west profile through Lothagam Peak. The key for the units and other features is shown in figure 2.17.

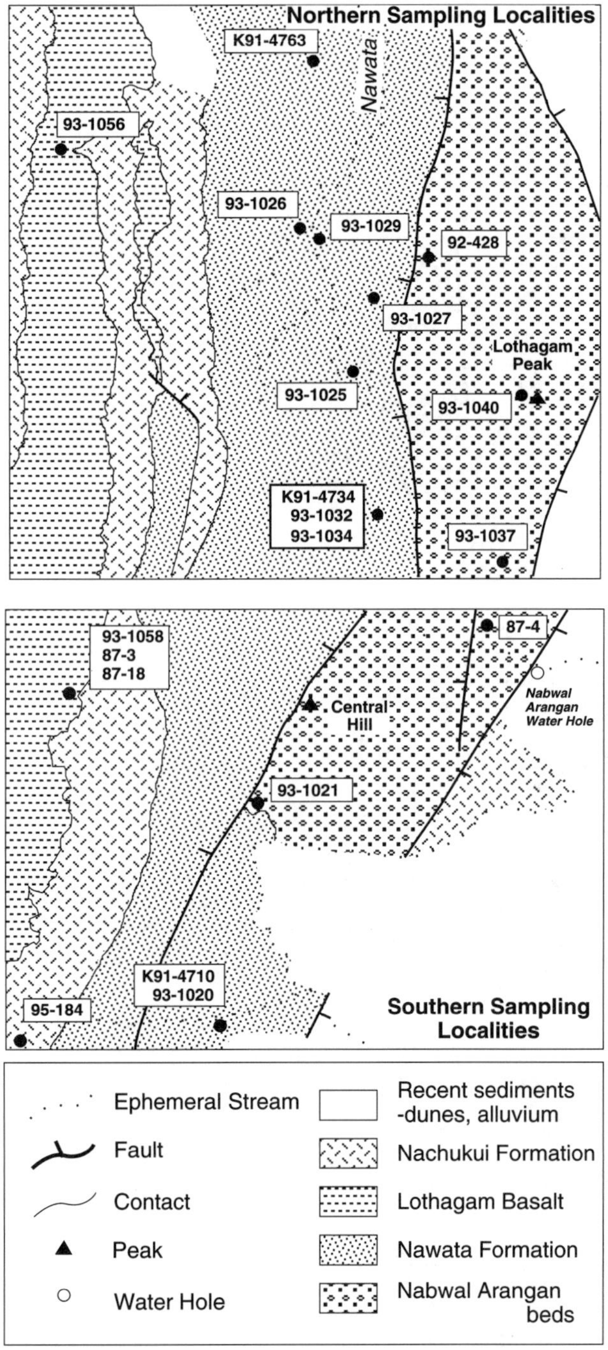

Figure 2.17 Sample locality maps showing provenance of material used for the isotopic dating reported here, together with the key for the geological maps.

Formation was deposited mainly in alluvial fan to fluvial environments.

The remaining ~400 m of section exposed in Lothagam overlies the Nawata Formation essentially conformably, perhaps disconformably, and is assigned to the Nachukui Formation of the Turkana Basin Omo Group of Harris et al. (1988) by Leakey et al. (1996) after Powers and Feibel (unpublished). The Apak Member, consisting of ~100 m of fluvial to lacustrine sediments comprising upward-fining cycles from sandstone to claystone, lies immediately above the Nawata Formation (figure 2.18). Overlying the Apak Member is the Lothagam Basalt, a unit about 50 m thick, forming the prominent ridge (Muruongori) on the western flanks of Lothagam (figures 2.16 and 2.18). Although originally described as a sill (Patterson et al. 1970), further study of the features and context of this basalt suggest that it was emplaced as a flow moving into soft sediments (Powers and Feibel, unpublished). Above the Lothagam Basalt occur the mainly lacustrine mudstones of the Muruongori Member, overlain by the sandstones and mudstones of the fluvial Kaiyumung Member of the Nachukui Formation, truncated by the present-day sand plain (Leakey et al. 1996).

The rich and diverse fauna from Lothagam comes mainly from the Nawata Formation and the Apak Member of the Nachukui Formation, and includes fish, reptiles, and a particularly good mammalian record (Patterson et al. 1970; Smart 1976; Leakey et al. 1996). The larger herbivores and their carnivore predators are especially well represented. Two hominoid teeth were found in the upper member of the Nawata Formation, and an important mandible fragment was recovered from the lowermost beds in the Apak Member (Patterson et al. 1970; Kramer 1986; White 1986; Hill et al. 1992; Leakey et al. 1996).

Methods

Sampling for isotopic dating was undertaken from lavas of the Nabwal Arangan beds and from the Lothagam Basalt, from the pumice clasts found associated with the analcimolitic units in the Nawata Formation, and from a localized occurrence in the Apak Member of the Nachukui Formation (figures 2.18, 2.19, and 2.20).

Only lava flows of fresh appearance were sampled, and each specimen subsequently was examined in thin sections under a petrographic microscope. Samples chosen for dating had fresh primary mineralogy with minimal alteration. Most samples from the Nabwal Arangan beds appear to be phonolites. They range from aphyric to phyric in purplish clinopyroxene and occasional kaersutite, with or without microphenocrysts of clinopyroxene, iron oxide, plagioclase, and olivine (often altered), set in an extremely fine grained but essentially holocrystalline groundmass of similar minerals, possibly also containing nepheline. The basalt at the top of the Nabwal Arangan beds (93-1021) is nearly aphyric, with some microphenocrysts of clinopyroxene and altered olivine. Although extremely fine grained, the rock appears to be well crystallized in clinopyroxene, iron oxide, a little plagioclase, and a feld-

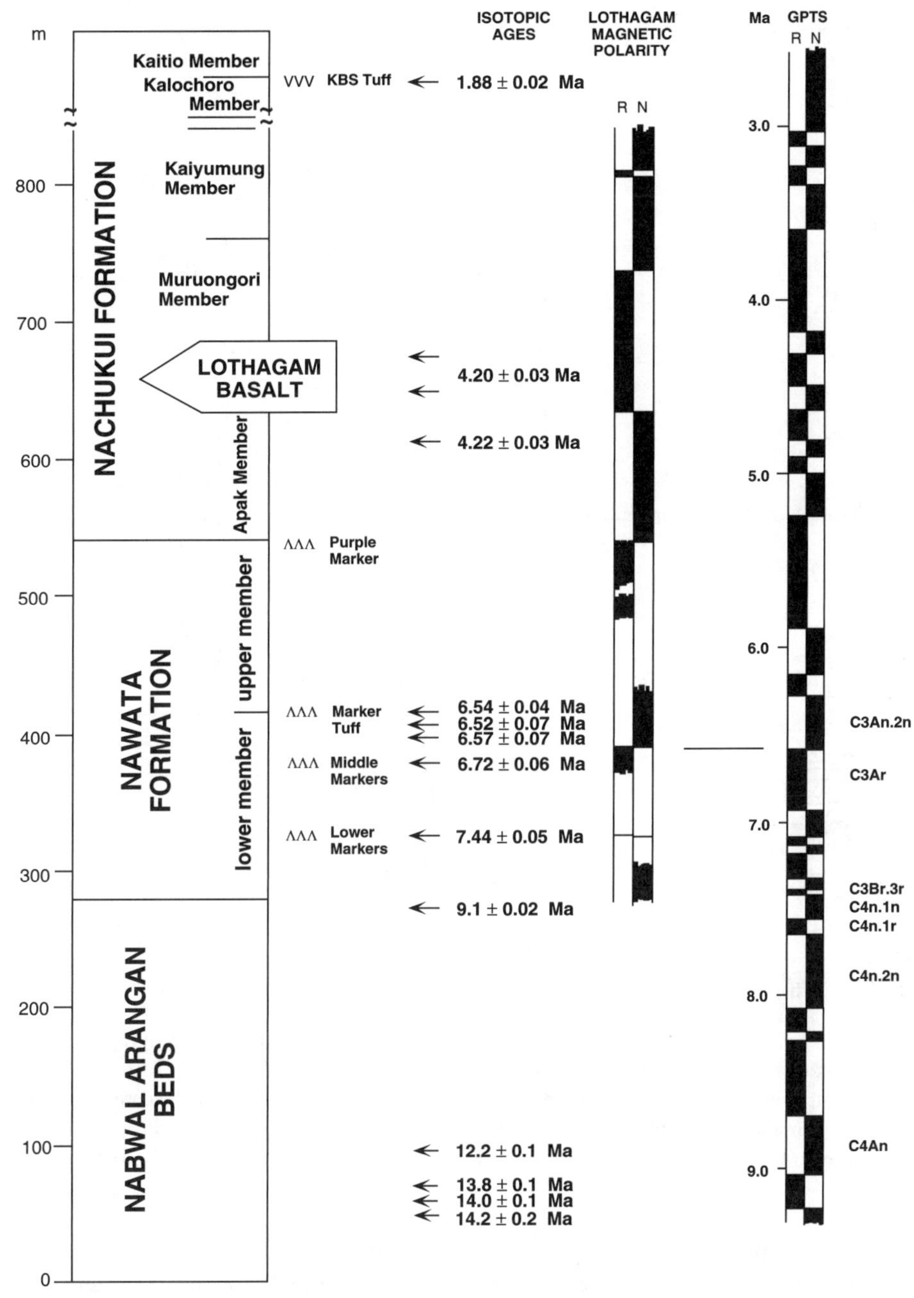

Figure 2.18 Schematic composite stratigraphic column for Lothagam, scale by thickness. Prominent marker units are identified and isotopic ages (simple mean and standard deviation of population) shown at the appropriate stratigraphic level. Magnetostratigraphy (MPS) of the Lothagam sequence, based on Powers (1980) and Feibel (unpublished), is shown, together with the geomagnetic polarity time scale (GPTS) of Cande and Kent (1995) on the right.

spathoid and is probably an undersaturated alkali basalt rather than a phonolite.

Fresh samples of Lothagam Basalt are all similar petrographically, as might be expected, as it appears to be a single, thick, lava flow. Samples are medium grained (average about 0.1 to 0.2 mm) in plagioclase (~50%), clinopyroxene (~20%), iron oxide, and olivine, together with about 20 percent of pale brown, intersertal glass. The primary minerals generally are unaltered, although olivine commonly shows incipient alteration. As the glass likely contains much of the potassium, its state of preservation is of some importance in regard to dating. In unweathered samples of the basalt, the glass is surprisingly fresh and isotropic, often with some small microlites within it. Small areas of brownish green clay occur in thin sections, but comprise only a few percent by volume. In one sample of Lothagam Basalt (93-1056), an aggregate about 1 mm across of anhedral clinopyroxene crystals, individually in the range of 0.2 to 0.5 mm in size, was found, but it is not clear whether this represents a glomerocryst or a xenocrystic aggregate.

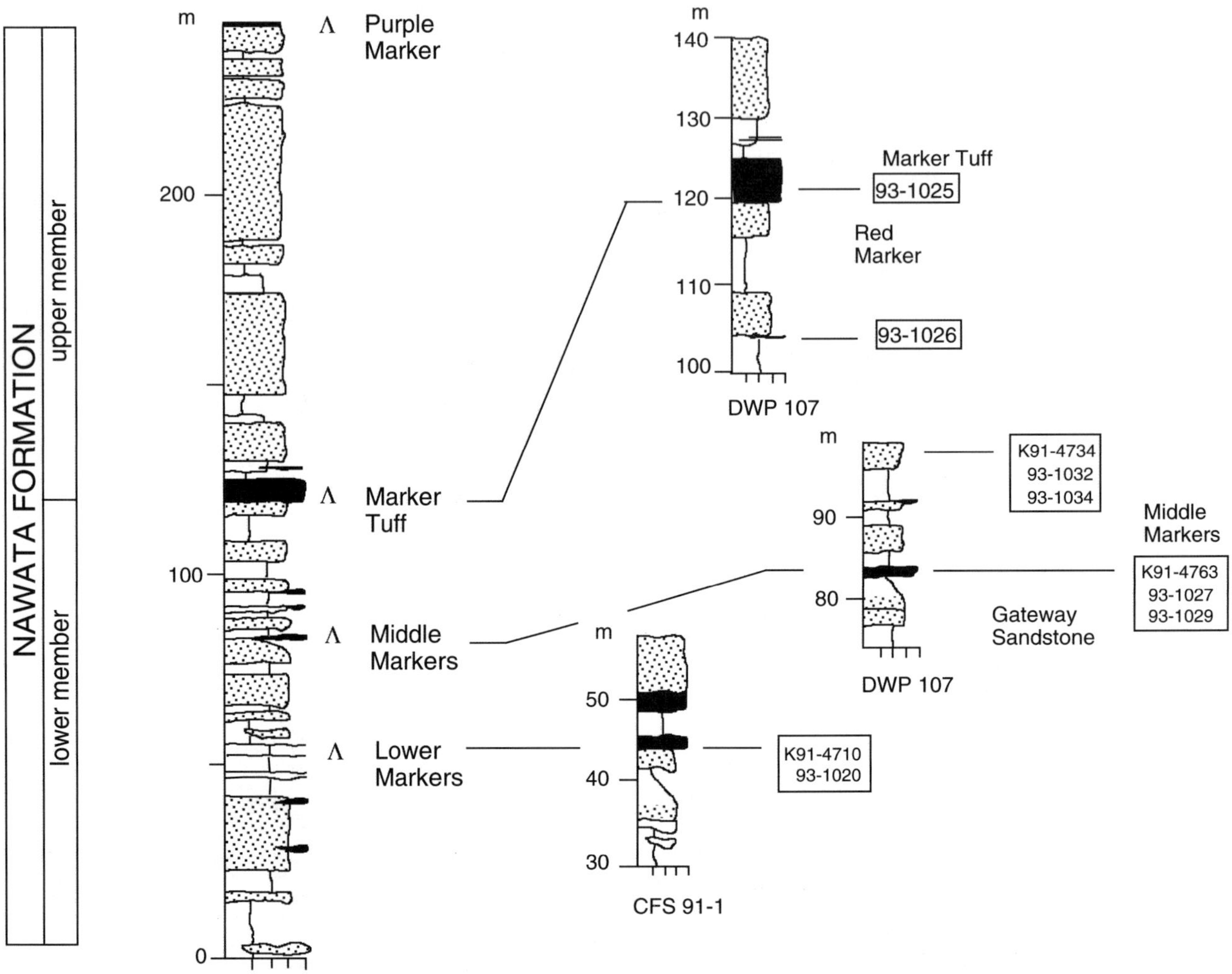

Figure 2.19 Detailed stratigraphic sections (after Powers and Feibel, unpublished), showing stratigraphic levels yielding samples used for isotopic dating. Sections labeled DWP are from Powers (1980), those labeled CSF are from Feibel (unpublished). Note change in scale between composite column (left) and detailed sections. Bars at the base of each section refer to the kind of material comprising the section above; one bar on the left signifies clay, the next to the right indicates silt, the next, sand, and the one on the right signifies gravel or tephra; each unit in the section is plotted according to this convention.

Samples considered suitable for whole rock K-Ar dating were crushed to a fragment size of 0.25 to 0.5 mm, a representative aliquot taken and further reduced to less than 0.15 mm to be reserved for potassium analysis by flame photometry. The coarser material was used for argon extraction. Plagioclase separates were prepared using heavy liquids and magnetic techniques from several samples of Lothagam Basalt, although this was difficult owing to the very small crystal size. Techniques of K-Ar measurement were similar to those described by McDougall and Schmincke (1977), and involved argon extraction in an ultrahigh vacuum (UHV) system by melting the sample in a molybdenum crucible by means of radiofrequency heating. Following addition of an ^{38}Ar tracer, the gas was purified and the argon was isotopically analyzed using an MS10 mass spectrometer operated in the static mode. The precision of a K-Ar age generally is about 1 percent standard deviation. Subsequent to the K-Ar measurements on the Lothagam Basalt, three of the whole rock samples and two plagioclase separates were chosen for ^{40}Ar-^{39}Ar step-heating analysis. Each sample was packed in its own aluminum container centrally within which was placed a smaller cylinder containing the fluence monitor, GA1550 biotite, of K-Ar age 97.9 Ma (McDougall and Roksandic 1974). The individual sample containers were packed into an irradiation canister, at either end of which a synthetic K-silicate glass sample was placed, all in a geometry similar to that shown in McDougall and Harrison (1988:figure 3.11), including 0.2 mm of cadmium shielding to minimize production of nucleogenic ^{40}Ar (Tetley et al. 1980). This irradiation (ANU 7) was for 48 h in facility X33 or X34 of HIFAR nuclear reactor, Lucas Heights, New South Wales, with three inversions of the sample canister

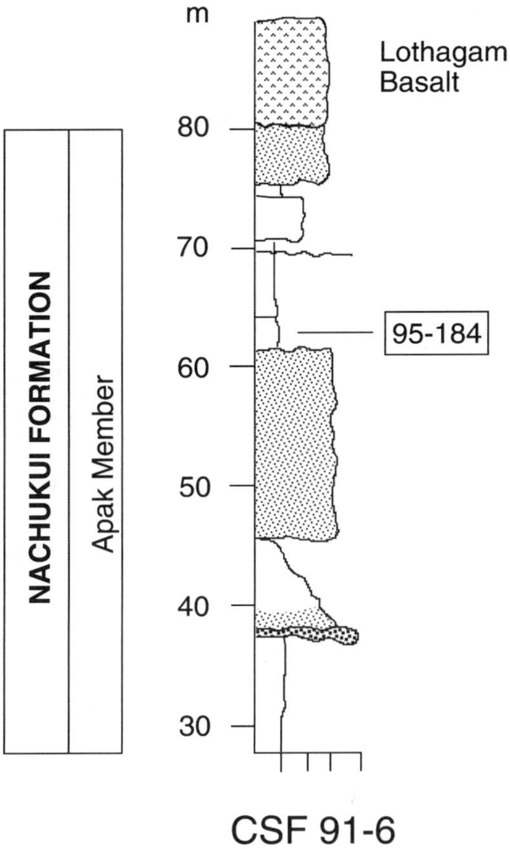

Figure 2.20 Stratigraphic section, CSF 91-6, through the Apak Member, Nachukui Formation, toward the southern end of outcrop, including the locality for sample 95-184.

end-for-end at exactly 12 h intervals to help minimize the neutron flux gradient.

Following irradiation, samples were placed in a UHV system, and the argon was released by fusion with a laser beam for the fluence monitor mineral, or by step heating for the whole rocks and plagioclase feldspars. The gases released were purified over SAES getters and then expanded directly into a VG3600 gas source mass spectrometer operated statically. Argon isotope measurements were made on ^{36}Ar, ^{37}Ar, ^{38}Ar, ^{39}Ar, and ^{40}Ar, collecting data through a Daly detector and photomultiplier system. Sensitivity of the VG3600 operated at 200 μA trap current and 4.4 kV accelerating potential was about 5×10^{-4} amps/torr for argon, or about 2.5×10^{-17} mol/mV on the Daly system at the gain chosen for this work over the course of these experiments. Mass discrimination was measured regularly using purified atmospheric argon from a gas pipette; the discrimination did not exceed 0.8 percent per amu over the period these measurements were made. For the step-heating experiments on the whole rock and plagioclase samples, between 100 and 150 mg of sample was loaded into the UHV system and then successively dropped into a tantalum crucible heated resistively, with temperature control through a thermocouple feedback system. A minimum of 13 steps, each of 15 minutes duration at temperature, were undertaken on each sample, progressively increasing the temperature from about 500°C to over 1400°C. All isotopic measurements of the purified argon were undertaken using the Daly collector in the VG3600 mass spectrometer.

In the ^{40}Ar-^{39}Ar dating, the correction factor $(^{40}Ar/^{39}Ar)_K$ was measured in each irradiation on K-silicate glass and ranged from 0.020 to 0.035. The Ca correction factors used were measured in other irradiations done over the same period. Values used in all calculations are: $(^{36}Ar/^{37}Ar)_{Ca} = 3.49\ (\pm 0.14) \times 10^{-4}$ and $(^{39}Ar/^{37}Ar)_{Ca} = 7.86\ (\pm 0.01) \times 10^{-4}$ (Spell et al. 1996).

Although altered, the pumice clasts collected from the analcimolitic tuffaceous horizons within the Nawata Formation contain remarkably fresh, limpid phenocrysts of alkali feldspar, ideal for single crystal ^{40}Ar-^{39}Ar dating purposes. Where possible, pumice clasts were collected from the same stratigraphic horizon at more than one locality. Pumice clasts often were only about 1 cm in diameter, but at a few localities they were up to 5 cm or more across. In the laboratory, following cleaning of the surface of the pumices, they were individually carefully crushed followed by separation of the alkali feldspar crystals greater than 0.5 mm in size by handpicking or by using heavy liquids. Feldspars from each pumice were maintained as a discrete sample. In the case of the Marker Tuff, the lowermost unit of the upper member, Nawata Formation, no pumice clasts were found. However, at one locality in Nawata Laga, crystals of alkali feldspar up to about 1 mm in size were present and, using tweezers, were picked out directly in the field from the altered tuffaceous matrix. The final feldspar concentrates normally were washed ultrasonically in 7 percent HF for five minutes to remove traces of altered glass and matrix.

Toward the southern limit of outcrop in Lothagam in the Apak Member of the Nachukui Formation, a lens about 20 cm thick was found locally containing small (5 to 30 mm), yellow, rounded clasts of what appeared to be altered pumices. These clasts occur in a darker colored sandy matrix. Although very friable, individual clasts were collected and cleaned of adhering matrix as far as possible in the field. Feldspar crystals were found in this material and separated as described above from two subsamples.

Some 20 to 30 crystals of feldspar, usually 0.5 to 1 mm maximum dimension, from each sample were packed in aluminum foil and successively placed in a silica glass tube 6 mm internal diameter and 33 mm long, interspersed with similar small packets of a sanidine fluence monitor (92-176, separated from Fish Canyon Tuff, Colorado) of nominal K-Ar age 27.9 Ma (Steven et al. 1967; Cebula et al. 1986). A synthetic

K-silicate glass was placed at either end of the silica glass tube, and the whole package was encased in 0.2 mm thick cadmium within an aluminum reactor vessel for irradiation in facility X33 or X34 in HIFAR nuclear reactor. Irradiations for reactor canisters K787 and G536 were for 72 h, whereas ANU23 was irradiated for 48 h. In each case the reactor vessel was inverted three times during the irradiation, as previously described; the neutron fluence gradient was found to be less than 2 percent along the reactor can for each irradiation. Following irradiation, individual alkali feldspar crystals were loaded into wells in a copper sample tray, installed in the UHV system, and baked overnight at 200°C. A focused argon-ion laser beam up to 10W power was used to fuse individual crystals. The gases released from each crystal were purified and then expanded directly into the VG3600 gas source mass spectrometer for argon isotopic analysis, as outlined previously. Mass spectrometer control and data acquisition were accomplished using a Macintosh computer and Noble software, which also enabled full data processing to be carried out immediately following isotopic analysis. The $^{40}Ar/^{39}Ar$ ratio measured on argon released from the Lothagam alkali feldspar crystals normally had a precision of 0.3 to 0.6 percent (standard deviation), and as the proportion of radiogenic argon commonly was high (>85%), uncertainties in individual calculated ^{40}Ar-^{39}Ar ages generally are similar.

The irradiation parameter, *J*, for each unknown was derived by interpolation from the argon isotopic measurements made on gas released from the sanidine fluence monitor crystals (92-176). A minimum of four crystals was measured for each level of fluence monitor in the silica glass tube, with a precision ranging from 0.1 to 0.4 percent (standard deviation of the population). The interpolated *J* parameter for each unknown had an estimated error of 0.2 to 0.4 percent standard deviation, included quadratically in the calculation of error in the age of each unknown. In the case of the Lothagam Basalt irradiation (ANU 7), a minimum of five individual crystals of the biotite fluence monitor (GA1550) from each level were fused in the vacuum system with the laser and the purified argon isotopically analyzed. The uncertainty in *J*, calculated from the standard deviation of the population for each level, ranged from 0.28 to 0.68 percent; the appropriate uncertainty was used in the subsequent age calculations.

Although the individual alkali feldspar crystals subjected to ^{40}Ar-^{39}Ar dating were successfully fused by the laser beam, the short interval that the laser coupled with the material, sometimes for only a few seconds, means that it is unlikely that the argon was quantitatively released in all cases. For alkali feldspar crystals that have not been significantly reheated or disturbed since their crystallization and cooling, the measured ^{40}Ar-^{39}Ar ages on gas released by laser fusion should accurately reflect the time elapsed since cooling. This is because it is expected that such crystals would yield a flat ^{40}Ar-^{39}Ar age spectrum, so that an age determined on only a portion of the gas is likely to be closely similar to that determined in the case of complete gas release. In support of this view, McDougall (1985) showed that feldspars from pumice clasts in similar environments in tuffaceous beds of the Turkana Basin gave essentially flat ^{40}Ar-^{39}Ar age spectra. The feldspars from the Nawata Formation are as fresh, clear, and unaltered as those from the Koobi Fora Formation, despite the greater alteration of the associated tuffs. Finally, as will be seen subsequently, the extremely good agreement between ^{40}Ar-^{39}Ar ages on individual crystals from each pumice suggests that the above line of argument is sound.

In this paper we use the numerical time scale of Harland et al. (1990) in which the Eocene-Oligocene boundary is estimated to be at 35.4 Ma, the Oligocene-Miocene boundary at 23.3 Ma, the Miocene-Pliocene boundary at 5.1 Ma, and the Pliocene-Pleistocene boundary at 1.64 Ma.

Results

Potassium-argon age data are listed in table 2.4, and ^{40}Ar-^{39}Ar single crystal results are summarized for each pumice and for each horizon in table 2.5; analytical data for one set of single crystal measurements are listed in table 2.6 as an example, with the bulk of the data to be found in appendices. Step-heating analytical results are summarized in table 2.7, with analytical data listed in appendices, deposited with the Geological Society of London Library and British Library at Boston Spa, W Yorkshire, U.K., as Supplementary Publication No. SUP 18131 (8 pages). All errors are given at the level of one standard deviation. In the case of the single crystal results from each pumice, the results are presented as an arithmetic mean and standard deviation of the population, as well as a weighted mean together with a standard error of the mean; the weighting is done according to the inverse of the variance. For a step-heating experiment, an integrated total fusion age was calculated by summation of the individual ages and errors according to the size of each step. A plateau was identified in an age spectrum when consecutive steps had concordant ages at the level of one standard deviation, and together comprised a significant proportion (>35%) of the ^{39}Ar released. Calculation of a plateau age was made by weighting each step age by the inverse of its variance to provide a weighted mean age. The results will be discussed in stratigraphic order from oldest to youngest.

Nabwal Arangan Beds

Measured whole rock K-Ar ages on three samples of phonolite and one sample of phonolitic basalt lie in the range 12.2 to 14.2 Ma (table 2.4). The oldest measured age, 14.2 ± 0.2 Ma, was obtained on sample 93-1037 from a large clast (>1 m) in a breccia bed cropping out on the prominent eastern ridge of Lothagam. This age might be expected to reflect the time of eruption of the original lava, with later incorporation into the breccia. Sample 93-1040 is from a virtually aphyric phonolitic lava from just below the summit of the northern peak (Lothagam) on the eastern ridge; it yields a similar age of 13.8 ± 0.1 Ma. A somewhat younger apparent age of 12.2 ± 0.1 Ma was found for a phonolite (87-4) collected from a breccia in the Nabwal Arangan gorge. A columnar jointed phonolitic basalt (92-428), apparently in the western bounding fault of the eastern horst block, yielded an age of 14.0 ± 0.1 Ma. These four results suggest that the phonolitic volcanism occurred over a restricted interval between about 14 and 12 Ma ago in the mid Miocene. In marked contrast, the highest basalt in the Nabwal Arangan beds as mapped on the southern flanks of Central Hill, immediately below the Nawata Formation, gives a K-Ar age on whole rock sample 93-1021 of 9.1 ± 0.2 Ma (table 2.4). This sample appears to be well crystallized, even if extremely fine grained, and relatively fresh, so that the age can be taken at face value, considerably younger than the phonolitic lavas. This raises the question as to whether there is a significant hiatus within the Nabwal Arangan beds, as previously suggested by Patterson et al. (1970) and Behrensmeyer (1976). Patterson et al. (1970) quoted K-Ar whole rock ages of 16.8 ± 0.5 and 8.31 ± 0.25 Ma for two basaltic rocks from what are now called the Nabwal Arangan beds, with the younger age probably from one of the basalts on the southern slopes of Central Hill. Unfortunately no details of localities or analytical data were given.

Nawata Formation

Alkali feldspar crystals were obtained from five tuffaceous horizons within the Nawata Formation (figures 2.18, 2.19, and 2.20). Four of the levels sampled are from within the lower member, and the stratigraphically highest sample was from the Marker Tuff, the basal unit of the upper member (Leakey et al. 1996). As previously mentioned, individual feldspar crystals were dated by the ^{40}Ar-^{39}Ar method; prior to discussing the data, some general comments about how the results are treated are in order. A very simple approach is adopted. Ages obtained on feldspars from each pumice sample are combined to obtain a simple arithmetic mean or average age, and the standard deviation of the population is calculated. Any age outside two standard deviations is then rejected as an outlier, and a new mean age and population standard deviation are calculated if necessary. This simple approach is adopted as generally the individual ages have similar uncertainties, so that unit weighting of each age would seem to be justified. We have chosen to utilize the standard deviation of the population, rather than the standard deviation of the mean, as the statistic to be used in discussion because it gives a clearer indication of the spread of the results and a more realistic uncertainty in the age. Nevertheless, the standard deviation of the mean can be calculated, if desired, simply by dividing the standard deviation of the population by $\sqrt{n}$, where n is the number of results obtained. Results are listed for feldspars from each pumice, and this average is then compared with results from other pumices from the same stratigraphic level. If concordance is found, then an overall mean age is calculated for the level in the same manner.

In single crystal ^{40}Ar-^{39}Ar dating it is common practice to calculate a weighted mean, with weighting of results according to the inverse of their variance; that is, so the more precise results are given greater weight. For comparison we have also calculated a weighted mean, shown in the summary table 2.5. The data for individual pumices or stratigraphic levels also have been assessed by means of probability plots (Kelley and Bluck 1989; Deino and Potts 1992), which are useful in determining the homogeneity of a given population. In addition, results from each pumice have been plotted in isotope correlation diagrams, and an age (as well as the indicated composition of the trapped argon) is derived from regression analysis (York 1969); results are listed in table 2.5. Generally the isochron ages differ little from the arithmetic or weighted mean values, showing that the trapped argon composition in most cases is indistinguishable from atmospheric argon. The MSWD is a goodness-of-fit parameter associated with the isochron analysis, and it has an expected value of 1.0 for a fit of the regression line to the data points to within experimental error. Generally it is accepted that an MSWD value greater than 2.5 indicates scatter of data about the line greater than can be accounted for by experimental error alone. Simple mean K/Ca ratios have been calculated for feldspars from each pumice and also are listed in table 2.5.

The lowermost unit sampled in the Nawata Formation was the Lower Marker, about 43 m above the base of the formation in the southern exposures (figure 2.19). This analcimolite, an altered tuffaceous unit about 2 m thick, contains flattened, altered pumice clasts up to ~2 cm in diameter, concentrated locally in lenses. Alkali feldspar was separated from four individual pumice clasts, collected on two different occasions from the same locality. A minimum of 6 and as many as 11 crystals were measured from each pumice clast, and the results will be discussed in some detail as a representative example. Eleven feldspars from pumice K91-4710(B) yielded a nearly concordant set of ages, with an arithmetic mean of 7.442 ± 0.047 Ma (table 2.5). The total spread in measured age is only 1.7 percent. Indeed, the remarkably good statistics for the weighted mean and the isochron age (table 2.5), together with their concordance, serve to emphasize that, in fact, there is a fair degree of homogeneity of the population and that the spread in apparent ages is small. Ten feldspar crystals measured from pumice 93-1020(A) form an essentially concordant population with a simple mean age of 7.470 ± 0.063 Ma. The calculated age on crystal 8 of 7.605 ± 0.053 Ma can be omitted because it is 2σ from the mean; the new mean of 7.455 ± 0.044 Ma that is calculated differs little from the initial mean. Results from the six crystals analyzed from pumice 93-1020(C) are close to concordant and yield a mean of 7.464 ± 0.077 Ma; the somewhat younger apparent age for crystal 4 compared with the rest of the population causes the population standard deviation to be just over 1 percent, but this result cannot be regarded as unduly anomalous as it remains within 2σ of the mean. The mean age for 9 crystals measured from pumice 93-1020(D) is 7.358 ± 0.112 Ma, but there are two anomalously low measured ages (on crystals 4 and 9), that may be eliminated on the 2σ criterion; the revised mean age is 7.411 ± 0.033 Ma. The average ages for the four pumice clasts lie within 1 percent of one another and are indistinguishable within the errors. Note that the weighted mean ages and those derived from the isochron approach all are essentially concordant. The overall mean age based on 33 ages is 7.443 ± 0.051 Ma, with a weighted mean age of 7.461 ± 0.004 Ma. The probability plot (figure 2.21a) incorporating these 33 results is dominated by a single peak but with some asymmetry reflecting several somewhat younger measured ages. Although there is considerable scatter in the K/Ca ratios measured on the feldspars, in part owing to the small size of the ^{37}Ar ion beams, the data are consistent with a single population indicated from the ^{40}Ar-^{39}Ar age measurements. The measured age is considered to reflect the time of cooling of the alkali feldspars immediately following explosive eruption. As tuffaceous deposits can be expected to be deposited very soon after eruption, the cooling age can reasonably be regarded as giving a very close approximation to the time of deposition in the Lothagam sedimentary sequence.

The Middle Marker lies just above the distinctive Gateway Sandstone (figure 2.19); it is estimated to be stratigraphically about 30 m above the Lower Marker. It is mainly an analcimolitized tuffaceous unit about 1 m thick which contains lenses of small, flattened pumice clasts up to a maximum of 5 cm in diameter. Alkali feldspars were separated from seven pumice clasts collected from three different localities in the same horizon up to about 1 km apart (figure 2.17) in the northern area in and around the Nawata drainage. A minimum of five individual crystals was dated by ^{40}Ar-^{39}Ar total fusion techniques from each pumice (table 2.5). Apart from one anomalously young age (5.70 ± 0.02 Ma) on crystal 5 from K91-4763(B) pumice, the results from each of the seven pumice clasts are essentially concordant. The arithmetic means of results from each of the pumice clasts range from a minimum of 6.681 ± 0.049 Ma (standard deviation of the population) to a maximum of 6.765 ± 0.028 Ma, a spread of about 1.3 percent, but with errors that virtually overlap at the one standard deviation level (table 2.5). Thus, the results are concordant for each pumice, for pumice clasts at each sampling locality, and between sample localities. The overall simple mean age is 6.720 ± 0.062 Ma giving unit weight to each of the 42 measurements, with a weighted mean age of 6.731 ± 0.003 Ma. The probability plot indicates a single broad peak with some tailing to slightly younger ages (figure 2.21b). The isochron ages derived for results from individual pumice clasts also are concordant, except for pumice 93-1027(A) for which all points plot so close to one another that the isochron approach is not particularly meaningful (table 2.5). Our best estimate for the age of the explosive volcanic eruption leading to the deposition of the Middle Marker tuffaceous horizon is the arithmetic mean age of 6.72 ± 0.06 Ma.

About 15 m above the Middle Marker and approximately 15 m below the Red Marker in the lower member of the Nawata Formation occurs another (unnamed) altered tuffaceous unit about 2 m thick, well exposed in the central area of Lothagam near the head of Nawata Laga (figures 2.18 and 2.19). Locally in the so-called primate area, this tuffaceous unit contains relatively abundant flattened, altered pumice clasts in which fresh alkali feldspar phenocrysts are sporadically present. Feldspars were separated from four pumice clasts (K91-4734(A), (B), 93-1032, 93-1034), collected on two different occasions from the same locality, and ^{40}Ar-^{39}Ar ages were determined on five to nine individual crystals from each pumice (table 2.5). For each of the four pumice clasts, the age populations are fairly homogeneous with no outliers; the mean ages range from 6.640 ± 0.022 Ma to 6.525 ± 0.066 Ma, an apparent range of ~1.8 percent, with overlapping errors at the level of two standard deviations. The isochron ages also are concordant, although that for 93-1032(B) is very uncertain owing to limited dispersion of points in the isotope correlation diagram. Assuming a homogeneous population,

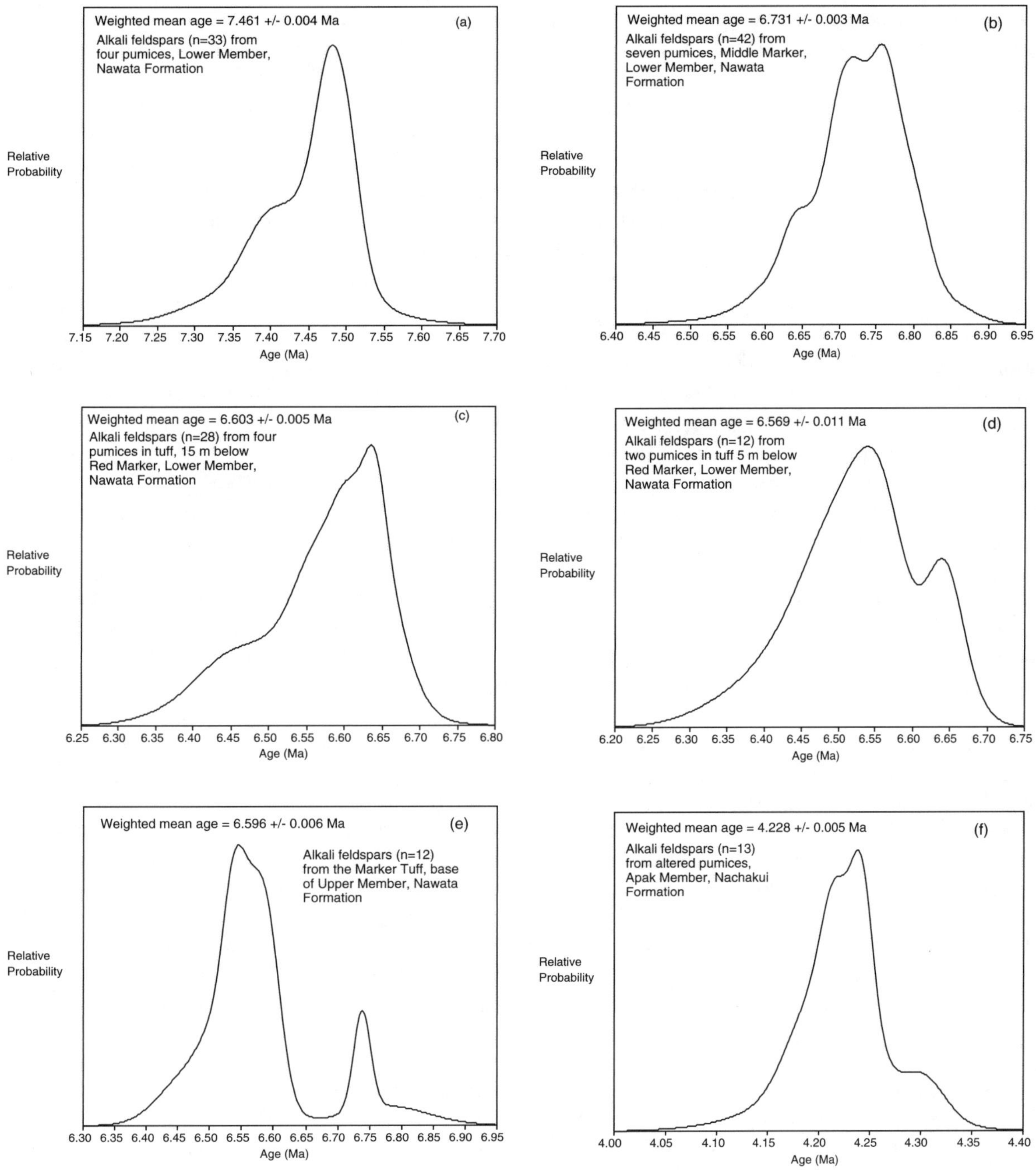

Figure 2.21 Probability plots (Deino and Potts 1992) showing combined results of single crystal alkali feldspar dating for individual horizons in the Nawata Formation (a–e) and from the Apak Member, Nachakui Formation (f).

supported by the distinctive and fairly uniform K/Ca ratio of 8.1 ± 3.3, a mean age of 6.566 ± 0.075 Ma is derived based on 28 measurements, or 6.603 ± 0.005 Ma if a weighted mean age is preferred. The probability plot shows a strong maximum with some scatter to lower ages (figure 2.21c). Again, the age derived from these feldspars is regarded as a good estimate for the time elapsed since explosive eruption and deposition. Note that the average age is only marginally younger than that found for the Middle Marker. However, the differences in the K/Ca ratios for the feldspars from the two horizons confirm that they are products of different eruptions.

At a locality in Nawata Laga, a rather silty horizon about 5 m stratigraphically below the Red Marker, and about 6 m above the stratigraphic level just discussed, was found to contain flattened pumice clasts ranging from about 2 cm to 6 cm in diameter, each containing some alkali feldspar crystals (figure 2.19). Feldspars were separated from two pumice clasts (93-1026(A) and 93-1026(C)), and single crystal ^{40}Ar-^{39}Ar age results are summarized in table 2.5. For both pumice clasts, six crystals were measured, and the age populations appear to be homogeneous and indistinguishable, with means of 6.511 ± 0.040 and 6.537 ± 0.097 Ma for the two pumice clasts. An overall mean of 6.523 ± 0.072 Ma is derived or 6.569 ± 0.012 Ma if a weighted mean is calculated; the ages obtained from the correlation plots are comparable (table 2.5). The probability plot (figure 2.21d) is essentially a single broad peak with a subsidiary peak reflecting two older ages at 6.64 Ma in 93-1026(C). These ages are indistinguishable from those on the pumice clasts from the unit about 6 m below and raise the possibility of being the product of the same eruptions reworked to a higher stratigraphic level. Note that the K/Ca ratios on the feldspars also are not distinguishable between the two horizons (table 2.5).

The stratigraphically highest horizon in the Nawata Formation for which dating was found to be possible was the Marker Tuff, the defined basal unit of the upper member, only a few meters above the Red Marker (figure 2.19). In the Nawata Laga this unit crops out well as an altered agglomeratic or laharic tuffaceous unit a few meters thick; single alkali feldspar crystals were handpicked from this unit. Some 12 individual crystals were dated (sample 93-1025, table 2.5), providing an overall mean age of 6.575 ± 0.094 Ma, but with a bimodal probability plot (figure 2.21e). Excluding the two ages, more than two standard deviations from the average yield a mean of 6.539 ± 0.044 Ma or a weighted mean age of 6.555 ± 0.007 Ma and a similar isochron age (table 2.5). This age again is indistinguishable from those derived from the two units last discussed. As the K/Ca ratios are very scattered for the feldspars from the Marker Tuff (table 2.5), this does not act as a discriminator.

The single crystal ^{40}Ar-^{39}Ar dating of alkali feldspars from the Nawata Formation documents a history extending from about 7.44 ± 0.05 Ma to 6.54 ± 0.04 Ma (about 0.9 Ma), for deposition of much of the lower member, based on results from four units from about 70 m of stratigraphic section. Thus an average sedimentation rate of about 80 mm/1,000 years is derived. The Marker Tuff at the base of the upper member yields ages on alkali feldspars indistinguishable from those found for tuffaceous horizons in the upper 10 to 15 m of the lower member, so that either there was more rapid deposition than previously, or possibly there was some reworking of volcanic material into higher stratigraphic levels.

Apak Member, Nachukui Formation

As previously discussed, alkali feldspar crystals were separated from what appeared to be altered pumice clasts in a sandy lens within the Apak Member. Stratigraphically this level lies about 35 m above the Purple Marker at the top of the Nawata Formation, and about 17 m below the Lothagam Basalt, as shown in section CSF 91-6 (figure 2.20) measured in the vicinity of the sample locality by Feibel (Powers and Feibel, unpublished).

Feldspar crystals were separated from two subsamples; 95-184A consisted of small intact clasts, and 95-184B was a subsample comprising clay clasts that had disintegrated during transportation.

Results of single crystal total fusion ^{40}Ar-^{39}Ar age measurements are given in table 2.6. For sample 95-184A, six alkali feldspar crystals were measured; a concordant set of results was obtained yielding an arithmetic mean of 4.23 ± 0.04 Ma. In the case of sample 95-184B, seven alkali feldspar crystals were measured, again with concordancy of results yielding a mean of 4.21 ± 0.03 Ma, indistinguishable from the results on the other subsample. The overall mean based upon all 13 single crystal measurements gives an age of 4.22 ± 0.03 Ma, with no outliers; the probability plot shows a dominant peak with a subsidiary minor peak at a slightly higher age owing to a single apparent age at 4.30 Ma (figure 2.21f). Nevertheless, the mean age is accepted as that of the eruption that produced the pumices. The concordancy of all results provides considerable confidence that we are dealing with a juvenile population of phenocrysts from an igneous environment. Clearly, the level of the Apak Member in which the pumice fragments occur can be no older than 4.22 Ma. It is reasonable to postulate that deposition of the pumiceous material occurred very soon after the explosive eruption that produced the pumice. Thus, this part of the Apak Member is considered to have a depositional age of 4.22 ± 0.03 Ma, or slightly younger.

Lothagam Basalt

The Lothagam Basalt is intercalated between the Apak and Muruongori Members of the Nachukui Formation (figure 2.18), after Powers and Feibel (unpublished) and Leakey et al. (1996). It is a unit up to about 50 m thick, forming the prominent northerly trending ridge (Muruongori) on the western flanks of Lothagam. We favor an extrusive origin for the Lothagam Basalt. The basalt shows good columnar jointing, and it commonly is strongly weathered, often spheroidally. However, relatively fresh, unweathered samples of the basalt were obtained from the summit of the ridge south of Apak and from a gorge incised into the basalt about 2 km north of Apak (figure 2.17). These samples are typical basalts as previously described. Patterson et al. (1970) reported an age of 3.71 ± 0.23 Ma for a whole rock sample from the basalt, but no analytical or locality data were given. The new K-Ar results are listed in table 2.4. Ages have been measured on a number of whole rock samples and on plagioclase separated from two of the samples. There is a considerable spread in the calculated ages from 4.6 to 6.1 Ma; the range in apparent age is very much larger than can be accounted for by analytical errors, which rarely exceed a few percentage points. Indeed, there is an even older measured K-Ar age of 9.1 Ma on one whole rock sample from the Lothagam Basalt; as this result is clearly anomalous and quite inconsistent with the age information on the underlying Nawata Formation, the analytical data are not listed in table 2.4. Possible explanations for the range in apparent age might include variable incomplete degassing at the time of eruption of preexisting radiogenic argon derived from the source regions of the basalt, unusual in subaerially erupted lavas, or incorporation of xenolithic material in the magma, although none was definitely identified; both explanations would lead to old apparent ages. Laboratory-induced fractionation of atmospheric argon in the sample during pumping in the vacuum system prior to argon extraction also can result in old apparent ages (cf. Baksi 1974; McDougall et al. 1976). A spread to younger ages is most likely caused by variable loss of radiogenic argon, quite possibly from the glass in the whole rock samples. Thus, there are plausible explanations for both old and young apparent K-Ar ages, so that it is very difficult to assign an age of emplacement of the basalt on the basis of these K-Ar results. It should be noted, however, that all the K-Ar ages on the Lothagam Basalt reported here (table 2.4) are inconsistent with the age of 4.22 ± 0.03 Ma measured on pumice from the underlying (older) Apak Member, discussed above.

In an attempt to resolve these inconsistencies, three whole rock and two plagioclase samples from the Lothagam Basalt were measured using the ^{40}Ar-^{39}Ar step-heating approach. An overall summary of the results is given in table 2.7. Note that the conventional K-Ar ages and the ^{40}Ar-^{39}Ar integrated total fusion ages agree closely in three cases, with slightly older apparent ages by the ^{40}Ar-^{39}Ar method in the remaining two cases. Of greater importance, however, is the nature of the age spectra, shown in figure 2.22.

Whole rock sample 87-3, from the Lothagam Basalt from the ridge south of Apak (figure 2.17), yields a very distinctive age spectrum (figure 2.22a), showing a monotonically decreasing age with progressive release of ^{39}Ar to about 46 percent, followed by the final eight steps, comprising 54 percent of the ^{39}Ar release, giving essentially concordant ages with a mean of 4.19 ± 0.03 Ma. This experiment is strongly suggestive of an excess argon component being released in the early stages of gas release (cf. McDougall and Harrison 1988), probably from the glass in the basalt, as the K/Ca ratios also monotonically decrease in an analogous manner to the age. The plateau age, 4.19 ± 0.03 Ma, can be regarded with confidence as a maximum age for the basalt, very much younger than the integrated total fusion age of 5.1 ± 0.1 Ma. Using the data from the steps considered to comprise the plateau in the age spectrum, isochron analysis yields a virtually identical age with a trapped argon component that is indistinguishable from atmospheric argon (table 2.7).

Plagioclase separated from sample 87-3 was also step-heated. The age spectrum shows initial high apparent ages, decreasing to yield an apparent plateau age of 4.58 ± 0.05 Ma over five steps between 17 and 70 percent ^{39}Ar release (figure 2.22b). In the five higher temperature steps, older apparent ages are found. A slightly younger apparent age is found if the same steps identified as forming the plateau in the age spectrum are regressed in the isotope correlation diagram (table 2.7). If the whole rock plateau age estimate is accepted as approximating the eruption age, then the plagioclase apparent age must be regarded as excessively old, presumably because of the presence of excess argon.

Whole rock sample 87-18 from the Lothagam Basalt within about 100 m of sample 87-3 yields a similar age spectrum to the latter sample. However, the plateau comprising four steps and about 37 percent of the ^{39}Ar release (figure 2.22c) gives an age of 4.32 ± 0.04 Ma, which is somewhat older than that found for 87-3 (table 2.7).

A whole rock sample, 93-1056, from the basalt in the steep valley cutting through the Lothagam Basalt ridge about 2 km north of Apak (figure 2.17), yielded an age spectrum with mainly decreasing apparent ages up to 47 percent ^{39}Ar release, with the remaining 53 percent of gas release yielding a concordant age at 4.20 ± 0.02 Ma (figure 2.22d); a similar result is obtained by isochron analysis (table 2.7). Plagioclase from the sample also shows monotonically decreasing ages over the first 16 percent of gas release, and then a plateau at 4.43 ± 0.04 Ma comprising 65 percent of the gas release (figure 2.22e).

Overall, these results are interpreted as confirming the ubiquitous occurrence of excess argon in the Lothagam Basalt on the basis of the generally monotonic decrease in age over the first 15 to 50 percent of gas release. Based upon good age plateaus in two of the whole rock age spectra yielding ages agreeing at 4.20 ± 0.03 Ma, and a third sample yielding a slightly older apparent plateau age, we suggest 4.20 Ma as a maximum age for the Lothagam Basalt. Plagioclase separated from two of these basalt samples yield somewhat older apparent plateau ages, interpreted as indicating the presence of excess argon even in the plateau segments of the age spectra. As plagioclase is an important

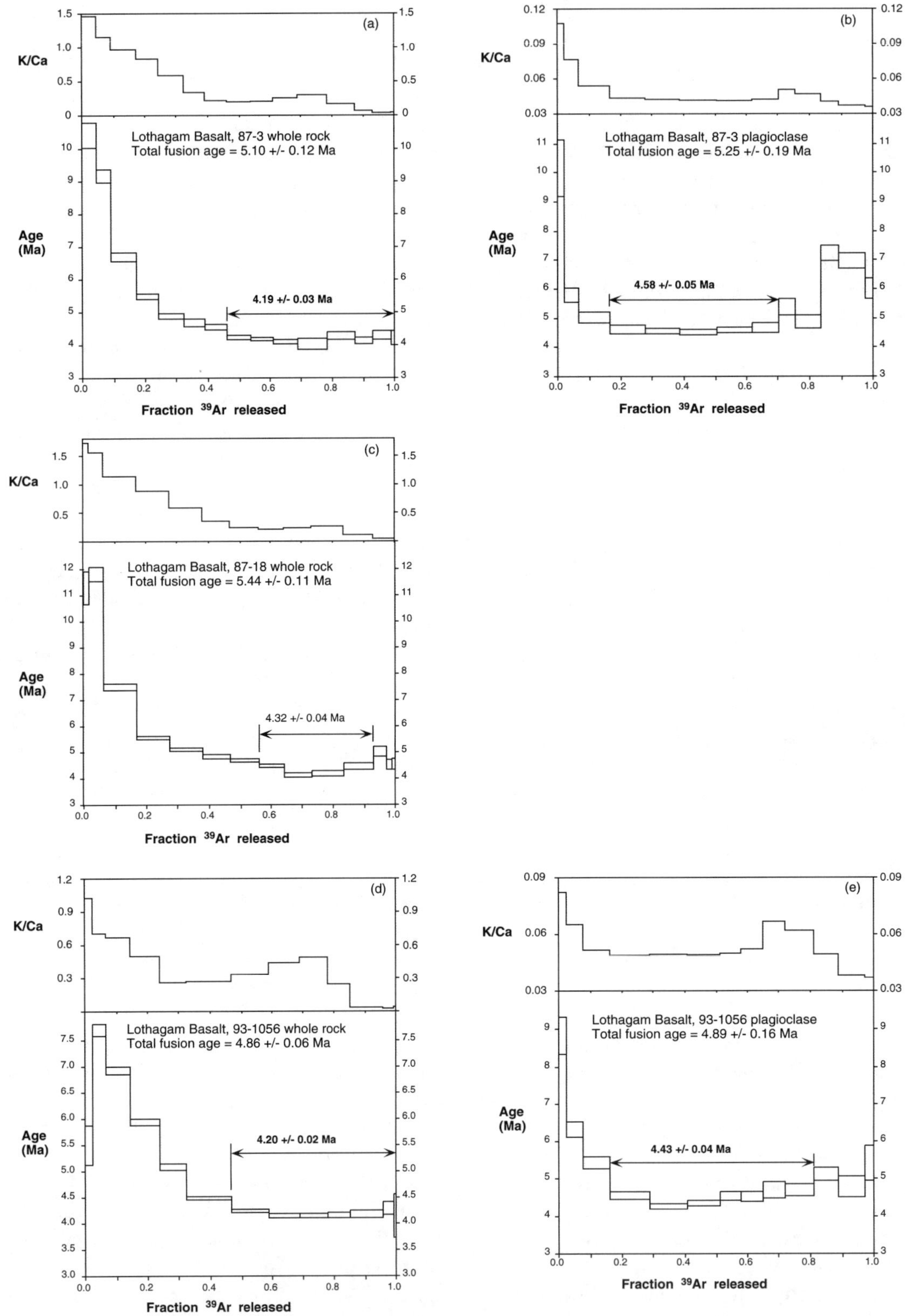

Figure 2.22 ^{40}Ar-^{39}Ar age spectra for whole rock basalts and plagioclase feldspars from the Lothagam Basalt. Also shown are the K/Ca plots derived from each step heating experiment.

component in the whole rocks also, caution must be exercised in accepting the whole rock plateau ages as other than maxima. Nevertheless, on present evidence we accept an age of 4.20 ± 0.03 Ma as a good estimate for the time of eruption of the Lothagam Basalt. Note that such an age is consistent with and indistinguishable from the age of 4.22 ± 0.03 Ma found for the pumiceous fragments in the Apak Member just below the Lothagam Basalt, so that we have considerable confidence in the age assignment.

Having derived an age for the Lothagam Basalt based upon the ^{40}Ar-^{39}Ar age spectra, we emphasize that the evidence for widespread but variable presence of excess argon in the Lothagam Basalt is very strong. As much of the excess argon appears to be released early in the step-heating experiments, with relatively high K/Ca ratios, an important carrier may well be the volcanic glass in the basalt. But the older apparent ages of the plagioclase separates suggest that they also are affected by the presence of excess argon. These results not only demonstrate the power of the ^{40}Ar-^{39}Ar step-heating technique in resolving such problems but also show how little we actually know about how excess argon is incorporated or occurs in subaerial basalts that are virtually free of obvious xenolithic or xeno-crystic contamination.

Discussion

In the broader perspective, Lothagam, an uplifted and tilted block, provides information on the geological history of this part of the Kenya Rift over about a 10 Ma interval commencing about 14 Ma ago in the mid Miocene (figure 2.18). As demonstrated by Morley et al. (1992), Lothagam is an uplifted segment of the much more extensive Kerio Basin, one of a series of half-graben basins that are characteristic of the northern Kenya Rift from Oligocene times. Our K-Ar dating results indicate that there was a period of phonolitic volcanism from ~14 to 12 Ma ago, together with deposition of proximal volcaniclastics, all included in the Nabwal Arangan beds. As an aside, a sequence of phonolitic volcanics and volcanically derived clastics of similar age occurs in the Lothidok Range, about 30 km north of Lothagam, in a rather similar tectonic and structural setting as Lothagam (Boschetto et al. 1992; Morley et al. 1992). Basalt at the top of the Nabwal Arangan beds at Lothagam yields an age of 9.1 ± 0.2 Ma, possibly indicating significant hiatuses in the recorded depositional history. The overlying conformable or near-conformable Nawata Formation indicates a rather quieter depositional regime as, apart from basal conglomerates, the formation is dominated by fluvial sandstones and mudstones. Interbedded tuffaceous beds, now analcimolites, together with some pumice clasts, indicate contemporaneous rhyolitic volcanism.

Dating of single crystals of alkali feldspar from pumice clasts shows that the lower member of the Nawata Formation was deposited over an interval from somewhat older than 7.4 Ma ago to 6.5 Ma ago in the Late Miocene at an average rate of ~80 mm/1,000 years. In the composite section (figure 2.18), the Lower Marker, the lowest dated unit, is estimated to be about 43 m above the base of the member. Extrapolating the sedimentation rate, we obtain an estimate for the age of initiation of deposition of the Nawata Formation of about 8.0 Ma.

Although there is good control on the age of the base of the upper member of the Nawata Formation from the 6.54 ± 0.04 Ma age on the Marker Tuff, the top of the upper member can only be directly constrained to be older than 4.2 Ma, the age obtained on units in the Nachukui Formation, higher in the sequence (figure 2.18). As an approximation, if we extrapolate the average sedimentation rate found for the lower member into the upper member of the Nawata Formation, we obtain an estimated age of 5.0 Ma for the top of the upper member, the Purple Marker, at about the Miocene-Pliocene boundary. Recognizing the difficulties inherent in such extrapolations, we must take into account the uncertainties of at least 0.2 Ma in the age estimates. Based upon the direct ages and the extrapolations, the Nawata Formation is considered to have been deposited over an interval of time extending over ~3.0 Ma from about 8.0 to 5.0 Ma ago in the latest Miocene, possibly extending into the earliest Pliocene.

The concordant ages on alkali feldspars from altered pumice clasts in the upper part of the Apak Member of the Nachukui Formation at 4.22 ± 0.03 Ma indicate deposition of this horizon at this time or shortly thereafter. The inferred age of 4.20 ± 0.03 Ma for the overlying Lothagam Basalt, based on ^{40}Ar-^{39}Ar step-heating results on whole rock samples, is consistent with the Apak result. We shall return to providing an age estimate for the base of the Apak Member subsequently.

Correlation of the Muruongori Member of the Nachukui Formation, overlying the Lothagam Basalt (figure 2.18), with the Lonyumun Member of the same formation (Leakey et al. 1996), suggests that the Muruongori Member is older than 3.92 ± 0.04 Ma, the age of the Moiti Tuff (Leakey et al. 1995), which immediately overlies the Lonyumun Member elsewhere in the Turkana Basin.

In order to attempt to obtain additional indirect age control, we shall now briefly consider the magnetostratigraphy available from Lothagam, summarized from Powers (1980) in figure 2.18, together with the Cande and Kent (1995) version of the geomagnetic polarity time scale (GPTS). Some additional palaeomagnetic results from Lothagam were obtained by Kamau (1994), and these broadly confirm the earlier measurements of Powers (1980). Leakey et al. (1996) proposed some correlations between the Lothagam magnetostratigraphy and the GPTS, but it will be evident from figure 2.18, that even with the new well-determined ages on parts of the Lothagam sequence, few unique or unequivocal correlations can be made. The fact that less than half the number of magnetozones has been recognized in the Lothagam sequence above the Nabwal Arangan beds, compared with the number shown for the GPTS for the age interval being discussed, emphasizes the problems involved in making secure correlations. Nevertheless, there appears to be a good match in the lower member of the Nawata Formation at a level just above the Middle Markers, where a reverse to normal polarity transition is tightly constrained (CSF, unpublished data), which may be correlated with the C3Ar-C3An.2n chron boundary of estimated age 6.57 Ma (Cande and Kent 1995). Across the Nawata Formation–Apak Member boundary, Powers (1980) showed a reverse to normal polarity change (figure 2.18). In the GPTS, the most likely correlate is the base of the C3n.4n subchron (the Thvera subchron) with estimated age of 5.23 Ma. This estimate is not very different from that derived above by extrapolation of sedimentation rate, which yielded an age of ~5.0 Ma for the Purple Marker at the top of the Nawata Formation.

However, these estimates of age for the base of the overlying Nachukui Formation represented by the Apak Member must be treated with caution as there may be a significant hiatus across the boundary between the Nawata and Nachukui Formations. Powers and Feibel (unpublished) note that the Purple Marker shows evidence for exposure on a land surface for an extended period of time. Leakey et al. (1996) reported differences in the fauna across the Purple Marker–Apak boundary, including a nearly complete turnover in the fossil pig species, perhaps also indicating a significant time break. On this basis, the normal polarity of the Apak Member could be correlated with any of the younger normal polarity subchrons in chron C3, the Gilbert Chron. Thus, the base of the Apak Member is older than 4.22 Ma and younger than 5.2 Ma, but in view of the arguments given above for a hiatus between the Apak and the Nawata, we suggest that the base is unlikely to be older than about 4.9 Ma and possibly somewhat younger. These same age limits apply to the hominoid mandibular fragment (KNM-LT 329) from Lothagam as it was found in sediments from the lowermost Apak Member (Leakey et al. 1996). The relationship of the hominoid mandible at Lothagam to the excellent hominid fossils from Kanapoi, assigned to the new species *Australopithecus anamensis* (Leakey et al. 1995), remains undetermined. Kanapoi is 65 km south of Lothagam, and the sequence in which most of the hominid fossils occur has been dated at between 4.17 ± 0.03 and 4.07 ± 0.02 Ma (Leakey et al. 1995, 1998), and therefore is probably correlative with the latest Apak and the Muruongori Members at Lothagam. The two hominoid teeth recovered from the upper member of the Nawata Formation (Leakey et al. 1996) can be bracketed between 6.54 Ma and about 5.0 Ma, latest Miocene to earliest Pliocene. It is worth noting that deposition of the upper member of the Nawata Formation covers the time interval correlated with the Messinian, the youngest Miocene stage.

The faunal turnover seen in the Lothagam sequence between the time of deposition of the Nawata Formation and the Nachukui Formation, reported by Leakey et al. (1996), is at or near the Miocene-Pliocene boundary at about 5.0 Ma. Although a depositional hiatus is considered likely across this boundary, it was also suggested by Leakey et al. (1996) that palaeoenvironmental factors were also important, with a change from riparian woodland to more open habitats.

Finally, the Lothagam block clearly has been uplifted subsequent to deposition of the Kaiyumung Member of the Nachukui Formation (figure 2.18). This is younger than 3.9 Ma, the estimated age for the top of the underlying Muruongori Member. The presence of even younger strata, including the KBS Tuff of age 1.88 ± 0.02 Ma (McDougall 1985) at Lothagam (see figure 2.18), further attests to youthful faulting. The main point being made is that major extensional faulting which resulted in the uplift of the Lothagam, probably owing to flexural isostatic response of the footwall to the major fault (Morley 1989; Morley et al. 1992), occurred in Pliocene or Pleistocene times. This faulting, principally along northerly trending lines, is characteristic of the Kenya Rift in the Turkana region (Morley et al. 1990, 1992; Dunkelman et al. 1989) and reflects the continued rifting to the present time from initiation about 30 Ma ago in the Oligocene.

Conclusions

New isotopic age determinations provide good chronostratigraphic control for faunal evolution and rift valley development within the Lothagam sequence, northern Kenya. A cluster of Middle Miocene ages (14.2–12.2 Ma) for the lower part of the Nabwal Arangan beds constrains a major episode of volcanic and volcaniclastic accumulation within the Kerio Valley half-graben. An age of 9.1 ± 0.2 Ma for a basalt flow in the uppermost part of the Nabwal Arangan beds provides a limit on the base of the overlying Nawata Formation. Fluvial strata of the lower Nawata Formation include four tephra layers with ages ranging from 7.44 ± 0.05 Ma to 6.52 ± 0.07 Ma (Late Miocene). The prominent Marker Tuff, separating the lower and upper members of the formation, yields an age of 6.54 ± 0.04 Ma and is indistinguishable from the subjacent dated level. Overlying strata of the upper member of the Nawata Formation did not yield datable materials, but extrapolated ages, based on average sedimentation rates in the lower member, suggest an age of about 5.0 Ma for the top of this unit. One datable horizon within the overlying Apak Member of the Nachukui Formation produced an age of 4.22 ± 0.03 Ma. Despite problematic K-Ar age results, the Lothagam Basalt, which caps the Apak Member, has yielded ^{40}Ar-^{39}Ar age spectra from which we infer an emplacement age of 4.20 ± 0.03 Ma.

The highly fossiliferous levels of the lower Nawata Formation are largely bracketed by the 7.4 Ma to 6.5 Ma dated tephras. Two hominoid teeth from the upper Nawata can be constrained to lie between 6.5 and 5 Ma based on extrapolated ages. The hominoid mandible KNM-LT 329 from the lower Apak Member can be placed between 4.2 and 5 Ma. This new age control on the Lothagam sequence places the fossil faunas in a tight framework relative to regional and global patterns of change in and around the Messinian Stage at the end of the Miocene. In addition, they establish Lothagam as an early extension of the long and relatively continuous record of Plio-Pleistocene rift valley evolution, documenting both changing landscapes and biotas, from the Turkana Basin.

Acknowledgments

We thank Dr. Meave G. Leakey and the National Museums of Kenya for their encouragement and for logistical support. Work by CSF at Lothagam was supported by grants from the National Science Foundation (BNS 90-07662) and the L.S.B. Leakey Foundation. Additional assistance was provided by the Koobi Fora Field School and Kenya Wildlife Service. Comments on drafts of this paper by M. G. Leakey and F. H. Brown were very helpful, as were the more formal reviews by F. H. Brown and R. Watkins, and the subject editor, R. Burgess. F. H. Brown initiated the attempts at K-Ar dating of the Lothagam Basalt reported in this paper, and introduced the first author to Lothagam.

Technical support in the geochronology laboratory at the Australian National University was provided by J. Mya for mineral separation, and by R. Maier and A. Doulgeris for the K-Ar and ^{40}Ar-^{39}Ar dating. Neutron irradiations were facilitated by the Australian Institute of Nuclear Science and Engineering and the Australian Nuclear Science and Technology Organization. C. Krayshek assisted with drafting of the figures. We thank B. Turrin for the synthetic K-silicate glass.

References Cited

Baksi, A. K. 1974. Isotopic fractionation of a loosely held atmospheric argon component in the Picture Gorge Basalts. *Earth and Planetary Science Letters* 21:431–438.

Behrensmeyer, A. K. 1976. Lothagam Hill, Kanapoi, and Ekora: A general summary of stratigraphy and faunas. In Y. Coppens, F. C. Howell, G. L. Isaac, and R. E. Leakey, eds., *Earliest Man and Environments in the Lake Rudolf Basin: Stratigraphy, Paleoecology, and Evolution,* pp. 163–170. Chicago: University of Chicago Press.

Boschetto, H. B., F. H. Brown, and I. McDougall. 1992. Stratigraphy of the Lothidok Range, northern Kenya, and K/Ar ages of its primates. *Journal of Human Evolution* 22:47–71.

Buck, W. R. 1988. Flexural rotation of normal faults. *Tectonics* 7:959–973.

Cande, S. C., and D. V. Kent. 1995. Revised calibration of the geomagnetic polarity timescale for the Late Cretaceous and Cenozoic. *Journal of Geophysical Research* 100:6093–6095.

Cebula, G. T., M. J. Kunk, H. H. Mehnert, C. W. Naeser, J. D. Obradovich, and J. F. Sutter. 1986. The Fish Canyon Tuff, a potential standard for the $^{40}Ar/^{39}Ar$ and fission track dating methods [abstract]. *Terra Cognita* 6:139.

Deino, A., and R. Potts. 1992. Age-probability spectra for examination of single-crystal $^{40}Ar/^{39}Ar$ dating results: Examples from Olorgesailie, southern Kenya Rift. *Quaternary International* 13/14:47–53.

Dunkelman, T. J., J. A. Karson, and B. R. Rosendahl. 1988. Structural style of the Turkana Rift, Kenya. *Geology* 16:258–261.

Dunkelman, T. J., B. R. Rosendahl, and J. A. Karson. 1989. Structure and stratigraphy of the Turkana Rift from seismic reflection data. *Journal of African Earth Sciences* 8:489–510.

Harland, W. B., R. L. Armstrong, A. V. Cox, L. E. Craig, A. G. Smith, and D. G. Smith. 1990. *A Geologic Time Scale 1989.* Cambridge: Cambridge University Press.

Harris, J. M., F. H. Brown, and M. G. Leakey. 1988. Stratigraphy and paleontology of Pliocene and Pleistocene localities west of Lake Turkana, Kenya. *Contributions in Science* 399:1–128.

Hill, A., S. Ward, and B. Brown. 1992. Anatomy and age of the Lothagam mandible. *Journal of Human Evolution* 22:439–451.

Kamau, R. K. 1994. Paleomagnetic study of the Plio-Pleistocene hominid-bearing strata in northern Kenya. M.S. thesis, University of Utah.

Kelley, S., and B. J. Bluck. 1989. Detrital mineral ages from the Southern Uplands using ^{40}Ar-^{39}Ar laser probe. *Journal of the Geological Society* (London) 146:401–403.

Kramer, A. 1986. Hominid-pongid distinctiveness in the Miocene-Pliocene fossil record: The Lothagam mandible. *American Journal of Physical Anthropology* 70:457–473.

Leakey, M. G., C. S. Feibel, I. McDougall, and A. Walker. 1995. New four-million-year-old hominid species from Kanapoi and Allia Bay, Kenya. *Nature* 376:565–571.

Leakey, M. G., C. S. Feibel, R. L. Bernor, J. M. Harris, T. E. Cerling, K. M. Stewart, G. W. Storrs, A. Walker, L. Werdelin, and A. J. Winkler. 1996. Lothagam: A record of faunal change in the Late Miocene of East Africa. *Journal of Vertebrate Paleontology* 16:556–570.

Leakey, M. G., C. S. Feibel, I. McDougall, C. Ward, and A. Walker. 1998. New specimens and confirmation of early age for *Australopithecus anamensis. Nature* 393:62–66.

McDougall, I. 1985. K-Ar and $^{40}Ar/^{39}Ar$ dating of the hominid-bearing Pliocene-Pleistocene sequence at Koobi Fora, Lake Turkana, northern Kenya. *Geological Society of America Bulletin* 96:159–175.

McDougall, I., and T. M. Harrison. 1988. *Geochronology and Thermochronology by the $^{40}Ar/^{39}Ar$ Method.* New York: Oxford University Press.

McDougall, I., and Z. Roksandic. 1974. Total fusion $^{40}Ar/^{39}Ar$ ages using HIFAR reactor. *Journal of the Geological Society of Australia* 21:81–89.

McDougall, I., and H.-U. Schmincke. 1977. Geochronology of Gran Canaria, Canary Islands: Age of shield building volcanism and other magmatic phases. *Bulletin Volcanologique* 40:57–77.

McDougall, I., and R. T. Watkins. 1988. Potassium-argon ages of volcanic rocks from northeast of Lake Turkana, northern Kenya. *Geological Magazine* 125:15–23.

McDougall, I., N. D. Watkins, and L. Kristjansson. 1976. Geochronology and paleomagnetism of a Miocene-Pliocene lava sequence at Bessastadaa, eastern Iceland. *American Journal of Science* 276:1078–1095.

Morley, C. K. 1989. Extension, detachment, and sedimentation in continental rifts (with particular reference to East Africa). *Tectonics* 8:1175–1192.

Morley, C. K., R. A. Nelson, T. L. Patton, and S. G. Munn. 1990. Transfer zones in the East African rift system and their relevance to hydrocarbon exploration in rifts. *American Association of Petroleum Geologists Bulletin* 74:1234–1253.

Morley, C. K., W. A. Wescott, D. M. Stone, R. M. Harper, S. T. Wigger, and F. M. Karanja. 1992. Tectonic evolution of the northern Kenya Rift. *Journal of the Geological Society* (London) 149:333–348.

Patterson, B., A. K. Behrensmeyer, and W. D. Sill. 1970. Geology and fauna of a new Pliocene locality in northwestern Kenya. *Nature* 226:918–921.

Powers, D. W. 1980. Geology of the Mio-Pliocene sediments of the lower Kerio River Valley. Ph.D. diss., Princeton University.

Smart, C. 1976. The Lothagam 1 fauna: Its phylogenetic, ecological and biogeographic significance. In Y. Coppens, F. C. Howell, G. L. Isaac, and R. E. Leakey, eds., *Earliest Man and Environments in the Lake Rudolf Basin: Stratigraphy, Paleoecology, and Evolution,* pp. 361–369. Chicago: University of Chicago Press.

Spell, T. L., I. McDougall, and A. P. Doulgeris. 1996. Cerro Toledo Rhyolite, Jemez Volcanic Field, New Mexico: $^{40}Ar/^{39}Ar$ geochronology of eruptions between two caldera-forming events. *Geological Society of America Bulletin* 108:1549–1566.

Steven, T. A., H. H. Mehnert, and J. D. Obradovich. 1967. *Age of Volcanic Activity in the San Juan Mountains, Colorado.* U.S. Geological Survey Professional Paper 575-D:47–55. Washington D.C.: Government Printing Office.

Tetley, N., I. McDougall, and H. R. Heydegger. 1980. Thermal neutron interferences in the $^{40}Ar/^{39}Ar$ dating technique. *Journal of Geophysical Research* 85:7201–7205.

White, T. D. 1986. *Australopithecus afarensis* and the Lothagam mandible. *Anthropos* (Brno) 23:79–90.

York, D. 1969. Least squares fitting of a straight line with correlated errors. *Earth and Planetary Science Letters* 5:320–324.

Zanettin, B., E. J. Visentin, G. Bellieni, E. M. Piccirillo, and F. Rita. 1983. Le volcanisme du bassin du Nord-Turkana (Kenya): Age, succession et évolution structurale. *Elf-Acquitaine Bulletin de Centres Recherches Exploration-Production* 7:249–255.

TABLE 2.4 Potassium-Argon Ages on Whole Rock Samples and on Plagioclase from Volcanic Rocks at Lothagam, Turkana Region, Northern Kenya

Lab No.	Field No.	Material	K (wt %)	$^{40}Ar^*$ (10^{-12} mol/g)	$^{40}Ar^*$ (%)	Calculated Age (Ma ± 1 s.d.)	Locality
Lothagam Basalt							
93-1056	93-1056	Basalt	0.857, 0.862	6.81	56.3	4.56 ± 0.05	Gorge, 2 km north of Apak; aerial photograph 775/118-071;
93-1056	93-1056	Plagioclase	0.327, 0.330	2.78	13.6	4.87 ± 0.09	2° 55′ 40″ N, 36° 02′ 52″ E
93-1058	93-1058	Basalt	0.751, 0.751	7.92	39.1	6.07 ± 0.06	Ridge, south of Apak; photo 774/106–117
87-3	K86-2899B	Basalt	1.014, 1.011	8.86	35.9	5.04 ± 0.06	Ridge, south of Apak, close to 93-1058
87-3	K86-2899B	Plagioclase	0.328, 0.331	2.69	20.7	4.70 ± 0.09	
87-18	K86-2898	Basalt	1.138, 1.141	10.59	35.7	5.35 ± 0.07	Ridge, south of Apak near 87-3
Nabwal Arangan beds							
93-1021	93-1021	Basalt	0.500, 0.510	8.01	66.9	9.12 ± 0.15	South flank Central Hill, photo 774/131–137; 2° 53′ 21″ N, 36° 03′ 10″ E
87-4	K86-2893	Phonolite	3.703, 3.697	78.6	73.5	12.2 ± 0.1	Clast from breccia in gorge
93-1040	93-1040	Phonolite	2.970, 2.977	71.3	84.3	13.8 ± 0.1	Ridge below northern peak, photo 775/194–119
93-1037	93-1037	Phonolite	2.648, 2.681	65.8	79.3	14.2 ± 0.2	Clast in breccia, on east ridge, photo 775/188–136
92-428	K90-4658	Basalt	1.701, 1.704	41.6	64.9	14.0 ± 0.1	Intrusive? into fault, photo 776/176–192

$\lambda_{e+e'} = 0.581 \times 10^{-10}$/year.

$\lambda_{\beta} = 4.962 \times 10^{-10}$/year.

$^{40}K/K = 1.167 \times 10^{-4}$ mol/mol.

$^{40}Ar^*$: radiogenic argon.

TABLE 2.5 Summary of ^{40}Ar-^{39}Ar Age Results on Single Crystals of Alkali Feldspar from Tuffaceous Horizons, Nawata Formation, and Apak Member, Nachukui Formation, Lothagam

Sample No.	No. of Crystals	K/Ca (±1 s.d.)	Arithmetic Mean Age (Ma ± 1 s.d.)	Weighted Mean Age (Ma ± 1 s.d.)	Isochron Analysis Age (Ma ± 1 s.d.)	Isochron Analysis $(^{40}Ar/^{36}Ar)_i$ (±1 s.d.)	Isochron Analysis MSWD	Stratigraphic Level
Apak Member, Nachukui Formation A								
95-184(A)	6	28.5 ± 11.3	4.232 ± 0.037	4.233 ± 0.006	4.220 ± 0.017	318.7 ± 7.5	1.2	
95-184(B)	7	43.2 ± 29.1	4.212 ± 0.029	4.218 ± 0.009	4.241 ± 0.018	263.4 ± 13.0	0.5	
All	13	36.4 ± 23.1	4.221 ± 0.033	4.228 ± 0.005	—	—	—	
Nawata Formation								
93-1025	12	37.1 ± 50.4	6.575 ± 0.094	6.596 ± 0.006	6.624 ± 0.094	265.7 ± 42.5	18.0	Marker Tuff
	10	40.8 ± 54.7	6.539 ± 0.044	6.555 ± 0.007	6.542 ± 0.041	308.6 ± 22.1	2.8	
93-1026(A)	6	6.5 ± 2.1	6.511 ± 0.040	6.525 ± 0.019	6.532 ± 0.051	291.8 ± 25.7	0.9	Tuff ~5 m below Red Marker
93-1026(C)	6	43.7 ± 17.0	6.537 ± 0.097	6.594 ± 0.015	6.574 ± 0.105	306.3 ± 20.0	5.6	
All	12	25.1 ± 22.6	6.523 ± 0.072	6.569 ± 0.012	—	—	—	
K91-4734(A)	7	9.5 ± 4.6	6.640 ± 0.022	6.641 ± 0.008	6.630 ± 0.022	299.7 ± 4.0	1.0	Tuff ~15 m below Red Marker
K91-4734(B)	9	6.6 ± 0.5	6.553 ± 0.072	6.593 ± 0.008	6.658 ± 0.055	211.7 ± 34.6	5.0	
93-1032(B)	5	7.9 ± 1.9	6.525 ± 0.066	6.537 ± 0.019	6.362 ± 0.386	377.9 ± 134.2	2.4	
93-1034	7	8.6 ± 4.5	6.537 ± 0.078	6.558 ± 0.017	6.619 ± 0.050	255.8 ± 27.3	2.0	
All	28	8.1 ± 3.3	6.566 ± 0.075	6.603 ± 0.005	—	—	—	
K91-4763(A)	5	26.0 ± 1.6	6.671 ± 0.048	6.681 ± 0.008	6.705 ± 0.046	239.8 ± 48.2	3.0	Middle Marker
K91-4763(B)	7	10.9 ± 2.9	6.744 ± 0.038	6.753 ± 0.007	6.779 ± 0.051	281.4 ± 14.5	4.9	
K91-4763(C)	5	12.0 ± 2.6	6.765 ± 0.028	6.768 ± 0.007	6.773 ± 0.047	277.6 ± 54.7	3.1	
93-1027(A)	7	13.6 ± 3.6	6.681 ± 0.049	6.686 ± 0.011	6.493 ± 0.017	479.6 ± 127.5	1.9	
93-1027(C)	6	13.9 ± 2.4	6.759 ± 0.033	6.762 ± 0.010	6.761 ± 0.016	297.0 ± 10.5	1.1	
93-1029(A)	6	15.7 ± 4.6	6.723 ± 0.042	6.715 ± 0.014	6.713 ± 0.017	296.9 ± 5.8	1.9	
93-1029(B)	6	21.2 ± 4.5	6.697 ± 0.109	6.740 ± 0.013	6.769 ± 0.049	281.3 ± 15.1	7.2	
All	42	15.9 ± 5.8	6.720 ± 0.062	6.731 ± 0.003	—	—	—	
K91-4710(B)	11	19.0 ± 3.6	7.442 ± 0.047	7.451 ± 0.006	7.440 ± 0.061	310.2 ± 24.6	6.6	Lower Marker
93-1020(A)	10	28.8 ± 4.9	7.470 ± 0.063	7.477 ± 0.008	7.470 ± 0.033	304.4 ± 16.7	1.8	
	9	28.8 ± 5.2	7.455 ± 0.044	7.474 ± 0.008	7.471 ± 0.025	299.0 ± 12.3	1.0	
93-1020(C)	6	18.8 ± 21.2	7.464 ± 0.077	7.487 ± 0.010	7.421 ± 0.208	351.7 ± 59.4	5.5	
93-1020(D)	9	48.4 ± 79.8	7.358 ± 0.112	7.407 ± 0.014	7.506 ± 0.042	221.6 ± 13.7	1.9	
	7	23.3 ± 9.3	7.411 ± 0.033	7.431 ± 0.016	7.465 ± 0.041	265.9 ± 27.9	0.3	
All	33	22.6 ± 10.7	7.443 ± 0.052	7.461 ± 0.004	—	—	—	

$\lambda = 5.543 \times 10^{-10}\ a^{-1}$; ages referenced to an age of 27.9 Ma for sanidine 92-176 from the Fish Canyon Tuff, Colorado.

TABLE 2.6 Results of ^{40}Ar-^{39}Ar Dating of Single Crystals of Alkali Feldspar from Altered Clasts in a Fluvial Sandstone in Apak Member, Nachukui Formation, About 17 m Below Lothagam Basalt

Sample	Mass (mg)	$^{36}Ar/^{39}Ar$ $\times 10^{-4}$	$^{37}Ar/^{39}Ar$ $\times 10^{-2}$	$^{40}Ar/^{39}Ar$	K/Ca	$^{40}Ar^*/^{39}Ar_K$ $\pm$ c.v.	$^{40}Ar^*$ (%)	^{39}Ar 10^{-14} mol	Calculated Age Ma $\pm$ 1 s.d.
95-184A alkali feldspar from altered clasts up to 10 mm in sandstone						$J = 0.001111 \pm 0.35\%$; ANU 23, level 9			
Crystal 1	1.5	1.705	2.957	2.177	17.8	2.104 ± 0.44%	96.7	2.04	4.214 ± 0.024
2	5.0	0.949	1.349	2.169	39.0	2.117 ± 0.23%	97.6	5.26	4.239 ± 0.018
4	1.6	1.167	4.600	2.158	11.4	2.102 ± 0.29%	97.4	2.84	4.210 ± 0.019
5	1.6	1.612	1.447	2.184	36.4	2.112 ± 0.56%	96.7	2.76	4.229 ± 0.028
6	1.3	1.448	1.479	2.163	35.6	2.097 ± 1.37%	96.9	2.17	4.198 ± 0.059
7	1.9	17.23	1.710	2.681	30.8	2.148 ± 0.48%	80.1	2.42	4.301 ± 0.026
								$\bar{x}$ 6	4.231 ± 0.037
95-184B alkali feldspar from altered clasts up to 10 mm in sandstone						$J = 0.001110 \pm 0.35\%$; ANU 23, level 10			
Crystal 1	1.6	6.848	1.422	2.351	37.0	2.123 ± 1.40%	90.3	2.35	4.248 ± 0.061
2	1.5	3.802	2.498	2.235	21.1	2.099 ± 0.91%	93.9	2.09	4.199 ± 0.041
3	0.9	2.633	1.495	2.209	35.2	2.106 ± 0.60%	95.3	1.31	4.213 ± 0.029
4	1.3	1.233	1.518	2.180	34.7	2.119 ± 0.31%	97.2	1.81	4.239 ± 0.020
5	1.2	7.712	0.577	2.342	91.2	2.089 ± 0.47%	89.2	1.88	4.178 ± 0.025
6	1.6	17.79	5.750	2.633	9.2	2.087 ± 1.63%	79.3	2.54	4.174 ± 0.069
7	1.1	2.604	0.711	2.218	74.0	2.115 ± 0.90%	95.4	1.93	4.231 ± 0.041
								$\bar{x}$ 7	4.212 ± 0.029

$\lambda = 5.543 \times 10^{-10}\ a^{-1}$. Fluence monitor sanidine 92-176 from Fish Canyon Tuff, Colorado, with nominal K-Ar age 27.9 Ma.

Correction factors: $(^{36}Ar/^{37}Ar)_{Ca} = 3.49 \times 10^{-4}$; $(^{39}Ar/^{37}Ar)_{Ca} = 7.86 \times 10^{-4}$; $(^{40}Ar/^{39}Ar)_K = 0.0256$ for 95-184A and 0.0265 for 95-184B. Overall mean age: $\bar{x}$ 13 = 4.221 ± 0.033 Ma.

Sensitivity ~3.2×10^{-17} mol/mV. 48 h irradiation in X33 or X34, HIFAR reactor, 0.2 mm Cd shielding used.

TABLE 2.7 Summary of ^{40}Ar-^{39}Ar Age Results on Samples from the Lothagam Basalt

Sample No.	K-Ar Age (Ma ± 1 s.d.)	Integrated Total Fusion Age $^{40}Ar/^{39}Ar$ (Ma ± 1 s.d.)	Plateau Age			$^{36}Ar/^{40}Ar$ versus $^{39}Ar/^{40}Ar$ Regression "Plateau" Points (Ma ± 1 s.d.)	$(^{40}Ar/^{36}Ar)_i$	MSWD
			(Ma ± 1 s.d.)	Steps	(% ^{39}Ar)			
87-3 WR	5.04 ± 0.06	5.10 ± 0.12	4.19 ± 0.03	8	54.0	4.20 ± 0.10	295.0 ± 3.6	0.67
87-3 Plagioclase	4.70 ± 0.09	5.25 ± 0.19	4.58 ± 0.05	5	53.7	4.36 ± 0.26	302.6 ± 7.8	0.08
93-1056 WR	4.56 ± 0.05	4.86 ± 0.06	4.20 ± 0.02	7	53.2	4.21 ± 0.05	294.2 ± 4.4	1.97
93-1056 Plagioclase	4.87 ± 0.09	4.89 ± 0.16	4.43 ± 0.04	7	64.8	4.07 ± 0.11	311.7 ± 4.6	0.66
87-18 WR	5.35 ± 0.07	5.43 ± 0.11	4.32 ± 0.04	4	37.1	4.43 ± 0.35	291.5 ± 12.7	6.10
All $\bar{x}$ 5	4.90 ± 0.31	5.11 ± 0.24	4.34 ± 0.16			4.25 ± 0.14		

$\lambda = 5.543 \times 10^{-10}\ a^{-1}$.

Irradiation ANU7/214, 48 h irradiation end day 258/1995.

Fluence monitor used, GA1550 Biotite, nominal K-Ar age 97.9 Ma.

WR: whole rock.

3

CRUSTACEA AND PISCES

3.1

Fossil Crabs (Crustacea, Decapoda, Brachyura) from Lothagam

Joel W. Martin and Sandra Trautwein

Remains of fossil crabs attributable to the family Potamonautidae have been recovered from the Nawata Formation and the Apak Member of the Nachukui Formation. Their occurrence in the Lothagam sequence is consistent with the presence of a well-oxygenated riverine system. More precise identification requires access to features not normally preserved in fossil crab material.

Freshwater crabs are a tremendously diverse assemblage of true (brachyuran) crabs known from Central and South America, Africa (including Madagascar), Australasia, southern Europe, and south and Southeast Asia. There are an estimated 900 species in the group, making them one of the most diverse assemblages of crabs. Species are known from cold, rapidly flowing mountain streams, tropical rainforest floors (where they may even be semiterrestrial or arboreal), warm lowland ponds and paddies, and just about any other freshwater environment (Ng 1988; Rodríguez 1982, 1992; Cumberlidge 1991; Cumberlidge and Sachs 1989a, 1989b). Their diversity, range, and size make them important ecologically, economically, and medically (as vectors of some tropical diseases). Although they were originally thought to comprise a single family (Potamidae), the group was treated as 11 families in three superfamilies by Bott (1970a, 1970b) and Pretzmann (1972, 1973). In turn, Bott's and Pretzmann's work has been questioned by more recent workers who employ cladistic methodology that is based on new morphological and molecular sequence data (e.g., Guinot et al. 1997; Cumberlidge 1999; Sternberg and Cumberlidge 1999; Abele et al. 1999). The taxonomy and phylogeny of the group are actively being revised on the basis of some of these new data. In this contribution, we follow the admittedly conservative classification of Martin and Davis (2001), where the freshwater crabs are composed of one family (Trichodactylidae) in the otherwise marine superfamily Portunoidea, one family (Psuedothelphusidae) in its own superfamily Pseudothelphusoidea, and six Old World freshwater families. The Old World families are partitioned among the superfamilies Gecarcinucoidea (families Gecarcinucidae and Parathelphusidae) and Potamoidea (families Deckiniidae, Platythelphusidae, Potamidae, and Potamonautidae (table 3.1). Many formerly recognized families have been synonymized in recent years (see discussion in Martin and Davis 2001).

Details of the timing of the invasion of freshwater by these crabs remain unclear. Hypotheses range from 11 independent unrelated invasions of freshwater by different groups of marine crabs during the Late Cretaceous to lower Tertiary (e.g., Bott 1970a, 1970b; Pretzmann 1973), to two lower Tertiary invasions (one in the Americas that resulted in the Trichodactylidae and another one elsewhere that led to all other families from some widespread marine ancestor; see Sternberg et al. 1998), to a single, much older (~200 Ma) colonization of the freshwater habitat; this resulted in freshwater crab monophyly (e.g., Rodríguez 1986; Ng et al. 1995).

In Africa, only the superfamily Potamoidea is known. (The superfamily Pseudothelphusoidea is restricted to Central and South America, as is the family Trichodactylidae of the otherwise marine superfamily Portunoidea; the two families of the Gecarcinucoidea are restricted to the Indian subcontinent, Southeast Asia, and Australasia). Of the four currently recognized potamoid families (Martin and Davis 2001), the Platythelphusidae are restricted to Lake Tanganyika, and the Potamidae are found in northwest Africa, southeastern Europe, and Asia (figure 3.1).

Thus, with some certainty we can say that the fossil crabs from Lothagam could only belong to one of two families: Potamonautidae (known only from sub-Saharan Africa plus the Nile in Egypt and from Madagascar) and Deckeniidae (known only from East Africa). Species identifications in these families often are based on the detailed structure of the male pleopods (among other features), such that even remarkably preserved fossils could not be identified further (that is, to the level of genus or species). We are assuming that all of the Lothagam fossils are members of the freshwater crab superfamily Potamoidea and that, based on the modern-day distribution of this family, they are probably members of the family Potamonautidae. The "almost complete lack of a fossil record for all groups of African freshwater crabs" (Sternberg and Cumberlidge 1999:493) makes comparisons with existing fossil material virtually impossible. In this report we give brief descriptions of fossil freshwater crabs from the Late Miocene hominid-bearing locality of Lothagam, Kenya, that were collected during the 1991 and 1992 field seasons. Catalog numbers for these specimens begin with the acronym KNMI-LT, which denotes invertebrate fossils from Lothagam in the collections of the National Museums of Kenya, Nairobi.

Figure 3.1 Two of the more complete fossil crab fragments from Lothagam. Top = KNMI-LT 23667, Upper Nawata, posterior two thirds of carapace, dorsal view, with carpus of right cheliped visible at upper right. Estimated size of entire crab is carapace width, 38.5 mm; carapace length (estimated because of incomplete frontal region), 33.1 mm. Bottom = KNMI-LT 24193, Lower Nawata, ventral view of different specimen with intact left cheliped (in outer view) and with portions of pereiopods two and three. Length of chela (base of propodus to tip of propodal finger), 28.4 mm; height (just proximal to articulation of dactylus and propodus), 11.5 mm.

Materials and Methods

We examined 32 specimens that represented parts of fossil crabs from Mio-Pliocene strata exposed at Lothagam. These samples derived from both members of the Nawata Formation and from the Apak Member of the Nachukui Formation; thus, they ranged in age from 4.2 to 7.4 Ma (McDougall and Feibel 1999). By far the majority of the specimens comprised extremities and midsections of the fingers of the chelipeds, including both dactylar and propodal finger pieces. Occasional larger specimens contained fragments of carapace, but none was complete enough to allow positive identification, even to the family level. Specimens selected for photography were lightly cleaned with a dry paintbrush. Observations and line illustrations were made with a Wild M5 APO stereomicroscope. Measurements were made with digital calipers and rounded to the nearest tenth of a millimeter.

Systematic Description

Superfamily Potamoidea

The superfamily Potamoidea contains four families, two of which—Deckiniidae and Potamonautidae—occur today in East Africa.

Family Potamonautidae

Most characters that serve to distinguish crabs of the family Deckiniidae from those of the family Potamonautidae involve details of the orbital margins and the fifth pereiopod dactylus (Sternberg and Cumberlidge 1999), features that are not preserved in any of the Lothagam fossils. The Deckiniidae contains only the genus *Deckenia,* which currently contains two species, *D. imitatrix* and *D. mitis* (Ng et al. 1995). The group is characterized by an "ovate carapace" caused by (or facilitating) greatly swollen branchial chambers, which "seems to be associated with terrestrial habits or life in stagnant, poorly oxygenated waters" (Ng et al. 1995:583). The few fragments of dorsal carapace (e.g., figure 3.1) revealed no signs of an expanded branchial region, and thus the genus *Deckinia* (and the family Deckiniidae) have been ruled out. Although we could not detect a clear epigastric crest or a postorbital crest that extends to the epibranchial tooth, both of which are reported to characterize species in the Potamonautidae (Sternberg and Cumberlidge 1999:505, 506), these regions of the carapace were very poorly preserved. By default, and assuming also that all fossilized pieces sent to us came from crabs with similar carapace structure, we have assigned all of the Lothagam fossils to the family Potamonautidae. Arguing against this placement is the fact that one chelipedal carpus was preserved (KNMI-LT 23667; figure 3.1, top), and it appeared to possess a single anteromedial spine, whereas potamonautids typically have two such spines (Sternberg and Cumberlidge 1999; Cumberlidge 1999).

Potamonautidae gen. and sp. indet.

(Figures 3.1, 3.2)

Lothagam Material

- *Lower Nawata:* 1, dactylus fragment; 24193, claw and part exoskeleton; 24194, claw and part exoskeleton; 24195, claw fragments; 24196, chela fragment; 24197, claw fragment; 25094, Rt. propodal finger; 25095, dactylus; 25100, claw fragment; 25101, chela fragment; 25102, chela fragments; 25415, exoskeleton and claw fragment; 25416, limb fragments.
- *Upper Nawata:* 23667, exoskeleton; 24188, claw fragment; 24190, claws; 24191, 2 claw fragments; 24192, chela fragment; 25087, claw fragment; 25088, claw fragments; 25089, exoskeleton; 25090, claw and exoskeleton fragments; 25091, claw; 25092, chela fragment; 25093, 3 claw fragments; 25096, chela fragments; 25097, propodal finger fragment; 25098, claw fragment; 25099, claw fragment; 25128, dactylus fragment.
- *Apak Member:* 24187, chela fragment; 24189, claw fragment.

KNMI-LT 23667 (figure 3.1, top) is a large specimen, consisting mostly of a badly fractured posterior two-thirds of carapace and part of the right cheliped. The carpus of the right cheliped is striking, with sharp anteromedial and smaller anterolateral spines. Greatest carapace width 38.5 mm; greatest carapace length (estimated because of deteriorated frontal region) 33.1 mm.

KNMI-LT 24187 comprises the middle portion of a chelipedal finger. Five teeth are visible, the middle being largest and approximately twice the height of the other four. Its curvature suggests this is the dactylus of the left chela or possibly the propodus of the right chela (less likely). Length 10.0 mm; greatest height (at basal tooth) 5.4 mm.

KNMI-LT 24188 consists of a small basal to three-fourths length of a left chela dactylus with six low, rounded teeth on the cutting surface. Length 9.9 mm; height 4.6 mm at base.

KNMI-LT 24189 is a small fragment of only the sclerotized portion (i.e., teeth and immediately adjacent area) of a right dactylus or left propodus. Nine teeth, of varying sizes, are visible along the cutting surface. Length 17.2 mm; greatest height (at approximate level of basalmost tooth) 3.5 mm.

KNMI-LT 24190 comprises the dactylus and most of the propodus of a right chela in "outer" view. Length 28.5 mm; height 12.3 mm.

KNMI-LT 24191 comprises two claw fragments. The larger (thicker) fragment (length 16.4 mm; greatest height 6.5 mm) appears to be the right propodal finger and bears a row of cutting teeth. These increase in size from 1 to 5; tooth 6 is small, tooth 7 is approximately equal to 5, and thereafter the teeth decrease in size toward the tip. The smaller (thinner) fragment (length 18.5 mm; height 5.1 mm) appears to be dactylar (slender, more curved), but this is not definite. Approximately eight teeth are visible on this fragment, the proximal four of which are larger than the distal four.

KNMI-LT 24192 (figure 3.2e) is a large (length 23.5 mm; height 7.9 mm at base), strongly curved dactylus of the right chela and is obviously from a crab that possessed a large "gape" when chela fingers were closed. The cutting surface has a row of three or four small, rounded teeth and has minute tubercles distal to the last tooth.

KNMI-LT 24193 (figure 3.1) represents an entire left chela, most of the cheliped, and the coxa of pereiopods 2 and (partial) 3. Part of the thoracic sternum is visible, showing cuticular punctae. Length of the entire chela (base of propodus to tip of propodal finger) 28.4 mm; estimated length of chela plus carpal segment 30.5 mm. Greatest height of chela (measured just proximal to point of articulation of dactylus) 11.5 mm.

KNMI-LT 24194 comprises parts of both fingers of the left chela; additionally, a small piece of the sternal plastron is visible. Length (of entire fossil) 16.8 mm; width of entire fossil 16.8 mm. Greatest length of chela fingers 13.5 mm; height of fingers (combined) 8.2 mm.

KNMI-LT 24195 contains 13 fragments of chelipedal fingers. Two of the fragments are quite large (length of largest 21.5 mm; height 9.9 mm measured from the bottom of the first tooth to the top of the finger).

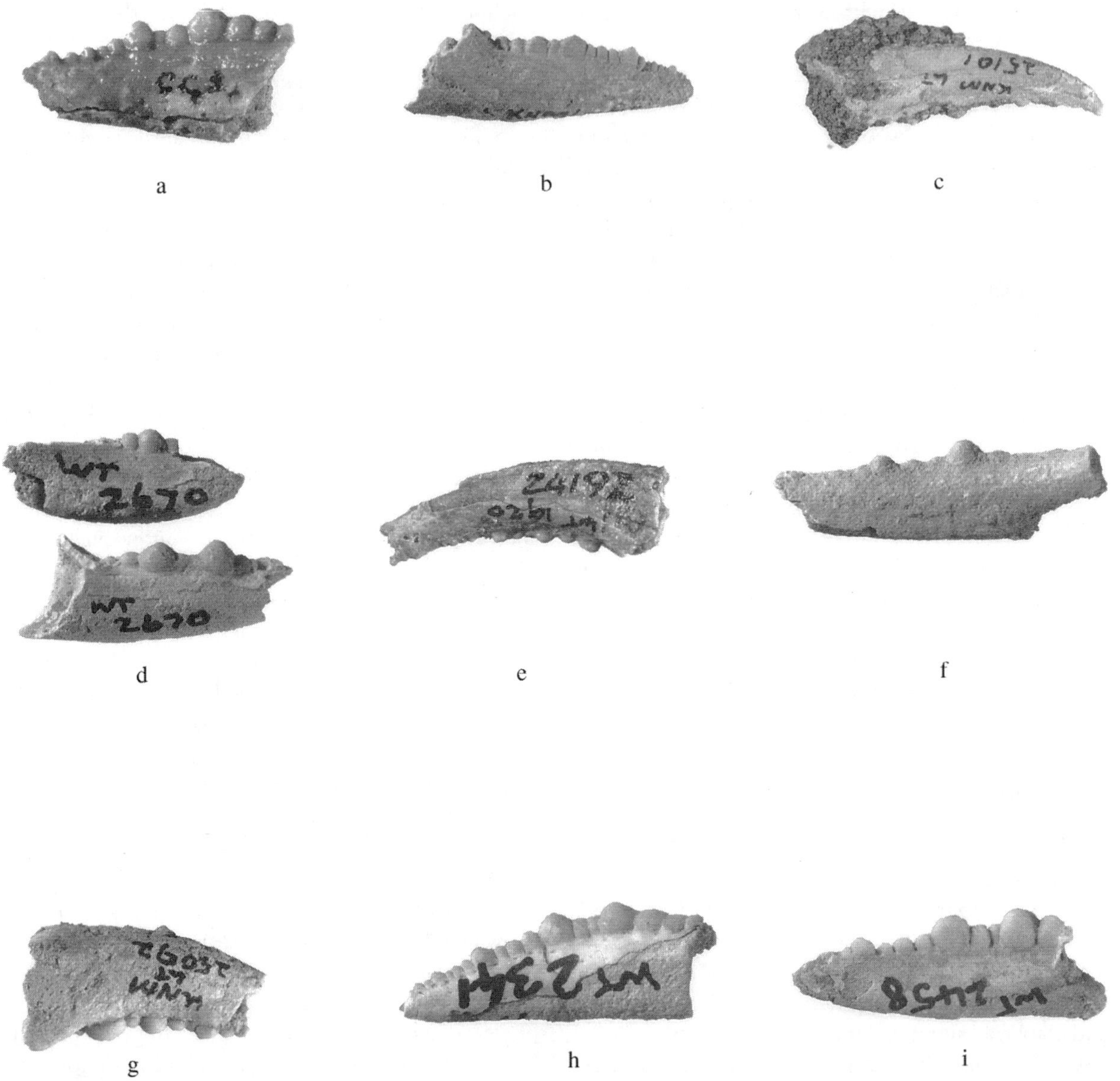

Figure 3.2 Photographs of representative fossil crab fragments from the Lower Nawata (a–d) and Upper Nawata (e–i). a = KNMI-LT 24196, length = 15.8 mm; b = KNMI-LT 25094, length = 23.2 mm; c = KNMI-LT 25101, length = 23.9 mm; d = KNMI-LT 25102 (two fragments), lengths = 19.7 mm and 17.2 mm; e = KNMI-LT 24192, length = 23.5 mm; f = KNMI-LT 25087, length = 18.0 mm; g = KNMI-LT 25092, length = 15.8 mm; h = KNMI-LT 25097, length = 13.3 mm; i = KNMI-LT 25099, length = 20.5 mm.

KNMI-LT 24196 (figure 3.2a) is a large piece of a chela tip (probably the propodal finger of the right chela), one of few that extends to almost the distal extremity. The cutting surface has 12 closely set teeth of varying sizes; teeth 4 and 7 (proximal to distal) are largest. Length 15.8 mm; height (at basalmost of two large teeth) 7.4 mm.

KNMI-LT 24197 is a minute tip of a claw with four or five low, well-worn teeth. Length 7.1 mm; height 3.0 mm.

KNMI-LT 25087 (figure 3.2f) probably represents a left dactylus. It consists of a long, thin claw fragment with two large teeth and one smaller one between them, also one small tooth distal to second large tooth. Length 18.0 mm; height 5.9 mm at basalmost large tooth.

KNMI-LT 25088 constitutes two fragments, both in poor condition. The larger (length 14.2 mm; height 3.5 mm at second basal tooth) consists of a row of eight teeth, sizes of which vary; teeth 2 and 4 (proximal to distal) are larger than the others. Smaller fragment length 11.2 mm; height at midpoint 3.4 mm.

KNMI-LT 25089 includes part of a fairly large female; the abdomen and part of the left chela propodus are fairly clear in ventral view. Greatest length of the left chela 28.1 mm; greatest height 9.9 mm. There is also a partial right dactylus (length 18.0 mm; height 8.0 mm at base).

KNMI-LT 25090 includes six individual fragments of chelipeds. The matrix is extremely hard, and possibly for this reason the specimens appear glossy, much more so than in other samples. All specimens are badly fragmented; very little information can be gleaned from them. One notable specimen is a nearly entire claw in "outer view," with the manus ("hand" of the propodus) broken open and with tips of both fingers missing. Length of entire chela (fingers and imprint of propodus) 29.3 mm; height (measured just proximal to articulation with dactylus) 12.3 mm. Other pieces are mostly claw and leg fragments.

KNMI-LT 25091 constitutes an extremely large (length 25.4 mm; height 11.0 mm at base) dactylus of a right chela; most of the cutting surface is obscured by adhering calcareous matrix.

KNMI-LT 25092 (figure 3.2g) represents the proximal three-fourths of a right chela dactylus. A large basal tooth is followed distally by smaller, then larger, then smaller teeth. Eight teeth are visible along the cutting surface. Length 15.8 mm; height 8.7 mm at base.

KNMI-LT 25093 comprises three claw fragments. The largest fragment (length 18.7 mm; height 6.4 mm at base) is gently curving and bears 12 teeth on the cutting surface; the teeth are more or less alternate in size, medium to small. A second fragment constitutes a very small (length 10.1 mm; height 4.2 mm) claw tip, either left dactyl or right propodal finger; 13 teeth are present on the cutting surface, with teeth 4 and 6 larger and slightly more acute than the others. The third fragment comprises a midsection of a claw tip. Length 11.9 mm; height 6.5 mm.

KNMI-LT 25094 (figure 3.2b) is a large (length 23.2 mm; height 8.5 mm at base), heavy, well-formed right propodal fixed finger, entire nearly to the distal extremity. The entire row of teeth on the cutting surface is visible; teeth 5 and 9 (proximal to distal) are markedly larger and more acute than the others.

KNMI-LT 25095 is a thick, blunt finger, clearly the dactylus of a right chela, worn, with few details discernible. Length 18.0 mm; height 8.4 mm at base.

KNMI-LT 25096 comprises four small fragments. The largest (length 20.6 mm; height 8.8 mm at base) is probably a right chela dactylus. Another is the base of a chela dactylus.

KNMI-LT 25097 (figure 3.2h) is a small but remarkably clean fragment of a right propodal finger; it is fragile and hollow. Eighteen teeth are visible along the cutting surface; teeth 4 and 7 (proximal to distal) are larger than the others, some of which are minute, especially toward the tip. Length 13.3 mm; height 5.0 mm.

KNMI-LT 25098 is a thick basal portion of right chela dactylus. The basal tooth is large and is followed by seven or more teeth (extremity of finger missing) of alternating sizes. Length 13.2 mm; height at base 7.8 mm.

KNMI-LT 25099 (figure 3.2i) is a large piece of what appears to be the right propodal finger of an obviously sizeable crab (length 20.5 mm; height 7.4 mm at second basalmost tooth). Approximately ten teeth are visible along the cutting edge; the basal-most two teeth and teeth 5 and 9 (proximal to distal) rise above the others.

KNMI-LT 25100 represents the partial dactylus of a right chela with a large basal tooth plus eight small teeth; lots of adhering matrix. Length 20.2 mm; height 10.0 mm.

KNMI-LT 25101 (figure 3.2c) is a large (length 23.9 mm; height 9.5 mm at base), curving dactylus of a left chela (possibly the propodus of a right chela, but very delicate if so). Cutting surface with eight teeth, varying in size but mostly small.

KNMI-LT 25102 comprises two specimens. The larger is a dactylus of a left chela (length 19.7 mm; height 7.9 mm at most basal tooth) (figure 3.2d, lower photograph); there are two large teeth, with smaller teeth proximally and distally (not between them). The smaller specimen (length 17.2 mm; height 6.6 mm) (figure 3.2d, upper photograph) is a right dactylus or left propodal finger, with a cluster of three tightly opposed teeth, of which the center one is the largest.

KNMI-LT 25128 comprises a midsection of a right chela dactylus; the cutting surface has eight visible teeth. Length 11.5 mm; height 7.8 mm at base.

KNMI-LT 25415 comprises two specimens. The (larger) first specimen is a rare "whole crab" fossil, showing the underside (mostly) and partial upper surface of the carapace, plus part of one cheliped. Some sternal plates and sutures are visible, but it is difficult to enumerate these as the specimen is in very poor condition. Estimated size of the entire crab is 27.8 mm carapace width. The second specimen is a very thin (probably dactylar) chela finger with approximately seven low, worn teeth. Length 12.4 mm; height 3.9 mm at base.

KNMI-LT 25416 includes a nearly complete left chela and a large carpal segment (possibly from a different crab). Length of nearly complete chela 21.5 mm; height 7.8 mm. Other material includes small pieces of carapace and a partial right cheliped dactylus (length 15.1 mm; height 9.5 mm at base). There are also many unidentifiable fragments.

KNMI-LTI 1 appears to be a partial left chela dactylus; seven teeth are visible near the base of the cutting surface. Length 17.6 mm; height 10.9 mm at base.

Paleoenvironmental Interpretation

Had these fossils proven to be members of the East African family Deckiniidae, some speculation on habitat might have been warranted because the inflated carapace of deckiniids may reflect a low oxygen environment (Ng 1988; Ng et al. 1995). The fact that they are apparently members of the Potamonautidae reveals less information, as potamonautids are known from a much wider range of habitats, virtually throughout sub-Saharan Africa and including the Nile River in Egypt, eastern Africa, South Africa (e.g., Stewart 1997; Cumberlidge 1999), and parts of Madagascar. In general, however, potamonautids are commonly referred to as "river crabs" because of their propensity for these habitats. Thus, we may assume that the Lothagam fossil crabs are indicative of a relatively well-oxygenated riverine system. Virtually all freshwater crabs, including the African potamonautids, are opportunistic scavengers and predators, and thus the presence of these crabs in Lothagam furnishes little information about cooccurring species.

Acknowledgments

We thank the government of Kenya and the trustees of the National Museums for permission to study these interesting specimens. We thank Neil Cumberlidge and Trisha Spears for sharing their thoughts on the origins, distributions, and classification of the modern freshwater crab families. This work was funded in part by a U.S. National Science Foundation grant to J. W. Martin and Dave Jacobs (DEB 9978193, PEET program in Systematic Biology).

References Cited

Abele, L. G., T. Spears, and N. Cumberlidge. 1999. Biogeography and phylogeny of freshwater crabs based on molecular evidence. In *Program and Abstracts, The Crustacean Society Summer Meeting,* p. 20. Lafayette, Louisiana, 26–30 May 1999.

Bott, R. 1970a. Betrachtungen über die Entwicklungsgeschichte und Verbreitung der Subwasser-Krabben nach der Sammlung des Naturhistorischen Museums in Genf/Schweiz. *Revue Suisse de Zoologie* 77:327–344.

Bott, R. 1970b. Die Subwasserkrabben von Europa, Asien, Australien und ihre Stammesgeschichte. Eine revision der Potamoidea und der Parathelphusoidea (Crustacea, Decapoda). *Abhandlungen der Senckenbergischen Naturforschenden Gesellschaft* 526:1–338.

Cumberlidge, N. 1991. *Sudanonautes kagorensis,* a new species of fresh-water crab (Decapoda: Potamoidea: Potamonautidae) from Nigeria. *Canadian Journal of Zoology* 69: 1938–1944.

Cumberlidge, N. 1999. *The Freshwater Crabs of West Africa: Family Potamonautidae.* Faune et Flore Tropicales No. 35. Paris: Institut de Recherche pour le Développement.

Cumberlidge, N., and R. Sachs. 1989a. A key to the crabs of Liberian fresh waters. *Zeitschrift für Angewandte Zoologie* 76:221–229.

Cumberlidge, N., and R. Sachs. 1989b. Three new subspecies of the West African freshwater crab *Liberonautes latidactylus* (de Man, 1903) from Liberia, with notes on their ecology. *Zeitschrift für Angewandte Zoologie* 76:425–439.

Guinot, D., B. G. M. Jamieson, and C. C. Tudge. 1997. Ultrastructure and relationships of spermatozoa of the freshwater crabs *Potamon fluviatile* and *Potamon ibericum* (Crustacea, Decapoda, Potamidae). *Journal of Zoology* (London) 241: 229–244.

Martin, J. W., and G. E. Davis. 2001. *An Updated Classification of the Recent Crustacea.* Science Series No. 39. Los Angeles: Natural History Museum of Los Angeles County.

McDougall, I., and C. S. Feibel. 1999. Numerical age control for the Miocene-Pliocene succession at Lothagam, a hominoid-bearing sequence in the northern Kenya Rift. *Journal of the Geological Society* (London) 156:731–745.

Ng, P. K. L. 1988. *The Freshwater Crabs of Peninsular Malaysia and Singapore.* Singapore: Department of Zoology, National University of Singapore, and Shing Lee Publishers.

Ng, P. K. L., Z. Stevcic, and G. Pretzmann. 1995. A revision of the family Deckiniidae Ortmann, 1897 (Crustacea: Decapoda: Brachyura: Potamoidea), with description of a new genus (Gecarcinucidae, Gecarcinucoidea) from the Seychelles, Indian Ocean. *Journal of Natural History* 29: 581–600.

Pretzmann, G. 1972. Die Pseudothelphusidae (Crustacea, Brachyura). *Zoologica* 42:1–182.

Pretzmann, G. 1973. Grundlagen und Ergebnisse der Systematik der Pseudothelphusidae. *Zeitschrift für Zoologische Systematik und Evolutionsforschung* 11:196–218.

Rodríguez, G. 1982. *Les crabes d'eau douces d'Amerique: Famille des Pseudothelphusidae.* Faune Tropicale 22. Paris: ORSTOM.

Rodríguez, G. 1986. Centers of radiation of fresh-water crabs in the neotropics. In R. H. Gore and K. L. Heck, eds., *Biogeography of the Crustacea,* pp. 51–67. Crustacean Issues 4. Rotterdam: Balkema.

Rodríguez, G. 1992. *The Freshwater Crabs of America: Family Trichodactylidae and Supplement to the Family Pseudothelphusidae.* Faune Tropicale 31. Paris: ORSTOM.

Sternberg, R. von, and N. Cumberlidge. 1999. A cladistic analysis of *Platythelphusa A.* Milne-Edwards, 1887, from Lake Tanganyika, East Africa (Decapoda: Potamoidea: Platythelphusidae) with comments on the phylogenetic position of the group. *Journal of Natural History* 3:493–511.

Sternberg, R. von, N. Cumberlidge, and G. Rodríguez. 1998. On the marine sistergroups of the freshwater crabs (Crustacea: Decapoda). *Journal of Zoological Systematics and Evolutionary Research* 37:19–38.

Stewart, B. A. 1997. Morphological and genetic differentiation between populations of river crabs (Decapoda: Potamonautidae) from the Western Cape, South Africa, with a taxonomic revision of *Gecarcinautes brincki. Zoological Journal of the Linnean Society* 199:1–21.

TABLE 3.1 Classification of Freshwater Crabs[a] and Their Present-day Distribution

Subphylum Crustacea
Order Decapoda
Suborder Pleocyemata
Infraorder Brachyura
Section Eubrachyura[b]
Subsection Heterotremata

Family	Location
Superfamily Portunoidea[c]	
Family Trichodactylidae	Central and South America
Superfamily Pseudothelphusoidea	
Family Pseudothelphusidae	Central and South America
Superfamily Gecarcinucoidea	
Family Gecarcinucidae	Indian subcontinent and Southeast Asia
Family Parathelphusidae	Indian subcontinent, Southeast Asia, and Australasia
Superfamily Potamoidea	
Family Deckiniidae	East Africa
Family Platythelphusidae	Lake Tanganyika
Family Potamidae	Northwest Africa, southeast Europe, the Middle East, the Himalayas, Southeast Asia, and China
Family Potamonautidae	Subsaharan Africa (plus the Nile in Egypt) and Madagascar

[a] Contains a total of three subsections, 20 superfamilies, and 61 families (Martin and Davis 2001).

[b] Other families of brachyuran crabs have species that can or must live in freshwater (e.g., *Metapaulius depressus* in the Grapsidae, *Uca subcylindrica* in the Ocypodidae, and others) but that are not confused with the potamon-like crabs (the former "Family Potamida").

[c] Contains two other families of crabs restricted to marine or estuarine waters.

3.2

Fossil Fish Remains from Mio-Pliocene Deposits at Lothagam, Kenya

Kathlyn M. Stewart

More than 7,000 fossil fish elements collected from Late Miocene and Pliocene strata at Lothagam show considerable change throughout the sequence from the lower member of the Nawata Formation to the Kaiyumung Member. The Nawata Formation fish fauna appears to be uniform, although more sites with fish fossils occur in the Lower Nawata than in the Upper. The Nawata Formation fauna contains small-sized fish and archaic genera. In the Apak Member of the Nachukui Formation, the archaic elements are lost or scarce and extant genera predominate. Both Nawata and Apak faunas are river-adapted. The superjacent Muruongori and Kaiyumung Members contain a predominantly lake fauna with several new taxa. *Sindacharax deserti*, *Semlikiichthys rhachirhinchus*, and *Tetraodon* sp. are new to the Turkana Basin but are also known from Mio-Pleistocene Egyptian and/or Western Rift deposits in Zaire/Uganda; they represent exchange of faunas with those regions through a newly opened hydrological connection. The freshwater puffer also makes its first appearance in the basin as a new species. Characids show considerable evolutionary change, with new species recognized from the Nawata Formation and from the Apak and Kaiyumung Members of the Nachukui Formation. The near absence of tilapiine cichlids throughout the Lothagam succession may signify a later immigration from Asia than previously thought.

More than 7,000 fossil fish elements were collected from the Lothagam deposits from 1991 to 1993 by a National Museums of Kenya team that included the author. Previous expeditions had noted the presence of fish (e.g., Smart 1976), and occasional surface collections had been made (e.g., Schwartz 1983) but before the National Museums of Kenya expeditions no systematic recovery of fish had been undertaken.

Fish elements were collected from the Lower and Upper Nawata, and from the Apak, Muruongori, and Kaiyumung Members of the Nachukui Formation. Elements from Holocene-aged deposits were rarely collected and are not reported here. In the 1991 and 1992 seasons, we recovered all teeth and bone elements; in the 1993 season, we recovered all teeth but only rare or unusual bone elements. Although they are certainly not comprehensive, these findings do give some approximation of the diversity and abundance of species throughout the deposits.

Elements were identified using both the extant fish collections of the osteology department of the National Museums of Kenya in Nairobi and the fossil fish collections, primarily those collected by H. Schwartz, in the palaeontology division. Although only elements from Lothagam are reported here, elements and teeth from the nearby Pliocene-aged Turkana Basin sites of Kanapoi, Ekora, South Turkwel, North Napudet, and Eshoa Kakurongori were also collected between 1991 and 1995; these other collections are mentioned below only if they contribute to identification of Lothagam elements.

Only the type specimens of the new fish species erected in this contribution have been provided with accession numbers; other specimens are referred to by their field numbers pending their formal accession into the collections of the National Museums of Kenya. The divisions on the scales provided in the illustrations are at millimeter intervals unless otherwise indicated.

Systematic Description

Order Protopteriformes
Family Protopteridae
Protopterus Owen, 1839
Protopterus sp. (lungfish)

Lothagam Material

- *Lower Nawata:* 1644, toothplate fragment; 1658, 11 lower toothplates, 2 upper toothplates; 1659, lower toothplate, 2 upper toothplate fragments; 1672, lower toothplate; 1710, 2 lower toothplates, 2 toothplate fragments; 1732, 2 toothplate fragments; 1733, 2 lower toothplates, toothplate fragment; 2301, upper toothplate; 2365 upper toothplate; 2383, 2 lower toothplates; 2385, upper toothplate; 2410, toothplate; 2413, upper toothplate.
- *Upper Nawata:* 1594, lower toothplate fragment; 1655, toothplate fragment; 1765, lower toothplate portion; 1766, upper toothplate; 1957, upper toothplate; 2223, toothplate.
- *Apak Member:* 1768, lower toothplate.
- *Kaiyumung Member:* 1850, 3 toothplates; 1852, toothplate; 1992, 2 toothplate fragments.

Protopterus toothplates are robust and preserve well as fossils. Unfortunately, they are not diagnostic below the level of genus. Lothagam *Protopterus* fossils represent individuals that ranged in size from an estimated total length of 10 cm to over 1 m.

Discussion

Protopterus toothplates were reasonably frequent at Nawata Formation sites. They were only occasionally present in later deposits, although they were abundant at the nearby Pliocene-aged site of Eshoa Kakurongori.

Lungfish are large eel-like fish, with one extant species obtaining lengths of almost 2 m. They have a largely molluscivorous diet. Their unique aestivation habits increase the chances of preservation as fossils: they make burrows in the mud, and at least one species aestivates for 7 to 8 months (Greenwood 1986). African lungfish elements are known from ?Eocene and Oligocene deposits in northern Africa (Lavocat 1955), Early Miocene deposits at Loperot (Van Couvering 1977), early-mid Miocene deposits at Rusinga and Karungu, Kenya (Greenwood 1951; Van Couvering 1977), Mio-Pliocene deposits in Sinda and the Lake Albert Basin (Greenwood and Howes 1975; Van Neer 1994) and in Manonga, Tanzania (Stewart 1997), Pliocene deposits in the Lake Edward-Albert Basin (Stewart 1990), and Plio-Pleistocene deposits in eastern Turkana (Schwartz 1983). Lungfish are no longer present in Lake Turkana, but they are widespread in Kenya, the Nile basin, and throughout the African continent.

Order Polypteriformes
Family Polypteridae
Polypterus Geoffroy Saint-Hilaire, 1802
Polypterus sp. (bichir)

Lothagam Material

- *Lower Nawata:* 1214, 5 cranial fragments; 1635, 2 scales; 1644, 16 scales; 1658, 2 cranial fragments, 3 trunk vertebrae centra, 201 scales, 2 dorsal spine fragments; 1659, trunk vertebra centrum, 20 scales; 1672, cranial fragment, 6 trunk vertebrae centra, 19 vertebrae centra, 129 scales, 2 dorsal spine fragments; 1710, cranial fragment, 6 trunk vertebrae centra, 2 caudal vertebrae centra, 63 scales, 2 dorsal spine fragments; 1733, 2 cranial fragments, 6 scales; 1751, 2 cranial fragments, trunk vertebra centrum, 5 vertebrae centra, 69 scales, dorsal spine; 1752, cranial fragment, trunk vertebra centrum, 2 vertebrae centra, 125 scales, 2 dorsal spine fragments; 1773, 4 scales; 1971, scale; 1987, vertebra centrum; 1988, scale; 1990, cranial fragment; 20 scales; 1996, 32 scales; 2312, scale; 2365, 18 scales; 2386, vertebra centrum, scale; 2412, 11 scales; 2413, 8 scales.
- *Upper Nawata:* 1594, cranial fragment, 3 vertebrae centra, 13 scales, 4 dorsal spine fragments; 1655, 18 scales, dorsal spine fragment; 1765, 10 scales; 1766, 4 scales; 1950, vertebra centrum, 17 scales; 1957, 146 scales, 3 dorsal spine fragments; 1977, vertebra centrum, 9 scales.
- *Apak Member:* 1760, 3 scales; 1849, 2 scales; 1942, 2 scales; 1944, 2 scales; 1948, vertebra centrum; 1960, 2 scales.
- *Muruongori Member:* 3153, 11 scales.
- *Kaiyumung Member:* 1850, 5 scales; 1851, scale; 1852, scale; 1994, 2 scales; 1999, 2 scales; 2000, 2 scales.

Polypterus elements are robust and preserve well, particularly the distinctive ganoid scales and the ganoine-covered spines and cranial fragments. Unfortunately, these elements are not useful for diagnosis below the level of genus.

Discussion

The modern family Polypteridae is represented by two genera: *Polypterus* and *Calamoichthys*, both of which are restricted to Africa. In the literature, most elements have been referred to the larger and today much more widely distributed genus *Polypterus,* or as Polypteridae. *Polypterus* remains were apparently abundant at the Nawata Formation sites.

Polypterus are long, slender fishes with a distinctive long dorsal fin that is divided by spines into portions that resemble sails; they have a lung-like organ to breathe air. Their diet is small fish and insects. Like lungfish, they have Paleozoic origins and are known from Cretaceous deposits in Egypt (Stromer 1916). Their Cenozoic record is long and includes Eocene deposits in Libya (Lavocat 1955), Miocene deposits in Rusinga and Loperot, Kenya (Greenwood 1951; Van Couvering 1977) and Bled ed Douarah, Tunisia (Greenwood 1973), Pliocene deposits at Wadi Natrun, Egypt (Greenwood 1972), and Plio-Pleistocene deposits at Koobi Fora (Schwartz 1983). *Polypterus* has never been recovered from the Western Rift sites. Two extant species are known from Lake Turkana—*P. senegalus* and *P. bichir. Polypterus* is widespread from Senegal to the Nile basin up to Lake Albert, as well as in the Zaire basin and Lake Tanganyika.

Order Osteoglossiformes
Family Osteoglossidae
Heterotis Ruppell, 1829
Heterotis sp.

Lothagam Material

- *Lower Nawata:* 1659, 2 ?lacrimal fragments, 2 opercular fragments, 42 cranial fragments, 6 trunk vertebrae centra; 1672, 2 opercular fragments, 6 cranial fragments; 1733, 3 cranial fragments; 1751, opercular fragment, 2 cranial fragments; 1996, opercular fragment, 2 cranial fragments, 2 trunk vertebrae centra; 2365, 2 opercular fragments; 2301, 2 opercular fragments; 2386, trunk vertebra centrum; 2414, 2 cranial fragments.
- *Upper Nawata:* 1594, opercular fragment, trunk vertebra; 1655, cranial fragment, 3 trunk vertebrae centra, 10 ?scales; 1658, 3 cranial fragments; 1957, ?dentary portion, ?lacrimal fragment, opercular fragment, trunk vertebra centrum.
- *Apak Member:* 1948, vertebra fragment; 2420, 2 vertebrae.
- *Muruongori Member:* 3153, 2 ?trunk vertebrae; 3153/3154, trunk vertebra centrum.

Only robust *Heterotis* elements were recovered, primarily opercula and vertebrae. All elements recovered were similar to those of extant species. The estimated length of the fossil fish falls within modern limits: up to 90 cm in total length. Skulls, which may have been useful for identifying evolutionary changes, were not recovered.

Discussion

Osteoglossidae are at present represented by one genus and species in Africa—*Heterotis niloticus. Heterotis* was apparently common in the Nawata Formation but rare at later horizons. It can attain a large size, up to a meter in length. Its diet is zooplankton, phytoplankton, and insects. Its fossil record is poor, with records only from Plio-Pleistocene deposits at Koobi Fora (Schwartz 1983). Extant *Heterotis* is known from the Omo and Kerio deltas in Lake Turkana and throughout the Gambia, Senegal, Volta, Niger, Chad, and Nile basins.

Order Mormyriformes
Family Mormyridae
Hyperopisus Gill, 1862
Hyperopisus sp.

Lothagam Material

- *Apak Member:* 1760, tooth; 1942, 2 teeth; 1944, 13 teeth.
- *Muruongori Member:* 3153/3154, 2 teeth.
- *Kaiyumung Member:* 1850, 10 teeth; 1851, tooth; 1852, tooth; 1994, 13 teeth; 1998, 5 teeth; 1999, tooth.

Hyperopisus teeth are distinctive—smooth and round in shape, with a flat attachment surface. They attach not to the jaws but to the parasphenoid and basihyal inside the mouth. Being robust, they preserve well. The estimated size range of the Lothagam teeth (1.5–2 mm in diameter) is similar to that from extant *Hyperopisus* that are 50–80 cm in total length. The Lothagam fossils appear to be within the size range of present-day specimens, unlike some Zaire specimens (Stewart 1990) and some later Turkana Basin specimens that were larger. Van Neer (1994) noted that *Hyperopisus* teeth may be confused with *Bunocharax* teeth, but certainly those recovered in the Upper Semliki deposits were easy to separate, by shape, size, texture, and fossilization. No teeth resembling *Bunocharax* were recovered at Lothagam.

Discussion

Hyperopisus elements are rare throughout the Lothagam sequence, first appearing in the Apak Member of the Nachukui Formation.

Hyperopisus and other mormyroids have muscles in the caudal peduncle that were modified to form a weak electromagnetic field with which they sense their surroundings (e.g., Beadle 1981). Waters of high salinity apparently interfere with this sensory ability and, although they are occasionally found at the Omo delta, mormyroids are generally absent from modern Lake Turkana and other bodies of water with high salinity. Fossil *Hyperopisus* teeth are known from Pliocene deposits of Wadi Natrun, Egypt (Greenwood 1972), Plio-Pleistocene deposits in the Lake Edward-Albert Basins (Greenwood and Howes 1975; Stewart 1990), Mio-Pleistocene Lake Albert-Edward Basin deposits (Van Neer 1994), and Plio-Pleistocene deposits at Koobi Fora (Schwartz 1983). Extant *Hyperopisus bebe* is known from the Omo delta of Lake Turkana and from the Senegal, Volta, Niger, Chad, and Nile basins.

Family Gymnarchidae
Gymnarchus Cuvier, 1829
Gymnarchus sp.
(Figure 3.3)

Lothagam Material

- *Lower Nawata:* 1644, 5 teeth; 1658, 24 teeth; 1659, 28 teeth; 1672, 7 teeth; 1710, 3 teeth; 1751, 3 teeth; 1752, 15 teeth; 1948, 2 teeth; 1987, 2 teeth; 1990, 8 teeth; 2365, 2 teeth; 2413, 3 teeth; 2416, tooth; 3150, 5 teeth.
- *Upper Nawata:* 1594, 3 teeth; 1655, tooth; 1765, 3 teeth; 1950, 2 teeth.
- *Apak Member:* 1760, tooth; 1942, 18 teeth; 1944, 14 teeth; 1948, 2 teeth; 1960, tooth.
- *Muruongori Member:* 3153/3154, 2 teeth.
- *Kaiyumung Member:* 1850, tooth; 1851, 3 teeth; 1852, 5 teeth; 1993, tooth; 1994, 40 teeth; 1998, 5 teeth; 1999, 7 teeth; 2000, 2 teeth; 2332, tooth.

The only *Gymnarchus* specimens from Lothagam are oral teeth that lined the premaxilla and dentary. These teeth are robust and distinctive both in fossil and extant animals, with a square, triangular or subtriangular shape and fine serration along the edges (figure 3.3). The appearance and size range appear similar to those of extant *Gymnarchus;* several teeth have a base of about 2–3 mm, which suggests an estimated total length of about 60 to 90 cm (within the range of extant specimens).

Discussion

Gymnarchus was present throughout the Lothagam sequence but was most abundant in the Nawata Forma-

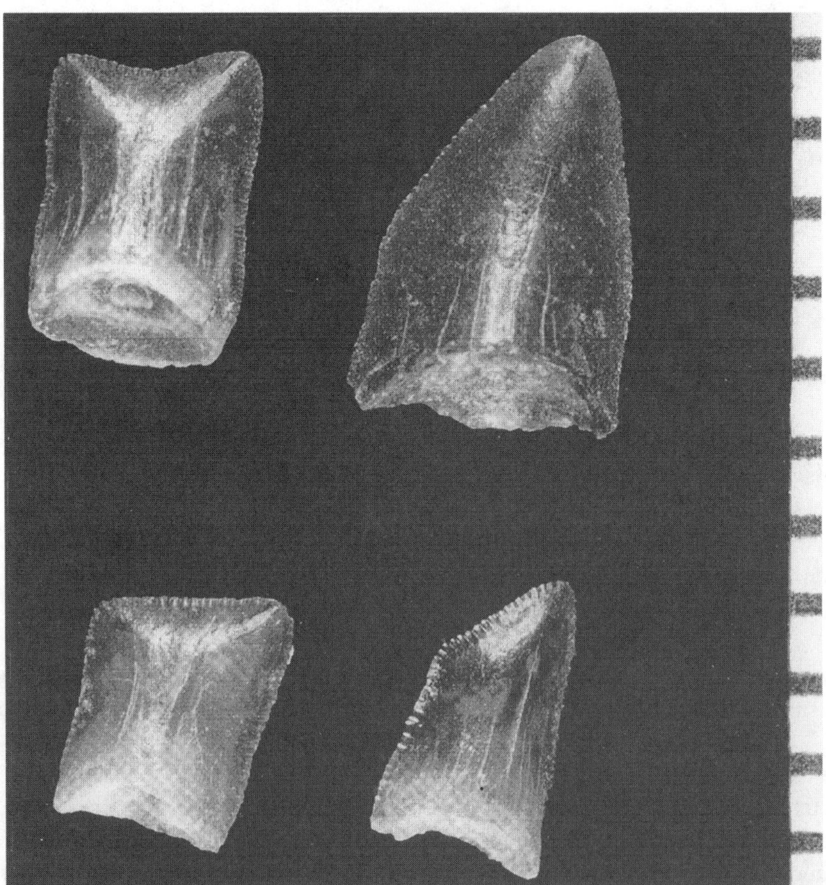

Figure 3.3 *Gymnarchus* teeth showing different shapes and serrated outlines.

tion. As in *Hyperopisus,* its caudal muscles are modified to create a weak electrosensory field with which to sense its surroundings. It is an eel-like fish, with a diet of fish and snails. It is reported from Miocene-Pleistocene deposits in the Lake Albert-Edward Basin (Van Neer 1994), Pliocene deposits in the Lake Edward-Albert Basin (Stewart 1990; Van Neer 1992), and Plio-Pleistocene deposits in eastern Turkana (Schwartz 1983). At present, *Gymnarchus niloticus* is known from the Omo delta in Lake Turkana and in the Gambia, Senegal, Niger, Volta, Chad, and Nile basins.

Order Mormyriformes
Mormyriformes indet.

Lothagam Material

- *Lower Nawata:* 1658, vertebra centrum fragment; 1672, 3 vertebrae centra; 1751, 16 vertebrae centra; 1752, vertebra centrum; 1971, caudal vertebra centrum; 1990, 3 trunk vertebrae centra; 2419, trunk vertebra.
- *Apak Member:* 1944, trunk vertebra centrum.
- *Muruongori Member:* 3153, trunk vertebra centrum.
- *Kaiyumung Member:* 1850, vertebra centrum.

Discussion

Mormyriform vertebrae are difficult to distinguish taxonomically, but the large size of several of these vertebrae makes them almost certainly *Gymnarchus.*

Order Cypriniformes
Family Cyprinidae
Labeo Cuvier, 1817
Labeo sp.

Lothagam Material

- *Lower Nawata:* 1659, tooth; 1751, tooth.
- *Apak Member:* 1942, 4 teeth; 1944, 2 teeth; 1948, 2 teeth; 1960, 3 teeth.
- *Muruongori Member:* 3154, tooth.
- *Kaiyumung Member:* 1835, tooth; 1850, 5 teeth; 1852, 17 teeth; 1994, tooth; 1995, 2 teeth; 1998, tooth; 2332, 2 teeth.

Labeo is represented by its distinctive flattish pharyngeal teeth, which preserve well in fossil accumulations. Other elements are generally not robust enough to preserve as fossils. The size of the Lothagam individuals is estimated to be up to 90 cm total length.

Discussion

Fossils of *Labeo* are rare throughout Lothagam but become more common in the Kaiyumung Member (Pliocene). *Labeo* has a toothless underhanging mouth; its diet is chiefly algae and benthic detritus. It can reach lengths of almost a meter. *Labeo* is only certainly known from Pliocene deposits in Wadi Natrun, Egypt (Greenwood 1972), eastern Turkana, Kenya (Schwartz 1983), and the Edward-Albert Basin deposits (Stewart 1990) and Pleistocene deposits in the Edward-Albert Rift (Van Neer 1994). A reported Miocene occurrence from western Uganda may be in error; the author states that certain Mio-Pliocene sites had Pleistocene-aged fossils mixed in (Van Neer 1994:90); *Labeo*-like teeth are also reported from the mid Miocene of Loperot (Van Couvering 1977). This distribution suggests that *Labeo,* an Asian taxon, probably migrated to the eastern part of the African continent during mid to late Miocene times, possibly in several episodes when geographic conditions were suitable. From there it moved to other sites in Africa. Extant *Labeo* is represented by one species—*L. horie*—in Lake Turkana, but the genus is widespread throughout the continent, including the Nile basin, western Africa, eastern Africa, and the Zaire and Zambezi basins.

Barbus Cuvier and Cloquet, 1816
Barbus sp.
(Figure 3.4)

Lothagam Material

- *Apak Member:* 1942, 2 teeth.
- *Kaiyumung Member:* 1850, 8 teeth; 1852, 44 teeth; 1993, 9 teeth; 1994, 4 teeth; 1995, 8 teeth; 1998, 3 teeth; 1999, 2 teeth; 2332, 6 teeth.

Barbus is represented only by its teeth, which attach to pharyngeal plates and are very robust. The teeth consist of both round, robust teeth with a molariform appearance and smaller, flatter, and less robust teeth (figure 3.4). Many of the large teeth lack a "dome" but are round with a large central concavity. The fossil teeth are very large when compared with those of extant *B. altianalis* or *B. bynni* specimens, which have an estimated total length of 50–70 cm. In his revision of the genus, Banister (1973) reported considerable variation of tooth size within species, making it difficult to assign the fossil to species.

Discussion

Barbus is a minnow-like fish, often reaching lengths of over a meter. Like *Labeo,* it has an underhanging

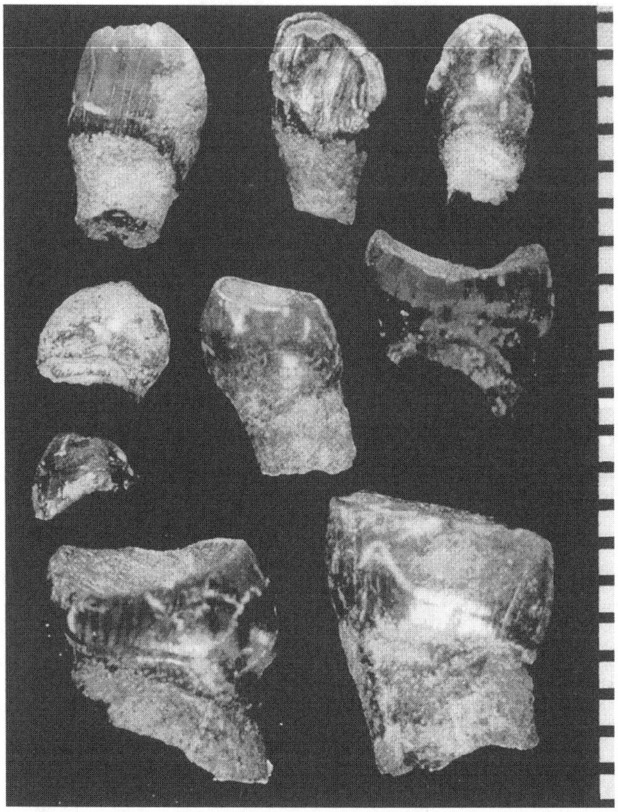

Figure 3.4 *Barbus* teeth showing different shapes and sizes.

mouth. Its diet includes mainly invertebrates, including mollusks, insects, and ostracods. Its teeth are rare at Lothagam, first appearing at Apak Member sites and becoming more common at Kaiyumung Member sites.

Its fossil record is known primarily from the Pliocene to Recent, in particular from Pliocene deposits in the Edward-Albert Rift, Zaire (Stewart 1990), Plio-Pleistocene deposits from Koobi Fora (Schwartz 1983), and Pleistocene deposits from the Edward-Albert Rift (Greenwood 1959; Van Neer 1994). Van Couvering (1977) notes that "*Barbus*-like" teeth are known from mid Miocene deposits in Kenya. Like *Labeo, Barbus* probably reached Africa from Asia during the Miocene. *Barbus* has radiated enormously in Africa, but today it is represented by only three species in Lake Turkana. It is known throughout the African continent, including western Africa, the Nile, Zambezi and Zaire basins, and the eastern and southern African systems.

Cyprinidae indet.

Lothagam Material

- *Lower Nawata:* 1644, vertebra centrum; 1659, 2 vertebrae centra; 1672, tooth.
- *Apak Member:* 1762, vertebra centrum; 1948, 2 teeth.
- *Muruongori Member:* 3153, 2 trunk vertebrae; 3153/3154, trunk vertebra.
- *Kaiyumung Member:* 1851, 2 vertebral centra; 1852, tooth fragment.

Order Characiformes
Family Citharinidae
Distichodus Muller and Troschel, 1845
Distichodus sp.

Lothagam Material

- *Lower Nawata:* 1659, tooth.
- *Apak Member:* 1944, tooth; 1998, tooth.

Distichodus teeth are distinctive, being conical with a bifurcate crown. The teeth are very small, but are usually the only elements that are preserved because the bones are delicate. The teeth recovered at Lothagam represent fish of up to a meter long, thus similar in size to extant specimens.

Discussion

Distichodus remains are not common at Lothagam, but this undoubtedly reflects their poor preservation and/or recovery rather than actual numbers.

Distichodus is a deep-bodied fish with a diet primarily of invertebrates. It can reach lengths of over a meter. It is known from Mio-Pliocene deposits in the Lake Albert-Edward Basins (Van Neer 1994), Pliocene deposits in the Lake Edward-Albert Basin (Stewart 1990), and Pleistocene deposits in eastern Turkana (Schwartz 1983). Extant *Distichodus* is known from Lake Turkana—*D. niloticus*—and from the Nile basin up to Lake Albert.

Family Characidae
Hydrocynus Cuvier, 1817
Hydrocynus sp. (tigerfish)

Lothagam Material

- *Lower Nawata:* 1635, 2 teeth; 1644, tooth; 1658, 7 teeth, tooth fragment; 1659, 5 teeth; 1672, tooth; 1710, 6 teeth; 1751, 3 teeth; 1752, 4 teeth; 1987, 3 teeth; 1990, 13 teeth; 2275, tooth.
- *Upper Nawata:* 1594, 2 teeth; 1655, 6 teeth; 1765, 3 teeth.
- *Apak Member:* 1847, 2 teeth; 1942, 27 teeth; 1944, 27 teeth; 1948, 2 teeth; 1960, 8 teeth, tooth fragment.
- *Muruongori Member:* 3153, 7 teeth; 3153/3154, 24 teeth.

- *Kaiyumung Member:* 1850, 5 teeth, 3 tooth fragments; 1851, 7 teeth; 1852, 63 teeth, 2 tooth fragments; 1993, 23 teeth, tooth fragment; 1994, 235 teeth, 9 tooth fragments; 1995, 29 teeth; 1998, 23 teeth, 5 tooth fragments; 1999, 34 teeth; 2000, 3 teeth; 2332, 46 teeth, 4 tooth fragments.

Hydrocynus is represented only by long, conical, and pointed teeth; its bones are usually too delicate to preserve, except in the largest specimens. A great size range of teeth was recovered, representing fish of between 10 cm to a meter in total length. The teeth are identical to those of present-day species.

Discussion

Hydrocynus was abundant throughout the Lothagam sequence, as it is in the modern Lake Turkana. It can reach lengths of about 70 cm, although some of the fossils represent larger fish. It is piscivorous. Its fossil record is based almost completely on teeth: from Mio-Pleistocene deposits in the Lake Albert-Edward Rift, Uganda (Van Neer 1994), Miocene deposits of Sinda, Zaire (Van Neer 1992), Pliocene deposits in Wadi Natrun, Egypt (Greenwood 1972) and the Lake Edward-Albert Rift, Zaire (Stewart 1990), and Plio-Pleistocene deposits in the Omo Valley (Arambourg 1947) and at Koobi Fora (Schwartz 1983). Today, *Hydrocynus* is represented by one species, *H. forskalii,* in Lake Turkana but a second species, *H. lineatus,* is present in the Omo River. *Hydrocynus* is widespread from Senegal to the Nile, including the Volta, Niger, and Chad basins.

Alestes Muller and Troschel, 1844
Alestes sp.

Lothagam Material

- *Lower Nawata:* 1710, first inner premaxillary tooth; 1752, 2 inner premaxillary teeth; 2410, tooth; 3154, 2 first inner premaxillary teeth.
- *Apak Member:* 1944, outer dentary tooth.
- *Kaiyumung Member:* 1850, outer dentary tooth, first inner premaxillary tooth; 1852; 4 outer dentary teeth, second inner premaxillary tooth, second inner premaxillary tooth, third/fourth inner premaxillary tooth; 1994, 2 outer dentary teeth, first inner premaxillary tooth; 1995, first outer dentary tooth.

Alestes is represented only by teeth, as its other elements are delicate. Its cusped teeth are unusual among fish. However, the teeth are small, and many teeth may have passed through the 1-mm screen used in the field. Identification of the teeth to species was not possible on outer teeth, and while one inner tooth showed some affinity to *A. dentex* and one to *A. stuhlmanni,* these similarities were not definitive enough for species designation. The teeth represent fish of between 40 and 50 cm in total length.

Discussion

The sparse representation of *Alestes* in the Lothagam deposits is probably due to both poor preservation and lack of recovery of very small teeth. *Alestes* is an open-water fish that can reach over 50 cm in length. Its diet is chiefly zooplankton and insects. The fossil record of *Alestes* is poor, with remains known from Plio-Pleistocene deposits in the Lake Edward-Albert Basin (Stewart 1990) and from Mio-Pleistocene deposits in the Lake Edward-Albert Rift (Van Neer 1994) and Manonga, Tanzania (Stewart 1997). Miocene teeth with affinities to *Alestes* are reported from Loperot and Mpesida, Kenya (Van Couvering 1977). Extant *Alestes* is represented by six species in Lake Turkana, including *A. baremose, A. dentex, A. nurse, A. macrolepidotus, A. ferox,* and *A. minutus. Alestes* is known from the Volta, Niger, and Chad basins to the Nile River and in the Zaire, Zambezi, and Limpopo basins.

Sindacharax Greenwood and Howes, 1975

A total of 3,688 teeth, occluded jaw, 3 partial premaxillae, 2 partial dentaries, and 1,410 tooth fragments were attributable to *Sindacharax.* The teeth described here were virtually all isolated specimens; therefore placement in the jaw was determined by analogy with the orientation of the teeth of the upper and lower jaws of *Sindacharax greenwoodi* (Stewart 1997) and by reference to two premaxillary jaw fragments with in situ teeth (described later in this contribution). However, because there is considerable variation in cusp patterns in known *Sindacharax* jaws, placement of these isolated teeth is tentative.

Sindacharax lothagamensis sp. nov.
(Figures 3.5–3.8)

Diagnosis

Holotype comprises medium robust right premaxilla fragment with two rows of teeth, with no interspace between the two rows, distinguished from *S. lepersonnei* by cusps forming ridges on inner premaxillary teeth rather than discrete cusps as in *S. lepersonnei.* Distinguished from *S. greenwoodi* by absence of interspace between tooth rows rather than wide separation of the inner and outer rows of teeth as in *S. greenwoodi,* and by lower plane for outer tooth attachment. Also distinguished from *S. greenwoodi* by cusp pattern of second premaxillary tooth; chiefly the squarish shape of the tooth and the lack of the ridged arc surrounding the dominant lingual cusp; in *S. lothagamensis,* the dominant cusp is flanked by two cusps on each side, and a short ridge anterior to the dominant cusp. Distinguished from *S. deserti* by absence in second inner tooth of raised circular ridge radiating from the dominant lingual cusp.

Holotype

Premaxillary fragment with second inner tooth (figure 3.5). KNM-LT 38264, collected by Sam N. Muteti and Peter Kiptalam in 1992 from site 1990, lower member of the Nawata Formation, Lothagam.

Paratypes

- *Lower Nawata:* 1644, first inner premaxillary tooth; 1658, 3 second inner premaxillary teeth; 1659, second inner premaxillary tooth; 1710, second inner premaxillary tooth; 1751, first inner premaxillary tooth; 1752, 3 second inner premaxillary teeth; 1990, 4 first inner premaxillary teeth, 8 second inner premaxillary teeth; 2413, first inner premaxillary tooth, 3 second inner premaxillary teeth; 38264, holotype (see above).
- *Upper Nawata:* 1950, second inner premaxillary tooth.
- *Apak Member:* 1942, 2 second inner premaxillary teeth; 1944, 3 second inner premaxillary teeth; 1948, first inner premaxillary tooth, second inner premaxillary tooth.

Figure 3.5 *Sindacharax lothagamensis* sp. nov., occlusal view of premaxilla and second inner tooth, holotype, KNM-LT 38264.

Etymology

Named after the site—Lothagam—where these elements were recovered.

The premaxilla fragment has one *in situ* second inner tooth and tooth attachment bases for three outer teeth and three inner teeth (figure 3.5). The outer tooth bases, although incomplete, appear oval and small, less than a fifth the size of the *in situ* tooth. The outer bases are contiguous with those of the inner teeth—that is, there is no interspace between the first and second rows of teeth. Similarly, the inner teeth bases abut each other, but as they are incomplete, nothing can be said about the shape or size of these teeth. The dental shelf of the premaxilla is unusual in that the outer tooth bases are on a lower plane than the inner teeth (figure 3.6). This is more similar to dental shelves in *Alestes,* while in other *Sindacharax* jaws the outer and inner tooth bases are on the same horizontal plane. The premaxilla itself is of medium robusticity and is certainly much more gracile than that of *S. greenwoodi.* No tooth crypts were identified. The second inner tooth is *in situ,* and is squarish in shape, with rounded corners (figure 3.5). It has a dominant lingually placed cusp that is flanked on each side by two smaller cusps. Between anterior flanking cusps is a short ridge of cusps. Anterior to this are three ridges composed of discrete, weakly ridged cusps.

In addition to the premaxilla fragment, 29 isolated second inner premaxillary teeth were identified. One was from an Upper Nawata site; the remainder were from the Lower Nawata. As in the *in situ* second tooth, these teeth generally have a squarish shape; in some there is slight narrowing at the lingual end. A dominant cusp lies lingually and is flanked on each side by one or two cusps that lie along the margins of the tooth. Anterior to the dominant cusp is a short lateral ridge that lies between the flanking cusps, and anterior to this ridge are one or more ridges that traverse the width of the tooth. These ridges are made up of one or more discrete cusps. The second inner teeth varied in length up to 5.5 mm, but most were under 3 mm in length.

Seven first inner premaxillary teeth were associated with the second inner teeth that were recovered from Lower Nawata sites. Because only second teeth belonging to *S. lothagamensis* were found at Lower Nawata sites, unlike some later sites where a mix of *Sindacharax* species were recovered, first inner teeth recovered in association with the second teeth were assumed to belong to the same species and are identified as *S. lothagamensis.* These teeth are long and narrow, longer than wide. The dominant cusp is at the lingual end of the

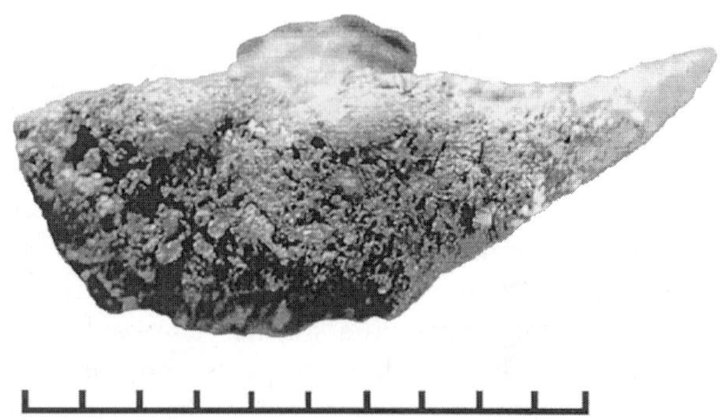

Figure 3.6 *Sindacharax lothagamensis* sp. nov., anterior view of premaxilla and outer tooth bases.

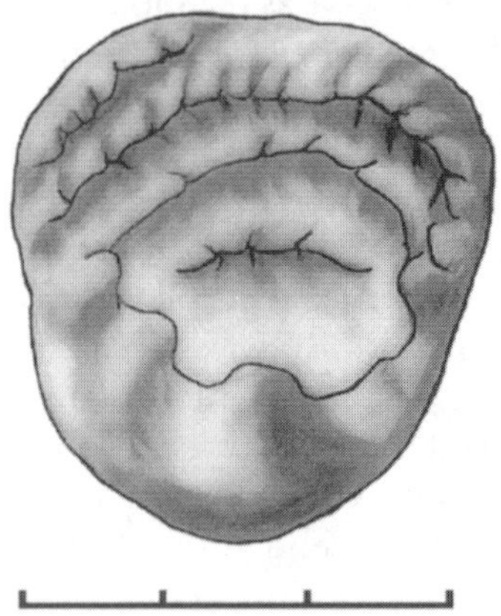

Figure 3.7 *Sindacharax lothagamensis* sp. nov., drawing of second inner premaxillary tooth (occlusal view).

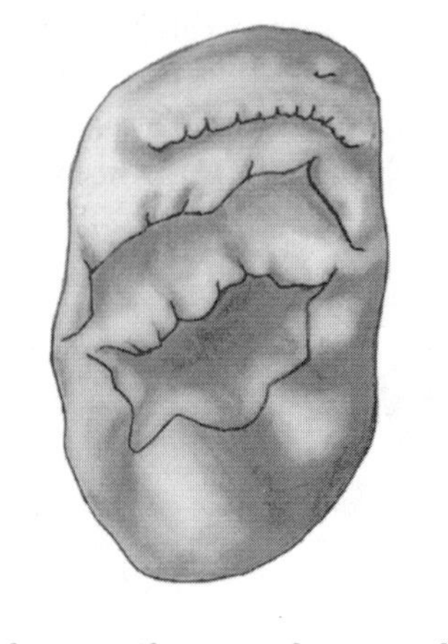

Figure 3.8 *Sindacharax lothagamensis* sp. nov., drawing of first inner premaxillary tooth (occlusal view).

tooth with two cusps veering in a diagonal line toward the presumed buccolabial side (figure 3.8). Anterior to the dominant cusp are one or more ridges that traverse the width of the tooth.

Presumed third and fourth teeth were found in association with the other inner teeth of *S. lothagamensis.* Because I could not distinguish the third from the fourth, I have classified them both as *Sindacharax* sp., but describe them separately here. The presumed third inner premaxillary teeth are long and thin with an oval shape. A dominant cusp flanked on one side by three or more lesser cusps makes up the presumed buccolingual margin of the tooth, while a short ridge made up of one or more cusps lies parallel along the opposing margin of the tooth.

On one specimen a short row of two cusps lies perpendicular to the main rows, at the anterior end of the tooth. Based on comparison with other known inner teeth, it is presumed that these are third inner teeth, although it is often difficult to distinguish third from fourth inner teeth.

The presumed fourth tooth is similar to the third inner tooth, being long and oval in shape. There is no second ridge on the opposing margin of the tooth, although one or two cusps may be in evidence.

All outer premaxillary teeth associated with *S. lothagamensis* were referred to *Sindacharax* sp. type A; all outer dentary teeth were referred to *Sindacharax* sp. type A (see the following discussion and under *Sindacharax* sp.).

Discussion

All teeth designated as *S. lothagamensis* were found as isolated teeth, with only the type specimen having a tooth attached to a premaxillary fragment. Only first and second inner premaxillary teeth were distinctive enough to classify to species; third and fourth inner and all outer teeth were classified as *Sindacharax* sp. *Sindacharax lothagamensis* was primarily confined to the Lower Nawata sites, although one specimen was found in an Upper Nawata site, and a few were found in Apak Member sites. In general, the teeth appeared to come from small individuals, in contrast to later *Sindacharax* specimens that could represent individuals with lengths of over a meter. These teeth bear greater resemblance to *Alestes* teeth than later *Sindacharax* in several features, including smaller size, less cusp ridging, attachment on a lower plane for premaxillary outer teeth, and less robusticity of premaxilla. However, they differ from *Alestes* in having cusps that form ridges on the inner teeth.

Sindacharax mutetii sp. nov.

(Figures 3.9–3.14)

Diagnosis

Second inner premaxillary tooth distinguished from *Sindacharax lepersonnei* by cusps forming ridges rather than discrete cusps as in *S. lepersonnei.* Distinguished from *S. lothagamensis* by lengthened cusp ridge anterior to the dominant cusp (truncated in *S. lothagamensis*), which traverses the width of the tooth; distinguished from *S. deserti* by absence of raised circular ridge radiating from the dominant lingual cusp; distinguished from *S. greenwoodi* by lack of the ridged arc that surrounds the dominant lingual cusp.

Lothagam Material

Holotype

A second inner premaxillary tooth (figure 3.9), KNM-LT 38265, collected by Sam N. Muteti and Peter Kiptalam in 1993 from Site 1944, in the Apak Member of the Nachukui Formation.

Paratypes

- *Apak Member:* 1760, 3 first inner premaxillary teeth; 1942, 7 first inner premaxillary teeth, 10 second inner premaxillary teeth, 7 third inner premaxillary teeth; 1944, 12 first inner premaxillary teeth, 18 second inner premaxillary teeth, 8 third inner premaxillary teeth; 1960, first inner premaxillary tooth, 3 second inner premaxillary teeth, 2 third inner premaxillary teeth; 2420, first inner premaxillary tooth, second inner premaxillary tooth.
- *Kaiyumung Member:* 1850, 2 second inner premaxillary teeth; 1852, 12 second inner premaxillary teeth; 1994, first inner premaxillary tooth; 2000, second inner premaxillary tooth.

Etymology

Named in honor of Sam Muteti, who helped collect the type specimen.

The holotype is a broadly oval second inner premaxillary tooth with transverse ridges in which the cusps are poorly defined (figures 3.9 and 3.10). A dominant cusp is positioned lingually, and a smaller cusp flanks each side. Anterior to these three cusps are one or more ridges (with very weakly defined cusps) that traverse the width of the tooth. Tooth size varies from small (length under 3 mm) to large (length 7.6 mm). *S. mutetii* is the only *Sindacharax* species recovered from Kanapoi, and teeth from that site serve as the species standard. A total of 89 teeth of *S. mutetii* recovered from Lothagam were referred to this species on the basis of similarity of size and overall similarity of cusp morphology.

The first inner teeth recovered had attachment bases exactly like that on the *Sindacharax* cf. *S. mutetii* premaxilla (see description that follows). These teeth are long and narrow, at least twice as long as wide (figures 3.11 and 3.12). They have a distinctive cusp pattern, with a dominant cusp at the lingual end of the tooth, and one cusp, not two as in *S. lothagamensis*, flanking it. Anterior to this second cusp is one or more ridges that traverse the width of the tooth. The ridges are comprised of very weakly defined cusps; on some specimens, the cusps cannot be distinguished. On a few teeth, these transverse ridges are interrupted in the middle.

The second inner premaxillary teeth recovered had the same broad oval shape of the type specimen, although some had a slightly more triangular shape. They had the same cusp pattern as the holotype.

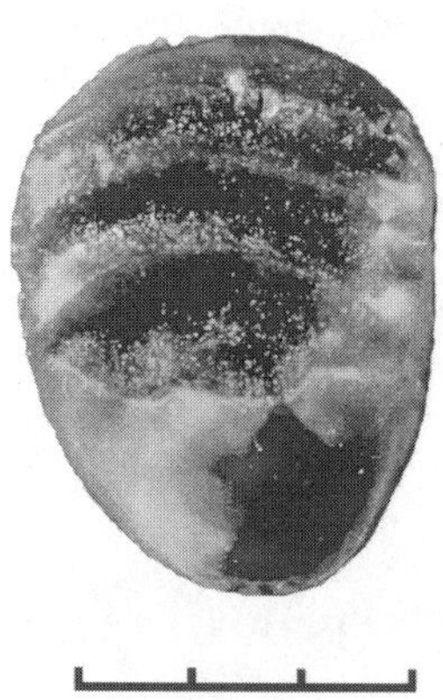

Figure 3.9

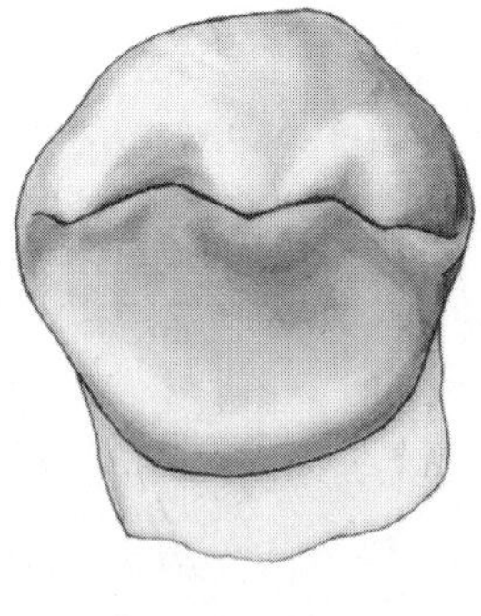

Figure 3.10

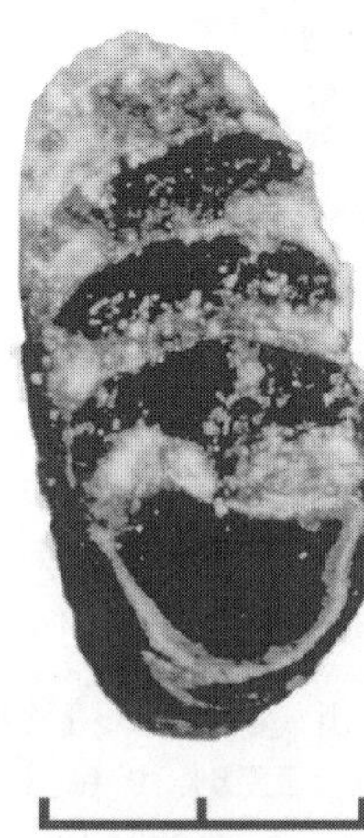

Figure 3.11

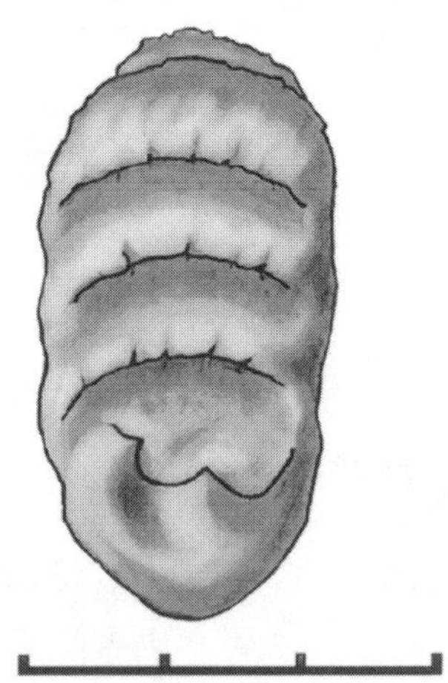

Figure 3.12

Figure 3.9 *Sindacharax mutetii* sp. nov., second inner premaxillary tooth (occlusal view), holotype, KNM-LT 38265.

Figure 3.10 *Sindacharax mutetii* sp. nov., drawing of second inner premaxillary tooth (occlusal view).

Figure 3.11 *Sindacharax mutetii* sp. nov., first inner premaxillary tooth (occlusal view).

Figure 3.12 *Sindacharax mutetii* sp. nov., drawing of first inner premaxillary tooth (occlusal view).

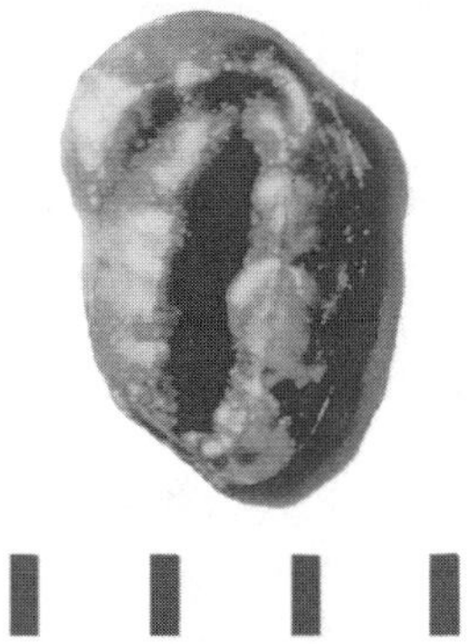

Figure 3.13

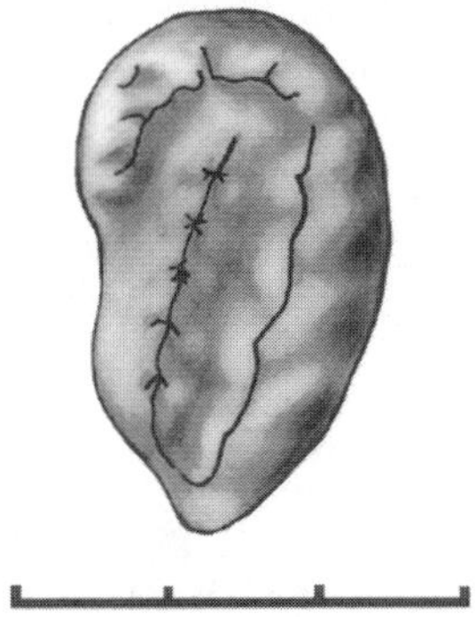

Figure 3.14

Figure 3.13 *Sindacharax mutetii* sp. nov., third inner premaxillary tooth (occlusal view).

Figure 3.14 *Sindacharax mutetii* sp. nov., drawing of third inner premaxillary tooth (occlusal view).

The third inner premaxillary teeth identified were oval in shape, with narrowing at the lingual end (figures 3.13 and 3.14). A dominant cusp is placed buccolabially, flanked by ridge-shaped cusps along the margin of the tooth. Anterior to these cusps are one or more ridges made up of very weakly defined cusps. A trench separates the dominant cusp and the anterior ridges. Although several isolated teeth are probably fourth inners by analogy with other *Sindacharax* jaws, this was not certain, and these have been classified as *Sindacharax* sp. However, I will describe them here. These teeth are long and narrow, at least twice as long as wide. There is a ridge made up of a dominant cusp with three or more flanking cusps along one margin of the long axis; a much weaker ridge may or may not be present along the opposing margin. The ridges are separated and do not curve into each other, as with the third inner teeth.

All outer premaxillary and dentary teeth associated with *S. mutetii* were classified as *Sindacharax* sp.

Discussion

Apak sites contain primarily *Sindacharax mutetii* teeth. Teeth from this species are also known from Pliocene-aged sites, particularly Kanapoi, where *S. mutetii* is abundant and the only *Sindacharax* species identified. The dominance of *S. mutetii* at the Apak sites and at Kanapoi may indicate a preferred habitat—Kanapoi is composed of fluviodeltaic sediments (Feibel, personal communication). *Sindacharax mutetii* teeth were larger than *S. lothagamensis*, and more abundant in the sites where both occur.

Sindacharax cf. *S. mutetii*
(Figures 3.15–3.18)

Lothagam Material

- *Apak Member:* 1942, right premaxilla with two *in situ* teeth.

Tentatively assigned to this species is a premaxilla with two *in situ* teeth. The teeth are similar to those of *S. mutetii*, but the cusp patterns are slightly different and preclude a definitive species designation. This is a large right premaxilla, about 90 percent complete, and it lacks only fragments of the posterior shelf (figure 3.15). It is robust, with the inner second and third teeth *in situ*, and attachment bases are visible for the first and second outer and the first and fourth inner teeth (the third outer tooth was presumed present; however, the base has been obscured). There is no interspace between outer and inner tooth rows, as the outer tooth bases abut the inner bases. The inner tooth bases abut each other. Both outer and inner teeth are on the same horizontal plane.

The premaxilla itself has an apparently truncated ascending arm. There are three clear hinges at the medial edge, for articulation with the left premaxilla (figure 3.16). The shelf is absent distal to the fourth tooth base. The anterior depth including the ascending arm is greater than for other *Sindacharax* premaxillae; although similar to *S. greenwoodi*; however, the arm is too incomplete for measurements (figure 3.17). This premaxilla, with that of *S. greenwoodi*, is the largest recovered from the Lothagam deposits. No tooth crypts

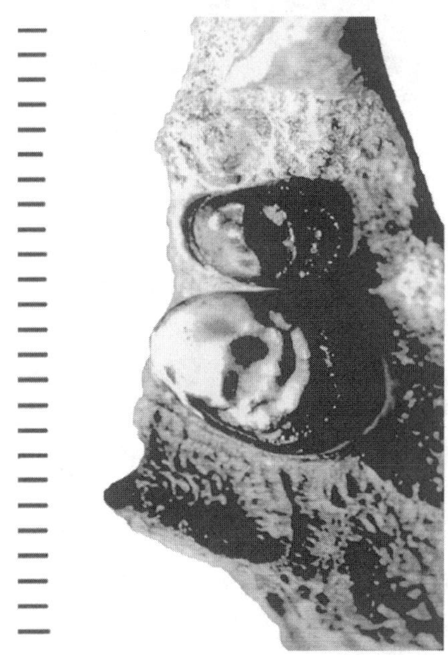

Figure 3.15 *Sindacharax* cf. *S. mutetii*, premaxilla and second, third inner teeth (occlusal view).

Figure 3.16 *Sindacharax* cf. *S. mutetii,* premaxilla (medial view) showing hinges.

Figure 3.17 *Sindacharax* cf. *S. mutetii,* premaxilla (anterior view).

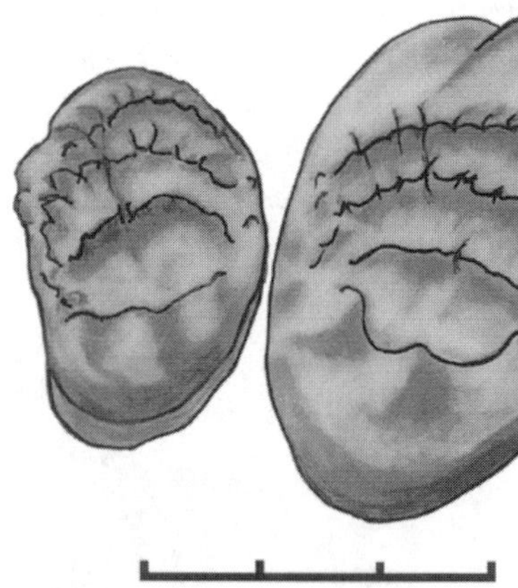

Figure 3.18 *Sindacharax* cf. *S. mutetii,* drawing of second and third inner premaxillary teeth (occlusal view).

were visible. The attachment bases for the outer teeth are oval and between two-thirds and identical size when compared to the inner second tooth. In *S. lothagamensis* the outer tooth bases are circular and less than one-fifth the size of the second inner tooth. The attachment base for the first inner tooth is very long and narrow, although it is somewhat shorter than that for the second inner tooth. The base of the fourth inner tooth is not clearly outlined.

The *in situ* second inner premaxillary tooth has a broad oval shape (figure 3.18). There is a dominant cusp lying lingually, flanked by one large cusp on either side and joined to each by a low ridge. Anterior to the dominant cusp are one or more ridges, slightly arced, which traverse the width of the tooth. These ridges are more "ridged" than in *S. lothagamensis,* in that the cusps are not as clearly defined. The flanking cusp on the left lateral side of the dominant cusp forms a rough semicircle with the first anterior ridge.

The *in situ* third inner premaxillary tooth has an oval shape and is roughly two-thirds the size of the second tooth (figure 3.18). Again, there is a dominant cusp at the lingual end that is flanked by two lesser cusps. The dominant cusp and two lesser cusps are joined together to form a low ridge, which then joins up with an anterior ridge to form an off-center oval ridge, with very weakly defined cusps. Anterior to this oval ridge are one or more ridges with weakly defined cusps that traverse the width of the tooth.

No first inner teeth were *in situ* on the premaxilla. However, the tooth attachment base indicates a very long, narrow tooth. The base for the fourth tooth was not clear enough to predict shape.

Discussion

It was decided not to make this premaxilla the holotype for *S. mutetii* because the cusp patterns on the teeth differed from the very consistent patterns of the other teeth ascribed to this species. Nevertheless, the cusp pattern on the inner second tooth was similar enough to that of the holotype to refer the premaxilla tentatively to this species, although no tooth similar to the third tooth has yet been found. However, because the premaxilla is very distinctive, both in its robusticity and size, this identification will remain tentative until further jaw material definitely attributable to *S. mutetii* is recovered.

Sindacharax howesi sp. nov.
(Figures 3.19–3.21)

Diagnosis

Second inner premaxillary tooth, distinguished from *S. lepersonnei* by presence of transverse ridges rather than discrete cusps as on latter; distinguished from *S. greenwoodi* by lack of a ridge encircling the lingual cusp as in *S. greenwoodi;* distinguished from *S. deserti* by shape (triangular in *S. deserti*) and by lack of a raised circular ridge radiating from the dominant lingual cusp as in *S. deserti;* distinguished from *S. lothagamensis* by shape of teeth and dominant cusp flanked by one, not two cusps as in *S. lothagamensis;* distinguished from *S. mutetii* by truncated ridge anterior to dominant cusp.

Lothagam Material

Holotype

KNM-LT 32866, a second inner premaxillary tooth (figure 3.19), collected by Sam N. Muteti in 1992 from site 1994 in the north Kaiyumung Member deposits, Lothagam.

Paratypes

- *Kaiyumung Member:* 1850, first inner premaxillary tooth; 23 second inner premaxillary teeth; 1852, 53 first inner premaxillary teeth, 113 second inner premaxillary teeth; 1993, first inner premaxillary tooth, 4 second inner premaxillary teeth; 1994, 138 first inner premaxillary teeth, 191 second inner premaxillary teeth; 1995, 10 first inner premaxillary teeth, 27 second inner premaxillary teeth; 2000, 10 first inner premaxillary teeth, 15 second inner premaxillary teeth; 2332, 14 first inner premaxillary teeth, 50 second inner premaxillary teeth.

Etymology

Named after Gordon Howes, who with P. H. Greenwood named the genus *Sindacharax* (Greenwood and Howes 1975).

The second inner premaxillary tooth (figures 3.19 and 3.20) is a broad oval in shape, narrowing at the lingual end, but not as shoe-shaped as in some *S. greenwoodi.* It is relatively large compared to *S. lothagamensis* and *S. deserti: S. howesi* second inner teeth range up to 8.8 mm in length. There is a dominant cusp at the lingual end of the tooth, and it is flanked on each side by a smaller cusp; a low ridge joins the three cusps. Anterior to the main cusp is a ridge made up of distinguishable cusps; the ridge lies between the two flanking cusps. The ridge and three cusps form a loose circle that is not as clearly defined or as small and circular as in *S. deserti.* Anterior to this loose circle are one or more ridges with poorly defined cusps; these ridges traverse the width of the tooth.

Figure 3.19 *Sindacharax howesi* sp. nov., first inner premaxillary tooth (occlusal view).

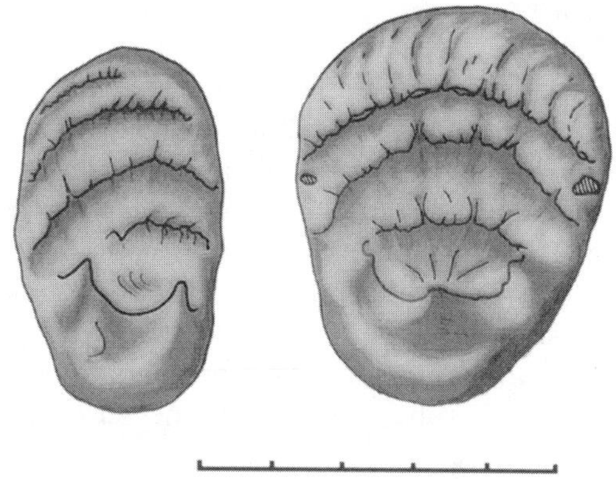

Figure 3.20 *Sindacharax howesi* sp. nov., drawing of first and second inner premaxillary teeth (occlusal view).

Figure 3.21 *Sindacharax howesi* sp. nov., second inner premaxillary tooth, holotype KNM-LT 32866 (occlusal view).

The first inner premaxillary teeth (figures 3.20 and 3.21) have an oval shape, although none has the wide oval shape seen in *S. deserti;* some have a longer, narrow shape. The dominant cusp is mediolingual; it forms a short ridge to a smaller cusp that lies on the distal margin. A second short lateral ridge lies anterior to the dominant cusp. The two short ridges do not unite to form a semicircle but are separated. Anterior to the short ridge are one or more longer ridges that traverse the width of the tooth. Numerous third and fourth inner teeth that did not differ significantly from those of *S. mutetii* have been assigned to *Sindacharax* sp.

Both premaxillary and dentary outer teeth were numerous but did not differ distinctly from those of *S. lothagamensis, S. mutetii,* or *S. greenwoodi.* Outer premaxillary teeth were assigned to *Sindacharax* sp. types A, B, and C; outer dentary teeth were assigned to *Sindacharax* sp. type A.

Discussion

These teeth were very common in the north Kaiyumung Member deposits. They attain a much larger size than *Sindacharax* teeth from earlier deposits. Their cusp patterns bear some similarity to those of *S. lothagamensis*, although the cusp patterns of *S. howesi* are more ridged and are on average much larger. There may be some evolutionary relationship between the two species. These teeth are most common in northern Kaiyumung Member deposits, but occasionally they also occur in later deposits.

Sindacharax deserti Greenwood and Howes, 1975
(Figures 3.22–3.24)

Lothagam Material

- *Muruongori Member:* 3151/3152, second inner premaxillary tooth; 3153, first inner premaxillary tooth, 2 second inner premaxillary teeth, third inner premaxillary tooth; 3154, 7 first inner premaxillary teeth, 20 second inner premaxillary teeth, 2 third inner premaxillary teeth.
- *Kaiyumung Member:* 1852, 2 first inner premaxillary teeth, 8 second inner premaxillary teeth; 1994, 6 first inner premaxillary teeth, 30 second inner premaxillary teeth; 1998, second inner premaxillary tooth; 2000, first inner premaxillary tooth, second inner premaxillary tooth; 2332, 3 second inner premaxillary teeth.

Because these teeth were misassigned in their original description (Greenwood 1972; see also Greenwood 1976a), I will describe them here. Eight of 17 first inner premaxillary teeth have an oval shape, but with widening at the labial end and narrowing at the lingual (figure 3.22). Nine of the 17 have the long and narrow shape of *S. lothagamensis* and *S. mutetii*. In all teeth, the dominant cusp is lingually placed. A low ridge joins it with a smaller cusp that is located on the opposite tooth margin from the dominant cusp. This low ridge continues to join up with another small cusp that is located just anterior to the dominant cusp. The ridge and the cusps thereby form a semicircle; smaller cusps may be incorporated in the semicircle. One or more ridges traverse the width of the tooth anterior to the semicircle.

The second inner premaxillary teeth are exactly as pictured in Greenwood (1972: figures 2a, 2b; 3a, 3b; see also figure 3.22). The complete circle that forms anterior to the main lingual cusp but includes the cusp distinguishes them. This circle is completely united and ridged, with only the main cusp and two weakly defined cusps apparent in the circle.

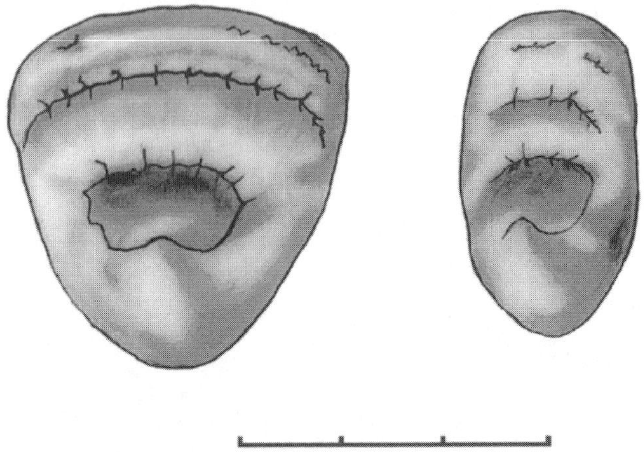

Figure 3.22 *Sindacharax deserti*, drawing of second (at left) and first (at right) inner premaxillary teeth (occlusal view).

The third inner premaxillary teeth have not been described before. They have two shapes: oval with a pointed end, or a long oval. With both, the dominant cusp is at the lingual end flanked on one side by one smaller cusp and on the other side by a ridge of smaller cusps that radiate along the outer tooth margin (figures 3.23 and 3.24). Paralleling the ridge, with a deep channel between, are one or more truncated ridges of small cusps, unlike the long ridge of *S. lothagamensis* or *S. mutetii*. In a few specimens, one or more short ridges occur anterior and perpendicular to the other ridges.

One fourth inner premaxillary tooth was possibly identified, but because it is a lone specimen, it is only tentatively considered a fourth tooth, and hence it is classified as *Sindacharax* sp. It had a long oval shape, with a cusped ridge along each of the long margins of

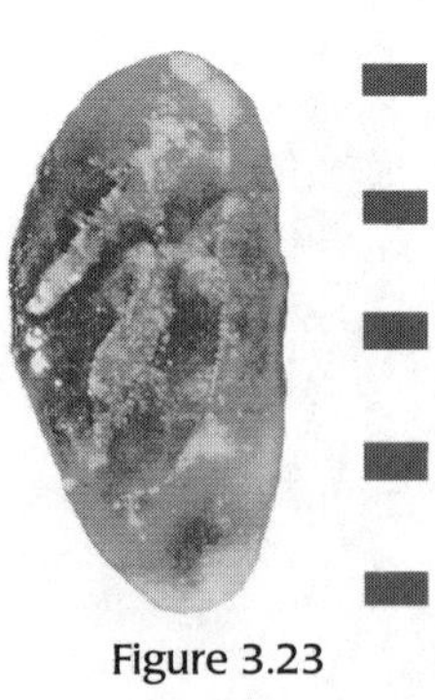

Figure 3.23

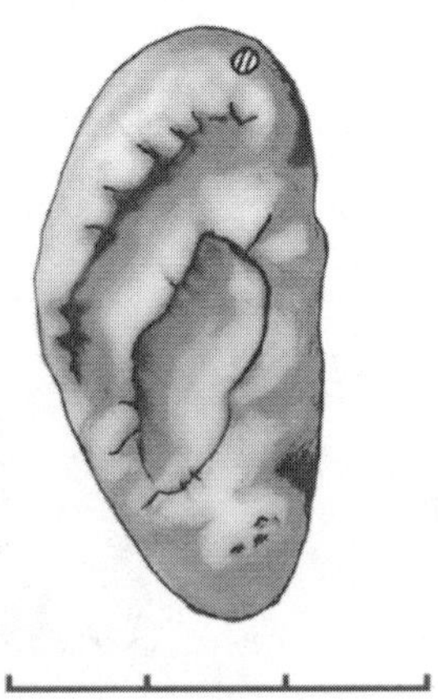

Figure 3.24

Figure 3.23 *Sindacharax deserti*, third inner premaxillary tooth (occlusal view).

Figure 3.24 *Sindacharax deserti*, drawing of third inner premaxillary tooth (occlusal view).

the tooth, and a channel in between the two ridges. One ridge comprised four cusps; the other six smaller cusps.

The outer premaxillary teeth were indistinguishable from those of other species and were ascribed to *Sindacharax* sp. types A and B.

The outer dentary teeth were indistinguishable from those of *S. lothagamensis* and *S. mutetii* and were assigned as *Sindacharax* sp. type A.

Discussion

Teeth assigned to *S. deserti* are very different in appearance from those of *S. lothagamensis* and *S. mutetii.* They are much rounder and often have a distinctive fossilization color and texture. Teeth from *S. deserti* are known primarily from Muruongori Member sites and from the Pliocene-aged site of Ekora, where the majority of *Sindacharax* teeth are *S. deserti.* The Ekora teeth have a much greater size range than those at Lothagam, which are uniformly small (second inner premaxillary teeth are under 4 mm in length). These teeth have little precedent in the Lothagam deposits; they appear suddenly in the Muruongori Member and in the deposits at Ekora, but they are rare in later deposits. Because they are known from the Pliocene-aged Wadi Natrun sites (Egypt), some connection with the Nile basin must be assumed in Muruongori times.

Sindacharax greenwoodi Stewart, 1997

Lothagam Material

- *Muruongori Member:* 3153, 4 first inner premaxillary teeth, 6 second inner premaxillary teeth; 3154, 2 first inner premaxillary teeth.
- *Kaiyumung Member:* 1835, 2 first inner premaxillary teeth, 3 second inner premaxillary teeth; 1850, 2 partial, occluded premaxillae and dentaries from same individual with 17 *in situ* teeth; 35 first inner premaxillary teeth, 35 second inner premaxillary teeth; 1852, 36 first inner premaxillary teeth, 34 second inner premaxillary teeth; 1993, first inner premaxillary tooth, 3 second inner premaxillary teeth; 1994, 70 first inner premaxillary teeth, 40 second inner premaxillary teeth; 1995, 11 second inner premaxillary teeth; 1998, 17 first inner premaxillary teeth, 13 second inner premaxillary teeth; 1999, 21 first inner premaxillary teeth, 41 second inner premaxillary teeth; 2000, 6 first inner premaxillary teeth, 7 second inner premaxillary teeth; 2332, 23 first inner premaxillary teeth, 17 second inner premaxillary teeth.

A complete jaw (two occluded premaxillae and dentaries with 17 *in situ* teeth) and a total of 427 isolated teeth were assigned to this species. The occluded jaw was described and named by Stewart (1997) and forms the basis for identification of the isolated teeth.

The first inner premaxillary teeth are identical to those of the type (Stewart 1997), although several are much fresher, with clear cusped ridges. There is a considerable size range in teeth, although none is larger than the type specimen.

The second inner teeth are also identical to the type (Stewart 1997), although many were less worn and had fresher, more clearly cusped ridges than the type. Again, there is a considerable size range in teeth but none is larger than the type.

Third and presumed fourth teeth recovered showed a variety of cusp patterns and were classified as *Sindacharax* sp.

Many outer premaxillary and dentary teeth were similar to the type, but because these overlapped with later, Pleistocene-aged *Sindacharax* teeth (personal observation), they were all assigned to *Sindacharax* sp.

Discussion

This species has been described and reported as *Sindacharax greenwoodi* based on a partial, occluded upper and lower jaw with right upper and lower teeth *in situ,* recovered from site 1850 in the southern Kaiyumung Member deposits (Stewart 1997). This species is first seen in the Muruongori Member sites, but rarely, and only as very small teeth. In Kaiyumung Member deposits the teeth are many times larger and far more numerous, comprising the majority of teeth recovered.

Teeth of *S. greenwoodi* are known from Plio-Pleistocene deposits around the Turkana Basin, including the Shungura deposits at Omo (Greenwood 1976a; Stewart 1997), and possibly from the Lake Edward-Albert Basin deposits (personal observation).

Sindacharax sp.

(Figures 3.25–3.32)

A partial, fairly robust left premaxilla from site 1998 in the southern Kaiyumung deposits (figure 3.25) with no teeth but with partial attachment outlines for first, second, third, and fourth inner teeth and two outer teeth can only be identified as *Sindacharax* sp. The inner and outer tooth bases are virtually contiguous, excluding it from *S. greenwoodi.* Without teeth, it is not possible to further assign species.

A total of 166 third and 27 fourth inner premaxillary teeth were assigned only as *Sindacharax* sp., due to apparent similarity between the species, which may, however, be a factor of the small number of these teeth recovered overall.

Figure 3.25 *Sindacharax* sp., premaxilla (occlusal view).

Third and Fourth Inner Premaxillary Teeth

Lothagam Material

- *Lower Nawata:* 1710, third inner tooth; 1752, third inner tooth, fourth inner tooth.
- *Apak Member:* 1760, third inner tooth; 1944, 2 fourth inner teeth.
- *Muruongori Member:* 3153, fourth inner tooth.
- *Kaiyumung Member:* 1850, 16 third inner teeth, fourth inner tooth; 1852, 28 third inner teeth, 4 fourth inner teeth; 1994, 102 third inner teeth, 17 fourth inner teeth; 1995, 3 third inner teeth; 1999, 7 third inner teeth, fourth inner tooth; 2000, third inner tooth; 2332, 6 third inner teeth.

Many third and fourth inner premaxillary teeth could not be assigned to species, as they were very similar throughout the deposits. Often the third and fourth teeth could not be distinguished from each other, as only one fourth tooth is preserved *in situ* on the jaw (*S. greenwoodi* type specimen) and even this is very worn; thus they are assigned here.

Outer Teeth

All outer premaxillary teeth and outer dentary teeth were identified only as *Sindacharax* sp., for two reasons. First, because only the *S. greenwoodi* jaw had outer premaxillary and dentary teeth *in situ* it was not possible to associate outer teeth with inner teeth of the other species, unless identical teeth were exclusively associated in the same members. Second, recovery of hundreds of outer teeth indicates considerable similarity in shape and cusp pattern, making species differentiation impossible. There are several distinct types, however, and I have classified the outer teeth into these types, with an indication of the species with which they were most consistently associated.

Outer Premaxillary Teeth: Type A

Lothagam Material

- *Lower Nawata:* 1644, tooth; 1659, tooth; 1752, 2 teeth; 1990, 4 teeth; 2413, 2 teeth; 2419, tooth; 3150, 2 teeth.
- *Apak Member:* 1942, 7 teeth; 1944, 15 teeth; 2420, tooth.
- *Muruongori Member:* 3153, 2 teeth; 3154, 5 teeth.
- *Kaiyumung Member:* 1850, 4 teeth; 1852, 15 teeth; 1993, 2 teeth; 1994, 121 teeth; 1995, 8 teeth; 1999, 5 teeth; 2000, tooth; 2332, 4 teeth.

Discussion

Type A teeth consist of one dominant and two much smaller flanking cusps that slope into a short platform on one side but have a steep shelf on the other side (figures 3.26 and 3.27). The platform is uncusped. They

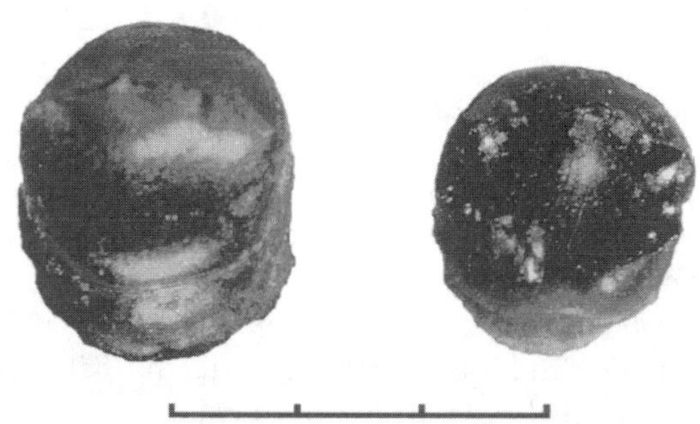

Figure 3.26 *Sindacharax* sp., Type A (at left) and Type B (at right) outer premaxillary teeth (occlusal view).

Figure 3.27 *Sindacharax* sp., drawings of Type A (at left) and Type B (at right) outer premaxillary teeth (occlusal view).

have a round or oval attachment base. These teeth are found primarily in the Nawata Formation and the Apak and Muruongori Members, and they are the only outer teeth associated with *S. lothagamensis*. Those associated with *S. mutetii* often have elongated attachment bases.

Outer Premaxillary Teeth: Type B

Lothagam Material

- *Muruongori Member:* 3153, 3 teeth; 3154, 12 teeth.
- *Kaiyumung Member:* 1835, tooth; 1850, 5 teeth; 1852, 42 teeth; 1994, 136 teeth; 1995, 2 teeth; 1998, 5 teeth; 2000, 3 teeth; 2332, 13 teeth.

Discussion

Type B teeth are similar to Type A but have one or more discrete cusps at the base of the platform (figures 3.26 and 3.27). Their attachment base is round or a roundish oval. These teeth first appear in the Muruongori Member and are especially common in the Kaiyumung north deposits, associated with *S. howesi*.

Outer Premaxillary Teeth: Type C

Lothagam Material

- *Muruongori Member:* 3153, 5 teeth.
- *Kaiyumung Member:* 1835, 2 teeth; 1850, 67 teeth; 1852, 41 teeth; 1993, tooth; 1994, 135 teeth; 1995, 4 teeth; 1998, 25 teeth; 1999, 56 teeth; 2000, 4 teeth; 2332, 14 teeth.

Discussion

These teeth again have a dominant cusp flanked by two smaller cusps, which drop into a short platform that has one or two rows of discrete cusps (pictured in Stewart 1997:figures 2A and 3A, first and third outer teeth). The base is an elongated oval. These appear in the Kaiyumung north deposits and are common particularly in Kaiyumung south deposits, associated with *S. greenwoodi*. They comprise the first and third outer premaxillary teeth of the *S. greenwoodi* type specimen.

Outer Premaxillary Teeth: Type D

Lothagam Material

- *Kaiyumung Member:* 1835, tooth; 1850, 49 teeth; 1852, 16 teeth; 1994, 23 teeth; 1998, 37 teeth; 1999, 50 teeth; 2332, 9 teeth.

Discussion

These teeth differ considerably from the previous types in that they are flat, with no steep shelves. There is a central cusp, surrounded by one or two concentric circles of discrete cusps (pictured in Stewart 1997:figure 3.2A and 3A, second outer tooth). These teeth do not appear until the Kaiyumung south deposits and are associated with *S. greenwoodi* teeth. This tooth is identical to the second outer premaxillary tooth of the *S. greenwoodi* type specimen.

Outer Dentary Teeth

The varieties described next are based on outer dentary teeth, probably mainly first, second, and third teeth; fourth teeth are much smaller, and few have been recovered. The first tooth in both types is usually truncated posteriorly, to accommodate the inner tooth (figure 3.28). There is considerable wear visible on most dentary teeth, and it is often difficult to describe any morphology on the teeth.

Outer Dentary Teeth Variety A

Lothagam Material

- *Lower Nawata:* 1658, 5 teeth; 1659, tooth; 1710, 3 teeth; 1733, tooth; 1735, tooth; 1752, 12 teeth; 1990, 16 teeth; 2413, 2 teeth; 3150, 2 teeth.
- *Apak Member:* 1760, 6 teeth; 1942, 14 teeth; 1944, 34 teeth; 1948, 2 teeth; 1960, 4 teeth; 2420, 4 teeth.
- *Muruongori Member:* 3151/3152, 3 teeth; 3153, 20 teeth; 3154, 29 teeth; 3153/3154, 2 teeth.
- *Kaiyumung Member:* 1835, tooth; 1850, 68 teeth; 1852, 139 teeth; 1992, tooth; 1993, 12 teeth; 1994, 464 teeth; 1995, 21 teeth; 1998, 20 teeth; 1999, 26 teeth; 2000, 27 teeth; 2332, 81 teeth.

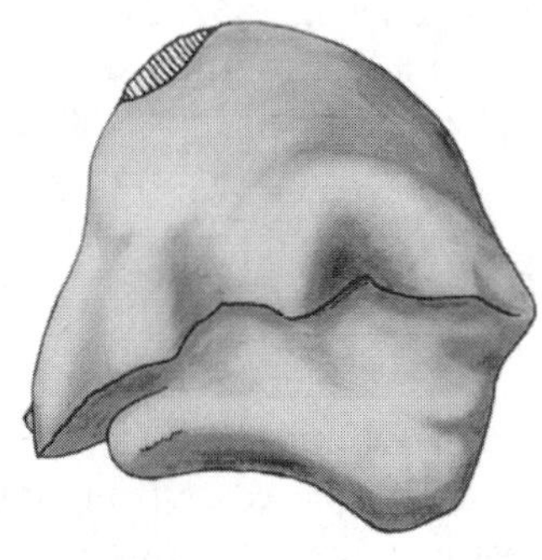

Figure 3.28 *Sindacharax* sp., drawing of Type A first outer dentary tooth (occlusal view).

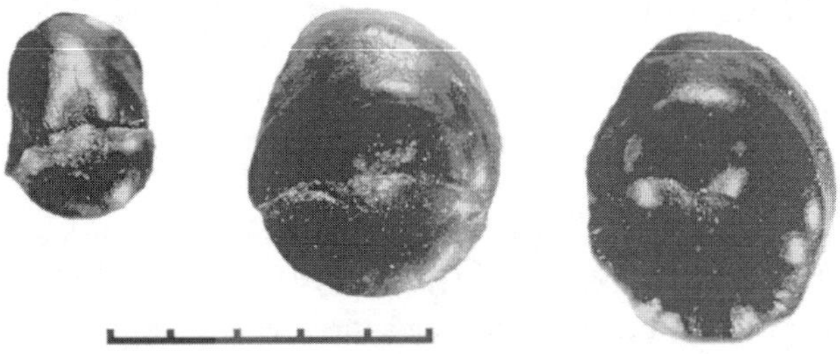

Figure 3.29 *Sindacharax* sp., Type A (two teeth at left) and Type B (right) outer dentary teeth (occlusal view).

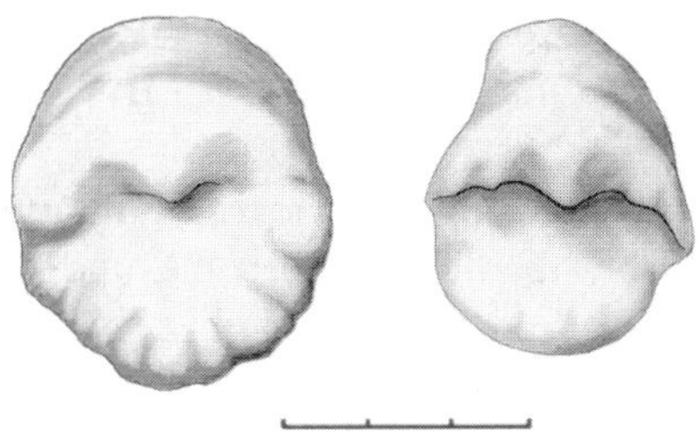

Figure 3.30 *Sindacharax* sp., drawings of outer dentary teeth (occlusal view); Type A (right), Type B (left).

These teeth also have a dominant cusp, with a prominent point, flanked by two smaller cusps, which form a shelf on one side, more elongated and less steep than that of the premaxillary teeth (figures 3.29 and 3.30). On the other side the cusps slope into a platform that is broader than that of the premaxillary teeth.

The platform is usually uncusped but may be weakly cusped. The attachment base is much more elongated than in most premaxillary teeth. These teeth are found throughout the sequence, but primarily in the Nawata and early Nachukui formations.

Outer Dentary Teeth Variety B

Lothagam Material

- *Muruongori Member:* 3153, 2 teeth.
- *Kaiyumung Member:* 1835, 2 teeth; 1850, 26 teeth, dentary with 2 teeth; 1852, 21 teeth; 1993, 2 teeth; 1994, 6 teeth; 1995, tooth; 1998, 18 teeth; 1999, 7 teeth, dentary with 3 teeth; 2000, 2 teeth; 2332, 6 teeth.

Discussion

These teeth are generally larger and rounder than variety A in overall size and shape (figures 3.29 and 3.30). They have a centrally placed cusp, but it is much smaller than that in variety A, and the ridges leading away from the cusp are vestigial. These are very flat teeth; with wear, the cusp is not visible. There are often weak cusps posterior to the main cusp, on the posterior part of the shelf. The attachment bases are elongated. These teeth appear in the Muruongori Member, but are rare until the Kaiyumung Member southern deposits.

The teeth of the two partial dentary specimens from sites 1850 and 1999 in the southern Kaiyumung deposits are too worn to name to species. The dentary from site 1999 has the first and second outer teeth *in situ,* and although these are very worn, they are evident as variety B (figure 3.31). The dentary from 1850 has the first, second, and third teeth *in situ* and an attachment base for the fourth. The teeth are very worn but appear to be variety B (figure 3.32).

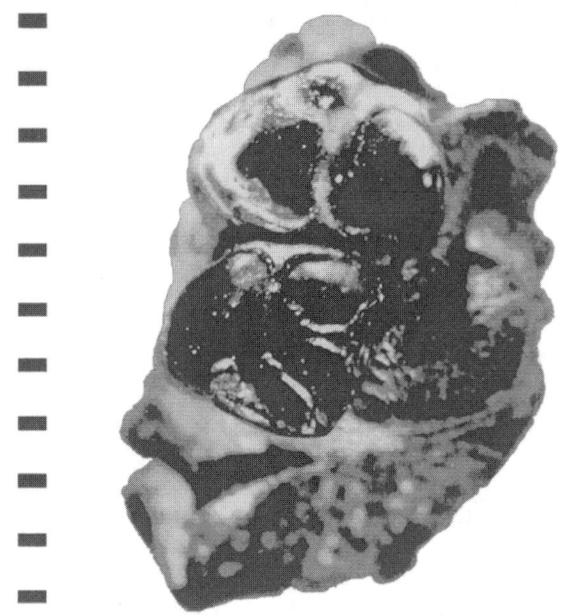

Figure 3.31 *Sindacharax* sp., dentary and two teeth (occlusal view).

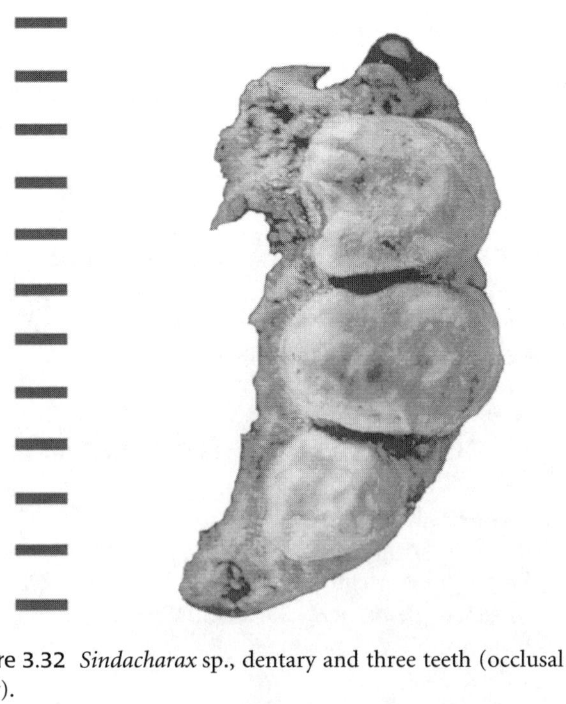

Figure 3.32 *Sindacharax* sp., dentary and three teeth (occlusal view).

Inner Dentary Teeth

Lothagam Material

- *Apak Member:* 1944, 6 teeth; 1948, tooth; 1960, tooth.
- *Muruongori Member:* 3153, 2 teeth; 3153/3154, tooth.
- *Kaiyumung Member:* 1850, 7 teeth; 1852, 13 teeth; 1994, 56 teeth; 1998, 2 teeth; 1999, 4 teeth; 2000, tooth; 2332, 2 teeth.

These teeth are very similar in both extant *Alestes* and fossil *Sindacharax.* They are the only inner dentary teeth and are positioned posterior to the first outer dentary tooth, which usually has a notch for the adjacent inner dentary tooth. The inner dentary teeth are often very worn, but pristine teeth are small and round in shape, with a single elongated centrally placed cusp.

Worn and/or Fragmented Teeth

Lothagam Material

- *Lower Nawata:* 1710, tooth.
- *Apak Member:* 1760, 2 teeth; 1942, 21 teeth; 1944, 33 teeth; 1948, tooth; 1960, tooth.
- *Muruongori Member:* 3153, 16 teeth; 3154, 3 teeth; 3153/3154, tooth.
- *Kaiyumung Member:* 1835, 2 teeth; 1850, 225 teeth; 1852, 236 teeth; 1992, 2 teeth; 1993, 22 teeth; 1994, 426 teeth; 1995, 40 teeth; 1998, 32 teeth; 1999, 194 teeth; 2000, 53 teeth; 2332, 99 teeth.

All worn and/or fragmented teeth not assigned to species are listed here.

Discussion

Several problems exist with the interpretation of the *Sindacharax* material from Lothagam. The genus was erected by Greenwood and Howes (1975) based on isolated teeth from Western Rift sites in Zaire; the paucity of jaws with *in situ* teeth has made classification and taxonomy very difficult. Hundreds of teeth have since been recovered and reported, but varying tooth shapes and cusp patterns, and lack of characoid relatives with similar tooth morphology, have led to considerable confusion in the literature (Greenwood and Howes 1975; Greenwood 1976a; Van Neer 1992; Stewart 1997).

One occluded jaw, several premaxillary and dentary fragments, and thousands of isolated teeth recovered from the Lothagam and nearby sites confirm that there is considerable individual variation in *Sindacharax* jaw and tooth morphology, and this variation hinders a coherent classification. Further, because of some reworking of the Lothagam deposits during the high lake levels of the Early Holocene, there is some mixing of fossils. However, most sites had internal consistency in representation of species.

At least five species were present in the Late Miocene to Pliocene-aged deposits at Lothagam—two previously reported species and three new species. Whereas there is considerable overlap between the outer teeth and the inner third and fourth teeth of the species, the first and second inner teeth were distinctive and consistently represented within the members, thereby allowing easy separation. Each species was primarily and consistently identified with a stratigraphic unit or geographic locale within such a unit: *Sindacharax lothagamensis* with the Nawata Formation; *S. mutetii* with the Apak Member deposits (and at the site of Kanapoi); *S. howesi* with the north Kaiyumung Member deposits; *S. deserti* with the Muruongori Member deposits (and the site of Ekora); and *S. greenwoodi* (and probably later *Sindacharax* species represented by very worn teeth) with the southern Kaiyumung Member deposits.

Sindacharax lothagamensis, primarily from Lower Nawata sites, is sparsely represented. It was a smaller fish than later *Sindacharax* species, judging from the size of the teeth. The teeth are more *Alestes*-like in shape and morphology than those of later species, particularly the outer dentary and premaxillary teeth and the third and fourth inner premaxillary teeth. Because of its similarities to many extant *Alestes* species, *S. lothagamensis* was possibly less specialized in diet than later *Sindacharax* species that show more tooth specialization.

By Apak Member times, *Sindacharax mutetii* had evolved: it was larger in size than *S. lothagamensis* and with greater ridging on its inner premaxillary teeth. However *S. mutetii* shared many similarities with *S. lothagamensis,* in particular outer tooth morphology. *Sindacharax mutetii* is at present known only from the Lake Turkana Basin deposits. It is the only *Sindacharax* species recovered from the Pliocene-aged Kanapoi deposits, and it is rare in Lothagam deposits later than the Apak Member. A robust premaxilla with teeth displaying slightly different cusp patterns is tentatively assigned to this species.

A completely different species of *Sindacharax*—*S. deserti*—appeared in the Muruongori Member deposits. Teeth of this species were also small at Lothagam, though larger at Ekora, and with a completely different shape and cusp morphology from earlier species. The ridges in the inner premaxillary teeth were so defined that individual cusps were almost not discernable. Further the teeth were of a different shape, in particular the second inner premaxillary teeth. The outer premaxillary teeth differed in that many possessed small cusps. *S. deserti* is also known from deposits in Egypt (Wadi Natrun), and possibly Lake Edward, Zaire. As mentioned, it is also the dominant *Sindacharax* species recovered from the Pliocene site of Ekora, southeast of Lothagam. While *S. deserti* appears in later Lothagam deposits, it is rare.

The northern Kaiyumung deposits are dominated by *S. howesi*, which is larger than most earlier species of *Sindacharax*, based on the length of the second inner premaxillary teeth (up to 9 mm in length). The inner teeth of *S. howesi* have lost the extreme ridging of *S. deserti*, with individual cusps weakly defined. The second inner teeth acquired more of a "shoe-shape" that is also seen in later, Pleistocene-aged *Sindacharax*. Most outer premaxillary teeth have cusps, as do some dentary teeth. This species is also known from Allia Bay deposits (personal observation).

The southern Kaiyumung exposures yield fossils from several different units, but *S. greenwoodi* is dominant at all sites. The type specimen, an occluded jaw, has been reported elsewhere (Stewart 1997), and many isolated teeth were recovered. These fish were also larger in size than were earlier species. Teeth from this species have been recovered elsewhere in the Turkana Basin, including from the Omo Shungura deposits (Greenwood 1976a; Stewart 1997) and Allia Bay (personal observation).

Many narrow, shoe-shaped second inner premaxillary teeth were recovered from the Kaiyumung Member, primarily the southern deposits. Virtually all showed considerable weathering. These are similar to teeth known from later, Pleistocene deposits (teeth collected by Craig Feibel and described by Rose Difley of the University of Utah [unpublished ms.]) and may represent reworking of deposits by the Holocene high lake. They are here identified as *Sindacharax* sp.

More jaw elements with *in situ* teeth are needed to establish the evolutionary relationships of *Sindacharax;* with present material consisting almost solely of isolated teeth, only taxonomy can be undertaken. It has been suggested that cusped ridges are a derived character for the Characidae (Greenwood and Howes 1975), which could make *Sindacharax* a sister group of the South American serrasalmines. Further study should clarify relationships between the African and South American groups; it has been suggested that *Sindacharax* may be the last of an old world serrasalmine lineage (Greenwood and Howes 1975) that probably had a Gondwanaland origin.

A little known extant species of *Alestes*—*A. stuhlmanni*—has ridged cusps on its teeth and possibly should be reclassified as *Sindacharax* (see Stewart 1997 for further discussion).

Characidae indet.

Lothagam Material

- *Lower Nawata:* 1752, 2 vertebrae centra.
- *Apak Member:* 1658, 2 trunk vertebrae centra; 1942, 2 trunk vertebrae centra; 1944, 2 trunk vertebrae centra; caudal vertebra centrum, vertebra centrum.
- *Muruongori Member:* 3153, caudal vertebra centrum; 3154, trunk vertebra.
- *Kaiyumung Member:* 1994, 3 caudal vertebrae centra.

Order Siluriformes
Family Bagridae
Bagrus Bosc, 1816
aff. *Bagrus* sp.

Lothagam Material

- *Lower Nawata:* 1710, trunk vertebra centrum, pectoral spine portion; 1987, caudal centrum.
- *Upper Nawata:* 1594, frontal fragment; 1950, trunk vertebra centrum.
- *Apak Member:* 1853, vertebra centrum; 1942, posterior skull (basioccipital, parasphenoid, supraoccipital, frontal fragment); 1948, posterior portion skull.
- *Kaiyumung Member:* 1851, vertebra centrum.

Elements of *Bagrus* are not as robust as other bagrid genera and are not as well represented in fossil deposits. Skull fragments are distinctive by their striated appearance, unlike the "bumpy" crania of other bagrid and clariid fish.

Discussion

Recently Van Neer (1994) identified a new bagrid genus and species (*Nkondobagrus longirostris*) from Late Miocene–Pliocene deposits in the Albert Rift, Uganda, based solely on anterior portions of neuro-crania. There were many similarities to skulls of the genus *Bagrus.* Because anterior elements of bagrid neurocrania were not recovered from Lothagam, it was not possible to tell if *Nkondobagrus* was present; elements similar to *Bagrus* were classified only by affinity to the modern genus, with which they compared most favorably. The size range of the Lothagam specimens was within the extant range, from about 20 cm up to about 1 m.

Bagrus fossils are known from Late Miocene deposits at Chalouf, Egypt (Priem 1914), possibly from Pliocene Wadi Natrun (Greenwood 1972), Pliocene Lake Edward-Albert rift (Greenwood 1959), and the Plio-Pleistocene deposits of the Omo Valley (Arambourg 1947) and Koobi Fora (Schwartz 1983). Extant *Bagrus* is a deepwater catfish, with a diet primarily of fish. It is common in Lake Turkana where it is represented by *B. docmac* and *B. bayad.* It is also found from Senegal to the Nile, including the Volta, Niger, and Chad basins.

Clarotes Kner, 1855
Clarotes sp.

Lothagam Material

- *Lower Nawata:* 1659, basioccipital portion, 5 cranial fragments, 2 trunk vertebrae centra; 1710, parasphenoid portion, basioccipital portion, 2 cleithra fragments, pectoral spine fragment; 1732, pectoral spine fragment; 1733, posttemporal fragment; 1751, anterior portion skull (dermethmoid, prefrontals, frontals, anterior parasphenoid), prefrontal, supraoccipital portion, dentary portion, quadrate, cleithrum fragment, posttemporal; 1990, cranial fragment, 2 pectoral spine fragments; 2365, supraoccipital/basioccipital articulated; 2409, partial posterior skull fragment (supraoccipital, basi-occipital), 2 frontals, supraoccipital, basioccipital, articular, quadrate, 2 pectoral spine fragments, 3 weberians, 4 vertebrae centra; 2412, cleithrum fragment, pectoral spine fragment.
- *Upper Nawata:* 1594, frontal portion, sphenotic fragment, articular portion, quadrate fragment, 2 cranial spine fragments, 2 pectoral spine fragments, opercular fragment, 2 cleithra fragments, weberian, 2 trunk vertebrae centra, 3 caudal vertebrae centra; 1755, pectoral spine; 1765, cleithrum fragment; 1969, 2 cleithra/coracoid fragments.
- *Apak Member:* 1849, cranial fragment, pectoral spine fragment; 1942, cranial fragment.
- *Kaiyumung Member:* 1851, pectoral spine.

Clarotes has a robust cranial shield, and its elements preserve well. The cranial elements and pectoral spines are also distinctive, with a striated bumpy texture. Because *Clarotes* can achieve lengths up to 1.5 m, their elements are often abundant in fossil deposits. Elements from the whole skeleton were common in the Lothagam deposits.

Discussion

Van Neer (1994) recently identified a new species of bagrid, *Chrysichthys macrotis,* from Late Miocene-Pliocene deposits in the Albert Rift, Uganda, based primarily on complete and nearly complete skulls. The most commonly recovered postcranial elements—pectoral spines, dorsal spines, articulars, and vertebrae—bear considerable similarity to living *Clarotes* and *Chrysichthys* elements, so that identification based on these elements alone even to genus is usually not possible (Van Neer 1994:105ff.). I considered this new species when investigating the Lothagam fossils. However, *Clarotes* has a strong Plio-Pleistocene presence in the Lake Turkana Basin (e.g., Schwartz 1983), and much of the Lothagam bagrid material resembles extant skulls of this genus. Further, the Lothagam *Clarotes* material contains several individuals that are estimated to be much larger than the Ugandan *Chrysichthys macrotis*—up to 1.5 m in length—and their skulls appeared broader than the new *Chrysichthys* fossils. These factors, combined with the similarity to extant *Clarotes* material, made it apparent that the fossils were *Clarotes.*

Clarotes fossils are known from Miocene deposits at Sinda, Zaire (Greenwood and Howes 1975) and Bled ed Douarah, Tunisia (Greenwood 1973), Pliocene deposits at Wadi Natrun, Egypt (Greenwood 1972), Plio-Pleistocene deposits in the Lakes Edward-Albert Basins (Stewart 1990; Greenwood 1959; Greenwood and Howes 1975), and Plio-Pleistocene deposits in the Omo Valley (Arambourg 1947) and Koobi Fora (Schwartz 1983). One species of *Clarotes, C. laticeps,* is present in the Omo River, but not in Lake Turkana, and is widespread throughout the Nile, Senegal, and Niger systems and in rivers in eastern Africa.

Bagridae indet.

Lothagam Material

- *Lower Nawata:* 1658, trunk centrum, pectoral spine fragment; 1710, 2 pectoral spine fragments; 1733, pectoral spine fragment; 1751, pectoral spine fragment; 1971, articular fragment, 5 trunk vertebrae centra, pectoral spine fragment; 1990, cranial spine fragment; 2417, caudal vertebra centrum.
- *Upper Nawata:* 1594, basioccipital fragment, 4 trunk vertebrae centra, caudal vertebra centrum; 1734, trunk vertebra centrum; 1951, 4 caudal vertebrae centra; 1957, pectoral spine fragment.
- *Apak Member:* 1760, trunk vertebra centrum; 1942, caudal vertebra centrum.
- *Kaiyumung Member:* 1851, pectoral spine; 1994, 2 trunk vertebrae centra; 1995, trunk vertebra centrum.

Discussion

Van Neer's (1994) report on Miocene-Pleistocene fossil fish from Uganda documents the frequency of bagrid catfish remains (including a new genus and species with similarities to *Bagrus*) and a new species with many elements similar to those of genera *Clarotes* and *Chrysichthys*. Bagrids were not common at Lothagam and do not appear to have radiated as they did in the Western Rift localities, although they are more common in later Pliocene-Pleistocene Turkana Basin deposits (remains of large *Clarotes* individuals are common in Pliocene-aged South Turkwel sites [personal observation], and Plio-Pleistocene Turkana Basin deposits [Schwartz 1983]).

Family Schilbeidae
Schilbe Oken, 1817
?*Schilbe* sp.

Lothagam Material

- *Upper Nawata:* 1594, caudal vertebra.

This vertebra showed most affinity with *Schilbe*, but as it is the only element attributed to this genus in all Lothagam deposits, the identification should be viewed with caution.

Discussion

Schilbe is present only at one Upper Nawata site. An unpublished report lists the genus from Miocene Chiando Uyoma deposits in Kenya (Schwartz 1983). *Schilbe uranoscopus* is present in modern Lake Turkana, while extant *Schilbe* is widespread in systems throughout the African continent.

Family Clariidae
Clarias Scopoli, 1777
Heterobranchus Geoffroy Saint-Hilaire, 1809
Clarias sp. or *Heterobranchus* sp.

Lothagam Material

- *Lower Nawata:* 1635, dermethmoid portion, parasphenoid fragment, supraoccipital, complete and partial articular; 1644, pectoral spine fragment; 1658, dermethmoid portion, prefrontal, supraorbital fragment, 5 supraoccipital portions, 5 pectoral spine fragments, 3 trunk vertebrae centra; 1659, dermethmoid fragment, supraoccipital fragment, 26 cranial fragments, pectoral spine, trunk vertebra centrum; 1672, dermethmoid fragment, supraorbital portion, 3 supraoccipital portions, complete and 3 portions articulars, cerato- and epihyal portion, 2 pectoral spine fragments, caudal vertebra centrum, trunk vertebra centrum, 3 caudal vertebrae centra; 1710, anterior skull (dermethmoid, prefrontals, anterior frontals), 2 dermethmoid portions, 2 complete prefrontals, complete and portion frontals, posttemporal, 2 articular fragments, 2 urohyals, operculum, 4 pectoral spine fragments, 2 trunk vertebrae centra, 2 caudal vertebrae centra; 1732, dermethmoid, articular, trunk vertebra centrum, caudal vertebra centrum; 1733, dermethmoid portion, 4 frontal fragments, 3 supraoccipital portions, ventral hypohyal, pectoral spine fragment, caudal vertebrae centra; 1751, 8 dermethmoid portions, frontal fragment, supraorbital portion, postorbital fragment, 2 pterygoid fragments, 10 supraoccipital portions, posttemporal portion, urohyal, ventral hypohyal portion, complete and partial articulars, operculum, cleithrum portion, weberian portion, 4 trunk vertebrae centra, caudal vertebra centrum; 1752, pterygoid portion, supraoccipital fragment, 119 cranial fragments, pectoral spine fragment; 1773, articular portion, 6 cranial fragments; 1781, pectoral spine fragment; 1990, supraoccipital fragment; 1996, anterior skull portion (frontals, prefrontals, dermethmoid, vomer), anterior skull fragment (prefrontal, dermethmoid, vomer), 2 dermethmoids, 3 complete and 3 partial frontals, sphenotic, 2 pterotics, supraoccipital portion, 66 cranial fragments, dentary portion, articular portion,1 pectoral spine fragment, weberian; 2301, anterior skull portion, frontal, supraoccipital portion; 2312, 3 dermethmoid portions; 2365, complete and 2 partial dermethmoids, complete and 2 partial prefrontals, 5 supraoccipital portions, 128 cranial fragments, 2 trunk vertebrae centra; 2386, prefrontal, 2 pterotics, supraoccipital portion, posttemporal; 2413, posterior skull fragment, supraoccipital portion, 70 cranial fragments.
- *Upper Nawata:* 1594, dermethmoid fragment, articular, cleithrum fragment, 3 cranial fragments, caudal vertebra centrum; 1655, 5 dermethmoid portions, 3 prefrontal portions, frontal fragment, complete and fragment supraoccipital, 2 complete and 5 fragmented articulars, 40 cranial fragments, complete and 4 fragmented pectoral spines, trunk vertebra centrum, caudal vertebra centrum; 1756, dermethmoid portion, cranial fragment, 2 pectoral spine fragments; 1765, medioposterior half skull (frontals, pterygoids, left sphenotic, dermosphenotics, right operculum, cleithra, supraoccipital, basioccipital), supraoccipital portion; 1766, dermethmoid portion, prefrontal, supraoccipital, basioccipital, pectoral spine fragment, caudal vertebra centrum; 1950, epihyal, 3 cranial fragments, 2 trunk vertebrae centra; 1957, skull fragment (right vomer, dermethmoid), 14 complete dermethmoids and 33 fragments, 9 complete and 25 partial prefrontals, 8 complete frontals and 7 fragments, complete supraorbital and 19

fragments, postorbital, 4 dermosphenotics, 28 sphenotics, 16 pterotics, 13 parasphenoid portions, 39 supraoccipital portions, 11 posttemporals, 2 complete cerato- and epihyals, 3 ceratohyals, 6 ceratohyal portions, 6 epihyals, 2 urohyals, 6 premaxillae, 12 dentary portions, 8 complete and 48 partial articulars, 9 complete and quadrate fragment, operculum portion, 44 cleithra/coracoid fragments, 906 cranial fragments, complete and 25 pectoral spine fragments, 3 weberians, vertebra centrum; 1977, cleithrum portion, 9 cranial fragments; 1988, supraoccipital fragment, articular portion, pectoral spine fragment, 7 caudal vertebrae centra.

- *Apak Member:* 1759, medial skull (frontal, parasphenoid, prootic, supraoccipital); 1760, pterotic fragment, 11 cranial fragments; 1942, 2 frontal fragments, supraoccipital fragment, cranial fragment; 1944, cranial fragment, trunk vertebra centrum; 1948, trunk centrum vertebra; 1959, frontal fragment.
- *Kaiyumung Member:* 1850, articular, vertebra centrum; 1852, 1 caudal vertebra centrum.

Elements of *Clarias* and *Heterobranchus* are very similar in morphology, although a few elements (e.g., the palatine) can be distinguished between the two. *Heterobranchus* is extremely rare today and almost nonexistent at any fossil deposit where separation of genera was possible. Although I have classified fossils as either *Clarias* or *Heterobranchus,* the vast majority are attributable to *Clarias.* Clariid elements are very robust, particularly the cranial elements, and are common fossils. At Lothagam, elements from all parts of the skeleton were represented, particularly the robust cranial fragments. As has been noted from other fossil sites (personal observation) vertebrae were not as abundant as cranial remains. The Lothagam remains represent fish in a variety of size ranges from <10 cm to about 1 m in total length, with the majority between 40 and 50 cm in length. Several partial or almost-complete skulls were recovered for use in systematic study of fossil clariids by future researchers.

Discussion

Clarias is a shallow water species with an accessory air breathing organ, allowing it to survive out of water for up to 18 hours. Its diet is varied, including insects, plankton, and fish. It can reach lengths of almost 2 meters. Clariid remains are common throughout the Lothagam deposits and in late Cenozoic deposits of Africa. Only tentative identifications are reported in the mid Miocene from Bled ed Douarah (Tunisia) (Greenwood 1973) and Ngorora, Kenya (Schwartz 1983). Definite clariid remains are known from Miocene deposits in Sinda, Zaire (Van Neer 1992) and Chalouf, Egypt (Priem 1914), Mio-Pliocene deposits in Manonga, Tanzania (Stewart 1997), Mio-Pleistocene deposits in the Albert-Edward Rift (Van Neer 1994), Pliocene deposits in Wadi Natrun, Egypt (Greenwood 1972), and Plio-Pleistocene deposits at Koobi Fora (Schwartz 1983). Extant *Clarias* is represented by *C. lazera* in Lake Turkana. *Clarias* is widespread throughout Africa, including the Nile, Zaire, and Zambezi basins.

Heterobranchus has a similar appearance to *Clarias* and can achieve similar size. It may be more sensitive to high salinity values than *Clarias.* At present, it is represented in Lake Turkana by one species—*H. longifilis*—but this species is rare in the lake. Like *Clarias, Heterobranchus* is widespread throughout the major river basins of Africa. It has no fossil record.

Family Mochokidae
Synodontis Cuvier, 1817
Synodontis sp.

Lothagam Material

- *Lower Nawata:* 1658, 2 pectoral spine fragments; 1659, supraoccipital, 2 cleithra fragments, 2 trunk vertebrae centra, weberian portion; 1710, cleithrum fragment, cranial fragment; 1732, medial skull portion (parietals, sphenotics, pterotics, posttemporals, nuchal plate), 1733, anterior skull portion (dermethmoid, frontals, prefrontals), medial skull portion (frontals, parietals, sphenotics, pterotics, posttemporals, nuchal plate, supraoccipital), 2 supraoccipital fragments, operculum fragment, cleithrum portion; 1735, medial skull portion (prefrontals, frontals, parietals, sphenotics, pterotics, supraoccipital, nuchal plate, weberian); 1751, cleithrum portion, supraoccipital portion; 1989, cleithrum.
- *Upper Nawata:* 1594, parietal fragment, cleithrum fragment, 2 cranial fragments.
- *Apak Member:* 1758, supraoccipital portion; 1942, medial skull fragment; 2 cranial fragments; 1968, cleithrum; 1756, 2 cleithra fragments, pectoral spine fragment; 1949, complete skull; 1950, pectoral spine fragment.
- *Muruongori Member:* 3153, pectoral spine fragment.
- *Kaiyumung Member:* 1835, anterior skull fragment; 1994, 2 teeth; 1998, 3 teeth; 1999, tooth.

Synodontis remains were present throughout the sequence, although not abundant except for a concentration in the Lower Nawata. Elements from all parts of the body were recovered, including several skulls, and isolated teeth. As with most other catfish, the robust cranial shield preserves especially well in fossil deposits.

Discussion

It is possible to distinguish species by the humeral process on the cleithrum and by dentary teeth (Greenwood 1959), but with 112 extant *Synodontis* species described (Poll 1971) and several closely related genera, it was not possible to obtain the necessary dry comparative material to distinguish these. Greenwood (1959) was able to distinguish *S. schall* from *S. frontosus* using the humeral process on cleithra, and from his discussion, the Lothagam cleithra are more similar to *S. frontosus*. Several skulls and partial skulls were recovered and will be useful for further investigation by future researchers.

Synodontis remains were found throughout the Lothagam deposits. *Synodontis* is a small catfish, well protected by a cranial shield and two heavily serrated pectoral spines and one cranial spine. Its diet is varied, including insects, plankton, mollusks, and fish. Fossil *Synodontis* is known from Miocene deposits at Rusinga and Chianda, Kenya (Greenwood 1951; Van Couvering 1977), Moghara and Chalouf, Egypt (Priem 1920), and Bled ed Douarah, Tunisia (Greenwood 1973), Mio-Pleistocene deposits in Lake Albert-Edward Rift (Greenwood and Howes 1975; Van Neer 1992, 1994), Pliocene deposits in the Lake Edward-Albert Rift (Stewart 1990) and Wadi Natrun (Greenwood 1972), and Plio-Pleistocene deposits at Koobi Fora (Schwartz 1983). There are two species of *Synodontis* in modern Lake Turkana—*S. schall* and *S. frontosus*. Extant *Synodontis* is widespread in systems throughout the African continent.

Siluriformes indet.

Lothagam Material

- *Lower Nawata:* 1635, cleithrum fragment, 5 cranial fragments; 1644, 5 cranial fragments; 1655, 49 cranial fragments, 3 pectoral spine fragments; 1658, cleithrum fragment, 124 cranial fragments, cranial spine fragment, 10 pectoral spine fragments, 5 spine fragments; 1659, 3 cranial fragments, 3 pectoral spine fragments; 1672, 2 parasphenoid fragments, basioccipital fragment, operculum fragment, cleithrum fragment, 142 cranial fragments, 3 pectoral spine fragments, caudal centrum fragment; 1710, dermethmoid fragment, parasphenoid fragment, articular fragment, 5 cleithra fragments, 16 pectoral spine fragments, 140 cranial fragments, 2 trunk vertebrae fragments, caudal vertebra fragment, ray fragment; 1732, basioccipital fragment, 12 cranial fragments; 1733, cleithrum fragment, 29 cranial fragments; 1751, 2 cleithra fragments, 117 cranial fragments, 19 pectoral spine fragments, weberian, trunk centrum fragment; 1752, 40 cranial fragments, 2 pectoral spine fragments; 1971, cleithrum fragment, weberian portion; 2275, trunk vertebra fragment; 2410, 20 cranial fragments; 2411, articular portion; 2412, 2 cleithra portions; 2419, pectoral spine fragment.
- *Upper Nawata:* 1594, prefrontal fragment, frontal fragment, 7 cleithra fragments, 40 cranial fragments, 26 pectoral spine fragments, trunk vertebra fragment, 2 caudal vertebrae fragments; 1756, cranial fragment; 1765, weberian fragment, pectoral spine fragment; 1766, 15 cranial fragments, pectoral spine fragment; 1950, 4 pectoral spine fragments; 1957, weberian fragment, 19 vertebrae fragments; 1977, skull fragment, parasphenoid fragment, 5 pectoral spine fragments, 3 weberian fragments, trunk vertebra fragment, 2 vertebrae fragments.
- *Apak Member:* 1760, 7 cranial fragments, 2 pectoral spine fragments; 1769, cranial fragment; 1849, pectoral spine fragment; 1942, 2 cranial fragments, pectoral spine fragment, cranial spine fragment; 2 weberian fragments, caudal vertebra fragment; 1944, quadrate fragment, pectoral spine fragment, 2 vertebrae fragments; 2420, pectoral spine fragment.
- *Muruongori Member:* 3153, 4 pectoral spine fragments.
- *Kaiyumung Member:* 1850, 3 cleithra fragments, 24 cranial fragments; 1852, cranial fragment, pectoral spine fragment; 1994, pectoral spine fragment; 1995, cranial fragment; 1998, 2 pectoral spine fragments, vertebra centrum; 1999, pectoral spine fragment, trunk vertebra centrum.

Order Perciformes
Suborder Percoidei
Family Latidae Jordan, 1923
Lates Cuvier, 1828
Lates niloticus Linnaeus, 1758
(Figure 3.33)

Lothagam Material

- *Lower Nawata:* 1971, basioccipital portion, 2 trunk vertebrae centra.
- *Upper Nawata:* 1765, basioccipital fragment, dentary fragment, cleithrum fragment, trunk vertebra centrum, caudal vertebra centrum.
- *Apak Member:* 1617, dentary, caudal vertebra centrum; 1759, 2 trunk vertebrae centra; 1760, 2 caudal vertebrae centra; 1761, premaxilla fragment, trunk vertebra centrum, caudal vertebra centrum; 1762, posterior portion skull (basioccipital, exoccipitals,

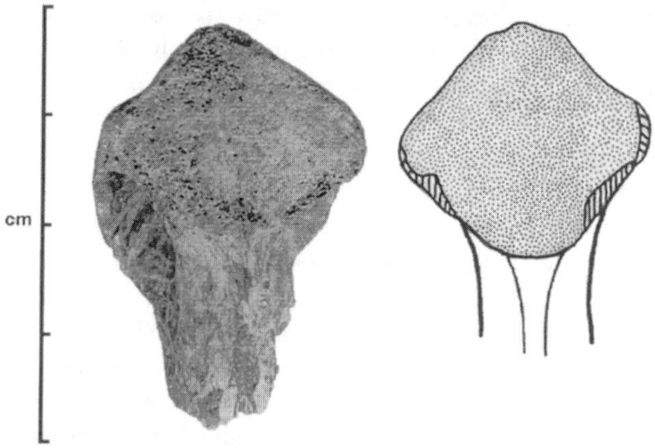

Figure 3.33 *Lates niloticus,* vomer tooth patch in photo and in drawing.

epiotics, prootics, opisthotics, parasphenoid, frontals), trunk vertebra centrum; 1763, trunk vertebra centrum; 1849, parasphenoid fragment, dermethmoid; 1941, 2 trunk vertebrae centra; 1942, angular, dentary fragment, articular fragment, dentary fragment, 7 trunk vertebrae centra, 3 caudal vertebrae centra; 1943, articular fragment, trunk vertebra centrum; 1944, basioccipital fragment, dentary fragment, trunk vertebra centrum; 1960, trunk vertebra centrum, caudal vertebra centrum; 1972, posterior skull fragment; 1973, basioccipital fragment; 1978, trunk vertebra centrum.

- *Kaiyumung Member:* 1835, basioccipital, dentary portion, 2 trunk vertebrae centra; 1850, 2 vomers, 2 anterior premaxillae, anterior dentary, 2 trunk vertebrae centra, caudal vertebra centrum; 1852, 2 anterior trunk vertebrae; 1993, trunk vertebra centrum, dentary fragment; 1994, 3 vomers, vertebra centrum; 1999, vomer; 2000, vomer.

Many elements of this species were found throughout the Lothagam deposits, often in association with *Semlikiichthys* cf. *S. rhachirhinchus* (which until 1999 was named *Lates rhachirhinchus*). With the exception of the vomer, all elements were identical to the extant *L. niloticus* elements. The vomers, as discussed below, were similar to the extant specimens, but with a different shaped dentigerous area and a more sharply inclined vomerine crest.

Discussion

The elements ascribed to *Lates niloticus* were identical to those of extant specimens. Only the vomers showed slight differences, in shape of dentigerous surface (which is in any case variable in extant specimens) and in the angle of the vomerine ridge. The size range of the fish was similar to those of today, up to 2 m in length. Because elements of both *L. niloticus* and *Semlikiichthys* cf. *S. rhachirhinchus* were found in association, it seems that the species coexisted. Elements of both species were also found in the Pleistocene Lakes Edward-Albert Basin deposits of the Western Rift.

Fossil *Lates niloticus* elements are common in African deposits because of their robusticity and size. They are known from Eocene deposits in Fayum, Egypt (Weiler 1929), Miocene deposits from Rusinga, Kenya (Greenwood 1951), Gebel Zelten and Cyrenaica, Libya, (Arambourg and Magnier 1961), Moghara and Chalouf, Egypt (Priem 1920), Bled ed Douarah, Tunisia (Greenwood 1973), Plio-Pleistocene deposits from the Lake Edward-Albert Basins (*L. niloticus* and *S. rhachirhinchus*) (Greenwood 1959; Greenwood and Howes 1975; Stewart 1990; Van Neer 1994), an unpublished report from Marsabit Road, Kenya (Schwartz 1983), the lower Omo Valley (Arambourg 1947) and Koobi Fora (Schwartz 1983), and Pliocene deposits from Manonga, Tanzania (Stewart 1997), and Wadi Natrun, Egypt (Greenwood 1972).

Extant *Lates* is known from Lake Turkana (*L. niloticus* and *L. longispinis*) and is widespread throughout northern, eastern, and western Africa from Senegal to and including the Nile and the Zaire basins.

Family *incertae sedis*
Semlikiichthys Otero and Gayet, 1999
Semlikiichthys rhachirhinchus (Greenwood and Howes, 1975)
Semlikiichthys cf. *S. rhachirhinchus*

(Figures 3.34–3.41)

Lothagam Material

- *Lower Nawata:* 1990, posterior caudal; 2385, first vertebra; 2412, first vertebra; 2419, anterior abdominal vertebra, 2 vertebrae fragments.
- *Upper Nawata:* 1977, anterior 3 premaxilla, trunk vertebra centrum; 1979, mid abdominal vertebra.
- *Apak Member:* 1942, trunk centrum; 1960, basioccipital fragment, posterior abdominal vertebra, vertebra fragment; 1974, first vertebra.
- *Muruongori Member:* 3153, first vertebra; 3154, first vertebra, anterior abdominal vertebrae, vertebra fragment.
- *Kaiyumung Member:* 1835, 2 first vertebrae, 4 anterior abdominal vertebrae, 7 mid abdominal vertebrae, 3 posterior abdominal vertebrae, 4 anterior caudal vertebrae, 2 caudal vertebrae; 1850, posterior skull, articulated basioccipital with 2 exoccipitals, 2 anterior dentaries, 2 angulo-articular fragments, maxilla fragment, 25 first vertebrae, 18 anterior ab-

dominal vertebrae, 11 mid-abdominal vertebrae, 10 posterior abdominal vertebrae, 8 anterior caudal vertebrae, 7 other caudal vertebrae; 1851, 40 trunk vertebrae centra, 15 caudal vertebrae centra; 1852, 3 first vertebrae, 2 anterior abdominal vertebrae, 7 mid-abdominal vertebrae, 2 posterior abdominal vertebra, 2 anterior caudal vertebra, 2 caudal vertebrae, trunk vertebra centrum; 1994, 2 anterior dentaries, 2 first vertebra, 5 anterior abdominal vertebrae, mid-abdominal vertebra, 15 trunk vertebrae, posterior abdominal vertebra, anterior caudal vertebra, caudal vertebra, 1995, 2 first vertebrae, anterior abdominal vertebra, mid-abdominal vertebra, trunk vertebra centrum; 1998, 2 anterior premaxillae, 7 first vertebrae, 2 anterior abdominal vertebrae, 3 mid-abdominal vertebrae, posterior abdominal vertebra, anterior caudal vertebra, caudal vertebra, trunk vertebra centrum, 1999, first vertebra, 2 anterior abdominal vertebrae, 6 mid-abdominal vertebrae, 3 posterior abdominal vertebrae, 3 caudal vertebrae, trunk vertebra centrum, 2 vertebrae fragments; 2000, basioccipital, dentary fragment, 3 first vertebrae, anterior abdominal vertebra, mid-abdominal vertebra, posterior abdominal vertebra, 8 abdominal vertebrae centra; 2332, anterior abdominal vertebra, mid-abdominal vertebra.

Three hundred elements of *Semlikiichthys* cf. *S. rhachirhinchus* were recovered from the Lothagam deposits. Because of the difficulty in naming these elements, they are discussed in detail. *Semlikiichthys* is a new genus erected to describe fossils formerly ascribed to *Lates rhachirhinchus*; they are now *S. rhachirhinchus* (Otero and Gayet 1999).

Skull

A posterior half skull, KNM-LT 1850, is one of the most diagnostic elements recovered and is treated here in some detail. The specimen consists of the posterior frontals, posterior pterosphenoids, posterior sphenotics, prootics, parietals, intercalars, partial pterotics, right epiotic, exoccipitals, basioccipital, supraoccipital, and posterior parasphenoid (figure 3.34). The right lateral portions of the skull are too damaged to recognize features. In the following discussion, the skull will be compared with extant *Lates niloticus* skulls, descriptions and drawings of other extant *Lates* species (Greenwood 1976b), and with the fossil *S. rhachirhinchus* elements recovered from Lake Albert Basin Miocene-Pliocene deposits in Zaire (Greenwood and Howes 1975).

In comparison with extant *L. niloticus* skulls, the Lothagam fossil showed narrowing of the parietal/supraoccipital region in dorsal view, and especially the otic region as seen in ventral view, similar to *S. rhachirhinchus*. Other parts of the skull were not present to judge overall narrowing. Similar to that in *S. rhachirhinchus*, the supraoccipital crest is short in length, extending anteriorly only to the posterior prootic, unlike the extension to the anterior sphenotic in *L. niloticus* and other extant species. In two *S. rhachirhinchus* specimens this crest extended slightly more anteriorly than that in the Lothagam fossil. The crest is incomplete posteriorly, so the height cannot be compared.

Similar to *S. rhachirhinchus* and unlike all extant *Lates* species, in KNM-LT 1850 the posttemporal fossa is a shallow pit with a bony floor, not the deep trench covered with a membrane that is seen in extant species. The pterosphenoid is broken anteriorly in the Lothagam fossil but clearly extends forward like *L. niloticus* and *S. rhachirhinchus*, and unlike *L. mariae* and *L. macrolepis*. Although somewhat damaged, the prootic of the Lothagam fossil seems similar to that of *S. rhachirhinchus* in being more anteriorly extensive than in extant species. Also similar to *S. rhachirhinchus*, the sphenotic of KNM-LT 1850 has a deep fossa anterior and dorsal to the sphenotic/prootic suture, unlike extant *L. niloticus* and other extant *Lates* where no such fossa exists. Unlike extant *Lates* specimens, but similar to *S. rhachirhinchus*, the lateral commissure on the prootic is very wide. However, in the Lothagam fossil, a nerve foramen opens just posterior to the trigeminofacialis chamber, which is not seen in *L. niloticus* or in *S. rhachirhinchus* but appears to be present in *L. stappersi* and possibly *L. microlepis*, both members of the subgenus *Luciolates* (Greenwood 1976b). The facets for the hyomandibular are similar in all specimens.

In the occipital region, on the exoccipital, KNM-LT 1850 differs from *S. rhachirhinchus*, but is similar to *L. niloticus* in having separate, circular openings to the vagus and the more anterior glossopharyngeal nerves, while in *S. rhachirhinchus* they are contained in the same long, deep groove (Greenwood and Howes 1975:81). From drawings in Greenwood (1976b), *L. cal-*

Figure 3.34 *Semlikiichthys* cf. *S. rhachirhinchus*, posterior skull (left lateral view).

Figure 3.35 *Semlikiichthys* cf. *S. rhachirhinchus*, ventral skull showing grooved parasphenoid.

carifer, L. macrophthalmus, L. longispinis, and *L. augustifrons* have separated openings, while *L. microlepis, L. mariae*, and *L. stappersi* have a long grooved opening. As in *S. rhachirhinchus*, the exoccipital facets are round to oval, rather than kidney shaped as found in extant species. Slightly dorsal to the opening for the occipital-spinal nerve is another foramen, which is not seen in *L. niloticus* or in drawings of *S. rhachirhinchus*, although it just may not be preserved in the latter. The basioccipital of KNM-LT 1850 is similar to that of *S. rhachirhinchus* but different from *L. niloticus* in that the facets (for Baudelot's ligament) are more ventrally placed in the fossils—in some cases so they are almost contiguous. In the Lothagam fossil and in *S. rhachirhinchus* the parasphenoid is narrower, without the widening seen in *L. niloticus* where it meets the prootic. As in *S. rhachirhinchus*, but unlike all extant specimens, the groove along the ventral surface of the parasphenoid (figure 3.35) stretches from where it meets the basioccipital to a line dropped down from the lateral commissure. Although little of the fossil anterior parasphenoid is present, similar to *S. rhachirhinchus* but unlike extant *Lates*, the anterior parasphenoid is rounded (figure 3.35).

Basioccipital

Two basioccipital elements were recovered, in addition to the one articulated to KNM-LT 1850; one is articulated with two exoccipitals and a first vertebra. These specimens differ from extant *Lates* but are similar to *S. rhachirhinchus* in having facets for Baudelot's ligament in a more ventral position, with their edges almost contiguous on the ventral surface of the basioccipital, while in the other specimens the facets are positioned laterally, on the lateral sides of the basioccipital. In the Lothagam fossils the facets have a lip on their edges, but this lip is not usually noticeable in the extant species.

Vomer

A total of nine vomers were recovered from the Lothagam and Kanapoi sites (because of the diagnostic value of this element, non-Lothagam-derived vomers are also considered in this description). Because of the overall similarity of the Lothagam fossils to those described from Zaire (Greenwood and Howes 1975; Van Neer 1992), and because *S. rhachirhinchus* was named for the unique vomer, this element has special importance in naming the Lothagam material. The vomer of *S. rhachirhinchus* is unique in possessing a forward projecting spine, furrows that separate the spine from the body of the vomer, and a tooth patch that is delimited by a raised shelf from the body of the vomer itself (Greenwood and Howes 1975).

Six of the nine Lothagam specimens had the anterior region preserved enough to observe if a spine existed; no spine was evident, nor were the furrows separating the spine from the vomer. Eight of the nine specimens had enough of the tooth patch preserved to observe that the patch was part of the vomer body; it was not separated by a distinct shelf. Eight of the nine vomers were very similar to those of extant *L. niloticus*, with two differences. In all Lothagam specimens, the vomerine tooth patch had a distinctive shape not seen in five extant *L. niloticus* specimens observed (figure 3.36), or in 18 specimens of *L. niloticus* pictured (Van Neer 1987: figure 2). Moreover, the median crest, which forms a spine in *S. rhachirhinchus*, inclines dorsally and posteriorly more steeply in the Lothagam fossils than in *L. niloticus*. In all other features, eight of the nine Lothagam vomers closely resemble the extant *L. niloticus* vomers.

However, one large vomer from Kanapoi differs considerably from the other eight fossil vomers. I am discussing it in the context of the Lothagam material, because its interpretation affects the naming of the Lothagam specimens. It is unfortunately somewhat damaged, so the presence of a spine cannot be discerned on its anterior surface (figure 3.37). It does have furrows on either side of the damaged area, similar to the *S.*

Figure 3.36 *Semlikiichthys* cf. *S. rhachirhinchus*, vomer (dorsal view).

Figure 3.37 *Semlikiichthys* cf. *S. rhachirhinchus,* vomer (ventral view).

rhachirhinchus vomer and unlike all other fossil and extant *Lates* vomers, and it is clear that there was a protrusion of some sort between the furrows (figure 3.37). The ventral "occlusal" surface of the vomer is differently shaped from the other Lothagam fossils, in having a lateral projection on each side, like horns (figure 3.38). Further, the "occlusal" surface is not tooth-bearing as in other Lothagam fossils, but is bony anteriorly, with a large, flat, oval-shaped surface medially from which a plate-like structure appears to have broken or worn off (figure 3.38), again similar to *S. rhachirhinchus.* This whole structure has a posterior extension that is not present in other fossils but which is similar to that of *S. rhachirhinchus* as pictured in Greenwood and Howes (1975). Further, there are lateral extensions dorsal to the "occlusal" surface, which again are not present on the other fossils and which more closely approach *S. rhachirhinchus.* In short, this specimen is almost certainly a damaged *S. rhachirhinchus* vomer, with the remnants of a spine.

Figure 3.38 *Semlikiichthys* cf. *S. rhachirhinchus* and *L. niloticus* premaxillae compared (anterior views).

Premaxilla

Several anterior portions of premaxillae were recovered. These elements appear exactly like those pictured and described for *S. rhachirhinchus* (Greenwood and Howes 1975:86ff.). Compared to extant *Lates* premaxillae (figures 3.39 and 3.40), the Zaire and Lothagam fossils lack the medial extension of the dentigerous shelf, the ascending process unites with the articular process rather than being separated, there is a large medially-located foramen absent in extant species, and the fossil articular process inclines medially and lies in a different plane from the ascending process.

Dentary

Only anterior portions of dentaries were recovered. Surprisingly, two types were represented in addition to dentaries ascribed to *L. niloticus.* The first type included two specimens that were very similar to those of *L. niloticus,* but, unlike the extant species and similar to *S. rhachirhinchus* (Greenwood and Howes 1975), they had a slightly narrower dentigerous surface as well as enlarged lateral line sensory openings. The second type included three specimens that differed considerably from *L. niloticus* and also from *S. rhachirhinchus* (Greenwood and Howes 1975). This type has a flat, two-dimensional, squared-off appearance when viewed from the anterior, unlike extant *Lates* and *S. rhachirhinchus* dentaries, which are curved from top to bottom. Viewed dorsally, the Lothagam dentaries curve only slightly backward, unlike the extant *Lates* specimens, which curve back quite abruptly into the anguloarticulars. These second types also have an extremely narrow dentigerous surface (much narrower than extant *Lates* or *S. rhachirhinchus*) and larger lateral line sensory canal openings. This dentigerous surface occasionally has an anterior "lip" that does not occur in other specimens. The ventral surface is also unique in that the medial surface curves up into a flange, along which at least two of the lateral line sensory canal openings lie; this differs from both the extant specimens and *S. rhachirhinchus.* In fact, the very narrow dentigerous surface and ventral flange shows some similarity with cichlid dentaries, particularly *Tilapia,* although the size is greater than for extant cichlids.

For this reason, this second type of dentary is classified as Perciformes indeterminate until further material is recovered.

Figure 3.39 *Semlikiichthys* cf. *S. rhachirhinchus* and *L. niloticus* premaxillae compared (posterior views).

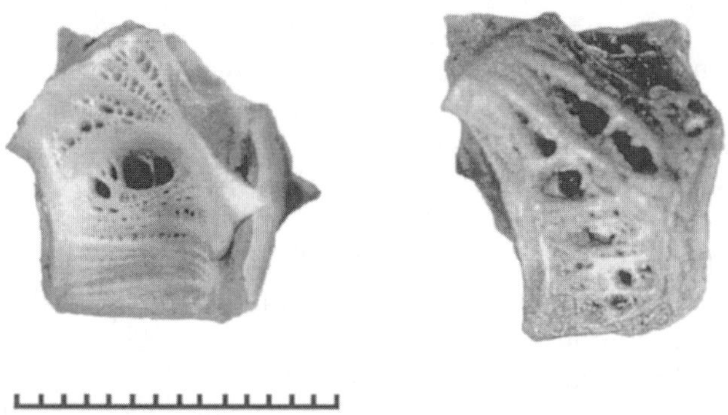

Figure 3.40 *Semlikiichthys* cf. *S. rhachirhinchus* and *L. niloticus* first vertebrae compared (left lateral views).

Angulo-articular

Several of the angulo-articulars were recovered, usually the anterior 50 percent, with retroarticular attached. These differ primarily in the shape of the latero-sensory groove, which is short in extant *Lates*, long in *S. rhachirhinchus*, and divided into two in the Lothagam fossils by a small bridge. This bridge is narrow, and it is possible that it may have worn away in the *Semlikiichthys* cf. *S. rhachirhinchus* specimens. The Lothagam articular differs from extant *Lates* specimens but is similar to *S. rhachirhinchus* as it has a more bowl-shaped facet that curves up and around posteriorly into a hook shape. The ascending arm is also narrower in the fossil than in extant *Lates*. The articular in the Lothagam and Zaire fossils appears to be relatively smaller than in extant specimens and attaches more ventrally to the angular than does that of extant *L. niloticus* specimens.

While some elements of the Lothagam fossil represent fish of over 1 m in length, many were from smaller fish, probably under 60 cm in total length.

Vertebrae

A total of 51 first vertebrae were recovered from the Lothagam deposits. These first vertebrae are compact and robust and by far the most abundant element represented of *Semlikiichthys* cf. *S. rhachirhinchus*. The first vertebra is identical to the first vertebra of *S. rhachirhinchus* described and figured by Greenwood and Howes (1975:figure 16) and photographed by Van Neer (1992:plate 3, figure 10), but differs from extant *Lates* primarily by having a tapering centrum and upswept exoccipital facets (figures 3.40 and 3.41a). The fossil exoccipital facets also differ from extant species by their smaller surface area, as well as being wider than long; in extant *Lates*, the exoccipital facets are kidney shaped and are longer than wide.

The remaining vertebrae differ considerably from the extant *Lates* vertebrae. While Greenwood and Howes (1975) assigned positions in the column for *S. rhachirhinchus* vertebrae, I find these uncertain until an intact

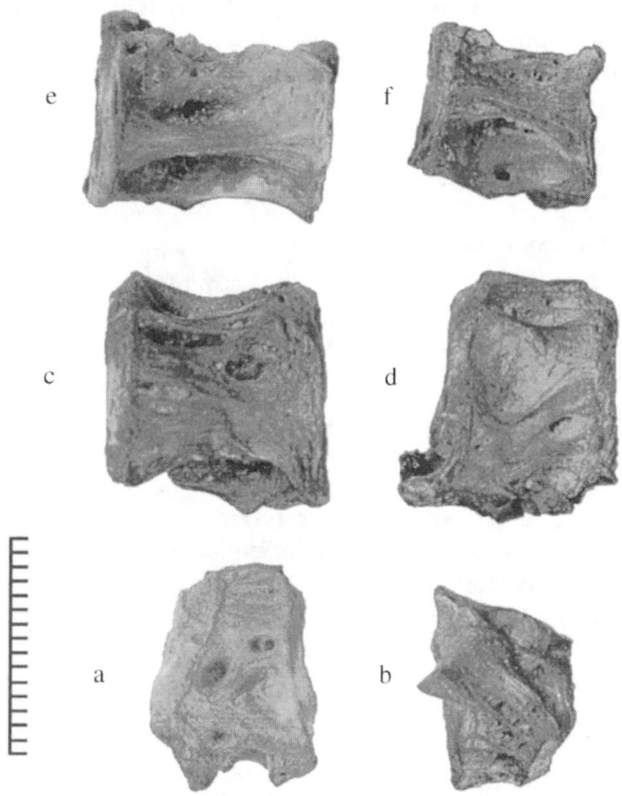

Figure 3.41 *Semlikiichthys* cf. *S. rhachirhinchus* vertebrae: (a) first vertebra, (b) anterior trunk, (c) middle trunk, (d) posterior trunk, (e) anterior caudal, (f) caudal.

vertebral column is recovered, and have therefore described their position generally, determined primarily using the location of the rib facet and the arrangement of the trabeculae.

Anterior trunk vertebrae

The anterior trunk vertebrae are probably second or third vertebrae, primarily based on the compressed body and the more loosely organized arrangement of the trabeculae in the Lothagam examples than in the extant specimens (figure 3.41b). The anterior centrum is circular in outline, while the posterior centrum is more oval. Unlike extant vertebrae, the Lothagram vertebrae have a groove along the ventral surface and the pleural rib facet is not as well defined. The arrangement of trabeculae is similar to that of *S. rhachirhinchus* as pictured by Greenwood and Howes (1975:figure 17).

Middle trunk vertebrae

The middle trunk vertebrae are probably fourth or fifth vertebrae, based on the placement of the rib facet. Some have more compressed bodies and slightly more medially placed pleural rib facets and may be fourth vertebrae (figure 3.41c). Some have a pronounced groove on the ventral surface and may be fifth vertebrae, based on comparison with the extant specimens. The trabeculae are more loosely organized, and the basapophyses are better developed than for extant specimens. These specimens resemble those pictured by Greenwood and Howes (1975:figures 18 and 19).

Posterior abdominal vertebrae

The posterior abdominal vertebrae are difficult to place, although four or five are likely sixth or seventh in position, based on very small developing transverse processes (figure 3.41d). These vertebrae differ from those of extant *L. niloticus* because they lack trabeculae below the main process and have the existing trabeculae more loosely organized. The rib facet is smaller than in extant specimens, and more anteriorly placed. These specimens resemble those of Greenwood and Howes, although because their ventral view is missing, the placement of the facet is unknown (1975:figure 20a–b).

Anterior caudal vertebrae

These vertebrae were determined to be anterior caudal because of the more ventral placement of the trabeculae; in posterior caudals, the trabeculae are centrally placed (figure 3.41e). The fossil vertebrae are not dissimilar from extant specimens, although they are more elongated and slender, and the core of trabeculae is also narrower. These were not figured by Greenwood and Howes (1975).

Other caudal vertebrae

It is not possible to assign a more specific position to the remaining vertebrae. The trabeculae are centrally arranged, as in the extant specimens, but they are narrower (figure 3.41f). The vertebrae themselves are more elongated and narrow than in the extant *L. niloticus* specimens, and they are similar to the figures of Greenwood and Howes (1975:figure 20c).

Discussion

The preceding description makes clear the overwhelming number of shared characters between the Lothagam *Lates* elements and the *S. rhachirhinchus* neurocranial and other elements including vertebrae, and the many shared differences with extant *Lates* specimens. However, I am comparing them with, rather than referring them to, *Semlikiichthys rhachirhinchus* because of the lack of a clearly spined vomer (notwithstanding the damaged Kanapoi vomer). It is hoped that future fieldwork will uncover spined vomers, for a more definitive classification.

Semlikiichthys cf. *S rhachirhinchus* is known from Mio-Pleistocene deposits at Sinda, Zaire (Greenwood and Howes 1975; Van Neer 1992). Its size ranged up to 2 m in length.

Percoidei indet.

Lothagam Material

- *Lower Nawata:* 1752, trunk vertebra, caudal vertebra centrum; 1809, cranial fragment; 1989, cleithrum fragment; 1990, 2 caudal vertebrae centra, 2 vertebrae fragments; 2385, trunk vertebra centrum; 2407, caudal vertebra centrum; 2412, 2 trunk vertebrae centra; 2419, anterior quadrate.
- *Upper Nawata:* 1570, trunk centrum; 1594, caudal vertebra centrum; 1979, parasphenoid fragment, 6 cleithra fragments.
- *Apak Member:* 1757, cleithrum fragment; 1759, trunk vertebra centrum; 1769, premaxilla fragment, trunk vertebra fragment; 1848, trunk vertebra cen-

trum; 1849, articular fragment, 6 trunk vertebrae centra, 2 caudal vertebrae centra; 1942, basioccipital, 3 articular fragments, quadrate, basihyal portion, 4 trunk vertebrae centra, 7 caudal vertebrae centra, 3 vertebrae fragments, 2 dorsal spine fragments, pterygiophore fragment; 1944, vomer fragment, parasphenoid fragment, trunk vertebra centrum, vertebra fragment, pterygiophore fragment; 1948, trunk vertebra centrum; 1958, anterior skull portion (vomer, lateral ethmoids), posterior skull portion (basioccipital, exoccipitals, parasphenoid portion); 1960, skull fragment, trunk vertebra centrum, vertebra fragment, 1976, caudal centrum.

- *Muruongori Member:* 3153, dorsal spine, 2 dorsal spine fragments; 3154, quadrate fragment; trunk vertebra centrum.
- *Kaiyumung Member:* 1835, 11 vertebrae centra; 1850, premaxilla fragment, 47 vertebrae centra; 1851, 2 articular fragments; 1992, angular, caudal vertebra centrum.

Classification of the Lothagam Percoidei fossils was a conundrum, in that while many elements recovered were clearly referable to *L. niloticus,* most were virtually identical to the *Semlikiichthys rhachirhinchus* fossils pictured by Greenwood and Howes (1975) and Van Neer (1992), and seen by the author at the Tervuren museum. However, the element that was most dissimilar was the vomer, which in most specimens lacked the characteristic spine for which the Zaire fossils were named ("rhachirhinchus" means loosely "snout with spine"). Although eight recovered vomers were identical to each other and very similar to *L. niloticus,* the ninth was substantially different. It may have had an anterior spine, as well as the dentigerous plate typical of *S. rhachirhinchus.* My inclination was to name the non-*L. niloticus* Lothagam fossils as *S. rhachirhinchus,* but this was impossible without a clearly spined vomer from which the name is derived. Given the great similarities of the elements to *S. rhachirhinchus,* along with the presence of a vomer with a probable spine, the fossils are referred to as *Semlikiichthys* cf. *S. rhachirhinchus.*

Assuming that this identification is correct, two hypotheses can be suggested for the presence of *S. rhachirhinchus* in the Turkana Basin. Either an ancestral Percoidei population exploited both the Lake Albert and Turkana Basins, giving rise to two distinct and endemic species with numerous, very similar adaptations, or *S. rhachirhinchus* migrated between the Lake Albert Basin (where it is known from Early Miocene to Pleistocene deposits [Greenwood and Howes 1975]) and the Turkana Basin.

Evidence for similar endemic Percoidei populations is seen in modern Lakes Turkana (*L. longispinis*) and Albert (*L. macrophthalmus*). Each lake has a small endemic species of *Lates* that inhabits deep water and shows apparently derived features of enlarged eyes and elongated third dorsal spines, as well as a tendency for fourth spines on the preoperculum (Greenwood 1976b; Beadle 1981). Two theories are possible for this occurrence. During the high lake levels of the Early Holocene, both lakes were colonized by fauna from the Nile river, including *Lates niloticus,* and these two species may have evolved separately from the parent Nile population. Alternatively, Greenwood (1976b) suggests that a second *Lates* species colonized the two lakes with *Lates niloticus,* from which the populations evolved. This seems less likely, as no bones indicative of an hypothesized parent species have been recovered.

While the modern example shows that endemic *Lates* populations can evolve with similar adaptations, I find it unlikely that the numerous unique specializations shared by *Semlikiichthys rhachirhinchus* and *Semlikiichthys* cf. *S. rhachirhinchus* could be generated in two physically separated populations, particularly when *S. rhachirhinchus* evolved in a deep lake, while the Lothagam species would have evolved in a river, hence necessitating different habitat adaptations. Pre-Holocene exchange of faunas between the Lakes Albert/Edward Basin and Lake Turkana have been postulated before, and the presence of *S. rhachirhinchus* in both basins seems to provide support for this suggestion. Further support for exchange of faunas within the Nile-linked systems is found in a palatine, similar to that of *S. rhachirhinchus,* recovered in the Wadi Natrun Pliocene deposits (Greenwood 1972; Greenwood and Howes 1975).

In a brief discussion of the relationships of *S. rhachirhinchus,* Greenwood and Howes (1975; Greenwood 1976b) suggested that extant *Lates* could be split into two lineages, and that *S. rhachirhinchus* shared two derived characters with the Lake Tanganyika lineage and none with the *L. niloticus* lineage. However, they questioned whether these characters reflect convergence or actual phyletic affinity. A recent doctoral dissertation (Otero 1997) and publication (Otero and Gayet 1999) have presented a revision of *Lates,* putting the Zaire fossils into a new genus and putting other *Lates* species (as well as *Psammoperca* and *Eolates*) into the family Latidae.

The Percoidei specimens were found throughout the Lothagam deposits, but particularly in the Muruongori Member deposits and later. While five specimens of *Semlikiichthys* cf. *S. rhachirhinchus* were found in the Nawata Formation deposits, it cannot be excluded that they were mixed in from later deposits. Large numbers of this species do not appear until the Kaiyumung Member, but it is possible it had a minor presence in the earlier deposits.

Family Cichlidae
Tribe Tilapiini
Tilapiini indet.

Lothagam Material

- *Muruongori Member:* 3152, 2 first anal pterygiophores; 3153/3154, caudal vertebra.
- *Kaiyumung Member:* 1850, 51 vertebrae centra; 1852, trunk vertebra centrum.

Cichlids are poorly represented at Lothagam and only by two groups of elements: the distinctive first anal pterygiophores and vertebrae centra. However, some elements may be inadvertently included in the category Perciformes, as they appear to be indistinguishable from *Lates* elements.

Discussion

The paucity of cichlid material at Lothagam except in the latest sites is puzzling. Although the majority of cichlid elements are delicate, the dorsal/anal spines are robust and usually preserve well as fossils. Cichlid spines can usually be distinguished from *Lates* elements, so their absence apparently means that cichlids were either extremely rare or absent from all but the latest Lothagam deposits. Cichlids are divided into the tribes Haplochromini and Tilapiini (e.g., Trewavas 1983), which are usually distinguishable by key characters, particularly the structure of the basal apophysis on the skull. All cichlids represented in Lothagam deposits are tilapiines.

Fossil cichlids are known from ?Eocene and Oligocene deposits in Tanzania and Somalia (Van Couvering 1982), Miocene deposits from Lamatina, Uganda, and from Kirimun, Ngorora, Mpesida, Loperot, and Rusinga, Kenya (Van Couvering 1982), Pliocene deposits in the Lake Edward-Albert rift (Stewart 1990) and Manonga, Tanzania (Stewart 1997), Pleistocene deposits from the Lake Albert Basin (Van Neer 1994), and Plio-Pleistocene deposits in the lower Omo Valley (Arambourg 1947) and Koobi Fora (Schwartz 1983). Seven species of cichlids are known from modern Lake Turkana, and the family is widespread throughout systems in the African continent.

Perciformes indet.

Lothagam Material

- *Lower Nawata:* 1710, trunk vertebra centrum.
- *Upper Nawata:* 1594, basioccipital fragment, 4 dorsal spine fragments.
- *Apak Member:* 1764, operculum fragment; 1942, dorsal spine fragment; 1960, dorsal spine fragment, pterygiophore fragment; 2420, 5 dorsal spine fragments.
- *Muruongori Member:* 3151, pelvic spine fragment, dorsal spine fragment; 3152, pelvic spine fragment, dorsal spine fragment; 3153, dorsal spine fragment; 3154, dorsal spine fragment.
- *Kaiyumung Member:* 1850, 2 dentaries, 67 spine fragments; 1851, 2 dorsal spines; 1852, 7 dorsal spine fragments; 1994, 2 dentaries, quadrate fragment; 5 dorsal spine fragments; 1998, quadrate fragment.

It is often difficult to distinguish incomplete dorsal, anal, and pectoral spines and pterygiophores of *Lates* versus cichlids. For this reason they are here listed as indeterminate perciforms.

As described above, three anterior dentary elements are referred only as Perciformes indeterminate. They seem transitional between extant *Lates* and cichlid dentaries, and as no other elements with such transitional characters were recovered, they cannot be further identified except as perciform.

Order Tetraodontiformes
Family Tetraodontidae
Tetraodon Linnaeus, 1758
Tetraodon sp. (puffer fish)
(Figures 3.42, 3.43)

Lothagam Material

- *Kaiyumung Member:* 1850, 14 toothplates, ?caudal vertebra centrum; 1851, 3 toothplates; 1852, 4 toothplates; 1994, 2 toothplates, toothplate fragment, 2 caudal vertebrae centra; 1998, 2 toothplates; 1999, toothplate, 2 caudal vertebrae centra.

Tetraodon postcranial elements are small and delicate, so they do not preserve well in fossil sites, with the occasional exception of vertebrae. However, the jaws, in particular the toothplates, are robust and do preserve well, and they are often the only remains preserved of puffer fish. Toothplates from the Kaiyumung Member sites differ from one recovered from the Muruongori Member.

Tetraodon sp. nov.
(Figures 3.42, 3.43)

Lothagam Material

- *Muruongori Member:* 3163, articular, dentary and toothplate articulated.

Figure 3.42 *Tetraodon* sp. nov. dentary, articular, and toothplate (left) compared with *Tetraodon* sp. dentary (dorsal aspect).

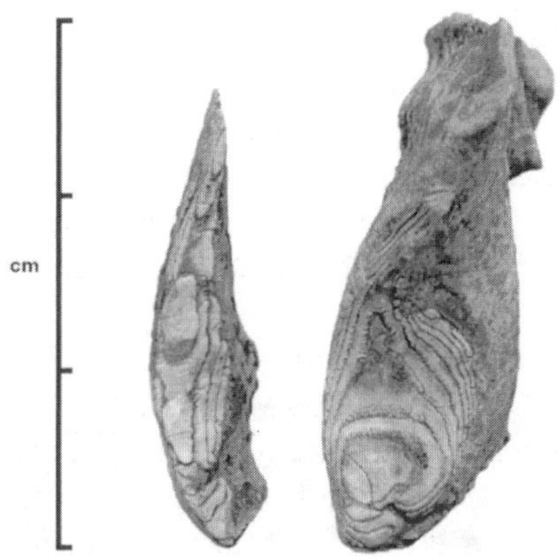

Figure 3.43 *Tetraodon* sp. nov. toothplate (left) compared with *Tetraodon* sp. toothplate (right) (medial aspect).

An articulated articular, dentary, and toothplate has an occlusal surface that differs considerably from later tetraodontid toothplates, with the Muruongori Member plate being much wider and more robust (figures 3.42 and 3.43, at bottom) than the later, Kaiyumung Member, ones (figures 3.42 and 3.43, at top).

Discussion

Wide toothplates characteristic of the Muruongori Member puffer are also found at the contemporaneous Ekora site deposits, but not elsewhere. No toothplates characteristic of the later (Kaiyumung) puffers were found in the Muruongori Member or Ekora deposits. Lack of sufficient comparative material means that the new species cannot be fully described or named at this time.

The differences between tetraodontid cranial material from the Muruongori and Kaiyumung Members are significant enough to put the Muruongori material into a new species. More elements are needed to better describe it. The question also arises of the origin of the puffers from the Lothagam succession.

The only published fossil records of *Tetraodon* are from Pleistocene deposits in the Lake Albert-Edward Rift (Van Neer 1994) and from Plio-Pleistocene deposits at Koobi Fora (Schwartz 1983). Tetraodontid material previously reported from mid Miocene sites is believed to be of Pleistocene origin (Van Neer 1994:90, 117).

The first appearance of Lothagam puffer fossils at a Muruongori Member site coincides with that of *Sindacharax deserti.* In the Pliocene, some exchange of fish fauna must have occurred between the Turkana Basin and the Nile River and/or the Albert-Edward Basin. This occurrence is almost certainly the earliest record of freshwater puffers in Africa.

Tetraodon fahaka is present in modern Lake Turkana, whereas the genus *Tetraodon* is known from the Nile, Senegal, and Niger systems, as well as the Zaire basin and Lake Tanganyika.

Paleoecology

While changes seen in assemblages from the Lothagam sequence have evolutionary and biogeographical importance, they also mark paleoenvironmental and paleoecological changes. In the lower and upper members of the Nawata Formation sites, *Protopterus, Polypterus, Heterotis, Gymnarchus,* and the catfish *Clarias* were common; all these are fish that live in shallow, swampy water that has considerable vegetation for spawning and for feeding. The waters must have been fresh, as *Polypterus* is intolerant of even slightly saline water. The presence of *Hydrocynus,* and to a lesser extent large *Lates,* signifies the presence of open waters; *Lates* also requires well-oxygenated, oxygenated, or well-mixed waters. The assemblage suggests a large, slow-moving river with numerous well-vegetated back swamps and bays that were well-oxygenated and not brackish.

In the Apak Member of the Nachukui Formation, there is a considerable change in represented fish taxa. Species that dominated the Nawata Formation—*Protopterus, Polypterus,* and *Heterotis*—were rare or absent in the Apak Member, and the diversity of taxa declined. *Lates, Sindacharax,* and *Hydrocynus* became more common, and this suggests a faster flowing river with fewer vegetated backwaters than in Nawata times. Muruongori Member outcrops were limited, and few fossils were recovered; these were mainly teeth, so no accurate statement of environment can be made. While *Sindacharax* is dominant, teeth of *Hydrocynus* and *Gymnarchus* were also recovered, as well as *Polypterus* scales. Some *Lates* and *Tetraodon* elements were also found.

A considerable change occurred at the Kaiyumung Member sites. Although the exposures were fewer, and less diversity of fossils was recovered, elements of the new species of *Sindacharax* and *Lates* were abundant. Unfortunately, these taxa are of limited use for interpreting ecology and environment because both groups are now extinct. In the southern Kaiyumung deposits, however, there is considerable diversity, with new or previously rare taxa—including *Barbus, Labeo, Tetraodon,* and the new *Sindacharax* species—becoming more common. A diversity of catfish is also evident. The overall size of individuals is much larger than in earlier units, and estimated lengths of fish such as *Lates* and *Sindacharax* are much greater. The fauna represents a great diversity of habitats and trophic groups and is consistent with the presence of a large lake.

Discussion and Summary

The fish fauna shows considerable change from the Lower Nawata to the northern Kaiyumung deposits of the Nachukui Formation, a time span estimated to be from about 7.4 Ma to about 2.7 Ma (Leakey et al. 1996; McDougall and Feibel 1999). The Nawata Formation fish fauna appears to be uniform throughout this formation, although Lower Nawata sites with fish fossils are considerably more common and have greater numbers of specimens than have the Upper Nawata sites. Therefore, the apparent uniformity of the fauna may be an artifact of poor recovery. The Nawata Formation fish assemblage consists of relatively small-sized individuals, many of which represent archaic but still extant fish genera; where enough elements were recovered, these are similar to extant species. A previously unrecognized species of the extinct *Sindacharax* is smaller and more similar to its sister group *Alestes* than to later *Sindacharax* species. The most common fish taxa inhabit shallow, swampy, well-vegetated bays and probably indicate a large, slow-moving river with backwaters.

The Apak Member fauna differs from the Nawata fauna due to loss or scarcity of the archaic elements. *Sindacharax, Lates,* and *Hydrocynus* are best represented, which suggests a faster moving river with fewer vegetated bays. A new species of *Sindacharax*—*S. mutetii*—appears, and it is larger and more robust than *S. lothagamensis.* Enigmatically, *S. mutetii* is the only species of *Sindacharax* represented in numerous teeth recovered from the nearby Pliocene-aged Kanapoi site. Kanapoi deposits are described as fluvial in origin (Feibel, personal communications), as are Apak deposits, and this species of *Sindacharax* may only be represented in rivers.

While relatively few fossils were recovered from Muruongori Member deposits (about 4 Ma), three new taxa appear. The dominant species—*Sindacharax deserti*—is quite different from earlier *Sindacharax* teeth and is also found in Nile system deposits of a similar age (Greenwood 1972). This species of *Sindacharax* is likely an immigrant to the Lonyumun lake in which the Muruongori Member deposits accumulated. The presence of this species in both Egyptian deposits of similar age and the Turkana Basin suggests an exchange of fauna between the systems.

Also present in the Muruongori Member is *Semlikiichthys* cf. *S. rhachirhinchus,* otherwise known only from Mio-Pleistocene Lake Albert Basin deposits. Elements from this species were recovered from the Nawata Formation and Apak Member sites, but they were rare and may have been mixed in from later sites. In the Muruongori and Kaiyumung Members, elements of this new *Semlikiichthys* are abundant. Its presence at Lothagam may be due to exchange between the Albert/Edward and Turkana Basins. Because this species is known earlier in the Albert Basin (Early Miocene), it is likely to have originated there and migrated to Turkana. A probable element of this species is also reported from Pliocene Egyptian sites.

A third new fish group—*Tetraodon* sp. nov.—is present in the Muruongori Member. This genus has a poor fossil record in Africa, with its only other record being in the Pleistocene of the Lake Albert Basin. The toothplates recovered from Muruongori Member sites are quite different from those in later deposits, as well as from the extant puffer in Lake Turkana, and this indicates the presence of a new species. Further analysis and comparison with other extant and fossil *Tetraodon* elements is needed to adequately name and describe this species. This is probably the earliest record of freshwater puffers in Africa.

A new, probably endemic species of *Sindacharax*—*S. howesi*—appears in the northern Kaiyumung deposits, and another species—*S. greenwoodi*—appears in the later, southern deposits. The northern Kaiyumung deposits show little diversity in fish, but this undoubtedly reflects incomplete fossil recovery. The southern deposits represent a much more diverse and larger (in size) fauna than was seen previously. While considerable mixing of deposits is evident in several southern Kaiyumung sites, some sites do show internal consistency of taxa. These latter sites document considerable increase in size of previously represented taxa, as well as the appearance of *S. greenwoodi,* and much greater abundance of *Barbus, Labeo, Hydrocynus,* and *Semlikiichthys rhachirhincus.* Similarly-aged deposits from South Turkwel sites (ca 3.5 Ma; Ward et al. 1999; F. Brown, personal communication) also show a substantial increase in total lengths of fauna over earlier deposits, particularly *Hyperopisus, Sindacharax, Clarotes,* and *Lates.*

The ecological composition of the Nawata Formation assemblage is of small-sized fish that are dominated by piscivores and herbivores. In the Muruongori and Kaiyumung Members the species are represented by much larger individuals, with a dominance of molluscivorous and piscivorous taxa. By later Pliocene and Pleistocene times, based on collections from eastern Turkana deposits (table 3.2), the taxa are even more diverse and large-sized with increasing presence of large catfish (table 3.2), and the fauna is not dissimilar from that of the modern lake. No elements of *Semlikiichthys rhachirhinchus* or *Tetraodon* sp. nov. are reported in the Pleistocene collections (Schwartz 1983), nor did I find evidence for them in the collections at the National Museums of Kenya. A ray makes a brief appearance in the early Pleistocene (table 3.2; Schwartz 1983; Feibel 1988), as does *Auchenoglanis;* however the latter is rare, and its absence at Lothagam may well be an artifact of recovery.

At no time does the composition of the Lothagam fish assemblage reflect the overwhelming dominance of large molluscivores seen in the Pliocene Western Rift fauna (e.g., Stewart 1990); instead, it remains diverse throughout. In comparison with the Pleistocene and modern faunas, the near absence of cichlids and the paucity of citharinids and cyprinids at Lothagam is surprising, although representatives of the latter two groups have delicate bones and do not preserve well. The depauperate (tilapiine) cichlid fauna at Lothagam is enigmatic but has also been noted at Mio-Pleistocene deposits in the Western Rift (Greenwood and Howes 1975; Stewart 1990) and to some extent in Pliocene Egypt deposits (Weiler 1929; Greenwood 1972). While haplochromine cichlid remains are well documented from the Early Cenozoic in Africa (e.g., Van Couvering 1982), tilapiine cichlid remains are rare to absent until the Pleistocene in known African deposits. This absence may signify a later immigration to Africa from Asia than previously thought.

Of considerable interest are the evolutionary changes in the characid fauna, particularly *Sindacharax,* and in the centropomid fauna (*Lates* and *Semlikiichthys*). Unfortunately, few *Alestes* elements were collected; nevertheless, analysis is continuing on the relationships between *Alestes* and *Sindacharax* and on developments in characids since Gondwana times (Stewart and Murray, in preparation). New work on *Lates* (e.g., Otero and Gayet 1999) reexamines relationships among *Lates* and *Semlikiichthys* species.

Also of considerable interest are the biogeographic implications of the Lothagam fauna. The appearance in Muruongori Member sites of three new species—*Sindacharax deserti, Semlikiichthys* cf. *S. rhachirhinchus,* and *Tetraodon* sp.—all three of which are known from Mio-Pleistocene Western Rift sites, and *Sindacharax deserti* and *Semlikiichthys rhachirhinchus* from Pliocene Egyptian sites, signifies faunal exchange between these regions. Such exchanges have implications for migrations of other aquatic and terrestrial faunas in the Pliocene.

In sum, the Lothagam fossils document change from a later Miocene fish fauna dominated by small-sized archaic taxa that are primarily piscivorous and herbivorous and that prefer shallow, swampy vegetated habitats to a larger-sized, more diverse Pliocene assemblage with a greater molluscivore component and a more open water, well-oxygenated habitat. However, the Lothagam fauna at no time includes the large component of molluscivores seen in the Western Rift Pliocene faunas. The appearance in the Muruongori Member of three new species—*Sindacharax deserti, Semlikiichthys* cf. *S. rhachirhincus,* and *Tetraodon* sp.—denotes an exchange of fauna with the Western Rift and/or northern proto-Nile faunas in the Pliocene. The near absence of tilapiine cichlids is enigmatic in the Lothagam deposits and may represent a later immigration from Asia than was previously thought.

This contribution provides an exhaustive treatment of the Lothagam fish fauna, but much more work needs to be done, particularly on the siluriforms. Diagnostic skeletal elements were collected from all represented taxa in the Lothagam deposits, and they are now housed in the National Museum in Nairobi, awaiting further analysis.

Acknowledgments

My gratitude to Dr. Meave Leakey, who invited me to join the Turkana Basin project and study the fossil fish from Lothagam and who provided support in the field. My gratitude also to Sam N. Muteti, who worked tirelessly to collect fossil fish elements throughout the Lothagam deposits. Thanks also to Peter Kiptalam, Justus Edung, and all members of the National Museums of Kenya fossil team for their help in collecting fossils. Special thanks to Kamoya Kimeu for leadership and support in the field. Thanks to the paleontology staff at the National Museum in Nairobi for assistance in the lab. Special thanks to Donna Naughton for illustrations and photography and for long hours counting fish teeth. My thanks to the Social Sciences and Humanities Research Council of Canada for transport assistance in 1991 and to the Canadian Museum of Nature Research Advisory Council for transport and field assistance in subsequent years. Thanks as well to the editors—Meave Leakey and John Harris—of this volume.

References Cited

Arambourg, C. 1947. *Mission scientifique de l'Omo (1932–1933).* Fasc. 3. Paris: Muséum National d'Histoire Naturelle.

Arambourg, C., and P. Manier. 1961. Gisement de vertébrés dans le bassin tertiare de Syrte (Libye). *Comptes Rendus de l'Académie des Sciences* (Paris) 252:1181–1183.

Banister, K. E. 1973. A revision of the large *Barbus* of East and Central Africa. *Bulletin of the British Museum (Natural History), Zoology* 26:1–148.

Beadle, L. C. 1981. *The Inland Waters of Tropical Africa.* London: Longman.

Feibel, C. S. 1988. Palaeoenvironments of the Koobi Fora Formation, Turkana Basin, northern Kenya. Ph.D. diss., University of Utah.

Greenwood, P. H. 1951. Fish remains from Miocene deposits of Rusinga Island and Kavirondo Province, Kenya. *Annals and Magazine of Natural History* 12:1192–1201.

Greenwood, P. H. 1959. Quaternary fish-fossils. *Exploration du Parc National Albert, Mission J. de Heinzelin de Braucourt (1950)* Fasc. 4:1–80.

Greenwood, P. H. 1972. New fish fossils from the Pliocene of Wadi Natrun, Egypt. *Journal of Zoology* (London) 168: 503–519.

Greenwood, P. H. 1973. Fish fossils from the Late Miocene of Tunisia. *Notes du Service Géologique de Tunisie* 37:41–72.

Greenwood, P. H. 1976a. Notes on *Sindacharax* Greenwood and Howes, 1975, a genus of fossil African characid fishes. *Revue de Zoologie Africaine* 90:1–13.

Greenwood, P. H. 1976b. A review of the family Centropomidae (Pisces, Perciformes). *Bulletin of the British Museum (Natural History), Zoology* 29:1–81.

Greenwood, P. H. 1986. The natural history of African lungfishes. *Journal of Morphology,* Supplement 1:163–179.

Greenwood, P. H., and G. J. Howes. 1975. Neogene fossil fishes from the Lake Albert–Lake Edward rift (Zaire). *Bulletin of the British Museum (Natural History), Geology* 26:69–127.

Lavocat, R. 1955. Découverte de Dipneustes du genre *Protopterus* dans le Tertiare ancien de Tomaguilelt (Soudan français). *Comptes Rendus de l'Académie des Sciences* (Paris) 240:1915–1917.

Leakey, M. G., C. S. Feibel, R. L. Bernor, J. M. Harris, T. E. Cerling, K. M. Stewart, G. W. Storrs, A. Walker, L. Werdelin, and A. J. Winkler. 1996. Lothagam: A record of faunal change in the Late Miocene of East Africa. *Journal of Vertebrate Paleontology* 16:556–570.

Martin, J. W., and G. E. Davis. 2001. *An Updated Classification of the Recent Crustacea.* Science Series No. 39. Los Angeles: Natural History Museum of Los Angeles County.

McDougall, I., and C. S. Feibel. 1999. Numerical age control for the Miocene-Pliocene succession at Lothagam, a hominoid-bearing sequence in the northern Kenya Rift. *Journal of the Geological Society* (London) 156:731–745.

Otero, O. 1997. Paléoichthyofaune de l'Oligo-Miocène de la Plaque arabique, approches phylogénétique, paléoenvironnementale et paléobiogéographique. Ph.D. diss., University of Lyon.

Otero, O., and M. Gayet. 1999. *Semlikiichthys* (Perciformes Incertae sedis), genre nouveau, et position systematique nouvelle pour *Lates rhachirhynchus* Greenwood and Howes, 1975, du Plio-Pléistocène africain. *Cybium* 23:13–27.

Poll, M. 1971. Révision des *Synodontis* africains (famille Mochocidae). *Annales du Musée Royal du Congo Belge, Tervuren, Sciences Zoologiques* 191:1–497.

Priem, R. 1914. Sur les poissons fossiles et en particulier des Silurides du Tertiaire supérieur et des couches récentes d'Afrique. *Bulletin de la Société Géologique de France* 21:1–13.

Priem, R. 1920. Poissons fossiles du Miocène de l'Egypte. In R. Fourtau, ed., *Contribution à l'étude des vertébrés Miocènes de l'Egypte,* pp. 12–15. Cairo: Government Press.

Schwartz, H. L. 1983. Paleoecology of Late Cenozoic fish from the Turkana Basin, northern Kenya. Ph.D. diss., University of California, Santa Cruz.

Smart, C. 1976. The Lothagam I fauna: Its phylogenetic, ecological and biogeographic significance. In Y. Coppens, F. C. Howell, G. L. Isaac, and R. E. Leakey, eds., *Earliest Man and Environments in the Lake Rudolf Basin: Stratigraphy, Paleoecology, and Evolution,* pp. 361–370. Chicago: University of Chicago Press.

Stewart, K. M. 1990. Fossil fish from the Upper Semliki. In N. T. Boaz, ed., *Evolution of Environments and Hominidae in the African Western Rift Valley,* pp. 141–162. Memoir No. 1. Martinsville: Virginia Museum of Natural History.

Stewart, K. M. 1997. A new species of *Sindacharax* (Teleostei: Characidae) from Lothagam, Kenya, and some implications for the genus. *Journal of Vertebrate Paleontology* 17:34–38.

Stromer, E. 1916. Die Entdeckung und die Bedeutung der Land- und Süsswasserbewohnenden Wirbeltiere in Tertiär und der Kreide Ägyptens. *Zeitschrift der Deutschen Geologischen Gesellschaft* 68:397–425.

Trewavas, E. 1983. *Tilapiine Fishes.* London: British Museum (Natural History).

Van Couvering, J. A. H. 1977. Early records of fresh-water fishes in Africa. *Copeia* 1:163–166.

Van Couvering, J. A. H. 1982. *Fossil Cichlid Fish of Africa.* Special Papers in Palaeontology 29. London: Palaeontological Association.

Van Neer, W. 1987. A study on the variability of the skeleton of *Lates niloticus* (Linnaeus, 1758) in view of the validity of *Lates maliensis* Gayet, 1983. *Cybium* 11:411–425.

Van Neer, W. 1992. New late Tertiary fish fossils from the Sinda region, eastern Zaire. *African Study Monographs,* Supplementary issue 17:27–47.

Van Neer, W. 1994. Cenozoic fish fossils from the Albertine Rift Valley in Uganda. In B. Senut and M. Pickford, eds., *Geology and Palaeobiology of the Albertine Rift Valley, Uganda-Zaire.* Vol. 2. *Palaeobiology/Paléobiologie,* pp. 89–127. Occasional Publication No. 29. Orléans: Centre International pour la Formation et les Echanges Géologiques.

Ward, C. V., M. G. Leakey, B. Brown, F. Brown, J. Harris, and A. Walker. 1999. South Turkwel: A new Pliocene hominid site in Kenya. *Journal of Human Evolution* 36:69–95.

Weiler, W. 1929. Die mittel- und obereocäne Fischfauna Ägyptens mit besonderer Berücksichtigung der Teleostomi. *Abhandlungen der Bayerische Akademie der Wissenschaften* 1:1–57.

TABLE 3.2 Fish Taxa from the Nawata Formation and the Apak, Muruongori, and Kaiyumung Members and from Plio-Pleistocene Deposits at Koobi Fora

Taxa	Nawata[a]	Apak[a]	Muruongori[a]	Kaiyumung[a]	Koobi Fora[b]
Myliobatiformes	—	—	—	—	+
Protopterus sp.	+	+	—	+	+
Polypterus sp.	+	+	+	+	+
Heterotis sp.	+	+	+	—	+
Hyperopisus sp.	—	+	+	+	+
Gymnarchus sp.	+	+	+	+	+
Labeo sp.	?+	+	+	+	+
Barbus sp.	—	+	—	+	+
Distichodus sp.	+	+	—	—	+
Citharinus sp.	—	—	—	—	+
Hydrocynus sp.	+	+	+	+	+
Alestes sp.	+	+	—	+	—
Sindacharax lothagamensis	+	+	—	—	—
S. mutetii	—	+	—	+	—
S. cf. *mutetii*	—	+	—	—	—
S. howesi	—	—	—	+	—
S. deserti	—	—	+	+	—
S. greenwoodi	—	—	+	+	—
Sindacharax sp.	+	+	+	+	+
Bagrus sp.	—	—	—	—	+
aff. *Bagrus* sp.	+	+	—	+	—
Clarotes sp.	+	+	—	+	+
Auchenoglanis sp.	—	—	—	—	+
?Schilbe sp.	+	—	—	—	—
Clarius/Heterobranchus	+	+	—	+	+
Synodontis sp.	+	+	+	+	+
Lates niloticus	+	+	—	+	—
Semlikiichthys cf. *S. rachirhinchus*	+	+	+	+	—
Cichlidae	—	—	+	+	+
Tetraodon sp. nov.	—	—	+	—	—
Tetraodon sp.	—	—	—	+	+

Note: Absence may only reflect incomplete fossil recovery.

[a] This report.

[b] Schwartz 1983; Feibel 1988.

4

REPTILIA AND AVES

4.1

Fossil Turtles from Lothagam

Roger C. Wood

Fossil turtles from Lothagam were abundant and diverse in the lower and upper members of the Nawata Formation and Apak Member of the Nachukui Formation. Three families (Pelomedusidae, Testudinidae, and Trionychidae) and at least six different species are represented. Remains of the pelomedusid *Turkanemys pattersoni* gen. and sp. nov. outnumber those of all the other identifiable Lothagam chelonians combined. Other components of the fauna found only at Lothagam are the pelomedusid *Kenyemys williamsi* and the trionychid *Cycloderma debroinae.* Remains of two additional types of trionychid turtles and of a giant tortoise have also been recovered. The Lothagam turtles represent a mixture of extinct and modern forms, including the earliest known occurrence of the living species *Cycloderma frenatum* and the last known continental African representative (*Turkanemys pattersoni*) of a lineage that survives today only as the Madagascan species *Erymnochelys madagascariensis.* Except for the giant tortoise, all the Lothagam turtles clearly represent highly aquatic species, an observation that is consistent with interpreting Lothagam's sediments as being laid down in a riverine environment bordered by gallery forests. The exceptional abundance of *Turkanemys pattersoni* may result from annual congregations of this species at communal nesting beaches.

When Professor Bryan Patterson's Harvard expedition first discovered the Lothagam fossil locality during the latter part of the summer of 1966, it was immediately apparent that fossil turtles were an abundant and often well preserved component of the fauna. Subsequent collections confirmed that Lothagam has an unusually diverse chelonian fauna as well. Some of the Lothagam fossil turtle material has already been described (Meylan et al. 1990; Wood 1976, 1983). The purpose of this contribution is to formally document the single most abundant and best preserved of all the Lothagam fossil chelonians, a new genus and species of pelomedusid (side-necked) turtle. In addition, in keeping with the intent of this volume, an overview of the entire fossil chelonian fauna from Lothagam will also be provided. Abbreviations used in this contribution and terminology for the bones and scutes of the shell (figures 4.1 and 4.2) follow those of Loveridge and Williams (1957). KNM is the standard abbreviation for National Museums of Kenya, Nairobi.

Systematic Description

In addition to summarizing what is known about previously described fossil turtles from Lothagam, this section includes the description of a new genus and species of pelomedusid (side-necked) turtles, which is of particular interest for several reasons. It is, in terms of numbers of individual specimens, by far the most abundant of the fossil turtles from Lothagam. Moreover, a nearly complete specimen represents one of the best (if not the very best) preserved fossil turtles yet to be described anywhere in Africa. Finally, it is the last continental African representative of a side-necked turtle lineage, extending back to the Early Tertiary of Egypt, which is today represented only by *Erymnochelys madagascariensis* from Madagascar.

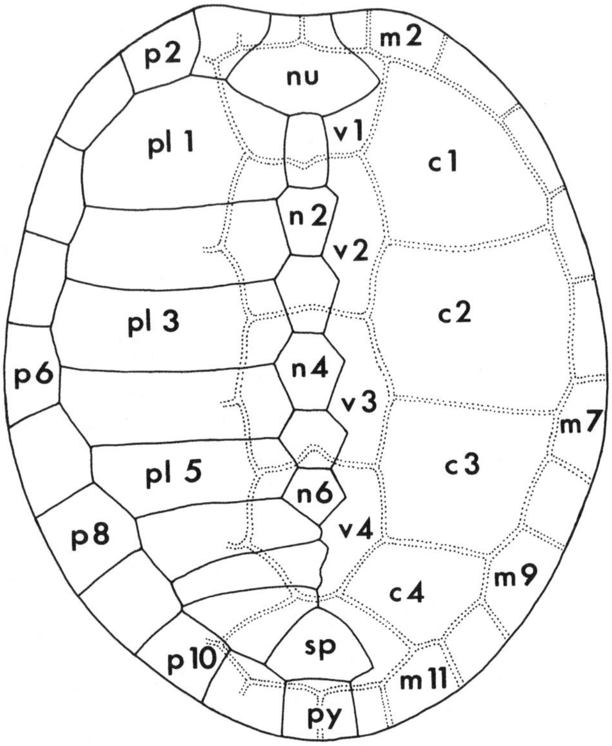

Figure 4.1 Bone and scute terminology for a typical pelomedusid carapace. Abbreviations for bones: n = neural; nu = nuchal; p = peripheral; pl = pleural; py = pygal; sp = suprapygal. Abbreviations for scales: c = costal; m = marginal; v = vertebral.

Order Testudines Linnaeus, 1758
Suborder Pleurodira Cope, 1898
Family Pelomedusidae (?)
Turkanemys gen. nov.

Diagnosis

Shell differs from all other African pelomedusid species by virtue of having trapezoidally shaped first vertebral scute and in tendency of nuchal bone to be proportionally broader than in other species. Shell differs from all South American species of *Podocnemis* (to which many African fossil pelomedusids have been assigned in the past as a matter of convenience) in having six rather than seven neural bones and a triangular rather than pentagonal intergular with gulars meeting in the midline behind it.

Skull differs from *Erymnochelys madagascariensis* as follows: general proportions relatively narrow and elongate rather than moderately broad and squat; deeply emarginated in cheek region, preventing quadrate from meeting jugal; interparietal scute trapeziform, with broad contact between parietal scutes in midline behind it; prominent triturating ridges on palatal surface of maxilla; nares higher than broad; external opening of tympanic cavity more elongate anteroposteriorly; mandible with well-developed median triturating ridges inside masticating troughs; very broad biting surface at mandibular symphysis.

Differs from South American skulls of *Podocnemis* in lacking anteroposterior midline depression between orbits; more lateral orientation of orbits and nares higher than broad (features that are shared with *Peltocephalus dumerilianus*). Differs from the skull of *Peltocephalus dumerilianus* in: presence of two, rather than one, triturating ridges; same general proportional differences as with *E. madagascariensis;* deep emargination of cheek region; size and shape of interparietal scute.

Differs from all other pelomedusid species in structure of cervical series, with articular surfaces being intermediate in shape between saddle joints of typical podocnemines and procoelous condition of pelomedusines and *E. madagascariensis.*

Type species

Turkanemys pattersoni sp. nov.

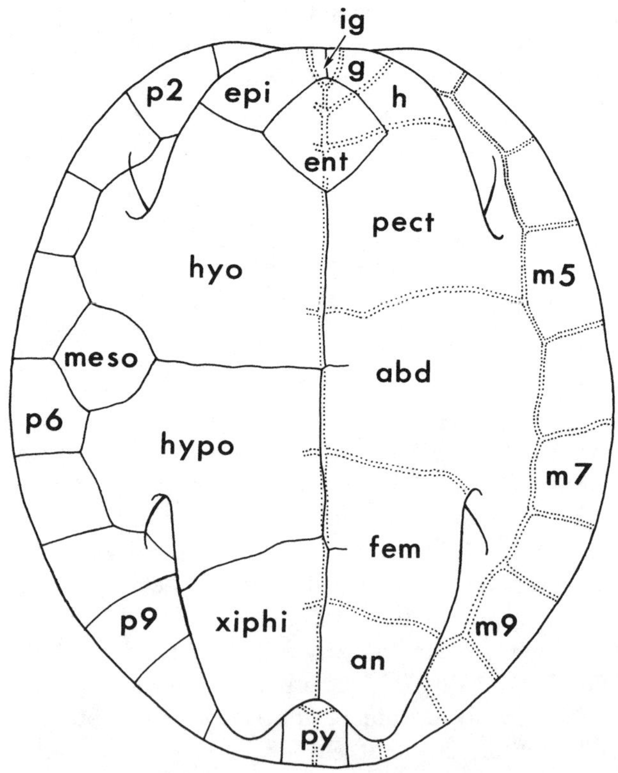

Figure 4.2 Bone and scute terminology for a typical pelomedusid plastron. Abbreviations for bones: ent = entoplastron; epi = epiplastron; hyo = hyoplastron; hypo = hypoplastron; meso = mesoplastron; p = peripheral; py = pygal; xiphi = xiphiplastron. Abbreviations for scales: abd = abdominal; an = anal; fem = femoral; g = gular; h = humeral; ig = intergular; m = marginal; pect = pectoral.

Etymology

Named after the Turkana District of northwestern Kenya where remains of this taxon have been collected, the Turkana people who live there, and the nearby Rift Valley lake of the same name.

Distribution

Miocene and Pliocene of the Turkana District, northwestern Kenya.

Remarks

Until recent years, many African pelomedusids when first described were routinely (and, in retrospect, erroneously) referred to as the extant South American genus *Podocnemis.* However, as the fossil record of African pelomedusids continues to improve, it has become increasingly clear that *Podocnemis sensu strictu* never occurred in Africa. As they became better known, it has been possible to reassign the supposed African representatives of *Podocnemis* to other genera (e.g., "*Podocnemis*" *antiqua* = *Shweboemys antiqua;* Wood 1971).

Reference to the South American pelomedusids *Podocnemis* and *Peltocephalus* was therefore made in the diagnosis to underscore the fact that *Turkanemys* is not a representative of any known South American lineage.

The proportions of the first vertebral scute on the carapace have been used here as a diagnostic character for *Turkanemys.* It should be noted that there are specimens of both "*Podocnemis*" *fajumensis* and "*Podocnemis*" *aegyptiaca* whose ratio of width to length for the first vertebral slightly overlaps the lower limit of the width/length ratios found in specimens of *Turkanemys* (table 4.1). Nevertheless, there is a marked tendency for the width/length ratio in *Turkanemys* to be greater than in any of the other African fossil pelomedusids with similar shells that were originally described as *Podocnemis.*

Turkanemys pattersoni sp. nov.

(Figures 4.3–4.11; tables 4.1–4.3)

Diagnosis

As for the genus.

Holotype

KNM-LT 569, a complete plastron and nearly complete carapace, an entire skull with associated mandibles and hyoids, as well as parts of the axial and appendicular

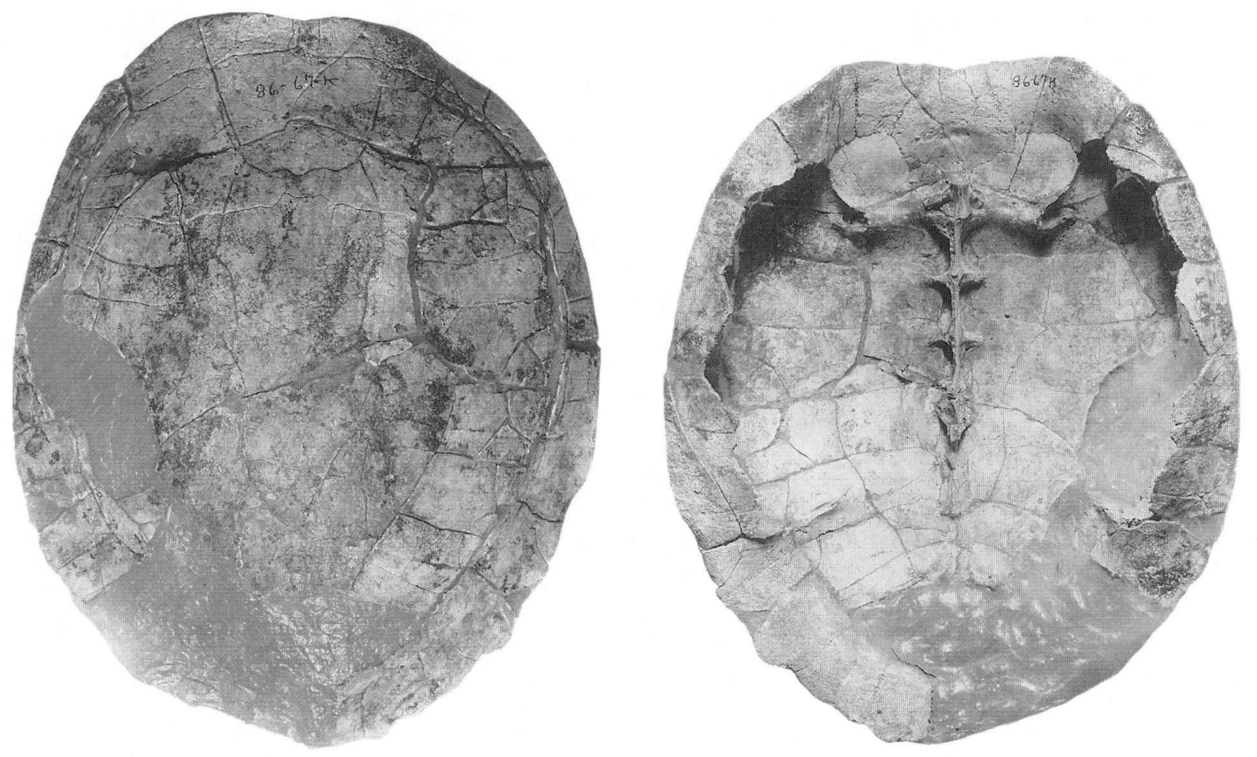

A B

Figure 4.3 *Turkanemys pattersoni* gen. and sp. nov. carapace, KNM-LT 569: A = dorsal view, B = visceral surface. Note facets on the first pair of pleurals for the strongly developed axial buttresses. Midline length = 37.8 cm.

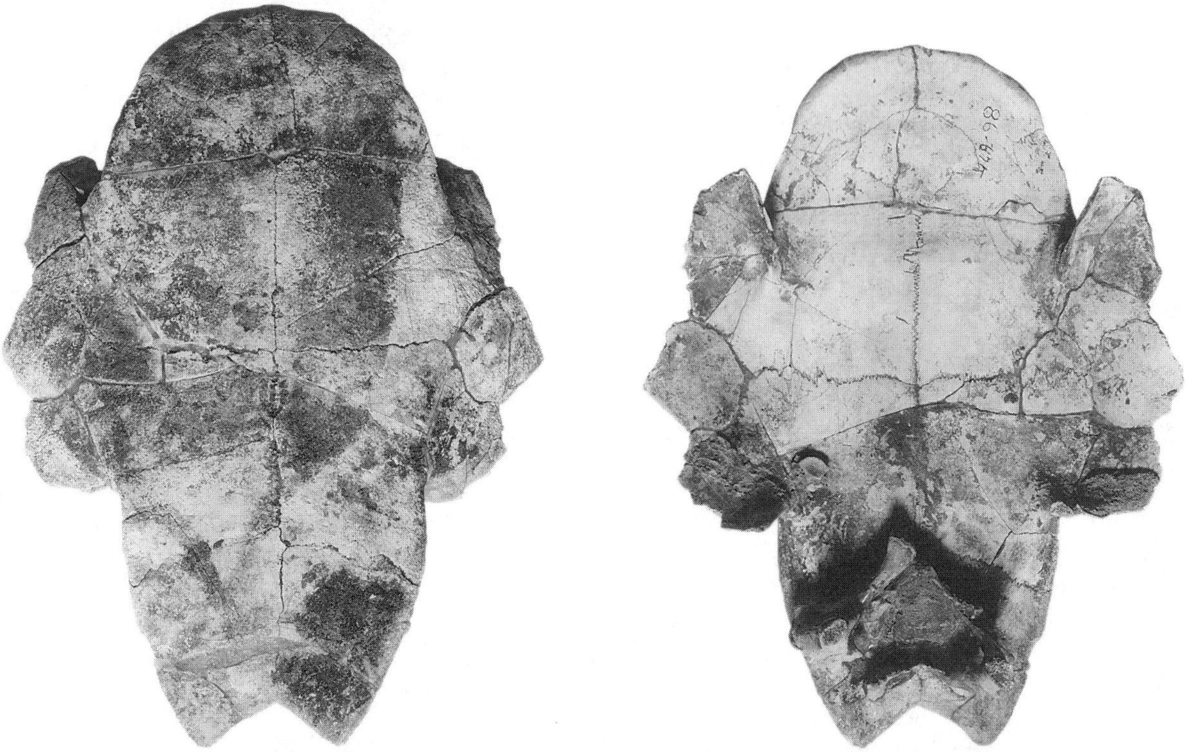

A B

Figure 4.4 *Turkanemys pattersoni* gen. and sp. nov. plastron, KNM-LT 569: A = ventral view, B = visceral surface. The pelvic girdle remains fused to the inner surfaces of the xiphiplastron. Midline length = 33.0 cm.

skeleton (a complete but poorly preserved pelvis, head of right and proximal one-half of left humerus, right and left femora, tibiae, and fibulae, left calcaneum, astragalus, fifth metatarsal, and six phalanges); from the lower member of the Nawata Formation.

Etymology

Named after my mentor, friend, and companion on many field trips to remote areas, the late Professor Bryan Patterson. Pat (as he was universally known) was the leader of five paleontological expeditions from Harvard University's Museum of Comparative Zoology to northern Kenya between 1963 and 1967. These resulted in the discovery of several important fossil localities (most notably Kanapoi and Lothagam) that have yielded an abundance of fossil vertebrates, including the new species here described. Pat also discovered the magnificently preserved type specimen of *Turkanemys pattersoni* sp. nov. in the summer of 1967.

Distribution

Southwestern Turkana District, Kenya, in Late Miocene horizons at Lothagam (lower and upper members of the Nawata Formation); Early Pliocene Kanapoi Formation; and Mio-Pliocene sediments west of Ekora (Patterson et al. 1970; Behrensmeyer 1976).

Lothagam Material

- *Lower Nawata:* 429, carapace fragments; 432, carapace and plastron fragments; 430, carapace fragments; 434, plastron and carapace fragments; 440, nearly complete plastron and carapace fragments; 569, holotype; 23178, carapace and plastron; 26482, complete carapace and plastron; 26531, carapace and plastron fragments; 26532, plastron fragments and humerus.
- *Upper Nawata:* 426, plastra fragments; 427, complete carapace and partial plastron; 428, complete carapace and plastron; 445, partial carapace and plastron; 451, anterior fragment plastron; 452, partial plastron; 453, fragmentary plastron; 454, partial plastron and carapace fragments; 556, carapace fragment; 566, partial carapace and plastron; 570, fragmented carapace and plastron; 571, carapace and plastron; 23981, complete carapace and plastron fragment; 23984, nearly complete carapace and plastron; 23986, complete carapace and plastron; 23989, almost complete skull;

23992, mandible; 24025, half carapace and plastron; 26473, plastron; 26474, partial plastron; 26475, carapace and plastron fragments; 26477, carapace and plastron; 26478, carapace fragments; 26479, 3 plastron fragments; 26480, partial carapace and plastron; 26487, mandible; 26494, cranium; 26496, carapace fragment with carnivore tooth mark; 26497, plastron fragment; 26498, broken carapace; 26507, half carapace and plastron; 26511, Rt. mandible; 26512, complete plastron and carapace fragment; 28771, partial cranium.

- *Apak Member:* 438, partial plastron and fragment of carapace; 568, partial carapace and complete plastron; 23985, nearly complete carapace and plastron; 26470, complete carapace; 26471, carapace and plastron fragments; 26516, complete plastron.
- *Kaiyumung Member:* 441, neural; 26523, carapace.
- *Horizon indet.:* 431, complete carapace and plastron; 565, carapace and plastron; 8739, partial skull; 23048, complete skull; 24073, plastron.

Kanapoi Material

- *Kanapoi Formation:* KP 435, partial internal mold and many unattached bone fragments of a relatively small individual; KP 436, partial carapace and plastron; KP 437, almost complete right xiphiplastron; KP 451, fragmentary specimen including many

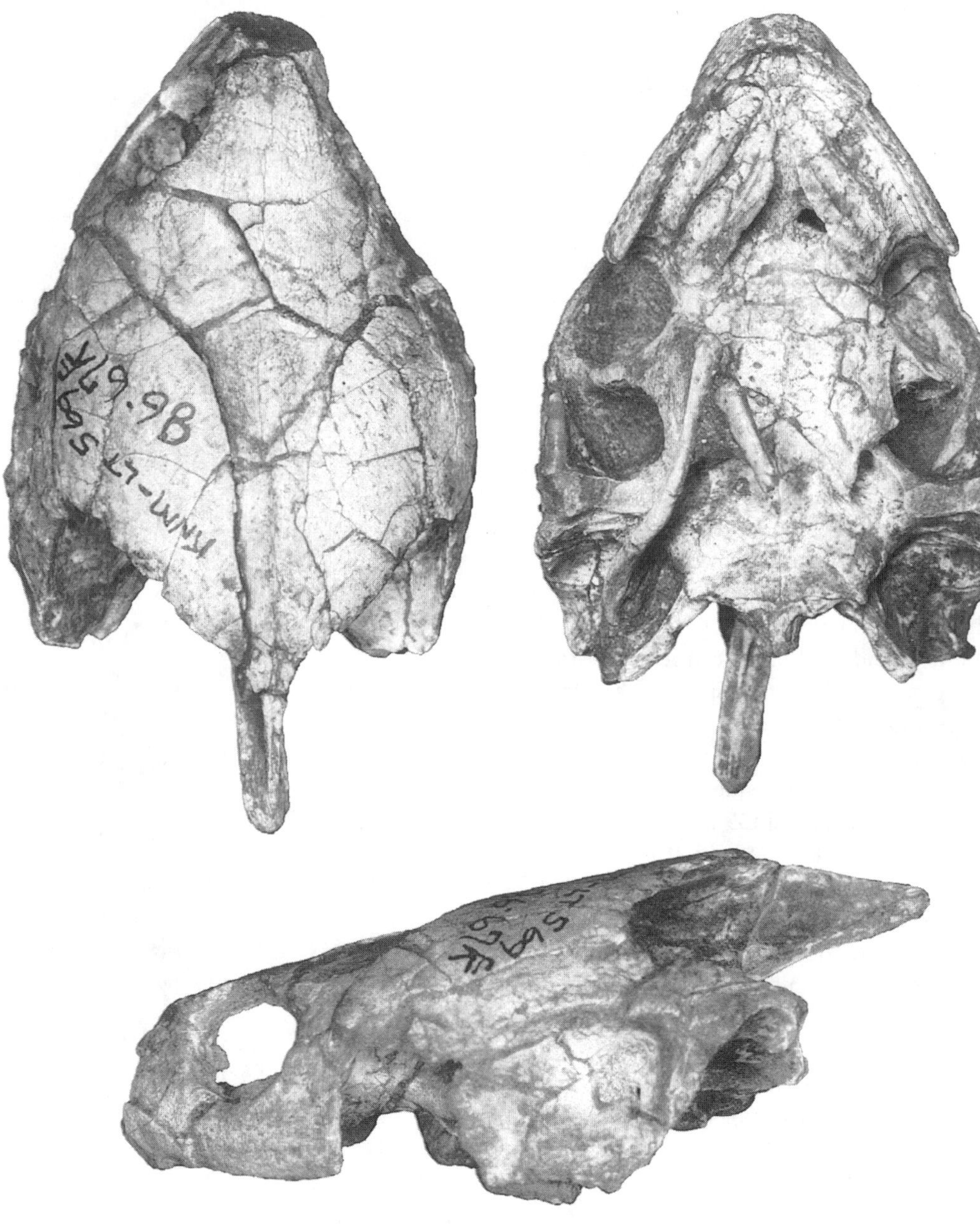

Figure 4.5 *Turkanemys pattersoni* gen. and sp. nov. cranium, KNM-LT 569: top left = dorsal view; top right = ventral view; bottom = lateral view. Snout–condyle length = 68.3 mm; maximum width = 49.2 mm.

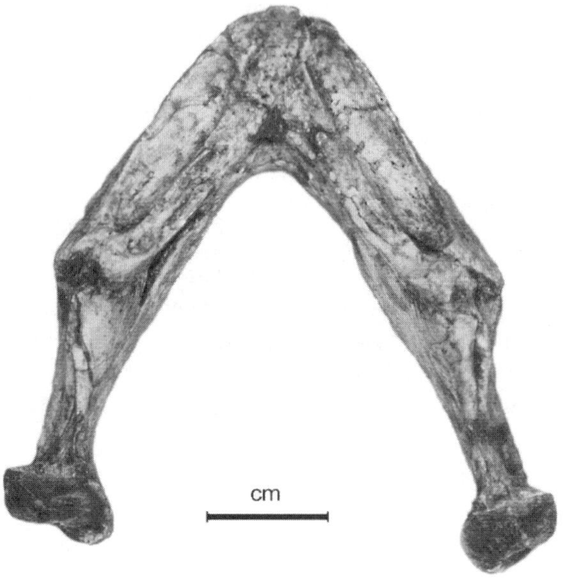

Figure 4.6 *Turkanemys pattersoni* gen. and sp. nov. mandible, KNM-LT 569 (dorsal view).

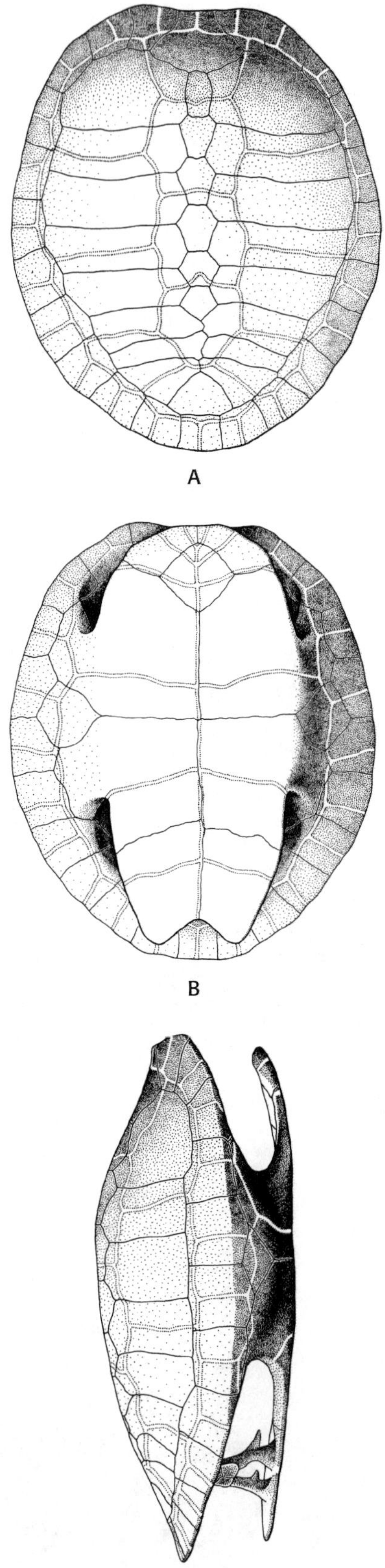

Figure 4.7 Reconstruction of the shell of *Turkanemys pattersoni* gen. and sp. nov.: A = dorsal view; B = ventral view; C = right lateral view.

pieces of carapace and plastron; KP 562, partial carapace and somewhat more complete plastron.

The type of *Turkanemys* is the best-preserved fossil turtle yet to have been discovered in Africa. The shell is virtually undistorted and nearly complete. No parts of the plastron are wanting, and only the following elements are missing from the carapace: the pygal, the suprapygal, the eighth pair of pleurals as well as the lateral ends of the third through seventh pleurals of the left side, the ninth through eleventh left peripherals and also parts of the fifth and sixth ones of the same side. But these areas are preserved in other specimens so that it is possible to reconstruct the entire shell with complete confidence (figure 4.7).

Removal of the matrix from inside the shell of the type revealed the presence of a skull, a mandible, and much of the rest of the skeleton. So perfect is the preservation of the skull that even the bony components of the hyoids are present in their proper anatomical position. Compaction has, however, caused a slight amount of breakage and crushing on the roof of the skull and some distortion of the snout region. Extending back from the occipital condyle is a series of eight articulated cervical vertebrae. Both shoulder girdles and most of the pelvis are intact. Parts of the two humeri, all of both femora, a tibia, a fibula, and some assorted tarsals, metatarsals, and phalanges have also been preserved. Even a few caudal vertebrae are present. The following description is for the most part based on this specimen.

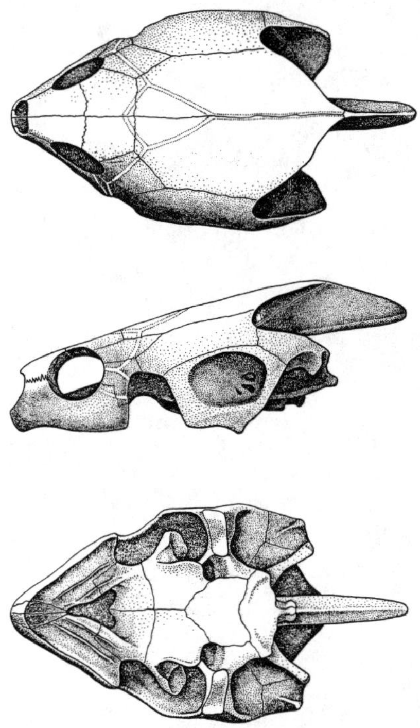

Figure 4.8 Reconstruction of the cranium of *Turkanemys pattersoni* gen. and sp. nov.: top = dorsal view; middle = left lateral view; bottom = ventral view.

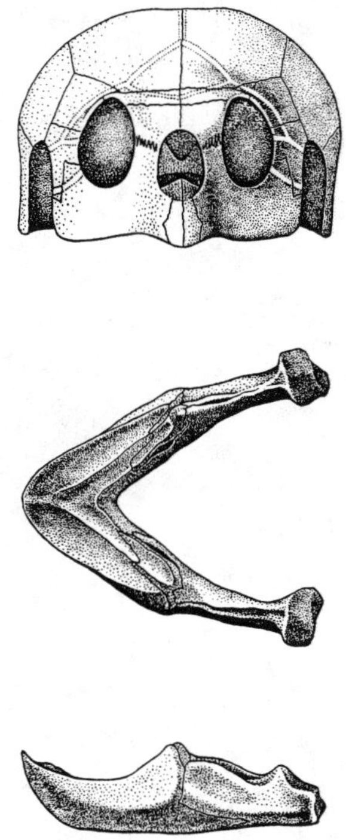

Figure 4.9 Reconstruction of the cranium and mandible of *Turkanemys pattersoni* gen. and sp. nov.: top = anterior view of cranium; middle = dorsal view of mandible; bottom = left lateral view of mandible.

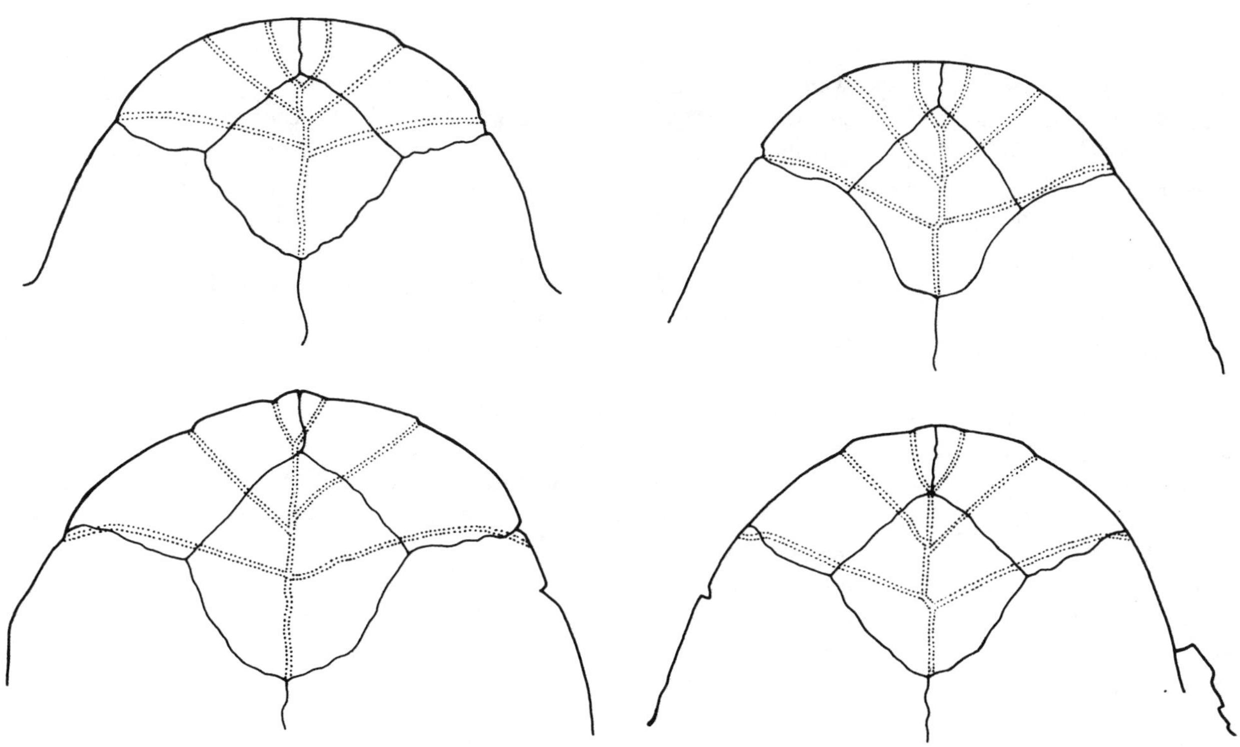

Figure 4.10 Variations in shape of the entoplastron and scute configurations on the anterior lobes of representative specimens of *Turkanemys pattersoni* gen. and sp. nov. The specimens in this figure are not drawn to scale.

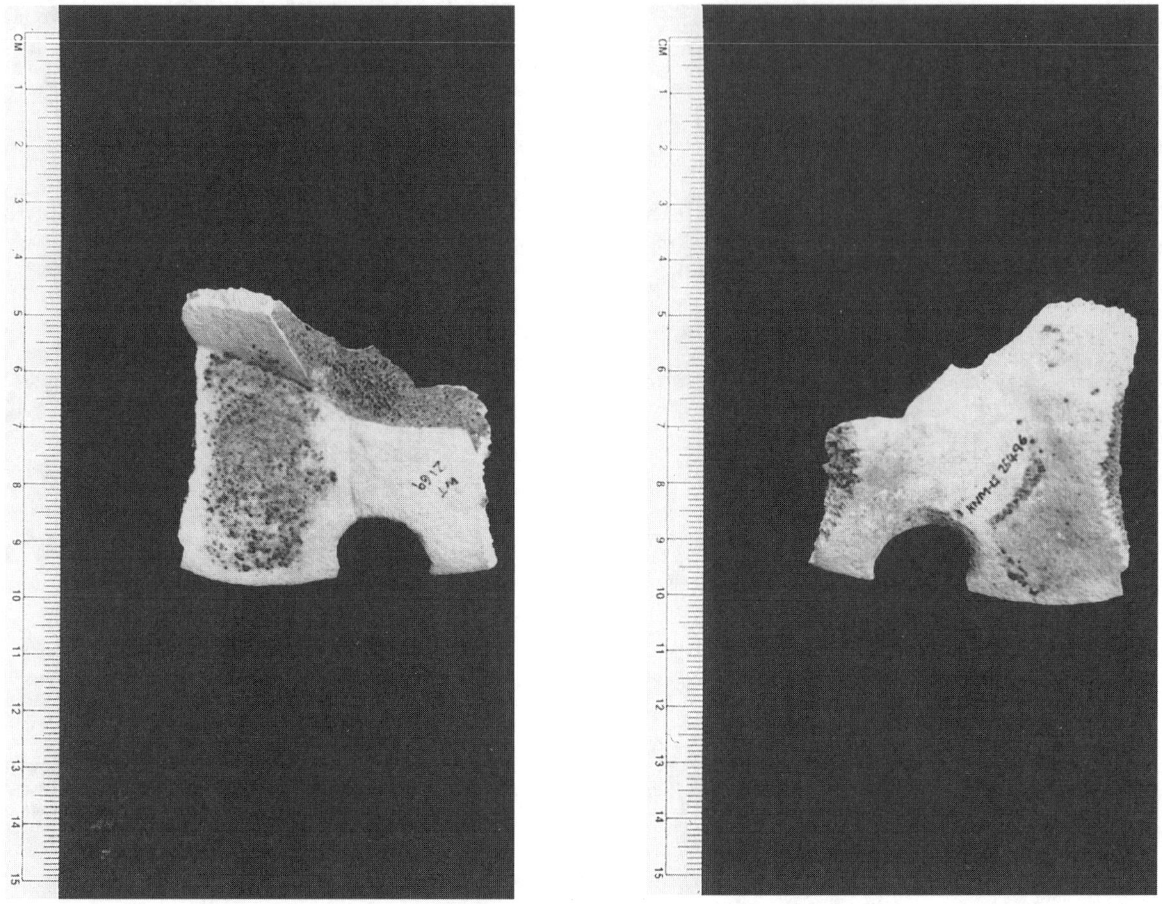

Figure 4.11 *Turkanemys pattersoni* gen. and sp. nov., KNM-LT 26496. Isolated peripheral bone (probably the eighth from the right side of the carapace) showing probable puncture wound from a crocodile tooth: right = external view; left = internal view.

The carapace of *T. pattersoni* is fairly large (table 4.2), depressed and oval in outline. In every specimen examined, the nuchal bone is always considerably broader than long, its width being in one case (LT 110) nearly one and one-half times greater than its length (table 4.1). At the front margin the nuchal is rather narrow, but its anterolateral sides curve outwards toward the rear so that the greatest width is more than double that at the anterior border. As is true for all the other African fossil species originally attributed to *Podocnemis,* there are six neurals arranged in a continuous series. The first of these is fusiform and abuts directly against the posterior end of the nuchal bone. Then follow four hexagonal neurals whose proportions gradually change from front to rear. The second, third, and fourth are all longer than broad and have their anterolateral sides shorter than their posterolateral ones. The fifth neural forms a nearly symmetrical hexagon with equal dimensions on all sides. The sixth and last neural is pentagonal and slightly wider than long. Eight pairs of pleurals flank either side of the neurals. The seventh pair becomes progressively narrower as the distance away from the midline increases; as far as I have been able to determine, somewhat similarly shaped seventh pleurals are found only in *Erymnochelys madagascariensis.* Part of the sixth pair and all of the seventh and eighth pairs of pleurals meet in the midline to separate the sixth neural from the suprapygal, which is roughly triangular in outline. The pygal is a narrow, somewhat trapezoidal bone. Eleven pairs of peripherals fringe the circumference of the carapace. The first four dorsal vertebrae, as well as part of the fifth, have been preserved in the type. No differences from the corresponding elements of *E. madagascariensis* are apparent. Iliac scars extend over parts of the inner sides of the seventh and eighth pairs of pleurals.

As is the case in all pelomedusids, there were five vertebral scutes extending along the midline of the carapace from front to rear. The first vertebral narrowed posteriorly and was trapezoidal in outline, a feature evidently unique to this species. The remaining vertebrals were all hexagonal, although their proportions varied greatly; the second and third were elongated with the antero- and posterolateral sides of the fourth and the anterolateral sides of the fifth being the longest. This results in a pronounced constriction in the width of the vertebral series at the junction between these scutes. The greatest width of the fifth (last) vertebral exceeds its greatest length whereas the reverse situation prevails for all the preceding ones.

Four pairs of costals were arranged on either side of the vertebrals. In typical pleurodiran fashion, these scutes overlapped onto the upper surfaces of the peripheral bones. No cervical scute was present, and the midline union of the first pair of marginals prevented the first vertebral from reaching the front margin of the carapace. As is usual in pelomedusids, there were 12 pairs of marginals.

The posterior lobe of the plastron is somewhat longer than the anterior one, and the portion between the axial and inguinal notches is longer than either lobe (figures 4.4 and 4.7B); except for a slight upturning at the front end of the anterior lobe, and in some cases an almost imperceptible downwarping along the lateral edges of the posterior lobe, the bottom surface of the plastron appears to have been essentially flat. As in most pelomedusids, some variability in the shape of the entoplastron is common. Different individuals may have diamond-shaped or almost pentagonal entoplastra, or some form intermediate between these extremes (figure 4.10). Usually, on the visceral surface at the front of the entoplastron, and often extending forward onto the epiplastra, there is a small, more or less oval swelling that presumably is related in some way to the connection of the neck musculature. Otherwise, the thickness of the anterior plastral lobe is surprisingly uniform. Sutures between the hyo- and hypoplastra bisect the middle of the bridge and terminate laterally in a junction with the medial mesoplastral sutures.

The mesoplastra are small, hexagonal elements wedged between the outer ends of the hyo- and hypoplastra at the base of the bridge. The anal notch is broad and V-shaped, and the xiphiplastra terminate in blunt points. The position of the pelvic scars does not differ from that of *E. madagascariensis.* The axial and inguinal buttresses of the bridge are well developed. Furrows on the outer surface of the anterior lobe show that the intergular was a small, triangular scute. Whether or not the intergular extended back onto the entoplastron is variable within the sample (figure 4.10).

Behind the intergular, the trapeziform gulars always met in the midline. Generally, the lateral portions of the humero-pectoral sulcus either coincided with or lay somewhat in front of the suture between the epiplastra and hyoplastra. The pectoral-abdominal groove traverses the bridge in front of the mesoplastra and does not cut across any part of them. The positions of the remaining sulci are shown in figure 4.7B. On the external surface of many specimens a network of anastomosing vermiculations can be seen. The plastral formula is pectoral > abdominal > femoral > anal > intergular > humeral > gular.

The skull of *T. pattersoni* (figures 4.5, 4.8, and 4.9) is rather elongate, being nearly one and one-half times as long as it is wide (snout–condyle length, 68.3 mm; width above tympanic cavities, 49.2 mm). The temporal region is moderately emarginated dorsally and deeply emarginated ventrally. Both the nares and orbits are slightly exposed in dorsal view. The interorbital width is equal to the height of the orbits. A relatively small, trapeziform interparietal scute covered the anteromedian surfaces of the parietal bones and its anterior tip extended far enough forward to cover a very small portion of the frontals. Behind it, the paired parietal scutes met extensively along the midline. There appear to have been small subocular scutes behind the orbits, separating the median frontal scute from the maxillary scutes. The jugal is prevented from meeting the parietal by the intervention of a large postorbital.

The tympanic cavity is entirely enclosed by the quadrate. In contrast to *E. madagascariensis,* in which the shape of the funnel leading into the columellar foramen is essentially circular externally, this opening in *T. pattersoni* is oval, its long axis being aligned anteroposteriorly. A very shallow precolumellar foramen and a post-otic antrum are nearly central in position within the tympanic cavity, whereas they are situated at the posterior end of the cavity in *E. madagascariensis.*

Because of lateral distortion in the snout region, it is difficult to reconstruct the shape of the nares. As preserved, they are higher than wide. Preservation is not sufficiently good to be able to determine if there were dorsal and ventral median projections of bone that partially separated the nasal openings into two equal portions, as is the case for most of the South American species of *Podocnemis.*

On the ventral surface of the skull, paired triturating ridges are present on each side of the palate. The medial set arises from the anterior part of the palatines and extends forward across the maxillaries but does not unite at the midline on the posterior portion of the premaxillae, thus forming a single V-shaped ridge. Between this pair of ridges and the tomia is a second set of ridges that are essentially restricted to the surface of the maxillae and do not join each other anteriorly. There is a deep premaxillary fossa for the reception of the strong hook on the mandibular symphysis. The maxillae do not meet in the midline behind the premaxillae and, contrary to the condition in all specimens of Recent species of *Podocnemis* that I have examined, there is no U-shaped depression at the posterior end of the palatal symphysis. Prominent ectopterygoid processes project into the subtemporal fossae. The carotid canals were quite large as is always the case in pelomedusids. The contact between basioccipital and quadrate is also characteristic of this family. On the ventral surface of the basioccipitals, only a very faint trace of a semicircular precondylar fossa is present. A pair of sigmoid shafts, remains of the bony portion of the hyoids, is also preserved in essentially the correct anatomical position beneath the basicranium.

The mandible differs from that of *E. madagascariensis* in several important respects. The biting surface at the symphysis is very broad, and within the masticating troughs on either side is a single, well-developed triturating ridge. Although these ridges extend far forward, they do not quite meet in the midline. Whereas the masticating trough in *E. madagascariensis* is narrowest at the mandibular symphysis and continuously broadens toward the crest of the coronoid process, the reverse is true for *T. pattersoni;* the widest part of the biting surface occurs at the mandibular symphysis, and this becomes progressively narrower toward the rear so that part of the lateral face of the dentary becomes exposed in dorsal view.

All eight cervical vertebrae have been preserved. I have not examined the shape of the connections between atlas and axis, cervicals 3 and 4 and cervicals 7 and 8, since these were separated. The articular pattern for the remainder of the cervicals does not conform to either of the two characteristic pelomedusid patterns described by Williams (1950:515 and appendix 1), nor, in fact, does it conform to the cervical pattern of any kind of previously described chelonian. Those articulations that are visible appear to be intermediate in structure between the procoelus centra of modern African pelomedusids (*Pelomedusa* and *Pelusios,* as well as *Erymnochelys* from Madagascar) and the saddle joints of typical South American pelomedusids (*Podocnemis* and *Peltocephalus*).

A number of caudal vertebrae have been preserved. Judging from their dorsoventral flatness and relatively small size, I suspect that these vertebrae are all from the middle and terminal sections of the tail rather than from its base. They do not appear to differ in any way from those of living pelomedusids that I have examined.

Preserved portions of the girdles, limb, and foot bones reveal no differences from *E. madagascariensis.*

Discussion

Living pelomedusid turtles are all freshwater aquatic forms. There is growing consensus, however, that many Late Mesozoic and Early Tertiary pelomedusids were adapted to marine environments (de Broin and Werner 1998; Wood 1984). There is even evidence of a modest radiation of tortoise-like, presumably terrestrial pelomedusids in Africa during the Tertiary that survived into the Pleistocene of Olduvai (Auffenberg 1981; Wood 1971). But the streamlined, hydrodynamically efficient shell of *Turkanemys pattersoni,* roughly comparable in its proportions to the shells of modern pelomedusids from Africa, South America, and Madagascar, coupled with the associated Lothagam fauna, leaves no doubt that this species was a freshwater aquatic turtle.

A comprehensive cladistic analysis of all fossil pelomedusids is currently being prepared for publication elsewhere. Consequently, a rigorous phylogenetic analysis of *Turkanemys* will not be undertaken here. Nonetheless, some general comments about its taxonomic significance seem worthwhile.

Throughout the better part of the Tertiary there was a lineage of side-necked (pelomedusid) turtles in Africa whose sole surviving representative, *Erymnochelys madagascariensis,* is found today only on the neighboring island of Madagascar. Until recently, the extinct African forms were known only from shells. All of these were originally assigned to the genus *Podocnemis,* which was long used as a catch-all name for almost any fossil pelomedusid from anywhere in the world (e.g., "*Podocnemis*" *alabamae* from the coastal plains of the eastern United States; "*Podocnemis*" *indica* from India; "*Podocnemis*" *andrewsi* from Europe).

Today, the accepted view is that *Podocnemis sensu strictu* should be restricted in its usage to a variety of living and fossil species confined to South America. The sole supposed living Old World representative that had been traditionally called *Podocnemis* has been reallocated to *Erymnochelys.* Furthermore, one of the extinct African taxa initially assigned to *Podocnemis*—"*P.*" *antiqua* from the Fayum of Egypt—has been recognized as a representative of the genus *Shweboemys* (Wood 1970).

Remaining, nonetheless, is a group of African fossil pelomedusids—"*Podocnemis*" *fajumensis,* "*P.*" *aegyptiaca,* and "*P.*" *bramlyi* (all from the Early to mid Tertiary of Egypt), as well as *Turkanemys* and several other forms not yet formally described from the Miocene of East and Central Africa—which seem to represent a coherent assemblage of related taxa. What they all have in common with each other (and with the extant *Erymnochelys madagascariensis*) is a strong and enduring similarity of shell morphology. Shared features include moderate size; thick bones; strong buttresses between carapace and plastron; moderately arched carapace; six neural bones of identical shape and arrangement; virtually identical plastral proportions; and a standardized, nearly invariant scute configuration on both the carapace and plastron. Evidently, this type of shell has undergone remarkable little change over a very long period of time (Late Eocene to the present).

Because of this stability—almost stasis—in shell structure, it is difficult to make inferences about relationships among these clearly related pelomedusids (which can for convenience be designated as members of the "*Erymnochelys* lineage") on the basis of shell morphology alone. Fortunately, specimens have now been recovered from three different fossil localities in East and Central Africa that, for the first time, provide *Erymnochelys*-type shells associated with skulls. This relative bonanza of new material offers some glimmers of insight into the evolutionary history of the lineage. These separate discoveries—including the *Turkanemys* assemblage here described—offer the prospect of a wealth of valuable new information. The other material includes a well-preserved skeleton from the Miocene of Rusinga Island, Kenya (currently under study by Larry Witmer) and a large collection of fragmentary material from the Miocene of the Western Rift Valley in Zaire (Hirayama 1992; Wood and Gawlas in preparation). Though lacking any complete specimens, this latter assemblage does, nonetheless, enable confident reconstruction of a complete composite specimen shell and much of the skull.

What is beginning to emerge from all this newly available material is quite interesting. While shell morphology has been remarkably conservative, skull structure has varied appreciably. The several partial skulls from Zaire are nearly indistinguishable, insofar as they have been preserved, from those of the modern *E. madagascariensis.* Moreover, the shells of these two taxa are virtually identical. The striking similarity of the fossil to the living species makes the former as ideal a candidate as one could ever hope to find in the paleontological record for a direct ancestor of the latter (Wood and Gawlas in preparation). This reinforces the impression, already afforded by shell morphology, of a conservative evolutionary history for *Erymnochelys.* It also, for the first time, furnishes direct evidence of the African ancestry for the surviving Madagascan species of *Erymnochelys.*

In contrast, the skull of *Turkanemys* is radically different from that of *E. madagascariensis.* For the most part, its various features are strongly reminiscent of many of the living South American species of *Podocnemis,* which can best be interpreted as a striking example of parallelism.

Because the shell of *Turkanemys,* nevertheless, is virtually indistinguishable from that of *Erymnochelys madagascariensis* in terms of both shape and size, a rather close relationship between these genera is likely. The maximum recorded carapace length for *E. madagascariensis* is 43.3 cm (Williams 1954a); the largest carapace of *Turkanemys pattersoni* (table 4.2) measures 44.5 cm in length. Further hindering any analysis is the unique structure of the cervical articulations, which do not really suggest close kinship to any known pelomedusid.

Turkanemys, in sum, exhibits a peculiar combination of African, South American, and even *sui generis* characters. This novel pastiche of anatomical features clearly indicates that *Turkanemys* is not a direct ancestor of *E. madagascariensis.* Instead, it appears to be a sterile offshoot from the main lineage (represented by the more or less contemporaneous Western Rift Valley material) that ultimately gave rise to the sole living representative of the genus.

Turkanemys pattersoni is found not only at Lothagam but also at the nearby and somewhat younger fossil localities of Kanapoi and Ekora, representing a time differential of as much as three to five million years. Not surprisingly, given the conservative history of shell morphology in the "*Erymnochelys* lineage," shells from these three localities are indistinguishable from one another.

Turkanemys pattersoni is the only pelomedusid found in the Pliocene deposits at Kanapoi and Ekora and is the latest surviving representative of the *Erymnochelys* lineage on continental Africa so far known. Thus, it would seem that *E. madagascariensis* and *T. pattersoni* are terminal species of divergent phyletic lines that must have separated no later than the Middle Miocene—in view of the fact that *T. pattersoni* is found abundantly in the Late Miocene beds at Lothagam—and probably not before the Early Miocene—since *Podocnemis aegyptiaca* represents an acceptable structural ancestor for both species.

The material referred to *Turkanemys pattersoni* represents the largest, best preserved population sample of any African fossil turtle yet described. Not unexpectedly, some variability can be detected in certain characters, including whether or not the intergular extended back onto the front of the entoplastron; the shape and proportions of the entoplastron; and the shape of the anal notch. Analysis of these variations, however, reveals that there are no consistent differences between the Kanapoi and Lothagam samples such as might indicate the presence of separate species at these two localities of somewhat differing ages. Furthermore, within these subsamples, it is not possible to distinguish more than one taxonomic group by any combination of characters. In fact, the kinds of variation observed in this species are not greater than those I have found in individuals in populations of living African pelomedusids.

One other specimen (LT 126), an oddly symmetrical fragment from the posterior end of a carapace (Wood 1976), deserves special mention. In terms of bone thickness, shell curvature, shape of the suprapygal, and general size, it matches comparable features of *Turkanemys.* However, there is no suprapygal bone, and the adjacent posteriormost peripheral bones are both triangular rather than rectangular. Moreover, the rim is uniformly thickened, suggesting that this peculiar shell morphology does not represent damage sustained (and subsequently healed) during its lifetime by some predator or accident. It could, perhaps, represent some kind of birth defect, but I have never seen anything remotely resembling this osteological arrangement in any of the several thousand turtle hatchlings or adults that I have examined. It is conceivable, therefore, that this peculiar specimen might represent a new taxon with a decidedly unusual carapace structure. Given how little has been preserved of the entire specimen, however, designating it as a new taxon would clearly be inappropriate. Instead, it seems best to regard this odd specimen for the time being as an aberrant representative of *Turkanemys.*

Nearly all the specimens of *Turkanemys pattersoni* are of similar size, presumably representing adults, and consequently nothing can be said regarding ontogenetic changes. Nor can the sex ratio be determined on the basis of existing evidence. No depressions, often a sexually dimorphic character indicative of males, occur on the posterior plastral lobes of any specimens. Perhaps all of the individuals in the hypodigm are females, or, as a more probable alternative, in this particular species males may not have had plastral indentations.

In many living turtles, males also have considerably longer tails than do females. Presumably, an osteological correlative of this character would be that in specimens of comparable sizes and belonging to the same species, there would be two nonoverlapping sizes of the anal notch, males being characterized by the larger ones and females by the smaller ones. However, Wood and Diaz de Gamero (1971) have shown that in pelomedusids this may not necessarily be true, depending on the species being considered. In any case, although a few specimens of *Turkanemys pattersoni* have notably small anal notches and several others have notably large ones, most are of an intermediate shape and size. The intergrading continuum of anal notch shapes makes it impossible to determine sex ratios in this species on the basis of this character also.

In many chelonian species, adult size is a sexually dimorphic character, with one sex routinely attaining a larger size than the other. (In some of these dimorphic species, females are larger than males, while in others the reverse is the case.) But the uniformity in size of all the *Turkanemys* shell material precludes any useful inferences about sex ratio based on differences in adult size.

An isolated peripheral bone (LT 26496, probably the eighth on the right side; figure 4.11) from an adult shell of *Turkanemys* affords an interesting glimpse of an encounter with a potential predator. At the outer edge of this bone is a semicircular notch. Its size and shape strongly suggest that it represents a puncture wound from a crocodile tooth. Because the bone around the rim of this tooth puncture is smooth, it appears that the turtle survived this particular crocodile attack long enough for the bone to heal. Miocene crocodiles do not seem to have been too discriminating in their taste for turtles, as a similar puncture wound has been reported in a trionychid ("soft shelled" turtle) carapace from the Sahabi fauna of Libya (Wood 1987).

Kenyemys williamsi Wood, 1983

(Figure 4.12; table 4.1)

Diagnosis

Differs from all other known pelomedusids by the following combination of characters: (1) a series of elongate tuberosities forming an interrupted keel extending along the midline rearward from the dorsal surface of the second neural bone; (2) six neural bones forming a continuous series, the anterior end of the first abutting directly against the rear margin of the nuchal bone and the sixth one being heptagonal; (3) outer corners of nuchal bone extending beyond lateral margins of first vertebral scute; (4) pentagonal shape of first vertebral scute; (5) only eighth and posterior part of seventh pair of pleural bones meet at midline of carapace; (6) anterior plastral lobe truncated; (7) triangular intergular scute not overlapping anterior end of entoplastron and only partially separating the gular scutes along the midline axis of the plastron.

Holotype

KNM-LT 567, a nearly complete but somewhat crushed shell from the lower member of the Nawata Formation at Lothagam. The catalogue number of KNM-LT 127 formerly cited for this specimen (Wood 1983) was erroneous.

Lothagam Material

- *Lower Nawata:* 567, holotype; 23990, ridged neural; 26521, ridged neural; 26528, ridged neural; 26529, ridged neural; 26530, ridged neural; 38296, ridged neural.
- *Apak Member:* 23991, ridged neural.

A detailed description is provided by Wood (1983).

Discussion

Kenyemys williamsi is the only pelomedusid other than *Turkanemys pattersoni* known from Lothagam. In contrast to *Turkanemys,* which is far more abundantly represented at Lothagam, *Kenyemys* is a rare component with only a handful of known specimens. Aside from a single complete but somewhat crushed carapace, nearly

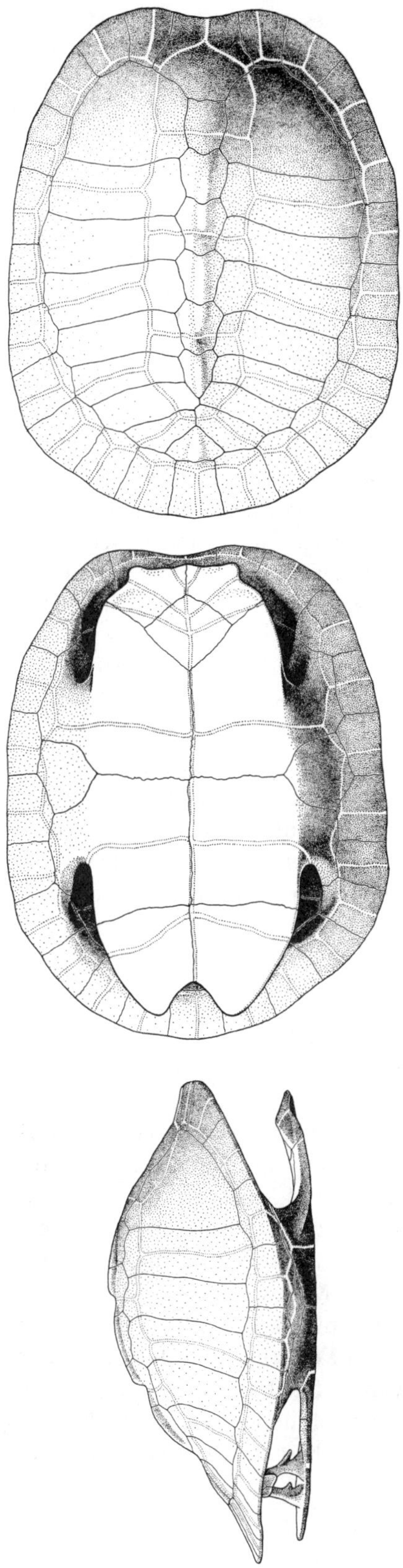

Figure 4.12 Reconstruction of the shell of *Kenyemys williamsi*, KNM-LT 567: top = dorsal view; middle = ventral view; bottom = lateral view. Estimated carapace midline length = 32 cm.

all of the other specimens assigned to *Kenyemys* are represented by distinctively ridged isolated neural bones.

Kenyemys is readily distinguished from all other fossil and recent pelomedusids. Compared to *Turkanemys*, *Kenyemys* is somewhat smaller in size (estimated straight line carapace length is 32 cm for the latter versus similar measurements ranging between 37.8 and 44.5 cm for the former). The general proportions of the shell differ, too. The lateral margins of the carapace of *Kenyemys* are essentially straight and parallel to each other, whereas the equivalent part of the carapace in *Turkanemys* is rounded. Moreover, the carapace of *Kenyemys* is more highly arched than that of *Turkanemys*. Its convexity is reminiscent of the carapace curvature that typifies the modern African pelomedusid *Pelusios sinuatus*, the largest in size, proportionately thickest boned, and most highly arched of the several species included within this genus.

Many specimens of *Turkanemys* were recovered from sandstones, whereas the type of *Kenyemys* was found in a limey clay. This might indicate that *Kenyemys* typically occupied a different kind of habitat from that favored by *Turkanemys*. Perhaps *Kenyemys* had habitat preferences (such as oxbow lakes) somewhat comparable to those of the modern African side-necked turtle *Pelomedusa* which (at least in parts of East and Central Africa) tends to occur in calm water environments that are rarely, if ever, represented in the fossil record, such as isolated water holes and ephemeral pools which seasonally dry up (Wood 1973).

No likely precursors of *Kenyemys* have yet been identified in the African fossil record, nor have any descendents so far been recognized.

Family Trionychidae (Fitzinger, 1826)

Commonly referred to as "soft-shelled" turtles, trionychids have a unique shell morphology. The carapace is disc-shaped in outline and nearly flat. Peripheral bones around the rim of the carapace are entirely absent save for one genus, *Lissemys*, in which a reduced number of posterior peripherals have been retained. The margins of the carapace are considerably extended by a ring of thick, leathery skin. The outer surfaces of both the carapace and plastron bones are moderately to strongly textured with irregularly-shaped tiny grooves, ridges, and/or tubercles. Unlike all other turtles except the highly aberrant leatherbacks (dermochelyids), enlarged scales do not cover the external surfaces of the shell. These are instead replaced by a layer of skin.

The plastron, too, is peculiar, being composed of a series of loosely articulated and variably shaped bony elements that correspond in number and position, but not in shape, to the plastral bones of more conventional turtles. There are no bony bridges connecting the carapace to the plastron between the front and hind limbs. The eighth cervical vertebra has a unique articulation with the first dorsal vertebra.

In short, fossil trionychids can be readily recognized even on the basis of small fragments of their shells. However, sorting out one trionychid taxon from another with any degree of confidence can be extremely frustrating unless substantial parts of the carapace and/or plastron have been preserved. Fortunately, several fairly complete trionychid carapaces have been collected at Lothagam. These have enabled the unequivocal identification of two soft-shelled species, *Cycloderma frenatum* and *Cycloderma debroinae.* In addition, a diagnostic fragment has permitted recognition of the presence of a third kind of trionychid in the Lothagam fauna, although not enough has been preserved to characterize it in detail (Meylan et al. 1990).

Cycloderma frenatum Peters, 1854
(Figure 4.13)

Diagnosis

Shell differs from that of other species of the genus in having a reduced entoplastral callosity, in often having a thin sheet of bone present between the two pairs of costiform processes on the ventral surface of the nuchal bone, and in having distal widths of first and eighth pleural bones greater than those of the other pleurals.

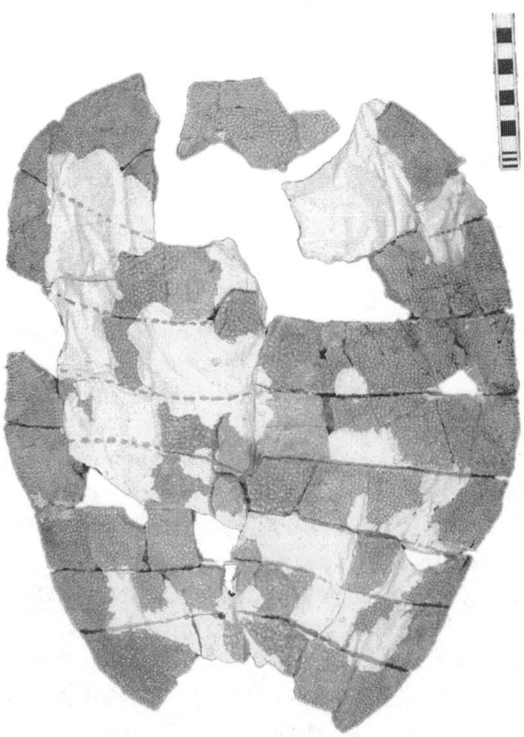

Figure 4.13 Restoration of the carapace of *Cycloderma frenatum,* KNM-LT 17197: dorsal view. Approximate midline length = 45 cm.

Lothagam Material

- *Upper Nawata:* 17197, parts of at least two adults including a nearly complete carapace (figure 4.13).
- *Nawata Formation, Horizon indet.:* 17199, plastral remains of more than one individual.

The midline length of the nearly complete carapace, when restored, would have been approximately 45 cm. Its oval disc is slightly indented along the rim on both sides at the sutural junction between the fifth and sixth pleural bones. Parts or all of the eight pairs of pleurals have been preserved. The lateral borders of the first pair of pleurals are substantially longer than those of the second through seventh pleurals. The eighth pleurals have both sustained extensive damage posteriorly. If restored, the distal widths of the eighth pleurals would clearly be longer than those of any of the other bones in the pleural series. Lateral portions of both sides of the typically broad nuchal bone are missing, the damage being greatest on the left side. What does remain of the nuchal exhibits the morphology typical of modern specimens of *C. frenatum.* Portions of only three neurals are present, probably representing the third, fifth, and sixth. Damage to these and the absence of the remainder makes it impossible to reconstruct the neural series with any degree of confidence.

Discussion

Cycloderma frenatum is an extant species whose modern distribution encompasses the southern half of Tanzania, the northern half of Mozambique, and all of Malawi. Fossil representatives of this species have been described from several localities, including the Pliocene Chiwondo beds of Malawi, the Plio-Pleistocene of Omo (southern Ethiopia), and the Plio-Pleistocene of Koobi Fora (northern Kenya). Except for the Chiwondo beds of Malawi, these fossil localities all lie outside the current boundaries of the species. The Lothagam specimen represents the earliest known occurrence of *Cycloderma frenatum.*

Cycloderma debrionae Meylan et al., 1990

Diagnosis

Differs from other species of the genus in having an almost vestigial, irregularly shaped entoplastral callosity, in having distal width of second pair of pleurals greater than that of first pair, and in lacking ischial projections of the pelvis into the thyroid fenestra.

Lothagam Material

- *Upper Nawata:* 17200 (holotype), virtually complete skeleton.

This is an exceptionally complete specimen, which represents most of the skeleton. The carapace lacks only the eighth right pleural and a distal portion of the second left pleural. All plastral elements are represented from one side or the other, including the right epiplastron, most of the entoplastron, the left hyo-hypoplastron, and the left xiphiplastron. The pectoral and pelvic girdles are both intact, and substantial portions of the limbs have been preserved as well. Elements of both the cervical and caudal vertebrae were recovered, as well as the condylar region of the skull and parts of the hyoid apparatus. A detailed description of this skeleton is provided by Meylan et al. (1990).

Discussion

The type (and only known specimen) of *Cycloderma debrionae* is the most completely preserved soft-shelled turtle yet known from the African fossil record. Aside from the remarkably complete type specimen of *Turkanemys pattersoni,* it is also the best preserved of all the Lothagam fossil turtles. When I collected this specimen in the field it was largely disarticulated. Not until the specimen was prepared back in the laboratory at Harvard did its exceptional preservation become apparent.

Four species of *Cycloderma* have been described. Two of these—*C. debroinae* and *C. victoriae*—are known only as fossils, each from a single locality and represented by only a single specimen. Of the two living species, *C. aubryi* has no fossil record, whereas *C. frenatum* has been recognized at four separate fossil localities (Lothagam being the oldest of these); all but one—the Chiwondo beds of Malawi—are outside the current range of this species.

Taking into consideration both extinct and living species, Lothagam is the only place so far known where two species of *Cycloderma* occur sympatrically. Given how little is known about the ecology and behavior of the extant species of *Cycloderma,* however, it would be premature to speculate on the possible significance of this fact.

Phylogenetic analysis (Meylan et al. 1990) suggests *C. debroinae* and *C. victoriae* are sister species, both being more closely related to *C. frenatum* than is the only other living species of *Cycloderma, C. aubryi.* At present, however, far too little is known to permit a clear understanding of the evolutionary history of this genus.

Trionychinae indet.

Lothagam Material

- *Horizon indet.:* 17201, right pleural.

In Africa, modern trionychids can be readily separated into two distinct subfamilies, the Cyclanobinae (within which *Cycloderma* is placed) and the Trionychinae. Cyclanobines are endemic to sub-Saharan Africa, whereas trionychines have a nearly global distribution when their fossil and recent distributions are taken into account. All African fossil soft-shelled turtles so far known can be included within one or the other of these subfamilies.

Several osteological characters are consistently useful for distinguishing between representatives of these two subfamilies. Cyclanorbines invariably have two neural bones that intervene between the first pair of pleurals. In contrast, trionychines have only a single elongate neural that separates these two bones. In cyclanorbines, the short sides of the typically hexagonal neural bones (often described as coffin-shaped) invariably face posterolaterally, whereas in trionychines the neural series always contains at least one reversal of neural orientation. Some neurals have their short sides directed anterolaterally, while in others the short sides face posterolaterally. Moreover, the hyo- and hypoplastra are always fused in posthatchling cyclanorbines, but these same bones are never fused in trionychines.

With these characteristics in mind, it is possible to identify a single Lothagam specimen (LT 17201) as an indisputable representative of the subfamily Trionychinae. This specimen is a nearly complete first right pleural bone. A moderately straight neural suture along the medial border indicates the likelihood that a single elongate neural was situated along the midline of the carapace between the left and right first pleural bones. Another trionychine feature is the position of the second rib, fused to the underside of the first pleural, which curves forward laterally so that it would have underlain the outer end of the nuchal bone's visceral surface.

Though a rather tenuous indication of the presence of trionychines in the Late Miocene sediments at Lothagam, this single bone nonetheless represents the earliest record of this subfamily in the Lake Turkana Basin. Unfortunately, it cannot be located in the collections at the National Museums of Kenya in Nairobi, and there is no record of its having been returned to Nairobi after study in North America. A single trionychine species,

Trionyx triunguis, is the sole surviving soft-shelled turtle species in Lake Turkana today.

Trionychidae indet.

Lothagam Material

- *Lower Nawata:* 26515, carapace fragments.
- *Upper Nawata:* 23993, hyo-hypoplastron; 26505, carapace fragments; 26522, carapace fragments; 26524, carapace fragments.
- *Horizon indet:* 25124, cervical vertebra

Several additional specimens from Lothagam indisputably represent the remains of soft-shelled turtles. These are so fragmentary, in some cases, as to be otherwise indeterminate. Included in this category are three specimens (LT 26505, 26522, and 26524) catalogued as "carapace fragments" from the Upper Nawata. Two other specimens—one (LT 23583) collected in 1967 and the other (LT 23993) in 1991—are both catalogued as hyo-hypoplastra and are therefore, if correctly identified, presumably the remains of cyclanorbines.

Family Testudinidae Gray, 1825
cf. *Geochelone*

Lothagam Material

- *Lower Nawata:* 23978, carapace and plastron fragments; 26485, carapace and plastron fragments; 26514, carapace and plastron fragments; 26518, carapace and plastron fragments; 26527, squashed carapace and plastron fragments; 26533, carapace fragments.
- *Upper Nawata:* 23583, hyo-hypoplastron; 26307, Lt. tibia; 26469, carapace fragments; 26491, carapace and plastron fragments; 26493, posterior carapace and plastron fragments; 26517, carapace and plastron fragments; 26525, carapace fragments.
- *Apak Member:* 24070, carapace, plastron and distal phalanx; 26433, carapace fragments; 26472, carapace fragment; 26488, posterior carapace and carapace and plastron fragments; 26489, carapace and plastron fragments; 26490, carapace fragments, 26534, distal and proximal humerus.

Discussion

Fossil tortoises are the only component of the chelonian fauna from Lothagam that have not yet been studied in detail. The specimens referred here to *Geochelone* all represent the remains of giant tortoises. Whether these document the presence of more than a single taxon cannot yet be stated with any degree of certainty. Meylan and Auffenberg (1986a) described two contemporaneous species of fossil tortoises (*Geochelone laetoliensis* and *G. brachygularis*) from the somewhat younger Laetolil Beds of northern Tanzania, so the possibility cannot be dismissed out of hand that more than one tortoise taxon may be eventually recognized at Lothagam.

Many giant tortoises, both extinct and extant, and from many different continents, have at one time or another been referred to as *Geochelone.* This genus has served as a catch-all taxon and cannot currently be rigorously diagnosed (Meylan and Auffenberg 1986b). Nonetheless, as a matter of convenience, it seems reasonable, at least for the present, to assign all the Lothagam fossil tortoise remains to *Geochelone* until such time as they can be properly studied in order to clearly signify the presence of giant fossil tortoises at this locality. Particularly in view of the fact that *Geochelone* is the most abundant fossil tortoise in Africa (Meylan and Auffenberg 1986b), this seems a reasonable working hypothesis.

Giant fossil tortoises are a ubiquitous component of all the major East African Mio-Pliocene fossil vertebrate localities including Laetoli (Meylan and Auffenberg 1986a), Omo (Arambourg 1947), and Olduvai (Auffenberg 1981). Though not yet formally described, monumentally large tortoises are also known from Kanapoi and Koobi Fora, where one impressive specimen has been left *in situ* as an exhibit with a shelter built over it. Elsewhere in Africa, *Geochelone* has been reported from the Late Eocene of Egypt (Andrews 1906), the Early Miocene of western Kenya (Andrews 1914), the Middle Miocene of Namibia (Stromer 1926), and the [Late?] Miocene of Libya (Wood 1987). Oddly, no giant tortoises have yet been reported from the fossil-rich localities in the Afar region of Ethiopia.

In any case, giant fossil tortoises were in the past broadly dispersed across a continent from which they are now absent. Africa today has the greatest diversity of living tortoise species of any continent, with South Africa (and particularly its Cape region) having more taxonomic diversity (five genera and 11 or perhaps 12 species) than anywhere else on the continent (Boycott and Bourquin 1988). But modern African tortoises are all small- to moderate-sized forms. Giant tortoises today survive only on the island of Aldabra in the Indian Ocean and on many of the larger islands of the Galapagos Archipelago in the eastern Pacific. What caused the disappearance of these behemoths from Africa (and other continents) remains largely a matter of conjecture.

Overview of the Lothagam Turtle Fauna

Lothagam fossil turtles are abundant, diverse and by and large, notably well preserved. The remains of three different chelonian families (pelomedusids, trionychids, and testudinids) have been found at this locality. These represent six different taxa, at least two of which (*Kenyemys williamsi* and *Cycloderma debroinae*) are known from this site alone. *Turkanemys pattersoni* is found elsewhere only at the nearby and slightly younger localities of Kanapoi and Ekora. *Cycloderma frenatum* is a modern species also represented in the fossil record to one extent or another in the lower Omo Valley, at Koobi Fora, and in the Chiwondo Beds of Malawi. Remains of one additional type of trionychid have also been recovered here, but they cannot yet be adequately diagnosed.

Whether or not the tortoises at Lothagam represent one or more new species, or a taxon that has already been described, remains to be determined. Taken as a whole, the Lothagam turtle fauna is somewhat transitional in nature, combining extinct genera (the pelomedusids), extinct species of living genera (*Cycloderma debroinae* and perhaps the tortoises), and modern ones (*Cycloderma frenatum* and perhaps some or all of the other trionychids as well).

Such taxonomic diversity is rivaled in Africa only by the Late Eocene and Early Oligocene chelonian faunas of the Fayum Depression in Egypt (Andrews 1906). No other East or Central African fossil locality equals Lothagam in terms of diversity, abundance, or quality of preservation.

From Laetoli, only two species of tortoises have been reported. *Geochelone brachygularis* is known from half a dozen complete or nearly complete shells, while evidence of *G. laetoliensis* is based only on fragments from six different specimens. Both of these tortoises are endemic to this locality.

From the lower Omo Valley of Ethiopia, fossil representatives of three living species have been described on the basis of reasonably good material: *Cycloderma frenatum, Pelusios sinuatus,* (*Sternothaerus rudolphi* of Arambourg 1947), and *Pelusios adansonii* (de Broin 1969, 1979). In addition, Arambourg (1947) mentioned that remains of large tortoises are not uncommon, though preserved only as broken fragments that have never been described. Thus, four different turtle taxa can be distinguished at this locality.

Four chelonian species have also been identified from Olduvai. Two of these (*Pelusios sinuatus* and *Taisternon microsulcae*) are pelomedusids, while the other two are tortoises (*Geochelone pardalis*, a modern species, and *Geochelone* "species"). *Pelusios sinatus* is a living species that is abundantly represented by several essentially complete shells and very large quantities of disarticulated shell fragments. In contrast, *Latisternon microsulcae* was described as a new genus and species based only on two isolated bones, a left epiplastron (the type specimen), and a nuchal bone not associated with it. Both of the tortoises were recognized on the basis of shell fragments alone. As at Omo, therefore, four different kinds of turtles have been reported. But only one of them (*Pelusios sinuatus*) can be well characterized. *P. sinuatus* is the only chelonian taxon common to both Omo and Olduvai.

Koobi Fora is comparable to Lothagam in terms of its chelonian taxonomic diversity, but not in terms of the quantity or quality of preservation of its fossil turtle remains. The modern side-necked turtles *Pelusios adansonii* and *Pelusios sinuatus* are represented by several largely complete shells. The presence of two soft-shelled turtle species, *Cycloderma elegans* and *Cycloderma frenatum*, can also be reasonably well documented. Both of these, too, are extant species. Two additional taxa of soft-shelled turtles—*Cyclanorbis senegalensis* (another living species) and a trionychine—have also been reported as components of the Koobi Fora turtle fauna (Meylan et al. 1990). Each of these, however, is represented only by two or three shell fragments. The undoubted presence at Koobi Fora of the genus *Trionyx*, and perhaps even the living species *T. triunguis*, is established by a fairly complete shell (Wood 1979). As noted previously, giant tortoise remains have been discovered though not yet described at Koobi Fora.

Fossil turtles have undoubtedly been recovered from at least some of the various Ethiopian Rift Valley localities (e.g., de Heinzelin et al. 1999), but none have so far been described, so comparisons with these sites are not yet possible. Miocene vertebrate localities elsewhere in Africa have yielded some interesting chelonian material, but none of these occurrences begins to compare with Lothagam in terms of diversity and abundance. At Kachuku in western Kenya, remains of a soft-shelled turtle (*Cycloderma victoriae*) and a giant tortoise, along with a juvenile pelomedusid reminiscent of *Erymnochelys* or *Turkanemys*, have been discovered (Andrews 1914). The nearby islands of Rusinga and Maboko in Lake Victoria have each yielded a single complete shell that is virtually indistinguishable from that of *Turkanemys*. Associated with the Rusinga shell is an excellent skull (currently under study by L. Witmer). The Early Miocene sediments of Rusinga have also provided the oldest record of the living genus *Pelusios, P. rusingae* (Williams 1954b).

Mid- to Late Tertiary fossil localities in the Western Rift Valley have produced, for the most part, only fragmentary remains of turtles (de Broin and Gmira 1994; Meylan 1990; Swinton 1926). The one notable exception is provided by pelomedusid remains that have been collected from the Sinda Basin (Western Rift Valley of

Zaire). Most of these are disarticulated, unassociated individual bones, but several partial and nearly complete shells have also been found. These indicate the presence of a form whose shell is very similar to that of the living Madagascan species *Erymnochelys madagascariensis* (Hirayama 1992). Several partial skulls further reinforce this striking similarity (Wood and Gawlas in preparation). Also recovered from here is a smattering of trionychid and testudinid fragments that are sufficient to establish the presence of these families, but are otherwise undiagnostic.

Further afield in Africa, the Sahabi Formation of Libya has produced both soft-shelled turtles (*Trionyx triunguis*) and tortoises (*Geochelone*), neither in great abundance (Wood 1987). A complete carapace, partial skull, and diagnostic plastral fragments leave little doubt that the Sahabi soft-shelled turtle is the earliest confirmed record of *Trionyx.*

As noted previously, the Lothagam turtles represent a mixture of extinct and modern forms. *Turkanemys pattersoni* is the last known representative on the continent of Africa of a lineage stretching back to the Early Tertiary of North Africa. Earlier occurrences of this lineage include the Fayum of Egypt, Rusinga and Maboko Islands in Kenya, and the Sinda Basin of Zaire. The sole surviving representative of this group is *Erymnochelys madagascariensis* on Madagascar.

Giant tortoises survived in Africa until more recent times, being known from Laetoli, Koobi Fora, Omo, and Olduvai. But they, too, eventually disappeared from the African fauna and today survive only on isolated oceanic islands.

Soft-shelled turtles appear to have been as diverse at Lothagam as anywhere in Africa today, with at least three contemporaneous species (*Cycloderma debroinae, C. frenatum,* and a trionychine undiagnosable at the generic level). The Lothagam occurrence of *C. frenatum* is the earliest known for this modern species, which, however, does not encompass the Lake Turkana Basin within its present range. A single trionychine species, *Trionyx triunguis,* is the only living soft-shelled turtle in Lake Turkana today. This species has at present a widespread geographic distribution across sub-Saharan Africa and along the Nile River drainage. It is the only member of the subfamily Trionychinae found in Africa today. The occurrence of trionychines at Lothagam is not surprising, given the fact that they are also known in the Miocene of Sahabi, Libya. Not enough is known about the habitat of the modern African soft-shelled turtles to be able to explain why trionychines eventually supplanted cyclanorbines in the Lake Turkana Basin.

The paleoenvironment of Lothagam has been interpreted as primarily riverine with adjacent swamps, ponds, and fringing gallery forests (Leakey et al. 1996). The fossil turtles of Lothagam are consistent with this scenario. Pelomedusids and trionychids are aquatic forms. The giant tortoises were certainly terrestrial. Both of the modern tortoise species in northern South America (*Geochelone carbonaria* and *G. denticulata*) reportedly prefer forested areas (Pritchard and Trebbau 1994). Perhaps the Lothagam tortoises may also have been primarily denizens of the gallery forests that lined the riverbanks and surrounding backwater areas. The proximity of this type of habitat to bodies of water where sediments were routinely accumulating may well account for the relative abundance of *Geochelone* specimens recovered from Lothagam.

Soft-shelled turtles are much less common than either pelomedusids or testudinids at Lothagam. Their remains are more common in the Upper Nawata than in other members. Tortoises are equally abundant in the Upper and Lower Nawata and in the Apak Member.

Perhaps the single most remarkable component of the Lothagam fossil turtle fauna is *Turkanemys pattersoni.* This pelomedusid represents the largest and best-preserved sample of any fossil turtle species yet discovered anywhere in Africa. The reproductive behavior of a living South American pelomedusid may help us understand the phenomenal abundance of *T. pattersoni.* In the Amazon and Orinoco River systems, large populations of the giant side-necked river turtle *Podocnemis expansa* are known to concentrate during the nesting season (which coincides with the height of the annual dry season) around certain extensively exposed sandbar islands. So intensive is the nesting activity at these sites that it is not unusual for previously laid nests to be inadvertently dug up by the subsequent nesters. With the onset of the rainy season, these concentrated populations become widely dispersed.

The characteristics of the *T. pattersoni* sample at Lothagam fit this model. These specimens may well represent females who perished during the annual nesting season. All the shells are of large and relatively uniform size, indicating a population of adults. Large numbers of this species may have been irresistibly attracted to a favored nesting site in the ancient river that deposited the Lothagam sediments. Rapid burial of skeletons in the waters adjacent to this presumed nesting site would account for the generally excellent preservation of so many specimens. If this interpretation of the evidence is correct, the Lothagam sediments may have accumulated over a series of sequential dry seasons.

Acknowledgments

I am grateful to the late Professor Bryan Patterson for affording me the opportunity to be a member of Harvard University paleontological expeditions to northern Kenya in the summers of 1965, 1966, and 1967. These expeditions were funded by National Science Foundation grants to Professor Patterson (GA 425 and GA 1188). I also wish to thank Professor E. E. Williams of Harvard's Museum of Comparative Zoology for his advice on and encouragement of my studies of African turtles.

Two grants from the National Geographic Society enabled me to study living pelomedusid turtles in East Africa for seven months in 1967–1968 and subsequently to visit European natural history museums for the purpose of examining both fossil and recent turtles from Africa. Additional grant support has been provided by faculty research grants from Richard Stockton College of New Jersey.

I am greatly indebted to the editors of this volume for their invitation to contribute to it and for their long-suffering patience. They have been most helpful in furnishing information about fossil turtle specimens in the collections of the National Museums of Kenya in Nairobi. I would also like to thank Bob Campbell for the photograph in Figures 4.11 and 4.13. Mary Muungu, Kyalo Manthi, Samual Ngui, and Ngala Jillani are thanked for assistance in locating and measuring Lothagam specimens. Dr. Monte McCrossin kindly provided me with photographs of the Maboko Island fossil pelomedusid shell. In addition, I am most appreciative to the curators (too numerous to mention individually) at the many museums I have visited, not only for their gracious hospitality but also for access to and information about the collections under their care.

I am much obliged to my colleagues Drs. Gene Gaffney and Peter Pritchard for critically reviewing this manuscript. Preparation of the type specimen of *Turkanemys pattersoni, Kenyemys williamsi,* and *Cycloderma debroinae* was skillfully carried out by Arnie Lewis. Illustrations of shell reconstructions for both *Turkanemys pattersoni* and *Kenyemys williamsi* were prepared by Lazlo Meszoly. Photographs of the skull of *T. pattersoni* were generously furnished by Dr. Gene Gaffney. I also thank Diane Baxter-Daly and Caralyn Zehnder for their considerable help with preparation of this manuscript.

References Cited

Andrews, C. W. 1906. *A Descriptive Catalog of the Tertiary Vertebrata of the Fayum, Egypt.* London: British Museum (Natural History).

Andrews, C. W. 1914. On the lower Miocene vertebrates from British East Africa, collected by Dr. Felix Oswald. *Journal of the Geological Society* (London) 70:163–186.

Arambourg, C. 1947. Contribution à l'étude géologique et paléontologique du bassin du lac Rudolphe et de la Basse Vallée de l'Omo. In *Mission scientifique de l'Omo (1932–1933).* Vol. 1, fasc. 3. *Paléontologie,* pp. 231–562. Paris: Muséum National d'Histoire Naturelle.

Auffenberg, W. 1981. The fossil turtles of Olduvai Gorge, Tanzania, Africa. *Copeia* 3:509–522.

Behrensmeyer, A. K. 1976. Lothagam Hill, Kanapoi, and Ekora: A general summary of stratigraphy and fauna. In Y. Coppens, F. C. Howell, G. L. Isaac, and R. E. Leakey, eds., *Earliest Man and Environments in the Lake Rudolf Basin: Stratigraphy, Paleoecology, and Evolution,* pp. 163–172. Chicago: University of Chicago Press.

Boycott, R. C., and O. Bourquin. 1988. *The South African Tortoise Book: A Guide to South African Tortoises, Terrapins and Turtles.* Johannesburg: Southern Book Publishers.

de Broin, F. L. 1969. Sur la présence d'une tortue, *Pelusios sinuatus* (A. Smith) au Villafranchian inférieur du Tchad. *Bulletin de la Société Géologique de France* 7:909–916.

de Broin, F. L. 1979. Chéloniens du Miocène d'Afrique orientale. *Bulletin de la Société Géologique de France* 21:323–327.

de Broin, F. L., and S. Gmira. 1994. Les chélonians dulçiaquicoles du rift occidental, Ouganda. In B. Senut and M. Pickford, eds., *Geology and Palaeobiology of the Albertine Rift Valley, Uganda-Zaire.* Vol. 2. *Palaeobiology/Paléobiologie,* pp. 157–186. Occasional Publication No. 29. Orléans: Centre International pour la Formation et les Echanges Géologiques.

de Broin, F. L., and C. Werner. 1998. New Late Cretaceous turtles from the western desert, Egypt. *Annales de Paléontologie* 84:131–214.

de Heinzelin, J. J., D. Clark, T. White, W. Hart, P. Renne, G. WoldeGabriel, Y. Beyene, and E. Vrba. 1999. Environment and behavior of 2.5-million-year-old Bouri hominids. *Science* 284:625–629.

Hirayama, R. 1992. Fossil turtles from the Neogene strata in the Sinda Basin, eastern Zaire. *African Study Monographs,* Supplementary issue 17:49–65.

Leakey, M. G., C. S. Feibel, R. L. Bernor, J. M. Harris, T. E. Cerling, K. M. Stewart, G. W. Storrs, A. Walker, L. Werdelin, and A. J. Winkler. 1996. Lothagam: A record of faunal change in the Late Miocene of East Africa. *Journal of Vertebrate Paleontology* 16:556–570.

Loveridge, A., and E. E. Williams. 1957. Revision of the African tortoises and turtles of the suborder Cryptodira. *Bulletin of the Museum of Comparative Zoology* 115:163–557.

Meylan, P. A. 1990. Fossil turtles from the Upper Semliki, Zaire. In N. T. Boaz, ed., *Evolution of Environments and Hominidae in the African Western Rift Valley,* pp. 163–170. Memoir No. 1. Martinsville: Virginia Museum of Natural History.

Meylan, P. A., and W. Auffenberg. 1986a. New land tortoises (Testudines: Testudinidae) from the Miocene of Africa. *Zoological Journal of the Linnean Society* 86:279–307.

Meylan, P. A., and W. Auffenberg. 1986b. The cheloniams of the Laetolil Beds. In M. D. Leakey and J. M. Harris, eds., *Laetoli: A Pliocene Site in Northern Tanzania,* pp. 62–78. Oxford: Clarendon Press.

Meylan, P. A., B. S. Weig, and R. C. Wood. 1990. Fossil soft-shelled turtles (family Trionychidae) of the Lake Turkana Basin, Africa. *Copeia* 2:508–528.

Patterson, B., A. K. Behrensmeyer, and W. D. Sill. 1970. Geology of a new Pliocene locality in northwestern Kenya. *Nature* 256:279–284.

Pritchard, P. C. H., and P. Trebbau. 1984. *The Turtles of Venezuela.* Contributions to Herpetology 2. Athens, Ohio: Society for the Study of Amphibians and Reptiles.

Stromer, E. V. 1926. Rest von land- und susswasserbewohnender Wirbeltiere aus dem Diamentfeldern Deutsch Sudwestafrikas. In E. Kaiser, ed., *Die Diamentenwuste Sudwestafrikas,* pp. 139–142. Berlin: Reiner.

Swinton, W. E. 1926. Part II: Fossil Reptilia. In E. J. Wayland, ed., *The Geology and Palaeontology of the Kaiso Bone Beds,* pp. 37–44. Occasional Papers 2. Kampala: Geological Survey of Uganda.

Williams, E. E. 1950. Variation and selection in the cervical central articulations of living turtles. *Bulletin of the American Museum of Natural History* 94:505–562.

Williams, E. E. 1954a. A key and description of the living species of the genus *Podocnemis* (sensu Boulenger) (Testudines, Pelomedusidae). *Bulletin of the Museum of Comparative Zoology* 111:279–295.

Williams, E. E. 1954b. A new Miocene species of *Pelusios* and the evolution of that genus. *Breviora* 25:1–7.

Wood, R. C. 1970. A review of the fossil Pelomedusidae (Testudines, Pleurodira) of Asia. *Breviora* 357:1–24.

Wood, R. C. 1971. The fossil Pelomedusidae (Testudines, Pleurodira) of Africa. Ph.D. diss., Harvard University.

Wood, R. C. 1973. A possible correlation between the ecology of living Africa pelomedusid turtles and their relative abundance in the fossil record. *Copeia* 3:627–629.

Wood, R. C. 1976. An enigmatic chelonian fragment from the Pliocene of Kenya. *Copeia* 3:589–591.

Wood, R. C. 1979. First record of a fossil trionychid skull from Africa. *Herpetologica* 35:360–364.

Wood, R. C. 1983. *Kenymys williamsi,* a fossil pelomedusid turtle from the Pliocene of Kenya. In A. J. G. Rhodin and K. Miyata, eds., *Advances in Herpetology and Evolutionary Biology: Essays in Honor of Ernest E. Williams,* pp. 74–85. Cambridge, Mass.: Museum of Comparative Zoology, Harvard University.

Wood, R. C. 1984. Evolution of the pelomedusid turtles. *Studia Geologia Salmanticensia* 1:269–282.

Wood, R. C. 1987. Fossil turtles from the Sahabi Formation. In N. T. Boaz, A. El-Arnauti, A. W. Gaziry, J. de Heinzelin, and D. D. Boaz, eds., *Neogene Paleontology and Geology of Sahabi,* pp. 107–112. New York: Liss.

Wood, R. C., and M. L. Diaz de Gamero. 1971. *Podocnemis venezuelensis,* a new fossil pelomedusid (Testudines, Pleurodira) from the Pliocene of Venezuela and a review of the history of *Podocnemis* in South America. *Breviora* 376:1–23.

TABLE 4.1 Comparisons of the Proportions of the Nuchal Bones (Maximum Width vs. Maximum Length) in Living and Fossil Examples of the *Erymnochelys* Lineage

Species	Specimen No.[a]	Width[c] (mm)	Length[c] (mm)	W/L ratio (%)	Average W/L Ratio for Each Species (%)
E. madagascariensis	MCZH 5198	48.0	48.0	100.0	100
Pod. fajumensis	AMNH 5087	35.0	35.2	99.0	—
Pod. fajumensis	YPM 6202	43.4	40.6	107.0	111
Pod. fajumensis	YPM 6203	36.6	28.9	127.0	—
Pod. aegyptiaca[b]	CGM, no catalog number	45.0	44.0	102.0	115
Pod. sp. cf. *P. aegyptiaca*	BU 6416	77.9	61.2	127.0	—
K. williamsi	KNM-LT 127	62.0	51.0	122.0	122
T. pattersoni	KNM-LT 426	(101.0)	71.3	142.0	—
T. pattersoni	KNM-LT 428	82.2	(60.0)	137.0	—
T. pattersoni	KNM-LT 431	90.0	70.0	129.0	131
T. pattersoni	KNM-LT 569 (type)	75.6	56.9	133.0	—
T. pattersoni	KNM-LT 23981	82.0	64.1	116.0	—
T. pattersoni	KNM-LT 23985	82.0	(70.5)	131.0	—

[a] MCZH = Museum of Comparative Zoology, Harvard University; AMNH = American Museum of Natural History; YPM = Peabody Museum, Yale University; CGM = Cairo Geological Museum, Egypt; BU = Department of Geology, University of Bristol; KNM-LT = specimen from Lothagam in National Museums of Kenya collections.

[b] For this specimen, the dimensions are approximate and have been estimated using measurements taken from Fourtau 1920: figure 21.

[c] (measurement) = estimated measurement

TABLE 4.2 Shell Measurements for Specimens of *Turkanemys pattersoni* gen. and sp. nov.

	Length (cm)	
Specimen No.	Carapace	Plastron
KNM-LT 428	40.2	36.0
KNM-LT 431	45.0	(41.0)
KNM-LT 438	—	40.3
KNM-LT 565	—	40.0
KNM-LT 568	—	38.0
KNM-LT 569	37.8	33.0
KNM-LT 571	—	40.7
KNM-LT 23178	(40.0)	36.6
KNM-LT 23981	—	37.5
KNM-LT 23984	—	39.9
KNM-LT 23984	46.0	41.0
KNM-LT 23470	45.0	39.0
KNM-LT 23512	—	38.7
KNM-LT 23516	—	(38.0)

TABLE 4.3 Comparisons of Various Features of the Skulls of *Erymnochelys madagascariensis, Turkanemys pattersoni* gen. and sp. nov., and the Living South American Species of *Podocnemis* (Exclusive of the Somewhat Aberrant *Peltocephalus dumerilianus,* Long Considered a Member of *Podocnemis*)

Character	***E. madagascariensis***	***T. pattersoni***	**South American Forms**
Extent of emargination in cheek region	Slight	Considerable	Considerable
General proportions of skull	Moderately broad and squat	Narrow and elongate	Narrow and elongate
Quadrate meets jugal	Yes	No	No
Shape of interparietal scute	Triangular	Trapeziform	Variable
Midline depression between orbits	No	No	Yes
Hook at median symphysis of upper jaw	Yes	No	No
Number of triturating ridges on maxillary	1	2	1–3
Maxillae meet at midline behind premaxillae on palatal surface	No	No	Yes, only for some species
U-shaped depression at posterior end of palatal symphysis	Yes	No	Yes
Narial opening higher than broad	No	Yes	No
Median triturating ridge present on biting surface of mandible	No	Yes	Yes, for most species
Broad biting surface at mandibular symphysis	No	Yes	Intermediate
Dimensions of mandibular masticating trough	Narrowest at mandibular symphysis and broadens continuously toward crest of coronoid process	Widest at mandibular symphysis and narrows continuously toward crest of coronoid process	Of approximately uniform breadth along entire length

4.2

Late Miocene–Early Pliocene Crocodilian Fauna of Lothagam, Southwest Turkana Basin, Kenya

Glenn W. Storrs

A remarkably diverse fauna of crocodilians comprised of four genera and five species documents a high degree of ecological niche partitioning in the Late Miocene–Early Pliocene aquatic paleoenvironment of Lothagam, Kenya. The first records of the extant species *Crocodylus niloticus* and *C. cataphractus* are from Lothagam. Until the Quaternary, the dominant crocodilian species at Lothagam and elsewhere in eastern Africa was the giant brevirostrine, here established as a new genus (*Rimasuchus lloydi*). A new species of *Eogavialis* from Lothagam is distinct from the earliest known gavialids (from the Fayum Basin Paleogene of Egypt) but plesiomorphic relative to *Gavialis* of the Indian subcontinent. The Lothagam *Eogavialis* specimens (and others newly recognized from the Early Miocene of Loperot, southwest Turkana Basin) are some of the few records of undoubted gavialids from the Miocene and to date the only ones from East Africa. Well-preserved specimens of the distinctive longirostrine *Euthecodon* at Lothagam confirm a relationship of this taxon with the Crocodylidae.

The Late Miocene–Early Pliocene East African locality of Lothagam in the southwestern portion of the Turkana Basin, first exploited by Patterson and others (Patterson et al. 1970; Behrensmeyer 1976; Smart 1976; Leakey et al. 1996), preserves an important and unusually diverse assemblage of Neogene eusuchian-grade crocodilians (Crocodylia *sensu* Benton and Clark 1988). Previous studies of the fossil crocodilians of East Africa have focused on those from the Early Miocene of the Victoria Basin, for example Rusinga Island (Tchernov and Van Couvering 1978), and on those of the Plio-Pleistocene of the north and eastern Turkana Basin (Tchernov 1976, 1986). However, many of the specimens in these studies were not described in detail. The Lothagam sample provides a new opportunity to elucidate the anatomy, taxonomic diversity, and evolutionary history of East African Crocodylia.

Despite their general abundance as fossils, the history and relationships of East African crocodilians are poorly known. A substantial part of this situation can be blamed, not so much on the lack of good quality specimens, as on the historical collecting bias of most previous workers that favored mammalian, and particularly primate, faunas. Another reason for our ignorance is the remarkable conservatism of eusuchian crocodilians in general, hence the few identified characters suitable for the construction of rigorous phylogenetic hypotheses. Indeed, much observed variation in African Neogene crocodilians is of proportion, generally in the length and breadth of the rostrum. Even where such proportional differences are important, they are difficult to quantify in a meaningful way. Whereas Dodson (1975), Kälin (1933), Kramer and Medem (1956), and Iordansky (1973) have produced cranial indices for living crocodilians, and Tchernov (1976, 1986) and Pickford (1996) have followed with data for African fossils, the results are ontogenetically and functionally dependent, and are thus rarely useful for phylogenetic analysis.

Current sample sizes for East African fossil taxa also provide little opportunity for insight into individual and ontogenetic variation within and between populations. As in living crocodilians, the sutural relationships of the cranial bones may vary. So too may the number of teeth in the jaws. Tooth number has certainly increased independently as a response to rostral elongation in a variety of lineages as suggested, in part, by Tchernov (1986). Like extant forms, the postcrania of most species, where known, are little different from one another (although known gavialids differ considerably from other taxa). All of these problems have hampered historical studies.

Two recent morphological studies of crown group Crocodylia that provide discrete cranial characters for analysis are those of Brochu (1997) and Norell (1989). Only a single study that included East African crocodilian fossils within a quantitative cladistic framework has been undertaken (Brochu 1997), but East African material was not its primary focus. Although fossil material of most African crocodilians is too limited for detailed study of variation, an attempt can be made to document characters that may be significant for the Lothagam fossils. In particular, some discussion of Norell's (1989) characters and examination of the relevant portions of Brochu's (1997) analysis is made here.

Sediments of the Lothagam Group, estimated to range from 5 to 7 million years old on the basis of biostratigraphic evidence (Leakey et al. 1996), have now been dated radiometrically and determined to have been deposited during the Late Miocene and Early Pliocene (McDougall and Feibel 1999). Within the Lothagam Group, fossil crocodiles have mostly been recovered from the lower and upper members of the Nawata Formation, but others are known from the basal Apak Member of the overlying Nachukui Formation, and several important specimens have originated within the younger Kaiyumung Member. These deposits represent fluvial, floodplain, and associated lacustrine paleoenvironments in the drainage area of ancient Lake Turkana (Feibel this volume; Leakey et al. 1996). The fauna and flora were diverse and abundant. The crocodilian materials, like most of the fossils from the site, are generally in a good state of preservation, and some are quite remarkable. The material studied here is housed in the National Museums of Kenya, Palaeontology Division, Nairobi (KNM), unless otherwise indicated (ER = Koobi Fora accession series; LP = Loperot accession series; LT = Lothagam accession series).

Systematic Description

Whereas a single crocodilian species (*Crocodylus niloticus*) now dominates East Africa (with a small relict population of *C. cataphractus* at Lake Tanganyika), at least five species occupied the region during the Late Miocene as demonstrated by the Lothagam locality. At least three species coexisted during the Early Pliocene. All appear to have been competitively excluded from one another through the occupation of specialized ecological niches. Such niche partitioning was apparently facilitated by the extremely rich range of prey. The five crocodilian species are readily distinguished on the basis of skull morphology (the few associated postcrania are undiagnostic).

Crocodylus Laurenti, 1768
Crocodylus niloticus Laurenti, 1768
(Figures 4.14a, 4.15–4.17)

Diagnosis

Moderate- to large-sized extant crocodylid with generalized rostrum of moderate proportion, median nasal promontorium, typically 14 maxillary and 15 mandibular teeth. Anterior nuchal osteoderms well developed.

Lothagam Material

- *Lower Nawata:* 23108, skull.
- *Upper Nawata:* 24027, cranial fragments; 24029, Rt. mandible fragment; 26618, skull.
- *Apak Member:* 24146, Rt. mandible.

Figure 4.14a Restoration of brevirostrine crocodiles from Lothagam by Mauricio Antón: top = *Crocodylus niloticus;* center = *Crocodylus cataphractus;* bottom = *Rimasuchus lloydi* gen. nov.

Figure 4.14b Restoration of longirostrine crocodiles from Lothagam by Mauricio Antón: top = *Eogavialis andrewsi* sp. nov.; bottom = *Euthecodon brumpti.*

Lothagam provides the earliest known record of *C. niloticus,* the dominant crocodilian of contemporary East Africa. To date, it is known only from the lower and upper members of the Nawata Formation and the Apak Member of the Nachukui Formation. In spite of the familiar status and well-documented anatomy of *C. niloticus,* little attention has been paid to the fossil history of this species. Tchernov (1976) suggested that the fossil *C. niloticus* had relatively shorter snouts than those of extant populations. However, a nearly perfect, undistorted skull of *C. niloticus,* LT 23108 from the Lower Nawata, is virtually indistinguishable from modern representatives, and its rostrum falls well within the range of variation of recent examples (figures 4.15 and 4.16). This fossil preserves the distinctive

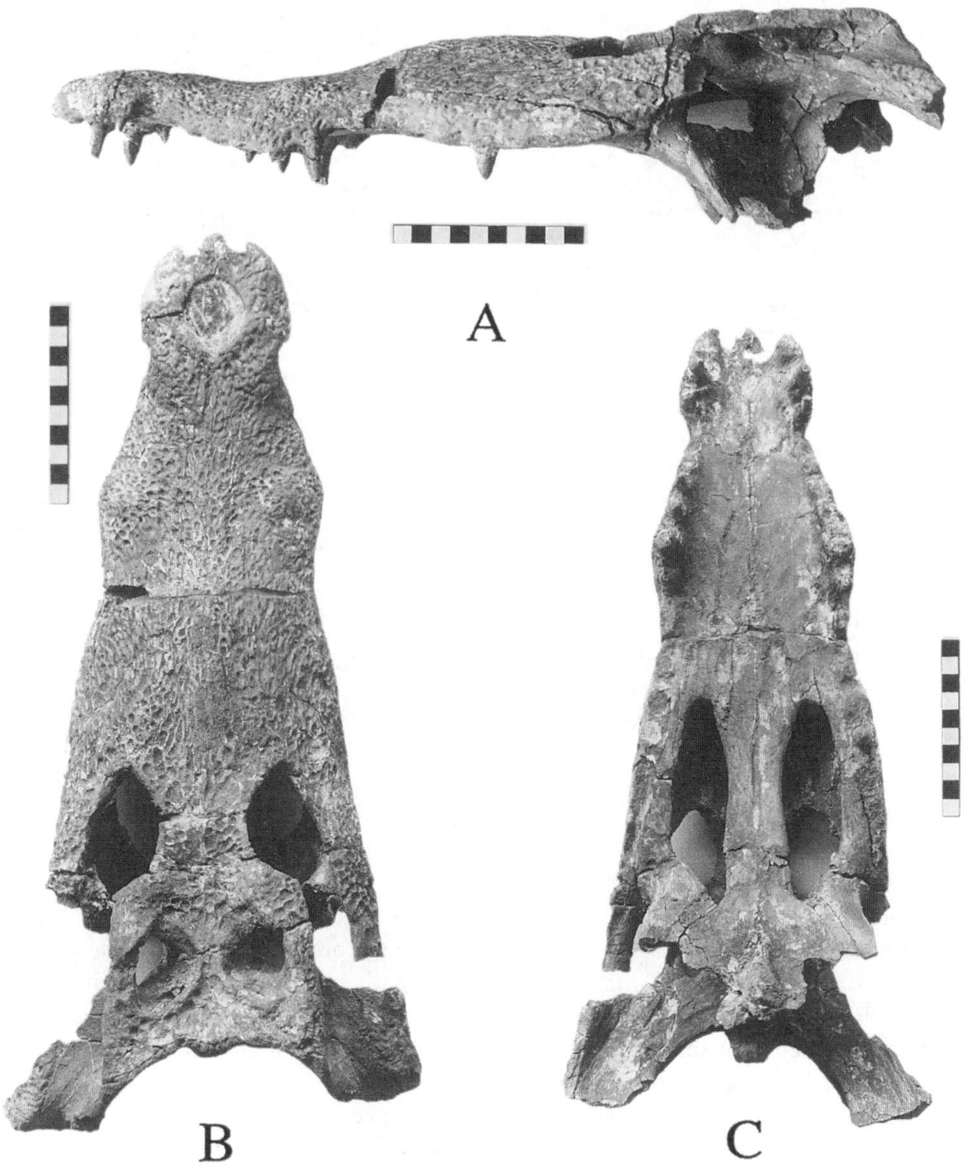

Figure 4.15 Skull of *Crocodylus niloticus,* KNM-LT 23108: A = left lateral aspect; B = dorsal aspect; C = ventral aspect. Scale bar equals 100 mm.

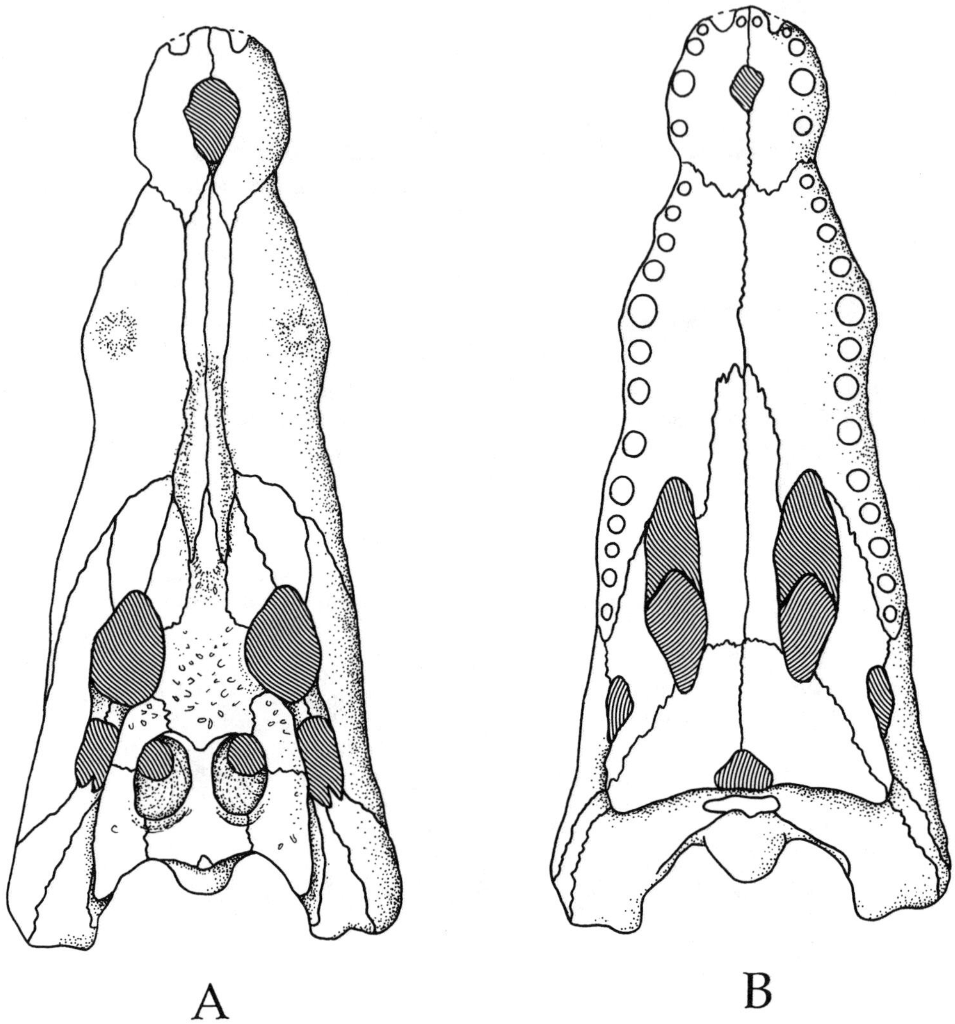

Figure 4.16 Skull of *Crocodylus niloticus,* KNM-LT 23108, restored: A = dorsal aspect; B = ventral aspect.

longitudinal nasal ridge ("preorbital promontorium" of Hecht 1987) of living *C. niloticus.* This rounded ridge gives the animal the appearance in profile of having a "Roman nose" and is lacking from other Lothagam *Crocodylus* species. LT 26618, from the Upper Nawata, is another very good *C. niloticus* skull missing only the anterior half of the rostrum; it also exhibits the prominent nasal ridge. The Early Miocene *C. pigotti* of Rusinga Island (Tchernov and Van Couvering 1978), lacks a nasal promontorium and is flat-faced by comparison. The promontorium appears to be a derived feature that links *C. niloticus* with New World species of *Crocodylus* (*C. acutus, C. intermedius, C. moreletii, C. rhombifer*). It is not present in any out-group taxon (although perhaps it is independently derived in some examples of *Euthecodon*). Brochu (1997) links *C. niloticus* and *C. rhombifer* as the most derived taxa in his analysis, although not all extant taxa were included. Subsequent work (Brochu personal communication 1999) links *C. niloticus* with all New World *Crocodylus.*

Hecht (1987) similarly discusses the significance of the nasal promontorium, which was constructed from a thickening of the nasals, frontals, and adjacent antorbital elements, with regard to the extant New World species *C. rhombifer, C. moreletii,* and the relatively longirostrine *C. acutus.* Hecht (1987) noted its presence also in *C. "checchiai"* from the Early Pliocene of Sahabi, Libya (Boaz 1982; Hecht 1987; Maccagno 1948, 1952). While Hecht (1987) acknowledged a nasal promontorium in *C. niloticus,* he failed to appreciate the frequency of its development, stating it to be "rare." Although the promontoria of *C. "checchiai"* and LT 23108 are slightly more pronounced than those of most modern *C. niloticus,* no other features distinguish them from the extant population. In fact, other, sometimes larger, specimens in the Lothagam collection display less well pronounced promontoria, and a certain amount of intraspecific or ontogenetic variation in this character is to be expected.

Kälin (1933) describes a wide degree of variation in rostrum shape and sculpturing in *C. niloticus* (*vulgaris*).

Some apparent examples of *C. niloticus* from Lothagam (e.g., LT 24027) are very large, approaching the dimensions of "*Crocodylus*" *lloydi,* discussed below (e.g., LT 26465), and have rather broad rostra, although these still fall within the range of variation for extant *C. niloticus,* especially for large body size. Tchernov (1986) may have mistaken some very large examples of *C. niloticus* for his concept of relatively narrow-snouted variants of "*Crocodylus*" *lloydi.* This could partially explain the reported absence of *C. niloticus* at Koobi Fora (Feibel et al. 1991; Tchernov 1976, 1986).

LT 23108 is a relatively small skull, about 400 mm from the tip of the snout to the end of the supraoccipital (~435 mm to the posterior ends of the quadrates) and approximately 135 mm across the skull at a position just in front of the orbits. Only minor sutural variation, normal among individuals in a population (e.g., Kälin 1933), distinguishes LT 23108 from a typical example of extant *C. niloticus.* For example, the anteriormost tips of the nasals do not enter the external nares on the dorsal surface of the fossil, but are overgrown by the premaxillae. The nasals are slightly constricted transversely at mid length, a condition sometimes seen in extant *C. niloticus.* The lack of lateral skull table ridges or "horns" on the dorsolateral corners of the squamosals in the Lothagam fossil, correlated with large size in living *C. niloticus,* probably indicates relative youth.

The positions of the palatal sutures, a trait given great weight by Mook (1921), Tchernov (1986), and Tchernov and Van Couvering (1978), are essentially identical in *C. "checchiai,"* LT 23108, LT 26618, and extant *C. niloticus.* Although Tchernov (1986) and Tchernov and Van Couvering (1978) emphasize differences in palatal suture pattern and relative tooth position between various East African crocodilian taxa, there is more variation in these patterns in extant *C. niloticus* than they recognize. The range of variation in the latter easily incorporates the pattern found in LT 23108. Sutures may fall at or between teeth and thus Tchernov's (1986) strict tooth/suture position chart (see also Tchernov and Van Couvering 1978) cannot apply. A sample of only 15 extant *C. niloticus* in the Kenyan National Museum shows that the maxillary/palatine suture may vary widely from a position in front of the 6th maxillary tooth to the back of the 7th. Thus, the position of this suture is not a useful character for differentiation of species.

Figure 4.17 Right mandibular rami: A = *Crocodylus niloticus,* KNM-LT 24146, lateral aspect; B = recent *Crocodylus niloticus,* lateral aspect; C = *Rimasuchus lloydi,* gen. nov., KNM-LT 22966, lateral aspect. Scale bars equal 100 mm.

There is more constancy in the premaxillary/maxillary palatal suture of *C. niloticus* that normally extends to between the 1st and 2nd maxillary teeth, as in LT 23108 (*contra* Tchernov and Van Couvering 1978 and Tchernov 1986). Even the forward limit of the palatal fenestrae is variable in *C. niloticus* and may lie at the back of the 8th maxillary tooth, although it is generally alongside the 9th; it lies level with the 9th maxillary tooth in LT 23108. For further discussion of the significance of sutural position in Recent taxa, see Kälin (1933), who examined a large sample of *C. niloticus* (*vulgaris*) for individual variation.

Clear anterior occlusion pits or inferior cavities have been worn in the premaxillae for the reception of the 1st mandibular teeth of LT 23108. These pierced the upper jaw just as they do in many modern *C. niloticus* individuals (Kälin 1933). Such perforations are not always present, but those of LT 23108 are more pronounced than in most other *C. niloticus* specimens from Lothagam in becoming notches; the perforations in the specimen identified by Hecht (1987) as *C. "checchiai"* are identical to those of LT 23108. Tchernov (1986) states that the occlusion pits of *"Crocodylus" lloydi* are not directed dorsoventrally as in *C. niloticus* but are more anteroposterior in orientation, presumably a result of its considerably foreshortened premaxillae. The 1st premaxillary teeth appear to have been lost in LT 23108 as in the specimen of *C. "checchiai"* described by Hecht (1987). The small 1st or 2nd teeth of extant *Crocodylus* may be alternately forced out of the mouth during ontogeny. Earlier in life, there would surely have been five teeth in each premaxilla of LT 23108 as in virtually all other crocodilians (but not in *Euthecodon;* see the following discussion).

A prominent maxillary boss or protuberance on the dorsal surface of the snout occurs near each large 5th maxillary tooth as in virtually all species of *Crocodylus* including *C. niloticus* and *C. "checchiai," C. cataphractus,* and *C. pigotti.* In LT 23108, as in extant *C. niloticus,* the boss is positioned between the 5th and 6th teeth. The tooth row margins of the maxillae of LT 23108 are festooned in typical *Crocodylus* fashion. *Crocodylus "checchiai"* apparently had 14 teeth in each maxilla (although Hecht 1987 suggests that it has "about 15"), while LT 23108 had 13 or 14. Extant *C. niloticus* generally have 13 maxillary teeth, although 14 may sometimes be present. LT 23108, LT 26618, and *C. "checchiai"* are surely part of the lineage leading to extant *C. niloticus.* While perhaps genetically distinct from the living population, as it is impossible to distinguish between them morphologically, all are best considered to be conspecific with *C. niloticus.* Buffetaut (1984, 1985) considers *C. pigotti* to be a suitable "ancestor" to *C. niloticus.*

A right mandibular ramus of *C. niloticus* from Lothagam (LT 24146), like the upper tooth row of LT 23108, displays prominent festoons and is little different from extant examples (figure 4.17). There are four mandibular teeth in the symphysis, and it is estimated that 15 teeth were present in the ramus (the coronoid eminence and posterior end of the tooth row are broken), as in extant *C. niloticus.* Tchernov (1976) proposed that a higher mandible was characteristic of fossil *C. niloticus* specimens, but this is not apparent in LT 24146.

Crocodylus cataphractus Cuvier, 1824

(Figures 4.14a, 4.18, 4.19)

Diagnosis

Moderate-sized extant crocodylid with slightly constricted "piscivorous" rostrum with gently concave profile, no preorbital ridges, typically 13 maxillary and 15 or 16 mandibular teeth. Anterior nuchal osteoderms continuous with dorsal series.

Lothagam Material

- *Lower Nawata:* 23104, cranium and mandible.

A single, well-preserved specimen of *C. cataphractus,* LT 23104 from the Lower Nawata, demonstrates the unequivocal presence of this species at Lothagam and (like the *C. niloticus* specimens) represents the earliest occurrence of this species in the fossil record (Leakey et al. 1996). The specimen (figures 4.18 and 4.19) is of moderate size (approximately 510 mm in total length, about 470 mm from the tip of the rostrum to the end of the supraoccipital, and nearly 160 mm broad across the front of the orbits).

The fossil is closely similar to its extant counterpart, and little additional description is required. Both are easily recognized by a moderately longirostrine morphotype and a concave rostral profile with an upturned rostral tip. The region immediately anterior to the orbits is broad and flat; the face then slopes ventrally to the approximate level of the 6th or 7th maxillary teeth whence it is recurved upward. There is no nasal promontorium. The external nares lie upon a raised premaxillary platform, well above the lowest part of the dorsal surface of the snout. This morphology contrasts markedly with the rostrum of *C. niloticus* which, as a result of its nasal ridge, is broadly convex in profile with external nares that are not significantly raised relative to the remainder of the snout. The nasals, prefrontals, and lachrymals are longer and

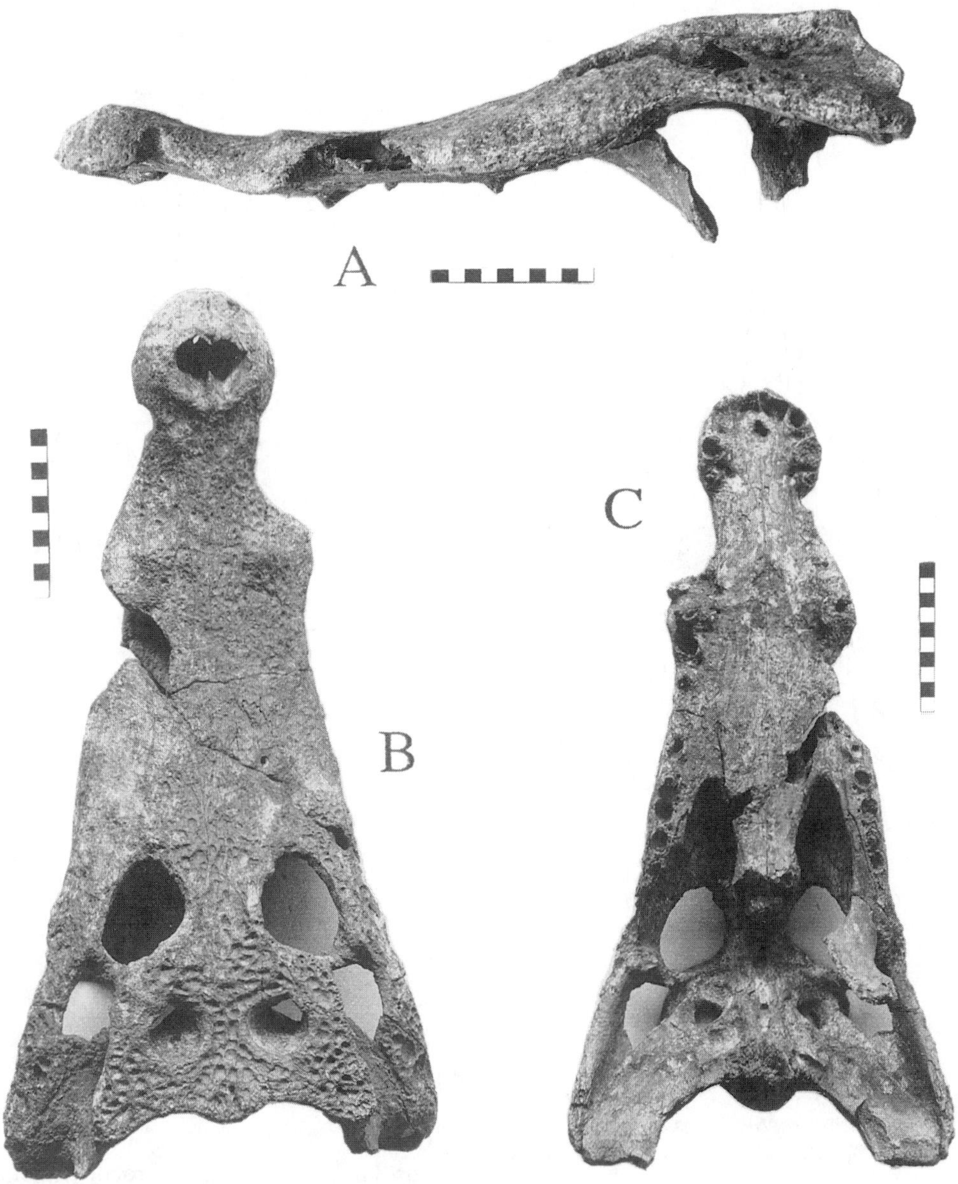

Figure 4.18 Skull of *Crocodylus cataphractus*, KNM-LT 23104: A = left lateral aspect; B = dorsal aspect; C = ventral aspect. Scale bars equal 100 mm.

relatively more slender than in *C. niloticus*, and the lachrymals extend well anterior of the prefrontals *contra* the Nile crocodile. The skull table of *C. cataphractus*, LT 23104, is relatively much broader than in *C. niloticus*, and both the supratemporal fenestrae and orbits are more circular. Prominent maxillary bosses are present in *C. cataphractus*, as in many *Crocodylus* species. These, however, lie directly above the 5th maxillary tooth in the Lothagam specimen, whereas in the Nile crocodile and others they generally occur between the 5th and 6th teeth. The small 2nd premaxillary tooth is absent in LT 23104, at least on one side, as is often the case in *Crocodylus*. Only moderate festooning of the tooth row margin is present.

Crocodylus cataphractus, LT 23104, possesses 13 maxillary teeth, as do extant representatives, but its rostrum is rather shorter and somewhat broader than that of the living form, with interesting implications, as discussed later in this contribution. The anterior edge of the palatal fenestra lies at the level of the 10th maxillary tooth. Significantly, the fenestra ends at the 11th tooth position in a Plio-Pleistocene *C. cataphractus* from Koobi Fora (KNM-ER 929) and frequently between the 11th and 12th in Recent examples (Tchernov 1986). Bearing in mind the reservations just noted about variation in palatal suture positions, the palatines of the Lothagam fossil extend anteriorly to the level of the 8th maxillary tooth, between the

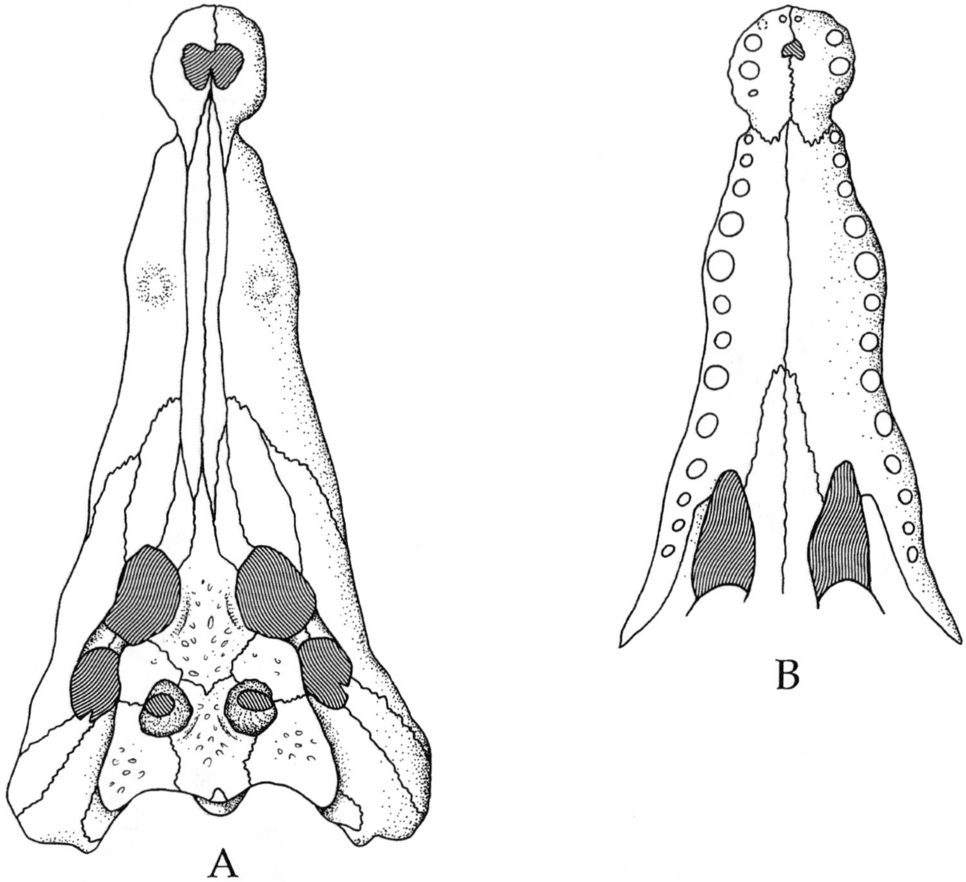

Figure 4.19 Skull of *Crocodylus cataphractus*, KNM-LT 23104, restored: A = dorsal aspect; B = ventral aspect.

8th and 9th in the Plio-Pleistocene *C. cataphractus* and often to the level of the 9th in the extant form. Furthermore, Tchernov (1986) suggests that fewer teeth lie adjacent to the palatal fenestrae in individuals of the putative *C. cataphractus* lineage as they approach the Recent, a gradual phyletic shift moving the teeth forward into an elongated rostrum. In the Plio-Pleistocene animal from Koobi Fora, 2.5 teeth lay alongside the fenestra, whereas two teeth are present today. In the Lothagam specimen, 3.5 teeth lie opposite the fenestra. Tchernov (1986) believed that intertooth distances should increase as the *C. cataphractus* lineage approaches the present, but this is difficult to confirm with the limited sample sizes available.

Much of the lower jaw is preserved, the right ramus better than the left. The mandible is relatively longer, shallower, and more slender than in *C. niloticus*, and it lacks significant festooning. The coronoid eminences are damaged but were apparently lower than in *C. niloticus*. The shallow mandibular symphysis of LT 23104 extends back to the 6th mandibular tooth, while in the Plio-Pleistocene *C. cataphractus* from Koobi Fora (KNM-ER 929), it reaches the 7th, and ends between the 7th and 8th in extant examples (Tchernov 1986).

Rimasuchus gen. nov.

Diagnosis

A very large, brevirostrine crocodylid characterized by premaxillae that are broader than long with a relatively straight premaxillae/maxillae palatal suture, deep "canine" occlusal notch, slight dorsal maxillary boss, closely spaced anterior dentary teeth, broadly diverging mandibular rami, and prominent dentary festoon. Particularly distinguished from *Crocodylus niloticus* by the lack of a nasal promontorium.

Rimasuchus lloydi (Fourtau, 1920)

(Figures 4.14a, 4.20–4.23)

Etymology

Latin *rima* = crack; genus named after the East African Rift Valley where most specimens have been found.

Holotype

CGM 15597, incomplete cranium and mandible collected by Lt. Col. Arthur H. Lloyd and conserved at the Cairo Geological Museum, Egypt. A topotype specimen in the Natural History Museum, London (a nearly complete skull from the Miocene of Wadi Moghara, Egypt, BMNH uncatalogued) provides suitable comparative material.

Lothagam Material

- *Lower Nawata:* 22966, Rt. mandible; 23151, cranial and postcranial fragments; 24026, Rt. mandible; 24068, mandible fragments.
- *Upper Nawata:* 24038, mandibular fragment; 24069, mandible fragments; 24072, mandible fragment; 24074, premaxilla fragment; 24080, cranial fragments; 24166, cranial and vertebrae fragments; 24645, cranium and mandible; 26464, Rt. mandibular symphysis; 28651, mandible fragments.
- *Upper Nawata or Apak Member:* 421, skull.
- *Kaiyumung Member:* 23676, mandible fragments; 24058, mandible fragments; 24060, Rt. mandible; 26305, fragmented skull.

Euthecodon brumpti, discussed later in this contribution, and *Rimasuchus lloydi* are easily the most common (as well as the largest) fossil crocodilians in the Turkana Basin. *Rimasuchus lloydi,* hitherto referred to the genus *Crocodylus,* was briefly described by Fourtau (1920) and Müller (1927) and later in more detail by Tchernov (1986) and Pickford (1996). *R. lloydi* is a very large, broad-snouted form that frequently reached an estimated 7 m or more in length. By comparison with *Crocodylus niloticus, R. lloydi* is notably brevirostrine. Younger, hence smaller, individuals of *R. lloydi* are also represented in the Lothagam fauna.

No complete skull of *R. lloydi* has been collected from Lothagam, so some reliance has been placed on the slightly younger (Plio-Pleistocene) Koobi Fora fauna in the present description. No significant morphological differences have been noted between the two populations. A large skull and partial skeleton (LT 421) collected by Patterson's team (Patterson et al. 1970) and noted by Tchernov (1986) as having the shortest and broadest snout known of any *R. lloydi* specimen, is from approximately 0.8 km south of Lothagam Hill. It is almost certainly younger than other specimens from Lothagam and may be equivalent in geological age to specimens from Koobi Fora. However, there is a good, although fragmented, skull from Lothagam proper, LT

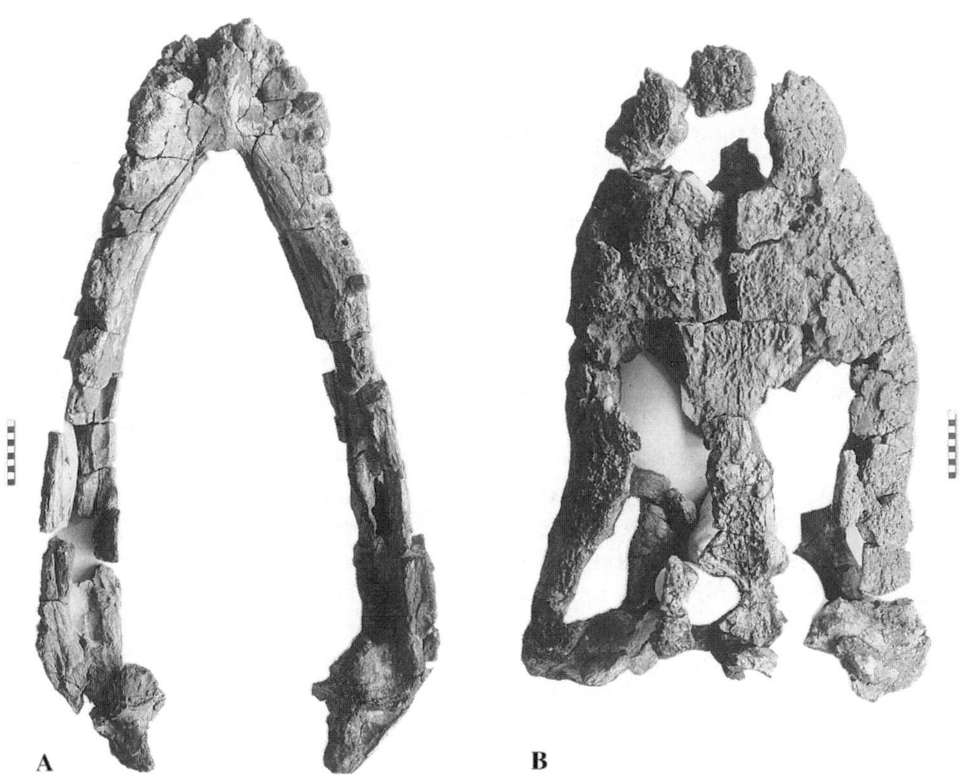

Figure 4.20 Mandible and partial skull of *Rimasuchus lloydi* gen. nov., KNM-LT 26305: A = mandible in dorsal aspect; B = skull in dorsal aspect. Scale bars equal 100 mm.

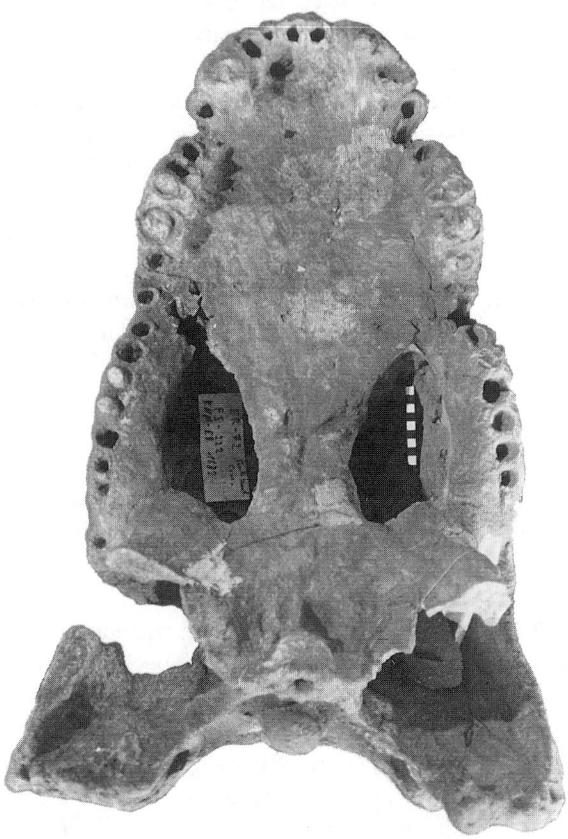

Figure 4.21 Palate of *Rimasuchus lloydi* gen. nov., KNM-ER 1682. Scale bar equals 100 mm.

26305 (from the Kaiyumung Member of the Nachukui Formation), that represents the largest known example of *R. lloydi* (figure 4.20). The estimated skull length of this specimen, from the snout tip to the back of the quadrates is approximately 970 mm, and it is about 830 mm to the back of the supraoccipital. The total length of the mandible of LT 26305 is nearly 1.08 m; it is approximately 580 mm wide at its maximum point. The mandibular symphysis is about 170 mm long.

The most distinctive features of *R. lloydi* are its relatively short and broad premaxillae, and the short but deep mandibular symphysis. As opposed to *Crocodylus niloticus* and *C. cataphractus*, the combined premaxillae of *R. lloydi* are noticeably broader on the palate than they are long (figure 4.23). The palatal premaxillary/maxillary suture (clearly seen in LT 26305) is straighter transversely in *R. lloydi* than in *C. niloticus* (Tchernov 1986). The posterior extension of the premaxilla on the dorsal surface of the skull is also relatively short, and the distances from the tip of the snout to the external nares and from the tip to the incisive foramen are often small. This morphology is apparently derived, as no close potential outgroup possesses such short premaxillae.

The "canine notches" or occlusal grooves between the premaxillae and maxillae for reception of the 4th mandibular teeth are relatively much shorter anteroposteriorly in *R. lloydi* than in *Crocodylus niloticus*, and they are usually deeper transversely (figure 4.21). The occlusion pit for the 1st mandibular tooth does not commonly pierce the roof of the premaxilla. Typically, five premaxillary and 13 to 14 maxillary teeth occur in *R. lloydi*, as in *C. niloticus*. Five premaxillary and 14 maxillary teeth are present in the very large LT 26305, the 4th premaxillary and 5th maxillary teeth being the largest. Maxillary bosses above and between the 5th and 6th maxillary teeth are not strongly expressed in *R. lloydi* and are relatively smaller than in *C. niloticus*. Although this may represent a scaling factor, even the largest extant *C. niloticus* possess prominent bosses (Kälin 1933). Unlike the derived condition in *C. niloticus*, the preorbital area is flat in all examples of *R. lloydi*—that is, there is no nasal promontorium, as noted by Tchernov (1986).

Like the skull table of *Crocodylus cataphractus*, that of *R. lloydi* is relatively broader than in all but the largest *C. niloticus*. Squamosal "horns" have not been observed. Tchernov (1986) states that the lateral borders of the supratemporal fenestrae run more obliquely in *R. lloydi* than in *C. niloticus*, but while generally true, the relationship is somewhat variable. The exaggerated "up-rolled" orbital edges of *R. lloydi* noted by Tchernov (1986) are a size-dependent feature, and big individuals of extant *C. niloticus* develop these as well (most small crocodilians also possess this feature to some degree).

The lower jaw and mandibular symphysis of *R. lloydi* are also characteristic, as noted by Tchernov (1986: plates 4 and 6). The symphysis is relatively shorter in *R. lloydi* than in *Crocodylus niloticus* but incorporates more teeth, on average; the symphysis is frequently relatively deeper and stouter in *R. lloydi* as well. Tchernov (1986) states, incorrectly, that there are only three to four teeth in the symphysial portion of the mandible, whereas there are never fewer than four and are more usually five. The length of the symphysis varies within species, as shown by extant *C. niloticus* where four to four and one-half teeth normally lie in the symphysial region, but very large individuals may also incorporate five (personal observation). A line at the back of the symphysis normal to the sagittal plane of the jaw (i.e., normal to the symphysial suture) defines the number of included teeth. This number is partially a function of the angle at which the jaw rami meet; the angle is much more acute in *C. niloticus* than in *R. lloydi*, hence fewer teeth generally lie in that portion of the dentary framed by the symphysis of *C. niloticus* (note: this character may be difficult to judge from a single ramus without projection of a mirror image). The mandibular rami diverge much more broadly in *R. lloydi* than in any other Lothagam crocodilian.

Figure 4.22 Right mandibular rami: A = Recent *Crocodylus niloticus*, dorsal aspect; B = *Rimasuchus lloydi* gen. nov., KNM-LT 22966, dorsal aspect. Scale bar equals 100 mm.

LT 22966 (figure 4.22) is a good example of an isolated right mandibular ramus of a medium-sized *R. lloydi* from the Lower Nawata; only the retroarticular process and the anterior tip of the symphysis are missing. The external mandibular fenestra is, *contra* Tchernov (1986), not more elongate than that found in *C. niloticus,* although the shape and relative size of crocodilian mandibular fenestrae vary ontogenetically. The jaw is relatively short, deep, and robust. There are 15 mandibular teeth in LT 22966, as in *Crocodylus niloticus. Rimasuchus lloydi* teeth are typically fat and become even more bulbous toward the rear of the jaw. Like most crocodilian teeth, they are bicarinate (front and back). The teeth of *R. lloydi* are generally blunter than those of *C. niloticus* and, according to Tchernov (1986), are rarely sharp. However, crown sharpness is related to tooth size in living crocodilians. The 4th tooth is the largest, with the 1st having only a slightly smaller diameter.

LT 22966 shows that *R. lloydi* possesses a particularly marked dentary eminence or festoon that is crowned by the large 10th and 11th mandibular teeth. The elevation of this festoon is much more marked than in *C. niloticus;* whereas the 11th to 15th mandibular teeth of *C. niloticus* lie at essentially the same horizontal plane, the last several teeth of *R. lloydi* lie well below the level of the 11th tooth. The teeth of *R. lloydi* are generally more closely spaced throughout the ramus than those of *C. niloticus,* especially so within the symphysis. Although there is a slight occlusal groove diastema between the 2nd and 3rd teeth and also between the 8th and 9th teeth in LT 22966, as in *C. niloticus,* the largest *R. lloydi* specimen, LT 26305, has closely spaced teeth. There is no anterior diastema in LT 26305, and only a slight occlusal space occurs after the 7th tooth. Seemingly, occlusal groove spacing was lessened with age and increased overall size.

LT 23151 is a small individual of *R. lloydi* from the Lower Nawata at Lothagam that includes some associated postcranial material. The best preserved elements are a fine, matrix-free atlas/axis complex, the axis with an obvious hypapophysis. These, and the other post-crania, are unremarkable and typically crocodylid in form.

Although most of his *R. lloydi* characters are related to proportion and scaling, Tchernov (1986) does not discuss the role of ontogeny in their manifestation nor the possibility of variable ontogenetic sampling in his

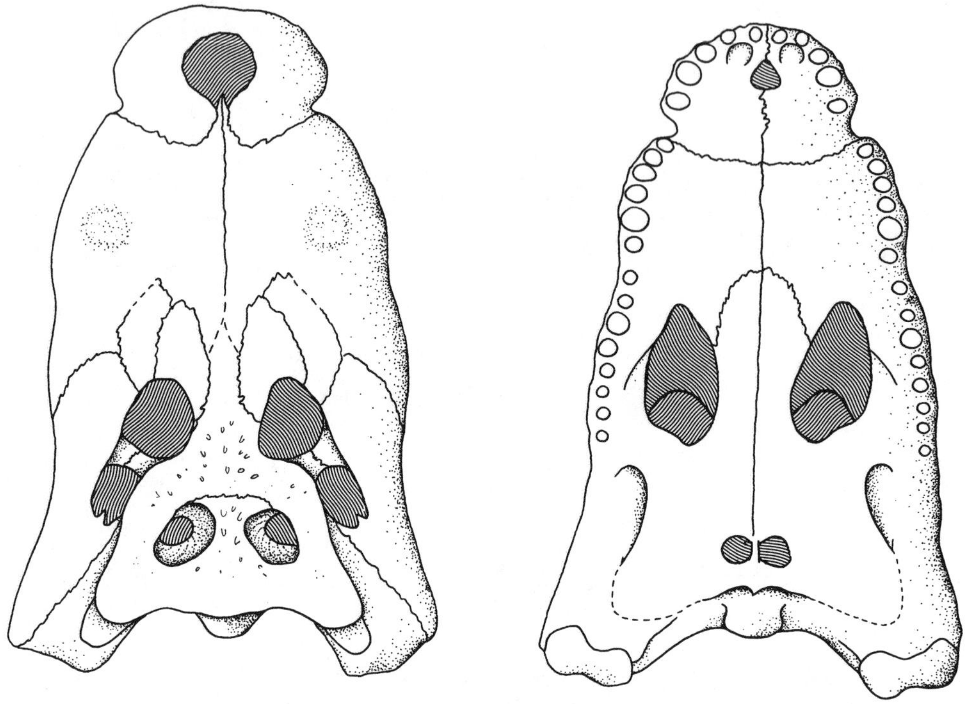

Figure 4.23 Composite skull of *Rimasuchus lloydi* gen. nov., based on KNM-LT 00421 and 26305, and on KNM-ER 1682 and 2275; certain sutures are indiscernable: A = dorsal aspect; B = ventral aspect.

material. Much more study of these factors is required, even among living taxa. Additionally, Tchernov (1986) relies heavily on the condition of palatal suture patterns (which are probably variable and furthermore unclear in all but a few cases). Because of these difficulties, some of Tchernov's (1986) putative *R. lloydi* specimens are likely to be large examples of *C. niloticus.* Further work will be required to sort out the complete samples of these two taxa.

At the latest, *Rimasuchus lloydi* appears in the Lower Miocene. At Lothagam, it is known with certainty from the lower and upper members of the Nawata Formation and from the Kaiyumung Member of the Nachukui Formation, whence came the very large LT 26305. As noted by Tchernov (1986), *Rimasuchus lloydi* bears at least a superficial resemblance to "*C.*" *megarhinus* from the Paleogene of the Fayum Basin (Andrews 1905, 1906), although the premaxillae are relatively much longer in the latter. "*Crocodylus*" *megarhinus,* however, does not appear to be closely related to *R. lloydi.* Brochu (1997) separates the two taxa by two clade nodes. Nevertheless, "*C.*" *megarhinus* represents a relatively brevirostrine morphotype among Early Tertiary crocodiles that could represent a functional precursor to *R. lloydi.* As such, *R. lloydi* may represent an archaic element of the Lothagam fauna. Brochu (1997) also distances *R. lloydi* from *Crocodylus* by a node that also defines, in part, *Euthecodon* and *Osteolaemus,* suggesting that *R. lloydi* is not congeneric with *Crocodylus.* This interpretation is accepted here, and a new generic name is employed for the large, broad-nosed fossil crocodilian of East Africa.

Eogavialis Buffetaut, 1982

Hecht and Malone (1972) reviewed this enigmatic genus, first described (as *Tomistoma*) by Andrews (1901), and concluded that it belonged within a broadened concept of *Gavialis.* Langston (1965) and Sill (1970) had also questioned the tomistomine relationships of Andrews's (1901, 1905, 1906) material. Buffetaut (1982), however, provided the new generic name *Eogavialis* for the obviously plesiomorphic specimens from the Paleogene of the Fayum Basin, Egypt, although he did not elaborate on the morphology or significance of *Eogavialis. Eogavialis* had previously been the object of much debate, as it bears a superficial resemblance to the extant longirostrine crocodylid, *Tomistoma schlegelii,* the Malay false gavial (Andrews 1901, 1905, 1906; Buffetaut 1978, 1982; Hecht and Malone 1972; Joleaud 1930; Müller 1927; Sill 1968). Kälin (1955), Langston (1965), and Antunes (1987) provided lists of characters that distinguish *Gavialis* from *Tomistoma,* but no distinction was made by them between apomorphic and plesiomorphic features. Tchernov (1986) also failed to recognize such distinctions and continued to place *Eogavialis* material within *Tomistoma,* apparently on the bases of historical convention and overall similarity of skull proportion, rostrum shape, and tooth number. The most obvious plesiomorphy of *Eogavialis* is its retention of a premaxilla/nasal contact, a primitive character shared with *Tomistoma,* and thus the focus of some confusion. This contact is present in most outgroup taxa and, as a symplesiomorphy, is useless in demonstrating relationship between *Eogavialis* and *Tomistoma.*

Gavialids, however, are easily distinguished from crocodylids (including *Tomistoma*) and alligatorids by the large crista that runs across the midpoint of the jugal bar of the postorbital in the gavials. This crista is much reduced in other Crocodylia, but it is very prominent in *Eogavialis andrewsi.* Gavialids possess the synapomorphies of a rectangular skull table (versus trapezoidal in other crocodilians), large circular supratemporal fenestrae (although circularization of these fenestrae may also increase ontogenetically: Joffe 1967), a robust and subvertical postorbital bar, ventral end of the postorbital bar not inset relative to a dorsal lamina of the jugal, pterygoid bullae, constricted antorbital area, and subcircular orbits with everted orbital rims (Brochu 1997, Norell 1989). *Eogavialis* possesses all of these characters, as shown by both the Kenyan material and the very well preserved skull of *E. africanus* at Yale (YPM 6263) that was examined by Hecht and Malone (1972) (personal observation). *Gavialis* and *Eogavialis* also share a sloping occipital plate (as opposed to the subvertical plate of crocodylids and alligatorids), although the polarity of this character is equivocal.

Eogavialis andrewsi sp. nov.
(Figures 4.14b, 4.25–4.28)

Diagnosis

A longirostrine gavialid crocodilian differing from *Gavialis* by virtue of the plesiomorphic retention of a premaxilla/nasal contact and lesser contrast between the elongate rostrum and facial region. Distinguished from Paleogene representatives of *Eogavialis,* i.e., *E. africanus* (and, if distinct, *E. gavialoides*), by a rostrum approximately 85–90 percent as broad and 5–10 percent longer, a broader skull table, wider separation (by nearly a factor of two) of the supratemporal fenestrae, more everted anterior orbital rims, and a more marked constriction of the rostrum at the premaxilla/maxilla suture. Differs from *Gryposuchus* in having less eversion to the orbital rims, frontals which are largely excluded from the supratemporal fenestrae by thin splints of the parietals, and deeply pitted sculpturing of the skull roof.

Holotype

KNM-LT 22943, a largely undeformed skull lacking pterygoids and ectopterygoids, most of the palatines save for their anterior tips, and virtually all of both jugals, from the lower member of the Nawata Formation, Lothagam.

Etymology

Named after Charles William Andrews, who was the first to study and describe African gavialids (from the Fayum of Egypt).

Lothagam Material

- *Lower Nawata:* LT 22943, holotype; 23088 fragmentary cranium and mandibles with numerous incomplete postcranial elements including appendicular bones, vertebrae, ribs and osteodermal scutes.

The hypodigm potentially also includes two undescribed skulls of *Eogavialis* cf. *E. andrewsi* from the lower Miocene of Loperot, southwest Turkana Basin, KNM-LP 23295 and KNM-LP 28830.

Eogavialis was not recognized from the Turkana Basin until the recent work at Lothagam (Leakey et al. 1996). As detailed in the preceding text, it is clearly a member of the Gavialidae, which is presumed to be the most primitive family of extant crocodilians (Brochu 1997, Norell 1989). *Eogavialis andrewsi* lies in a functionally (perhaps morphologically) intermediate position between the modern gavial, *Gavialis gangeticus,* and the more plesiomorphic gavialid(s) of the Fayum Basin Paleogene. In spite of its superficially similar rostrum, it bears no close relationship to *Tomistoma schlegelii* and lacks important crocodylid (and hence tomistomine) synapomorphies. For example, the tomistomine character of exposure of the vomer on the palate (Iordansky 1973) is notably lacking. The rostrum of *E. andrewsi* is more slender than rostra of the early Fayum specimens and approximates the narrow snout of *Gavialis.* However, *E. andrewsi* lacks the extreme facial constriction just anterior to the orbits seen in extant *Gavialis,* yet it is more slender-faced than either *Tomistoma* or *E. africanus.* The superficial similarity of the preorbital region of *E. andrewsi* with some tomistomines (e.g., the Ugandan *Tomistoma coppensi;* Pickford 1994) highlights the need for evolutionary novelties in the reconstruction of crocodilian phylogenetic relationships. No one, for example, would consider *Rimasuchus lloydi* to be an alligatorid merely because their rostrum length to breadth ratios are similar; such proportions are functionally mediated.

This species is clearly not part of the *Gavialis* clade, but to date can be placed neither as a plesiomorphic sister to *Gavialis* nor in a clade containing *Eogavialis africanus* on the basis of unequivocal synapomorphies based on discrete characters. Brochu (1997) positions *Gryposuchus*, a South American gavial (Buffetaut 1982, Langston 1965, Langston and Gasparini 1997), between *Eogavialis* and *Gavialis*, but it too is distinct from the Lothagam taxon. The currently unresolved position of the Lothagam specimen (figure 4.24) dictates the conservative use of the generic name *Eogavialis* rather than the premature creation of a new genus.

The *Eogavialis andrewsi* holotype (figures 4.25 and 4.26) is approximately 770 mm long (700 mm to the back of the broken supraoccipital, and 750 mm to the end of the occipital condyle) and an estimated 150 mm across at the front of the orbits. It is 160 mm across the skull table at the middle of the temporal fenestrae, with an estimated 180 mm total width across the center of the frontals. It has an extremely elongate rostrum relative to the Lothagam crocodilians detailed so far (540 mm long from the front of the orbits). As noted, the nasals reach the premaxillae in the holotype, the restriction of the nasals in *Gavialis* being a (presumably post-Miocene) derivation. However, as in *Gavialis* (but less than in *Tomistoma*), the premaxillae of *Eogavialis* have only a short dorsal exposure: they reach to only behind the 1st maxillary tooth. The elongate, splint-like lachrymals reach forward to a point in front of the 10th maxillary tooth positions and contact the nasals as in *Gavialis* and crocodylids. The facial profile of *Eogavialis andrewsi* is concave in the manner of *C. cataphractus* (i.e., no promontorium), although the external nares were not raised.

An obvious and prominent diastema, approximately 40 mm long, occurs between the last (5th) premaxillary tooth and the 1st maxillary tooth in *Eogavialis andrewsi*.

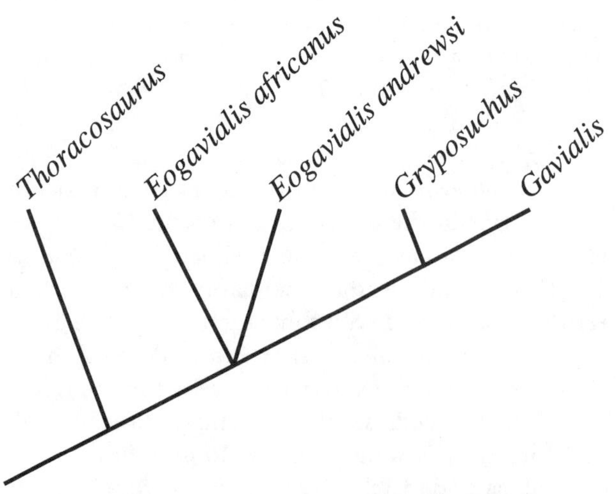

Figure 4.24 Cladogram of proposed phylogenetic relationships of *Gavialidae*, modified from Brochu (1997).

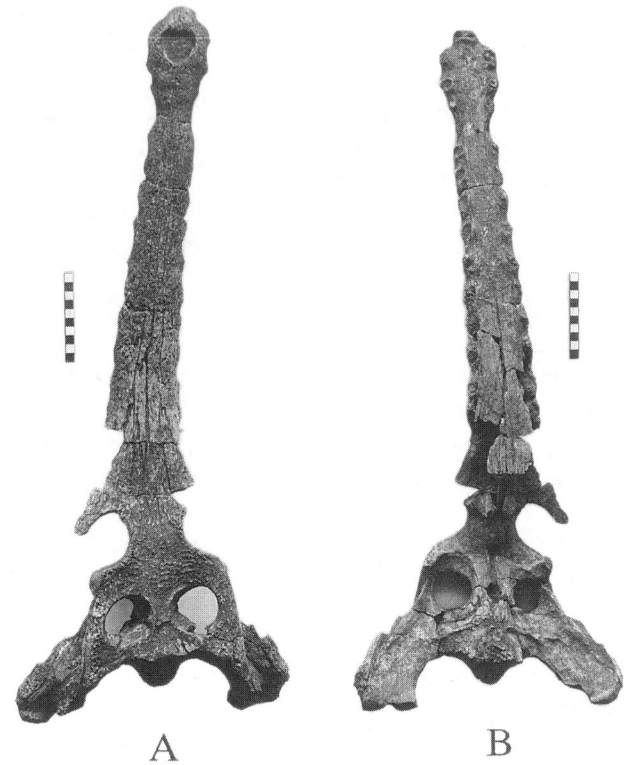

Figure 4.25 Skull of *Eogavialis andrewsi* sp. nov., KNM-LT 22943: A = dorsal aspect; B = ventral aspect. Scale bars equal 100 mm.

Gavialis gangeticus has lost this diastema. However, both *E. andrewsi* and *Gavialis* possess deep occlusal notches on the anterior tips of the premaxillae to accommodate their long 1st dentary teeth, contrary to most crocodylids where the 1st dentaries are received in occlusion pits and are not visible dorsally unless they pierce the premaxillae. At least 10 tooth positions are preserved in the maxillae of the *E. andrewsi* holotype, LT 22943 (the posterior ends of the maxillae are lost, and more teeth were originally present). No intact teeth are preserved in place (the broken bases of several remain, however). A single, loose, associated tooth crown (25 mm long) is bicarinate, sharp, and acutely conical in typical gavial fashion.

As in *Gavialis* and the Fayum form(s) but unlike in *Tomistoma* the interorbital bar of *Eogavialis andrewsi* is wide. The Lothagam *Eogavialis* also has a broader parietal roof than does *Tomistoma,* but this roof is narrower than in *Gavialis* (and also *C. cataphractus*). Tchernov (1986) reproduced a plate (no. 4, figure 4) of a cranium from Loperot (mistakenly labeled as from Kanapoi) that he identified as *C. cataphractus* (KNM-LP 23295, from the Early Miocene, perhaps 17 million years old). Another skull from Loperot, KNM-LP 28830, is a slightly smaller example and conspecific. These specimens are not *C. cataphractus,* however, but clearly belong to the Gavialidae and are presumed to be

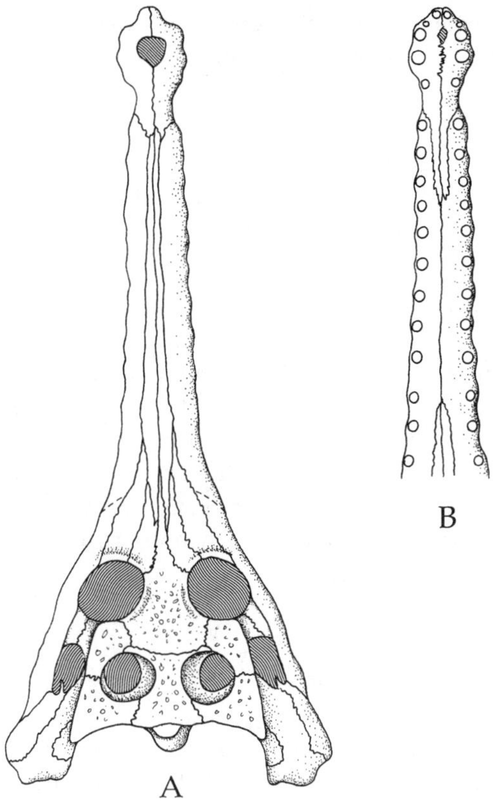

Figure 4.26 Skull of *Eogavialis andrewsi* sp. nov., KNM-LT 22943, restored: A = dorsal aspect; B = anterior palate.

Eogavialis. Although the premaxilla/nasal contact, or lack thereof, cannot be observed, they agree in all important ways with the *Eogavialis* skull from Lothagam, except that they are smaller (approximately 95 and 80 mm across immediately anterior to the orbits, respectively). They are likely immature individuals. The Loperot skulls have wider areas between the supratemporal fenestrae than the Lothagam specimen, which also has relatively more circular fenestrae and a broader interorbital area. During living crocodilian ontogeny, the fenestrae widen more quickly than the braincase expands (Dodson 1975; Iordansky 1973; Kälin 1933). An only slightly everted anterior orbital rim is apparent in the Lothagam *Eogavialis*, but this is a preservational artifact. The rims are damaged or lacking, and the small *Eogavialis* skulls from Loperot clearly exhibit both anterior and lateral eversion, although this eversion is less than that seen in *Gavialis*.

The postorbital bars are broken in the holotype but would have been stout and oriented subvertically, as evidenced by their transversely sectioned remains. These sections are subcircular—not thin and laterally compressed as in crocodylids—although neither are they as massive as in *Gavialis*. This morphology is apparently primitive, but the polarity of this character is not entirely clear (Norell 1989). The right postorbital bar retains part of the anterior crista but this is very well displayed on the Loperot skulls. It is also evident on a pair of loose postorbitals associated with LT 23088 (figure 4.27). A pronounced anterolateral corner to the dorsal surface of the postorbital produces the characteristically gavialid subrectangular skull table. This rectangle is less well defined than in *Gavialis*, but is again more pronounced than in the Paleogene *Eogavialis* of the Fayum. The nuchal process of the supraoccipital, broken off in the Lothagam specimen, is nevertheless pronounced in *Eogavialis* (e.g., LP 23295, figure 4.28) (among East African crocodilians, this process is largest in *Eogavialis*, whereas the smallest nuchal process occurs in *C. niloticus*).

The anterior edges of the quadratojugals are damaged in the holotype of *E. andrewsi*, but KNM-LP 23295 retains the complete jugal bar and shows the primitive character of a large anterior process of the quadratojugal overlapping the medial surface of the jugal. According to Norell (1989), this feature is not present in most crocodylids, and he suggests that there is some variation in its presence in *C. niloticus*. *Contra* Norell (1989), a substantial anterior process exists in at least the Lothagam *C. cataphractus*. KNM-LP 28830 also shows the quadratojugal process on its left side, as well as a small posterior ectopterygoid extension onto the medial surface of the jugal, the lack of which is noted by Norell (1989) to be derived for crocodylids, including *Tomistoma*. There does not appear to be a posterior process of the postorbital on to the jugal in *E. andrewsi* as, according to Norell (1989), there should be in gavialids. Crocodylids and some primitive alligatorids have lost this process, but its lack here suggests that it may not be as useful a character as believed by Norell (1989).

Only the rostral part of the palate is preserved in the holotype, LT 22943. The anterior ends of the palatines extend to between the levels of the 9th and 10th maxillary teeth. The premaxillae extend caudal to the 3rd maxillary tooth. All three Kenyan specimens of *Eogavialis* lack their pterygoids and at least the posterior ends of palatines, thus no indications of the palatal fenestrae or of the presence of pterygoid bullae remain; these are presumed to be similar to those of *Eogavialis africanus*.

Figure 4.27 Free postorbitals of *Eogavialis andrewsi* sp. nov., KNM-LT 23088: A = left postorbital; B = right postorbital. Scale bar equals 50 mm.

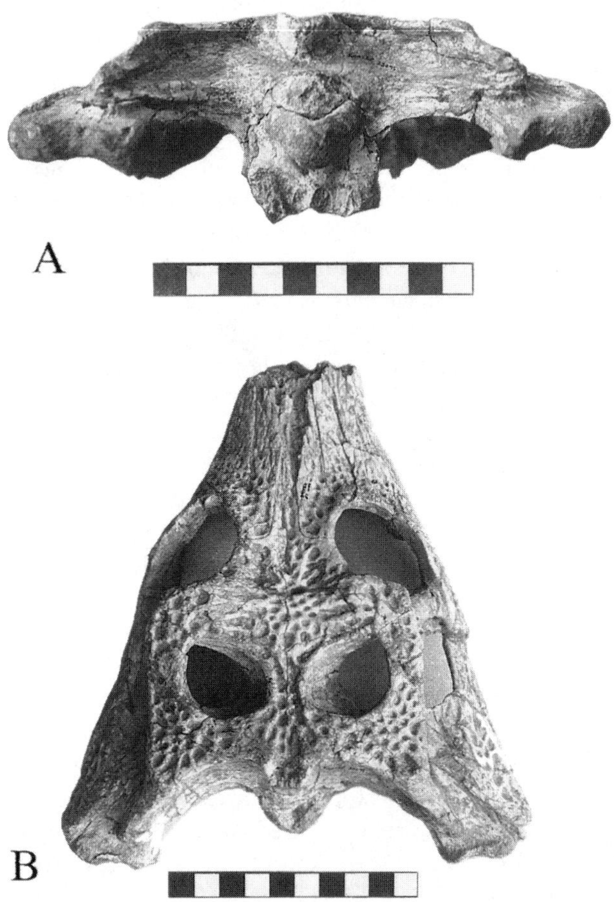

A

B

Figure 4.28 Skull, lacking rostrum, of *Eogavialis* cf *E. andrewsi*, KNM-LP 23295, from Loperot. A = occipital aspect; B = dorsal aspect. Scale bars equal 100 mm.

The braincase relationships of *Eogavialis andrewsi* are difficult to determine accurately but appear unremarkable. The foramen ovale is prominent, but whether or not the prootic is broadly exposed at its posterior margin is not discernible in the present state of preparation. Such a condition is to be expected, however, as it occurs in both *Gavialis* and *Tomistoma* (Norell 1989). The basioccipital tuberosities are primitive and not notably different from those of crocodylids (figure 4.28). They are not pendulous as in *Gavialis* (Hecht and Malone 1972).

LT 23088 retains part of the dentaries and some postcranial elements. The mandible has an extremely long symphysis, but the alveoli are not produced into salients as in *Euthecodon*, as discussed later in this contribution. It is not possible to tell if the splenial perforation or foramen illustrated by Norell (1989) is present in *Eogavialis* because these bones are missing from the mandible. The atlas and the few other vertebral fragments of this specimen are too abraded or fragmentary to determine the presence of either a small axial diapophysis, anterior cervical hypapophyses (particularly the 2nd postatlantal), or the shape of the axial neural spine. These features are characteristic in *Gavialis*. The few preserved scutes and other bones are unremarkable.

Euthecodon Fourtau, 1920
Euthecodon brumpti (Joleaud, 1920)

(Figures 4.14b, 4.29–4.31)

Diagnosis

Very large eusuchian with extremely elongate and narrow rostrum with deeply scalloped dental margins, premaxillae and nasals attenuated, prominent narial ridge, moderate premaxillary/maxillary diastema, four premaxillary teeth, skull table small and nearly square, occiput vertical, long mandibular symphysis, teeth isodont and slender.

Lothagam Material

- *Lower Nawata:* 22956, mandible fragment; 23177, partial cranium; 24066, cranial and mandible fragments.
- *Upper Nawata:* 24030, mandible fragments and scutes; 24037, mandible fragments; 24064, cranium and mandible fragments; 24065, mandible fragment; 24067, mandible fragment; 24077, mandible fragments; 24083, rostrum; 26456, cranial fragments; 26457, mandible fragments; 26458, premaxillae; 26460, mandible fragment; 26462, mandible fragment; 28650, mandibular symphysis.
- *Apak Member:* 24036, mandible fragment; 24040, mandible fragments and premaxillae.
- *Kaiyumung Member:* 24075, mandible fragments; 24076, mandible and maxilla fragment; 26306, cranium and mandible.)

Remains of *Euthecodon* are extremely common at Lothagam and elsewhere in the Turkana Basin, but the taxon has been described in only a superficial manner. The present material includes some of the best specimens yet collected and also one of the largest examples known. One of the better preserved specimens, LT 26306, must have been an enormous animal approximately 10 m long. Its skull is approximately 1.52 m long by 270 mm (greatest width) across the frontals, and 300 mm deep at the occiput from the skull table to the ventral tip of the pterygoid flange.

This species was first described, from the Pliocene Omo Group of Ethiopia, by Joleaud (1920) as a species of *Tomistoma* (subgenus *Euthecodon*). The type species for the genus, *Euthecodon nitriae* Fortau 1920, from the Lower Pliocene of Wadi Natrun, Egypt (Fortau 1920),

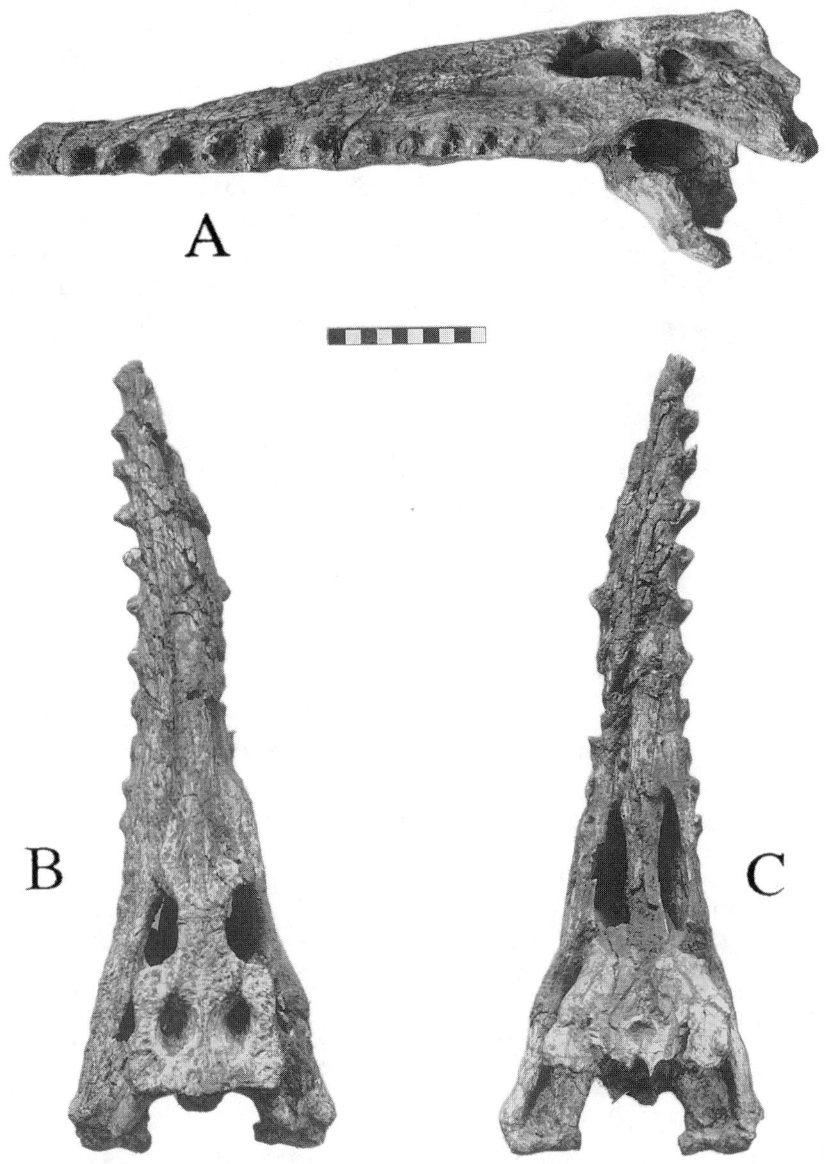

Figure 4.29 Skull, lacking anterior portion of rostrum, of *Euthecodon brumpti*, KNM-LT 23177: A = left lateral aspect; B = dorsal aspect; C = ventral aspect. Scale bar equals 100 mm.

has been considered distinct—largely on the basis of general skull size, tooth number, and rostral proportion—by numerous workers (Ginsburg and Buffetaut 1978; Kälin 1955; Steel 1973; Tchernov 1976, 1986). Tchernov's (1976, 1986) identification of the Turkana Basin material as *E. brumpti* is accepted here pending further study of *E. nitriae*.

An Early Miocene, shorter snouted, North African species of *Euthecodon, E. arambourgi,* occurs at Gebel Zelten, Libya (Arambourg and Magnier 1961; Buffetaut 1985; Ginsburg and Buffetaut 1978; Savage and Hamilton 1973; Tchernov 1986). An indeterminate species of *Euthecodon* occurs in the Early Pliocene, Sahabi Formation of Libya (Hecht 1987), and Early Miocene Victoria Basin rocks of Rusinga Island have yielded a single mandibular fragment of *Euthecodon* (Tchernov and Van Couvering 1978). Aoki (1992) and Pickford (1994) note fragmentary *Euthecodon* material from Albertine Rift sediments of the Congo, as does Buffetaut (1979) from the Lower Miocene of Ombo, Kenya. The relationships of these fossils to *E. brumpti* are poorly understood.

Tchernov (1976, 1986) considered *Euthecodon* to be a tomistomine and assumed it was a direct offshoot of "*Tomistoma*" (actually *Eogavialis*) of the Paleogene. Ginsburg and Buffetaut (1978) and Kälin (1955) also allied *Euthecodon* with the tomistomines. Tchernov (1976) states that *Euthecodon* and "*Tomistoma*" share

Figure 4.30 Occiput of *Euthecodon brumpti*, KNM-LT 26306. Scale bar equals 50 mm.

many characters, and lists (Tchernov 1986) those found in extant *Tomistoma* as given by Kälin (1955), Langston (1965), and Hecht and Malone (1972). However, none of these shared characters is unequivocally derived. While there are no recognized synapomorphies of *Euthecodon* and Recent *Tomistoma*, *Euthecodon* bears a potential relationship with Crocodylidae. Brochu (1997) allies *Euthecodon* with *Rimasuchus* and *Osteolaemus*, and Brochu and Storrs (1995) ally it with the large, "horned" *"Crocodylus" robustus* of Madagascar. *Euthecodon* is certainly not closely related to gavialids.

Euthecodon is most readily identified by its characteristic rostrum with deeply scalloped margins that delineate salients between the teeth, hence the generic name (figures 4.30 and 4.31). The tooth crowns project from the margins of the mouth, alternate in occlusal sequence, and produce an intermeshing net of sharp tines, apparently for capturing fish. The rostrum is extremely long and narrow—piscivorous rostra presumably have this shape to minimize drag during sideways sweeps through schools of fish, while maximizing the radius of attack; struggling fish produce little torsional stress on crocodilian snouts, so broadly buttressed rostra are unnecessary in piscivorous taxa.

A short but marked nasal promontorium is present in *Euthecodon brumpti*, immediately anterior to the orbits. It is very well developed in LT 26306. Preliminary analysis suggests that the promontorium is independently derived from that of *C. niloticus*. The promontorium of *E. brumpti* is narrow and longitudinally oriented and is bounded posteroventrally by a distinct antorbital depression; it extends anteriorly to the level of the 6th or 7th maxillary tooth. The snout of *Euthecodon* slopes evenly from the promontorium toward the premaxillae until, near the level of the premaxilla/maxilla suture, the tip is redirected upward. As a result, the external nares lie on a raised premaxillary pedestal, much as they do in *C. cataphractus*. The posterodorsal extensions of the premaxillae reach to the level of the 6th maxillary teeth. Like *Eogavialis*, *Euthecodon* never exhibits inferior occlusal notches or piercings of the premaxillae, but the 1st mandibular teeth always occlude outside of the premaxillae via distinct anterior grooves.

The nasals retain their plesiomorphic contact with the premaxillae in *E. brumpti* (*contra* Tchernov 1986), and, as shown in a Koobi Fora specimen with particularly clear sutures (KNM-ER 1778), the nasals may sometimes fuse to form a single median element, although the nasals are certainly not fused in one Lothagam rostrum, LT 24083. The nasals, and indeed most other bones, are also fused in the very large LT 26306. The lachrymals are apparently long and splintlike, reaching to the front of the promontorium in moderately sized animals such as LT 23177, and about midway along it in LT 26306. The lachrymals contact the nasals as in gavialids and crocodylids (but not alligatorids).

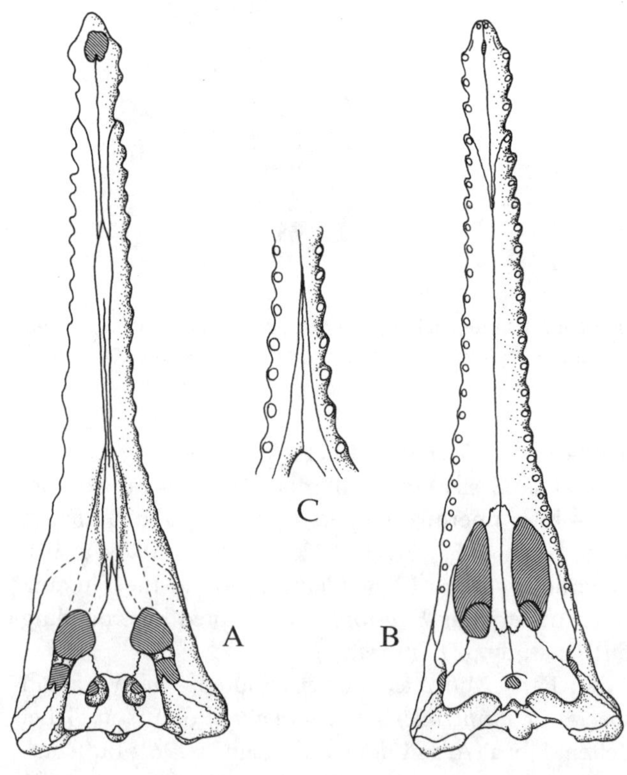

Figure 4.31 Composite skull of *Euthecodon brumpti;* the sutural positions are based largely on KNM-ER 1778: A = dorsal aspect; B = ventral aspect; C = posterior part of mandibular symphysis.

Another notable feature of *Euthecodon* is its relatively small skull table, when contrasted with that of other longirostrine forms: it is about as long as it is broad and is roughly square in plan; the skull table of *Eogavialis,* in contrast, is rather broader than it is long. Similarly, the supratemporal fenestrae of *Euthecodon* are narrowly oval or lozenge-shaped, not circular as in the gavialids. Squamosal ridges or horns may develop in larger, older individuals, and very large "horns" are present in LT 26306. There is a very small nuchal process on the supraoccipital in *Euthecodon.* The occiput above the basioccipital is vertically oriented as in crocodylids and alligatorids (Tarsitano et al. 1989), in contrast to the sloping occiput of gavialids.

The articular area of the quadrate is narrow in *Euthecodon* relative to that of typical crocodilians, and the intercondylar notch may be pronounced. Additionally, the innermost condyle may have a pointed or angular, rather than a rounded, medial corner. The shape of the quadrate in crocodilians is frequently a diagnostic character (Langston 1975, Norell and Storrs 1989). The lower temporal fenestrae of *Euthecodon* are relatively small and are subquadrate in form, not ovate or subtriangular as is usual in crocodilians. *Euthecodon* exhibits a quadratojugal spine (LT 23177 retains its vestige on the left side, LT 26306 the base of the right), although this delicate feature is almost always broken off in fossil crocodilians. The presence of this spine, however, is a primitive character; it is lost only in the alligatorids in posthatching stages. The postorbital bar has no crista, and here again *Euthecodon* is distinct from gavialids. The postorbital bar itself is relatively delicate and laterally compressed in *Euthecodon,* and it bends laterally to join the jugal.

The teeth of *Euthecodon* are relatively isodont and slender. They are bicarinate and extremely sharp, as is characteristic of piscivorous dentition. The largest Lothagam *Euthecodon,* LT 26306, has 21 upper teeth and 20 lower teeth. This contrasts with 24 to 25 upper teeth (usually 24), and 21 to 22 mandibular teeth in the several Plio-Pleistocene *Euthecodon* fossils from Koobi Fora. Specimens from both faunas have only four teeth in each premaxilla, as opposed to the usual five in many crocodilians. Probably the second premaxillary tooth has been lost; *E. arambourgi* Ginsburg and Buffetaut 1978, from the Lower Miocene of Gebel Zelten, Libya, has five premaxillary teeth with a small 2nd tooth. Tchernov (1986) uses this character as a basis for a phylogenetic link with *Tomistoma,* but the 2nd tooth is small in *Crocodylus* and gavials as well, and at least *Crocodylus* may also lose this tooth in early ontogeny. The palatal fenestra reaches anteriorly to the level of the 15th maxillary tooth in LT 26306 and to the 18th in the Plio-Pleistocene Koobi Fora form. However, a smaller example of *Euthecodon* from Lothagam, LT 24083, shows the fenestra reaching to between the 17th and 18th maxillary teeth. The significance of this variation is unknown.

The mandibular symphysis lies in front of the 17th mandibular tooth in the largest Lothagam *Euthecodon,* LT 26306. It is at the level of, or just behind, the 19th tooth in the Koobi Fora animals. The splenials of *E. brumpti* are incorporated into the mandibular symphysis to the level of the 13th mandibular teeth (figure 4.31). It is not known if the splenial is imperforate (following Norell 1989). There is relatively little surficial sculpturing on the symphysial part of the *Euthecodon* mandible; rostral sculpturing is also slight.

Euthecodon is known at Lothagam from the lower and upper members of the Nawata Formation and from the Apak and Kaiyumung Members of the Nachukui Formation. The largest example, LT 26306, is from the Kaiyumung Member (as was the largest *Rimasuchus lloydi*).

Discussion

No East African crocodilians are known prior to the Neogene; therefore all Miocene occurrences, such as that at Lothagam, assume special significance. The recognition of five distinct crocodilian taxa at Lothagam in the Miocene and Pliocene is somewhat surprising, however, given the generally conservative morphology of the group and the lack of diversity among East African crocodiles today. Such high diversity for African Neogene crocodilians is so far unique to Lothagam. The well-preserved Lothagam fauna, with its special stratigraphic and geographic positions, provides a new opportunity for consideration of the evolutionary history and behavior of African Crocodylia.

The presence of four genera and five species attests to a relatively high degree of niche partitioning in the aquatic predator realm for the Late Miocene–Early Pliocene of East Africa. This supports an interpretation of the area as one of high productivity and favorable living conditions over a substantial period of time (Leakey et al. 1996). Large numbers of a high diversity of fish including *Lates,* the Nile perch; *Polypterus,* the bichir; *Protopterus,* the African lungfish; *Gymnarchus,* the electric fish; and others as detailed by Stewart (this volume) provided plentiful food for crocodilians. At least two of the crocodilian taxa, *Euthecodon* and *Eogavialis,* exhibit extreme specialization for piscivory, while the sharp-snouted crocodile, *Crocodylus cataphractus* is only a little less specialized in this regard. *Crocodylus* at Lothagam displays two different average ratios for rostrum length to breadth—the relatively narrow snouted and seemingly more piscivorous *C. cataphractus,* and the broader (generalist?) *C. niloticus.* Living *C. cataphractus* is highly aquatic and feeds predominantly on fish, amphibians, birds, and crustaceans (Alderton 1991; Ross and Magnusson 1989; Steel 1989). Extant *C. niloticus* will attack and consume any animal that it can catch, including large mammals; juveniles rely more heavily on fish (Alderton 1991). At the opposite extreme falls the broad-faced, probable large mammal hunter, *Rimasuchus lloydi.*

Two of the crocodilian species from Lothagam are extant, although only *Crocodylus niloticus* is found in the area today. Lothagam, however, confirms the once widespread occurrence, and now reduced range, of *C. cataphractus.* The now dominant position of *C. niloticus* in East Africa seemingly reflects the currently arid conditions of the East African Rift Valley. Extant *C. cataphractus* is most often found in open waters (Alderton 1991). *Crocodylus niloticus* was, as today, widely distributed throughout East Africa during the Neogene, occurring in the Miocene Lothagam succession and in the Pliocene of both the Omo Basin, Ethiopia, and Kanapoi, Kenya (Feibel et al. 1991; Tchernov 1976, 1986). *Crocodylus "checchiai"* of the Lower Pliocene, Sahabi Formation of Libya (Qasr el Sahabi) is here considered synonymous with *C. niloticus.* Tchernov (1976) suggests, however, that *C. niloticus,* while common today, is relatively rare in the fossil record. Indeed, it has not been positively identified at Koobi Fora. As discussed earlier in this contribution, however, confusion over the definition of *C. niloticus* may account for at least part of this situation. Alternatively, a severe historical collecting bias against fossil crocodilians may have played a role.

Crocodylus cataphractus, once widespread in East Africa, is now restricted to the Ujiji River and Lake Tanganyika, as well as to the Gabon, Senegal, and Congo River drainage basins of West and Central Africa (Groombridge 1987; Steel 1989; Tchernov 1976). The species has been recorded from fossils in the Pliocene of Wadi Natrun, Egypt, and from the Plio-Pleistocene of the Omo Basin, Ethiopia, and of Koobi Fora, Kenya (Arambourg 1947; Feibel et al. 1991; Joleaud 1930; Tchernov 1976, 1986). A possibly allied form ("*Mecistops*") has been recorded from the Congo (Aoki 1992). However, *C. cataphractus* has been the least commonly found crocodilian in the Turkana Basin. Living *C. cataphractus* lead rather solitary lives and are never found at high densities (Steel 1989). The apparent displacement of *C. cataphractus* by *C. niloticus* in East Africa since the Plio-Pleistocene can be attributed to the latter's greater tolerance of arid conditions, such as those now prevalent in Kenya; *C. niloticus* appears adapted to more euryhaline conditions than its Congo Basin cousin. The living Nile crocodile frequently experiences periodic droughts and the relatively high salinity that characterizes the variable (sometimes ephemeral) water bodies of the modern East African Rift system.

Certainly, the Late Miocene and Early Pliocene in East Africa were characterized by much greater availability of water and by more permanent large water bodies than are found today. It is assumed that *Euthecodon,* as an obligatory piscivore (Tchernov 1986), was restricted to these water bodies (Hecht 1987). The same was probably true of *Eogavialis* (living *Gavialis* is primarily aquatic: Alderton 1991; Steel 1989) and the extinction of these two longirostrine taxa by the mid Pleistocene at the latest may be attributable to the loss of suitable permanent water sources. *Lates,* which is associated with these taxa at Lothagam (Stewart this volume; Leakey et al. 1996), today requires well-oxygenated waters in large permanent stands (Hecht 1987), and *Polypterus* is intolerant of any saline influence (Leakey et al. 1996).

Euthecodon brumpti appears restricted to the greater Turkana drainage basin, and perhaps lacked the overland travel capacity of *Crocodylus.* It is extremely common in the Lothagam and Plio-Pleistocene Omo groups and in the Pliocene Kanapoi Formation, only becoming extinct in the last million years. According to Tchernov (1976), these *E. brumpti* populations are indistinguishable. *Euthecodon* has not, however, been found in the Lake Baringo Basin or other southeasterly exposures (Leakey et al. 1996; Tchernov 1976), suggesting a disjunction of water bodies between these regions and the Turkana drainage. Tchernov (1976) and Tchernov and Van Couvering (1978) suggest a Pleistocene link between the ancestral Nile Basin and Turkana for an exchange of crocodilians, especially *Euthecodon* and *Rimasuchus lloydi.*

Rimasuchus lloydi was apparently very common over much of East Africa from the Lower Miocene until at least the mid Quaternary, often occurring in close association with *Crocodylus niloticus.* Specimens of *Rimasuchus* have been noted by Tchernov (1976, 1986) in the Miocene of Egypt (Moghara), Kenya (Baringo, Lukeino, Lothagam), Libya (Gebel Zelten), Saudi Arabia (Al-Sarrar), the Sinai Peninsula (Erg-el-Ahmar) and southern Tunisia; the Pliocene of Ethiopia (Omo), Kenya (Koobi Fora, Kanapoi), and Uganda (Kaiso); and the Quaternary of Ethiopia (Omo), Kenya (Koobi Fora), Sudan (Abu Huggar), and Tanzania (Olduvai). Pickford (1994) noted additional specimens from the Plio-Pleistocene of Uganda. The earliest specimens of the species have been thought to be from the Lower Miocene of North Africa, while its first appearance in East Africa is Late Miocene. However, Pickford's (1996) report of this species from the Lower Miocene of southern Africa suggests that much of the biogeographic history of *Rimasuchus* remains unknown.

The relatively brevirostrine habitus of *Rimasuchus lloydi* suggests a formidable nearshore predator of large mammals. The stout dentition would have allowed the breaking and crushing of large bones, and the short, broad snout could withstand the significant torsional forces imparted by the struggles of large prey items. Tchernov (1976, 1986) suggests variability of the brevirostrine condition among several populations of *R. lloydi* around Lake Turkana; however, he fails to comment on the potential role of ontogenetic variation in rostral index within his limited sample sizes. As noted earlier in this contribution, the relative lengths and breadths of crocodilian snouts are clearly dependent on an individual's ontogeny (Dodson 1975; Iordansky 1973; Kälin 1933). The potential confusion between identifications of *R. lloydi* and *C. niloticus* has also been noted. The more generalized "semiaquatic" *C. niloticus*, in which mature individuals change diet from piscivorous to largely carnivorous, may have been able to coopt part of the "semiterrestrial" predator niche of *R. lloydi* after the Quaternary extinction of the latter, thereby increasing its own numbers. The large, contemporary fossil *C. niloticus* probably competed directly with *Rimasuchus. Rimasuchus*, like *C. cataphractus*, may have lost out to *C. niloticus* as a result of increasing aridity in eastern Africa.

Crocodylus is the most common Old World crocodilian today, but its ancestry and the phylogenetic relationships of its component species remain uncertain. In some areas of the global record, the genus is severely split; in others it serves as a "wastebasket taxon." Few rigorously objective studies of its phylogeny have been undertaken. Tchernov (1986) postulates a direct, gradualistic lineage from Rimasuchus ("*Crocodylus*") *lloydi* to *C. niloticus*, but his hypothesis is entirely ad hoc. Tchernov (1976, 1986) and Tchernov and Van Couvering (1978) are correct that the rostrum has been the focus of much phylogenetic change within crocodilians. However, a well-documented account of species variation for East African crocodilians does not exist, weakening any gradual hypothesis. As noted above, Brochu (1997), using analysis of synapomorphies and including several living species of *Crocodylus*, has separated *R. ("C.") lloydi* and *C. niloticus* into distinct genera. A close relationship between *C. niloticus* and *C. cataphractus* is also problematic. Brochu's (1997) analysis separates them by two clade nodes.

A phyletic elongation of the *Crocodylus cataphractus* rostrum, as suggested by Tchernov (1986), is more plausible in light of the new Lothagam specimen. It has a shorter, broader rostrum, with fewer teeth, than does the modern population. Furthermore, the specimen from the Plio-Pleistocene of Koobi Fora is intermediate in morphology between the Lothagam fossil and modern examples. Nevertheless, this suggested phyletic change is difficult to test, as only single specimens are available from Lothagam and Koobi Fora. Furthermore, tentative acceptance of this possibility does not presuppose the rapid transformation of *"C." articeps* of the Fayum Paleogene (Andrews 1905, 1906) into *C. cataphractus* as proposed by Tchernov (1986). *"Crocodylus" articeps* almost certainly bears no close relationship with *Crocodylus*. Tchernov's (1986) hypothesis lacks methodological rigor because, like his lineage hypotheses for *Euthecodon* and *C. niloticus*, it is based only on stratigraphic/geographic position and an assumption of continually increasing snout length as a trend toward piscivory. Tchernov and Van Couvering (1978) also state that phyletic shortening of the crocodilian rostrum has never occurred but give no supporting evidence for this position. Their position is contrary to the most recent phylogenetic analysis (Brochu 1997). Palatal suture/tooth relationships as used by Tchernov (1986) and Tchernov and Van Couvering (1978) may have some validity for species determination, but there is clearly some overlap between taxa.

Lothagam represents a new geographic record for the Gavialidae; never before have gavials been recorded in East Africa, and it would be very interesting to know when they made their last appearance there. *Eogavialis* has not yet been recovered from rocks younger than the Lower Nawata, although the other four Lothagam taxa occur in the Plio-Pleistocene Omo Group rocks of the Turkana Basin. By the Late Pliocene, *Gavialis* had already appeared on the Indian subcontinent in the form of *G. browni* and *G. lewisi* of the Siwalik Group (Mook 1932 and Lull 1944, respectively). *Gavialis gangeticus* of India, Pakistan, and Bangladesh is the only living representative of the family. *Eogavialis andrewsi* surely represents a relict of Paleogene faunas that, along with numerous other taxa noted by Leakey et al. (1996), may have been impacted by the end-Miocene extinction. Previously, Buffetaut (1985) believed all African gavialids to be extinct after the Early Miocene. *Eogavialis* may well have been extinct by Koobi Fora, or even Nachukui time, but past bias against collecting African fossil crocodilians leaves this question open. Although ancestral relationships cannot be demonstrated conclusively, Lothagam provides in *Eogavialis andrewsi* a plesiomorphic taxon with the potential for relationship to later gavialids such as *Gavialis*. As documented here, *E. andrewsi* displays some proportional and other advances over *Eogavialis africanus* from the Paleogene of the Fayum Basin, Egypt. However, although *Eogavialis* displays an apparent narrowing of the rostrum over time, significant rostral elongation is not obvious.

The claim of ever-increasing rostral elongation in all African crocodilians (Tchernov 1976, 1986), although not necessarily incorrect, is not supported by empirical evidence or demonstrated synapomorphies. At the very least, it begs the question of the origin of the *Rimasuchus* brevirostrine snout, as outgroups to the Eusuchia frequently exhibit relatively longer rostra than does *R.*

lloydi. It could even be argued that *Crocodylus niloticus*, which *contra* Tchernov (1976) did not have a demonstrably broader snout in fossil populations, has lost maxillary teeth phyletically while developing a more generalist habitus.

Acknowledgments

I thank Meave Leakey for her invitation to undertake study of the Lothagam crocodilians, John Harris for his editorial efforts, and Michael Benton for his aid and encouragement. I am also grateful to Robin Storrs for her generous help and indulgence. Reviews of an earlier version of this contribution were provided by Michael Benton, Christopher Brochu, Wann Langston, and the late Robert Savage. This work was partially supported by the Cincinnati Museum Center, the University of Bristol Department of Geology, the National Museums of Kenya, and a generous travel grant from the University of Bristol Alumni Foundation.

References Cited

Alderton, D. 1991. *Crocodiles and Alligators of the World.* London: Blandford.

Andrews, C. W. 1901. Preliminary note on some recently discovered extinct vertebrates from Egypt (Part II). *Geological Magazine* 4:436–444.

Andrews, C. W. 1905. Notes on some new Crocodilia from the Eocene of Egypt. *Geological Magazine,* n.s., 2:481–484.

Andrews, C. W. 1906. *A Descriptive Catalogue of Tertiary Vertebrata of the Fayûm, Egypt.* London: British Museum (Natural History).

Antunes, M. T. 1987. Affinities and taxonomical status of Miocene longirostrine crocodilians from western Europe with remarks on phylogeny, paleoecology, and distribution [in Portuguese]. *Comunicações dos Serviços Geológicos Portugal* 73:49–58.

Aoki, R. 1992. Fossil crocodilians from the Late Tertiary strata in the Sinda Basin, eastern Zaire. *African Study Monographs,* Supplementary issue 17:67–85.

Arambourg, C. 1947. Contribution à l'étude géologique et paléontologique du Bassin de lac Rudolphe et de la Basse Vallée de l'Omo. In *Mission scientifique de l'Omo (1932–1933).* Vol. 1, fasc. 3. *Paléontologie,* pp. 231–562. Paris: Muséum National d'Histoire Naturelle.

Arambourg, C., and P. Magnier. 1961. Gisement de vertébrés dans le bassin tertiare de Syrte (Libye). *Comptes Rendus de l'Académie des Sciences* (Paris) 252:1181–1183.

Behrensmeyer, A. K. 1976. Lothagam Hill, Kanapoi, and Ekora: A general summary of stratigraphy and fauna. In Y. Coppens, F. C. Howell, G. L. Isaac, and R. E. Leakey, eds., *Earliest Man and Environments in the Lake Rudolf Basin: Stratigraphy, Paleoecology, and Evolution,* pp. 163–172. Chicago: University of Chicago Press.

Benton, M. J., and J. M. Clark. 1988. Archosaur phylogeny and the relationships of the Crocodylia. In M. J. Benton, ed., *The Phylogeny and Classification of the Tetrapods,* vol. 1, pp. 295–338. Oxford: Clarendon Press.

Boaz, D. D. 1982. Preliminary assessment of taphonomy and paleoecology at Sahabi. *Garyounis Science Bulletin* 4:109–121.

Brochu, C. 1997. Morphology, fossils, divergence timing, and the phylogenetic relationships of *Gavialis. Systematic Biology* 46:479–522.

Brochu, C., and G. W. Storrs. 1995. The giant dwarf crocodile: A reappraisal of *"Crocodylus" robustus* from the Quaternary of Madagascar [abstract]. In B. D. Patterson, S. M. Goodman, and J. L. Sedlock, eds., *Environmental Change in Madagascar,* p. 6. Chicago: Field Museum.

Buffetaut, E. 1978. Sur l'histoire phylogénetique et biogéographique des Gavialidae (Crocodylia, Eusuchia). *Comptes Rendus de l'Académie des Sciences* (Paris) 287:911–914.

Buffetaut, E. 1979. Présence du crocodilien *Euthecodon* dans le Miocène inférieur d'Ombo (golfe de Kavirondo, Kenya). *Bulletin de la Société Géologique de France* 21:321–322.

Buffetaut, E. 1982. Systématique, origine et évolution des Gavialidae sud-américains. *Geobios* 6:127–140.

Buffetaut, E. 1984. On the occurrence of *Crocodylus pigotti* in the Miocene of Saudi Arabia, with remarks on the origin of the Nile crocodile. *Neues Jahrbuch für Geologie und Paläontologie Monatshefte* 9:513–520.

Buffetaut, E. 1985. Zoogeographical history of African crocodilians since the Triassic. In Karl L. Schuchmann, ed., *Proceedings of the International Symposium on African Vertebrates: Systematics, Phylogeny and Evolutionary Ecology,* pp. 453–469. Bonn: Zoologisches Forschungsinstitut und Museum Alexander Koenig.

Dodson, P. 1975. Functional and ecological significance of relative growth in *Alligator. Journal of Zoology* (London) 175: 315–355.

Feibel, C. S., J. M. Harris, and F. H. Brown. 1991. Paleoenvironmental context for the Late Neogene of the Turkana Basin. In J. M. Harris, ed., *Koobi Fora Research Project.* Vol. 3. *The Fossil Ungulates: Geology, Fossil Artiodactyls, and Palaeoenvironments,* pp. 321–370. Oxford: Clarendon Press.

Fourtau, R., ed. 1920. *Contribution à l'étude de vertébrés Miocènes de l'Egypt.* Cairo: Government Press.

Ginsburg, L., and E. Buffetaut. 1978. *Euthecodon arambourgi* n. sp. et l'évolution du genre *Euthecodon,* crocodilien du Néogène d'Afrique. *Géologie Méditerranéenne* 5:291–302.

Groombridge, B. 1987. The distribution and status of world crocodilians. In G. J. W. Webb, S. C. Manolis, and P. J. Whitehead, eds., *Wildlife Management: Crocodiles and Alligators,* pp. 427–444. Chipping Norton, Australia: Surrey Beatty.

Hecht, M. K. 1987. Fossil snakes and crocodilians from the Sahabi Formation of Libya. In N. T. Boaz, A. El-Arnauti, A. W. Gaziry, J. de Heinzelin, and D. D. Boaz, eds., *Neogene Paleontology and Geology of Sahabi,* pp. 101–106. New York: Liss.

Hecht, M. K., and B. Malone. 1972. On the early history of the gavialid crocodilians. *Herpetologica* 28:281–284.

Iordansky, N. N. 1973. The skull of the Crocodilia. In C. Gans and T. S. Parsons, eds., *The Biology of the Reptilia,* vol. 4, pp. 201–262. London: Academic Press.

Joffe, J. 1967. The "dwarf"crocodiles of the Purbeck Formation, Dorset: A reappraisal. *Palaeontology* 10:629–639.

Joleaud, L. 1920. Sur la présence d'un Gavialide du genre *Tom-*

istoma dans le Pliocène d'eau douce de l'Ethiopie. *Comptes Rendus de l'Académie des Sciences* (Paris) 70:816–818.

Joleaud, L 1930. Les crocodiliens du Pliocène d'eau douce de l'Omo (Ethiopie). *Centenaire de la Société Géologique de France, Livre Jubilaire 1830–1930* 1930:411–423.

Kälin, J. A. 1933. Beiträge zur vergleichenden Osteologie des Crocodilidenschädels. *Zoologischer Jahrbücher* 57:535–714.

Kälin, J. 1955. Crocodilia. In J. Piveteau, ed., *Traité de paléontologie,* vol. 5, pp. 695–784. Paris: Masson.

Kramer, G., and F. Medem. 1956. Über wachstumsbedingte Proportionsänderungen bei Krokodilen. *Zoologischer Jahrbucher* 66:62–74.

Langston, W., Jr. 1965. *Fossil Crocodilians from Colombia and the Cenozoic History of the Crocodilia in South America.* Publications in the Geological Sciences 52. Los Angeles: University of California.

Langston, W., Jr. 1975. Ziphodont crocodiles: *Pristichampsus vorax* (Troxell), new combination, from the Eocene of North America. *Fieldiana Geology* 33:291–314.

Langston, W., Jr., and Z. Gasparini. 1997. Crocodilians, *Gryposuchus,* and the South American gavials. In R. F. Kay, R. H. Madden, R. L. Cifelli, and J. J. Flynn, eds., *Vertebrate Paleontology in the Neotropics: The Miocene Fauna of La Venta, Colombia,* pp. 113–154. Washington, D.C.: Smithsonian Institution

Leakey, M. G., C. S. Feibel, R. L. Bernor, J. M. Harris, T. E. Cerling, K. M. Stewart, G. W. Storrs, A. Walker, L. Werdelin, and A. J. Winkler. 1996. Lothagam: A record of faunal change in the Late Miocene of East Africa. *Journal of Vertebrate Paleontology* 16:556–570.

Lull, R. S. 1944. Fossil gavials from North India. *American Journal of Science* 242:417–430.

Maccagno, A. M. 1948. Descrizione di una nuova specie di "*Crocodilus*" del giacimento di Sahabi (Sirtica). *Atti della Reale Accademia Nazionale dei Lincei, Memorie* 1:61–96.

Maccagno, A. M. 1952. I coccodrilli di Sahabi. *Rendiconti Accademia Nazionale dei XL* 3:71–117.

McDougall, I., and C. S. Feibel. 1999. Numerical age control for the Miocene-Pliocene succession at Lothagam, a hominoid-bearing sequence in the northern Kenya Rift. *Journal of the Geological Society* (London) 156:731–745.

Mook, C. C. 1921. Skull characters of the Recent Crocodilia with notes on the affinities of the Recent genera. *Bulletin of the American Museum of Natural History* 44:126–268.

Mook, C. C. 1932. A new species of fossil gavial from the Siwalik beds. *American Museum Novitates* 514:1–5.

Müller, L. 1927. Ergebnisse der Forschungsreisen Prof. E. Stromers in den Wusten Ägyptens. V. Tertiäre Wirbeltiere. 1. Beitrage zur Kenntnis der Krokodilier des ägyptischen Tertiärs. *Abhandlungen der Mathematisch-naturwissenschaftlichen Abteilung der Königlichen Bayerischen Akademie der Wissenschaften* 31:1–96.

Norell, M. A. 1989. The higher level relationships of the extant Crocodylia. *Journal of Herpetology* 23:325–335.

Norell, M. A., and G. W. Storrs. 1989. A review of the type fossil crocodilians in the Yale Peabody Museum. *Postilla* 203: 1–28.

Patterson, B., A. K. Behrensmeyer, and W. D. Sill. 1970. Geology and fauna of a new Pliocene locality in northwestern Kenya. *Nature* 226:918–921.

Pickford, M. 1994. Late Cenozoic crocodiles (Reptilia: Crocodylidae) from the Western Rift, Uganda. In B. Senut and M. Pickford, eds., *Geology and Palaeobiology of the Albertine Rift Valley, Uganda-Zaire.* Vol. 2. *Paleobiology/Paléobiologie,* pp. 137–155. Occasional Publication No. 29. Orléans: Centre International pour la Formation et les Echanges Géologiques.

Pickford, M. 1996. Fossil crocodiles (*Crocodylus lloydi*) from the lower and Middle Miocene of southern Africa. *Annales de Paléontologie (Vert.-Invert.)* 82:235–250.

Ross, C. A., and W. E. Magnusson. 1989. Living crocodilians. In C. A. Ross, ed., *Crocodiles and Alligators,* pp. 58–73. Oxford: Facts on File.

Savage, R. J. G., and W. R. Hamilton. 1973. Introduction to the Miocene mammal faunas of Gebel Zelten, Libya. *Bulletin of the British Museum (Natural History)* 22:515–527.

Sill, W. D. 1968. The zoogeography of the Crocodilia. *Copeia* 1968:76–88.

Sill, W. D. 1970. Nota preliminar sobre un nuevo gavial del Plioceno de Venezuela y una discusión de los gaviales sudamericanos. *Ameghiniana* 7:151–159.

Smart, C. 1976. The Lothagam 1 fauna: Its phylogenetic, ecological and biogeographic significance. In Y. Coppens, F. C. Howell, G. L. Isaac, and R. E. Leakey, eds., *Earliest Man and Environments in the Lake Rudolf Basin: Stratigraphy, Paleoecology, and Evolution,* pp. 361–369. Chicago: University of Chicago Press.

Steel, R. 1973. Crocodylia. In O. Kuhn, ed., *Handbuch der Paläoherpetologie,* vol. 16, pp. 1–116. Stuttgart: Fischer.

Steel, R. 1989. *Crocodiles.* London: Helm.

Tarsitano, S. F., E. Frey, and J. Riess. 1989. The evolution of the Crocodilia: A conflict between morphological and biochemical data. *American Zoologist* 29:843–856.

Tchernov, E. 1976. Crocodylidae from the Pliocene/Pleistocene formations of the Rudolf Basin. In Y. Coppens, F. C. Howell, G. L. Isaac, and R. E. Leakey, eds., *Earliest Man and Environments in the Lake Rudolf Basin: Stratigraphy, Paleoecology, and Evolution,* pp. 370–378. Chicago: University of Chicago Press.

Tchernov, E. 1986. *Evolution of the Crocodiles in East and North Africa.* Cahiers de Paléontologie. Paris: Centre National de la Recherche Scientifique.

Tchernov, E., and J. Van Couvering. 1978. New crocodiles from the Early Miocene of Kenya. *Palaeontology* 21:857–867.

4.3

Lothagam Birds

John M. Harris and Meave G. Leakey

Most of the thirty-six avian postcranial specimens are from the Nawata Formation (Lower Nawata 22, Upper Nawata 12). Almost all represent waterfowl, but a bustard (*Eupodotis* sp.), an owl (Strigidae indet.), and a small stork (Ciconiidae indet.) were recovered from the Lower Nawata, and a Marabou stork (cf. *Leptoptilus* sp. indet.) from the Upper Nawata. From the Apak Member, a large Marabou (*Leptoptilus* sp. indet.) and an indeterminate partial ulna were recovered. Fragments of a single ratite egg from the Lower Nawata show an aepyornithoid pore pattern. Many fragments of *Struthio* eggshell were recovered from the Nawata Formation and Apak Member, but only one from the Kaiyumung Member. An abrupt decrease in the size of the *Struthio* pore basins occurs above the Lower Nawata.

Only two avian specimens were collected by the Harvard University expeditions but remains of more than 30 individuals were retrieved by the recent National Museums of Kenya expeditions. The majority of bones recovered are those of waterfowl. Fragments of ratite eggshell have also been collected and appear to represent two, perhaps three, different species.

Systematic Description

Family Aepyornithidae
Genus and species indet.
(Figure 4.32)

Lothagam Material

- *Lower Nawata:* 25085, many shell fragments.

A single specimen comprising many shell fragments shows the aepyornithoid pore pattern described by Sauer (1972) with dagger point pores, sting pores, and linear grooves. The shell is relatively thin. This specimen is tentatively referred to the Aepyornithidae.

Family Struthionidae
Struthio Linnaeus
cf. *Struthio* sp. indet.
(Figure 4.33)

Lothagam Material

- *Lower Nawata:* 24964, shell fragment; 24965, shell fragments; 24966, shell fragment; 24967, shell fragments; 24968, shell fragment; 24969, shell fragment; 24970, shell fragment; 24971, shell fragment; 24972, shell fragment; 28665, egg shell; 25084, shell fragment; 26568, shell fragment; 26569, egg shell.
- *Upper Nawata:* 24973, shell fragments; 24974, shell fragment; 24977, shell fragment; 25075, shell fragment; 25077, shell fragment; 25079, shell fragment; 25080, shell fragment; 25082, shell fragments; 26566, shell; 26567, shell; 28660, shell fragment; 28666, shell fragments.
- *Nawata Formation:* 28663, shell fragments; 28775, shell fragments.
- *Apak Member:* 24975, shell fragment; 24976, shell fragment; 25074, shell fragment; 25076, shell fragment; 25078, shell fragment; 25081, shell fragments; 25083, shell fragment; 25104, shell fragments; 26071, shell fragments; 26072, shell fragments; 28721, shell; 28723, eggshell.

Figure 4.32 Lower Nawata eggshell fragments, KNM-LT 25085, showing the aepyornithoid pore pattern.

- *Kaiyumung Member:* 25086, shell fragment.
- *Horizon indet:* 28664, shell fragment.

The majority of the eggshell fragments from Lothagam show struthious pore patterns that indicate that ostriches were present throughout the section. Most of the shell fragments assigned to *Struthio* sp. are superficially similar to those from extant ostrich shells. However, there is an abrupt decrease in the size of the pore basins that coincides with the junction of the Lower and Upper Nawata. The large, widely spaced pore basins of the Lower Nawata shell fragments have a mean diameter of 4.9 mm (range = 2.06–6.95 mm), whereas those from the Upper Nawata have a mean of 1.2 mm (range 0.7–1.94 mm). The pore basins of fragments found in the Marker Tuff are intermediate in size. Decrease in pore basin size may reflect decrease in overall egg size and/or decrease in water vapor conductance as required by changing environmental conditions (Tullett and Board 1977). Change in pore basin size could also reflect a taxonomic difference (Tyler and Fowler 1979).

Fossil ratite eggshell is commonly found in Neogene deposits in Africa and Eurasia (Andrews 1911; Burchak-Abramovich and Vekua 1971; Mikhailov 1988; Mikhailov and Kurochkin 1988; Sauer 1966, 1979; Sauer and Roth 1972; Sauer and Sauer 1978; Whybrow and Hill 1999). Sauer (1972) described two types of pore pattern typical of ratite eggshell.

In the aepyornithoid pattern, displayed by *Aeypyornis* from Madagascar, the surface of the shell is characterized by numerous small and irregular longitudinal lines, bent and forked grooves, and small pits. Sauer recognized three types of pores: "dagger point" (small short slit-like grooves), "sting pores" (small circular pits), and "linear grooves" (longer linear depressions). These conspicuous pore openings are oriented parallel with the long axis of the egg. The struthious pattern, characteristic of extant *Struthio,* has the shell surface characterized by small circular pore openings, clusters of pores, and clusters of irregular pore grooves randomly distributed over the surface of the eggshell with no particular alignment with the axis of the egg. The grooves form irregular reticulate or rosette patterns; like the clustered pores, they are mostly located in pits that may be interconnected by furrows. Sauer (1979) documented that in the Miocene and Pliocene an extinct heavily built ostrich with short and thick feet (Type A) produced relatively large ovate-shaped eggs with an aepyornithoid pore pattern. This form existed side by side with slender-built long-footed ostriches (Type S) that produced eggs with a struthious pore pattern. He noted that the resemblance of the pore pattern of the ovate Type A eggs to those of aepyornithoid eggs from Madagascar implies homology but does not necessarily indicate taxonomic affinity.

Fossil eggshells from Eurasia show a transition from the aepyornithoid Type A pattern in the Miocene to the struthious pattern in the Pliocene, and in southern Morocco the change from the aepyornithoid to struthious pore patterns occurs in Mio-Pliocene deposits (Sauer 1979). The many intermediate shell types from Morocco were interpreted to suggest that evolutionary transition between the two types may represent millions of years. Similar chronoclinal changes in eggshell patterns have been observed in different branches of the

Eurasian Ratitae (Mikhailov 1988; Mikhailov and Kurochkin 1988).

The occurrence of two types of ratite shell pore pattern in the Lower Nawata is thus consistent with evidence elsewhere for two contemporaneous types of ostrich—Type A and Type S—in Miocene deposits. The abrupt change in the pore basin diameter of the struthious type egg shells from Lothagam has not been documented elsewhere, but shells with similar large pore basins are found in the Baynunah Formation, Abu Dhabi (Andrew Hill, personal communication).

Figure 4.33 Lothagam eggshell fragments showing the struthioid pore pattern: top left = KNM-LT 24966 from the Lower Nawata; top right = KNM-LT 26072 from the Apak Member; bottom left = KNM-LT 24973 from the Upper Nawata on the Marker Tuff; bottom right = KNM-LT 25077 from the Upper Nawata.

Family Pelecanidae
Pelecanus Linnaeus
Pelecanus sp.
(Figure 4.34)

Lothagam Material

- *Upper Nawata:* 24018, complete Lt. radius, Lt. distal ulna, Rt. distal ulna, Rt. distal ulna and shaft, Lt. distal tarsometatarsus and shaft, Rt. proximal shaft tarsometatarsus, complete pedal phalanx; 25111, Lt. proximal femur.

The pelican family is represented by a large proximal left femur (25111) and associated skeletal elements (24018) from the Upper Nawata. These are of comparable size and shape to those of *Pelecanus rufescens.*

Family Phalacrocoracidae
Phalacrocorax Brisson, 1760
Phalacrocorax carbo (Linnaeus)

Lothagam Material

- *Galana Boi:* 25122, Rt. distal and proximal humerus with shaft, Lt. distal humerus with shaft, complete Lt. ulna.

These associated wing bones are almost certainly from the Galana Boi and are indistinguishable from those of the extant White-necked Cormorant.

Family Anhingidae
Anhinga Brisson, 1760
Anhinga rufa (Lacépède and Daudin, 1802)
Anhinga cf. *A. rufa*

Lothagam Material

- *Lower Nawata:* 23080, distal Rt. tibiotarsus with shaft, distal Lt. ulna with shaft; 25118, distal Lt. tarsometatarsus with shaft; 25120, proximal Rt. ulna; 37096, fragment Rt. ulna shaft; 37101, fragment Lt. ulna shaft; 37103, humeral end Rt. coracoid, Rt. distal humerus, proximal shaft fragment Rt. humerus.
- *Upper Nawata:* 24016, Lt. distal immature tarsometatarsus; 25107, proximal and distal Rt. and Lt. humerus, proximal Lt. and Rt. ulna, distal Rt. ulna, Rt. sternal coracoid, Lt. humeral coracoid; 25121, proximal Rt. humerus.

The commonest waterfowl species at Lothagam is a darter of comparable size to the extant African Darter (*Anhinga rufa*).

Family Ardeidae
Genus and species indet.

Lothagam Material

- *Lower Nawata:* 23973, ungual phalanx; 37098, proximal Rt. tibiotarsus.

Figure 4.34 *Pelecanus* sp. limb bones, KNM-LT 24018: top = left radius; middle = distal right ulna; bottom = right tarsometatarsus.

- *Upper Nawata:* 25110, distal Lt. tibiotarsus; 25115, distal Lt. femur; 37102, humeral end Rt. coracoid with corpus.

The ardeid material includes the remains of herons or egrets that are represented by two leg bones from the Upper Nawata and, perhaps, an ungual phalanx from the Lower Nawata.

Family Ciconiidae
Leptoptilos (Lesson, 1831)
Leptoptilos crumeniferus (Lesson, 1831)
Leptoptilos cf. *L. crumeniferus*
(Figure 4.35)

Lothagam Material

- *Upper Nawata:* 24015, distal Lt. tarsometatarsus shaft, proximal Lt. tarsometatarsus with shaft, Lt. proximal first phalanx (II or IV), Lt. first manus phalanx (III).

A specimen from the Upper Nawata represents a stork of a size comparable to that of the extant Marabou.

Leptoptilos sp. indet.
(Figure 4.36)

Lothagam Material

- *Apak:* 25106, Lt. proximal tibiotarsus.

This Apak Member specimen represents a stork similar to but larger than the extant Marabou. Skeletal elements of the fossil Marabou *Leptoptilos* sp. from the 11.5 Ma Ngorora Formation (Hill and Walker 1979) are slightly larger than those of 24015, although the first phalanx is of similar size. The Apak tibiotarsus, 25106, is markedly larger than that of the Ngorora specimen. Today, only one species of Marabou is known in Africa, where it is found in association with herds of ungulates and also at sites of human habitation where it may encounter carrion. Its presence is not a good indicator of environment.

Figure 4.35 *Leptoptilos* cf. *L. crumeniferus:* proximal left tarsometatarsus, KNM-LT 24015.

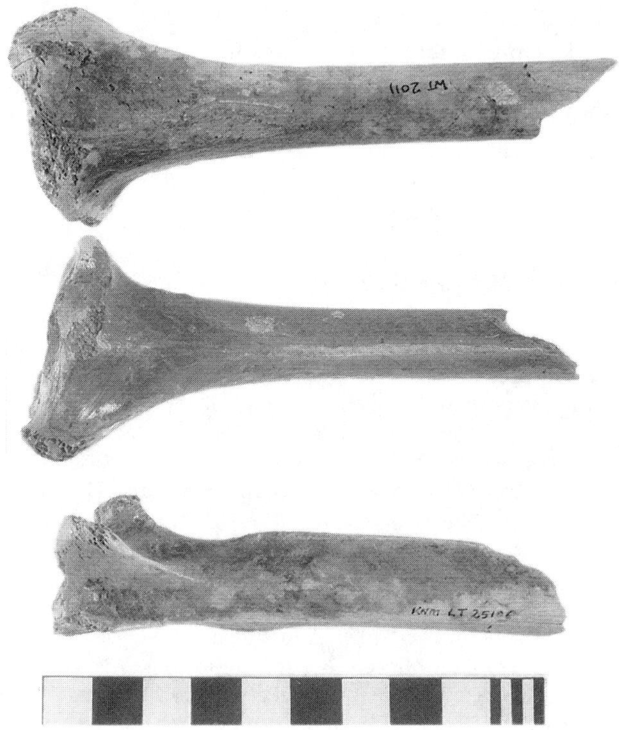

Figure 4.36 *Leptoptilos* sp. indet.: left proximal tibiotarsus, KNM-LT 25016: top = medial view; middle = lateral view; bottom = mediolateral view.

Family Ciconiidae
Genus and species indet.

Lothagam Material

- *Lower Nawata:* 25105, distal Rt. carpometacarpus; 25117, distal Lt. tarsometatarsus; 37095 humeral end Rt. scapula.

Specimens from the Lower Nawata represent one or two small stork species.

Family Anatidae
Genus and species indet.
(Figure 4.37)

Lothagam Material

- *Lower Nawata:* 116, Rt. proximal carpometacarpus; 407, Rt. distal tarsometatarsus; 24011, Lt. proximal carpometacarpus; 25113, Rt. proximal carpometacarpus.
- *Upper Nawata:* 24012, Lt. carpometacarpus.

Two of four anatid specimens from the Lower Nawata are duck-sized (24011, 25113), and two probably represent geese (116, 407). A carpometacarpus (24012) from the Upper Nawata is about the same size as the large Spur-winged Goose.

Family Rallidae
Genus and species indet.

Lothagam Material

- *Lower Nawata:* 24017, Lt. proximal carpometacarpus, proximal Rt. radius, proximal pedal phalanx; 37097, distal Rt. humerus.
- *Upper Nawata:* 25109, distal Rt. tarsometatarsus.

Rails are present in both the Lower and Upper Nawata. The younger material is approximately the same size as the extant Red-knobbed Coot (*Fulica cristata*) whereas the Lower Nawata representative is about 20 percent smaller.

Family Otididae
Eupoditis Lesson, 1839
Eupodotis sp.

Lothagam Material

- *Upper Nawata:* 25116, distal Lt. tibiotarsus.

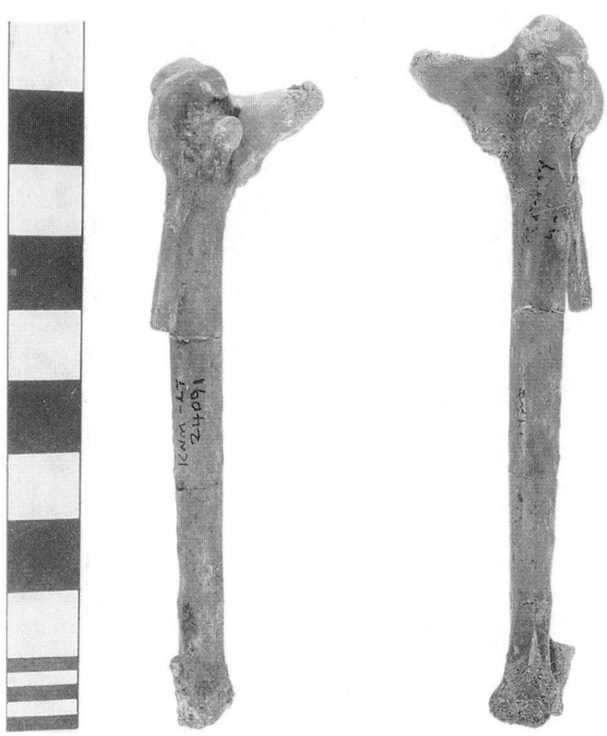

Figure 4.37 Anatidae gen. and sp. indet.: left carpometacarpus, KNM-LT 24091: left = lateral view; right = medial view.

One specimen, probably from the base of the Upper Nawata (25116), represents a medium-sized bustard.

Family Strigidae
Genus and species indet.

Lothagam Material

- *Lower Nawata:* 25114, Rt. proximal tibiotarsus and fibula, Lt. crushed proximal tibiotarsus, Lt. distal tibiotarsus, Rt. proximal tarsometatarsus, Rt. proximal ulna, Lt. distal ulna, Rt. proximal radius, Lt. distal femur fragment, Lt. distal humerus fragment.

Associated bones from the Lower Nawata represent a small species of owl.

Family indet.

Lothagam Material

- *Lower Nawata:* 23045, proximal and distal Rt. ulna, distal Rt. humerus, proximal Rt. femur, and fragment tibiotarsus shaft; 25119, distal Rt. radius; 25112, distal Rt. radius; 37099, humeral end Lt. coracoid.
- *Upper Nawata:* 24014, distal Lt. femur.
- *Apak Member:* 25108, proximal Lt. ulna.
- *Galana Boi:* 23585, distal radius; 25123, Lt. femur, Rt. tibia shaft.

Discussion

The presence of ratites and waterfowl in the Lothagam biota is not entirely unexpected. The brief summary provided here was not intended to be a comprehensive treatment but merely to document the existence of the Lothagam avifauna pending further investigation by appropriately qualified researchers.

Acknowledgments

We thank the government of Kenya and the trustees of the National Museums of Kenya for permission to work with this material. Diana Mattheisen provided provisional identifications of the fossil material and also provided helpful references. Kimball Garrett kindly made available comparative osteological material from the collections of the Natural History Museum of Los Angeles County. Alan Walker kindly assisted with identification and interpretation of the eggshell fragments.

References Cited

Andrews, C. W. 1911. Note on some fragments of the fossil eggshell of a large struthious bird from southern Algeria, with some remarks on some pieces of the eggshell of an ostrich from northern India. In H. Schalow, ed., *Verhandlungen des Fünften Internationalen Ornithologen-Kongresses, Berlin, Mai 30 bis 4 Juni, 1910,* pp. 169–174. Berlin: Deutsche Ornithologische Gesellschaft.

Burchak-Abramovich, N. I., and A. K. Vekua. 1971. The fossil ostrich from the Akchagil layers of Georgia. *Acta Zoologica Cracoviensia* 16:1–28.

Hill, A., and A. Walker. 1979. A fossil Marabou (Aves: Ciconiidae) from the Miocene Ngorora Formation, Baringo District, Kenya. *Netherlands Journal of Zoology* 29:215–220.

Mikhailov, K. E. 1988. The comparison of East European and Asian ostriches' Pliocene eggshells. *Fossil Reptiles and Birds of Mongolia: Transactions of the Joint Soviet-Mongolian Palaeontology Expedition* 34:65–72.

Mikhailov, K. E., and E. N. Kurochkin. 1988. The eggshells of Struthionoformes from the Palearctic and its position in the system of views on Ratitae evolution. *Fossil Reptiles and Birds of Mongolia: Transactions of the Joint Soviet-Mongolian Palaeontology Expedition* 34:43–65.

Sauer, E. G. F. 1966. Fossil egg shell fragments of a giant struthious bird (*Struthio oshanai,* sp. nov.) from Etosha Pan, South West Africa. *Cimbebasia* 14:1–52.

Sauer, E. G. F. 1972. Ratite eggshells and phylogenetic questions. *Bonner Zoologische Beiträge* 23:3–48.

Sauer, E. G. F. 1979. A Miocene ostrich from Anatolia. *Ibis* 121:494–501.

Sauer, E. G. F. 1976. Aepyornithoide eierschalen aus dem Miozan und Pliozan von Anatolien, Turkei. *Palaeontographica* A 153:62–115.

Sauer, E. G. F., and P. Roth. 1972. Ratite eggshells from Lanzarote, Canary Islands. *Science* 176:43–45.

Sauer, E. G. F., and E. M. Sauer. 1978. Ratite eggshell fragments from Mio-Pliocene continental sediments in the District of Ouarzazate, Morocco. *Palaeontographica* A 161:1–54.

Tullett, S. G., and R. G. Board. 1977. Determinants of avian egg shell porosity. *Journal of Zoology* (London) 183:203–211.

Tyler, C., and S. Fowler. 1979. The size, shape, and orientation of pore grooves in the egg shells of *Rhea* sp. *Journal of Zoology* (London) 187:283–298.

Whybrow, P. J., and A. Hill, eds. 1999. *Fossil Vertebrates of Arabia.* New Haven: Yale University Press.

5

LAGOMORPHA AND RODENTIA

Rodents and Lagomorphs from the Miocene and Pliocene of Lothagam, Northern Kenya

Alisa J. Winkler

At least 13 genera and 15 species of rodents and lagomorphs are reported from Lothagam from sediments dating from the Late Miocene through the Pliocene. The fauna includes the earliest African record of the Family Hystricidae (Old World porcupines), and one of the earliest African records of the Family Leporidae (rabbits and hares). Lothagam has yielded the extinct leporid *Alilepus,* previously described only from Eurasia and North America. A diverse thryonomyid fauna is present, including two taxa from the lower member of the Nawata Formation (*Paraphiomys chororensis, Paraulacodus* cf. *P. johanesi*), that had previously been described only from Chorora, Ethiopia. Younger deposits at Lothagam have yielded the derived extant cane rat, *Thryonomys.* The gerbil *Abudhabia* is reported for the first time from sub-Saharan Africa. An unnamed new genus and species of murid from the lower member of the Nawata Formation has affinity with *Myocricetodon magnus* from northern Africa. Murine rodents from Lothagam include a common extinct East African genus (*Saidomys*) and a new species of the genus *Karnimata, K. jacobsi. Karnimata* is poorly known from Africa but is better known from southern Asia. The giant squirrel, *Kubwaxerus,* in the Lower Nawata suggests the presence of forests. The occurrence of *Thryonomys* in younger sediments suggests more open habitats.

The fossiliferous deposits at Lothagam provide a rare glimpse into the faunas and environments of the poorly known Late Miocene of East Africa. These deposits are significant not only in yielding Late Miocene faunas but also in producing taxa from successively younger strata that document both faunal and environmental change through time (Leakey et al. 1996). Rodents and lagomorphs are uncommon from Lothagam. However, the Late Miocene sample (Lower and Upper Nawata) includes at least 12 taxa, some of which are known from fairly complete cranial and postcranial remains.

Excavations at Lothagam prior to the late 1980s yielded one rodent taxon, a new genus and species of giant ground squirrel, *Kubwaxerus pattersoni* (Cifelli et al. 1986; figure 5.1). Since Cifelli et al.'s study, additional remains of *Kubwaxerus,* as well as specimens of at least 19 other species of rodents and lagomorphs, have been recovered from Late Miocene through Holocene sediments by M. G. Leakey's field parties from 1989 to 1993 (table 5.1). Most specimens are from the lower (eight taxa) and upper (five taxa) members of the Nawata Formation (Late Miocene). Rodents and lagomorphs are rare from the Early Pliocene Apak (one taxon) and Kaiyumung (two taxa) Members of the Nachukui Formation. Preliminary sampling in the Holocene Galana Boi Formation has yielded a minimum of five taxa, which are listed in table 5.1, but are not discussed.

Most of the material described in this chapter was collected as surface finds. Some were also recovered by dry screening through 2 and 6 mm mesh. Wet screening through fine mesh, which enhances the recovery of rat- and mouse-sized animals, was not possible because of the scarcity of water. Thus, larger rodents and lagomorphs are better represented at Lothagam, and the generally more numerically abundant and speciose smaller-sized rodents, common in more completely sampled faunas, are underrepresented.

Figure 5.1 Restoration of *Kubwaxerus pattersoni* by Mauricio Antón. Reconstructed head and body length = 40 cm.

Larger specimens were measured with dial calipers. Smaller fossils were measured with a dissecting microscope fitted with a reticule. Geologic age of the specimens is based on radiometric dates from adjacent sediments provided by McDougall and Feibel (1999). Original specimens are housed in the Division of Palaeontology, National Museums of Kenya, Nairobi, under the prefix KNM-LT.

Systematic Description

Order Lagomorpha
Family Leporidae Gray, 1821
Alilepus Dice, 1931
Alilepus sp.

(Figure 5.2A–B)

Lothagam Material

- *Lower Nawata:* 23179, complete skull, Rt. anterior dentary (I, P_{3-4}), Rt. proximal ulna, Lt. and Rt. distal humeri, Rt. proximal femur, Lt. and Rt. distal femora, Lt. proximal ilium, two lumbar vertebrae, proximal forelimb phalanx; 22999, fragmentary and crushed skull and skeleton including maxillary fragments (Lt. and Rt. I^{1-2}, Rt. P_4 and M^{1-2}, and fragment with two cheek teeth), Lt. dentary (I, P_3–M_2), Rt. dentary (I, P_3–P_4 [trigonid only]), two vertebrae (one lumbar), and Rt. calcaneum. 23179 is from the Monkey Area, just below the Marker Tuff. 22999 is from the Northern Area, below the Marker Tuff. Both specimens are 6.54–6.57 Ma based on a radiometric date of 6.54 ± 0.04 for the Marker Tuff and 6.57 ± 0.07 for 5 m below the Red Marker (the next older radiometrically dated unit; McDougall and Feibel 1999). 23179 is likely closer to 6.54 Ma.

Formal description of the Lothagam *Alilepus* will be provided elsewhere (Winkler in preparation). The P_3s are illustrated (figure 5.2A–B) and briefly described here in order to establish the presence of this genus at Lothagam. Tooth terminology follows White (1991). Occlusal mesiobuccal length by labiolingual width measurements in millimeters are: 22999 right P_3 3.92 by 3.42; left P_3 4.00 by 3.17 and 23179 3.33 by 2.92. The enamel pattern of 22999 is more distinct than that of 23179. The latter specimen has sufficient occlusal and postmortem wear so that the thickness, much less the pattern, of the enamel is often difficult to distinguish. Overall, the morphology of the two specimens is comparable. The posterior internal reentrant (PIR) is as deep as the posterior external reentrant (PER). Both extend about halfway across the tooth. The isthmus between the trigonid and talonid is about two times thicker on 23179 than it is on 22999, and the lateral edge of the PIR points anteriorly on 23179. An anterior internal reentrant (AIR) is shallow to lacking, compared to the posterior internal reentrant. The anterior reentrant (AR) is missing. Thick enamel on the anterior edge of PER (TH) and thin enamel on the posterior edge of PER (TN) are smooth to slightly folded. The anterior external reentrant (AER) is relatively shallow (compared to PIR and PER) with smooth thin enamel.

Discussion

The genus *Alilepus* was erected by Dice (1929, 1931) based on the type species "*Lepus*" *annectens* Schlosser from the Late Miocene localities of Ertemte and Olan Chorea, Inner Mongolia, China (Schlosser 1924; Flynn et al. 1995). White (1991) revised the genus based on study of *Alilepus* from the Late Miocene to Pliocene of Eurasia and North America and diagnosed the genus as being of medium to large size with a fully modernized cranium and dentary. His diagnosis was also based on morphology of the P^2 (morphology uncertain on the one Lothagam specimen) and P_3. LT 23179 and 22999 (figure 5.2A–B) are assigned to *Alilepus* based on White's diagnosis of the P_3: PIR as deep or shallower than the PER, AIR shallower (usually missing) than the PIR, AR missing, TH smooth to slightly folded, and AER shallow with smooth thin enamel. In *Alilepus* the PIR is often (but not always) pinched off to form an enamel lake (White 1991). This was not observed on the *Alilepus* specimens from Kenya.

The *Alilepus* morphotype may be found as a variant within samples of P_3s of the Pliocene-Pleistocene genus *Serengetilagus* (Averianov personal communication 1999) and the Late Miocene–Pleistocene genus *Trischizolagus* (Averianov and Tesakov 1997). *Serengetilagus* is known from Lothagam in the Apak Member, Nachukui Formation. It is possible that the "*Alilepus*" specimens from Lothagam are really *Serengetilagus* (or *Trischizolagus;* the distinction between these two genera is controversial). However, since both of the known Nawata specimens have the *Alilepus* pattern, it is more likely that they pertain to *Alilepus.*

The presence of *Alilepus* in Late Miocene sediments at Lothagam is one of the earliest records of the Family Leporidae in Africa. It may be the only, or one of only two, African reports of the genus *Alilepus.* An undescribed isolated P_3 that may be attributable to *Alilepus* is known from the Lukeino Formation, Tugen Hills, Kenya, which is constrained by isotopic dates of 6.2 and 5.6 Ma (Hill 1999). The morphology of *Alilepus* is incompletely known: most remains are tooth and jaw fragments, and only one skull has been described

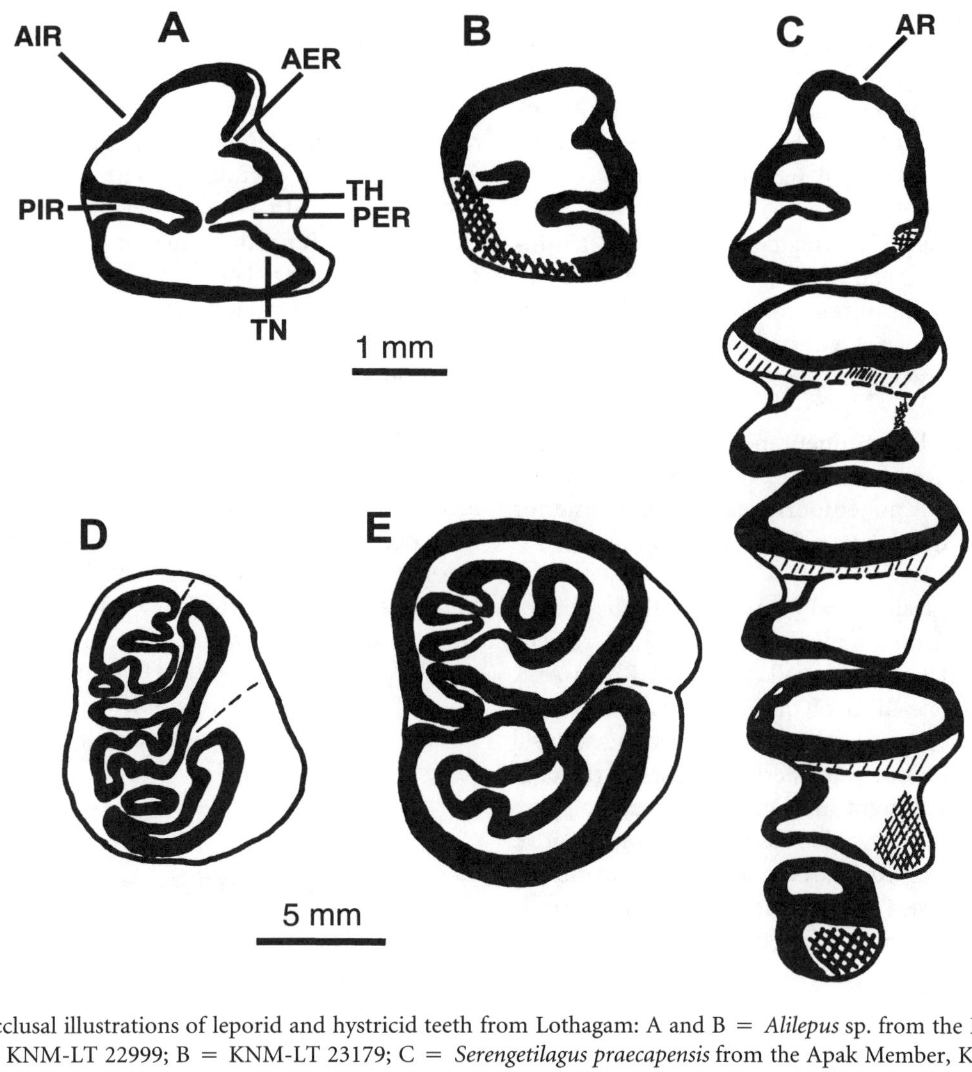

Figure 5.2 Occlusal illustrations of leporid and hystricid teeth from Lothagam: A and B = *Alilepus* sp. from the Lower Nawata, right P_3; A = KNM-LT 22999; B = KNM-LT 23179; C = *Serengetilagus praecapensis* from the Apak Member, KNM-LT 24963, left P_3-M_3. Leporid P_3 tooth terminology illustrated in A and C: AER = anterior external reentrant; AIR = anterior internal reentrant; AR = anterior reentrant; PER = posterior external reentrant; PIR = posterior internal reentrant; TH = thick enamel in PER; TN = thin enamel in PER. D = *Hystrix* sp. (small), from the Lower Nawata, KNM-LT 24948, right dP_4. E = *Hystrix* sp. (large), from the Kaiyumung Member, KNM-LT 23115, right M_1 or M_2. 1 mm bar scale applies to A, B, and C; 5 mm bar scale applies to D and E.

(White 1991). The more complete remains from Lothagam will add significantly to our knowledge of the functional morphology of this animal. In addition, these African records of *Alilepus* may provide insight into the origin of the endemic African leporids *Bunolagus* and *Pronolagus,* which have P_3s morphologically similar to those of *Alilepus* (Averianov personal communication 1999).

Serengetilagus Dietrich, 1941
Serengetilagus praecapensis Dietrich, 1941
(Figure 5.2C; table 5.2)

Lothagam Material

- *Apak Member:* 24963, incomplete Lt. dentary (broken incisor, P_3–M_3). This specimen is from the Central Area, above the Purple Marker. Its age is estimated to be older than 4.22 Ma, based on radiometric dates of 4.22 Ma ± .03 for 17 m below the Lothagam Basalt and 35 m above the Purple Marker, and 4.20 Ma ± .03 for the overlying Lothagam Basalt (McDougall and Feibel 1999).

The Apak lagomorph is a fragmentary dentary that includes the posterior end of the diastema and the tooth row (figure 5.2C; measurements in table 5.2). There is considerable evidence of weathering: the bone and teeth have many fine cracks, and there is exfoliation of a thin outer layer of bone.

There is a single prominent mental foramen located about 2 mm anterolateral to P_3. The beginning of the ascending ramus is present, originating at the posterior end of M_3. On the lingual side of the mandible the incisor ends at the level of P_3.

The incisor is trapezoidal in outline, with the base of the trapezoid oriented ventrolabially. The P_3 is crescentic in outline. It has very small AR and AIR. An AER extends about one-third of the way across the tooth. The PER runs approximately halfway across the occlusal surface. The anterior and posterior borders of the posterior external reentrant are lightly crenulated. There may be a small PIR, but there is breakage in this area. The P_4–M_2 are each composed of an elevated ovoid trigonid and a lower ovoid talonid. The small M_3 is formed by two flattened ovals (trigonid and talonid).

Discussion

Serengetilagus praecapensis was originally described by Dietrich (1941, 1942) based on specimens collected by Kohl-Larsen in 1938–1939 from Laetoli, Tanzania, and now housed in collections at the Museum of Natural History, Humboldt University, Berlin. MacInnes (1953) described additional remains of this species, probably recovered from the northeastern end of Lake Eyasi (?Pleistocene age; not from Laetoli; Davies 1987), and now in collections at the British Museum (Natural History), London. Additional remains of *S. praecapensis* collected from Laetoli in 1959 are in the collections of the National Museums of Kenya, Nairobi. This material has not been formally studied. The British Museum material was reexamined by Davies (1987), and he suggests that two different subspecies may be present, one in the Laetolil Beds (3.7–3.59 Ma) and one in the Upper Ndolanya Beds (3.0–2.5 Ma). Erbaeva and Angermann (1983) redescribed the material Dietrich had originally examined; they designated a lectotype and noted and illustrated that the P_3 of this species shows much morphologic variability. LT 24963 is assigned to *S. praecapensis* based on possession of a PER that crosses about one-half the occlusal surface, an AER, and an AR and AIR (although the latter two are rudimentary). The Lothagam specimen may have a PIR. Presence of this reentrant is unusual in *Serengetilagus* but is within the range of variation observed in this species (Erbaeva and Angermann 1983:figure 3). With few exceptions, the size of individual teeth and length of the toothrow of LT 24963 is within the range of variation observed in comparative material of *S. praecapensis* (table 5.2). Measurements out of range of comparative material include a slightly narrower P_3 and slightly wider M_1 and M_2. Location of the base of the incisor in LT 24963 is as observed in *S. praecapensis* (Erbaeva and Angermann 1983; Davies 1987).

The occurrence of *S. praecapensis* at Lothagam establishes the presence of this species in Kenya, in addition to its occurrence in Tanzania. *Serengetilagus* aff. *S. praecapensis* has been reported recently from Chad (circa 5 Ma based on biochronology; Brunet and M.P.F.T. 2000). I agree with Flynn and Bernor (1987) that a leporid P_3 (KW 138) from the Pliocene Kanam West locality, Kenya, is likely referable to *Serengetilagus.* The Lothagam *Serengetilagus* is one of the oldest records of the genus.

Family Hystricidae Burnett, 1830
Hystrix Linnaeus, 1758
Hystrix sp. (small)
(Figure 5.2D; table 5.3)

Lothagam Material

- *Lower Nawata:* 24948, isolated Rt. dP_4. The specimen comes from the Northern Area, low in the section, possibly >7.44 Ma based on an isotopic age for the

Lower Markers of 7.44 ± 0.05 (McDougall and Feibel 1999). The minimum age for the lower member of the Nawata Formation is 6.54 ± 0.04 for the Marker Tuff, which is a distinct boundary between the Upper and Lower Nawata (McDougall and Feibel 1999). Dates for the underlying Nabwal Arangan Beds range from 14.2 ± 0.02 Ma to 9.1 ± 0.02 Ma (McDougall and Feibel 1999).

This relatively small tooth (figure 5.2D) is elongate and highly compressed labiolingually, especially anterior to the labial sinusid (terminology of Denys 1987). There is light occlusal wear. The enamel is heavily crenulated. There are three roots: a larger anterior root at the midline and two smaller posterolabial and posterolingual roots.

Discussion

Measurements of the Lothagam specimen and comparative material are given in table 5.3. Specific identification of isolated porcupine teeth based on the occlusal pattern or size can be very difficult. The occlusal pattern changes with continued wear (Masini and Rook 1993; Sen and Kovatchev 1987). Porcupine teeth are known to vary in size with crown height (Masini and Rook 1993; Sen and Kovatchev 1987), so the significance of these differences in size (especially with small sample sizes) is uncertain, if comparisons are not made with specimens at a similar wear stage.

LT 24948 is larger than a dP_4 of the small extinct species *H. leakeyi* from Laetoli, Tanzania (Denys 1987). Crown heights of the Laetoli (6.3 mm) and Lothagam (6.65 mm) specimens are close. The Lothagam tooth is slightly smaller than four dP_4s of *Hystrix* cf. *H. makapanensis* from Olduvai Bed I, Tanzania (measurements from Sabatier 1978, in Denys 1987). Crown height for these specimens is not reported. One heavily worn dP_4 of the extant species *H. cristata* is smaller (crown height 3.12 mm). Denys (1987) does not report any dP_4s of the large fossil African genus *Xenohystrix*, which would be even larger than *H. makapanensis*. LT 24948 is likely too small to be *Xenohystrix*. Regarding the different species of *Hystrix* (see the following section), it is most parsimonious to assign the Lothagam tooth to *Hystrix* sp.

The fossil record of porcupines in Africa is limited. The earliest southern African record of porcupines is from the Early Pliocene at Langebaanweg, South Africa (Hendey 1981). Two taxa are recognized, but neither has yet been identified nor described. Greenwood (1955, 1958) described material from the Late Pliocene at Makapansgat, South Africa. This material was reanalyzed by Maguire (1976, 1978; in Collings et al. 1976), who also studied fossil porcupines from other South African localities. Three species are identified from fossil deposits in South Africa: (1) *Xenohystrix crassidens*, an extinct, large, relatively brachyodont form with distinct roots; (2) *Hystrix makapanensis*, which is hypsodont relative to *X. crassidens* and is intermediate in size between *X. crassidens* and the extant species *H. africaeaustralis*; and (3) *H. africaeaustralis*.

The early remains of hystricids from East Africa are generally better constrained chronologically by isotopic dates. Specimens from Hadar, Ethiopia (3.4–3.18 Ma; *X. crassidens*), have been described by Dietrich (1942), Sabatier (1978), and Denys (1987). Sabatier (1982) and Wesselman (1984) studied porcupines from the Omo Group, Ethiopia. The Omo specimens include *X. crassidens* (about 3 Ma), *H. makapanensis* (about 3–2 Ma), and the extant species *H. cristata* (about 3 Ma and younger). Specimens from Laetoli, Tanzania, have been described by Dietrich (1942), Sabatier (1978), and Denys (1987). Material from the Laetolil Beds (3.7–3.59 Ma), as summarized by Denys (1987), includes *X. crassidens*, *H.* cf. *makapanensis*, and a small extinct species, *Hystrix leakeyi*. *Hystrix* sp. indet. is questionably reported from the geologically younger Upper Ndolanya Beds at Laetoli (3–2.5 Ma; Denys 1987). *Hystrix makapanensis* is also known from Olduvai Bed I, Tanzania (1.75–1.65 Ma; Sabatier 1978).

Reports of Miocene-Pliocene hystricids from northern Africa are rare. *Hystrix* sp. is reported from the Late Miocene at Marceau, Algeria (Arambourg 1959), but the age of this material is uncertain.

Prior to recovery of the dP_4 from Lothagam, the oldest specimen of a porcupine from Africa was an isolated upper incisor (LU 974/615) of *Hystrix* sp. from the Lukeino Formation, Tugen Hills, central Kenya (Winkler 1990; 6.2–5.6 Ma: Hill 1999). This specimen is not identifiable at the species level. The oldest known hystricid is *Sivacanthion*, a primitive form from the Middle Miocene (lower Siwaliks) of northern India (Colbert 1933). The presence of a geologically older primitive taxon in southern Asia suggests an origin in, and subsequent dispersal from, southern Asia (Jacobs et al. 1985).

Hystrix sp. (large)

(Figure 5.2E; table 5.3)

Lothagam Material

- *Kaiyumung Member:* 23115, Rt. M_1 or M_2. The specimen derives from the extreme northern end of the Kaiyumung exposures. The Kaiyumung Member cannot be dated isotopically (McDougall and Feibel 1999); however, the northern exposures of the Kaiyumung Member likely date between about 4.0 and 3.5 Ma (M. G. Leakey personal communication; Leakey et al. 1996).

LT 23115 is illustrated in figure 5.2E. The tooth has moderate occlusal wear and is relatively brachyodont. Crown height is about 10 mm. This tooth is from a relatively large porcupine (table 5.3), most comparable in size to that of either an M_1 or M_2 of *Hystrix* cf. *H. makapanensis* from Olduvai Bed I, Tanzania (1.75–1.65 Ma; Sabatier 1978 presented in Denys 1987), or *H. makapanensis* from Makapansgat, South Africa (about 3–2.5 Ma; Greenwood 1955 presented in Denys 1987). Crown height for these specimens is not provided. LT 23115 is also comparable in size with the Late Miocene–Early Pliocene species *H. primigenia* from the circum-Mediterranean regions and the Near East (Masini and Rook 1993:table 2). Crown height is not indicated for these specimens.

Discussion

Hystrix makapanensis is generally considered to have hypsodont cheek teeth. Maguire (in Collings et al. 1975) describes *H. makapanensis* as less brachyodont than *Xenohystrix,* but less hypsodont than *H. africaeaustralis. Hystrix primigenia* is relatively brachyodont (Masini and Rook 1993). The porcupine molar from Lothagam is assigned conservatively to *Hystrix* sp. (large) for the following reasons: (1) there is only one tooth known of this taxon; (2) there are many problems with identification of hystricid teeth as discussed above; and (3) the relationships between African and European hystricids has not yet been established (except for *H. cristata;* Masini and Rook 1993).

Family Sciuridae Gray, 1821
Tribe Protoxerini Moore, 1959
Kubwaxerus Cifelli et al., 1986
Kubwaxerus pattersoni Cifelli et al., 1986

Lothagam Material

- *Lower Nawata:* 447 (holotype), anterior Lt. dentary (base I_1 and P_4–M_2 alveoli), proximal Rt. scapula, Rt. humerus, shaft and distal Lt. humerus, Lt. innominate, incomplete Rt. ischium, a distal metapodial, an ungual, Rt. cuboid, vertebrae and vertebral fragments, rib fragments (not all the fragments catalogued as LT 447 pertain to this individual); 23082, isolated Lt. M_3, proximal and distal Lt. femur, Lt. tibia, one thoracic and three cervical vertebrae, three distal phalanges, four proximal and four distal metapodials, Lt. and Rt. naviculars, proximal and distal Lt. calcaneum; 23087, Rt. anterior dentary (I_1–M_1); 24210, crushed skull fragment (Lt. P^4, M^{1-2}, Rt. M^1, Lt. $\underline{I}$), Lt. anterior dentary (I_1 and roots M_{1-3}), Rt. anterior dentary (I_1), Lt. and Rt. proximal femora, distal femur fragment, Rt. proximal humerus, Rt. proximal ilium, one caudal vertebra; 24951, Lt. anterior dentary (I_1, M_{1-2}); 24954, Lt. incomplete dentary (P_4–M_3).
- *Horizon indet.:* 10019, Lt. M^1 or M^2, anterior Lt. dentary with I_1, fragments of upper and lower incisors, proximal Rt. and Lt. ulnae, proximal Lt. radius, Rt. calcaneum, three thoracic, two lumbar, and three caudal vertebrae, two proximal metapodials, three incomplete phalanges.

The first specimen of *Kubwaxerus,* LT 10019, was collected in 1968. There are no field data for this specimen, but Cifelli et al. (1986) noted it was from "Member" B or C (former Lothagam stratigraphic terminology). In current terminology, "B" would range from the Lower Nawata into the Upper Nawata, and "C" would include some of the upper part of the Upper Nawata and some of the Apak Member. The holotype and all other specimens are from the Lower Nawata. The holotype, 23087, and 24951 are from the Northern Area. 24951 is from 15 m above the Gateway Sandstone. 24210 and 24954 are from the Monkey Area. 24954 is from below the Red Marker. Most of these specimens range in age from a minimum of 6.54 ± 0.02 Ma (age of the Marker Tuff) to a maximum of about 7.44 ± 0.05 Ma (age of the Lower Marker; McDougall and Feibel 1999). 24954 has a minimum age of about 6.57 Ma based on a date of 6.57 ± 0.05 Ma for 5 m below the Red Marker (McDougall and Feibel 1999).

Discussion

The original description of *K. pattersoni* (Cifelli et al. 1986) was based on LT 447 and 10019. Both these specimens include a good collection of postcrania, but essentially no dentition (see above). Further collecting at Lothagam has yielded five additional individuals, and the complete dentition (based on a composite from several individuals) is now known. Description and analysis of these dental remains and additional postcranial material will be presented elsewhere (Winkler in preparation).

Family Thryonomyidae Pocock, 1922
Paraphiomys Andrews, 1914
Paraphiomys chororensis Geraads, 1998
(Figure 5.4A; table 5.4)

Lothagam Material

- *Lower Nawata:* 22998, Rt. dentary (incisor, dP_4, M_{1-3}). The specimen derives from the Northern Area: 7.44–6.54 Ma.

The lingual side of the mandible is somewhat crushed, and the labial side is abraded. The anterior end of the jaw and the portion of the jaw posterior to M_3 are lacking. The incisor is roughly ellipsoidal in outline, with thick enamel that extends approximately one-third the way along the lingual and labial sides. There is a faint ridge running along the anterolabial face of the incisor.

Measurements of the cheek teeth are given in table 5.4. Tooth terminology for thryonomyids is illustrated in figure 5.3. The dP_4–M_2 show moderate occlusal wear, and the M_3 has light wear (figure 5.4A). Hypsodonty is moderate, comparable to that seen in species of *Paraulacodus* and *Paraphiomys,* but not as great as that of *Neosciuromys.* The dP_4 is proportionally the most elongate tooth. It has three transverse lophs: metalophid, hypolophid, and posterolophid. The anterior positions of the metaconid and entoconid result in the metalophid and hypolophid, respectively, being strongly oblique. The anterolingual sulcus is wider and deeper than the posterolingual sulcus. There is also a deep labial sulcus. This tooth has a very large anterior root. It is uncertain if the large lingual and posterolingual roots are fused in the midline.

The M_1 and M_2 are roughly trapezoidal in outline and have three transverse lophs. The metaconid is located anterior to the protoconid; thus the metalophid is strongly oblique. The hypolophid is transverse. A posterior arm of the protoconid is lacking. There is a distinct anterolabial cusp that is connected to the metalophid. The anterolabial cusp on the M_2 is about twice the size of that on the M_1. The valley between the anterolabial cusp and the metalophid opens labially. These teeth have a large anterior root. It cannot be determined if the lingual root and large posterolabial root are fused in the midline.

The M_3 is approximately triangular in outline, tapering posteriorly. Occlusal morphology of this tooth is similar to that of the M_1 and M_2. The anterolabial cusp is comparable in size to that of the M_2. The M_3 has large anterolabial and lingual roots, and a larger posterior root.

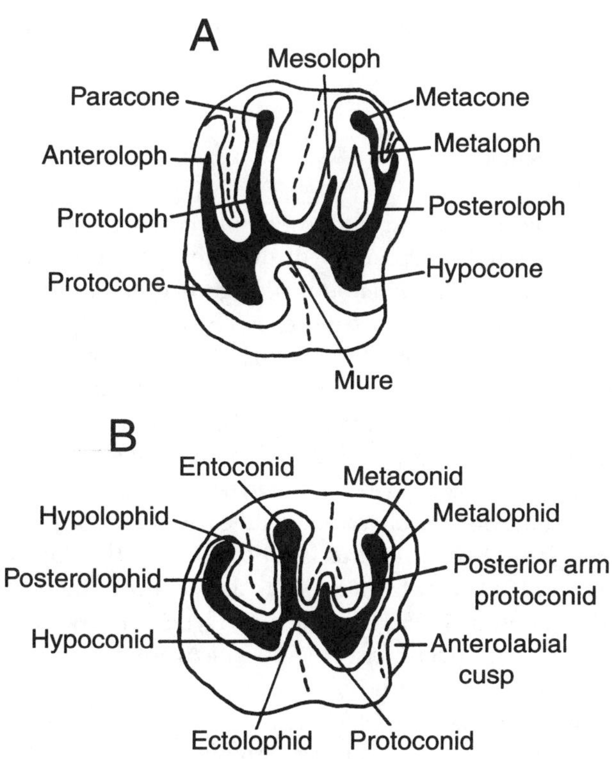

Figure 5.3 Tooth terminology for thryonomyids using *Paraphiomys pigotti* (an Early Miocene taxon): A = left M^2; B = right M_2 (from Winkler 1992).

Discussion

Paraphiomys chororensis was erected by Geraads (1998a) based on a sample of four isolated upper molars, five lower dentitions, and isolated lower molars and incisors from Chorora, Ethiopia (10.7–10.4 Ma). The species is diagnosed by the following characters: (1) similar in size to *P. stromeri hopwoodi* (Early Miocene, Songhor, Kenya; Lavocat 1973); (2) upper incisors lacking grooves; (3) upper molars with a constant moderately long mesoloph; (4) metaloph generally rejoining the mesoloph and rarely fusing with the posteroloph; and (5) lower molars lacking a posterior arm of the protoconid (derived). LT 22998 is assigned to *P. chororensis* based on similar size (table 5.4) and morphology, including cheek teeth with three lophs, absence of the posterior arm of the protoconid, anterolabial cusp attaching to the metalophid, and the valley between the anterolabial cusp and protoconid opening labially. The slightly longer molars (about 4–10 percent longer) of the Lothagam specimen may prove (with larger sample sizes of both populations) to be of taxonomic significance, but the difference is currently insufficient to assign LT 22998 to a different species.

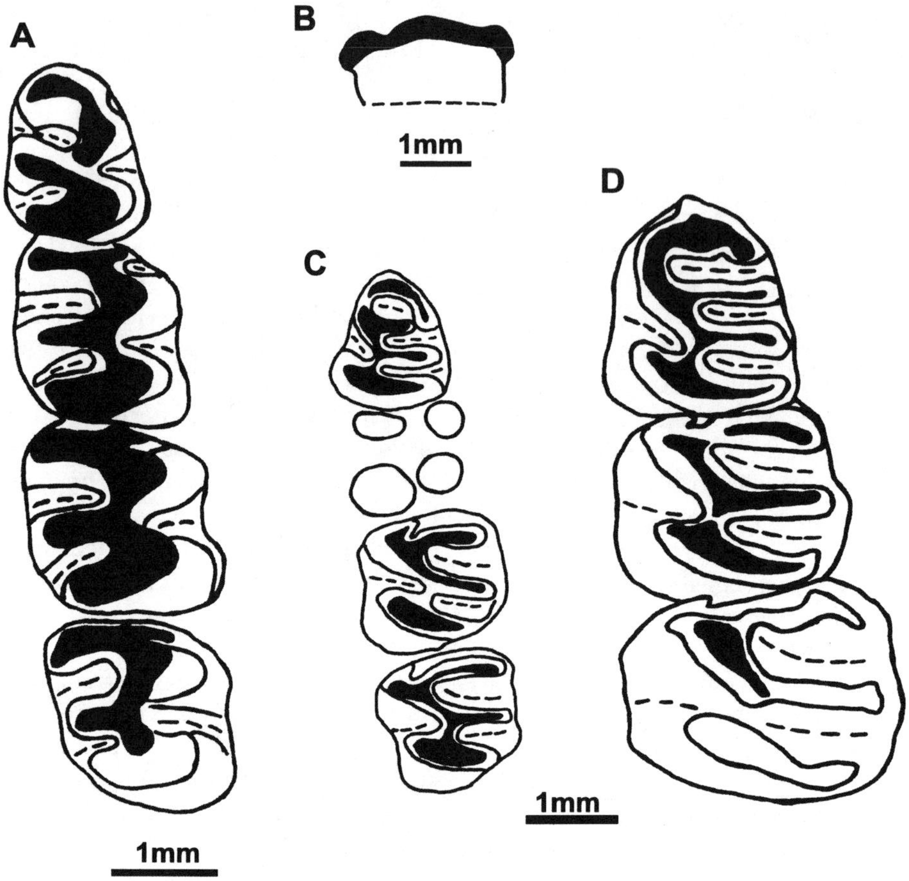

Figure 5.4 Occlusal illustrations of Lower Nawata thryonomyids and an extant *Thryonomys*. A = *Paraphiomys chororensis*, KNM-LT 22998, right dP_4, M_{1-3}; B = *Paraulacodus* cf. *P. johanesi*, KNM-LT 26542, upper incisor; C = *Thryonomys* sp. (small), KNM-LT 24950, left dP_4, M_{2-3}; D = extant *Thryonomys swinderianus*, left dP_4, M_{1-2}.

Paraulacodus Hinton, 1933
Paraulacodus johanesi Jaeger, Michaux, and Sabatier, 1980
Paraulacodus cf. *P. johanesi*
(Figure 5.4B; table 5.5)

Lothagam Material

- *Lower Nawata:* 26542, isolated upper incisor. The specimen derives from the Monkey Area, just below the Marker Tuff. Slightly more than 6.54 Ma based on a date of 6.54 ± 0.04 Ma for the Marker Tuff.

LT 26543 (figure 5.4B; table 5.5) is triangular in cross-section (complete antero-posterior height is unavailable). There are two distinct anterior grooves in the enamel, with a wide median ridge between them. The two grooves are approximately equal distance from the edges of the tooth. The medial groove is broader and shallower than the labial groove. Enamel extends equally and only slightly posteriorly along the labial and lingual sides of the tooth. The anterolabial and antero-lingual intersections are both angular.

Discussion

This specimen is assigned to *Paraulacodus* based on the presence and morphology of the two distinct grooves in the enamel. *Paraulacodus* is diagnosed in part by an upper incisor with two grooves (Hinton 1933; Black 1972). The extant and fossil genus *Thryonomys* has three grooves. The upper incisors of *Paraphiomys pigotti*, *P. stromeri*, and *P. chororensis* (Geraads 1998a) are smooth. The upper incisors of *Paraphiomys shipmani* (Denys and Jaeger 1992), *Apodecter* and *Neosciuromys* are unknown. *Paraulacodus* includes two described species. *Paraulacodus indicus* is reported from the Potwar Plateau, Pakistan, from deposits dating from 12.9–12.5 Ma (Hinton 1933; Black 1972; Flynn and Winkler 1994). Two upper incisors of *P. indicus* are known, one of which is relatively narrow and may be a juvenile (Flynn and Winkler 1994). *Paraulacodus johanesi* is described only from Chorora, Ethiopia (10.7–10.5 Ma; Jaeger et al. 1980; Geraads 1998a), from a sample that includes five upper incisors. This species is also reported, but not described, from the early Late Miocene at Berg Aukas, Namibia (Conroy et al. 1992).

Paraulacodus indicus and *P. johanesi* are similar in morphology and size. Differences include *P. johanesi* having a heavier dentary and upper incisor (shorter antero-posterior height), relatively wider lower incisors, and weaker anterolabial cusp on lower molars (Flynn and Winkler 1994; Jaeger et al. 1980). LT 26542 is comparable in morphology and close in width to both of the two described species (table 5.5). It is possible that the Lothagam tooth represents a species with a wider incisor, but a larger sample is needed to test this. Since the antero-posterior height cannot be compared, the Lothagam specimen is tentatively referred to *P. johanesi*, based on its geographic proximity (to Ethiopia) and the presence of another Chorora taxon in the Lothagam fauna.

Genus *Thryonomys* Fitzinger, 1867
Thryonomys gregorianus Thomas, 1894
Thryonomys cf. *T. gregorianus*
(Table 5.6)

Lothagam Material

- *Kaiyumung Member:* 23692, fragment of palate with Rt. and Lt. M^1–M^3. This specimen is from the Kaiyumung southern exposures, which may be slightly younger than the Kaiyumung northern exposures. The age is 4.0–3.5 Ma.

The bone of LT 23692 is badly abraded, and the teeth are cracked. Cracked portions of the left M^2 are displaced, so the width of the tooth cannot be determined. The M^1s have moderate occlusal wear, the M^2s have light wear, and the M^3s are essentially unworn (both are not fully erupted). Assignment of LT 23692 to *Thryonomys* is based on relatively wide and antero-posteriorly compressed cheek teeth that have three simple oblique lophs (a mesoloph is lacking; Winkler 1992; Flynn and Winkler 1994). The left M^2 has a metaloph, which would be subsumed with moderate wear. The morphology of the Lothagam specimen is within the range of variation observed in the two extant species, *T. gregorianus* and *T. swinderianus.* The teeth of LT 23692 are not as antero-posteriorly compressed (transverse lophs are more widely separated) as comparative adult specimens of *T. gregorianus* and *T. swinderianus.* However, the degree of compression is comparable with juvenile specimens of both extant species. *Thryonomys gregorianus* is generally smaller in size than *T. swinderianus* (although there is overlap, table 5.6). The two species can also be distinguished by soft tissue structure (e.g., tail length), upper incisor morphology, and some other cranial and dental characters (Kingdon 1974; Denys 1987). The palate from Lothagam is small and is closer in size to *T. gregorianus.* Incompleteness and lack of other diagnostic characters on the single fossil specimen preclude definitive assignment to *T. gregorianus.*

Discussion

The oldest records of *Thryonomys* are from the Early Pliocene (possibly Late Miocene). Two isolated upper cheek teeth of a small *Thryonomys* are known from the Wembere-Manonga Formation, Inolelo 1 locality, Tanzania (about 5–4 Ma; Winkler 1997). A sample of five specimens of a small *Thryonomys* (including one M^1) are reported from the Upper Ndolanya Beds, Laetoli, Tanzania (3.5–2.4 Ma; Denys 1987). Morphology of these specimens is similar to that of the Lothagam teeth, but the teeth differ in size. There is a single incomplete lower cheek tooth of another small *Thryonomys* from the Chemeron Formation, Tabarin locality, Kenya (Winkler 1990) about 4.5–4.4 Ma (Hill 1999). More recent fossil *Thryonomys* are discussed by Denys (1987), and *Thryonomys* sp. is also known from the Lusso Beds, Upper Semliki Valley, Zaire (2.3–2 Ma; Boaz et al. 1992).

Thryonomys sp. (small)
(Figure 5.4C)

Lothagam Material

- *Upper Nawata:* 26544, isolated Rt. lower incisor; 24200, Rt. dentary (M_{1-3}); 24202, Lt. dentary (incisor, dP_4, M_{1-2}); 24949, Lt. dentary (incisor, roots dP_4, M_{1-3}); 24950, Lt. dentary (incisor, dP_4, roots M_1, and M_{2-3}); 24956, Rt. dentary (incisor, M_{1-3}). The specimens derive from the Central Area (south), below the Purple Marker. The Purple Marker cannot be dated radiometrically, but an extrapolated date based on sedimentation rates and paleomagnetic data suggests the Purple Marker could be 5.23 Ma (McDougall and Feibel 1999).

Discussion

A representative specimen is illustrated in figure 5.4C, and can be compared to an extant (and much larger) specimen of *T. swinderianus* (figure 5.4D). The fossil taxon is assigned to *Thryonomys* based on a dP_4 with four lophs, lower molars with three lophs, an anterolabial cusp that is connected to the protoconid, the valley between the anterolabial cusp and protoconid opening lingually, and relatively wide cheek teeth. These

Lothagam thryonomyids are as small or smaller than the extant species *T. gregorianus* and may be assignable to a new species (Winkler in preparation).

Family Muridae Rochebrune, 1883
Subfamily Gerbillinae Alston, 1876
Abudhabia de Bruijn and Whybrow, 1994
Abudhabia sp.

(Figure 5.6C; table 5.7)

Lothagam Material

- *Upper Nawata:* 24211, isolated Rt. M^1. The specimens derive from the Central Area, below the Purple Marker. Estimated minimum age is 5.23 Ma based on sedimentation rates and paleomagnetic data, which suggests the Purple Marker could be 5.23 Ma (McDougall and Feibel 1999).

Tooth terminology is comparable to that for murines (figure 5.5). LT 24211 is illustrated in figure 5.6C. Measurements are given in table 5.7. This tooth is elongate and has relatively high cusps. There is light occlusal wear. The anteroconid is wide and medially located. It is single, but there are two weak grooves on its anterior face, and these grooves signify derivation from a tricuspid condition. A lingual cingulum is lacking, and a labial cingulum is faint. Wide deep valleys separate the anteroconid from the protoconid and metaconid, and the protoconid and metaconid from the hypoconid and entoconid. The protoconid is a little larger than, and is located only slightly posterior to, the metaconid. The protoconid and metaconid are distinct transversely elongate ovals. The hypoconid is somewhat smaller than, and is situated slightly posterior to, the entoconid. A distinct conical posterior cingulum is in the midline of the tooth, adjacent to the hypoconid. A faint cingulum extends labially from the posterior cingulum along the posterior side of the hypoconid. A short cingulum extends lingually from the posterior cingulum and is separated from the entoconid by a valley. There are large anterior and posterior roots, and a central rootlet.

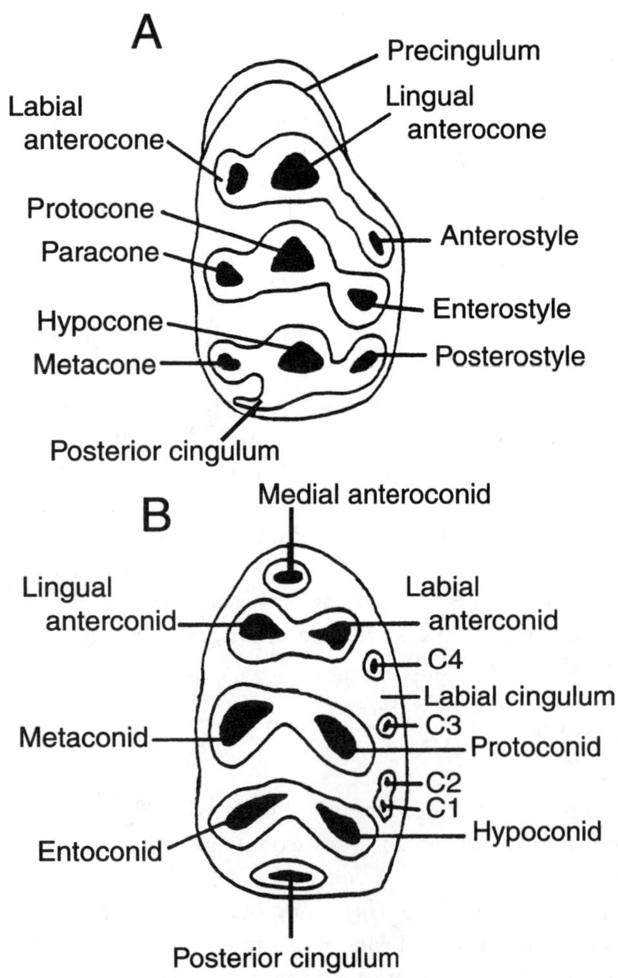

Figure 5.5 Tooth terminology for murine rodents: A = right M^1; B = right M_1 (after Jacobs 1978).

Discussion

LT 24211 is similar in morphology to *Abudhabia* and *Protatera,* both of which include Miocene-Pliocene gerbils that generally lack longitudinal crests. The genus *Abudhabia* was erected by Bruijn and Whybrow (1994) to include the species *A. baynunensis* and "*Protatera*" *kabulense* (Sen 1983). The diagnosis of the genus *Abudhabia* (Bruijn and Whybrow 1994) states that M^1 and M_1 have a postero-central cusp (= posterior cingulum); M^2 and M_2 have remnants of an anterior cingulum; the main cusps of M^1, M^2, and M_2 form transverse ridges; the main cusps of M_1 are alternating; the longitudinal crest is absent between cusp-pairs; the enterocone of upper molars is absent; and the upper incisor has one longitudinal groove.

Abudhabia baynunensis is known from ten isolated teeth from the Late Miocene of Abu Dhabi, United Arab Emirates. *Abudhabia kabulense* is based on 41 isolated teeth from the Early Pliocene of Pul-e Charkhi, Afghanistan. *Abudhabia* cf. *A. kabulense* is reported from the Early and Late Pliocene of India (localities at 4.5–3.5 Ma and 2.5 Ma; Patnaik 1997). Flynn and Jacobs (1999) describe *A. pakistanensis* from two maxillary fragments (lower dentition unknown) from the Late Miocene of Pakistan (8.6 Ma); they suggest that *Protatera yardangi,* from the Late Miocene of Sahabi, Libya (Munthe 1987), belongs in the genus *Abudhabia.*

Protatera was erected by Jaeger (1977) based on 33 isolated teeth of *P. algeriensis* from the Late Miocene at Amama 2, Algeria. *Protatera* was diagnosed as having cusps fused into transverse or oblique lophs (M^1 or M^2;

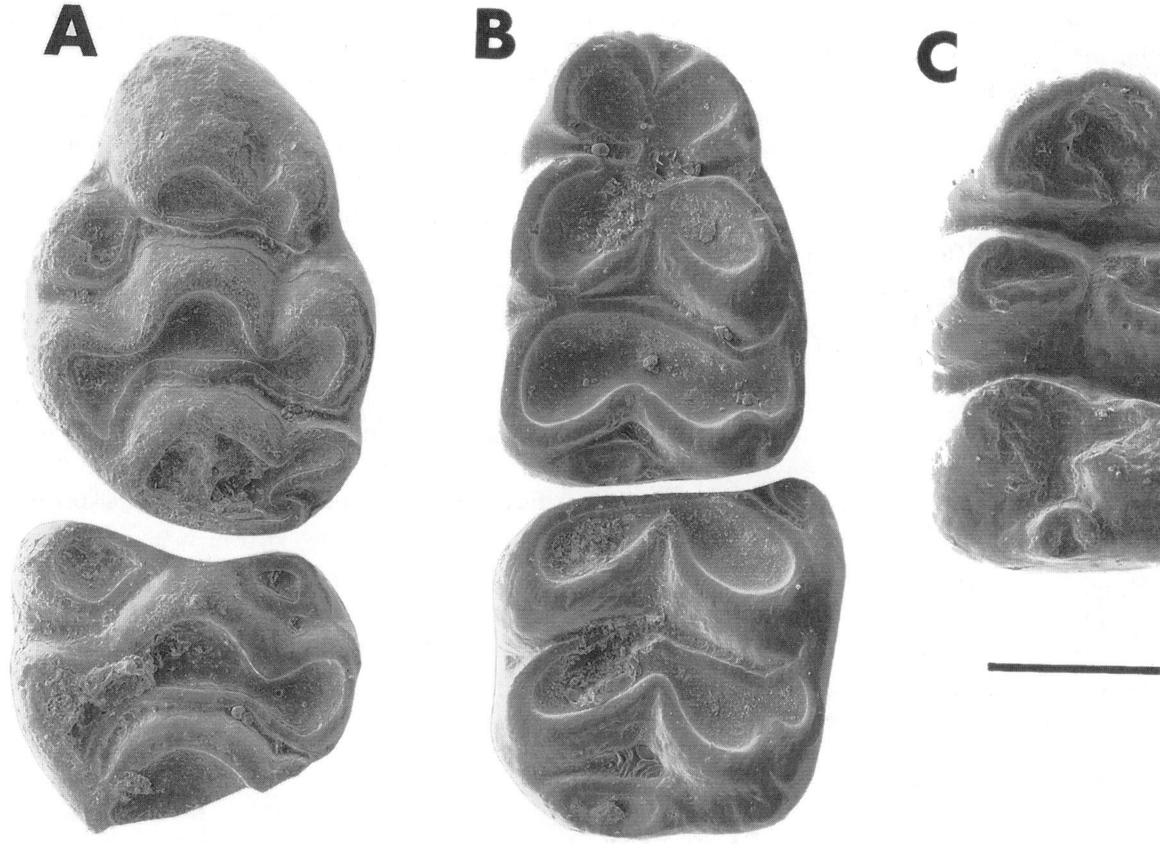

Figure 5.6 Scanning electron photomicrographs of *Karnimata jacobsi* sp. nov and *Abudhabia* sp. from Lothagam: A = *K. jacobsi*, Lower Nawata, KNM-LT 24208, left M^{1-2}, holotype; B = *K. jacobsi*, Lower Nawata, KNM-LT 26538, right M^{1-2}; C = *Abudhabia* sp., Upper Nawata, KNM-LT 24211, right M_1. Bar scale is 1 mm.

oblique on M^1), vestigial longitudinal crests, prelobe on anterocone of M^1 large and with a smooth anterior wall, anterocone transversely elongate, anteroconid of M_1 complex and including an anterior sinus, a central fovea, and a cingular tubercle, M_3 very reduced, and simple root pattern. *Protatera* also includes *P. almenarensis* from the Late Miocene of Spain (Agustí 1990; localities summarized in Geraads 1996), and *P. davidi* from the Miocene-Pliocene of Morocco (Geraads 1996). *Protatera* has prismatic cusps and stronger longitudinal crests than *Abudhabia* (Bruijn and Whybrow 1994); Flynn and Jacobs (1999) emphasize that *Protatera* species (*P. algeriensis* and *P. almenarensis*) have tall cusps that fuse into transverse lophs at an earlier wear stage than is observed in *Abudhabia*. LT 24211 has lower cusps that would not fuse into transverse crests until late in wear, a simple anteroconid, and a complete lack of longitudinal crests.

The Lothagam gerbil is assigned to *Abudhabia* based on an M_1 with a broad simple anteroconid (derived at an early wear stage from a tricuspid condition) and the presence of a posterior cingulum. The protoconid and metaconid of the Lothagam tooth are only slightly alternating, compared to illustrations of the more alternating cusps of *A. baynunensis* (Bruijn and Whybrow 1994), *A. kabulense* (Sen 1983), and "*P.*" *yardangi* (Munthe 1987). Orientation of the hypoconid and entoconid of 24211 is comparable to that of described species of *Abudhabia*. Like known species of *Abudhabia*, the Lothagam tooth lacks longitudinal crests. A weak anterior mure that projects posteriorly from the anteroconid is found in *A. baynunensis* and *A. kabulense*. Munthe (1987:140) notes that three of four M_1s of "*P.*" *yardangi* are worn, so that the anteroconid is in contact with the following row of cusps, but there is no discussion of an anterior mure intervening between these cusps (and this is not illustrated). There is also no hint of an anterior mure on the Lothagam gerbil.

The Lothagam M_1 is most similar to the corresponding tooth of "*P.*" *yardangi*. Similarities include size (table 5.7) and presence of a broad anteroconid that lacks a labial cingulum (observed in *A. baynunensis* and *A. kabulense*). LT 24211 may be derived relative to known species of *Abudhabia* because the protoconid and metaconid are less alternating and there is no trace of an anterior mure. A larger sample (including other tooth positions) of the Lothagam *Abudhabia* is needed for specific assignment, in particular to determine if this material represents a new species.

Subfamily Murinae Illiger, 1811
Karnimata Jacobs, 1978
Karnimata jacobsi sp. nov.

(Figures 5.6A–B, 5.7A–B; table 5.8)

Diagnosis

A murine with brachyodont, rounded cusps that are weakly connected in chevrons. Stephanodonty and a posterostyle are absent from M^1. The anterostyle on M^1 is slightly displaced posteriorly, and there is a distinct posterior cingulum (a primitive feature). The metacone is small relative to the hypocone (a derived condition), in contrast to the proportionally larger metacone of species such as *K. darwini*. The M^3 is elongate. The anterior two rows of cusps of M_1 wear to form an "x" shaped pattern. One of two known M_1s has a short anterior mure (primitive). M^1 has four roots (three large roots and a smaller root under the paracone), and is derived relative to all other species of *Karnimata* (although close to *K. huxleyi*), whose M^1's lack a distinct root under the paracone. The M^2 has three or four roots. *K. jacobsi* is larger in size than *K. darwini*, *K. huxleyi*, and *K. minima*, and comparable to, or larger than, *Karnimata* sp. (Jacobs 1978) and *K. intermedia*.

Holotype

KNM-LT 24208, incomplete left maxilla with M^1 and M^2 from the northern area (Gateway), WT 1733 (locality field number), Lothagam, west Lake Turkana, northern Kenya.

Etymology

Named after Louis L. Jacobs for his contributions to our understanding of murine rodents.

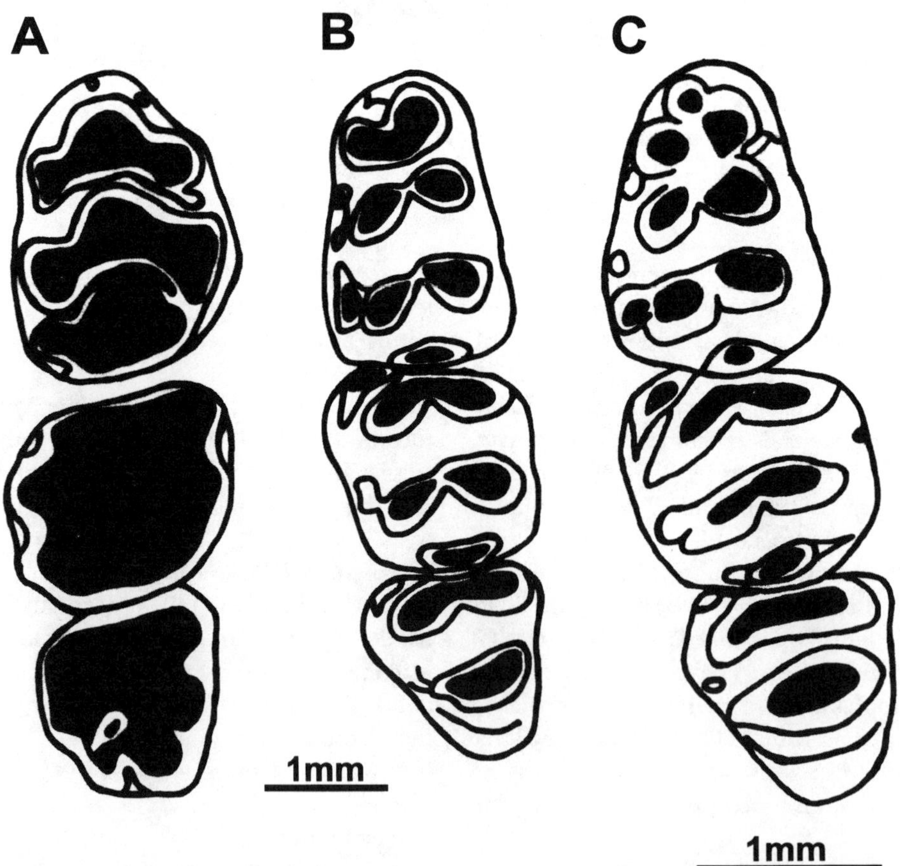

Figure 5.7 Occlusal illustrations of murine rodents from the Nawata Formation: A = *Karnimata jacobsi* sp. nov., Lower Nawata, KNM-LT 24962, right M^{1-3}; B = *Karnimata jacobsi* sp. nov., Lower Nawata, KNM-LT 24953, left M_{1-3}; C = *Saidomys* sp., Upper Nawata, KNM-LT 24201, left M_{1-3}.

Lothagam Material

- *Lower Nawata:* the holotype; 24962, anterior portion cranium (Lt. and Rt. M^{1-3}); 24953, Lt. dentary (incisor, M_{1-3}); 26538, Rt. dentary (incisor, M_{1-2}).

The holotype, LT 24208, is from slightly below the Marker Tuff and above the Gateway Sandstone; it is about 6.54 Ma based on a radiometric date of 6.54 ± 0.04 for the Marker Tuff (McDougall and Feibel 1999). 24962, 24953, and 26538 are from the northern area, Carnivore Site, from below the Red Marker, and are approximately 6.57 Ma based on a radiometric date of 6.57 Ma ± .07 Ma from 5 meters below the Red Marker (McDougall and Feibel 1999).

Cranium

LT 24962 includes much of the cranium anterior to M^3. The zygomatic arches and much of the left zygomatic plate are missing. The bone is badly weathered and often crushed. On the right side, the zygomatic plate is wide and slightly oblique to the mid-sagittal plane. Anterior palatine foramina extend posteriorly to the anterior border of M^1. If present, the posterior palatine foramina would have been small (none are observed). The infraorbital foramen is expanded (myomorphous condition). Upper incisors are relatively robust.

Upper dentition

Tooth terminology is illustrated in figure 5.5. The dentition of LT 24962 (figure 5.7A) shows heavy occlusal wear, while that of LT 24208 (figure 5.6A) has moderately heavy wear. Tooth measurements are given in table 5.8. Polarity of characters for both murines (*Karnimata* and *Saidomys*) described in this chapter uses *Antemus chinjiensis* (the oldest murine) as the outgroup (Jacobs 1977, 1978; Jacobs et al. 1989). The M^1 is relatively elongate—LT 24208 more so than LT 24962. A precingulum is lacking. The right M^1 of LT 24962 has two small prestyles. A large lingual anterocone is transversely aligned (or slightly anterior to) a labial anterocone that is about half its size (and labiolingually compressed). The labial anterocone has a short posterolabially projecting spur. An oval anterostyle is weakly connected to the lingual anterocone. The anterostyle is displaced slightly behind the lingual and labial anterocones. The anterostyle is a little larger than the labial anterocone. Neither the anterostyle nor the enterostyle is significantly compressed. There are no connections between the first and second rows of cusps. LT 24962 has an additional cuspule on the lingual margin of the tooth between the anterostyle and enterostyle. The paracone and enterostyle are both connected to the protocone. The paracone is only slightly posterior to the protocone and the enterostyle a little more posterior. The paracone and enterostyle are similar in size and smaller than the protocone. The enterostyle extends posteriorly to contact the hypocone. The largest cusp of the tooth is the bulbous, centrally placed hypocone, which is connected to a smaller, anteriorly placed, elongate metacone and a short, but distinct (on LT 24208) posterior cingulum. A faint labial cingulum connects the paracone and metacone on LT 24208. A posterostyle is lacking. There is no trace of stephanodonty, even with moderately heavy occlusal wear. Four roots are present: a large root under the lingual and labial anterocones, a small root under the paracone, a large root under the metacone and hypocone, and a large root (formed by two fused roots) under the anterostyle and enterostyle.

Heavy occlusal wear of the M^2 of LT 24962 has obscured the details of its morphology, but the outline of the tooth is similar to that of LT 24208. There is a large oval anterostyle and an oval labial anterocone that is about half the size of the anterostyle. Both the anterostyle and labial anterocone contact the protocone. The labial anterocone also has a cingular extension that contacts the paracone. The protocone, paracone, and enterostyle are developed as in M^1, except that the paracone contacts the metacone. The M^2 of LT 24208 is broken across the hypocone, so development of the metacone and posterior cingulum cannot be determined. Based on LT 24962, the posterior cingulum was probably short, but distinct (close in development to that of the M^1). The M^2 of LT 24208 has four main roots: under the anterostyle, the labial anterocone and paracone, the enterostyle, and the metacone and hypocone. On LT 24962 the roots under the anterostyle and enterostyle have fused.

Heavy occlusal wear of the only known M^3 of *K. jacobsi* has obscured much of the detail of cusp morphology. The tooth is relatively large and elongate. Development of the anterostyle is comparable to that of the M^2. Presence or absence of a labial anterocone is indeterminate. The protocone, paracone, and enterostyle have completely fused to form a crescent that is concave posteriorly. The hypocone and metacone have also fused to form a crescent that is concave posteriorly. Anteriorly, there are probably separate anterolabial and anterolingual roots. A large posterior root is also present.

Lower dentition

Two incomplete dentaries of *K. jacobsi* are known. LT 26538 (figure 5.6B) has heavy occlusal wear, and LT 24953 (figure 5.7B) has moderate wear. The dentaries are relatively gracile. Both dentaries are incomplete anteriorly and also posterior to M_3. The mental foramen is located about 1–1.5 mm anterolabial to the anterior end of M_1. There is a strong masseteric crest. The M_1 is elongate. There is a faint anterior cingulum between the labial and lingual anteroconids. These two cusps are separated from the next row of cusps by a deep valley that is bisected on LT 26538 by a short anterior mure. The anterior and middle rows of cusps wear to form an "x" shaped pattern. The transversely compressed oval posterior cingulum lies in the midline on LT 24953 but is displaced lingually on LT 26538. A labial cingulum extends from the labial anteroconid to the hypoconid. This cingulum includes a strong C1 and weak C3. LT 24953 also has a weak C2. The M_1 has large anterior and posterior roots and a labial rootlet under the protocone.

The M_2 is broad anteriorly and tapers posteriorly. There is a distinct labial anteroconid with a cingulum connecting it to the protoconid. Labial major cusps are larger than lingual major cusps on LT 26538. On LT 24953 the protoconid is slightly larger than the metaconid, but the entoconid is larger than the hypoconid. The posterior cingulum of M_2 is about twice the size of M_1, but the location of this cusp is similar to that of M_1. LT 26538 has a large C1. On LT 24953 C1 is relatively small and slightly more anterior in position. This specimen also has a weak C2. It is uncertain if there are two large roots (anterior and posterior) or four smaller roots.

The M_3 is triangular in outline. The M_3 of LT 26538 is represented only by roots, which include two partially fused anterior roots and a posterior root. LT 24953 has a low weak labial anteroconid. The protoconid and metaconid are fused into a crescent that is concave posteriorly. There is a large crescentic posterocentrally located hypoconid. A short low cingulum connects the protoconid and hypoconid. There is also a low weak posterior stylar shelf.

Discussion

This Lothagam murine is assigned to the extinct, primarily southern Asian taxon *Karnimata*. *Karnimata* was diagnosed by Jacobs (1978) as being brachyodont, with rounded distinct cusps that are weakly connected in chevrons, stephanodonty and a posterostyle absent, anterostyle relatively forward in position, short and narrow anterior mure usually present on M_1, M^2 with three or four large roots, and M_1 with three large roots. *Karnimata* includes *K. darwini* (type species; 8.35 Ma; Jacobs and Downs 1994) and *K. huxleyi* (5.7 Ma) from the Siwalik Beds of Pakistan. Jacobs (1978) also listed *Karnimata* sp. (8.4 Ma) from the Siwaliks of Pakistan. Brandy (1979, 1981) described *K. minima* and *K. intermedia* from the late Miocene of Afghanistan. Brandy additionally named *K. afghanensis* from Afghanistan, but this species has been reassigned to *Saidomys* (Sen 1983). *Karnimata intermedia* is also known from the Siwalik Beds of northern India (no older than 6.5 Ma; Flynn et al. 1990). Cheema et al. (1983) noted *Karnimata* sp. nov. from Jalapur, northern Pakistan (Early Vallesian, but specific age uncertain). In Africa, *Karnimata* is reported (but undescribed) from Berg Aukas, Namibia (early Late Miocene; Conroy et al. 1992). Jacobs et al. (1989) suggested that three isolated teeth referred to *Paraethomys* cf. *P. miocaenicus* from Algeria (Late Miocene; Jaeger 1977) may be more closely related to *Karnimata*. Mein et al. (1993) synonymized Jacob's (1978) type species, *K. darwini*, with *Progonomys woelferi*, but they did not name the genus in which the remaining species of *Karnimata* should be included.

Karnimata jacobsi is morphologically quite similar to *K. darwini*, the geologically oldest known species of *Karnimata*. However, *K. jacobsi* is significantly larger (table 5.8) and has a proportionally smaller (and usually more oblique) metacone relative to the hypocone on M_1. Reduction of the metacone relative to the hypocone is likely a derived character; greater obliquity of the metacone is also likely to be derived. *Karnimata jacobsi* also has a smaller labial anteroconid on M_2 and three roots on M_3 (derived; *K. darwini* has two).

Karnimata huxleyi is described as close in morphology to *K. darwini* (Jacobs 1978). *Karnimata huxleyi* is distinguished by its larger M_3 (table 5.8), larger M_3 with the labial anteroconid strongly reduced, and M_3 with three roots. Unlike *K. huxleyi* (and similar to *K. darwini*), *K. jacobsi* has an M^1 with a less rounded occlusal outline with the anterostyle more posterior in position, and a less reduced labial anteroconid on M_3. The M^3 of *K. huxleyi* is proportionally longer (87 percent of the length M2; n = 2) than that of *K. darwini* (69 percent; n = 1) and *K. jacobsi* (64 percent, n = 1). Jacobs (1978) suggested that *K. huxleyi* is derived relative to *K. darwini* because of the consistent possession of four (not two) roots on M^2, and the elongation of M^3. The M^1 of *K. huxleyi* (and *K. jacobsi*) may be derived relative to that of *K. darwini* by possessing a metacone that is relatively reduced in size compared to the hypocone. The M^1 of *K. huxleyi* is described (Jacobs 1978) as having three major roots plus a "fairly well developed accessory rootlet . . . approximately beneath the paracone." This root morphology is close to that of *K. jacobsi*.

Jacobs (1978) assigned two M^1s from the Late Miocene of Pakistan to *Karnimata* sp. These teeth are similar to *K. darwini* in overall morphology, but they differ from this species in their larger size (table 5.8). Brandy

(1979, 1981) considered these larger teeth to be distinct from those of *K. intermedia* (the latter slightly higher crowned and proportionally wider) and *K. minima* (the latter smaller in size and with a less elongate M^1). *Karnimata jacobsi* is distinct from *Karnimata* sp. in the former's larger size (table 5.8), more posterior placement of the anterostyle on M^1, and proportionally smaller metacone and posterior cingulum. In addition to three large roots, *Karnimata* sp. has a minute rootlet under the protocone, while *K. jacobsi* has a distinct root under the paracone.

Karnimata minima is known from five teeth (Brandy 1979, 1981). It is diagnosed as similar in size to *K. darwini* (table 5.8), M^1 amygdaloidal in outline, metacone oblique and antero-externally located, hypocone very large, and anterostyle and enterostyle only connected to the lingual anterocone and protocone, respectively, with extensive wear. Overall, the morphology of *K. minima* and *K. jacobsi* are similar. However, *Karnimata jacobsi* is larger in size (table 5.8), appears to have a less isolated enterostyle, and has a less distinct labial anteroconid. The M^2 of *K. minima* has four roots, and the M_2 has two transverse roots. The roots of M^1, M^3, and M_3 (M_1 has not been recovered) are not described.

Another large species of *Karnimata, K. intermedia* (table 5.8) is known from five teeth from Afghanistan (Brandy 1979, 1981) and tentatively from an M^2 and M^3 from India (Flynn et al. 1990). This species is diagnosed by its large size, distinct posterior cingulum on M^2, and very oblique metacone on M^1 and M^2. *Karnimata intermedia* has three major roots plus one very reduced (and centrally located) rootlet on M^1, four equally sized roots on M^2, and two on M_2. *Karnimata jacobsi* and *K. intermedia* are close in size, but *K. jacobsi* has a less oblique metacone on M^1 and M^2 (primitive) and a larger fourth root on M^1 (derived).

Cheema et al. (1983) considered five isolated teeth from near Jalapur, northern Pakistan, to belong to a new species of *Karnimata.* This species was considered distinct based on its smaller size, large anterostyle relative to the lingual anterocone on M^1, lower crown height, and cusps more rounded and weakly connected. Although the Jalapur specimens are smaller in size than *K. darwini* (table 5.8), based on published descriptions and illustrations their morphology appears to be within the range of variation of that species. The difference in size may be a result of regional variation within a species. A Jalapur M_3 is likely aberrant, or a different taxon. The tooth is rectangular in outline (versus triangular), lacks a posterior cingulum, and has three cusps in the posterior row (comparably sized hypoconid, entoconid, and probably a slightly smaller C1).

Jaeger (1977) referred three isolated teeth from Amama 2, Algeria, to *Paraethomys* cf. *P. miocaenicus.* Jacobs et al. (1989:170) suggested these teeth were more similar to *Karnimata* than *Paraethomys* in having the paracone separate from the metacone. Whatever genus these teeth should be assigned to, they are distinct from *K. jacobsi. Karnimata jacobsi* is significantly larger in size. The M^1 of *P.* cf. *P. miocaenicus* is 2.06 by 1.47 mm, M^2 1.46 by 1.40 mm, and M_3 is 1.06 by 1.07 mm. The metacone of the M^1 of *K. jacobsi* is proportionally smaller. *Karnimata jacobsi* is derived in root morphology, with an M^1 with four rather than three roots, and an M_3 with three rather than two roots.

Jacobs and Downs (1994) suggested that the Asian genus *Parapelomys* evolved from *Karnimata* through cladogenesis between 8.45 and 8.0 Ma. *Parapelomys* includes *P. robertsi* (type species from Pakistan, 5.7 Ma; Jacobs 1978), *P. charkhensis* (Brandy 1979, 1981; Sen 1983) and *P. orientalis* (Sen et al. 1979). *Parapelomys charkhensis* is known from Pul-e Charkhi, Afghanistan (Early Pliocene; Sen 1983). Sen et al. (1979) described *P. orientalis* from Hadji Rona, Afghanistan (Pliocene). *Parapelomys* sp. is reported from Pakistan at 7.1 Ma (Jacobs and Downs 1994).

Parapelomys differs from *Karnimata* in larger mean size and has rows of cusps more arcuate in appearance, a generally smaller posterior cingulum on M_1, and lacks the "x"-shaped wear pattern on M_1 (Jacobs 1978). The size of *Parapelomys* (Jacobs 1978:table 15) is comparable to the larger-sized species of *Karnimata* (i.e., *K. intermedia, K. jacobsi, Karnimata* sp. of Jacobs 1978). *Karnimata jacobsi* differs from *Parapelomys: K. jacobsi* has less arcuate rows of cusps and an "x"-shaped wear pattern on M_1 (both characters primitive); it may also have the anterostyle more posterior in position, and it lacks stephanodonty (primitive), which may be present in *P. robertsi.*

Few specimens are known of many of the species of *Karnimata* (i.e., *K. jacobsi, K. minimi, K. intermedia*), so the range of individual variation is unknown. This may compromise the validity of poorly represented species, if a larger sample is eventually recovered. As is currently known, however, *K. jacobsi* is derived relative to other species of *Karnimata* as it possesses an M^1 with a strong fourth root.

The presence of *Karnimata* in East African Late Miocene sediments has important paleobiogeographic implications. The genus has its origins in the Late Miocene (>8 Ma) of southern Asia and relatively quickly emigrated into Africa. In Africa, *Karnimata* is reported from the Late Miocene of south (Namibia), east (Kenya), and at least a closely related form from north Africa. As the Late Miocene of Africa becomes better known, it is likely that our knowledge of the African members of this genus will improve considerably.

Saidomys James and Slaughter, 1974
Saidomys sp.
(Figure 5.7C; table 5.9)

Lothagam Material

- *Upper Nawata:* 24201, Lt. dentary (incisor, M_{1-3}). The specimen derives from the Central Area (south), below the Purple Marker. Estimated minimum age is 5.23 Ma based on sedimentation rates and paleomagnetic data, which suggest the Purple Marker could be 5.23 Ma (McDougall and Feibel 1999).

The incisor is triangular in outline. Enamel is preserved along the anterior face of the tooth and extends posteriorly only slightly along the sides. The dentary is poorly preserved and extensively cracked. The specimen is delicate, and some of the matrix could not be removed; this remaining matrix made observation of root development difficult. The tip of the coronoid process, ascending ramus, and angular process are missing. A large oval mental foramen is situated along the labial edge of the diastema, about 1 mm anterior to the root of M_1. The strong masseteric crest extends to the anterior end of M_1. The ascending ramus originates about halfway along M_2 and obscures the posterior half of M_2 and M_3 from labial view.

The three cheek teeth have moderate occlusal wear (figure 5.7C). They are large and proportionally wide for their length (table 5.9). The M_1 has labial major cusps slightly smaller and a little more posteriorly situated than are the lingual cusps. The labial and lingual anteroconids, protoconid, and metaconid form an "x"-shaped wear pattern. However, the labial and lingual anteroconids do not connect to the protoconid or metaconid. A small spur extends posteriorly from the lingual anteroconid. The hypoconid and entoconid are connected anteriorly in the midline. There is a large circular medial anteroconid. A labial cingulum (including C4 and a small C2) extends from the posterior end of the labial anteroconid to the large C1. A large oval posterior cingulum is present on a posterior stylar shelf. Root development is partially obscured by bone, but the tooth has large anterior and posterior roots; a smaller root may also be present under the metaconid.

The M_2 is nearly trapezoidal in outline, but narrows posteriorly. The two main rows of cusps each form a wide inverted "v" shape, with the labial and lingual cusps connected anteriorly in the midline. The labial cusps are located slightly behind the lingual cusps. There is a large labial anteroconid: a narrow labial cingulum extends from it to contact the protoconid. A distinctive C1 is present, and it connects with the hypoconid and protoconid by a narrow labial cingulum. There is a small lingual stylar cusp between the metaconid and entoconid. The posterior cingulum is a large oval that lies on a posterior stylar shelf. Root development is obscured by matrix.

The M_3 is triangular in outline and relatively elongate. The protoconid and metaconid are fused and nearly transversely aligned. There is a large hypoconid that had fused with C1. The labial anteroconid is a small oval. A small C2 is present posterolabial to the protoconid. There is a strong posterior cingulum developed as a shelf, which extends labially along the hypoconid and C1. Large anterior and posterior roots are present, but detailed root development cannot be determined.

Discussion

Saidomys is a relatively common fossil murine from the Late Miocene–Late Pliocene of East Africa and the Early Pliocene of Afghanistan. Described species (summarized in Winkler 1997) include *S. natrunensis* from Egypt, *S. afarensis* from Ethiopia, *S. parvus* from Tanzania, and *S. afghanensis* from Afghanistan. A fifth unpublished species is known from the Early Pliocene of the Tugen Hills, central Kenya (Winkler 1990). There is an isolated M_1 of *Saidomys* sp. indet. reported from upper Member G, Shungura Formation, lower Omo Valley, Ethiopia (Late Pliocene; Wesselman 1984). *Saidomys* is also present in the Kanapoi fauna, northern Kenya (Early Pliocene; personal observation). Chaimanee (1998) described a new species of *Saidomys, S. siamensis,* from the Late Pliocene–Early Pleistocene of Thailand. Although close in morphology to *Saidomys, S. siamensis* likely belongs to a different genus. *Saidomys siamensis* is derived relative to all other species of *Saidomys* as it lacks a posterior cingulum on M^1. Chaimanee states that lack of a posterior cingulum is due to the younger geological age of *S. siamensis* compared to other species of *Saidomys.* Although possibly geologically younger, *S. siamensis* is more primitive than all other species of *Saidomys* in that it lacks a medial anteroconid on M_1. A medial anteroconid is present, and strong, on all other species of *Saidomys* (Winkler 1997).

The Lothagam specimen is assigned to *Saidomys* based on the following (Winkler 1997): (1) relatively large size; (2) M_1 with large medial anteroconid, large C1, and lacking an anterior mure; (3) M_2 with C1 reduced relative to M_1; and (4) strong posterior cingulum on M_1 and M_2. LT 24201 is a small *Saidomys.* The teeth are similar in size to *S. parvus* and the Tugen Hills taxon (table 5.9). *Saidomys parvus* is known from 14 isolated teeth from the Ibole Member, Wembere-Manonga Formation, north-central Tanzania. Fauna from the Ibole

Member is estimated to be 5.5–5.0 Ma based on biochronology using the entire fauna (Harrison and Baker 1997) and 5–4 Ma using only the rodents (Winkler 1997). Other than its small size, the only diagnostic character of *S. parvus* of relevance to the Lothagam specimen is that two of three M_1s of *S. parvus* have a small conical (rather than elongate) posterior cingulum. The third specimen of *S. parvus* has an elongate posterior cingulum. A conical posterior cingulum is the derived state for this character for *Saidomys* (Winkler 1997). Since the morphology of the posterior cingulum may be variable, it is impossible to say with confidence that an elongate posterior cingulum is diagnostic for the Lothagam *Saidomys*. No M_2 or M_3s of *S. parvus* have been recovered.

The Tugen Hills sample of *Saidomys* includes 15 isolated teeth from the Tabarin locality, Chemeron Formation (around 4.5–4.4 Ma; Hill 1999). This material likely represents a new species, similar in size to *S. parvus* but differing from it in other aspects of its morphology (Winkler 1997). Compared to the M_1s of *S. parvus,* the one M_1 from the Tugen Hills has a more elongate posterior cingulum (similar to that of LT 24201). The size and morphology of the Lothagam M_1 is closely comparable to that of the Tugen Hills tooth, including the presence of small C2 and C4. The five known M_1s of *S. parvus* have C4, but none has C2. Two *Saidomys* M_2s are known from the Tugen Hills. These teeth are similar in size and morphology to the Lothagam specimen.

The Lothagam M_3 differs significantly from the three Tugen Hills M_3s in that the former has a robust posterior cingulum along the hypocone and C1. The Tugen Hills teeth either lack a posterior cingulum ($n = 1$) or it is very reduced. *Saidomys natrunensis* may lack a posterior cingulum (but the specimens have heavy occlusal wear). A posterior cingulum is usually present, but miniscule to vestigial on M_3s of *S. afarensis* (Sabatier 1982). The robust posterior cingulum on the Lothagam tooth is within the range of variation observed for *S. afghanensis.* On M_3, a strongly developed posterior cingulum is the derived condition for murines (Jacobs et al. 1989:figure 5).

The Lothagam M_3 has a small but distinct labial anteroconid, similar to that of one of three Tugen Hills specimens (the labial anteroconid is reduced in the other two specimens). Development of the labial anteroconid of the Lothagam tooth is within the range of variation of *S. afghanensis.* The condition of the labial anteroconid cannot be confidently determined in *S. natrunensis,* but a cingulum, at least, appears present. In *S. afarensis,* the labial anteroconid is either small or nonexistent (Sabatier 1982). *Antemus* (Jacobs et al. 1989:figure 5) has a relatively strong labial anteroconid (primitive condition). The more reduced labial anteroconid on the M_3 of *S. afarensis* is derived relative to that of the other species.

The Lothagam *Saidomys* has a unique combination of characters in comparison with other known species. It is small—similar in size to *S. parvus* and the Tugen Hills species. However, the Lothagam taxon is more primitive than *S. parvus* as it possesses an M_1 (and M_2) with an elongate (versus conical) posterior cingulum. The M_3 of the Lothagam specimen is derived compared to the Tugen Hills species in that LT 24201 has a strongly developed posterior cingulum; development of the labial anteroconid on M_3 is, however, similar to or stronger than that of the Tugen Hills species. The Lothagam specimen is similar to *S. afghanensis* as they both possess an elongate posterior cingulum on M_1 and M_2 and have a strong posterior cingulum and small, but distinct, labial anteroconid on M_3. However, *S. afghanensis* is significantly larger (table 5.9). The Lothagam mandible may represent a new species of *Saidomys.* However, without an appreciation of the range of variation of diagnostic characters (e.g., development of the posterior cingulum), it is presumptive to name a new species on this single specimen.

Subfamily *incertae sedis*
Genus and species unknown
(Figure 5.8; table 5.10)

Lothagam Material

- *Lower Nawata:* 24203, fragmentary cranium (Lt. $\underline{I}$, Lt. M^1, Rt. M^{1-2}), Rt. dentary (M_{1-3}). The specimen derives from the Monkey Area, below the Red Marker. It is about 6.57 Ma based on a date of 6.57 ± 0.07 at five meters below the Red Marker.

Tooth terminology for this specimen is similar to that for murine rodents (figure 5.5). The teeth of LT 24203 are illustrated in figure 5.8, and measurements are in table 5.10. The skull is badly crushed. The zygoma appears to have been massive. Length of the palatine foramina cannot be determined. The upper incisor is slender and ungrooved.

The M^1 has three rows of conical cusps. There is a large anterocone, which has a faint anterior midline valley. The width of the anterocone is about two-thirds the width of the following row of cusps. There is an anterolingual inflection (a derived character; seen especially on the right M^1) lingual to the anterocone. The protocone and slightly larger paracone are oriented transverse to each other, as are the hypocone and metacone. The latter two cusps are similar in size (right side), or the metacone is slightly larger (left side). The middle and

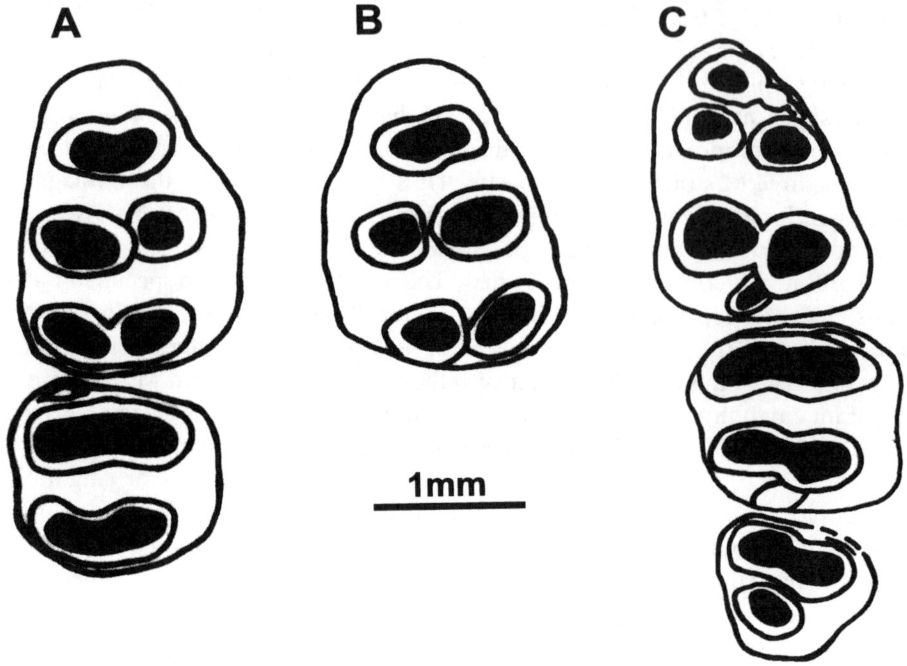

Figure 5.8 Occlusal illustration of Muridae, gen. and sp. nov., KNM-LT 24203, from the Lower Nawata: A = right M^{1-2}; B = left M^1; C = right M_{1-3}.

posterior rows of cusps are each joined posteromedially. Longitudinal crests and a posterior cingulum are lacking. There are three large roots: under the medial anterocone and anterior end of the paracone, under the protocone and hypocone, and under the metacone and posterior end of the paracone.

The M^2 is rectangular in outline. It has a small labial anterocone. The protocone and paracone are fused through occlusal wear to form a transverse loph, as is the metacone and hypocone. Longitudinal crests and a posterior cingulum are lacking. There is a large lingual root under the protocone and hypocone. Details of lateral root development are obscured by bone, but a large root under the paracone and metacone is likely.

The dentary is fractured and crushed. It is missing the anterior end and incisor, as well as the bone posterior to M_3. Overall, the dentary appears relatively robust and deep. Development of the masseteric crest is obscured by crushing, and the location of the mental foramen also cannot be determined.

The lower molars are relatively narrow. Major cusps are conical, and the labial cusps are displaced slightly posterior relative to lingual cusps. With occlusal wear, major labial and lingual cusps coalesce to form transverse lophs. The M_1 has a large anteroconid located lingual to the midline of the tooth. Extending posterolingually from the anteroconid is a short, low cingulum, which contacts the metaconid. Posterolabial to the anteroconid is a cingular cuspid, which is continuous with a cingulum that contacts the protoconid. The protoconid is slightly larger than the metaconid, and these two cusps are separate from each other. The hypoconid and entoconid are larger than the protoconid and metaconid, and the hypoconid is slightly smaller than the entoconid—these two cusps are in contact through occlusal wear. There is an elongate posterior cingulum oriented lingual to the midline of the tooth and in contact with the hypoconid. A large root is present under the medial anteroconid and metaconid. Posteriorly, one large or two smaller roots are present.

The M_2 is rectangular in outline. There is a weak, low anterolabial cingulum. The protoconid and metaconid, and the hypoconid and entoconid, respectively, have fused to form two transverse lophs. The posterior cingulum is elongate, in contact with the hypoconid, and located lingual to the midline of the tooth. A large labial root is present, but it is uncertain if there are two separate or one fused lingual root.

The M_3 is triangular in outline and little reduced. The protoconid and metaconid have coalesced to form a transverse loph. There is a large hypoconid situated just lingual to the midline of the tooth. A low faint anterolabial cingulum is present. A posterior cingulum is lacking. Roots are present under all major cusps, but it is unclear if roots are separate or fused.

Discussion

LT 24203 is currently not assigned to subfamily, but its affinities probably lie with the myocricetodontines, an extinct group known primarily from the Miocene of

Africa (e.g., Jaeger 1977; Tong 1989; Tong and Jaeger 1993), but also from southern Asia (e.g., Lindsay 1987). The Lothagam fossil is most similar to the poorly known taxon *Myocricetodon magnus* from Pataniak 6, Morocco (Late Middle Miocene; Jaeger 1977). *Myocricetodon magnus* is known only from a maxillary fragment with M^1, a mandibular fragment with M_2, and an isolated M_1. The species is diagnosed by the following characters: (1) size of the extant gerbil *Tatera;* (2) posterior palatine foramina little extended anteriorly and stopping at the anterior part of M^2; and (3) molars with round tubercles lacking arms. The generic assignment of *M. magnus* has been questioned by Bruijn and Whybrow (1994), but a new generic assignment was not proposed. Generic assignment was accepted by Tong (1989) in her extensive study of the origin and evolution of North African gerbils.

LT 24203 is similar to *M. magnus* in having molars with rounded major cusps that are essentially transverse to each other on M^1 and M_2 (and M_3 of the Lothagam specimen) and strongly oblique to each other on M_1. The M^1 and M_2 of both samples lack longitudinal crests (derived condition). The Lothagam M_1 lacks any trace of an anterior mure (derived condition). Although not noted in the description, the M_1 of *M. magnus* has a short anterior mure. A posterior cingulum is lacking on the M^1 (and M^2 of LT 24203). The M_1 and M_2 of LT 24203 have a low, but distinct, posterior cingulum. Although a faint posterior cingulum is present in the illustration of the M_1 of *M. magnus* (Jaeger 1977:plate II, figure 11), the text describes this cingulum as lacking (Jaeger 1977:36). The M_2 of *M. magnus* has distinct posterior and anterolabial cingula. The anterolabial cingulum is weak on the Lothagam M_2.

The M^1 of *M. magnus* is the tooth most comparable to the Lothagam taxon (except in the former's smaller size). Both are massive and wide, and both have three roots. They also have a transversely elongate medial anterocone and an anterolingual inflection (stronger on LT 24203). The major differences between the two taxa are in the lower dentition, as described above.

Diagnosis and detailed comparisons of the Lothagam murid will be provided elsewhere (Winkler in preparation). In comparison to *M. magnus,* LT 24203 is derived in being significantly larger in size (table 5.10), lacking any trace of an anterior mure, and having a weaker anterolabial cingulum on M_2.

Discussion and Conclusions

The Lothagam rodents and lagomorphs augment our limited knowledge of the composition of African Late Miocene faunal communities. The African later Pliocene is relatively better known (see, for example, Jaeger 1977; Hendey 1981; Wesselman 1984; Denys 1987; Munthe 1987), but still sparse. The Lothagam sample from the Miocene-Pliocene succession is small in the total number of individuals identifiable at least to family level (minimum 43), but diversity is high (13 genera, 15 species), which suggests a rich micromammal community. The presence of external age control in the form of radiometric dates (McDougall and Feibel 1999) significantly enhances the usefulness of the Lothagam fauna for evolutionary and paleobiogeographic studies.

An understanding of the Late Miocene record of African rodents and lagomorphs is crucial to understanding the evolutionary history of several groups that are found in Africa today but that likely had their origins elsewhere. This includes the Leporidae (rabbits and hares) and the Hystricidae (Old World porcupines). The earliest, well-dated African record of the hystricids is at Lothagam, where a specimen is recorded from low in the Lower Nawata, possibly >7.44 Ma. Leporids are first recorded from Lothagam at 6.57–6.54 Ma. The earliest African record of leporids may be from the Mpesida Beds in the Tugen Hills, from deposits dating between 7 and 6.2 Ma (Winkler 1990; Hill 1999). The Tugen Hills may also record *Alilepus* at 5.6–6.2 Ma (Hill 1999) from the Lukeino Formation.

Thryonomyids from the Lothagam sequence add to our knowledge of the evolution of the extant cane rat, *Thryonomys,* which is known exclusively from Africa. At least four species of thryonomyids are present in the Lothagam section (table 5.1), in sediments dating from the Late Miocene through the Holocene. Of particular significance from an evolutionary standpoint is the presence of the extinct genus *Paraulacodus* in the Lower Nawata, but *Thryonomys* in the Upper Nawata and younger deposits.

Hinton (1933) noted similarities between *Paraulacodus* and *Thryonomys.* Jaeger et al. (1980) suggested that the morphology of *Paraulacodus* was intermediate between that of Early Miocene thryonomyids and *Thryonomys.* Bruijn (1986), however, did not consider *Paraulacodus* (which he reassigned to *Neosciuromys*) to be closely related to *Thryonomys.* Cladistic analysis suggests that *Paraulacodus* is the sister-taxon to *Thryonomys* (Winkler 1992; Flynn and Winkler 1994). *Paraphiomys* is a more distantly related genus. Specimens of *Thryonomys* from the upper member, Nawata Formation, are relatively small and may represent a new species. A small, isolated P^4 from the Upper Nawata (not described here) appears close in morphology to *Paraulacodus* but is comparable in size to the Upper Nawata small *Thryonomys.* This P^4 is proportionally smaller than the upper incisor of *Paraulacodus* from the Lower Nawata. *Thryonomys* from the Early Pliocene Kaiyumung Member, Nachukui Formation, is tentatively assigned to an extant species. A specimen from the Holocene Galana Boi Formation is proportionally larger and definitely belongs to an extant species. Detailed study of the *Thryonomys* from the Upper Nawata will be important to determine if this taxon is not only smaller than extant species but also primitive in its morphology.

The Lothagam rodent and lagomorph fauna includes taxa that are unique to this locality and those with a wider paleobiogeographic distribution. The squirrel, *Kubwaxerus*, is known only from Lothagam. The new murid from the Lower Member, Nawata Formation, is also unique to Lothagam, although it is a sister-taxon to *Myocricetodon magnus* from North Africa. Other members of the fauna have affinity with fossil taxa from northern Africa (*Abudhabia*, *Serengetilagus praecapensis*, possibly *Karnimata*), East Africa (*Paraulacodus* cf. *P. johanesi*, *Paraphiomys chororensis*, *Thryonomys*, *Serengetilagus praecapensis*, *Alilepus*), Namibia (*Karnimata*, *Paraulacodus johanesi*), and Eurasia (*Alilepus*, *Karnimata*). *Hystrix*, *Alilepus*, and *Karnimata* are probably immigrants from Eurasia. The murine *Saidomys* is widely distributed in both time and space. It is known from the Late Miocene to Late Pliocene, and it has been recovered from Afghanistan and North and East Africa. The greatest similarity of the Lothagam fauna is with other East African localities. It is apparent, however, that during the Late Miocene–Early Pliocene faunal exchange was occurring between East, North, and southern Africa, as well as with Eurasia. This faunal exchange has been suggested by many authors—for example, in the circum-Mediterranean region (Benammi et al. 1996; Geraads 1998b) and within Africa and between Africa and Eurasia (Winkler 1994; Denys 1987).

Examination of the faunal list from Lothagam (table 5.1) reveals that different taxa are present in the different members. Are these taxonomic differences real, and a reflection of ecological change through time? Or are these differences simply indicative of sampling biases? Certainly, the sample of rodents and lagomorphs from Lothagam is small. Because of this, any conclusions based on comparing micromammal faunas through time, and using faunal change to imply ecologic change, are extremely tentative. There is the suggestion, however, of change from a more closed forest habitat of the Lower Nawata (based on the giant squirrel, *Kubwaxerus*; Cifelli et al. 1986) to more open habitats higher in the section (based on *Thryonomys*, which currently inhabits moist savannas to marshes and reed beds; Kingdon 1974).

Acknowledgments

I wish to thank M. G. Leakey for inviting me to study the Lothagam rodents and lagomorphs. This research benefited from discussions of Lothagam fauna, chronology, and stratigraphy with M. G. Leakey and J. M. Harris. Louis L. Jacobs kindly reviewed a draft of the manuscript. The assistance of B. Barnes with final preparation of the figures is greatly appreciated. Financial support from the L.S.B. Leakey Foundation is gratefully acknowledged.

References Cited

Agustí, J. 1990. The Miocene rodent succession in eastern Spain: A zoogeographical appraisal. In E. H. Lindsay, V. Fahlbusch, and P. Mein, eds., *European Neogene Mammal Chronology*, pp. 375–404. New York: Plenum Press.

Arambourg, C. 1959. *Vertébrés continentaux du Miocène supérieur de l'Afrique du nord.* Paléontologie Mémoires No. 4. Algiers: Carte Géologique d'Algérie.

Averianov, A. O., and A. S. Tesakov. 1997. Evolutionary trends in Mio-Pliocene Leporinae, based on *Trischizolagus* (Mammalia, Lagomorpha). *Paläontologische Zeitschrift* 71: 145–153.

Benammi, M., M. Calvo, M. Prévot, and J.-J. Jaeger. 1996. Magnetostratigraphy and paleontology of Aït Kandoula Basin (High Atlas, Morocco) and the African-European Late Miocene terrestrial faunal exchanges. *Earth and Planetary Science Letters* 145:15–29.

Black, C. C. 1972. Review of fossil rodents from the Neogene Siwalik Beds of India and Pakistan. *Palaeontology* 15: 238–266.

Boaz, N. T., R. L. Bernor, A. S. Brooks, H. B. S. Cooke, J. de Heinzelin, R. Dechamps, E. Delson, A. W. Gentry, J. W. K. Harris, P. Meylan, P. P. Pavlakis, W. J. Sanders, K. M. Stewart, J. Verniers, P. G. Williamson, and A. J. Winkler. 1992. A new evaluation of the significance of the late Neogene Lusso Beds, Upper Semliki Valley, Zaire. *Journal of Human Evolution* 22:505–517.

Brandy, L. D. 1979. Rongeurs nouveaux du Néogène d'Afghanistan. *Comptes Rendus des Séances de l'Académie des Sciences* (Paris), ser. D, 289:81–83.

Brandy, L. D. 1981. Rongeurs muroïdés du Néogène supérieur d'Afghanistan: Evolution, biogéographie, corrélations. *Palaeovertebrata* 11:133–179.

Bruijn, H. de. 1986. Is the presence of the African family Thryonomyidae in the Miocene deposits of Pakistan evidence for fauna exchange? *Palaeontology*, Proceedings B 89:125–134.

Bruijn, H. de, and P. J. Whybrow. 1994. A Late Miocene rodent fauna from the Baynunah Formation, Emirate of Abu Dhabi, United Arab Emirates. *Proceedings of the Koninklijke Nederlandse Akademie van Wetenschappen* 97:407–422.

Brunet, M., and M.P.F.T. 2000. Chad: Discovery of a vertebrate fauna close to the Mio-Pliocene boundary. *Journal of Vertebrate Paleontology* 20:205–209.

Chaimanee, Y. 1998. *Plio-Pleistocene Rodents of Thailand.* Thai Studies in Biodiversity No. 3. Bangkok: Biodiversity Research and Training Program, National Center for Genetic Engineering and Biotechnology.

Cheema, I. U., S. Sen, and L. J. Flynn. 1983. Early Vallesian small mammals from the Siwaliks of northern Pakistan. *Bul-*

letin du Muséum National d'Histoire Naturelle (Paris), 5th ser., 1:267–280.

Cifelli, R. L., A. K. Ibui, L. L. Jacobs, and R. W. Thorington Jr. 1986. A giant tree squirrel from the Late Miocene of Kenya. *Journal of Mammalogy* 67:274–283.

Colbert, E. H. 1933. Two new rodents from the lower Siwalik Beds of India. *American Museum Novitates* 633:6.

Collings, G. E., A. R. I. Cruickshank, J. M. Maguire, and R. M. Randall. 1975. Recent faunal studies at Makapansgat Limeworks, Transvaal, South Africa. *Annals of the South African Museum* 71:153–165.

Conroy, G. C., M. Pickford, B. Senut, J. Van Couvering, and P. Mein. 1992. *Otavipithecus namibiensis*, first Miocene hominoid from southern Africa. *Nature* 356:144–148.

Davies, C. 1987. Note on the fossil Lagomorpha from Laetoli. In M. D. Leakey and J. M. Harris, eds., *Laetoli: A Pliocene Site in Northern Tanzania*, pp. 190–193. Oxford: Clarendon Press.

Denys, C. 1987. Fossil rodents (other than Pedetidae) from Laetoli. In M. D. Leakey and J. M. Harris, eds., *Laetoli: A Pliocene Site in Northern Tanzania*, pp. 118–193. Oxford: Clarendon Press.

Denys, C., and J.-J. Jaeger. 1992. Rodents of the Miocene site of Fort Ternan (Kenya). First part: Phiomyids, bathyergids, sciurids and anomalurids. *Neues Jahrbuch für Geologie und Paläontologie, Abhandlungen* 185:63–84.

Dice, L. R. 1929. The phylogeny of the Leporidae, with description of a new genus. *Journal of Mammalogy* 10:340–344.

Dice, L. R. 1931. *Alilepus*, a new name to replace *Allolagus* Dice, preoccupied, and notes on several species of fossil hares. *Journal of Mammalogy* 12:159–160.

Dietrich, W. O. 1941. Die säugetierpaläontologische Ergebnisse der Kohl-Larsen'schen Expedition, 1937–1939 im nördlichen Deutsch-Ostafrika. *Centralblatt für Mineralogie, Geologie und Paläontologie* (Stuttgart) 1941B:217–223.

Dietrich, W. O. 1942. Ältestquartäre Säugetiere aus der südlichen Serengeti, Deutsch-Ostafrika. *Paläontographica* (Stuttgart) 94A:43–133.

Erbaeva, M. A., and R. Angermann. 1983. Das Originalmaterial von *Serengetilagus praecapensis* Dietrich, 1941: Ergänzende Beschreibung und vergleichende Diskussion. *Schriftenreihe für Geologische Wissenschaften* 19/20:39–60.

Flynn, L. J., and R. L. Bernor. 1987. Late Tertiary mammals from the Mongolian People's Republic. *American Museum Novitates* 2872:1–16.

Flynn, L. J., and L. L. Jacobs. 1999. Late Miocene small mammal faunal dynamics: The crossroads of the Arabian Peninsula. In P. Whybrow and A. Hill, eds., *Fossil Vertebrates of Arabia*, pp. 412–419. New Haven: Yale University Press.

Flynn, L. J., and A. J. Winkler. 1994. Dispersalist implications of *Paraulacodus indicus:* A South Asian rodent of African affinity. *Historical Biology* 9:223–235.

Flynn, L. J., A. Sahni, J.-J. Jaeger, B. Singh, and S. B. Bhatia. 1990. Additional fossil rodents from the Siwalik Beds of India. *Proceedings of the Koninklijke Nederlandse Akademie van Wetenschappen* 93:7–20.

Flynn, L. J., Z. Qiu, N. D. Opdyke, and R. H. Tedford. 1995. Ages of key fossil assemblages in the Late Neogene terrestrial record of northern China. In *Geochronology Time Scales and Global Stratigraphic Correlation*, pp. 365–373. SEPM Special Publication No. 54. Tulsa: Society for Sedimentary Geology.

Geraads, D. 1996. Rongeurs du Mio-Pliocène de Lissasfa (Casablanca, Maroc). *Geobios* 31:229–245.

Geraads, D. 1998a. Rongeurs du Miocène supérieur de Chorora (Ethiopie): Cricetidae, Rhizomyidae, Phiomyidae, Thryonomyidae, Sciuridae. *Palaeovertebrata* 27:203–216.

Geraads, D. 1998b. Biogeography of circum-Mediterranean Miocene-Pliocene rodents: A revision using factor analysis and parsimony analysis of endemicity. *Palaeogeography, Palaeoclimatology, Palaeoecology* 137:273–288.

Greenwood, M. 1955. Fossil Hystricoidea from the Makapan Valley, Transvaal. *Palaeontologia Africana* 3:77–85.

Greenwood, M. 1958. Fossil Hystricoidea from the Makapan Valley, Transvaal: *Hystrix makapanensis* nom. nov. for *Hystrix major* Greenwood. *Annals and Magazine of Natural History* 13:365.

Harrison, T., and E. Baker. 1997. Paleontology and biochronology of fossil localities in the Manonga Valley, Tanzania. In T. Harrison, ed., *Neogene Paleontology of the Manonga Valley, Tanzania: A Window into the Evolutionary History of East Africa*, pp. 361–393. New York: Plenum Press.

Hendey, Q. B. 1981. Palaeoecology of the Late Tertiary fossil occurrences in "E" Quarry, Langebaanweg, South Africa, and a reinterpretation of their geological context. *Annals of the South African Museum* 84:1–104.

Hill, A. 1999. The Baringo Basin, Kenya: From Bill Bishop to BPRP. In P. Andrews and P. Banham, eds., *Late Cenozoic Environments and Hominid Evolution: A Tribute to Bill Bishop*, pp. 85–97. London: Geological Society.

Hinton, M. A. C. 1933. Diagnoses of new genera and species of rodents from Indian Tertiary deposits. *Annals and Magazine of Natural History* 10:620–622.

Jacobs, L. L. 1977. A new genus of murid rodent from the Miocene of Pakistan and comments on the origin of the Muridae. *PaleoBios* 25:1–11.

Jacobs, L. L. 1978. *Fossil Rodents (Rhizomyidae and Muridae) from Neogene Siwalik Deposits, Pakistan.* Bulletin Series No. 52. Flagstaff: Museum of Northern Arizona Press.

Jacobs, L. L., and W. R. Downs. 1994. The evolution of murine rodents in Asia. In Y. Tomida, C. K. Li, and T. Setoguchi, eds., *Rodent and Lagomorph Families of Asian Origins and Diversification*, pp. 149–156. Monographs No. 8. Tokyo: National Science Museum.

Jacobs, L. L., L. J. Flynn, and W. R. Downs. 1989. Neogene rodents of southern Asia. In C. C. Black and M. R. Dawson, eds., *Papers on Fossil Rodents in Honor of Albert Elmer Wood*, pp. 157–177. Science Series No. 33. Los Angeles: Natural History Museum of Los Angeles County.

Jacobs, L. L., L. J. Flynn, and C.-K. Li. 1985. Comments on rodents from the Chinese Neogene. *Bulletin of the Geological Institutions of the University of Uppsala*, n.s., 11:59–78.

Jaeger, J.-J. 1977. Les rongeurs du Miocène moyen et supérieur du Maghreb. *Palaeovertebrata* 8:1–166.

Jaeger, J.-J., J. Michaux, and M. Sabatier. 1980. Premières données sur les rongeurs de la formation de Ch'orora (Ethiopie) d'âge Miocène supérieur. I: Thryonomyidés. *Palaeovertebrata, Mémoire Jubilaire R. Lavocat*, pp. 365–374.

Kingdon, J. 1974. *East African Mammals: An Atlas of Evolution in Africa.* Vol. 2, pt. B. *Hares and Rodents.* London: Academic Press.

Lavocat, R. 1973. *Les rongeurs du Miocène d'Afrique orientale. I: Miocène inférieur.* Mémoires et Travaux Ecole Pratique des

Hautes Etudes No. 1. Montpellier: Ecole Pratique des Hautes Etudes, Institut Montpellier.

Leakey, M. G., C. S. Feibel, R. L. Bernor, J. M. Harris, T. E. Cerling, K. M. Stewart, G. W. Storrs, A. Walker, L. Werdelin, and A. J. Winkler. 1996. Lothagam: A record of faunal change in the Late Miocene of East Africa. *Journal of Vertebrate Paleontology* 16:556–570.

Lindsay, E. H. 1987. Cricetid rodents from Siwalik deposits near Chinji Village. Part I: Megacricetodontinae, Myocricetodontinae, and Dendromurinae. *Palaeovertebrata* 18:95–154.

MacInnes, D. G. 1953. *The Miocene and Pleistocene Lagomorpha of East Africa.* Fossil Mammals of Africa No. 6. London: British Museum (Natural History).

Maguire, J. 1976. A taxonomic and ecological study of the living and fossil Hystricidae with particular reference to southern Africa. Ph.D. diss., University of Witwatersrand.

Maguire, J. 1978. Southern African fossil porcupines. *South African Journal of Science* 74:144.

Masini, F., and L. Rook. 1993. *Hystrix primigenia* (Mammalia, Rodentia) from the Late Messinian of the Monticino gypsum quarry (Faenza, Italy). *Bollettino della Società Paleontologica Italiana* 32:79–87.

McDougall, I., and C. S. Feibel. 1999. Numerical age control for the Miocene-Pliocene succession at Lothagam, a hominoid-bearing sequence in the northern Kenya Rift. *Journal of the Geological Society* (London) 156:731–745.

Mein, P., E. M. Suárez, and J. Agustí. 1993. *Progonomys* Schaub, 1938 and *Huerzelerimys* gen. nov. (Rodentia): Their evolution in western Europe. *Scripta Geologica* 103:41–64.

Munthe, J. 1987. Small-mammal fossils from the Pliocene Sahabi Formation of Libya. In N. T. Boaz, A. El-Arnauti, A. W. Gaziry, J. de Heinzelin, and D. D. Boaz, eds., *Neogene Paleontology and Geology of Sahabi,* pp. 135–144. New York: Liss.

Patnaik, R. 1997. New murids and gerbillids (Rodentia, Mammalia) from Pliocene Siwalik sediments of India. *Palaeovertebrata* 26:129–165.

Sabatier, M. 1978. Les rongeurs des sites à hominidés de Hadar et Melka Kunturé (Ethiopie). Thesis, Université Sciences et Techniques, Montpellier.

Sabatier, M. 1982. Les rongeurs du site Pliocène a hominidés de Hadar (Ethiopie). *Palaeovertebrata* 12:1–56.

Schlosser, M. 1924. Tertiary vertebrates from Mongolia. *Palaeontologica Sinica,* ser. C, fasc. 1, 1:1–119.

Sen, S. 1983. Rongeurs et lagomorphes du gisement Pliocène de Pul-e Charkhi, bassin de Kabul, Afghanistan. *Bulletin du Muséum National d'Histoire Naturelle* (Paris), 5th ser., 1:33–74.

Sen, S., and D. B. Kovatchev. 1987. The porcupine *Hystrix primigenia* (Wagner) from the Late Miocene of Bulgaria. *Palaeontology,* Proceedings B 90:317–323.

Sen, S., M. Brunet, and E. Heintz. 1979. Découverte de rongeurs "africains" dans le Pliocène d'Afghanistan (bassin de Sarobi), implications paléobiogéographiques et stratigraphiques. *Bulletin du Muséum National d'Histoire Naturelle* (Paris), 4th ser., 1:65–75.

Slaughter, B. H., and G. T. James. 1979. *Saidomys natrunensis,* an arvicanthine rodent from the Pliocene of Egypt. *Journal of Mammalogy* 60:421–425.

Tong, H. 1989. Origine et évolution des Gerbillidae (Mammalia, Rodentia) en Afrique du nord. *Bulletin de la Société Géologique de France* 155:1–120.

Tong, H., and J.-J. Jaeger. 1993. Muroid rodents from the Middle Miocene Fort Ternan locality (Kenya) and their contribution to the phylogeny of muroids. *Palaeontographica* A 229:51–73.

Wesselman, H. B. 1984. The Omo micromammals: Systematics and paleoecology of early man sites from Ethiopia. *Contributions to Vertebrate Evolution* 7:1–219.

White, J. A. 1991. North American Leporinae (Mammalia: Lagomorpha) from Late Miocene (Clarendonian) to latest Pliocene (Blancan). *Journal of Vertebrate Paleontology* 11:67–89.

Winkler, A. J. 1990. Systematics and biogeography of Neogene rodents from the Baringo District, Kenya. Ph.D. diss., Southern Methodist University.

Winkler, A. J. 1992. Systematics and biogeography of Middle Miocene rodents from the Muruyur Beds, Baringo District, Kenya. *Journal of Vertebrate Paleontology* 12:236–249.

Winkler, A. J. 1994. The Middle/Upper Miocene dispersal of major rodent groups between southern Asia and Africa. In Y. Tomida, C. K. Li, and T. Setoguchi, eds., *Rodent and Lagomorph Families of Asian Origins and Diversification,* pp. 173–184. Monographs No. 8. Tokyo: National Science Museum.

Winkler, A. J. 1997. Systematics, paleobiogeography, and paleoenvironmental significance of rodents from the Ibole Member, Manonga Valley, Tanzania. In T. Harrison, ed., *Neogene Paleontology of the Manonga Valley, Tanzania: A Window into the Evolutionary History of East Africa,* pp. 311–332. New York: Plenum Press.

Table Abbreviations

AMNH = American Museum of Natural History, New York

KNM = National Museums of Kenya, Nairobi

KW = Kanam West locality

LT = Lothagam localities

LU = Lukeino locality

TABLE 5.1 Composite Faunal List of Rodents and Lagomorphs and Their Stratigraphic Occurrence at Lothagam, Kenya

	Stratigraphic Occurrence				
	Nawata		Nachukui		
Taxon	**Lower**	**Upper**	**Apak**	**Kaiyumung**	**Galana Boi**
Order Lagomorpha					
Family Leporidae					
Alilepus sp.	X	—	—	—	—
Serengetilagus praecapensis	—	—	X	—	—
Leporidae sp.[a]	—	X	—	—	X
Order Rodentia					
Family Sciuridae					
Kubwaxerus pattersoni	X	—	—	—	—
Xerus rutilis[a]	—	—	—	—	X
Family Hystricidae					
Hystrix sp. (small)	X	—	—	—	—
Hystrix sp. (large)	—	—	—	X	—
Family Thryonomyidae					
Paraphiomys chororensis	X	—	—	—	—
Paraulacodus cf. *P. johanesi*	X	—	—	—	—
Thryonomys cf. *T. gregorianus*	—	—	—	X	—
Thryonomys sp. (small)	—	X	—	—	—
Thryonomys sp. (large)	—	—	—	—	X
Family Muridae					
Subfamily Gerbillinae					
Abudhabia sp.	—	X	—	—	—
Tatera sp.	—	—	—	—	X
Subfamily Murinae					
Karnimata jacobsi sp. nov.	X	—	—	—	—
Saidomys sp.	—	X	—	—	—
Taxon c[a]	—	X	—	—	—
Taxon d[a]	X	—	—	—	—
Arvicanthis sp.	—	—	—	—	X
Subfamily *incertae sedis*					
Gen. and sp. nov.	X	—	—	—	—

[a] Not discussed in the text.

TABLE 5.2 Occlusal Measurements (in mm) of the Lower Dentition of *Serengetilagus praecapensis* from Kenya and Tanzania

	Locality			
Measurement	Lothagam	Laetoli (Humboldt cln)[a]	Laetoli (KNM cln)	NE of Lake Eyasi (BM cln)[b]
P_3 length				
Number	1	54	9	1
Mean	—	3.27	3.36	3.25
Range	3.14	2.9–3.7	3.08–3.83	—
P_3 width				
Number	1	52	9	1
Mean	—	3.08	3.22	—
Range	2.57	2.7–3.5	2.71–3.5	3.0
P_4 length				
Number	1	47	9	1
Mean	—	2.88	2.84	—
Range	2.57	2.5–3.3	2.57–3.0	2.9
P_4 width				
Number	1	43	9	1
Mean	3.28	3.16	3.35	3.25
Range	—	2.5–3.6	3.0–3.58	—
M_1 length				
Number	1	—	8	1
Mean	—	—	2.85	—
Range	2.71	—	2.71–3.0	3.0
M_1 width				
Number	1	—	8	1
Mean	—	—	3.14	—
Range	3.28	—	3.0–3.17	3.0
M_2 length				
Number	1	—	3	1
Mean	2.71	—	2.83	2.8
Range	—	—	2.71–2.92	—
M_2 width				
Number	1	—	3	1
Mean	—	—	2.88	—
Range	3.14	—	2.86–2.92	3.1
M_3 length				
Number	1	—	2	1
Mean	1.71	—	—	2.0
Range	—	—	1.71–1.86	—
M_3 width				
Number	1	—	2	1
Mean	—	—	—	—
Range	1.71	—	1.86	1.75
Length toothrow				
Number	1	10	1	—
Mean	—	16.2	—	—
Range	15.28	14.6–17.4	16.24	—

[a] Measurements from Laetoli (Humboldt Collection) are from Erbaera and Augermann (1983).

[b] Measurements from northeast of Lake Eyasi (British Museum [BM] Collection) are from Macinnes (1953).

TABLE 5.3 Measurements (in mm) of *Hystrix* sp. from Lothagam, *H. leakeyi* from Laetoli (Denys 1987), *H.* cf. *H. makapanensis* from Olduvai Bed I, Tanzania (Sabatier 1978), *H. makapanensis* (Greenwood 1955) and *Xenohystrix crassidens* from Makapansgat, South Africa (Denys 1987), and *X. crassidens* from Hadar, Ethiopia (Sabatier 1978)

	***Hystrix* sp.**[a]	***H. leakeyi***	***H. cristata***	***H. cristata* Corbet and Jones**	***H. africaeaustralis* Corbet and Jones**	***H.* cf. *H. makapanensis***	***H. makapanensis***	***Xenohystrix crassidens***	***Xenohystrix crassidens***
Site	Lothagam	Laetoli	Various	—	—	Olduvai Bed I	Makapansgat	Makapansgat	Hadar
Length dP_4									
Number	1	1	1	—	—	4	—	—	—
Range	9.14	8.0	8.26*	—	—	9.8–11.1	—	—	—
Mean	—	—	—	—	—	10.2	—	—	—
Width dP_4									
Number	1	1	1	—	—	4	—	—	—
Range	6.65	5.7	5.63*	—	—	6.0–7.4	—	—	—
Mean	—	—	—	—	—	6.95	—	—	—
Length M_1									
Number	1 M_1 or M_2	2	1	—	—	5	—	1	1
Range	11.44	7.9–8.2	8.5	—	—	10.1–11.5	—	14.0	12.5
Mean	—	—	—	8.5	8.7	10.6	—	—	—
Width M_1									
Number	1 M_1 or M_2	2	1	—	—	5	1	1	1
Range	9.16	7.1	7.0	—	—	7.0–8.9	9.5	11.0	11.9
Mean	—	7.1	—	7.5	7.4	7.9	—	—	—
Length M_2									
Number	—	2	1	—	—	4	—	1	1
Range	—	8.7–8.8	9.0	—	—	10.4–11.6	—	14.5	14.0
Mean	—	—	—	10.0	9.0	11.1	—	—	—
Width M_2									
Number	—	2	1	—	—	4	—	1	1
Range	—	6.7–7.0	7.8	—	—	7.0–9.5	—	11.5	12.0
Mean	—	—	—	8.0	7.7	8.8	—	—	—

Note: Extant comparative material includes *H. cristata* (Shuler Museum Collection, SMU [Denys 1987]) and mean values from Corbet and Jones (1965), as cited in Denys (1987). Extant specimens of *H. africaeaustralis* are mean values from Corbet and Jones (1965), as cited in Denys (1987).

[a] dP_4 = *Hystrix* sp. (small); M_1 or M_2 = *Hystrix* sp. (large).

TABLE 5.4 Measurements (in mm) of *Paraphiomys chororensis* from Lothagam, Kenya, and Chorora, Ethiopia

Measurement and Locality	P_4	M_1	M_2	M_3
Length				
Lothagam				
Number	1	1	1	1
Mean	—	—	—	—
Observed range	2.42	2.58	2.58	2.67
Chorora				
Number	6	5	5	2
Mean	2.32	2.32	2.48	—
Observed range	2.20–2.47	2.13–2.44	2.38–2.56	2.23–2.56
Width				
Lothagam				
Number	1	1	1	1
Mean	—	—	—	—
Observed range	1.83	2.29	2.50	2.25
Chorora				
Number	7	5	5	2
Mean	1.91	2.22	2.44	—
Observed range	1.77–2.29	2.01–2.47	2.29–2.59	2.17–2.41

Source: Measurements from Geraads (1998a).

TABLE 5.5 Measurements (in mm) of the Antero-posterior Height and Transverse Width of Upper Incisors of *Paraulacodus johanesi* from Lothagam, Kenya (KNM-LT 26542), and Chorora, Ethiopia, and of *P. indicus* from the Potwar Plateau, Pakistan (GSI D281, YGSP 33105)

	Antero-posterior Height			Width		
Taxon Specimen No.	Number	Mean	Observed Range	Number	Mean	Observed Range
Paraulacodus johanesi						
KNM-LT 26542	—	—	—	1	—	3.08
Chorora specimens[a]	4	3.26	3.05–3.55	5	2.61	2.40–2.95
Chorora specimens[b]	—	3.35	3.02–3.63	—	2.53	2.40–2.76
Paraulacodus indicus						
GSI D281	1	—	3.08	1	—	2.17
YGSP 33105[c]	1	—	2.5[d]	1	—	2.5

[a] Measurements from Jaeger et al. (1980).

[b] Measurements from Geraads (1998a).

[c] Measurements from Flynn and Winkler (1994).

[d] Approximate.

TABLE 5.6 Measurements (in mm) of *Thryonomys* from the Nachukui Formation, Lothagam, Kenya; the Wembere-Manonga Formation, Inolelo 1, Tanzania; the Upper Ndolanya Beds, Laetoli, Tanzania and Extant Specimens from Various Localities

	Taxon and Locality				
	T.* cf. *T. gregorianus Lothagam	***Thryonomys* sp.** Inolelo 1[a]	***Thryonomys* sp. indet.** Laetoli[b]	***Thryonomys gregorianus***	***Thryonomys swinderianus***
				Extant, Various Localities	
M_1 length					
Number	2	1	1	11	7
Mean	—	—	—	4.1	4.3
Range	3.4	3.1	4.1	3.7–4.7	3.8–4.8
M_1 width					
Number	2	1	1	11	7
Mean	—	—	—	5.1	6.1
Range	3.6	3.8	4.8	4.5–6.1	5.3–6.6
M_2 length					
Number	1	1	—	8	7
Mean	—	—	—	4.3	4.9
Range	4.0	3.2	—	3.9–4.7	4.5–5.2
M_2 width					
Number	—	1	—	8	7
Mean	—	—	—	5.3	6.4
Range	—	3.4	—	4.9–5.6	5.3–6.9

[a] Winkler (1997).

[b] Denys (1987).

TABLE 5.7 Occlusal Measurements (in mm) of *Abudhabia* sp. from Lothagam, Kenya; "*Protatera*" *yardangi* from Sabahi, Libya; *A. baynunensis* from the Emirate of Abu Dhabi, United Arab Emirates (UAE); and *A. kabulense* from Pul-e Charkhi, Afghanistan

	Taxon and Locality			
Measurement	***Abudhabia* sp.** Kenya	**"*Protatera*" *yardangi*** Libya[a]	***Abudhabia baynunensis*** UAE[b]	***Abudhabia kabulense*** Afghanistan[c]
M_1 length				
Number	1	2	2	7
Mean	—	—	—	2.73
Observed range	2.56	2.40–2.50	2.22–2.33	2.60–2.85
M_1 width				
Number	1	3	2	7
Mean	—	1.60	—	1.89
Observed range	1.60	1.60	1.45–1.47	1.79–1.94

[a] Munthe (1987).

[b] de Bruijn and Whybrow (1994).

[c] Sen (1983).

Table 5.8 Occlusal Measurements (in mm) of *Karnimata jacobsi* sp. nov., from Lothagam, *K. darwini* (Jacobs 1978), *K. huxleyi* (Jacobs 1978), *K. minima* (Brandy 1981), *K. intermedia* (Brandy 1981), *K.* cf. *K. intermedia* (Flynn et al. 1990), *Karnimata* sp. (Jacobs 1978), and *Karnimata* sp. (Cheema et al. 1983)

							***Karnimata* sp.**	
	K. jacobsi	***K. darwini***	***K. huxleyi***	***K. minima***	***K. intermedia***	***K.* cf. *K. intermedia***	**Jacobs**	**Cheema et al.**
Length M^1								
Number	2	39	16	1	1	—	2	1
Range	2.52–2.64	1.95–2.38	1.80–2.22	2.29	2.77	—	2.50–2.60	1.90
Mean	—	2.14	1.99	—	—	—	—	—
Width M^1								
Number	2	39	16	1	1	—	2	1
Range	1.80	1.22–1.50	1.28–1.55	1.53	1.85	—	1.50–1.60	1.12
Mean	—	1.37	1.42	—	—	—	—	—
Length M^2								
Number	1	36	9	1	1	1	—	—
Range	2.52	1.32–1.62	1.25–1.48	1.71	2.02	2.02	—	—
Mean	—	1.46	1.37	—	—	—	—	—
Width M^2								
Number	2	36	9	1	1	1	—	—
Range	1.84–1.88	1.25–1.52	1.30–1.58	1.59	1.82	1.78	—	—
Mean	—	1.40	1.46	—	—	—	—	—
Length M^3								
Number	1	9	2	—	1	1	—	1
Range	1.60	0.88–1.12	1.18–1.20	—	1.36	1.53	—	1.00
Mean	—	1.01	—	—	—	—	—	—
Width M^3								
Number	1	9	2	—	1	1	—	1
Range	1.60	0.95–1.12	1.12–1.15	—	1.47	1.39	—	0.88
Mean	—	1.03	—	—	—	—	—	—

Length M_1								
Number	2	38	18	—	—	—	—	2
Range	2.32	1.72–2.15	1.55–1.98	—	—	—	—	1.56
Mean	—	1.98	1.80	—	—	—	—	—
Width M_1								
Number	2	38	18	—	—	—	—	2
Range	1.44–1.56	1.02–1.30	1.00–1.25	—	—	—	—	0.96
Mean	—	1.17	1.16	—	—	—	—	—
Length M_2								
Number	2	43	15	2	1	—	—	—
Range	1.72–1.84	1.38–1.72	1.30–1.68	1.50–1.68	1.83	—	—	—
Mean	—	1.52	1.44	—	—	—	—	—
Width M_2								
Number	2	43	15	2	1	—	—	—
Range	1.60–1.80	1.15–1.45	1.18–1.52	1.29–1.55	1.56	—	—	—
Mean	—	1.29	1.36	—	—	—	—	—
Length M_3								
Number	1	18	2	1	1	—	—	1
Range	1.48	1.18–1.40	1.32–1.42	1.32	1.53	—	—	1.08
Mean	—	1.26	—	—	—	—	—	—
Width M_3								
Number	1	18	2	1	1	—	—	1
Range	1.40	0.98–1.22	1.22–1.35	1.29	1.40	—	—	0.88
Mean	—	1.33	—	—	—	—	—	—

TABLE 5.9 Occlusal Measurements (in mm) of Lower Molars of *Saidomys* from Lothagam and Tabarin, Kenya; the Manonga Valley, Tanzania; Wadi Natrun, Egypt; Hadar, Ethiopia; and Pul-e Charkhi and Dawrankhel 14 and 15, Afghanistan

	Taxon and Locality					
	***Saidomys* sp. Kenya**[a]	***Saidomys* sp. nov. Kenya**[a]	***Saidomys parvus* Tanzania**[b]	***Saidomys natrunensis* Egypt**[c]	***Saidomys afarensis* Ethiopia**[d]	***Saidomys afghanensis* Afghanistan**[e]
M_1 length						
Number	1	1	3	2	60	6
Mean	—	—	2.52	—	2.97	2.91
Range	2.60	2.67	2.33–2.67	2.70–2.90	2.78–3.15	2.71–3.07
M_1 width						
Number	1	1	3	2	60	13
Mean	—	—	1.73	—	2.06	2.03
Range	1.80	1.83	1.67–1.79	2.00–2.10	1.85–2.27	1.90–2.19
M_2 length						
Number	1	2	—	2	55	11
Mean	—	—	—	—	2.23	2.27
Range	1.92	1.92–2.04	—	2.10–2.20	2.05–2.40	2.17–2.58
M_2 width						
Number	1	2	—	2	55	11
Mean	—	—	—	—	2.12	2.22
Range	2.00	1.88–2.04	—	2.20–2.40	1.97–2.30	2.14–2.32
M_3 length						
Number	1	3	—	2	47	10
Mean	—	—	—	—	2.07	2.19
Range	1.80	1.71–1.82	—	2.00–2.20	1.93–2.24	2.00–2.30
M_3 width						
Number	1	3	—	2	47	11
Mean	—	—	—	—	1.92	2.11
Range	1.68	1.59–1.71	—	2.10–2.20	1.75–2.14	2.05–2.09

[a] Winkler (1990).
[b] Winkler (1997).
[c] Slaughter and James (1979).
[d] Sabatier (1982).
[e] Sen (1983).

TABLE 5.10 Occlusal Length by Width Measurements (in mm) of Muridae, gen. and sp. nov., from the Upper Member, Nawata Formation, Lothagam, Kenya, and *Myocricetodon magnus* from Pataniak 6, Morocco

	Tooth				
Taxon	**M^1**	**M^2**	**M_1**	**M_2**	**M_3**
Muridae gen. and sp. nov.	4.25 × 3.00	2.42 × 2.67	3.67 × 2.75	2.42 × 2.67	2.00 × 1.92
	4.42 × 2.92				
Myocricetodon magnus	2.77 × 1.91	—	2.49 × 1.72	1.89 × 1.74	—

Source: Measurements from Jaeger (1977).

6

PRIMATES

6.1

Cercopithecidae from Lothagam

Meave G. Leakey, Mark F. Teaford, and Carol V. Ward

The Lothagam collection of Cercopithecidae constitutes the largest collection of this family known from the Late Miocene of Africa. The majority of the specimens derive from the Nawata Formation, where papionins make up 80 percent of the cercopithecid collection; three species of colobine and indeterminate species make up the remaining 20 percent. Cercopithecids are rare in the Apak Member of the Nachukui Formation—only three specimens are known—but are common again in the Kaiyumung Member where *Theropithecus* cf. *T. brumpti* is the dominant cercopithecid. The postcranial specimens show that both colobines and cercopithecines were semiterrestrial. There are very few characters that distinguish the two subfamilies at this time, suggesting that African colobines became increasingly arboreal only after the end of the Miocene. Studies of the molar microwear show that the colobines were eating foods much like those eaten by colobines today, whereas the lack of large pits on the molars indicates that the cercopithecines, in contrast to extant species, were not ingesting hard objects. Despite the fact that the fossil colobines and cercopithecines showed no significant differences in molar microwear, morphological features of the occlusal surface and the disparity in the size of the anterior dentition suggest that there were dietary differences. The colobines, with their small anterior dentition and prominent transverse loph(id)s, were probably eating seeds and some leaves, whereas the cercopithecines, with their large anterior dentition and large cheek teeth with less occlusal relief, were eating mainly fruits. Features of the cranium and deciduous dentition that are shared with *Victoriapithecus* suggest the Victoriapithecinae and Cercopithecinae belong within the same family.

Figure 6.1 Restoration of *Parapapio lothagamensis* sp. nov. by Mauricio Antón.

The origins of the Cercopithecidae are believed to date back to the earliest Miocene, although the oldest known fossil, an M^2 approximately 19 Ma old, is from Napak, Uganda (Pilbeam and Walker 1968). Other early occurrences known from East Africa are from slightly younger sites (~17 Ma) and include an M_3 from Ombo, Kenya (Le Gros Clark and Leakey 1951); a mandible fragment and an isolated molar from Loperot, Kenya (Szalay and Delson 1979); and 16 specimens including mandibular and maxillary fragments and isolated teeth from Buluk, northern Kenya (Leakey 1985). In North Africa, cercopithecids assigned to the genus *Prohylobates* are known from two Early Miocene localities: Wadi Moghara, Egypt (Simons 1969), and Gebel Zelten, Libya (Delson 1979). An M_3 has also been reported from the Middle Miocene at Ongoliba, Zaire (Hooijer 1963). In contrast to these sparse occurrences of early cercopithecids, at the Middle Miocene site on Maboko Island, Lake Victoria (15 Ma), cercopithecids are well represented (Benefit 1993, 1994; Benefit and McCrossin 1990).

In earlier studies (von Koenigswald 1969; Delson 1973; Simons and Delson 1978; Szalay and Delson 1979), the Miocene cercopithecids were regarded as belonging to a separate subfamily within the Cercopithecidae. With the increased sample from Maboko Island, detailed morphological comparisons led Benefit (1993) to identify a number of primitive dental traits unique to these early monkeys and at least three derived traits common to the extant subfamilies, Colobinae and Cercopithecinae, but exclusive of the Victoriapithecinae. She therefore proposed that the subfamily Victoriapithecinae be raised to the family rank (Benefit 1993). In contrast to other Miocene sites where hominoids are commonly found and cercopithecids are absent, at Maboko *Victoriapithecus* is the most commonly recovered primate (Benefit and McCrossin 1990). The latest victoriapithecines are found in the Ngorora Formation, Baringo, Kenya, at a site dated at 12.5 Ma (Hill 1999; Hill et al. 2002).

The earliest known colobines are *Microcolobus tugenensis* from Ngeringerowa, Tugen Hills, Kenya (Benefit and Pickford 1986), which is now dated at approximately 9 Ma (Hill 1999) and a single molar from the similar aged site at Nakali, Kenya (Benefit and Pickford 1986). The earliest known cercopithecines are found at Lothagam. The divergence of the two subfamilies thus occurred prior to the accumulation of the Ngeringerowa assemblage.

The Lothagam collection of Cercopithecidae constitutes the largest African Late Miocene collection of this family (168 specimens). Most derive from the lower member of the Nawata Formation where the papionins constitute 79 percent of the cranial collection (95 specimens). Papionins make up 64 percent of the 48 Upper Nawata cercopithecid specimens. The Nawata Formation taxa include one species of papionin and two species of colobine, one or both of which is represented by postcranial elements. Only three cercopithecid specimens, two colobine and one cercopithecine, were documented from the Apak Member of the Nachukui Formation, although a new species of *Cercopithecoides* may be from this member. Fourteen cercopithecid specimens were recovered from the youngest Lothagam deposits in the Kaiyumung Member. Most represent *Theropithecus* (12 specimens), with only one specimen each of an indeterminate colobine and an indeterminate species of *Parapapio*.

The first cercopithecids found at Lothagam were collected by Patterson's expedition in 1967. Three of these specimens—a mandible fragment, an isolated P_4, and a distal humerus—were initially identified as Papionin cf. *Parapapio* and cf. *Cercocebus* (Smart 1976), although the attribution of the respective specimens was not given. The most complete, the mandible LT 415 (173-67K) discovered in 1967, was briefly described and referred to Papionini gen. and sp. indet. A by Leakey and Leakey (1976). A single molar of *Theropithecus* collected in the same year, was, for some time, considered to be the earliest evidence of *Theropithecus* (Szalay and Delson 1979; Delson 1993). A colobine premolar remained unpublished. Additional material was recovered by Princeton University Expeditions in 1972 and 1973 (five specimens) and by a short survey led by Richard Leakey in 1980 (three more). The majority of the Lothagam collection was recovered by the National Museums of Kenya field expeditions between 1989 and 1993. All are housed in the collections of the National Museums of Kenya.

The collection provides the opportunity to assess the evolutionary status of these early monkeys and their relationship to the later cercopithecid radiation seen in the Pliocene and Pleistocene East African deposits.

Systematic Description

Order Primates Linnaeus, 1758
Infraorder Catarrhini E. Geoffroy, 1812
Family Cercopithecidae Gray, 1821
Subfamily Cercopithecinae Gray, 1821
Tribe Papionini Burnett, 1828
Genus *Parapapio* Jones, 1937

Diagnosis

A medium to large fossil papionin distinguished by the lateral profile of the muzzle dorsum, which forms a straight line or a smooth, slightly concave curve from nasion to rhinion or beyond to nasospinale. The supraorbital tori are usually weakly developed and do not project forward in either sex; ophryonic groove little developed or absent. There are no strong maxillary ridges or deep maxillary fossae, although in the larger individuals there is some hollowing. Fossae on the lateral mandibular faces are weakly excavated or absent (modified from Leakey and Delson 1987).

Type species

P. broomi Jones, 1937

Parapapio lothagamensis sp. nov.

(Figures 6.1–6.3, 6.4A–B, 6.4D, 6.5A, 6.6B, 6.6E, 6.7A–C; tables 6.1–6.6, 6.9)

Diagnosis

Distinguished from all other *Parapapio* species by its small size, long obliquely oriented mandibular symphysis, relatively broad P_3s and a dP_4 that lacks a distal transverse crest. The symphysial region is narrow, the lower incisors are positioned mesial to the canines, and the I_1s mesial to the I_2s. There is some sexual dimorphism in the size of the teeth. The molars have closely approximated cusps and flare toward the cervix, giving a high cusp width to crown base index. Features shared with *Victoriapithecus* exclusive of most extant colobines and cercopithecines include a high degree of molar flare, distally constricted M^3s with variable absence of distal shelf, the frequent occurrence of a metaconid on P_3, an obliquely oriented P_4, labiolingually wide M_1, the retention of a weakly developed crista obliqua on the dP^4, and, occasionally, a weakly developed hypoconulid on the dP^4.

Holotype

23091, male mandible lacking both rami and with Lt. I_2, /C, P_4–M_3; Rt. P_4–M_3, tip Rt. /C crown and Rt. I_2.

Locality

Nawata Formation, Lothagam.

Horizon

The holotype was collected in 1989, before aerial photographs had been obtained. Its exact provenance is therefore uncertain, but it is recorded as from the "central area of Lothagam" and thus is from either the Upper or the Lower Nawata.

Lothagam Material

- *Lower Nawata:* 114, Rt. mandible fragment (partial M_2, roots M_1); 115, Rt. P^4; 415, Rt. mandible fragment (M_{2-3}); 419, Rt. female maxilla fragment (C/–M^3); 420, fragment squashed crown Rt. male C/; 22971, Lt. mandible (M_{2-3}); 22972, Lt. maxilla fragment (P^4–M^3); 22973, Rt. maxilla fragment (M^{1-3}) and Lt. P_3; 22974, Lt. proximal femur; 23065, male edentulous Lt. mandible corpus and symphysis (roots Lt. and Rt. I_{1-2}, Lt. /C–M_1, broken M_{2-3}; 23066, Lt. M_3; 23070, Lt. M_1, Lt. M_3 trigonid and phalanx; 23074, Lt. distal humerus; 23075, Rt. distal humerus shaft; 23077, Rt. distal humerus; 23079, Rt. I_2, male Lt. /C, root Rt. /C, Rt. C/, root Lt. C/, Rt. maxilla fragment (M^2); 23081, Rt. talus, middle cuneiform and cuboid; 23086, Rt. distal tibia; 23090, female mandible fragment (Lt. /C–P_4) (Lt. M_3), (Lt. M_1), Lt. maxilla (M^{2-3}) and skull fragments; 23122, Lt. talus; 23124, juvenile Rt. mandible fragment (distal dP_3 and dP_4, P_3 and /C in crypt); 23163, Lt. I^1, Rt. M^1 or M^2; 23173, Rt. I_1; 23717, broken M_1 or M_2; 24097, Lt. female P_3; 24099, Rt. M_1; 24101, Rt. M^1, M^2; 24105, Lt. M_3; 24106, Rt. maxilla (very weathered M^{2-3}); 24108, Lt. mandible fragment (dP_4–M_1), worn dP_3; 24111, male facial parts, maxilla, premaxilla and nasals (Lt. P^{3-4}, Lt. and Rt. M^2, Lt. M^3, Rt. M^1); 24114, Lt. proximal humerus; 24120, mandible fragment (roots P_4–M_3); 24121, Lt. distal femur; 24122, Lt. mandible (M_1–M_3, broken P_4); 24125, Rt. calcaneus; 24134, Lt. I^2; 24135, Rt. mandible (M_{1-3}); 24138, Rt. dP^4; 24139, mandible (Lt. and Rt. P_4–M_3); 26187, male edentulous Rt. mandible (roots I_1–M_3); 26370, Rt. proximal scapula; 26371, Rt. M^2; 26373, Rt. M^1; 26384, Lt. M^2; 26385, Lt. proximal humerus, Rt. proximal Mt 2, Lt. triquetral; 26386, male distorted mandible symphysis (Rt. /C–P_4, Rt. and Lt. I_{1-2}, broken Lt. /C); 26388, Lt. male P_3; 26389, Rt. distal humerus shaft; 26391, Rt. mandible fragment (M_3); 26393, Lt. M^2; 26394, juvenile maxilla and mandible fragments, symphysis (Lt. and Rt. I_1 in crypt, roots d/C and dP_3), Lt. mandible fragment (M_2 erupting), Rt. mandible fragment (Rt. distal half M_1) half /dP; 26395, Rt. M_1 or M_2; 26398, Lt. broken M_3; 26400, Lt. M^2; 26405, Lt. I^1; 26406, Rt. I^1, Lt. I^2; 26402, Rt. calcaneus; 26403, Lt. proximal femur; 26404, Lt. proximal femur; 26409, Lt. M_3; 26410, Rt. distal humerus; 26579, Lt. mandible fragment (distal half dP_3), Rt. distal half dP_4, Lt. M_1, unerupted Rt. I_2; 26608, Rt. worn I^1; 26617, Rt. worn dP^4; 26619, Rt. dP^4, Rt. dI^2; 28575, Lt. calcaneus; 28576, Rt. maxilla (P^3–M^3); 28728, Lt. M_2; 28755, Lt. M_3; 28766, mandible symphysis, alveoli Lt. /C–Rt. /C; 28791, Rt. maxilla (P^3–M^1); 28792, Lt. female edentulous mandible (alveoli P_3–M_3), Rt. maxilla fragment (root I^2),

Lt. and Rt. I^1–P^3, Lt. P^4, M^2, Rt. M^3; 30238, Lt. M^3; 30263, Rt. M_3; 30608, Lt. I_1.

- *Upper Nawata:* 123, Rt. M^2; 23067, Rt. humerus lacking head; 23068, Lt. distal humerus and partial shaft; 24089, weathered dP_4; 24094, Lt. mandible (P_4–M_3); 24095, Lt. mandible (M_2); 24096, Rt. maxilla fragment (M^2), Lt. I^1; 24100, Lt. I^1; 24102, Rt. M^1; 24109, Lt. male /C, Rt. M^3; 24112, frontal; 24113, Rt. maxilla (dP^4–M^1, and P^3, P^4, M^2 in crypt); 24117, Lt. M_3; 24119, Rt. proximal ulna; 24123, Rt. distal humerus; 24127, worn M_1 or M_2; 24133, Lt. M_3; 24136, Lt. male mandible fragment (M_{2-3}) and edentulous symphysis; 24137, maxilla fragments (Rt. P^3–M^2, Lt. P^4, roots M^{1-3}), fragment Lt. P^3; 24140, Lt. mandible fragment (worn M_2, roots M_1 and M_3); 26366, Rt. female C/; 26367 M/; 26374, broken $M^{?1}$; 26375, Rt. proximal femur; 26376, Lt. proximal tibia; 26377, two tooth fragments; 28769, Rt. proximal humerus; 28781, Lt. and Rt. broken I^1, tooth fragments; 28783, edentulous mandible, Rt. M_3 and two M fragments; 28786, Lt. weathered M^3; 30606, Rt. male P_3; 36910, Rt. I_1, Lt. I_2.
- *Nawata Formation, horizon indet.:* 23091, (holotype); 23164, Lt. female C/.

Parapapio lothagamensis is the smallest papionin recognized. It is on average smaller than *P. ado* from Laetoli, although there is some overlap in the size of the teeth (table 6.5).

Skull

Cranium

LT 24112 (figure 6.2) preserves a portion of the frontal and interorbital septum, which is damaged in the region of glabella, thus exposing cancellous bone. There is no evidence of a frontal sinus. On the left side, a thin supraorbital torus with a distinct supraorbital notch and spine is preserved. Superiorly, the frontal extends posteriorly 27.4 mm from the supraorbital margin. There is a slight depression behind the tori but no evidence of temporal lines. The superior 21 mm of the narrow nasals are preserved, and although the morphology is unclear, these nasals appear to be bordered by extensions of the frontal with frontomaxillary squamae. On the left side, a small portion of maxilla appears to be preserved; if this is the correct interpretation, the frontal extends unusually far distally and the general morphology indicates a long rostrum. There is no evidence of the lachrymal or of the lachrymal fossa on either side of the nasal bridge, so that this morphology is similar to that of *Victoriapithecus* in which the narrow interorbital ex-

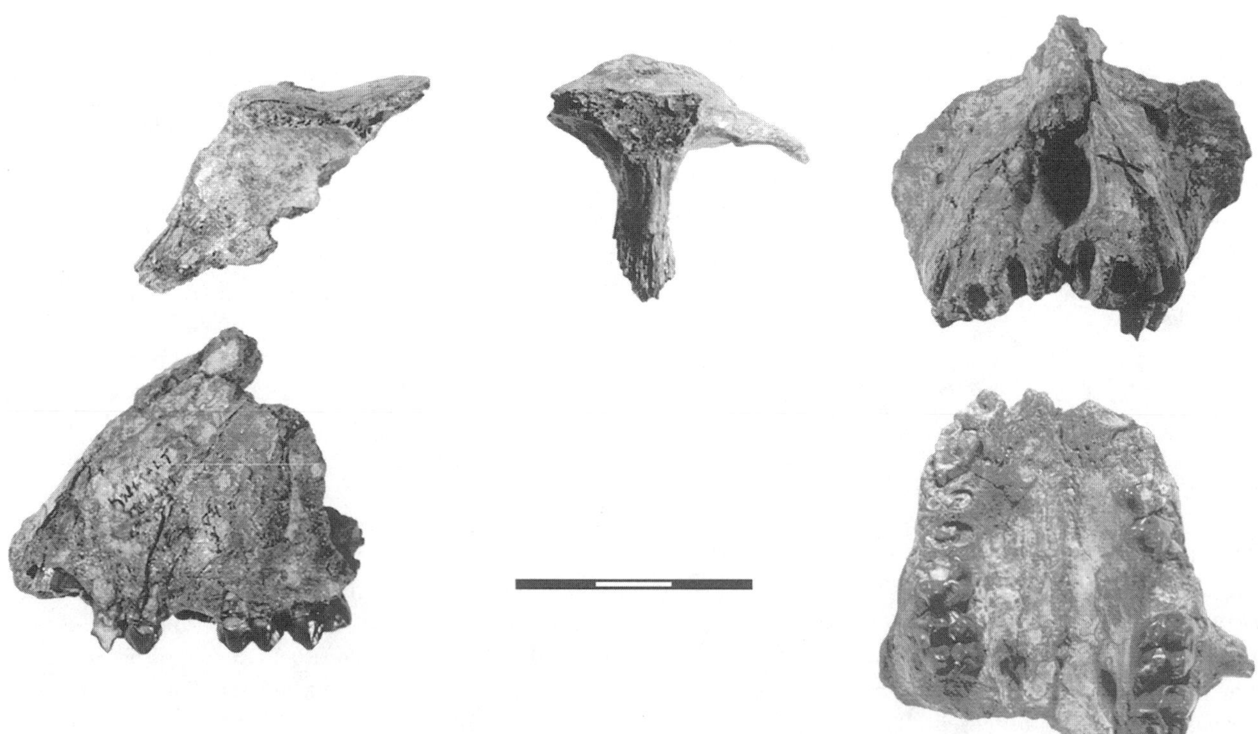

Figure 6.2 *Parapapio lothagamensis* sp. nov. KNM-LT 24111, maxilla from the Lower Nawata, and KNM-LT 24112, frontal from the Upper Nawata: left = left lateral view; center and top right = anterior view; lower right = occlusal view.

tends as a thin keel anterior to the orbits and the lachrymal fossa is positioned well posterior to the nasals. Characters that *Parapapio* shares with *Victoriapithecus* are seen in the steep angle that the bridge of the nose makes with the frontal and the apparently rather straight, very slightly concave muzzle dorsum in lateral profile.

Of two preserved maxillae and palates, LT 24111 and 24137, the former, a male, is the more complete; it includes the premaxillae, portions of the zygomatic and partial nasals, and most of the dentition (figure 6.2). Bone is lost at the incisor alveoli. Posteriorly, the bone is broken behind the M^3s, across the malar region of the zygomatic, and behind the lower orbital margins. A small portion of the zygomatic is preserved along the left zygomaticomaxillary suture. LT 24137 is a partial maxilla broken anteriorly at the level of the P^3s, posteriorly at the M^3s, and superiorly across the maxillary fossae. The right P^3 to M^3 and the left P^4 and partial P^3 are preserved. The remaining nasal profile of LT 24111 matches that of the frontal, LT 24112 described above, in the raised bridge of the nose and the steep angle of the nasals. In lateral profile, the angle of the muzzle dorsum is about 35° to the occlusal plane and only slightly convex. Lateral to the nasals, there is a slight depression on the maxilla, which inferiorly is inflated over the canine roots. There is only a very faint hint of a maxillary fossa. Four distinct infraorbital foramina on the right and three on the left border the frontomaxillary suture. Because the nasals are broken inferiorly, the shape of the nasal aperture is unclear, but the premaxilla is preserved and is about 4.5 mm wide along the left aperture margin. Judging by the position and size of the incisor alveoli, the premaxilla would have been prognathic with large strong central incisors that extended well anterior to the canines. The maxilla is deep below the orbits relative to the length of P^3–M^3, and the malar process of the zygomatic is also deep (~19 mm). The malar process departs above M^2 and close to the alveolar margin. The palate is U-shaped with I^1 positioned anterior to I^2, which is anterior to the canine. The preserved palate, LT 24111, is slightly distorted; the undistorted preserved portion of the palate LT 24137 shows a slight ridge at the midline. The relatively small incisive foramina of LT 24111 are damaged; although their outline is unclear, they can be seen to be small and positioned between the canines with the anterior border posterior to the canine mesial margin and the posterior border anterior to the canine distal margin. The maxillary-premaxillary suture runs across the center of the foramina. The posterior border of the palate is missing, but the clearly defined greater palatine foramen with a sharp medial edge is clearly visible on the left side.

Mandible

The best-preserved mandible is the type specimen, LT 23091 (figure 6.3). This male mandible is broken on both sides just distal to M_3 and is thus missing both rami. The worn left and right I_2, C, and P_4–M_3 are preserved, but only the roots of the remaining teeth are present. The mandible is quite deep compared to the size of the teeth (25.5 mm below M_2), and the alveolar plane is essentially parallel to the inferior border. The mandibular symphysis is distinctive in its unusually oblique orientation: in lateral profile the anterior aspect of the mandible makes an angle of approximately 25° to the axis of the occlusal plane of the cheek teeth, receding steeply from the alveolar margin to the inferior border, which extends to the level of M_1. The anterior aspect, which is smooth and with no trace of mental ridges or rugosities, is narrow and, toward the incisor alveoli, somewhat keeled. There is a centrally positioned foramen symphyseosum. Superiorly, the shallow postincisive plane extends posteriorly to the level of M_1, and there is a distinct genial fossa. In occlusal view, the U-shaped dental arcade is distinctive in being narrow anteriorly with prognathic incisors; the lateral incisors are placed mesial to the canines and distal to the central incisors. The incisor alveolar margins are in the same plane as those of the cheek teeth. The tooth rows diverge only slightly posteriorly. There is a hint of a fossa, and a large mental foramen is positioned below M_1 on the lower third of the lateral face. On the right side, this foramen is double with a second smaller foramen 3.8 mm immediately distal to it. On the left side, a second very small foramen is 3.5 mm immediately superior to the larger. The I_1 is relatively small, with oblique occlusal wear almost to the cervix on the labial face. The canine is worn along its distal face from the tip to the distal cuspule. Dentine is exposed continuously between the buccal and lingual cusps of P_4, M_1, and M_2 and the mesial cusps of M_3. Only small dentine pits appear on the hypoconid and hypoconulid.

Although there are four edentulous male mandible fragments (LT 23065, 24136, 26187, and 28766), which all show the same distinctive symphysial morphology seen in LT 23091, there are no comparable female symphyses.

Permanent Dentition

Both upper and lower permanent dentitions are well represented (tables 6.1–6.6).

Upper permanent incisors

There are eight I^1s from which measurements can be taken; all are worn with dentine exposed along the

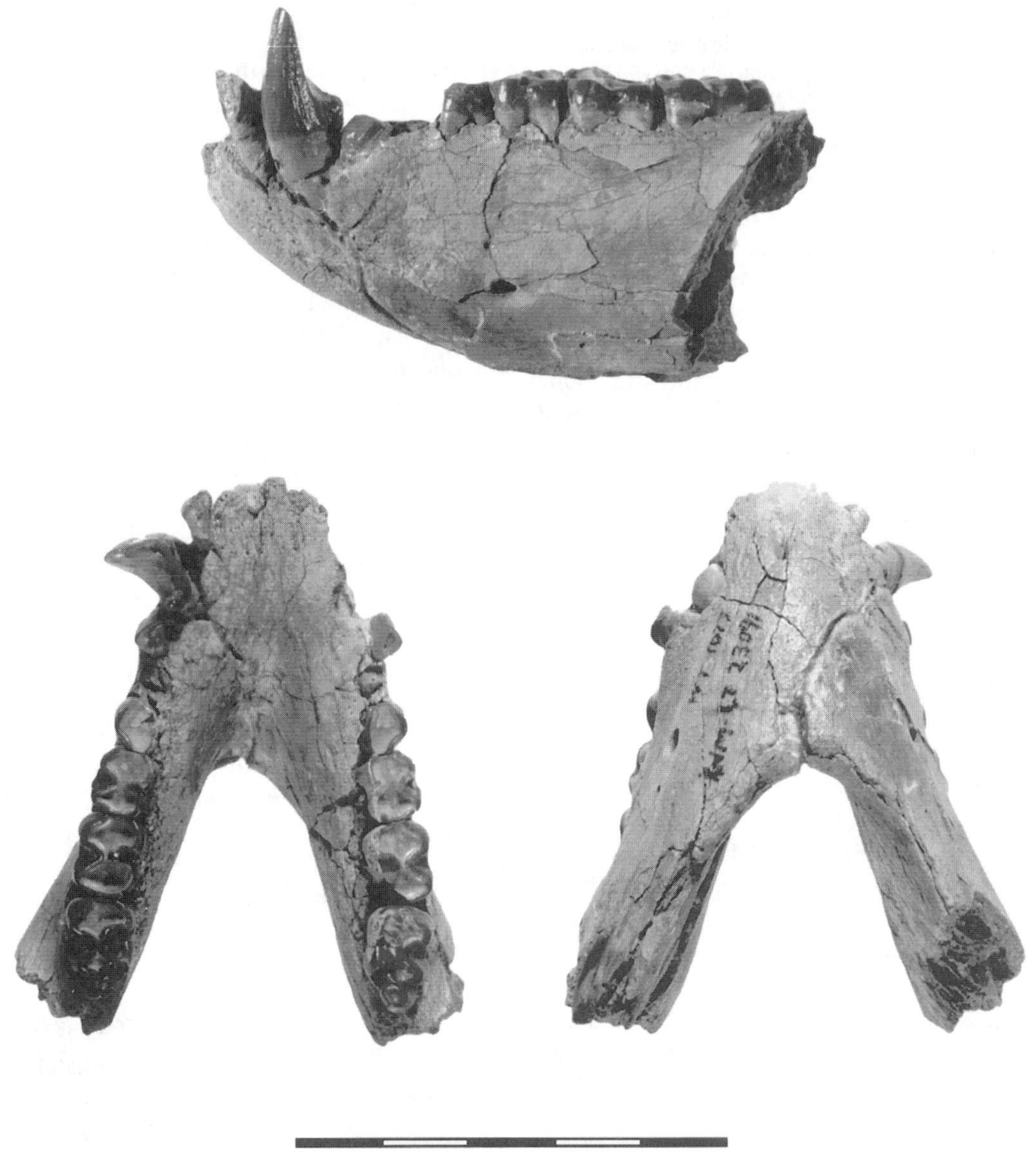

Figure 6.3 *Parapapio lothagamensis* sp. nov. KNM-LT 23091, mandible (holotype) from the Nawata Formation: top = lateral view; bottom left = occlusal view; bottom right = ventral view.

length of the occlusal surface. The least worn (LT 24096, 24100, 26406, and 28792) preserve approximately two-thirds of the crown. As is typical in the Papionini, the I^1s are large, mesiodistally broad, and high crowned. There is a distinct sulcus on the labial aspect, which forms a deep vertically oriented groove at its deepest point slightly mesial to the midline. A thin sharp ridge borders the sulcus distally and converges with a mesial, more rounded ridge approximately 2–2.5 mm from the crown root junction where a distinct cuspule may be present (LT 26405 and 28792). The triangular-shaped root is relatively short, measuring less than the height of the crown on all but the three most worn specimens. On the most complete specimens, the root to crown height ratio measured on the buccal face is 10.1/11.7 for 28792 and 9.5/11.5 for 26406. Two specimens are associated with I^2s (LT 26406, 28792): the I^2 is significantly smaller than the I^1 being shorter mesiodistally (75% and 68%, respectively) and also narrower buccolingually (75% and 85%, respectively).

Only three specimens with I^2s are preserved. The I^2s associated with I^1s mentioned above are moderately worn, with small areas of dentine exposed on the main cusp, but LT 24134 is too worn to show any significant morphology. On the labial aspect of the least worn specimen, LT 26406, there is a sulcus on the upper portion of the crown, bordered mesially and distally by thin slight ridges which converge 2.4 mm from the crown root junction. The sulcus disappears toward the crown tip where the labial surface is convex mesiodistally. The mesiodistally compressed root is slightly shorter than the worn crown height measured on the labial face (LT

28792, root/crown length = 9.6/9.5; 26404 root/crown length = 8.7/9.5) and slightly shorter than the I^1 root of 28792 (10.1 mm).

Upper permanent canines

One male and four female specimens preserve complete upper canines. The broken erupting male canines of the facial specimen described above, LT 24111, reveal the canine cross section close to the enamel junction (MD 9.7 mm, LL 6.2 mm). The male canine, LT 23079, is very worn and only a small part of the crown is preserved. The mesial and distal wear surfaces give the crown a triangular profile.

The female canines are smaller than those of males indicating a significant degree of sexual dimorphism in size as well as morphology. Three of the female specimens are of similar size, but one, LT 38792, the least worn, is slightly larger. This latter specimen has a small island of dentine exposed at the crown apex, and a thin strip of enamel is worn from this point along the mesial margin almost to the cervix. A distinct lingually facing groove is bordered by the mesial margin and is continuous with the lingual and the distal cingulum. All three aspects of the root have more or less developed furrows, that on the lingual face being the deepest.

Upper permanent premolars

Six P^3s are preserved. Of the best specimens, LT 24113 is in the crypt and unworn, LT 28792 has only small islands of dentine exposed on the two cusp tips, and LT 28576 is only moderately worn. The buccal cusp (paracone) is taller than the lingual (protocone), which is set mesial to it. The two cusps are set proximal to each other and separated by the shallowest point of a C-shaped fissure, which is continuous from the mesial to the distal margins. The P^3s are buccolingually broad compared to the mesiodistal length (mean LL/MD 128%, range 121–133%, n = 3), relative to those of extant monkeys (*Colobus, Cercopithecus, Cercocebus* and *Papio;* mean 105, range 75–128%) (Benefit 1993). *Victoriapithecus* has a similarly wide P^3 relative to length: mean LL/MD 126%, range 119–137%, n = 9 (Benefit 1993). The crown is also high (LT 28792; MD/paracone height = 80%). The P^3 shows significant flare from the proximally set cusp tips to the base of the crown. This is particularly marked on the lingual face so that when measured from the enamel line to the cusp tip the paracone and protocone 'heights' are similar, but when viewed from the mesial or distal aspect the less flared buccal cusp (paracone) is actually the higher. On the buccal face the enamel line dips steeply from the distal to the mesial corner. The lingual face is more narrow than the buccal face, and this gives the tooth a distinctly triangular outline in occlusal view.

Nine P^4s are preserved. The unerupted P^4 (LT 24113) is preserved in the crypt but is only partially visible. The least worn P^4s (LT 24111, 22972, and 28792) have only small circular islands of dentine exposed on each cusp. The morphology is similar to that of P^3, but because the lingual face is relatively broader, the occlusal outline is D-shaped rather than triangular. Like the P^3s, the P^4s are labiolingually broader (LL/MD = 143%, range 131–152%, n = 8) than those of extant monkeys (*Colobus, Cercopithecus, Cercocebus,* and *Papio;* mean 124%, range 110–131%; Benefit 1993) and more similar to those of *Victoriapithecus* (LL/MD = 139%, range 127–145%, n = 11; Benefit 1993). The buccal face and its enamel line is essentially symmetrical; when measured from the enamel line to the cusp tip, the steeply flaring lingual cusp (protocone) appears the 'higher' (see 28792), although, as for the P^3s, when viewed from the mesial or distal aspect, the buccal cusp (paracone) is actually the higher.

Upper permanent molars

Several specimens of upper molars include one or more molars associated with other cheek teeth (table 6.3). The upper molars are approximately as wide as they are long, although there is some variability. The mean MD/LLa index of M^1 is 103 percent (range 96–114%, n = 8); of M^2 is 98 percent (range 90–111%, n = 11), and of M^3 is 97 percent (range 84–108%, n = 9). *Victoriapithecus* has upper molars that are wider than they are long: for M^1, MD/LLa = 94 percent, range 75–110 percent, n = 37; for M^2, MD/LLa = 91 percent, range 80–111 percent, n = 64; for M^3, MD/LLa = 90 percent, range = 75–105 percent, n = 56 (Benefit 1993). Among *Parapapio* upper molars, Benefit (1993) found that only *P. broomi* M^1s shared the same pattern as *Victoriapithecus.* The *P. lothagamensis* upper molars all have closely approximated cusp tips, and both the buccal and lingual sides of the crown flare toward the cervix, although the lingual flare is the more exaggerated. The flare is most pronounced on M^2. The average index of the labiolingual distance between the mesial cusps/maximum labiolingual width on unworn molars is 34 percent for M^1, 37 percent for M^2, and 34 percent for M^3, suggesting even more closely approximated cusps than those of *Victoriapithecus* for which the same indices are 43 percent for M^1, 42 percent for M^2, and 52 percent for M^3. The crown is high (LT 28792; MD/paracone height = 80%). All three molars have a large mesial fovea, and M^1 and M^2 also have a large distal fovea. Two of the six preserved M^3s lack distal shelves, a feature frequently found in *Victoriapithecus* M^3s (Benefit 1993). The median lingual cleft deeply incises the lingual faces becoming shallower toward the cervix and extending close to the enamel line, whereas the median buccal cleft is shallow. Both the median lingual notch and the median buccal notch are shallow. The mesial lingual cleft is variably developed but is usually distinct on M^1 and M^2. The distal lingual cleft is faint or absent on all molars. The mesial and distal buccal clefts are not pronounced.

The relative mesiodistal lengths of the associated upper molars are $M^1 < M^2 > M^3$, with the M^3 usually close to the length of M^2. M^1 and M^2 distal widths are only moderately narrower than mesial widths—or in the case of M_1 sometimes slightly wider—and the cusps are of similar height. The M^3 is distinguished from M^1 and M^2 by its markedly reduced distal width and lower distal cusps. Of seven measurable M^1s, the mean index LLa/LLb was 102.1 percent with a range of 91.8–109.6 percent; for 12 M^2s, the mean was 107.3 percent and the range was 101.1–114.1 percent. In contrast, nine measurable M^3s show a mean index LLa/LLb of 126.9 percent with a range of 105.1–145.3 percent. The *Victoriapithecus* M^3 shows a disparity of mesial and buccal widths that is similar to that of the other molars. Equivalent indices for *Victoriapithecus* associated molars is as follows: for M^1 = 108 percent, range 102.4–113 percent; M^2 = 108 percent, range 102–117 percent; and M^3 = 121 percent, range 102–132 percent. As for *Victoriapithecus,* the M^3 mesial buccal cusp (paracone) projects farther buccally than does the distal buccal cusp (metacone), but the two lingual cusps are aligned, which gives the crown a skewed appearance. *Cercocebus albigena* has similarly unequal medial and distal widths and cusp heights (Benefit 1993). The M^1 of LT 28791, a maxillary fragment with the cortical bone missing and thus exposing the tooth roots, is long relative to width (MW/DW = 112). All maxillary specimens in which cortical bone has been lost (LT 419, 22973, and 28791) show the roots are long relative to the MD length.

Lower permanent incisors

Four specimens preserve I_1s, and all are worn. Except for LT 26386, a partial male mandible with canines, all are isolated. The least worn, LT 30608 and 23173, show some morphological detail. The occlusal face of each is worn obliquely from the high mesial corner, and their buccal faces are gently convex and high (12.9 mm and 13.8 mm, respectively) and would have been several mm higher when unworn. The lingual face, as in all papionins, lacks enamel, although there is some indication that there may have been some enamel development. Microscopic study of thin sections is needed to show if this is the case. The I_1 is markedly convex from the base to the tip. There is a slight sulcus bordered by faint but relatively sharp ridges along the mesial and distal margins. The crown profile from the lateral and mesial aspects is wedge shaped with the buccal face slightly convexly curved. The root is mesiodistally compressed and triangular in cross section.

The I_2s, which are all worn, include three males, LT 26386, LT 23079, and LT 23091 (associated with the well-preserved male mandible mentioned earlier), one female, LT 23090, and two of indeterminate sex, LT 26579 and 26910. LT 36910 is the least worn, and its small size suggests it is probably female. The I_2 occlusal face wears more obliquely than that of the I_1. As in the I_1s there may be some development of enamel or only very thin enamel on the lingual face. From the lingual aspect the distal face is almost vertical but the mesial face curves towards the midline away from the root. From lateral and mesial aspects, the tooth crown is wedge shaped and gently curved, concave buccally and convex mesially. The roots are mesiodistally compressed.

Lower permanent canines

Canines are preserved in the two male mandibles, LT 23091 and 26386. In addition, there is a complete right male canine associated with the root of the left broken at the cervix (LT 23079), an isolated male canine (LT 24109), and a worn female canine, associated with mandibular and maxillary fragments (LT 23090). The four male canines show various degrees of wear: LT 24109 and 26386 are slightly worn, and 23091 and LT 23079 are moderately worn. The unworn canines are high crowned, pointed, and mesiodistally compressed; from the lingual aspect, the axis is slightly sinusoidal, curving initially laterally and then mesially toward the tip. There is a slight cingulum on the mesial margin and above this a distinct mesial sulcus. The mesiodistally compressed root of LT 24109 is 15.3 mm long from its tip to the lowest point on the enamel line, which, on the mesial face is close to the lingual border. The crown of LT 24109 is slightly higher (16.3 mm) from the same point to the tip. The same measurements for LT 26386 are 17.5 mm and 16.2 mm and for LT 23079 18.0 and 17.0, respectively. The morphology of the female canine, LT 23090, is obscured by wear but is considerably smaller than any of the male canines (MD for the female compared to the mean for three males is 60% and LL is 68%).

Lower permanent premolars

Five P_3s are preserved, three male and two female. The male premolars are considerably larger than the female ones, with a long rootward extension of enamel (honing facet) on the mesiobuccal aspect of the buccal cusp (protoconid). The vertical protoconid height on the unworn male P_3 (LT 30606) is 52 percent (6.8 mm) of the height of the honing facet (13 mm). The protoconid is positioned at the midpoint between the mesial and distal margins as in *Victoriapithecus* and there is a small metaconid situated on the distolingual crest that defines the mesial border of the deep posterior fovea. Short metaconids were found on all *Victoriapithecus* P_3s except one (Benefit 1993). Metaconids occur at a low frequency in extant cercopithecids. The steep postprotocristid defines the buccal margin. The angle between the distolingual crest and the preprotocristid, which defines the honing facet, is close to a right angle. A cingulum borders the lingual face; that of LT 30606 is only developed anteriorly, while that of LT 26388 is clearly defined along the length of the border. Weathering obscures that of LT 26386. The female P_3s are considerably smaller than those of the males. Both are worn, LT 23090 moderately so and 24097 with only small dentine pits exposed on the protoconid tips. As in males, there is a small metaconid. The distolingual crest (postprotocristid) is clearly developed. The female P_3 differs from that of the male by its very short honing facet and distinct, relatively long and raised distal margin.

The preserved P_4s are all associated with mandible fragments. All are moderately worn and little can be said of the occlusal morphology. The least worn, LT 24094, shows a relatively large buccal cusp (protoconid), a high lingual cusp (metaconid), and deep well-defined anterior and posterior foveae. In several of these specimens, the enlarged buccal cusp skews the orientation of the P_4 relative to the molar row, so that the mesial portion of the tooth is buccal to the distal portion, and the mesial border of the tooth is oblique to the main axis of the tooth row. This is most clearly seen on LT 23091. The edentulous mandibles, LT 23065 and LT 26187, appear also to have had skewed P_4s; the P_4 mesial buccal root is situated well buccal to the distal buccal root. An obliquely oriented P_4 is a character typical of the Victoriapithecinae.

Lower permanent molars

The lower molars include several specimens in which one or more molars are associated with other cheek teeth and others that are unassociated (table 6.4). Because the dimensions of the smallest M_2 and the largest M_1 overlap, two specimens cannot be securely identified (LT 24127 and 26395). The lower molars are high crowned but with relatively little cusp relief and slight basal flare on the buccal face. Cusps are closely approximated, and the buccal cusps have clearly defined rounded buccal faces separated from the adjoining cusp by deep buccal notches.

The associated lower molars show relative mesiodistal lengths $M_1 < M_2 < M_3$, with the M_3 always the largest due to the pronounced hypoconulid. The mesial and distal widths of M_1 are on average almost equal, whereas the distal width of M_2 is on average less than the mesial width. The mesial and distal foveae of M_1 and M_2 are large and deep, with the distal foveae usually larger than the mesial. On unworn M_2s and M_3s, the height of the mesial lingual cusp (metaconid) is greater than that of the distal lingual cusp (entoconid) but sometimes for M_1 this may be reversed. On all three molars, the distal cusps (entoconid and hypoconid) are subequal. The hypoconulid is well developed, and a tuberculum sextum is frequently present between the hypoconulid and entoconid. The buccal face flares gently from the cusp tips to the cervix, whereas the lingual face is almost vertical. Although the shear crests are relatively short, the median buccal clefts are deep and extend a variable distance from the cervix; they clearly separate the mesial and distal cusps and usually have a rather squared-off base. Mesial and distal buccal grooves are clearly defined but variably developed.

Benefit (1993) noted that the M_1 crowns of *Victoriapithecus* have a squarer outline (mean MD/LLa = 116, range 102–132) than is observed in most other cercopithecoids, with the exception of *Colobus satanus* and *Cercocebus galeritus*. This is also true for *P. lothagamensis* (mean MD/LLa = 123.3, range = 113–134.6), which is comparable to *Victoriapithecus* in this feature.

There appears to be some difference related to sexual dimorphism in the size of the molars: the molars of females where the sex is known are smaller than those of males. The longest M_3, LT 24117, has a large hypoconulid and tuberculum sextum. Benefit (1993) found that the development of the hypoconulid of *Victoriapithecus* was related to sex, females having smaller hypoconulids than males.

Deciduous Dentition

Upper deciduous teeth

One right dI^2, LT 26619, has a small island of dentine exposed on the cusp tip, and was associated with a cracked and very lightly worn dP^3. A maxillary fragment (LT 24113) with M1 and the germs of P^3, P^4, and M^2 in the crypts preserves a moderately worn dP^4. Two isolated dP^4s (LT 24138 and 26617) are also preserved.

The dI^2 is asymmetrical with the cusp positioned mesially and a longer distal than mesial crest extending from it to the distal and mesial corners, respectively. The distal crest forms the buccal border of a shallow but broad sulcus and terminates at the distal corner, where it joins the cingulum. The cingulum continues mesially on the lingual face to the midpoint, where it curves toward the crown apex, thus defining the lingual and mesial margins of the sulcus. At the mesial corner the mesial crest terminates at a small but distinct cuspule. In occlusal view the crown is roughly triangular shaped, with the base along the buccal margin and the apex at the lingual corner.

The dP^4s are similar in proportion to the permanent first molars, and the mesial and distal cusps are set opposite each other. They differ in lacking completely developed distal transverse lophs, and in this feature they resemble the upper deciduous premolars of *Victoriapithecus.* The best preserved dP^4, LT 26619, has a diagonal crack running through the mesial buccal cusp (paracone), and this crack slightly obscures the mesial occlusal surface morphology. A transverse crest joins the closely approximated mesial cusps (paracone and protocone), whereas the distal cusps, which are positioned approximately opposite each other, are separated by a C-shaped fissure that originates from the distal fovea. An oblique crest (crista obliqua) passes from the metacone to the continuous lingual margin (composed of the prehypocrista and postprotocrista), joining it at the base of the lingual notch or, perhaps, just distal to the base. The trigon basin is thus separated from the distal fovea. Benefit (1994) found a true crista obliqua in 34 of 39 specimens (87%) of *Victoriapithecus* dP^4s. The cusp tips are in close proximity. The ratio of the width (2.09 mm) between the distal cusps (metacone and hypocone) to the distal transverse width (5.8 mm) is 36 percent, and that contrasting the distance between the buccal cusps (metacone and paracone) (2.33 mm) to the mesiodistal length (6.47 mm) is also 36 percent. Benefit (1994) found the mean value of the same index for *Victoriapithecus* to be 45 percent.

The wear on the other preserved dP^4s obscures the detailed occlusal morphology, but the least worn, LT 24113, has an oblique crest passing toward the median lingual notch, which is terminated by the distinct longitudinal fissure. The detailed lingual occlusal morphology is lost from wear. The slightly more worn LT 24138 has a true crista obliqua, which joins the lingual transverse border at the lingual notch and thus separates the trigon basin and the distal fovea. LT 26117 is too worn to show the occlusal morphology. LT 24113 is associated with an M^1 and an M^2 in a maxillary fragment; in this specimen the dP^4 is 85 percent as long and 86 percent as wide as the first molar and 70 percent as long and 74 percent as wide as the second molar.

Lower deciduous teeth

Several lower deciduous teeth were recovered, but these are mostly incomplete and do not include either incisors or canines. No mesial but several distal fragments of the dP_3 are preserved, along with several complete and partial dP_4s. A fragment of mandible, LT 23124, preserves the distal half of a worn dP_3 and a moderately worn dP_4; a slightly less worn second fragment, LT 24110, is smaller and is probably a distal half dP_3. Its width is close to the distal dP_3 in the mandible fragment LT 23124. LT 24108 includes a worn dP_4 and a permanent M_1; LT 26579 includes a less worn distal half dP_3 and an isolated fragment of the distal half of dP_4; LT 26394 preserves a d/P fragment associated with a right mandible fragment with a distal M_1; and LT 24110, a dP_4, lacks associated bone.

The distal dP_3 (LT 26579) is moderately worn. There is a small distal fovea, and the mesial sides of the mesial cusps indicate these were higher than the distal cusps. LT 26394 is very incomplete although relatively unworn, and it has a relatively large distal face. Nothing much can be said of the worn partial dP_3 (LT 23124), but it is associated with the best preserved dP_4, which is only moderately worn. This dP_4 is similar to the permanent M_1 except that it is narrower mesially than distally and is low crowned with thinner enamel. The mesial and distal foveae are also rather large relative to the mesiodistal length. The unworn distal half dP_4 fragment associated with a mandible fragment with a distal dP_3, LT 26579, has a very small island of dentine exposed on the distobuccal corner proximal to the interstitial facet that is similar to the hypoconulid of the *Victoriapithecus* dP_4. All dP_4 fragments have a large fovea and distinct lophs that connect the lingual and buccal cusps.

Postcranial Skeleton

The postcrania are only tentatively attributed here to *Parapapio lothagamensis.* As discussed previously and later in this contribution, the cercopithecid postcrania from Lothagam are similar morphologically, and thus difficult to attribute to a specific taxon. All specimens in this list are relatively large, corresponding roughly to the size of the *P. lothagamensis* craniodental remains, and display morphologies found only in cercopithecines among extant cercopithecids. Based on dental/postcranial proportions of several extant cercopithecid taxa, all measurable specimens appear to be too large to belong to the smallest colobine, Colobinae species A. They all appear to reflect habitual terrestrial locomotion, with no obvious adaptations to committed arboreality. Many of these morphologies are primitive and are also found in *Victoriapithecus* (Harrison 1989). The morphology of the skeletal elements is described collectively because the specimens are morphologically equivalent, although, of course, it is possible that they represent more than one taxon.

The single scapula fragment (LT 26370) includes a glenoid, and the bases of the coracoid process and part of the spine. It has a cercopithecine morphological pattern. The spine originates 12.8 mm from the superior margin of the glenoid surface, low for a cercopithecid and found only in extant cercopithecines (Larson 1995). The base of the spine is narrow and almost forms a foramen for the suprascapular nerve, rather than being smoothly concave as in extant colobines. The glenoid measures 17.2 mm SI by 12.9 mm AP, which is in the broad end of the range of all cercopithecids.

The humerus (figure 6.4A, B, and D) is also like that of many extant cercopithecines, with a posteriorly directed humeral head that is ovoid in outline, seen in LT 22769, which measures 17.3 mm by 15.3 mm. It tends to project proximally at or just below the greater tuberosity. This specimen also shows tuberosities that are disparate in size—15.2 mm for the greater and 9.0 mm for the lesser tubercle breadths—and separated by a broad (6.8 mm), flat bicipital groove. The shaft is posteriorly and medially inclined, with the large deltopectoral crest apparently located closer to the proximal than to the distal end of the bone, although this location cannot be determined in any specimen.

Distally, the medial epicondyle is directed posteriorly. The highly asymmetrical trochlea has almost no groove, and the capitulum is fairly flat. The olecranon fossa is deep, and there is a strong extension of the trochlear surface along its lateral margin. Proximal to this region, the margin of the olecranon fossa displays a concavity to accommodate the ulna, a feature found in cercopithecines but lacking in extant colobines given their larger olecranon fossae that permit a greater range of elbow extension than is typical for cercopithecines.

Ulnar morphology (figure 6.5A) is less well represented but reflects the asymmetrical trochlear morphology and anteriorly oriented radial notch characteristic of cercopithecids, and it is particularly pronounced in cercopithecines given their more terrestrial locomotor emphasis. There is a double radial notch facet and a straight or slightly posteromedially deflected olecranon process, both of which are primitive features found in *Victoriapithecus.* This olecranon morphology is also found in Lothagam colobines.

The femora (figure 6.6B and E) all have head articular surfaces that project onto the femoral neck. The femoral neck-shaft angles are low (105–115°), near the bottom or below the range of variation seen in extant African colobines. They have correspondingly low foveas and greater trochanters that project proximally above the level of the head. Distally, they all have narrower, deeper patellar grooves than found in extant colobines. They have anteroposteriorly deeper condyles, and the proximal tibial specimen (LT 26376) has a correspondingly anteroposteriorly deep plateau. The tibiae also all have talar joint surfaces that are roughly square in outline.

The fossil tali from Lothagam lack all diagnostic colobine morphologies identified by Strasser (1988) and entirely fit the cercopithecine pattern. The medial malleolar facet meets the posterior calcaneal facet, the head is set at a low angle, the distal facet is present, and the posterior groove for m. flexor tibialis is absent. The trochlea is also deeply grooved and narrow. The cuboid (LT 23081) has a large navicular facet, with subequal proximal and distal ectocuneiform facets; this is another cercopithecine feature (Strasser 1988). The medial cuneiform (LT 23081) displays a distal end with a lateral indentation and rounded plantar margin. The plantar tubercle is large. Strasser notes no features of the calcanea that distinguish colobines from cercopithecines. All of the Lothagam *P. lothagamensis* calcanea (figure 6.7A–C), however, are significantly larger than the two attributed to Colobinae, LT 30610 and LT 26392.

Remarks

These papionin specimens show some disparity in the size of the teeth, but it is unlikely that they represent more than one species. The specimens that can be sexed consistently show the females to be smaller than the males, which indicates that the collection most probably represents one sexually dimorphic species. In particular, two lower right central incisors—LT 23173 and LT 36910—show extreme size difference.

The *P. lothagamensis* postcrania display shoulder, elbow, hip, knee, ankle, and foot morphologies that reflect adaptations to movement predominantly in a sagittal plane with pronated hands. Such morphology is retained in extant cercopithecines, especially papionins. Thus the *P. lothagamensis* postcrania demonstrate adaptations to habitual terrestriality but retain primitive characters seen in *Victoriapithecus* (Harrison 1989). These *P. lothagamensis* specimens are as large or larger than any attributed to Colobinae from Lothagam.

Benefit's (1993, 1994) detailed morphological descriptions and analyses of the hundreds of *Victoriapithecus macinnesi* teeth from the 15 Ma deposits of Maboko Island distinguished those characters unique to *Victoriapithecus,* those shared by *Victoriapithecus* and cercopithecines, and those shared with colobines. This led her to conclude that the Victoriapithecinae probably gave rise to the two extant subfamilies and that the Victoriapithecinae should be given family rank. The number of features that *P. lothagamensis* shares with *Victoriapithecus* to the exclusion of the Colobinae and Cercopithecinae calls into question the justification for this suggestion.

Benefit (1993) listed seven features of the permanent dentition unique to *Victoriapithecus.* The first two of these are not found in *P. lothagamensis:*

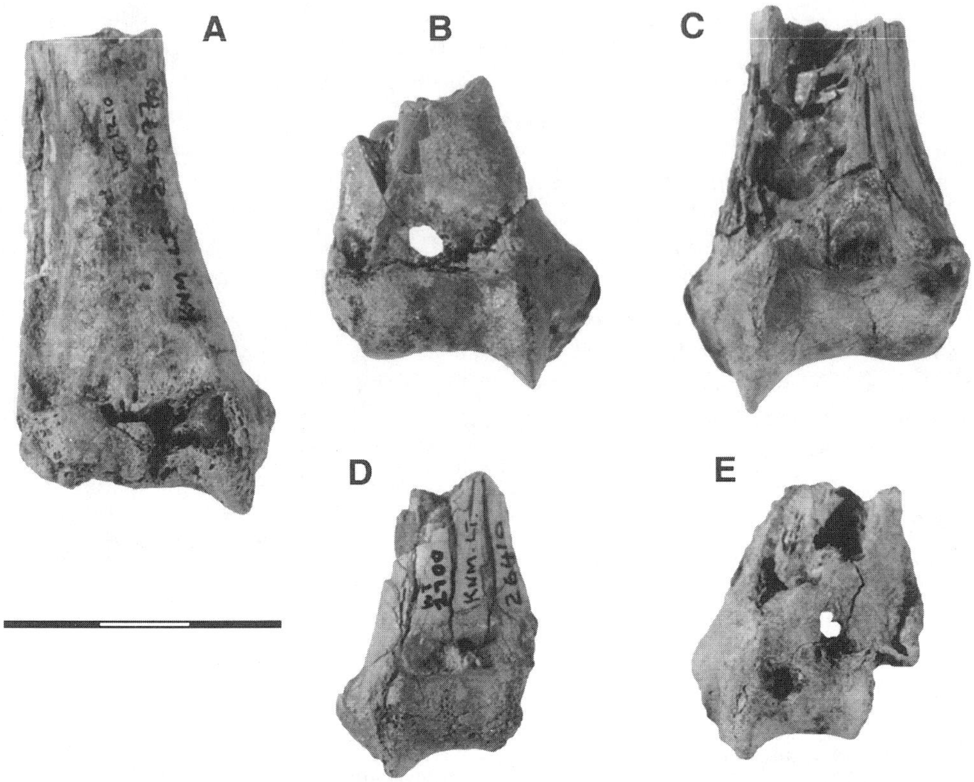

Figure 6.4 Anterior views of cercopithecid distal humeri: A = KNM-LT 23074, *Parapapio lothagamensis* sp. nov., Lower Nawata; B = KNM-LT 23077, *Parapapio lothagamensis* sp. nov., Lower Nawata; C = KNM-LT 416, Cercopithecinae cf. *Parapapio* sp. indet., Apak Member; D = KNM-LT 26410, *Parapapio lothagamensis* sp. nov., Lower Nawata; E = KNM-LT 26381, Cercopithecinae cf. *Parapapio* sp. indet., Apak Member.

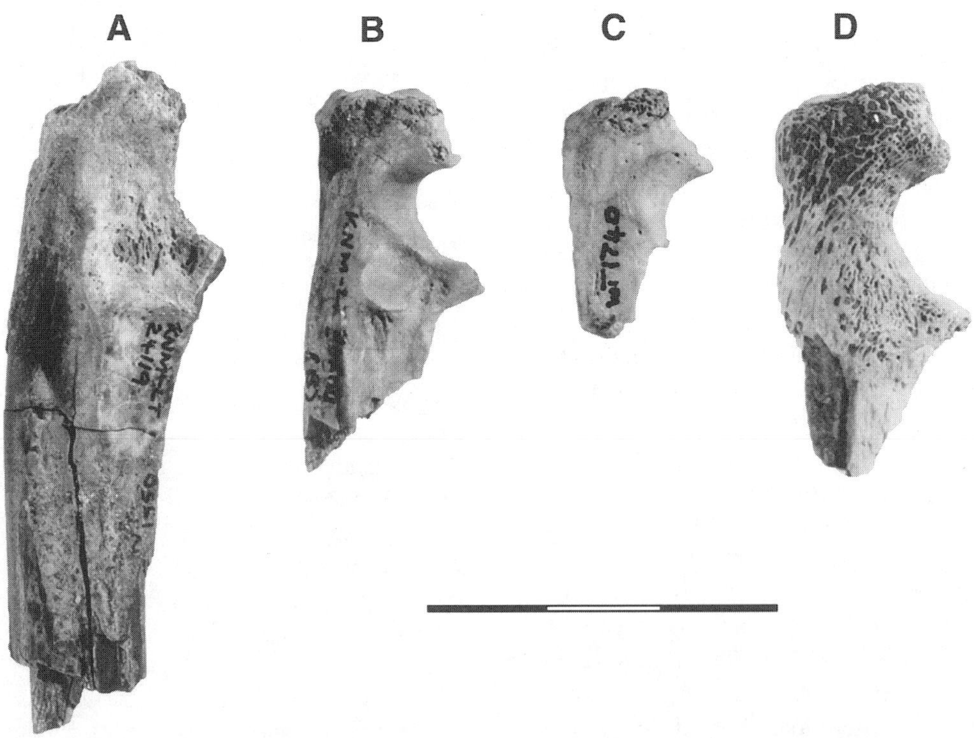

Figure 6.5 Cercopithecid proximal ulnae: A = KNM-LT 24119, *Parapapio lothagamensis* sp. nov., Upper Nawata, lateral view; B = KNM-LT 30609, Colobinae gen. and sp. indet. (small), Lower Nawata, lateral view; C = KNM-LT 24126, Colobinae gen. and sp. indet. (small), Lower Nawata, medial view; D = KNM-LT 26407, *Cercopithecidae* gen. and sp. indet., Lower Nawata, lateral view.

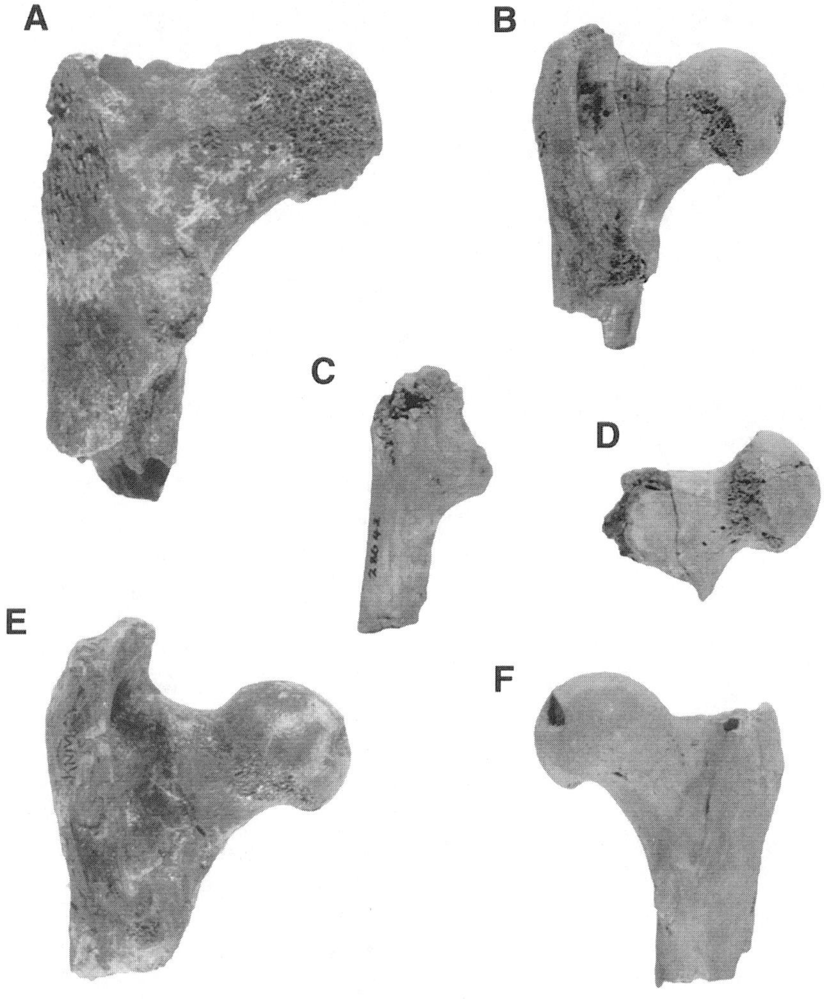

Figure 6.6 Cercopithecid proximal femora (posterior view): A = KNM-LT 24104, *Theropithecus* cf. *T. brumpti,* Kaiyumung Member; B = KNM-LT 26403, *Parapapio lothagamensis* sp. nov., Lower Nawata; C = KNM-LT 28642, Colobinae gen. and sp. indet. (small), Lower Nawata; D = KNM-LT 26390, Colobinae gen. and sp. indet. (small), Lower Nawata; E = KNM-LT 2974, *Parapapio lothagamensis* sp. nov., Lower Nawata; F = KNM-LT 28724, Cercopithecinae cf. *Parapapio* sp. indet., Apak Member.

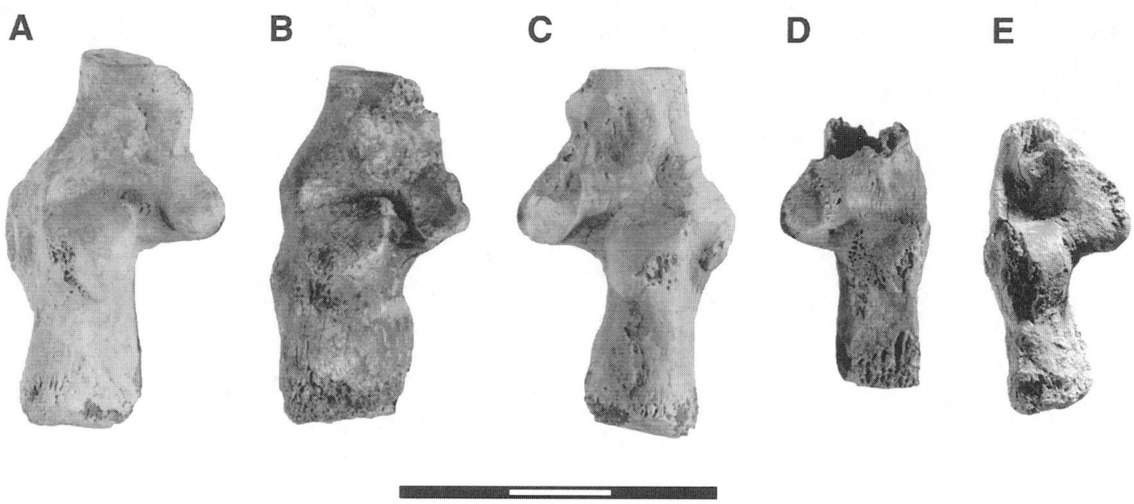

Figure 6.7 Cercopithecid calcanei (dorsal view): A = KNM-LT 28575, *Parapapio lothagamensis* sp. nov., Lower Nawata; B = KNM-LT 26402, *Parapapio lothagamensis* sp. nov., Lower Nawata; C = KNM-LT 24125, *Parapapio lothagamensis* sp. nov., Lower Nawata; D = KNM-LT 26392, Colobinae gen. and sp. indet. (small), Upper Nawata; E = KNM-LT 30610, Colobinae species B, Lower Nawata.

1. variable retention of crista obliqua on the upper molars
2. variable retention of M_1/M_2 hypoconulids

The remaining five characters are shared with *P. lothagamensis:*

3. skewed orientation of P_4 relative to the long axis of the molar row
4. presence of a P_3 metaconid
5. M_1s that are more square than comparable molars of extant cercopithecoids
6. M^3s distally constricted, with shorter and more closely approximated distal cusps than is observed for extant species
7. high degree of molar flare due to the closer proximity of cusp tips than is observed among extant monkeys

Subsequently, Benefit (1994) found several characters of the deciduous dentition that are not seen in *Victoriapithecus* but which are shared by the Colobinae and Cercopithecinae. Only the characters of the upper and lower P_4s of *P. lothagamensis* can be compared, but of the five characters of the dPs, distinguishing the two modern subfamilies from *Victoriapithecus,* two are not characteristic of *P. lothagamensis.* These features of the deciduous teeth in *P. lothagamensis*—a crista obliqua on the dP^4 and upper deciduous premolars that are not fully bilophodont and that lack distinct lophs between the hypocone and metacone—align it with the Miocene monkeys rather than with the extant subfamilies.

Benefit (1993) found only two dental features shared between the *Victoriapithecus* and the Colobinae to the exclusion of the Cercopithecinae. One of these, the short mesial shelves of the M_1 and M_2, is also shared with *Prohylobates,* so this feature appears to be variable within the Victoriapithecinae. The other, the wide upper premolars relative to their length, is also seen in *P. lothagamensis.* The new material from Lothagam displays several characters shared with the middle Miocene *Victoriapithecus* but not with extant Colobinae and Cercopithecinae, thus weakening the case presented by Benefit for elevating the Victoriapithecidae to family rank.

Cercopithecinae cf. *Parapapio* Species indet.

(Figures 6.4C, 6.4E, 6.6F; tables 6.7, 6.9)

Lothagam Material

- *Apak Member:* 416, Rt. distal humerus; 24110, half dP_3 or dP_4; 26381, Rt. distal humerus; 28724, Rt. proximal femur.
- *Kaiyumung Member:* 26369, Lt. maxilla (M^1–M^3).
- *Horizon indet.:* 448, female mandible with dentition, Rt. maxilla (P^3–M^2); 449, Lt. male maxilla fragment (C/–P^4). These specimens were collected in 1980 and are probably from the faulted northern area which it has not been possible to place securely in the Lothagam stratigraphy. It is possible that these strata correlate with the Apak Member.

Only three cranial specimens and a lower deciduous premolar fragment (LT 24110) are included here; the latter is too incomplete to identify further. LT 448 is a relatively complete female mandible and a right maxilla with moderately worn teeth. It is rather smaller than would be expected for a female of *P. lothagamensis,* but it has characteristic *Parapapio* dental morphology. The mandible, however, lacks the characteristic symphysial morphology of the common Nawata Formation species. LT 449 is a male maxillary and premaxillary fragment with C/–P^4. The canine is missing part of the distal portion of the crown close to the alveolus and part of the tip of the root. There is some wear on the lingual and mesial faces close to the crown tip. A deep mesial sulcus extends from the worn tip to the root. When viewed from the mesial aspect, the crown is distinctly curved: convex on the lingual and concave on the labial faces. It is likely that the deposits from which this specimen and LT 448 were recovered are considerably younger than the Nawata Formation and may even have been younger than the Apak Member. The difference in size between the two specimens is larger than might reasonably be explained by sexual dimorphism, suggesting that they may represent two species. Of comparable size to LT 449 is the left male maxillary fragment with M^{1-3} and part of the palate from the Kaiyumung Member, LT 26369. This can be readily identified as a medium-sized papionin from the low cusp relief of the partially worn M^2 and little-worn M^3. The long axis of the tooth row is bowed, although this feature appears to have been exaggerated by distortion. The molars have closely approximated cusps and considerable basal flare.

Theropithecus Geoffroy, 1843 *Theropithecus brumpti* (Arambourg, 1947) *Theropithecus* cf. *T. brumpti*

(Figure 6.6A; tables 6.8, 6.9)

Lothagam Material

- *Kaiyumung Member:* 417, Rt. M_2; 24104, Lt. proximal femur; 24128, Rt. M_3; 24129, Rt. P^4; 24130, fragment Rt. P_4; 26368, Lt. female C/, lacking crown tip; 26372, Lt. I^1; 26396, Rt. female C/; 26397, Lt. I^1;

26401, Lt. male P_3, lower molar fragments; 26615, /M; 37105, Rt. dP_4.

Eight specimens of this taxon have been recovered, the majority isolated teeth. The M_2, LT 417, was collected in 1967 by Bryan Patterson and described in detail by Delson (1993). It has long been considered to represent an early occurrence of this genus because the age of Lothagam-3 was erroneously believed to be between 4.0 and 4.5 Ma (Hill and Ward 1988).

Five upper teeth are recognizably *Theropithecus:* these include two upper incisors, LT 26372 and 26397; two female canines, LT 26396 and 26368; and a high-crowned, large right P^4, LT 24129. The incisors are small, being short mesiodistally and wide labiolingually. *Theropithecus* has relatively small incisors, compared to the size of the cheek teeth. The most complete canine, LT 26396, is little worn and broken lingually at the tip of the crown, but the morphology is typical of this genus. The second canine lacks much of the crown but preserves most of the root. The premolar is complete, unworn, and large, and again typical of *Theropithecus.* The proximal femur, LT 24104, is somewhat abraded but is large in size, with a short neck and apparent extension of the femoral head onto the neck.

Remarks

These specimens were all recovered from the southern exposures of the Kaiyumung Member. The faunal evidence suggests that they may be younger than those recovered from the northern Kaiyumung exposures. The five lower teeth include a left P_3 with a long honing facet, LT 26401; a right M_2, LT 417; and a right M_3, LT 24128; along with several partial molars and tooth fragments. Except for the P_3, these teeth are all worn, but they are typically *Theropithecus* in morphology. The long honing facet of the P_3 is characteristic of male *T. brumpti* rather than *T. darti* and *T. oswaldi.* The latter two species have reduced canines and shorter P_3 honing facets. The M_2, LT 417, is described by Delson (1993), who considers the morphology too simple to be attributable to *T. brumpti.* However, early *T. brumpti* from the Koobi Fora Formation show a less complex morphology than those from later deposits (Leakey 1993). These unworn Lothagam specimens compare well with KNM-ER 127, attributed to *T. brumpti,* from the Tulu Bor Member of the Koobi Fora Formation. The similarities between the Lothagam and Koobi Fora specimens, together with the long honing face of the P_3, another character of *T. brumpti,* indicate these specimens are correctly attributed to this species. The Kaiyumung thus provides an early record of *T. brumpti* known elsewhere in the Turkana Basin from similar-aged deposits in the Lokochot Member of the Koobi Fora Formation (3.36 and 3.5 Ma; Leakey 1993). *T. brumpti* is absent from the South Turkwel deposits, about 25 km to the north of Lothagam, which may be slightly older than the southern exposures of the Kaiyumung (Ward et al. 1999). *T. darti* first occurs in the Sidi Hakoma Member at Hadar (3.4 Ma; Eck 1993; Delson et al. 1993), and *T. quadratirostris* is known from the Usno Formation in the Turkana Basin (3.3 Ma; Delson et al. 1993), although Delson and Dean (1993) consider this species best assigned to *Papio* (*Dinopithecus*).

Subfamily Colobinae Jerdon, 1867
Genus *Cercopithecoides* Mollett, 1947

Diagnosis

Fairly large extinct colobines. Braincase large and rounded, muzzle relatively narrow, face wide, and orbits large. Frontal process of zygoma narrow, nasals moderately long, malar region shallow superioinferiorly, nasal aperture small and straight in lateral profile, postorbital constriction slight, postglabella sulcus present, and basioccipital wide. Temporal lines meet posteriorly and sagittal crest only slightly developed or absent, nuchal crests may be well developed in males, postglenoid process small. Mandibular body relatively shallow with marked lateral ridge (prominentia lateralis) and flat anterior surface. Gonial region small, ramus low and at an oblique angle to occlusal plane, superior edge of coronoid process approximately level with mandibular condyle. Premolars relatively small and P^3 lacking a protocone. Sexual dimorphism apparent in canines and P_3. Postcranial skeleton shows features typical of more terrestrial cercopithecids. Differs from *Libypithecus, Nasalis,* and *Rhinocolobus* in the short, rounded braincase and relatively shorter muzzle. Differs from all other colobines in the low, shallow mandible with short, oblique ramus (after Leakey 1982).

Cercopithecoides kerioensis sp. nov.
(Figure 6.8; tables 6.10, 6.11)

Diagnosis

A *Cercopithecoides* that, in the male of the species, differs from both *C. williamsi* and *C. kimeui* in its small size, relatively thin supraorbital tori, narrow internasal width, well-developed nuchal crests, and presence of a sagittal crest close to inion. The mandibular body is relatively shorter and deeper than that of either of the larger species; anteriorly the inferior margin is inflated and the foramen symphyseosum, absent.

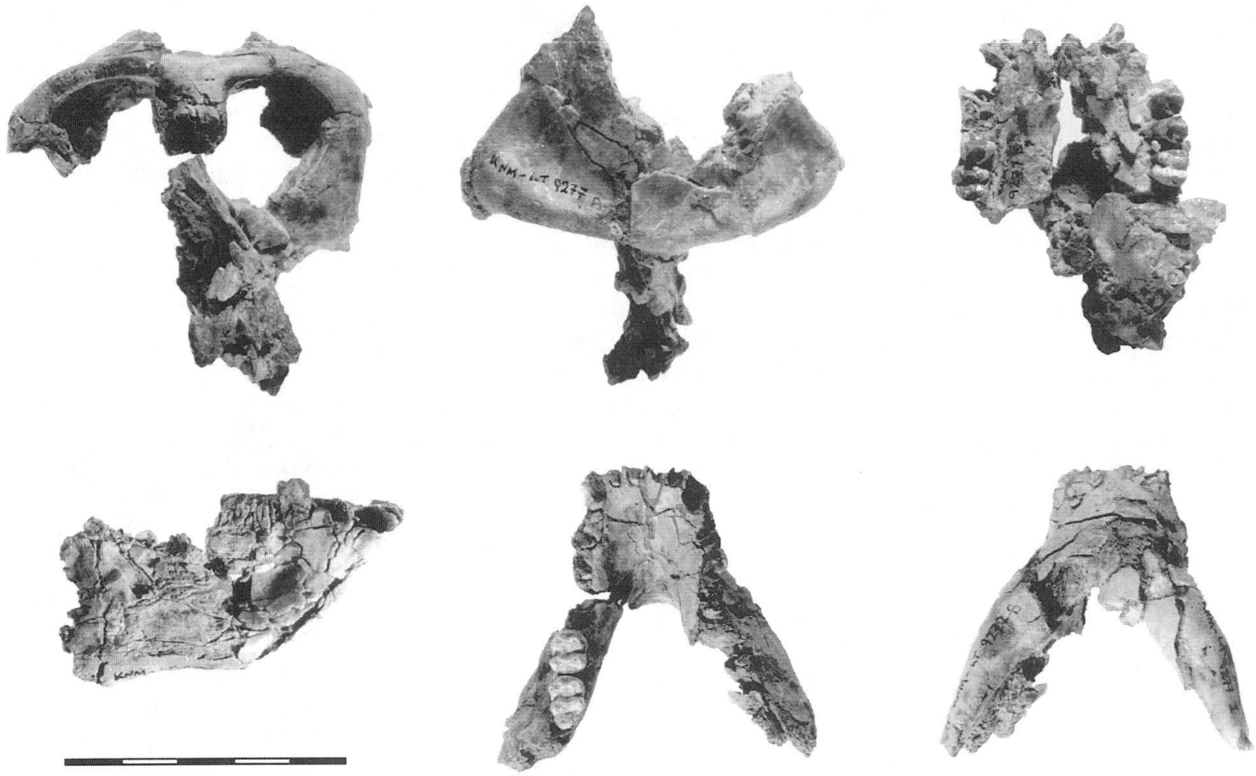

Figure 6.8 *Cercopithecoides kerioensis* sp. nov., KNM-LT 9277, holotype, horizon indet.: top left = cranium, anterior view; top center = cranium, dorsal view; top right = palate, occlusal view; bottom left = mandible, lateral view; bottom center = mandible, occlusal view; bottom right = mandible, inferior view.

Lothagam Material

Holotype

LT 9277, a partial male cranium and mandible of uncertain provenance.

Type locality

Lothagam, Kenya. The specimen was collected in 1980 and is probably from the faulted area in the northern section to the west of the River Nawata where the sediments may correlate with the Apak Member of the Nachukui Formation. It is possible, but less likely, that it is from the Kaiyumung Member.

Etymology

Named after the River Kerio to the east of Lothagam. The ancestral River Kerio deposited the sediments of the Apak Member (see Feibel, chapter 2.1 in this volume).

The single specimen of this species preserves parts of the cranium and mandible. The cranium includes the frontal, which was broken posteriorly anterior to the frontoparietal suture; partial left and right maxillae with the left P^3–M^1 and the right P^4–M^1; fragments of zygomatic and parietal; a left temporal fragment with the articular surface and postglenoid process; a left occipital fragment with petrous; the partial basioccipital; and a fragment of occipital at inion. The mandible, which preserves the right P_4 and left M_{2-3}, is broken posteriorly on the right side across the ramus and on the left side just posterior to M_3. Fragments of the right and left ramus preserve both condyles. All fragments are invaded by matrix-filled cracks, which have caused some distortion.

This relatively small species was smaller than *C. williamsi* but close in size to the extant *Colobus abyssinicus.* Although much of the preserved bone is distorted and crushed, morphological details are well preserved. In most features, this specimen is typical of *Cercopithecoides,* which is well represented in the cave deposits in South Africa and in East Africa in the Koobi Fora Formation to the east of Lake Turkana and at Olduvai Gorge (Leakey and Leakey 1976; Leakey 1982).

The largest preserved fragment of cranial vault is the frontal, which has relatively thinner supraorbital tori and a narrower interorbital distance than the other known *Cercopithecoides* species, although these features are known to be variable. There is a slight supraorbital sulcus, and, viewed from the anterior aspect, the supraorbital tori are only very slightly bow-shaped. The postorbital constriction is not marked because anteriorly the temporal lines originate laterally close to the frontozygomatic suture and curve only gently posteriorly. A fragment of the cranial vault at inion has well-developed nuchal crests, slight development of the sagittal crest (although this is broken), and a distinct crest that passes inferiorly from inion, dividing the occiput. It is likely that the temporal lines would have converged on the posterior portion of the vault, probably posterior to bregma. Unfortunately, the maxilla is crushed and distorted on both sides so that there is little evidence of the infraorbital foramina, which are generally well developed and multiple in this genus. The malar process of the zygomatic was relatively shallow dorsoventrally as in other species of *Cercopithecoides.* A fragment of the left temporal preserves the glenoid fossa, which presents a rather flat articular surface and a small postglenoid process. A fragment of the left occiput preserves the mastoid process and petrous. The basioccipital fragment is crushed and cracked.

The mandible, too, is rather crushed so that both the slenderness and depth of the body are exaggerated. However, even when undistorted, the body would have been deeper and the inferior margin more inflated than that of *C. williamsi.* Features characteristic of *Cercopithecoides* are the short post-incisive planum, the well-developed inferior transverse torus, the thick inferior margin of the body, and the rather square anterior face of the symphysial region. The foramen symphyseosum is absent, whereas it is normally present in *Cercopithecoides.* Both condyles are preserved; they are lightly built and narrow anteroposteriorly.

The maxilla preserves the left P^3–M^1 and right P^4–M^1, and the mandible preserves the right P_4 and left M_{2-3}. These teeth, which are only slightly worn, are similar to those of *C. williamsi* although they are relatively narrower buccolingually. The P^3 lacks a protocone, and the transverse lophs on all the cheek teeth are well developed. The cheek teeth are smaller than those of *C. williamsi* from eastern and southern Africa: the upper and lower P4s fall outside the range in their mesiodistal widths but just within the range in mesiodistal length (table 6.10), and the dimensions of the M^1s, M_2s, and M_3s fall outside the range of *C. williamsi.*

Remarks

Cercopithecoides was originally described by Mollett (1947) from a single damaged skull from Makapansgat, South Africa. Freedman (1957) described a second species, *C. molletti,* from Swartkrans, but this species was later synonymized with *C. williamsi* (Freedman 1960).

Additional crania, mandibles, and postcranial elements have been recovered in South Africa from Makapansgat, Sterkfontein, Taung, Graveyard, and Cooper's (Freedman 1957, 1960, 1965, 1970, 1976; Freedman and Brain 1972; Maier 1970). The recognition of both *C. williamsi* and a larger species *C. kimeui* in East Africa has shown this genus to have been widely dispersed geographically (Leakey and Leakey 1973; Leakey 1982).

Cercopithecoides was relatively common in the South African cave deposits where it was the only colobine. In the Turkana Basin, *Cercopithecoides* occurs in the Koobi Fora Formation together with diverse species of colobines, but it is curiously absent from similar-aged deposits in the Shungura and the Nachukui Formations (Leakey 1987; Harris et al. 1988) and from Laetoli (Leakey and Delson 1987), suggesting quite specific habitat preferences. The *Cercopithecoides* postcranial skeleton indicates a rather terrestrial habitat and the extreme wear of the dentition of many of the specimens may be due to the consumption of sand with the diet (Leakey 1982).

C. kerioensis not only represents a new species of *Cercopithecoides* but probably also the earliest known.

Colobinae species A
(Tables 6.10, 6.11)

Lothagam Material

- *Lower Nawata:* 24107, Lt. mandible (M_2, roots P_3–M_1); 26607, Lt. I_1; 36913, Rt. I_2, 37104, Lt. I_2
- *Upper Nawata:* 23078, Lt. male C/; 26383, Rt. male /C; 36912, Rt. I^2
- *Horizon indet.:* 418, Rt. P^4

This is a small colobine, smaller than the extant *Colobus abyssinicus,* with a gracile shallow mandible. The eight specimens all show colobine characters and are assigned to a single taxon largely because they are, respectively, from individuals of similar small size. There is no other criterion on which they can be said to be conspecific.

Upper permanent lateral incisors

The two right I^2 (LT 36912 and the rather damaged LT 37104) are lightly worn. On the lingual face a deep groove-shaped sulcus has its deepest point adjacent to the distinct mesial ridge and defines the mesial margin. The distal ridge is equally distinct but shorter; it joins the superior margin, which curves superiorly to meet it at a slight notch. The crown is low, and the mesiodistal length (4.1 mm) is only slightly greater than the buccolingual width.

Upper permanent canine

LT 23078 is a small upper male C/ lacking the tip of the crown that is significantly smaller than others assigned to *P. lothagamensis.* It is worn along the distolingual border from crown tip to cervix. It has a large root relative to the crown height and a deep mesial groove, which continues on to the root. It is too small to be comfortably assigned to *P. lothagamensis.*

Upper permanent premolars

The P^4, LT 418, is unworn and lacks roots. In occlusal view it is triangular in outline, with the base of the triangle the buccal face. The mesial margin is straight, in contrast to the rounded distal margin. The mesiodistal length is 85 percent of the buccolingual width. The cusps are high, and the tooth is relatively straight sided with some buccal flare on the lingual cusp (metacone). The buccal cusp (protocone) is taller than the slightly more mesially positioned metacone. A loph joins the two cusps transected by the shallowest point on a C-shaped fissure which is continuous from the mesial to the distal margins. Distinct mesial and distal shelves border the foveae mesially and distally, respectively.

Mandible

The shallow, gracile half mandible, LT 24107, is weathered, so it was probably not as lightly built as it appears now. The weathering has removed much of the cortical bone, particularly on the posterior inferior margin, and this weathering is probably the reason that the body does not deepen posteriorly in the typical colobine manner. The mandible is broken anteriorly and superiorly at the canine alveolus, but more inferiorly the anterior break reaches almost to the mid-line of the symphysis. The remaining canine alveolus is small, indicating that the individual was a female. Distally it is broken across the ramus behind the M_3 alveolus. The ramus ascends at, rather than posterior to, the M_3.

The mandible measures more than 14.3 mm below M_2 and more than 15.9 mm at the junction of P_4 and M_1. The mental foramen is difficult to discern, but it appears to have been located anteriorly just at the canine alveolus and the anterior break. The remaining bone at the symphysis preserves enough of the superior transverse torus to show that it was large. The single remaining tooth, the M_1, is lightly worn, with tiny islands of dentine exposed on the buccal cusps. The molar is straight sided with limited lateral flare and lingual cusps taller than buccal. The deep mesial fovea, which is situated buccally, is smaller than the more lingually positioned distal fovea. The cusps are tall, with the lingual portion of the lingual lophids sharp and passing directly buccally from the apex of the anteriorly directed lingual crests as is typical of the Colobinae. The median buccal cleft is well defined and deeply incises the crown. The base forms an open V-shaped notch. There is no evidence of either mesial or distal buccal clefts.

Lower permanent incisors

Left I_1, LT 26607, lacks roots and is incomplete and rather too worn to show any meaningful morphology. It is relatively low crowned and can thus be confidently assigned to the Colobinae. The right I_2, LT 36913, is slightly worn, and some enamel has been removed from the lingual face by weathering. It is a small and a relatively low crowned nondescript tooth with, in lingual view, the occlusal wear facet oriented obliquely inferiorly from the highest point on the mesial margin.

Lower permanent canine

The male lower right canine, LT 26383, is moderately worn. A facet extends superiorly from the distal heel at the base of the crown to wrap around the crown onto the lingual face close to the apex, where a small, obliquely oriented circular island of dentine is exposed. The wear facet close to the canine tip faces lingually; this is an unusual orientation. Below this wear facet, the lingual face of the crown is broken and the enamel is missing. Viewed from the mesial aspect, the root and crown of the tooth curve laterally so that the buccal face is gently concave and the lingual face is convex. Viewed from the buccal aspect, the mesial face of the crown is convex and the mesiodistally compressed root is relatively straight distally. The crown curves from the distal heel toward the tip.

Remarks

The material assigned to this small colobine is fragmentary and only tentatively assigned to a single taxon. The most complete specimen is the left mandible with M_2. The small size of the canine alveolus indicates that this specimen is female. If the upper male canine, LT 23078, and the lower male canine, LT 26383, are correctly assigned to this taxon, there was considerable sexual dimorphism in canine size.

Colobinae species B

(Figures 6.7E, 6.9; tables 6.9–6.12)

Lothagam Material

- *Lower Nawata:* 23064, male Lt. C/; 23162, Rt. mandible fragment (P_3, root /C), Lt. M_1; 23165, Lt. I_1; 23167, Lt. mandible fragment (M_1); 23166, Lt. dP^4; 26387, Lt. M_3 talonid; 26399, juvenile mandible (Lt.

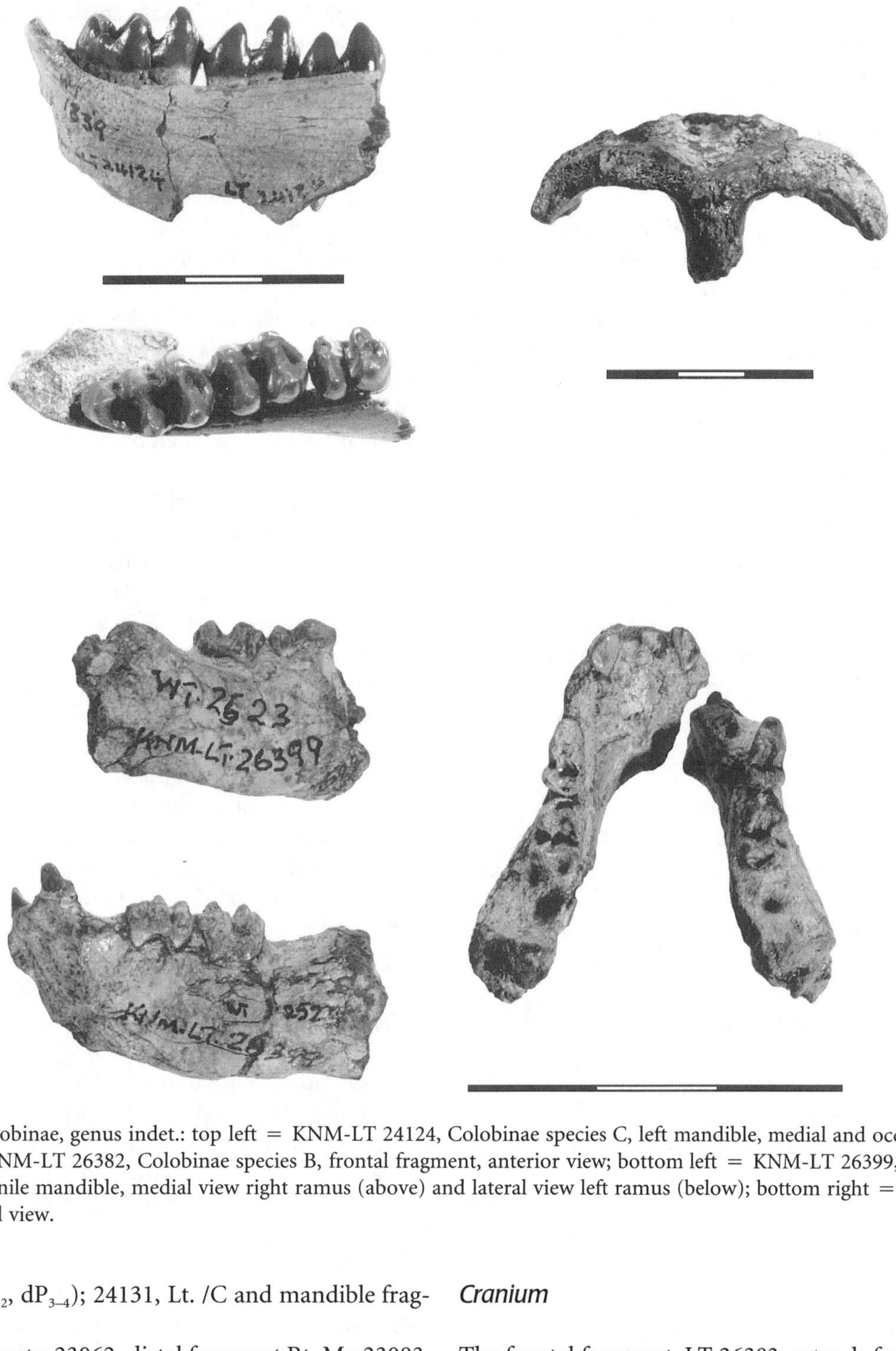

Figure 6.9 Colobinae, genus indet.: top left = KNM-LT 24124, Colobinae species C, left mandible, medial and occlusal views; top right = KNM-LT 26382, Colobinae species B, frontal fragment, anterior view; bottom left = KNM-LT 26399, Colobinae species B, juvenile mandible, medial view right ramus (above) and lateral view left ramus (below); bottom right = KNM-LT 26399, occlusal view.

and Rt. dI_2, dP_{3-4}); 24131, Lt. /C and mandible fragments.

- *Upper Nawata:* 23062, distal fragment Rt. M_3; 23083, Lt. M_3; 24098, M_2; 24116, Rt. male /C; 24132, Lt. M_1; 26382, frontal; 30610, Rt. P/, Rt. calcaneus; 36911, Rt. M^2.

This taxon is represented by 17 specimens, which are assigned to a single taxon largely on the basis of size. With the exception of the frontal fragment and the calcaneus, these specimens are all isolated or associated teeth. This species is slightly larger than the extant *Colobus abyssinicus.*

Cranium

The frontal fragment, LT 26382, extends from the midline at glabella laterally on both sides to the zyogomaticofrontal suture, posteriorly for 23 mm and anteriorly only a few millimeters. The supraorbital tori are thick (5.2 mm) and robust compared to the rather small orbits, as discerned from the curvature of the superior orbital margin, and there is little evidence of the supraorbital notch. Viewed frontally, the superior surface dips at the midline, and on the superior aspect there is a distinct postglabella sulcus. The temporal lines converge sharply behind the orbits but do not meet before the posterior break. The interorbital distance is wide (8.8 mm). Evidence of the frontonasal suture can be seen at the break.

Upper permanent dentition

One upper male canine, LT 23064, is assigned to this species. It is missing all but 5 mm of crown and is broad. It has a triangular, cross-sectional profile at the cervix. An upper premolar, LT 30610, is probably a P^4; it is lightly worn and shows the morphology clearly, although the enamel surface is damaged from weathering. The paracone is the taller of the two cusps and, as is typical of *Colobus* and *Victoriapithecus* (Benefit 1993), the paracone height (5.57 mm) is greater than the mesiodistal length (4.87 mm). Two lingually directed crests occur on the buccal cusp (paracone); the mesial of these is positioned opposite the lingual cusp (protocone). Only a very weak suggestion of a crest is visible on the protocone. The two cusps are separated by a C-shaped fissure, which curves around the protocone and shallows at the level of the protocone crest. The lingual and buccal faces flare toward the cervix.

Upper permanent molars

The upper molars are represented by one M^2, LT 36911, which is very lightly worn and has sharp lophs that pass lingually from the buccal cusps. There is a distinct mesial buccal cleft and the median buccal cleft is V-shaped. The mesial fovea is only slightly smaller than the distal fovea.

Lower permanent anterior dentition

A left I_1, LT 23165, is moderately worn and has lost enamel on the buccal face close to the cervix. In buccal view, the tip is worn obliquely from its highest point at the midline. The orientation of the wear facet on the lingual face is almost vertical. The crown is 4.52 mm long mesiodistally and slightly over 3.8 mm wide labiolingually. Two lower male, lightly worn canines are preserved: a left with the tip of the crown missing but complete root (LT 24131) and a right with complete crown and slightly damaged root (LT 24116). These male canines differ from those of *P. lothagamensis* in being longer mesiodistally and having relatively shorter strongly tapering roots.

A left P_3, LT 23162, is labiolingually broader (5.43 mm) than the mesiodistal length (5.55 mm). The buccal face of the protoconid extends almost vertically distally onto the mesial root, in contrast to the oblique extension of *P. lothagamensis.* The occlusal profile is a hypotenuse triangle with the hypotenuse along the broad distal margin. The occlusal surface is weathered, but there appears to have been a small metaconid. A cingulum is visible along the lingual margin. The posterior fovea was large, and the mesial fovea was only a small pit. This specimen is contained in a mandibular fragment next to an oval-shaped broken tooth root with a very small island of enamel on the buccal face. It is not clear what this represents.

Lower permanent molars

The lower molars show typical colobine morphology with prominent transverse lophids and parallel buccal and lingual faces. A right M_1 (LT 23162), associated with the P_3 described above, is moderately worn such that dentine is just exposed on the higher lingual cusps and more extensively on the buccal cusps. The cusps are high, with sharp lophids leading from the anteriorly directed lingual marginal crests. The median buccal cleft is well defined and flattens at its base. There is no evidence of either mesial or distal buccal clefts. A second left M_1 (LT 23167) in a fragment of mandible is less worn; a third M_1 that lacks roots (LT 24132) is unworn, although the cusps are slightly weathered. Both these specimens have distal buccolingual widths greater than their mesial widths. The buccal cusps are positioned distal to the opposing lingual cusps. This is particularly true of the distal buccal cusp (hypoconid) such that the distal lophid is obliquely oriented to the mesial lophid rather than parallel to it. In all M_1s, the steep distal face of the only slightly oblique mesial lophs are high.

A lightly worn M_2 (LT 24098) has similar obliquely oriented distal lophs. This tooth has a distinct cuspule at the base of the wide and deep buccal cleft. LT 23083, an M_3, has lost enamel on the distal face of the large hypoconulid. This tooth has high cusps, the lingual being the higher, a deep median buccal cleft, and distal buccal groove. The lingual cusps are positioned slightly mesial to the buccal cusps, and the lophs are thus slightly obliquely oriented but parallel to each other. One of two M_3 talonid fragments (LT 23062) has a distinct tuberculum sextum separated from the hypoconulid by a deep distal lingual cleft. The other fragment (LT 26387) is worn and like LT 23083 lacks a tuberculum sextum.

Upper deciduous tooth

A dP^4 (LT 23166) provides the only evidence of the upper deciduous teeth. It has dentine exposed on all the cusps and is low crowned but with sharp transverse lophs. The morphology is essentially that of the M^1 with lingually flared faces to the lingual cusps and the distal buccolingual width less than the mesial. The mesial fovea is similar in size to the distal fovea.

Lower deciduous teeth

A small, well-preserved juvenile mandible, LT 26399 (figure 6.9), has an almost complete set of very lightly worn deciduous teeth, lacking only the canines and I_1s. The two halves of the mandible are separate and broken through the right side of the symphysis at the canine alveolus. The bone is damaged at the break, so it is not possible to reconstruct the mandible without distortion. Posteriorly each fragment is broken across the ramus about 10 mm distal to dP_4. The larger piece includes the left and right dI_2, along with the left dP_3 and dP_4. The right piece includes only the right dP_3 and dP_4. The dI_2s are only just worn at the tips and apart from faint indications of a wear facet on the mesial margin of the dP_3, the cheek teeth are essentially unworn.

The anterior face of the symphysis is quite broad and flat, and it narrows inferiorly. The symphysis is 14 mm deep at the midline and is oriented obliquely, at an angle of about 45° to the occlusal plane. The postincisive planum extends distally to the level of the mesial border of dP_3. The depth of the mandible is 10 mm below the dP_4. The curved inferior margin deepens from the symphysis posteriorly until the level of dP_4 when it curves slightly superiorly and then deepens again below the ramus. The body thickens posteriorly to accommodate the M_1 germ.

The dI_2 is asymmetrical. The lingual face is V-shaped with a longer mesial than distal margin. A shallow lingual sulcus is bordered by two relatively thick ridges, which converge at the lingual heel. The margins, which form relatively thick ridges, border a slight sulcus with its deepest part toward the base of the tooth. The main cusp is positioned mesially so that the mesial ridge is shorter than the distal ridge. A short mesial margin passes to the mesially positioned main cusp. The long distal margin runs at an oblique angle from the cusp tip to the distal corner. The dI_2 is wider buccolingually (2.8 mm) than it is long mesiodistally (2.5 mm), and the crown leans mesially in lingual view so that the occlusal margin is positioned mesial to the root.

The dP_3 is an elongated tooth due to the very large mesial fovea, which is almost equal to half the length of the tooth (length mesial fovea 2.43 mm, length of tooth posterior to the mesial fovea 2.8 mm). The distal fovea is very small. The protoconid is the largest and tallest cusp followed by the subequal hypoconid and entoconid. The metaconid is the smallest cusp but of similar height to the entoconid; it is situated distal to the protoconid but very close to it. The mesial fovea is a shallow basin with the floor deepest mesially. The buccal margin of the fovea is clearly demarcated by the preprotocristid. A very faint crest can be seen running obliquely from the mesial buccal cusp (protoconid) to the center of the mesial fovea. Sharp transverse crests pass laterally from the mesial lingual cusp (metaconid) and distal lingual cusp (entoconid), clearly defining the mesial and distal limits of the central basin. There is a faint mesial buccal groove on the mesial face, and a wide mesial buccal cleft deeply incises the crown between the mesial (protoconid) and distal (hypoconid) buccal cusps. On the lingual face the lingual notch is moderately deep, but there is no evidence of a mesiolingual notch.

The dP_4 is similar in morphology to the permanent M_1 except that the mesial fovea is large and deep. It is slightly larger than the distal fovea. The tooth narrows distally, and the mesial buccolingual width is less than that of the distal. Sharp transverse lophids connect the mesial and distal cusps. A wide medial buccal cleft bisects the buccal face and deeply incises the crown; faint mesial and distal buccal grooves are just discernable.

Postcranium

The only associated postcranial element of Colobinae species B is LT 30610, a nearly complete calcaneus missing only its cuboid facet and lateral half of the proximal astragalar facet. It is small, measuring 24.8 mm in maximum length. It is the smallest calcaneus in the Lothagam sample, but it is closely matched in size by LT 26392, referred below to Colobinae gen. and sp. indet. The posterior portion of the LT 30610 calcaneus, from the anterior margin of the proximal astragalar facet to the proximal end of the bone, is 16.6 mm long. The proximal astragalar facet is short, as in other cercopithecids, and the middle and distal facets are separate from one another. This facet is tightly curved, measuring 7.9 mm in maximum length with a maximum height of 3.0 mm along a perpendicular line that connects these points.

Remarks

This material, too, is fragmentary. The most complete specimens are LT 26399, the juvenile mandible with deciduous teeth, and LT 26382, frontal fragment; both give some indication of cranial morphology and are clearly colobine. The dentition is typically colobine and of similar size to Colobinae species C described later in this contribution, but it is less hypsodont and morphologically distinct. The ulna, LT 30609, is similar in morphology to LT 24119, which is attributed to *P. lothagamensis,* but is much smaller in size. The calcaneus is small but otherwise reveals no features that would clearly diagnose it as colobine. The calcaneus is about 30 percent smaller than the LT 24125, 26402, and 28575, all probably papionin specimens. Since tooth crown dimensions of Colobinae species B and the most abundant papionin, *P. lothagamensis,* are roughly equivalent, this may (admittedly weakly) suggest that either Colobinae species B is megadont or *P. lothagamensis* is microdont. The size variation in this sample indicates that there was some size sexual dimorphism in this colobine, although with additional material it could prove to represent more than one species.

Colobinae species C

(Figure 6.9; tables 6.10, 6.11)

Lothagam Material

- *Apak Member:* 24124, Lt. mandible fragments, M_{1-3}; 30607, Lt. M^3.

A moderately worn M^3 (LT 30607) is constricted distally with the distal labiolingual width (5.7 mm) significantly less than the mesial (7.25 mm). The mesial fovea is positioned buccal to the midline, as is the circular distal fovea. Sharp transverse lophs, which are worn mesially and distally, link the mesial and distal cusps. The central basin is deep.

A second specimen (LT 2412), a mandible fragment with lightly worn M_1 to M_3, has little of the mandible preserved. Cortical bone is lost on the lateral side, thus exposing both roots of M_2 and the posterior roots of M_1. All three molars have exceptionally high pointed cusps and sharp, obliquely oriented transverse crests. The mesial lingual cusps (metaconids) of all three teeth are higher than the distal lingual cusps (entoconids), and both lingual cusps (metaconids and entoconids) of M_1 and M_2 and the mesial lingual cusp (metaconid) of M_3 are significantly higher than their opposite buccal cusps (protoconid and hypoconid), respectively. The distal pair of cusps of the M_3 is almost equal in height and size. The lingual transverse crests are steeply sloping and longer than the buccal. The mesial and distal lophids of all three teeth are oriented parallel to each other but obliquely to the main axis of the teeth. The buccal faces of all three teeth have deep median buccal notches, and the median buccal clefts incise the crowns almost to the cervix. The distal buccal groove of M_3 is also deep. On the lingual face, the median lingual cleft incises the occlusal surface almost to the cervix. The mesial and distal fovea of M_1 are approximately the same size, whereas the mesial fovea of M_2 is slightly smaller than the distal fovea and about the same size as that of M_3. The M_3 hypoconulid is large, and there is a distinct tuberculum sextum.

Remarks

This colobine is of similar size to Colobinae species B. It is distinguished by the unusually high occlusal relief, high cusps, long shearing crests, and the oblique orientation of the transverse crests relative to the long axis of the molar teeth. The dentition of this Apak Member colobine most closely resembles that of *Cercopithecoides*, but the material is insufficient to give a certain generic designation.

Colobinae genus and species indet. (small)

(Figures 6.5B–C, 6.6C–D, 6.7E; table 6.9)

Lothagam Material

- *Lower Nawata:* 24126, Rt. proximal ulna fragment; 26390, Lt. femur head and neck; 28642, Lt. proximal femur shaft; 30609, Rt. mandible (roots M_{2-3}), proximal Rt. ulna.
- *Upper Nawata:* 22976, Lt. proximal humerus and partial shaft; 26379, Rt. distal femur; 26392, Lt. calcaneus; 28652, distal tibia fragment.

The proximal ulna, LT 24126 (figure 6.5C) is preserved distally through the depth of the trochlear notch anteriorly and posteriorly at the level of the coronoid process. It is abraded along the posterolateral border of the olecranon process, and the posteromedial border is broken away to the midtrochlear level. It has a moderately asymmetrical trochlea and distinct proximal and distal trochlear joint surfaces connected by a narrow isthmus. It measures a minimum of 8.9 mm AP from the trochlear notch to posterior shaft. The olecranon is 9.1 mm tall from the proximal trochlear surface, which is 7.5 mm ML at its widest point. The posterior portion of the shaft is 5.3 mm wide.

The ulna fragment, LT 30609 (figure 6.5B), is preserved proximal to a point about 1 cm distal to the radial notch. Much of the margins of the olecranon and coronoid processes are abraded. The trochlear notch is fairly asymmetrical. The trochlear surface is almost continuous across the notch but has fairly distinct proximal and distal portions, common to extant cercopithecines but also seen in LT 24126. LT 30609 is the smallest one of the three in the Lothagam sample, although only slightly so. It measures a minimum of only 8.4 mm AP from the trochlear notch to posterior shaft, with an olecranon height of only 6.4 mm from the superior trochlear notch, a proximal trochlear ML width of 6.8 mm, and a trochlear notch length of only 14.2 mm—all roughly the size of an extant vervet monkey ulna. Its surface is largely abraded, but the contours are preserved.

LT 26390, a fragment of femoral head and neck, has abrasion along the margins of the head. The neck is fairly long and displays a small posterior tubercle. This specimen appears to belong to a colobine because the head does not seem to extend onto the surface of the neck at any point and there is only a mild anterior ridge along the greater trochanter for muscle attachment. The head measures 16.3 mm in diameter, falling near the middle of the size range in the Lothagam sample of five measurable femoral heads.

LT 28652 is attributed to the Colobinae based on its small size and joint shape. It is a tiny distal tibial fragment that is abraded along the region of the epiphyseal line. It measures only a maximum of 10.6 mm AP by 10.8 mm ML, with a malleolus measuring 5.7 mm in length. A trochlear surface that is shorter AP than ML is a colobine trait, typical of arboreal catarrhines, and differs from the square outline of extant papionins.

LT 22976, the proximal end and about one-fourth of the humerus shaft, is perfectly preserved except for abrasion along the recently fused epiphyseal line. It displays several typical colobine morphologies. The head is fairly round in contour and measures a maximum of 18.8 mm by a minimum of 17 mm. It projects proximally past the level of the tuberosities. The neck is broader ML than AP and measures 11.5 mm by 13.2 mm. The 6.4 mm wide bicipital groove is flat. The greater tuberosity is 16.9 mm wide, and the lesser one is 11.3 mm.

LT 26379 preserves the distal end of the femur and distalmost end of the shaft. It is small, with a bicondylar width of 20.4 mm. Its lateral condyle measures only 16.1 mm AP, however, which is shallower than typical of extant cercopithecines and other Lothagam distal femora. The patellar surface is deeper and narrower than most extant colobines and measures 11.7 mm ML by 9.7 mm SI with a depth of 1.7 mm.

The left calcaneus, LT 26392, is missing its cuboid facet, inferolateral margin, and edges of its heel process, and it is abraded along the margins of the posterior astragalar facet. It displays no diagnostic colobine features but is considerably smaller than most in the Lothagam sample. It is almost as small as LT 30610, the smallest calcaneus that is associated with a premolar and is attributed to Colobine species B. LT 26392 measures only 16.6 mm from the distal margin of the proximal astragalar facet proximally, and the proximal astragalar facet measures 10.9 mm and is curved to a maximum height of 3.0 mm.

Remarks

Isolated postcranial specimens are difficult to attribute to a particular taxon in the Lothagam sample because the bones show only minor morphological variation, most of which display predominantly terrestrial traits similar to those found in the earlier *Victoriapithecus.* All specimens show at least some functional characters that reflect the largely terrestrial evolutionary origin of cercopithecids (Harrison 1989). Proportions of tooth to postcranial dimensions of extant cercopithecid taxa generally subsume the variation in such proportions in the fossil sample, making it difficult to use size as a discriminating variable. The specimens listed here, however, all represent either extreme small size among similar elements attributed to *Parapapio lothagamensis* and cf. *Parapapio* sp. indet. or some morphology that is fairly typical of extant colobines.

The most conclusive taxonomic assignments among these specimens are LT 26390 and LT 22976, which have some distinctive colobine attributes. LT 26390 appears to have a femoral head adapted to a wider range of joint excursions than typical of most extant cercopithecines, with no extension onto the neck. LT 22976 has a fairly rounded, tall humeral head, which also indicates a wider range of habitual shoulder postures during locomotion than is typical of arboreal catarrhines today. Still, its morphology in this regard is not as well developed as most extant colobine taxa, and this lack of development reflects a presumed terrestrial heritage and possible partial terrestrial habitus.

The distal femur, LT 26379 demonstrates a mix of typical extant colobine features, with its fairly rounded lateral condylar margin and a primitive or more terrestrial narrow, deep patellar groove that differs from that of most extant colobines. A more rounded femoral condyle suggests habitual knee use during locomotion in a variety of flexion-extension postures, while a deep patellar groove implies relatively rapid running or leaping. Because the patellar morphology appears to be primitive for cercopithecids, the more extant colobine-like condylar shape suggests that this specimen does belong to a colobine.

Even so, assigning any of these specimens to Colobinae must be regarded as tentative until more associated cranial and postcranial fossils of the Lothagam taxa are recovered.

Colobinae genus and species indet. (large)

Lothagam Material

- *Kaiyumung Member:* 23680 Rt. M_2

A single isolated M_2 of a large colobine, LT 23680, is the only evidence of this subfamily in the Kaiyumung Member. The tooth is partially worn, and the surface of the enamel is weathered. The morphology is typically colobine, with high cusp relief, sharp transverse crests, small anterior fovea, and a vertically high median buccal cleft that deeply incises the crown. Large-bodied colobines make their first appearance in the fossil record in the Turkana Basin at about this time and are reported from the Usno Formation and Member A of the Shungura Formation (Leakey 1987), the Lokochot Member of the Koobi Fora Formation, and South Turkwel (Ward et al. 1999). This Kaiyumung molar represents a large-bodied colobine but is too incomplete to give specific or generic attribution.

Cercopithecidae genus and species indet.
(Figure 6.5D)

Lothagam Material

- *Lower Nawata:* 23072, Rt. juvenile tibia lacking epiphyses; 24103, tail vertebra in coprolite; 26407 Rt. proximal ulna.
- *Upper Nawata:* 23056, Lt. femur shaft (2 pieces); 23120, Lt. distal weathered juvenile humerus.
- *Apak:* 26380, Lt. distal femur.

These specimens are either poorly preserved or represent elements that are taxonomically undiagnostic.

Microwear

As in many previous microwear analyses (e.g., Strait 1993; Teaford et al. 1996; Ungar and Teaford 1996), the dental microwear analysis of the Lothagam fossil monkeys is based on scanning electron microscope (SEM) examinations of high-resolution casts of teeth. The scanning electron microscope is used because of its superior depth of focus (or depth of field), and the casts are used because they allow the specimens to be examined at higher accelerating voltages (in traditional, high vacuum, nonenvironmental SEMs), which, in turn, yields better images. The casts are extremely stable and yield excellent resolution of detail (below 0.1 micron) (Teaford and Oyen 1989). Casts are also useful because it is often impossible to examine rare museum specimens in analyses that require gold-coating of the specimen or analyses in high vacuums.

The study began with all of the available mandibular and maxillary first and second molars from the Lothagam cercopithecids. First and second molars were used because the distinction between them is difficult in isolated teeth. Only specimens that showed average-to-moderate amounts of wear were used in the analyses—that is, slightly worn specimens or those showing dentin exposures over more than two-thirds of the occlusal surface were not used (see the discussion that follows, of postmortem wear, to see the actual sample sizes used in the statistical analyses of microwear measurements). As in previous studies (e.g., Teaford 1994; Teaford et al. 1996), dental impressions were taken with Coltene-Whaledent's "President Jet Regular" polysiloxane, and positive casts were made using Ciba-Geigy's "Araldite" cold-cure epoxy. All casts were then sputter-coated with 200 angstroms of gold and examined in an Amray 1810 scanning electron microscope in secondary emissions mode at an accelerating voltage of 20 Kv.

Postmortem Wear and Sample Sizes

Several specimens had suffered severe damage due to postmortem wear. Usually, the wear was apparent over most of the tooth crown in patterns similar to those generated in Gordon's laboratory studies of the effects of wind-blown abrasives on teeth (Gordon 1984). Occasionally, specimens exhibited a combination of abrasive and erosive postmortem wear. The net effect was that 45 percent of the specimens were unsuitable for dental microwear analyses (table 6.13). While this may seem like a low percentage of usable specimens, the figure compares favorably with other dental microwear analyses of fossil monkeys—for example, the Plio-Pleistocene monkeys of Kenya (Teaford and Leakey 1992), where figures ranged from 38 to 75 percent.

Scanning Electron Microscopy and Microwear Measurements

Whenever possible, two micrographs were taken of each tooth. All micrographs were taken at a magnification of 500, taking every precaution to maintain a standard working distance and to minimize stage tilt (see Pastor 1993 for further discussions of stage tilt). All micrographs were taken of the crushing surfaces (Phase II facets) defined by Kay (1977). Micrographs were scanned into a computer and then digitized using Peter Ungar's semiautomated analysis package, Microware 3.0. The program uses a 4:1 ratio of the length to width of each feature as the cutoff for determining which features are pits and scratches. For each micrograph, it yields measures of average pit and scratch width, percentage of pits, number of microwear features, and a measure of the homogeneity of scratch orientation. The last measurement merely reflects whether scratch orientation is homogeneous or heterogeneous. Thus it is not dependent on tooth or micrograph orientation (for a further discussion of this measurement, see Ungar 1994).

Statistical Analyses

Using averages for the microwear measurements for each specimen, sample t-tests were run for comparisons of the molar microwear patterns exhibited by the Lothagam colobines and cercopithecines. In all comparisons, Lilliefors and Bartlett's tests were run to check for normal distributions and equal variances in the samples. When necessary, the microwear data were either rank- or log-transformed to meet the assumptions of parametric statistics (Conover and Iman 1981; Zar 1984).

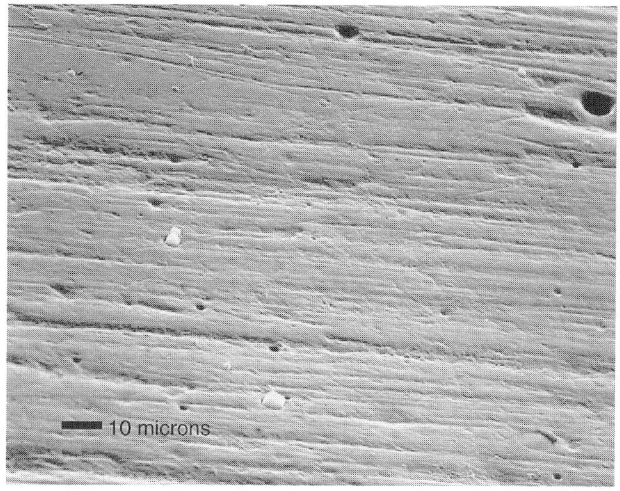

Figure 6.11 Molar microwear on basin facet of Colobinae species C, KNM-LT 24124.

Results

There were no significant differences in molar microwear between the colobines and cercopithecines from Lothagam (tables 6.14 and 6.15). However, certain comparisons (e.g., the amount of microwear per micrograph) were undoubtedly hampered by the small sample size for the colobines. Another complication was provided by the single specimen of Colobinae species C, LT 24124. This showed a unique variety of microwear patterns for the Lothagam sample, with its molar cusp tip facets showing heavy pitting and its basin facets showing fine scratches (figures 6.10 and 6.11). While the percentage of pits for these specimens is superficially similar to those published for extant cercopithecines (see, for example, Teaford 1988), these similarities are probably misleading artifacts of a change in measurement techniques. Microwear measurements in the earlier studies were computed using a digitizer employing little or no magnification of the micrographs. As already noted, the Lothagam samples were measured using Ungar's semiautomated computer technique, which effectively magnifies the micrograph by enlarging it to fit a computer screen. The net effect is that more small features, especially small pits, are measured, thus increasing the number of features and the percentage of pits for each micrograph. Despite these methodological differences, the most striking find of this study is that the fossil cercopithecines resembled the fossil colobines rather than extant cercopithecines in their microwear patterns (figures 6.12, 6.13, and 6.14). This was largely due to a notable absence of large pits on the teeth of the cercopithecines, in contrast to their extant counterparts. See the following discussion of diet for the implications of these results.

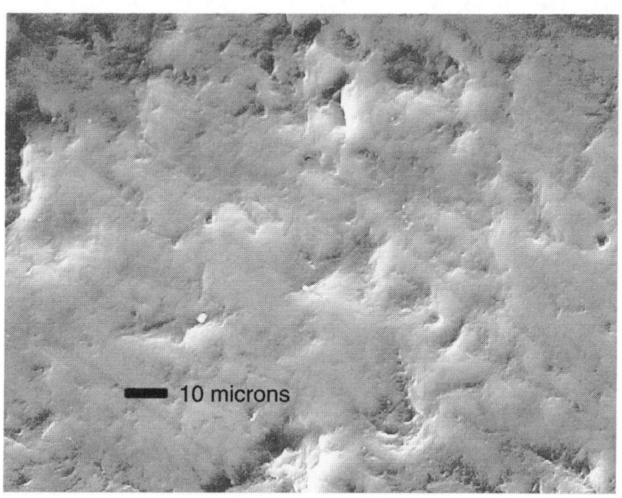

Figure 6.10 Molar microwear on cusp tip facet of Colobinae species C, KNM-LT 24124.

Discussion

As the largest collection of early African cercopithecids, the Lothagam material provides important new insights relevant to our understanding of the evolution, biostratigraphy, and geographic distribution of the Cercopithecinae and Colobinae. The presence of three cercopithecid species in the Nawata Formation, *P. lothagamensis,* and two species of colobines, show that the two extant subfamilies, the Colobinae and Cercopithecinae, were well established in the Late Miocene. The collection from the Nawata Formation represents the earliest evidence of the Pliocene and Pleistocene radiation of African cercopithecids, which culminated in the evolution of several species of *Parapapio* and *Papio,* four species of *Theropithecus,* three genera and at least five species of large bodied colobines, and several species of smaller colobines.

Although Late Miocene and Early Pliocene cercopithecids are not well known, the Lothagam cercopithecid distribution can be compared to that from other sites. Only a few Late Miocene African sites of comparable age to the Nawata Formation at Lothagam are known that preserve cercopithecids. In Kenya, Lukeino (6.2–5.6 Ma) in the Baringo Basin (Hill 1999), has not

Figure 6.12 Molar microwear on crushing facet of extant *Colobus* (OM 3110).

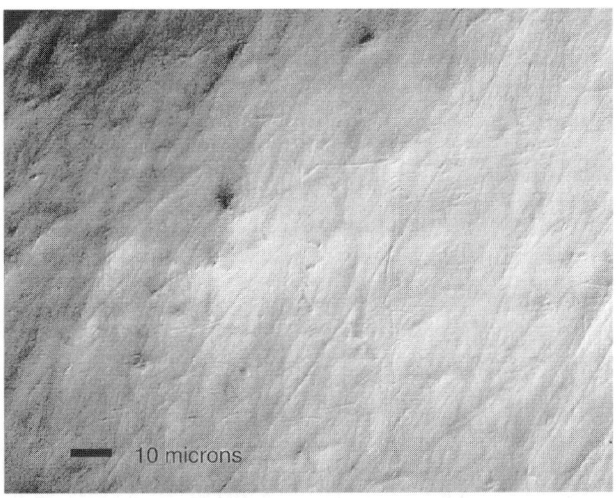

Figure 6.13 Molar microwear on crushing facet of extant *Papio* (OM 7266).

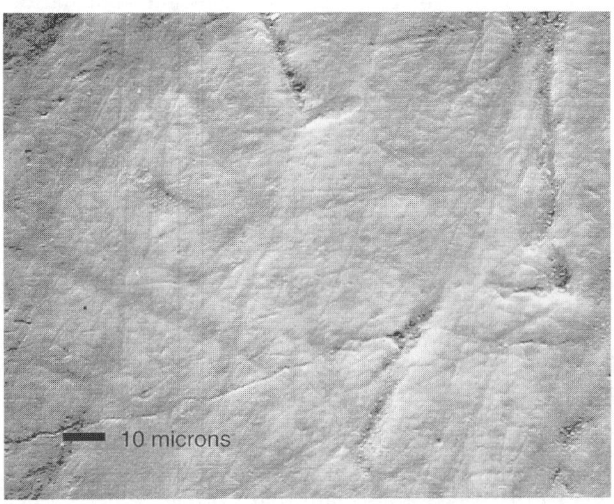

Figure 6.14 Molar microwear on crushing facet of *Parapapio lothagamensis* sp. nov., KNM-LT 24139.

yet yielded any cercopithecines, but Mpesida (~6.5 Ma) has evidence of a colobine of extant size (Hill 1999). Recently a new site, Lemudongo, near Narok in Kenya, which on biostratigraphic and preliminary Ka/Ar dating is probably ~6 Ma, has yielded a cercopithecid fauna that is unique in its exclusively colobine representation (work in progress with Stanley Ambrose). More recently still, WoldeGabriel et al. (2001) report cercopithecids from the Kusaralee Member of the Sangatole Formation (~5.2 Ma) in the Central Awash Comples of Ethiopia. In North Africa, two cercopithecid taxa, referred to cf. *Macaca* sp. and Colobinae gen. and sp. indet., respectively (Meikle 1987), have been recovered at Sahabi, Libya (~5 Ma; Bernor and Pavlakis 1987). At Wadi Natrun in Egypt, the colobine *Libypithecus markgrafi* is known from a single relatively complete cranium (Szalay and Delson 1979). The paucity of the Late Miocene African fossil record is due, at least in part, to lack of suitably aged sites.

The Early Pliocene is only slightly better known and again there is relatively little with which to compare the few cercopithecines recovered from the Apak Member, which probably dates from just over 5 Ma to 4.2 Ma. The sediments in the Manonga Valley in Tanzania (Harrison 1997) are of an equivalent age, but only one tooth, a weathered and abraded canine of a large cercopithecid, was recovered from the Kiloleli Member (~4.5–4.0 Ma; Harrison and Baker 1997). Aramis in Ethiopia (4.3–4.5 Ma) has yielded a large collection of cercopithecids (WoldeGabriel et al. 1994) which is at present under study (Delson and Frost personal communication). In South Africa, two isolated molars were reported from Langebaanweg (~5–4 Ma; Grine and Hendey 1981).

Sites equivalent to the middle Pliocene deposits of the Kaiyumung Member (~3.5 Ma) are better known and have good cercopithecid representation. East African sites include several in the Turkana Basin: to the west of Lake Turkana, South Turkwel (Ward et al. 1999), and the Lomekwi and Kataboi Members of the Nachukui Formation (Harris et al. 1988); to the east of the Lake, the Lokochot and Tulu Bor Members of the Koobi Fora Formation (Brown and Feibel 1991); and to the north, the Usno Formation and Members A and B of the Shungura Formation (Brown et al. 1985). Sites in Ethiopia of this age also include the rich fossiliferous deposits at Hadar (3.3–3.0 Ma; Johanson et al. 1982; Walter and Aronson 1993) and in Tanzania, Laetoli (3.5–3.6 Ma; Drake and Curtis 1987).

Species Distribution Within Lothagam

Papionines are numerically the most abundant of the Lothagam cercopithecids (107 specimens). Only one papionin species, *Parapapio lothagamensis,* is encountered in the Nawata Formation comprising 109 specimens, 76 of them from the Lower Nawata. Only 24 postcranial specimens are known of this species. *P. lothagamensis* represents 79 percent of the cercopithecid fauna in the Lower Nawata and 64 percent in the Upper Nawata. Two specimens are of unknown provenance. Surprisingly, only one papionin, cf. *Parapapio* sp. indet., was found in the Apak Member deposits, but *Theropithecus* cf. *T. brumpti* and cf. *Parapapio* sp. indet. were recovered from the Kaiyumung Member.

Throughout the Nawata Formation, colobines are numerically less abundant (29 specimens) than the papionins but they are more diverse as they include at least two species. Colobines comprise a higher proportion of the Upper Nawata sample than of the Lower Nawata, 29 and 33 percent, respectively. Fewer than 20 percent of the postcranial specimens can be attributed to colobines, which is consistent with the ratio of craniodental remains. In the Nawata Formation, one sexually dimorphic small species is known from seven specimens and a larger species is known from 16 specimens. In the Apak Member, two specimens of a distinctive colobine with high-crowned molars were recovered (Colobinae species C), and a new species of *Cercopithecoides, C. kerioensis,* is probably also from this member. One specimen of a large colobine was recovered from the Kaiyumung Member and represents early evidence (also seen at South Turkwel) of the radiation of the large-bodied colobines that are so diverse in slightly later deposits.

Morphological Comparisons

Recent discoveries of *Victoriapithecus macinnesi* from Maboko Island, Kenya (Benefit and McCrossin 1997) have provided new insights into the cercopithecid ancestral morphotype. In particular, a complete skull with a macaque-like face and moderately long snout, a deep malar region, and narrow interorbital septum demonstrates that many characters of the long-faced Papionini are primitive (Benefit and McCrossin 1997). Moreover, while several features of the *Victoriapithecus* dental morphology (such as the variable occurrence of a crista obliqua on the upper molars and of a hypoconulid on the lower molars) are primitive retentions of features shared with apes, others (including the relatively large anterior dentition, the high-flaring crowns of the molar teeth with little cusp relief, the closely approximated cusps, and the short-shearing crests) are derived features shared with the later papionins. Benefit (1993) found only two characters of *Victoriapithecus* shared with the Colobinae exclusive of the Papionini, and one of these is also observed in *P. lothagamensis.*

Despite the derived papionin affinities, however, the Victoriapithecinae are generally believed to have given rise to both colobines and cercopithecines sometime in the Late Miocene prior to the appearance of the earliest known colobine, *Microcolobus tugenensis* (between 10.5 and 8.5 Ma; Benefit and Pickford 1986). A diminutive colobine, *M. tugenensis,* was slightly smaller than the smallest extant colobine, the olive colobus, *Procolobus verus.* The number of shared derived features common to both *P. lothagamensis* and *V. macinnesi* weakens the case for separating the Victoriapithecinae from the Cercopithecidae at the family level.

Similarities in mandibular morphology and dentition are seen between *P. lothagamensis* and the early papionin attributed to cf. *Macaca* from Sahabi (Meikle 1987). The axis of the symphysis of the Lothagam specimens and of Sahabi IP28A is approximately 25 degrees to the occlusal plane. In later papionins the axis is much steeper. A lower central incisor from Sahabi, 105P16A, like those from Lothagam and all later papionins, lacks enamel or the enamel is very thin on the lingual face. A distinct line on both the mesial and distal faces defines the lingual extent of the thick labial enamel. It is likely that the Sahabi specimens should be attributed to *Parapapio* rather than to cf. *Macaca.* Comparative dental measurements show that the Sahabi specimens are smaller and just outside the range of *P. lothagamensis.* Specimens of cf. *Macaca* are recorded from Marceau in Algeria and from Wadi el Natrun in Egypt, but these specimens are too fragmentary for accurate taxonomic designation.

Parapapio is described from Laetoli (Leakey and Delson 1987) where *P. ado* is well represented at 3.5 Ma. This species lacks the primitive features that *P. lothagamensis* shares with *Victoriapithecus,* and the cheek teeth are larger although the ranges overlap. *Parapapio* is also common at the 4.4 Ma site at Aramis in Ethiopia and at Kanapoi (4.1 Ma) in Kenya. Excavations at site 261-1 at Allia Bay, Kenya (3.9 Ma), have produced numerous isolated teeth that appear to represent more than one species of papionin (Coffing et al. 1994). The material from these sites is currently under study.

Two teeth from Langebaanweg (Grine and Hendey 1981) are the only Late Miocene monkeys reported from South Africa. Four species of *Parapapio* are known from the South African Pliocene: the small *P. jonesi,* the intermediate-sized *P. broomi* and *P. antiquus,* and the large *P. whitei.* There appears to be significant sexual dimorphism in this sample, and size ranges overlap so that taxonomic identification from fragmentary specimens and isolated teeth is difficult. The primitive features of *P. lothagamensis* distinguish it from the South African *Parapapio.*

The appearance and dominance of *Theropithecus* cf. *T. brumpti* in the Kaiyumung Member is consistent with its occurrence at other sites of this age in the Turkana Basin. From the Omo Group deposits, *T. brumpti* is the most common cercopithecine between 3.5 and approximately 2.5 Ma; this reflects the presence of relatively closed riparian woodland or forest (Eck 1987). At Hadar, dated between 3.3 and 3.0 Ma, *T. darti*, the presumed immediate ancestor of *T. oswaldi*, is common, and this may reflect a more open grassy habitat (Eck 1993). At Laetoli, *Theropithecus* is absent, and here the common cercopithecine is *Parapapio* (Leakey and Delson 1987). This may indicate a more open environment, one too open for *T. brumpti*, and at this time too early for *T. darti*, which first appears in the Sidi Hakoma Member (3.3 Ma) at Hadar (Eck 1993). Alternatively, 3.6–3.5 Ma ago, Laetoli may have been outside the geographic range of *Theropithecus*. Kanapoi (4.1 Ma; Leakey et al. 1995) to the west of Lake Turkana lacks *Theropithecus* probably because of its early age.

Although taxonomically diverse, colobine specimens are often numerically less common in the fossil record than cercopithecines, and their remains are often fragmentary, which obscures their taxonomic identity. Few of comparable age to the Nawata Formation are known. Two colobine teeth are described from Sahabi, an M^3 and dP_4 (Meikle 1987), but they are too fragmentary to provide a definitive taxonomic identification although in size they are a little larger than the Colobinae species B from the Nawata Formation. Colobine isolated teeth are also known from the Late Miocene sites in North Africa at Marceau, Algeria, and Wadi el Natrun, Egypt. A complete skull of *Libypithecus markgrafi* was recovered from the latter, but the remaining specimens are fragmentary. At Aramis, Ethiopia, at a time equivalent in age to the Apak Member, colobines are more common than cercopithecines, and they are the only cercopithecid represented at the Late Miocene site of Lemudongo, Kenya, that is currently under investigation by Stanley Ambrose. Colobines are also encountered at Kanapoi (Leakey et al. 1995) and site 261-1 at Allia Bay, but they are not common (Coffing et al. 1994).

Locomotion

In contrast to the dietary differences between the Lothagam colobines and cercopithecines, their post-cranial anatomy reveals a notable lack of distinction between these groups. Extant colobines are readily distinguishable from cercopithecines in most areas of the skeleton, reflecting a greater commitment to an arboreal lifestyle. These adaptations have been documented by a variety of researchers—for example, Anemone (1993), Birchette (1982), Ciochon (1986), Conroy (1976), Fleagle (1976), Jolly (1965), Krentz (1993), Larson (1993, 1995), Rose (1993), Strasser (1988), and Su and Jablonski (1998). Extant colobine limbs are generally adapted to habitual loading in a greater variety of limb postures, thus reflecting the variation in the size, location, and compliance of supports in an arboreal setting. Their joints reflect adaptation to loading in a variety of postures, with rounder scapular glenoids; higher, more globular humeral heads; higher, rounder femoral heads; knee joints that are less expanded anteroposteriorly; and more mobile joints within the foot. Among other features, they have a more distal insertion of the deltoid muscle on the humerus; a shorter, less retroflexed olecranon process; a lower femoral greater trochanter and a shorter calcaneal heel process; and longer, more curved phalanges. Cercopithecines contrast with colobines in these features, reflecting a greater commitment to terrestriality, which provides for a more predictable pattern of loading of the limbs than does moving in the trees. Exceptions to this rule are occasionally found in the genus *Macaca*, which includes some of the most arboreal species of cercopithecines.

The Lothagam cercopithecids all appear to be quite similar in the preserved parts of their postcranial skeletons, to the point that they are difficult to distinguish from one another. Only size can be used in some cases to assign these fossils to particular taxa, and then only tentatively. All Lothagam cercopithecids resemble extant cercopithecines more closely than they do extant colobines in skeletal adaptation, as they exhibit semiterrestrial traits. Many of these traits are primitive for cercopithecids, as they are also found in *Victoriapithecus* (Harrison 1989).

There is no evidence of extreme arboreal cercopithecid adaptations in the fossil record, from the time of first appearance of cercopithecids in the Early Miocene, until the Late Pliocene. The first good evidence for arboreal locomotion is with the large-bodied colobines, a radiation that apparently took place after 4 Ma. The most arboreal of these is *Rhinocolobus turkanensis*, which first appears in Member A of the Shungura Formation (~3.0 Ma; Leakey 1982). *Paracolobus chemeroni* from the Chemeron Formation, Baringo, believed to be 3.3 to 3.2 Ma (Birchette 1981), shows a greater degree of arboreality than earlier colobines, but it still may have been at least partly terrestrial. *Theropithecus brumpti* also shows some arboreal features of its postcranial skeleton (Krentz 1993), although it appears to have been largely terrestrial.

The apparently recent appearance of arboreal adaptations in colobines raises interesting taxonomic and evolutionary issues. Extant African colobines resemble Asian colobines postcranially in many ways, and these traits have generally been viewed as synapomorphies that unite the Colobinae (Strasser and Delson 1987). The divergence time of these lineages, however, has recently been estimated to have been as early as 10 Ma (Goodman et al. 1998). If this is the case, then the postcranial similarities among African and Asian colobine lineages must represent homoplasies and reflect a shared arboreal habitus rather than phylogenetic heritage.

It is interesting to note that it is around the Miocene-Pliocene transition that Asian hominoids disappeared from the fossil record in Asia. Also, at about this time, the extant African hominoid lineages diversified (review in Stewart and Disotell 1998), all of which are at least partly terrestrial. One can speculate that this may have opened up arboreal primate niches in Africa and Asia, which cercopithecids then invaded (Ward 1998). The timing of these transitions is not precise, and so such hypotheses must remain only speculative at the present time.

Diet

The dental morphology of the Lothagam cercopithecids is of interest in view of recent studies and interpretations of cercopithecid diets. In order to predict diets of fossil cercopithecids, Benefit (1999) measured molar features shown to be functionally correlated among extant monkeys. By averaging estimates from regression equations that correlate diet with shear crest length, flare, and cusp relief, she predicted the proportions of leaves and fruits in the annual diets of extant and extinct monkeys. The Middle Miocene *Victoriapithecus* was found to have a predicted diet of 79 percent fruit and 7 percent leaves which is more frugivorous than the most frugivorous of extant monkeys, the Tana River mangabey *Cercocebus galeritus* (Benefit 1999). The dental morphology and several derived features of the cranium, such as the absence of the maxillary sinus, appear to be adaptations that provide strength against high occlusal forces. Benefit (1999) suggests that these features show that the victoriapithecids were undoubtedly frugivores; this interpretation is in agreement with previous molar microwear analyses (Lucas and Teaford 1994; Teaford et al. 1996) and in contrast to the long-held view that bilophodonty evolved as an adaptation for leaf eating and that the earliest monkeys were folivores (Napier 1970).

Some support for this change in perception of the adaptive role of early cercopithecids comes from Lucas and Teaford (1994), who assessed the physical properties of food in terms of its varied fracture behavior. By relating three modes of fracture to the different design features of wedges and blades, and then assessing the mechanical properties of colobine foods, they have shown that colobine teeth are adapted to process both seeds and leaves. It should be emphasized, however, that the seeds in question are those with tough, pliant coverings, not the hard, brittle objects often envisioned as seeds (cf. Happel 1988). This confusion is due, in part, to the mechanical complexity of fruits and seeds. Quite simply, not all seeds are alike. Moving outward, from the plant embryo and its food supply through the seed coatings and any layers of surrounding fruit, one finds a myriad of physical properties in a variety of plants, including tough pliant innermost layers (e.g., *Milletia atropurpurea*) and hard, fibrous outer shells (e.g., *Mezzetia parviflora*) (Lucas 1994). Moreover, the properties of these layers can change throughout the life of a seed (Kinzey and Norconk 1990; Williamson and Lucas 1995). With this complexity in mind, the colobine molar should be viewed as a multipurpose tool: in effect, a pair of wedges (the transverse symmetrical lophs) bordered by blades (the bladelike shearing crests along the buccal and lingual sides of the tooth) that can be used to process a variety of foods, ranging from those that are easily fragmented (such as young leaves and flowers) to those that are significantly tougher (such as mature leaves and flexible seed coats). For example, large seeds containing difficult-to-fragment storage tissues and covered by relatively thin flexible testa (e.g., *Milletia atropurpurea*) could be fractured by the wedges, whereas tough leaf tissues, such as the thick-walled veins of *Calophyllum inophyllum* (Lucas et al. 1991) could be fractured by the bordering blades. Seed storage tissues are richer sources of energy than are leaves, so that an animal that can make efficient use of this resource would have an advantage over one that could not. As the colobine dentition is also adapted to process leaves, colobines would also have access to an easily obtained source of protein, and one that might be less affected by seasonal changes than are fruits. In comparison, the thick-enameled cercopithecine molars with their blunt, closely-approximated cusps and relatively short transverse lophs and shearing crests are adapted to process a different range of foods, those that are soft and easily fragmented like the flesh of many fruits, as well as those that are hard and brittle such as the nut of *Balanites glabra* that is found inside a hard shell (Altmann 1998).

The range of usable food items, together with their physical properties, must always be kept in mind to understand the origin and evolution of bilophodonty in primates. In short, many primate foods (e.g., fruits, young leaves, and flowers) are surprisingly easy to process (Lucas and Teaford 1994). This is evident from ongoing work with *Alouatta palliata,* where tooth wear generally has little effect on food processing ability (Teaford et al. 1999). This implies that primate teeth are very well suited for the tasks they perform—so much so that they may only be seriously tested, physically, by occasional crucial items in the diet (Rosenberger and Kinzey 1976). With this in mind, primate bilophodont molars, with relatively low occlusal relief and thin enamel, may well have originated as an efficient means of processing a variety of foods. Only the hardest and toughest foods could not be processed by such teeth. Subsequent changes in occlusal relief and the use of lophs and blades would then have allowed a gradual incorporation of tougher foods into the diets of prehistoric colobines and harder foods into the diets of prehistoric cercopithecines.

The differences in diet of the two cercopithecid subfamilies are also reflected in the relative size of the anterior and posterior dentitions (Lucas 1989). Colobines, which use their teeth to break and destroy seeds, do most of the work inside the mouth with their cheek teeth. Colobines have large posterior teeth and relatively small incisors. In contrast, cercopithecines, which have cheek pouches and can store harvested fruits prior to their separating the fruit skin from the fruit flesh, do much of this separation with the anterior dentition (removing the fruit skins), holding the fruit in their hands. The fruit flesh is then removed from the seeds inside the mouth and ingested, whereas the hard inner seeds are spat out. Cercopithecines have both large anterior teeth and large cheek teeth.

The *Victoriapithecus* dental morphology with its similarities to that of the Papionini may be the ancestral adaptation to frugivory. However, among the large collection of dentitions from Maboko, there is no indication of colobine adaptations. This could be either because ancestral colobines had not yet diverged from a common victoriapithecine ancestor or because primitive colobines simply weren't present at Maboko for ecological reasons. It must be remembered that the time of emergence of the colobine dentition in the evolutionary record is unclear. The preserved molar teeth of *Microcolobus tugenensis,* the earliest known colobine, show the typical colobine morphology (Benefit and Pickford 1986). In Europe, the larger *Mesopithecus pentilici* was common at Pikermi and less common at other sites all dating between 8.5 and 6 Ma. There is thus no evidence earlier than *Microcolobus* for the divergence of the two subfamilies and the appearance of Lucas and Teaford's colobine seed-eating adaptation.

Previous studies of molar microwear (Teaford and Leakey 1992; Teaford et al. 1996) indicate that the case for seed eating in ancestral colobines is not straightforward. Colobines from the early hominid sites of East Africa show molar microwear patterns like those of extant colobines—that is, they have a preponderance of scratches and small pits on the enamel surfaces (figure 6.12). However, fossil cercopithecines from the same sites also show molar microwear patterns like those of extant colobines. This indicates that prehistoric colobines were probably feeding on foods much like those eaten by extant colobines. Fossil cercopithecines, by contrast, were not ingesting hard objects, as evidenced by the lack of large pits on their molars. As noted by Lucas and Teaford (1994) the "hard objects" in question could be either hard food items or large-grained abrasives on food. In either case, the Plio-Pleistocene cercopithecines of East Africa were rarely ingesting them. Molar microwear analyses of the Lothagam fossil monkeys show the same pattern as in the more recent Plio-Pleistocene samples. In other words, the colobines and cercopithecines are statistically indistinguishable (table 6.15).

One might legitimately ask why the cercopithecines were not ingesting hard objects. Two explanations spring to mind. First, they might not have been as terrestrial as some of their extant counterparts (e.g., *Papio*), thus avoiding many of the large-grained abrasives that often cling to foods like roots and tubers; this would run counter to analyses of their postcrania that suggest a semiterrestrial lifestyle, however. Second, they might not have been forced to use hard objects as "fallback foods," or at least not to ingest them in processing fallback foods. In other words, they might have always had ready access to softer, easier to process foods, unlike, for example, extant savanna baboons that are often forced to rely on grass corms when preferred foods are not available (Altmann 1998). The idea of fallback foods is not new to primatology, and a number of investigators have documented seasonal changes in diet in which species are forced to eat lower quality foods when prime foods are not available (e.g., Conklin-Brittain et al. 1998; Rudran 1978; Wrangham et al. 1998). The idea is new to studies of primate paleobiology, however, where dietary interpretations have traditionally focused on broad diet categories that have been determined by the most commonly eaten foods. If critical food items are indeed crucial to the survival and reproduction of individuals, they may well be crucial for the evolution of morphological differences (Lambert et al. 1999; Teaford et al. 1999) such as the dental differences between early colobines and cercopithecines. Interestingly, the idea that the Lothagam cercopithecines had ready access to soft, easily processed foods finds support in Benefit's recent analyses (1999) of shear crest lengths, which suggest that East African *Parapapio* was more frugivorous than its South African counterpart.

Conclusions

The apparently greater similarity in locomotor adaptation between Lothagam colobines and cercopithecines suggests that these monkeys were distinguished primarily by their different dietary preferences, colobines eating seeds and some leaves and cercopithecines eating largely fruits. If this is the case, the occurrence of colobines in fossil faunas might be less indicative of paleoecology than previously supposed. Monkeys, particularly colobines, when present in fossil faunas, are conventionally taken to indicate forest or closed woodland environments (WoldeGabriel et al. 1994). If the scenario outlined in the discussion in this contribution is correct, it may be very misleading.

The study of the Narok colobines will be of interest in this context. If colobine dentitions are adapted primarily to seed eating, the largely folivorous diet of extant African colobines and their specialized sacculated stomachs may be a relatively recent adaptation, and the seed-eating "specializations" of species like *Colobus satanus* may include some of the more primitive colobine adaptations. The Pliocene and Pleistocene colobines were more likely primarily seed eaters living in more open country and eating leaves only seasonally when seeds were unavailable. The semiterrestrial rather than arboreal adaptations of the post-cranial anatomy of the fossil colobines support this suggestion. In essence, then, mature leaves may have been the original fallback food of colobines. The prevalence of arboreal monkeys in Africa today is a result of recent radiations of both arboreal forest-dependant colobines and the largely forest-inhabiting guenons. This modern prevalence of forest and closed woodland cercopithecid habitats has led to our prejudiced views as to the habitats of past cercopithecid species. The Lothagam fossils are changing these perspectives.

The cercopithecid fossil record presents a timely reminder that caution should be exercised when interpreting past ecologies using the presence of taxa which today are associated with specific habitats. The graminivorous *Theropithecus gelada* is today restricted to the grassy highlands in Ethiopia. In the past, although *T. oswaldi* was a committed open country gramnivore, studies of the postcranial anatomy (Krentz 1993), the dental microwear (Teaford 1993), and the dental morphology of another species, *T. brumpti*, have shown this to be an arboreal forest living species, possibly largely folivorous. The specialized arboreal folivorous Colobinae known today are in sharp contrast to their fossil record of a long evolutionary history of semiterrestrial seed eaters.

Acknowledgments

We thank the government of Kenya and the governors of the National Museums of Kenya. We particularly thank the field crew whose sharp eyes spotted not only the larger more complete specimens but also the many tiny monkey teeth in the collection; the collection managers who spent long hours accessioning the collection; and the preparators for their skill and patience in cleaning the material. Special thanks go to Christopher Kiarie, Benson Kyongo, Kyalo Manthi, Joseph Mutaba, Samuel Ngui, and Mary Muungu; their constant cheerful help and support has been invaluable. We also thank Bob Campbell, who took several of the photographs and loaned his Sprint Scan scanner. The microwear studies were supported by NSF grants 8803570, 8904327, 9118876, and 9601766.

References Cited

Altmann, S. A. 1998. *Foraging for Survival: Yearling Baboons in Africa.* Chicago: University of Chicago Press.

Anemone, R. L. 1993. Functional anatomy of the hip and thigh of primates. In D. L. Gebo, ed., *Postcranial Adaptation in Nonhuman Primates,* pp. 150–174. DeKalb: Northern Illinois University Press.

Benefit, B. R. 1993. The permanent dentition and phylogenetic position of *Victoriapithecus* from Maboko Island, Kenya. *Journal of Human Evolution* 25:83–172.

Benefit, B. R. 1994. Phylogenetic, paleodemographic, and taphonomic implications of *Victoriapithecus* deciduous teeth from Maboko, Kenya. *American Journal of Physical Anthropology* 95:277–331.

Benefit, B. R. 1999. Biogeography, dietary specialization and the diversification of African Plio-Pleistocene monkeys. In T. Bromage and F. Shrenk, eds., *African Biogeography, Climatic Change and Early Hominid Evolution,* pp. 172–188. Oxford: Oxford University Press.

Benefit, B. R., and M. L. McCrossin. 1990. Diet, species diversity and distribution of African fossil baboons. *Kroeber Anthropological Society Papers* 71/72:77–94.

Benefit, B. R., and M. L. McCrossin. 1997. Earliest known Old World monkey skull. *Nature* 388:368–371.

Benefit, B. R., and M. Pickford. 1986. Miocene fossil cercopithecoids from Kenya. *American Journal of Physical Anthropology* 69:441–464.

Bernor, R. L., and P. P. Pavlakis. 1987. Zoogeographic relationships of the Sahabi large mammal fauna (early Pliocene, Libya). In N. T. Boaz, A. El-Arnauti, A. W. Gaziry, J. de Heinzelin, and D. D. Boaz, eds., *Neogene Paleontology and Geology of Sahabi,* pp. 349–384. New York: Liss.

Birchette, M. G. 1982. The postcranial skeleton of *Paracolobus chemeroni.* Ph.D. diss., Harvard University.

Brown, F., and C. S. Feibel. 1991. Stratigraphy, depositional environments and palaeogeography of the Koobi Fora Formation. In J. M. Harris, ed., *Koobi Fora Research Project.* Vol. 3. *The Fossil Ungulates: Geology, Fossil Artiodactyls, and Palaeoenvironments,* pp. 1–30. Oxford: Clarendon Press.

Brown, F. H., I. McDougall, I. Davies, and R. Maier. 1985. An intergrated Plio-Pleistocene chronology for the Turkana Basin. In E. Delson, ed., *Ancestors: The Hard Evidence,* pp. 83–90. New York: Liss.

Ciochon, R. L. 1986. The evolution of the cercopithecoid primate forelimb with special reference to African Plio-Pleistocene species. Ph.D. diss., University of California, Berkeley.

Coffing, K., C. Feibel, M. Leakey, and A. Walker. 1994. Four-million-year-old hominids from East Lake Turkana, Kenya. *American Journal of Physical Anthropology* 93:55–65.

Conklin-Brittain, N. L., R. W. Wrangham, and K. D. Hunt. 1998. Dietary response of chimpanzees and cercopithecines to seasonal variation in fruit abundance. II. Macronutrients. *International Journal of Primatolology* 19:971–998.

Conover, W. J., and R. L. Iman. 1981. Rank transformations as a bridge between parametric and nonparametric statistics. *American Statistician* 35:124–129.

Conroy, G. C. 1976. Primate postcranial remains from the Fayum province, Egypt, UAR. Ph.D. diss., Yale University.

Delson, E. 1973. Fossil colobine monkeys of the circum-Mediterranean region and the evolutionary history of the Cercopithecidae (Primates, Mammalia). Ph.D. diss., Columbia University.

Delson, E. 1979. *Prohylobates* from the early Miocene of Libya: A new species and its implications for cercopithecid origins. *Geobios* 12:725–733.

Delson, E. 1993. *Theropithecus* fossils from Africa and India and the taxonomy of the genus. In N. Jablonski, ed., Theropithecus: *The Rise and Fall of a Primate Genus,* pp. 157–189. Cambridge: Cambridge University Press.

Delson, E., and D. Dean. 1993. Are *Papio baringensis* R. Leakey, 1969, and *P. quadratirostris* Iwamoto, 1982, species of *Papio* or *Theropithecus*? In N. Jablonski, ed., Theropithecus: *The Rise and Fall of a Primate Genus,* pp. 125–156. Cambridge: Cambridge University Press.

Delson, E., G. G. Eck, M. G. Leakey, and N. G. Jablonski. 1993. Appendix I: A partial catalogue of fossil remains of *Theropithecus.* In N. Jablonski, ed., Theropithecus: *The Rise and Fall of a Primate Genus,* pp. 499–526. Cambridge: Cambridge University Press.

Drake, R., and G. H. Curtis. 1987. K-Ar geochronology of the Laetoli fossil localities. In M. D. Leakey and J. M. Harris, eds., *Laetoli: A Pliocene Site in Northern Tanzania,* pp. 48–52. Oxford: Clarendon Press.

Eck, G. G. 1987. The skull of *Theropithecus brumpti* compared with those of other species of the genus *Theropithecus.* In Y. Coppens and F. C. Howell, eds., *Les faunes Plio-Pléistocène de la Basse Vallée de l'Omo (Ethiopie).* Vol. 3. *Cercopithecidae de la Formations de Shungura,* pp. 1–122. Paris: Centre National de la Recherche Scientifique.

Eck, G. G. 1993. *Theropithecus darti* from the Hadar Formation, Ethiopia. In N. Jablonski, ed., Theropithecus: *The Rise and Fall of a Primate Genus,* pp. 15–83. Cambridge: Cambridge University Press.

Fleagle, J. G. 1976. Locomotor behavior and skeletal anatomy of sympatric Malaysian leaf monkeys (*Presbytis obscura* and *Presbytis melalophus*). *Yearbook of Physical Anthropology* 20:440–463.

Freedman, L. 1957. The fossil Cercopithecoidea of South Africa. *Annals of the Transvaal Museum* 23:121–262.

Freedman, L. 1960. Some new fossil cercopithecoid specimens from Makapansgat, South Africa. *Palaeontologia Africana* 7:7–45.

Freedman, L. 1965. Fossil and subfossil primates from the limestone deposits at Taung, Bolt's Farm, and Witkrans, South Africa. *Palaeontologia Africana* 9:19–48.

Freedman, L. 1970. A new check list of fossil Cercopithecoidea of South Africa. *Palaeontologia Africana* 13:109–110.

Freedman, L. 1976. South African fossil Cercopithecoidea: A reassessment including a description of new material from Makapansgat, Sterkfontein and Taung. *Journal of Human Evolution* 5:297–315.

Freedman, L., and C. K. Brain. 1972. Fossil cercopithecoid remains from the Kromdraai australopithecine site (Mammalia: Primates). *Annals of the Transvaal Museum* 28:1–16.

Goodman, M., C. A. Porter, J. Czelusniak, S. L. Page, H. Schneider, J. Shoshani, G. Gunnell, and C. P. Groves. 1998. Toward a phylogenetic classification of primates based on DNA evidence complemented by fossil evidence. *Molecular Phylogenetics and Evolution* 9:585–598.

Gordon, K. D. 1984. Taphonomy of dental microwear, II. *American Journal of Physical Anthropology* 63:164–165.

Grine, F. E., and Q. B. Hendey. 1981. Earliest primate remains from South Africa. *South African Journal of Science* 77:374–376.

Happel, R. 1988. Seed-eating by West African cercopithecines, with reference to the possible evolution of bilophodont molars. *American Journal of Physical Anthropology* 75:303–327.

Harris, J. M., F. H. Brown, and M. G. Leakey. 1988. Stratigraphy and paleontology of Pliocene and Pleistocene localities west of Lake Turkana, Kenya. *Contributions in Science* 399:1–128.

Harrison, T. 1989. New postcranial remains of *Victoriapithecus* from the Middle Miocene of Kenya. *Journal of Human Evolution* 18:3–54.

Harrison, T., ed. 1997. *Neogene Paleontology of the Manonga Valley, Tanzania: A Window into the Evolutionary History of East Africa.* New York: Plenum Press.

Harrison, T., and E. Baker. 1997. Palaeontology and biochronology of fossil localities in the Manonga Valley, Tanzania. In T. Harrison, ed., *Neogene Paleontology of the Manonga Valley, Tanzania: A Window into the Evolutionary History of East Africa,* pp. 361–393. New York: Plenum Press.

Hill, A. 1999. The Baringo Basin, Kenya: From Bishop to BPRP. In P. Andrews and P. Banham, eds., *Late Cenozoic Environments and Hominid Evolution: A Tribute to Bill Bishop,* pp. 85–97. London: Geological Society.

Hill, A., M. G. Leakey, J. D. Kingston, and S. Ward. In press. New cercopithecoids and a hominoid from 12.5 Ma in the Tugen Hills succession, Kenya [Special issue: Paleoanthropological Research in the Tugen Hills, Kenya]. *Journal of Human Evolution.*

Hill, A., and S. Ward. 1988. Origin of the Hominidae: The record of African large hominoid evolution between 14 My and 4 My. *Yearbook of Physical Anthropology* 31:49–83.

Hooijer, D. A. 1963. Miocene Mammalia of the Congo. *Annales Musée Royal de l'Afrique Central,* 8th ser., 46:1–71.

Johanson, D. C., M. Taieb, and Y. Coppens. 1982. Pliocene hominids from the Hadar Formation, Ethiopia (1973–1977): Stratigraphic, chronologic, and palaeoenvironmental contexts, with notes on hominid morphology and systematics. *American Journal of Physical Anthropology* 57:373–402.

Jolly, C. J. 1965. Origins and specializations of the longfaced cercopithecoidea. Ph.D. diss., University of London.

Kay, R. F. 1977. The evolution of molar occlusion in the Cercopithecidae and early catarrhines. *American Journal of Physical Anthropology* 46:327–352.

Kinzey, W. G., and M. A. Norconk. 1990. Hardness as a basis of fruit choice in two sympatric primates. *American Journal of Physical Anthropology* 81:5–15.

Krentz, H. B. 1993. Postcranial anatomy of extant and extinct species of *Theropithecus.* In N. G. Jablonski, ed., Theropithecus: *The Rise and Fall of a Primate Genus,* pp. 383–422. Cambridge: Cambridge University Press.

Lambert, J. E., C. A. Chapman, R. W. Wrangham, and N. L. Conklin-Brittain. 1999. The hardness of cercopithecine foods: Implications for the critical function of enamel thickness in exploiting fallback foods. *American Journal of Physical Anthropology,* Supplement 28:1–178.

Larson, S. G. 1993. Functional morphology of the shoulder in primates. In D. L. Gebo, ed., *Postcranial Adaptation in Nonhuman Primates,* pp. 45–69. DeKalb: Northern Illinois University Press.

Larson, S. G. 1995. New characters for the functional interpretation of primate scapulae and proximal humeri. *American Journal of Physical Anthropology* 98:13–35.

Leakey, M. G. 1982. Extinct large colobines from the Plio-Pleistocene of Africa. *American Journal of Physical Anthropology* 58:153–172.

Leakey, M. G. 1985. Early Miocene cercopithecids from Buluk, northern Kenya. *Folia Primatologica* 44:1–14.

Leakey, M. G. 1987. Colobinae (Mammalia, Primates) from the Omo Valley, Ethiopia. In Y. Coppens and F. C. Howell, eds., *Les faunes Plio-Pléistocène de la Basse Vallée de l'Omo (Ethiopie).* Vol. 3. *Cercopithecidae de la Formations de Shungura,* pp. 147–169. Paris: Centre National de la Recherche Scientifique.

Leakey, M. G. 1993. Evolution of *Theropithecus* in the Turkana Basin. In N. Jablonski, ed., Theropithecus: *The Rise and Fall of a Primate Genus,* pp. 85–124. Cambridge: Cambridge University Press.

Leakey, M. G., and E. Delson. 1987. Fossil Cercopithecidae from the Laetoli beds. In M. D. Leakey and J. M. Harris, eds., *Laetoli: A Pliocene Site in Northern Tanzania,* pp. 91–107. Oxford: Clarendon Press.

Leakey, M. G., and R. E. Leakey. 1973. New large Pleistocene Colobinae (Mammalia: Primates) from East Africa. In L. S. B. Leakey, R. J. G. Savage, and S. C. Coryndon, eds., *Fossil Vertebrates of Africa,* vol. 3, pp. 121–138. London: Academic Press.

Leakey, M. G., and R. E. F. Leakey. 1976. Further Cercopithecinae (Mammalia, Primates) from the Plio-Pleistocene of East Africa. In R. J. G. Savage and S. C. Coryndon, eds., *Fossil Vertebrates of Africa,* vol. 4, pp. 121–146. London: Academic Press.

Leakey, M. G., C. S. Feibel, I. McDougall, and A. Walker. 1995. New four-million-year-old hominid species from Kanapoi and Allia Bay, Kenya. *Nature* 376:565–571.

Le Gros Clark, W. E., and L. S. B. Leakey. 1951. *The Miocene Hominoidea of East Africa.* Fossil Mammals of Africa No. 1. London: British Museum (Natural History).

Lucas, P. W. 1989. A new theory relating seed processing by primates to their relative tooth sizes. In L. H. Schmitt, L. Freedman, and N. W. Bruce, eds., *The Growing Scope of Human Biology,* pp. 37–49. Proceedings of the Australasian Society for Human Biology, 2. Perth: Centre for Human Biology, University of Western Australia.

Lucas, P. W. 1994. Categorization of food items for oral processing. In D. J. Chivers and P. Lander, eds., *The Digestive System of Mammals,* pp. 197–218. Cambridge: Cambridge University Press.

Lucas, P. W., and M. F. Teaford. 1994. Functional morphology of colobine teeth. In A. G. Davies and J. F. Oates, eds., *Colobine Monkeys: Their Ecology, Behaviour and Evolution,* pp. 173–203. Cambridge: Cambridge University Press.

Lucas, P. W., M. F. Choong, H. T. W. Tan, M. I. Turner, and A. J. Berrick. 1991. The fracture toughness of the leaf of the dicotyledon *Calophyllum inophyllum* L. (Guttiferae). *Philosophical Transactions of the Royal Society of London,* ser. B, 334:95–106.

Maier, Q. 1970. New Cercopithecoidea from the lower Pleistocene cave deposits of the Makapansgat Limeworks, South Africa. *Palaeontologia Africana* 13:69–108.

Meikle, W. E. 1987. Fossil Cercopithecoidea from the Sahabi Formation. In N. T. Boaz, A. El-Arnauti, A. W. Gaziry, J. de Heinzelin, and D. D. Boaz, eds., *Neogene Paleontology and Geology of Sahabi,* pp. 119–127. New York: Liss.

Mollett, V. D. 1947. Fossil mammals of the Makapan Valley, Potgietersrust. I. Primates. *South African Journal of Science* 43:295–303.

Napier, J. R. 1970. Paleoecology and catarrhine evolution. In J. R. Napier and P. H. Napier, eds., *Old World Monkeys: Ecology, Systematics and Behaviour,* pp. 53–95. New York: Academic Press.

Pastor, R. F. 1993. Dental microwear among prehistoric inhabitants of the Indian subcontinent: A quantitative and comparative analysis. Ph.D. diss., University of Oregon.

Pilbeam, D., and Walker, A. C. 1968. Fossil monkeys from the Miocene of Napak, N.E. Uganda. *Nature* 270:657–660.

Rose, M. D. 1993. Functional anatomy of the elbow and forearm in primates. In D. L. Gebo, ed., *Postcranial Adaptation in Nonhuman Primates,* pp. 70–95. DeKalb: Northern Illinois University Press.

Rosenberger, A. L., and W. G. Kinzey. 1976. Functional patterns of molar occlusion in platyrrhine primates. *American Journal of Physical Anthropology* 45:281–298.

Rudran, R. 1978. Socioecology of the Blue Monkeys (*Cercopithecus mitis stuhlmanni*) of the Kibale Forest, Uganda. In G. G. Montgomery, ed., *The Ecology of Arboreal Folivores,* pp. 483–503. Washington, D.C.: Smithsonian Institution Press.

Simons, E. 1969. A Miocene monkey (*Prohylobates*) from northern Egypt. *Nature* 223:687–689.

Simons, E., and E. Delson. 1978. Cercopithecidae and Parapithecidae. In V. J. Maglio and H. B. S. Cooke, eds., *Evolution of African Mammals,* pp. 100–119. Cambridge, Mass.: Harvard University Press.

Smart, C. 1976. The Lothagam 1 fauna: Its phylogenetic, ecological and biogeographic significance. In Y. Coppens, F. C. Howell, G. L. Isaac, and R. E. Leakey, eds., *Earliest Man and Environments in the Lake Rudolf Basin: Stratigraphy, Paleoecology, and Evolution,* pp. 361–369. Chicago: University of Chicago Press.

Stewart, C. B., and T. R. Disotell. 1998. Primate evolution in and out of Africa. *Current Biology* 8:R582–588.

Strait, S. G. 1993. Molar microwear in extant small-bodied faunivorous mammals: An analysis of feature density and pit frequency. *American Journal of Physical Anthropology* 92:63–79.

Strasser, E. 1988. Pedal evidence for the origin and diversification of cercopithecid clades. *Journal of Human Evolution* 17:225–245.

Strasser, E., and E. Delson. 1987. Cladistic analysis of cercopithecid relationships. *Journal of Human Evolution* 16:81–99.

Su, D., and N. G. Jablonski. 1998. Locomotion and forelimb morphology in snub-nosed monkeys (Colobinae: Rhinopithecus). *American Journal of Physical Anthropology,* Supplement 26:213.

Szalay, F. S., and E. Delson. 1979. *Evolutionary History of the Primates.* New York: Academic Press.

Teaford, M. F. 1988. A review of dental microwear and diet in modern mammals. *Scanning Microscopy* 2:1149–1166.

Teaford, M. F. 1993. Dental microwear and diet in extant and

extinct *Theropithecus:* Preliminary analyses. In N. G. Jablonski, ed., Theropithecus: *The Rise and Fall of a Primate Genus,* pp. 331–349. Cambridge: Cambridge University Press.
Teaford, M. F. 1994. Dental microwear and dental function. *Evolutionary Anthropology* 3:17–30.
Teaford, M. F., and M. G. Leakey. 1992. Dental microwear and diet in Plio-Pleistocene cercopithecoids from Kenya. *American Journal of Physical Anthropology,* Supplement 14:160–161
Teaford, M. F., and O. J. Oyen. 1989. Live primates and dental replication: New problems and new techniques. *American Journal of Physical Anthropology* 80:73–81.
Teaford, M. F., M. C. Maas, and E. L. Simons. 1996a. Dental microwear and microstructure in early Oligocene Fayum primates: Implications for diet. *American Journal of Physical Anthropology* 101:527–543.
Teaford, M. F., P. S. Ungar, R. F. Kay, and M. G. Leakey. 1996b. The evolution of diet in Old World monkeys. *American Journal of Physical Anthropology,* Supplement 22:1–227.
Teaford, M. F., N. Yamashita, K. Aldridge, and K. E. Glander. 1999. Food processing in different social groups of *Alouatta palliata. American Journal of Physical Anthropology,* Supplement 28:263–264.
Ungar, P. S. 1994. Incisor microwear of Sumatran anthropoid primates. *American Journal of Physical Anthropology* 94:339–363.
Ungar, P. S., and M. F. Teaford. 1996. A preliminary examination of non-occlusal dental microwear in anthropoids: Implications for the study of fossil primates. *American Journal of Physical Anthropology* 100:101–113.
von Koenigswald, G. H. R. 1969. Miocene Cercopithecoidea and Oreopithecoidea of East Africa. In L. S. B. Leakey, ed., *Fossil Vertebrates of Africa,* vol. 1, pp. 39–52. London: Academic Press.
Walter, R. C., and J. L. Aronson. 1993. Age and source of the Sidi Hakoma Tuff, Hadar Formation, Ethiopia. *Journal of Human Evolution* 25:229–240.
Ward, C. V. 1998. Comments. *Current Biology* 8:R746.
Ward, C. V., M. G. Leakey, B. Brown, F. Brown, J. Harris, and A. Walker. 1999. South Turkwel: A new Pliocene site in Kenya. *Journal of Human Evolution* 36:69–95.
Williamson, L., and P. W. Lucas. 1995. The effect of moisture content on the mechanical properties of seed shell. *Journal of Materials Science* 30:162–166.
WoldeGabriel, G., Y. Haile-Selassie, P. R. Renne, W. K. Hart, S. H. Ambrose, B. Asfaw, G. Heiken, and T. White. 2001. Geology and palaeontology of the Late Miocene Middle Awash Valley, Afar Rift, Ethiopia. *Nature* 412:175–178.
WoldeGabriel, G., T. D. White, G. Suwa, P. Renne, J. de Heinzelin, W. K. Hart, and G. Heiken. 1994. Ecological and temporal placement of early Pliocene hominids at Aramis, Ethiopia. *Nature* 371:330–333.
Wrangham, R. W., N.-L Conklin-Brittain, and K. D. Hunt. 1998. Dietary response of chimpanzees and cercopithecines to seasonal variations in fruit abundance. I. Antifeedants. *International Journal of Primatology* 19:949–970.
Zar, J. H. 1984. *Biostatistical Analysis.* Englewood Cliffs, N.J.: Prentice Hall.

Table Abbreviations

General

A = Apak
Acc. No. = Accession number
K = Kaiyumung
KNM = Kenya National Museum
LN = Lower Nawata
LT = Lothagam
N = Nawata
UN = Upper Nawata
unkn. = horizon unknown

Anatomical

AP = anteroposterior
Dist = distal
LL = labiolingual
LLa = anterior labiolingual
LLp = posterior labiolingual
Lt. = left
Mand = mandible
Max = maxilla
max = maximum
MD = mesiodistal
ML = mediolateral
min = minimum
PD = proximodistal
Prox = proximal
Rt. = right
SI = superoinferior

Scapula

1. craniocaudal diameter of glenoid fossa
2. ML diameter of glenoid fossa
3. craniocaudal distance from base of spine to cranial margin of glenoid fossa

Humerus

1. maximum diameter of head
2. minimum diameter of head
3. transverse breadth of greater tuberosity
4. transverse breadth of lesser tuberosity
5. width of bicipital groove
6. proximal shaft AP
7. proximal shaft ML
8. proximal projection of greater tuberosity above humeral head

9. distance between center of deltoid tuberosity and proximal margin of head
10. midshaft AP
11. midshaft ML
12. distance from deltoid tuberosity to depth of trochlear groove
13. bicondylar width
14. ML width of distal articular surfaces taken anteriorly
15. ML width of capitulum
16. ML width of trochlea
17. width of medial pillar bordering olecranon fossa
18. width of lateral pillar bordering olecranon fossa
19. retroflexion angle of medial epicondyle in distal view from ML axis

Ulna

1. minimum AP distance from trochlear notch to posterior shaft
2. PD height of olecranon from trochlear notch
3. ML width of proximal portion of trochlear notch
4. ML width of posterior shaft adjacent to trochlear notch
5. PD length of trochlear notch
6. ML breadth of radial notch
7. AP diameter of shaft 2.5 cm distal to radial notch
8. ML diameter of shaft 2.5 cm distal to radial notch

Femur

1. SI diameter of head
2. ML depth of head
3. PD distance between highest point on head and lowest point on neck
4. PD distance between highest point on greater trochanter and lowest point on neck
5. neck-shaft angle
6. proximal shaft AP diameter
7. proximal shaft ML diameter
8. bicondylar width
9. AP diameter of lateral condyle
10. patellar surface ML width
11. patellar surface PD length
12. depth of patellar groove

Tibia

1. plateau AP diameter
2. plateau ML diameter
3. distance from anterior margin of plateau to center of tuberosity
4. midshaft AP diameter
5. midshaft ML diameter
6. distal epiphysis AP diameter
7. distal epiphysis ML diameter
8. PD length of malleolus

Astragalus

1. maximum PD length parallel to trochlear groove
2. PD length of trochlea at its midpoint
3. ML width of trochlea at its midpoint
4. maximum diameter of head
5. minimum diameter of head
6. depth of trochlear groove

Calcaneus

1. maximum PD length
2. anterior PD length from center of posterior astragalar facet to cuboid facet
3. posterior PD length from center of posterior astragalar facet to heel process
4. posterior astragalar facet PD length
5. posterior astragalar facet DP maximum height of arc in medial view

Table 6.1 Measurements (in mm) of the Upper Anterior Dentition and Premolars of *Parapapio lothagamensis*

			I^1		I^2		C/ Male		C/ Female		P^3			P^4		
Acc. No.	Side	Level	MD	LL	MD	LL	MD	LL	MD	LL	MD	LL	LL/MD	MD	LL	LL/MD
115	Rt.	LN	—	—	—	—	—	—	—	—	—	—	—	4.6	7.0	—
419 female	Rt.	LN	—	—	—	—	—	—	6.0	5.7	5.1	6.7	131.4	4.9	7.2	146.9
22972	Lt.	LN	—	—	—	—	—	—	—	—	—	—	—	5.3	7.7	145.3
23090 female	—	LN	—	—	—	—	—	—	—	—	—	—	—	—	—	—
23163	Rt.	LN	7.5	6.3	—	—	—	—	—	—	—	—	—	—	—	—
23164 female	Lt.	Unkn.	—	—	—	—	—	—	6.0	5.2	—	—	—	—	—	—
24096	Lt.	UN	7.9	6.1	—	—	—	—	—	—	—	—	—	—	—	—
24100	Lt.	UN	7.3	6.4	—	—	—	—	—	—	—	—	—	—	—	—
24111 male	Rt.	LN	—	—	—	—	>9.5	>6.4	—	—	—	—	—	—	—	—
24111 male	Lt.	LN	—	—	—	—	>9.1	>6.4	—	—	>4.9	—	—	5.0	7.0	—
24113	Rt.	UN	—	—	—	—	—	—	—	—	—	—	—	5.0	—	—
24134	Rt.	LN	—	—	4.6	~4.8	—	—	—	—	—	—	—	—	—	—
24137	Lt.	UN	—	—	—	—	—	—	—	—	—	—	—	4.9	6.4	130.6
24137	Rt.	UN	—	—	—	—	—	—	—	—	4.6	>5.0	—	4.9	6.4	130.6
26366 female	Rt.	UN	—	—	—	—	—	—	5.9	5.7	—	—	—	—	—	—
26405	Lt.	LN	7.7	—	—	—	—	—	—	—	—	—	—	—	—	—
26406	Rt./Lt.	LN	6.8	6.1	5.1	4.6	—	—	—	—	—	—	—	—	—	—
26608	Rt.	LN	>6.0	>5.3	—	—	—	—	—	—	—	—	—	—	—	—
28576	Rt.	LN	—	—	—	—	—	—	—	—	6.1	7.4	121.3	5.6	—	—
28781	—	UN	—	6.4	—	—	—	—	—	—	—	—	—	—	—	—
28791	Rt.	LN	—	—	—	—	—	—	—	—	>5.5	—	—	5.3	8.1	152.8
28792 female	Lt.	LN	7.3	6.4	5.0	5.4	—	—	6.6	6.5	—	—	—	5.5	8.0	145.5
28792 female	Rt.	LN	7.4	—	5.3	5.4	—	—	6.6	6.5	5.7	7.6	133.3	5.4	8.1	150.0
Mean	—	—	7.4	6.3	5.0	5.1	—	—	6.2	5.9	5.4	7.2	128.7	5.1	7.3	143.1
Max	—	—	7.9	6.7	5.3	5.4	—	—	6.6	6.5	6.1	7.6	133.3	5.6	8.1	152.8
Min	—	—	6.8	6.1	4.6	4.6	—	—	5.9	5.2	4.6	6.7	121.3	4.6	6.4	130.6

TABLE 6.2 Measurements (in mm) of the Lower Anterior Dentition and Premolars of *Parapapio lothagamensis*

			I_1		I_2		/C Male		/C Female		P_3 Male		P_3 Female		P_4	
Acc. No.	Side	Level	MD	LL	MD	LL	MD	LL	MD	LL	MD	LL	MD	LL	MD	LL
23079 male	Lt.	LN	—	—	4.6	>6.2	6.0	>9.8	—	—	—	—	—	—	—	—
23090 female	—	LN	—	—	—	—	—	—	3.3	5.9	—	—	3.9	5.0	—	—
23091 male	Lt.	N	—	—	3.8	5.5	5.3	8.7	—	—	—	5.5	—	—	—	5.7 > 5.3
23091 male	Rt.	N	—	—	—	—	—	—	—	—	—	—	—	—	5.4	5.5
23173	—	—	5.4	6.5	—	—	—	—	—	—	—	—	—	—	—	—
24094	Lt.	UN	—	—	—	—	—	—	—	—	—	—	—	—	4.7	5.2
24097 female	Lt.	LN	—	—	—	—	—	—	—	—	—	—	3.5	5.9	—	—
24109 male	—	UN	—	—	—	—	5.4	8.5	—	—	—	—	—	—	—	—
24139	Rt.	LN	—	—	—	—	—	—	—	—	—	—	—	—	5.4	5.9
26386 male	Lt.	LN	4.8	>6.1	4.5	>5.7	5.5	8.6	—	—	—	—	—	—	—	—
26386 male	Rt.	LN	4.7	>5.7	4.2	>6.0	<5.8	~9.5	—	—	4.5	6.5	—	—	5.2	6.0
26388	Lt.	LN	—	—	—	—	—	—	—	—	5.0	7.4	—	—	—	—
26579	Rt.	LN	—	—	3.8	3.7	—	—	—	—	—	—	—	—	—	—
30606 male	Rt.	UN	—	—	—	—	—	—	—	—	5.7	7.1	—	—	—	—
30608	Lt.	LN	5.8	>5.3	—	—	—	—	—	—	—	—	—	—	—	—
30608	Lt.	LN	5.1	>6.0	—	—	—	—	—	—	—	—	—	—	—	—
36910	Rt./Lt.	UN	4.4	5.3	3.9	5.0	—	—	—	—	—	—	—	—	—	—
Mean	—	—	5.0	5.9	4.1	4.7	5.6	8.6	3.3	5.9	5.1	6.6	3.7	5.5	5.2	5.7
Max	—	—	5.8	6.5	4.6	5.5	6.0	8.7	3.3	5.9	5.7	7.4	3.9	5.9	5.4	6.0
Min	—	—	4.4	5.3	3.8	3.7	5.3	8.5	3.3	5.9	4.5	5.5	3.5	5.0	4.7	5.2

Table 6.3 Measurements (in mm) of the Upper Molars of *Parapapio lothagamensis*

			M^1					M^2					M^3				
Acc. No.	Side	Level	MD	LLa	LLp	LLa/LLp	MD/LLa	MD	LLa	LLp	LLa/LLp	MD/LLa	MD	LLa	LLp	LLa/LLp	MD/LLa
123	Rt.	UN	—	—	—	—	—	9.5	8.8	8.7	101.1	108.0	—	—	—	—	—
419 female	Rt.	LN	7.7	8.0	7.3	109.6	96.3	8.7	9.1	8.1	112.2	95.6	8.3	8.3	6.1	136.07	100.0
22972	Lt.	LN	7.7	8.0	7.5	106.7	96.3	8.7	9.7	8.5	114.1	89.7	8.5	9.4	7.0	134.29	90.4
22973	Rt.	LN	8.3	8.0	8.1	98.8	103.8	9.3	9.7	9.6	101.0	95.9	9.2	9.7	9.0	107.78	94.8
23079 male	Rt.	LN	—	—	—	—	—	10.4	10.2	9.3	—	—	—	—	—	—	—
23090 female	—	LN	—	—	—	—	—	8.7	9.3	—	—	—	8.6	8.4	6.5	129.23	102.4
23163	Rt.	LN	—	—	—	—	—	9.4	~8.7	~7.9	—	—	—	—	—	—	—
24101	Rt.	LN	8.9	7.8	8.5	91.8	114.1	10.2	>10.8	9.8	—	—	—	—	—	—	—
24109	Rt.	UN	—	—	—	—	—	—	—	—	—	—	9.9	9.2	7.4	124.32	107.6
24111 male	Lt.	LN	—	—	—	—	—	8.9	9.1	8.5	107.1	97.8	8.8	8.2	7.8	105.13	107.3
24111	Rt.	LN	>7.4	—	7.4	—	—	8.7	9.3	8.8	105.7	93.5	—	—	—	—	—
24134	Rt.	LN	—	—	—	—	—	—	—	—	—	—	—	—	—	—	—
26371	Rt.	LN	—	—	—	—	—	~8.3	8.1	7.1	114.1	~1.0	—	—	—	—	—
26373	Rt.	LN	7.3	7.3	6.7	—	100.0	—	—	—	—	—	—	—	—	—	—
26384	Lt.	LN	—	—	—	—	—	8.9	—	8.9	—	—	—	—	—	—	—
26393	Lt.	LN	—	—	—	—	—	8.3	8.6	8.1	106.2	96.5	—	—	—	—	—
26400	Lt.	LN	—	—	—	—	—	9.0	9.8	8.8	111.4	91.8	—	—	—	—	—
28576	Rt.	LN	8.0	—	—	—	—	8.6	—	—	—	—	8.5	10.1	7.6	132.89	84.2
28791	Rt.	LN	8.4	8.6	8.0	107.5	97.7	—	—	—	—	—	—	—	—	—	—
28792 female	Lt.	LN	—	—	—	—	—	9.4	10.3	9.5	108.4	91.3	—	—	—	—	—
28792 female	Rt.	LN	—	—	—	—	—	—	—	—	—	—	8.5	9.3	6.4	145.31	91.4
24096	Lt.	UN	—	—	—	—	—	10.7	10.2	9.9	103.0	104.9	—	—	—	—	—
24102	Rt.	UN	8.6	7.6	7.7	98.7	113.2	—	—	—	—	—	—	—	—	—	—
24113	Rt.	UN	7.4	7.4	7.3	101.4	100.0	9.2	8.3	8.0	103.8	110.8	—	—	—	—	—
24137	Rt.	UN	>5.4	—	—	—	—	>7.6	—	—	—	—	—	—	—	—	—
26374	—	UN	7.6	—	~6.8	—	—	—	—	—	—	—	—	—	—	—	—
28786	Lt.	UN	—	—	—	—	—	—	—	—	—	—	9.9	>9.9	>7.5	—	—
30238	Lt.	LN	—	—	—	—	—	—	—	—	—	—	8.3	8.9	7.0	127.14	93.3
Mean	—	—	8.0	7.8	7.6	102.1	102.6	9.2	9.3	8.8	107.3	97.8	8.9	9.1	7.2	126.9	97.3
Max	—	—	8.9	8.6	8.5	109.6	114.1	110.8	9.9	9.9	114.1	110.8	9.9	10.1	9.0	145.3	107.6
Min	—	—	7.3	7.3	6.7	91.8	96.3	8.3	8.1	7.1	101.0	89.7	8.3	8.2	6.1	105.1	84.2

TABLE 6.4 Measurements (in mm) of the Lower Molars of *Parapapio lothagamensis*

			M_1					M_2					M_3				
Acc. No.	Side	Level	MD	LLa	LLp	LLa/LLp	MD/LLa	MD	LLa	LLp	LLa/LLp	MD/LLa	MD	LLa	LLp	LLa/LLp	MD/LLa
415	Rt.	LN	—	—	—	—	—	8.2	7.4	6.9	107.2	110.8	11.0	7.8	7.0	111.4	141.0
22971	Lt.	LN	—	—	—	—	—	8.8	8.4	7.8	107.7	104.8	—	~8.2	—	—	—
23066	—	LN	—	—	—	—	—	—	—	—	—	—	12.5	7.9	7.7	102.6	158.2
23070	—	LN	7.4	6.0	5.8	103.4	123.3	—	—	—	—	—	—	—	6.2	—	—
23090 female	—	LN	7.0	6.0	5.9	101.7	116.7	—	—	—	—	—	10.5	7.5	6.6	113.6	140.0
23091 male	Lt.	LN	7.6	6.5	6.7	97.0	116.9	9.1	7.9	7.8	101.3	115.2	11.6	8.0	7.4	108.1	145.0
23091 male	Rt.	LN	7.7	6.8	7.0	97.1	113.2	9.4	8.2	—	—	114.6	12.4	7.8	—	—	159.0
23717	—	LN	—	—	—	—	—	—	—	>7.1	—	—	—	—	—	—	—
24094	Lt.	UN	7.0	5.2	5.1	102.0	134.6	8.1	6.5	5.9	110.2	124.6	10.4	6.7	5.9	113.6	155.2
24095	—	UN	—	—	—	—	—	9.1	7.4	7.0	105.7	123.0	—	—	—	—	—
24099	Rt.	LN	7.5	5.8	6.4	90.6	129.3	—	—	—	—	—	—	—	—	—	—
24105	Lt.	LN	—	—	—	—	—	—	—	—	—	—	12.5	9.2	7.6	121.1	135.9
24108	Lt.	LN	8.4	6.7	7.0	95.7	125.4	—	—	—	—	—	—	—	—	—	—
24117	Lt.	UN	—	—	—	—	—	—	—	—	—	—	13.5	8.4	7.9	106.3	160.7
24122	—	LN	7.1	—	—	—	—	9.5	7.4	6.6	112.1	128.4	11.0	7.0	6.0	116.7	157.1
24127	—	UN	8.4	6.6	6.9	95.7	127.3	—	—	—	—	—	—	—	—	—	—
24133	Lt.	UN	—	—	—	—	—	—	—	—	—	—	11.9	8.9	7.6	117.1	133.7
24135	Rt.	LN	6.8	5.7	5.8	98.3	119.3	8.3	7.4	7.2	102.8	112.2	10.4	7.4	—	—	—
24136 male	Lt.	UN	—	—	—	—	—	—	—	—	—	—	12.4	7.8	6.8	114.7	159.0
24139	Lt.	LN	—	—	—	—	—	—	—	—	—	—	10.5	—	—	—	—
24139	Rt.	LN	7.5	—	—	—	—	9.2	8.1	—	—	113.6	11.1	8.1	7.1	114.1	137.0
24140	—	UN	—	—	—	—	—	~8.6	~7.0	~7.2	—	—	—	—	—	—	—
26391	Rt.	LN	—	—	—	—	—	—	—	—	—	—	10.7	7.2	6.4	112.5	148.6
26395	Rt.	LN	—	—	—	—	—	8.5	7.1	6.9	102.9	119.7	—	—	—	—	—
26398	Lt.	LN	—	—	—	—	—	—	—	—	—	—	—	—	6.7	—	—
26409	Lt.	LN	—	—	—	—	—	—	—	—	—	—	11.2	7.3	6.4	114.1	153.4
26579	Rt.	LN	8.9	7.0	6.9	—	127.1	—	—	—	—	—	—	—	—	—	—
28728	Lt.	LN	—	—	—	—	—	10.3	9.0	8.4	107.1	114.4	—	—	—	—	—
28755	Lt.	LN	—	—	—	—	—	—	—	—	—	—	9.7	7.3	6.3	115.9	132.9
28783	—	UN	—	—	—	—	—	—	—	—	—	—	>10.6	8.0	6.9	115.9	—
30263	—	LN	—	—	—	—	—	—	—	—	—	—	>10.3	6.7	6.0	111.7	—
Mean	—	—	7.6	6.2	6.4	97.9	123.3	9.0	7.7	7.2	106.3	116.5	11.4	7.7	6.8	113.1	147.8
Max	—	—	8.9	7.0	7.0	103.4	134.6	10.3	9.0	8.4	112.1	128.4	13.5	9.2	7.9	121.1	160.7
Min	—	—	6.8	5.2	5.1	90.6	113.2	8.1	6.5	5.9	101.3	104.8	9.7	6.7	5.9	102.6	132.9

TABLE 6.5 Measurements (in mm) of the Dentition of *Parapapio lothagamensis* Compared with Those of *P. ado*

	C/ Male		C/ Female		P_3		P_4	
Upper-anterior	**MD**	**LL**	**MD**	**LL**	**MD**	**LL**	**MD**	**LL**
P. lothagamensis								
Mean	10.60	7.90	6.3	6.0	5.4	7.2	5.1	7.4
Max	10.60	7.90	6.6	6.5	6.1	7.6	5.6	8.1
Min	10.60	7.90	5.9	5.2	4.6	6.7	4.6	6.4
P. ado								
Mean	11.25	8.90	6.35	5.3	6.07	7.1	6.8	8.1
Max	12.00	9.20	6.4	5.9	6.5	7.9	7.5	8.3
Min	10.50	8.60	6.3	4.7	5.4	6.5	6.1	7.8

	I_1		/C Male		P_3 Male		P_3 Female		P_4	
Lower-anterior	**MD**	**LL**	**MD**	**LL**	**MD**	**LL**	**MD**	**LL**	**MD**	**LL**
P. lothagamensis										
Mean	5.0	5.3	5.6	8.6	6.6	5.1	5.5	3.7	5.7	5.2
Max	5.8	5.3	6.0	8.7	7.4	5.7	5.9	3.9	6.0	0.4
Min	4.4	5.3	5.3	8.5	5.5	4.5	5.0	3.5	5.2	4.7
P. ado										
Mean	4.9	4.8	6.3	10.2	—	5.4	7.3	4.4	7.1	5.9
Max	5.1	5.3	6.8	10.6	—	5.8	10.9	5.0	8.5	6.5
Min	4.5	3.8	5.8	10.0	—	5.0	6.1	3.6	5.8	5.1

	M^1			M^2			M^3		
Upper Molars	**MD**	**LLa**	**LLa/MD**	**MD**	**LLa**	**LLa/MD**	**MD**	**LLa**	**LLa/MD**
P. lothagamensis									
Mean	8.0	7.8	0.9	9.2	9.3	1.0	8.9	9.1	1.0
Max	8.9	8.6	1.1	10.7	10.3	1.1	9.9	10.1	1.2
Min	7.3	7.3	1.0	8.3	8.1	0.9	8.3	8.2	0.9
P. ado									
Mean	9.0	8.1	0.9	11.2	10.4	0.9	10.8	10.2	0.9
Max	9.9	8.4	1.0	12.3	12.0	1.1	11.8	11.3	1.0
Min	8.0	7.5	0.9	9.9	8.5	0.9	9.4	8.2	0.9

	M_1			M_2			M_3		
Lower Molars	**MD**	**LLa**	**LLa/MD**	**MD**	**LLa**	**LLa/MD**	**MD**	**LLa**	**LLa/MD**
P. lothagamensis									
Mean	7.6	6.2	0.8	9.0	7.7	0.9	11.4	7.7	0.7
Max	8.9	7.0	0.9	10.3	9.0	1.0	13.5	9.2	0.8
Min	6.8	5.2	0.7	8.1	6.5	0.8	9.7	6.7	0.6
P. ado									
Mean	9.1	7.0	0.8	10.8	9.0	0.8	13.9	8.9	0.7
Max	10.3	7.9	0.8	12.5	9.8	0.9	15.3	10.8	0.7
Min	7.7	6.3	0.7	9.6	7.3	0.8	12.5	7.8	0.6

P. lothagamensis

Deciduous dP_4	**MD**	**LLa**
Mean	6.8	5.6
Max	7.0	5.8
Min	6.6	5.1

P. ado

Deciduous dP_4	**MD**	**LLa**
Mean	8.0	5.6
Max	8.4	6.0
Min	7.6	5.4

TABLE 6.6 Measurements (in mm) of the Deciduous Dentition of *Parapapio lothagamensis*

			dC/		dP^3			dP^4		
Acc. No.	Side	Level	MD	LL	MD	LLa	LLp	MD	LLa	LLp
24113	Rt.	UN	—	—	—	—	—	6.4	6.2	>5.7
24138	Rt.	LN	—	—	—	—	—	6.6	6.4	6.1
26617	Rt.	LN	—	—	—	—	—	6.4	6.1	6.0
26619	Rt.	LN	4.5	3.3	—	—	—	6.4	—	5.7
			d/C		dP_3			dP_4		
			MD	LL	MD	LLa	LLp	MD	LLa	LLp
23124	Rt.	LN	—	—	—	—	4.4	6.6	5.1	5.5
24108	Lt.	LN	—	—	—	—	—	7	5.8	~6.2
26579	Rt./Lt.	LN	—	—	—	—	4.1	—	—	5.3

TABLE 6.7 Measurements (in mm) of the Dentition of cf. *Parapapio* sp. indet.

Anterior Dentition

			I^1		I^2		C/ Male		C/ Female		P^3			P^4		
Acc. No.	Side	Level	MD	LL	MD	LL	MD	LL	MD	LL	MD	LL	LL/MD	MD	LL	LL/MD
448 female	Rt.	LN	—	—	—	—	—	—	—	—	—	—	—	>4.8	—	—
			I_1		I_2		/C Male		/C Female		P_3			P_4		
			MD	LL	MD	LL	MD	LL	MD	LL	MD	LL	LL/MD	MD	LL	LL/MD
448 female	Lt.	LN	>4.6	>5.3	—	—	—	—	—	—	—	—	>4.8	>5.1	—	—
448 female	Rt.	LN	>5.0	>5.7	>4.8	>5.1	—	—	>6.0	>3.7	>5.2	>4.4	>4.8	>5.1	—	—
449 male	Lt.	Unkn.	—	—	—	—	10.6	7.9	—	—	—	7.2	—	6.5	7.7	1.18

Posterior Dentition

			M^1			M^2			M^3		
Acc. No.	Side	Level	MD	LLa	LLp	MD	LLa	LLp	MD	LLa	LLp
448 female	Rt.	Unkn.	>6.8	>7.7	>7.7	>9.1	>9.0	>8.9	—	—	—
26369	—	K	8.9	—	—	~11	—	9.6	10.7	10.3	8.7
			M_1			M_2			M_3		
			MD	LLa	LLp	MD	LLa	LLp	MD	LLa	LLp
448 female	Lt.	LN	>6.9	>5.7	>6.1	>8.5	>7.8	>7.2	>11.2	—	>6.1
448 female	Rt.	LN	>6.5	>5.9	>6.1	>8.3	>7.5	>7.0	>10.8	>7.2	>6.1

Deciduous Dentition

			dP_3		
Acc. No.	Side	Level	MD	LLa	LLp
24110	—	A	—	4.9	—

TABLE 6.8 Measurements (in mm) of the Dentition of *Theropithecus* cf. *T. brumpti* from Lothagam

Anterior Dentition

		I^1		C/ Male		P^3	
Acc. No.	Level	MD	LL	MD	LL	MD	LL
24129	K	—	—	—	—	9.5	10.9
26368 female	K	—	—	9.5	8.9	—	—
26372	K	8.0	6.0	—	—	—	—
26396 female	K	—	—	9.4	8.6	—	—
26397	K	~7.4	6.4	—	—	—	—

		P_3	
		MD	LL
26401 male	K	11.0	7.7

Posterior Dentition

		M_2			M_3	
Acc. No.	Level	MD	LLa	LLp	MD	LLa
417	K	15.0	11.3	10.6	—	—
24128	K	—	—	—	~18.5	~12.2

Deciduous Dentition

		dP_4		
Acc. No.	Level	MD	LLa	LLp
37105	K	12	8.2	8.3

Table 6.9 Lothagam Cercopithecoid Postcranial Measurements

Acc. No.	Element	1	2	3	4	5	6	7	8	9	10	11	12	13	14	15	16	17	18	19
Parapapio lothagamensis																				
26370	Scapula	17.1	12.8	12.8	—	—	—	—	—	—	—	—	—	—	—	—	—	—	—	—
23067	Humerus	—	—	—	—	—	—	—	—	—	9.5	11.3	100	—	18.4	6.8	—	4.8	7.8	—
23074	Humerus	—	—	—	—	—	—	—	—	—	—	—	—	30.9	20.7	8.3	11.5	6.4	9.9	70
23077	Humerus	—	—	—	—	—	—	—	—	—	—	—	—	27.5	19.1	—	—	5.1	7	75
24114	Humerus	—	—	—	—	—	~14.7	—	—	—	—	—	—	—	—	—	—	~14	—	—
24123	Humerus	—	—	—	—	—	—	—	—	—	—	—	100.7	—	—	—	8.1	—	—	40
26385	Humerus	—	—	—	—	—	14.4	15	—	72	—	—	—	—	—	—	—	—	—	—
26410	Humerus	—	—	—	—	—	—	—	—	—	—	—	—	—	—	—	~7.5	10.8	—	55
28769	Humerus	17.3	15.3	15.2	9	6.8	—	—	0.9	—	—	—	—	—	—	—	—	—	—	—
24119	Ulna	12.7	—	—	—	—	11.8	13	7	—	—	—	—	—	—	—	—	—	—	—
22974	Femur	18.1	11.5	4	10.5	105	—	—	—	—	—	—	—	—	—	—	—	—	—	—
24121	Femur	—	—	—	—	—	—	—	29.3	24.8	~15	18.1	2.5	—	—	—	—	—	—	—
26375	Femur	—	—	—	8.6	—	—	—	—	—	—	—	—	—	—	—	—	—	—	—
26403	Femur	16.5	9.9	2.9	5	110	—	—	—	—	—	—	—	—	—	—	—	—	—	—
26404	Femur	15.7	8.2	3.9	—	110	—	—	—	—	—	—	—	—	—	—	—	—	—	—
23086	Tibia	—	—	—	11	8.8	12	9.3	—	—	—	—	—	—	—	—	—	—	—	—
26376	Tibia	27.2	~31	24.4	~21	~13	—	—	—	—	—	—	—	—	—	—	—	—	—	—
23081	Talus	23.4	13.1	10.8	1	—	11.5	—	—	—	—	—	—	—	—	—	—	—	—	—
23122	Talus	(>21)	12.2	10.5	—	—	1.4	—	—	—	—	—	—	—	—	—	—	—	—	—
24125	Calcaneum	35.9	14.2	22.6	9.6	3.5	—	—	—	—	—	—	—	—	—	—	—	—	—	—
26402	Calcaneum	34.5	12	21.2	11.1	4.1	—	—	—	—	—	—	—	—	—	—	—	—	—	—
28575	Calcaneum	36.7	12.6	23.3	10.9	4.4	—	—	—	—	—	—	—	—	—	—	—	—	—	—

Cercopithecinae cf. *Parapapio* sp. indet.																				
416	Humerus	—	—	—	—	—	—	—	—	—	—	—	—	29.6	19.2	7.2	9.6	6.7	7.4	45
26381	Humerus	—	—	—	—	—	—	—	—	—	—	—	—	—	—	—	8.9	8.7	13.2	—
28724	Femur	16.8	11.3	6.8	—	112	—	—	—	—	—	—	—	—	—	—	—	—	—	—
Theropithecus brumpti																				
24104	Femur	~21	—	~5	—	115	—	—	—	—	—	—	—	—	—	—	—	—	—	—
Colobinae species B																				
30610	Calcaneum	—	—	16.6	7.9	3	—	—	—	—	—	—	—	—	—	—	—	—	—	—
Colobinae gen. and sp. indet. (small)																				
22976	Humerus	18.8	17.4	16.9	11.3	6.4	11.5	13.2	[−1.9]	—	—	—	—	—	—	—	—	—	—	—
24126	Ulna	8.9	9.1	7.5	5.3	—	—	—	—	—	—	—	—	—	—	—	—	—	—	—
30609	Ulna	8.4	6.4	6.8	5.8	8.7	10.1	—	6.5	—	—	—	—	—	—	—	—	—	—	—
26379	Femur	—	—	—	—	—	—	—	20.4	16.1	11.7	9.7	1.7	—	—	—	—	—	—	—
26390	Femur	16.3	—	—	—	—	—	—	—	—	—	—	—	—	—	—	—	—	—	—
28642	Femur	—	—	—	—	—	10.2	9.1	—	—	—	—	—	—	—	—	—	—	—	—
26392	Calcaneum	—	—	17.1	7.6	3.2	—	—	—	—	—	—	—	—	—	—	—	—	—	—
Cercopithecidae gen. and sp. indet.																				
26407	Ulna	9.7	8.5	9.2	5.8	14.2	10.1	—	—	—	—	—	—	—	—	—	—	—	—	—
23056	Femur	—	—	—	—	—	11.7	8.8	—	—	—	—	—	—	—	—	—	—	—	—
26380	Femur	—	—	—	—	—	—	—	21.2	18.7	~9	12.5	1.5	—	—	—	—	—	—	—
23072	Tibia	—	—	—	9.4	7.1	—	—	—	—	—	—	—	—	—	—	—	—	—	—

TABLE 6.10 Measurements (in mm) of Colobinae Anterior Dentition and Premolars

			I^1		I^2		C/ Male		P^3		P^4	
Acc. No.	**Side**	**Level**	**MD**	**LL**	**MD**	**LL**	**MD**	**LL**	**MD**	**LL**	**MD**	**LL**
Cercopithecoides kerioensis												
9277	Lt.	—	—	—	—	—	—	—	—	6.2	5.2	6.6
9277	Rt.	—	—	—	—	—	—	—	—	—	5.3	6.7
C. williamsi												
Range: East and South Africa[a]	—	—	—	—	—	—	—	—	—	—	5.2–6.8	7.6–8.7
Colobinae species A												
418	Rt.	Unkn.	—	—	—	—	—	—	—	—	4.6	5.5
23078 male	—	UN	—	—	—	—	7.2	5.2	—	—	—	—
26607	Lt.	LN	3.55	—	—	—	—	—	4.8	—	—	—
36912	Lt.	UN	—	—	4.2	4.0	—	—	—	—	—	—
Colobinae species B												
23064 male	Lt.	LN	—	—	—	—	10.6	8.0	—	—	—	—
30610	Rt.	UN	—	—	—	—	—	—	—	7.4	—	—
36911	Lt.	UN	—	—	—	—	—	—	—	—	—	—
Colobinae species C												
30607	Lt.	A	—	—	—	—	—	—	—	—	—	—

			I_1		I_2		C/ Male		P_3		P_4	
Acc. No.	**Side**	**Level**	**MD**	**LL**	**MD**	**LL**	**MD**	**LL**	**MD**	**LL**	**MD**	**LL**
Cercopithecoides kerioensis												
9277	—	—	—	—	—	—	—	—	—	—	6.5	4.6
C. williamsi												
Range: East and South Africa[a]	—	—	—	—	—	—	—	6.2–9.0	5.2–6.9	—	—	—
Colobinae species A												
26383 male	Rt.	LN	—	—	—	—	4.5	6.5	—	—	—	—
36913	—	LN	—	—	3.6	3.1	—	—	5.6	—	—	—
Colobinae species B												
23162	Lt.	LN	—	—	—	—	—	—	—	5.9	—	—
23165	Rt.	LN	4.6	—	—	—	—	—	—	—	—	—
24116 male	Rt.	UN	—	—	—	—	6.2	10.3	—	—	—	—
24131 male	Lt.	LN	—	—	—	—	6.1	10.1	—	—	—	—

[a] Freedman (1957).

TABLE 6.11 Measurements (in mm) of Colobinae Molars

Acc. No.	Side	Level	M^1 MD	M^1 LLa	M^1 LLp	M^2 MD	M^2 LLa	M^2 LLp	M^3 MD	M^3 LLa	M^3 LLp
Cercopithecoides kerioensis											
9277	Lt.	—	6.6	7.0	6.5	—	—	—	—	—	—
9277	Rt.	—	6.6	7.1	6.6	—	—	—	—	—	—
C. williamsi											
Range: East and South Africa[a]	—	—	7.8–9.5	8.4–9.6	7.8–9.0	—	—	—	—	—	—
Colobinae species B											
36911	Lt.	UN	—	—	—	9.4	8.9	8.6	—	—	—
Cercopithecoides kerioensis											
9277	Lt.	—	—	—	—	7.9	6.6	6.7	9.6	6.3	6.0
C. williamsi											
Range: East and South Africa[a]	—	—	8.5–12.5	6.8–8.7	6.8–9.7	10.9–14.0	7.7–8.8	7.1–8.9	—	—	—
Colobinae species A											
24107	Lt.	LN	—	—	—	6.6	5.0	5.1	—	—	—
Colobinae species B											
23062	Rt.	UN	—	—	—	—	—	—		—	6.5
23083	Lt.	UN	—	—	—	—	—	—	>10.5	6.9	6.5
23162	Lt.	LN	7.0	5.3	5.9	—	—	—	—	—	—
23167	Lt.	LN	7.3	5.1	5.5	—	—	—	—	—	—
24098 male	Rt.	UN	—	—	—	9.8	7.3	7.3	—	—	—
24132	Lt.	UN	7.4	5.8	5.6	—	—	—	—	—	—
26387	Lt.	LN	—	—	—	—	—	—	—	—	6.2
Colobinae species C											
24124	Lt.	A	7.2	5.8	6.4	8.0	6.9	6.7	11	6.9	—

[a] Freedman (1957).

TABLE 6.12 Measurements (in mm) of Colobinae Deciduous Dentition

Acc. No.	Side	Level	dI^2 MD	dI^2 LL	dP^3 MD	dP^3 LLa	dP^3 LLp	dP^4 MD	dP^4 LLa	dP^4 LLp
Colobinae species B										
23166	Lt.	LN	—	—	6.1	5.6	5.00	—	—	—
			dI$_2$ MD	dI$_2$ LL	dP$_3$ MD	dP$_3$ LLa	dP$_3$ LLp	dP$_4$ MD	dP$_4$ LLa	dP$_4$ LLp
Colobinae species B										
26399	Lt.	UN	2.2	2.9	5.6	3.2	3.40	5.8	4.1	4.3
26399	Rt.	UN	2.2	2.6	5.6	3.1	3.40	5.9	4.0	4.4

TABLE 6.13 Cercopithecid Teeth Examined for Microwear

	Parapapio	Colobinae
Specimens used in this study	KNM-LT 415, 23079, 23091, 24095, 24096, 24101, 24108, 24111, 24139, 26393, 28576	KNM-LT 23162, 24098, 24107, 24124
Specimens unsuitable for molar microwear due to postmortem wear	KNM-LT 22971, 23090, 24094, 24099, 24102, 26373, 26374, 26395, 26400, 28791, 28792	KNM-LT 23167, 24132

TABLE 6.14 Cercopithecid Molar Microwear Measurements (Means)

Accession No.	Mean No. Features/Micrograph	Mean Pit Width (μm)	Mean Scratch Width (μm)	Mean Vector of Scratch Orientation	% Pits
Parapapio					
415	144	2.01	0.86	0.430	50.0
23079	114.5	2.41	0.88	0.194	45.5
23091	196	2.81	0.85	0.602	32.9
24095	430	1.60	0.85	0.611	38.6
24096	176	2.17	0.92	0.640	52.1
24101	338	1.91	0.97	0.587	25.2
24108	461.5	1.73	0.80	0.526	48.9
24111	255	2.53	1.09	0.558	52.1
24139	335	1.72	0.88	0.308	38.3
26393	202	2.41	0.96	0.223	35.9
28576	147	2.25	0.88	0.588	34.1
Colobinae					
23162	258.5	2.19	0.99	0.652	48.8
24098	223.5	1.74	0.80	0.634	29.5
24107	684	1.64	0.82	0.433	30.6
24124	354	1.95	0.86	0.615	45.4

TABLE 6.15 Statistics for Molar Microwear Measurements from Lothagam Fossil Monkeys (Means + Standard Error)

	Mean No. Features/Micrograph	Mean Pit Width (mm)	Mean Scratch Width (mm)	Mean Vector of Scratch Orientation	% Pits
Parapapio	254 ± 36	2.14 ± .12	0.90 ± .02	.479 ± .05	40.3 ± 3.0
Colobinae	380 ± 105	1.88 ± .12	0.87 ± .04	.583 ± .05	38.6 ± 5.0

6.2

The Lothagam Hominids

Meave G. Leakey and Alan C. Walker

Hominids are rare at Lothagam; only seven specimens are attributed to the Hominidae, and only three of these are from the Late Miocene deposits. These three specimens represent populations from close to the time of the divergence between the lineages of the Hominidae that gave rise to extant chimpanzees and bonobos on the one hand and extant humans on the other. The primitive aspects of two isolated teeth from the top of the upper member of the Nawata Formation indicate that they could equally represent an early hominin or the ancestral morphotype of both lineages. The mandible recovered in 1967 from the base of the Apak Member shows close affinities with *Australopithecus anamensis* mandibles but, without comparative material from earlier populations, it is not possible to give an attribution more secure than Hominidae indeterminate. Four isolated teeth from the Kaiyumung Member show closest individual matches with specimens from Laetoli and from Hadar and are attributed to *Australopithecus* cf. *A. afarensis.*

In spite of increasing interest in recent years in the earliest stages of human evolution, fossils that document this crucial time are frustratingly few (Hill and Ward 1988). Recent discoveries at Aramis in Ethiopia (White et al. 1994) and Kanapoi in Kenya (Leakey and Walker 1997; Leakey et al. 1998) have provided good samples of two new species of early hominins aged between 4.4 and 4 Ma. The ca. 6 Ma hominids from Kenya (Senut et al. 2001) whose hominid status is debated and the recently announced 5.8–5.2 Ma hominids from the Middle Awash, Ethiopia (Haile-Selassie 2001) are important new additions to the fossil record. The Tabarin mandible and Chemeron humerus from the Baringo Basin in Kenya are dated at 4.5 to 4.4 Ma (Hill 1999).

Molecular estimates for the time of divergence between chimpanzees and hominins have varied over the years but the most recent estimate is 5.5 Ma (Kumar and Hedges 1998, based on a nuclear gene molecular clock calibrated by the diapsid-synapsid divergence time). Lothagam thus represents a site with the potential for providing further human fossils from this important time interval. Disappointingly, only three hominid specimens were recovered from the earlier sediments: the mandible found by Bryan Patterson in 1967 and two isolated teeth.

These specimens, which derive from the upper member of the Nawata Formation and the base of the Apak Member of the Nachukui Formation, are close to this splitting time and could represent (1) the last common ancestor of chimpanzees, bonobos, and hominins, or (2) the earliest ancestor of chimpanzees and bonobos, or (3) the earliest hominins. The Upper Nawata accumulated between 6.54 ± 0.07 Ma and the time of deposition of the Purple Marker. Based on paleomagnetism and sedimentation rates, the Purple Marker has an estimated age of 5.23 Ma but there is evidence for substantial time loss in the section at about this time and this age is by no means certain (C. S. Feibel personal communication). The two isolated hominid teeth are from the upper part of the Upper Nawata and therefore closer to 5.23 Ma than to 6.54 Ma. For more discussion of the age of the Upper Nawata, see McDougall and Feibel (1999).

The mandible, LT 329, comes from the lowermost part of the Apak Member, less than 3 m above the Purple Marker. The one date obtained for this member is for a pumiceous bed that lies 17 m below the Lothagam Basalt but 35 m above the Purple Marker. The date of 4.22 ± 0.03 Ma represents the minimum age of the mandible. The real age of the fossil is likely to be nearer

5.0 than 4.22 Ma because the mandible was recovered only just above the Purple Marker. However, estimates based on sedimentation rates are likely to have high errors associated with them (McDougall and Feibel 1999).

The rarity of fossil hominids from the Late Miocene sediments at Lothagam probably indicates that hominoids were not common constituents of the regional biota. Other mammals are well represented, and their remains include well-preserved specimens of carnivores that are not usually found in abundance; this is another reason to think that hominid primates were living in low densities and that their relative paucity is not due to taphonomic filtering.

Four specimens from the Kaiyumung Member of the Nachukui Formation are much younger, possibly close to 3.5 Ma. The underlying Muruongori Member represents a mostly lacustrine phase and is probably laterally equivalent to the Lonyumun Member of the Koobi Fora Formation. The upper boundary of the Lonyumun Member is the base of the Moiti tuff, which has been dated at about 3.9 Ma (WoldeGabriel et al. 1994; Leakey et al. 1995), and so the Kaiyumung Member is younger than this.

Systematic Description

Superfamily Hominoidea Gray, 1825
Family Hominidae Gray, 1825
Hominidae indet.

(Figures 6.15, 6.16d–h, 6.17–6.19; table 6.16)

Lothagam Material

- *Upper Nawata:* 22930, Lt. M^3; 25935, Rt. I_1.
- *Apak Member:* 329, Rt. mandibular fragment with M_1.

Figure 6.15 Restoration of Hominidae indeterminate from Lothagam by Mauricio Antón.

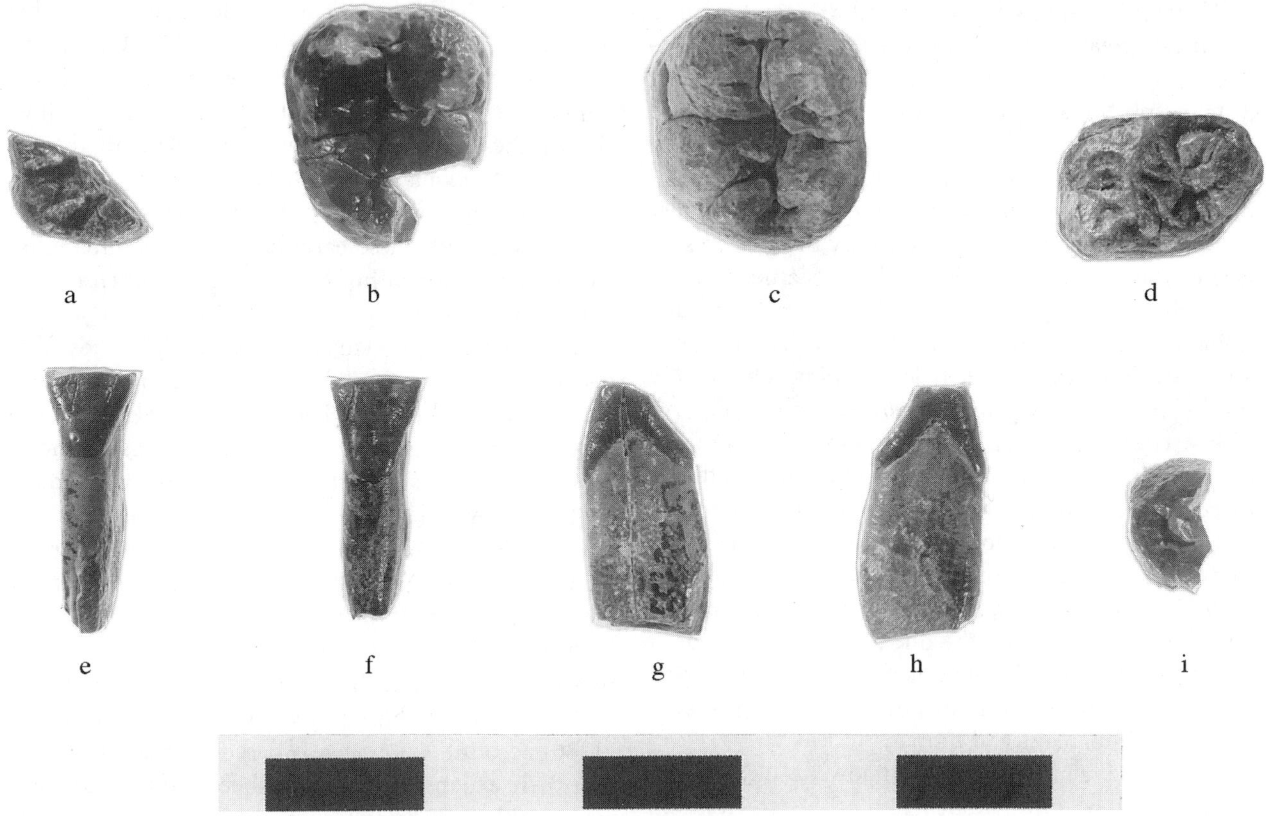

Figure 6.16 Lothagam hominoid teeth. Top row: occlusal views of KNM-LT 23181 (a), KNM-LT 23182 (b), KNM-LT 23183 (c), and KNM-LT 22930 (d), Bottom row: labial (e), lingual (f), mesial (g), and distal (h) views of KNM-LT 25935, and occlusal view of KNM-LT 25936 (i). The scale is in centimeters.

These three specimens are described in order from the oldest to the youngest. Comparisons cannot yet be made with *Ardipithecus ramidus* specimens, but it is anticipated that those will, once fully published, help in understanding the early Lothagam specimens. It is unfortunate that the Upper Nawata teeth are not particularly diagnostic; hominid lower incisors are fairly uniform in shape, and upper third molars are quite variable in most primates.

LT 22930, a left M^3 (figure 6.16d), consists of a weathered crown that is missing the distobuccal enamel and, in some places, some cervical enamel. Not much of the crown is missing, however, as the dentine cap is clearly seen. The crown is compressed mesiodistally (the mesiodistal diameter is greater than 9.7 mm and is estimated at 10.0 mm; the buccolingual diameter is 13.7 mm). The distal cusps were evidently much smaller than the two mesial ones. Places where the distolingual enamel is broken away show that the enamel is relatively thin. It might be that the enamel is thin over the whole crown, but this would have to be determined by sectioning or ultra-high-resolution computed X-ray tomography. The unequal-sized cusps are, from largest to smallest, mesiolingual, mesiobuccal, distolingual, and distobuccal. All of the cusps have relatively fine radial grooves running down their slopes from their tips. These are especially clear on the mesial and lingual slopes of the mesiolingual cusp. The crown as a whole is low (estimated to be ~6 mm from the cervix to the tips of the mesial cusps). Both mesial cusps slope markedly toward the cervix, the mesiolingual especially so. The mesial fovea is ~6 mm wide and reaches nearly all the way across the mesial side of the mesiobuccal cusp. The central basin of the tooth is centered about 5.3 mm from the mesial margin and about 8.3 mm from the lingual margin of the tooth. Because of weathering, it is difficult to say if the tooth was in wear, but it was probably not. Parts of the three roots are preserved for about 6 mm above the cervix. The lingual root is subelliptical in section, and the superior break is just above the level where the buccal and lingual roots separated. The break has exposed a small circular pulp cavity. At the break the root is 6.3 mm mesiodistally by 5.2 mm buccolingually. The two buccal roots are both compressed mesiodistally. The mesial one is 7.5 mm buccolingually by 2.9 mm mesiodistally. The distal one is 5.6 mm buccolingually by 3.3 mm mesiodistally.

Although mesiodistally shortened M^3s can be found in African great apes, they are usually a product of reducing the size of the tooth, not abbreviation of the distal cusps. Molars similar to the Lothagam one can be found in the sample of *Australopithecus anamensis* from Kanapoi, e.g., KP 30498D and E (Leakey et al. 1998). But these teeth, apart from strong mesial wear, are taller and have thicker enamel, as judged by the broken surface of the crown of 30498D and the enamel sections seen where the cervix of 30498E is damaged. The distolingual cusps are also less clearly defined, and the lingual surface of the crown is less sloping. One of the specimens of *Australopithecus afarensis* from Hadar, AL 333x-1 (Johanson et al. 1982), has similar-sized mesial cusps but much larger distal cusps and a crown that is twice as high as that of the Lothagam tooth.

In summary, this tooth is a lower-crowned and thinner-enameled version of some third molars of early *Australopithecus* species. As such it may have come from a very early hominin on the one hand, or, as Ungar et al. (1994) suggested for the Lukeino molar, it may represent the condition in the ancestral morphotype shared by both chimpanzees and humans.

LT 25935 is a heavily worn right lower central incisor (figure 6.16e–h). Only about 7 mm of the crown is left. Its maximum diameters just above the cervix are 7.4 mm labiolingually and 4.9 mm mesiodistally. At the occlusal plane these diameters are 3.7 mm and 5.9 mm, respectively. The occlusal surface is rectangular in outline, with a narrow, raised mesial ribbon of 1 mm thick enamel. The enamel is hardly raised above the dentine level at the distal margin. The labial surface is slightly damaged by weathering but has a hollowing of the center of the surface that shows as an irregular longitudinal groove. The lingual surface is slightly concave from occlusal surface to cervix and shows a small cervical swelling. Weathering has obscured the details of the interstitial facets, but a small facet is just visible on the lingual corner of the distal surface. About 10 mm of the root is preserved. It is strongly compressed mesiodistally, being 4.0 mm by 8.0 mm about halfway down its preserved length.

It is difficult to say much conclusively about this worn tooth, but it appears unlikely that it could have been as mesiodistally broad as those of living African apes. It is much more similar to the incisors of early *Australopithecus*. For instance, it is an almost perfect match in size and in shape for the right I_1 of the type specimen of *A. anamensis* (Leakey et al. 1995), although, because it is less worn than that specimen, it was probably lower crowned. Like the previous specimen, this worn tooth could be from an early hominin, unless the ancestral morphotype of both chimpanzees and humans more closely resembled early *Australopithecus* rather than African great apes.

LT 329, the Lothagam mandible (figure 6.17), has been a subject of analysis by White (1977, 1986), Kramer (1986), and Hill and Ward (1988). There is also an unpublished 1971 manuscript by Patterson and Howells. All these accounts were written well before the discovery of *A. anamensis* and *Ardipithecus ramidus* and so suffer, through no fault of the authors, from lack of comparison with these early hominins. The mandible has been mentioned numerous times in other articles, books and reviews.

This specimen is a small part of the right body from just anterior to the mental foramen to the notch for the facial vessels posteriorly. The ramus is broken off at its root, and the upper part of the body contains the mesial roots and the mesial part of the alveolus for the distal root of M_3. A large crack runs through the length of the body so that the lower half to one-third was separated from the upper part. Another matrix-filled vertical crack was present at the time of discovery, but together with the large horizontal crack, it was opened and the matrix was removed shortly after discovery (Patterson and Howells unpublished manuscript 1971). Kramer's (1986) account was evidently based on a cast made before these expansion cracks had been reduced. The first molar, although heavily worn, is present, along with most of the roots of the other molars.

Details of the morphology of the specimen are provided by White (1977, 1986), Kramer (1986), and Hill and Ward (1988), and a multivariate analysis was attempted by Corruccini and McHenry (1980). A summary of their observations, with our own emphasis and our own comparative observations added, is as follows.

The body is relatively wide, especially in the alveolar region. It is 20.4 mm wide by 31.4 mm high at M_2 (White 1986), with the widest part at the alveolar margin. The extramolar sulcus is narrow and placed high near the alveolar margin. The ramus takes off from an anterior position between M_2 and M_3. The area around the mental foramen is hollowed. Just anterior to M_1, the lateral surface of the alveolar bone swings laterally. White (1986) thought that this might be the configuration in life, but cautioned that it could be the result of plastic deformation. Now that we have more comparative material of early species of *Australopithecus* that show this feature (Leakey et al. 1998), we think that it is probably not distorted. The mental foramen itself is placed below the mesial part of M_1, high on the body, and opening directly anteriorly. Lingual subalveolar hollowing is very pronounced, giving a triangular cross-section under the last molars. Figure 6.18 shows cross-sections of the body of this mandible and various hominids at the level of M_2. Both superior and inferior tori were present, as judged by their lateral traces, and the

Figure 6.17 Lothagam mandible, KNM-LT 329, in lateral (top), medial (middle), and occlusal (bottom) views. The scale is in centimeters.

inferior torus was relatively low. The base of the body is thin, especially posteriorly.

The root morphology has been discussed by Hill and Ward (1988). A "serrate" pattern is seen in which the distal molar roots are inclined progressively more toward the buccal side. The molar roots can be seen in X-rays and are of roughly equal lengths. The distal root of the third molar is angled sharply distally. Figure 6.19 is a tracing from a lateral X-ray of the specimen.

The M_1 crown is complete, but worn. It is nearly square in outline (12.7 mm mesiodistally by 12.9 mm buccolingually in the worn state: 13.2 mm by 12.9 mm when tooth loss by interstitial wear is estimated). There are three deep, contiguous dentine exposures on the two buccal cusps and the distal cusp. The mesiolingual cusp has a tiny dentine exposure, while the mesiodistal cusp has none. As Corrucini and McHenry (1980) and Kramer (1986) noted, the mesiolingual and distolingual cusps are the largest, followed in order by distobuccal, mesiobuccal, and distal cusps. White (1986) suggested, based purely on his observations and comparisons, and not on sectioning or computed X-ray tomography studies, that the buccal enamel in LT 329 was thinner than in similarly worn *A. afarensis* teeth. White (1986) also

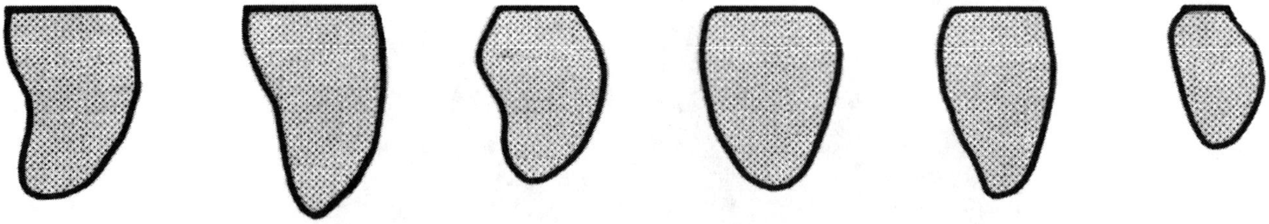

Figure 6.18 Vertical cross-sections through early hominid mandibles at the level of M_2 compared with the Lothagam mandible KNM-LT 329. From left to right: KNM-LT 329, *Australopithecus anamensis* KNM-KP 29281, *A. anamensis* KNM-KP 31713, *Australopithecus afarensis* LH 4, *A. afarensis* AL 198-1, and *Pan paniscus.*

pointed out the mesially positioned hypoconulid, which helps give the tooth its square outline, and the greater width of the talonid relative to the trigonid.

Many, but not all, of the features of this specimen are probably primitive for hominins and many of them, as noted by previous authors, are also found in *A. afarensis* specimens. For the first time we can now compare the Lothagam hominin with mandibles of *A. anamensis* from between 4.2 and 4.1 Ma. There are two specimens from Kanapoi complete enough to make a comparison, the holotype of *A. anamensis,* KP 29281, and the slightly smaller KP 31713. Based on size and particularly relative canine size, these are both presumed to be females. It is unfortunate that the large, presumed male mandible, KP 29287 is not complete enough at the base to enable us to make a complete comparison with a large *A. anamensis* mandible. In practically all respects, the Lothagam mandible finds a better match with these specimens than with any of *A. afarensis.* Even AL 198-1 (White and Johanson 1982), which several authors note is most similar among that hypodigm to KMN-LT 329, is more different than any *A. anamensis* specimen. It should be noted in passing that this specimen had a

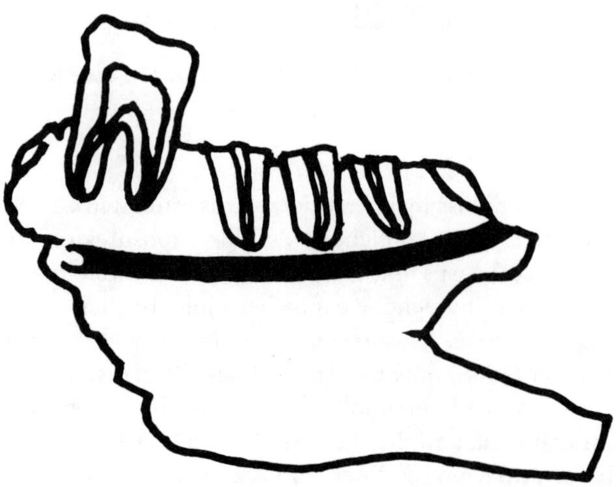

Figure 6.19 Tracing from a lateral X-ray of KNM-LT 329 to show the molar roots. The black line represents the mandibular canal.

fourth molar, which might have contributed to its tall thin body. The two small mandibles from Kanapoi have very strong subalveolar hollowing (figure 6.18). They, like the Lothagam mandible, have marked superior and inferior tori, and the latter is low in all three cases. They have broad molar alveolar regions, and the hypoconulid is positioned mesially on M_1. Their mental foramina open directly forward but are not placed as relatively high on the body as is the Lothagam one. They are positioned about halfway up the body, whereas in *A. afarensis* they tend to be positioned somewhat below the halfway level of the body. The area around the foramen is hollowed where it can be seen (in the most complete specimen), and the lateral alveolar bone at the level of the premolars flares laterally toward the canine jugum, in part because of the lateral position of the premolars relative to the first molar. None of the Kanapoi mandibles is complete enough to say whether the base was as thin posterior to the facial vessel notch as it is in LT 329, but their cross-sectional outlines are more similar than any *A. afarensis* ones, in which the lingual surface is flat, not hollowed, under M_2. In fact, we would find only tooth enamel thickness, and that unquantified, that would tell us that the Lothagam mandible would not fit well in the sample from Kanapoi and Allia Bay. The 4.4 Ma specimens of *Ardipithecus ramidus* from Aramis (White et al. 1994, 1995) have not yet been published in enough detail to say whether they are even closer in morphology to the Lothagam specimen, but in the matter of enamel thickness they would seem to be closer. White et al. (1994) report that, based on broken teeth, enamel thickness is "intermediate between chimpanzees and *A. afarensis/africanus*/early *Homo* conditions." Leakey et al. (1995) recorded from broken specimens that enamel thickness in *A. anamensis* was as thick as that in *A. afarensis,* and so the *Ar. ramidus* material might prove to match Lothagam better than *A. anamensis.* The only published mandible of *Ar. ramidus* (ARA-VP-1/129) is the most anterior part of one of an infant, and so cannot help here. An adult mandible of *Ardipithecus* has been found associated with a partial skeleton (White et al. 1995), but to date nothing has been published about it.

It might be thought, following this discussion, that the Lothagam mandible can be relatively securely placed as an early hominin, with strong affinities to both *Ar. ramidus* and *A. anamensis.* But this requires that mandibles from even earlier populations (which might include the last common ancestor of chimpanzees and humans) have a different morphology from these later populations. We have no fossils of the last common ancestor, so extrapolating the morphology of modern African apes back far into the past is likely to be misleading. After all, these three living species have had as much time to evolve, over 5 Ma, as humans have. For the moment, the Lothagam specimen is attributed to "Hominidea indeterminate."

Subfamily Hominidae Gray, 1825
Australopithecus Dart, 1925
Australopithecus cf. *A. afarensis*
(Figures 6.16a–c, 6.16i)

Lothagam Material

- *Kaiyumung Member:* 23181 fragmentary Rt. dM^2 germ; 23182, Rt. M_3 crown; 23183, Lt. M_2; 25936 P/ fragment.

There are four specimens from the Kaiyumung Member that are probably about 3.5 Ma in age. All are considered to belong in *Australopithecus* cf. *A. afarensis* as they have their closest individual matches with specimens from that species.

LT 23181 is a small fragment of tooth germ (figure 6.16a). It is probably the distal third of a right dM^2 and is reasonably matched in size and shape by the same part in Laetoli Hominid 21 (White 1980) but has fewer small accessory cuspules in the central basin.

LT 23182 is most of a lower molar crown (figure 6.16b). It is likely to be an M_3 rather than an M_2 because of its lower crown height, relatively elongated occlusal outline, and backward-sloping roots. The distobuccal cusp is missing, and a small wedge of the distolingual crown—together with a piece of the mesial part of the broken distal root—is also missing. Only about 8 mm of the mesial platelike root is preserved. The crown, which is slightly worn and more so buccally than lingually, is 15.5 mm (estimated) mesiodistally and 13.2 mm buccolingually. It has suffered some slight chemical weathering and has damage to the cervix. There is a small (5.5 mm by 3.5 mm) interstitial facet on the mesial face. The straight 4.2 mm long mesial fovea is met at right angles by a central groove that runs distally to the posterior fovea. Accessory grooves run into the central basin on the distobuccal part of the mesiolingual cusp, into the lingual part of the distobuccal cusp, and into the posterior fovea. Mild lingual grooves and a stronger buccal groove mark the outer faces of the tooth. The best comparison in the *A. afarensis* hypodigm is with the M_3 of Laetoli Hominid 4 (Leakey et al. 1976).

LT 23183 is a lower left molar, complete with roots (figure 6.16c). It is probably an M_2 because of its relatively rectangular shape, straight roots, and high crown. The whole surface has suffered some chemical weathering and has a frosted appearance. The roots are closed apically, and there should have been some wear, at least on the buccal cusps, but the weathering makes it difficult to be certain of this. The crown is very high and 15.2 mm mesiodistally by 13.6 mm buccolingually. The mesiolingual cusp is the tallest and largest, followed in size by the mesiobuccal, the distobuccal, the distolingual, and the distal cusps. There is a 4.7 mm wide mesial fovea and a 3.5 mm wide distal fovea. Accessory grooves run down from the main cusps into the central basin, but these are somewhat obscured by the surface erosion. There is a weak distal buccal groove and a stronger mesial buccal groove that ends in a protostylid shelf. The mesial root is 16.0 mm long, hollowed down its long axis both mesially and distally and is bifid at the apex. The 16.4 mm long distal root is oval in section and is hollowed only mesially. It has a single apex. There is a good match for this tooth in AL. 145-35 from Hadar (Johanson et al. 1982), although that tooth has more wear, especially buccally.

LT 25936 is a fragment of worn premolar and a small piece of root (figure 6.16i). It is probably a buccal corner of an upper or lower premolar, but little useful can be said about it.

Discussion

Several points can be made about the Lothagam hominid fragments. First, despite their fragmentary nature, the earlier fossils represent populations from around the time of the divergence between ancestral chimpanzees and bonobos and ancestral humans. Also they appear to be primitive relative to later hominin species in practically all respects. They suggest by their relative rarity that hominids were not an abundant part of the Lothagam fauna during the accumulation of the Nawata Formation, and this is interesting in view of the increase in the hominin fossil record in East Africa after about 4.5 Ma. The later hominins from Lothagam, from the Kaiyumung Member, are most like specimens of *Australopithecus afarensis.* This species is relatively common at sites in Tanzania, Kenya, and Ethiopia from sediments of the same age, and the Lothagam sample, though very small, probably adds another locality to the widespread geographic range of that species.

Acknowledgments

We thank the government of Kenya and the governors of the National Museums of Kenya. Thanks are due to the many sponsors of the field and laboratory research, the collection managers and the preparators at the National Museums of Kenya, and the field crew led by K. Kimeu, particularly Sila Dominic and Joseph Mutaba, who found the Upper Nawata isolated lower incisor and M^3, respectively. Thanks also to Kamoya Kimeu, Mwongela Muoka, Samuel Ngui, and Kathy Stewart, who discovered the Kaiyumung isolated teeth. W. W. Howells, G. Suwa, and T. D. White provided valuable information, and B. Brown and S. Ward helped with figures 6.18 and 6.19. C. S. Feibel helped us understand the complications of the age determinations.

References Cited

Corruccini, R., and H. M. McHenry. 1980. Cladometric analysis of Pliocene hominoids. *Journal of Human Evolution* 9:209–221.

Eckhardt, R. B. 1977. Hominid origins: The Lothagam problem. *Current Anthropology* 18:356.

Haile-Selassie, Y. 2001. Late Miocene hominids from the Middle Awash, Ethiopia. *Nature* 412:178–181.

Hill, A. 1999. The Baringo Basin, Kenya: From Bill Bishop to BPRP. In P. Andrews and P. Banham, eds., *Late Cenozoic Environments and Hominid Evolution: A Tribute to Bill Bishop*, pp. 85–97. London: Geological Society.

Hill, A., and S. Ward. 1988. Origin of the Hominidae: The record of African large hominoid evolution between 14 My and 4 My. *Yearbook of Physical Anthropology* 31:49–83.

Howell, F. C. 1978. Hominidae. In V. J. Maglio and H. B. S. Cooke, eds., *Evolution of African Mammals*, pp. 154–248. Cambridge, Mass.: Harvard University Press.

Johanson, D. C., and B. Edgar. 1996. *From Lucy to Language.* New York: Simon and Shuster.

Johanson, D. C., T. D. White, and Y. Coppens. 1982. Dental remains from the Hadar Formation, Ethiopia: 1974–1977 collections. *American Journal of Physical Anthropology* 57: 545–603.

Kramer, A. 1986. Hominid-pongid distinctiveness in the Miocene-Pliocene fossil record: The Lothagam mandible. *American Journal of Physical Anthropology* 70:457–473.

Kumar, S., and S. B. Hedges. 1998. A molecular timescale for vertebrate evolution. *Nature* 392:917–920.

Leakey, R. E. F. 1973. Australopithecines and hominines: A summary of the evidence from the Early Pleistocene of eastern Africa. *Symposium of the Zoological Society of London* 33:53–69.

Leakey, M., and A. Walker. 1997. Early hominid fossils from Africa. *Scientific American* 276:60–65.

Leakey, M. D., R. L. Hay, G. H. Curtis, R. E. Drake, M. K. Jackes, and T. D. White. 1976. Fossil hominids from the Laetolil Beds at Laetoli. *Nature* 262:460–466.

Leakey, M. G., C. S. Feibel, I. McDougall, and A. Walker. 1995. New four-million-year-old hominid species from Kanapoi and Allia Bay, Kenya. *Nature* 376:565–571.

Leakey, M. G., C. S. Feibel, R. L. Bernor, J. M. Harris, T. E. Cerling, K. M. Stewart, G. W. Storrs, A. Walker, L. Werdelin, and A. J. Winkler. 1996. Lothagam: A record of faunal change in the Late Miocene of East Africa. *Journal of Vertebrate Paleontology* 16:556–570.

Leakey, M. G., C. S. Feibel, I. McDougall, C. Ward, and A. Walker. 1998. New specimens and confirmation of an early age for *Australopithecus anamensis. Nature* 393:62–66.

McDougall, I., and C. S. Feibel. 1999. Numerical age control for the Miocene-Pliocene succession at Lothagam, a hominoid-bearing sequence in the northern Kenya Rift. *Journal of the Geological Society* (London) 156:731–745.

Patterson, B., A. K. Behrensmeyer, and W. D. Sill. 1970. Geology and fauna of a new Pliocene locality in northwestern Kenya. *Nature* 226:918–921.

Pilbeam, D. 1972. *The Ascent of Man.* New York: Macmillan.

Senut, B., M. Pickford, D. Gommery, P. Mein, C. Cheboi, and Y. Coppens. 2001. First hominid from the Miocene (Lukeino Formation, Kenya). *Comptes Rendus de l'Académie des Sciences* (Paris) 332:137–144.

Simons, E. L. 1978. Diversity among the hominids: Vertebrate paleontologist's viewpoint. In C. J. Jolly, ed., *Early Hominids of Africa*, pp. 543–566. London: Duckworth.

Szalay, F. S., and E. Delson. 1979. *Evolutionary History of the Primates.* New York: Academic Press.

Tobias, P. V. 1978. The South African australopithecines in time and hominid phylogeny, with special reference to the dating and affinities of the Taung skull. In C. J. Jolly, ed., *Early Hominids of Africa*, pp. 45–84. London: Duckworth.

Ungar, P. S., A. Walker, and K. Coffing. 1994. Reanalysis of the Lukeino molar. *American Journal of Physical Anthropology* 94:165–1173.

White, T. D. 1977. The anterior mandibular corpus of early African Hominidae: Functional significance of shape and size. Ph.D. diss., University of Michigan.

White, T. D. 1980. Additional fossil hominids from Laetoli, Tanzania: 1976–1979 specimens. *American Journal of Physical Anthropology* 46:197–230.

White, T. D. 1986. *Australopithecus afarensis* and the Lothagam mandible. *Anthropos* (Brno) 23:79–90.

White, T. D., and D. C. Johanson. 1982. Pliocene hominid mandibles from the Hadar Formation, Ethiopia: 1974–1977 collection. *American Journal of Physical Anthropology* 57: 501–544.

White, T. D., G. Suwa, and B. Asfaw. 1994. *Australopithecus ramidus*, a new species of early hominid from Aramis, Ethiopia. *Nature* 371:307–312.

White, T. D., G. Suwa, and B. Asfaw. 1995. Corrigendum: *Australopithecus ramidus*, a new species of early hominid from Aramis, Ethiopia. *Nature* 375:88.

Wolpoff, M. H. 1982. *Ramapithecus* and hominid origins. *Current Anthropology* 23:501–522.

Wolpoff, M. H. 1999. *Paleoanthropology.* Boston: McGraw-Hill.

WoldeGabriel, G., T. D. White, G. Suwa, P. Renne, J. de Heinzelin, W. K. Hart, and G. Heiken. 1994. Ecological and temporal placement of Early Pliocene hominids at Aramis, Ethiopia. *Nature* 371:330–333.

TABLE 6.16 Sequential Identifications of Lothagam Mandible KNM-LT 329

Author(s)	**Year**	**Attribution**
Patterson et al.	1970	*Australopithecus* sp. cf. *A. africanus*
Pilbeam	1972	*A. africanus*
Leakey	1973	Possible late-surviving *Ramapithecus*
Eckhardt	1977	Hominoid resembling modern African pongids
White	1977	Hominoid
Howell	1978	Possibly *Australopithecus*
Simons	1978	Hominid
Tobias	1978	Species close to, or identical with, *A. africanus*
Szalay and Delson	1979	Homininae *incertae sedis*
Wolpoff	1982	Either ramapithecine or early hominid
Corruccini and McHenry	1980	Hominidae indet.
White	1986	Hominoidea indet.
Kramer	1986	*Australopithecus* sp. cf. *A. afarensis*
Hill and Ward	1988	*Australopithecus afarensis*
Leakey et al.	1996	Hominoid
Johanson and Edgar	1996	Consistent with Hominidae but too fragmentary to assign to extinct pongid or hominid
Wolpoff	1999	Hominid

7

CARNIVORA

Mio-Pliocene Carnivora from Lothagam, Kenya

Lars Werdelin

Lothagam is a key site for understanding the evolution of the African Plio-Pleistocene carnivore fauna. The extensive Carnivora collection includes at least 21 taxa, with 15 represented in the Lower Nawata, 9 in the Upper Nawata, 4 in the Apak Member, and 3 in the Kaiyumung Member. Represented families comprise Amphicyonidae (two species), Mustelidae (four species), Viverridae (five species), Hyaenidae (four species), Felidae (five species), and Canidae (one species). The mustelid material includes a new giant-sized form with hypercarnivorous adaptations, also a possible ancestor of the living honey badger, *Mellivora capensis,* and a new species of the enhydrine genus *Vishnuonyx.* A new cursorial species of the hyaenid *Ictitherium* is represented by a complete skeleton. Another partial skeleton represents a new genus and species of saber-toothed felid that is related to *Homotherium.* The Lothagam biota includes the youngest record of Amphicyonidae and the first appearances of modern Mellivorinae, *Viverra, Genetta,* the *Hyaena* lineage, and *Dinofelis.* There are indications of several biogeographic dispersals from Eurasia but only tenuous connections to the Middle Miocene carnivorans of Africa. The carnivoran fauna of Lothagam shows similarities with both the Late Miocene assemblage from Sahabi, Libya, and the earliest Pliocene fauna from Langebaanweg, South Africa.

In contrast to the extensive investigations of Late Miocene carnivores of Eurasia (summarized in Werdelin 1996b; Werdelin and Solounias 1996), very little has to date been written about Late Miocene carnivores of Africa. This is partly because of a lack of appropriate sites and material, but mainly because carnivores from this time period have been studied relatively little compared to other taxa that have had a greater perceived significance to efforts at dating and paleoecological analysis. Thus, despite several publications on other faunal elements at Lothagam (see references in Leakey et al. 1996), the only record of carnivores collected during the 1960s expeditions comprised part of the faunal list of Lothagam-1 by Smart (1976):

??*Civettictis*
aff. *Euryboas*
Subfamily Felinae (large, primitive form)
Subfamily Machairodontinae

In comparison, the material and faunal list presented in this chapter reflects the substantial additions to this fauna that have been contributed by the National Museums of Kenya expeditions.

Other Late Miocene carnivores from Africa are almost equally poorly known, except for the fauna from Sahabi described by Howell (1987), which includes material belonging to Ursidae, Viverridae, Hyaenidae, and Felidae. Sahabi is broadly correlative with MN 13 (latest Miocene; ca. 6–4.8 Ma) in Eurasia, meaning that it correlates approximately with the upper member of the Nawata Formation of Lothagam (dated 6.54–~5.5 Ma) and that the two sites can be expected to have carnivore taxa in common. The same is true of the slightly younger (earliest Pliocene) Langebaanweg fauna from South Africa (Hendey 1974). Other described carnivores from the Late Miocene of Africa include the few specimens from the somewhat earlier (ca. 8–10 Ma) Namurungule Formation, Samburu Hills (Nakaya et al. 1984); the single record of *Hyperhyaena leakeyi* from Nakali, Kenya, also older than Lothagam (ca. 10–11 Ma; Aguirre and Leakey 1974); and material from two North African Miocene sites—Beni Mellal in Morocco (Ginsburg 1977) and Bled Douarah in Tunisia (Kurtén 1976).

Figure 7.1 Restoration of *Ekorus ekakeran* sp. nov. by Mauricio Antón. Shoulder height = 60 cm.

Other carnivore occurrences in the Late Miocene of Africa are recorded only in the form of citations in faunal lists (e.g., Hill 1995; Nakaya et al. 1984; Pickford 1978; and others).

In this contribution I provide identifications and descriptions of the majority of carnivore specimens from Lothagam, including those from the 1960s collections. Because the material includes two complete skeletons and one partial one, in the interests of brevity most descriptions are necessarily preliminary. In some cases, fairly extensive discussions of affinities and of other African records are provided; in others, this aspect is skimmed over as it would require too much analysis to fit within the framework of this descriptive paper. The Late Miocene–Pleistocene African carnivore record is currently being revised on a taxon by taxon basis (generally at the generic level) in collaboration with Dr. Margaret E. Lewis. A paper on *Dinofelis* has recently been published (Werdelin and Lewis 2001), and others will follow. More extensive discussion of the affinities and evolutionary significance of the Lothagam carnivores will be presented in those papers.

Systematic Description

Family Amphicyonidae

The Amphicyonidae is an extinct family of arctoid carnivores with an extensive Northern Hemisphere fossil record from the Eocene to the Late Miocene. This record has recently been revised and reviewed by Hunt (e.g., 1996a) and Viranta (1996). In contrast, the African record of Amphicyonidae is sparse (reviewed in Savage 1978). Recently, material of very large amphicyonids has been described from Arrisdrift, Namibia (Morales et al. 1998). As will be made clear later in this chapter, the Lothagam records are as young or younger than any Northern Hemisphere records of amphicyonids, and hence these records are of great significance to studies of this family, one of the few carnivoran families to become extinct in the Neogene.

Amphicyonidae species A

(Figure 7.2A–F; table 7.1)

Lothagam Material

- *Lower Nawata:* 23049, Lt. M^2; 23073, distal parts Lt. Mc V and ?Mc IV; 23051, damaged proximal articulation of Lt. radius.

Specimen 23049 (figure 7.2A–B) is a large, square tooth. Buccally, there is a large, low paracone and an equally large and still lower metacone. Buccal to these two cusps there is a bulging cingulum shelf. Both buccal cusps are worn, the metacone more so than the paracone. The protocone is also large and worn down to the level of the lingual cingulum, which is broad and runs from a substantial, anteromedially positioned paraconule, around the lingual margin of the tooth, to end at a large, posteromedially positioned hypocone. The enamel of the lingual cingulum is worn down considerably anterolingual to the protocone and has a tendency to beading on the posterolingual margin. There is a small gap between the buccal and lingual cingula, both anteriorly and posteriorly. The paraconule is joined to the buccal cingulum just anterolingual to the paracone by a low but distinct paracrista that continues lingually past the paraconule to the protocone. Aside from the aforementioned wear, the tooth crown is well preserved. There are three roots, one lingual and two buccal, all of which are broken.

The metapodial fragments are most remarkable for their width, while at the same time they are not dorsoventrally compressed. The Mc V (figure 7.2C–D) in particular has a very wide lateral shelf and a strongly asymmetric articular surface for the first phalanx. The radius fragment (figure 7.2E–F) has an oval articular fovea with a deep articular circumference and a slight lip overlying the shaft.

Discussion

The square outline, prominent lingual cusps, and posterolingual enamel beading of the upper molar is diagnostic of the Amphicyonidae. However, the amphicyonid M^2 is not very diagnostic at lower taxonomic levels and the specimen cannot be attributed to any particular amphicyonid taxon. The Lothagam M^2 clearly represents a large species, as it is similar in size to the M^2 of *Amphicyon major* of Europe, which has an estimated body mass of 100–300 kg (Viranta 1996). This specimen is quite different from the upper molars of *Ysengrinia,* as reported from Arrisdrift by Morales et al. (1998:figure 4). However, Morales et al. also report the presence of *Amphicyon giganteus* at Arrisdrift, and the molar reported here is consistent with assignment to that taxon. The postcranial fragments are also large. The width of the distal parts of the metapodials and the oval shape of the articular fovea of the radius differentiate them from Ursidae and ally them with the Amphicyonidae. The Mc V is about the same size as an Mc V of *Amphicyon giganteus* figured by Ginsburg and Antunes (1968). Thus, all these specimens belong to an amphicyonid of very large size and, although they cannot with absolute certainty be referred to the same taxon, the probability that there would be two such taxa at Lothagam seems remote. The estimated age of the lower member of the Nawata Formation (McDougall and Feibel 1999) makes this one of the youngest records of the Amphicyonidae known thus far.

Amphicyonidae species B

(Figure 7.3)

Lothagam Material

- *Upper Nawata:* 23944, partial Rt. horizontal ramus (roots P_4, broken M_1, alveolus for M_2 and anterior part of the M_3 alveolus).

The long and relatively slender ramus is deepest and thickest beneath the M_1 talonid. The ramus is broken anteriorly at the anterior end of the anterior root of P_4 and posteriorly just at the point where the horizontal ramus begins to ascend to the coronoid process. The latter break continues posteroventrally, and the ventral half of the ramus is about 30 mm longer than the dorsal half. Posterior to M_2 the ramus becomes noticeably

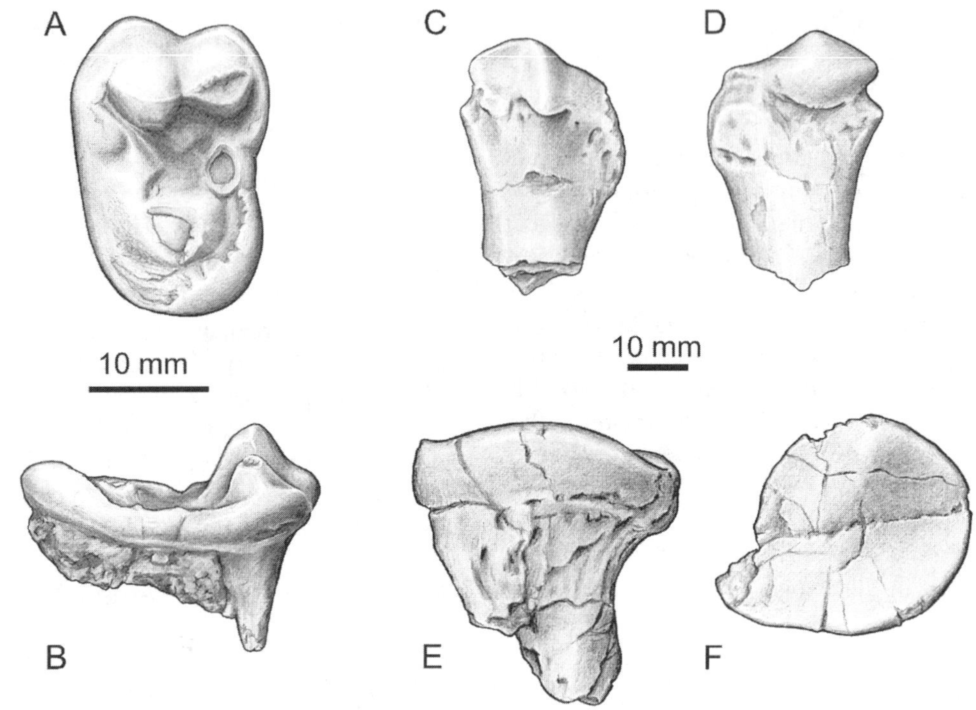

Figure 7.2 Specimens of Amphicyonidae sp. A: A = KNM-LT 23049, Lt. M^2, occlusal view; B = KNM-LT 23049, posterior view; C = KNM-LT 23073, distal left Mc V, ventral view; D = KNM-LT 23073, dorsal view; E = KNM-LT 23051, damaged proximal part of left radius, caudal view; F = KNM-LT 23051, proximal view.

thinner. The anterior end of the masseteric fossa can be felt rather than seen and lies posterior to the tooth row. The lower fourth premolar was a relatively short and probably quite slender tooth. The lower carnassial is short, with the talonid making up about one-third of the total length of the tooth. The trigonid is relatively short for the size of the animal. Neither trigonid cusp can be clearly distinguished because of a strong, nearly horizontal wear facet that runs from the anterior end of the trigonid to the posterior end of the talonid. The talonid has a single, well-developed and probably trenchant cusp. The metaconid is well developed and confluent with the posterolingual cingulum through a cristid. There is no buccal cingulum, only a slight marginal swelling at the talonid. The M_2 was, to judge from the alveoli, a broad tooth of about 14 mm length. The M_3 alveolus is much smaller than either alveolus of M_2.

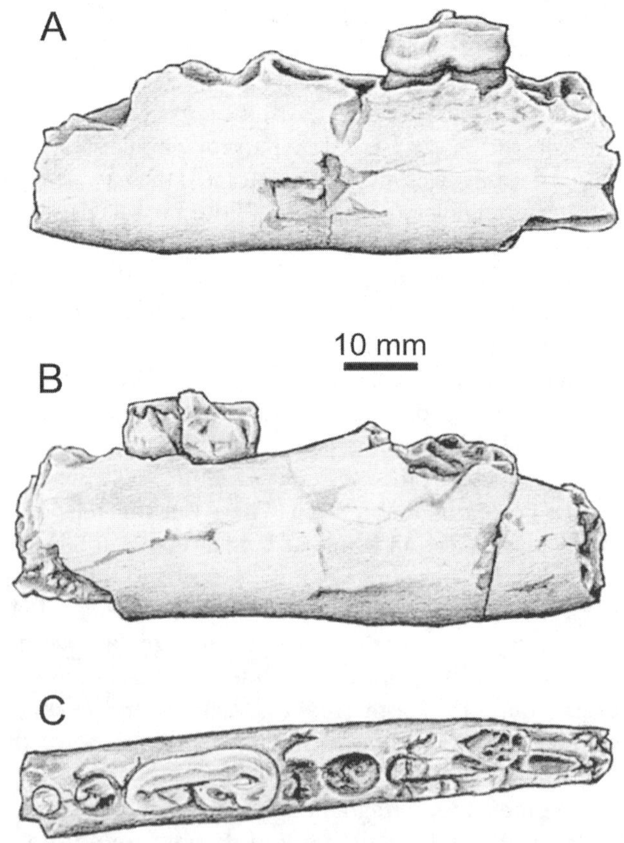

Figure 7.3 Partial right mandibular ramus of Amphicyonidae sp. B, KNM-LT 23944: A = buccal view; B = lingual view; C = occlusal view.

Discussion

The presence of M_3 limits the possible taxonomic attribution of this specimen to two families: Amphicyonidae and Canidae. Of these, the narrow, single cusped talonid of M_1 and the relatively large and long M_2 strongly argue for assignment to the Amphicyonidae. This referral is corroborated to some extent by the fact that the oldest record of canids in Africa is from the supposedly slightly younger Langebaanweg site (Hendey

1974), although the difference in age is slight enough that in itself it would not be sufficient for placing this specimen in the Amphicyonidae. The specimen has no known counterpart elsewhere and its exact phylogenetic affinities cannot be determined at present. It belonged to a considerably smaller taxon than the amphicyonid specimens described above and almost certainly represents a new genus and species. Unfortunately, the material available is not sufficient for unconditional taxonomic attribution, but, if the assignment is correct, this specimen represents the youngest record of the Amphicyonidae in the Old World and probably the youngest anywhere (Hunt 1996a, 1996c).

Family Mustelidae

The Mustelidae has an extensive but poorly understood fossil record, and nearly all subfamilies and genera are in need of revision (Werdelin 1996a, 1996b). Two areas of mustelid evolution are currently of particular interest. One concerns the evolution of aquatic mustelids of the subfamily Lutrinae (Willemsen 1992). The origin of this radiation in the Miocene is still obscure. Another aspect of mustelid evolution that has been neglected is the parallel evolution of mustelids of gigantic size in several lineages (Werdelin 1996a). Mustelids of gigantic size (more than twice the mass of the largest living forms, which reach ca. 30–35 kg) are not uncommon in the Miocene, with examples known from Eurasia, North America, and Africa (see discussion below). Why this distinction between modern and fossil mustelids should exist is unknown, as are the ecological parameters that circumscribe these taxa. The Lothagam fauna presents significant new evidence to address these and other issues in mustelid evolution.

Subfamily indet.
Genus *Ekorus* gen. nov.

Diagnosis

Mustelidae of gigantic size. Dentition highly modified for slicing, with narrow canines and slender premolars. Lower carnassial lacking metaconid, talonid reduced to a single, tall cusp placed directly posterior to the trigonid blade. Lower second molar small and peg-shaped. Upper first molar very reduced and much broader than long. Upper second molar lost. Appendicular skeleton modified, with relatively long limbs but short, broad, semi-plantigrade feet. Humerus lacking entepicondylar foramen. Vertebral column slender. Tail long.

Etymology

After *ekor,* the Turkana word for "badger," in reference to the possible relationship between this taxon and the honey badger *Mellivora.*

Hypodigm

Type species only.

Ekorus ekakeran sp. nov.
(Figures 7.1, 7.4–7.9; tables 7.2, 7.3)

Diagnosis

As for genus.

Holotype

KNM-LT 23125, a nearly complete skeleton with cranium, mandible, and nearly all of the postcranium from the lower member of the Nawata Formation.

Etymology

After *ekakeran,* the Turkana word for "runner," in reference to the relatively cursorial (for a mustelid) features of the holotype skeleton.

Lothagam Material

- *Lower Nawata:* holotype; 23951, Rt. M_1; 23956, two mandible fragments, including a ?P_1.

The following description is based entirely on the holotype skeleton as the other two specimens are very small and contribute no additional information about the taxon.

The cranium (figure 7.4) has been laterally flattened and crushed so that very few characteristic features can be distinguished. The crushing has also resulted in some shearing, so the right side is positioned somewhat dorsal to the left side. As is the case in all mustelids, the splanchnocranium is markedly shortened, while the neurocranium is long and low. The incisive foramina are large and long. The zygomatic arches (preserved separately; figure 7.4C) are robust, especially in their posterior part. The occipital crest originates on the anterior part of the frontals, near the zygomatic process, and continues to the posterior end of the skull. It is low throughout, nowhere being more than 10 mm in height. The nuchal crests appear to have been prominent and to form a narrow "U" at the midline, where they meet the occipital crest. The occipital condyles are large. The paroccipital processes are long and narrow. The mandibular fossa and retroarticular process would not have locked the mandible in place, unlike in many other mustelids. The external auditory meatus is large and

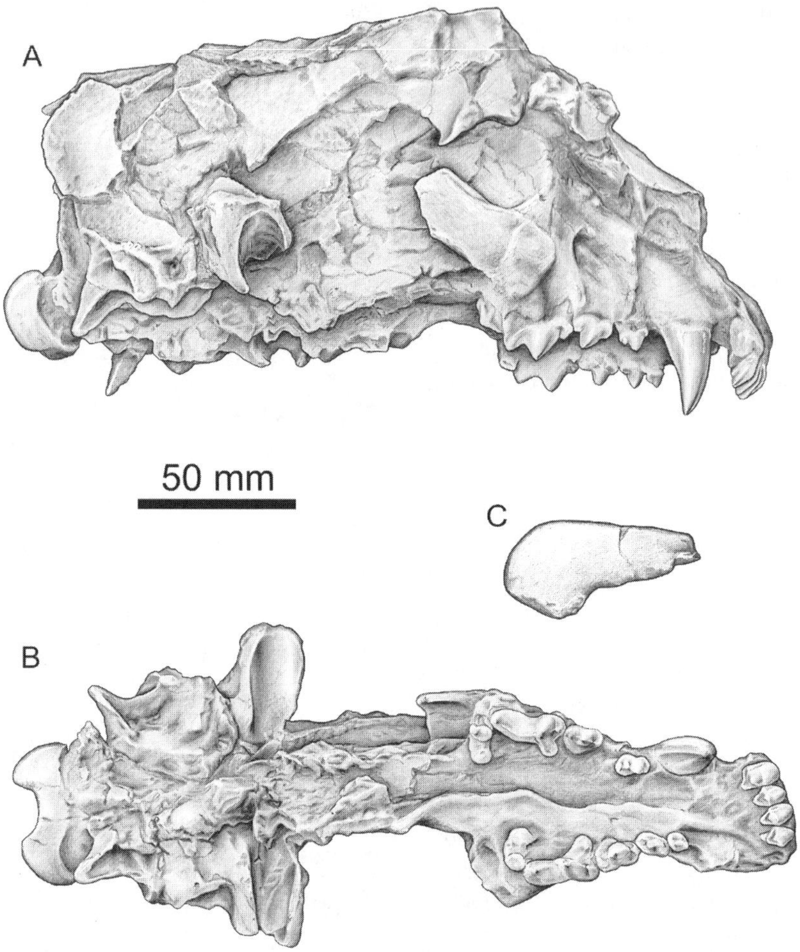

Figure 7.4 *Ekorus ekakeran* gen. and sp. nov. holotype cranium, KNM-LT 23125: A = right lateral view; B = ventral view; C = right zygomatic arch, right lateral view.

was in all likelihood surrounded by a bony tube, which is partly preserved on the right side.

The mandible (figure 7.5) is complete but is laterally flattened and the right ramus is somewhat distorted while the left coronoid process is incomplete, lacking the apex. A part of the right ramus anterior to the canine has also been broken off. The horizontal rami are deep and robust; their ventral margin is nearly straight, unlike in many carnivores, in which the ventral margin is more or less convex. There is a single large mental foramen located in the middle of the ramus beneath the juncture of P_2 and P_3. The rami did not become separated at the symphysis during postdepositional compression, and this indicates that the symphysis probably was tightly fused. The masseteric fossa is very large but relatively shallow. Its anterior margin is located just posterior to the root of M_2. The coronoid process is low and anteroposteriorly long. The condyles are dorsoventrally large.

The dentition (figures 7.6 and 7.7) is preserved in its entirety except for the left I_2, though some teeth have fallen out of their alveoli and are preserved separately. The tooth rows are straight, with no or only minor imbrication. There is some overlap between the left P_4 and M_1, as the anterior end of M_1 is set slightly lingual to the posterior end of P_4. The upper incisors are set in a straight line. The same is presumably true of the lower incisors, but this cannot be verified on the specimen, as the anterior end of the mandible is damaged. The first upper incisor has a broad crown with three distinct cusps. The central cusp is the largest, while the medial and lateral accessory cusps are about equal in size. The central cusp has a wear facet on its lingual face. The base of I^1 is long and narrow and has only a slight trace of a lingual cingulum shelf.

The second upper incisor is larger than the first incisor. It has the same general cusp arrangement, though in this tooth the lateral cusp is larger than the medial one. The wear facet on the central cusp faces buccolaterally. The base of I^2 is long and narrow. In this case, there is clear development of a lingual cingulum shelf.

The third upper incisor (figure 7.6C) is a large, trenchant tooth. The main cusp is caniniform. There is no medial or lateral accessory cusp, but a basal cingulum runs along the entire medial side of the tooth. There is a buccolateral wear facet that runs from the tip of the cusp to the buccolateral side, where there is a small swelling at the base of the enamel.

The upper canine is short and slightly more transversely compressed than in modern pantherine felids; it is also somewhat recurved. There is a salient ridge on its posterior margin, and this ridge is bounded posterolaterally by a shallow groove, while a similar groove is present anteromedially. On the anterior face there is a pronounced wear facet that runs slightly obliquely from medial to lateral, beginning about one-third of the distance from base to tip.

The first upper premolar is preserved in place on the left side, while the right side is damaged in this area and its P^1 is preserved separately. The tooth is square in apical view. It is a small tooth whose three small cusps are all located on the buccal side. The main cusp is set just in front of the anteroposterior midline of the tooth. The anterior accessory cusp is very small and closely appressed to the main cusp. The posterior accessory cusp is larger and is separated from the main cusp by a shallow valley. The lingual side of P^1 is formed into a broad cingulum shelf.

The second upper premolar is longer and more slender than the P^1 and P^3. This premolar has a large, trenchant main cusp that is conical in lingual view, with slightly convex anterior and posterior margins. The anterior accessory cusp is very small and appressed to the main cusp. The posterior accessory cusp is also small, though it is slightly larger than the anterior, and is separated from the main cusp by a shallow valley. Directly behind the posterior accessory cusp there is a very small cingulum cusp. The posterolingual part of the tooth bulges out into a small shelf.

The third upper premolar is three-rooted and nearly triangular in occlusal view. The main cusp is high and trenchant but relatively short. The anterior and posterior cusps are about equal in size, less than one third of the height of the main cusp, and both are separated from it by shallow valleys. There is a broad lingual shelf formed above the internal root of P^3.

The upper carnassial is relatively long and slender for a mustelid carnassial. There is a large parastyle with a fully developed cusp. The protocone is low, and its anterior margin lies just behind the anterior margin of the parastyle. The protocone has two distinct cusplets, one set anteriorly and one set lingually, of which the

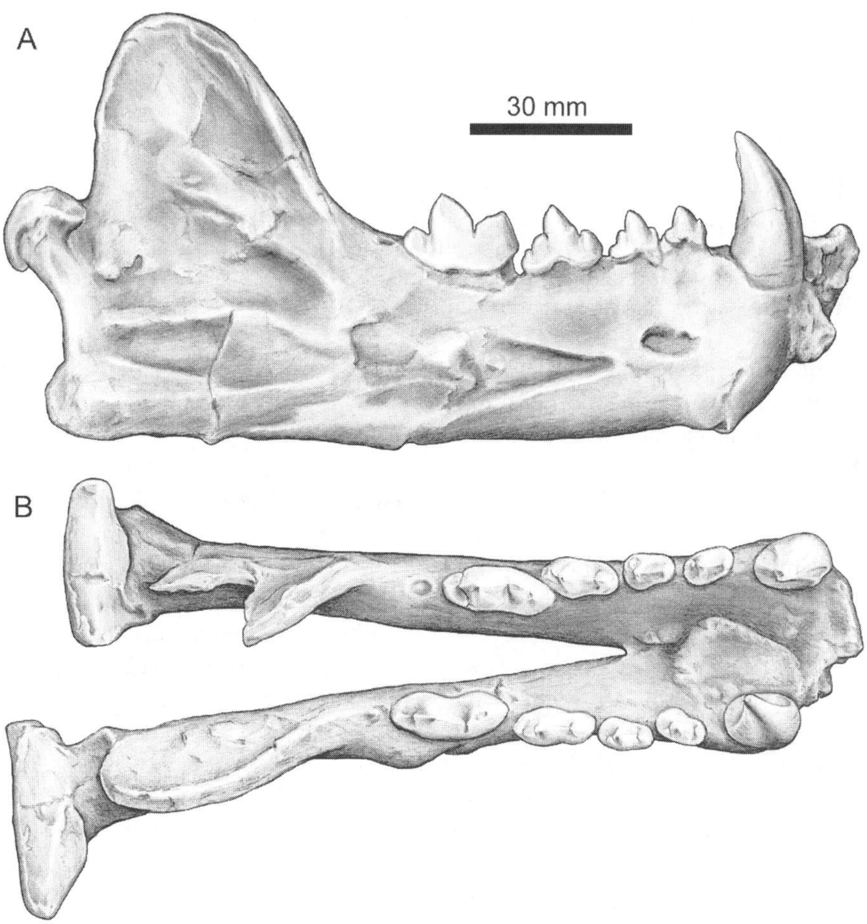

Figure 7.5 *Ekorus ekakeran* gen. and sp. nov. holotype mandible, KNM-LT 23125: A = right lateral view; B = occlusal view.

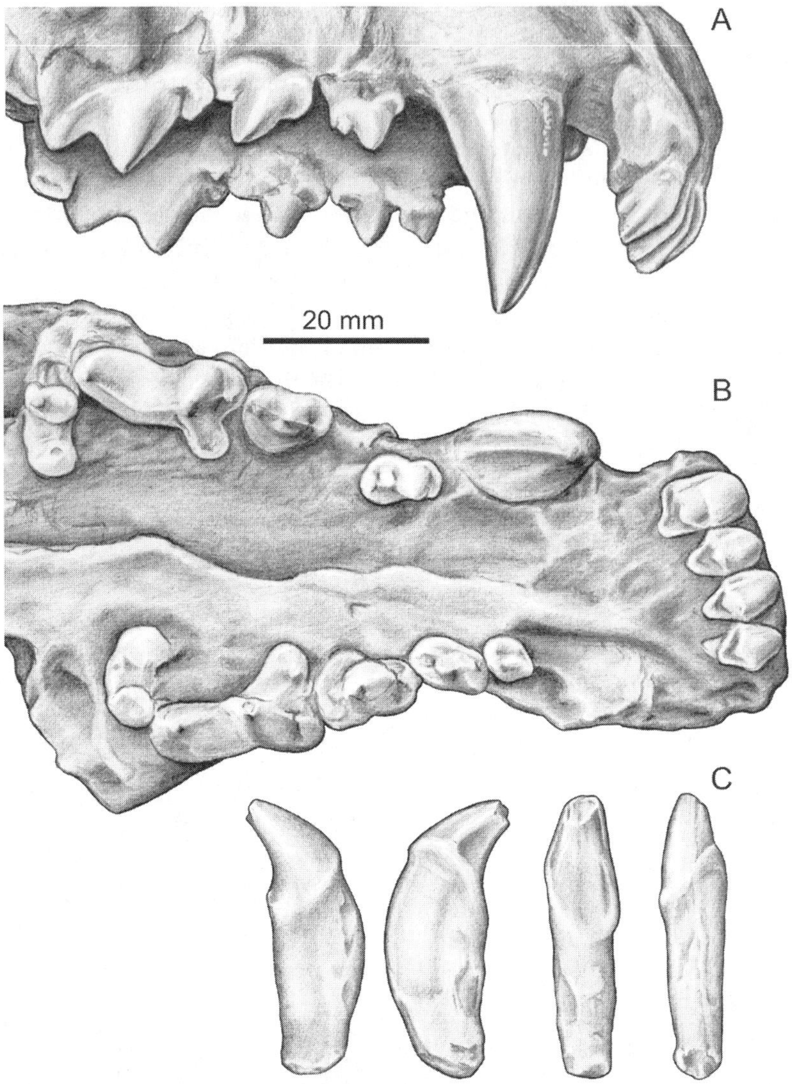

Figure 7.6 *Ekorus ekakeran* gen. and sp. nov. holotype upper dentition, KNM-LT 23125: A = right lateral view; B = occlusal view; C = left I^3 in (left to right) lateral, medial, posterior, and anterior view.

former is much the larger. The paracone and metastyle are formed into a cutting blade. As in most other mustelids, this specimen has no carnassial notch. The paracone is much higher than the metastyle. Unlike those in felids and hyaenids, the metastyle blade is not horizontal but forms a clear, posterior metacone. The tooth as a whole is slightly arched, with the parastyle and metacone ends set somewhat buccal to the paracone.

The first upper molar is rectangular in occlusal view, as it is anteroposteriorly equally long throughout its width. It has two buccal cusps, the paracone and metacone, which are low but distinctly developed. The protocone is large and bulbous and equal in size to the paracone and metacone together. The tooth is strongly dorsoventrally arched.

The first lower incisor (figure 7.7E) is a small, narrow, single-cusped tooth. It shows a wear facet for the upper incisor on its anterior face. The apex is angled slightly ventrally from medial to lateral. There is no buccal cusp, only a very small buccal shelf. The second lower incisor (figure 7.7F) is only slightly larger than I_1 and narrow. The apex shows incipient separation into medial and lateral cusps. Both of these cusplets show horizontal wear facets. There is no visible buccal cusp or shelf. The third lower incisor (figure 7.7G) is somewhat larger than I_1 and I_2. It has a broad crown with two cusps, a large medial cusp, and a smaller and lower lateral one. There is a wear facet for I^3 on the anteromedial side of the tooth. The base is anteroposteriorly long and transversely narrow, with only a slight hint of a buccal shelf.

The lower canine is short, stout, and markedly recurved. It has a shallow groove anteromedially and a low salient ridge posteriorly. A marked wear facet runs obliquely from lateral to medial along the posterior face of the tooth, starting at the base and extending almost to the tip of the crown.

The first lower premolar (figure 7.7D) is a small, narrow tooth with three distinct cusps of which the central, main cusp is the tallest. The anterior accessory cusp is set well away from the main cusp and is separated from it by a shallow valley, while the posterior accessory cusp is closely appressed to the main cusp.

The second lower premolar has a tall main cusp that is nearly round in apical view. There is a marked anterior accessory cusp that is closely appressed to the main cusp. A posterior accessory cusp is also present. It is slightly larger but lower than the anterior accessory cusp. There is a very small cusp on the posterior cingulum directly behind the posterior accessory cusp. The cingulum is expanded to a broad posterolingual shelf.

The third lower premolar is tall and stout. It has a tall main cusp that is only slightly longer than it is wide. The anterior and posterior accessory cusps are subequal in size (the anterior is very slightly the larger) and are appressed to the main cusp. There is a minute posterior cingulum cusp, while posterolingually the cingulum is expanded into a broad shelf.

The fourth lower premolar is tall, long, and slender. It has a tall and narrow main cusp. All the cusps are set in a single anteroposterior file. The anterior accessory cusp is large and separated from the main cusp by a shallow valley. The posterior accessory cusp is slightly smaller than the anterior. The valley separating it from the main cusp is smaller than that between the anterior accessory cusp and the main cusp. There is a relatively large cingulum cusp situated directly behind the posterior accessory cusp. Unlike P_2 and P_3, P_4 does not have a posterolingual cingulum shelf.

The lower carnassial is long and slender. The paraconid is somewhat lower than the protoconid but is broader and longer. The trigonid as a whole forms a trenchant blade and has a prominent carnassial notch. The talonid consists of a single, tall cusp placed directly behind and in line with the trigonid cusps. The talonid is separated from the protoconid by a shallow valley. There is no metaconid.

The second lower molar (figure 7.7C) is nearly round in apical view, being only slightly longer than it is broad.

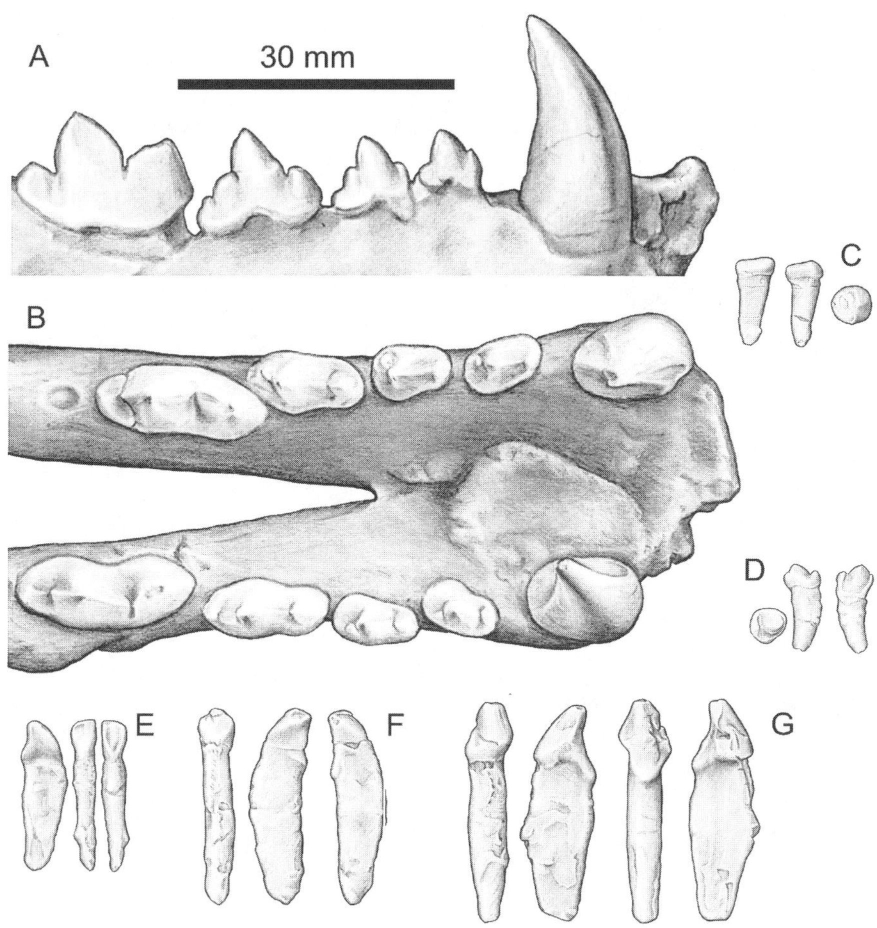

Figure 7.7 *Ekorus ekakeran* gen. and sp. nov. holotype lower dentition, KNM-LT 23125: A = right lateral view; B = occlusal view; C = ?right M_2 in (left to right) lingual, buccal, and occlusal views; D = right P_1 in (left to right) occlusal, buccal, and lingual views; E = right I_1 in (left to right) medial, anterior, and posterior views; F = left I_2 in (left to right) anterior, lateral, and medial views; G = left I_3 in (left to right) anterior, lateral, posterior, and medial views.

It lacks distinct cusps but has a low ridge running in the midline along the entire length of the crown. This is presumably a vestige of one or both of the trigonid cusps.

Both scapulae are preserved. The right scapula is complete, except for the middle part of the anterior edge of the supraspinous fossa, and is criss-crossed by numerous cracks with some resultant distortion. The left scapula is broken about halfway along the spine and is missing most of the supraspinous and infraspinous fossae. The glenoid cavity is oval and smoothly concave. There is no medial constriction delineating the part that leads to the coracoid process. The supraspinous fossa is larger than the infraspinous. The spine is high and extends nearly the whole length of the scapula. The acromion is large and lies above the margin of the glenoid cavity. Due to breakage, no distinctive features can be identified on the medial side of the scapula.

Both humeri are complete (figure 7.8A–B), lacking only small fragments of the shafts. The head is longer anteroposteriorly than transversely and is very round, with the articular surface extending well down on both the medial and lateral sides. The greater tubercle is very large and long, making the proximal end of the humerus anteroposteriorly very long as well. The greater tubercle extends proximally well beyond the head. The lesser tubercle is robust but low, and it blends imperceptibly into the articular surface of the head. The tricipital line is prominent and forms a ridge with a slight overhang to the caudal side. This ridge extends distally about two fifths of the distance down the shaft, which is wide anteroposteriorly but quite compressed mediolaterally. The lateral epicondylar crest is very prominent and originates about two thirds of the way down the shaft from the proximal end. The distal articulation is narrower than that in other mustelids. The lateral epicondyle is large, but both trochleae are narrow, which gives them a truncated appearance. There is no supratrochlear foramen, nor is there, unlike the condition in most mustelids, an entepicondylar foramen or any vestige of a bony bar.

Both radii are present and complete (figure 7.8C), lacking only small fragments of the shafts. The head is remarkable in being almost rectangular, with the lateral edge only slightly longer than the medial. The articular circumference overhangs the neck slightly on the medial side. The radial tuberosity is small. The shaft is nearly triangular, with the laterocaudal side slightly concave while the cranial side is slightly arched. The shaft as a whole is very straight in the mediolateral direction and slightly curved craniocaudally. The distal articulation is relatively narrow and oval in outline. The grooves for the extensor digitalis communis, extensor carpi radialis, and abductor pollicis longus are not very distinct. The

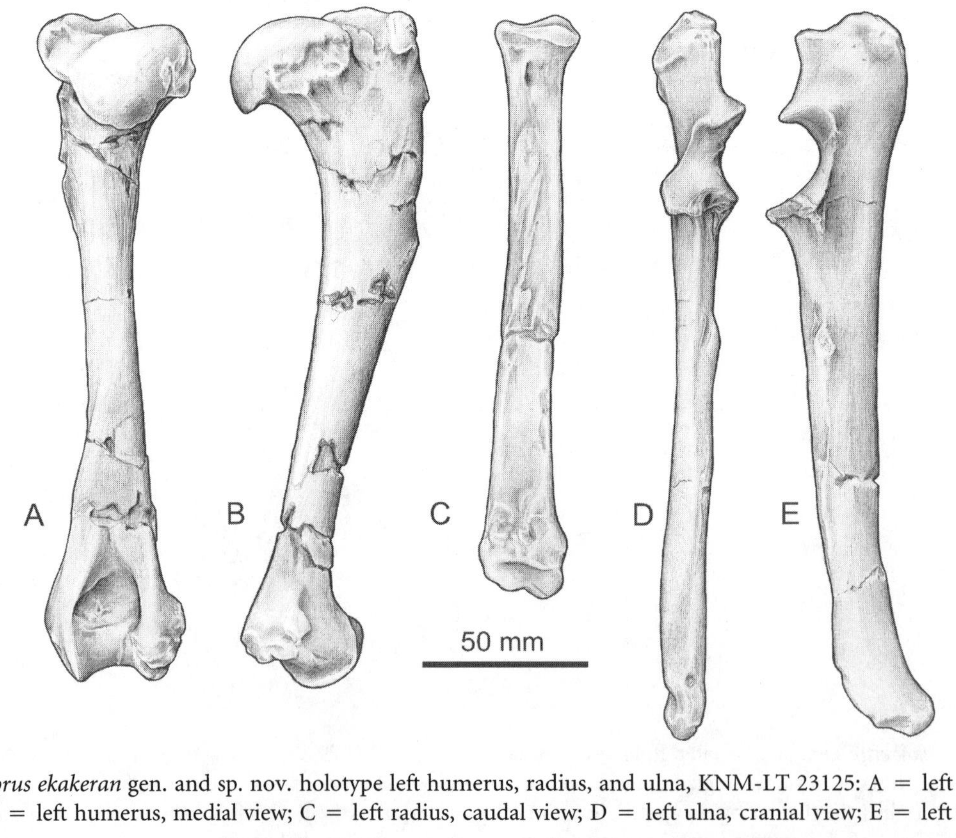

Figure 7.8 *Ekorus ekakeran* gen. and sp. nov. holotype left humerus, radius, and ulna, KNM-LT 23125: A = left humerus, caudal view; B = left humerus, medial view; C = left radius, caudal view; D = left ulna, cranial view; E = left ulna, lateral view.

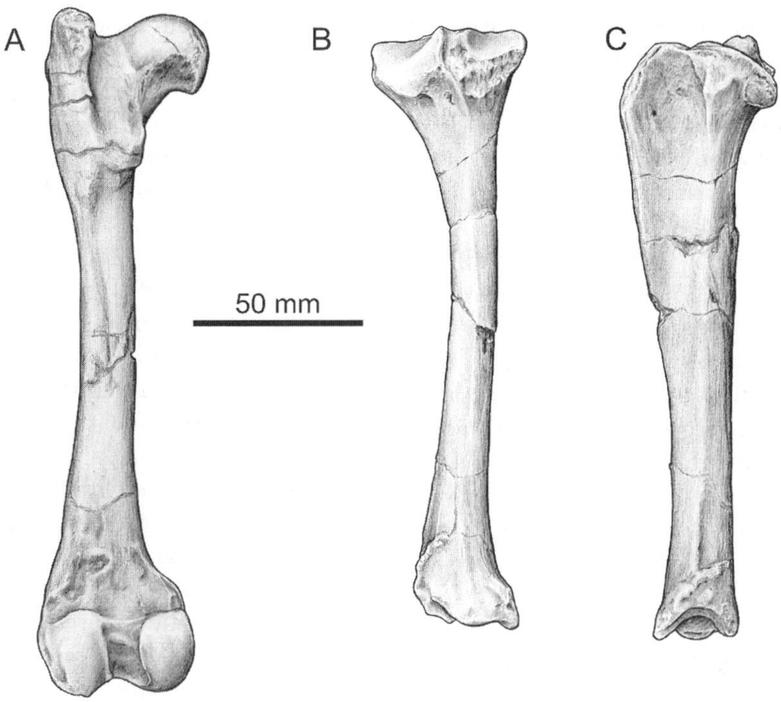

Figure 7.9 *Ekorus ekakeran* gen. and sp. nov. holotype left femur and tibia, KNM-LT 23125: A = left femur, caudal view; B = left tibia, caudal view; C = left tibia, medial view.

latter, however, can be seen to be set at about 75° to the long axis of the shaft, and this feature suggests a very broad foot with a semi-opposable first digit.

Both ulnae are present and complete (figure 7.8D–E), lacking only fragments of the shafts. The ulna is very stout. The olecranon is substantial. Its cranial side does not bear a distinct triceps groove. The attachment area for the triceps is quite large on the caudal side, however. The trochlea is relatively shallow, as is the articular surface for the radius. The shaft is mediolaterally compressed but craniocaudally long. There is a prominent attachment area for the abductor pollicis longus. The distal end of the ulna and the styloid process are very large and robust.

Both femora are present and complete (figure 7.9A). The left femur is almost undistorted and lacks only a small fragment of the shaft. In the right femur the head has been pushed laterally into the trochanteric fossa, damaging the greater trochanter. The femur is a robust bone. The head and greater trochanter are about equal in height. The head is large. The neck is proximodistally long but craniocaudally relatively narrow. The greater trochanter is relatively small, as is the trochanteric fossa. The latter is quite deep, however. The lesser trochanter is small but extends quite far out from the shaft. The shaft is long and is broader craniocaudally than mediolaterally. The distal articulation is narrow. The surfaces for the attachment of the gastrocnemius are prominent. The lateral and medial condyles are subequal in size. The patellar groove is relatively narrow and shallow. Both patellae are present and complete. In cranial view, the patella forms a long triangle. The femoral articular surface is large and forms a low arch; it covers about three-quarters of the caudal surface of the patella.

Both tibiae are present (figure 7.9B–C). The left tibia is nearly complete, lacking only some minor pieces of the shaft and a piece of the proximomedial condyle. The right tibia is represented by the proximal and distal parts and lacks about 30 mm in the middle of the shaft. The cranial intercondylar area is high. The medial and lateral condyles are subequal in size and very flat. The popliteal notch is deep. The cranial border is broad and long. The distal part of the shaft is triangular in cross section. The distal articular surface is small but robust. The medial malleolus is short and broad.

Only the proximal and distal parts of the left fibula and the distal part of the right fibula are available. Neither the proximal nor the distal articulation was broad and flaring. The 25 mm of the left fibular shaft that are preserved are flat and narrow.

Both os coxae are present. The right is complete except for a part of the iliac blade, the pubis, and part of the ischium. The left os coxae lacks the pubis, about half of the ilium, and the ventral part of the ischium. The iliac blade is narrow and long. The gluteal surface of the ilium is deep and bounded dorsally by a low crest.

The auricular surface is wider than it is long. The lunate surfaces are deep and wide, and the acetabular fossae are correspondingly small. The ischiatic spine is small, but the ischiatic tuberosity is large and robust. Judging by the angles of the ilium and ischium, the pelvis, when complete and articulated, must have been relatively narrow.

Nearly all of the vertebral column is present. The only elements missing are the sacrum (present as a compressed mass of bone) and about 5–6 terminal caudal vertebrae. Most vertebrae are broken, and some are represented only by the centrum. However, it is possible to get an estimate of the absolute and relative lengths of the divisions of the vertebral column.

The atlas is short and not very wide. The lateral vertebral foramen is large and opens almost wholly laterally. The alar notch is surrounded by bone and made into a foramen. The atlas wings are short and do not extend caudally beyond the centrum. The transverse foramen is not present, but instead there are vestigial foramina in the form of deep pits on the ventral face of the atlas. The vertebral foramen is round. The caudal articular foveae are shallow.

The axis is short and high. The dens is short and robust, while the cranial articular surface is wide and relatively flat. The spinous process is moderately high cranially and rises to form a peak at the caudal end. This peak also flares slightly laterally to a width of about 10 mm at the apex. The caudal articular surfaces are mildly flaring and angled about 20° from the horizontal.

The third cervical vertebra is short and square in dorsal view. The spinous process has been broken, but it must have been low. The left transverse process is also broken. Both cranial and caudal articular surfaces are slightly flaring and angled about 20° from the horizontal. There is a transverse foramen present.

The fourth cervical vertebra is preserved in two pieces, having been broken at the pedicles. The right transverse process is missing. The spinous process is short and low. The transverse process is large and extends backward beyond the centrum. The cranial and caudal articular surfaces are large and slightly flaring. The cranial is angled about 20° and the caudal about 40° from the horizontal. A transverse foramen is present.

The fifth cervical vertebra is missing the cranial articular surfaces, the left caudal articular surface, the spinous process, and both transverse processes. This leaves a relatively nondescript bone with the centrum, right caudal articular surface, and pedicles. We can, however, note that the spinous process was craniocaudally short.

The sixth cervical vertebra is short and high, with a short spinous process that is slightly damaged. The left caudal articular surface and both transverse processes are missing or distorted. The spinous process is higher than that of C4 but lower than that of C7 (that of C5 is missing). The transverse process must have been large and the transverse foramen is present.

The seventh cervical vertebra is complete, apart from missing a section of the right transverse process. The spinous process is broken dorsally but is higher than that of C6. The caudal articular surfaces are small and do not flare laterally. The transverse process is robust and extends anteriorly rather than posteriorly. There is no transverse foramen, but on the left side between the pedicle and transverse process there are pits for a vestigial (or, alternatively, an incipient) foramen.

The first three thoracic vertebrae are represented by the centra and spinous processes. The fourth thoracic vertebra is represented by the centrum, including part of the mammillary process, and the proximal part of the spinous process. The fifth thoracic vertebra is relatively complete, although the spinous process has been pushed ventrally into the vertebral foramen, thus damaging the mammillary processes and the transverse processes, the right of which is lost. The spinous process is missing its distal part; the centrum is considerably wider than high (as measured between the caudal costal foveae). The sixth thoracic vertebra is preserved much as T5, but the spinous process is not pushed as far ventrally into the vertebral foramen. The transverse processes are large and robust. The distal part of the spinous process is broken and missing.

The seventh thoracic vertebra is relatively complete. The right transverse process has been pushed into the centrum, and the spinous process is broken, whereupon the distal part has been displaced about 4 mm to the right of the proximal part. The transverse processes of this vertebra are large and knoblike. The eighth thoracic vertebrae is complete, but it has been sheared ventrally and to the right. When this shearing occurred, the spinous process was broken off and the distal part reattached about 6 mm down and to the right of the proximal part. The ninth thoracic vertebra is represented by the centrum and the spinous process, including the left transverse process and right mammillary process, while the tenth thoracic vertebra is complete but has been crushed so that all except for the left transverse process, the distal half of the spinous process, and the centrum form a jumbled mass of fragments. The eleventh thoracic vertebra is complete, but the transverse processes are damaged and part of the right pedicle is broken off. The caudal articular surfaces are set high on the spinous process. The transverse process is mediolaterally short and craniocaudally elongated. The twelfth thoracic vertebra is represented by the centrum only. This vertebra or possibly T13 is the anticlinal vertebra.

The thirteenth thoracic vertebra is complete except for the left mammillary process and the spinous process. It is a large and long vertebra with an extended caudal articular process and a short accessory process. The fourteenth thoracic vertebra is complete except for a part of the left cranial articular process. The accessory process is larger than that of the preceding vertebra, and the vertebra as a whole is considerably larger than T13.

The first lumbar vertebra is large and long, with a robust spinous process. The accessory process is small. The vertebra is complete, except for missing the right mammillary process. The right mammillary process of L2 is attached to the caudal articular process of L1. The second lumbar vertebra lacks the caudal parts of the pedicles. The distal part of the spinous process has been pushed into the proximal part. The third lumbar vertebra is represented by the centrum and the mammillary processes, the base of the spinous process, and part of the right caudal articular process. The fourth lumbar vertebra is missing the cranial part of the pedicles, the mammillary processes, the spinous process, and the left caudal articular process. The fifth and sixth lumbar vertebrae are represented only by the centra.

The sacrum is represented by two compressed masses of bony fragments. No distinguishing features can be recovered from these fragments.

Nineteen caudal vertebrae are preserved. Judging by the size and morphology of the caudalmost of them, there may originally have been five or six additional caudal vertebrae, but hardly more. Most of the caudal vertebrae are represented by centra only, but the third, fourth, and eighth preserve one or both mammillary processes. The ninth caudal vertebra is the last to have a cranial articular surface. After the fifteenth caudal vertebra, the mammillary processes become indistinct.

The left manus is nearly complete. The positively identified elements include Mc I–V, proximal phalanges I and III–V (the latter lacking the distal part), middle phalanges III–IV, and ungual phalanges III–IV. The foot as a whole is short and broad. The right manus is less complete but does include Mc I and the proximal parts of Mc II–III, as well as a number of proximal, middle, and ungual phalanges.

The scapholunar is broad and relatively flat, with little arching of the radial articular surface. It has a small, deep facet on the cuneiform side. The articular surfaces for the unciform and magnum are broad and separated by a shallow groove set at 45° to the long axis of the scapholunar. The sulcus for the flexor carpi radialis is shallow. The cuneiform is a slender, flat bone with a large articular surface for the unciform, which is available from the right side only. It is a blocky bone, with a process that extends dorsolaterally. The magnum is an elongated bone with a wide dorsal end and a high palmar end. The trapezium is a small, subtriangular wedge of bone with substantial proximal and distal articular surfaces. The pisiform is relatively large; it has a dorsoventrally high and transversely narrow proximal end with a V-shaped articular surface for the cuneiform, and the distal end is nearly round.

The first metacarpal is a long, slender bone with a large base and head. The shaft is straight. The proximal phalanx of digit I is long and slender and is slightly curved on the lateral side. It has a large head for the articulation with the ungual phalanx, which is missing but presumably was substantial.

The second metacarpal is relatively short and wide; it is about one third longer than that of Mc I but twice as wide in the shaft. The base and head are both strongly expanded.

The third metacarpal is about one fifth longer than Mc II, while the shaft is about the same width. The head and base are both wide. The proximal phalanx of digit III is robust and has a relatively narrow head. The middle phalanx is short and wide, with a head that is wider than the base. The ungual phalanx has a short and wide ungual process, which suggests a short, broad claw. The ungual crest is high and narrow.

The fourth metacarpal is slightly longer than the third but has about the same shaft width. The proximal phalanx is longer and more slender than the corresponding bone in digit III. The middle phalanx is broad and flat and a little larger than that of digit III. The ungual phalanx is similar to that of digit III, but larger.

The fifth metacarpal is slightly longer than the second and somewhat more robust, with a wide base and head. The facet for the cuneiform is quite large and set laterally. The proximal phalanx is missing the head. This metacarpal is quite slender, compared to the corresponding bone of digits III and IV, and has some slight curvature on the lateral side.

The astragalus is very square in general outline. The head is short and wide, while the neck is short and robust. The head is angled medially relative to the body. The body is as wide as it is long. The trochlear notch is shallow. The medial calcanear articulation extends along the neck all the way to the head but does not reach the trochlea. The lateral articulation with the calcaneum is broad. The lateral process is short. The trochlea does not reach the plantar side of the astragalus.

The calcaneum head is short and round, being slightly wider transversely than anteroposteriorly. The gastrocnemial groove on the head is shallow. The tuber is short and considerably deeper than it is wide. The body is also short and wide. The sustentaculum extends considerably medially. The base is angled toward the medial side relative to the long axis of the bone.

The left pes is nearly complete, with the positively identified elements including Mt II–V, proximal phalanges II–V, middle phalanges II–IV, and ungual phalanges II and IV. The right pes is less complete but does include the proximal parts of Mt II–IV, as well as a number of proximal, middle, and ungual phalanges. As a whole, the pes is relatively short and wide, although it is not as extreme in this regard as the manus.

The navicular has a wide articulation for the astragalus on its proximal face. On the distal face there are two articulations, one each for the second and third tarsals. The latter is the larger and is set slightly more toward the plantar side than is the former. The plantar process is set at the lateral edge of the central tarsal. The cuboid is robust. The proximal articulation for the calcaneum is wider transversely than anteroposteriorly, and it is angled slightly toward the lateral side. The sulcus for the tendon of the musculus peroneus longus is set low on the plantar side and extends the entire length of that side; it is nowhere very deep. The distal face has a large, concave articular surface for Mt IV and a much smaller and slightly convex one for Mt V.

The ectocuneiform is robust and wedge-shaped, with a proximal articulation for the central tarsal and a distal one for Mt III.

Only the left mesocuneiform is present. It is quite robust, with articulations proximally for the navicular and distally for Mt II.

The entocuneiform is a narrow but relatively deep wedge of bone. It has a proximal articulation for the central tarsal and a distal one for Mt I, which can therefore be inferred to be present, although it has not been recovered.

The second metatarsal is relatively long and slender. The head is transversely narrow but anteroposteriorly long, while the base is broad. The proximal phalanx is narrow in the middle of the body but more robust proximally and distally. The middle phalanx is short and dorsoventrally flattened.

The third metatarsal is robust and about one fifth longer than Mt II. Its base articulates with the tarsus about 5 mm distal to the Mt IV articulation with the carpus. As a whole, Mt II–III are set more distally than Mt IV–V. The proximal phalanx of digit III is relatively slender, with a broad base and head. The middle phalanx is broad and flat, and the head and base are considerably widened. The ungual phalanx is very robust; its base is triangular, with a high and broad ungual crest. The ungual process is broad and blunt, more so than in the manus, suggesting a broad, blunt claw.

The fourth metatarsal is robust. It is slightly longer than Mt III. The shaft is broken but has been restored so that the bone is complete. The proximal phalanx is wide, with a marked medial prominence at about two thirds of the length of the body from the proximal end. The middle phalanx is broad and dorsoventrally flattened. The ungual phalanx is like that of digit III, but slightly narrower.

The fifth metatarsal is long and slender. The shaft is curved in the lateral direction, making the pes slightly splayed. The proximal phalanx is also slender, but it has some slight curvature on the lateral side.

Discussion

The relationships of this form are extremely difficult to establish, and only a few preliminary notes will be provided here. A number of lineages of mustelids of gigantic size have been recorded from several continents (Werdelin 1996a). The earliest of these are from the Early Miocene of North America (the genera *Megalictis* and *Aelurocyon*, recently synonymized by Hunt and Skolnick 1996). These are relatively primitive forms with short limbs. Comparison of measurements herein and in Hunt and Skolnick (1996:table 2) show that *Megalictis* and *Ekorus* have about the same lower carnassial length, but the limb bones of the latter are more than 20 percent longer and the indices of distal to proximal limb elements are quite different (table 7.3). These North American forms clearly have no relationship with the present material.

In the later Miocene of Eurasia, several genera show possible relationships with *Ekorus*—*Ischyrictis, Laphictis, Hadrictis, Eomellivora,* and *Perunium.* The last two genera have recently been synonymized, which results in the single valid species *Eomellivora wimani* (Wolsan and Semenov 1996). It appears likely that *Hadrictis* also should be synonymized with *Eomellivora* (cf. Orlov 1948; Pia 1939; Zdansky 1924). In a previous publication (Leakey et al. 1996), it was noted that *Ekorus* has some traits in common with *Eomellivora.* However, most of these (such as the reduction of the molars) are much more extreme in *Ekorus,* and this fact, together with differences in the shape of the upper canines and premolars, indicates that we are dealing with a genus distinct from *Eomellivora,* though perhaps in the same lineage.

Ginsburg and Morales (1992) recently suggested that *Eomellivora* is derived from the older *Ischyrictis,* and it is certainly possible that *Ekorus* represents a continuation of this lineage or has a separate derivation from *Ischyrictis.* Alternatively, *Ekorus* may represent a derivation from the somewhat more feloid-like genus *Hoplictis* (Ginsburg and Morales 1992; Viret 1951). In the absence of any older African taxon to tie it to, and in view of the highly derived nature of the skull, dentition, and postcranium of *Ekorus,* it is not as yet possible to provide a definitive answer to the question of its phylogenetic affinities.

Subfamily Mellivorinae
Genus *Erokomellivora* gen. nov.

Diagnosis

Small-sized Mellivorinae with slender mandibular horizontal ramus. Premolars long and slender. Lower carnassial low and long with small metaconid. The M_2 present and single-rooted.

Etymology

After *eroko,* the Turkana word for "prior," in reference to the possible ancestral status of this form relative to the genus *Mellivora.*

Hypodigm

Type species only.

Erokomellivora lothagamensis sp. nov.
(Figure 7.10; table 7.4)

Diagnosis

As for genus, only species.

Holotype

KNM-LT 23926, a left horizontal ramus with posterior root of P_3, roots of P_4, complete M_1, and alveolus for M_2 from the upper member of the Nawata Formation.

Referred material

Type specimen only.

Etymology

After the locality.

Of the lower fourth premolar, only the anterior root and a part of the posterior end of the tooth are preserved, but these parts indicate that this tooth was relatively long and slender. The lower carnassial has heavy wear on the entire crown. The cusps are relatively low, and the shearing blade is short. The carnassial notch was probably not very deep, as it is almost worn down. The paraconid and protoconid are subequal in height. There is a very small metaconid. The talonid is a simple basin with a single, small terminal cusp. As judged from the alveolus, the M_2 was a small, single-rooted tooth. The ramus is broken but was relatively narrow and deep. The anterior end of the masseteric fossa ends posterior to the tooth row.

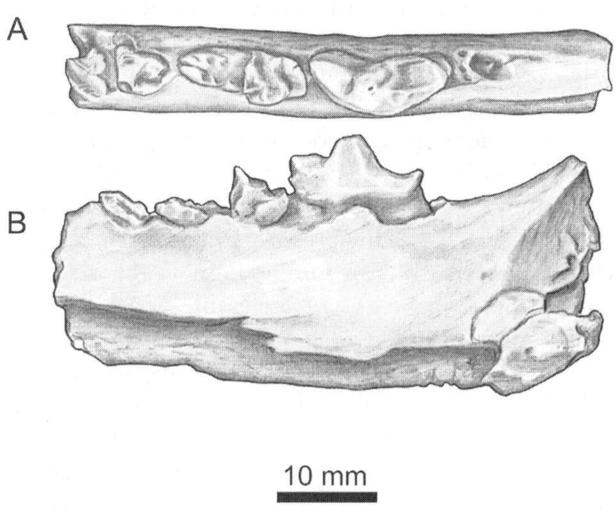

Figure 7.10 *Erokomellivora lothagamensis* gen. and sp. nov. holotype, partial left mandibular ramus, KNM-LT 23926: A = occlusal view; B = lateral view.

Discussion

This specimen matches *Mellivora benfieldi* from Langebaanweg in size and in most morphological particulars (Hendey 1974, 1978b). It differs from that taxon in the considerably longer P_4, the more obliquely set and shorter M_1 trigonid, and, above all, the retention of M_2. The last tooth is absent in *Mellivora benfieldi* and the modern *Mellivora capensis.* It is present in the much larger and more primitive *Eomellivora* species, while the condition in *Promellivora punjabiensis* cannot be determined because of the state of preservation of the type and only specimen of that taxon.

The characters that distinguish the Lothagam specimen from *M. benfieldi* are all primitive characters, and structurally at least we may here see a possible ancestor of *Mellivora.*

Mellivorinae gen. and sp. indet.
(Table 7.5)

Lothagam Material

- *Lower Nawata:* 23160, left lower ramus fragment (posterior part of ?P_4).
- *Upper Nawata:* 25130, proximal Rt. humerus.

The ramus beneath the preserved premolar of 23160 is relatively shallow and broad. The ?P_4 is short and wide and is dominated by the main cusp. The posterior edge of this cusp is crest-like, and this crest extends to the posterior end of the tooth. Posteriorly, the tooth forms a short shelf. There is no posterior accessory cusp, but the cingulum is drawn up into a low cingulum cusp at the extreme posterior margin of the tooth.

The humerus, 25130, has a head that is transversely narrow and anteroposteriorly long. Its articular surface reaches relatively far caudally. The greater tubercle is small and low, not reaching further proximally than the head. The lesser tubercle is low but broad and this, together with a relatively narrow shaft, makes the proximal articulation wide in relation to the shaft.

Discussion

The ?P_4 of the ramus fragment resembles P_4 in modern *Mellivora capensis.* A comparison with *Erokomellivora* described above shows the P_4 to be much longer in that form; if the two are conspecific, the tooth preserved in 23160 must be P_3, which does not accord well with the morphology. At present there is no way to determine whether or not these specimens are conspecific, and 23160 is therefore provisionally assigned to Mellivorinae gen. and sp. indet. The humerus 25130 is generally mellivorine in appearance, but given the possibility just noted that there may be more than one mellivorine species in the material, it is here also assigned to Mellivorinae gen. and sp. indet.

Subfamily Lutrinae
Genus *Vishnuonyx* Pilgrim, 1932
Vishnuonyx angololensis n. sp.
(Figure 7.11; table 7.6)

Diagnosis

A species of *Vishnuonyx* larger than the genoholotype of *V. chinjiensis.* Protocone shelf broader and hypocone better developed. Metastyle relatively longer, making up about 50 percent of the anteroposterior length of the upper carnassial.

Holotype

KNM-LT 23948, a complete right P^4 from the lower member of the Nawata Formation.

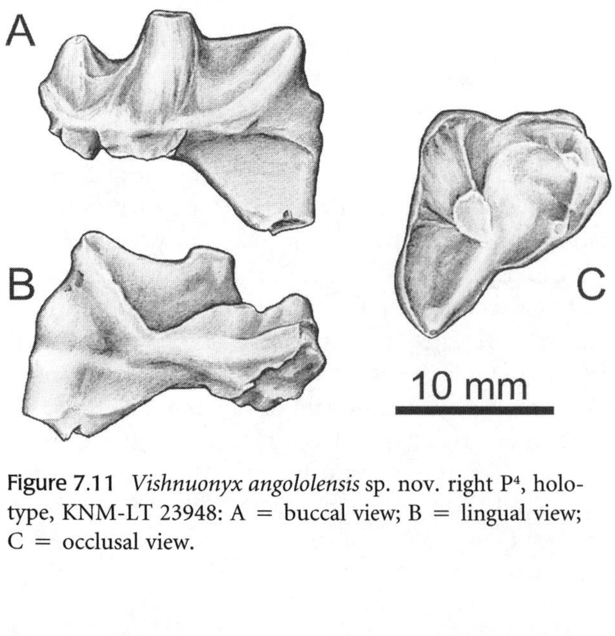

Figure 7.11 *Vishnuonyx angololensis* sp. nov. right P^4, holotype, KNM-LT 23948: A = buccal view; B = lingual view; C = occlusal view.

Etymology

After *angolol,* the Turkana word for "river," in reference to the presumed fluviatile environment this taxon inhabited.

This short, broad upper carnassial lacks a parastyle but has a strong ridge that connects the paracone to the anterior end of the tooth. The paracone is relatively low and has an elliptic cross section; the main axis of the ellipse angles lingually about 30° from the buccal edge of the tooth. There is no carnassial notch. The paracone and metastyle meet at the lowest point of the carnassial blade at an angle of about 70°, and the angle between the metastyle and the buccal edge of the tooth is therefore about 20°. The metastyle is short and low, without any distinct cusp in evidence. The protocone is large and set far anteriorly, reaching further anteriorly than the anterobuccal end of the tooth. A substantial lingual shelf separates the paracone and protocone; this shelf is demarcated anteriorly by a low ridge that runs from the anterolingual side of the paracone to the buccal side of the protocone. Posterior to the protocone, a lower but well-developed hypocone is separated from the protocone by a low ridge that continues posterior to the hypocone to the posterior end of the lingual shelf of the tooth. The tooth is surrounded by a stout cingulum, except at the anterobuccal side of the paracone and the posterior end of the metastyle, where the cingulum merges with the crown.

Discussion

This tooth shows a curious mixture of characters of the different tribes of Lutrinae (Willemsen 1992). It has the short lingual shelf of the Lutrini, the carnassial shear of some Aonyxini, such as *Cyrtaonyx,* and the hypocone of the Enhydrini. The closest comparison would seem to be with *Vishnuonyx chinjiensis* from the Siwaliks (Pilgrim 1932), in which the combination of short lingual shelf and developed hypocone is also present. This probably early Late Miocene form was considered by Willemsen (1992) to be the oldest Enhydrini, and it is not implausible that the present specimen represents a species derived from an immigration of the older Siwalik form or one of its derivatives into Africa. Though clearly more primitive in its characters than all other Enhydrini except *Vishnuonyx,* the Lothagam form is unlikely to be ancestral to later Enhydrini because it is only slightly younger than the much more derived *Enhydriodon africanus,* known from, for example, Langebaanweg (Hendey 1978b). Nevertheless, the Lothagam *Vishnuonyx* is closely related to a smaller, and possibly more primitive, *Vishnuonyx* from the Lukeino Formation (Pickford 1975, 1978).

Family Viverridae

The evolution of the family Viverridae is concentrated in Eurasia and Africa. Aspects of the early evolution of the family have recently been discussed by Hunt (1998), but the early evolution of the extant genera of this family, such as *Genetta* and *Viverra,* is poorly understood, and the Lothagam fauna can contribute significantly to improving this situation.

Subfamily Viverrinae
Genus *Viverra* Linnaeus, 1758
Viverra cf. *V. leakeyi* Petter, 1963

(Figure 7.12; table 7.7)

Lothagam Material

- *Lower Nawata:* 25413, right upper carnassial attached to a small piece of maxilla.

The upper carnassial 25413 has a high and conical paracone that slopes anteriorly to a very short shelf and minute parastyle. The protocone is almost triangular in occlusal view. It is set slightly anterior to the anteriormost point of the parastyle and is separated from it by an indentation of the anterior tooth margin. The protocone is separated from the paracone by a deep and broad valley. From the protocone the lingual side of the tooth curves gently back to the metastyle without any noticeable indentation. The buccal margin of the tooth is also straight, but it narrows slightly at the level of the paracone. The metastyle is short and low and set at about 20° to the long axis of the tooth. Basal cingula are present anterior to the paracone and on the lingual side posterior to the protocone; these cingula extend to the posterior extremity of the tooth.

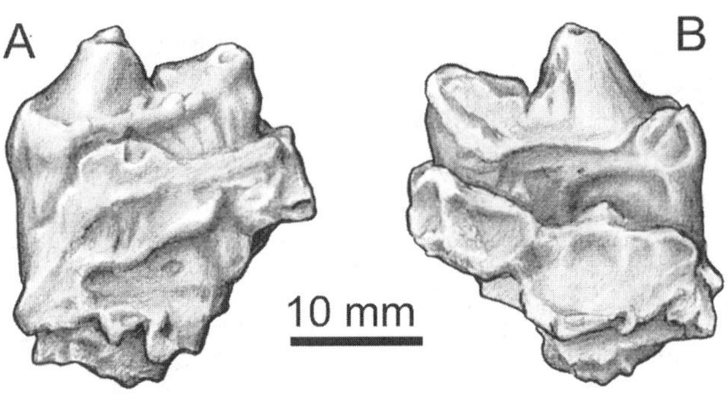

Figure 7.12 *Viverra* cf. *V. leakeyi,* right P^4, KNM-LT 25413: A = buccal view; B = lingual view.

Discussion

This tooth is similar in most respects to the P^4 of *Viverra leakeyi* from Langebaanweg and Laetoli (Hendey 1974; Hunt 1996b; Petter 1963, 1987). It differs in that the protocone is less anteriorly placed, the metastyle is shorter and more angled relative to the long axis of the tooth, and the buccal margin is less curved. It also differs from the Laetoli specimen in having a reduced parastyle, which is seen in some specimens from Langebaanweg (Hendey 1974). It is also larger than any of the specimens from Langebaanweg or Laetoli hitherto described, and in this feature matches *Vishnuictis durandi* from the Siwaliks of India (Lydekker 1884; Pilgrim 1932).

Hunt (1996b) has recently shown that although both have typical viverrid bullae, *Viverra leakeyi* and the species of *Vishnuictis* (*V. salmontanus* and *V. durandi*) differ in features of the skull that indicate different ancestries and different adaptations, with *Viverra leakeyi* being more similar to modern *Viverra* in skull shape than is *Vishnuictis.* The dentition of the latter is also more hypercarnivorous.

Other taxa potentially related to *Viverra leakeyi* are the two species assigned to the genus *Megaviverra* (*M. carpathorum* and *M.* (?) *appenina*) by Kretzoi and Fejfar (1982). Except for the more angled metastyle, the reduced parastyle, and the slightly larger size, the Lothagam P^4 is indistinguishable from the P^4 of *M. carpathorum.* Kretzoi and Fejfar (1982) suggested on the basis of the then available material that *Viverra leakeyi* should be considered congeneric with *Megaviverra* species. Since no *Megaviverra* skull is known, comparisons with *V. leakeyi* are not possible in this regard. However, in *Megaviverra* species (as well as in *Vishnuictis* species) the P^4 parastyle seems more strongly developed than that of any *V. leakeyi* specimen. In this feature the latter is similar to modern *Viverra* species. The polarity of this feature is uncertain, but it indicates that some caution should be used in assigning *V. leakeyi* to *Megaviverra.*

In summary, the Lothagam specimen is undoubtedly most similar to *Viverra leakeyi* from Langebaanweg and Laetoli and, despite the somewhat greater size, is likely to belong to that species. *Viverra leakeyi* differs from *Megaviverra* and *Vishnuictis* in a number of features, most noticeably the development of the parastyle on P^4 and the structure of the rostrum. In both these features *V. leakeyi* is more similar to extant *Viverra.* The polarities of these features have yet to be analyzed; however, it appears unlikely that *V. leakeyi* should belong to either of the other genera mentioned and may thus be a member of the genus *Viverra sensu stricto,* despite its geographic distribution well outside the modern range of that genus.

Viverrinae sp. indet.
(Figure 7.13; table 7.8)

Lothagam Material

- *Lower Nawata:* 23032, Rt. horizontal ramus (alveoli for /C, P_1, broken P_2, alveolus for P_3, roots of P_4, complete M_1, and alveolus of M_2).

The alveolus for the lower canine indicates that it is more or less circular in cross section and quite robust. The first lower premolar is small and single-rooted, while the second lower premolar is double-rooted. The latter is a high, narrow tooth that, as far as can be determined, lacks anterior and posterior accessory cusps. The third lower premolar appears to have been relatively long, almost as long as P_4. Both of these teeth are slender. The lower carnassial has a paraconid that is intermediate in height between protoconid (tallest cusp) and metaconid (lowest). None of the cusps is particularly tall. The distance between the protoconid and metaconid is slightly less than between either and the paraconid. The talonid is long and low, with a substantial and worn hypoconid that is separated from the protoconid by a broad valley. There is a small but distinct hypoconulid and a very low and indistinct entoconid that is separated from the metaconid by a valley that houses a very small entoconulid. The second lower molar was small and elongated. All the cusps of M_1 are worn at the tips. In addition, the paraconid and protoconid show some wear down the buccal side, but there is no distinct shear facet developed.

The ramus is slender. The symphysis is strong but low, with a distinct posteroventral bulge. The symphysis ends between the roots of P_2. The masseteric fossa is indistinct but appears shallow and ends level with the posterior end of M_2.

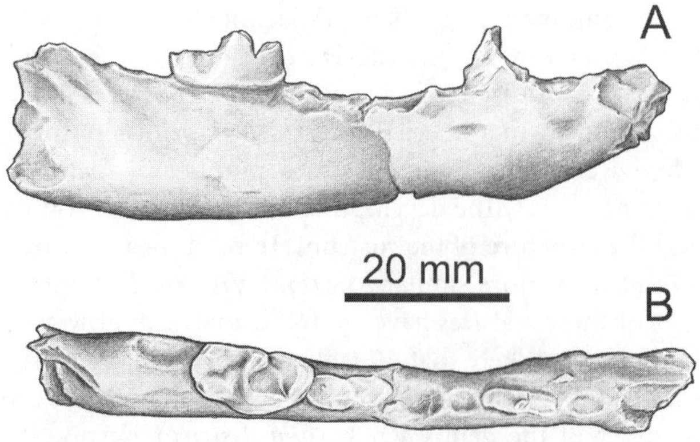

Figure 7.13 Viverrinae sp. indet., partial right mandibular ramus, KNM-LT 23032: A = lateral view; B = occlusal view.

Discussion

This specimen is similar in morphology to, but considerably smaller than, the *Viverra leakeyi* material from Langebaanweg (Hendey 1974). The difference is greater than 20 percent and seems unlikely to represent sexual dimorphism within a single species, although such dimorphism cannot be ruled out. However, a clear distinguishing feature is that, judging by the alveolus, the M_2 of the present form appears larger and, in particular, longer than the M_1 of *V. leakeyi* from Langebaanweg. The present specimen thus appears to represent a species of viverrine distinct from known forms.

Viverridae gen. and sp. indet. (large species)
(Table 7.9)

Lothagam Material

- *Lower Nawata:* 25411, lower ramus fragments (P_4).
- *Upper Nawata:* 25132, lower ramus fragments (broken P_4).

Both specimens are extremely poorly preserved. Both retain P_4, which is long and narrow. There is no anterior accessory cusp. Instead, the anterior part of the tooth slopes gently and somewhat concavely up to a relatively low and short main cusp. The posterior edge descends more steeply to the posteriormost end of the tooth, where there is a low cingulum cusp. The ramus is relatively deep and narrow, but otherwise it is too poorly preserved to be informative.

Discussion

The closest comparison would seem to be with *Viverra leakeyi.* The lower premolars of this taxon are poorly known, but data from Hendey (1974) indicate that the width/length ratios are about the same, with the Lothagam specimens just a little more slender than those from Langebaanweg. The Lothagam specimens are also larger, which would be in keeping with the relative sizes of the P^4s of *Viverra leakeyi* from Lothagam and Langebaanweg. However, these specimens are too fragmented and damaged for this attribution to be more than very tentative, and they are here left as Viverridae gen. and sp. indet.

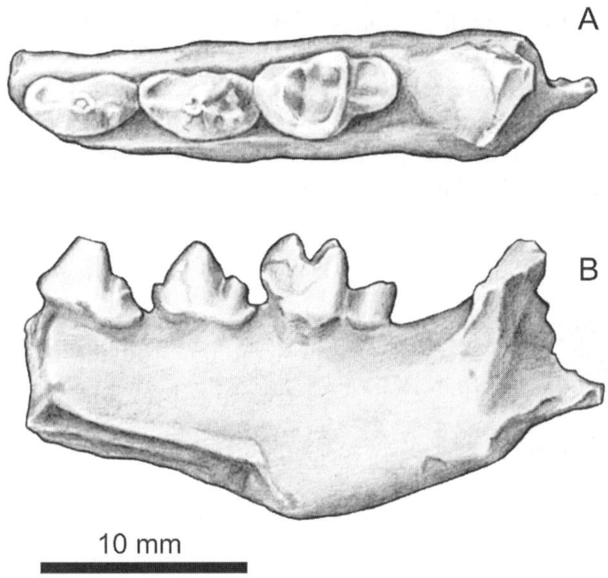

Figure 7.14 cf. *Genetta* sp. A, partial left mandibular ramus, KNM-LT 25409: A = occlusal view; B = lateral view.

Genus *Genetta* G. Cuvier, 1816
cf. *Genetta* sp. A
(Figure 7.14; table 7.10)

Lothagam Material

- *Lower Nawata:* 25409, partial Lt. horizontal ramus (P_3–M_1); 23031, Lt. ramus fragments (broken/C, P_3, broken P_4, M_1).

Specimen 25409 has some horizontal wear on the tips of the tooth crowns. The third lower premolar has a very small anterior accessory cusp that is situated anterolingually and is closely appressed to the main cusp. The main cusp is conical, with more or less symmetric anterior and posterior crests. Its apex is over the anterior root. The posterior accessory cusp is low and free of the main cusp, although there is only a very small notch between them.

The fourth lower premolar has a distinct though small anterior accessory cusp that is set in the anterolingual corner of the tooth. The main cusp is asymmetric with the apex over the anterior root. The posterior accessory cusp is very distinct and set off from the main cusp by a wide valley. A posterior cingulum cusp is individuated slightly lingual to the line formed by the main cusp and posterior accessory cusp.

The paraconid is the tallest cusp of the lower carnassial, and there is a clear anterior paracristid with a small wear facet. This paracristid is set at about 60° to the main axis of the ramus. The anteriormost part of the paracristid has a noticeable cingulum swelling that is not manifest elsewhere in the tooth. The carnassial notch descends abruptly from the paraconid and climbs more gradually to a protoconid that is slightly lower than the paraconid but slightly higher than the metaconid. The metaconid is only slightly lower than the protoconid. These two cusps are placed close to each other and about equidistant from the paraconid. The talonid is much lower than the trigonid; it is short and has two distinct cusps. Buccally there is a tall hypoconid that is joined to the protoconid by a low cristid. Lingually there is an entoconid that is lower than the hypoconid. The entoconid mostly takes the form of a cristid at the posterior extremity of the tooth.

The ramus is relatively robust, and the masseteric fossa is short and ends anteriorly behind the tooth row.

Specimen 23031 is less complete but matches 25409 in all respects in which they can be compared. To judge from the remaining fragments, the lower canine was relatively robust with a round cross section.

Discussion

The general appearance of the lower carnassial indicates that this material may be assigned to the genus *Genetta*. As such, it would be about the size of *Genetta servalina*, though it is unlikely, given its age, to represent any of the extant species of *Genetta*. However, there are many small viverrids and herpestids with similar morphologies of the lower teeth, and the referral to *Genetta* must remain tentative. If it is *Genetta*, this find would represent the earliest occurrence of the genus, with the next known occurrence from Kanapoi at ca. 4.1 Ma.

cf. *Genetta* sp. B
(Figure 7.15; table 7.11)

Lothagam Material

- *Upper Nawata:* 23945, Lt. lower horizontal ramus (P_4, alveoli for /C, P_2–P_3, and M_1–M_2).

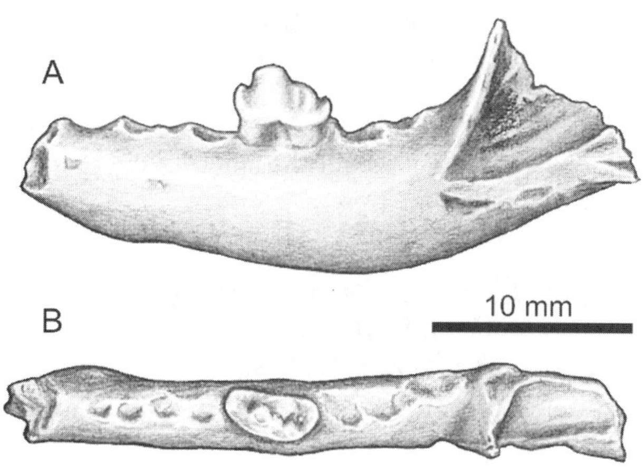

Figure 7.15 cf. *Genetta* sp. B, partial left mandibular ramus, KNM-LT 23945: A = lateral view; B = occlusal view

The fourth lower premolar has an anterior accessory cusp that is set well off from and anterolingual to the main cusp. The anteriormost part of the tooth is broken (though no more than a fraction of a millimeter is missing). The main cusp is situated above the anterior root. The posterior accessory cusp is substantial and is more closely appressed to the main cusp than is the anterior accessory cusp. There is a cingulum crest set lingual to the main axis of the tooth.

The ramus is slender, and its deepest point lies beneath the posterior root of M_1. There are two mental foramina. The anteriormost of these is set about 1.5 mm below the alveolar line just anterior to the anterior root of P_2. The posterior mental foramen is set slightly lower, below the anterior root of P_3. The masseteric fossa is deep and short, and it ends level with the posterior part of the M_2 root. The symphysis is robust and ends at the posterior root of P_2. The coronoid process rises nearly vertically from the horizontal ramus.

Discussion

This specimen is not by itself diagnostic, but it is very similar to those described here as cf. *Genetta* sp. A. However, it is some 15 to 20 percent smaller, which would make it smaller than any living species of the genus. The two forms may represent the two sexes of the same species, but in view of the difference in age I have here preferred to maintain them as separate taxa.

Family Hyaenidae

Among the families of Feloidea, the Hyaenidae has perhaps the best documented fossil record, although it is still very imperfectly known (Werdelin and Solounias 1991, 1996; Werdelin and Turner 1996). This family is nevertheless sufficiently well known to investigate community patterns of evolution in the fossil record (Werdelin 1996c). However, there is a lacuna in our knowledge of this group in the Late Miocene of Africa, and the Lothagam material will be instrumental in plugging this gap, allowing for an investigation of whether the evolutionary pattern within the family is the same in Africa as in Eurasia.

Genus *Ictitherium* Roth and Wagner, 1854
Ictitherium ebu sp. nov.
(Figures 7.16–7.20; table 7.12)

Diagnosis

A species of *Ictitherium* with shortened M_1 talonid, long and slender premolars, and somewhat reduced upper and lower molars. Bones of both thoracic and pelvic limb extremely long and slender with limb indices indicative of cursorial abilities. Unreduced metacarpal I on a digit bearing two phalanges, the terminal of which bears a claw. Metatarsal I possibly present. Third cervical vertebra large and long.

Holotype

KNM-LT 23145, a nearly complete skeleton with skull, mandible, and most of the postcranium, from the lower member of the Nawata Formation.

Etymology

After *ebu,* the Turkana word for "hyena."

Lothagam Material

- *Lower Nawata:* the holotype; 23047, partial Rt. horizontal ramus (M_1) and a distal Rt. humerus fragment.
- *Apak Member:* 10031, partial Rt. horizontal ramus (M_1 talonid and root of M_2).

The following description will be based almost exclusively on the holotype, as the referred specimens are small and fragmentary and serve mainly to confirm observations made on the holotype.

The cranium (figure 7.16A–B) is nearly complete, lacking only the entire left and posterior four-fifths of the right zygomatic arch (though a part of the left zygomatic arch is preserved separately). Both pterygoid hamuli are broken off. The ventral parts of the left and right bulla are broken open, and the bulla cavity is filled with sediment. A part of the side wall of the left parietal is missing, as well as a small piece near the juncture of the frontal, parietal, and basisphenoid on the right side. There are numerous cracks in the cranial bones, especially in the maxillary, palatal, and parietal regions. The skull is slightly laterally compressed, and the right side has been raised a few mm relative to the left side. Overall, however, the skull is only slightly distorted.

The premaxillae are long and narrow. The caudal part of the suture between the premaxilla and maxilla has been obliterated on both sides. Ventrally, the premaxillary-maxillary suture can be observed on the left side. It meets the palatine fissure at about two-thirds of the length of that fissure from its anterior end. In this feature it is apparently more derived than *I. viverrinum,* the only other member of the genus where this character is recorded (Werdelin and Solounias 1991). It cannot be determined whether the premaxillae meet the frontals dorsally or whether the nasals intervene, but

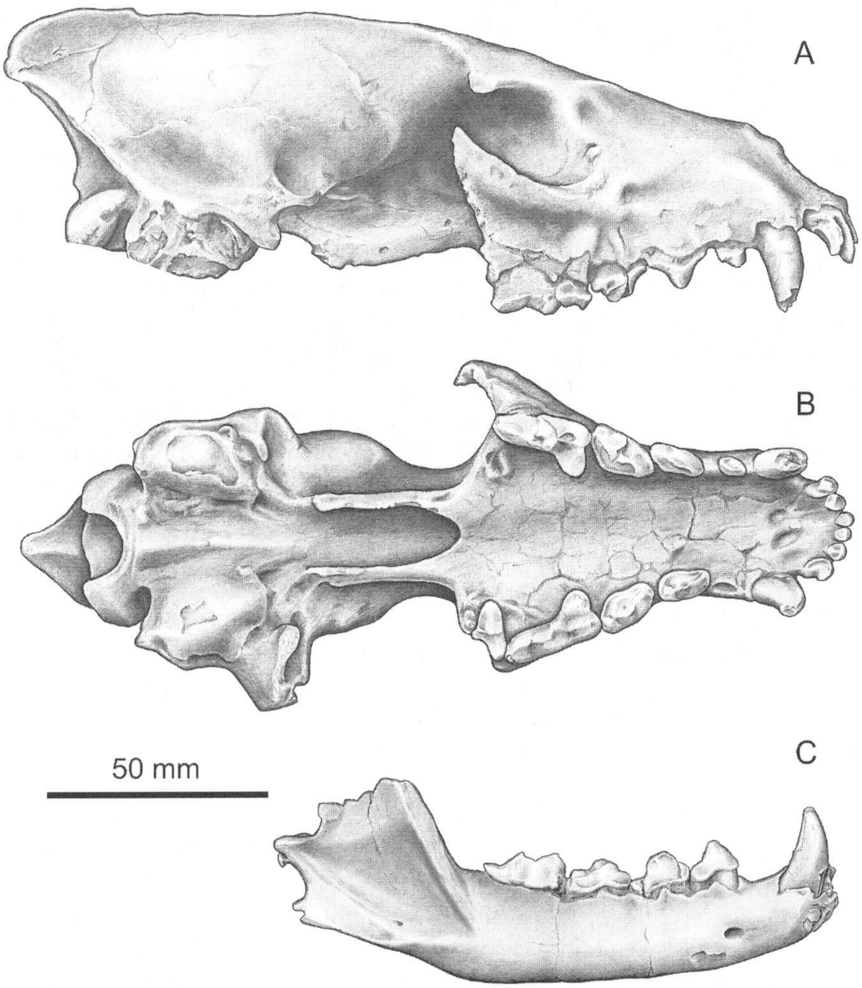

Figure 7.16 *Ictitherium ebu* sp. nov. holotype, KNM-LT 23145: A = cranium, right lateral view; B = cranium, ventral view; C = right mandibular ramus, lateral view.

the distance between the premaxillae and frontals must in any case have been very small.

The skull is narrow across the maxillae, which are relatively low and long. Posteriorly they reach back to nearly half the antero-posterior length of the orbit. The maxillae bulge out markedly for the roots of the canines. Posterior to this bulge the area above P^1–P^3 is flat. The infraorbital foramen is single, large, round, and situated above the main cusp of P^3. Above P^4 the maxillae are strongly indented, and on the right side the maxilla has been worn through to the anterior root of P^4. This phenomenon is seen among some extant carnivores as well (e.g., some felids; cf. Salles 1992).

The palatine is mediolaterally arched, with the median part lying clearly dorsal to the lateral parts, though this arching could be somewhat exaggerated due to the slight lateral compression of the skull. The orbital wing of the palatine is indistinct, inasmuch as the sutures are mostly obliterated. However, the orbital shelf of the palatine is broad and slightly vaulted. The palatine does not extend far dorsally on the zygomatic arch, unlike in modern hyaenids, but as in most other Miocene representatives of the family. The palatine foramina cannot be clearly distinguished, although the sphenopalatine and caudal palatine foramina appear to emanate into a large common pit in a manner similar to that of the modern brown hyena, *Parahyaena brunnea* (Werdelin and Solounias 1991). A distinct ridge runs from the anterior part of the orbital wing of the palatine and posterad, ending on the presphenoid, between the orbital fissure and the rostral alar canal.

The nasals are long and narrow. They do not extend as far posteriorly as the maxillae, their posteriormost point being approximately level with the anterior end of the orbita. From this point they flare out gradually in the anterior direction. This flaring increases slightly anterior to the point at which the distance between the maxillae is smallest—that is, at the anteriormost point of the frontals. This point is situated dorsal to the posterior accessory cusp of P^2.

As noted, the frontals either reach the premaxillae or are separated from them by a very short gap. Posterior to the anterior end of the orbits, the frontals rise gently throughout their length without presenting any bulges. The posteriormost extent of the frontals is just anterior to the mandibular fossa. The zygomatic processes are broad and short. Beginning at the zygomatic processes, a lyriform crest in the form of a low ridge runs posteromedially to the posterior end of the frontals, at which point the left and right halves of the crest are 10–15 mm apart (the exact position of the frontal-parietal suture is difficult to establish). Farther posteriorly, these ridges join up to form the sagittal crest. Laterally, the frontals form the anterior part of a moderately expanded braincase. There is a pit approximately at the juncture of the frontal, parietal, and basisphenoid. This pit is of unknown significance, but it matches one seen in KNM-OM 5314, a juvenile individual of *Crocuta crocuta,* the extant spotted hyena. Some pathology may be implied.

The extent of the lacrimal can only be approximated. It is clearly short, not reaching the palatine. In this feature it is like the living hyenas *C. crocuta, P. brunnea,* and *Hyaena hyaena* but unlike the aardwolf, *Proteles cristatus,* in which the lacrimal has a suture with the palatine, excluding the frontal from the maxillary (Werdelin and Solounias 1991). The lacrimal fossa is large, taking up some 50 percent of the area of the lacrimal. There appear to be two foramina for inferior oblique muscles, as is the case in *H. hyaena.* The exact positions of these foramina cannot be determined, as the sutures separating the frontal, the orbital wing of the palatine, and the orbital wing of the maxillary are indistinct.

The extent of the presphenoid cannot be determined at all, as both its demarcating sutures and the ethmoid foramina have been obliterated. The basisphenoid is ventrally flat. Its orbital wing curves around the braincase to meet the frontal and parietal as is normal in hyaenids.

Like the basisphenoid, the basioccipital is flat and has low ridges that run laterally and medially. The laterally situated ridges and the roots of the pterygoid hamuli bound the anterior lacerate foramen and petrotympanic fissure. The median ridge of the basioccipital is distinct but narrow. This basioccipital structure is like that of most hyaenids and unlike that of *C. crocuta* and its relatives, which have a basioccipital with a median fissure and lateral low and broad bulges (Qiu 1987; Werdelin and Solounias 1991).

The parietals are unexceptional in surrounding a moderately expanded braincase. The low ridges that originated at the zygomatic processes of the frontals meet about halfway along the parietals to form a small but well-developed sagittal crest. The sagittal crest of the supraoccipital is drawn out quite far caudally, extending some 15 mm beyond the posteriormost extent of the occipital condyles. The nuchal crests are also well developed and form a narrow "U" where they meet the sagittal crest.

The exoccipitals are small and harbor slender occipital condyles that are set relatively far apart, making the foramen magnum relatively large for the size of the skull compared with, for example, the condition in *C. crocuta.* The jugular processes are short and do not extend as far ventrally as the auditory bullae do.

The temporals are long and low. They do not present any distinguishing features of note, especially as the zygomatic processes of the temporals are broken on both sides. The retroarticular process is narrow and wraps around the condyloid process to a greater extent than is seen in most hyaenids. It does not lock the mandible in place, however. The auditory bullae are both broken. What remains shows general features of primitive hyenas (Hunt 1974; Hunt and Solounias 1991).

The right mandibular ramus (figure 7.16C) is complete except for the dorsal part of the coronoid process. The left ramus is slightly damaged anterior to the canine and is broken about 15 mm from the anterior end of the masseteric fossa. A piece estimated to be about 15 mm long is missing, while a piece including the left condyloid process is preserved separately.

The mandible is long and slender. Its deepest point is beneath the talonid of M_1. An incisive foramen lies beneath I_2, and a single mental foramen lies beneath the anterior root of P_2. This is unlike, for example, *Ictitherium viverrinum* from Eurasia, which invariably has more than one mental foramen (Werdelin 1988a; Werdelin and Solounias 1991). The masseteric fossa is quite deep but short, and it ends anteriorly at approximately the midpoint of M_2. When articulated, the right and left rami form a narrow "V." The symphysis is somewhat rugose anteriorly and smooth posteriorly, suggesting some degree of mobility.

The dentition is nearly complete. In the upper jaw it comprises I^1–I^3 on both sides, both canines, the right P^1, P^2–P^4 on both sides, and the left M^1–M^2. Of these, the tips of the right I^3 and both canines are broken, while the buccal sides of the right P^3 and P^4 are damaged. In the lower jaw the dentition available comprises the right canine and root of the left canine, the left P_1, the right P_2 and root of the left P_2, and P_3–M_1 on both sides. Of these, the right P_3 and left M_1 are damaged. All the teeth are worn. The wear on the cheek teeth increases considerably from anterior to posterior.

The upper incisors are set in a gentle arc with the size of the teeth increasing from medial to lateral. The first upper incisors are small, peg-like teeth. They are too worn to determine whether there was an internal accessory cusp, but given the anteroposterior length of the base of I^1 as preserved, it is likely that there was at least a significant internal swelling if not an actual cusp.

The second upper incisors are also small and peglike. They are not as worn as the I^1, and there is an indication on the right I^2 that there may have been a low posterior accessory cusp on this tooth. The third upper incisors are considerably larger than I^1 and I^2 and are somewhat caniniform. There is no internal (posterior) accessory cusp, but there is a strong basal cingulum on the posteromedial side of the tooth. The I^3 is separated from the canine by a diastema of about 5 mm length. The wear facets on I^1 and I^2 run more or less from anteroventral to posterodorsal at an angle of about 30°. In addition, I^1 has wear on the anterior face, so that overall the tooth has a wear facet that runs from the anterior enamel-dentine juncture, arches over the crown of the tooth, and continues down on the posterior side to the root. The anterior part of this facet runs at a steeper angle (about 70°) than the posterior part. The wear on I^3 is similar to that on I^1, but is not as great. It is oriented relative to the sagittal plane in the same way as the wear on I^1–I^2, but since I^3 itself is oriented differently relative to the sagittal plane, the wear facet on this tooth is mainly restricted to the buccolateral face.

The upper canines are short, narrow, and gently recurved. They have a distinct wear facet on the anterior face. The tip of the left canine has worn smooth, which indicates that it was either worn down in life or broken off and then worn. The canine is separated from the P^1 by a diastema of about 3 mm.

The first upper premolar is small and oval in occlusal outline. There is no anterior accessory cusp, and the main cusp rises directly from the anteriormost point of the tooth. There is a low, blunt posterior accessory cusp and a slight swelling at the base of the enamel on the lingual side. The wear facet on P^1 starts on the apex of the main cusp and continues down the tooth (i.e., dorsad relative to the skull) at an angle of about 30°, to end on the apex of the posterior accessory cusp.

The second upper premolar is a long and slender tooth. There is no anterior accessory cusp, but a ridge runs from a small swelling at the anteriormost point of the base of the enamel to the apex of the main cusp. The main cusp is narrow; it is a simple cone in buccal view. A weakly developed crest runs from the apex of the main cusp to the posterior accessory cusp. The latter is well developed, low and broad. There is a well-developed cingulum that runs from the anterobuccal corner of the tooth, around the anterior basal swelling, and along the lingual side of the tooth to its posteriormost point, immediately posterior to the posterior accessory cusp. This cingulum is better developed than in other members of the genus *Ictitherium* and, indeed, than in the majority of stem-group hyaenids (Werdelin and Solounias 1991). The apex of the main cusp shows some slight wear; otherwise P^2 is unworn.

The third upper premolar is very similar to the second but is larger and relatively broader. It is somewhat imbricated and overlaps P^2 anteriorly and P^4 posteriorly. There is no anterior accessory cusp, but there is a broad cusp that runs from an anterolingual basal swelling to the apex of the main cusp. The main cusp is robust and triangular; it is connected to the posterior accessory cusp by a short crest. The posterior accessory cusp is well formed, low, and slightly longer than it is wide, and it is moderately appressed to the main cusp. The basal cingulum runs from the buccal side approximately level with the wear facet on the main cusp, around the anterior end of the tooth and the anterolingual basal swelling, and along the entire lingual side of the tooth to the posterolingual corner, where the tooth abuts P^4. The apex of the main cusp is worn to a greater extent than in P^2. This wear facet is inclined slightly from anterior (ventralmost point) to posterior (dorsalmost) and from buccal (ventral) to lingual (dorsal). On the left side it has begun to encompass the crest that unites the main cusp with the posterior accessory cusp.

The upper carnassial is a short, slender tooth. It has a robust, round parastyle with a salient crest that runs from the anterior end to the apex of the cusp and down into the valley that separates the parastyle from the paracone. The protocone is large and triangular and set about level with the anterior end of the parastyle. On the lingual side the protocone merges gradually and without any abrupt break into the paracone. The paracone is the tallest and longest cusp of P^4. It is more or less triangular in buccal view. The metastyle is short and narrow, which is one of the features that identify this form as belonging to the genus *Ictitherium*. The upper carnassial is heavily worn, and extensive shear facets run along the entire buccal side of the tooth. The valley between the parastyle and protocone has been worn smooth, and the protocone itself has been worn concave. The ridge that connects the paracone and protocone has been made more prominent by wear. Posterior to this ridge, the wear facet is slightly concave. At the midpoint of the paracone, the wear facet turns sagittally and in its continuation runs parallel to the main axis of the tooth. All cusps of the upper carnassial have been worn down considerably from their estimated original height. In terms of wear stage, the carnassials match Kurtén's (1953:figure 23) wear stage 5 of *Hyaenotherium wongii*, though the protocone is more heavily worn than is the norm for that wear stage.

The first upper molar is relatively large, with a metastyle wing that has not been significantly reduced. These are also features of *Ictitherium*. In dorsal view the tooth is strongly concave. All the cusps have been worn smooth, making it impossible to determine their exact positions. As can be seen from a comparison with illustrations in Zdansky (1924:plates XIII–XIV) the M^1 is more reduced than in *Ictitherium viverrinum* but less so than in *Hyaenotherium wongii.* Metrically it matches species that in M^1 reduction are coded 0 (primitive) by Werdelin and Solounias (1991).

The second upper molar is a small, transversely oval tooth that is worn smooth, thus providing no detailed morphological information. Like the M^1 it is more reduced than in *Ictitherium viverrinum* but less than in *Hyaenotherium wongii.*

The lower canine is short and slender and somewhat laterally compressed. The long axis of the lower canine lies at an angle of about 45° to the long axis of the mandible. The canine is also angled significantly from anteromedial to posterolateral. Even when the symphysis is articulated, the long axes of the two lower canines do not align with the sagittal plane. Instead, their continuation would meet posteriorly at an angle of about 30°. This angle is greater than in, for example, *Thalassictis robusta* and other primitive hyaenids, but it is more similar to other species of *Ictitherium.* The lower canine is worn apically and, due to wear on the posterobuccal side, is almost triangular in occlusal view. A low ridge runs from the apex of the canine and two thirds of the distance to the enamel-dentine juncture on the anteromedial side. The posterobuccal side has a wear facet with a similar dorsoventral extent. The lower canine is separated from the first lower premolar by a diastema of about 7 mm.

The first lower premolar, which is present only on the left side, is a small, peglike tooth that has been worn down to a flat surface; it shows no morphological detail. This specimen is similar in size to the homologous tooth in *Ictitherium viverrinum* and *Hyaenotherium wongii* (Werdelin 1988a, 1988b).

The second lower premolar is long and slender. In occlusal view the tooth broadens gradually from anterior to posterior to the posterior cingulum shelf, which is the broadest part of the tooth. There is no anterior accessory cusp. A strong salient ridge runs from the anterior end of the tooth to the apex of the main cusp, defining the anterior face of this cusp, which is very slightly convex. The main cusp is an equilateral triangle in lingual view and has a posterior ridge that runs down to the posterior cingulum shelf. This shelf is broad and flat, with only a very slight hint of a posterior accessory cusp. The tooth is slightly worn apically, and there is also some wear on the posterior cingulum shelf directly posterior to the main cusp.

The third lower premolar is long and slender. It is nearly rectangular in occlusal outline, and the broadest part of the tooth is near the posterior end of the main cusp. There is no anterior accessory cusp. A low and narrow ridge runs from the anterior end of the tooth to the apex of the main cusp, defining the anterior face of this cusp, which is slightly concave. The main cusp is an equilateral triangle in lingual view. There is a very small posterior accessory cusp on the buccal side of the posterior cingulum shelf. An indistinct cingulum runs around the anterior and posterior ends of the tooth. The apex of the main cusp is worn, and this wear facet extends down the posterior face of the cusp, ending near the valley between the cusp and the posterior cingulum shelf.

The fourth lower premolar is long and anteriorly narrow. Posteriorly, the cingulum shelf and cusp cause this tooth to broaden considerably. There is a small and low anterior accessory cusp. The main cusp is short and broad. There was probably a substantial posterior accessory cusp, but it has been nearly obliterated by wear. The buccal cingulum cusp is long and narrow; when unworn this cusp was probably separated from the main cusp by a valley, but again wear has obliterated most of this feature. A marked cingulum runs the length of the buccal side of the tooth. The most significant feature of this tooth is the wear pattern. The wear facet starts on the anterior half of the main cusp. It is dorsoventrally aligned, with the posterior part being more ventral than the anterior. In other words, the wear facet runs down the posterior side of the main cusp and onto the cingulum shelf, which is worn nearly flat. The posteriormost few mm of P_4 are broken off on both sides, probably also as a result of the extensive wear on this tooth. Manipulation of the skull and mandible together shows that this wear facet is due to its specific alignment with the protocone of P^4 which, as noted, is also extensively worn.

The lower carnassial of the holotype is heavily worn. The paraconid and protoconid are about equal in length, while the talonid is long and wide as in other *Ictitherium.* The wear on this specimen is too extensive for other features to be seen. However, in specimens LT 10031 and LT 23047 the metaconid is tall and robust and set at about the level of the posterior end of the protoconid. The talonid is at least two-cusped, with an entoconid and hypoconid that are about subequal in height. The entoconid is separated from the metaconid by a valley, while the hypoconid is connected to the protoconid by a low ridge that continues from the hypoconid posteriorly to the posterior end of the tooth. At this point there may be a small hypoconulid, but the available specimens are too worn for the presence or absence of this feature to be determined. As noted, the carnassials of the holotype are extensively worn. The trigonids are worn to such an extent that the shearing blade has been almost obliterated. The metaconid is entirely missing on the left side, while on the right side the posterior half of the metaconid has been worn away. The talonid on both sides has been worn flat.

None of the specimens retains the second lower molar. Judging from the alveolus and root, this tooth was small and simple, rather smaller than the average for

Ictitherium viverrinum. It is set at the lingual margin of the ramus directly behind M_1.

In general, the wear pattern of the dentition of the holotype is very interesting, increasing as it does from anterior to posterior. The anterior premolars are relatively unworn, while P_4 and M_1 (and P^4 and M^{1-2}) would seem to have a limited remaining functional life span. This contrasts markedly with extant hyenas, indicating that a study of the evolution of dental wear and occlusal patterns in hyenas would provide valuable information on dietary shifts. Such a study is beyond the scope of this contribution, however.

The left scapula is complete except for the anterodorsal and posterodorsal parts of the supraspinous fossa and the caudal angle of the infraspinous fossa, as well as a portion of the infraglenoid tubercle. The acromion and lateral and distalmost parts of the spine are also missing. The supraspinous fossa is distorted in that the caudal portions have been pushed medially, causing the bone to fold over itself in its posterodorsal part. The right scapula is represented only by the glenoid and a portion of the blade as far posteriorly as 15 mm of the (broken) spine. The glenoid cavity is oval, and the supraglenoid tubercle is well developed. Despite the breakage and distortion, it is evident that the supraspinous fossa was broader than the infraspinous. This is in contrast to modern hyenas in which the infraspinous fossa is much the larger. Judging by the skeleton figured by Gaudry (1862–1867:plate X), *I. viverrinum* from Pikermi matches the Lothagam form in this feature. In this respect *Ictitherium* more closely resembles canids. The spine is relatively high, as is also true of modern hyenas.

Both humeri are present and complete (figure 7.17A–B). They are both crisscrossed by numerous small cracks, and there is some minor damage to the greater tubercle of the right humerus. The left humerus is slightly distorted by cracks in the condyle, below the neck, and in mid-shaft. This has made the left humerus slightly longer than the right. The right humerus seems undistorted. The humeri are long and slender, relatively more so than in any other known hyaenid, fossil or recent. The greater tubercle is long and narrow and raised some 5 mm above the proximalmost point of the head. The head itself is small and very round, both craniocaudally and mediolaterally. In this it differs from the humerus of *Crocuta crocuta,* which is much flatter craniocaudally. This is due to the greater indentation be-

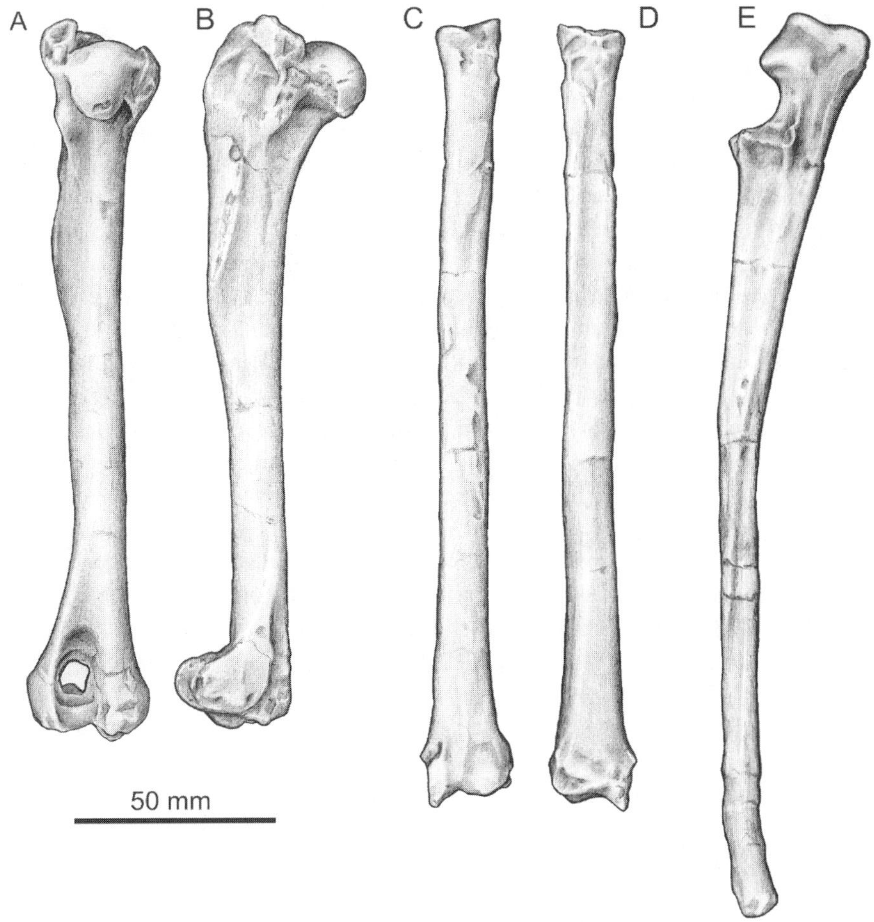

Figure 7.17 *Ictitherium ebu* sp. nov. holotype, KNM-LT 23145: A = left humerus, caudal view; B = lateral view; C = left radius, cranial view; D = caudal view; E = left ulna, lateral view.

tween the humeral head and the greater tubercle in *I. ebu*. It may be noted that in this feature *I. ebu* also differs from material from Samos tentatively referred to *H. wongii* (Lewis and Werdelin unpublished observations). The head is bounded medially by a prominent lesser tubercle that rises a little higher proximally than the head itself. The tricipital line is strong; at about 10 mm from its proximal terminus it forms a pronounced tubercle. The shaft distal to the deltoid tuberosity is slender and relatively straight. The olecranon fossa is narrow and deep, and the supratrochlear foramen is higher than it is wide. The trochlea is slender, much as in other Miocene hyenas. As in other hyenas the entepicondylar foramen is missing, and only minute vestiges of its bony bar remain.

Both radii are present (figure 7.17C–D). Both show numerous cracks and breaks throughout the shaft. These cracks are particularly prevalent on the right radius, and, due to the effects of these cracks, the right radius is about 6 mm longer than the left. The proximal articular surface is oval, with only a sight protrusion on the lateral side. The articular circumference does not form a marked lip overhanging the neck, unlike in *Crocuta,* where this lip is prominent all around the head and especially laterally. The radial tuberosity is small and forms a bump that demarcates the separation of the neck from the shaft. The shaft is extremely long, straight, and slender, much more so than in any other known hyaenid. Though *I. ebu* is a small animal, of some 10–15 kg estimated body mass based on regression analyses of M_1 length (Van Valkenburgh 1990), the radii are nearly as long as those of *Crocuta,* an animal of some 5–6 times the body mass. The distal articulation is narrow. The groove for the abductor pollicis longus is deep, but the grooves for the extensor carpi radialis and extensor digitalis communis are relatively indistinct. The ulnar notch is small and low. The styloid process of the right ulna and the right pisiform are still attached to the distolateral part of the right radius.

Both ulnae (figure 7.17E) are present. The left ulna is complete, though cracked in places, while the right ulna lacks the distal half, except for the styloid process, which as noted above is attached to the right radius. The ulna, like the humerus and radius, is extremely long and slender. The olecranon is narrow and straight, with only slight lateral curvature. The anconeal process is short and does not extend as far cranially as does the lateral part of the coronoid process. The radial notch is small and narrow. There is a marked pit below the radial notch on both left and right ulnae, much as in modern hyenas. The shaft is very narrow. The styloid process is anteroposteriorly slightly broader than the narrowest part of the shaft and is transversely very narrow.

Both femora are present and nearly complete (figure 7.18A–B). The left femur is broken and cracked in many places, and the distal articular area with condyles has been displaced a few mm craniad. In addition, the left femur as reconstructed is artificially straight and due to these postdepositional effects is about 10 mm longer than the right femur. The right femur is missing some small pieces of the shaft. The presence of a prominent, 3 mm high ridge on the laterocaudal side about half way along the shaft indicates that the right femur has been compressed in that area. The right femur also has a slightly damaged greater trochanter.

The head is small and very round. The neck is narrow and long and is 50 percent longer proximodistally than craniocaudally. The greater trochanter is not very prominent and does not exceed the head in height. The trochanteric fossa is relatively deep and in the right femur is separated by a crest into two distinct fossules. The lesser trochanter is small. The shaft is long and narrow and the femur as a whole significantly longer than the humerus. The lateral supracondylar tuberosity is very large, while the medial one is almost invisible. The distal articular area is small and narrow with the lateral condyle slightly larger than the medial. The trochlea is narrow and round in cross section.

The ?right patella is present. In cranial view it is a 17 mm long ovoid. The base and apex are about equally wide. There are two small articular areas on the caudal face.

Both tibiae are present (figure 7.18C–H), though both are broken. The right tibia is cracked in many places and is missing a small piece out of the shaft some 155 mm from the proximal end of the tibia. By comparison with the distal part of the left tibia, the missing part is unlikely to have been more than 10 mm in length. The left tibia is damaged proximally. The medial condyle is broken caudally and medially, and the lateral condyle is broken laterally. The tibial tuberosity is broken proximally and is missing a piece at about its midpoint. Like the right, the left tibia is broken in many places and a small piece is missing about two fifths of the way down the shaft, as measured from the proximal end. As in the case of the right tibia, however, the missing piece is unlikely to have been more than a few mm in length. The proximal articulation is narrow, with the lateral condyle slightly larger than the medial. The tibial tuberosity is low, not reaching the articular area. It is angled only slightly from medial toward lateral. The distal articulation is also small. The lateral articular surface is broader but shorter than the medial. The medial malleolus is short and does not reach further distally than the distocaudal projection of the medial articular surface.

Neither fibula is preserved intact and isolated, but the proximal 20 mm of the right fibula and the distal half of the left fibula are attached to their respective tibiae. These fragments show the fibula to be very slen-

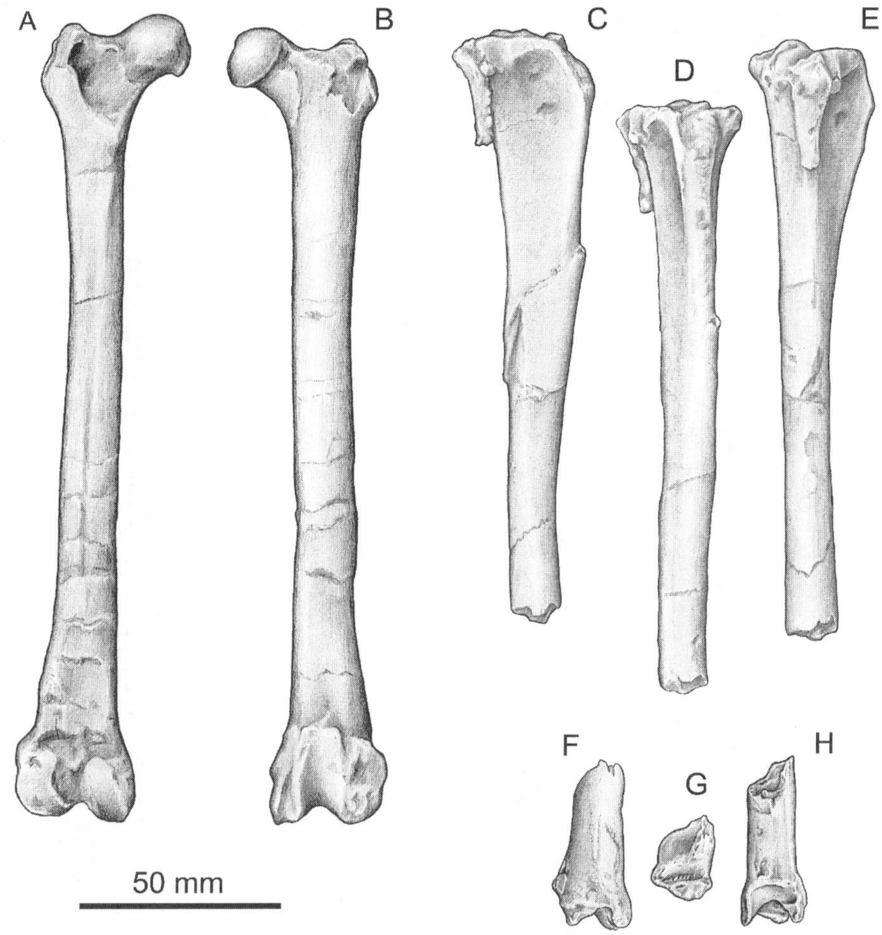

Figure 7.18 *Ictitherium ebu* sp. nov. holotype, KNM-LT 23145: A = left femur, caudal view; B = cranial view. C = proximal right tibia, lateral view; D = cranial view; E = craniolateral view (E shows the proximal fibula attached to the tibia); F = distal right tibia, cranial view; G = distal view; H = lateral view.

der, only some 2 mm wide at the narrowest point. The head is short and wide, as is the lateral malleolus. The groove of the lateral malleolus is very prominent.

A small part of the right os coxae is preserved; it incorporates the anterodorsal part of the acetabular fossa and the posteroventral part of the ilium. The ilium is about 20 mm wide at this point. In addition, another very small part of the right ilium is preserved. The two fragments do not articulate with each other. Given their respective sizes and shapes, however, only some 5 mm are likely to have separated the two.

The atlas is missing the left wing and the lateral border of the right wing. It is significantly arched and the vertebral foramen is higher than it is wide. The lateral vertebral foramina are large and there is no alar notch. The extension of the wings caudal to the body of the atlas makes up more than half of the total length of this vertebra. This is in marked contrast to *Crocuta,* in which this caudal extension makes up only about 40 percent of the total length of the atlas, but is more in line with the condition in striped and brown hyenas (Buckland-Wright 1969).

The third cervical vertebra (C3) is preserved in its entirety except for the spinous process, which is broken off at the base (figure 7.19A–E). It is relatively very long and differs considerably from the C3 of *Crocuta.* The cranial articular surfaces for the axis are small and point straight anteriorly without lateral flaring. They are inclined about 45° from horizontal. The body is long, and the anterior and posterior articulations lean dorsoventrally about 30° from horizontal. The ridge that connects the cranial and caudal articular surfaces is narrow and long. The vertebral foramen is small, and at this point it is wider than it is high. The transverse processes are narrow and long, and they extend caudally about as far as the caudalmost point of the body. They flare outward about 10 mm to either side of the caudal articular surfaces.

The fourth cervical vertebra (C4) is represented only by the body, which equals that of C3 in length.

The fifth cervical vertebra (C5) lacks the right cranial articulation and the anterior and posterior parts of the right transverse process, as well as the anteriormost part of the left transverse process, the tip of the

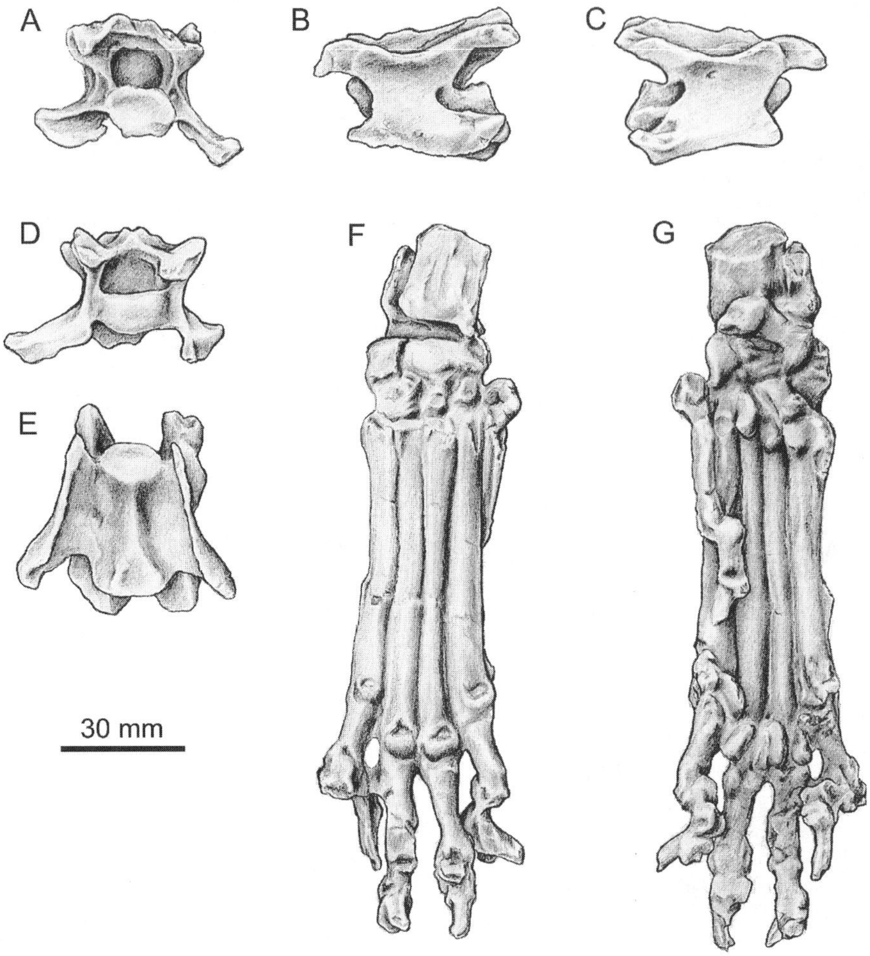

Figure 7.19 *Ictitherium ebu* sp. nov. holotype, KNM-LT 23145: A = third cervical vertebra, cranial view; B = left lateral view, C = right lateral view, D = caudal view; E = ventral view; F = right manus, dorsal view; G = ventral view.

spinous process, and a part of the cranial end of the body. This vertebra is almost cubic in general form, being subequal in length, width, and height. The cranial articulations flare outward and constitute the broadest part of C5. The caudal articular surfaces are less flaring. The cranial and caudal articular surfaces are inclined about 25° from the horizontal plane. The ridge that connects the cranial and caudal articular processes is broad and robust. The transverse processes are caudally very broad and massive. The accessory processes are robust and extend laterally slightly beyond the transverse processes.

The seventh cervical vertebra (C7) is preserved complete, except for the tip of the spinous process and the caudalmost part of the right transverse process. The cranial articular surfaces flare laterally. The arch is short, and the caudal articular surfaces are only slightly extended laterally. The transverse processes are short and robust and constitute the widest part of the vertebra. The body is relatively long.

The second thoracic vertebra (T2) is present, lacking only the transverse processes and the tip of the spinous process. This vertebra is short and has a high spinous process that is inclined caudally about 20° from the vertical. At this point, the vertebral canal is wider than high.

An additional thoracic vertebra is present isolated in the material. Judging by its general aspect it may be the vertebra directly behind the anticlinal thoracic vertebra. Since the vertebral count has not been determined, this vertebra cannot be given a number in the series. It lacks the transverse processes and the caudal articular surfaces. This vertebra is relatively short, contrasting it with the long cervicals. The cranial articular surfaces flare outward and are inclined about 70° from the horizontal plane. The spinous process is long and low and slightly inclined cranially.

The remaining thoracic vertebra (except the first) are contained within a fused mass, together with a number of ribs. It was decided not to prepare these out of the matrix. Unfortunately, the condition of this mass of bones makes it impossible to determine the full thoracic vertebral count of the specimen.

All four feet were found fully articulated. The front feet are both nearly complete. The right manus lacks most of the tarsals and some ungual phalanges, while the left manus is broken across the metapodials and lacks the distal moiety.

The right manus was left intact as found, but the bones of the left manus were prepared out individually (figure 7.19F–G). The scapholunar is a small, compact bone. The articular surface for the radius bears a distinct ridge that is positioned almost mediolaterally, with a 30° angle from dorsal to ventral. The sulcus for the flexor carpi radialis is broad and shallow. The sesamoid face is large. The cuneiform is a small, ovoid bone that is only slightly vaulted. The trapezoid and trapezium are small, while the magnum and unciform are rather large, with the latter being larger than the scapholunar. Both pisiforms are present, though the right is attached to the right radius as was noted earlier. The pisiform is large, as in other primitive hyenas. It is similar in shape to the pisiform of most carnivores in that the main axes of the proximal and distal ends are at right angles to each other.

Metacarpal I is long and slender. Unlike in modern hyenas, in which Mc I is reduced, digit I in the present form was fully functional and carried two phalanges, the second of which bore a claw. All the other metacarpals are long and slender, in keeping with the rest of the extremities. In articulation they form a narrow, long foot in which Mc III extends slightly beyond Mc IV. This pattern is the same as that seen in Chinese Miocene material referred to *Ictitherium viverrinum* (Lewis and Werdelin unpublished observations), while in most other hyenas these metacarpals either extend equally far distally or Mc IV extends further, the latter being the case in extant hyenas. All four toes have three phalanges, of which the third bears a substantial claw.

The right astragalus is missing part of the head and the lateral side. The left astragalus is more complete, although the neck is broken and has been restored and part of the head is lost. Like that of most hyenas, the astragalus is short and broad. The head is very wide but anteroposteriorly short. The lateral articular surface for the calcaneum is larger than the medial. The trochlear groove is broad and shallow.

The right calcaneum is missing the tuber, while the left is complete. The articular surfaces for the astragalus are broad and short, and they do not reach the distal extremity of the bone. The tuber is narrow, while the head is broad and nearly round, as it is only slightly deeper than wide. The distal articulation is slightly angled from the horizontal. The sulcus for the digital flexors is very deep and broad and is demarcated on either side by a distinct ridge.

Both hind feet have been prepared out of articulation. The left has all tarsals but lacks everything distal to the midpoint of the metatarsal shafts. The right hind paw lacks the tarsals but has a complete set of metatarsals and all phalanges except the third phalanges of digits II and III. The left navicular is wide, thin, and flat. It presents a smoothly concave surface for the astragalus articulation. The plantar process is small and extends dorsally beyond the articular surface for the astragalus. The entocuneiform is long and narrow and mediolaterally flattened. It has a distinct articular area for the navicular and also one for the first metatarsal. This is the first indication of the presence of Mt I in primitive hyenas. This bone has, however, not itself been preserved. The mesocuneiform is a small wedge of bone that has distinct articular surfaces for the navicular and Mt II. The ectocuneiform is high and narrow. The plantar process is long and high and very narrow. The cuboid is a high and narrow bone. The sulcus for the musculus peroneus longus is continuous and deeply intended in the plantar face of this bone.

The metatarsals are all long and slender. Unlike in modern hyenas, the manus and pes are about equal in length and width. The Mt III extends slightly further distally than Mt IV, which is true also of modern hyenas.

Discussion

The referral of this form to *Ictitherium sensu stricto* of Werdelin and Solounias (1991) can be justified on the basis of several dental characters, notably the relatively unreduced M_2 and M^1 and the short P^4 metastyle. Several other characters, though not as compelling, indicate the same generic attribution. The premolars of the Lothagam form are relatively long and wide, compared to those of *I. viverrinum*, a feature it shares with other *Ictitherium* larger than *I. viverrinum*, such as *I. pannonicum* and *I. ibericum* (Werdelin 1988a; Werdelin and Solounias 1991). In one feature, however, the length of the M_1 talonid, it differs markedly from other *Ictitherium*. The talonid is relatively reduced, as in species of other primitive hyaenid genera such as *Thalassictis* and *Hyaenotherium*. In general, *Thalassictis* and *Hyaenotherium*, and other more derived Miocene hyaenid genera, are more hypercarnivorous than *Ictitherium* species, and one of the characteristics that define this feature is the length of the M_1 talonid. The Lothagam *Ictitherium* seems to have evolved a more hypercarnivorous mode of existence in parallel with the aforementioned genera.

The most characteristic feature of the Lothagam *Ictitherium*, the extremely long and slender limbs, would tend to confirm this interpretation of its ecology, given that long, slender limbs can be viewed as an adaptation to cursoriality. However, since this is not necessarily the case, such an interpretation must still be considered somewhat premature. More detailed analysis of the ecomorphology of the Lothagam *Ictitherium* in the future

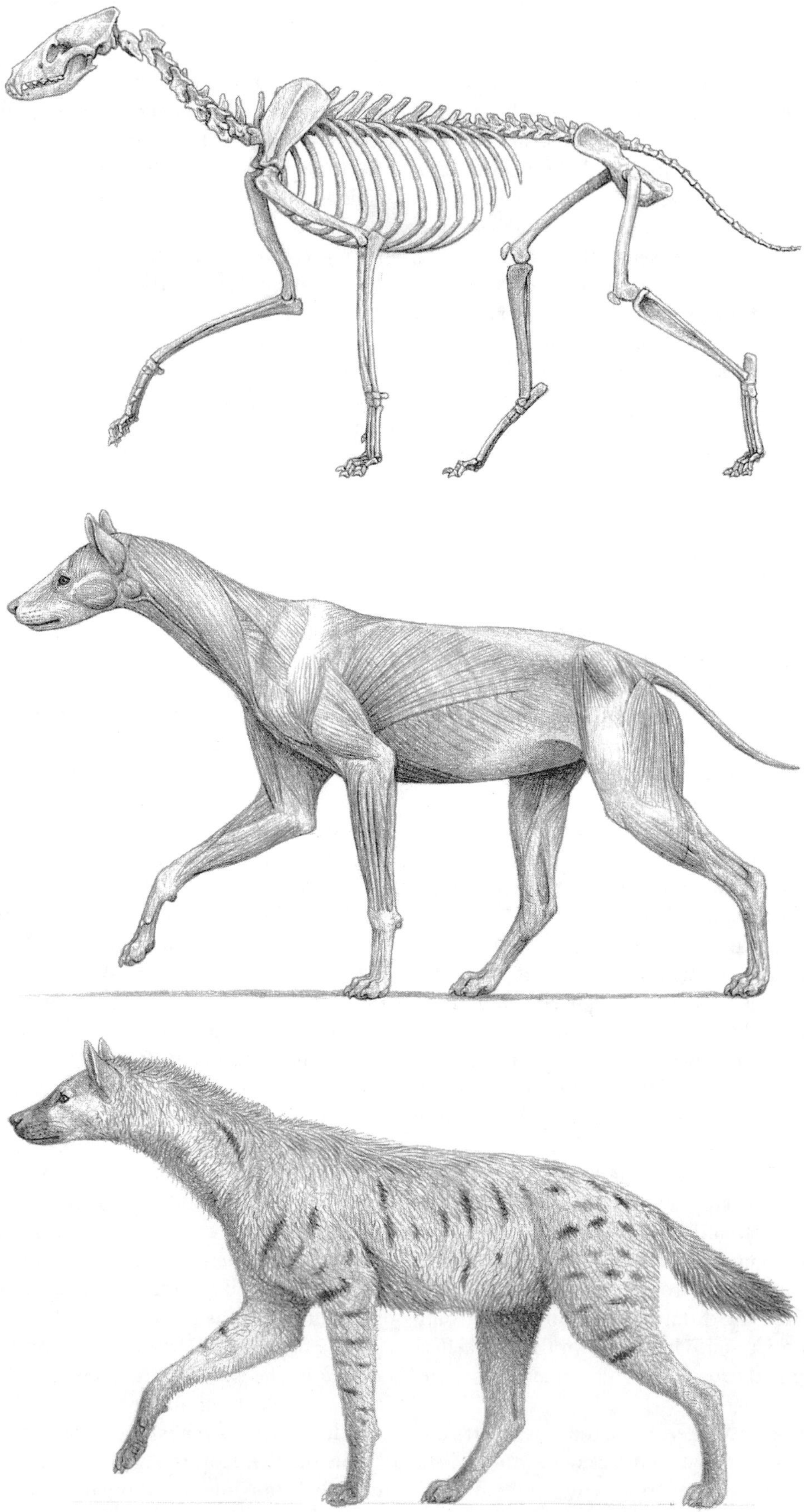

Figure 7.20 Restoration of *Ictitherium ebu* sp. nov. by Mauricio Antón. Shoulder height = 60 cm.

may settle this issue. Nevertheless, the postcranial skeleton certainly sets this form apart from its congeners insofar as their postcranial skeleton is known. This, together with some features of the dentition, confirm that we are dealing with a new species, which I am here calling *Ictitherium ebu.* Given that the slightly earlier record of this genus from Samburu Hills (Nakaya et al. 1984) should more correctly be referred to *Protictitherium, I. ebu* represents the first definite record of the genus in Africa (cf. Werdelin and Turner 1996). The Apak record is temporally approximately correlative with the latest *Ictitherium* in Eurasia, where the genus ranges from ca. 11.5 Ma to ca. 5.2 Ma.

Genus *Hyaenictitherium* Kretzoi, 1938
Hyaenictitherium cf. *H. parvum*
(Figure 7.21A–B; table 7.13)

Lothagam Material

- *Lower Nawata:* 23013, Lt. /C.
- *Upper Nawata:* 23937, Rt. horizontal ramus fragment (/C); 10032, partial Rt. horizontal ramus (P_2–P_4, root /C, and alveolus of P_1).

The description to follow will be based exclusively on KNM-LT 10032 (figure 7.21A–B) since the other specimens are hardly diagnostic and are only tentatively assigned to this taxon.

Judging by its alveolus, the first lower premolar was a small tooth set anterior and slightly lingual to P_2. The second lower premolar has a very small anterior accessory cusp, a large main cusp with a slightly convex anterior margin, and a small but distinct posterior accessory cusp that is set directly in line with the main cusp. An incipient cingulum in the form of a swelling is found around the posterior end of the tooth. The third lower premolar has a minute anterior accessory cusp, a high but narrow main cusp with a convex anterior margin, and a small posterior accessory cusp. The cingulum shelf is well developed on either side of the posterior accessory cusp. The fourth lower premolar is long and slender. A small but distinct anterior accessory cusp, a high, narrow main cusp, and a large and trenchant but low posterior accessory cusp, are all present. The lingual cingulum of P_4 is formed into a crest that runs from the groove between the main cusp and posterior accessory cusp to the posterior end of the tooth. The ramus is deep and relatively robust. A single large mental foramen is located beneath the middle of P_2.

Discussion

The dentition of this form is very similar to that of *Hyaenictitherium namaquensis* from Langebaanweg. The morphology of P_4 with its posterior cingulum crest is especially characteristic of the genus *Hyaenictitherium* (Werdelin et al. 1994). However, specimen LT 10032 is smaller than the known range of *H. namaquensis* would allow (Werdelin et al. 1994), and, in view of the known variability of Miocene hyaenid dentitions (Werdelin 1988a, 1988b), the Lothagam form is likely to represent a taxon that is distinct from *Hyaenictitherium namaquensis.* The same criterion differentiates the Lothagam form from *H. hyaenoides* and *H. pilgrimi,* although *H. parvum* is similar in size to LT 10032. The Lothagam specimen differs from *H. parvum* as illustrated by Khomenko (1914:plate I) and Semenov (1989:figure 37) in the more hypsodont P_2, the more convex anterior margins of P_2–P_3, and the less prominent anterior accessory cusp on P_4. These differences, together with the geographic distance that separates the respective localities, suggest but do not prove that the Lothagam material represents a new species of *Hyaenictitherium.* In Eurasia, *Hyaenictitherium* ranges from ca. 10.6 Ma to 5.2 Ma, with *H. parvum* present from 7.4 Ma to 5.2 Ma, thus providing a temporal match for the Lothagam specimens.

Genus *Hyaenictis* Gaudry, 1861
cf. *Hyaenictis* sp.
(Figure 7.21C–G; tables 7.14, 7.15)

Lothagam Material

- *Lower Nawata:* 23057, Rt. lower horizontal ramus fragments (root /C, roots and posterior crown of P_2, roots of P_3, partial P_4, and M_1 missing nearly all of the buccal side); 23033, partial Rt. lower ramus (worn and damaged P_4, the roots of P_2–P_3 and possibly the alveolus for P_1) (figures 7.21C–D); 23089, partial Rt. Mt III (figure 7.21G).
- *Upper Nawata:* 28772, distal radius fragment.
- *Apak Member:* 23930, distal fragment Rt. radius (figure 7.21E–F).

The postcranial specimens are only tentatively referred to the same taxon as the mandibular and dental specimens.

Specimen 23057 appears to lack a P_1, although due to the fragmentary nature of the specimen this cannot be definitely determined. The second lower premolar has a low main cusp and a low but distinct posterior accessory cusp. The posterior end of this tooth is very

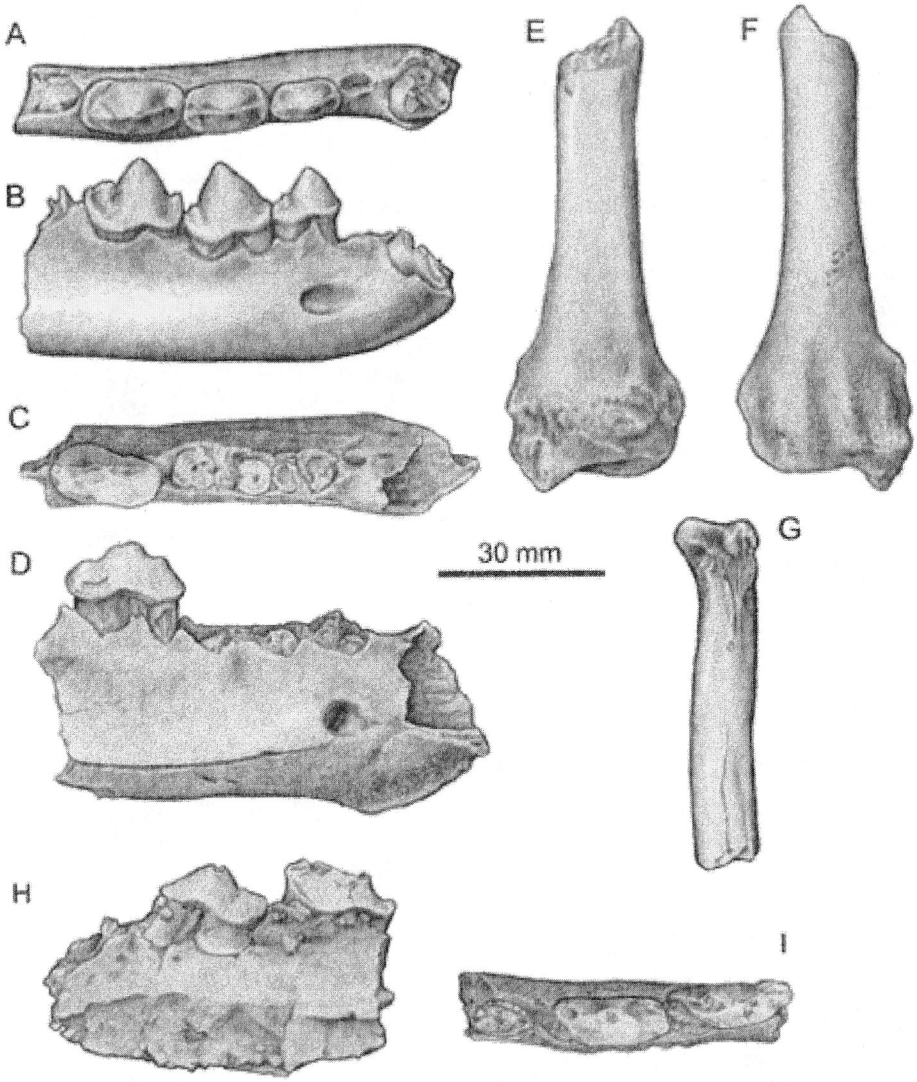

Figure 7.21 *Hyaenictitherium* cf. *H. parvum* partial right mandibular ramus, KNM-LT 10032: A = occlusal view; B = lateral view; C = *Hyaenictis* sp., KNM-LT 23033, partial right mandibular ramus, occlusal view; D = lateral view; E = *Hyaenictis* sp., KNM-LT 23930, right distal radius, cranial view; F = caudal view; G = *Hyaenictis* sp., KNM-LT 23089, partial right Mt III, lateral view; H = *Ikelohyaena* cf. *I. abronia* partial left mandibular ramus, KNM-LT 23947, lateral view; I = occlusal view.

square. The third lower premolar seems to have been large, long, and relatively slender. The fourth lower premolar is a slender tooth. The anterior accessory cusp is small but distinct and is clearly separated from the main cusp. The main cusp is high and narrow with a convex anterior border. The posterior accessory cusp is large, and there is a posterolingual cingulum crest that is much less developed than in *Hyaenictitherium* and has enamel with very slight beading. The lower carnassial has a large anterior cingulum. The paraconid is larger than the protoconid. The metaconid is large but low. The talonid was probably two-cusped with a high entoconid and lower hypoconid.

In 23033 (figure 7.21C–D) the fourth lower premolar has a low but distinct anterior accessory cusp that rises rapidly from the base of the crown and is set off from the main cusp by a shallow valley. The main cusp is small and conical. The remainder of the tooth has been severely damaged, presumably by taphonomic processes that have shorn off some of the enamel and rendered most other features indistinguishable. The third lower premolar was long and apparently slender, while the second lower premolar was relatively short and compact. Anterobuccal to the roots of the P_2 there is a small pit that may have housed a first lower premolar. However, this pit does not quite match the alveolus of P_1 in other hyaenids in either morphology or position, so the presence or absence of P_1 in this specimen must be considered moot. The ramus is robust but not massive. The symphysis is large with a distinct "chin" and ends at the level of the anterior root of P_2. A single large mental foramen lies ventral to P_2.

Specimens 23930 (figure 7.21E–F) and 28772 are nearly identical in size and morphology. Both are broken, worn, and abraded, and few morphological features can be distinguished on either. What can be seen indicates that they should be assigned to the Hyaenidae. They are closely similar to but larger than material referred to *Hyaenictitherium* from China (Lewis and Werdelin, unpublished observations). Judging purely by size, they most likely belong to the cf. *Hyaenictis* sp. described above, but it cannot be entirely precluded that they belong to a taxon not identified on the basis of craniodental material.

The partial right Mt III, 23089 (figure 7.21G), lacks the distal end. It has a broad articulation for the magnum and is otherwise also quite robust. It is larger than available material from China and Greece (Pikermi and Samos), except for that definitely referred to *Adcrocuta eximia.* On the basis of size, this specimen is most plausibly assigned to cf. *Hyaenictis* sp.

Discussion

A number of attributes of 23057—such as the shape of P_2, the development of the posterior cingulum cusp on P_4, and the development of the M_1 talonid, as well as the size of the specimen—indicate that it belongs to the genus *Hyaenictis.* A character that argues against this attribution is the presence of a metaconid on M_1, a feature that is not present in *Hyaenictis* as conceived by Werdelin et al. (1994). However, in the Langebaanweg material there are two specimens, PQ-L21788 and PQ-L21792, that are close to the Langebaanweg *Hyaenictis, H. hendeyi,* in all respects except for the presence of a metaconid in M_1 and the absence of M_2. These specimens were left as indeterminate by Werdelin et al. (1994). The presence of specimen 23057 at Lothagam, which shows the same sort of character combination (although the presence or absence of M_2 is unknown in this specimen), might argue for *Hyaenictis* being polymorphic in this respect and for the presence of a species of *Hyaenictis* at Lothagam. Specimen 23033 is of the general size and appearance of *Hyaenictis* and probably represents the same taxon as 23057.

Postcranial material of hyaenids not directly associated with craniodental material is at present almost impossible to identify to species or even genus. This is mainly due to the very limited amount of work done on fossil hyaenid postcrania up to the present time. As noted, it cannot be precluded that one or more of the above-mentioned specimens pertain to a taxon or taxa not present in the craniodental material. However, it is more likely that they do belong to those taxa identified on the basis of craniodental material, and these postcranial fragments do not provide any reason to increase the number of hyaenid taxa present in the Lothagam fauna beyond the four already discussed.

Genus *Ikelohyaena* Werdelin and Solounias, 1991
Ikelohyaena cf. *I. abronia*
(Figure 7.21H–I; tables 7.16, 7.17)

Lothagam Material

- *Lower Nawata:* 23947, partial Lt. horizontal ramus (P_4–M_1 and posterior root of P_3, severely abraded and possibly etched) (figure 7.21H–I).
- *Apak Member:* 25127, broken proximal fragment and a distal fragment Rt. tibia—referred very tentatively to this taxon.

The fourth lower premolar is broken where the anterior accessory cusp should be and therefore the existence of this cusp cannot be positively determined. However, the appearance of the break gives the impression that there was a protruding cusp that has been broken off and there is therefore every reason to believe that an anterior accessory cusp was in fact present. The main cusp is conical and leans slightly posterad. Its anterior margin is very slightly convex. The posterior accessory cusp is small and there is a small posterior cingulum crest. The whole tooth has suffered extensive postmortem damage and it is not possible to make out any additional features with certainty. The lower carnassial is even more worn and damaged than P_4. The paraconid is longer and wider than the protoconid. Both cusps are broken and worn so their relative heights cannot be determined. The talonid is short and narrow, although the latter feature is exaggerated due to the poor state of preservation. Its cusps are entirely worn away. The metaconid is broken off, but its former presence is attested to by the presence of a polished area of breakage on the protoconid. The posterior root of M_1 appears somewhat pathological in that its basal 7 mm are curved from vertical to posterolabial. The ramus seems to have been relatively deep but not very broad.

Discussion

The combination of characters such as the posterior cingulum crest on P_4 and the proportions of M_1 and presence of a metaconid in this tooth indicate that this material belongs to the Hyaenidae. Comparison with the other identified hyaenid taxa from Lothagam indicates that 23947 is too large to belong to *I. ebu* and too small to belong to a species of *Hyaenictis*. It is more or less the correct size for *Hyaenictitherium* cf. *H. parvum*, but the morphology of P_4 even in this worn state differs substantially from that of 10032, which above has been assigned to that taxon, while the morphology of M_1 in 23947 differs considerably in its proportions from the M_1 of, for example, *H. namaquensis*. However, all the described features, as well as the measurements, are an excellent match for *Ikelohyaena abronia* from Langebaanweg (Hendey 1978a:figures 1a–b, 2a–b; Werdelin et al. 1994) and Laetoli (Barry 1987; Werdelin and Solounias 1991). The low, slender P_4 with low accessory cusps, the M_1 with a short talonid and metaconid present, and the deep but relatively narrow ramus are all features of *I. abronia*. The teeth are relatively long and narrow for that taxon (Werdelin et al. 1994:appendix table 1), but part of the narrowness may be due to the abraded nature of 23947. Due to the nature of the specimen, it is impossible to make a definite referral to species and I therefore assign 23947 to *Ikelohyaena* cf. *I. abronia*.

Family Felidae

The Felidae is another family with an extensive and well-studied fossil record and an extremely broad geographic distribution. The family is known in the fossil state from all continents except Australia and Antarctica, with the bulk of the fossil record being composed of representatives of the sabertooth cats (subfamily Machairodontinae). The main contribution of the Lothagam material to the study of the Felidae is in the information it provides about a key time period in the evolution of those felid taxa that were to become dominant in the African Plio-Pleistocene, such as *Dinofelis* and *Homotherium*.

Subfamily Machairodontinae
Genus *Lokotunjailurus* gen. nov.

Diagnosis

Felidae of large size. Mandibular horizontal ramus slender, ascending ramus relatively tall. Upper canine strongly laterally compressed, but not extremely tall. Serrations present on both anterior and posterior edges. The P^2 present but small and peglike. Upper carnassial long and slender with completely reduced protocone. The P_3 small and single-rooted, though not peglike in appearance. Lower carnassial long, slender, and low. Metaconid-talonid complex absent. Appendicular skeleton relatively slender, lacking extreme machairodont features. First digit of manus very robust, with extremely large claw, more than twice the size of the claws on the other digits.

Hypodigm

Type species only.

Etymology

After *lokotunj*, the Turkana word for "cat."

Lokotunjailurus emageritus sp. nov.
(Figures 7.22–7.30; tables 7.18–7.22)

Diagnosis

As for genus, only species.

Holotype

KNM-LT 26178, a partial skeleton from the lower member of the Nawata Formation that includes cranial fragments, left and right mandibular rami, left and right I^1, right I^2, left I^3, right P^2, a partial left P^4, right /C, right P_4, left and right M_1, left and right scapulae, left and part of the right humerus, left and right radii, left ulna, right scapholunar, right magnum, unciform, trapezium, and trapezoid, right pisiform, right Mc I, left and right Mc II, right Mc III, left and right Mc IV, right Mc V, most proximal, middle, and ungual phalanges of the manus, left femur, right tibia, ?right fibula, right astragalus, left and right calcanea, left and right cuboid, left and right navicular, left and right ectocuneiform, left Mt II, left Mt III, left Mt IV, left Mt V, most proximal, middle, and ungual phalanges of the pes, atlas, C4, C5, C6, C7, T1, T2, T3, ?T4, ?T5, ?T6, ?T7, ?T8, ?T9, ?L4, 3 caudal vertebrae, nine ribs of the left side, eight ribs of the right side, several sternebrae and numerous other smaller fragments (figures 7.23A–C, 7.24–7.28, and 7.29A–B). Measurements are given in table 7.18.

Etymology

After *emagerit*, the Turkana word for "claw," in reference to the disproportionately large claw on the first digit of the manus.

Lothagam Material

- *Lower Nawata:* the holotype, 23053, C/ root and crown fragments; 23055, root C/; 25405, Rt. P^4 (figure 7.23F–G); 23950, Lt. mandibular ramus (roots

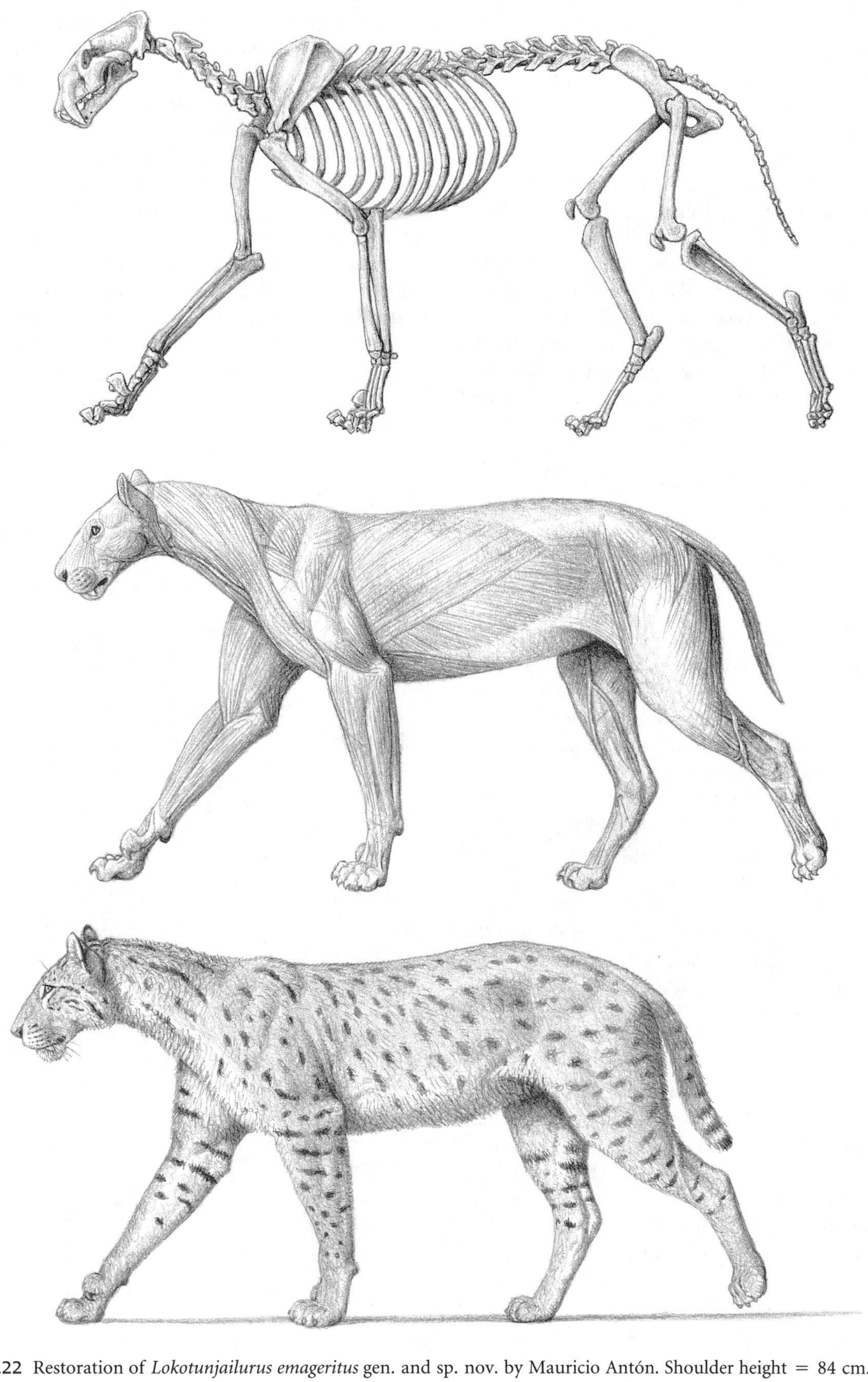

Figure 7.22 Restoration of *Lokotunjailurus emageritus* gen. and sp. nov. by Mauricio Antón. Shoulder height = 84 cm.

of P_4 and M_1); 23060, Lt. M1 (figure 7.23D–E); 25408, proximal Lt. Mc V; 23037, associated posterior limb fragments including proximal femur, calcaneum, a phalanx, tibia (figure 7.29M–N); 23084, Rt. os coxae; 23036, Lt. femur lacking proximal articulation; 22879, distal Lt. tibia; 25410, of proximal Lt. femur fragment; 23940, Rt. calcaneum (figure 7.29K–L); 550, navicular (figure 7.29I–J); 23061, distal metapodial; 23586, Rt. Mt IV (figure 7.29G–H); 25399, middle phalanx lacking distal end; 30308, astragalus.

- *Upper Nawata:* 23941, Lt. maxilla fragment (P^3); 23929, associated postcranial fragments including Lt. and Rt. os coxae, axis, C3, L1; 23928, proximal Lt. ulna; 23949, Lt. astragalus; 23935, Lt. astragalus (figure 7.29E–F); 23931, Rt. astragalus (figure 7.29C–D); 23924, Rt. astragalus; 23955, distal metapodial; 24165, partial axis; 23050, Rt. C/ (figure 7.23H–I); 25133, proximal Rt. Mc II; 25406, distal humerus fragment); 23953, distal metapodial; 28773, distal femur fragments; 30309, Rt. Mt III–Mt IV and distal end of Mt V.
- *Nawata Formation Horizon indet.:* 446, proximal and distal Rt. femur; 23939, Lt. astragalus.

Most specimens referred to the present taxon duplicate elements found in the holotype and morphologically are practically identical to them. The only specimens that cannot be directly or indirectly compared with elements present in the holotype are the upper canines, P^3 and P^4. These are assigned to *Lokotunjailurus emageritus* because they are the right size and they are unquestionably machairodont and because it is unlikely that there would be two distinct, equally large species present in the material. Measurements are given in tables 7.19 to 7.22. The following description will in the main be based on the holotype. When information from other specimens is included, this will be specifically stated.

Cranium 26178 has unfortunately been reduced to a number of small fragments. There are three somewhat larger pieces: a portion of the right maxilla and jugal with a fragment of the orbit and lacrimal foramen and the anterior two-thirds of the zygomatic arch, a part of the left and right premaxilla with alveoli for I^1–I^3 and part of the canine alveolus, and a part of the occiput with the posterior part of the occipital and dorsal parts of the nuchal crests.

The infraorbital foramen is large and located just anteroventral to the orbit. The lacrimal foramen is also large. The zygomatic arch is not very robust in the preserved part. The premaxillae seem to have been slightly procumbent, with a short shelf in front of the nasal opening. The occiput is high and narrow and has prominent occipital and nuchal crests. There are large, dorsally situated pits for the attachment of the nuchal musculature. The one on the left side is longer than that on the right.

The mandible (figure 7.23A–C) is long and slender. The symphysial rugosity has a high relief. It runs from the alveolus of the canine to the ventral extremity of the anterior part of the ramus but is anteroposteriorly relatively short. The incisors and canines are set in a gently curving row. A marked ridge runs from the posterior end of the canine alveolus to the anterior end of the alveolus for P_3. There is only a very small vertical antero-ventral projection, but the anteriormost part of the mandible is mediolaterally angled, which creates a shallow fossa on the anterolateral part of the ramus. In this fossa is set a groove that runs posterad into the anterior mental foramen, the opening of which is directed anterolaterally and situated about halfway along

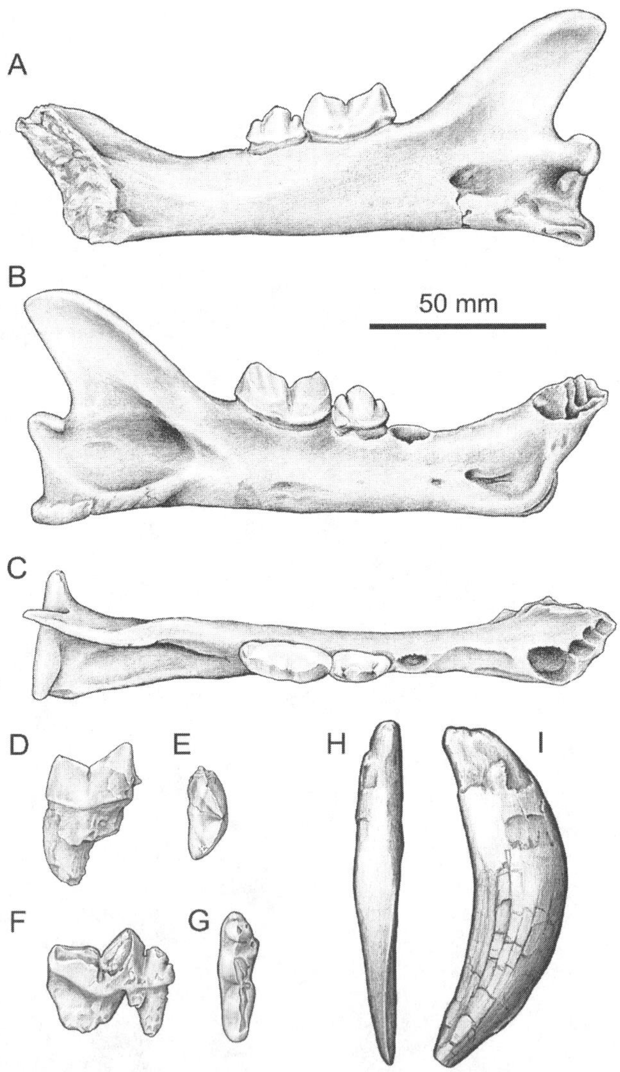

Figure 7.23 *Lokotunjailurus emageritus* gen. and sp. nov., holotype right mandibular ramus, KNM-LT 26178: A = medial view; B = lateral view; C = occlusal view; D = KNM-LT 23060, left M_1 lingual view; E = occlusal view; F = KNM-LT 25405, right P^4, lingual view; G = occlusal view; H = KNM-LT 254051, right C/, anterior view; I = lateral view.

the diastema between /C and P_3. A second, smaller mental foramen is located posterior and slightly ventral to the first, below the anterior end of the alveolus for P_3. The mandible is shallowest just anterior to P_3, then deepens gradually until it reaches the ascending ramus of the coronoid process. The masseteric fossa is small but relatively deep; it extends anteriorly almost to the posterior end of M_1 and is bounded dorsally by a shelf that extends partway up the coronoid process and ventrally by a prominent lateral masseteric shelf. The angular process is relatively large and broad. On the medial side it is formed into a small (5 mm wide) shelf. The condyle is broad but shallow. The coronoid process is intermediate in size between that of pantherines and that of later sabertooths such as *Megantereon*. It does not extend as far posteriorly as the condyle.

Two probable upper first incisors are present in the material. The I^1 is a small tooth with a narrow and deep root. The crown is worn apically. The wear facet angles about 30° from lateral to medial and slightly from anterior to posterior. There is no lingual cusp or shelf.

A probable right upper second incisor is present. It has a deep root that is wider anteriorly than posteriorly. A wear facet starts on the buccal side just below the apex, crosses over the apex, and extends down on the lingual side nearly to the root. There is no lingual cusp or shelf, but there is a slight swelling at the base of the enamel.

The left upper third incisor is present in the material. This tooth is more robust than I^1 or I^2 and has a short, deep, and narrow root. The medial side of the tooth is flat, while the lateral is convex. A wear facet starts on the buccal side halfway down the crown, crosses the apex, and extends down the lingual side to the root. There is a slight lingual swelling at the base of the enamel and a small medial cusplet that is relatively long in the buccolingual direction.

Upper canine 23050 (figure 7.23H–I) is strongly laterally compressed but not very hypsodont. The crown is complete, but the enamel is formed into a mosaic of smaller pieces through a series of minor longitudinal and transverse cracks. The crown is distinctly recurved, and there are small serrations on both the anterior and posterior margins. The crown makes up slightly more than 50 percent of the total height of the tooth.

The second upper premolar is strongly reduced and is oval in apical view. There are two cusps. The larger cusp is placed centrally in the tooth and has a wear facet that extends down its lingual side. The posterior accessory cusp is prominent, somewhat trenchant, and set well away from the main cusp; it has no wear facet. Anterior to the main cusp there is a slight swelling, which indicates a vestigial anterior accessory cusp.

The third upper premolar (23941) has a large anterior accessory cusp that is set away from the main cusp and curved slightly back on it. The anterior cusp is set slightly lingual to the main axis of the tooth. The main cusp is directed somewhat posteriorly but is not very tall. The posterior accessory cusp is about equal in size to the anterior accessory cusp. There is also a small posterior cingulum cusp present behind the posterior accessory cusp.

Upper carnassial 25405 (figure 7.23F–G) is long and slender. There is a small preparastyle at the anteriormost extremity, just anterior to a tall, trenchant parastyle. The paracone is very tall and slender and, together with the long and low metastyle, forms an extensive cutting blade. The protocone is completely reduced; its former existence is merely attested to by the presence of a reduced anterolingual root and a slight lingual swelling. The anterobuccal root is short and straight, while the posterior root extends from the middle of the paracone to the posterior end of the tooth. Neither root is deep. A weak anterobuccal cingulum lies just behind the preparastyle and extends as a basal swelling to the anterior end of the metastyle. There is no discernible lingual cingulum. The preparastyle is unworn, but the parastyle has a small wear facet on the apex, descending down some 2 mm on the lingual face. The paracone and metastyle are extensively worn from the middle of the paracone to the posterior end of the metastyle. In this region the enamel has worn away, exposing the dentine and leaving a very sharp cutting blade at the apex of the tooth. At the carnassial notch this wear facet reaches the base of the enamel. On the paracone it has worn slightly more than, and on the metastyle slightly less than, halfway down the enamel.

The third lower premolar is not represented in the sample, but in the holotype mandible the alveolus indicates that this tooth was small and single-rooted, though considerably longer than wide and thus probably not peglike in shape. The right fourth lower premolar is present *in situ* in the mandible. It is a well-developed tooth with a tall main cusp and prominent anterior and posterior accessory cusps. The accessory cusps are set well off from the main cusp and are about equal in size. The main cusp is worn from the apex down on to the posterobuccal side. There is also a wear facet on the posterobuccal side below the posterior accessory cusp.

The lower carnassial is long and relatively low. The paraconid is somewhat lower, shorter, and broader than the protoconid. Both lower carnassials of the holotype are heavily worn on the buccal side, while an isolated M_1 (23060; figure 7.23D–E) is unworn. On the paraconid the wear facet is smooth, but on the protoconid there is incipient vertical ridging of the wear facet, which is evident from a slight indentation of the facet in the middle of the protoconid. A small vertical groove on both the lingual and the buccal sides separates the protoconid from the posteriormost part of the tooth. Posterior to this groove there is a slight swelling at the base of the enamel indicative of a vestigial talonid. A basal swelling of the enamel occurs around the entire crown. The anterior root is present in 23060; it is somewhat taller than the crown and very robust.

The left scapula is missing a small part of the dorsal border of the supraspinous fossa but is otherwise undistorted (figure 7.24). The right scapula is complete but cracked in several places. The supraspinous fossa is more or less semicircular in shape and larger than the infraspinous fossa. The latter is triangular in overall shape. The caudal angle juts out slightly from the border. The spine is low and long. It curves over to the caudal side for almost its entire extent. The acromion is large. The glenoid cavity is oval. The supraglenoid tubercle is robust and sends off the coracoid process in an anteroventral direction. Overall, this process is small and slender. Beneath the spine the glenoid cavity is concave.

The complete left humerus and the head and distal condyle of the right humerus are represented (figure 7.25A–B). The head is large. It is strongly arched anteroposteriorly, less so mediolaterally. The greater tubercle is relatively small and has a concave face, although this concavity is slightly exaggerated because the proximal part of the shaft was compressed. However, the greater tubercle is relatively long in the anteroposterior direction, and the lesser tubercle is robust but low. As noted, the shaft is damaged distal to the greater tubercle. This is true also on the medial side, where the shaft has completely collapsed inward. The distal part of the shaft and condyle are undistorted, apart from a crack just above the olecranon fossa, which causes some slight (about 1 mm) displacement of the proximal and distal halves of the entepicondylar bar relative to each other. There is no supratrochlear foramen. The olecranon fossa is large and shallow. There is a smaller fossa between the olecranon fossa and the medial epicondyle. The latter is prominent and provides massive attachment for the carpus and digital extensors. The entepicondylar foramen opens proximocaudally rather than directly caudally. Both radii are preserved (figure 7.25C–D).

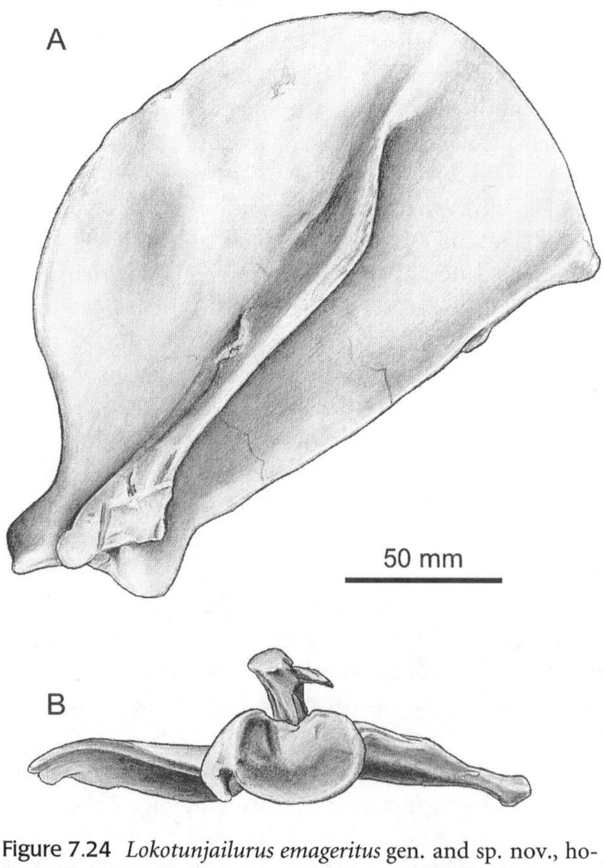

Figure 7.24 *Lokotunjailurus emageritus* gen. and sp. nov., holotype left scapula, KNM-LT 26178: A = lateral view; B = ventral view.

In both, the caudal surface of the shaft has been pushed craniad, causing the shaft to collapse inward. The articular fovea is shallowly concave. The articular circumference is very broad and has a marked medial lip overhanging the shaft. The distal articulation is broad and high. The ulnar notch is large. The grooves for the extensor digitis communis, extensor carpi radialis, and abductor pollicis longus are shallow but very broad. The carpal articular surface is broad and very narrow. It extends some way up the caudal side of the radius.

The left ulna (figure 7.26) is cracked below the ulnar tuberosity and halfway down the shaft but is undistorted. The olecranon is low but very broad; it forms a prominent overhang on the cranial side. A large pit lies proximomedial to the trochlea. The triceps groove is broad and deep. The coronoid articular surface is directed markedly craniocaudally. The shaft is robust and has a long, prominent rugosity for the abductor pollicis longus. The styloid process is very robust and flares out strongly from the shaft.

The left femur is represented by its proximal third. It is broken at the trochanteric fossa and beneath the head. The head has been restored in position. The right femur is represented by the proximal quarter and most of the distal three quarters. These pieces do not articulate, but there can only be a maximum of 30 mm of the shaft missing between them. The head is large and round. The greater trochanter is craniocaudally long and does not exceed the head in height. The trochanteric fossa is wide and deep. The lesser trochanter is formed into a process that juts out considerably (nearly 10 mm) from the shaft. The lateral supracondylar tuberosity is large, considerably more so than the medial. The lateral condyle is larger than the medial. The patellar groove is broad and shallow.

Specimen 25410 is the proximal end of a left femur that shows poor preservation with many cracks and

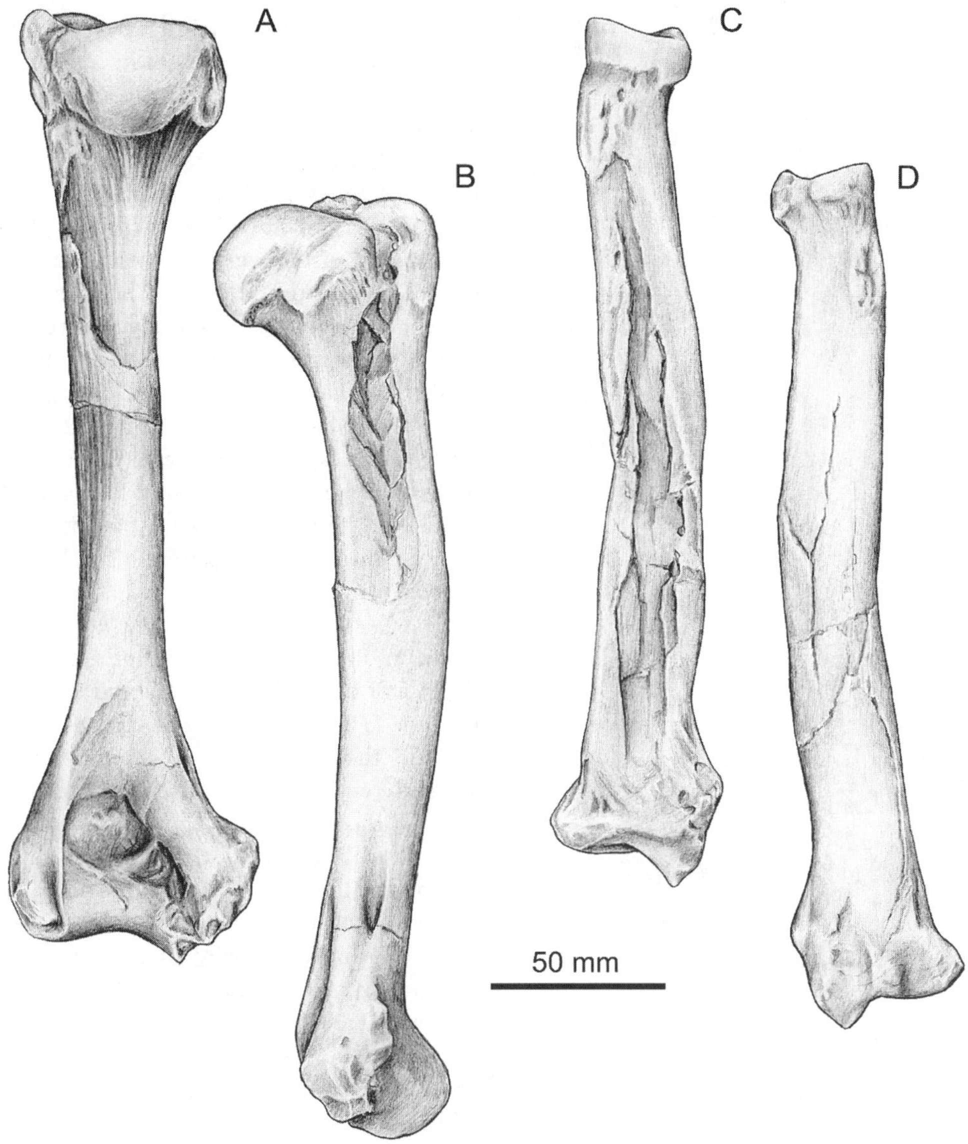

Figure 7.25 *Lokotunjailurus emageritus* gen. and sp. nov., holotype, left humerus and left radius, KNM-LT 26178: A = left humerus, caudal view; B = medial view; C = left radius, caudal view; D = cranial view.

breaks. The head is round and set on a long and anteroposteriorly compressed neck. The greater trochanter is very prominent and knoblike. The trochanteric fossa is deep. The articular surface for the vastus lateralis and intermedius is bounded proximally by a prominent groove.

The right tibia (figure 7.27A–B) is complete, but the shaft has been broken and restored in two places and the proximomedial part of the shaft has collapsed inward. The intercondyloid areas are very wide and shallow. The tibial tuberosity is low but very broad. The attachment areas for the flexor hallucis longus, flexor digitis longus, and popliteus are very marked on the caudal surface and are separated by low ridges. The medial malleolus is very robust, with a deep groove for the tendon of the tibialis caudalis muscle.

The ?right fibula (figure 7.27C–D) is complete except for missing a small fragment from the shaft. Both the head and the lateral malleolus are extremely broad, while the shaft is long and very slender. The shaft has a triangular cross section with the base of the triangle facing the tibia.

Specimen 23084 appears to be a subadult os coxae. It has a modestly developed ischiatic tuberosity an small ischiatic spine. The obdurator foramen forms a broad oval with the long axis directed from anterolateral to posteromedial. The acetabular fossa is deep. The iliopubic eminence is low. The caudal dorsal iliac spine is not well defined in this specimen.

Only a limited number of vertebrae, mostly from the cervical and thoracic regions, are available. The atlas is very broad and has a large vertebral foramen that is

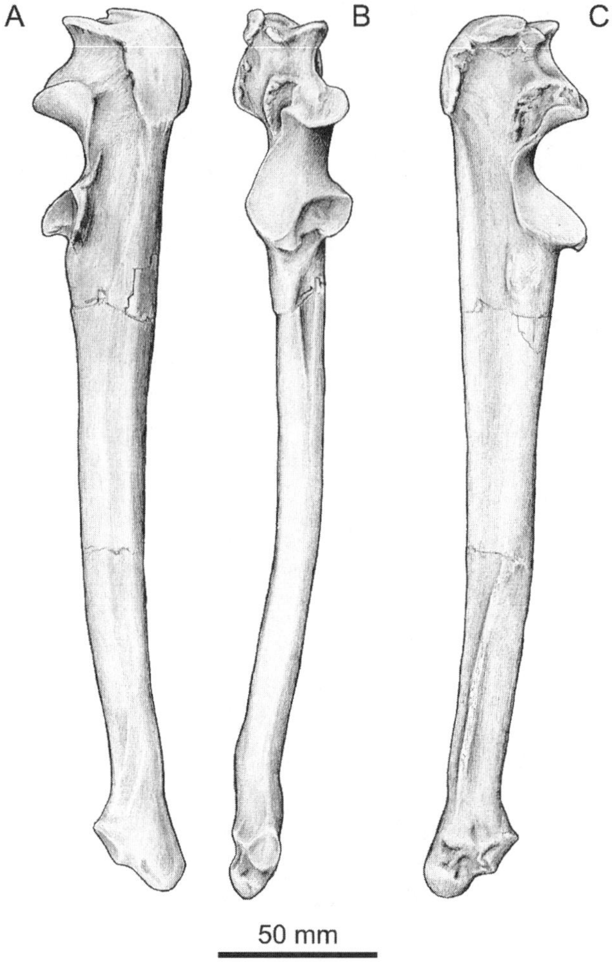

Figure 7.26 *Lokotunjailurus emageritus* gen. and sp. nov., holotype left ulna, KNM-LT 26178: A = lateral view; B = cranial view; C = medial view.

almost square in cross section. The anterior part of the left wing and most of the right are missing. The lateral vertebral foramen opens wholly laterally. The alar notch opens anterolaterally, as it is anteriorly overgrown by the lip of the articular fovea. The body is relatively short, while the wings are long and have a marked distolateral angle. There is no transverse foramen.

The fourth cervical vertebra is present in the holotype. It is large and robust. The neural spine is broken, as are the transverse processes, while the arch is sheared to the left, as seen in anterior view. The transverse processes of C4 are broad. The sixth cervical vertebra has a broken spine and transverse processes and is sheared to the left and ventrally. There is a piece missing from the arch to the right of the neural spine. The C6 transverse foramen is large, and the transverse processes have a long base. The seventh cervical vertebra lacks the neural spine and part of the right side of the arch, as well as the left transverse process. The C7 centrum is shorter than that of C6.

The first thoracic vertebra is nearly complete, lacking only the neural spine. The base of the neural spine is robust, and the transverse processes are large and robust with large costal foveae. The second thoracic vertebra is represented by the centrum only, while the third thoracic vertebra is missing the right transverse process and part of the right side of the centrum. The neural spine is long and wholly vertical, with no discernible caudal lean. The centrum is short. The transverse process is large and has a large costal fovea. The fourth thoracic vertebra is missing the left transverse process and part of the left side of the centrum. The spine is long and leans somewhat caudad. The transverse process is robust, with a large costal fovea. The ?fifth thoracic vertebra is complete but is slightly compressed on the right side. The spine is high and becomes craniocaudally shorter toward the dorsal end. The transverse process is robust.

The ?sixth thoracic vertebra is represented by the centrum only. The ?seventh thoracic vertebra lacks the distal half of the neural spine and the right transverse process. The left transverse process is robust. The ?eighth thoracic vertebra is complete but slightly sheared ventrally and to the left. The spine leans caudally to a considerable degree. The transverse process is robust. The ?ninth thoracic vertebra is complete except for the distal part of the neural spine, and it is slightly sheared ventrally and to the right. The spine leans still more caudally than does that of ?T7, and the transverse processes are short and robust. The ?tenth thoracic vertebra is complete, although the pedicle of the right transverse process is slightly damaged. The neural spine is shorter than that of ?T8, with strong caudal lean. The transverse process is short and robust.

The ?fourth lumbar vertebra is represented by the centrum and part of the left transverse process. The latter has a very long base and is flat. The centrum has a strong ventral keel.

Most of the right manus was found articulated and has been left in that condition (figure 7.28). It includes the scapholunar, cuneiform, trapezium, magnum, unciform, pisiform, metacarpals I–V, proximal phalanges I and III, and ungual phalanx I. Most of these are articulated as in life, but the scapholunar has been displaced laterally, the pisiform has become attached to the distal end of the ungual phalanx I, the unciform is attached to Mc V, and the proximal phalanx III is displaced and attached by its shaft to Mc III. The most outstanding feature of this specimen is the huge first digit, and especially the ungual phalanx, which, as we shall see, is several times larger than the other ungual phalanges of the manus. This leads to interesting functional considerations that will be explored elsewhere. Aside from this, the manus is unremarkable, although

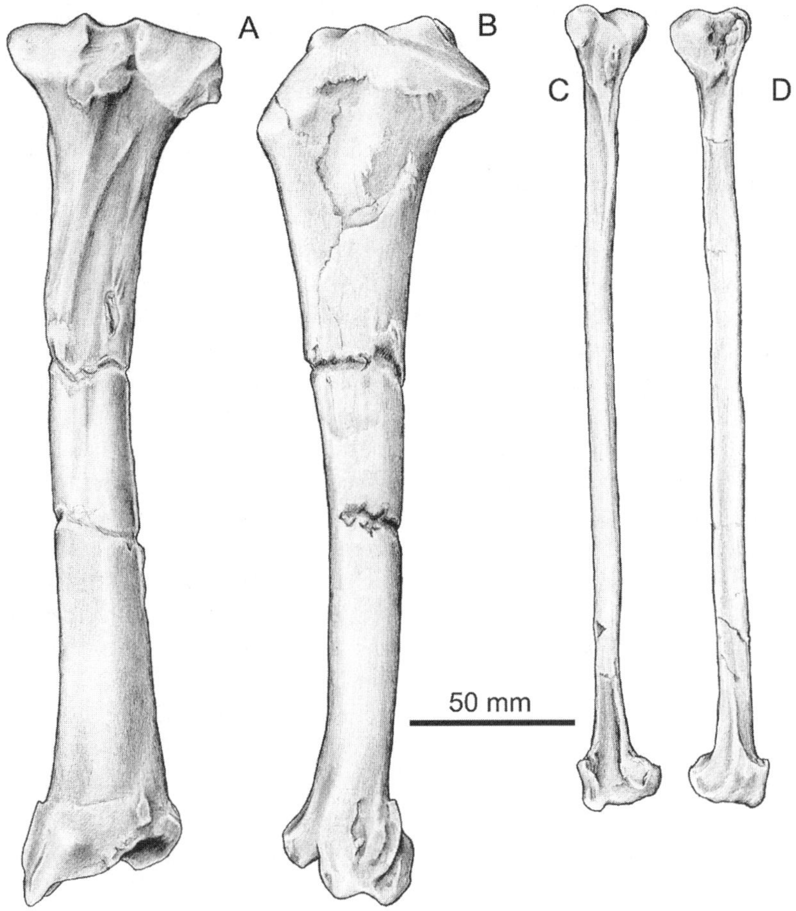

Figure 7.27 *Lokotunjailurus emageritus* gen. and sp. nov., holotype KNM-LT 26178: A = right tibia, caudal view; B = medial view; C = ?right fibula, lateral view; D = medial view.

we may note that Mc II is proximally set well above Mc III–V and that Mc III extends further distally than Mc IV when in articulation.

The scapholunar is very large and robust. The articular surface for the radius is long and narrow and has strong dorsolateral to ventromedial arching, as well as a large medial facet for the pisiform.

The cuneiform, unciform, magnum, and trapezium are articulated with the right manus and present no particular distinguishing characteristics that are visible on the specimen.

The pisiform is relatively narrow with broad proximal and distal processes set at 90° to each other in the normal fashion.

The first metacarpal is a large bone with a prominent, blunt dorsal keel. The ventral side is broad and flat.

The proximal phalanx of digit I is broad but short, a square bone that is dorsoventrally somewhat flattened. The medial basal articulation is smaller than the lateral. The articulation for the ungual phalanx is wide and flat.

The ungual phalanx of digit I has already been mentioned. It is a huge bone, some 60 mm long by 45 mm deep. The articular surfaces for the extensor and flexor muscles are in proportion to the size of the specimen. The isolated left ungual phalanx is also present. The ungual process is completely missing on the right side and nearly completely on the left. What little that can be discerned on the left side indicates that the ungual process must have been very narrow, more so relative to the ungual processes of the other digits than in modern cats.

Both second metacarpals are present, the right in the articulated right manus. The base of Mc II is very wide and has a large overlap with the proximal end of Mc III. These two metacarpals abut for a large proportion of their length, thus creating a single strong unit. The shaft of Mc II is triangular in cross section; its dorsal side is flat and the ventral (palmar) side is keeled. The shaft is slightly curved toward medial. The head is large and somewhat asymmetric, with the medial side narrow but longer than the lateral side.

The right proximal phalanx of digit II has been tentatively identified. It is large and robust and has a slight lateral curvature of the shaft. The proximal articular surface is somewhat asymmetric, as it is deeper on the lateral side.

Both middle phalanges of the manus digit II have been identified. The shaft is slender and has a marked dorsal keel and strong curvature on the lateral side. The head is low and broad.

The ungual phalanx of digit II has been identified from both sides. It is long and high. The articular surfaces for the extensors and flexors are large. The ungual process is broken on both sides, but what is left indicates that it was high and very narrow.

The third metacarpal is present only in the articulated right manus. The shaft is more rounded and less triangular than that of Mc II; it is dorsally quite flat. The head is almost symmetric.

The proximal phalanx of the right digit III is present in the articulated manus, while the left has been recovered isolated. It is a long, robust, nearly symmetric bone. The base is very wide, although the head is narrow. The shaft is somewhat dorsally arched.

The middle and ungual phalanges of manus digit III have been tentatively identified. The middle phalanx is relatively small and slender, and the shaft has a marked curvature on the lateral side. The head is broad and low. The ungual phalanx is a robust bone that has large attachment areas for the flexors and extensors. The tip of the ungual process is broken; the remainder is long, curved, and narrow, though not as narrow as in digit II.

The fourth metacarpal is known from both sides, the right articulated with the right manus and the left isolated. It is a long and slender bone that has a relatively narrow shaft and blunt keels ventrally and dorsomedially. The articulation at the base is almost semicircular, and there is a large lateral articular surface for Mc III. The head is broad and symmetric.

The proximal phalanx of manus digit IV has been tentatively identified from both sides. It is long and slender and has a moderate curvature on the lateral side. The base is taller than wide. There is slight dorsal arching. The head is broad and low.

The middle phalanx of manus digit IV is not known, but the ungual phalanx has been tentatively identified on both sides. It has large articular surfaces for the extensors and flexors. The ungual process is broken but was very narrow.

The right fifth metacarpal, present in the articulated manus, is relatively short and slender. The proximal articulation is strongly arched. The head is strongly asymmetric, as the medial side is much larger than the lateral.

The middle phalanx of manus digit V has been tentatively identified on both sides. It is short and flat and has a marked curvature on the lateral side and a broad, flat head.

The astragalus (figure 7.29A–F) is represented by several specimens in the available material, including the right astragalus from the holotype skeleton. It is a short, square bone. The head is transversely wide and dorsoventrally narrow. It is set on a short, broad neck. The articular surface of the head is confluent with the medial articulation for the calcaneum. This articulation almost reaches the trochlea. The lateral calcanear articulation is broad and evenly concave. The trochlea reaches a short distance onto the plantar face of the astragalus and is strongly angled from medial to lateral.

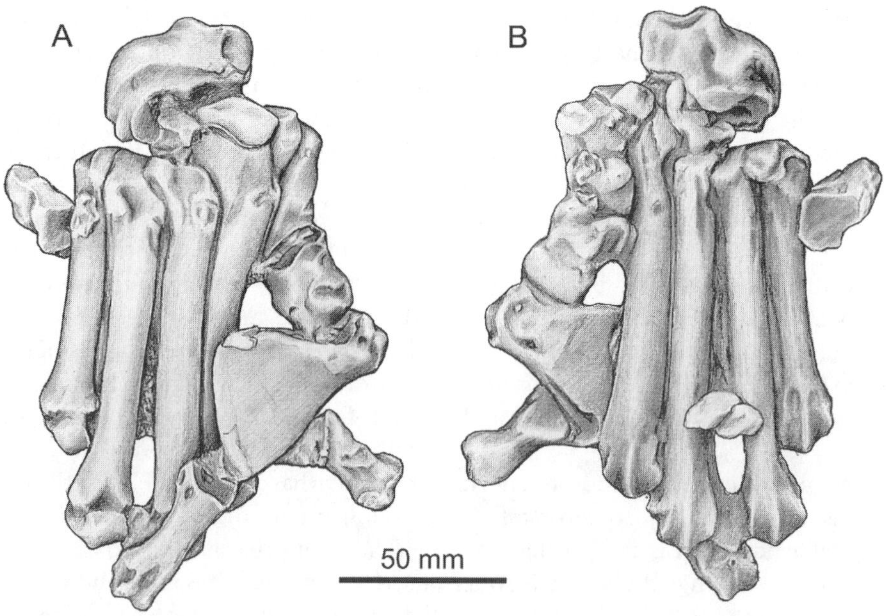

Figure 7.28 *Lokotunjailurus emageritus* gen. and sp. nov., holotype right manus, KNM-LT 26178: A = dorsal view; B = ventral view.

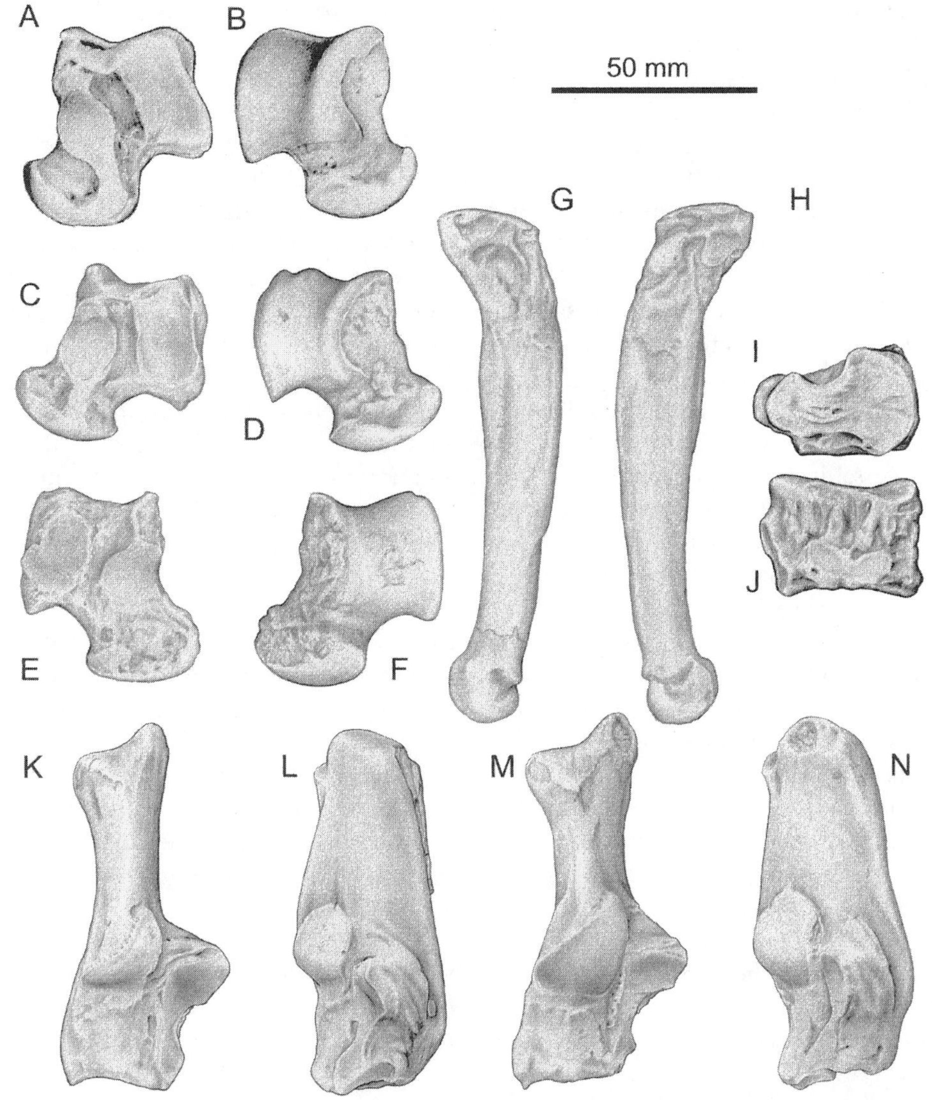

Figure 7.29 *Lokotunjailurus emageritus* gen. and sp. nov., tarsals and metatarsal: A = KNM-LT 26178, holotype, right astragalus, ventral view; B = dorsal view; C = KNM-LT 23931, right astragalus, ventral view; D = dorsal view; E = KNM-LT 23935, left astragalus, ventral view; F = dorsal view; G = KNMT-LT 23586 right Mt IV, lateral view; H = medial view; I = KNM-LT 550, navicular, proximal view; J = lateral view; K = KNM-LT 23940, right calcaneum, dorsal view; L = medial view; M = KNM-LT 23037, right calcaneum, dorsal view; N = medial view.

Several calcanea are present in the material (figure 7.29K–N), including both the left and the right calcaneum from the holotype skeleton, though the right is somewhat damaged. The calcaneum is long and robust and has a dorsoventrally long and mediolaterally narrow tuber. The medial process of the tuber is very tall, and the sulcus for the biceps and gastrocnemius muscles is broad. The groove for the flexors is large. The medial articular surface for the astragalus is long, reaching from the sustentaculum to the base of the calcaneum. The lateral articulation for the astragalus is smaller. It reaches some 10 mm up on the lateral side of the tuber, while it is quite short distally, ending about 25 mm from the base. The base is set at right angles to the main axis of the calcaneum.

The pes, especially the tarsus, is not as complete as the manus but is still represented by most elements. The navicular is broad and short (figure 7.29I–J). The articular facets for the medial, central, and lateral cuneiforms are distinct on the distal surface. There is no plantar process.

The cuboid is robust. The sulcus for the musculus peritoneus longus tendon is deep and wide and set low on the plantar face. The articulation with the calcaneum is nearly flat, while the articulations for Mt IV–V are set at oblique angles to each other.

The ectocuneiform is a robust bone with a large, proximally situated plantar process. There is a long, low articular surface on the lateral side for the cuboid and a smaller, more distally placed one on the medial side for the medial cuneiform.

The second metatarsal is present in the material. It is long and slender and has a laterally compressed shaft. The shaft has a marked keel on the dorsal side. The head is distinctly asymmetric, as the medial side is larger but shorter than the lateral.

All three phalanges of pes digit II have been tentatively identified. The proximal phalanx is long and slender and has a marked curvature of the shaft. The proximal articulation is asymmetric. The middle phalanx is narrow and low and has a marked dorsal keel. The shaft has pronounced curvature on the lateral side. The head is broad and low. The ungual phalanx is short and has a relatively robust ungual crest. The ungual process is curved and narrow.

The third metatarsal is a long and very broad bone in which the shaft has a flattened cross section. The plantar tubercle is long. The shaft only flares out slightly proximal to the sesamoid fossa. The head is relatively flat, in comparison with the heads of the metacarpals.

All three phalanges of pes digit III have been tentatively identified. The proximal phalanx is long and relatively flat and has some dorsal arching. The middle phalanx is low and flat but has no arching, although it displays moderate curvature on the lateral side. The ungual phalanx has large attachment areas for the flexors and extensors, and it bears a curved ungual process that is less narrow than that of the digits of the manus.

The fourth metatarsal (figure 7.29G–H) is long and has a narrow shaft that is arched dorsally and curved slightly in the lateral direction. The shaft has low proximodorsal and distoventral keels. The head is high and short.

All three phalanges of pes digit IV have been tentatively identified. The proximal phalanx is long and flat and has a straight, dorsoventrally compressed shaft that has some slight dorsal arching. The middle phalanx has a low, flat shaft but has no arching; it is asymmetric and has a dorsolateral keel, but it shows no curvature. The ungual phalanx has large articular surfaces for the extensors and flexors. The ungual process is broken.

The fifth metatarsal is long and slender and has a mediolaterally compressed shaft that curves somewhat toward lateral. The base is asymmetric, as the lateral side is shorter but wider than the medial. The head is short and tall.

All three phalanges of pes digit V have been tentatively identified. The proximal phalanx is long and slender and has a shaft that is somewhat curved on the lateral side; this shaft is not arched. The middle phalanx is short and flat and has a straight, unarched shaft that bears a dorsolateral keel. The ungual phalanx has relatively small attachment areas for the extensors and flexors and a short, slightly curved, narrow ungual process.

Discussion

This taxon resembles species referred to the genus *Machairodus* in many respects. However, it differs in some significant characteristics of the dentition in ways that relate the Lothagam form to *Homotherium.* Sotnikova (1991) has shown that there is a trend from early (early Late Miocene) forms of *Machairodus,* such as *Machairodus aphanistus* and *M. copei,* over later (middle to late Late Miocene) *Machairodus,* such as *M. giganteus* and *M. palanderi* (often referred to *Amphimachairodus;* Beaumont 1978), via the latest Miocene *M. kurteni* to Pliocene *Homotherium.* Features of this trend are, above all, the reduction of the P^4 protocone, reduction of P_3, and narrowing of M_1. This trend is illustrated for the P_4 protocone in figure 7.30, which also shows that *Lokotunjailurus* has achieved the *Homotherium* grade in this feature.

Another feature in which *Lokotunjailurus* differs from *Machairodus* and approaches some *Homotherium* is the single-rooted P_3. No known *Machairodus,* including the dentally derived *M. kurteni* (Sotnikova 1991), has a single-rooted P_3. By contrast, early *Homotherium* from Laetoli (Barry 1987), West Turkana (Harris et al. 1988), and Eshoa Kakurongori (unpublished data) appear to variably have one or two roots on P_3, which suggests that *Lokotunjailurus* may not be in the direct ancestry of *Homotherium.*

Thus, despite the absence of typical *Homotherium* features in the postcranial skeleton, mandible and most of the dentition including the canines, the above listed features indicate that *Lokotunjailurus* is closer to *Ho-*

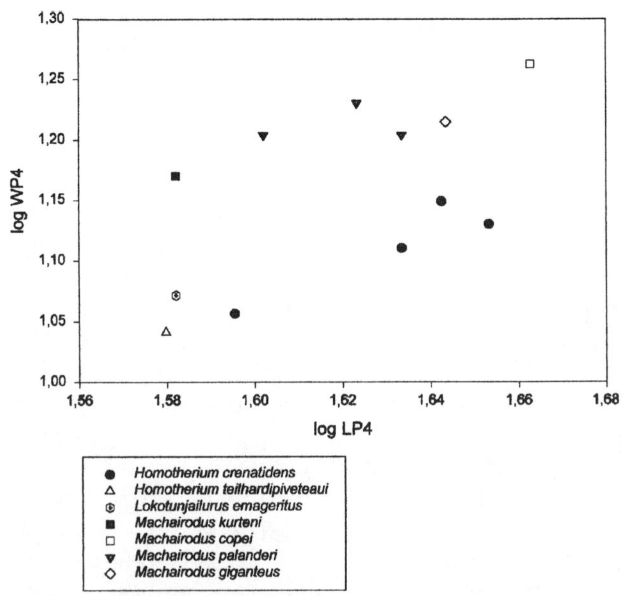

Figure 7.30 Bivariate diagram showing relationship between length of P^4 and anterior width (at protocone) of P^4 for various machairodont taxa as labeled.

motherium than any other known machairodont. The two can tentatively be considered to be sister taxa.

Genus *Dinofelis* Zdansky, 1924
Dinofelis sp.
(Figure 7.31; tables 7.23, 7.24)

Lothagam Material

- *Lower Nawata:* 24041, distal shaft Rt. humerus; 30310, proximal Lt. radius; 23942, partial Lt. C/ (figure 7.31C–D); 561, proximal Lt. femur shaft.
- *Upper Nawata:* 127, Rt. horizontal ramus (roots of /C, P_3, P_4; figure 7.31A–B); 25397, posterior Rt. P_4 (figure 7.31E–F); 25401, proximal Rt. femur shaft; 25398, proximal Rt. radius; 25137, distal Lt. tibia; 23934, proximal Lt. radius and ulna; 25407, proximal Lt. Mc V; 23933, proximal Rt. tibia.
- *Apak Member:* 23696, Rt. calcaneum (figure 7.29G–H); 25129, shaft and distal radius fragments; 28711, Rt. horizontal ramus (partial alveolus for /C, roots P_3, P_4); 25126, proximal Lt. tibia; 25403, phalanx I of the left manus digit 1; 30305, Lt. horizontal ramus (alveoli for /C, P_3, anterior root and posterior alveolus of P_4, anterior alveolus of M_1).
- *Horizon indet.:* 23932, Lt. calcaneum.

The three partial mandibles 127, 28711, and 30305 are very similar to each other. The best preserved is 127 (figure 7.31A–B). The ramus is slender and narrow, and it flares anteriorly on both the lateral and medial sides to form a broad anterior chin. Most of the space in the anterior dental arcade is taken up by the canine, the alveolus for which is large though somewhat mediolaterally compressed. The large canine means that there is not enough space for a regularly curved incisor arcade. Instead, I_2 is set at a level posterior to I_1 and I_3, quite unlike the condition in *Machairodus* spp. (Sotnikova 1991). A long postcanine diastema is topped dorsally by a narrow crest that runs all the way to the alveolus for P_3. Two mental foramina, one below the middle of the diastema and one below the anterior root of P_3, lie at about the same distance from the ventral

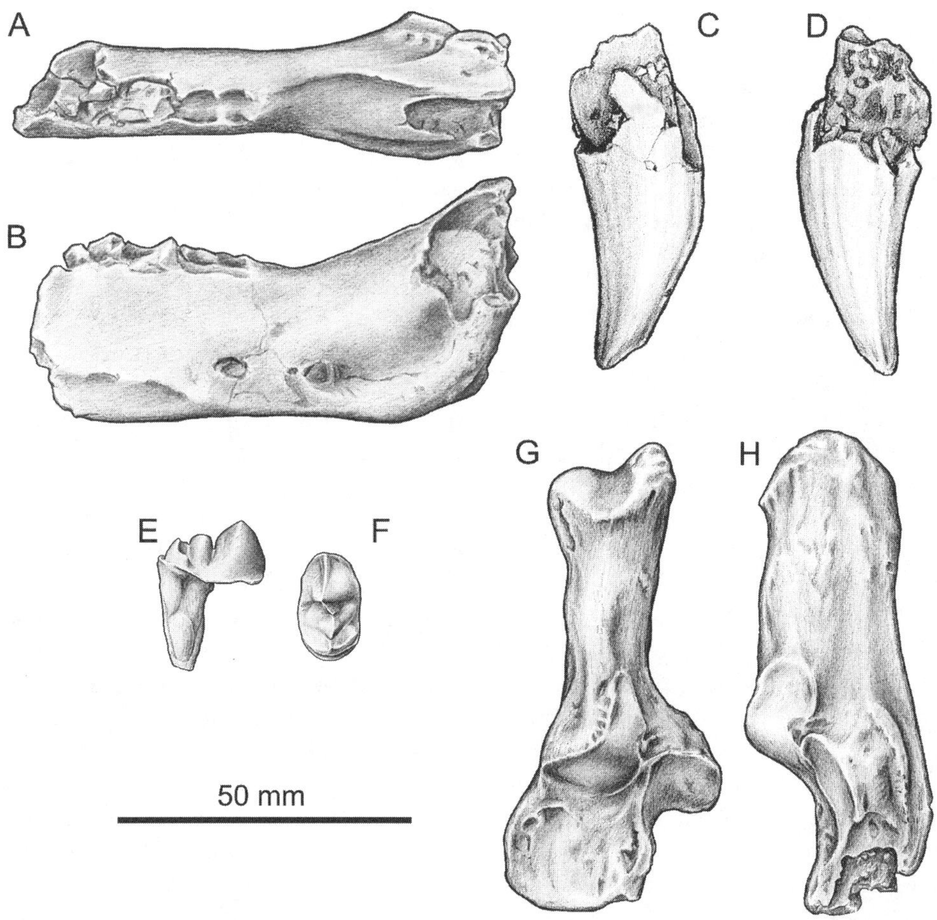

Figure 7.31 *Dinofelis* sp. specimens: A = KNM-LT 127, partial right mandibular ramus, occlusal view; B = right lateral view; C = KNM-LT 23942, partial left C/, medial view; D = lateral view; E = KNM-LT 25397, partial right P^4, buccal view; F = occlusal view; G = right calcaneum, dorsal view; H = medial view.

margin of the ramus. This ventral margin is absolutely straight in the part preserved in all specimens to the extent that, when placed on a flat surface, the entire ventral margin is in contact with the surface. The ramus begins to broaden mediolaterally at the level of P_3 and continues to broaden gradually in its entire preserved length. The symphysis is tall and very short, and it has deep relief, which indicates a strengthening against rotational forces in the sagittal plane. The third lower premolar was slender and two-rooted, as judged from its alveoli. The fourth lower premolar was considerably larger than the third.

Upper canine 23942 (figure 7.31C–D) is broken and somewhat abraded, and the enamel is flaked off some distance from the enamel-dentine juncture. Thus, neither the height of the crown nor its width/length ratio can be accurately determined, though the tooth is clearly mediolaterally compressed and the ratio can be estimated as something less than 0.65. The posterior cutting edge is distinctly developed and set directly at the posterior margin of the tooth, while the anterior edge is set slightly to the medial side. Neither edge shows any sign of serrations.

Specimen 25397 represents the posterior two thirds of a P_4 (figure 7.31E–F). The tooth is relatively narrow in its preserved part, and the widest part is level with the posterior accessory cusp. The main cusp is trenchant and has a leaf-shaped lateral outline and sharp anterior and posterior cutting edges. The large posterior accessory cusp is likewise leaf-shaped and has a sharp cutting edge. The posterior shelf is formed into a sharp crest that runs on both buccal and lingual sides of a median cusp that is separated from the posterior accessory cusp by a narrow notch. Posterior to the crests there is a short posterior cingulum shelf.

Specimen 23041 is a fragment of the distal part of the shaft of the left humerus, broken distally just proximal to the distal articulation. An entepicondylar foramen was present, showing this to be a humerus of a large felid, but otherwise there are no characteristic features of note. It is placed in *Dinofelis* sp. on the basis of size.

Specimens 25398, 30310, and 25129 include parts of the radius. The two former show the proximal end and are very similar in morphology. The articular circumference is deep on the caudal side and is formed into a distinct anterior lip, which overlies the lateral side of the shaft. The articular fovea is shallowly concave. There is a very large radial tuberosity, while the attachment for the abductor pollicis longus (seen in 30310) is relatively deeply indented. 25129 is a fragment of the shaft and distal part of a radius. It shows typical characters of a large felid.

Specimen 25407 is a proximal part of a left Mc V. It shows no distinguishing features apart from being the fifth metacarpal of a large felid, while specimen 25403 is the first phalanx of digit 1 of a large felid. These specimens are attributed to *Dinofelis* sp. on the basis of their size, which is considerably less than that of the homologous elements in *Lokotunjailurus emageritus.*

Two proximal parts of felid femoral shafts, 561 and 25401, may belong to this species, but they are not diagnostic.

Three tibia fragments are represented, 23933 and 25126, right and left proximal fragments, and 25137, a distal left tibia fragment. All three are damaged and abraded and are here placed in *Dinofelis* sp. on the basis of their size match with the other limb bones.

Specimens 23932 and 23696 are left and right calcanea (figure 7.31G–H). The latter has a damaged distal articular surface. The lateral and medial processes of the tuber are prominent and flank a deep groove for the gastrocnemius. The tuber is long and transversely flattened. The medial articular surface for the astragalus is round and has a narrow distal extension that reaches the distal articular surface. This extension is much narrower than in *Lokotunjailurus emageritus.* The lateral articular surface of the astragalus is narrow and reaches some way up the tuber. The flexion halfway down this articular surface is much greater than in *L. emageritus,* thus indicating a tighter articulation with the astragalus. The area between the articular surfaces does not form a groove as in large machairodontines. The distal articular surface is set obliquely relative to the longitudinal axis of the calcaneum.

Discussion

This material has been described in greater detail elsewhere (Werdelin and Lewis 2001). The craniodental material can with some assurance be referred to *Dinofelis,* while the postcranial material is for the most part assigned to this genus through relative size comparisons with the other Lothagam felid material. Compared to other *Dinofelis,* the Lothagam form is relatively small and slender of build, with generally more primitive felid morphology than seen in later species of the genus. An interesting feature is that the first digit of the manus appears to have been relatively large in this taxon, and this condition is similar to, but much less extreme than, the situation in *Lokotunjailurus emageritus,* as discussed earlier in this chapter. Although the material here assigned to *Dinofelis* sp. is not inconsiderable and almost certainly represents a new taxon, I refrain from naming it because of the dearth of properly diagnostic craniodental material, which has made comparisons between this form and other *Dinofelis* difficult. The Lothagam form represents the earliest known *Dinofelis* from Africa.

Genus *Metailurus* Zdansky, 1924
cf. *Metailurus* sp.
(Figure 7.32; table 7.25)

Lothagam Material

- *Lower Nawata:* 23059, Lt. femur.
- *Upper Nawata:* 10030, Rt. C/ (figure 7.32); 23118, Rt. astragalus; 25131, distal tibia; 23927, proximal and distal Rt. tibia; 25404, proximal Lt. Mc III; 28571, Rt. Mt III.

Right upper canine 10030 (figure 7.32) is complete except for missing about 5 mm or so of the apex. It is slightly abraded, and some enamel has flaked off the medial side. The tooth is small and mediolaterally compressed, and it has a width/length ratio of about 0.65. The anterior cutting crest is obliterated by a narrow vertical wear facet that extends from the enamel-dentine juncture to the point at which the tip of the tooth is broken off. At its widest, this facet is about 2 mm. The posterior cutting crest is prominent and shows no sign of serrations. The root is longer and wider than the crown, and it is relatively straight until about 10 mm before it meets the crown, where it begins to curve. Maximum curvature is reached where the root and crown meet. The crown continues to curve apically, but this curvature is slight and gradually decreasing.

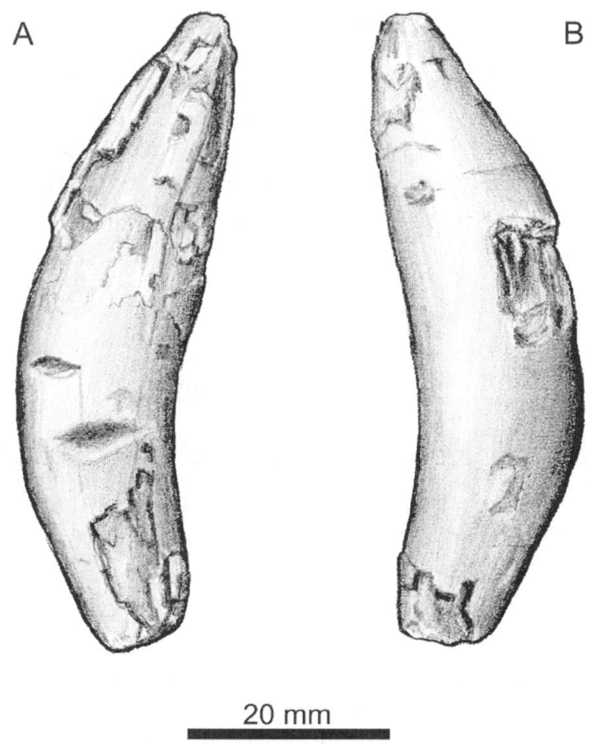

Figure 7.32 cf. *Metailurus* sp., right C/, KNM-LT 10030: A = lateral view; B = medial view.

Radius fragment 23934 is very similar to the radius of *Dinofelis* sp. but is smaller and the articular surface is more angled relative to the shaft than in that taxon. In addition, the proportions of the articular rim differ between the two forms.

The ulna of 23934 has a robust olecranon and a deep groove for the triceps attachment. The attachment for the flexor carpi ulnaris is bounded laterally by a broad ridge. The anconeus attachment area is also bounded by a ridge on the caudal side, although this ridge is less prominent than the aforementioned one. Further distally on the shaft, the attachments for extensor pollicis longus and abductor pollicis longus are distinctly visible. The trochlear notch is deep, while the radial notch is quite shallow, though high.

Proximal left Mc IV, 25404, is typical of a medium- to large-sized felid and has a very deep articular facet for Mc III. Other than this, it presents no special characteristics and is placed here mainly on the basis of size.

Specimen 23059 is a left femur that lacks everything proximal to a point just distal to the base of the lesser trochanter and with a damaged cranial half of the distal articulation. Due to the missing parts, the specimen presents few distinguishing characters. The shaft is narrow and gently curved. In cross section, the proximal part is slightly transversely flattened, while the distal part is nearly round. The lateral and medial condyles are nearly equal in size, though breakage makes their exact sizes difficult to determine. The attachment area for the adductor magnus et brevis is distinct, while the attachment for the medial head of the gastrocnemius is deeply indented.

The tibia is represented by two specimens, 23927 and 25131, of which only the former is well enough preserved to present useful characters. This specimen consists of the proximal and distal parts of the right tibia; an undetermined length of the shaft is missing. The proximal tibia is very similar to that of *Lokotunjailurus emageritus*. The medial condyle is relatively flat, and the lateral condyle is more curved, although both extend about equally far down the caudal side of the bone. The intercondyloid area is shallow, though the intercondyloid eminence is higher than in *L. emageritus*. The proximal part of the shaft is triangular in cross section. The attachment area for the flexor hallucis longus is very distinct, but those for the caudal tibial and flexor digitorum longus are less so. In the distal fragment the groove for the tibialis caudalis is deep. Specimen 25131 is very abraded; it is similar in size and, as far as can be determined, morphology to 23927.

Right astragalus, 23118, is very similar to that of the much larger *Lokotunjailurus emageritus*, though 23118 is relatively flatter and has a narrower neck. The head is transversely long and anteroposteriorly narrow. Medially the head extends some distance beyond the neck, which is short and is angled from lateral to medial. The medial articular surface for the calcaneum is oval and sits on the neck abutting the head; it does not extend down to the trochlea. The lateral articular surface for the calcaneum is broad and extends medially, causing the groove between the articular surfaces to angle from medial to lateral at a greater angle than the main axis of the articular surface itself. This is in contrast to *L. emageritus*, in which these two features run in parallel. The lateral and medial sides of the trochlea are about equally broad, with the lateral trochlea extending further dorsally than the medial (in *L. emageritus* the reverse is the case).

Third metatarsal, 28571, is that of a medium-sized felid and matches in size the other postcranial elements here listed but presents no distinguishing characteristics of its own.

Discussion

The assignment of this material to *Metailurus*, a Eurasian machairodont genus, is highly tentative. The upper canine seen in figure 7.31A–B is of about the size of the upper canine of a small *Metailurus major* (Zdansky 1924) and the postcranial material is more or less consistent with this size. Unfortunately, the appendicular skeleton of *Metailurus* is largely unknown, hence no direct comparisons can be made. The only other referral of African material to *Metailurus* is by Petter (1973), who referred a partially erupted upper canine from Olduvai Gorge to this genus. The Olduvai Gorge specimen is of about the same size as the Lothagam one, but the age difference tends to preclude a direct relationship between them. In addition, the Olduvai Gorge specimen is much younger than any Eurasian specimen of *Metailurus*, the genus becoming extinct in Eurasia at the end of the Miocene. The Lothagam specimen matches the Eurasian material better in time.

Specimens from the Kaiyumung Member, Nachukui Formation

In this treatment of carnivore material from Lothagam, I have decided to present the specimens from the Kaiyumung Member of the Nachukui Formation separately from the rest of the material. This is because there is both a temporal and a taxonomic gap between this material and the rest. As seen in figure 7.33, considerable taxonomic overlap occurs between the carnivores of the Nawata Formation and the Apak Member of the Nachukui Formation. However, this overlap is lost in dealing with the Kaiyumung Member material, which more resembles material from sites such as Hadar (Ethiopia) and South Turkwel (Kenya) that date from around 3.5 Ma or less. This is not to say that the Kaiyumung Member is temporally equivalent with these faunas, but it certainly postdates the Apak Member by a good margin, and this at a time when there was massive turnover in the carnivore guild. Thus, presenting the Kaiyumung material together with the rest may lead to some confusion so it is here presented separately instead.

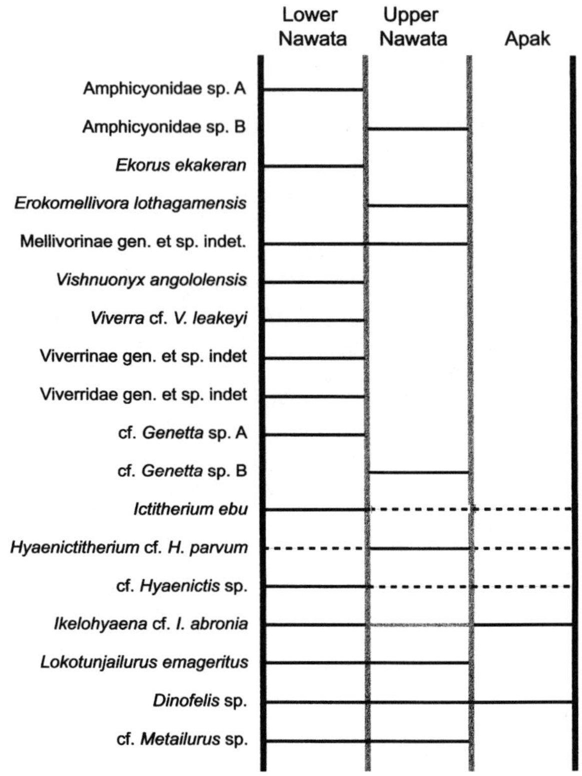

Figure 7.33 Diagram showing representation of the carnivore taxa identified at Lothagam by member, excluding the Kaiymung specimens. The horizontal dashed line indicates uncertain presence; gray bars indicate range-through taxa.

Family Felidae
Subfamily Machairodontinae
Genus *Dinofelis*
Dinofelis aronoki Werdelin and Lewis, 2001

Referred Material

- 23111, Lt. radius.

Specimen 23111 is very large and robust, and it has a more enlarged bicipital tuberosity than do the later *Dinofelis*. The shaft is flat on the distal, posterior surface, as in all *Dinofelis* and in contrast to *Homotherium*. The grooves for the extensor tendons are eroded, and the medial edge of the distal end is broken. The carpal articulation is very large relative to the size of the specimen, and this enlargement suggests a relatively large manus. The medial edge of the large styloid process is in line with the shaft, as in all *Dinofelis*. The dimensions of this specimen are similar to the distal radius from Hadar, AL 363-20, and the two specimens are similar in age.

Discussion

This specimen obviously represents a different species from the material of *Dinofelis* sp. as reported earlier in this chapter. It is figured and discussed in more detail in Werdelin and Lewis (2001), in which paper its taxonomic affinities are established.

Subfamily Felinae
Genus *Leptailurus* or *Caracal* sp.

Referred Material

- 23116, Lt. and Rt. ramus fragments (partial Lt. P_4, Rt. P_4–M_1).

The specimen represents a small-sized felid. The P_4 has a tall and pointed main cusp and low accessory cusps. The anterior accessory cusp is well separated from the main cusp.

Discussion

In size, this specimen matches a serval or caracal. Small fossil cats are poorly known in general and in Africa in particular, and without a revision of the totality of the material it is not possible to assign this specimen to a specific taxon.

Family Canidae
Genus *Canis*
cf. *Canis* sp.

Referred Material

- KNM-LT 25402, ?Lt. ?lower canine.

The specimen is relatively worn. The crown is slender and pointed. Other features are poorly visible.

Discussion

The specimen is quite nondescript and is only tentatively referred to the Canidae.

General Discussion of the Carnivoran Fauna

This general discussion will briefly consider a range of issues, from distribution to biogeographic connections. With respect to distribution, there are distinct differences between the various geological members. The material from the lower member of the Nawata Formation includes a number of species of small to medium size, such as *Vishnuonyx angololensis* and several viverrids, which are less common in the upper member of the Nawata Formation or in the Apak Member of the Nachukui Formation. Three possible reasons for this difference may be suggested. The first, but perhaps the least likely in the present context, is that it is a real reflection of a decrease in carnivore diversity during the time period covered. This hypothesis is gainsaid by the fact that it is apparently only smaller species that are absent in the Upper Nawata and Apak Member. This size bias rather suggests one or both of two alternative hypotheses. The first is that the differences represent a preservational bias against smaller taxa in the Upper Nawata and Apak Member. The nonexclusive alternative is that the differences reflect collecting effort. There are fewer taxa in the Upper Nawata than in the Lower Nawata and fewer still in the Apak, and this pattern correlates with the total number of specimens recovered. Since it is well known that the number of taxa increases with the size of the available sample, this may be the reason for the differences between the various geological members.

Whatever the cause, it should thus be recalled in the following discussion that the Lower Nawata appears reasonably representative as far as carnivore taxa are concerned, while the Upper Nawata and Apak are clearly depauperate in smaller carnivores, while the larger carnivores appear to remain relatively well represented (and for the most part identical to those from the Lower Nawata). The representation of carnivores in the various members is summarized in figure 7.33. As far as the 1960s collections and the faunal list of Smart (1976) are concerned, the specimen referred by him to ??*Civettictis* is here assigned to *Ictitherium ebu* (a very viverrid-like hyaenid), the aff. *Euryboas* specimen to *Hyaenictitherium* cf. *H. parvum* (a not dissimilar but more primitive hyaenid), and the specimens of subfamily Felinae and subfamily Machairodontinae are here distributed between *Lokotunjailurus emageritus* and *Dinofelis* sp., both now considered to belong to the Machairodontinae, although the latter taxon shows convergence with Felinae (Werdelin and Lewis 2001).

Paleoecological Indications

Carnivores are of limited utility in paleoecological reconstruction, but some indications may be provided by the Lothagam carnivores. The persistence of amphicyonids, which are generally considered forest-dwelling (Viranta 1996), is an indication that the environment included large enough expanses of forest to accommodate large to very large forest-dwelling carnivores. The felids present also indicate the existence of closed habitats, although in this case we need not necessarily be dealing with forest per se, but rather with closed woodland. However, several of the taxa, including the two complete skeletons, indicate more open environments in the vicinity. Although *Ekorus* is a mustelid and would not be expected to inhabit open grassland environments, the postcranial skeleton of this form indicates a degree of cursoriality not seen in other mustelids, which suggests a degree of adaptation to open environments. The same is true of *Ictitherium ebu.* Congeners in Eurasia inhabit a variety of habitats and are well known in "Pikermian" environments, which may be considered semi-open, or at least not fully closed. *I. ebu,* however, shows adaptations to extreme cursoriality in its postcranial skeleton, thus suggesting the presence of environments that are more open than those of, for example, Pikermi or Samos in Greece. The presence of *Vishnuonyx,* a specialized aquatic form, indicates the existence of permanent water, which is not surprising. The smaller carnivores are all viverrids and indicate the availability of arboreal or semi-arboreal habitats.

Thus, the environment around Lothagam during the period represented in the present collection appears on the basis of the carnivores to have included a mixture of very closed and very open habitats. The interesting thing here is that the species indicating this mixture are fairly extreme in both directions (amphicyonids with regard to closed habitats and *I. ebu* with regard to open ones), and it is unlikely that there would be a single environment that these taxa would inhabit on a regular basis, with the exception of their common need for water.

Biogeographic Connections

The biogeographic relationships implied by the taxa present in the Lothagam collections are mainly with Eurasia. Most taxa in the fauna have close relatives in the Miocene of Eurasia. These will be dealt with here on a taxon by taxon basis.

Amphicyonidae sp. A is clearly closely related to the genus *Amphicyon* itself and may belong to that genus. In particular, affinities between this form and *A. major* and *A. giganteus* may be sought. The affinities of *Ekorus ekakeran* are difficult to ascertain, as noted in the preceding discussion. However, it still remains likely that this taxon belongs in the same group as the European large-sized mustelids of the Miocene, so that its affinities also lie to the north. *Vishnuonyx* is known from a more primitive form in the Siwalik Hills of India, and this suggests that there might be a direct connection between Africa and the Indian subcontinent, bypassing Europe, from which *Vishnuonyx* is not known (Willemsen 1992). *Viverra* is a Eurasian genus with limited representation in Africa, one of the earliest occurrences being at Lothagam. *Ictitherium* is a Eurasian genus, with the Lothagam record being the only one from sub-Saharan Africa (the *Ictitherium* discussed by Nakaya et al. 1984 should in all likelihood be assigned to the more primitive genus *Protictitherium;* cf. Werdelin and Turner 1996). The hyaenid genera *Hyaenictitherium* and *Hyaenictis* have wide distributions in Eurasia and Africa, but in both cases the more primitive taxa are Eurasian, and the African representatives are likely to be derived offshoots of immigrants from Eurasia. The same is true of *Lokotunjailurus* and cf. *Metailurus.*

Taxa with more distant relationships to Eurasia are *Erokomellivora, Genetta, Ikelohyaena,* and *Dinofelis. Mellivora* may be of African origin, and the distance between its possible ancestor *Erokomellivora* and the next most distant possible relatives is great, thus opening the possibility for a relatively long evolutionary history in the Miocene of Africa for this lineage. The same is true of *Genetta*—an African endemic—with the exception of the extant *G. genetta,* which has a range that extends into Europe. The evolutionary history of this genus is not known but may lie entirely within Africa. *Ikelohyaena* is only known from Africa (at Langebaanweg, Laetoli, and Hadar, in addition to Lothagam), and while the group to which it belongs originated in Eurasia, it is not clear how long a history this specific lineage had in Africa before the appearance of *I.* cf. *I. abronia* at Lothagam. *Dinofelis* is closely related to *Metailurus,* but it cannot be determined whether the latter is the ancestor of *Dinofelis* or whether they shared a common ancestor within the *Paramachairodus* group. In either case, the Lower Nawata record of *Dinofelis* is the oldest of the genus.

The biogeographic relationships of Amphicyonidae sp. B are entirely unknown, both because of the limited material and because medium-sized amphicyonids are rare in the Late Miocene (Viranta 1996).

In summary, biogeographic relationships of the Lothagam carnivores may indicate several dispersal events, one from India, possibly in the early Late Miocene, and two from Eurasia, probably in the early and middle Late Miocene. The first Eurasian dispersal included the ancestors of the *Mellivora, Genetta,* and *Ikelohyaena/Hyaena* lineages and the second dispersal the forebears of the other Lothagam taxa. Direct connections with Middle Miocene African carnivores appear unlikely.

Evolutionary Significance

In this section I will consider the relationships between the Lothagam carnivore faunas and those of Sahabi and Langebaanweg, the only other reasonably sized carnivore faunas in the Late Miocene to earliest Pliocene of Africa. I will also consider the record of first and last appearances of carnivores at Lothagam and discuss the significance of the Lothagam fauna in the context of the subsequent evolution of the descendants of its constituent taxa.

There are some similarities between the carnivore faunas of Sahabi (Howell 1987) and Lothagam, but also obvious differences. The similarities include the presence of one or several species of *Viverra* (not likely to be identical at the species level, though), the presence of species of *Hyaenictitherium* (not the same species, as already noted), and the presence of both a large and a smaller machairodont. Differences include the presence of ursids at Sahabi rather than amphicyonids as at Lothagam and, more important, the presence at Sahabi of several large, bone-cracking percrocutids and hyaenids, which are entirely absent from Lothagam.

The latter absence is worth considering more fully. In Eurasia, the species *Adcrocuta eximia* is an almost ubiquitous member of the fauna from MN 10 to the end of the Miocene (ca. 10.6–5.2 Ma). This species is also represented in the only well-sampled carnivore fauna of North Africa—that is, Sahabi (Howell 1987:figure 3). However, neither this species nor any other true bone-cracking hyaena is known from the Miocene of sub-Saharan Africa: the material assigned to *Percrocuta australis* (Hendey 1974) and later *Adcrocuta australis* (Hendey 1978a) now being referred to *Chasmaporthetes* (Werdelin and Solounias 1990; Werdelin et al. 1994). In addition, Sahabi records the presence of the youngest known percrocutid, *P. senyureki*. Percrocutids were very rare in the Late Miocene of Eurasia, but they did reach sub-Saharan Africa in the form of *Percrocuta tobieni* from Kabarsero (Ngorora Formation, Members C, D, and E) and *Hyperhyaena leakeyi* from Nakali. These records are from the Middle Miocene or earlier part of the Late Miocene (Vallesian equivalent) and are thus considerably older than the Lothagam fauna. In sub-Saharan Africa the Percrocutidae appear to have become extinct in the Late Miocene, before the time interval recorded at Lothagam. At Lothagam, as at Langebaanweg, bone-cracking hyenas are represented by the transitional and relatively unspecialized *I. abronia* or a close relative. The absence of larger bone-cracking hyenas during the last few million years of the Miocene of Africa is surprising and is not addressed until the appearance in early Pliocene sites such as Kanapoi of hyenas of modern aspect.

Similarities between Langebaanweg and Lothagam are more extensive than between Lothagam and Sahabi, and the differences are more easily explained. The differences include the presence of phocids and canids at Langebaanweg, although phocids were absent from Lothagam for simple ecological reasons and canids were absent because canids had not yet reached Africa from North America in the time represented by the Nawata Formation at Lothagam. In addition, there are no herpestids known from Lothagam, which may be an artifact of sampling. Similarities between the two faunas include the presence of an Enhydrini, *Vishnuonyx* at Lothagam and the more derived *Enhydriodon* at Langebaanweg as befits its slightly younger age. *Viverra leakeyi* (or a related form) is present at both sites, as is *Genetta* sp. (although probably not the same species). The Hyaenidae are similar at the two sites, with the presence of the genera *Hyaenictitherium, Hyaenictis,* and *Ikelohyaena* at both sites, while Lothagam has the primitive *Ictitherium* and Langebaanweg the derived *Chasmaporthetes* (Werdelin et al. 1994). Both faunas include a large machairodont, which may turn out to be the same species, although this requires further study. Also present at both sites are species of *Dinofelis* (not the same one; cf. Werdelin and Lewis 2001) and a metailurine, cf. *Metailurus* sp. at Lothagam and *"Felis" obscura* at Langebaanweg. Langebaanweg also includes a small felid, not present at Lothagam until the Kaiyumung and a significant addition to the fauna, *Homotherium.*

The similarities between Lothagam and Langebaanweg are so extensive and the differences between them so clearly due to geographic and temporal position that this aids in determining the stratigraphic position of Langebaanweg. Judging by the carnivores, this fauna is no older than the Apak Member of the Nachukui Formation, dated at about 5.0–4.20 Ma (McDougall and Feibel 1999) and hence older than Kanapoi, the base of which is dated to 4.17 Ma and which has a much more modern carnivore fauna (Leakey et al. 1995, 1998).

From an evolutionary perspective, Lothagam has a pivotal role due to the several both first and last appearances in the carnivore fauna and the key phylogenetic role played by some taxa. Last appearances include Amphicyonidae (globally) and, since they are the only records from Africa, large-sized Mustelidae and *Ictitherium.* First appearances are more extensive and important and include Mellivorinae of modern type, *Viverra, Genetta, Ikelohyaena,* and *Dinofelis.* Most of these are the earliest representatives of taxa with a significant subsequent presence in Africa.

Several of the first appearances and some other taxa have a phylogenetically basal position relative to later African members of their respective clades. This is true of *Erokomellivora* relative to modern *Mellivora, Vishnuonyx angololensis* relative to *Enhydriodon, Ikelohyaena* relative to *Hyaena, Lokotunjailurus* relative to *Homotherium,* and *Dinofelis* sp. relative to later members of the genus. Thus, the Lothagam fauna has a pivotal role in the study of the evolution of Plio-Pleistocene carnivores of Africa.

Acknowledgements

I would like to thank the government of Kenya, the Kenya National Museums, and, in particular, Dr. Meave Leakey for the invitation to study this material. Thanks also to crews in the field and laboratory who have worked so hard at collecting and curating the specimens. Dr. Meave Leakey and Dr. Margaret Lewis provided valuable information and comments during the course of this work. All drawings of specimens were made by Mr. Björn Lindsten. My research was financed by the Swedish Natural Science Research Council.

References Cited

Aguirre, E., and P. Leakey. 1974. Nakali: Nueva fauna del *Hipparion* de Rift Valley de Kenya. *Estudios Geológicos* 30:219–227.

Barry, J. C. 1987. The large carnivores from the Laetoli region of Tanzania. In M. D. Leakey and J. M. Harris, eds., *Laetoli: A Pliocene Site in Northern Tanzania,* pp. 235–258. Oxford: Clarendon Press.

Beaumont, G. de. 1978. Notes complémentaires sur quelques félidés (Carnivores). *Archives des Sciences* (Geneva) 31:219–227.

Buckland-Wright, J. C. 1969. Craniological observations on *Hyaena* and *Crocuta. Journal of Zoology* (London) 159:17–29.

Gaudry, A. 1862–1867. *Animaux fossiles et géologie de l'Attique.* Paris: Savy.

Ginsburg, L. 1977. Les carnivores du Miocène de Beni Mellal (Maroc). *Géologie Méditeranéenne* 4:225–240.

Ginsburg, L., and M. T. Antunes. 1968. *Amphicyon giganteus,* carnassier géant du Miocène. *Annales de Paléontologie* 54:3–32.

Ginsburg, L., and J. Morales. 1992. Contribution à la connaissance des mustélidés (Carnivora, Mammalia) du Miocène d'Europe *Trochictis* et *Ischyrictis,* genres affines et genres nouveaux. *Comptes Rendus de l'Académie des Sciences* (Paris) 315:111–116.

Harris, J. M., F. H. Brown, and M. G. Leakey. 1988. Stratigraphy and paleontology of Pliocene and Pleistocene localities west of Lake Turkana, Kenya. *Contributions in Science* 399:1–128.

Hendey, Q. B. 1974. The late Cenozoic Carnivora of the southwestern Cape Province. *Annals of the South African Museum* 63:1–369.

Hendey, Q. B. 1978a. Late Tertiary Hyaenidae from Langebaanweg, South Africa, and their relevance to the phylogeny of the family. *Annals of the South African Museum* 76:265–297.

Hendey, Q. B. 1978b. Late Tertiary Mustelidae (Mammalia, Carnivora) from Langebaanweg, South Africa. *Annals of the South African Museum* 76:329–357.

Hill, A. 1995. Faunal and environmental change in the Neogene of East Africa: Evidence from the Tugen Hills sequence, Baringo District, Kenya. In E. S. Vrba, G. H. Denton, T. C. Partridge, and L. H. Burckle, eds., *Paleoclimate and Evolution, with Emphasis on Human Origins,* pp. 178–193. New Haven: Yale University Press.

Howell, F. C. 1987. Preliminary observations on Carnivora from the Sahabi Formation (Libya). In N. T. Boaz, A. El-Arnauti, A. W. Gaziry, J. de Heinzelin, and D. D. Boaz, eds., *Neogene Paleontology and Geology of Sahabi,* pp. 153–181. New York: Liss.

Hunt, R. M., Jr. 1974. The auditory bulla in Carnivora: An anatomical basis for reappraisal of carnivore evolution. *Journal of Morphology* 143:21–76.

Hunt, R. M., Jr. 1996a. Amphicyonidae. In D. R. Prothero and R. J. Emry, eds., *The Terrestrial Eocene-Oligocene Transition in North America,* pp. 476–485. Cambridge: Cambridge University Press.

Hunt, R. M., Jr. 1996b. Basicranial anatomy of the giant viverrid from "E" Quarry, Langebaanweg, South Africa. In K. M. Stewart and K. L. Seymour, eds., *Palaeoecology and Palaeoenvironments of Late Cenozoic Mammals: Tributes to the Career of C. S. (Rufus) Churcher,* pp. 588–597. Toronto: University of Toronto Press.

Hunt, R. M., Jr. 1996c. Biogeography of the order Carnivora. In J. L. Gittleman, ed., *Carnivore Behavior, Ecology, and Evolution,* vol. 2, pp. 485–541. Ithaca, N.Y.: Cornell University Press.

Hunt, R. M., Jr. 1998. Evolution of the aeluroid Carnivora: Diversity of the earliest aeluroids from Eurasia (Quercy, Hsanda-Gol) and the origin of felids. *American Museum Novitates* 3252:1–65.

Hunt, R. M., Jr., and R. Skolnick. 1996. The giant mustelid *Megalictis* from the Early Miocene carnivore dens at Agate Fossil Beds National Monument, Nebraska: Earliest evidence of dimorphism in New World Mustelidae (Carnivora, Mammalia). *Contributions to Geology* 31:35–48.

Hunt, R. M., Jr., and N. Solounias. 1991. Evolution of the aeluroid Carnivora: Hyaenid affinities of the Miocene carnivoran *Tungurictis spocki* from Inner Mongolia. *American Museum Novitates* 3030:1–25.

Khomenko, I. P. 1914. La faune méiotique du village Taraklia du district de Bendery. Fissipedia. Rodentia. Rhinoceratinae. Equinae. Suidae. Proboscidae. *Trudy Bessarabskoe Obshchestvo Estestvoispytatelei* (Kishinev) 5:1–55.

Kretzoi, M., and O. Fejfar. 1982. Viverriden (Carnivora, Mammalia) im europäischen Altpleistozän. *Zeitschrift für Geologische Wissenschaften* 10:979–995.

Kurtén, B. 1953. On the variation and population dynamics of fossil and recent mammal populations. *Acta Zoologica Fennica* 76:1–122.

Kurtén, B. 1976. Fossil Carnivora from the Late Tertiary of Bled Douarah and Cherichira, Tunisia. *Notes du Service Géologique de Tunisie* 42:177–214.

Leakey, M. G., C. S. Feibel, R. L. Bernor, J. M. Harris, T. E. Cerling, K. M. Stewart, G. W. Storrs, A. Walker, L. Werdelin, and A. J. Winkler. 1996. Lothagam: A record of faunal change in the Late Miocene of East Africa. *Journal of Vertebrate Paleontology* 16:556–570.

Leakey, M. G., C. S. Feibel, I. McDougall, and A. Walker. 1995. New four-million-year-old hominid species from Kanapoi and Allia Bay, Kenya. *Nature* 376:565–571.

Leakey, M. G., C. S. Feibel, I. McDougall, C. Ward, and A. Walker. 1998. New specimens and confirmation of an early age for *Australopithecus anamensis. Nature* 393:62–66.

Lydekker, R. 1884. Siwalik and Narbada Carnivora. *Paleontologica Indica,* 10th ser., 2:178–351.

McDougall, I., and C. S. Feibel. 1999. Numerical age control for the Miocene-Pliocene succession at Lothagam, a hominoid-bearing sequence in the northern Kenya Rift. *Journal of the Geological Society* (London) 156:731–745.

Morales, J., M. Pickford, D. Soria, and S. Fraile. 1998. New carnivores from the basal Middle Miocene of Arrisdrift, Namibia. *Eclogae Geologicae Helvetiae* 91:27–40.

Nakaya, H., M. Pickford, Y. Nakano, and H. Ischida. 1984. The Late Miocene large mammal fauna from the Namurungule Formation, Samburu Hills, northern Kenya. *African Study Monographs,* Supplementary issue 2:87–131.

Orlov, Y. A. 1948. *Perunium ursogulo* Orlov, a new gigantic extinct mustelid (A contribution to the morphology of the skull and brain and to the phylogeny of Mustelidae). *Acta Zoologica* 29:63–105.

Petter, G. 1963. Etude de quelques Viverridés (Mammiféres, Carnivores) du Pliocène inférieur de Tanganyika (Afrique orientale). *Bulletin de la Société Géologique de France,* 7th ser., 5:267–274.

Petter, G. 1973. Carnivores Pléistocènes du ravin d'Olduvai. In L. B. S. Leakey, R. J. G. Savage, and S. C. Coryndon, eds., *Fossil Vertebrates of Africa,* vol. 3, pp. 43–100. London: Academic Press.

Petter, G. 1987. Small carnivores (Viverridae, Mustelidae, Canidae) from Laetoli. In M. D. Leakey and J. M. Harris, eds., *Laetoli: A Pliocene Site in Northern Tanzania,* pp. 194–234. Oxford: Clarendon Press.

Pia, J. 1939. Ein riesiger Honingsdachs (Mellivorine) aus dem Unterpliozän von Wien. *Annalen des Naturhistoriches Museums* (Vienna) 50:537–583.

Pickford, M. 1975. Late Miocene sediments and fossils from the Northern Kenya Rift Valley. *Nature* 256:279–284.

Pickford, M. H. L. 1978. Stratigraphy and mammalian palaeontology of the Late Miocene Lukeino Formation, Kenya. In W. W. Bishop, ed., *Geological Background to Fossil Man,* pp. 263–278. Edinburgh: Scottish Academic Press.

Pilgrim, G. E. 1932. The fossil Carnivora of India. *Paleontologica Indica,* n.s., 18:1–232.

Qiu Zhanxiang. 1987. Die Hyaeniden aus dem Ruscinium und Villafranchium Chinas. *Münchner Geowissenschaftliche Abhandlungen,* A 9:1–108.

Salles, L. 1992. Felid phylogenetics: Extant taxa and skull morphology. *American Museum Novitates* 3047:1–67.

Savage, R. J. G. 1978. Carnivora. In V. J. Maglio and H. B. S. Cooke, eds., *Evolution of African Mammals,* pp. 249–267. Cambridge, Mass.: Harvard University Press.

Semenov, Y. A. 1989. *Iktiterii i morfologicheski skhodnye hieny neogena SSSR.* Kiev: Naukova Dumka.

Smart, C. 1976. The Lothagam 1 fauna: Its phylogenetic, ecological and biogeographic significance. In Y. Coppens, F. C. Howell, G. L. Isaac, and R. E. Leakey, eds., *Earliest Man and Environments in the Lake Rudolf Basin: Stratigraphy, Paleoecology, and Evolution,* pp. 361–369. Chicago: University of Chicago Press.

Sotnikova, M. V. 1991. A new species of *Machairodus* from the Late Miocene Kalmakpai locality in eastern Kazakhstan (USSR). *Annales Zoologici Fennici* 28:361–368.

Van Valkenburgh, B. 1990. Skeletal and dental predictors of body mass in carnivores. In J. Damuth and B. J. MacFadden, eds., *Body Size in Mammalian Paleobiology: Estimation and Biological Implications,* pp. 181–205. Cambridge: Cambridge University Press.

Viranta, S. 1996. European Miocene Amphicyonidae: Taxonomy, systematics and ecology. *Acta Zoologica Fennica* 204:1–61.

Viret, J. 1951. Catalogue critique de la faune des mammifères Miocènes de La Grive Saint-Alban (Isère). *Nouvelles Archives du Muséum d'Histoire Naturelle* (Lyon) 3:1–104.

Werdelin, L. 1988a. Studies of fossil hyaenids: The genera *Ictitherium* Roth and Wagner and *Sinictitherium* Kretzoi, and a new species of *Ictitherium. Zoological Journal of the Linnean Society* 93:93–105.

Werdelin, L. 1988b. Studies of fossil hyaenids: The genera *Thalassictis* Gervais ex Nordmann, *Palhyaena* Gervais, *Hyaenictitherium* Kretzoi, *Lycyaena* Hensel and *Palinhyaena* Qiu, Huang and Guo. *Zoological Journal of the Linnean Society* 92:211–265.

Werdelin, L. 1996a. Carnivoran ecomorphology: A phylogenetic perspective. In J. L. Gittleman, ed., *Carnivore Behavior, Ecology, and Evolution,* vol. 2, pp. 582–624. Ithaca, N.Y.: Cornell University Press.

Werdelin, L. 1996b. Carnivores, exclusive of Hyaenidae, from the later Miocene of Europe and western Asia. In R. L. Bernor, V. Fahlbusch, and H.-W. Mittmann, eds., *The Evolution of Western Eurasian Miocene Mammal Faunas,* pp. 271–289. New York: Columbia University Press.

Werdelin, L. 1996c. Community-wide character displacement in the lower carnassials of Late Miocene hyaenas. *Lethaia* 29:97–106.

Werdelin, L., and M. Lewis. 2001. A revision of the genus *Dinofelis* (Mammalia, Felidae). *Zoological Journal of the Linnean Society* 132:147:258.

Werdelin, L., and N. Solounias. 1990. Studies of fossil hyaenids: The genus *Adcrocuta* and the interrelationships of some hyaenid taxa. *Zoological Journal of the Linnean Society* 98:363–386.

Werdelin, L., and N. Solounias. 1991. *The Hyaenidae: Taxonomy, Systematics and Evolution.* Fossils and Strata No. 30. Oslo: Universitetsforlaget.

Werdelin, L., and N. Solounias. 1996. The evolutionary history of hyenas in Europe and western Asia during the Miocene. In R. L. Bernor, V. Fahlbusch, and H.-W. Mittmann, eds., *The Evolution of Western Eurasian Neogene Mammal Faunas,* pp. 290–306. New York: Columbia University Press.

Werdelin, L., and A. Turner. 1996. The fossil and living Hyaenidae of Africa: Present status. In K. M. Stewart and K. L. Seymour, eds., *Palaeoecology and Palaeoenvironments of Late Cenozoic Mammals: Tributes to the Career of C. S. (Rufus) Churcher,* pp. 635–659. Toronto: University of Toronto Press.

Werdelin, L., A. Turner, and N. Solounias. 1994. Studies of fossil hyaenids: The genera *Hyaenictis* Gaudry, and

Chasmaporthetes Hay, with a reconsideration of the Hyaenidae of Langebaanweg, South Africa. *Zoological Journal of the Linnean Society* 111:197–217.

Willemsen, G. 1992. A revision of the Pliocene and Quaternary Lutrinae from Europe. *Scripta Geologica* 101:1–115.

Wolsan, M., and Y. A. Semenov. 1996. A revision of the Late Miocene mustelid carnivoran *Eomellivora. Acta Zoologica Cracoviensia* 39:593–604.

Zdansky, O. 1924. Jungtertiäre Carnivoren Chinas. *Paleontologia Sinica*, ser. C, 2:1–149.

Table Abbreviations

Cranial Measurements

BL = basilar length of skull
CB = width across occipital condyles
CBL = condylobasal length of skull
IOB = intraorbital width
PL = length palate
POP = width skull at postorbital processes
POC = minimum width skull at postorbital constriction
P-P = width skull across the posterior end of the upper carnassials

Dental Measurements

L = mesiodistal length
LpP^4 = length of protocone of P^4
LmP^4 = length of metastyle of P^4
LpP_4 = length of main cusp of P_4
LtM_1 = length of trigonid of M_1
MandD at P_4–M_1 = depth horizontal ramus between P_4 and M_1
MandD behind M_1 = depth horizontal ramus behind M_1
M^2L (buccal) = buccal length of M^2
W = buccolingual width
WaP^4 = anterior width (including protocone) of P^4
$WblP^4$ = minimum width blade of P^4
WP_4 (middle) = width at middle of P_4
WP_4 (posterior) = width posterior end of P_4
Wtal M_1 = width talonid of M_1

Postcranial Measurements

a-p = anteroposterior
Astragalus TotH = maximum height astragalus
Calcaneum SustentaculumW = width of calcaneum at sustentaculum
Calcaneum TotH = maximum height of calcaneum
Digit X Phalanx X = phalanx X of digit X
Dist = distal
L = anatomical length
min = minimum
Prox = proximal
Scapula Glenoid fossa L = maximum dimension glenoid fossa of scapula
Scapula Glenoid fossa W = minimum dimension glenoid fossa of scapula
transv = transverse
Ulna RadNotchW = width radial notch of ulna
W = maximum width

Vertebra Measurements

Axis L (excluding dens) = length axis excluding dens
Atlas W ant = anterior width of atlas
Caudal X CentrumL = length centrum of caudal vertebra
XCX or TX CentrumL = length centrum of cervical or thoracic vertebra X
CX or TX W ant = anterior width cervical or thoracic vertebra X
CX or TX H ant = anterior height cervical or thoracic vertebra X
CX or TX W post = posterior width cervical or thoracic vertebra X
CX or TX H post = posterior height cervical or thoracic vertebra X
LX CentrumL = length centrum of lumbar vertebra X

Other Abbreviations

a = approximate
AMNH = American Museum of Natural History
F:AM = Frick Collection, American Museum of Natural History
FMNH = Field Museum of Natural History
KNM = National Museums of Kenya

TABLE 7.1 Measurements (in mm) of Material Referred to Amphicyonidae species A (Large Species)

Parameter	KNM-LT 23049	KNM-LT 23073	KNM-LT 23051
M^2L (buccal)	18.2	—	—
M^2W	25.5	—	—
?Mc IV DistW (transv)	—	24.9	—
?Mc IV DistW (a-p)	—	17.8	—
Mc V DistW (transv)	—	27.8	—
Mc V DistW (a-p)	—	19.1	—
Radius ProxW (min)	—	—	a34

TABLE 7.2 Measurements (in mm) of the Holotype of *Ekorus ekakeran*, KNM-LT 23125

Parameter	Measurement	Parameter	Measurement
LI^3	10.5	Scapula Glenoid fossa L	44.5
LC/	14.1	Scapula Glenoid fossa W	25.9
WC/	10.9	Femur L	221.0
LP^1	4.7	Femur ProxW	56.6
WP^1	5.5	Femur HeadW	29.2
LP^2	9.5	Femur ShaftW	a19.4
WP^2	6.2	Femur DistW	49.9
LP^3	12.1	Tibia L	198.0
WP^3	8.3	Tibia ProxW (transv)	48.2
LP^4	22.1	Tibia DistW (max)	36.7
WaP^4	12.8	Tibia DistW (min)	24.8
$WblP^4$	7.5	Astragalus HeadW	20.7
LpP^4	10.1	Astragalus Neck W	14.0
LmP^4	8.7	Astragalus BodyW	22.3
LM^1	6.7	Astragalus TotH	33.6
WM^1	13.9	Calcaneum HeadW (transv)	20.2
L/C	13.6	Calcaneum HeadW (a-p)	17.7
W/C	10.6	Calcaneum SustentaculumW	25.7
LP_2	8.9	Calcaneum TotH	61.0
WP_2	6.1	Atlas L	22.8
LP_3	10.4	Atlas W	99.0
WP_3	6.9	Axis L (excluding dens)	41.2
LP_4	14.9	C3 CentrumL	27.5
WP_4	7.2	C4 CentrumL	26.8
LpP_4	5.6	C5 CentrumL	26.6
LM_1	20.9	C6 CentrumL	27.0
WM_1	8.4	C7 CentrumL	27.9
LtM_1	16.1	T1 CentrumL	24.6
MandD behind M_1	35.6	T2 CentrumL	25.8
P_2-M_1 L	54.8	T3 CentrumL	26.1
Humerus L	204.0	T4 CentrumL	27.3
Humerus HeadW	32.0	T5 CentrumL	25.0
Humerus ProxW	48.4	T6 CentrumL	25.5
Humerus ShaftW	16.8	T7 CentrumL	25.1
Humerus DistW (max)	44.0	T8 CentrumL	25.9
Radius L	175.0	T9 CentrumL	25.5
Radius ProxW (min)	14.5	T10 CentrumL	a28.0
Radius ProxW (max)	23.4	T11 CentrumL	27.9
Radius DistW (max)	28.3	T12 CentrumL	27.4
Radius DistW (min)	21.4	T13 CentrumL	29.6
Ulna L	225.0	T14 CentrumL	31.6
Ulna OlecranonW	32.5	L1 CentrumL	32.6

TABLE 7.2 Measurements (in mm) of the Holotype of *Ekorus ekakeran*, KNM-LT 23125 (Continued)

Parameter	Measurement	Parameter	Measurement
L2 CentrumL	32.2	Mc II ProxW (a-p)	16.2
L3 CentrumL	34.4	Mc II DistW (transv)	12.4
L4 CentrumL	35.8	Mc III L	66.2
L5 CentrumL	36.3	Mc III ProxW (transv)	12.9
L6 CentrumL	33.8	Mc III ProxW (a-p)	18.8
Caudal 1 CentrumL	19.5	Mc III DistW (transv)	13.9
Caudal 2 CentrumL	20.9	Mc IV L	68.9
Caudal 3 CentrumL	19.2	Mc IV ProxW (transv)	12.5
Caudal 4 CentrumL	19.5	Mc IV ProxW (a-p)	17.3
Caudal 5 CentrumL	19.1	Mc IV DistW (transv)	13.8
Caudal 6 CentrumL	19.2	Mc V L	56.5
Caudal 7 CentrumL	20.8	Mc V ProxW (transv)	12.5
Caudal 8 CentrumL	22.9	Mc V ProxW (a-p)	15.7
Caudal 9 CentrumL	24.6	Mc V DistW (transv)	12.4
Caudal 10 CentrumL	26.2	Mt II L	61.8
Caudal 11 CentrumL	27.0	Mt II ProxW (transv)	9.8
Caudal 12 CentrumL	29.3	Mt II ProxW (a-p)	18.3
Caudal 13 CentrumL	29.3	Mt II DistW (transv)	12.1
Caudal 14 CentrumL	28.8	Mt III L	71.6
Caudal 15 CentrumL	27.8	Mt III ProxW (transv)	12.9
Caudal 16 CentrumL	26.5	Mt III ProxW (a-p)	20.6
Caudal 17 CentrumL	24.7	Mt III DistW (transv)	14.2
Caudal 18 CentrumL	23.1	Mt IV L	76.7
Caudal 19 CentrumL	19.1	Mt IV ProxW (transv)	11.0
Mc I L	36.9	Mt IV ProxW (a-p)	20.5
Mc I ProxW (transv)	8.9	Mt IV DistW (transv)	13.8
Mc I ProxW (a-p)	10.4	Mt V L	64.0
Mc I DistW (transv)	9.8	Mt V ProxW (transv)	10.9
Mc II L	54.2	Mt V ProxW (a-p)	15.3
Mc II ProxW (transv)	11.3	Mt V DistW (transv)	11.2

TABLE 7.3 Selected Measurements (in mm) of Some Large-Size Mustelidae

	LM_1	Humerus L	Radius L	Femur L	Tibia L	Radius/Humerus	Tibia/Femur
AMNH 12881	—	192.0	—	—	172.4	—	—
F:AM 54079	21.7	—	—	185.0	150.2	—	0.812
F:AM 25430	20.0	—	—	191.0	161.0	—	0.843
FMNH P12154	—	178.0	142.0	201.0	169.0	0.798	0.841
KNM-LT 23125	20.9	204.0	175.0	221.0	198.0	0.858	0.896

TABLE 7.4 Measurements (in mm) of the Holotype of *Erokomellivora lothagamensis*

Parameter	KNM-LT 23926
LM^1	13.3
WM^1	6.3
LtM^1	9.4
MandD behind M^1	19.0

TABLE 7.5 Measurements (in mm) of Material Referred to Mellivorinae gen. and sp. indet.

Parameter	KNM-LT 23160	KNM-LT 25130
$L?P_4$	a8.4	—
Humerus ProxW (transv)	—	26.9
Humerus ProxW (a-p)	—	30.4

TABLE 7.6 Measurements (in mm) of the Holotype of *Vishnuonyx angololensis* sp. nov.

Parameter	KNM-LT 23948
LP^4	15.3
WaP^4	12.9
$WblP^4$	6.3
LpP^4	7.5
LmP^4	6.3
L between protocone and hypocone	5.2

TABLE 7.7 Measurements (in mm) of Material Referred to *Viverra* cf. *V. leakeyi*

Parameter	KNM-LT 25413
LP^4	20.0
WaP^4	12.8
$WblP^4$	7.3
LpP^4	8.0
LmP^4	8.7

TABLE 7.8 Measurements (in mm) of Material Referred to Viverrinae sp. indet.

Parameter	KNM-LT 23032
LP_2	a8.6
LP^4	a11.1
LM^1	13.6
WM^1	7.0
LtM^1	8.9
MandD at P_4–M_1	15.2

TABLE 7.9 Measurements (in mm) of Material Referred to Viverridae gen. and sp. indet. (Large Species)

Parameter	KNM-LT 25132
LP_4	14.2

TABLE 7.10 Measurements (in mm) of Material Referred to cf. *Genetta* sp. A

Parameter	KNM-LT 25409	KNM-LT 23031
LP_3	5.3	—
WP_3	2.8	—
LP_4	5.9	—
WP_4	3.1	—
LpP_4	3.0	—
LM_1	—	6.5
WM_1	—	3.9
LtM_1	5.3	4.3
WM_1	3.9	—
MandD at P_4–M_1	a8.9	—

TABLE 7.11 Measurements (in mm) of Material Referred to cf. *Genetta* sp. B

Parameter	KNM-LT 23945
LP_4	a5.1
WP_4	2.6
LpP_4	2.3
MandD at P_4–M_1	8.0

TABLE 7.12 Measurements (in mm) of the Holotype and Other Material Referred to *Ictitherium ebu* sp. nov.

Parameter	KNM-LT 23145	KNM-LT 23047	KNM-LT 10031
LI^3	6.7	—	—
L/C	10.7	—	—
W/C	6.7	—	—
LP^1	7.3	—	—
WP^1	4.5	—	—
LP^2	12.8	—	—
WP^2	6.3	—	—
LP^3	16.1	—	—
WP^3	9.0	—	—
LP^4	a23.6	—	—
WaP^4	13.1	—	—
$WblP^4$	7.8	—	—
LpP^4	8.8	—	—
LmP^4	a7.8	—	—
LM^1	8.7	—	—
WM^1	15.4	—	—
LM^2	4.2	—	—
WM^2	6.5	—	—
PL	93	—	—
BL	175	—	—
CBL	185	—	—
P-P	a59	—	—
IOB	a30	—	—
POP	a50	—	—
POC	a31	—	—
CB	38	—	—
L/C	9.1	—	—
W/C	6.9	—	—
LP_2	12.0	—	—
WP_2	6.1	—	—
LP_3	14.9	—	—
WP_3	6.9	—	—
LP_4	—	—	—
WP_4	8.2	—	—
LM_1	18.7	—	—
WM_1	7.7	7.4	—
LtM_1	14.0	—	—
$WtalM_1$	6.7	6.3	6.5
C1 CentrumL	19	—	—
Femur L	208	—	—
Femur ProxW	37.3	—	—
Femur HeadW	17.2	—	—

TABLE 7.12 Measurements (in mm) of the Holotype and Other Material Referred to *Ictitherium ebu* sp. nov. (Continued)

Parameter	KNM-LT 23145	KNM-LT 23047	KNM-LT 10031
Femur ShaftW	14.9	—	—
Femur DistW	30.4	—	—
Tibia L	a201	—	—
Tibia ProxW (transv.)	33.5	—	—
Tibia ShaftW	12.9	—	—
Tibia DistW (max)	22.7	—	—
Tibia DistW (min.)	16.0	—	—
Humerus L	190	—	—
Humerus HeadW	25.0	—	—
Humerus ProxW (a-p)	42.7	—	—
Humerus ShaftW	14.9	—	—
Humerus DistW (transv)	32.2	32.5	—
Humerus DistW (a-p)	—	25.0	—
Radius L	209	—	—
Radius ProxW (min.)	13.0	—	—
Radius ProxW (max.)	16.6	—	—
Radius ShaftW	12.5	—	—
Radius DistW (max.)	21.0	—	—
Ulna L	230	—	—
Ulna OlecranonW	22.7	—	—
Ulna RadNotchW	18.4	—	—
Mc I L	30.9	—	—
Mc II L	72.0	—	—
Mc II ProxW (a-p)	10.3	—	—
Mc II ProxW (transv)	7.5	—	—
Mc II DistW (transv)	10.4	—	—
Mc III L	83.1	—	—
Mc III ProxW (a-p)	11.1	—	—
Mc III ProxW (transv)	9.5	—	—
Mc III DistW (transv)	9.0	—	—
Mc IV L	82.0	—	—
Mc IV ProxW (a-p)	13.0	—	—
Mc IV ProxW (transv)	8.1	—	—
Mc IV DistW (transv)	8.7	—	—
Mc V L	70.4	—	—
Mc V ProxW (a-p)	11.2	—	—
Mc V ProxW (dist)	11.8	—	—
Mc V DistW (transv)	9.9	—	—
Digit II Phalanx 1 L	23.7	—	—
Digit III Phalanx 1 L	28.1	—	—
Digit IV Phalanx 1 L	26.7	—	—

continued

TABLE 7.12 Measurements (in mm) of the Holotype and Other Material Referred to *Ictitherium ebu* sp. nov. (Continued)

Parameter	KNM-LT 23145	KNM-LT 23047	KNM-LT 10031
Digit V Phalanx 1 L	24.5	—	—
Digit III Phalanx 2 L	13.4	—	—
Digit IV Phalanx 2 L	13.3	—	—
Digit V Phalanx 2 L	12.5	—	—
Digit IV Phalanx 3L	11.0	—	—
Digit V Phalanx 3 L	14.3	—	—
Astragalus HeadW	15.4	—	—
Astragalus NeckW	11.6	—	—
Astragalus BodyW	15.1	—	—
Astragalus TotH	25.6	—	—
Calcaneum HeadW (max.)	15.1	—	—
Calcaneum HeadW (min.)	12.0	—	—
Calcaneum SustentaculumW	17.8	—	—
Calcaneum TotH	44.3	—	—
Mt II L	80.1	—	—
Mt II ProxW (a-p)	—	—	—
Mt II ProxW (transv)	5.2	—	—
Mt II DistW (transv)	—	—	—
Mt III L	a92	—	—
Mt III ProxW (a-p)	10.3	—	—
Mt III ProxW (transv)	8.4	—	—
Mt III DistW (transv)	—	—	—
Mt IV L	a97	—	—
Mt IV ProxW (a-p)	10.9	—	—
Mt IV ProxW (transv)	11.7	—	—
Mt IV DistW (transv)	—	—	—
Mt V L	77.6	—	—
Mt V ProxW (a-p)	13.3	—	—
Mt V ProxW (transv)	7.4	—	—
Mt V DistW (transv)	8.8	—	—
Digit II Phalanx 1 L	22.5	—	—
Digit III Phalanx 1 L	25.6	—	—
Digit IV Phalanx 1 L	26.0	—	—
Digit V Phalanx 1 L	20.4	—	—
Digit II Phalanx 2 L	11.4	—	—
Digit III Phalanx 2 L	12.7	—	—
Digit IV Phalanx 2 L	16.1	—	—
Digit V Phalanx 2 L	8.9	—	—
Digit II Phalanx 3 L	6.9	—	—
Digit III Phalanx 3 L	13.0	—	—
Digit V Phalanx 3 L	9.6	—	—

TABLE 7.13 Measurements (in mm) of Material Referred to *Hyaenictherium* cf. *H. parvum*

Parameter	KNM-LT 10032	KNM-LT 23937	KNM-LT 23013
LP_2	13.2	—	—
WP_2	7.1	—	—
LP_3	16.7	—	—
WP_3	9.0	—	—
LP_4	18.9	—	—
WP_4	9.4	—	—
LpP_4	9.9	—	—
L/C	—	a14.5	a12.5
W/C	—	a11.2	9.4

TABLE 7.14 Measurements (in mm) of Dental Material Referred to cf. *Hyaenictis* sp.

Parameter	KNM-LT 23057	KNM-LT 23033
WP_2	8.4	—
LP_4	20.7	>20.9
WP_4	9.8	>10.9
LpP_4	—	a9.5
LM_1	a24.5	—
LtM_1	19.0	—
WM_1	a11.0	—

TABLE 7.15 Measurements (in mm) of Postcranial Material Referred to cf. *Hyaenictis* sp.

Parameter	KNM-LT 23930	KNM-LT 23089
Radius DistW (max)	32.7	—
McIII ProxW (transv)	—	14.5

TABLE 7.16 Measurements (in mm) of Dental Material Referred to *Ikelohyaena* cf. *I. abronia*

Parameter	KNM-LT 23947
LP_4	a19.6
WP_4	9.1
LpP_4	9.7
LM_1	21.6
WM_1	8.3
LtM_1	16.7

TABLE 7.17 Measurements (in mm) of Postcranial Material Referred to *Ikelohyaena* cf. *I. abronia*

Parameter	KNM-LT 25127
Tibia DistW (max)	28.8
Tibia DistW (min)	20.5

TABLE 7.18 Measurements (in mm) of the Holotype of *Lokotunjailurus emageritus* gen. and sp. nov., KNM-LT 26178

Parameter	Measurement	Parameter	Measurement
L mandible	170.3	Astragalus BodyW	29.9
D at diastema	23.6	Femur HeadW	33.2
MandD behind M_1	33.19	Femur ProxW	71.5
L/C	13.5	Femur TrochanterW	36.7
W/C	10.0	Scapula Glenoid fossa L	55.0
LI_1	6.2	Scapula Glenoid fossa W	30.3
WI_1	4.9	Mc I L	a36.4
LI_2	7.4	Mc I ProxW (transv)	a22.6
WI_2	7.1	Mc I ProxW (a-p)	a51.1
LI_3	10.6	Mc III L	106.2
WI_3	9.1	Mc III ProxW (transv)	20.7
LP_3	8.1	Mc III ProxW (a-p)	24.5
WP_3	5.2	Mc III DistW (transv)	18.8
LP_4	17.9	Mc III DistW (a-p)	19.5
WP_4	8.5	Mc IV L	111.6
LpP_4	8.4	Mc IV ProxW (transv)	16.1
LM_1	27.8	Mc IV ProxW (a-p)	20.8
WM_1	10.9	Mc IV DistW (transv)	17.9
Radius ProxW (max)	33.4	Mc IV DistW (a-p)	18.4
Radius ProxW (min)	24.5	Mc V L	88.0
Radius DistW (transv)	44.8	Mt II L	106.3
Radius DistW (a-p)	30.3	Mt II ProxW (transv)	12.4
Ulna ProxW (a-p)	37.7	Mt II ProxW (a-p)	26.6
Ulna Olecranon fossa H (max)	34.9	Mt II DistW (transv)	17.6
Humerus ProxW (transv)	61.1	Mt II DistW (a-p)	17.5
Humerus ProxW (a-p)	65.6	Mt III L	117.8
Humerus DistW (transv)	74.8	Mt III ProxW (transv)	21.7
Humerus DistW (a-p)	42.2	Mt III ProxW (a-p)	28.2
Tibia ProxW (transv)	67.8	Mt III DistW (transv)	19.5
Tibia DistW (transv)	48.7	Mt III DistW (a-p)	18.3
Tibia DistW (a-p)	29.8	Mt IV L	116.2
Fibula ProxW (max)	24.9	Mt IV ProxW (transv)	13.3
Fibula DistW (max)	26.2	Mt IV ProxW (a-p)	23.6
Calcaneum HeadW (transv)	26.5	Mt IV DistW (transv)	16.9
Calcaneum HeadW (a-p)	26.3	Mt IV DistW (a-p)	17.6
Calcaneum NeckW (transv)	15.5	Mt V L	103.6
Calcaneum NeckW (a-p)	27.2	Mt V ProxW (transv)	16.3
Calcaneum SustentaculumW	37.0	Mt V ProxW (a-p)	17.3
Calcaneum DistW (a-p)	24.0	Mt V DistW (transv)	15.4
Calcaneum DistW (transv)	29.9	Mt V DistW (a-p)	15.2
Astragalus HeadW (transv)	29.2	Manus digit 1, phalanx I L	a31.7
Astragalus HeadW (a-p)	19.5	Manus digit 1, phalanx I ProxW (transv)	a22.6
Astragalus NeckW (transv)	23.8		

TABLE 7.18 Measurements (in mm) of the Holotype of *Lokotunjailurus emageritus* gen. and sp. nov., KNM-LT 26178 (Continued)

Parameter	Measurement
Manus digit 1, phalanx I, DistW (transv)	21.7
Manus digit 1, phalanx II, ProxW (transv)	21.0
Manus digit 1, phalanx II, H	61.7
Manus digit 1, phalanx II, L	43.4
Manus digit 2, phalanx I, L	41.7
Manus digit 2, phalanx I, ProxW (transv)	17.6
Manus digit 2, phalanx I, DistW (transv)	14.5
Manus digit 2, phalanx II, L	27.3
Manus digit 2, phalanx II, ProxW (transv)	14.8
Manus digit 2, phalanx II, DistW (transv)	16.0
Manus digit 2, phalanx III, ProxW (transv)	13.1
Manus digit 2, phalanx III, H	38.6
Manus digit 2, phalanx III, L	20.7
Manus digit 3, phalanx I, L	43.7
Manus digit 3, phalanx I, ProxW (transv)	17.6
Manus digit 3, phalanx I, DistW (transv)	14.0
Manus digit 3, phalanx II, L	25.9
Manus digit 3, phalanx II, ProxW (transv)	14.9
Manus digit 3, phalanx II, DistW (transv)	14.3
Manus digit 3, phalanx III, ProxW (transv)	11.0
Manus digit 3, phalanx I, H	21.7
Manus digit 4, phalanx I, L	40.4
Manus digit 4, phalanx I, ProxW (transv)	16.8
Manus digit 4, phalanx I, DistW (transv)	13.1
Manus digit 4, phalanx III, ProxW (transv)	9.7
Manus digit 4, phalanx III, H	25.1
Pes digit 2, phalanx I, L	38.2
Pes digit 2, phalanx I, ProxW (transv)	16.3
Pes digit 2, phalanx I, DistW (transv)	14.4
Pes digit 2, phalanx II, L	28.5
Pes digit 2, phalanx II, ProxW (transv)	15.1
Pes digit 2, phalanx II, DistW (transv)	14.1
Pes digit 2, phalanx III, ProxW (transv)	9.6
Pes digit 2, phalanx III, H	17.3
Pes digit 3, phalanx I, L	41.3
Pes digit 3, phalanx I, ProxW (transv)	17.7
Pes digit 3, phalanx I, DistW (transv)	14.3
Pes digit 3, phalanx II, L	22.7
Pres digit 3, phalanx II, ProxW (transv)	14.7
Pes digit 3, phalanx II, DistW (transv)	15.0
Pes digit 3, phalanx III, ProxW (transv)	9.9
Pes digit 3, phalanx III, H	21.0
Pes digit 4, phalanx I, L	38.8
Pes digit 4, phalanx I, ProxW (transv)	16.6
Pes digit 4, phalanx I, DistW (transv)	13.7
Pes digit 4, phalanx II, L	24.6
Pes digit 4, phalanx II, ProxW (transv)	14.4
Pes digit 4, phalanx II, DistW (transv)	13.5
Pes digit 4, phalanx III, ProxW (transv)	8.6
Pes digit 4, phalanx III, H	18.7
Pes digit 5, phalanx I, L	35.3
Pes digit 5, phalanx I, ProxW (transv)	13.8

continued

TABLE 7.18 Measurements (in mm) of the Holotype of *Lokotunjailurus emageritus* gen. and sp. nov., KNM-LT 26178 (Continued)

Parameter	Measurement	Parameter	Measurements
Pes digit 5, phalanx I, DistW	12.3	T2 W post	36.0
(transv)		T2 H post	19.4
Pes digit 5, phalanx II, L	19.4	T2 L	25.2
Pes digit 5, phalanx II, ProxW	13.0	T3 H ant	a19.0
(transv)		T3 W post	a35.2
Pes digit 5, phalanx II, DistW	12.3	T3 H post	19.7
(transv)		T3 L	25.8
Pes digit 5, phalanx III, ProxW	8.5	T4? W ant	23.4
(transv)		T4? H ant	20.5
Pes digit 5, phalanx I, H	17.5	T4? W post	36.7
Atlas W ant	62.5	T4? H post	20.8
Atlas WingL	61.9	T4? L	27.5
Atlas W	a112.0	T5? W ant	24.3
C4 W ant	28.2	T5? H ant	20.3
C4 H ant	21.3	T5? W post	37.4
C4 W post	25.1	T5? H post	20.4
C4 H post	21.1	T5? L	27.8
C4 CentrumL	28.0	T6? W ant	22.3
C5 W ant	25.5	T6? H ant	20.1
C5 H ant	20.8	T6? W post	34.0
C5 W post	25.7	T6? H post	20.2
C5 H post	18.4	T6? L	27.0
C5 L	25.8	T7? W ant	24.1
C6 W ant	28.4	T7? H ant	20.3
C6 H ant	20.1	T7? W post	36.3
C6 W post	24.8	T7? H post	20.1
C6 H post	19.0	T7? L	28.2
C6 L	27.2	T8? W ant	24.5
C7 H ant	18.1	T8? H ant	20.4
C7 H post	18.6	T8? W post	38.6
C7 L	26.4	T8? H post	20.8
T1 W ant	23.4	T8? L	28.4
T1 H ant	19.2	T9? W ant	22.6
T1 W post	31.1	T9? H ant	20.4
T1 H post	19.1	T9? W post	36.7
T1 L	25.9	T9? H post	19.9
T2 H ant	a18.7	T9? L	28.0

TABLE 7.19 Measurements (in mm) of Dental Material Referred to *Lokotunjailurus emageritus* gen. and sp. nov.

Parameter	KNM-LT 23050	KNM-LT 23941	KNM-LT 25405	KNM-LT 23950	KNM-LT 23060
LC/	26.9	—	—	—	—
WC/	12.3	—	—	—	—
HC/	a56	—	—	—	—
LP^3	—	14.4	—	—	—
WP^3	—	7.0	—	—	—
LP^4	—	—	38.2	—	—
WaP^4	—	—	11.8	—	—
$WblP_4$	—	—	10.7	—	—
LpP_4	—	—	13.3	—	—
LmP_4	—	—	16.3	—	—
LP_4	—	—	—	a20.3	—
LM_1	—	—	—	a26	a29
WM_1	—	—	—	—	13.3

TABLE 7.20 Measurements (in mm) of Humeri and Femora Referred to *Lokotunjailurus emageritus* gen. and sp. nov.

Parameter	KNM-LT 25406	KNM-LT 23036	KNM-LT 446	KNM-LT 25410
Humerus DistW (min)	86	—	—	—
Femur HeadW	—	—	34.6	—
Femur ProxW	—	—	73.1	62.3
Femur DistW	—	67	64.6	—

TABLE 7.21 Measurements (in mm) of Astragali Referred to *Lokotunjailurus emageritus* gen. and sp. nov.

Parameter	KNM-LT 23949	KNM-LT 23935	KNM-LT 23939	KNM-LT 23931	KNM-LT 23924
Astragalus HeadW	29.6	32.3	29.9	29.1	30.5
Astragalus NeckW	24.7	25.5	—	22.4	—
Astragalus BodyW	28.0	32.5	—	27.0	30.9
Astragalus TotH	46.3	a48.6	49.5	45.3	49.0

TABLE 7.22 Measurements (in mm) of Calcanea and Metatarsals Referred to *Lokotunjailurus emageritus* gen. and sp. nov.

Parameter	KNM-LT 23037	KNM-LT 23940	KNM-LT 23568
Calcaneum HeadW (transv)	26.4	23.8	—
Calcaneum HeadW (a-p)	—	26.0	—
Calcaneum SustentaculumW	37.2	35.9	—
Calcaneum TotH	92.9	91.4	—
Mt IV ProxW (transv)	—	—	13.5
Mt IV ProxW (a-p)	—	—	25.0
Mt IV DistW (transv)	—	—	18.3
Mt IV DistW (a-p)	—	—	17.5
Mt IV L	—	—	128.4

TABLE 7.23 Measurements (in mm) of Dental Material Referred to *Dinofelis* sp.

Variable	KNM-LT 25397
WP_4 (middle)	a8.4
WP_4 (posterior)	9.5

TABLE 7.24 Measurements (in mm) of Postcranial Material Referred to *Dinofelis* sp.

Variable	KNM-LT 25398	KNM-LT 25407	KNM-LT 25408	KNM-LT 25137	KNM-LT 30310	KNM-LT 23932	KNM-LT 23696
Ulna OlecranonW	—	—	—	—	—	—	—
Radius ProxW (max)	28.7	—	—	—	29.5	—	—
Radius ProxW (min)	21.1	—	—	—	20.8	—	—
Mc V ProxW (transv)	—	12.6	13.1	—	—	—	—
Mc V ProxW (a-p)	—	19.0	21.5	—	—	—	—
Femur ProxW	—	—	—	—	—	—	—
Tibia DistW (min)	—	—	—	44.0	—	—	—
Calcaneum HeadW (transv)	—	—	—	—	—	21.6	21.6
Calcaneum HeadW (a-p)	—	—	—	—	—	23.4	25.0
Calcaneum SustentaculumW	—	—	—	—	—	32.5	32.5
Calcaneum TotH	—	—	—	—	—	81.5	a84

TABLE 7.25 Measurements (in mm) of Material Referred to cf. *Metailurus* sp.

Parameter	KNM-LT 10030	KNM-LT 25397	KNM-LT 35404	KNM-LT 23118	KNM-LT 23059	KNM-LT 23927	KNM-LT 23934
LC/	17.0	—	—	—	—	—	—
WC/	9.9	—	—	—	—	—	—
WP_4	—	9.5	—	—	—	—	—
Mc III ProxW (transv)	—	—	10.1	—	—	—	—
Mc III ProxW (a-p)	—	—	16.5	—	—	—	—
Astragalus HeadW	—	—	—	19.4	—	—	—
Astragalus NeckW	—	—	—	13.7	—	—	—
Astragalus BodyW	—	—	—	20.8	—	—	—
Astragalus TotH	—	—	—	31.9	—	—	—
Femur DistW	—	—	—	—	39.9	—	—
Tibia ProxW (transv)	—	—	—	—	—	41.4	—
Tibia DistW (max)	—	—	—	—	—	30.8	—
Tibia DistW (min)	—	—	—	—	—	19.3	—
Ulna OlecranonW	—	—	—	—	—	—	27.2
Radius ProxW (max)	—	—	—	—	—	—	26.2
Radius ProxW (min)	—	—	—	—	—	—	19.1

8

PROBOSCIDEA AND TUBULIDENTATA

8.1

Elephantoidea from Lothagam

Pascal Tassy

Ten elephantoid taxa are described from the Late Miocene–Early Pliocene strata at Lothagam. Six are identified or tentatively identified at the species level. From the Lower Nawata come *Anancus kenyensis, Stegotetrabelodon orbus, Primelephas gomphotheroides,* and an elephantid (previously described as *Stegotetrabelodon orbus* or *Primelephas gomphotheroides*) that is also present in the Apak Member. Associated in the Upper Nawata are *Anancus kenyensis,* a trilophodont gomphothere, *Stegotetrabelodon orbus,* and an unidentified elephantid (species A). A new early *Elephas* species, also from the Upper Nawata and represented by a juvenile mandible, demonstrates that Elephantinae differentiation occurred during the Late Miocene. From the Apak Member of the Nachukui Formation have been recovered *Anancus kenyensis, Stegotetrabelodon orbus, Elephas* cf. *E. ekorensis, Loxodonta* ?aff. *L. exoptata,* and unidentified elephantids, *Elephas* species A and B.

Figure 8.1 Restoration of *Stegotetrabelodon orbus* by Mauricio Antón. Shoulder height estimated at about 3 meters.

Lothagam proboscideans provide a key to understanding the early differentiation of elephantids and were originally studied by Vincent J. Maglio in the 1970s (Maglio 1970, 1973; Maglio and Ricca 1977). Most of Maglio's material was recovered from the Lothagam region during the late 1960s by a team led by Brian Patterson of Harvard University. Maglio described two primitive elephantids from Lothagam, *Stegotetrabelodon orbus* and *Primelephas gomphotheroides,* in 1970. In the same paper, Maglio described a primitive member of the loxodont elephantine lineage, *Loxodonta adaurora,* from the Early Pliocene locality of Kanapoi and from Lothagam. Maglio thus established that elephantines originated in the Late Miocene and differentiated during the Early Pliocene.

"The Lothagam sequence was subdivided by Behrensmeyer (1976) into three units: Lothagam 1 and Lothagam 2, both of Miocene age, and Lothagam 3 of Pliocene age. Member 1B of Lothagam 1 yielded four proboscideans—*Deinotherium, Anancus,* and the elephantids *Primelephas* and *Stegotetrabelodon*—while Member 1C lacked *Stegotetrabelodon orbus* (Maglio 1973). The elephant *Loxodonta adaurora* was described in the Pliocene Lothagam 3 faunal unit (Maglio 1973).

The stegotetrabelodont was subsequently recognized from Mpesida in the Late Miocene of the Baringo Basin, Kenya (Tassy 1986:106: *Stegotetrabelodon* sp., "vraisemblablement *St. orbus*"), and in the lower Adu-Asa Formation of the Awash valley, Ethiopia (Kalb and Mebrate 1993:40). The species *Primelephas gomphotheroides,* first conceived as the ancestor of the three elephantine lineages (*Loxodonta, Elephas,* and *Mammuthus*), has also been described from the Lukeino Formation of the Baringo Basin (Tassy 1986), from the Adu-Asa Formation of the Awash Valley (Kalb and Mebrate 1993:50), from the lower Oluka Formation of Uganda as "cf. *P. gomphotheroides*" (Tassy 1994), and from the Manonga-Wembere Formation of Manonga Valley, Tanzania (Sanders 1997).

Fieldwork at Lothagam by National Museums of Kenya Expeditions from 1989 to 1993 provided a new stratigraphic framework (Leakey et al. 1996; McDougall and Feibel 1999) and, relevant to elephantoid systematics, several new fossils that allowed revision of the status of the previously described species and new hypotheses about the timing of elephantine differentiation. New fossils and new age documentation for many of the older specimens modify the previously estimated range of the taxa. *Primelephas gomphotheroides* is now restricted to the Lower Nawata (ca. 7–6.7 Ma). *Stegotetrabelodon orbus* is recognized from both the Lower and Upper Nawata and extends into the Apak Member of the Nachukui Formation. The gomphothere *Anancus kenyensis,* although rare, ranges from the Lower Nawata to the upper part of the Apak Member (ca. 7–4.2 Ma). Interestingly, fossils from the Upper Nawata and Apak show a greater diversity than previously thought. At least four elephantids are associated in this interval of time: *S. orbus* (Lower Nawata to Apak), and three elephantines: a new species of the *Elephas* clade (Upper Nawata), a taxon described here as "Elephantidae, gen. and sp. indet" (Lower Nawata and Apak) and another as "Elephantidae, gen. and sp. *incertae sedis* A" (Upper Nawata to Apak). From higher in the Apak Member are described *Elephas* cf. *E. ekorensis, Loxodonta* sp. indet. (?aff. *L. exoptata*), and "Elephantidae, gen. and sp. *incertae sedis* B" (middle Apak). Last but not least, a gomphothere that is clearly different from *Anancus* is present at the top of the Upper Nawata (~5 Ma).

Abbreviations

KNM = National Museums of Kenya, Nairobi
HI = height index (= 100H/W)
L = length
W = width (in parentheses, number of the plate where the width was measured)

Systematic Description

Proboscidea Illiger, 1811
Elephantoidea Gray, 1821
Gomphotheriidae Cabrera, 1929

This paraphyletic family consists of "bunodont mastodons"—that is, stem elephantids, the definition of which is open to discussion (i.e., Shoshani 1996; Tassy 1996b). Two gomphotheres are recognized at Lothagam: the previously described *Anancus kenyensis* and a probable trilophodont gomphothere not recorded previously from the Late Miocene of East Africa.

Anancus Aymard in Dhorlac, 1855
Anancus kenyensis (MacInnes, 1942)

(Figure 8.2 [1–4]; table 8.1)

Lothagam Material

- *Lower Nawata:* 346, Rt. P^4.
- *Upper Nawata:* 340, a Rt. maxilla with Rt. $M^{1\text{ or }2}$; 23781, Lt. dP_4.
- *Apak Member:* 341, associated portions of Rt. M_3, Lt. M_3, and Lt. M^3; 23790, anterior portion Lt. $M^{1\text{ or }2}$; 28567, Lt. dP^4.
- *Horizon indet:* 361, portion Rt. M_1; 383, portion Rt. M^1.

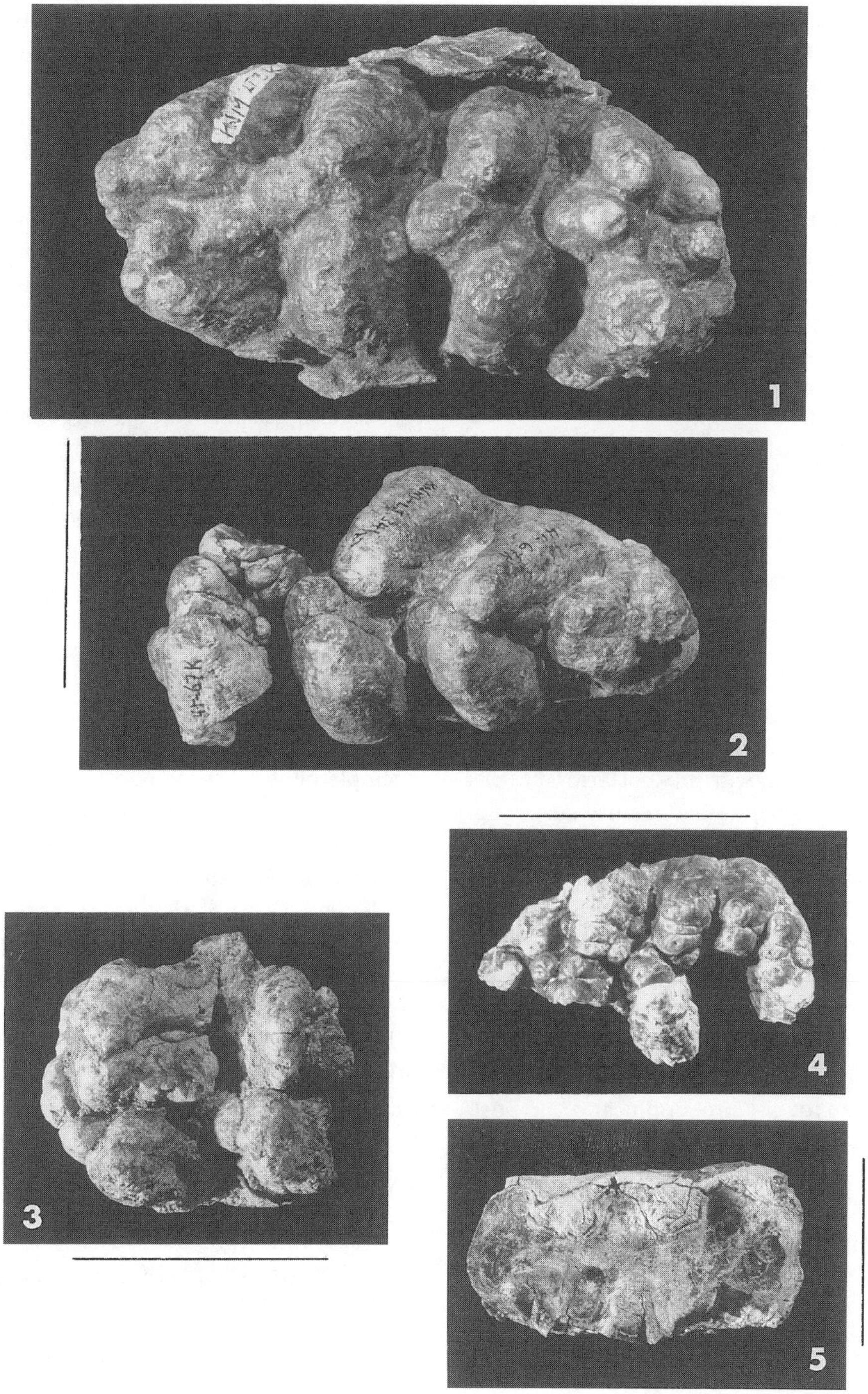

Figure 8.2 Gomphotheriid dentitions: 1 = KNM-LT 340, *Anancus kenyensis,* right maxilla fragment with right $M^{1 \text{ or } 2}$ (Apak Member), occlusal view; 2 = KNM-LT 341, *Anancus kenyensis,* associated portions of right M_3 (Apak Member), occlusal view; 3 = KNM-LT 23790, *Anancus kenyensis,* anterior portion left $M^{1 \text{ or } 2}$ (upper Apak Member), occlusal view; 4 = KNM-LT 28567, *Anancus kenyensis,* left dP^4 (middle Apak Member); 5 = KNM-LT 26324, Gomphotheriidae, gen. and sp. indet., right M_1 (Apak Member), occlusal view. Scales = 5 cm.

LT 28567, a left dP^4 from the middle Apak Member is broken, and the anterior part is missing (figure 8.2 [4]). This tooth displays derived characters. Supplementary accessory cusps are present. The postcingulum is inflated and consists of an outlined fifth loph and two posterior cuspules connected to the last loph. This arrangement (outlined fifth loph with primitive postcingulum) can be consistent with either a tetralophodont or pentalophodont grade: a dP^4 with a complicated posterior area can be associated with more primitive intermediate molars (M^{1-2}).

LT 23781, a left dP_4 (Upper Nawata), has four lophids with no trace of a fifth. The crown is heavily eroded and does not display distinctive traits.

An isolated premolar from the old collection (LT 346) was identified by Tassy (1986:95) as a right $P^{?4}$ of *Anancus kenyensis.* This tooth comes from the Lower Nawata. Its bunodonty is indicated by the round shape of the wear facets. It has four main cusps, and the pretrite cusps are enlarged. Pretrite and posttrite half lophs are separated. The postcingulum is inflated. It is very reminiscent of another premolar, from the Lukeino Formation, also allocated to *Anancus kenyensis* (Tassy 1986:94, plate 13, figure 5). The dislocation of the lophs gives an anancoid pattern to these teeth. However, no premolar of *Anancus* has yet been found to be associated with anancoid molars.

The anterior portion of upper left first or second molar (23790; Upper Apak) shows a complete trefoil on the second loph, a feature characteristic of a complex crown pattern (figure 8.2 [3]). Complex crowns previously recovered from undetermined horizons at Lothagam include 361 (portion of right M_1, rather worn) and 383 (portion of right M^1 with two posterior pretrite conules on the first loph, big posterior posttrite conule on the second loph). However, a partial right maxilla from the Upper Nawata (LT 340) displays a very simple right first or second molar with no posterior pretrite conule on the second loph and no accessory cusps (figure 8.2 [1]); cement is present, although not plentiful. This tooth is remarkably low crowned and narrow.

LT 341 (Apak Member) constitutes three fragments of upper and lower third molars from the same individual. The right M_3 (341A) comprises fragments of the four posterior lophids (figure 8.2 [2]). The left M_3 (341C) consists of the posterior area (last lophid and postcingulum), and the left M^3 (341B) constitutes the posterior half of the crown (three lophs and the postcingulum). The associated portions of the M^3 show many posttrite accessory cusps: three posttrite conules on the most anterior preserved loph (probably the third loph), one big anterior pretrite conule on the same loph, and two on the posterior loph. The postcingulum is complicated and forms two transverse lines. The successive lophs show no anancoid pattern, a condition reminiscent of the *Anancus* specimens described from Lukeino (KNM-LU 532, Tassy 1986: plate 13, figure 2). The M_3s have supplementary posttrite cusps. The enamel is thick (6.8 mm on the right lower molar). Cement is present but not plentiful on 341A–B. These teeth appear distinctively smaller and narrower than those described from Lukeino, but their weathering precludes precise measurements.

Discussion

In the late 1970s, Smart (1976) cited "a primitive form" of the genus *Anancus* at Lothagam; this was interpreted as *Anancus kenyensis* by Coppens et al. (1978). Tassy recognized two morphs among molars of *Anancus kenyensis:* the primitive "*kenyensis* morph" based on the holotype of *A. kenyensis* from Kanam (MacInnes 1942) and the derived "*petrocchi* morph" based on the sample from Sahabi (Libya) described by Petrocchi (1954). In East Africa the *kenyensis* morph was recognized at Kanam, Lukeino, Mpesida, and Lothagam, whereas the *petrocchi* morph was described from Sahabi (Libya), Kanapoi, Aterir, and the Chemeron Formation of Baringo Basin. However, both morphs are known to be associated at Lothagam, Lukeino, Chemeron, and Kanam, and this association suggests that evolutionary implications of these morphs may not be meaningful without supporting biostratigraphic evidence.

Three specimens were recovered during the 1989–1993 expeditions: 23790, portion of $M^{1 \text{ or } 2}$; 23781, left dP_4; and 28567, left dP^4. These clarify neither the status of the species nor the respective evolutionary level of Late Miocene and Early Pliocene molars. However, they do provide an opportunity to define the distinctive traits of the two morphs. Simple molars (*kenyensis* morph) display plesiomorphic characters such as lack of supplementary cusps and tetralophodonty of intermediate molars; in complex molars (*petrocchii* morph) accessory cusps are associated with pentalophodonty. The partial upper molar LT 23790 (upper Apak) has supplementary cusps that are not present in the tetralophodont $M^{1 \text{ or } 2}$ LT 340 from the Upper Nawata. The dP_4 (LT 23781; Upper Nawata) is too worn and weathered to show any character except the lack of a fifth lophid. The dP^4 (LT 28567; middle Apak) is complex with accessory cusps and an incipient fifth loph. This combination would fit with the *petrochii* morph, but on this tooth the cingular cusps are connected to the fifth loph. This last trait is more primitive than the latest known molars of *Anancus* where the postcingulum is fully separated from the last loph (M^2 from Sagantole Formation, Beeryada beds, Ethiopia, described by Kalb and Mebrate 1993:39).

Furthermore, it seems erroneous to expect clear-cut association of the pentalophodont grade with supplementary accessory cusps. In the important population of *Anancus* from the Early Pliocene of Dorkovo (Bulgaria), rare pentalophodont M_2s are found among much more common tetralophodont ones (Metz-Muller 1997). The holotype of *Anancus kenyensis* from Kanam shows the complex morph (supplementary accessory cusps) with no fifth loph. An intermediate form from Nkondo in Uganda has a pentalophodont M^2 without postcingulum.

Thus, without a large collection of teeth from one site, it is difficult to precisely determine the evolutionary level of *Anancus* molars. What can be safely concluded is that the molars from the Apak Member show distinct complex traits. But these molars are not among the most complicated of *Anancus* (Kanapoi, Chemeron, and Aterir in Kenya; the Beearyada beds in Ethiopia). At Kanapoi (4.2 Ma; Leakey et al. 1996) two M_2s (KNM-KP 384 and KNM-KP 410) are nearly identical and combine all derived traits: accessory conules, fifth lophid, strong anancoidy. Moreover no tooth at Lothagam matches those found at Sagantole, Ethiopia (Kalb and Mebrate 1993:37), or Baards Quarry at Langebaanweg (Hendey 1978:10), all of which show an extremely serrated enamel and very thin loph(id)s and are perhaps the ultimate evolutionary level of *Anancus* in eastern and southern Africa at around 4 Ma. Based on all the available morphological evidence, all teeth from Lothagam, including those from the Apak Member, would be older than 4 Ma.

Gomphotheriidae gen. and sp. indet.

(Figures 8.2 [5], 8.3; table 8.2)

Lothagam Material

- *Upper Nawata:* 26324, Rt. M_1.

This lower molar is heavily worn (figures 8.2 [5] and 8.3). The crown seen in occlusal view has the primitive shape of Miocene gomphothere molars. It is trilophodont. The third lophid is enlarged, and the wear patterns of the lophids are typical of bunodont teeth. Pretrite and posttrite halves of the lophids are well separated, and the posttrite half-lophids are located posteriorly to the pretrite half-lophids. Because the tooth is heavily worn, the central conules are contiguous with the lophids and all three lophids are confluent. Dentine indentations in the enamel of the posterior side of the second posttrite half-lophid indicate the presence of two small conules. Although the third lophid of the tooth is damaged, it is seen to be separated labially from the postcingulum. The posterior side of the tooth is preserved at the level of the cervix. Consequently, even if it could be deduced that the postcingulum was inflated, there is no room for a fourth lophid.

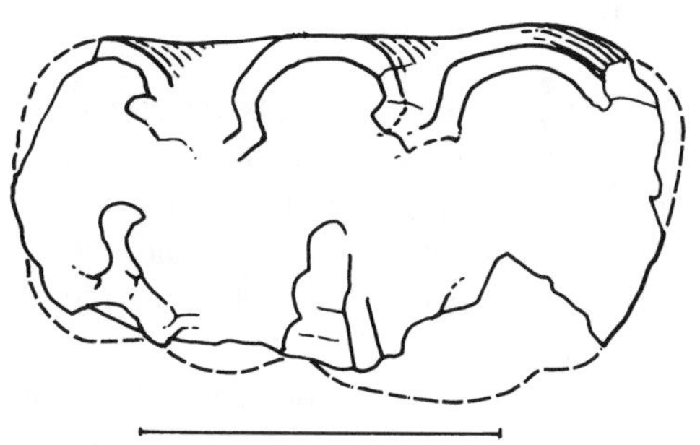

Figure 8.3 Gomphotheriidae gen. and sp. indet., right M_1, KNM-LT 26324 (Upper Nawata), occlusal view. Scale = 5 cm.

Discussion

The geographic and stratigraphic provenance of this tooth is unequivocal. According to M. G. Leakey's files (personal communication 1997) it was excavated in 1992 from the "lower white bed of the Purple Marker," a horizon that denotes the top of the Upper Nawata (Leakey et al. 1996:557). This tooth belongs to a trilophodont gomphothere because no fourth lophid can be reconstructed from what remains of the posterior area of its crown. When LT 26324 is compared with a tetralophodont M_1 or dP_4 of *Anancus kenyensis* such as LT 23781, it clearly cannot belong to *Anancus kenyensis* because no trilophodont dP_4 is known in *Anancus.* Morphological variation seen in the dP_4 of *Anancus* affects the size of the postcingulum but not the presence or absence of a fourth lophid.

This M_1 has a generalized morphology that could belong to any trilophodont Miocene species. For example, if it came from Middle Miocene strata in East Africa it could well belong to the trilophodont amebelodontid *Protanancus macinnesi* Arambourg 1945. However, Alengerr, a Middle Miocene locality, is probably the youngest locality to yield *P. macinnesi* (Tassy 1986:57). Thus *P. macinnesi* is not found in the latest Miocene. The only other Late Miocene trilophodont species from East Africa is *Choerolophodon ngorora* (Maglio 1974), whose latest occurrence at Ngorora and Nakali dates back to at least 10 Ma (Tassy 1986:69), thereby predating the upper member of the Nachukui Formation by about 4 million years. Moreover, although very worn, it is certain that the M_1 from the Upper Nawata cannot belong to a choerolophodont because it has neither wrinkled enamel nor a choerolophodont wear pattern.

Other than *Anancus,* the only gomphothere known from the latest Miocene of Africa (ca. 7–5 Ma) is *Amebelodon cyrenaicus* from Sahabi, Libya (Gaziry 1987). I consider this a junior synonym of *Mastodon grandincisivus* Schlesinger 1917 from Maragah in Iran, which is also present in the contemporaneous locality of Jebel Barakah in Abu Dhabi (see discussion by Tassy 1999). LT 26324 could well belong to this species and would be its first record from East Africa. The M_2 of this species is known to have reached the tetralophodont grade, but an association of trilophodont M_1 (and dP_4) with tetralophodont M_2 is not unexpected in amebelodontid species such as that from Libya. This condition is also displayed, for example, by *Platybelodon grangeri* from the Middle Miocene of Mongolia and China. Thus even in well-prospected areas, unexpected discoveries such as this worn small molar of a hitherto unrepresented species may still occur.

Elephantidae Gray, 1821
Stegotetrabelodon Petrocchi, 1941
Stegotetrabelodon orbus Maglio, 1970
(Figures 8.1, 8.4, 8.5 [1]; tables 8.3–8.7)

Holotype

KNM-LT 354, mandible with Lt. M_2 and M_3, Lt. and Rt. lower tusks, and associated Rt. M^2 and partial Lt. M^3.

Type locality

Lothagam, Apak Member of the Nachukui Formation.

Lothagam Material

- *Lower Nawata:* 343, ?Rt. P^4; 344, Lt. and Rt. M_3; 347, Lt. M^3; 349, Rt. M_3; 360, Lt. M^3; 23791, Lt. M_3; 26318, cranium with Rt. upper tusk, Rt. and Lt. M^3; 26334, anterior portion of Lt. $M_{1\ or\ 2}$ and isolated plate of indeterminate molar.
- *Upper Nawata:* 352, Lt. M_3.
- *Upper Nawata* or *Apak Member:* 355, two portions of Rt. and Lt. M_3; 359, Rt. M^2, Lt. M^3, Rt. and Lt. M_3s.
- *Apak Member:* 354, holotype (see above); 363, partial Rt. M^3.
- *Horizon indet.:* 365, Rt. dP_3; 366, Rt. M^3; 367, Rt. M^3.

The description and measurements of the original collection from Lothagam were published by Maglio (1970, 1973). Of the material previously described (Maglio 1970; 1973:18; Maglio and Ricca 1977:12), one specimen (LT 316, M^3) was not seen, one described as M^2 (LT 366) is here considered M^3, and one specimen (LT 350, two partial mandibles with right and left lower M_2, erroneously listed by Maglio as "LT 342") is not identified as *S. orbus* (see section "Elephantidae, gen. and sp. indet."). Accordingly, the range of variation of M^2 and M^3 is reevaluated (tables 8.3–8.6). In this section I emphasize two new specimens (LT 26318 and LT 26334) recovered between 1989 and 1993 from the lower part of the Nawata Formation.

Partial cranium LT 26318 is from an adult. The alveoli of M^2s are partly resorbed, and the three anterior plates of both right and left M^3s are partly worn (figure 8.4 [3]). The dorsal part of the skull is missing, and the basicranium is weathered and compressed under the palate so that the condyles are situated close to the choanae. Although the precise height of the maxilla cannot be measured, the area of the maxilla between the processus zygomaticus and the alveolar arch is large. This trait (plus the size of the M^3s) is consistent with an identification of this cranium as male (Tassy 1996a). The preserved portion of the right upper tusk, partly in the premaxilla, is straight, but the portion is too short (400 mm) to give a precise idea of the orientation of the entire tusk. The M^3s are large, with seven widely spaced thick plates, each made of five main cusps (figure 8.4 [4–5]). The laminar frequency is low (2.66–2.86). The cement is plentiful but does not embed the rear half of the teeth. The first three plates are worn. The first and second plates display a digitation that indicates the presence of a posterior pretrite conule in the first two interlophs. Conules are either lacking on the other plates or are confined to the base of the posterior side of the plates and hidden by a coating of cement. The cusps forming the plates are nearly equal in size. The plates are moderately high (HI = 73.7 on the fifth plate). In labial view, the anterior plates are anteriorly inclined, whereas the last two plates (sixth and seventh) are posteriorly inclined.

LT 26334A comprises the anterior three plates of a left $M_{1\ or\ 2}$ (figure 8.4 [6–7]). A big, round conule is present at the base of the posterior sides of the first and second plates. A smaller conule is present on the third plate. The shape of the plates is asymmetrical: the second is concave-concave, and the third is convex-convex.

LT 23791 is a left M_3 with seven plates and a distinct postcingulum (figure 8.4 [1]). The first five plates are worn. A posttrite conule is present on the posterior side of fourth and fifth plates, but no pretrite is present. This unusual feature is probably attributable to individual variation. The tooth is small, its size less than 80 percent of the holotype—a difference that is consistent with sexual dimorphism seen in proboscideans, although it can also be explained by size variation between populations.

Discussion

There is no major difference between specimens from the Lower Nawata and from higher in the succession. Only three specimens—all from the old collection—come from the later horizons: the holotype (M^3 and associated M_3s) from the Apak Member plus LT 352 (M_3) from the Upper Nawata and LT 355 (posterior portions of associated right and left M_3) that is either from the Upper Nawata or the Apak Member. In particular, the teeth of the holotype match well with the M^3s of cranium LT 26318 from the Lower Nawata. Size, proportions, height, lamellar frequency, and plate mor-

Figure 8.4 *Stegotetrabelodon orbus* dentitions: 1 = KNM-LT 23791, left M_3 (Lower Nawata), occlusal view; 2 = KNM-LT 366, right M^3 (horizon indet.), occlusal view; 3 = KNM-LT 26318, palate with right and left M^3 (Lower Nawata), occlusal view; 4 = KNM-LT 26318, left M^3 (Lower Nawata), occlusal view (same individual as 3); 5 = KNM-LT 26318, left M^3 (Lower Nawata), labial view (same individual as 3 and 4); 6 = KNM-LT 26334A, associated anterior portion of left $M_{1 \text{ or } 2}$ (Lower Nawata), occlusal view; 7 = KNM-LT 26334A, associated anterior portion of left $M_{1 \text{ or } 2}$ (Lower Nawata), lingual view (same individual as 6). Scales = 5 cm, except for figure 3.

phology are nearly identical in the molars of these two individuals that are perhaps separated by two million years. The variation in size and morphology displayed in the Lower Nawata remains unchanged in specimens from higher in the sequence. The M^3s have either seven plates (LT 26318 and LT 360 from Lower Nawata, and LT 367, horizon unknown) or six (LT 347 and LT 359 from the Lower and Upper Nawata, respectively, and LT 366 of unknown horizon). LT 347 and LT 366 are atypical. LT 347 is large (L = 210.0, W = 105.9 + [3])

and is of typical M^3 size, but its posterior end is abbreviated so that the shape of the crown is closer to that of a M^2 in occlusal view. However, the holotype M^2 has five plates and a postcingulum connected to the fifth plate, whereas in LT 347 the postcingulum is well separated from the sixth plate and forms the posterior end of the crown. LT 366 (figure 8.4 [2]) is small (L = 173.1, W = 86.0 taken at the third plate). Maglio (1973:18) interpreted LT 366 as an M^2. This tooth is indeed short (only slightly longer than the M^2 of the holotype (LT 354, L = 157.8, W = 92.4 taken at the fourth plate, FL = 3.25). But other characters match better with those of M^3s. The posterior end of this tooth is narrow, and the posterior root is subdivided. The wear pattern is asymmetrical: more pronounced lingually on the first two plates (normal condition) and more pronounced labially on the posterior plates. When present in elephants, such asymmetry of wear is usually characteristic of the M^3. A second example of unusual asymmetry in wear is provided by the cranium LT 26318: the labial half of the second plate of the M^3s is more worn than the lingual half. If LT 366 is an M^3, it is indeed a small tooth, probably that of a female, whereas large M^3s such as LT 367 belong to males (figure 8.5 [1]).

Specimen LT 363 (Apak Member) was allocated to *P. gomphotheroides* by Maglio (1973:21) as "partial right M^3 and mandibular symphysis." It is here allocated to *S. orbus*. The partial M^3 has plates made of big, round cusps closer to those of *S. orbus*. The mandibular symphysis does not belong to the same individual but is that of a suoid.

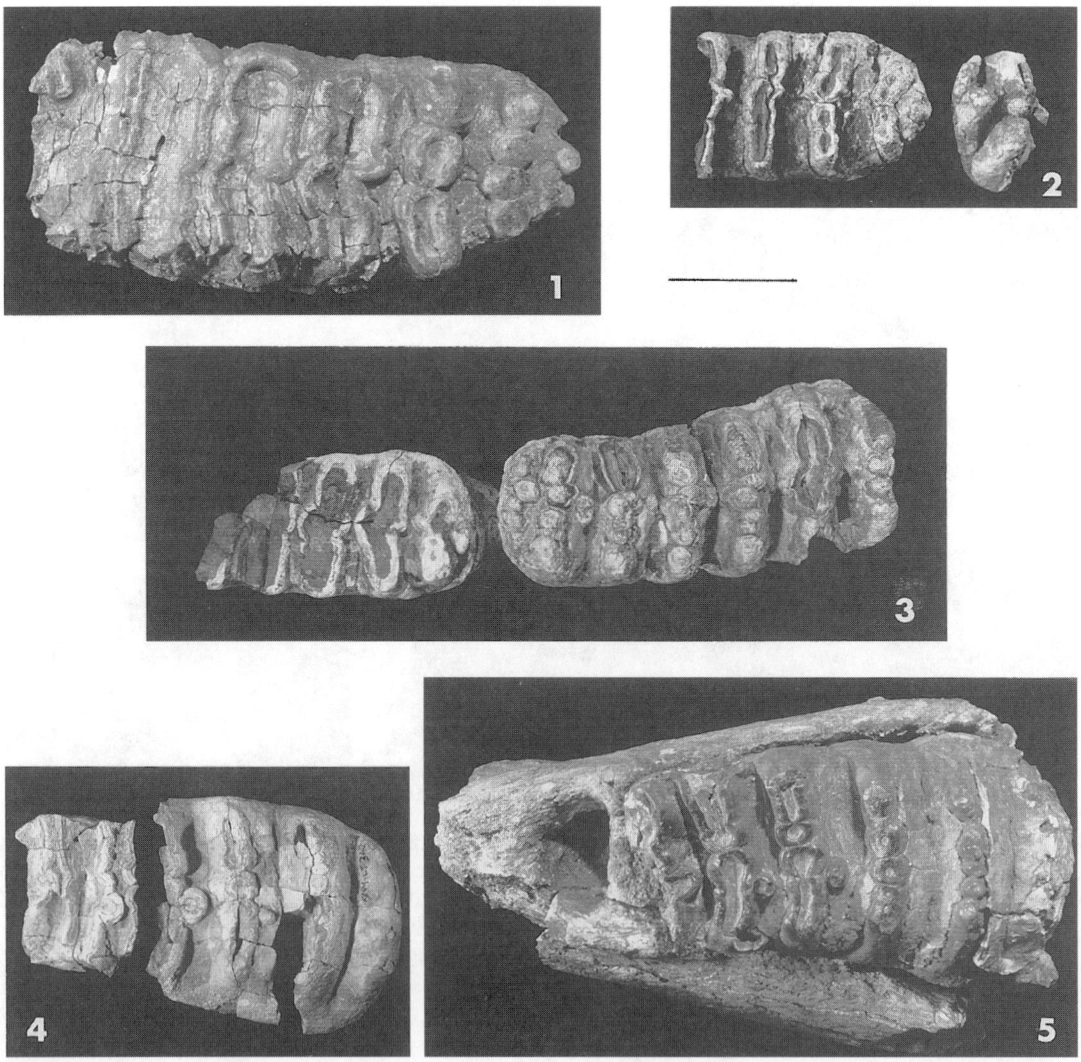

Figure 8.5 Elephantidae dentitions: 1 = KNM-LT 367, *Stegotetrabelodon orbus,* right M^3 (horizon indet.), occlusal view; 2 = KNM-LT 375, *Primelephas gomphotheroides,* right M_1 and first plate of M_2 (horizon indet.), occlusal view; 3 = KNM-LT 358, *Primelephas gomphotheroides,* associated right M_1 and M_2 (Lower Nawata), occlusal view; 4 = KNM-LT 342, Elephantidae gen. and sp. indet., left M_2 (?Lower Nawata), occlusal view; 5 = KNM-LT 350, Elephantidae gen. and sp. indet., left mandible fragment with M_2 (Upper Nawata/?Apak), occlusal view. Scale = 5 cm.

Sexual dimorphism is also seen on M_3s. Although it has seven plates, LT 23791 (figure 8.4 [1]) from Lower Nawata is smaller than the other known M_3s from Lothagam such as LT 359 (with seven plates and a reduced postcingulum; L = 259.0, W = 108.0[3]) or the holotype (with seven plates and a huge postcingulum forming an incipient eighth plate L = 266.0, W = 87.9[3]).

So-called intermediate molars (dP4, M1, M2) of *Stegotetrabelodon orbus* are rare. The holotype with associated upper and lower second molars is the only example from Lothagam. The partial germ LT 26334A, allocated to M_1 or M_2, shows an asymmetry in the shape of the plates: the second is concave-concave, and the third is convex-convex. This asymmetry is found in lower molars of *S. orbus* (e.g., LT 354 and LT 359).

Two lower M_3s (LT 349 from Lower Nawata and LT 352 from Upper Nawata), allocated to *S. orbus* by Maglio, contradict some previous hypotheses on evolutionary characters and trends. According to Maglio (1973:18), a diagnostic feature of *S. orbus* is the presence of posterior columns (= conules) only behind the first two plates. However, LT 349 and LT 352 display conules on plates posterior to the second plate.

Kalb and Mebrate (1993) and Kalb et al. (1996) emphasized the importance of the shape of the plates in erecting evolutionary hypotheses for elephantid molars where convex-convex plates is the primitive condition and concave-concave is the derived condition. In this feature, most of the molars of *S. orbus* are primitive, including LT 352 from the Apak Member. Some variation is seen: a concave-concave shape is present on the second plate of the holotype or on a second and third plate associated with the straight fourth plate in LT 359. However, LT 349 from the Lower Nawata is more derived. Although fragmentary, the first four plates of this tooth are concave-concave. Curiously, the primitive condition observed on LT 352 is associated with a derived trait, which makes this tooth unique in the collection of *S. orbus*. Numerous apical digitations are seen on the third and fourth plates (seven digitations on the fourth plate) and are associated with an anteroposterior compression of the upper part of the plates.

Variation exists, and associations of character states are sometimes self-contradictory. As a consequence, no clear-cut evolutionary separation can be observed between the molars found in Lower Nawata and those documented as Apak or "Upper Nawata or ?Apak." Morphological differences are few and merely due to individual variation. Size differences are conceivably due to sexual dimorphism. In this respect the holotype mandible would be that of a male. The relative shortening of the symphysis, compared to that of *Stegotetrabelodon syrticus* from Libya (also a male), would be characteristic of the species *S. orbus*. The hypothesis for sexual dimorphism in one species explaining the size differences of the symphyseal rostrum and tusks between the mandibles from Sahabi and Lothagam can be ruled out. The species *S. orbus* is valid and is clearly distinguishable by derived traits from the roughly contemporaneous (Late Miocene–Early Pliocene) species *S. syrticus* from Sahabi, Libya (Petrocchi 1954; Maglio 1973), and from Abu Dhabi (Tassy 1999).

Primelephas Maglio, 1970
Primelephas gomphotheroides Maglio, 1970
(Figures 8.5 [2–3], 8.6 [5]; tables 8.6–8.9)

Holotype

KNM-LT 351, associated Rt. and Lt. M^3, and fragment of palate.

Type locality and horizon

Lothagam, Nawata Formation, Lower Member.

Lothagam Material

- *Lower Nawata:* 351, holotype (see above); 358, Rt. and Lt. M^2, Rt. M_2, portion of Lt. M_2, Rt. and Lt. M_1, portions of mandibular ramus and symphysis, portion of lower tusk; 23782, Lt. P^4.
- *Horizon indet.:* 375, Rt. M_1 and first plate of M_2.

The only new specimen allocated to *Primelephas gomphotheroides* is an isolated premolar, probably left P^4 (table 8.8 and figure 8.6 [5]). This tooth is broken anteriorly and consists of the rear part of the first loph, the second loph, and a strong postcingulum. The first loph is made of two separate halves from the two round main cusps. The second loph and postcingulum are linear. Although larger, this tooth is reminiscent of the left $P^{?4}$ (KNM-LU 730) from Lukeino (Tassy 1986:111) and of the P^4s of a cranium from the Late Miocene of Uganda, allocated to *P. gomphotheroides* (Tassy 1994: plate 3, figure 3).

Discussion

No new important specimens of *Primelephas gomphotheroides* were recovered from Lothagam by the National Museum expeditions, but deciduous teeth possibly belonging to this species are described in the section labeled *Stegotetrabelodon* or *Primelephas*. Maglio (1973:21) assumed that "about fifteen additional specimens" belong to *P. gomphotheroides*, from Lothagam,

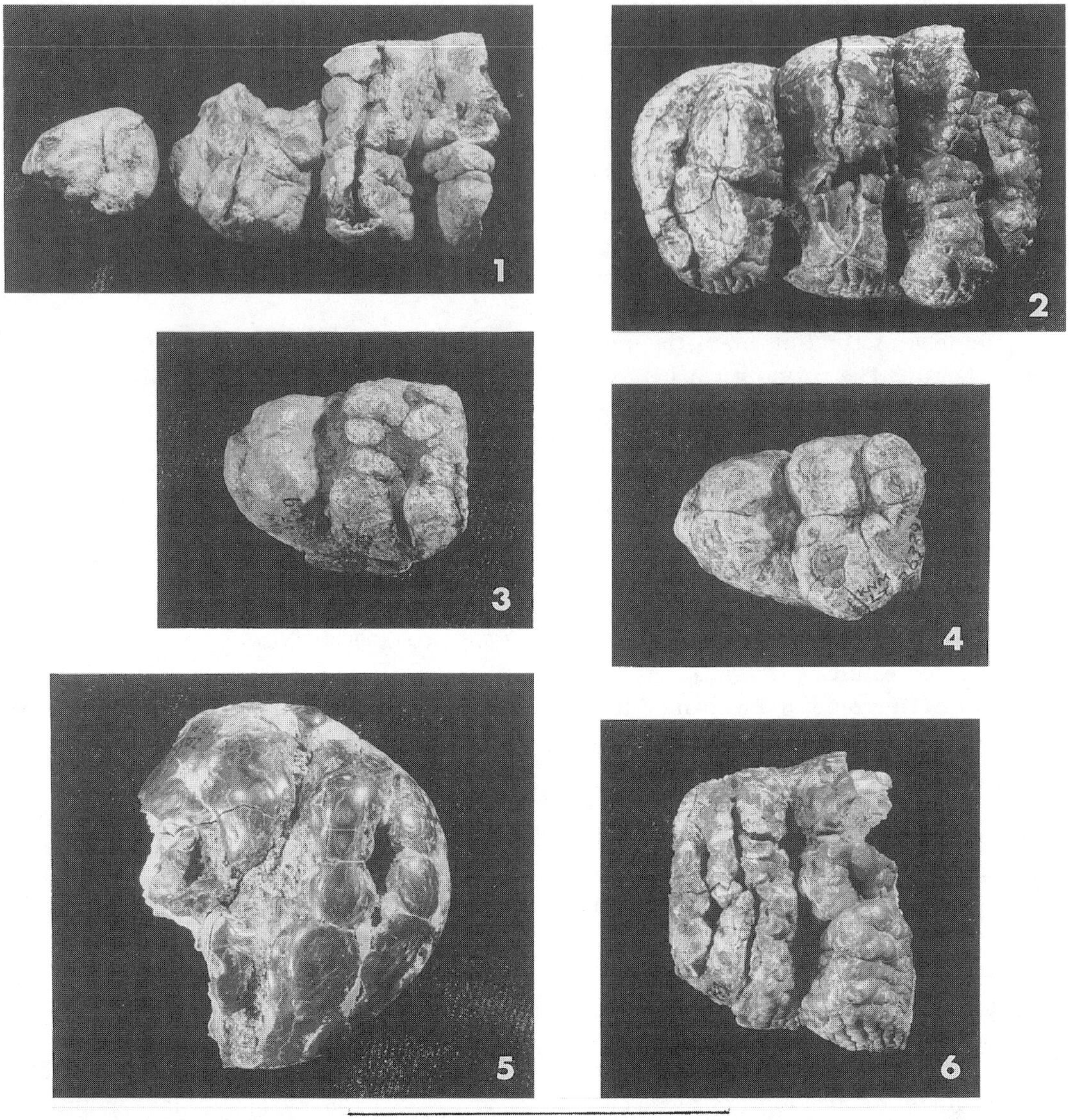

Figure 8.6 Elephantidae dentitions: 1 = KNM-LT 26332, *Stegotetrabelodon* or *Primelephas*, associated left dP_2–dP_3 (Lower Nawata), occlusal view; 2 = KNM-LT 22866, *Stegotetrabelodon* or *Primelephas*, left dP^3 (Lower Nawata), occlusal view; 3 = KNM-LT 26329, *Stegotetrabelodon* or *Primelephas*, ?right P_3 (Lower Nawata), occlusal view; 4 = KNM-LT 26339, *Stegotetrabelodon* or *Primelephas*, left P_3 (Lower Nawata), occlusal view; 5 = KNM-LT 23782, *Primelephas gomphotheroides*, ?left P^4 (Lower Nawata), occlusal view; 6 = KNM-LT 405, Elephantidae gen. and sp. *incertae sedis* A, partial right dP^3 (Upper Nawata), occlusal view. Scale = 5 cm.

Chemeron Formation, and Kanam East (Kenya) and from the Kaiso Formation (Uganda). However, from Lothagam I retain only four individuals: the holotype and LT 358 (nine associated specimens that belong to the same individual) from the Lower Nawata; LT 375 (M_1 of unknown horizon not included in the hypodigm by Maglio); and LT 23782—the newly found isolated broken P^4 also from the Lower Nawata. Another specimen listed by Maglio, LT 363 (partial right M^3 with mandibular symphysis) is here allocated to *S. orbus* (the symphysis being that of a suoid). Including LT 375 (the M_1) in the hypodigm implies some variation in molar morphology.

LT 358 is crucial for understanding *Primelephas gomphotheroides* and keeping the species separate from *S. orbus* (figure 8.5 [3]). This individual comprises associated upper and lower intermediate molars and mandibular symphysis and tusk. The relative states of wear of the M_1s and upper and lower second molars confirm that these teeth belong to only one individual and, consequently, that the broken mandible and tusk also belong to the same individual. Erupting lower and upper second molars show a slight wear on the anterior cingulum and first plate (but no dentine is seen). The symphysis and tusk were described in detail by Maglio and Ricca (1977:25, plate 2 [8]) under the field number 290

67-K. It was assumed from this specimen that *P. gomphotheroides* had lower tusks, which is a primitive feature among the Elephantinae. But there is no anatomical evidence to support the association of the portion of tusk and the mandibular fragment. The reconstruction proposed by Maglio can be refuted; nonetheless, this does not mean that the tusk LT 358 does not belong to that individual. Although a trough is preserved in the symphyseal area, extending posterior to the symphyseal border, Maglio and Ricca's (1977) description is not accurate. Because of its position in the mandible, this trough, into which the incisor was put by Maglio and Ricca, cannot be the alveolus but is instead the mandibular canal. In fact, that portion of the symphysis where the alveolus for the incisor should be is not preserved on this specimen. Consequently, it is not possible to prove or disprove the association of the tusk and the mandibular fragment. There is also no connection between the symphysis and the portion of ramus where the mandibular canal is clearly seen. The preserved part of the tusk, 95 mm long, must be close to the origin because of the large size of the pulp canal (the diameter of which is 37.6 mm). These proportions are too small for an upper tusk associated with erupting M^2s and match best with these of a lower tusk of a stegotetrabelodont. However, the molars of LT 358 do not belong to *S. orbus.* Because there is no reason to assume that the different specimens labeled LT 358 were not really associated, Maglio's conclusion that *P. gomphotheroides* had lower tusks is consistent although, again, it should be emphasized that there is no anatomical evidence to support Maglio's reconstruction. Now, as twenty-five years ago, a complete mandible is needed to confirm retention, orientation, and size of the lower tusks in primitive elephantines.

Direct association of second and third molars is not known in *P. gomphotheroides,* but the allocation of LT 358 to this species is based on several coherent dental characters of the intermediate molars (shape of crown, number and shape of plates) that confirm Maglio and Ricca's (1977) description. The association of M_1 and M_2 (figure 8.5 [3]) is important to identify the evolutionary level of *P. gomphotheroides.* The M_1s have five plates and a small postcingulum connected to the fifth plate. Both anterior and posterior columns are present. Plates 3–5 are convex-convex; plate 2 is straight-concave, and the posterior side of plate 1 is concave. The left M_2 consists of two broken plates. The right M_2 is very long; it is broken posteriorly and had at least six plates. It becomes wider posteriorly, a primitive feature more accentuated than on M_1s where maximal width is taken on the fourth plate. The plates are convex-convex, and anterior and posterior columns are present on the six preserved plates. The M^2s (right M^2 figured by Maglio 1973:plate 3, figure 3; left M^2 figured by Tassy 1986:plate 14, figure 2) have an incipient sixth plate that is narrow but well separated from the fifth plate (on the right M^2, an additional cingular cuspule is present on the posterior end), so that the associated M_2 could well have had seven plates. The plates are convex-convex. This character is more primitive here than in the holotype left M_3 (LT 351), where plates 2–4 are concave-concave, plates 5 and 6 are convex-convex (primitive), and plates 7 and 8 are concave-concave. This variation is perhaps significant for the evolutionary level of *P. gomphotheroides* with incipient elephantine derived traits in the shape of the plates.

On LT 358, plates are more numerous, more spaced, and more subdivided apically than in *S. orbus.* Although the differences are not very profound, they bring LT 358 closer to the holotype M^3s of *P. gomphotheroides* (especially the M^2s) than to the collection of molars of *S. orbus.*

LT 375 consists of a right M_1 and anterior portion (anterior cingulum and first plate) of right M_2 (figure 8.5 [2]). The anterior end of M_1 is broken, and four plates and a narrow postcingulum are preserved. The first anterior preserved plate has a concave posterior side, the second preserved plate is convex-straight, and the posterior two plates are convex-convex. A straight/concave shape was noticed on the first two plates of the M_1 of LT 358. The anterior end of the M_2 of LT 375 is less derived than that of the M_2 of LT 358: the anterior cingulum is more bulbous, and the first plate is less rectilinear. In contrast, the M_1 of LT 375 has more widely spaced plates than in other molars of *P. gomphotheroides.* If the partial M_2 could be allocated to *S. orbus,* the M_1 with thinner plates and more reduced columns cannot. The allocation of LT 375 to *P. gomphotheroides* is perhaps a better bet.

Stegotetrabelodon orbus or *Primelephas gomphotheroides*

(Figure 8.6 [1–4]; table 8.10)

Lothagam Material

- *Lower Nawata:* 434, Rt. dP^4; 22866, Lt. dP^3; 26325, partial dP^3; 26326, Lt. $dP^{?2}$; 26329, ?Rt. P_3; 26332, associated Lt. dP_2–dP_3; 26339, Lt. P_3.

LT 26326, a dP^2, consists of four main cusps, each subdivided into at least two cusps arranged in a linear fashion. Pretrite and posttrite half-lophs are separated by a deep median sulcus. The postcingulum is inflated and tends to form two half-lophs separated by a median sulcus. This tooth resembles KNM-LU 699 from the Lukeino Formation allocated to "cf. *Primelephas*" by Tassy (1986: plate 9, figure 5).

LT 26332 is an anterior left dP_2 associated with the anterior portion of a dP_3 (figure 8.6 [1]). The small dP_2 is complete. The primitive pattern of the crown is that of an elephantoid. The anterior cusp is inflated. The two posterior cusps are smaller; the lingual cusps, slightly worn, are transversely enlarged, whereas the smaller round labial cusp is connected to a postcingulum cuspule and the labial cusp. This tooth resembles that of Miocene gomphotheres rather than that of elephantids, such as the dP_2 from the Lukeino Formation described by Tassy (1986:111, figure 43) as "cf. *Primelephas*." The associated dP_3 is broken posteriorly. The first lophid lacks the pretrite (lingual) cusp. The second and third lophids are anteroposteriorly compressed (or plate-like), with numerous apical digitations. Conules are present in the two preserved interlophids. In the second interlophid, the posterior central conule of second lophid is weak and the anterior central conule of the third lophid is larger. The lophids are linear, much more so than those of the dP_3 of *S. orbus* (LT 365; Maglio and Ricca 1977: plate 1, figure 8; Tassy 1986: plate 14, figure 5), thus making this tooth clearly elephantine. The association of a small gomphothere-like dP_2 with an elephantine dP_3 can only be explained by an important variation of the morphology of dP_2.

LT 22866 and LT 26325, two dP^3s, share the plate-like loph pattern of elephantids. The complete left dP^3 (LT 22866; figure 8.6 [2]) is trilophodont with a linear, low postcingulum. A contact between the posttrite half-loph 1 and the pretrite half-loph 2 can be seen; this is a derived character for gomphotheres that persists in early elephantids (Tassy 1986) but is lost in later elephants. A contact also exists between a tiny posttrite cuspule on the posterior side of the second loph and pretrite cuspules on the anterior side of the third loph, but no anancoid dislocation is seen. The incomplete dP^3, LT 26325, consists of the posterior lingual region. Its inflated postcingulum is more complicated and less linear than that of LT 22866.

LT 434, a right dP^4 from the earlier collection (Lothagam 1B), consists of four entirely worn lophs. It is short. The wear pattern of the lophs reflects the plate shape of elephantids.

LT 26339 and LT 26329, left and right small lower third premolars, are nearly trilophodont and very reminiscent of the P_3s of tetralophodont gomphotheres. LT 26339 (figure 8.6 [4]) has a strong third lophid that is better developed than that in LT 26329 (figure 8.6 [3]). Both have a distinct thin postcingulum connected to the posttrite half of the third lophid. The first lophid is made of two cusps, and the second is made of four. The third lophid is low but distinct from both the second loph and the postcingulum; it is more linear on LT 26339 and more strongly connected to the pretrite half of second lophid on LT 26329. These two gomphothere-like premolars with no anancoid trait can only be interpreted as primitive elephantid premolars. They confirm the hypothesis of persistence of premolars in both *Stegotetrabelodon orbus* and *Primelephas gomphotheroides*, first based on P_4s from Lothagam and Lukeino (Maglio 1973; Tassy 1986).

Discussion

As the species *Stegotrabelodon orbus* is more common than *Primelephas gomphotheroides*, the specimens described in this section belong perhaps to the former, but there are no anatomical characters that unequivocally support this view. Because the two known P_3s are slightly different and because these differences lie in the respective states of elephantid features, the more derived LT 26339 could be allocated to *P. gomphotheroides* and LT 26329 to *S. orbus*. The same approach for dP_3 would determine LT 22866 as *P. gomphotheroides* and LT 26325 as *S. orbus*. This is mere speculation, however. Associated material, with deciduous teeth, premolars and molars, is the only clue to a safe systematic identification.

Elephas Linneaus, 1758
Elephas nawataensis sp. nov.
(Figures 8.7, 8.8 [1–4]; tables 8.11, 8.12)

Holotype

KNM-LT 23783, mandible with Rt. and Lt. dP_4s and M_1s, and various mandibular fragments including the tip of the symphysis.

Type locality and horizon

Lothagam, upper member of Nawata Formation.

Lothagam Material

- *Upper Nawata:* 23783, holotype (see above); 26327, portion of left $M_{1 \text{ or } 2}$.

Diagnosis

Elephantid with a brevirostrine mandible without lower tusks. In lateral view, ventral border of the symphysis convex and not bent downward. dP_4s have x6x plates with slight posterior widening. M_1s narrow with x5x or x6 plates. No protruding enamel loops on dP_4 and no columns on M_1, except on first plate. Low laminar frequency of M_1 (3.5–4.4). Enamel of M_1 thick (3.9 mm).

Crown of M_1 low (HI 74.3). Plates 1–4 of M_1 with both convex-convex and concave-concave orientations (plates 1–4 concave-concave, plates 5–6 [or 5x] convex-convex).

The mandible LT 23783 (figure 8.7) belongs to a young individual (dP_4 in wear, M_1 erupting, unworn). It is fairly well preserved and not distorted, although the symphysis is broken in the middle without contact between the horizontal rami and the tip of the symphyseal rostrum. Both ascending branches are lacking. Even if the total length of the rostrum cannot be directly deduced, the mandible is of the brevirostrine type. From its preserved parts, the symphyseal rostrum was probably short. In occlusal view the horizontal rami are very convergent so that there is space only for a narrow rostrum well individualized from the rami. The preserved parts of the symphysis are low. All these traits are those of a brevirostrine mandible. The symphyseal rostrum lacks any trace of incisor. Its tip is smoothly spatulate.

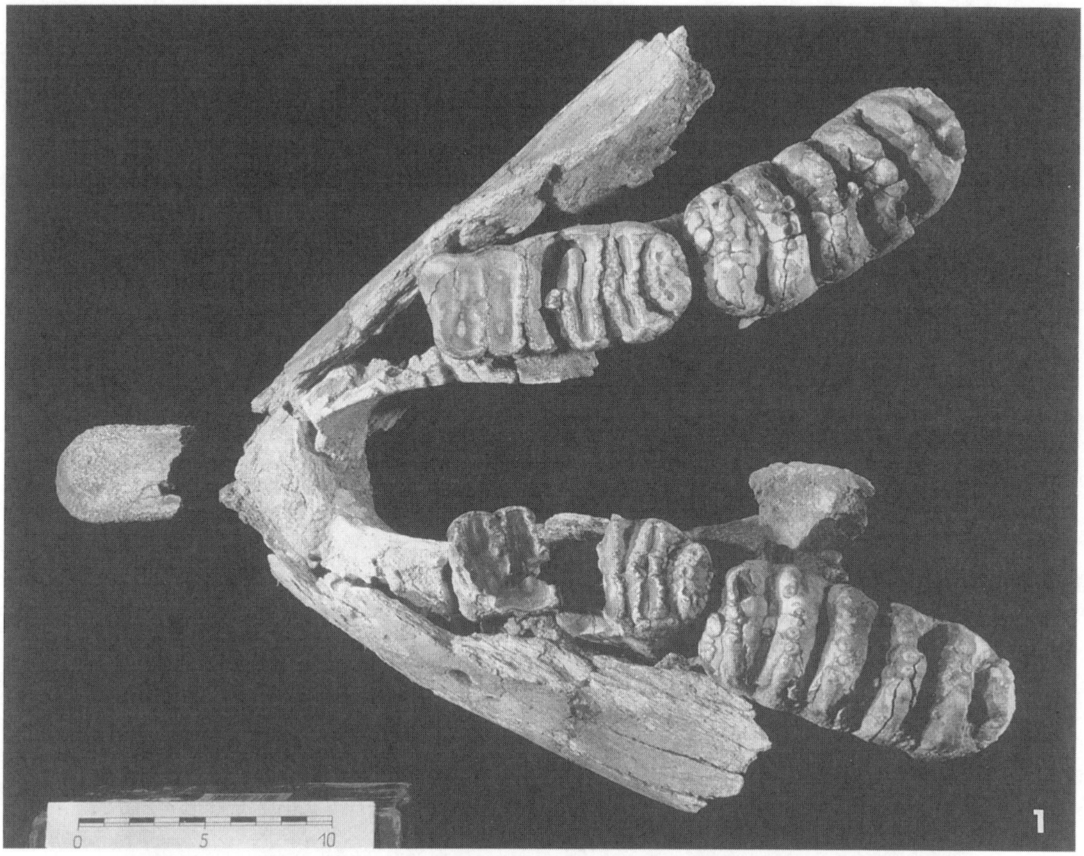

Figure 8.7 *Elephas nawataensis* sp. nov., holotype. KNM-LT 23783: 1 = mandible with right and left dP_4 and M_1 (Upper Nawata), occlusal view; 2 = lateral view of left ramus without M_1. Scales in cm.

The symphyseal trough is not deep. In lateral view the ventral angle of the rostrum is weak. This angulation is anteriorly placed, not at the posterior border of the symphysis, and this is a derived trait. On the contrary, in lateral view (figure 8.7 [2]), the ventral border of the ramus is convex at the level of the posterior symphyseal border; this trait is associated with a brevirostrine mandible that has a true elephantine beak. The mandibular canal is well preserved on this mandible. The small anterior mandibular foramen is anterior to the symphyseal posterior border. The large posterior mandibular foramen is situated at the level of the first plate of dP_4—that is, at the anterior end of the dental arch.

The dP_4s are one plate longer than M_1s. They have six plates and a postcingulum. The number of plates of M_1 can be expressed in alternate ways: M_1s either have five plates and a strong, well-separated postcingulum or six plates and no postcingulum. In either case, M_1s have a less derived laminar formula than dP_4 has.

In occlusal view, the crown of dP_4 is narrow, with an attenuated posterior widening at the level of the fifth plate; this is a derived condition. The dP_4s are fully worn, the occlusal plane is concave, and the first three plates are excavated. The enamel band is much wrinkled (a trait of deciduous teeth). The wear facets indicate the presence of weak columns (= conules), mostly integrated in the wrinkling made by the digitations of the plates. Consequently, the enamel band is rather straight without a loop. These columns appear on the posterior face of the first plate (both right and left sides), on the posterior face of the second plate (left side), and on the anterior face of the fifth plate (right side). In occlusal view, the plates vary from straight-concave (plates 1 and 2) to convex-convex (plates 3–6).

The erupting M_1s (figure 8.8 [1–3]) are narrow without posterior widening; this is a derived condition. As already mentioned, the last plate (sixth plate or postcingulum) is fully separated from the fifth plate. It is made of two main cusps (right side) or three (left side). Cement fills the interlophs and covers entirely the posterior face of the crown. The most anterior plates are made of numerous cusps (eight on the second loph of the left side), while the posterior plates are less derived (only four cusps). A pretrite column (= conule) is seen in the first interloph of both right and left M_1s, on the posterior face of the first plate. No other conule is seen. In occlusal view, the anterior plates (1–4) are concave-concave (a derived condition) and the posterior (plates 5 and 6 or postcingulum) are convex-convex.

LT 26327 is the posterior half of a partial $M_{1 \text{ or } 2}$ (figure 8.8 [4]). As in LT 23783, the crown is narrow, not enlarged posteriorly, plate ?4 is straight-concave, and plate ?5 is convex-convex. The main difference from LT 23783 is that the last plate is more robust and made of four cusps.

Discussion

The mandible LT 23783 clearly demonstrates the presence of an elephantid previously unrecognized from Lothagam. The association of several derived character states indicates that this elephantid is different from *Stegotetrabelodon orbus* present in the Upper Nawata Formation and from *Primelephas gomphotheroides* now recognized only in the Lower Nawata. The new species is also different from elephantid species found in the Nachukui Formation at Lothagam and at Kanapoi.

Lack of lower tusks precludes close relationship with *S. orbus.* The preserved parts of the mandibular symphysis indicate a brevirostry more pronounced than that of *P. gomphotheroides.* The symphyseal rostrum is more gracile than in *Loxodonta adaurora* from Kanapoi, and, unlike in *L. adaurora,* it is not bent downward. This mandible belongs to a young individual (erupting M_1s), while that of *L. adaurora* is known from an adult with M_3s in wear (Maglio 1973:23, plate 5, figures 2–3) but possible variation due to different growth stages can be excluded. Unpublished mandibles from Kanapoi of *L. adaurora* of the same individual age as LT 23783 show the same angulation of the ventral border of the symphysis as that seen on the adult (observations courtesy of M. Leakey and J. Harris personal communication). Definitely, this young mandible from Nawata is closer to mandibles of the *Elephas/Mammuthus* clade, such as *M. subplanifrons* from Langebaanweg, South Africa (Maglio 1973:plate 15, figure 2) or *Elephas recki shungurensis* from Shungura, Ethiopia (Beden 1983:75, figures 3–14; plate 3-3, figure 66; 1987a).

The plates of dP_4 display a wrinkled linear enamel band similar to those of *Elephas ekorensis* and *E. recki;* the lack of protruding loop makes them different from *P. gomphotheroides* and *L. adaurora.* The lack of columns in the interlophids posterior to plate 2 of the M_1 is another character that distinguishes *P. gomphotheroides* from *L. adaurora.* The concave-concave contour of plates 1–4 of M_1 is a derived trait found in Pliocene species of *Loxodonta* and *Elephas* but not seen in *P. gomphotheroides.* The convex-convex structure seen in figure 8.7 is a more primitive condition than that found in Pliocene elephantids.

This mandible does not belong to the most primitive species of *Elephas* previously known—*Elephas ekorensis* from Ekora and Kanapoi (Maglio 1973:33–34). All characters of M_1 are more primitive than those of *E. ekorensis.* The M_1s have only six plates and not eight, and they are shorter. The laminar frequency is lower (3.6–4.4 v. 4.3–6.2). The enamel is thicker (3.9 mm v. 2.1–3.0 mm). The crown is lower (74.6 mm v. 92.1 mm): the height index (74.3) is slightly higher than that of *P. gomphotheroides* (63.5–71.1) and much lower

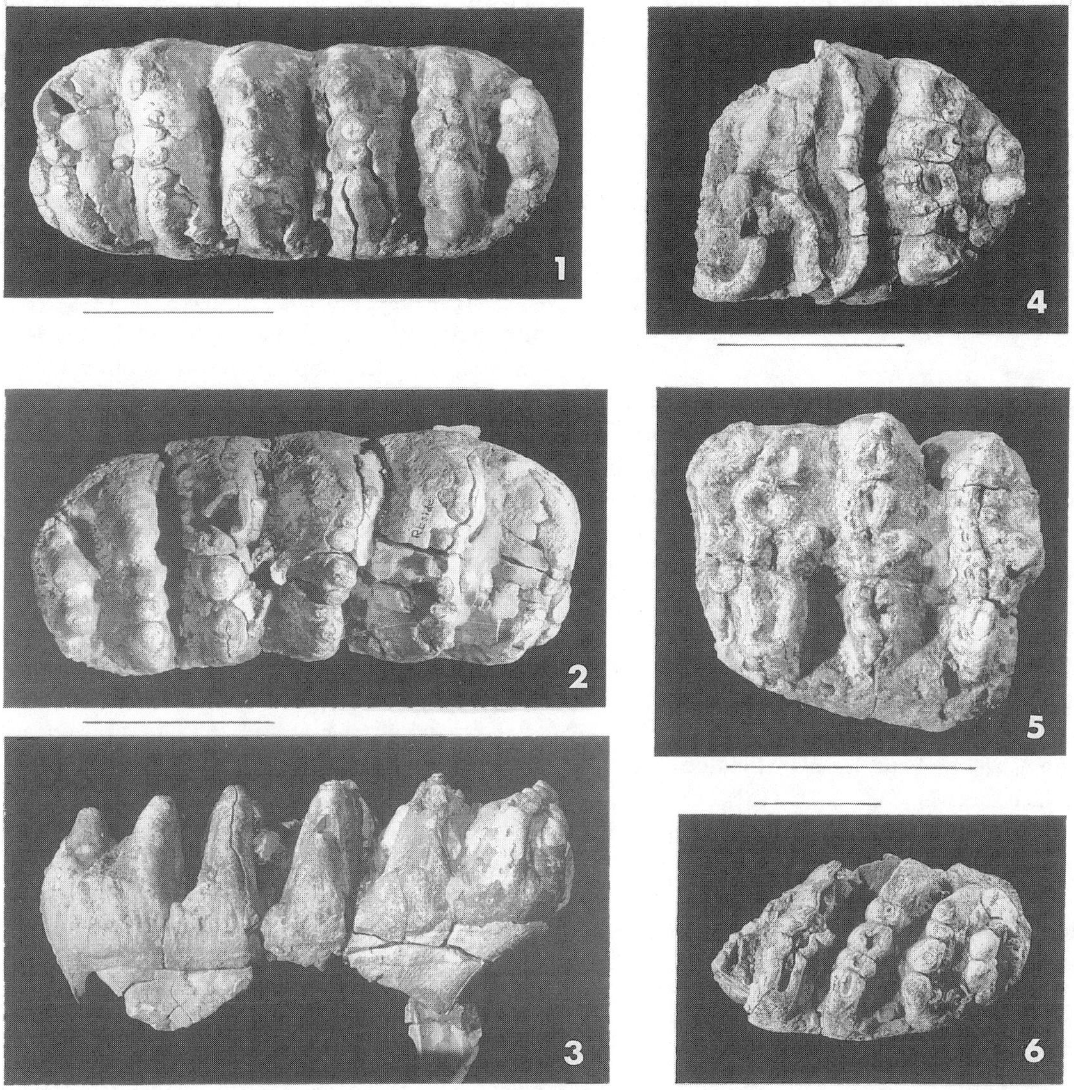

Figure 8.8 Elephantidae dentitions: 1 = KNM-LT 23783, *Elephas nawataensis* sp. nov., left M_1 of holotype mandible (Upper Nawata), occlusal view; 2 = KNM-LT 23783, right M_1, occlusal view; 3 = KNM-LT 23783, right M_1, labial view; 4 = KNM-LT 26327, *Elephas nawataensis* sp. nov., portion of left $M_{1\ \mathrm{or}\ 2}$ (Upper Nawata), occlusal view; 5 = KNM-LT 23795B, *Elephas* cf. *E. ekorensis*, partial right M^1 (Apak Member), occlusal view; 6 = KNM-LT 26323, Elephantidae gen. and sp. *incertae sedis* B, portion of right M_3 (middle Apak Member), occlusal view. Scales = 5 cm.

than in *E. ekorensis* (174.0). The posterior plates of M_1s are made of four main cusps, which is also a primitive condition. The cusps of LT 26327 are bigger and less gracile than those of LT 23783 so that the primitive morphology of this tooth is more accentuated.

In conclusion, LT 23783 belongs to a previously undescribed species and is attributable to *Elephas* rather than *Loxodonta*. The shape of the mandible is different from that of *S. orbus* and *P. gomphotheroides*. Dental characters are more derived than those in *S. orbus* and *P. gomphotheroides*, but they are more primitive than those in *E. ekorensis*. This mandible is allocated to a primitive species of the genus *Elephas*, the first described from Miocene strata, *Elephas nawataensis*, sp. nov. Morphologically, this species represents an intermediate stage between *P. gomphotheroides* and *E. ekorensis* and can be seen to be a possible forerunner of *E. ekorensis*. As the upper member of the Nawata Formation is dated between 6.24 Ma and 5.5 Ma, it is intermediate in time between *P. gomphotheroides* (Lower Nawata) and *E. ekorensis* (Ekora, Kanapoi). Yet, as early stages of *Mammuthus* evolution are still poorly known in East Africa, only further discoveries and investigations will yield firm diagnostic data of early species of *Elephas* and *Mammuthus*.

Elephantidae gen. and sp. indet.
(Figure 8.5 [4–5]; table 8.13)

Lothagam Material

- *Lower Nawata:* 342, Rt. and Lt. M_2; 26331, anterior part of Rt. $M_{?2 \text{ or } 3}$.
- *Upper Nawata:* 350, partial mandible with Rt. and Lt. M_2.
- *Horizon indet.:* 376, isolated plate of ?Rt. M_3.

Most specimens allocated to this taxon are from the early collection. One fragment of a lower molar from the Lower Nawata (LT 26331) is new. One was previously described with different interpretations (LT 350), and the two other specimens had not been published (LT 342 and LT 376).

Maglio (1973:plate 2, figure 4) described LT 350 under the number LT 342, and this mistake was perpetuated by later authors (Maglio and Ricca 1977; Tassy 1986; Kalb and Mebrate 1993; Froehlich and Kalb 1995). Maglio was aware of both specimens LT 342 (field number 57-67K) and LT 350 (field number 120-67K); the fact that they are nearly identical probably explains Maglio's mislabeling. In this contribution I adhere to the original accession numbers—that is, LT 350 is Maglio's LT 342, whereas LT 342 is a different specimen. The reader should be aware that Maglio's (1973) figure 4 of plate 2 is the right ramus with M_2 of LT 350, but I suspect that Maglio's (1973:18) listing of "LT 342, complete left lower M_2, and partial right lower M_2" is indeed LT 342 and not the specimen figured under this number in his plate 4.

The horizontal rami and M_2s comprising LT 350 were previously allocated to *Stegotetrabelodon orbus* by Maglio (1973:plate 2, figure 4) and Maglio and Ricca (1977:plate l, figure 7). Later they were allocated to *Primelephas gomphotheroides* (Froehlich and Kalb 1995: 390; Kalb and Mebrate 1993:44; Tassy 1986:113), but this identification is no longer supported here. They belong to a taxon clearly distinct from these two species.

The mandible LT 350 consists only of portions of right and left rami; it is missing those parts anterior to the alveoli of M_1 and behind the level of the origin of the ascending branch. At its posterior edge, the cross section of the ramus is different from that of the holotype mandible of *Stegotetrabelodon orbus;* it is both smaller (height = 112, width = 135) and less deep (the height of *S. orbus* is circa 160–180). The "rounder" cross section of LT 350 is more characteristic of a brevirostrine mandible.

The M_2s of LT 350 (left M_2; figure 8.5 [5]) were originally described by Maglio (1970, 1973). The right and left M_2s of LT 342 (left M_2; figure 8.5 [4]) are nearly identical in size, morphology, and state of wear—the first four plates being worn. There are, in all, five plates and a large postcingulum that forms the posterior side of the molars and is completely separated from the fifth plate. The crown is enlarged posteriorly at the level of the fifth plate, which is a primitive trait noted by Maglio (1973:19). Columns (= central conules) are prominent on the posterior face of plates 1–3 (LT 350 and right M_2 of LT 342) and of plates 1–4 (left M_2 of LT 342). The wear facet of the columns of the first (LT 350) and the first and second interlophid (LT 342) is that of a strong, rounded loop. There is no column (or even enamel thickening) on the anterior faces of the plates. The cement fills the interlophids and embeds the posterior cingulum, except in the right M_2 of LT 350. The plates are made of numerous cusps (apical digitations), and all are concave-concave in occlusal view.

Allocated to the same taxon are LT 376, an isolated plate, and LT 26331, the anterior part of a right lower molar. LT 26331 from the Lower Nawata consists of a very worn precingulum and first plate, worn second plate, and part of third plate. The plates are concave-concave, and the enamel is thick (4.4 mm measured on the pretrite half of third plate). The wear facet of the posterior loop of the second plate is confluent with the third loop; this means that the pretrite half of the third plate is inflated at its base, a character that cannot be seen on less worn teeth.

Discussion

The two individuals represented by LT 342 and LT 350 differ from both *Stegotetrabelodon orbus* and *Primelephas gomphotheroides.* Characters compatible with their allocation to *S. orbus* are primitive—M_2s with five plates and posterior enlargement. The posterior enlargement is also retained in *P. gomphotheroides.* Derived characters previously used to exclude LT 350 from the original hypodigm of *S. orbus* as conceived by Maglio (1973) are (1) the numerous digital apications (Tassy 1986) and (2) concave-concave plates (Kalb and Mebrate 1993). All plates of these lower M_2s are concave-concave, including the last one; in some lower molars of *S. orbus* the two anterior plates are concave-concave, as is the second plate of the only known M_2 of *P. gomphotheroides.* The M_2s are much wider than those of *P. gomphotheroides,* but, unlike in the right M_2 LT 358 of *P. gomphotheroides,* no anterior columns are present. The plates are thinner and more slender than in *P. gomphotheroides,* and their labial and lingual sides are less conical. The enamel is thinner (4.0–4.1 vs. 5.4–6.9 in *P. gomphotheroides*). All these traits distinguish these M_2s also from *S. orbus.* Except for the wide posterior end of the M_2s, all characters displayed by LT 350 and LT 342 are more derived than those of *P. gomphotheroides.* However, the laminar frequency is comparable to that of *P. gomphotheroides* (3.4).

The lower M_3 (LT 352) was described earlier in this contribution as a peculiar example of *S. orbus* (plates thinner with more numerous apical digitations). The association of this M_3 with the M_2s described here is unlikely because of the primitive convex-convex shape of the plates of LT 352.

The morphology of these M_2s is seen in no other species, and the narrow M_1s of *Elephas nawataensis* do not match with the M_2s described here. Because of their singular morphology, it seems very unlikely that LT 342 and LT 350 are variants of *S. orbus, P. gomphotheroides,* or *Elephas nawataensis.* They appear to belong to a different species, but more precise identification requires a larger sample and other teeth. The stratigraphic range of this unidentified species is uncertain. LT 342 is questionably assigned to the Lower Nawata (but could well be Upper Nawata, according to M. Leakey, personal communication), and LT 350 is from the Upper Nawata. Only LT 26331 is undoubtedly from the Lower Nawata, but the allocation of this worn fragment to the same taxon is tentative. More specimens are needed to assess the stratigraphic distribution of this elusive species. Although a precise systematic identification is not possible from the present material, it is sufficient to illustrate the important diversity of elephantids as early as the Late Miocene.

Elephas cf. *E. ekorensis* Maglio, 1970
(Figures 8.8 [5], 8.9 [1–2]; tables 8.14, 8.15)

Lothagam Material

- *Apak Member:* 23795, partial Rt. M^1; 26320, Lt. hemimandible with dP_4 and M_1.

The left hemimandible LT 26320 belongs to a young individual with dP_4 in wear and unworn M_1 erupting (figure 8.9 [1–2]). This mandible is brevirostrine, although the symphyseal rostrum is broken. The largest mandibular foramen is situated at the anterior border of dP_4. The ascending ramus is high, and this morphology conforms to a cranium with a strongly elevated basicranium. The top of the ascending branch is broken and lacking both coronoid apophysis and condyle.

The left dP_4 has seven plates and a reduced postcingulum (only one cuspule), all worn. Plates 1–4 possess a distinct pretrite column (= conule), the wear of which forms a lanceolate figure. On plates 2 and 3, a weak anterior column is partly merged in the folded figures of the enamel band. The crown is moderately enlarged at the level of plates 5 and 6. The posterior column of plates 5 and 6 is reduced and nearly merged in the folded figures of the enamel band. The seventh plate is narrow and consists of three main cusps. All plates are concave-concave, except the seventh which is convex-convex.

The left M_1 is not entirely visible. It has at least seven plates with the probable formula of $\times 7 \times$. At least the first five plates are concave-concave in occlusal view. The plates are made of five main cusps, and the highest cusps are the two cusps on each side of the remnant median sulcus. Pretrite columns are present in the first four interlophids: the most distinct is posterior to the second plate, posterior to the third plate, and anterior to the fourth plate. Although the precise height cannot be measured, the plates are moderately high and slightly wider than tall (HI = circa 83 calculated on the second plate). In lateral view, the plates are forwardly inclined with a concave anterior face.

LT 23795B, a right M^1 (figure 8.8 [5]) is tentatively allocated to the same taxon. It consists of the first three plates. The shape of the plates is elephantine with a moderate folding of the enamel band. Each plate possesses a posterior pretrite column. Seven apical digitations are seen on the top of the third plate, which is less worn than the others. Cement is preserved in the interlophs.

Discussion

This elephantine mandible is different from that of *Stegotetrabelodon orbus* or *Primelephas gomphotheroides* from the Nawata Formation. By its brevirostry and shape of the rami in the symphyseal area, it is closer to Pliocene elephantines. It does not belong to *Loxodonta* because neither the plates of dP_4 nor those of M_1 exhibit characteristic loxodont median loops. Relatively derived elephantine character states, such as the folding of the enamel band due to wear, make it closer to *Elephas.* Yet, the taxon is not *Elephas nawataensis,* the early *Elephas* from the Nawata Formation. Many traits of the molars of LT 26320 are more derived than those seen on the holotype of *E. nawataensis,* the juvenile mandible LT 23783 of the same ontogenetic age. Both dP_4 and M_1 have more plates, a higher laminar index, and taller height index. The morphology of M_1 in particular reflects more derived traits generally found in Pliocene species, namely taller plates, more closely spaced, with cusps less conical and more compressed anteroposteriorly, and with concave anterior sides in lateral view. All these characters match better with the later species *E. ekorensis* and *E. recki,* although *E. recki* with much more derived molars must be excluded. But LT 26320 is not absolutely identical to *E. ekorensis,* as

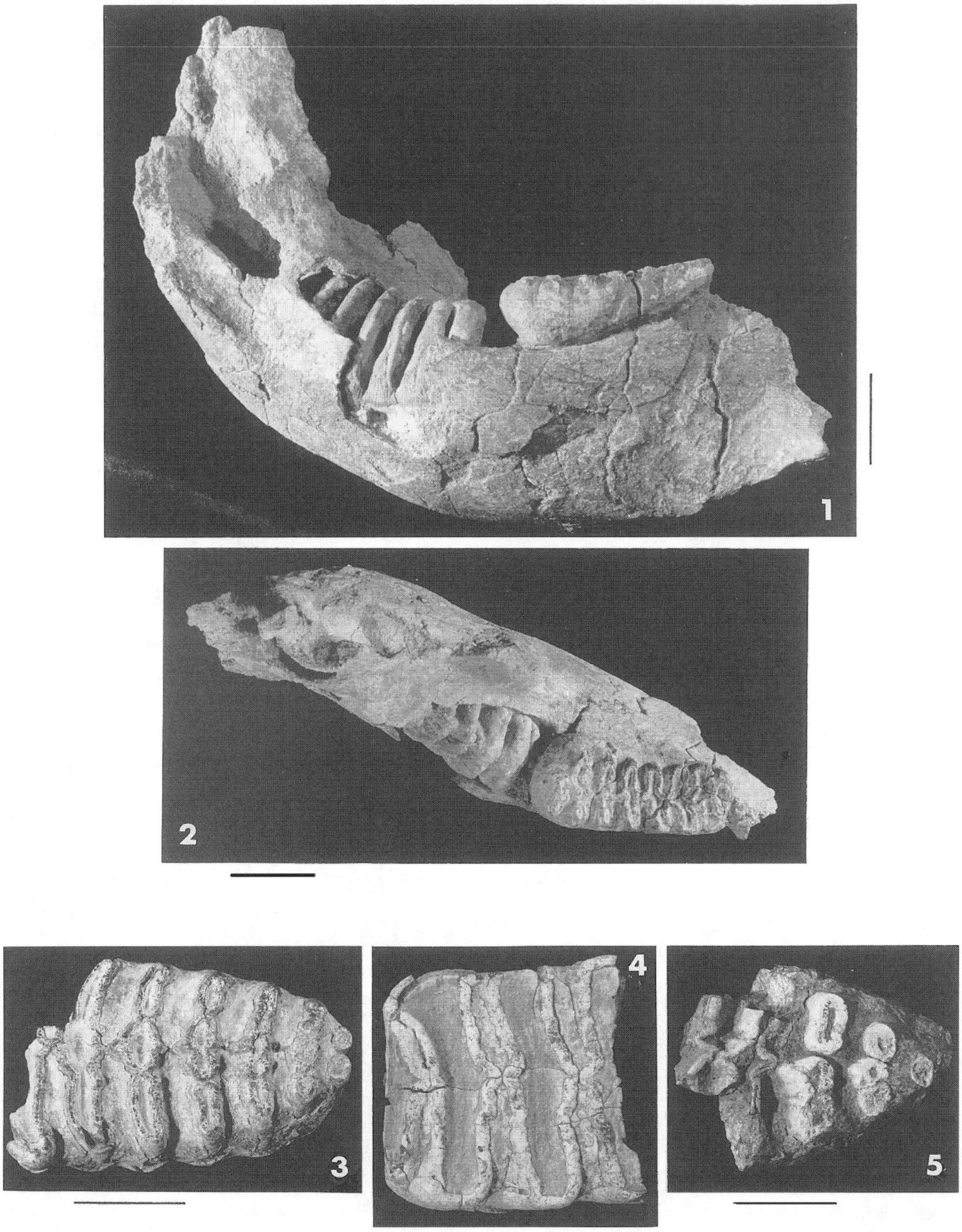

Figure 8.9 *Elephas* and *Loxodonta* dentitions: 1 = KNM-LT 26320, *Elephas* cf. *E. ekorensis*, left mandible with dP_4 and M_1 (upper Apak Member), lingual view; 2 = KNM-LT 26320, occlusal view; 3 = KNM-LT 23794, *Loxodonta* sp. indet. (?aff. *L. exoptata*), right M^3 broken anteriorly (Apak Member), occlusal view; 4 = KNM-LT 26321, *Loxodonta* sp. indet. (?aff. *L. exoptata*), anterior part of right M_3, (middle Apak Member), occlusal view; 5 = KNM-LT 23786, *Loxodonta* sp. indet. (?aff. *L. exoptata*), posterior part of left M_3, (Apak Member). Scales = 5 cm.

described from Ekora and Kanapoi, at about 4.2 Ma. The differences noticed between the Lothagam material described here and (1) the original hypodigm described by Maglio (1973:33–34) or (2) an unpublished specimen of *E. ekorensis* from Kanapoi (courtesy M. Leakey and J. Harris) are all related to more primitive character states displayed by the material from Apak. The dP_4 of LT 26320 has still a primitive shape (posterior enlargement). The M_1 is lower crowned and it is still less high than wide, unlike the proportions in *E. ekorensis.* On the basis of these two traits alone, it is not possible to accept a priori an individual variation and consider that LT 26320 is a bona fide *E. ekorensis.* These differences can probably be attributed to the more primitive evolutionary stage of the Apak material. They demonstrate that if the material from upper Apak belongs to the same lineage as *E. ekorensis,* it is necessarily older than 4.2 Ma.

Loxodonta F. Cuvier, 1825 (anonymous emendation 1827) *Loxodonta* sp. indet. (?aff. *L. exoptata*)

(Figure 8.9 [3–5]; table 8.16)

Lothagam Material

- *Apak Member:* 23786, posterior Lt. M_3; 23794, Rt. M^3 fragment; 26321, anterior part Rt. M_3.

The preserved portion of LT 23794, a right M^3 (figure 8.9 [3]), consists of six convex-convex plates. All plates except the postcingulum (two cusps) show wear. The wear figures (the dentine is visible on the first five preserved plates) display an incipient loxodont loop. The median area of the plates is inflated with a strong anterior column but posterior columns are weak. One consequence of this asymmetrical development of central columns is the convex shape of the posterior side of the plates. The laminar frequency is low. The enamel is rather thick and not much folded.

The preserved posterior portion of LT 23786, left M_3 (figure 8.9 [5]), consists of three and a half plates and the postcingulum made of one cusp. The median area of the plates is inflated and the wear figures make a nearly symmetrical loxodont loop. The second preserved plate is concave-concave. More posterior plates belong to the very end of the tooth; they are narrow and convex-straight and convex-convex. The enamel is thick.

The anterior part of LT 26321, right M3 (figure 8.9 [4]), is worn to the base and can only be tentatively compared to the other M_3s described above. The shape of the plates in occlusal view is nearly rectangular. They are concave-concave; the anterior side of the third plate is straight. The median part of the plates is inflated. The central columns are prominent. In the interlophids, facing anterior and posterior loops are in contact.

Discussion

These three partial molars indicate that a loxodont species very likely existed in the Apak Member of the Nachukui Formation. Its status remains obscure. The M^3 is different from either *Loxodonta exoptata* from Koobi Fora and Laetoli (Beden 1987b) or "*Loxodonta* sp. indet (Lukeino stage)" from Lukeino and Uganda (Tassy 1994). The M^3s of both taxa, like the modern *L. africana,* have convex-concave plates. The concave shape of the posterior plates is partly due to the enlargement of the posterior loop, not realized on LT 23794. Yet, the wear facets of LT 23794 can be found on some specimens of the collection of *Loxodonta exoptata* from Laetoli described by Beden (1987b:plate 8-3, figures 18 and 23: M^2, fourth plate of M^3). The M^3 described here is also different from *L. adaurora* from Kanapoi (Maglio 1970, 1973). *L. adaurora* is an atypical loxodont because the plates of the molars are nearly straight, without inflation medially, although a loxodont sinus can be present (though not on every tooth). In contrast, the M_3 from Apak exhibits the normal concave-concave shape of *Loxodonta.*

All in all, these partial molars seem more closely related to a group formed by *Loxodonta* sp. "Lukeino stage" and *L. exoptata* than to *L. adaurora.* However, three partial molars are not enough for an unequivocal systematic conclusion. Yet, because these molars are clearly different from those allocated to the genus *Elephas* and also found in the Apak Member, it can be concluded that two contemporaneous species representing the modern elephant genera *Loxodonta* and *Elephas* were present at Lothagam between ca. 4 and 5 million years.

Elephantidae gen. and sp. *incertae sedis* A

(Figure 8.6 [6]; table 8.17)

Lothagam Material

- *Upper Nawata:* 405, partial Rt. dP^3.
- *Apak Member:* 23785, portion of Rt. M^3.

The two preserved (anterior) plates of LT 405, a right dP^3 (figure 8.6 [6]), are thin and linear. The anterior cingulum is plate-like. Central columns are present on the posterior sides of the first and second plate and on the anterior side of the second plate.

LT 23785, a partial M^3, consists of a posterior half molar with five plates preserved and a postcingulum made of two cusps. It is an unworn tooth embedded in cement and only the top of the two anterior plates can be observed. The most anterior preserved plate shows six apical cones. This tooth has a low laminar index. It is moderately tall.

Discussion

There is no conclusive evidence that the dP^3 and M^3 described here belong to the same taxon, but they share derived traits compared to *Stegotetrabelodon orbus* and *Primelephas gomphotheroides,* and should probably be allocated to a more evolved elephantid. The dP^3 shows a more elephantine condition compared to the smaller dP^3 (LT 22866) allocated to *Stegotetrabelodon* or *Primelephas* (see above): the plates are thinner, more linear; the anterior cingulum is more plate-like. The M^3 is taller than the M^3s of *Stegotetrabelodon orbus* and *Primelephas gomphotheroides,* and belongs to an elephantine with a low laminar index. Because only a small part of this specimen is not covered in cement, it is not possible to allocate it more precisely to an early species of *Elephas* or *Loxodonta* or to what has been described here as "Elephantidae gen. and sp. indet."

Elephantidae gen. and sp. *incertae sedis* B

(Figure 8.8 [6]; table 8.18)

Lothagam Material

- *Apak Member:* 26323, portion of Rt. M_3.

LT 26323 comprises the rear portion of a partial M_3 (figure 8.8 [6]). The three preserved plates are linear and made up of four or five main cusps. They are straight-straight. The most anterior preserved plate is worn, and the wear figure shows the existence of a partly damaged posterior central column. No other central column is seen. The postcingulum is bulbous, made of five cups in a circle.

Discussion

This partial tooth exhibits no distinctive traits associated with any of the different taxa described elsewhere in this contribution. The plates are too linear and thin for *Stegotetrabelodon orbus;* the resemblance to *Primelephas gomphotheroides* is greater, although the plates are also thinner. The laminar frequency is also higher (3.75) than that of the holotype lower M_3 of *P. gomphotheroides* (3.4), and the enamel is thinner (3.9 mm vs. 4.9 mm). These derived characters make this tooth closer to elephantines such as *Elephas nawataensis, Elephas* cf. *E. ekorensis,* or "Elephantidae gen. and sp. indet." But the unlikely association of an M_3 with straight-straight plates and an M_2 with concave-concave plates such as these of "Elephantidae gen. and sp. indet." (LT 342 and LT 350) still has to be found in a single specimen. This partial tooth could then belong to a primitive member of the *Elephas* lineage.

General Conclusions

Recent fieldwork at Lothagam by Meave Leakey and colleagues proves that the elephantoid diversity in the Lothagam area is greater than previously thought (figure 8.10). The presence of *Anancus, Stegotetrabelodon,* and *Primelephas* is confirmed. In addition to these "classical" taxa, a primitive species of the extant genus *Elephas*—*Elephas nawataensis* sp. nov.—is described from the Late Miocene strata of the Nawata Formation. Another species, previously erroneously allocated either to *Stegotetrabelodon* or to *Primelephas* and here called "genus and species indet.," is also present in the Nawata Formation and Upper Nawata/?Apak. *Loxodonta adaurora* appears at Lothagam in the Kaiyumung Member of the Nachukui Formation, above the Lothagam basalt.

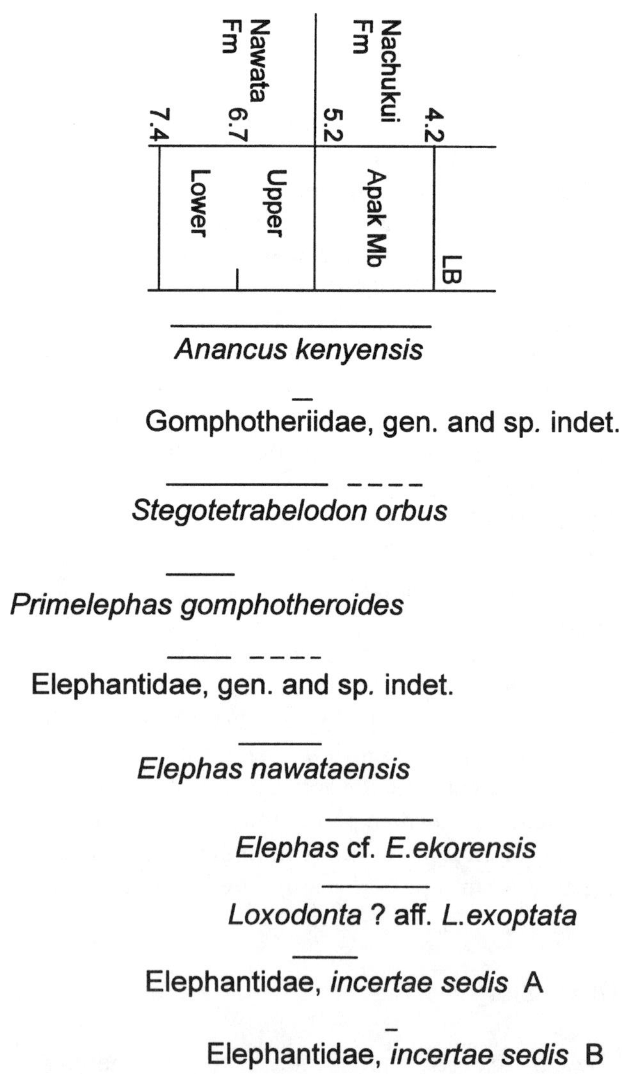

Figure 8.10 Stratigraphic extension of the elephantoid taxa recognized in the Nawata Formation and Apak Member of Nachukui Formation: LB = Lothagam basalt; dash = possible but uncertain extension. Stratigraphy taken from Leakey et al. (1996) and McDougall and Feibel (1999).

This provenance equates with that of the skull LT 353 described from Lothagam 3 by Maglio (1973:23) and of which only one M^3 survived transportation from Harvard University to Nairobi after Maglio's study. An indeterminate species of the genus *Loxodonta,* differing from *L. adaurora* and compared here to *L. exoptata,* is described from the Apak Member of the Nachukui Formation. From the same unit, a species of the genus *Elephas* intermediate in molar morphology between *Elephas nawataensis* and *Elephas ekorensis* is here assigned to "*Elephas* cf. *E. ekorensis.*"

A consequence of these discoveries is that the differentiation of elephantines is now seen to precede 4 Ma and to have occurred in the Late Miocene. At Lothagam, *Elephas* occurs in the upper part of the Nawata Formation ca. 6.7–5.2 Ma. *Loxodonta* occurs in the Apak Member of the Nachukui Formation, ca. 5.5–4.2 Ma. This species of loxodont is likely to be an intermediate between "*Loxodonta* sp. Lukeino stage" (described from Lukeino [ca. 6 Ma] and in the Nkondo Formation of Uganda [ca. 6–4 Ma]) and *Loxodonta exoptata* described from Laetoli. Thus two loxodont lineages were contemporaneous in the late Miocene: this taxon and the *L. adaurora* lineage. Although the morphological evidence is scanty, the Apak loxodont fits with the previous scheme that considers *Loxodonta adaurora* as an offshoot of the loxodont lineage (Beden 1983; Kalb and Mebrate 1993).

The taxon here called "Elephantidae, gen. and sp. indet.," from the Nawata Formation and Upper Nawata/?Apak (ca. 8–5 Ma), certainly played a role in the differentiation of elephantines. Because this taxon is contemporaneous with both *Stegotetrabelodon orbus* and *Primelephas gomphotheroides,* the hypothesis that *P. gomphotheroides* was the stem species of elephantines is clearly an oversimplification. Despite all the important fieldwork done in the Lothagam area, with its bounty of new specimens representing new taxa, our knowledge today of *Primelephas gomphotheroides* is nearly the same as it was in the 1970s. At Lothagam the species is restricted to four individuals only, certainly a frustrating state of affairs for students of elephantine evolution.

If the origin and differentiation of modern genera of elephants occurred earlier than previously thought, it is also worth noticing that trilophodont gomphotheres persisted in East Africa up to the latest Miocene. An isolated trilophodont M_1 from the uppermost Nawata Formation (ca. 5.2 Ma) is the only evidence for this statement but is enough to contrast elephantoid biodiversity in the Late Miocene at Lothagam with that of Sahabi, which is the only other African locality with a trilophodont gomphothere ca. 6–5 Ma.

New discoveries at Lothagam do not include stegodonts. Tassy (1994), who described numerous stegodont molars from Uganda, interpreted the absence of this taxon at Lothagam to emphasize the separation between the biota of the Eastern Rift and Western Rift. However Sanders (1999) has recently described the discovery of *Stegodon* at Mpesida in Kenya.

After recent fieldwork by the National Museums of Kenya expeditions from 1989 to 1993, Lothagam is confirmed as the type area of three elephantoid species: *Stegotetrabelodon orbus, Primelephas gomphotheroides,* and the new *Elephas nawataensis.* Nevertheless, it should be remembered that the fossil evidence for elephantid evolution is mainly based on molar morphology. Often when associated character states given by different organs are found together, we are forced to alter previous hypotheses. Consequently, we will continue to await new information from future discoveries.

Acknowledgments

I am especially eager to thank Meave Leakey, former head of paleontology at the National Museums of Kenya, for having given me the opportunity to study the Lothagam material. My research in the museum in 1997 benefits from her invaluable help and goodwill. I also thank her and John M. Harris for letting me look at the unpublished material from Kanapoi. My stay at Nairobi was financed by the CNRS (UMR 8569, dir. P. Janvier). Many thanks to Robert Campbell, who is responsible for the fine photographs of this article and to F. Pilard, D. Serrette, and D. Visset for their artistic help.

References Cited

Beden, M. 1983. Family Elephantidae. In J. M. Harris, ed., *Koobi Fora Research Project.* Vol. 2. *The Fossil Ungulates: Proboscidea, Perissodactyla, and Suidae,* pp. 40–129. Oxford: Clarendon Press.

Beden, M. 1987a. Les Eléphantidés (Mammalia, Proboscidea). In Y. Coppens and F. C. Howell, eds., *Les faunes Plio-Pléistocène de la Basse Vallée de l'Omo (Ethiopie).* Vol. 2. *Les Eléphantidés, Proboscidea (Mammalia),* pp. 1–162. Paris: Centre National de la Recherche Scientifique.

Beden, M. 1987b. Fossil Elephantidae from Laetoli. In M. D. Leakey and J. M. Harris, eds., *Laetoli: A Pliocene Site in Northern Tanzania,* pp. 259–294. Oxford: Clarendon Press.

Behrensmeyer, A. K. 1976. Lothagam Hill, Kanapoi, and Ekora: A general summary of stratigraphy and faunas. In Y. Coppens, F. C. Howell, G. L. Isaac, and R. E. Leakey, eds., *Earliest Man and Environments in the Lake Rudolf Basin: Stratigraphy, Paleoecology, and Evolution,* pp. 163–170. Chicago: University of Chicago Press.

Coppens, Y., V. J. Maglio, C. T. Madden, and M. Beden. 1978. Proboscidea. In V. J. Maglio and H. B. S. Cooke, eds., *Evolution of African Mammals,* pp. 336–367. Cambridge, Mass.: Harvard University Press.

Froehlich, D. J., and J. E. Kalb. 1995. Internal reconstruction of

elephantid molars: Applications for functional anatomy and systematics. *Paleobiology* 21:379–392.
Gaziry, A. W. 1987. Remains of Proboscidea from the Early Pliocene of Sahabi, Libya. In N. T. Boaz, A. El-Arnauti, A. W. Gaziry, J. de Heinzelin, and D. D. Boaz, eds., *Neogene Paleontology and Geology of Sahabi,* pp. 183–203. New York: Liss.
Hendey, Q. B. 1978. The age of the fossils from Baard's Quarry, Langebaanweg, South Africa. *Annals of the South African Museum* 75:1–24.
Kalb, J. E., and A. Mebrate. 1993. Fossil elephantoids from the hominid-bearing Awash Group, Middle Awash Valley, Afar Depression, Ethiopia. *Transactions of the American Philosophical Society,* 83:xiv, 1–114.
Kalb, J. E., D. J. Froehlich, and G. L. Bell. 1996. Phylogeny of African and Eurasian Elephantidae of the late Neogene. In J. Shoshani and P. Tassy, eds., *The Proboscidea: Evolution and Palaeoecology of Elephants and Their Relatives,* pp. 101–116. Oxford: Oxford University Press.
Leakey, M. G., C. S. Feibel, R. L. Bernor, J. M. Harris, T. E Cerling, K. M. Stewart, G. W. Storrs, A. Walker, L. Werdelin, and A. J. Winkler. 1996. Lothagam: A record of faunal change in the Late Miocene of East Africa. *Journal of Vertebrate Paleontology* 16:556–570.
MacInnes, D. G. 1942. Miocene and post-Miocene Proboscidea from East Africa. *Transactions of the Zoological Society of London* 25:33–106.
Maglio, V. J. 1970. Four new species of Elephantidae from the Plio-Pleistocene of northwestern Kenya. *Breviora* 341:1–43.
Maglio, V. J. 1973. Origin and evolution of the Elephantidae. *Transactions of the American Philosophical Society,* n.s., 63:1–149.
Maglio, V. J. 1974. A new proboscidean from the Late Miocene of Kenya. *Palaeontology* 17:699–705.
Maglio, V. J., and A. B. Ricca. 1977. Dental and skeletal morphology of the earliest elephants. *Verhandelingen der koninklijke Nederlandse Akademie van Wetenschappen* 29:1–51.
McDougall, I., and C. S. Feibel. 1999. Numerical age control for the Miocene-Pliocene succession at Lothagam, a hominoid-bearing sequence in the northern Kenya Rift. *Journal of the Geological Society* (London) 156:731–745.
Metz-Muller, F. 1997. A mandible of *Anancus arvernensis* (Proboscidea, Mammalia, Pliocene) with pentalophodont M2's: Significance of the pentalophodont grade in *Anancus. Neues Jahrbuch für Geologie und Paläontologie Monatsschrift* 12:709–726.
Petrocchi, C. 1954. I proboscidati di Sahabi. *Rendiconti Accademia Nazionale dei XL* 4:1–66.
Sanders, W. J. 1997. Fossil Proboscidea from the Wembere-Manonga Formation, Manonga Valley, Tanzania. In T. Harrison, ed., *Neogene Paleontology of the Manonga Valley, Tanzania: A Window into the Evolutionary History of East Africa,* pp. 266–310. New York: Plenum Press.
Sanders, W. J. 1999. Oldest record of *Stegodon* (Mammalia, Proboscidea). *Journal of Vertebrate Paleontology* 19:793–797.
Shoshani, J. 1996. Para- or monophyly of the gomphotheres and their position within Proboscidea. In J. Shoshani and P. Tassy, eds., *The Proboscidea: Evolution and Palaeoecology of Elephants and Their Relatives,* pp. 149–177. Oxford: Oxford University Press.
Smart, C. 1976. The Lothagam I fauna: Its phylogenetic, ecological and biogeographic significance. In Y. Coppens, F. C. Howell, G. L. Isaac, and R. E. Leakey, eds., *Earliest Man and Environments in the Lake Rudolf Basin: Stratigraphy, Paleoecology, and Evolution,* pp. 361–369. Chicago: University of Chicago Press.
Tassy, P. 1986. *Nouveaux Elephantoidea (Mammalia) dans le Miocène du Kenya.* Cahiers de Paléontologie, Travaux de Paléontologie Est-Africaine. Paris: Centre National de la Recherche Scientifique.
Tassy, P. 1994. Les Proboscidiens (Mammalia) fossiles du Rift occidental, Ouganda. In B. Senut and M. Pickford, eds., *Geology and Palaeobiology of the Albertine Rift Valley, Uganda-Zaire.* Vol. 2. *Palaeobiology/Paléobiologie,* pp. 215–255. Occasional Publication No. 29. Orléans: Centre International pour la Formation et les Echanges Géologiques.
Tassy, P. 1996a. Growth and sexual dimorphism among Miocene elephantoids: The example of *Gomphotherium angustidens.* In J. Shoshani and P. Tassy, eds., *The Proboscidea: Evolution and Palaeoecology of Elephants and Their Relatives,* pp. 92–100. Oxford: Oxford University Press.
Tassy, P. 1996b. Who is who among the Proboscidea? In J. Shoshani and P. Tassy, eds., *The Proboscidea: Evolution and Palaeoecology of Elephants and Their Relatives,* pp. 39–48. Oxford: Oxford University Press.
Tassy P. 1999. Elephantids (Mammalia) from the Emirate of Abu Dhabi, United Arab Emirates: Palaeobiogeographic implications. In P. J. Whybrow and A. Hill, eds., *Fossil Vertebrates of Arabia,* pp. 209–233. New Haven: Yale University Press.

Table Abbreviations

– = overestimated measurement
+ = incomplete measurement
[ ] = indices and measurements in brackets indicate an estimation or reconstruction
E = enamel thickness
H = maximum crown height (in parentheses the plate or loph measured)
HI = height index (= 100H/W)
I = index of robustness (= 100W/L)
KNM = National Museums of Kenya, Nairobi
L = length
LF = laminar frequency
P = number of plates; x prefix = anterior cingulum; x suffix = posterior cingulum; + prefix = anterior portion missing; + suffix = posterior portion missing
po = posttrite half loph (po1, 2, etc., first posttrite half loph, second posttrite half loph, etc.)
pr = pretrite half loph (pr1, 2, etc., first pretrite half loph, second pretrite half loph, etc.)
W = width (in parentheses the plate or loph measured)

TABLE 8.1 Measurements (in mm) for Dentition of *Anancus kenyensis* from Lothagam

	L	W	H	I	HI	E
dP^4						
KNM-LT 28567	e69.0	44.4 (3)	34.1 (po3)	—	(76.8) (3)	—
P^4						
KNM-LT 346	45.2	44.3	—	91.2	—	—
M^1						
KNM-LT 383	—	52.0 (1)	38 (po1)	—	—	—
$M^{1\text{ or }2}$						
KNM-LT 340	133.0	65 (1)	48.1 (pr1)	(48.9)	74.0 (1)	—
KNM-LT 23790	—	63+ (1)	53.4 (pr1)	—	84.8– (1)	—
M^3						
KNM-LT 341B	—	(69.6) (3)	51.7 (pr3)	—	(74.3) (3)	—
dP_4						
KNM-LT 23781	84.6	42+ (3–4)	—	—	—	—
M_1						
KNM-LT 361	—	57.5 (4)	—	—	—	—
M_3						
KNM-LT 341A	—	62+ (3)	—	—	—	6.8 (pr?3)
KNM-LT 341C	—	(57) (5)	—	—	—	—

TABLE 8.2 Measurements (in mm) of M_1 of Gomphotheriidae gen. and sp. indet. from Lothagam

	L	W	I	E
KNM-LT 26324	89.6	e56 (3)	e62.5	3.2 (po3)

TABLE 8.3 Measurements (in mm) of the Cranium of *Stegotetrabelodon orbus* (KNM-LT 26318) from Lothagam

Parameter	**Measurement**
Palatal length, from the anterior alveoli to the choanae	250.0
Internal maximal width of the palate	79.0
External maximal width of the palate	270.0
Internal width of the palate taken at the anterior grinding teeth	43.8
Minimal palatal width between the interalveolar cristae (= maxillary ridges)	41.0

Source: Parameters from Tassy (1996b:94).

TABLE 8.4 Measurements (in mm) of Second and Third Molars of *Stegotetrabelodon orbus* from Lothagam

	P	L	W	I	LF	H	HI	E
M^2								
KNM-LT 354	×5×	157.8	92.4 (4)	58.55	3.25	—	—	—
M^3								
KNM-LT 347	6×	210.0	105.9+ (3)	50.4+	3.2	—	—	—
KNM-LT 354	+3×	—	103.2 (3)	—	2.85	73.0 (3)	70.7 (3)	—
KNM-LT 359 r	×6	—	107.9 (2)	—	—	77.2 (4)	64.0 (2)	—
KNM-LT 366	6	173.1	86.0 (3)	49.7	3.25	—	—	—
KNM-LT 367	×6×	234.6+	103.4 (4)	44.1–	2.8+	—	—	—
KNM-LT 26318r	×7	241.8	110.0 (3)	45.5	2.7	—	—	—
KNM-LT 26318 l	×7	244.7	107.4 (2–3)	43.4	—	64.5 (5)	73.7 (5)	—
$M_{1\ or\ 2}$								
KNM-LT 26334	×3+	—	62.8 (2)	—	4.0	44.6 (2)	71.0 (2)	—
M_2								
KNM-LT 354	6×	165.0	89.0 (5)	53.9	3.6	—	—	—
M_3								
KNM-LT 344 r	+2×	—	—	—	—	—	—	—
KNM-LT 344l	+5×	—	—	—	3.7	—	—	5.2–6.5
KNM-LT 349	(?+)7×	233+	78.3 (2)	33.6–	3.4	—	—	—
KNM-LT 352	?×8×	e260.0	102.4 (?4)	e39.4	—	60.0+ (5)	60.6+ (5)	5.0 (pr2)
KNM-LT 354	×8	266.0	87.9 (3)	33.0	2.5	78.7 (3)	89.5 (3)	—
KNM-LT 355 r	—	—	83.6+	—	—	—	—	5.4 (pr)
KNM-LT 355 l	—	—	81.0+	—	—	—	—	5.2 (po)
KNM-LT 359 r	×7×	263.0	108.5 (3)	41.2	2.9	72.7 (3)	67.0 (3)	—
KNM-LT 359 l	×7×	259.0	108.0 (3)	41.7	3.0	68.0 (3)	62.9 (3)	—
KNM-LT 23791	×7×	211.3	87.8 (4)	41.5	3.65	47.2 (7)	74.4 (7)	5.4

TABLE 8.5 Comparative Measurements of M^2 of *Stegotetrabelodon*

	P	L	W	H	LF	ET	HI
S. orbus (Lothagam)	5	157.8	92.4	—	3.25	—	—
S. orbus (Adu-Asa)[a]	5	145.0	e85.0	45.0	4.0	—	54.0
S. syrticus (Sahabi)[b]	4–5	158.0–172.0	96.0	49.0	—	—	—
S. syrticus (Shuwaihat)[c]	4–5	174.4–176.9	101.4	50.2	2.9–3.0	—	49.5

[a] Kalb and Mebrate (1993).

[b] Gaziry (1987).

[c] Tassy (1999).

TABLE 8.6 Comparative Measurements (in mm) of M^3 in the Genera *Stegotetrabelodon* and *Primelephas*

	P	L	W	H	LF	ET	HI
S. orbus (Lothagam)	6–7	173.1–244.7	86.0–110.0	64.5–77.2	2.7–3.25	5.5–7.0	64.0–73.7
S. syrticus (Sahabi)[a]	6	232.0–242.0	109.8–126.1	73.0–80.1	2.8–3.2	6.0–7.1	65.6–67.2
S. syrticus (Shuwaihat)[b]	5+	—	e103	59.7	2.9	—	60.3
P. gomphotheroides (Lothagam)	7	211.6	96.6	52.5 (4)	3.1	—	57.1

[a] Maglio (1973).

[b] Tassy (1999).

TABLE 8.7 Comparative Measurements (in mm) of M_3 in the Genera *Stegotetrabelodon* and *Primelephas*

	P	L	W	H	LF	ET	HI
S. orbus (Lothagam)	7–8	211.3–266.0	78.3–108.4	68.0–78.7	2.5–3.7	5.0–6.5	62.9–89.5
S. syrticus (Sahabi)[a]	7	280.0–317.4	115.0–123.4	57.0–74.1	2.6–3.1	5.8–6.0	60.1
S. syrticus (Shuwaiat)[b]	8	277.0	109.1	62.0	3.0	6.6	56.4
Stegotetrabelodon sp. (Uganda)[c]	6–8	241.6+	94.0–102.0	81.0	3.0	5.4–6.8	79.0
P. gomphotheroides (Lothagam)	8	249.5+	92.6	58.8 (5)	3.4	5.6	64.3

[a] Modified from Maglio (1973).

[b] Tassy (1999).

[c] Tassy (1995).

TABLE 8.8 Measurements (in mm) of ?left P^4 of *Primelephas gomphotheroides* from Lothagam

	L	W	I	H	HI	E
KNM-LT 23782	43+	52.1	121.2–	22+	42.7+	2.6

TABLE 8.9 Measurements (in mm) of Molars of *Primelephas gomphotheroides* from Lothagam

	P	L	W	I	LF	H	HI	E
M^2								
KNM-LT 358 r	—	170.8	87.7 (4)	51.3	3.5	59.5 (4)	67.8 (4)	—
KNM-LT 358 l	×6	167.1	85.6 (4)	51.2	3.4	58.4 (4)	68.2 (4)	—
M^3								
KNM-LT 351	×7×	208.0	96.6 (1)	46.4	3.1	55.4 (4)	60.1 (4)	5.7 (po2)
M_1								
KNM-LT 358 r	×5×	106.0+	60.3 (4)	56.9–	4.4	—	—	—
KNM-LT 358 l	—	—	59.0 (4)	—	4.65	—	—	3.0 (pr2)
KNM-LT 375	+4×	94.0+	62.6	66.6–	4.0	—	—	—
M_2								
KNM-LT 358 r	×6+	170.0+	e80 (5)	e47–	3.65	54.7 (4)	74.3 (4)	5.4–6.2 (pr5-po6)
KNM-LT 358 l	—	—	79.4 (6)	—	—	53.7 (6)	67.6 (6)	6.9 (pre6)
M_3								
KNM-LT 351	×8×	249.0	92.7 (3)	37.2	3.4	67.2 (5)	73.8 (5)	4.9 (pr2)

TABLE 8.10 Measurements (in mm) of Premolars and Deciduous Premolars Allocated to *Stegotetrabelodon* or *Primelephas*

	P	L	W	I	H	HI	E
dP?2							
KNM-LT 26326	+2×	—	22.7	—	—	—	—
dP^3							
KNM-LT 22866	×3×	55.5	39.5 (3)	71.2	25.4 (3)	64.3 (3)	1.5 (3)
dP^4							
KNM-LT 434	4	64.7+	41.1 (3)	63.5–	—	—	—
dP_2							
KNM-LT 26332	2×	17.3	14.7 (2)	85.0	—	—	—
dP_3							
KNM-LT 26332	×3+	45.7+	31.0 (3)	67.8–	21.0 (3)	67.7 (3)	—
P_3							
KNM-LT 26329	×3×	32.8	25.7 (2,3)	78.3	15+ (2)	58.4+ (2)	—
KNM-LT 26339	×3×	33.8	25.0 (2)	74.0	—	—	—

TABLE 8.11 Measurements (in mm) of the Mandible of *Elephas nawataensis* sp. nov. (KNM-LT 23783) from Lothagam, Upper Nawata

Parameter	**Measurement**
Mandibular width taken at the root of ascending rami	335
Width of the horizontal ramus taken at the root of the ascending branch	113 (left side)
Width of the horizontal ramus, taken at the anterior of dP_3	50.3
Width of the horizontal ramus, taken at the anterior of dP_4	64
Posterior symphyseal width	159.1
Anterior symphyseal width	45.1
Height of symphyseal rostrum (taken at 60 mm of the tip)	25.8
Internal width between the dP_4s	63.2
Height of the horizontal ramus taken at the root of the ascending branch	e103 (left side)

Source: Parameters from Tassy (1996b:95).

TABLE 8.12 Measurements (in mm) of Molars of *Elephas nawataensis* sp. nov. from Lothagam

	P	L	W	I	LF	H	HI	E
dP_4								
KNM-LT 23783 (rt)	×6×	111.1	51.9 (4)	46.7	—	—	—	—
KNM-LT 23783 (lt)	×6×	—	50.3 (5)	—	—	—	—	2.1 (po4)
M_1								
KNM-LT 23783 (rt)	×6×	145.3	63.0 (2)	43.2	—	47.0 (2)	74.6 (2)	—
KNM-LT 23783 (lt)	×6×	138.8	61.8 (4)	44.5	—	45.6 (3)	74.4 (3)	—
$M_{1\ or\ 2}$								
KNM-LT 26327	+3×	—	69.5 (?4)	—	3.4	—	—	4.3

TABLE 8.13 Measurements (in mm) of Molars of Elephantidae gen. and sp. indet. from Lothagam

	P	L	W	I	LF	H	HI	E
M_2								
KNM-LT 342 (rt)	+3×	—	95.4 (5)	—	—	56.6 (5)	59.3	—
KNM-LT 342 (lt)	×6	—	96.1 (5)	—	3.0	—	—	4.1 (pr3)
KNM-LT 350 (rt)	×6	171.1	103.3 (5)	60.4	3.4	—	—	4.0 (po2)
KNM-LT 350 (lt)	×6	173.2+	101.9 (5)	58.8–	3.3	—	—	4.0 (po2)
$M_{?2 \text{ or } 3}$								
KNM-LT 26331	×3+	—	67.4 (2)	—	—	—	—	4.4 (pr)

TABLE 8.14 Measurements (in mm) of the Hemimandible of *Elephas* cf. *E. ekorensis*, KNM-LT 26320

Parameter	**Measurement**
Width of the horizontal ramus taken at the root of the ascending branch	99.4
Height of the horizontal ramus taken at the level of the fifth interlophid of dP_4	113.0
Height of the horizontal ramus taken at the root of the ascending branch	105.2

Source: Parameters from Tassy (1996b:95).

TABLE 8.15 Measurements (in mm) of *Elephas* cf. *E. ekorensis* from the Apak Member of the Nachukui Formation

	P	L	W	I	LF	H	HI
M^1							
KNM-LT 23795B	×3+	—	67.0 (3)	—	4.85	—	—
dP_4							
KNM-LT 26320	×7×	120.4	58.6 (5)	48.7	6.3	—	—
M_1							
KNM-LT 26320	×7×	(143)	c60 (2)	c41.9	5.3	(53)	(83.3)

TABLE 8.16 Measurements (in mm) of *Loxodonta* sp. indet. (?aff. *L. exoptata*) from the Nachukui Formation

	P	L	W	I	LF	H	HI	E
M^3								
KNM-LT 23794	+6×	—	95.4	—	4.3	—	—	3.6–4
M_3								
KNM-LT 23786	+3×	—	—	—	—	—	—	4.9 (po)
KNM-LT 26321	×3+	—	72.4 (3)	—	—	—	—	3.8 (pr3)

TABLE 8.17 Measurements (in mm) of the Upper Molar of Elephantidae gen. and sp. *incertae sedis* A from Lothagam

	P	L	W	I	LF	H	HI
dP^3							
KNM-LT 405	×2+	—	41.2 (2)	—	—	20.8+ (2)	—
M^3							
KNM-LT 23785	+5×	168.5+	85.1	—	—	72.8	88.1

TABLE 8.18 Measurements (in mm) of M_3 of Elephantidae gen. and sp. *incertae sedis* B from Nachukui Formation

	P	L	W	I	LF	H	HI	E
KNM-LT 26323	+3×	126+	e81	—	3.75	—	—	3.9

8.2

Deinotheres from the Lothagam Succession

John M. Harris

Deinotheres were rare components of the Lothagam biota, but deinothere enamel is distinctive and *Deinotherium bozasi* is documented from all four terrestrial members of the succession.

Figure 8.11 Restoration of *Deinotherium bozasi* by Mauricio Antón. Shoulder height estimated at about 3 meters.

Deinotheres were evidently restricted in distribution to Neogene and Quaternary localities of the Old World and appear to have originated in Africa as a sister group of the Elephantoidea. The earliest known representatives of the family, from Early Miocene localities in Kenya, had attained elephantine size but were characterized by dental and postcranial features that differentiated them unequivocally from the elephantoids. The dentition, in particular, is fundamentally diagnostic, and the combination of characters that is peculiar to the deinotheres—absence of upper tusks, downward curvature of the lower tusks, bifunctional and (mostly) bilophodont cheek teeth that are retained throughout the life of the individual—readily distinguish representatives of this family from other proboscideans.

Prodeinotherium hobleyi is characteristic of Early Miocene faunas from East and North Africa, and similar deinotheres became established in Europe (*P. bavaricum*) and Asia (*P. pentapotamiae*) by the onset of the Late Miocene. Toward the end of the Miocene, larger and more progressive forms representing the genus *Deinotherium* became established in Eurasia, initially coexisting with, and eventually replacing the smaller *Prodeinotherium* species. In Africa, replacement of *P. hobleyi* with *D. bozasi* reportedly occurred during the time interval represented by the Namurungule Formation of the Samburu Hills in Kenya (Nakaya et al. 1984). Deinotheres became extinct in Eurasia during the Pliocene but were still present in East Africa during the Early Pleistocene, and they occurred as late as Shungura Member L in the northern Lake Turkana Basin (Beden 1979).

Deinotheres are represented throughout the Lothagam succession. They are not common at any interval, and few specimens have been collected, but deinothere enamel is readily recognizable and has been collected for isotopic analysis, as well as for voucher specimens for the collections of the National Museums of Kenya. There is no indication that *P. hobleyi* might have persisted from the Namurungule Formation into the lowest part of the Nawata sequence.

Systematic Description

Family Deinotheriidae
Deinotherium Kaup, 1829
Deinotherium bozasi Arambourg, 1934

(Figure 8.11; table 8.19)

Diagnosis

Species of *Deinotherium* with teeth of similar size to those of *D. giganteum.* Skull rostrum steeply turned down anteriorly (as in *Prodeinotherium hobleyi*), with narrower external nares and rostral trough than in *D. giganteum*; preorbital swelling greatly reduced and sited just in advance of P^3; occiput steeply inclined; nasals with anterior median projection. Mandibular symphysis flexed at 90°. Upper premolars lacking the subsidiary styles of those of *D. giganteum.*

Lothagam Material

- *Lower Nawata:* 26346, tooth fragments.
- *Upper Nawata:* 26344, tooth fragment.
- *Apak Member:* 23806, molar fragments; 26345, tooth fragments.
- *Kaiyumung Member:* 23677, Rt. dP_4 fragment.
- *Horizon indet.:* 356, RM^1 and molar fragments.

Only two partial teeth were recovered from Lothagam—a lower deciduous fourth premolar fragment from the Kaiyumung Member and an incomplete right upper molar recovered in 1967 and hence of unknown provenance. The upper molar seemed to be slightly smaller than a comparable specimen from Koobi Fora (Harris 1983) but was not complete enough to afford detailed comparison. Other enamel fragments were consistent with derivation from *D. bozasi* rather than *P. hobleyi.*

Discussion

The bilophodont brachyodont teeth of deinotheres were superficially similar to those of tapirs and were admirably suited for processing soft vegetation. It is interesting to note, therefore, that whereas all the elephantoid proboscideans sampled from the Lothagam assemblages adopted a C_4 grazing diet during the course of accumulation of the Nawata Formation, the deinotheres persisted as C_3 browsers. Indeed, deinotheres and giraffine giraffes were the only large African mammals to have retained a diet of C_3 browse throughout their known history. However, whereas the $\delta^{18}O$ content of giraffid tooth enamel indicates these mammals derive(d) much of their water from their food, the more negative $\delta^{18}O$ values from deinothere tooth enamel suggests that deinotheres drank from local water sources.

Acknowledgments

I thank the government of Kenya and museum trustees of the National Museums of Kenya for permission to study the Lothagam deinothere material. I also thank the curatorial and preparation staff of the palaeontology division of the National Museums of Kenya, Nairobi, for making the material available for study.

References Cited

Arambourg, C. 1934. Le *Dinotherium* des gisements de l'Omo. *Comptes Rendus de la Société Géologique de France* 1934: 86–87.

Beden, M. 1979. Les éléphants (*Elephas* et *Loxodonta*) d'Afrique orientale: Systematique, phylogénie, intérêt biochronologique. Ph.D. diss., University of Poitiers.

Harris, J. M. 1983. Family Deinotheriidae. In J. M. Harris, ed., *Koobi Fora Research Project.* Vol. 2. *The Fossil Ungulates: Proboscidea, Perissodactyla, and Suidae,* pp. 22–39. Oxford: Clarendon Press.

Kaup, J. J. 1829. *Deinotherium giganteum. Isis* 4:401–404.

Nakaya, H., M. Pickford, Y. Nakano, and H. Nishida. 1984. The Late Miocene large mammalian fauna from the Namurungule Formation, Samburu Hills, northern Kenya. *African Study Monographs,* Supplementary issue 2:87–131.

TABLE 8.19 Measurements (in mm) of M^1 of *Deinotherium bozasi*

LT 356	
Length	91.5+
Protoloph tr	70.1+
Metaloph tr	76.7+
Tritoloph tr	—

8.3

Fossil Aardvarks from the Lothagam Beds

Simon A. H. Milledge

Two species of aardvark are represented at Lothagam, both restricted to the Nawata Formation. The smaller, *Leptorycteropus guilielmi*, known from a partial skeleton and an isolated partial femur, has a more generalized morphology than the extant aardvark but shows primitive fossorial adaptations. The larger, *Orycteropus* sp., is found in the Lower and Upper Nawata and is about one fifth smaller than the extant aardvark, *O. afer*. It displays some morphological differences of the postcranium when compared with the extant species, but its affinities are with the genus *Orycteropus*. The Lothagam aardvarks, although a rare component of the fossil assemblage, demonstrate that two aardvark species coexisted in the Lower Nawata.

The fossil tubulidentates from Lothagam represent two different species, a small and generalized form and a larger form, both restricted to the Nawata Formation. The smaller *Leptorycteropus guilielmi*, was described by Patterson (1975) from a partial skeleton (LT 419) recovered in 1967. *L. guilielmi* was about half the size of the extant *Orycteropus afer* and had a short facial region and slender limb bones with relatively small areas of muscle attachment. The genus is distinguished by primitive fossorial features that make it the most generalized member of the Orycteropodidae. The larger species was evidently about one fifth smaller than the extant aardvark. It is represented by three specimens, all postcranial, collected between 1989 and 1996. All are isolated elements, and most are incomplete. The material is sufficiently similar to extant *O. afer* to be assigned to the genus *Orycteropus*, but additional material is needed for species assignation.

Systematic Description

Order Tubulidentata
Family Orycteropodidae Bonaparte, 1850
Leptorycteropus Patterson, 1975
Leptorycteropus guilielmi Patterson, 1975
(Figure 8.12)

Diagnosis

I?/?, C1/1, P4/4, M3/3. Canines oval in section, larger than premolars; canines and premolars with peripheral ring of cement: known molars comparable in size and structure to those of *Orycteropus*, large relative size of animal. Maxillaries not extending forward to form elongate snout, not notched anteriorly, ventral surface of palatal portion flat, not grooved medially; anterior portion of jugal wider, relatively shallower than in *Orycteropus*, forming, with maxillary, a short blunt descending process; dorsal portion of frontoparietal suture more oblique than in *Orycteropus;* symphysis of mandible extending back to level of anterior end of P_2. Centra of presacral vertebrae generally wider relative to depth than in *Orycteropus*, sacrals five. Limb bones in general more slenderly constructed than in *Orycteropus*, especially in shafts; much narrower across distal extremities than in *Myorycteropus*. Humerus with deltoid crest merging into shaft distally, not forming distinct terminal projection, no large V-shaped deltopectoral area, entepicondyle extending as far distally as trochlea; anteroposterior diameter of distal end of radius short relative to transverse. Pectineal process of ilium much less projecting than in *Orycteropus;* medial surface of ischium very concave dorsal to obturator foramen. Hind leg bones more slenderly constructed than in *Orycteropus* (Patterson 1975).

Lothagam Material

- *Lower Nawata:* 419 (holotype), maxillary and mandibular fragments (including seven teeth), 18 vertebrae and partial humeri, radius, ulna, metacarpal II,

Figure 8.12 Restoration of *Leptorycteropus guilielmi* by Mauricio Antón. Shoulder height = 24 cm.

pelvis, femora, tibiae and fibula; 28573, distal Lt. femur with distal half shaft and Lt. femur head.

Patterson's (1975) description of the type specimen is comprehensive and need only be summarized here. *Leptorycteropus guilielmi* was about half the size of *Orycteropus afer;* it had a short facial region and slender limb bones with smaller areas of muscle attachment. *Leptorycteropus* had a flat, bony palate without median groove; no prolongation of maxillaries to form a long snout; limb bones generally more slender (especially shafts); humeri with less distinct terminal projection and no large V-shaped deltopectoral area; pectineal process of ilium less projecting and medial surface of ischium very concave dorsal to obturator foramen; and centrum of sacral vertebrae wider relative to depth—all different from extant *Orycteropus afer.* Patterson (1975) justified assigning the Lothagam specimen to a new genus and species on the basis of its primitive fossorial features making it the most generalized member of the Orycteropodidae.

The only additional specimen is a distal femur that lacks the medial condyle and has a distorted lateral condyle (LT 28573). Although slightly larger, this femur is similar to that of the type which Patterson describes as "similar in every feature to that of *Orycteropus* except in being relatively more slender in the shaft above the distal articular area." The total length of the fragment is 63.8 mm. Comparison with LT 419 shows this fragment's shaft is broken just proximal to the midpoint. This specimen also includes the femur head, which is not preserved in the type specimen. The head measures 16.6 mm anteroposteriorly and 16.0 mm mesiodistally. The lateral condyle is distorted.

Orycteropus Geoffroy, 1795
Orycteropus sp.
(Table 8.20)

Lothagam Material

- *Lower Nawata:* LT 28641, left astragalus, partial left tibia, partial distal metapodial, caudal vertebra, lumbar vertebra, and sacral fragment.
- *Upper Nawata:* LT 25136, partial right calcaneum; LT 37755, partial right ulna.

The complete left astragalus LT 28641 is well preserved with minimal surface damage. The dimensions of the Lothagam specimen are on average smaller (about 72%) than those of extant *O. afer,* and the major morphological differences are the longer neck with a relatively narrower head to the navicular facet (table 8.20). Fossil aardvark astragali representing the genus *Orycteropus* are well represented in the collections of the National Museums of Kenya and show three broad evolutionary trends.

First, they have an increased size. *O. minutus* (Songhor) from the Lower Miocene (19 Ma) has dimensions averaging 49 percent those of *O. afer.* At 9–12 Ma, *O. chemeldoi* (Ngorora Formation, Kenya) is somewhat larger (59%), with specimens of *Orycteropus* sp. from Lukeino (6.5 Ma), Lothagam (6–7 Ma), and Laetoli

(3.5 Ma) having dimensions 88 percent, 73 percent, and 84 percent, respectively.

Second, while the tibial facet (trochlear) dimensions have remained fairly constant, the neck of the navicular facet becomes shorter.

Third, the navicular facet in the later specimens is less oval and more rounded. *O. mauritanicus* from the Middle Miocene of Algeria (Patterson 1975) has an astragalus that is narrow with a relatively long neck, and these primitive features are consistent with its age. The calcanear facets are similar in all specimens except in that of the Lower Miocene *O. minutus,* where they are reduced. These evolutionary trends can be summarized as a gradual increase in robustness of the astragali, involving a shortening of the neck and more rounded navicular facet. The *Myorycteropus* astragali (Rusinga Island; Lower Miocene 18 Ma), while from a separate genus, seem to fit into the scheme of progression in *Orycteropus* astragalus evolution. They are slightly larger than *O. minutus* but retain the primitive condition of an overall elongated shape with long neck and narrow head. These similarities in the *Myorycteropus* and *Orycteropus* astragulae may reflect their common ancestry, and are in contrast to the astragalus from *Plesiorycteropus,* which is quite different from that of *Orycteropus,* indicating a more distant ancestry for this genus. As described by Patterson (1975:215), the astragalus of *Plesiorycteropus* is "short proximodistally, neck short and inclined medially, facet for internal malleolus of tibia large, foramen lacking."

The partial right calcaneum LT 25136 is missing the very distal extremity, including the medial and lateral processes. Again, the dimensions are around 76 percent of those of the extant *O. afer,* from which the major structural differences are the smaller size and less developed navicular facet. *Orycteropus afer* and the Lothagam *Orycteropus* sp. clearly differ in the joints between the calcaneum, astragalus, navicular, and cuboid—differences that testify to the evolution of a more robust and rigid hind foot. A similar situation is found in the evolution of the human foot, whereby movement is restricted distal to the metatarsals, with all proximal bones being locked to give a strong pivotal mechanism. In aardvarks, however, the functional advantage lies in digging rather than walking.

The proximal right ulna LT 37755 is missing its distal half and the tip of the olecranon, and it has damage to the semilunar notch and dorsal edge. The specimen is smaller (87%) than the ulna of *O. afer* and is less robust.

A fragment of left tibia, LT 28641, comprises a section of shaft from the central region and includes the distal emergence of the tibial crest. The tibia is smaller in size and more angular than that of extant *O. afer;* it also has a more pronounced interosseous crest. This fragment is the only element that can be compared with its anatomical equivalent from the contemporary *Leptorycteropus* from Lothagam. The *Leptorycteropus* tibia is smaller (70%) than that of *Orycteropus* sp., but they are morphologically similar.

The remaining elements of LT 28641 include a partial lumbar vertebra (missing spinous, caudal, and both transverse processes), a partial caudal vertebra (split along its midline and missing one transverse process and one mammillary process, and having damage to the other), a fragment of distal metapodial, and a fragment of sacrum (encompassing lateral sacral crest and one sacral foramen). All are smaller than extant *O. afer* equivalents. Two other postcranial specimens are too fragmentary to warrant detailed description.

Discussion

Patterson (1975) recognized four genera of orycteropodids: *Plesiorycteropus* from recent deposits in Madagascar (a genus isolated both geographically and anatomically), *Myorycteropus* from the Miocene of Kenya (most fossorially specialized), *Leptorycteropus* from the Late Miocene of Kenya (most generalized, with primitive fossorial features), and *Orycteropus* from Africa, Asia, and Europe (the only extant genus). In contrast, Pickford (1975) placed all the Early Miocene orycteropodids on the extant genus *Orycteropus.*

Because *Plesiorycteropus madagascariensis* is so morphologically distinct from the other three genera, Patterson (1975) suggested that two subfamilies, the Orycteropodinae and Plesiorycteropodinae, might be warranted. It is likely that *Plesiorycteropus* became geographically isolated in the Eocene, maintaining similarities to *Orycteropus* in the axial skeleton and pelvis, but evolving differences in limb and foot bones and masticatory apparatus. Patterson (1975:217) interpreted the plesiorycteropodines as a splinter group or bud from, and not a sister group of, the Orycteropodinae.

The earliest members of the order Tubulidentata, *Orycteropus minutus* (19 Ma) from Songhor (Pickford 1975) and *Myorycteropus africanus* (18 Ma) from Rusinga, Mfwangano, and Songhor (McInnes 1956; Pickford 1975) derive from Kenyan localities and provide circumstantial evidence that the order evolved in Africa. Considering the similarities between the skeletons of aardvarks and condylarths, one may have evolved from the other (Patterson 1975). Thereafter, an early radiation of the primitive aardvarks is represented by *Plesiorycteropus* in Madagascar. Between 9 and 9.5 Ma, the *Orycteropus* lineage spread into Europe and Asia. The only extant species, *O. afer,* is confined to sub-Saharan Africa.

The affinities between the genera *Orycteropus, Leptorycteropus,* and *Myorycteropus* are less clear, with as yet insufficient material to substantiate accurate phylogenetic relationships. *Myorycteropus* was originally described as a single species, *M. africanus,* from 18 Ma deposits at Rusinga Island, Kenya (McInnes 1956). This aardvark was half the size of extant *O. afer,* and while the dentition is not dissimilar, the postcranial skeleton of *M. africanus* displays highly fossorially specializations. Both McInnes (1956) and Patterson (1975) suggested the extreme fossorial adaptations of the skeleton of the Early Miocene *Myorycteropus* precluded it from the ancestry of the extant *O. afer,* believing that the fossorial specializations of *Orycteropus* evolved later and to less an extent. Patterson felt it more likely that the *Leptorycteropus* and *Orycteropus* lineages have a closer common ancestor than either has to *Myorycteropus.* Such an ancestor could have been a generalized fossorial aardvark, and this condition may have pertained through the *Leptorycteropus* lineage but was slowly replaced by greater fossorial adaptations in the *Orycteropus* lineage. In contrast, Pickford (1975) in describing additional Early Miocene specimens, attributed all this material to the extant genus *Orycteropus* and suggested that the fossorial specializations of these Early Miocene orycteropodids indicated that antbear evolution has proceeded very slowly. Although the status of *Myorycteropus* is unclear, the Lothagam material shows that at around 6 to 7 million years ago, both the *Orycteropus* and *Leptorycteropus* lineages lived contemporaneously in Kenya.

The available fossil material of *Orycteropus* is more abundant than that for the other genera, deriving from at least ten sites in Africa, Europe, and Asia. Pickford's *O. minutus* is reported as the earliest (and smallest) *Orycteropus* to date, from the 19 Ma deposits at Songhor, Kenya (Pickford 1975). Pickford also describes another species, *O. chemeldoi,* from Late Miocene (9–12 Ma) sediments in the Ngorora Formation, Kenya. The holotype material for *O. minutus* comprises articulated right metacarpals II and III, thus demonstrating a unique small size; that for *O. chemeldoi* is a left mandibular ramus with uniquely narrow teeth (P_2 to M_3). Slightly younger than *O. chemeldoi* is *O. mauritanicus* from Oued el Hammam, Algeria, followed by a series of fossils from Africa and Eurasia referred to as the *O. gaudryi* group by Pickford (1975). These probably reached Eurasia between 9.5 and 9 Ma. In Kenya, the 6.5 Ma Lukeino Formation yielded two specimens—a left talus of a large orycteropodid and a small phalanx of a tiny one—which suggests that two species of *Orycteropus* were living contemporaneously. *Orycteropus* sp. from Lothagam (6–7 Ma), Kanapoi (4.2 Ma) (unpublished material), and Laetoli (3.5 Ma) (Leakey 1987) were larger than other fossil aardvarks, but all were smaller and less robust than the extant *O. afer.* The earliest undoubted *O. afer* specimens are from Early Pleistocene deposits in Kenya at Koobi Fora, Rusinga Island, and Kanjera (Pickford 1975). Evolutionary traits displayed by *Orycteropus* fossils include increasing body size, enlargement of olfactory area, lengthening of facial region, increase in robusticity of limb bones, reduction of metacarpal V and fifth digit of the pes, and a relative increase in the size of the manus (Patterson 1975). To these may be added the gradual increase in robusticity of the astragali. These changes within the *Orycteropus* lineage have led to the extant *O. afer*—a committed myrmecophage with a well-established fossorial lifestyle.

Table Abbreviations

Max = maximum
Acc = accession
Nav. = navicular

References Cited

Leakey, M. G. 1987. Fossil aardvarks from the Laetoli Beds. In M. D. Leakey and J. M. Harris, eds., *Laetoli: A Pliocene Site in Northern Tanzania,* pp. 297–300. Oxford: Clarendon Press.

McInnes, D. G. 1956. *Fossil Tubulidentata from East Africa.* Fossil Mammals of Africa No. 10. London: British Museum (Natural History).

Patterson, B. 1975. The fossil aardvarks (Mammalia; Tubulidentata). *Bulletin of the Museum of Comparative Zoology* 147:185–237.

Pickford, M. 1975. New Fossil Orycteropodidae (Mammalia, Tubulidentata) from East Africa: *Orycteropus minutus* sp. nov. and *Orycteropus chemoldoi* sp. nov. *Netherlands Journal of Zoology* 25:57–88.

TABLE 8.20 Measurements (in mm) of Fossil Aardvark Material from Kenya and of Extant *Orycteropus afer*

Location	Age	Taxon	Acc. No.							
Radius					**Max Distal Width**	**Max Distal Breadth**	**Max Proximal Width**	**Max Width Tibial Facet (C)**	**Max Height Astragalus (D)**	**Neck Proportions (D/C)**
Modern	—	*O. afer*	—	Mean (n = 25)	32.63	23.27	19.89	—	—	—
				Range	30.51–35.16	21.72–26.04	18.18–21.80	—	—	—
Laetoli	—	*Orycteropus* sp.	LAET1813	—	24.78	18.51	14.38	—	—	—
Ulna					**Notch to Posterior Edge**	**Max Width Notch**	**Ulna Thickness Behind Notch**			
Modern	—	*O. afer*	—	Mean (n = 4)	21.12	39.55	6.51			
				Range	20.10–23.92	36.58–41.70	5.55–8.16			
Laetoli	3.5	*Orycteropus* sp.	**	—	15.58	31.09	6.43			
Lothagam	6.0–7	*Orycteropus* sp.	LT37755	—	18.88	33.67	6.01			
Tibia					**Max Distal Width**	**Max Distal Breadth**	**Max Proximal Width**	**Max Width Tibial Facet (C)**	**Max Height Astragalus (D)**	**Neck Proportions (D/C)**
Modern	—	*O. afer*	—	Mean (n = 27)	27.02	—	—	—	—	—
				Range	25.05–28.71	—	—	—	—	—
Laetoli	—	*Orycteropus* sp.	LAET3010		20.96	—	—	—	—	—

continued

TABLE 8.20 Measurements (in mm) of Fossil Aardvark Material from Kenya and of Extant *Orycteropus afer* (Continued)

Location	Age	Taxon	Acc. No.							
Astragalus					**Max Width Nav. Facet (A)**	**Max Breadth Nav. Facet (B)**	**Facet Proportions (A/B)**	**Max Width Tibial Facet (C)**	**Max Height Astragalus (D)**	**Neck Proportions (D/C)**
Modern	—	*O. afer*	—	Mean (n = 9)	17.22	14.98	1.15	23.77	38.48	1.62
				Range	16.49–18.75	14.36–15.84	—	21.21–25.93	35.85–42.19	—
Laetoli	3.5	*Orycteropus* sp.	LAET	—	15.13	13.31	1.13	19.58	31.1	1.59
Lukeino	6.5	*Orycteropus* sp.	LU	—	—	—	—	21.79	34.1	1.56
Lothagam	6.0–7	*Orycteropus* sp.	LT28641	—	12.55	10.4	1.21	17.16	30.04	1.75
Rusinga	18	*M. africanus*	—	Mean (n = 4)	11.63	7.68	1.51	13.62	24.04	1.77
				Range	10.44–12.82	6.97–8.38	—	12.18–14.28	21.66–26.42	—
Songhor	—	*O. minutus*	**	—	9.03	6.41	1.41	10.92	20.64	1.89
Calcaneum					**Max Width Nav. Facet**	**Max Width Shaft Below Nav. Facet**				
Modern	—	*O. afer*	—	Mean (n = 8)	43.3	26.25				
				Range	41.96–45.97	24.65–29.3				
Lothagam	—	*Orycteropus* sp.	LT25136	—	32.36	20.09				
MtIV					**Max Distal Width**	**Max Distal Breadth**	**Max Proximal Width**	**Max Width Tibial Facet (C)**	**Max Height Astragalus (D)**	**Neck Proportions (D/C)**
Modern	—	*O. afer*	—	Mean (n = 23)	—	—	20.79	—	—	—
				Range	—	—	19.28–22.83	—	—	—

9

PERISSODACTYLA

9.1

Lothagam Rhinocerotidae

John M. Harris and Meave G. Leakey

Three rhino species are represented in the Lothagam succession. *Brachypotherium lewisi* is the common rhino from the Nawata Formation, which in recent years has yielded additional dental and postcranial remains of this species. *Ceratotherium praecox* is present in both the Lower and Upper Nawata, and *Diceros bicornis* occurs in the Upper Nawata. Both *Ceratotherium* and *Diceros* occur in the Apak Member, but only *C. praecox* has been recovered from the Kaiyumung Member. Isotopic analysis of tooth enamel suggests that *B. lewisi* was primarily a browser and that *C. praecox* started exploiting C_4 grasses in the Upper Nawata.

The 1967 collection of rhinoceroses from Lothagam comprised a score of specimens, about half of which were believed to have derived from Lothagam Member 1B and the remainder from Member 1C. They were described by Hooijer and Patterson (1972), who recognized two new species—*Brachypotherium lewisi* and *Ceratotherium praecox*, the latter being better known from Kanapoi (Hooijer and Patterson 1972) and Langebaanweg (Hooijer 1972). The material was of interest in that it contained the largest and latest representative of the *Brachypotherium* lineage, and this evidently coexisted with the earliest known individual of *Ceratotherium*. These specimens were described in great detail by Hooijer and Patterson and need not be redescribed here. The only other Lothagam rhino specimen predating the 1990s collections is an incomplete upper molar of *Ceratotherium* (KNM-LT 396) that was recovered in 1972 and is now believed to be from surficial Holocene (Galana Boi) beds.

The Lothagam rhino hypodigm was more than doubled during the early 1990s by field parties from the National Museums of Kenya. During this project, the (sparse) field documentation of the 1967 specimens was reexamined and, where possible, the provenance of the earlier material was confirmed or refined. The new information helped clarify the distribution of *Brachypotherium*, now known to be mainly restricted to the Nawata Formation, and documented the presence of a third genus—*Diceros*. Teeth of the three genera may be distinguished by the criteria provided in table 9.1.

Using the criteria from table 9.1, it now appears evident that three of Hooijer and Patterson's original identifications were erroneous. Two incomplete lower dentitions, LT 82 from the Upper Nawata and LT 83 from the Apak Member, are now seen to belong to *Ceratotherium praecox*, whereas LT 84, an immature mandible from the Upper Nawata, is better identified as *Diceros*.

Systematic Description

Brachypotherium lewisi Hooijer and Patterson, 1972
(Figures 9.1–9.4; tables 9.2, 9.3)

Diagnosis

Size very large: condylobasal length of skull over 70 cm, antero-transverse diameters of M^{1-2} some 90 mm as opposed to 70 mm in *B. brachypus* (Lartet) or *B. snowi* (Fourtau) from Miocene of Europe and Egypt, respectively. Nasals hornless, slender, not very long, deepest point of nasomaxillary notch above P^4; anterior border of orbit above front of M^2, frontals flat and hornless, inferior squamosal processes united below subaural

Figure 9.1 Restoration of *Brachypotherium lewisi* by Mauricio Antón. Shoulder height = 150 cm.

channel. Upper incisors very large, upper cheek teeth brachyodont, ectoloph flattened behind paracone style, antecrochet moderate, protocone constriction slight, external cingula often present. Lower canines present, brachyodont cheek teeth with external groove between anterior and posterior lophids usually flattened out, external cingula often developed. Trochanter tertius of femur strongly developed.

Holotype

KNM-LT 88, cranium (Lt. and Rt. P^2–M^3) from the Lower Nawata.

Lothagam Material

- *Lower Nawata:* holotype; 80, Rt. P^3; 87, Rt. P^1; 94, crushed cranium (Lt. M^2, Lt. I^1, Rt. M^2, partial M^1 and M^3); 81, partial lower molar; 86, lower molar fragments; 93, Rt. M^3; 99, Lt. and Rt. P^2; 100, partial Lt. P^2 and P^3; 22874, Rt. M^2; 22961, Rt. P^2; 23800, Lt. Mc III; 23960, lower or upper I^1, I^2 and M fragments; 23963, Lt. P^2; 23964, Lt. Mt II. Lt. Mt IV; 23965, upper molar fragments; 24290, Lt. M^3, tooth fragments; 26280, broken Lt. $M_{?2}$; 26286, Rt. Mc II; 26300, distal Lt. Mc II or IV; 28735, Lt. and Rt. lower molar fragments.
- *Upper Nawata:* 85, partial Lt. I^1; 91, Lt. mandible (P_1–M_3); partial atlas; 95, Lt. juvenile mandible fragment and symphysis (P_2 and M_2, fragment M_1); 22872, Lt. upper P^2; 23962, Rt. P_1; 23967, Rt. P_1; 26279, partial Lt. mandible (M_3); 26281, proximal Rt. Mt III; 26301, sixth cervical vertebra; 26312, Lt. Mc IV.
- *Apak Member:* 90, mandible (Rt. P_2–M_3, Lt. P_2–M_2; 97, distal Lt. femur including third trochanter.
- *Horizon indet.:* 12686, Lt. humerus

Brachypotherium species comprise a group of large and heavy-bodied rhinoceroses characterized by the presence of upper tusks (I^1) and short limbs and feet (Hamilton 1973); they exploited riparian and forested habitats (Guérin 1980). *Brachypotherium lewisi* was a large hornless species that constituted the termination of the *Brachypotherium* lineage, outlasting *B. goldfussi* from the Pontian of Europe (Hooijer 1978). Initially recognized from Lothagam, the species is also represented at the Kenyan localities of Ngorora (Hooijer 1971) and

Mpesida (Hooijer 1973), as well as at Sahabi in Libya (Hooijer 1978). Twenty new specimens have been recovered from Lothagam by the National Museums of Kenya expeditions, mostly isolated teeth and limb bones (figure 9.2)—all of which were from the Nawata Formation and most from its lower member. The 1967 expedition, however, collected two specimens (LT 90, 97) that appear to be from the Apak Member.

A posterior cervical vertebra (C6?), LT 26301, lacks all but one of the articular processes but is identified as *Brachypotherium* because of its large size and slight differences from cervical vertebrae of the two extant species (such as two small protruding rugosities on the ventral surface). The articular facets for the ribs are relatively large. The maximum ventral anteroposterior length of 105 mm and the dorsal length of 88 mm are larger than in either extant species. Both anterior and posterior articular facets of the centrum are almost circular in profile; mediolateral and dorsoventral measurements of the strongly concave anterior articular surface are 81 by 80 mm; those for the strongly convex posterior facet are 80 by 75 mm.

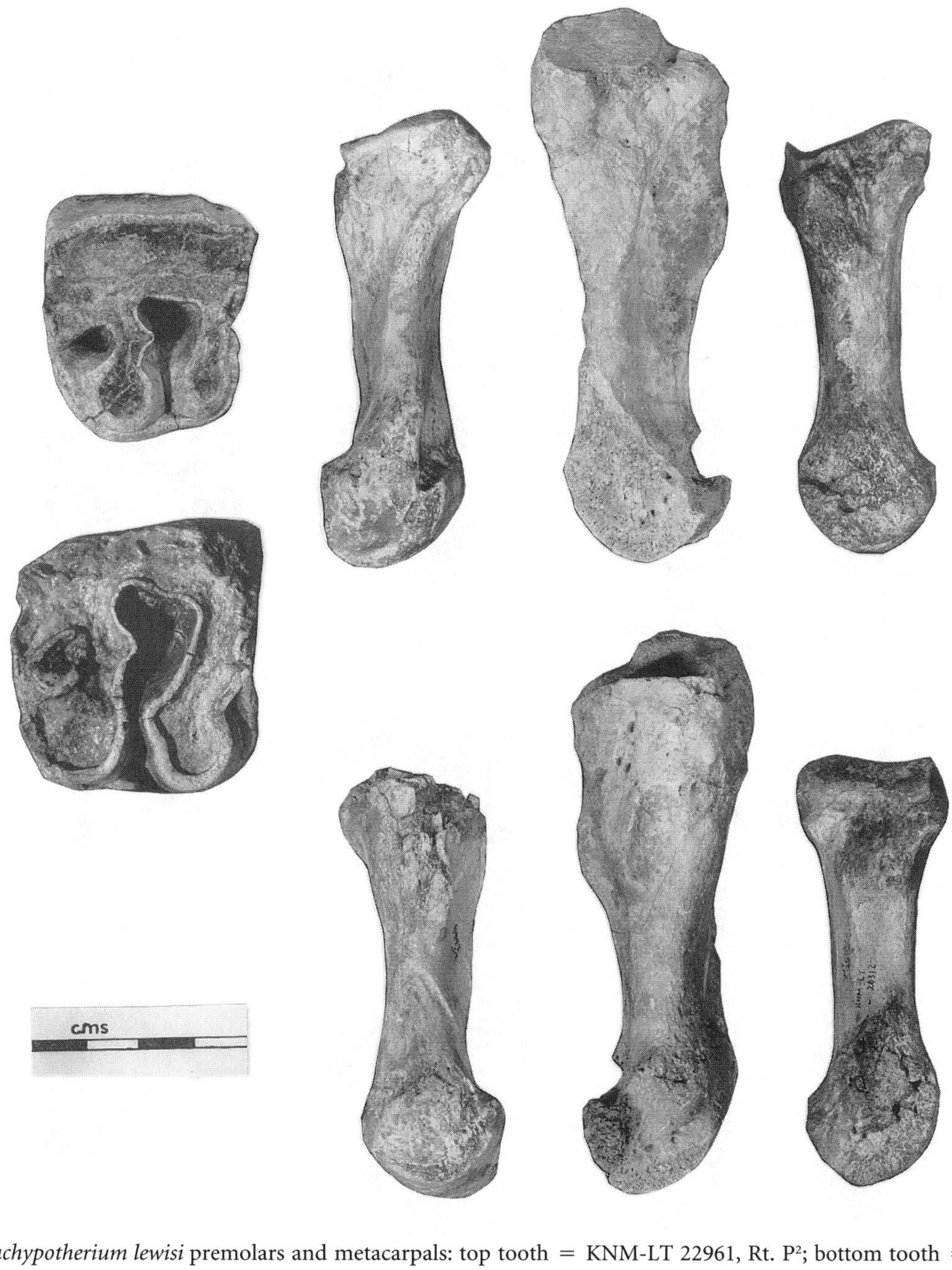

Figure 9.2 *Brachypotherium lewisi* premolars and metacarpals: top tooth = KNM-LT 22961, Rt. P^2; bottom tooth = KNM-LT 80, Rt. P^3. Left metacarpal = KNM-LT 26286, Rt. metacarpal II; center metacarpal = KNM-LT 23800, Lt. metacarpal III, right metacarpal = KNM-LT 26312, Lt. metacarpal IV; all three quarters of their natural size. Top row metacarpals = lateral view; bottom row = medial view.

A weathered left humerus attributable to *Brachypotherium*, LT 12686, lacks the lateral portion of the distal epiphysis, so that details of the relative proportions of the lateral epicondyle and radial articular surface have been lost. Bone is also lost proximally from the greater tuberosity. As in both extant rhino species, the medial epicondyle is only lightly developed. The proximal portion of the diaphysis is more strongly compressed than in the modern species. The greater tuberosity is rugose but rather less so than is typical of *Ceratotherium simum*. There is a sharp waisting of the shaft distal to the tuberosity that is evident in both anterior and posterior view.

Brachypotherium metacarpals of lengths comparable to those of modern species are proportionately stouter and more robust. The second and fourth metacarpals are distinctly "waisted" in the mid-portion of the diaphysis, a feature also present but less pronounced in Mc III (figure 9.3).

Metacarpal II of *Brachypotherium* is more symmetrical and has relatively broader epiphyses and a more waisted diaphysis than is characteristic of the modern

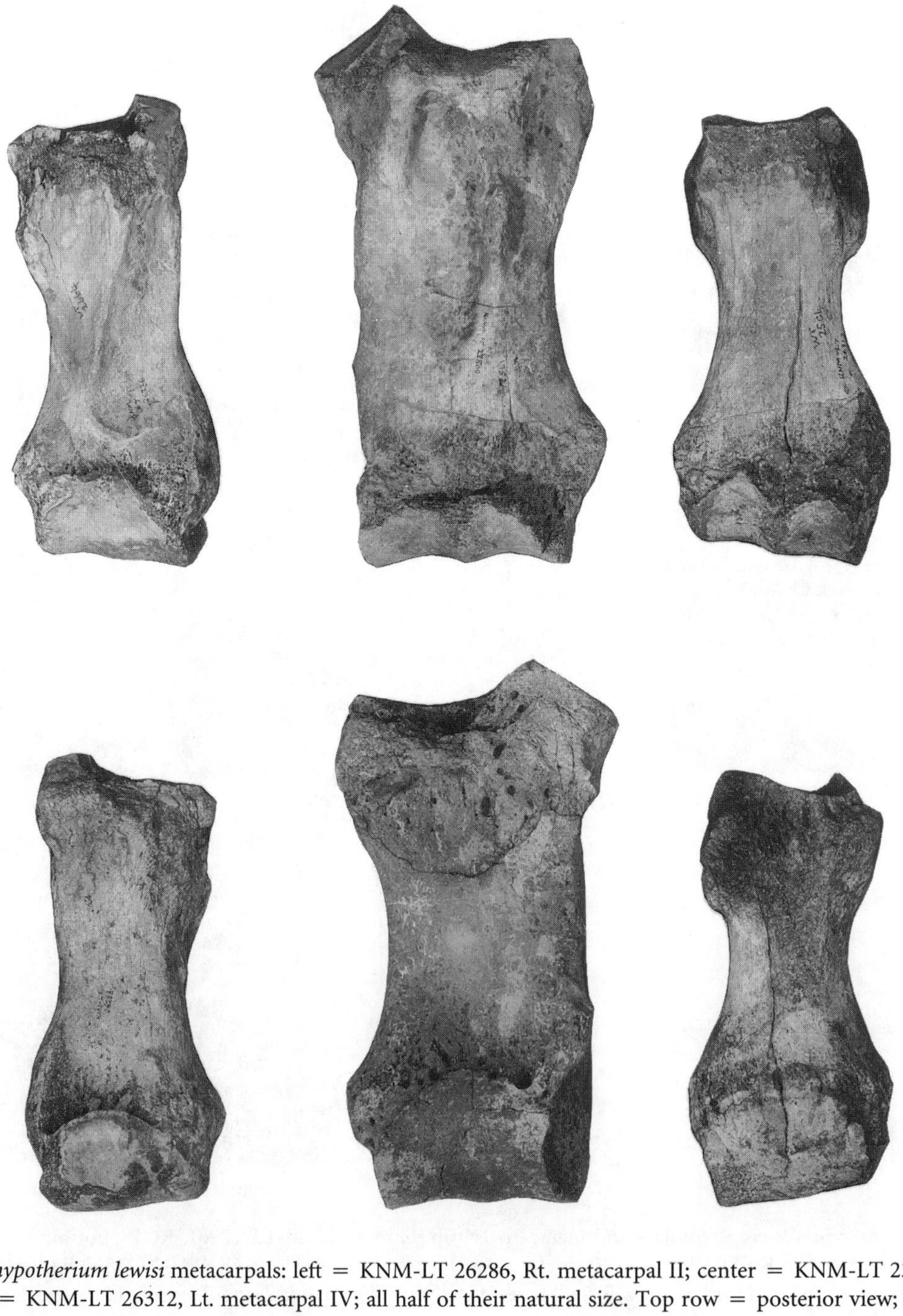

Figure 9.3 *Brachypotherium lewisi* metacarpals: left = KNM-LT 26286, Rt. metacarpal II; center = KNM-LT 23800, Lt. metacarpal III; right = KNM-LT 26312, Lt. metacarpal IV; all half of their natural size. Top row = posterior view; bottom row = anterior view.

species. Laterally, as in both modern species, the narrow facet for articulation with Mc III extends the entire length of the lateral surface. Proximally, in contrast to the modern species, the trapezoid facet is large and almost square, being almost as wide mediolaterally as it is deep anteroposteriorly.

Metacarpal III has a very anteroposteriorly compressed diaphysis but with stout carinate ridges that extend distally from the magnum facet along the lateral and medial margins to enclose a deeply excavated proximal hollow. The medial of these two ridges is the better developed. Proximally, the posterior portion of the large magnum facet is convex anteroposteriorly, whereas the anterior portion, although broken, appears to have been concave. Laterally, there is a clearly defined anterior facet for articulation with the unciform, and this is proportionately larger and more steeply angled than the concave facet of the extant species. The broken, proportionately small facet for articulation with Mc IV is located at the anterolateral corner of the magnum facet. There is no posterior facet for articulation with Mc IV, in contrast to the large circular posterior facets present for this purpose in both extant species. Medially, the small triangular-shaped anteromedial facet for articulation with Mc II is proportionately larger than in *Diceros* and comparable in relative size to that of *C. simum.*

Metacarpal IV has a distal epiphysis that is stouter and more symmetrical than in both extant species. The proximal epiphysis is also more symmetrical, with the proximal triangular facet for the trapezoid being much broader mediolaterally. Medially, the narrow facet for articulation with Mc III extends almost the length of the medial surface, although its posterior extent cannot be determined because of bone damage. There is no posterior facet for the third metacarpal, in contrast to both extant species where this facet is large and circular. There is a small, narrow, obliquely oriented facet for articulation with the magnum between the larger trapezoid facet and that for Mc III.

The metatarsals of *Brachypotherium* (figure 9.4) are much broader than those of *Ceratotherium simum.* Metatarsal II has a waisted diaphysis (though it is less pronounced than in the metacarpals), whereas Mt IV narrows proximally. The preserved portion of the proximal shaft of Mt III is markedly flattened from back to front and broadens distally. The proximal epiphyses of all three metatarsals are of comparable size to the extant species, although, except for Mt IV, they are mediolaterally broader, but the preserved distal epiphyses are much stouter.

Metatarsal II has very robust epiphyses and an elongated rugosity on the posterior face of the diaphysis. The proximal facets for the middle and external cuneiform are broader than in either extant species, and that for the internal cuneiform is circular and is angled obliquely to the diaphysis axis rather than parallel to it as in *D. bicornis* and *C. simum.* There is a large posterior internal cuneiform facet as in *Ceratotherium* (this facet is absent in *Diceros*).

Metatarsal III is represented only by its proximal portion, which is wider mediolaterally but shallower back to front than in *C. simum* and has a distinctly L-shaped proximal internal cuneiform facet. Laterally, there is only a very small anterior Mt IV facet (in contrast to both modern species); the posterior Mt IV facet, although present and large in the modern species, is absent. Medially, the posterior Mt II facet is absent. Because bone is missing in the area of the anterior Mt II facet, it is not possible to assess whether this facet was present in Mt III; however there is no corresponding facet on Mt II.

Metatarsal IV has a distal epiphysis that is larger than that of the extant species, and the proximal and distal epiphyses are in the plane of the axis of the shaft, which lacks torsion. In both modern species there is some torsion, so that the two epiphyses are offset. Proximally, the cuboid facet profile is broader posteriorly than in *C. simum* and extends to the posterior edge of the epiphysis. The medial margin is almost at a right angle to the dorsal margin, whereas in the modern species the angle between the two margins is closer to 45°. In contrast, the lateral margin meets the anterior margin at a right angle in the modern species but is closer to 45° in the fossil. Laterally, the anterior facet for Mt III is more elongated back to front than that of the modern species, and there is no posterior Mt III facet, although this facet is well developed in both modern species.

In general, the *Brachypotherium* metapodials are broader and without the tightly fitting joints (characteristic of the modern species) that occur both between the adjoining proximal epiphyses and between the proximal epiphyses and the podials.

A few *Brachypotherium* lower teeth were retrieved from the Namurungule Formation; these were evidently of similar size to *B. heinzeli* and thus smaller than those of *B. lewisi* (Nakaya et al. 1984).

Ceratotherium praecox Hooijer and Patterson, 1972

(Tables 9.4, 9.5)

Diagnosis

Skull differing from *C. simum* (Burchell) in greater concavity of skull roof, cranium less extended posteriorly, occiput more vertically inclined; cheek teeth not as hypsodont, lophs and lophids not markedly oblique, anterointernal corners of upper teeth not rounded, no

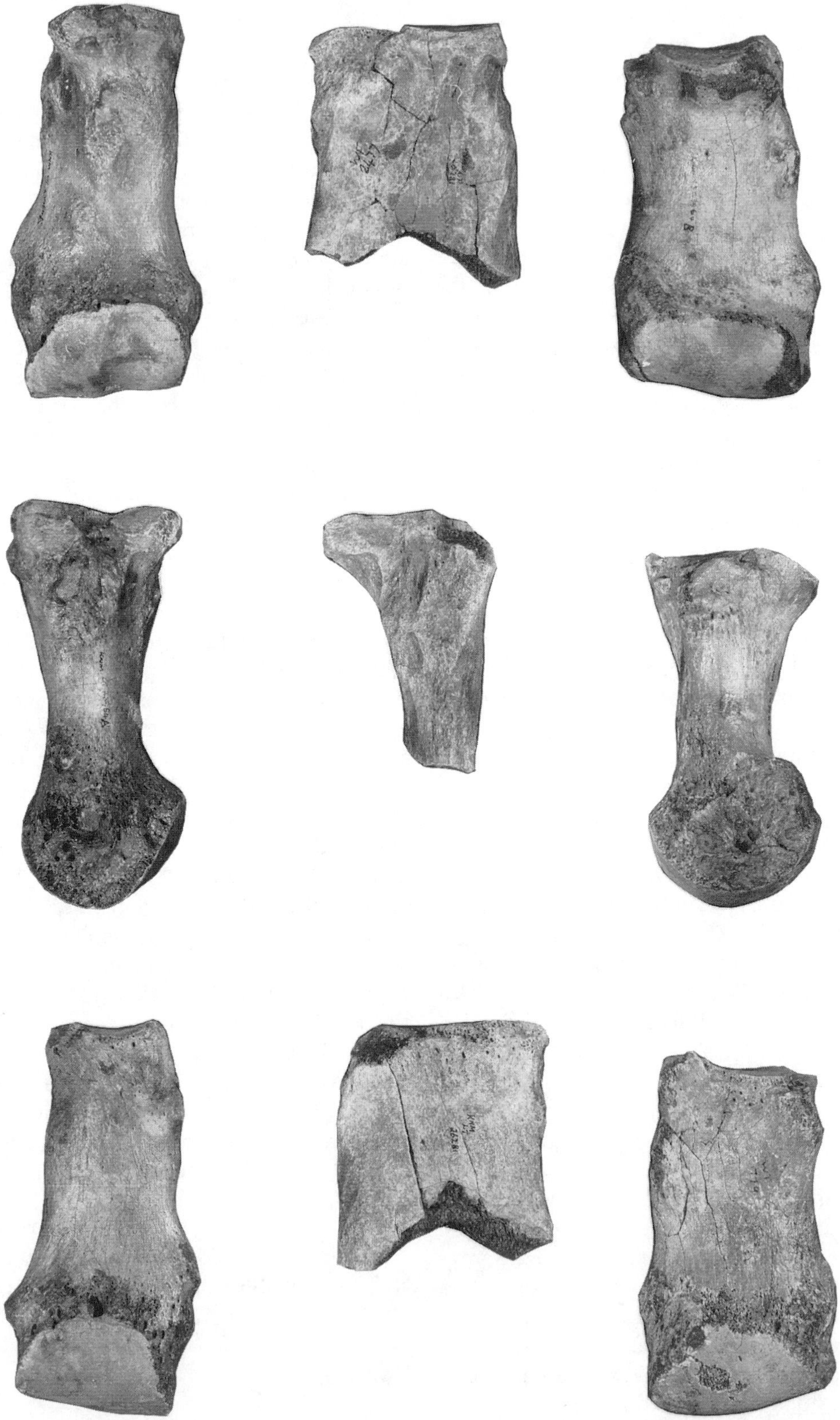

Figure 9.4 *Brachypotherium lewisi* metatarsals: left = KNM-LT 23964A, Lt. metatarsal II; center = KNM-LT 26281, Rt. metatarsal III; right = KNM-LT 26964B, Lt. metatarsal IV; all half natural size. Top row = posterior view; center row = lateral view; bottom row = anterior view.

medifossettes in P^2–M^2 and no fossettids in lower cheek teeth, internal cingula in upper cheek teeth variable.

Lothagam Material

- *Lower Nawata:* 89, Rt. M^2; 26289, Rt. ulna.
- *Upper Nawata:* 82, Rt. M_1; 23772, Lt. P^2 or P^3, and tooth fragments; 23970, partial cranium; 23972, Lt. astragalus; 26278, Rt. $M_{?2}$.
- *Apak Member:* 83, P_{2-4} and partial molars; 23966, Lt. calcaneum; 26296, sixth cervical vertebra.
- *Kaiyumung Member:* 23968, Lt. mandible fragment (P_2); 23969, Rt. maxilla (P^{1-2}); 26283, Rt. mandible (P_2–M_3); 26284, proximal phalanx III.
- *Galana Boi:* 96, Rt. P^4.

Nine new white rhino specimens were added to the Lothagam hypodigm, most from the upper part of the sequence. The record of *Ceratotherium* from the Lower Nawata is sparse, comprising an ulna collected recently and an upper molar collected in 1967, but the provenance data appear secure. According to Hooijer (1972), *C. praecox* is directly ancestral to the extant white rhino but differs from it cranially by having a more concave dorsal surface, a less extended posterior portion, a less posteriorly inclined occiput, and a less thick nuchal crest. The sole cranial specimen from Lothagam, LT 23970, derives from the Upper Nawata. The most readily recognizable parts constitute portions of the left and right zygomatic arches, the nasal boss, and fragments of the rear portion of the cranial vault. The zygomatic fragments are comparable in size to equivalent portions of an extant white rhino skull (OM 2184) from the osteology collections of the National Museums of Kenya. The nasal boss is closely comparable in shape but a little larger than that of the same modern white rhino specimen and appreciably larger than that of *Diceros bicornis* fossil cranium LT 23971 from the Apak Member. The occipital fragment includes much of the right and part of the left nuchal crest. In contrast to that of the extant white rhino, the nuchal crest of the fossil specimen extends laterally rather than posterolaterally. The upper part of the occiput also appears steeply angled backward as in the extant *C. simum* and in contrast to both fossil and extant examples of *D. bicornis.* However, more complete white rhino crania from Kanapoi and Kosia (and hence about 4 Ma in age) have occiputs that are less steeply angled than in extant crania. In addition, the dorsal surface of the rear portion of the cranial vault is almost horizontal, whereas that of fossil and recent examples of the black rhino rise steeply upward in front of the nuchal crest. The mandibular symphysis is narrower than in *C. simum* and more similar to that of *Diceros.* The cheek teeth are more hypsodont than those of *D. bicornis* but decidedly less so than those of *C. simum.* The transverse lophs of the upper cheek teeth are less backwardly inclined in *C. praecox* than in *C. simum,* and present a superficial resemblance to those of *Diceros.*

Several postcranial elements attributable to *Ceratotherium praecox* have been recovered. A posterior cervical vertebra, LT 26296, is almost complete and very similar to the C6 of *C. simum.* Its maximum ventral anteroposterior length is 93 mm, and its dorsal length is approximately 82 mm. Both articular surfaces of the centrum are almost circular in profile; dorsoventral and mediolateral measurements of the strongly concave anterior articular surface are 66 by 64 mm whereas those of the strongly convex posterior surface are 63 by 62 mm.

A right ulna, LT 26289, is almost indistinguishable from that of the extant *C. simum* except for its larger size and a larger degree of retroflexion of the diaphysis that gives it a more strongly concave curvature of the posterior surface. A left astragalus (LT 23972), a left calcaneum (LT 23966), and a proximal phalanx (LT 26284) are all indistinguishable morphologically from those of the extant species.

Postcranially, *C. simum* is larger than *D. bicornis* and has more massive metapodials, but otherwise the limb elements of the two extant species are remarkably similar; the abundant postcranial elements of *C. praecox* from Langebaanweg are rather larger than those of extant white rhinos (Hooijer 1972), but *C. praecox* foot elements from Lothagam are comparable in size to those of *C. simum,* and the limb bones fall within the range of variation of the living species.

Diceros bicornis (Linnaeus)

(Figure 9.5; table 9.6)

Lothagam Material

- *Upper Nawata:* 84, Rt. juvenile mandible (dP_1–M_1); 23665, Rt. M^1; 23961, Lt. $P^{2 \text{ or } 3}$; 26285, /M, Lt. P_3.
- *Apak Member:* 23971, partial cranium, Rt. M^{2-3}, Lt. M^3; 28563, cranium; 28762, Lt. M_1.

The earliest representatives of the genus *Diceros* are *D. pachygnathus* (Wagner) from the Pontian of Samos and Maragha and *D. douariensis* Guérin from the Late Miocene of Tunisia (Hooijer 1978). However, the presence of teeth indistinguishable from those of the extant *D. bicornis* in the Upper Nawata appear to represent the oldest specimens of the extant species. As previously observed by Guérin (1987), remains of the black rhinoceros have now been recovered from numerous localities of Late Pliocene and Pleistocene age in Africa

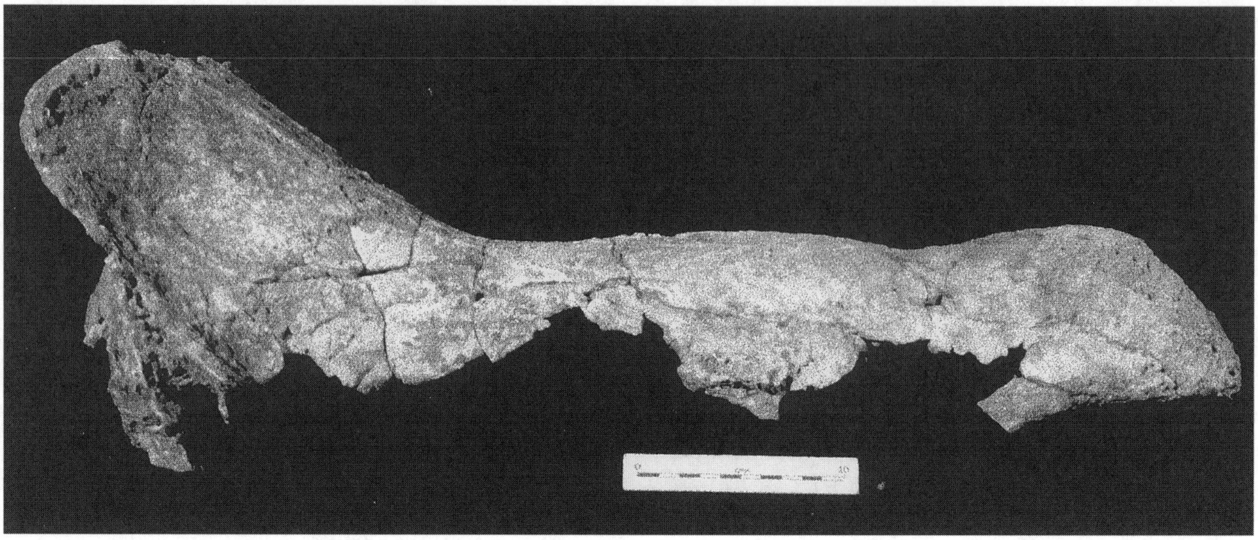

Figure 9.5 *Diceros bicornis* partial cranium KNM-LT 23971, right lateral view.

but not in sufficient quantities to document major anatomical (and hence taxonomic) differences from the extant species. The Lothagam hypodigm comprises two partial crania (KNM-LT 23971 and 28563), an immature mandible (LT 84), and several isolated teeth. The more complete of the two partial crania, LT 28563, is comparable in size to that of an extant black rhino, but it is missing much of the surface bone, which thus precludes a detailed description. The specimen is evidently an elderly adult because the associated teeth are worn almost to their roots. The rear of the cranial vault seems to rise more steeply than in extant black rhinos, but this may be because much of the surface bone is missing from the anterior part of the cranium. The nuchal fossa is more deeply excavated than in extant specimens; the paroccipital process is stouter, but the occiput is similarly vertically oriented. The other specimen, LT 23971, preserves the dorsal surface of the cranium from the nasals to the nuchal crest. It is identical in shape to extant black rhinos but is larger and with proportionately stouter nasals; however, its associated teeth are of identical size to extant representatives. In contrast, the crania of *Ceratotherium* are longer, and the cranial vault rises less abruptly anterior to the nuchal crest. The teeth appear identical in both size and morphology to those of extant representatives. Despite their larger size, it seems prudent to interpret the Lothagam crania as belonging to the extant species until such time that more complete material provides support for taxonomic differentiation.

Discussion

The lack of detailed locality information for the original (1967) collection is unfortunate because most of the better specimens were collected at that time. The provenance of the more recently collected material is, however, well substantiated. Of these, all of the recently collected *Brachypotherium* specimens are restricted to the Nawata Formation, whereas *Diceros* specimens hail from the Upper Nawata and Apak, and *Ceratotherium* comes from the Upper Nawata and Kaiyumung Members. Thus the change in the Lothagam rhino assemblage from that characteristic of the Miocene to essentially modern forms took place in the upper part of the Nawata Formation. Thenius (1955) proposed that *Ceratotherium* split off from *Diceros* stock somewhere in the Pliocene. Hooijer (1972) regarded *C. praecox* as little removed from the point of divergence. The presence of both the earliest representatives of *C. praecox* and *Diceros* in the Upper Nawata extends the timing of the dichotomy back into the Late Miocene.

The isotopic data are interesting in this respect. Some samples were taken from accessioned and identified specimens and some from rhino tooth fragments recovered at the outcrop. The *Brachypotherium* teeth sampled from the Nawata Formation (LT 86, 95, 100) were evidently C_3 browsers, but the sole *Brachypotherium* tooth from the Apak Member (LT 90) indicated a diet of mixed C_3 and C_4 vegetation. In contrast, a *Ceratotherium* individual from the Lower Nawata (LT 89) evidently browsed on C_3 vegetation, but others from the Upper Nawata (LT 23772) and Kaiyumung Members (LT 26283) were C_4 grazers. Of the unidentified rhino enamel fragments, those from the Lower Nawata were all C_3 browsers, whereas one sample from the Apak Member was from a C_3 browser (perhaps *Diceros?*) but five others were C_4 grazers (perhaps *Ceratotherium?*). The implications appear to be that there was a distinct ecological change between the lower and upper members of the Nawata Formation and that the diet of the white rhinos reflected this change.

Isotopic analyses of mammalian enamel from the Samburu Hills indicate that C_4 vegetation was present at the time that sequence accumulated and that some of the rhinoceroses were exploiting it. However, none of the rhino genera reported by Nakaya (1994) from the Samburu Hills assemblages (*Paradiceros, Chilotheridium, Kenyatherium,* iranotheriine) are present in the Lothagam sequence. It seems entirely possible, therefore, that at the Samburu Hills C_4 vegetation formed a significant portion of the biomass somewhat earlier than at Lothagam and was exploited there by taxa that are not represented in the Lothagam assemblages.

Acknowledgments

We thank the government of Kenya and the museum trustees of the National Museums of Kenya for permission to study the Lothagam rhino material. We also thank the curatorial and preparation staff of the palaeontology division of the National Museum of Kenya, Nairobi, for their assistance in making the material available for study. Bob Campbell assisted with the photography.

References Cited

Guérin, C. 1980. Les rhinocéros (Mammalia, Perissodactyla) du Miocène terminal au Pléistocène supérieur en Europe occidentale: Comparaison avec les espèces actuelles. *Documentation de Laboratoire de Géologie, Lyon* 79, fasc. 2:423–783.

Guérin, C. 1987. Fossil Rhinocerotidae (Mammalia, Perissodactyla) from Laetoli. In M. D. Leakey and J. M. Harris, eds., *Laetoli: A Pliocene Site in Northern Tanzania,* pp. 320–348. Oxford: Clarendon Press.

Hamilton, W. R. 1973. North African Lower Miocene rhinoceroses. *Bulletin of the British Museum (Natural History), Geology* 24:351–395.

Hooijer, D. A. 1971. A new rhinoceros from the late Miocene of Loperot, Turkana District, Kenya. *Bulletin of the Museum of Comparative Zoology* 142:339–392.

Hooijer, D. A. 1972. A late Pliocene rhinoceros from Langabaanweg, Cape Province. *Annals of the South African Museum* 29:151–191.

Hooijer, D. A. 1973. Additional Miocene to Pleistocene rhinoceroses of Africa. *Zoologische Mededelingen* (Leiden) 46:149–178.

Hooijer, D. A. 1978. Rhinocerotidae. In V. J. Maglio and H. B. S. Cooke, eds., *Evolution of African Mammals,* pp. 371–378. Cambridge, Mass.: Harvard University Press.

Hooijer, D. A., and B. Patterson. 1972. Rhinoceroses from the Pliocene of northwestern Kenya. *Bulletin of the Museum of Comparative Zoology* 144:1–26.

Nakaya, H. 1994. Faunal change of Late Miocene Africa and Eurasia: Mammalian fauna from the Namurungule Formation, Samburu Hills, northern Kenya. *African Study Monographs,* Supplementary issue 20:1–112.

Nakaya, H., M. Pickford, Y. Nakano, and H. Ishida. 1984. The late Miocene large mammal fauna from the Namurungule Formation, Samburu Hills, northern Kenya. *African Study Monographs,* Supplementary issue 2:87–131.

Thenius, E. 1955. Zur Kentniss der unterpliozänen *Diceros*-Arten (Mammalia, Rhinocerotidae). *Annalen der Naturhistorisches Museum* (Vienna) 60:202–211.

Table Abbreviations

ant = anterior
ap = anteroposterior length
dist = distal
dt = maximum depth
dv = dorsoventral depth
E = estimated measurement
lat = lateral
Lr = lower
Lt = left
lt = length
med = medial
post = posterior
prox = proximal
Rt = right
tc = tuber calcis
tr = transverse width
Ur = upper

TABLE 9.1 Distinguishing Features of the Lothagam Rhino Genera

Brachypotherium	***Diceros***	***Ceratotherium***
Teeth large and brachyodont	Teeth smaller and brachyodont	Teeth hypsodont
Thick enamel	Enamel of intermediate thickness	Thin enamel
Large upper, small lower incisors	No incisors	Moderate upper incisors
Labial edge of upper tooth oriented at wide angle to tooth row axis	Labial edge of upper teeth at lesser angle to tooth row axis	Labial edge of upper teeth at lesser angle to tooth row axis
Ectoloph very wide	Ectoloph narrow	Ectoloph narrow
Paracone style present	Paracone style present	No paracone style
Metacone style absent	Metacone style present	No metacone style
Wear surface at metacone forms ridge (consistent with transverse chewing)	Wear surface at metacone forms ridge	No metacone ridge (consistent with sagittal chewing)
Protolophs and metalophs transversely oriented	Protolophs and metalophs more diagonally orientated	Protolophs and metalophs most posteriorly orientated
Ectoloph behind metacone not reflected buccally	Ectoloph behind metacone reflected buccally	Ectoloph behind metacone not reflected buccally
Crotchet weak, never meeting protoloph	Crotchet stronger	Crotchet strong and crista present (but not connected as in extant white rhinos)

TABLE 9.2 Measurements (in mm) of Teeth of *Brachypotherium lewisi*

	LT 22872 Ur Nawata	LT 93 Lr Nawata	LT 24290 Lr Nawata	LT 99 Lr Nawata	LT 22961 Lr Nawata	LT 22874 Lr Nawata
P^2 ap	42.42	—	—	37.95	37.04	—
ant tr	46.42	—	—	46.08	43.02	—
post tr	50.29	—	—	49.0	45.0	—
M^2 ap	—	—	—	—	—	63.59
ant tr	—	—	—	—	—	—
post tr	—	—	—	—	—	65.78
M^3 ap	—	—	61.57	—	—	—
ant tr	—	—	70.25	—	—	—
post tr	—	79.4	76.84	—	—	—

	LT 94 Rt Lr Nawata	LT 94 Lt Lr Nawata	LT 88 Lt Lr Nawata	LT 88 Rt Lr Nawata	LT 23963 Lr Nawata
P^2 ap	—	—	39.67	38.89	40.88
ant tr	—	—	41.94	45.31	39.4
post tr	—	—	47.5	51.59	45.34
P^3 ap	—	—	42.17	50.04	—
ant tr	—	—	73.1	71.92	—
post tr	—	—	71.27	69.03	—
P^4 ap	—	—	44.71	56.95	—
ant tr	—	—	85.91	88.59	—
post tr	—	—	80.59	77.74	—
M^1 ap	—	—	63.55	65.25	—
ant tr	—	—	92.45	91.75	—
post tr	—	—	77.56	82.25	—
M^2 ap	76.28	80.58	75.81	76.07	—
ant tr	88.56	90.2	90.34	89.84	—
post tr	73.19	71.22	77.06	77.91	—
M^3 ap	—	—	73.58	70.32	—
ant tr	—	—	79.22	76.18	—
post tr	—	—	84.72	79.0	—

	LT 23962 Ur Nawata	LT 23967 Ur Nawata	LT 90 Lt Apak	LT 90 Rt Apak
P_1 ap	25.5	26.0	—	—
ant tr	13.93	15.15	—	—
post tr	20.39	20.45	—	—
P_2 ap	—	—	32.28	33.1
ant tr	—	—	19.64	19.45
post tr	—	—	22.82	26.14E

continued

TABLE 9.2 Measurements (in mm) of Teeth of *Brachypotherium lewisi* (Continued)

	LT 23962 Ur Nawata	LT 23967 Ur Nawata	LT 90 Lt Apak	LT 90 Rt Apak
P_3 ap	—	—	38.22	41.67
ant tr	—	—	24.97	25.94
post tr	—	—	30.4	33.09
P_4 ap	—	—	49.37	49.29
ant tr	—	—	31.04	32.96
post tr	—	—	36.17	35.92
M_1 ap	—	—	—	50.72
ant tr	—	—	—	—
post tr	—	—	—	—
M_2 ap	—	—	—	48.7
ant tr	—	—	—	—
post tr	—	—	—	—

	LT 26279 Ur Nawata	LT 81 Lr Nawata	LT 26280 Lr Nawata	LT 91 Ur Nawata	LT 28735 Lr Nawata
P_1 ap	—	—	—	20.23	—
ant tr	—	—	—	—	—
post tr	—	—	—	—	—
P_2 ap	—	—	—	28.57	—
ant tr	—	—	—	20.8	—
post tr	—	—	—	24.52	—
P_3 ap	—	—	—	36.2	—
ant tr	—	—	—	27.05	—
post tr	—	—	—	31.37	—
P_4 ap	—	—	—	45.17	—
ant tr	—	—	—	32.86	—
post tr	—	—	—	38.27	—
M_1 ap	—	—	—	50.12	—
ant tr	—	—	—	37.52	—
post tr	—	—	—	41.96E	—
M_2 ap	—	—	—	59.3E	—
ant tr	—	—	—	—	—
post tr	—	—	45.96	41.99E	—
M_3ap	72.03	—	—	54	—
ant tr	38.45	—	—	35.24	37.4
post tr	—	—	—	34.92	—

TABLE 9.3 Postcranial Measurements of *Brachypotherium lewisi*

	lt	prox ap	prox tr	dist ap	dist tr
Lt. Mt II 23964A	111	45	37	43	44
Rt. Mt III 26281	—	—	54	—	—
Lt. Mt IV 23964B	101	47	44	47	50
Rt. Mc II 26286	129	40	34	30	37
Lt. Mc III 23800	170	40	70	31	47
Dist Mc II or IV 26300	—	—	—	31	41
Lt. Mc IV 26312	140	46	54	43	59

TABLE 9.4 Measurements (in mm) of Teeth of *Ceratotherium praecox*

	LT 89 Lr Nawata	LT 23969 Kaiyumung	LT 23772 Ur Nawata
P^1 ap	—	22.67	—
ant tr	—	13.67	—
post tr	—	17.69E	—
P^2 ap	—	—	—
ant tr	—	—	—
post tr	—	—	—
P^3 ap	—	—	—
ant tr	—	—	35.87
post tr	—	—	48.89
M^2 ap	58.0	—	—
ant tr	63.78	—	—
post tr	—	—	—

	LT 26278 Ur Nawata	LT 26283 Kaiyumung	LT 23968 Kaiyumung	LT 82 Ur Nawata	LT 83 Apak
P_2 ap	—	30.33	—	—	—
ant tr	—	17.69	—	—	—
post tr	—	21.2	—	—	—
P_3 ap	—	37.55	43*	—	34.66
ant tr	—	25.2	—	—	—
post tr	—	—	—	—	23.43
P_4 ap	—	—	—	—	43.67
ant tr	—	—	—	—	27.09
post tr	—	—	—	—	—
M_1 ap	—	—	—	37.9	—
ant tr	—	—	—	22.79	—
post tr	—	—	—	25.93	—
M_2 ap	60.5	58.88	—	—	—
ant tr	45.0	42.43	—	—	30.34
post tr	46.0	42.66	—	—	—

TABLE 9.5 Postcranial Measurements of *Ceratotherium praecox*

	Astragalus	lat lt	med lt	prox tr	dist tr	dt
Upper Nawata	LT23972	865	82	85	77	65

	Calcaneum	lt	prox dv	prox tr	tc dv	tc tr
Apak Member	LT23966	133	72	89	70	55

	Proximal phalanx III	lt	prox ap	prox tr	dist ap	dist tr
Kaiyumung Member	LT26284	33	29	47	23	41

Table 9.6 Measurements (in mm) of Teeth of *Diceros bicornis*

	LT 84 Ur Nawata	LT 2397 Lt Apak	LT 23665 Ur Nawata	LT 23961 Ur Nawata	LT 28563 Apak
P^1 ap	—	—	—	—	—
ant tr	12.5	—	—	—	—
post tr	—	—	—	—	—
P^2 ap	24.57	—	—	31.45	35.12
ant tr	—	—	—	19.65	—
post tr	17.68	—	—	23.41	—
P^3 ap	37.77	—	—	—	43.79
ant tr	21.1	—	—	—	—
post tr	23.72	—	—	—	—
P^4 ap	41.26	—	—	—	49.76
ant tr	24.85	—	—	—	—
post tr	25.84	—	—	—	—
M^1 ap	50.22	—	53.21	—	—
ant tr	25.59	—	53.45	—	—
post tr	~26.0	—	53.32	—	—
M^2 ap	—	—	—	—	65.73
ant tr	—	—	—	—	—
post tr	—	—	—	—	63.72
M^3 ap	—	47.08	—	—	—
ant tr	—	56.12	—	—	—
post tr	—	58.09	—	—	—

	LT 23971 Rt Apak
M^1 ap	48.66
ant tr	60.7
post tr	58.57
M^3 ap	48.14
ant tr	55.2
post tr	57.48

	LT 28762 Apak
M_1 ap	48.22
ant tr	—
post tr	27.77

9.2

Systematics and Evolutionary Biology of the Late Miocene and Early Pliocene Hipparionine Equids from Lothagam, Kenya

Raymond L. Bernor and John M. Harris

Three hipparion species are present in the Lower Nawata: the large *Eurygnathohippus turkanense,* a new small species of *Eurygnathohippus* that was hitherto erroneously labeled *E. sitifense,* and the rare *Hippotherium* cf. *H. primigenium.* The *Hippotherium* specimens may be the last representatives of a lineage that entered Africa during the early Late Miocene. *Eurygnathohippus turkanense* is evidently closely related to *Sivalhippus perimense* from the Siwaliks and represents a lineage that entered Africa during the middle Late Miocene. The two *Eurygnathohippus* species occur also in the Upper Nawata, and very similar hipparions persist in the Apak and Kaiyumung Members. The presence of ectostylids in the Lothagam *Eurygnathohippus* species testifies to their close relationship with later African hipparions and to their separation from Eurasian stocks.

African hipparions had a complex history of migration, dispersion, adaptation, and evolutionary radiation (Bernor and Lipscomb 1991, 1995; Bernor and Armour-Chelu 1999). The first Old World hipparions occur near the base of the Late Miocene (Steininger et al. 1996). The most primitive taxon currently recognized, *Hippotherium primigenium,* would appear to be most closely related to the late Middle–early Late Miocene North American taxon *Cormohipparion occidentale s.l.,* or its predecessor, *Cormohipparion quinni* (Woodburne 1996). The migration from North America to Eurasia via Beringia is correlated with the terminal Serravallian regression (Bernor et al. 1988, 1989). Recently, there has been much systematic reevaluation of Old World hipparions in general (Bernor et al. 1989, 1996c) and of specific regions including East Asia (Qiu et al. 1987; Bernor et al. 1990), South Asia (Bernor and Hussain 1985), southeastern Europe to southwestern Asia (Bernor 1985), and Central Europe (Bernor et al. 1988, 1993a, 1993b). Reviews of hipparionine first occurrences include those for Indo-Pakistan (Pilbeam et al. 1996), Turkey (Kappelman et al. 1996), and western Eurasia and Africa (Sen 1996; Swisher 1996; Woodburne et al. 1996). These investigations indicate that the "*Hippotherium* Datum" (*sensu* Woodburne et al. 1996) is most accurately placed between 11 and 10.5 Ma. Moreover, whereas the first occurrence of Old World hipparions was erroneously viewed as having heralded the nearly instantaneous replacement of earlier Miocene forest environments with open country savanna-like environments (Berggren and Van Couvering 1974), it is now known that the evolution of Eurasian and African ecosystems from forested to open country "savanna mosaic woodlands" occurred diachronously and differentially in their degree through the Middle to Late Miocene interval (Bernor 1983, 1984; Bernor et al. 1979, 1996a, 1996b; Fortelius et al. 1996; Leakey et al. 1996).

The earliest African hipparions are of early Late Miocene age. Our current understanding of African hipparion evolution suggests that the founding population was derived from the Eurasian "*Hippotherium*" *primigenium* group. There would appear to be some regional differentiation, with a more derived form "*Hippotherium*" *africanum* occurring in North Africa and, at least on entry, a more primitive form, *Hippotherium* cf. *H. primigenium* in East Africa. The occurrence of a legitimate *Hippotherium* cf. *H. primigenium* is documented at Chorora, Ethiopia, between 10.5 and 9.3 Ma (constants corrected by Swisher in Woodburne et al. 1996 from 10.8 to 10.5 Ma; also, Bernor et al. 1989; Woodburne et al. 1996; Leakey et al. 1996). Kenyan records of this lineage include various localities in the Baringo Basin (Bernor personal observation) and Nakali (= *"Hipparion" africanum* of Aguirre and Alberdi 1974), but this material is meager in comparison to the Eurasian record of *Hippotherium primigenium*. By the end of the Late Miocene, East African members of this lineage are even more sparse—only a phalanx and two tibiae from the Lower Nawata representing the last plausible records. Post-Miocene referrals to *Hippotherium primigenium* would appear to be based on symplesiomorphic characters and, as such, are suspect.

The second founding population, the "*Sivalhippus*" Complex, would appear to have entered Africa no later than 8 Ma from South Asia. An evolutionary relationship between the large Lothagam species *Eurygnathohippus turkanense* and the Siwalik form *"Sivalhippus" perimense* has been documented previously (Bernor and Lipscomb 1991, 1995). Early members of the *Eurygnathohippus* clade variably express small ectostylids in the permanent mandibular dentition. The Nawata members of Lothagam include both a small and large species of *Eurygnathohippus*, but their frequency of ectostylids is low. In the Apak Member, ca. 5 Ma, the frequency increases substantially. Large and small hipparions (though not necessarily these or directly related species) persist together at least until the later Pliocene Omo Shungura G levels. A larger form, referred to the *Eurygnathohippus "ethiopicus"—E. cornelianus* lineage, then continues up to the Middle Pleistocene, ca. 0.6 Ma. A smaller Pleistocene form may occur in the Middle Awash and at Olduvai, but its morphology and chronological range are presently poorly known (Bernor personal observation).

The Lothagam assemblage is a critical East African fauna because it is derived from one of the longer and better chronostratigraphically controlled sequences in East Africa (Leakey et al. 1996; McDougall and Feibel 1999; Feibel this volume:chapter 2.1). This stratigraphic circumstance is rare in Africa, where it is matched only by the Middle Awash sequence in Ethiopia (Renne et al. 1999). The hipparion assemblage is of particular importance from the standpoints of its evolutionary and biogeographic relationships. The Lothagam hipparions occur during a time in which Eurasian hipparions undergo an explosive adaptive radiation and biogeographic extension that ends in the latest Miocene with a striking diversity crash. Equid workers have reported nothing comparable to this in Africa, and it would appear that East African hipparion evolution was decoupled from that of Eurasia by the end of the Miocene.

This contribution describes the diversity of hipparion species at Lothagam and their evolutionary and biogeographic relationships to critical hipparion species from broadly contemporaneous localities such as Sahabi in Libya (Bernor et al. 1987), Samos in Greece (Woodburne and Bernor 1980; Bernor et al. 1980), Maramena in Greece (Eisenmann and Sondaar 1995), and the Potwar Plateau of Pakistan (MacFadden and Woodburne 1982; Bernor and Hussain 1985). The analysis undertaken here marshals the continuous and character state variables pertinent to Old World hipparion evolution. The morphological diversity exhibited by the Lothagam hipparions is evaluated, their taxonomy is revised, hypotheses of their evolutionary and biogeographical relationships are formulated, and their probable ecological preferences are discussed within the limits imposed by the available material.

Materials and Methods

A number of equid researchers use both continuous and discrete variables to identify taxa and interpret phylogenetic relationships. The continuous variables used here follow those established in the 1981 American Museum of Natural History workshop on hipparion research published and illustrated by Eisenmann et al. (1988) and later augmented by Bernor et al. (1997) with measurements for some less common postcranial elements and the maxillary and mandibular cheek teeth. These measurements have been used by a number of investigators, including Bernor (1985) on the Maragheh hipparionines; Qiu et al. (1987) and Bernor et al. (1990) on separate suites of Chinese hipparionines; on Central European populations of *Hippotherium primigenium* from Austria (Bernor et al. 1993a), Hungary (Bernor et al. 1993b), and Germany (at Höwenegg: Bernor et al. 1997; at Dorn: Dörkheim, in Bernor and Franzen 1997); on African hipparions (Bernor and Armour-Chelu 1997, 1999); and in an overall review of Old World hipparion evolution (Bernor et al. 1989).

The character state analysis implements the use of 49 multistate characters of the skull, mandible, and dentition, which have been progressively refined through many of the studies cited above, and by Bernor and Lipscomb (1991, 1995), as well as those by colleagues who have studied North American hipparion evolution, including Skinner and MacFadden (1977), Woodburne and Bernor (1980), Woodburne et al. (1981), MacFadden (1984), Webb and Hulbert (1986), Hulbert (1988), Hulbert and MacFadden (1991), and Woodburne (1996). With one exception, these are all discrete variables. Utilization of postcranial character states awaits further comparative study across outgroup and sister-taxa.

In various studies, Eisenmann has used ratio diagrams to evaluate differences in hipparion metapodial proportions as a basis for recognizing taxa and their evolutionary relationships (for a comprehensive summary, see Eisenmann 1995). This procedure was followed by Bernor et al. (in press) on the large assemblage of hipparions from Sinap. Here, ratio diagrams of metapodials and of proximal phalanges from the third digit of the manus are used to assist in taxonomic decisions and to better interpret functional and evolutionary trajectories of hipparion locomotory systems. Because postcranial morphology may well be subject to a great deal of homoplasy, ratio diagrams are particularly useful for incorporation into broader analytical research designs that consider other anatomical regions.

Central European *Hippotherium primigenium* populations from Höwenegg (southern Germany, 10.3 Ma.; Swisher, 1996; Woodburne et al. 1996; Bernor et al. 1997) and Eppelsheim (western Germany, ca. 10.5 Ma; Bernor et al. 1996) provide standards with which to compare the Lothagam hipparion series and other pertinent hipparion populations from western Eurasia and Africa (table 9.7). The Höwenegg sample is particularly useful for postcranial comparisons because of the material's abundance and completeness (14 partial and complete skeletons). The Eppelsheim sample is better for maxillary and mandibular cheek tooth comparisons because the teeth are most often found without the associated jaws (allowing height measurements) and are more numerous than in the Höwenegg population. Also discussed in this contribution are hipparion populations from the Potwar Plateau of Pakistan (ca. 8.0 Ma; Pilbeam et al. 1996), the Kenyan sites of Ekora (ca. 4.0 Ma; Powers 1980) and Ngorora (Pickford 1978), the Greek sites of Maramena (ca. 5.5 Ma; Schmidt-Kittler 1995, but age estimate by Bernor based on correlation with Sahabi) and Samos (7.4–8.5 Ma; Solounias 1981), the Baynunah Formation of Arabia (ca. 6.0 Ma; Bernor estimate), and an unnamed site from Uganda (ca. 9.0 Ma; Pickford et al. 1993).

Hooijer and Maglio (1974) recognized three taxa from Lothagam: *Hipparion turkanense, Hipparion primigenium,* and *Hipparion* cf. *H. sitifense.* These three taxa were not clearly tied to a lithostratigraphic context, and the identification of *Hipparion* cf. *H. sitifense* was erroneous. Newly recovered material tied to the Lothagam chronostratigraphy (McDougall and Feibel 1999) documents the rare representation of *Hippotherium* cf. *H. primigenium* from the Lower Nawata; the common occurrence of *Eurygnathohippus turkanense* and *Eurygnathohippus feibeli* sp. nov. (binomen replaces *Hipparion* cf. *H. sitifense*) from the Nawata Formation, and the presence of large and small *Eurygnathohippus* specimens, probably related to *E. turkanense* and *E. feibeli,* respectively, in the Apak and Kaiyumung Members.

Abbreviations and Conventions

Locality Abbreviations

AL = Afar Locality. Most localities are at Hadar, Ethiopia, and collections are part of the National Natural History Museum of Ethiopia

AMNH = American Museum of Natural History, New York

BMNH = British Museum of Natural History, London (presently the Natural History Museum, London)

GSI = Geological Survey of India

HLMD = Hessisches Landesmuseum, Darmstadt

KNM-LT = National Museums of Kenya, Nairobi and Lothagam

MNHN = Museum National d'Histoire Naturelle, Paris

NHMW = Naturhistorisches Museum, Vienna

SMNK = Staatliches Museum für Naturkunde, Karlsrhue

Hipparion Definitions

1. Hipparionine or hipparion: horses with an isolated protocone on maxillary premolar and molar teeth and, as far as known, tridactyl feet, including species of the following genera: *Cormohipparion, Neohipparion, Nannippus, Pseudhipparion, Hippotherium, Cremohipparion, Hipparion, "Sivalhippus," Eurygnathohippus* (= senior synonym of "*Stylohipparion*"), *Proboscidipparion, "Plesiohipparion."* Characterizations of these taxa can be found in MacFadden (1984), Bernor and Hussain (1985), Webb and Hulbert (1986), Hulbert (1988), Qiu et al. (1987), Bernor et al. (1988, 1989, 1990, 1996, 1997), Woodburne (1989), Hulbert and MacFadden (1991), and Bernor and Armour-Chelu (1997, 1999).
2. "*Hipparion*": several distinct and separate lineages of Old World hipparionine horses formerly but no longer considered to be referable to the genus *Hipparion* (Woodburne and Bernor 1980; Bernor et al. 1980; MacFadden and Woodburne 1982; Bernor and Hussain 1985; Bernor 1985; Bernor et al. 1988; Bernor et al. 1989).
3. *Hipparion s.s.:* A discrete lineage of horses with the facial fossa positioned high on the face (MacFadden 1980, 1984; Woodburne and Bernor 1980; Woodburne et al. 1981; MacFadden and Woodburne 1982; Bernor and Hussain 1985; Bernor 1985; Bernor et al. 1987; Bernor et al. 1989; Woodburne 1989). The posterior pocket becomes reduced and eventually lost, and then confluent with the adjacent facial surface (includes Group 3 of Woodburne and Bernor 1980). No North American species of *Hipparion s.s.* is recognized; Bernor et al. (1989) have argued that

any morphologic similarity between North American "*Hipparion*" *s.s.* and *Hipparion s.s.* is due to homoplasy.

Measurements

Measurement numbers in the figures (M1, M2, M3, etc.) refer to those published by Eisenmann et al. (1988; and rounded to 0.1 mm) for the skulls and postcrania (also in Bernor et al. 1997), whereas tooth measurement numbers refer to those published by Bernor et al. (1997) and Bernor and Franzen (1997).

Anatomical Descriptions

The osteological nomenclature and the enumeration and/or lettering of the figures, have been adapted from Nickel et al. (1986). Getty (1982) was also consulted for morphological identification and comparison.

POB = Preorbital bar
POF = Preorbital fossa
Mc III = Metacarpal of the third (central) digit
Mt III = Metatarsal of the third (central) digit
A1PH3 = Proximal phalanx of the central digit of the manus

Systematic Description

Order Perissodactyla Owen, 1848
Suborder Hippomorpha Wood, 1937
Superfamily Equoidea Hay, 1902
Family Equidae Gray, 1821
Subfamily Equinae Steinmann and Doderlein, 1890
Eurygnathohippus Van Hoepen, 1930

All African hipparions of the genus *Eurygnathohippus* are united by the synapomorphy of ectostylids on the permanent cheek teeth. Eurasian and North American hipparions do not have this character except very rarely in extremely worn hipparion teeth from the Dinotheriensandes. *Stylohipparion* is the junior synonym of *Eurygnathohippus* by year priority.

Eurygnathohippus turkanense (Hooijer and Maglio, 1973)
(Figures 9.6, 9.7, 9.10–9.14; tables 9.8A–D, 9.9–9.11)

Synonymy

Hipparion turkanense Hooijer and Maglio, 1973:311
Hipparion turkanense Hooijer and Maglio, 1974:8
Not: *Hipparion turkanense* Hooijer, 1975:19–22
Eurygnathohippus turkanense Leakey et al., 1996:561

Diagnosis

A large species of hipparionine equid with a short, broad snout posterior to a line intersecting the two I^{3}s. Preorbital bar nearly indistinguishable due to strong reduction of the preorbital fossa; nasomaxillary fossa lost; orbital surface of lacrimal bone with a distinct foramen; preorbital fossa vestigial with no clearly distinguished outline, lacking a posterior rim, very shallow in its medial depth, lacking internal pits, and lacking a peripheral outline and anterior rim; infraorbital foramen inferior to the preorbital fossa; buccinator fossa distinct from the canine fossa and not pocketed; malar fossa absent; nasal notch incised just anterior to P^2. Maxillary cheek teeth with dP^1 strongly reduced; maxillary cheek teeth believed to be moderately curved; maximum crown height about 65 mm; fossette ornamentation moderately complex; posterior wall of postfossette mostly distinct; pli caballin usually double; hypoglyph moderately deeply to more deeply incised; protocone subtriangular shaped and lingually flattened; protoconal spur generally absent; protocone more lingually placed than hypocone in both premolar and molar teeth. There is no mandible associated with the type cranium, but the larger mandibular teeth from the Upper and Lower Nawata have the following characteristics: P$_2$ paraconid elongate; metaconid and metastylid generally rounded to elongate; metastylid spurs usually lacking on premolars and molars; ectoflexids usually separate metaconid and metastylid on premolars but not molars; pli caballinid usually absent, rarely single or complex; protostylid often not expressed on occlusal surface, but when it occurs it has a loop-like shape projected posterolabially; ectostylids are variable, being either absent or weakly developed, short in height and small in their length and width dimensions; linguaflexids are generally V-shaped on the premolars and have a deeper and broader U-shape on the molars; preflexids and postflexids have enamel margins that vary in their complexity; protoconid enamel band is usually rounded. Postcrania are relatively massive for a hipparionine; metapodials are short and robust with broad proximal and distal articular surfaces; anterior first phalanges III are also stoutly built with broad proximal and posterior distal surfaces (modified from Hooijer and Maglio 1973).

Holotype

KNM-LT 136, a female cranium with the maxillary dentition (figure 9.7) from the lower member of the Nawata Formation, Lothagam.

Figure 9.6 Restoration of *Eurygnathohippus turkanense* by Mauricio Antón. Shoulder height = 120 cm.

Age

Late Miocene, lower and upper members of the Nawata Formation, Lothagam, Kenya.

Lothagam Material

- *Lower Nawata:* holotype; 142, Lt. P^3; 154, Rt. P^3; 155, Lt. P^3, Lt. P^4; 167, Lt. P^3; 22867, upper teeth (Lt. and Rt. P^{2-4}, Lt. M^3); 22869, Rt. mandible and Lt. astragalus; 23107, cranium (Lt. P^4–M^3); 23994, Rt. I^1; 25430, Lt. P_4; 25433, Lt. astragalus; 25434, Lt. M_1; 25435, Rt. M_1, Rt. M_2; 25437, Lt. M_1, Lt. M_3; 25438, Lt. P_4; 25440, Rt. anterior second phalanx III, Lt. Mt III; 25441, Rt. M^2, Lt. M^2, Rt. M^3, Lt. M^3, Rt. P^2, Lt. P^3, Lt. P^4; 25442, Lt. M_3; 25443, Rt. M^1; 25446, Lt. I^2; 25456, Lt. P_3; 25457, Rt. M_1, Lt. M_3, Lt. P_3; 25458, Rt. M_3; 25459, Lt. M_1; 25460, Rt. P^3; 25462, Rt. Mt III; 25464, Lt. M^1; 26163, Rt. M^3; 26293, Lt. P^4, Rt. M^1; 38556, Lt. P3, Lt. $\bar{M}$
- *Upper Nawata:* 168, Lt. P^3; 22871, Lt. Mc III; 23685, 25455, Lt. P^3; Rt. P^4; 25467, Lt. M_3; 25470, Rt. Mt III; 25474, Rt. P_3; 25475, Lt. M_1; 25478, Rt. femur, Rt. tibia; 25481, Lt. Mt III; 25488, Rt. P_3; 25939, Lt. M_2; 25940, Rt. anterior first phalanx III; 26166, Lt./C; 26282, Lt. anterior second phalanx III; 26294, Lt. anterior first phalanx III.

Hooijer and Maglio (1973) initially described *"Hipparion" turkanense* as the earliest hipparion south of the Sahara. Comparison with subsequently collected *Hippotherium* cf. *H. primigenium* from Chorora, Ethiopia (Bernor and Armour-Chelu 1999), as well as with a small sample of hipparions from the Tugen Hills, demonstrates that *Eurygnathohippus turkanense* was neither the earliest nor the most primitive hipparion from sub-Saharan Africa. Analysis of the skull, dental, and postcranial anatomy of *E. turkanense* all suggest a close phylogenetic relationship with the Indo-Pakistan species *"Sivalhippus" perimense* (*sensu* Bernor and Hussain 1985). The Lothagam form is derived relative to the Siwalik form, particularly in the reduction of the preorbital fossa and the increased robustness of the distal appendicular skeleton.

Table 9.9 lists the character state distributions for *Eurygnathohippus turkanense* skulls KNM-LT 136 and LT 23107 and for *Eurygnathohippus* aff. *E. feibeli* skull KNM-EK 4. For *E. turkanense,* the following characters are of phylogenetic significance. The preorbital fossa,

Figure 9.7 *Eurygnathohippus turkanense* holotype cranium, KNM-LT 136: top = left lateral view; bottom = occlusal view.

placed far anterior on the face and high on the maxilla, is well separated from the lacrimal bone. The preorbital fossa is vestigial with a shallow depth, no peripheral border outline, and no anterior rim. These features reflect a maximum tooth crown height that was greater than most other Late Miocene Old World hipparions. Thus, *E. turkanense* closely resembles the Indo-Pakistan species "*Sivalhippus*" *perimense,* as characterized by Bernor and Hussain (1985), Bernor et al. (1989), and Bernor and Lipscomb (1991, 1995). The nasal notch is primitively incised just anterior to P^2, again like "*Sivalhippus*" *perimense* and in striking contrast to the deeply incised nasal notches of more derived members of the "*Sivalhippus*" Complex such as *Proboscidipparion pater* and *Proboscidipparion sinense.*

Table 9.10 lists the character state distributions for Lothagam upper teeth. Upper cheek teeth of *Eurygnathohippus turkanense* from the Lower Nawata differ from *E. feibeli* in the following features: cheek tooth curvature tends to be less; there is a modest incidence of the posterior wall of the postfossette not being distinctly separated from the posterior enamel wall of the tooth; there is a higher proportion of teeth that have complicated plications of the pre- and postfossettes; there is a tendency for the protocone to be more triangular-shaped. Also, parastyles and mesostyles are thicker in *E. turkanense* than in *E. feibeli,* and parastyle may be inflated in earlier stages of wear.

Table 9.11 lists the character state distributions for Lothagam lower hipparion teeth. The Lower Nawata sample of *E. turkanense* differs in the following morphological features from *Eurygnathohippus feibeli:* premolar and molar metaconids may be rounded, elongate, angular, and irregularly shaped; premolar and molar metastylids are similarly variable; premolar ectoflexid variably separates or does not separate metaconid and metastylid; pli caballinid likewise varies from being complex, rudimentary, or single and absent; when expressed, protostylid is an open loop extending posteriorly; none of the Lower Nawata specimens of *E. tur-*

kanense have ectostylids expressed on the occlusal surface; premolar ectoflexids vary from being shallow to V-shaped; molar linguaflexids are often deeper than those of *E. feibeli* from this level, achieving a deep, broad U-shape.

Upper Nawata lower teeth of *Eurygnathohippus turkanense* do not differ in metaconid, metastylid, and ectoflexid morphology from the Lower Nawata sample. Protostylid morphology is also very similar. Ectostylids may be observed in Upper Nawata representatives but are weakly developed and variable in their expression. Molar linguaflexids have deeper U-shapes in some individuals.

Only two specimens from the Apak Member are large enough to be assigned to *E. turkanense;* most equids from that unit are assigned to *Eurygnathohippus* sp. The most important feature of the Apak Member hipparion sample is that ectostylids occur with a greater frequency than in the subjacent unit (7 out of 17, or 41 percent of the time). However, the ectostylids are still small and have very short antero-posterior and mediolateral dimensions. Bernor and Armour-Chelu (1999) have documented that African horses exhibit an increase in ectostylid height and length over time. The two lower cheek teeth from the Kaiyumung Member both have relatively well-developed ectostylids.

Eurygnathohippus feibeli sp. nov.

(Figures 9.8–9.14; tables 9.10–9.12A–C)

Synonymy

Hipparion cf. *H. sitifense* Hooijer and Maglio 1974:20, 23

Eurygnathohippus "sitifense" Leakey et al. 1966:561

Diagnosis

A small hipparionine equid with gracile limbs. Metacarpal III is elongate and slender and with midshaft depth substantially greater than width. Anterior first phalanx III is long with very narrow midshaft width. Maxillary cheek teeth with thin parastyle and mesostyle; labiolingually moderately curved to straight; maximum crown height believed to be between 50 and 60 mm; mostly moderate to simple complexity of the pre- and postfossettes; posterior wall of postfossette mostly separated from posterior wall of the tooth; pli caballin mostly single or poorly defined double; hypoglyph variable with wear; protocone tending to be elongate and compressed; protoconal spur usually absent but may

Figure 9.8 Restoration of *Eurygnathohippus feibeli* sp. nov. by Mauricio Antón. Shoulder height = 110 cm.

appear as a small, vestigial structure; premolar and molar protocone placed lingually to hypocone. Mandibular cheek teeth having premolar metaconid/metastylid mostly rounded, molar metaconid/metastylid mostly rounded to elongate; metastylid spur absent; ectoflexid not separating metaconid/metastylid in premolars, variably separating metaconid/metastylid in molars; pli caballinid mostly absent; when expressed, protostylid is most often presented as a posteriorly directed, open loop; ectostylids are variably expressed and when present are diminutive structures which do not rise high on the labial side of the tooth; premolar and molar linguaflexid shallow V-shape; preflexid and postflexid enamel margins generally simple; protoconid enamel band rounded.

Holotype

KNM-LT 139, a partial right forelimb including fragmentary radius, metacarpal III, anterior first phalanx III, anterior second phalanx III, partial metacarpal II, 1st, 2nd and 3rd phalanx II and partial metacarpal IV. This specimen has been figured by Hooijer and Maglio (1974:plate 5, figure 7).

Etymology

Named in honor of Dr. Craig Feibel in recognition for his contributions to African Neogene geology and paleontology.

Type locality

Upper member of the Nawata Formation.

Age

Late Miocene, lower and upper members, Nawata Formation.

Geographic range

Kenya and possibly Ethiopia.

Lothagam Material

- *Lower Nawata:* 140, Lt. M^3; 22868, Lt. M^1, Lt. P^4; 23995, Rt. P_2; 25431, Rt. P^4; 25436, Rt. M^1, Lt. P^2, Rt. P^4; 25439, Lt. M^3, Rt. P^4; 25449, Lt. M^3, Lt. P^3; 25450, Rt. Mc III; 25451, Rt. Mt III; 25453, Rt. astragalus; 25454, Rt. C, Lt. P^2; 25461, Rt. M^3; 25463, Lt. M^3.
- *Upper Nawata:* 139, holotype; 141, Rt. M^1, Lt. P^3, Rt. P^4; 163, Rt. P^4, Rt. M^1, Rt. M^3, 392, Rt. P_4; 23687, Lt. M^2; 23770, Rt. dP^3; 23999, Rt. dP^4; Rt. M^1; 25468, Lt. M^2; 25469, Lt. P_4, Rt. P^4; 25472, Rt. anterior first phalanx III, Rt. anterior second phalanx III; 25473, Lt. P^4; 25476, Rt. P_2, Lt. M_1, Rt. M_3, Lt. M_3; 25477, Rt. dP_3, Lt. M_1, Lt. M^1; 25479, Rt. M_3; 25480, Lt. M^2; 25482, Rt. M^3; 25484, Lt. P^3; 25485, Lt. M^3; 25486, Lt. M^1, Rt. P^4; 25938, Rt. astragalus; 25941, Lt. astragalus; 25942, Rt. P^4; 26582, Lt. Mc III; 26621, Rt. tibia.

Small horses are not uncommon at African Late Miocene to Pleistocene localities. Hooijer (1975:22–26) found evidence of a small hipparion from as late as the Mid Pliocene horizons of Shungura Members B, C, E, F, and G. There is also a very small hipparion from the Middle Awash in 1 Ma horizons (T. D. White personal communication). While size can be a viable diagnostic criterion for hipparionine species, there is no compelling reason to unite all Late Miocene–Mid-Pleistocene East African small hipparions into a single species, or even a single lineage of small species.

The small horse in the Lothagam sequence is readily distinguishable from *E. turkanense.* Hooijer and Maglio (1974) assigned the small Lothagam form to the nomen *Hipparion* cf. *H. sitifense* Pomel, but *Eurygnathohippus feibeli* cannot be attributed to the North African *"Hipparion" sitifense* Pomel, 1897. According to Eisenmann (personal communication), Pomel never designated a type specimen for *"Hipparion" sitifense* from St. Arnaud Cemetery, Algeria (Pomel 1897), and the entire type series has been lost. Pomel did figure a single, worn upper cheek tooth, but this figure contains no species-significant character. *"Hipparion" sitifense* should thus be considered a nomen nudum.

The Upper Nawata distal forelimb holotype of *E. feibeli* (figure 9.9) is an important specimen because of its completeness and ease of comparison to a broad suite of other taxa. Bivariate plots for the third metacarpal (figure 9.10) and ratio diagrams (figure 9.11) demonstrate substantial differences in relative size and proportions between *E. feibeli* and *E. turkanense.* There is no question of their specific difference. The third metacarpal of *E. feibeli* deviates very little from that of Höwenegg *Hippotherium primigenium* in its measurements, with its greatest departures being in maximum length and mid-shaft width. It is very close in its dimensions to a third metacarpal (BMNH M2650) from the Siwalik Hills, a specimen that is part of the hypodigm of the type series of *Hipparion antelopinum.*

Analysis of the anterior first phalanx III (figures 9.12 and 9.13) clearly distinguishes *E. feibeli* from *E. turkanense* and *Hippotherium* cf. *H. primigenium.* However, *E. feibeli* compares most closely in this element with the Sahabi small hipparion (ISP32P25B), both of which have distinctly shorter length dimensions than Siwalik *Hipparion antelopinum.* Also, while the Lothagam-Sahabi grouping tracks the one Samos specimen in fig-

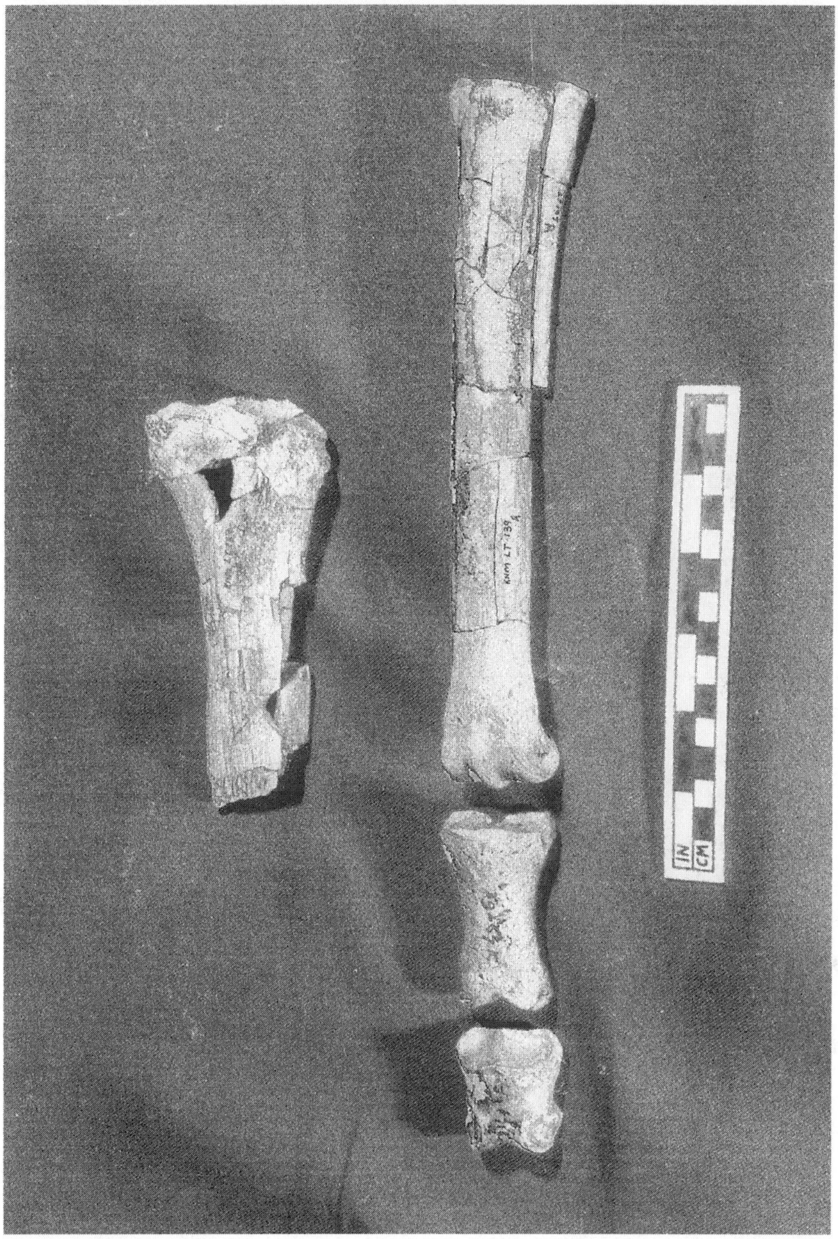

Figure 9.9 *Eurygnathohippus feibeli* sp. nov. partial right forelimb (holotype), KNM-LT 139, anterior view.

ure 9.11C, the Samos specimen is much smaller. The interspecific pattern demonstrated by the third metacarpal does not occur in the proximal phalanges.

A number of small maxillary and mandibular cheek teeth from the Upper and Lower Nawata are assigned to *E. feibeli*. There are relatively few discrete characters that clearly distinguish the small horse from the larger *E. turkanense*. Size and thin, bladelike parastyle and mesostyle morphology are the most reliable maxillary cheek tooth characters.

Upper cheek teeth of *Eurygnathohippus feibeli* from the Lower Nawata are moderately curved to straight, with maximum observed crown height between 40 and 60 mm. Occlusal morphology varies from somewhat complex to simple, probably due largely to stage of wear. The posterior wall of the postfossette is always distinct. Pli caballin morphology is predominantly single, being weakly double in some individuals. The hypoglyph varies from being very deeply incised to shallow, again largely due to stage of wear. The protocone morphology generally has an elongate, elliptical shape but becomes rounded in late wear; protocone is isolated from the protoloph except in late wear; protocone spur is usually not present; premolar and molar protocones are placed lingual to the hypocone. Teeth assigned to *Eurygnathohippus feibeli* also usually have very thin, bladelike parastyles and mesostyles, whereas those of *E. turkanense* are expanded on their worn, occlusal surface until later wear. Upper cheek teeth of *E. feibeli* from the Upper Nawata are essentially identical to those from the

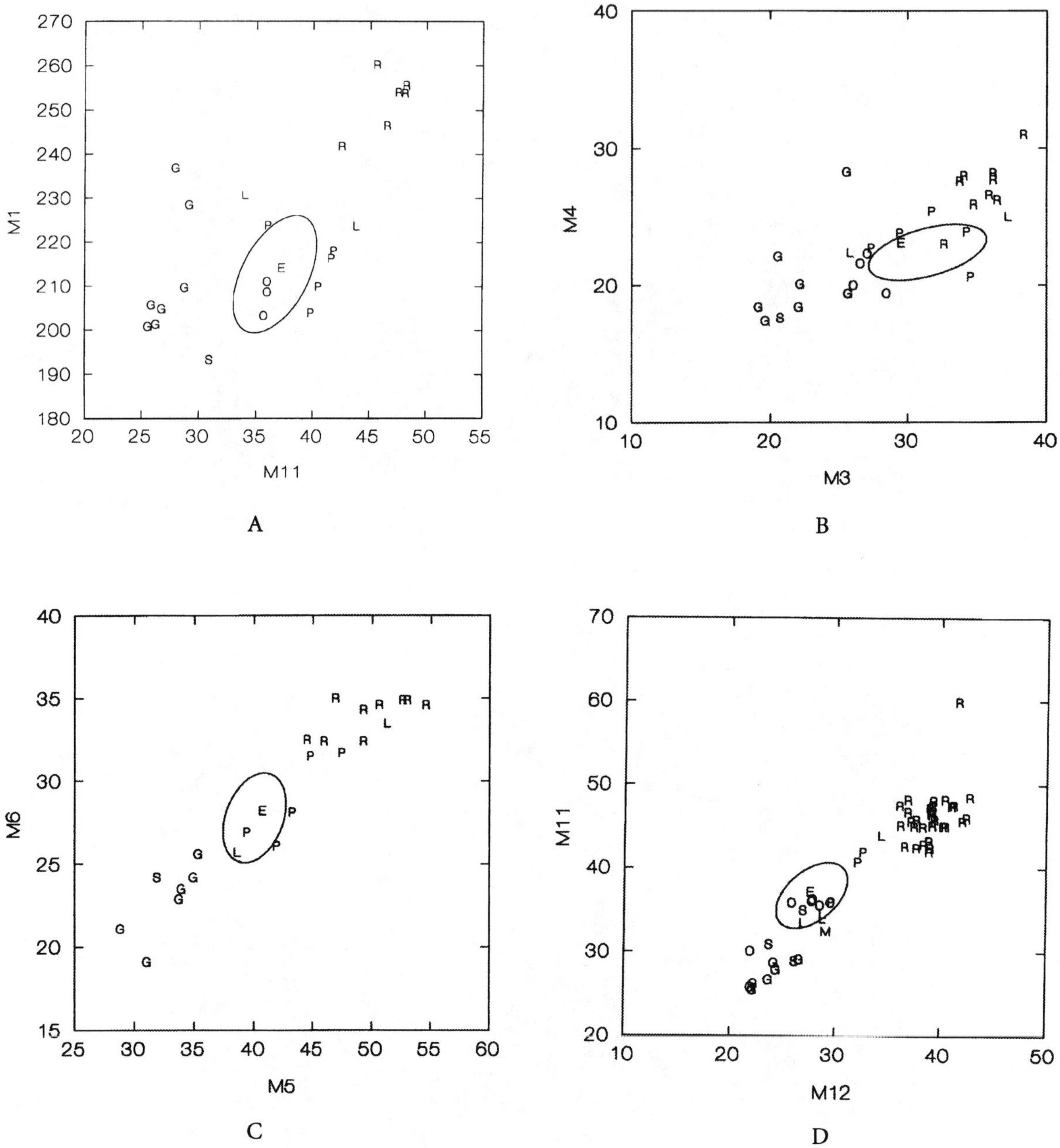

Figure 9.10 Bivariate plots of hipparion third metacarpals: A = maximum length (M1) versus distal articular width (M11); B = mid-shaft craniocaudal depth (M4) versus mid-shaft width (M3); C = proximal articular depth (M6) versus proximal articular width (M5); D = distal articular width (M11) versus distal maximum depth of keel (M12). In these and other bivariate plots, the ellipse represents Höwenegg specimens; for site abbreviations, see table 9.7.

Lower Nawata. Ectostylids may be observed in Upper Nawata representatives but are weakly developed and variable in their expression.

A small sample of hipparion lower cheek teeth from Ngorora (ca. 8.5 Ma), is characterized by the following: elongate P_2 paraconid; rounded premolar and molar metaconid; premolar metastylid rounded or absent; molar metastylid rounded; molar metastylid spur absent; premolar ectoflexid does not separate metaconid/metastylid, molar ectoflexid separates metaconid/metastylid; pli caballinids absent; protostylid mostly has the shape of a loop on the occlusal surface, extending posteriorly; ectostylids are absent; V-shaped premolar linguaflexids but linguaflexid U-shaped in the sole molar (M_1); preflexids mostly complex in their plication pattern, while postflexids are mostly complex or very complex; the postflexid does not have the anteriormost portion bent sharply lingually; the protoconid enamel band is always rounded. In all of these characters, the Ngorora hipparion compares most closely with Central European *Hippotherium primigenium.* This does not mean, however, that this taxon is referable to *H. primigenium,* because these are primitive characters for Old World hipparionine horses in general.

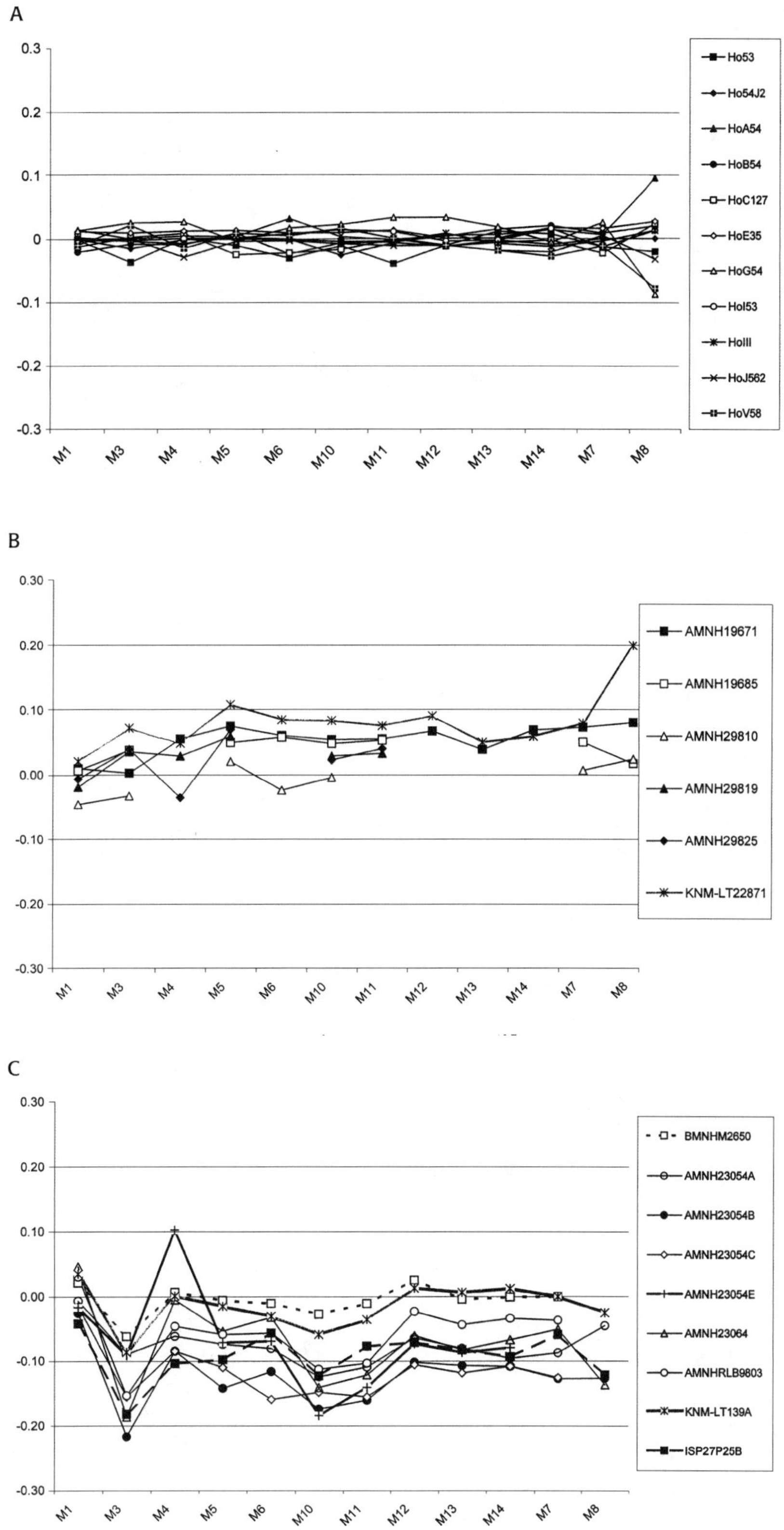

Figure 9.11 Hipparion third metacarpal ratio diagrams: A = Höwenegg standard; B = *Eurygnathohippus turkanense* and *"Sivalhippus" perimense;* C = Lothagam, Sahabi, Samos, and Siwalik small hipparions.

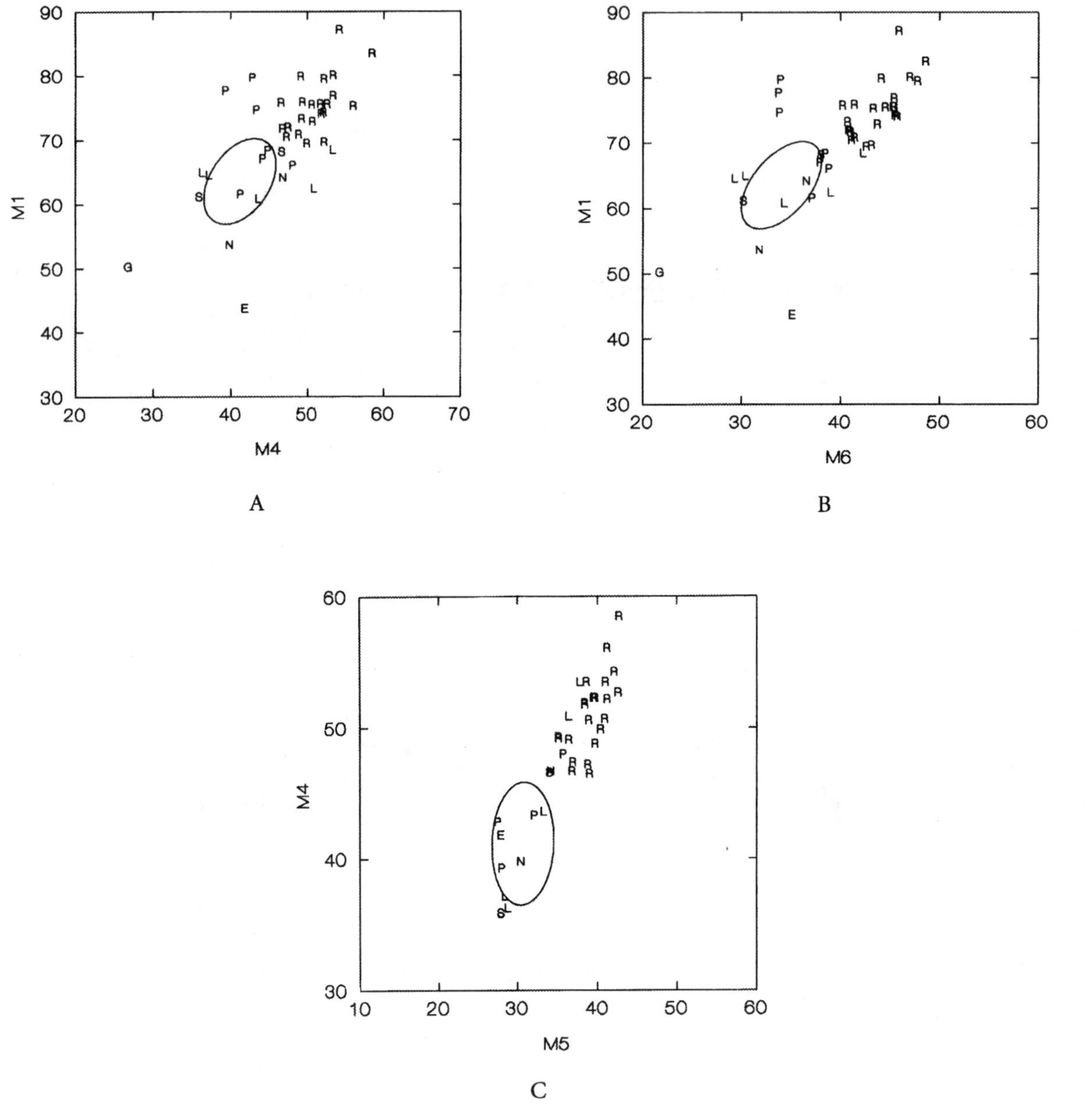

Figure 9.12 Bivariate plots of hipparion anterior first phalanx III (A1PH3): A = maximum length (M1) versus proximal width (M4); B = maximum length (M1) versus distal width at tuberosities (M6); C = proximal width (M4) versus proximal articular width (M5).

The Lower Nawata sample of *Eurygnathohippus feibeli* differs from the Ngorora sample by a number of characters, including the following: elongated or irregular-shaped premolar metaconid; elongated molar metaconid; in the molars metaconid and metastylid are not separated by ectoflexid (probably an ontogenetic bias); mostly simple premolar and molar preflexid and postflexid margins. Lower cheek teeth from the Upper Nawata sample of *Eurygnathohippus feibeli* do not differ significantly from the Lower Nawata *E. feibeli* sample in premolar and molar metaconid and metastylid and ectoflexid morphology. In the larger Upper Nawata sample the protostylid is slightly variable in the expression of the loop, as it is vertically placed or extends posterolabially. Although poorly developed and variable, ectostylids are present in adult individuals; premolar and molar ectoflexids are variable.

The Ekora cranium (KNM-EK 4) is a very important specimen, comprising a partial cranium with a deciduous dentition (dP^{2-4}) and M^1 clearly exposed in its crypt (Hooijer and Maglio 1974:13–15, plate 4, figures 1–3; plate 5, figure 1). Hooijer and Maglio (1974) assigned this specimen to "*Hipparion*" *primigenium* largely because of the size and depth of the preorbital fossa. While the preorbital fossa is large for a sub-Saharan African Pliocene hipparion, it is not as well developed as *Hippotherium primigenium s.s.* (Bernor et al. 1988, 1989; Bernor et al. 1997). Moreover, the overall

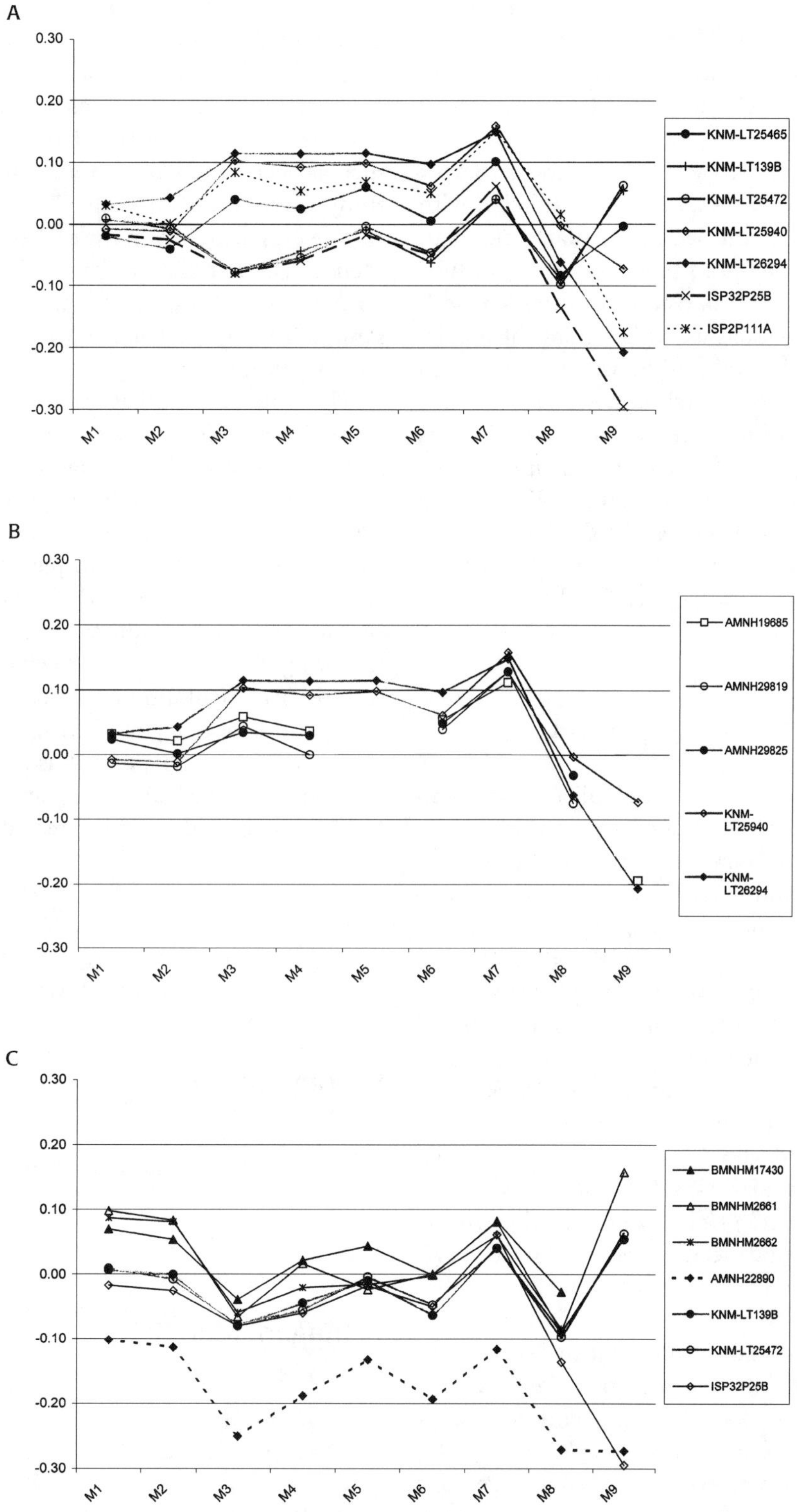

Figure 9.13 Hipparion anterior first phalanx III (A1PH3) ratio diagrams: A = Lothagam and Sahabi; B = Lothagam *Eurygnathohippus turkanense* and Siwalik *"Sivalhippus" perimense;* C = small, slender-limbed hipparions from Lothagam, Sahabi, the Siwalik Hills, and Samos.

cranial size is less than that of *Hippotherium primigenium s.s.,* while the measured crown height of the unerupted M^1 (= 55.0 mm; not 64 mm as reported by Hooijer and Maglio 1974) is greater than any known *Hippotherium primigenium s.s.* by a factor of 10 percent (Bernor and Franzen 1997; Bernor and Armour-Chelu 1999).

The Ekora specimen, KNM-EK 4, contrasts sharply with the two skulls of *E. turkanense.* It is smaller, with a shorter preorbital bar and lacrimal situated closer to the preorbital fossa. The preorbital fossa is large, subtriangular in shape, anteroposteriorly oriented with moderate posterior pocketing, with significant medial depth, and has a peripheral outline that is moderately well delineated. The nasal notch is not preserved, but would not have been highly retracted. KNM-EK 4 is clearly a different species from *E. turkanense,* but its phylogenetic relationships are uncertain. The cranial size and cheek tooth morphology of EK 4 suggest an affinity with Lower and Upper Nawata *Eurygnathohippus feibeli* and it is provisionally referred to *Eurygnathohippus* aff. *E. feibeli.*

Eurygnathohippus sp. indet.

Hooijer and Maglio (1973, 1974), Hooijer (1975), and Hooijer and Churcher (1985) cited an extended hypodigm for *"Hipparion" turkanense* that included Late Miocene (Mpesida) and Early Pliocene horizons (Kanapoi, Ekora, Mursi Formation of the Omo Group). This was not based on similarities in cranial and postcranial characters but, rather, on symplesiomorphic characters of the cheek teeth. The single most consistent dental character in this extended hypodigm was the variable occurrence of the ectostylids on the lower cheek teeth, a character that is primitive for the *Eurygnathohippus* clade. Hooijer and Maglio (1973, 1974) believed that there was a second, similarly sized hipparion which they referred to *"Hipparion" primigenium,* but they seemingly mistook ectostylid occurrence as a species-level difference: one species had an ectostylid, while the other did not.

An extended application of the nomen *"Hipparion" turkanense* has crept into the literature (see the very useful compendium of Late Miocene–Pliocene mammal species of East Africa; Behrensmeyer et al. 1997) and has become overextended to include any larger Late Miocene–Early Pliocene hipparion without consistently well-developed ectostylids. *Eurygnathohippus turkanense* is actually best characterized by its large size, lack of a preorbital fossa, short and broad snout, and relatively short and stoutly built limbs. No African equid assemblage that matches the Nawata skull and postcranial sample of *E. turkanense* has yet been reported (Bernor and Armour-Chelu 1999). The hypodigm is therefore restricted to those Lower and Upper Nawata horizons, and we recommend that the nomina *Eurygnathohippus* sp. indet (large) and *Eurygnathohippus* sp. indet. (small) be provisionally used for specimens from younger Lothagam horizons and other Early Pliocene East African localities until more diagnostic material such as occurs at Hadar is found and made available for study (Bernor and Armour-Chelu 1999).

The hipparions from the earliest Pliocene sites (Apak Member; Aterir, Ekora) differ little from the Nawata Formation samples, but the Aterir *Eurygnathohippus* sp. sample has a greater incidence of complex enamel ornamentation.

The youngest Lothagam hipparion sample comes from the Kaiyumung Member of the Nachukui Formation, which is close in age to the Basal Member of the Hadar succession. What appear to be large and small *Eurygnathohippus* populations may represent a single, size-variable species; the sample size for any tooth class is too small to test these differences. The Kaiyumung Member populations differ from older samples by the following: the cheek tooth crowns tend to be straighter; fossette ornamentation is somewhat more complex; pli caballin morphology is virtually always single (an advanced character); hypoglyph is generally less deeply incised (an advanced character); and the small form tends to have mediolaterally compressed protocones.

Eurygnathohippus sp. indet. (large)

(Tables 9.10, 9.11, 9.13)

Synonymy

Hipparion primigenium (partim) Hooijer and Maglio 1974:18

Lothagam Material

- *Apak Member:* 144, Lt. P^4; 23684, Rt. M_{1-3}; 23686, Lt. M_1; 24001, Rt. /C, Rt. I_1, Rt. I_2, Rt. I_3, Rt. M_1, Lt. M_1, Rt. M_2, Lt. M_2, Rt. M_3, Rt. P_2, Lt. P_2, Rt. P_3, Lt. P_4, Rt. P_4; 25943, Rt. M_1, Rt. M_2, Lt. M_3, Lt. P_2, Rt. P_3; 25945, P^4; 26292, Lt. dP_3; 26295, Rt. dP^3, Rt. dP^4, Rt. M^1, Rt. P^3, Rt. P^4.
- *Kaiyumung Member:* 164, Rt. P^4; 23998, Lt. P^4; 24052, Rt. P^3; 25947, Rt. M_1; 25948, Rt. P_4; 25949, Lt. M^3, Rt. M^3; 26164, Lt. P^2; 26165, Lt. dP^3.
- *Horizon indet.:* 8736, Lt. femur.

Eurygnathohippus sp. indet. (small)

(Tables 9.10, 9.11, 9.14)

Synonymy

Hipparion cf. *H. sitifense* Hooijer and Maglio 1974:23, 25

Lothagam Material

- *Apak Member:* 146, Lt. M^1; 23996, Lt. P^4, Lt. M_1; 24000, Lt. M_1; 25944, Rt. P^4; 25946, Rt. M_2.
- *Kaiyumung Member:* 23689, Rt. M^1; Rt. M^2; Rt. M^3; Rt. P^4.

Hippotherium von Meyer, 1829
Hippotherium primigenium von Meyer, 1829

Diagnosis

Orbital surface of the lacrimal bone with a sharply reduced or absent foramen; nasal notch retracted well anterior to P^2; preorbital fossa subtriangular, anteroventrally orientated, pocketed deeply posteriorly, medially deep, with a strongly delineated peripheral border outline, and a distinctly defined anterior rim; when present dP^1 and dP_1 vestigial to absent; maxillary cheek teeth moderately curved to straight, unworn specimens with a maximum crown height of approximately 50 mm, pre- and post-fossette enamel plications complex and maintain several deeply amplified plis until well beyond middle stage of wear, posterior wall of postfossette is always distinct, pli caballins mostly double, hypoglyph deeply incised well beyond the middle stage of wear, protocone commonly lingually flattened and labially rounded in middle adult stage of wear and clearly isolated from the protoloph until very late stage of wear, protocone spur very rare to absent, protocone more lingually placed than the hypocone, P^2 anterostyle/paraconid elongate; mandibular premolars and molars with rounded metaconids and metastylids, premolar ectoflexids do not separate metaconid and metastylid, although they do so in the molars, lower cheek teeth frequently with complex pli caballinids, protostylids commonly present in middle adult stage of wear, and expressed as enclosed circular structures but never strong and columnar, ectostylids virtually always absent, linguaflexids usually shallow to slightly V-shaped on the premolars, more V-shaped on the molars, pre- and postflexids with complex margins; protoconid enamel band rounded; metapodials robustly built but lengthened compared to *Cormohipparion.*

Hippotherium cf. *H. primigenium*

(Figures 9.14, 9.15; tables 9.10, 9.15)

Synonymy

cf. "*Hippotherium*" *primigenium* Leakey et al. 1966:561

Lothagam Material

- *Lower Nawata:* 160, Lt. tibia; 25447, Rt. M^3; 25448, Lt. tibia; 25465, Rt. anterior first phalanx III.

There is an elusive third species represented mainly by postcrania from the Lower Nawata that is assigned to *Hippotherium* cf. *H. primigenium.* This is best represented by KNM-LT 25465, an anterior first phalanx III (figure 9.14) but also probably the two tibiae, KNM-LT 160 and KNM-LT 25448. The anterior first phalanx III has proportions that are virtually identical to *Hippotherium primigenium,* and the tibiae fall clearly within the range of the same taxon.

KNM-LT 25447 differs from upper molars attributed to *Eurygnathohippus feibeli* and *E. turkanense* in the following features: occlusal pattern simple, hypoglyph shallowly incised; protocone is lingually flattened and labially rounded (a character consistently found in Potwar Plateau "*Sivalhippus*" *perimense*). These features may reflect state of wear; thus this tooth may represent either of the *Eurygnathohippus* species from the Nawata Formation, or, alternatively, it could represent the rare and elusive medium-size *Hippotherium* cf. *H. primigenium* that is represented by postcranial material.

Whether it was *Hippotherium primigenium* that populated sub-Saharan Africa, or a form closer to North American *Cormohipparion occidentale,* or the North African *Hippotherium africanum,* the initial immigrant was a primitive Old World species. Evidently, there was an early Late Miocene primitive group of *Hippotherium primigenium*/"*Hipparion*" *africanum*–like species in East Africa represented at Nakali (Aguirre and Alberdi 1974; Aguirre and Leakey 1974), in the Baringo Basin of Kenya and in the Chorora Formation of Ethiopia (Bernor personal observation). Likewise, the Ugandan early Late Miocene has evidence of a large derivative of a *Hippotherium primigenium*–like taxon in "*Hipparion*" *megadon* (Eisenmann 1994). The *Hippotherium* lineage was apparently replaced in the medial Late Miocene of sub-Saharan Africa by species of *Eurygnathohippus.*

Metric Analysis

Metric analysis was undertaken for those items most commonly used for species distinction and comparison

Figure 9.14 Hipparion anterior first phalanges III, anterior view: left = KNM-LT 139, *Eurygnathohippus feibeli* sp. nov. left A1PH3; center = KNM-LT 25465, *Hippotherium* cf. *H. primigenium* right A1PH3; right = KNM-LT 26294, *Eurygnathohippus turkanense,* left A1PH3.

by equid workers: the skull, astragalus, third metacarpal, anterior first phalanx III (central digit), and third metatarsal. Plots of the cheek teeth did not discriminate species; most of the analyzed sample fell within the Eppelsheim ellipse. As a result, we neither figure nor discuss cheek tooth size variables here (but see Bernor and Franzen 1998; Bernor et al. 1997; Bernor et al. in press). Calcanea are often useful for species discrimination but are absent from the Lothagam sample.

Analysis of Skull Continuous Variables

Figure 9.15A contrasts preorbital bar length versus P^2–M^3 length. The *Eurygnathohippus turkanense* holotype cranium and that of "*S.*" *perimense* (AMNH 19466) have relatively long preorbital bars compared with those of *Hippotherium primigenium* from Höwenegg, "*Hippotherium*" *africanum* from Algeria, and "*Sivalhippus*" *theobaldi* or *Hipparion antelopinum* from the Potwar Plateau. The Bou Hanifia sample of "*Hippotherium*" *africanum* retains a relatively long preorbital bar but has a marginally shorter tooth row. The preorbital fossa in the Höwenegg *Hippotherium* sample is longer (figure 9.15B), dorsoventrally deeper (figure 9.15C), and more developed; its ventral border more closely approaches the facial maxillary crest (figure 9.15C) than does that of *Eurygnathohippus turkanense* or "*Sivalhippus*" *perimense.* Note that *Eurygnathohippus* aff. *E. feibeli* from Ekora plots more closely to the *H. primigenium* sample (figure 9.15C). *Hippotherium* species have longer snouts than *E. turkanense* or "*Sivalhippus*" *perimense* (figure 9.15D). *Eurygnathohippus turkanense* also has a wider gape (figure 9.15E) than *Hippotherium* or the remainder of the sample. *Hippotherium primigenium* was adapted to living in subtropical forest/woodland environments (Bernor et al. 1988) and probably incorporated a large proportion of browse into its diet (Bernor et al. 1997; Bernor and Armour-Chelu 1999). Its snout proportions were probably slightly more elongate and narrow than the primitive condition for Old World hipparions (Bernor et al. in press).

Central European *Hippotherium primigenium* representatives (NHMW 4229 and HoA) have similar cheek tooth length proportions to those of *Cormohipparion occidentale* and *Cormohipparion* sp. nov. (figure 9.16A), but *H. primigenium* has a relatively longer snout than *C. occidentale.* Bernor et al. (in press) have argued that the new Turkish *Cormohipparion* species and the North American *C. occidentale* had more of a grazing regimen than *H. primigenium,* which clearly incorporated a major proportion of browse into its diet (Hayek et al. 1991; Bernor and Armour-Chelu 1999).

Where comparable dimensions are known, both "*Sivalhippus*" *perimense* and *Eurygnathohippus turkanense* are larger than *Cormohipparion occidentale* or *Cormohipparion* sp. nov., except in the strongly reduced preorbital fossa length dimensions. Decrease in size of

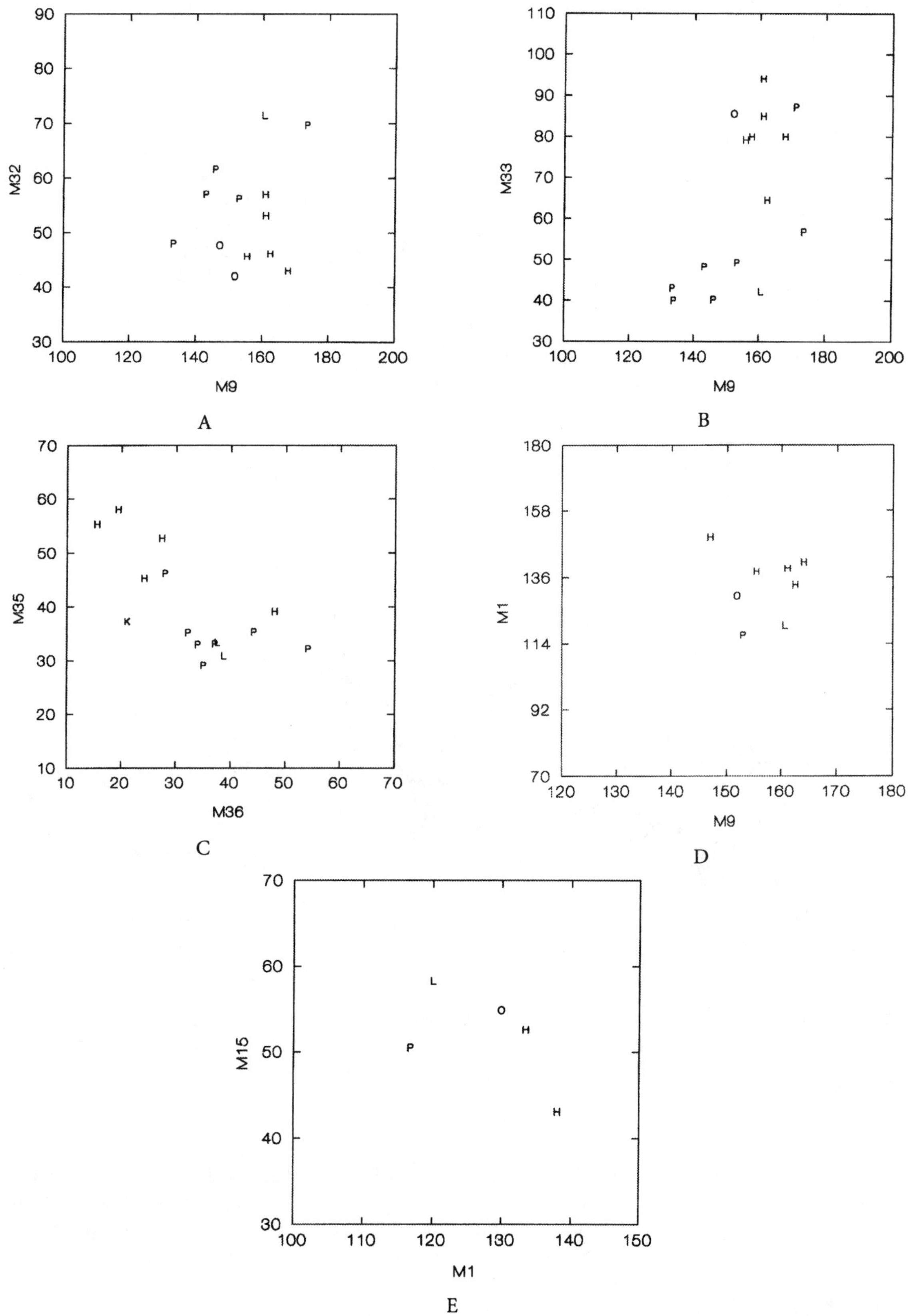

Figure 9.15 Bivariate plots for hipparion crania: A = length of preorbital bar [POB] (M32) versus length of P^2–M^3 (M9); B = length of preorbital fossa [POF] (M33) versus length of P^2–M^3 (M9); C = dorsoventral height of POF (M35) versus distance from ventral rim of POF to facial-maxillary crest (M36); D = muzzle length from prosthion to the middle of the line connecting the anterior borders of P^2 (M1) versus length of P^2–M^3 (M9); E = muzzle breadth between the posterior borders of the I^3s (M15) versus muzzle length from prosthion to the middle of the line connecting the anterior borders of P^2 (M1).

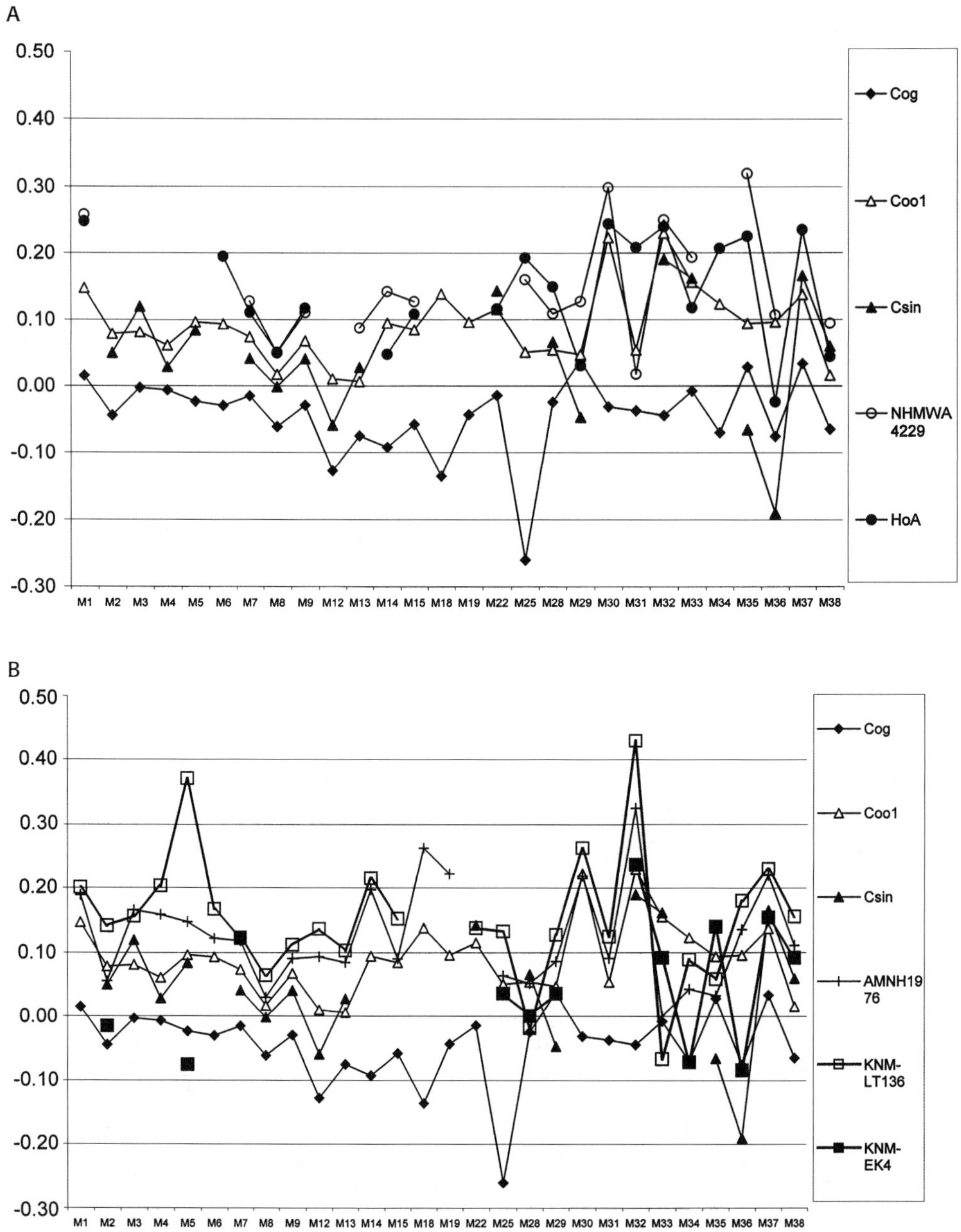

Figure 9.16 Hipparion skull ratio diagrams (log10) with *Cormohipparion quinni* as standard: A = *Cormohipparion–Hippotherium* evolutionary series; B = *Cormohipparion–"Sivalhippus"–Eurygnathohippus* evolutionary series. Data for North American Cormohipparion (*C. goorisi, C. occidentale 1,* and *C. quinni*), from Woodburne (1996), for Turkish *Cormohipparion* n. sp. (Sinap, basal MN 9, ca. 10.6–10.8 Ma) from Bernor et al. (in press).

the preorbital fossa is accompanied by its placement far anteriorly on the face, and this placement elongates the preorbital bar. These characters are significant at the superspecific rank (Woodburne et al. 1980; Bernor et al. 1989). Thus, *E. turkanense* is most similar to "*S.*" *perimense* (Bernor and Lipscomb 1995; Bernor and Armour-Chelu 1999). *Eurygnathohippus turkanense* and *Eurygnathohippus* aff. *E. feibeli* contrast most strongly in their palatal and postpalatal lengths with *Eurygnathohippus* aff. *E. feibeli,* being relatively short in both these dimensions (figure 9.16B). *Eurygnathohippus* aff. *E. feibeli* has a relatively shorter preorbital bar, longer preorbital fossa, greater preorbital fossa depth, and shorter distance between the ventral rim of the preorbital fossa and maxillary crest.

Forelimb

Radius

A single fragmentary radius from the Upper Nawata, part of the holotype of *E. feibeli* sp. nov. (LT 139), is

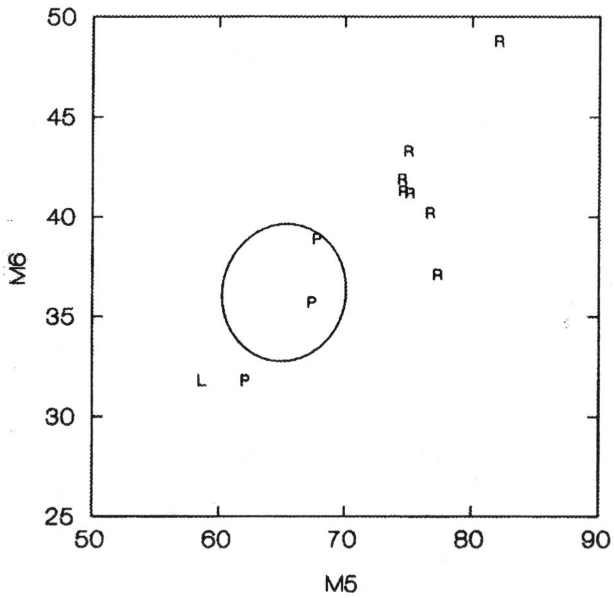

Figure 9.17 Bivariate plot for hipparion radius: proximal articular depth (M6) versus proximal articular width (M5).

substantially smaller than the Höwenegg sample and corresponds most closely in size to a specimen (BMNH M2657) from the Siwalik Hills (figure 9.17). All Hadar radii are substantially larger than the rest of the sample. Postcranial measurements have been found to be more stable than dental ones (Bernor et al. 1997), and this suggests that the differences found here are significant at the species level.

Metacarpal III

Two Lothagam third metacarpals have fundamentally different proportions. That of *E. feibeli* (LT 139) is relatively elongate and narrow, whereas that of *E. turkanense* (LT 22871) is shorter and much broader (figure 9.10A). Both of these specimens are derived from the Upper Nawata. There is great variability in the comparative sample. The Potwar Plateau sample sorts similarly to the Lothagam sample, but the variance is less. *Eurygnathohippus turkanense* and *E. feibeli* may also be distinguished on the basis of mid-shaft depth versus width (figure 9.10B), proximal articular depth versus width (figure 9.10C), and distal articular width versus distal maximum depth of the keel (figure 9.10D). Thus, variation in the third metacarpal supports the presence of two species at Lothagam: *E. feibeli* with a relatively elongate and narrow Mc III and *E. turkanense* with a relatively short and broad Mc III. *Eurygnathohippus feibeli* would probably be better adapted for open country cursorial behavior, while *E. turkanense* was probably adapted to a closed woodland setting where sustained running was less important.

Anterior first phalanx III

Three different-sized hipparion species are suggested by variation in anterior first phalanges. That of *E. feibeli* from the Upper Nawata (LT 139B and 25472) is relatively narrow proximally. "*Hippotherium*" cf. *H. primigenium* from the Lower Nawata (LT 25465) falls within the Höwenegg ellipse. *Eurygnathohippus turkanense* (LT 25940 of unknown provenance) is much wider proximally. Of two specimens from Sahabi (figure 9.12A), one plots with the small Lothagam form and the other at the upper end of the Höwenegg ellipse but with decidedly narrower proximal breadths. Individuals from Hadar are larger than those from Höwenegg, whereas *E. turkanense* plots at the higher end of the Hadar range for width dimensions but below, or in the lowermost part, of that range for length. The Potwar Plateau sample includes specimens within the Höwenegg ellipse and others above it within the length range, but below the width range of the Hadar sample (figure 9.12A).

The width of the distal tuberosities reflects proximal width (figure 9.12B), and this fact renders further support for the hypothesis that there are three hipparion species in the Nawata members. The separation of two Potwar Plateau populations is clearer from the distal width, probably corresponding to the smaller, more lightly built *Hipparion antelopinum* and larger, more heavily built *"Sivalhippus" perimense.* Proximal breadth versus proximal depth (figure 9.12C) again segregates three Lothagam species along the lines previously discussed.

Forelimb Log 10 Ratio Diagrams

Eisenmann (in Eisenmann and Sondaar 1995 and elsewhere) has shown the utility of Log[10] "Simpson's ratio diagrams" for depicting limb proportion differences. Bernor et al. (in press) have found these diagrams to be most effective when they are combined with analysis of bivariate plots. The same kind of comparison is employed here. However, whereas Eisenmann has used *Cremohipparion mediterraneum* as a standard, we used the mean values for the Höwenegg sample as the standard for the current analyses.

Metacarpal III

Figure 9.11A represents individual Höwenegg Mc III dimensions plotted against the mean of Höwenegg for each dimension. In Log 10 terms, this plot depicts individual variability around the Höwenegg mean. Of all the measurements, distal maximum depth is the only one that is highly variable (Bernor et al. 1997; Bernor and Franzen 1997).

Figure 9.11B plots *Eurygnathohippus turkanense* (KNM-LT 22871) with Potwar Plateau specimens of *"Sivalhippus" perimense* (AMNH numbers). *Eurygnathohippus turkanense* appears to be the most heavily built form, and lacks a great length dimension. Interestingly, while not as heavily built, *"S." perimense* tracks *E. turkanense* closely in nearly all its dimensions: the proportions are similar, and there is a similar tendency for the Mc III to be relatively heavily built. This similarity serves to further corroborate the phylogenetic relationship between these two taxa that was based on discrete skull characters (Bernor et al. 1989, 1996; Bernor and Lipscomb 1991, 1995; Bernor and Armour-Chelu 1999).

Figure 9.11C contrasts the small, slender built horses from Lothagam, Sahabi, Samos, and the Siwaliks. *Eurygnathohippus feibeli* (holotype KNM-LT 139A) plots close to the Höwenegg mean, except for its elevated length and sharply reduced mid-shaft width and, to a lesser extent, its proximal and distal width dimensions. *Eurygnathohippus feibeli* is very close in its morphology to BMNH M2650 which is assignable to *Hipparion antelopinum* (Bernor and Hussain 1985). The Sahabi small horse (ISP 27P25B) has a profile that is different from that of *E. feibeli:* it is shorter, with a much more slender shaft, and it has narrower proximal and distal epiphyses. The AMNH sample of *Cremohipparion matthewi-nikosi* is fairly homogeneous in itself except for AMNH 23054E, which has a very deep mid-shaft dimension. The Sahabi form is closest in its dimensions and proportions to the Samos sample.

Anterior first phalanx III

The Höwenegg sample of proximal phalanges from the (manus) third digit displays substantial variability for distal articular width, distal articular depth, and minimal length of the trigonum phalangis, the last being a notoriously variable measurement (Bernor et al. 1997).

Figure 9.13A directly compares the Lothagam and Sahabi phalanges. The *E. feibeli* holotype (LT 139B) compares very closely with Lothagam LT 25472 and Sahabi ISP 32P25B. *E. turkanense* is represented by two specimens, LT 25940 and LT 26294, with a series of measurements similar to those of the more heavily built form of Sahabi horse, *"Hipparion"* sp. (ISP2P111A). A third Lothagam morph, LT 25465 (*"Hippotherium"* cf. *H. primigenium*), is intermediate in size between the large and small Lothagam species, and it does not track either one particularly closely, except in proximal and distal width dimensions (which reflect body mass).

Figure 9.13B directly contrasts *Eurygnathohippus turkanense* with *"Sivalhippus" perimense* from the Potwar Plateau. The two populations are most dissimilar in their varying mid-shaft and proximal width dimensions and closest in their length, distal width, and distal articular depth measurements. *E. turkanense* is clearly more heavily built than *"S." perimense.*

Figure 9.13C plots the slender limbed hipparions from Lothagam, Sahabi, the Siwalik Hills (BMNH collection), and Samos. KNM-LT 25472 matches *Eurygnathohippus feibeli* (LT 139B) in all of its plotted points. Likewise, *E. feibeli* plots very closely to the Sahabi specimen, ISP 32P25B. Compared to *E. feibeli,* the Siwalik specimens of *Hipparion antelopinum* (BMNH numbers) have substantially greater length and proximal articular width dimensions, similar mid-shaft dimensions, and overlapping proximal articular depth and distal width and depth dimensions. The single Samos specimen, AMNH 22890, is much smaller and more slightly built than the remainder of the sample.

The anterior first phalanx III shows a closer morphological relationship between *E. feibeli* and the small Sahabi horse than is exhibited by the third metacarpal. This may reflect an actual phylogenetic relationship or may be due to a functional convergence.

Hindlimb

Tibia

There are only two complete hipparion tibiae in the Lothagam assemblage, both from the Upper Nawata, but they differ substantially in their size (figure 9.18A). The gracile LT 26621 falls within the Höwenegg range for maximum length but is more slenderly built in its distal maximum breadth. As such, it is referable to *E. feibeli.* The more robust LT 25478 (*E. turkanense*) is virtually the same length but is much broader. There is one comparable Hadar specimen that plots just outside the uppermost range of the Höwenegg hipparion and is more slender than *E. turkanense.*

A plot of distal maximum depth versus distal maximum width (figure 9.18B) provides somewhat less separation. The two Upper Nawata specimens plot outside and at either extreme of the Höwenegg ellipse. However, a third specimen from the Lower Nawata (LT 160) plots within the Höwenegg ellipse; this specimen may belong to the elusive third Lower Nawata taxon, referred here to *Hippotherium* cf. *H. primigenium,* which was previously known from a single first anterior phalanx III. Two specimens from Hadar plot at the uppermost extreme and just outside the Höwenegg ellipse,

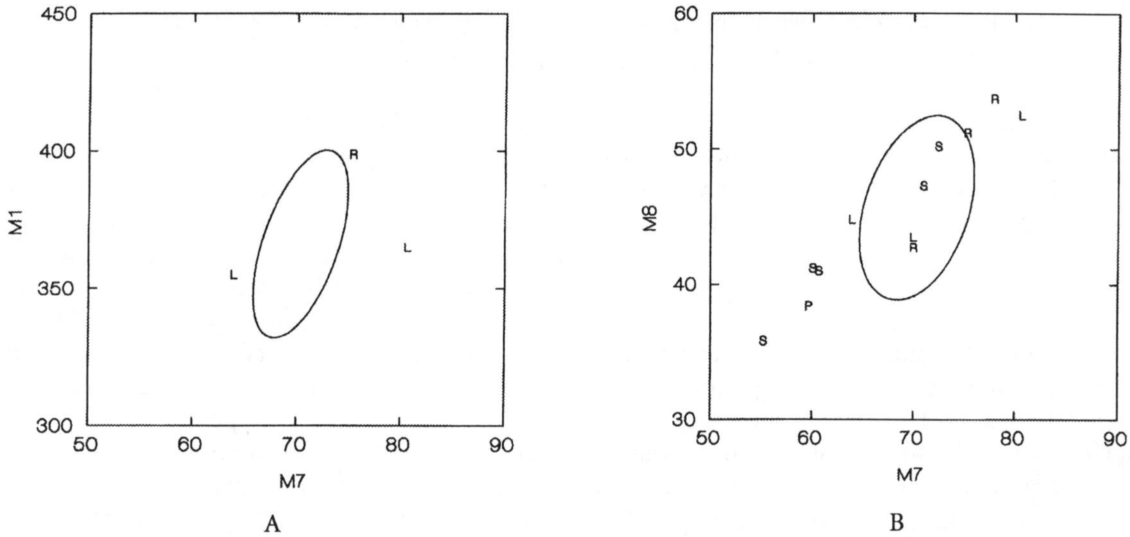

Figure 9.18 Bivariate plots for hipparion tibia: A = maximum length (M1) versus maximum width (M7); B = distal maximum depth (M8) versus maximum width (M7).

whereas a third plots in the lower range of the ellipse. Sahabi and the Potwar Plateau have individuals that are substantially smaller than the Höwenegg standard, principally in distal articular width. These undoubtedly represent small species of either *Cremohipparion* or *Hipparion s.s.*

Astragalus

Two data sets are plotted for the astragalus: maximum medial depth versus maximum diameter of the medial condyle (figure 9.19A) and distal articular depth versus distal articular width (figure 9.19B). In figure 9.19A, the large Lower Nawata individual (LT 25433) is assigned to *E. turkanense,* whereas the three smaller individuals are assigned to *E. feibeli.* Hadar material and individuals from the Potwar Plateau plot in the upper portion or outside the Höwenegg ellipse; they are larger horses. There is a large cluster of individuals from Sahabi, Maramena, the Potwar Plateau, and Bou Hanifia that plot well below the Höwenegg ellipse. These are all substantially smaller species than *Hippotherium primigenium* except, perhaps, for the Bou Hanifia specimen that may simply be a young individual.

The graph of distal articular depth versus distal articular width (figure 9.19B) pertains to a major load-bearing joint and, as such, is a reasonable proxy for body mass. Two Lower Nawata individuals and three Potwar Plateau individuals have higher values for distal articular width than the Höwenegg standard. In

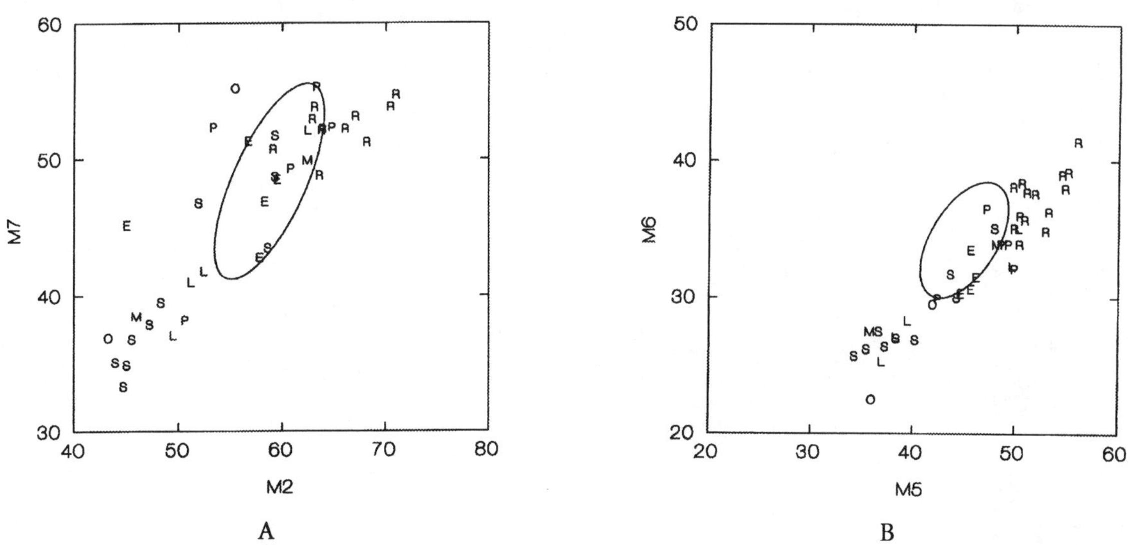

Figure 9.19 Bivariate plots for hipparion astragalus: A = maximum medial depth (M7) versus maximum diameter of the medial condyle (M2); B = distal articular depth (M6) versus distal articular width (M5).

contrast, one Lower Nawata and two Upper Nawata individuals are smaller than those from Höwenegg, grouping with individuals from Sahabi and Maramena. The Potwar Plateau is not represented by any really small astragali, but one specimen plots at the lowermost extreme of the Höwenegg ellipse. The one Ngorora specimen plots just inside the Höwenegg ellipse.

Third metatarsal

Figure 9.20A plots maximum length versus distal articular width of the third metatarsal. A large complete *E. turkanense* Mt III from the Upper Nawata (LT 25470) plots well outside the Höwenegg ellipse. An even larger Hadar hipparion Mt III dwarfs the rest of the sample, with a maximum length that approaches 300 mm. The only other Old World hipparion with comparable dimensions is the Pliocene Chinese species, *Proboscidipparion sinense* (Qiu et al. 1987), which also is a member of the "*Sivalhippus*" Complex. The Potwar Plateau sample includes individuals that vary greatly in both length and width, and this variation suggests that two or more species may be present in this sample. Sahabi has two individuals of comparable maximum length, but one is much narrower distally. Samos is here represented solely by slender-limbed forms, but these, too, exhibit considerable length variability.

Figure 9.20B plots Mt III mid-shaft depth versus mid-shaft width. Variation within the Lothagam sample is not as great for these parameters: an Upper Nawata individual plots at the upper extent of the Höwenegg ellipse, and a Lower Nawata individual plots at the lower extent of the ellipse. Likewise, the Potwar Plateau sample exhibits a modest dispersion within and around the periphery of the Höwenegg ellipse. All the Hadar metatarsals are larger than the Höwenegg sample, whereas those from Sahabi, Samos, and Maramena are smaller. Bou Hanifia "*Hippotherium*" *africanum* is slightly smaller than the Höwenegg standard range.

Figure 9.20C plots proximal articular depth versus proximal articular width. The proximal articular surface is a major load-bearing joint and a reasonable proxy for body mass. Two *Eurygnathohippus turkanense* individuals (LT 5462, Lower Nawata, and 25470, Upper Nawata) plot well above the Höwenegg ellipse and in the upper part of the Hadar sample. A single individual of *E. feibeli* (LT 25451) plots within the Höwenegg ellipse. The Potwar Plateau sample is scattered within, above, and below the Höwenegg ellipse and represents as many as three species. The very large individual plotting close

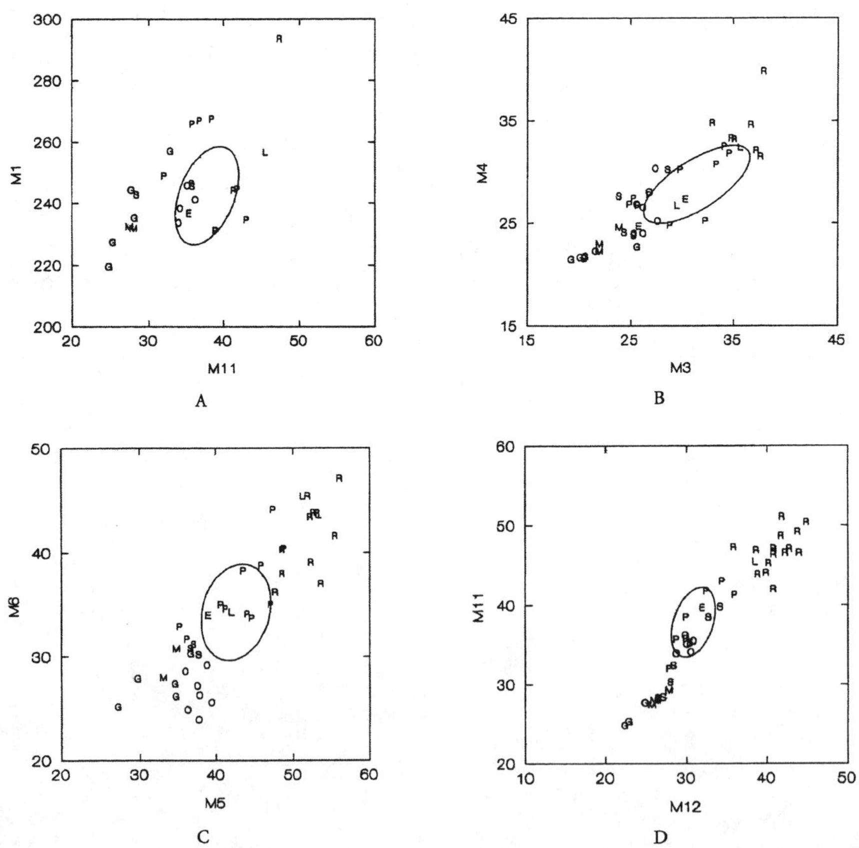

Figure 9.20 Bivariate plot for hipparion third metatarsals: A = maximum length (M1) versus distal articular width (M11); B = mid-shaft craniocaudal depth (M4) versus midshaft width (M3); C = proximal articular depth (M6) versus proximal articular width (M5); D = distal articular width (M11) versus distal maximum depth of keel (M12).

to the large Lothagam species may be the rare form *"Sivalhippus" theobaldi* (Bernor and Hussain 1985). All the Bou Hanifia sample is smaller than the Höwenegg range, suggesting that *"Hippotherium" africanum* had a smaller body mass than the Höwenegg hipparion. The Sahabi and Maramena samples are likewise smaller than the Höwenegg sample and close to the smaller Potwar Plateau specimens. Samos has the smallest specimens, and two Samos individuals have very narrow proximal articular widths.

In the graph representing distal articular width versus distal maximum depth of the keel (figure 9.20D), LT 25470 from the Upper Nawata plots well above the Höwenegg standard and within the Hadar sample distribution. The Potwar Plateau sample includes specimens whose range exceeds that of the Höwenegg sample. Sahabi includes two groupings of individuals, one smaller than the Höwenegg sample and another at the upper extreme. Maramena falls in the uppermost range of the Samos sample and overlaps extensively with the Sahabi sample.

Hind Limb Log 10 Ratio Diagrams

Metatarsal III

Figure 9.21 contrasts third metatarsals of the larger hipparion species in our sample: *Eurygnathohippus turkanense, "Sivalhippus" perimense* (AMNH specimens), and *Eurygnathohippus afarense* (Hadar AL numbers). *Eurygnathohippus turkanense* is remarkable for its relatively short, heavy build. *Eurygnathohippus afarense* is clearly a longer-limbed form that has values comparable to *E. turkanense* in mid-shaft, proximal facet width, and distal articular measurements. Although *E. turkanense* was shorter limbed than the Hadar horse(s), it probably had a comparable body mass. The Potwar Plateau sample is more heterogeneous, particularly for mid-shaft depth, than the Lothagam and Hadar samples but there appears to be one morph (i.e., AMNH 26953 and 29811) that closely tracks *E. turkanense.*

Discussion

The Lower and Upper Nawata have yielded most of the Lothagam hipparion specimens. There are three definable taxa from the Nawata members: the large *Eurygnathohippus turkanense* with massively built limbs, the small *Eurygnathohippus feibeli* sp. nov. with lightly built limbs, and an intermediate-sized species assigned to *Hippotherium* cf. *H. primigenium.* The last species is only certainly known from the Lower Nawata. Samples of the younger Apak and Kaiyumung Member hipparions include both large and small horses with some detectable advances in cheek tooth morphology, but with insufficient diagnostic features to characterize species. This younger Lothagam material is assigned to either *Eurygnathohippus* sp. indet. (large) or *Eurygnathohippus* sp. indet. (small).

Eurygnathohippus turkanense shares a close phyletic affinity with the Indo-Pakistan Late Miocene species *"Sivalhippus" perimense* (*sensu* Bernor and Hussain 1985). This affinity is supported by discrete cranial characters and continuous cranial and postcranial variables. *Eurygnathohippus feibeli* is related to *E. turkanense* by virtue of the mutual occurrence of ectostylids on the permanent cheek teeth. Comparison of *E. feibeli* to small horses from the Late Miocene of Sahabi (Libya),

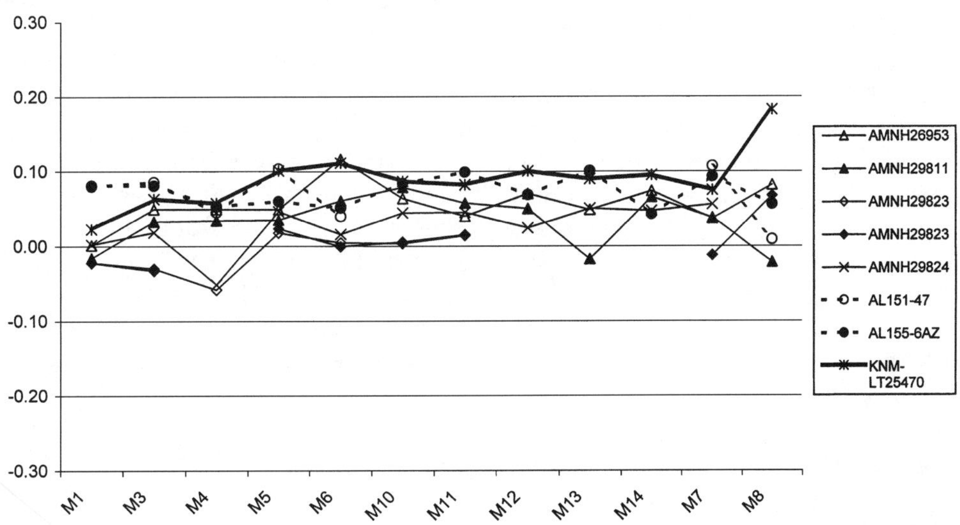

Figure 9.21 Ratio diagram for hipparion third metatarsals: *Eurygnathohippus turkanense, Eurygnathohippus afarense,* and *"Sivalhippus" perimense.*

Samos and Maramena (Greece), and the Siwaliks suggests that *E. feibeli* probably represents an independent small horse lineage. It would appear that the Sahabi small hipparion is identical in third metatarsal morphology to the Maramena hipparion; both forms are derived from terminal Miocene (MN13) age horizons. The Sahabi and Maramena hipparions appear to have phylogenetic affinities with the Samos hipparions rather than with *E. feibeli.* However, the morphology of the anterior first phalanx III supports a relationship between *E. feibeli* and a small horse from Sahabi. Because *E. feibeli* has ectostylids in permanent cheek teeth and a unique Mc III (which have not yet been demonstrated in the Sahabi small hipparion), we believe that the similarities between Sahabi and Lothagam phalanges may be homoplasious and related to cursoriality: a proportionally longer and narrower build that facilitated sustained, open country running.

An unexpected morphological similarity was found between the third metacarpal (LT 139) of *E. feibeli* and a specimen (BMNH M2650) from the type assemblage of the Indo-Pakistan species *Hipparion antelopinum.* Whereas the *H. antelopinum* material in question is of uncertain provenance, and it is uncertain that the cranial and postcranial material are associated, the preservation and size relationships of the cranial and postcranial material suggest that they were quarried from the same locality and belong to a single species. The cranial material of *Hipparion antelopinum* differs significantly from the Ekora skull in its very strong reduction of the preorbital fossa. Moreover, no Siwalik lower permanent cheek teeth are known to preserve ectostylids. These cranial and dental observations are incongruent with the similarity of metapodial proportions reported here and again deny a probable phyletic relationship between *Hipparion antelopinum* and *Eurygnathohippus feibeli.*

The rare, intermediate-sized species from the Lower Nawata is best interpreted as representing a late survivor of a more archaic Late Miocene hipparion, *Hippotherium* cf. *H. primigenium.* This hipparion is very similar in morphology to Central European *Hippotherium primigenium* (Bernor et al. 1997; Bernor and Franzen 1997) and to North African *"Hippotherium" africanum*; this similarity suggests that the Lower Nawata specimen may be a holdover from the first sub-Saharan African hipparion immigration. There is currently no evidence that this species survives at younger age horizons in sub-Saharan Africa.

Eurygnathohippus persisted after the Miocene and underwent a modest evolutionary radiation across Africa (Bernor and Armour-Chelu 1999). There is no good evidence of *E. turkanense* other than in the Lower and Upper Nawata. Because *E. turkanense* is extremely derived in its massive metapodial proportions, it is unlikely that it is related to the large, longer limbed Pliocene horses such as *E. hasumense, E. afarense,* and *E. "ethiopicus." Eurygnathohippus turkanense* was probably adapted to closed woodland settings and was not capable of sustained cursorial locomotion. *Eurygnathohippus feibeli* was a more lightly built form, with elongate distal limb elements that were probably adapted to more open country settings and were capable of more sustained cursorial locomotion. However, the length versus width proportions of the *E. feibeli* Mc III were not as advanced as those of contemporaneous eastern Mediterranean taxa; thus *E. feibeli* was probably less cursorial than these.

There have yet been no studies of enamel microwear on the Lothagam hipparions. However, carbon-isotope studies by Cerling et al. (this volume:chapter 12.2) suggest that both common Nawata hipparions were C_4 grazers. Bernor and Armour-Chelu (1999) have reviewed ecomorphological characteristics of African hipparions and concluded that by the beginning of the Pliocene, species of the *Eurygnathohippus* clade were strongly preferential grazers. At about the 2.5 Ma. horizon there was a marked increase in cheek tooth crown height, and by 1.7 Ma. there was a hipparion with even greater crown height (approaching 90 mm maximum), accompanied by incisor hypertrophy; this combination convincingly suggests that the most advanced members of the *Eurygnathohippus* clade were as dedicated to grazing as the white rhinoceros and wildebeest are today. It is likewise clear that the *Eurygnathohippus* clade was biogeographically restricted to Africa.

We are currently unable to address the phylogeny of Late Miocene–Pleistocene East African hipparions. This is due principally to the uneven distribution of material, or at least of studied material. We now have a good understanding of the Lothagam hipparion record and comparative material from Bou Hanifia (Algeria; ca. 9.5 Ma) and Sahabi (Libya; ca. 5.5 Ma). There is a critical gap, then, between about 5.5 Ma and the Hadar and equivalent-aged fauna of about 3.4 Ma; there is just too little material to evaluate conclusively (but note that the Apak mandible shows that ectostylids increase in frequency and height). Hadar has an excellent sample, yet to be published in toto, from 3.4 to 2.9 Ma. This sample overlaps with a modest hipparion sample from the Omo succession and localities adjacent to Lake Turkana. These collections show that a new, longer-legged hipparion existed in the Middle Pliocene. An emerging Middle Awash Late Pliocene hipparion fauna (Bernor personal observation), and the Olduvai Bed I and Bed II hipparion sample (ca. 1.9–1.7 Ma) include a modest-sized, but important record of the latest phases of hipparion evolution.

What we can say at this point in time, however, is that Lothagam's large hipparion, *E. turkanense,* was evidently replaced in the Pliocene by a new large hipparion with elongate limbs and the clear capability for sustained running. We do not know whether it was derived from an as yet unknown form, a form related to the South African Late Miocene form *E.* cf. *baardi* (Bernor and Armour-Chelu 1999), or a sister taxon of *E. feibeli.* We do not know whether all Plio-Pleistocene East African small hipparions are related to *E. feibeli* or really whether they are even monophyletic. We clearly need more complete material throughout the Late Miocene–Pleistocene record to pose any reasonable hypotheses of phylogenetic relationships.

A cautionary note: the similarities of the intermediate-sized hipparion from Lothagam to *Hippotherium primigenium* are based entirely on synplesiomorphic characters. It is entirely possible that these specimens, as well as other African taxa that hitherto have been referred to *Hippotherium,* instead represent one or more lineages derived from *Cormohipparion* immigrant stock.

Conclusions

The Lothagam assemblage fills a critical gap in our knowledge of African Late Miocene hipparion evolution. Our understanding of the early radiation of Old World hipparions is currently undergoing a great deal of change. North American *Cormohipparion* is now recognized in western Asia, and its relationship to the genera *Hippotherium, Hipparion s.s., "Sivalhippus,"* and *Eurygnathohippus* is ambiguous. The retention of a large, albeit medially shallower preorbital fossa in the Ekora *Eurygnathohippus* aff. *E. feibeli* suggests relationships within the *Cormohipparion-Hipparion-Hippotherium* trichotomy, while *E. turkanense's* reduced preorbital fossa and limb proportions suggest an alliance between *E. turkanense* and *S. perimense.* The occurrence of ectostylids in Lothagam *E. turkanense* and *E. feibeli* suggests a phylogenetic relationship exclusive of Eurasian hipparions and inclusive with Plio-Pleistocene African hipparions. Details of those relationships require a documentation that employs and analyses both continuous and discrete variables.

Acknowledgments

We thank Dr. Meave Leakey for her invitation to study the Lothagam hipparion assemblage. Support for this research was made possible by grants to R. B. from the National Science Foundation, the National Geographic Society, the L.S.B. Leakey Foundation, and the Alexander Von Humboldt Foundation.

References Cited

Aguirre, E., and M.-T. Alberdi. 1974. Hipparion remains from the northern part of the Rift Valley (Kenya). *Proceedings of the Koninklijke Nederlandse Akademie van Wetenschappen,* ser. B, 77:146–157.

Aguirre, E., and P. Leakey. 1974. Nakali: Nueva fauna de *Hipparion* del Rift Valley de Kenya. *Estudios Geológicos* 30:219–227.

Behrensmeyer, A. K., N. E. Todd, R. Potts, and G. E. McBrinn. 1997. Late Pliocene faunal turnover in the Turkana Basin, Kenya and Ethiopia. *Science* 278:1589–1594.

Berggren, W. A., and J. A. Van Couvering. 1974. The Late Neogene biostratigraphy, geochronology and paleoclimatology of the last 15 million years in marine and continental sequences. *Palaeogeography, Palaeoclimatology, Palaeoecology* 16:1–216.

Bernor, R. L. 1983. Geochronology and zoogeographic relationships of Miocene Hominoidea. In R. L. Ciochon and R. Corruccini, eds., *New Interpretations of Ape and Human Ancestry,* pp. 21–64. New York: Plenum Press.

Bernor, R. L. 1984. A zoogeographic theater and biochronologic play: The time/biofacies phenomena of Eurasian and African Miocene mammal provinces. *Paleobiology Continentale* 14:121–142.

Bernor, R. L. 1985. Systematic and evolutionary relationships of the hipparionine horses from Maragheh, Iran (Late Miocene, Turolian age). *Palaeovertebrata* 15:173–269.

Bernor, R. L., and M. Armour-Chelu. 1997. Later Neogene hipparions from the Manonga Valley, Tanzania. In T. Harrison, ed., *Neogene Paleontology of the Manonga Valley, Tanzania: A Window into the Evolutionary History of East Africa,* pp. 219–264. New York: Plenum Press.

Bernor, R. L., and M. Armour-Chelu. 1999. Toward an evolutionary history of African hipparionine horses. In T. Bromage and F. Schrenk, eds., *African Biogeography and Human Evolution,* pp. 189–215. Oxford: Oxford University Press.

Bernor, R. L., and J. Franzen. 1997. The equids (Mammalia, Perissodactyla) from the Late Miocene (Early Turolian) of Dorn-Dürkheim 1 (Germany, Rheinhessen). *Courier Forschungs-Institut Senckenberg* 1997:117–186.

Bernor, R. L., and S. T. Hussain. 1985. An assessment of the systematic, phylogenetic and biogeographic relationships of Siwalik hipparionine horses. *Journal of Vertebrate Paleontology* 5:32–87.

Bernor, R. L., and D. Lipscomb. 1991. The systematic position of *Plesiohipparion* aff. *huangheense* (Equidae, Hipparionini) from Glyazi, Turkey. *Mitteilung der Bayerischen Staatssammlung fr Palontologie und historisches Geologie* 31:107–123.

Bernor, R. L., and D. Lipscomb. 1995. A consideration of Old World hipparionine horse phylogeny and global abiotic processes. In E. Vrba, G. H. Denton, T. C. Partridge, and L. H. Burckle, eds., *Paleoclimate and Evolution, with Emphasis on Human Origins,* pp. 164–177. New Haven: Yale University Press.

Bernor, R. L., P. J. Andrews, N. Solounias, and J. A. H. Van Couvering. 1979. The evolution of Pontian mammal faunas: Some zoogeographic, paleoecologic and chronostratigraphic considerations. *Annals Geologica Pays Hellenica.* 1979:81–89 [Seventh International Congress on Mediterranean Neogene, Athens].

Bernor, R. L., V. Fahlbusch, P. Andrews, H. de Bruijn, M. Fortelius, F. Rögl, F. Steininger, and L. Werdelin. 1996a. The evolution of western Eurasian Neogene mammal faunas: A chronologic, systematic, biogeographic, and paleoenvironmental synthesis. In R. L. Bernor, V. Fahlbusch, and H.-W. Mittmann, eds., *The Evolution of Western Eurasian Later Neogene Faunas,* pp. 449–470. New York: Columbia University Press.

Bernor, R. L., V. Fahlbusch, and H.-W. Mittmann, eds. 1996b. *The Evolution of Western Eurasian Later Neogene Faunas.* New York: Columbia University Press.

Bernor, R. L., K. Heissig, and H. Tobien. 1987. Early Pliocene Perissodactyla from Sahabi, Libya. In N. T. Boaz, A. El-Arnauti, A. W. Gaziry, J. de Heinzelin, and D. D. Boaz, eds., *Neogene Paleontology and Geology of Sahabi,* pp. 233–254. New York: Liss.

Bernor, R. L., G. Koufos, M. O. Woodburne, and M. Fortelius. 1996c. The evolutionary history and biochronology of European and southwestern Asian Late Miocene and Pliocene hipparionine horses. In R. L. Bernor, V. Fahlbusch, and H.-W. Mittmann, eds., *The Evolution of Western Eurasian Later Neogene Faunas,* pp. 307–338. New York: Columbia University Press.

Bernor, R. L., J. Kovar-Eder, D. Lipscomb, F. Rögl, S. Sen, and H. Tobien. 1988. Systematics, stratigraphic and paleoenvironmental contexts of first-appearing *Hipparion* in the Vienna Basin, Austria. *Journal of Vertebrate Paleontology* 8:427–452.

Bernor, R. L., M. Kretzoi, H.-W. Mittmann, and H. Tobien. 1993a. A preliminary systematic assessment of the Rudabanya hipparions. *Mitteilungen der Bayerischen Staatssammlung für Paleontologie und historisches Geologie* 33:1–20.

Bernor, R. L., H.-W. Mittmann, and F. Rgl. 1993b. The Götzendorf hipparion. *Annalen der Naturhistorisches Museum* (Vienna) 95A:101–120.

Bernor, R. L., Z. Qiu, and L.-A. Hayek. 1990. Systematic revision of Chinese hipparion species described by Sefve, 1927. *American Museum Novitates* 2984:1–60.

Bernor, R. L., R. Scott, M. Fortelius, J. Kappelmann, M. Armour-Chelu, and S. Sen. In press. Systematics and evolution of the Late Miocene hipparions from Sinap, Turkey. In M. Fortelius, J. Kappelmann, S. Sen, and R. L. Bernor, eds., *The Sinap Formation, Turkey.* New York: Columbia University Press.

Bernor, R. L., H. Tobien, L.-A. Hayek, and H.-W. Mittmann. 1997. *Hippotherium primigenium* (Equidae, Mammalia) from the Late Miocene of Höwenegg (Hegau, Germany). *Andrias* 10:1–231.

Bernor, R. L., H. Tobien, and M. O. Woodburne. 1989. Patterns of Old World hipparionine evolutionary diversification and biogeographic extension. In E. H. Lindsay, V. Fahlbusch, and P. Mein, eds., *Topics on European Mammalian Chronology,* pp. 263–319. New York: Plenum Press.

Bernor, R. L., M. O. Woodburne, and J. A. Van Couvering. 1980. A contribution to the chronology of some Old World faunas based on hipparionine horses. *Geobios* 13:35–59.

Eisenmann, V. 1994. Equidae of the Albertine Rift Valley, Uganda. *CIFEG Occasional Publication* 29:289–307.

Eisenmann, V. 1995. What metapodial morphology has to say about some Miocene hipparions. In E. S. Vrba, G. H. Denton, T. C. Partridge, and L. H. Burkle, eds., *Paleoclimate and Evolution, with Emphasis on Human Origins,* pp. 148–163. New Haven: Yale University Press.

Eisenmann, V., and P. Y. Sondaar. 1995. The hipparions (Equidae, Perissodactyla, Mammalia). *Münchner Geowissenschaftliche Abhandlungen* 28:137–142.

Eisenmann, V., M.-T. Alberdi, C. de Giuli, and U. Staesche. 1988. Studying fossil horses. Pt. 1. Methodology. In M. O. Woodburne and P. Y. Sondaar, eds., *Collected Papers After the "New York International Hipparion Conference, 1981,"* pp. 1–71. Leiden: Brill.

Fortelius, M., L. Werdelin, P. Andrews, R. L. Bernor, A. Gentry, L. Humphrey, W. Mittmann, and S. Viratana. 1996. Provinciality, diversity, turnover and paleoecology in land mammal faunas of the later Miocene of western Eurasia. In R. L. Bernor, V. Fahlbusch, and H.-W. Mittmann, eds., *The Evolution of Western Eurasian Later Neogene Faunas,* pp. 414–448. New York: Columbia University Press.

Getty, R. 1982. *The Anatomy of Domestic Animals.* Philadelphia: Saunders.

Hayek, L.-A., R. L. Bernor, N. Solounias, and P. Steigerwald. 1991. Preliminary studies of hipparionine horse diet as measured by tooth microwear. In A. Forsten, M. Fortelius, and L. Werdelin, eds., Björn Kurten: A Memorial Volume. *Annales Zoologici Fennici* 28:187–200.

Hooijer, D. A. 1975. Miocene to Pleistocene hipparions of Kenya, Tanzania and Ethiopia. *Zoologische Verhandelingen* 142:1–80.

Hooijer, D. A., and R. Churcher. 1985. Perissodactyla of the Omo group deposits, American collections. In Y. Coppens and F. C. Howell, eds., *Les faunes Plio-Pléistocène de la Basse Vallée de l'Omo (Éthiopie).* Vol. 1. *Perissodactyles, Artiodactyles (Bovidae),* pp. 99–117. Paris: Centre National de la Recherche Scientifique.

Hooijer, D. A., and V. J. Maglio. 1973. The earliest hipparion south of the Sahara in the Late Miocene of Kenya. *Proceedings of the Koninklijke Nederlandse Akademie van Wetenschappen,* ser. B., 76:311–319.

Hooijer, D. A., and V. J. Maglio. 1974. Hipparions from the Late Miocene and Pliocene of northwestern Kenya. *Zoologische Verhandelingen* (Leiden) 134:3–34.

Hulbert, R. C. 1988. *Cormohipparion* and *Hipparion* (Mammalia, Perissodactyla, Equidae) from the Late Neogene of Florida. *Bulletin of the Florida State Museum of Biological Sciences* 33:229–338.

Hulbert, R., and B. J. MacFadden. 1991. Morphological transformation and cladogenesis at the base of the adaptive radiation of Miocene hypsodont horses. *American Museum Novitates* 3000:1–61.

Kappelmann, J., S. Sen, M. Fortelius, A. Duncan, B. Alpagut, J. Crabaugh, A. Gentry, J. P. Lunkka, F. McDowell, N. Solounias, S. Viranta, and L. Werdelin. 1996. Chronology and biostratigraphy of the Miocene Sinap Formation of central Turkey. In R. L. Bernor, V. Fahlbusch, and H.-W. Mittmann, eds., *The Evolution of Western Eurasian Later Neogene Faunas,* pp. 78–95. New York: Columbia University Press.

Leakey, M. G., C. S. Feibel, R. L. Bernor, J. M. Harris, T. E. Cerling, K. M. Stewart, G. W. Storrs, A. Walker, and L. Werdelin. 1996. Lothagam: A record of faunal change in the Late Miocene of East Africa. *Journal of Vertebrate Paleontology* 16:556–570.

MacFadden, B. J. 1980. The Miocene horse *Hipparion* from North America and from the type locality in southern France. *Palaeontology* 23:617–635.

MacFadden, B. J. 1984. *Systematics and Phylogeny of* Hipparion, Neohipparion, Nannippus *and* Cormohipparion *(Mammalia, Equidae) from the Miocene and Pliocene of the New World.* Bulletin of the American Museum of Natural History No. 179. New York: American Museum of Natural History.

MacFadden, B. J., and M. O. Woodburne. 1982. Systematics of the Neogene Siwalik hipparions (Mammalia, Equidae) based on cranial and dental morphology. *Journal of Vertebrate Paleontology* 2:185–218.

McDougall, I., and C. S. Feibel. 1999. Numerical age control for the Miocene-Pliocene succession at Lothagam, a hominoid-bearing sequence in the northern Kenya Rift. *Journal of the Geological Society* (London) 156:731–745.

Nickel, R., A. Schummer, and E. Seiferle. 1986. *The Anatomy of the Domestic Animals.* Vol. 1. *The Locomotor System.* Berlin: Schumberg.

Pickford, M. 1978. Geology, palaeoenvironments, and vertebrate faunas of the Mid-Miocene Ngorora Formation, Kenya. In W. W. Bishop, ed., *Geological Background to Fossil Man,* pp. 237–262. Edinburgh: Scottish Academic Press.

Pickford, M., B. Senut, and D. Hadoto, eds. 1993. *Geology and Palaeobiology of the Albertine Rift Valley, Uganda-Zaire.* Vol. 1. *Geology.* Occasional Publication No. 24. Orléans: Centre International pour la Formation et les Echanges Géologiques.

Pilbeam, D., M. Morgan, J. Barry, and L. Flynn. 1996. European MN units and the Siwalik faunal sequence of Pakistan. In R. L. Bernor, V. Fahlbusch, and H.-W. Mittmann, eds., *The Evolution of Western Eurasian Later Neogene Faunas,* pp. 96–105. New York: Columbia University Press.

Pomel, [N.] A. 1897. Les Equidé. *Carte Géologie Algérie* 12:1–44.

Powers, D. 1980. Geology of Mio-Pliocene sediments of the lower Kerio River Valley, Kenya. Ph.D. diss., Princeton University.

Qiu, Z., W. Huang, and Z. Guo. 1987. Chinese hipparionines from the Yushe Basin. *Palaeontologica Sinica,* ser. C, 175:1–250.

Renne, P. R., G. WoldeGabriel, W. K. Hart, G. Heiken, and T. D. White. 1999. Chronostratigraphy of the Miocene-Pliocene Sagntole Formation, Middle Awash Valley, Afar Rift, Ethiopia. *Geological Society of America Bulletin* 111:869–885.

Schmidt-Kittler, N. 1995. The vertebrate locality Maramena (Macedonia, Greece) at the Turolian-Ruscinian boundary (Neogene). *Münchner Geowissenschaftliche Abhandlungen* 28:1–180.

Sen, S. 1996. Present state of magnetostratigraphic studies in the continental Neogene of Europe and Anatolia. In R. L. Bernor, V. Fahlbusch, and H.-W. Mittmann, eds., *The Evolution of Western Eurasian Later Neogene Faunas,* pp. 56–63. New York: Columbia University Press.

Skinner, M. F., and B. J. MacFadden. 1977. Earliest known *Hipparion* from Holarctica. *Nature* 265:523–533.

Solounias, N. 1981. *The Turolian Fauna from the Island of Samos, Greece.* Contributions to Vertebrate Evolution 6. Basel: Karger.

Steininger, F. F., W. A. Berggren, D. V. Kent, R. L. Bernor, S. Sen, and J. Agustí. 1996. Circum-Mediterranean Neogene (Miocene and Pliocene) marine-continental chronologic correlations of European mammal units. In R. L. Bernor, V. Fahlbusch, and H.-W. Mittmann, eds., *The Evolution of Western Eurasian Later Neogene Faunas,* pp. 7–46. New York: Columbia University Press.

Swisher, C. C., III. 1996. New $^{40}Ar/^{39}Ar$ dates and their contribution toward a revised chronology for the Late Miocene nonmarine of Europe and West Asia. In R. L. Bernor, V. Fahlbusch, and H.-W. Mittmann, eds., *The Evolution of Western Eurasian Later Neogene Faunas,* pp. 64–77. New York: Columbia University Press.

Webb, D., and R. C. Hulbert. 1986. Systematics and evolution of *Pseudhipparion* (Mammalia, Equidae) from the Late Neogene of the Gulf Coastal Plain and the Great Plains. *Contributions to Geology* 3:237–272.

Woodburne, M. O. 1989. Hipparion horses: A pattern of endemic evolution and intercontinental dispersal. In D. R. Prothero and R. M. Schoch, eds., *The Evolution of Perissodactyls,* pp. 197–233. New York: Oxford University Press.

Woodburne, M. O. 1996. Reappraisal of the *Cormohipparion* from the Valentine Formation, Nebraska. *American Museum Novitates* 3163:1–56.

Woodburne, M. O., and R. L. Bernor. 1980. On superspecific groups of some Old World hipparionine horses. *Journal of Paleontology* 54:1319–1348.

Woodburne, M. O., R. L. Bernor, and C. S. Swisher III. 1996. An appraisal of the stratigraphic and phylogeneteic bases for the "*Hipparion* Datum" in the Old World. In R. L. Bernor, V. Fahlbusch, and H.-W. Mittmann, eds., *The Evolution of Western Eurasian Later Neogene Faunas,* pp. 124–136. New York: Columbia University Press.

Woodburne, M. O., B. J. MacFadden, and M. Skinner. 1981. The North American "*Hipparion*" Datum, and implications for the Neogene of the Old World. *Geobios* 14:1–32.

TABLE 9.7 *Hipparion* Assemblages Analyzed and Referred to in This Work

Locality	Symbol	Geologic Age (Bernor Estimate)
Baynunah Fm, Abu Dhabi	B	6.0
Bou Hanifia, Algeria	O	10.5
Ekora, Kenya	K	4.0
Epplesheim, Germany	E	10.5
Hadar, Ethiopia	R	2.9–3.4
Höwenegg, Germany	H	10.3
Lothagam, Kenya	L	3.5–6.5
Maramena, Greece	M	5.5
Ngorora, Kenya	N	9.5
Potwar Plateau, Pakistan	P	8.0
Sahabi, Libya	S	5.5
Samos, Greece	G	7.4–8.5

TABLE 9.8A Cranial Measurements of *Eurygnathohippus turkanense*

Measurement[a]	LT 23107 Lr Nawata	LT 136 Ur Nawata
Muzzle length	—	120.1
Palatal length	—	128.8
Vomerine length	97.0	113.1
Post-vomerine length	106.7	117.4
Post-palatal length	—	346.8
Basilar length	—	464.8
Premolar length	—	89.2
Molar length	64.5	74.4
Upper cheek teeth length	—	160.7
Choanal length	59.0	60.5
Minimal breadth of the choanae	35.2	32.2
Maximal breadth of choanae	44.2	44.3
Palatal breadth at P^4–M^1	81.8	68.9
Minimal muzzle breadth	—	43.9
Muzzle breadth	—	58.3
Length of fossa temporalis	70.4	80.8
Length basion to foramen ethmoidalis	107.2	—
Occipital breadth	—	58.4
Basioccipital breadth	93.8	98.9
Occipital height	—	67.4
Anterior ocular line	—	349.0
Posterior ocular line	—	239.0
Facial height	—	94.0
Cranial height	110.1	125.6
Anteroposterior orbital diameter	57.7	48.5
Dorsoventral orbital diameter	51.4	56.1
Length naso-incisival notch	—	128.1
Cheek length	—	185.9
Distance from orbit to preorbital fossa	70.5	71.5
Maximal length preorbital fossa	44.9	42.2
Length preorbital fossa to infraorbital foramen	—	55.9
Height preorbital fossa	33.4	30.9
Distance preorbital fossa to facial crest	37.6	38.8
Height infraorbital foramen above alveolar border	66.6	51.3
Height preorbital fossa above alveolar border	83.1	80.2

[a] Parameters from Eisenmann et al. (1988).

Table 9.8b Upper Teeth Measurements of *Eurygnathohippus turkanense*

Acc. No.	Member	Side	Specimen	M1	M2	M3	M4	M5	M6	M7	M8	M9	M10	M11
LT 23994	Lr Nawata	Lt.	I^1	17.4	10.4	7.8	10.7	—	—	—	—	—	—	—
LT 136	Ur Nawata	Rt.	I^1	13.1	—	12.3	—	—	—	—	—	—	—	—
LT 25446	Lr Nawata	Rt.	I^2	16.2	10.7	11.1	10.9	—	—	—	—	—	—	—
LT 136	Ur Nawata	Rt.	I^2	13.6	—	11.0	—	—	—	—	—	—	—	—
LT 136	Ur Nawata	Rt.	I^3	13.7	—	8.4	—	—	—	—	—	—	—	—
LT 136	Ur Nawata	Rt.	C	7	—	4.3	—	—	—	—	—	—	—	—
LT 26295	Apak	Lt.	dP^3	—	—	—	—	—	—	—	—	—	—	—
LT 26295	Apak	Lt.	dP^4	26.3	—	22.0	—	—	—	—	—	—	8.7	5
LT 26292	Apak	Rt.	dP^3	27.1	25.2	12.5	—	—	—	—	—	—	—	—
LT 22867	Lr Nawata	Lt.	P^2	29.9	27.8	21.3	19.6	20.6	2	4	2	1	7.7	4.4
LT 22867	Lr Nawata	Rt.	P^2	29.5	28.1	21.1	20.5	21.3	1	6	2	1	7.2	4.5
LT 25441	Lr Nawata	Lt.	P^2	29.9	27.9	21.4	21.1	10.1	2	4	3	1	7.8	5.3
LT 136	Ur Nawata	Rt.	P^2	36.4	—	25.0	—	—	—	3	2	1	0	—
LT 26295	Apak	Lt.	P^3	27.6	22.0	—	—	54.3	—	—	—	—	—	—
LT 142	Lr Nawata	Rt.	P^3	27.4	24.0	23.4	22.2	31.3	5	7	8	5	7.2	4.3
LT 154	Lr Nawata	Lt.	P^3	31.2	26.3	29.5	25.2	44.5	2	7	4	1	8.1	4.5
LT 155	Lr Nawata	Rt.	P^3	28.0	22.9	23.9	18.8	50.2	3	—	2	—	5.6	—
LT 167	Lr Nawata	Rt.	P^3	30.1	26.1	27.4	26.3	46.7	4	8	3	2	7.9	5
LT 22867	Lr Nawata	Rt.	P^3	25.4	21.1	24.8	23.9	24.2	1	6	4	1	9.5	4.2
LT 22867	Lr Nawata	Lt.	P^3	23.5	20.2	23.7	23.0	32.7	2	5	4	1	8.7	4.2
LT 25441	Lr Nawata	Rt.	P^3	23.0	19.8	25.3	25.2	9	1	5	5	2	7.7	5.2
LT 25455	Lr Nawata	Rt.	P^3	—	—	24.5	24.0	35.2	—	7	4	3	9.3	3.7
LT 25460	Lr Nawata	Lt.	P^3	24.5	21.4	23.8	22.8	27.8	1	4	3	1	9.0	4.2
LT 136	Ur Nawata	Rt.	P^3	27.3	—	27.4	—	—	1	5	3	1	0	—
LT 168	Ur Nawata	Rt.	P^3	24.8	21.6	23.6	21.7	38.9	3	7	5	1	6.4	3.6
LT 144	Apak	Rt.	P^4	23.1	22.0	24.6	23.0	32.3	3	7	3	2	8.8	4.1
LT 25945	Apak	—	P^4	28.3	24.0	26.3	23.5	46.0	—	—	—	—	7.4	—
LT 26295	Apak	Lt.	P^4	25.8	24.0	—	—	50.9	—	—	—	—	—	—
LT 155	Lr Nawata	Rt.	P^4	—	—	23.9	22.8	29.6	3	1	3	1	7.1	4.2
LT 22867	Lr Nawata	Lt.	P^4	22.3	21.3	22.4	22.2	27.7	1	4	3	2	8.7	4
LT 22867	Lr Nawata	Rt.	P^4	22.0	20.3	22.6	22.1	25.4	1	4	2	1	9.2	4.3
LT 23107	Lr Nawata	Rt.	P^4	21.0	20.0	24.7	—	10.1	0	2	0	0	7.8	5.2
LT 25441	Lr Nawata	Rt.	P^4	23.6	22.6	23.5	23.4	9.2	2	2	4	2	7.3	5.4
LT 26293A	Lr Nawata	Rt.	P^4	28.7	23.1	27.4	26.2	47.1	2	8	6	2	9.6	4.9
LT 136	Ur Nawata	Rt.	P^4	25.9	—	28.4	—	—	1	7	3	1	0	—
LT 23685	Ur Nawata	Lt.	P^4	21.9	20.6	24.4	22.3	34.4	6	7	6	2	7.2	3.9
LT 26295	Apak	Lt.	M^1	25.5	18.5	22.7	22.8	57.9	3	3	1	0	7.2	3.8
LT 23107	Lr Nawata	Rt.	M^1	—	20.4	25.7	—	5.2	—	—	—	—	9	6
LT 25443	Lr Nawata	Lt.	M^1	22.2	18.7	22.1	21.1	33.6	1	6	4	3	6	4.6
LT 25464	Lr Nawata	Rt.	M^1	27.7	20.9	23.1	24.2	57.7	6	7	7	1	8.3	4
LT 26293B	Lr Nawata	Lt.	M^1	24.1	22.4	—	—	50.0	3	8	4	3	—	—
LT 136	Ur Nawata	Rt.	M^1	22.7	—	26.7	—	—	1	5	4	1	—	—

TABLE 9.8B Upper Teeth Measurements of *Eurygnathohippus turkanense* (Continued)

Acc. No.	Member	Side	Specimen	M1	M2	M3	M4	M5	M6	M7	M8	M9	M10	M11
LT 23107	Lr Nawata	Rt.	M^2	20.5	20.5	24.1	—	3.2	0	0	0	0	8.9	5.9
LT 25441	Lr Nawata	Lt.	M^2	21.4	—	—	—	10.8	1	2	6	1	8.9	5.5
LT 25441	Lr Nawata	Rt.	M^2	—	—	23.4	22.7	10.0	1	6	5	1	8.5	5.4
LT 136	Ur Nawata	Rt.	M^2	23.3	—	28.1	—	—	1	5	5	1	—	—
LT 22867	Lr Nawata	Rt.	M^3	21.2	23	18.5	18.8	34.6	1	6	5	3	6.1	3.2
LT 23107	Lr Nawata	Rt.	M^3	21	20.2	21.6	—	6.4	0	3	1	0	8	5.8
LT 25441	Lr Nawata	Lt.	M^3	22.3	22.8	20.3	20.5	7.9	1	2	4	2	6.7	4.5
LT 25441	Lr Nawata	Rt.	M^3	22.4	21.2	20.4	19.7	10.9	1	2	4	2	7.7	4.3
LT 26163	Lr Nawata	Lt.	M^3	26.7	25.1	23.2	22.1	14.9	1	4	3	3	8.2	5.7
LT 136	Ur Nawata	Rt.	M^3	28.8	—	24.8	—	—	1	6	5	1	—	—
LT 163	Ur Nawata	Lt.	M^3	24.6	—	—	—	54.1	—	—	—	—	—	—

M1 = occlusal length.
M2 = length at 10 mm above roots.
M3 = occlusal width at mesostyle/protocone.
M4 = width at 10 mm above roots.
M5 = crown height along mesostyle.
M6 = number of plications on anterior face of prefossette.
M7 = number of plications on posterior face prefossette.
M8 = number plications on anterior face postfossette.
M9 = number plications on posterior face postfossette.
M10 = protocone length.
M11 = protocone width.
Source: Measurements from Bernor et al. (1997).

TABLE 9.8C Lower Teeth Measurements of *Eurygnathohippus turkanese*

Acc. No.	Member	Side	Specimen	M1	M2	M3	M4	M5	M6	M7	M8	M9	M10	M11	M12
LT 24001	Apak	Lt.	I_1	14.4	8.5	—	—	—	—	—	—	—	—	—	—
LT 24001	Apak	Lt.	I_2	15.1	8.6	—	—	—	—	—	—	—	—	—	—
LT 24001	Apak	Lt.	I_3	15.2	8.7	—	—	—	—	—	—	—	—	—	—
LT 24001	Apak	Lt.	/C	11.5	7.1	—	—	—	—	—	—	—	—	—	—
LT 26116	Lr Nawata	Rt.	/C	13	9.9	—	—	—	—	—	—	—	—	—	—
LT 23864	Ur Nawata	3	P_1	—	—	—	—	—	—	—	—	—	—	—	—
LT 24001	Apak	Lt.	P_2	29	—	11.3	7.7	10.4	10.1	—	9.2	11.3	—	—	—
LT 24001	Apak	Rt.	P_2	28.9	27.7	10.8	8.8	12	10.7	12.2	9.1	11.9	31.2	—	—
LT 25943	Apak	Rt.	P_2	28.2	27.6	12.2	7.5	11.1	14.1	13	11.1	12.5	31.4	—	—
LT 38556	Lr Nawata	Lt.	P_2	27.9	25.6	11.3	5.5	9.4	14.6	15.4	10.5	11.8	21.4	—	—
LT 24001	Apak	Lt.	P_3	25.9	—	15.3	9.2	12.7	12.1	—	11.6	12.4	—	—	—
LT 25943	Apak	Lt.	P_3	26.6	25.7	14.9	8.1	13.7	15.7	14.8	12.3	12.6	38.9	37.2	—
LT 25446	Lr Nawata	Rt.	P_3	26.4	23.9	13.8	—	—	12.5	12	12.7	14	39.8	37.5	—
LT 25457	Lr Nawata	Rt.	P_3	28	24.8	16.2	9.4	15.3	16.6	14.7	12.2	15.1	35.3	35.5	—
LT 38556	Lr Nawata	Lt.	P_3	25.3	25.6	15.4	8.2	9.7	16.5	16.7	14.2	13.2	12.6	11.8	—
LT 25474	Ur Nawata	Lt.	P_3	26	23	13.9	9.6	11.9	15.7	12.7	12.4	13.8	20.5	17.5	—
LT 25488	Ur Nawata	Lt.	P_3	25.1	21.8	14.2	9.1	12.9	11.6	12.2	11.1	12	38.9	36.9	—
LT 24001	Apak	Rt.	P_4	26	23.1	15.1	9.5	12	13.1	13	12.5	12.5	43.3	44.7	38.8
LT 24001	Apak	Lt.	P_4	24.4	—	14.5	8.2	11.2	12.6	—	12.1	12.1	—	—	—
LT 25430	Lr Nawata	Rt.	P_4	25.6	23	15.7	8.8	12.8	13.9	14	13.7	13.5	33.6	32.9	—
LT 25438	Lr Nawata	Rt.	P_4	27.6	24.3	14.4	8.9	13.4	11.8	12.5	—	—	55.5	—	—
LT 22869	Apak	Lt.	M_1	21.3	21.8	13.1	8.1	9.4	10.3	—	11.9	11.9	48.4	43.5	—
LT 23686	Apak	Rt.	M_1	23.5	21.3	13.9	7.1	10.5	10.9	13.6	9.1	8.8	53.5	49.1	—
LT 24001	Apak	Lt.	M_1	23.8	—	14.1	7.8	12.1	9.6	—	10.7	9.9	—	—	—
LT 24001	Apak	Rt.	M_1	23.3	21.7	13.9	6.7	9.6	9.7	13.5	10.5	9.8	40.8	41.4	31.6
LT 25943	Apak	Lt.	M_1	24.6	22.2	14.5	7.1	9.7	11.1	12.2	11.4	10.3	44.3	45.2	—
LT 25434	Lr Nawata	Rt.	M_1	25.2	22.4	13.1	7	10.2	15.2	14.7	12.1	12.1	44.6	43.4	—
LT 25435	Lr Nawata	Lt.	M_1	26.4	—	11.9	10.4	10.5	—	—	—	10.4	54.7	—	—
LT 25437	Lr Nawata	Rt.	M_1	22.8	20.5	12.4	6.7	7.9	9.1	12.7	12.1	11.4	28.4	26.1	—
LT 25457	Lr Nawata	Lt.	M_1	25	21.6	14.6	7.3	10.3	11.7	12.4	11.4	11	40.2	37	—

LT 25459	Lr Nawata	Rt.	M_1	26.6	23.6	—	—	—	—	11.5	—	—	52.7	—	—
LT 23684	Ur Nawata	Lt.	M_1	21.9	—	13.3	7.8	7.2	12.9	—	11.1	9.8	24.8	23.5	—
LT 25475	Ur Nawata	Rt.	M_1	26.1	—	12.6	7.3	10.2	9.4	—	8.2	7.7	55.3	55.8	—
LT 22869	Apak	Lt.	M_2	22.4	21	13.5	7.1	9.1	10.2	—	11.3	11.2	—	—	—
LT 24001	Apak	Lt.	M_2	24.1	—	13.1	7.8	10.6	9.1	—	9	8.9	—	—	—
LT 24001	Apak	Rt.	M_2	24.2	23.3	14.3	8.1	11.2	13.1	—	12	12.4	45.5	47.4	35
LT 25943	Apak	Lt.	M_2	24.8	22.5	12.5	7.3	10	12.9	12.3	9.4	9.4	48.1	—	—
LT 25435	Lr Nawata	Lt.	M_2	24.5	—	15.2	6.8	9.5	—	—	—	—	—	—	—
LT 23684	Ur Nawata	Lt.	M_2	21.9	—	12.3	8.3	7.1	11.5	—	3.9	9.1	—	—	—
LT 25939	Ur Nawata	Rt.	M_2	23.6	21.4	12.9	6.6	9.4	8.2	11.2	11.3	10.3	37.6	36.7	—
LT 24001	Apak	Lt.	M_3	23.8	—	10.2	6.6	9.5	7.9	—	7.8	7.2	—	—	—
LT 25943	Apak	Rt.	M_3	23.7	26.2	10.4	6	7.8	8.2	11.6	7.5	7.2	46.9	—	—
LT 25437	Lr Nawata	Rt.	M_3	27.5	30.3	10.9	5.9	7.4	11.7	11.5	10.4	11.6	31.8	28.8	—
LT 25442	Lr Nawata	Rt.	M_3	24.7	30	—	—	—	10.3	10.4	—	—	47.5	49.4	—
LT 25457	Lr Nawata	Rt.	M_3	25.2	26.6	11.1	7.5	8.1	9.2	11.1	9.8	7.7	27.8	—	—
LT 25458	Lr Nawata	Lt.	M_3	25.8	27.7	11.2	5.9	7.6	8.9	10.0	10.2	9.4	30	29.9	—
LT 23684	Ur Nawata	Lt.	M_3	24.3	—	11.1	7.4	7.8	11.7	—	8.9	8	—	—	—
LT 23684	Ur Nawata	Rt.	M_3	24	26.9	11.3	7.8	7.5	9.1	—	8.4	6.9	35.1	—	—
LT 25467	Ur Nawata	Rt.	M_3	24.2	27.7	11.1	6.7	8.7	10.4	11.2	8.4	8.2	35.1	35.4	—

M1 = occlusal length.
M2 = occlusal width.
M3 = length of metaconid-metastylid.
M4 = length of the prefossettid.
M5 = length of the postfossettid.
M6 = width of tooth across plane of ectoflexid/linguaflexid.
M7 = width 10 mm above root.
M8 = width across plane of metaconid and enamel band labial to protoconid.
M9 = width across plane of metastylid and enamel band labial to hypoconid.
M10 = crown height on mesial face of tooth.
M11 = length ectostylid.
M12 = width ectostylid.
Source: Measurements from Bernor et al. (1997:41).

TABLE 9.8D Postcranial Measurements of *Eurygnathohippus turkanense*

Acc. No.	Member	Side	Specimen	M1	M2	M3	M4	M5	M6	M7	M8	M9	M10[a]
LT 8736	Indet.	Rt.	Femur	—	—	47.7	52.0	—	—	—	—	61.8	—
LT 25478A	Ur Nawata	Lt.	Femur	411.0	398.0	44.5	54.1	133.6	98.7	106.5	131.3	65.4	63.7

Acc. No.	Member	Side	Specimen	M1	M2	M3	M4	M5	M6	M7	M8	M9[b]
LT 25478B	Ur Nawata	Lt.	Tibia	365.0	347.0	54.2	35.3	110.5	97.4	80.7	52.5	64.3

Acc. No.	Member	Side	Specimen	M1	M2	M3	M4	M5	M6	M7[c]
LT 22869	Lr Nawata	Rt.	Astrag	—	—	29.6	62.7	49.7	32.3	48.6
LT 25433	Lr Nawata	Rt.	Astrag	62.1	62.5	32.4	60.0	50.3	35.1	52.1

Acc. No.	Member	Side	Specimen	M1	M2	M3	M4	M5	M6	M7	M8	M9	M10	M11	M12	M13	M14[d]
LT 25440	Lr Nawata	Rt.	Mt III	—	—	—	—	—	—	—	—	—	—	31.8	29.9	31.4	—
LT 25462	Lr Nawata	Lt.	Mt III	—	—	—	—	51.2	45.4	46.2	14.7	8.7	—	—	—	—	—
LT 22871	Ur Nawata	Rt.	Mc III	223.7	216.0	37.2	25.0	51.3	33.5	41.5	18.4	10.3	47.6	43.9	34.4	27.4	29.8
LT 25470	Ur Nawata	Lt.	Mt III	256.7	247.8	35.7	32.4	53.3	43.6	46.8	15.5	6.4	48.8	45.5	38.6	31.1	33.6
LT 25481	Ur Nawata	Rt.	Mt III	—	—	—	—	—	—	—	—	—	38.6	—	30.1	24.7	26.7

				M1	M2	M3	M4	M5	M6	M7	M8	M9	M10	M11	M12	M13[e]
LT 26294	Indet.	Rt.	a1ph3	68.5	66.3	38.0	53.5	37.8	42.3	39.4	20.4	16.9	46.8	46.6	18.1	20.2
LT 25940	Ur Nawata	Lt.	a1ph3	62.5	58.6	37.0	50.9	36.4	39.0	40.3	23.4	23.0	34.2	36.6	21.2	20.7

				M1	M2	M3	M4	M5	M6[f]
LT 25440	Lr Nawata	Lt.	a2ph3	46.4	35.0	37.4	45.8	30.4	41.8
LT 26282	Ur Nawata	Rt.	a2ph3	41.2	30.9	39.4	44.0	29.1	40.3

[a] M1 = maximum length; M2 = length head to lateral condyle; M3 = minimum breadth (oblique); M4 = diameter at M3; M5 = maximum proximal breadth; M6 = maximum proximal depth; M7 = maximum distal breadth; M8 = maximum distal depth; M9 = maximum breadth trochlea; M10 = maximum depth head.

[b] M1 = maximum length; M2 = medial length; M3 = minimum breadth; M4 = minimum depth diaphysis; M5 = maximum proximal breadth; M6 = maximum proximal depth; M7 = maximum distal breadth; M8 = maximum distal depth; M9 = length fossa digitalis.

[c] M1 = maximum length; M2 = maximum diameter medial condyle; M3 = trochlea breadth; M4 = maximum breadth; M5 = distal articular breadth; M6 = distal articular depth; M7 = maximum medial depth.

[d] M1 = maximum length; M2 = medial length; M3 = minimum breadth; M4 = depth diaphysis at M3; M5 = proximal articular breadth; M6 = proximal articular depth; M7 = maximum diameter third tarsal articular facet; M8 = diameter fourth tarsal facet; M9 = diameter second tarsal facet; M10 = maximum distal supraarticular breadth; M11 = maximum distal articular breadth; M12 = maximum distal depth of keel; M13 = minimum depth distal lateral condyle; M14 = maximum depth distal medial condyle; M15 = angle of dorso-volar development of keel.

[e] M1 = maximum length; M2 = anterior length; M3 = minimum breadth; M4 = proximal breadth; M5 = proximal; breadth; M6 = distal breadth tuberosities; M7 = distal articular breadth; M8 = distal articular depth; M9 = minimum length trigonum phalangis; M10 = medial supratuberosity length; M11 = lateral supratuberosity length; M12 = medial infratuberosity length; M13 = lateral infratuberosity length.

[f] M1 = maximum length; M2 = anterior length; M3 = minimum breadth; M4 = maximum proximal breadth; M5 = maximum proximal depth; M6 = maximum distal articular breadth.

Source: Parameters from Eisenmann et al. (1988).

TABLE 9.9 Character State Distributions for Lothagam and Ekora Equid Crania

	C1	C2	C3	C4	C5	C6	C7	C8
KNM-LT 136	H	C	A	I	D	C–	A	D
KNM-LT 23107	H	C	—	I	D	C–	A	D
KNM-EK 4	C	C	A	E	B	A	A	B

	C9	C10	C11	C12	C13	C14	C15	C16
KNM-LT 136	B	B	A	A	A	A	C–	B
KNM-LT 23107	B	B	—	—	—	A	—	A
KNM-EK 4	B	B	A	A	A	A	—	A

C1: Relationship of lacrimal to the preorbital fossa: C = preorbital bar (POB) long with the anterior edge of the lacrimal placed more than half the distance from the anterior orbital rim to the posterior rim of the fossa; H = POB absent.

C2: Nasolacrimal fossa: C = nasomaxillary fossa absent (lost), leaving only nasolacrimal portion (when a POF is present).

C3: Orbital surface of lacrimal bone: A = with foramen.

C4: Preorbital fossa morphology: E = subtriangularly shaped and anteroposteriorly oriented; I = vestigial without C-shape outline, or absent.

C5: Fossa posterior pocketing: D = absent, no rim but a remnant depression; E absent.

C6: Fossa medial depth: A = deep, greater than 15 mm. in deepest place; C = shallow depth, less than 10 mm in deepest place.

C7: Preorbital fossa medial wall morphology: A = without internal pits.

C8: Fossa peripheral border outline: B = moderately delineated around periphery; D = absent with a remnant depression.

C9: Anterior rim morphology: B = absent.

C10: Placement of infraorbital foramen: B = inferior to, or encroaching on, anteroventral border of the preorbital fossa.

C11: Confluence of buccinator and canine fossae: A = present.

C12: Buccinator fossa: A = not pocketed posteriorly.

C13: Caninus (= intermediate) fossa: A = absent.

C14: Malar fossa: A = absent.

C15: Nasal notch position: C = at or near the anterior border of P^2.

C16: Presence of dP^1 (16U) or dP_1 (16L): A = persistent and functional; B = reduced and nonfunctional.

TABLE 9.10 Character State Distributions for Hipparion Upper Teeth from Lothagam and Selected Kenyan Sites

	Unit	Side	Tooth	C17	C18	C19	C20	C21	C22	C23	C24	C25	C26	C27	C28
Eurygnathohippus turkanense															
LT 136	Lr Nawata	Rt.	P^2	—	C	B/C	B	A	D	E	B	C	B	—	—
LT 136	Lr Nawata	Rt.	P^3	—	C	B/C	B	A	D	E	B	C	B	—	—
LT 136	Lr Nawata	Rt.	P^4	—	C	B/C	B	B	C	E	B	C	B	—	—
LT 136	Lr Nawata	Rt.	M^1	—	C	B/C	B	A	C	E/H	B	C	—	B	—
LT 136	Lr Nawata	Rt.	M^2	—	C	B/C	B	A	C	E/H	B	C	—	B	—
LT 136	Lr Nawata	Rt.	M^3	—	C	B/C	B	A	C	H	B	C	—	B	—
LT 25441	Ur Nawata	Lt.	P^2	—	—	C	B	B	D	E	—	—	—	—	—
LT 142	Ur Nawata	Rt.	P^3	C	C	B	B	B	C	E	B	C	B	—	—
LT 154	Ur Nawata	Lt.	P^3	C	C	A/B	A	A	A	E	B	C	C	—	—
LT 155	Ur Nawata	Lt.	P^3	C	—	B	B	B	B	F	B	C	B	—	—
LT 167	Ur Nawata	Rt.	P^3	C	—	A	B	A+	C	C/F	B	C	B	—	—
LT 168	Ur Nawata	Rt.	P^3	B/C	C	B	B	A	B	E	B	C	B	—	—
LT 25441	Ur Nawata	Rt.	P^3	—	—	C	B	B	D	E	—	—	—	—	—
LT 155	Ur Nawata	Rt.	P^4	C	—	B	B	B	C	E	B	C	B	—	—
LT 23107	Ur Nawata	Rt.	P^4	—	C	D	B	—	—	H	B	C	B	—	—
LT 25441	Ur Nawata	Rt.	P^4	—	—	C	B	B	D	E	—	—	—	—	—
LT 26293A	Ur Nawata	Rt.	P^4	C	?	A	B	A	C	I	B	C	B	—	—
LT 23107	Ur Nawata	Rt.	M^1	—	—	—	—	—	—	H	B	C	—	C	—
LT 25464	Ur Nawata	Rt.	M^1	C	?D	A	A	A+	C	J	A	C	—	B	—
LT 26293B	Ur Nawata	Lt.	M^1	C	?	A	B	—	—	—	—	—	—	—	—
LT 23106	Ur Nawata	Rt.	M^2	—	—	—	—	—	—	E	B	C	—	B	—
LT 25441	Ur Nawata	Lt.	M^2	—	—	C	B	B	D	E	—	—	—	—	—
LT 25441	Ur Nawata	Rt.	M^2	—	—	C	B	B	D	E	—	—	—	—	—
LT 25441	Ur Nawata	Lt.	M^3	—	—	C	B	B	C	E	—	—	—	—	—
LT 25441	Ur Nawata	Rt.	M^3	—	—	C	B	B	B	E	—	—	—	—	—
LT 23107	Ur Nawata	Rt.	M^3	—	—	—	B	—	—	H	B	C	—	B	—
LT 26163	Ur Nawata	Lt.	M^3	—	—	B	B	—	B	E	B	C	—	A	—

continued

TABLE 9.10 Character State Distributions for Hipparion Upper Teeth from Lothagam and Selected Kenyan Sites (Continued)

	Unit	Side	Tooth	C17	C18	C19	C20	C21	C22	C23	C24	C25	C26	C27	C28
Eurygnathohippus feibeli															
LT 23770	Lr Nawata	Rt.	dP^3	C	A	—	—	—	—	—	—	—	—	—	—
LT 23999	Lr Nawata	Lt.	dP^4	B	C	B	B	B	B	F	B	C	—	B	—
LT 25484	Lr Nawata	Rt.	P^3	B/C	B	B	B	A	B	E/G	B	C	B	—	—
LT 163	Lr Nawata	Lt.	P^4	B/C	C	B	B	B	—	F	—	—	B	—	—
LT 23685	Lr Nawata	Lt.	P^4	B/C	—	A	B	B	C	C!	B	C	B	—	—
LT 25473	Lr Nawata	Rt.	P^4	C	—	B	B	B	C	C	B	C	B	—	—
LT 25469	Lr Nawata	Rt.	P^4	B/C	—	A	B	B	C	C	B	B	—	B	—
LT 25486	Lr Nawata	Lt.	P^4	B	C	—	—	—	—	—	—	—	B	—	—
LT 163	Lr Nawata	Lt.	M^1	B/C	C	C	B	—	—	—	—	—	—	—	—
LT 23999	Lr Nawata	Lt.	M^1	B	C	B	B	B	B	F	B	C	—	B	—
LT 25477	Lr Nawata	Rt.	M^1	C	—	B	B	B+	C	C	B	C	—	B	—
LT 25486	Lr Nawata	Rt.	M^1	B	C	C	A	B	B	F	B	C	—	B	—
LT 23687	Lr Nawata	Rt.	M^2	B/C	C	B	B	A	C	F	B	B	—	B	—
LT 25468	Lr Nawata	Rt.	M^2	C	C	—	B	—	A	—	—	—	—	—	—
LT 25480	Lr Nawata	Rt.	M^2	B/C	C	C	C	—	B	—	F	B	B	—	—
LT 163	Lr Nawata	Lt.	M^3	B/C	C	—	—	—	—	—	—	—	—	—	—
LT 25482	Lr Nawata	Lt.	M^3	B/C	C	C	B	B	A	F	B	C	—	A	—
LT 25485	Lr Nawata	Lt.	M^3	C	C	C	A	B	A	F	B	C	—	A	—
LT 22867	Ur Nawata	Rt.	P^2	B/C	C	B	B	B	D	E	B	C	B	—	—
LT 22867	Ur Nawata	Rt.	P^2	B/C	C	B	B	B	D	E	B	C	B	—	—
LT 25436	Ur Nawata	Rt.	P^2	B/C	C	—	—	—	—	—	—	—	—	—	—
LT 25443	Ur Nawata	Rt.	P^2	B/C	C	B	B	A	C	G	B	C	B	—	—
LT 25454	Ur Nawata	Rt.	P^2	B/C	—	B/C	B	A	C	C	B	C	B	—	—
LT 22867	Ur Nawata	Rt.	P^3	B/C	C	B	B	B	C	E	B	C	B	—	—
LT 22867	Ur Nawata	Rt.	P^3	B/C	C	B	B	B	—	F	B	C	B	—	—
LT 25449	Ur Nawata	Rt.	P^3	B/C	C	B/C	B	—	B	C/J	B	C	B	—	—
LT 25460	Ur Nawata	Lt.	P^3	B/C	—	B/C	B	B	D	E	B	C	B	—	—
LT 22867	Ur Nawata	Rt.	P^4	B/C	C	B	B	B	C	F	B	C	B	—	—

LT 22867	Ur Nawata	Rt.	P^4	B/C	C	B	B	B	C	F	B	C	—	B	—
LT 22868	Ur Nawata	Rt.	P^4	C	C	A/B	B	B	C	F	B	C	B	—	—
LT 25431	Ur Nawata	Lt.	P^4	B/C	—	A	B	B	B	G	B	C	B	—	—
LT 25436	Ur Nawata	Lt.	P^4	B/C	C	B	B	—	—	—	—	—	—	—	—
LT 25439	Ur Nawata	Lt.	P^4	B/C	—	A	B	A	C	F	A	C	B	—	—
LT 25942	Ur Nawata	Rt.	P^4	B/C	C	B/C	B	B	A	F	B	C	—	—	—
LT 22868	Ur Nawata	Rt.	M^1	C	C	—	—	—	—	—	—	—	—	—	—
LT 25436	Ur Nawata	Lt.	M^1	B/C	C	—	—	—	—	—	—	—	—	—	—
LT 25443	Ur Nawata	Lt.	M^1	B/C	C	B	B	A	C	G	B	C	—	B	—
LT 25449	Ur Nawata	Rt.	M^3	B/C	C	B/C	B	—	—	—	—	—	—	—	—
LT 140	Ur Nawata	Rt.	M^3	B/C	C	C	B	A	B	F	B	C	—	B	—
LT 22867	Ur Nawata	Rt.	M^3	B/C	C	B	B	B	C	F	B	C	B	—	—
LT 25439	Ur Nawata	Rt.	M^3	B/C	—	B	B	A	A	F	A	C	—	B	—
LT 25461	Ur Nawata	Lt.	M^3	B/C	C	A	B	B+	A	E/C	B	A	—	B	—
Eurygnathohippus sp. indet. (large)															
LT 26295	Apak	Lt.	dP^3	—	—	—	—	—	—	C	A	C	B	—	—
LT 26295	Apak	Lt.	dP^4	—	—	C	B	—	C	C	B	C	B	—	—
LT 26295	Apak	Lt.	P^3	B	C	—	—	—	—	—	—	—	—	—	—
LT 26295	Apak	Lt.	P^4	B	C	—	—	—	—	—	—	—	—	—	—
LT 144	Apak	Rt.	P^4	B/C	—	A	B	A	B	C/E	B	C	B	—	—
LT 26295	Apak	Lt.	M^1	B	C	B	B	B	C	F	B	C	—	B	—
Eurygnathohippus sp. indet. (small)															
LT 26165	Kaiyumung	Rt.	dP^3	C	A	C	B	B	A	C	B	C	B	—	—
LT 26164	Kaiyumung	Rt.	P^2	B	C	B	B	A	B	D/E	B	C	B	—	A+
LT 24052	Kaiyumung	Lt.	P^3	C	C	A	B	—	—	—	—	—	—	—	—
LT 164	Kaiyumung	Lt.	P^4	B	C	B	B	—	C	F	B	C	B	—	—
LT 23998	Kaiyumung	Rt.	P^4	C	D	A	B	B	C	F	B	C	B	—	—
LT 25949	Kaiyumung	Rt.	M^3	C	C	A	B	—	—	B	C	—	B	—	—
LT 25949	Kaiyumung	Lt.	M^3	C	C	B	B	—	—	—	—	—	—	—	—
LT 23996	Apak	Rt.	P^4	C	—	B	B	—	C	F	A	C	B	—	—

continued

Table 9.10 Character State Distributions for Hipparion Upper Teeth from Lothagam and Selected Kenyan Sites (Continued)

	Unit	Side	Tooth	C17	C18	C19	C20	C21	C22	C23	C24	C25	C26	C27	C28
Eurygnathohippus sp. indet. (small)															
LT 25944	Apak	Lt.	P^4	B/C	D	C	B	A	C	F	A	—	B	—	—
LT 146	Apak	Rt.	M^1	C	C	B	B	B	B	F	B	C	—	B	—
LT 23689	Kaiyumung	Lt.	P^4	C	C	A	B	B	—	—	B	C	B	—	—
LT 23689	Kaiyumung	Lt.	M^1	C	C	A	B	B	C	F	B	C	—	B	—
LT 23689	Kaiyumung	Lt.	M^2	C	C	A	B	B	C	F	B	C	—	B	—
LT 23689	Kaiyumung	Lt.	M^3	C	C	A	B	B	C	F	B	C	—	B	—
Eurygnathohippus aff. *E. feibeli*															
KNM-EK4	Ekora	Rt.	dP^2	—	A	B	B	A	A	G	C	B	B	B	A
KNM-EK4	Ekora	Rt.	dP^3	—	A	B	B	A	B	G	C	C	B	B	—
KNM-EK4	Ekora	Rt.	dP^4	—	A	B	B	A	A	C	C	C	B	B	—
KNM-EK4	Ekora	Rt.	M^1	C	C	A/B	B	A	B	C	B	C	—	B	—
Eurygnathohippus sp. indet.															
LT 141	Apak	Rt.	P^3	B/C	C	B/C	B	B	A	F	B	—	—	—	—
LT 141	Apak	Lt.	P^4	B/C	C	B/C	B	B	B	F	B	—	—	—	—
LT 141	Apak	Lt.	M^1	B/C	C	B/C	B	—	B	—	—	—	—	—	—
KNM-AT150	Aterir	Lt.	P^2	B/C	C	A	B	A+	C	E/D	B	C	B	—	A

KNM-AT151	Aterir	Rt.	P^2	B/C	A	B	A	C	D	B	C	B	—	—	—
KNM-AT150	Aterir	Lt.	P^3	B/C	C	A	B	B	C	E/D	B	C	B	—	—
KNM-AT151	Aterir	Rt.	P^3	B/C	C	A	B	B	C	E	B	C	B	—	—
KNM-AT150	Aterir	Lt.	P^4	B/C	C	A	B	B	C	E/D	B	C	B	—	—
KNM-AT150	Aterir	Lt.	M^1	B/C	C	A	B	B	C	E/D	B	C	B	—	—
Hippotherium cf. *H. primigenium*															
LT 25447	Ur Nawata	Lt.	M^3	B	C	—	—	—	—	—	—	—	—	—	—

C17: Curvature of maxillary cheek teeth: A = very curved; B = moderately curved; C = straight.

C18: Maximum cheek tooth crown height: A = <30 mm; B = 30–40 mm; C = 40–60 mm; D = 60–75 mm; E = 75+ mm.

C19: Maxillary cheek tooth fossette ornamentation: A = complex, with several deeply amplified plications; B = moderately complex with fewer, more shortly amplified, thinly banded plications; C = simple complexity with few, shortly amplified plications; D = generally no plis; E = very complex.

C20: Posterior wall of postfossette: A = may not be distinct; B = always distinct.

C21: Pli caballin morphology: A = double; B = single or occasionally poorly defined double; C = complex; D = plis not well formed.

C22: Hypoglyph: A = hypocone frequently encircled by hypoglyph; B = deeply incised, infrequently encircled hypocone; C = moderately deeply incised; D = shallowly incised.

C23: Protocone shape: A = round q-shape; B = oval q-shape; C = oval; D = elongate-oval; E = lingually flattened-labially rounded; F = compressed or ovate; G = rounded; H = triangular; I = triangular-elongate; J = lenticular; K = triangular with rounded corners.

C24: Isolation of protocone: A = connected to protoloph; B = isolated from protoloph.

C25: Protoconal spur: A = elongate, strongly present; B = reduced, but usually present; C = very rare to absent.

C26: Premolar protocone/hypocone alignment: A = anteroposteriorly aligned; B = protocone more lingually placed.

C27: Molar protocone/hypocone alignment: A = anteroposteriorly aligned; B = protocone more lingually placed.

C28: P^2 anterostyle (28U)/paraconid (28L): A = elongate; B = short and rounded.

TABLE 9.11 Character State Distributions for Hipparion Lower Teeth from Lothagam and Ngorora

	Unit	Side	Tooth	C28	C29	C30	C31	C32	C33	C34	C35	C36	C37	C38	C39	C40	C41	C42	C43	C44	C45	C46	C47	C48	C49
Eurygnathohippus turkanense																									
LT 26166	Lr Nawata	Rt.	/C	—	—	—	—	—	—	—	—	—	—	—	—	—	—	—	—	—	—	—	—	—	—
LT 25457B	Lr Nawata	Lt.	P_2	—	—	—	—	A	—	A	B	—	—	B	—	C	B	—	B	B	—	A	A	A	A
LT 25446	Lr Nawata	Rt.	P_3	—	—	—	—	B	—	B	B	—	—	A	—	C	B	C	B	A	—	—	—	A	A
LT 25457A	Lr Nawata	Rt.	P_3	—	—	—	—	B	—	E	B	—	—	A	—	A	B	—	B	A	—	B	A	A	A
LT 25457B	Lr Nawata	Lt.	P_3	—	—	—	—	A	—	C	B	—	—	B	—	C	D	E	B	B	—	A	A	A	A
LT 25488	Lr Nawata	Lt.	P_3	—	—	—	—	D	—	B/C	B	—	—	A	—	B	B	C	B	B	—	B	B	A	A
LT 25430	Lr Nawata	Rt.	P_4	—	—	—	—	A/C	—	C	B	—	—	A	—	C	B	C	B	B+	—	A	A	A	A
LT 25438	Lr Nawata	Rt.	P_4	—	—	—	—	—	—	—	—	—	—	A	—	A	D	E	B	—	—	B	B	A	—
LT 22869	Lr Nawata	Lt.	M_1	—	—	—	—	—	A/C	—	—	C	B	—	B	C	F	E	B	—	D–	B–	A	A	A
LT 25434	Lr Nawata	Rt.	M_1	—	—	—	—	—	A	—	—	A/C	B	—	B	C	D	E	B	—	B+	A	A	A	A
LT 25435	Lr Nawata	Lt.	M_1	—	—	—	—	—	D	—	—	E	A	—	A	—	—	—	B	—	A	A	A	A	—
LT 25457A	Lr Nawata	Lt.	M_1	—	—	—	—	—	B	—	—	E	B	—	B	B+	D	E	B	—	D	A	A	A	A
LT 25459	Lr Nawata	Rt.	M_1	—	—	—	—	—	—	—	—	—	—	—	—	—	D	E	—	—	—	—	—	—	—
LT 22869	Lr Nawata	Lt.	M_2	—	—	—	—	—	A/C	—	—	C	B	—	B	B	E	B	B	—	D–	A	A	A	A
LT 25435	Lr Nawata	Lt.	M_2	—	—	—	—	—	D	—	—	E	A	—	B	—	—	—	B	—	A	B–	B–	A	—
LT 25458	Lr Nawata	Lt.	M_2	—	—	—	—	—	B	—	—	B	B	—	B	B	D	E	B	—	D	A	A	A	A
LT 25434	Lr Nawata	Rt.	M_3	—	—	—	—	—	A	—	—	E/C	B	—	B	C	D	E	B	—	B+	A	A	A	A
LT 25442	Lr Nawata	Rt.	M_3	—	—	—	—	—	—	—	—	—	—	—	—	—	D	E	—	—	—	—	—	—	—
LT 25457A	Lr Nawata	Rt.	M_3	—	—	—	—	—	B	—	—	B	B	—	B	C	B	—	B	—	D	A	A	A	B
LT 25474	Ur Nawata	Lt.	P_3	—	—	—	—	A	—	C	B	—	—	A	—	B	D	E	B	B+	—	A	A	A	B–
LT 23684	Ur Nawata	Lt.	M_1	—	—	—	—	—	A	—	—	C	B	—	B	C	E	E	A	—	D	B	B	B	A
LT 25434	Ur Nawata	Rt.	M_1	—	—	—	—	—	A/C	—	—	C	B	—	A	B–	—	—	B	—	B	A	A	A	A
LT 25475	Ur Nawata	Rt.	M_1	—	—	—	—	—	B	—	—	B/C	B	—	B	C	B	C	B	—	B	B	A	A	B
LT 23684	Ur Nawata	Lt.	M_2	—	—	—	—	—	A/C	—	—	C	B	—	B	C	A	C	A	—	D	A	B	B	A
LT 25939	Ur Nawata	Rt.	M_2	—	—	—	—	—	A	—	—	E	B	—	A	C	A	C	B	—	B+	A	A	A	A
LT 23684	Ur Nawata	Lt.	M_3	—	—	—	—	—	B/C	—	—	B	B	—	B	C	F	C	B	—	B+	A	B	B	A
LT 23684	Ur Nawata	Rt.	M_3	—	—	—	—	—	B	—	—	A	B	—	A	B	E	C	B	—	B+	A	B	B	A
LT 25467	Ur Nawata	Rt.	M_3	—	—	—	—	—	B	—	—	B	B	—	A	C	B	C	B	—	B	A	A	A	A

LT 23686	Apak	Rt.	M_1	—	—	—	—	—	B	—	—	E	B	—	B	C	—	—	B	—	D	B–	A	A	A
LT 24000	Apak	Rt.	M_1	—	—	—	—	—	B	—	—	E	B	—	B	B	—	—	B	—	B+	B	A	A	—
Eurygnathohippus feibeli																									
LT 23995	Lr Nawata	Lt.	P_2	—	—	—	—	B	—	A	B	—	—	A	—	C	—	—	B	B	—	A	A	A	A
LT 25463	Lr Nawata	Rt.	M_3	—	—	—	—	—	B	—	—	B/C	—	—	A	C	B	C	B	—	B+	B–	A	A	A
LT 25479	Lr Nawata	Lt.	M_3	—	—	—	—	—	—	—	—	—	—	—	—	—	—	—	—	—	—	—	—	—	—
LT 25477	Ur Nawata	Lt.	dP_3	—	—	—	—	B	—	A	B	—	—	B	—	B	D	E	A	E	—	A	A	A	A
LT 25476	Ur Nawata	Lt.	P_2	—	—	—	—	A	—	A	B	—	—	A	—	C	B	C	B	B	—	A	A	A	A
LT 25469	Ur Nawata	Rt.	P_4	—	—	—	—	—	—	B	B	—	—	A	—	B–	—	—	A	—	—	A	A	A	A
LT 392	Ur Nawata	Lt.	P_4	—	—	—	—	—	—	E	B	—	—	—	—	—	A	C	B	B+	—	A	A	A	A
LT 25476	Ur Nawata	Rt.	M_1	—	—	—	—	—	A	—	—	A	B	—	B	C	D	E	A	—	D	A	A	A	A
LT 25477	Ur Nawata	Lt.	M_1	—	—	—	—	—	—	—	—	—	—	—	—	—	B	C	A	—	—	—	—	—	—
LT 392	Ur Nawata	Lt.	M_1	—	—	—	—	—	D	—	—	E	B	—	A	C	—	—	B	—	B+	B	A	A	A
LT 25476	Ur Nawata	Lt.	M_3	—	—	—	—	—	B	—	—	B	B	A	A	C	F	C	B	—	B+	A	A	A	A
LT 25476	Ur Nawata	Rt.	M_3	—	—	—	—	—	B	—	—	B	B	—	A	C	—	—	B	—	B	A	A	A	A
Eurygnathohippus sp. indet. (small)																									
LT 23996	Apak	Rt.	M_1	—	—	—	—	—	A/C	—	—	A/C	B	—	B	C	A	C	?A	—	B+	A	A	A	A
LT 25946	Apak	Lt.	M_2	—	—	—	—	—	A/C	—	—	C	B	—	A	C	—	—	B	—	B+	A	A	A	A
Eurygnathohippus sp. indet. (large)																									
LT 24001	Apak	Lt.	I_1	—	B	A	—	—	—	—	—	—	—	—	—	—	—	—	—	—	—	—	—	—	—
LT 24001	Apak	Lt.	I_2	—	B	A	—	—	—	—	—	—	—	—	—	—	—	—	—	—	—	—	—	—	—
LT 24001	Apak	Lt.	I_3	—	A	A	A/B	—	—	—	—	—	—	—	—	—	—	—	—	—	—	—	—	—	—
LT 24001	Apak	Lt.	/C	—	—	—	—	—	—	—	—	—	—	—	—	—	—	—	—	—	—	—	—	—	—
LT 26292	Apak	Rt.	dP_3	—	—	—	—	—	—	—	—	—	—	—	—	—	—	—	—	—	—	—	—	—	—
LT 24001	Apak	Lt.	P_2	A+	—	—	—	B	—	A	B	—	—	A–	—	C	B	—	B	A	—	A	A	A	A
LT 24001	Apak	Rt.	P_2	A+	—	—	—	B	—	A	—	C	—	A–	—	B	B	—	B	B	—	A	B	A	B
LT 25943	Apak	Rt.	P_2	—	—	—	—	A	—	A	B	—	—	A	—	B	B	—	B	B	—	A	B	B	—
LT 25943	Apak	Lt.	P_3	—	—	—	—	A	—	A/C	B	—	—	A	—	C	A	C	B	B	—	B	B	B	B
LT 24001	Apak	Lt.	P_3	—	—	—	—	B	—	E	B	—	—	A–	—	B	F	C	A	B	—	B–	B	A	A
LT 24001	Apak	Lt.	P_4	—	—	—	—	A	—	C	A	—	—	A–	—	B	BF	C	A	B	—	B–	B	A	B
LT 24001	Apak	Rt.	P_4	—	—	—	—	A	—	C	—	C	—	A–	—	B	F	C	A	D–	—	B	B	A	A

continued

Table 9.11 Character State Distributions for Hipparion Lower Teeth from Lothagam and Ngorora (Continued)

	Unit	Side	Tooth	C28	C29	C30	C31	C32	C33	C34	C35	C36	C37	C38	C39	C40	C41	C42	C43	C44	C45	C46	C47	C48	C49
Eurygnathohippus sp. indet. (large)																									
LT 25943	Apak	Lt.	M_1	—	—	—	—	—	B	—	—	A/C	B	—	B	C	—	—	B	—	D	A	A	B	—
LT 24001	Apak	Lt.	M_1	—	—	—	—	—	A/C	—	—	C	B	—	B	B	A–	C	?	—	D–	B–	A	A	A
LT 24001	Apak	Rt.	M_1	—	—	—	—	—	A/C	—	C	—	B	—	B	C	A	C	A	—	D–	B–	A	A	A
LT 25943	Apak	Lt.	M_2	—	—	—	—	—	B	—	—	C	B	—	B	C	B	—	B	—	D–	A	B–	B	B–
LT 24001	Apak	Lt.	M_2	—	—	—	—	—	B	—	—	C	B	—	A!	B	A	C	A	—	D–	B–	B	A	B
LT 24001	Apak	Rt.	M_2	—	—	—	—	—	A/C	—	C	—	B	—	A!	B	A	C	A	—	D–	A	B	A	B–
LT 25943	Apak	Rt.	M_3	—	—	—	—	—	A	—	—	A/C	B	—	A	C	B	—	B	—	A	B	A	B	B–
LT 24001	Apak	Lt.	M_3	—	—	—	—	—	B	—	—	E	A	—	A!	B	F	C	?	—	D–	B–	B	A	B
LT 25948	Kai'mung	Lt.	P_4	—	—	—	—	CF	—	CF	B	—	—	—	B	B	A	C	A	E	—	A	A	A	B
LT 24947	Kai'mung	Lt.	M_1	—	—	—	—	—	A	—	—	C	B	A	—	B	—	—	A	—	B+	A	A	A	B
KNM-BN 1117	Ngorora	Lt.	P_2	—	—	—	—	A	—	A	B	—	—	A	—	C	B	—	B	B	—	A	A/B	A	A
KNM-BN 1117	Ngorora	Lt.	P_3	—	—	—	—	A	—	A	B	—	—	A	—	C	D	E	B	B	—	B	B/C	A	A
KNM-BN 1117	Ngorora	Rt.	P_3	—	—	—	—	A	—	A	B	—	—	A	—	C	D	E	B	B	—	B	B/C	A	A
KNM-BN 1117	Ngorora	Lt.	P_4	—	—	—	—	A	—	A	B	—	—	A	—	C	D	E	B	B	—	B	B	A	A
KNM-BN 1117	Ngorora	Rt.	P_4	—	—	—	—	A	—	A	B	—	—	A	—	C	D	E	B	B	—	B	B/C	A	A
KNM-BN 1117	Ngorora	Lt.	M_1	—	—	—	—	—	A	—	—	A	B	—	B	C	D	E	B	—	C	B	A	A	A

C28: P_2 anterostyle (28U)/paraconid (28L) : A = elongate; B = short and rounded.
C29: Mandibular incisor morphology: A = not grooved; B = grooved.
C30: Mandibular incisor curvature: A = curved; B = straight.

C31: 1_3 lateral aspect: A = elongate, not labiolingually constricted; B = very elongate, labiolingually constricted distally; C = atrophied.

C32: Premolar metaconid: A = rounded; B = elongated; C = angular on distal surface; D = irregular shaped; E = square shaped; F = pointed.

C33: Molar metaconid: A = rounded; B = elongated; C = angular on distal surface; D = irregular shaped; E = square shaped; F = pointed.

C34: Premolar metastylid: A = rounded; B = elongate; C = angular on proximal surface; D = irregular shaped; E = square shaped; F = pointed.

C35: Premolar metastylid spur: A = present; B = absent.

C36: Molar metastylid: A = rounded; B = elongate; C = angular on proximal surface; D = irregular shaped; E = square shaped; F = pointed.

C37: Molar metastylid spur: A = present; B = absent.

C38: Premolar ectoflexid: A = does not separate metaconid and metastylid; B = separates metaconid and metastylid.

C39: Molar ectoflexid: A = does not separate metaconid and metastylid; B = separates metaconid and metastylid; C = converges with preflexid and postflexid to abut against metaconid and metastylid.

C40: Pli caballinid: A = complex; B = rudimentary or single; C = absent.

C41: Protostylid: A = present on occlusal surface often as an enclosed enamel ring; B = absent on occlusal surface, but may be on side of crown buried in cement; C = strong, columnar; D = a loop; E = a small, poorly developed loop; F = a small, pointed projection continuous with the buccal cingulum.

C42: Protostylid orientation: A = courses obliquely to anterior surface of tooth; B = less oblique coursing, placed on anterior surface of tooth; C = vertically placed, lies flush with protoconid enamel band; D = vertically placed, lying lateral to protoconid band; E = open loop extending posterolabially.

C43: Ectostylids: A = present; B = absent.

C44: Premolar linguaflexid: A = shallow; B = deeper, V-shaped; C = shallow, U-shaped; D = deep, broad U-shape; E = very broad and deep.

C45: Molar linguaflexid: A = shallow; B = V-shaped; C = shallow, U-shaped; D = deep, broad U-shape; E = very broad and deep.

C46: Preflexid morphology: A = simple margins; B = complex margins; C = very complex.

C47: Postflexid morphology: A = simple margins; B = complex margins; C = very complex.

C48: Postflexid invades metaconid/metastylid junction by anteriormost portion bending sharply lingually: A = no; B = yes.

C49: Protoconid enamel band morphology: A = rounded; B = flattened.

Table 9.12a Upper Teeth Measurements of *Eurygnathohippus feibeli*

Acc. No.	Member	Side	Specimen	M1	M2	M3	M4	M5	M6	M7	M8	M9	M10	M11
LT 23770	Ur Nawata	Lt.	dP^3	29	26.6	20	21	23.9	—	—	—	—	8.8	—
LT 23999	Ur Nawata	Lt.	dP^4	25.8	—	—	—	9.4	1	9	1	0	8	7.5
LT 25436	Lr Nawata	Rt.	P^2	31.1	30.5	20.3	20.6	46.2	—	—	—	—	7.8	—
LT 25454	Lr Nawata	Rt.	P^2	32.4	27.7	21.5	19	34.5	3	3	4	2	7.5	4
LT 25449	Lr Nawata	Rt.	P^3	—	—	—	—	37	1	—	2	2	—	—
LT 25484	Ur Nawata	Rt.	P^3	24.3	22	23.7	21.9	38.9	2	5	4	1	6.4	3.7
LT 22868	Lr Nawata	Rt.	P^4	24.5	19.5	21.6	22.8	44.9	1	—	3	1	8.1	3.4
LT 25431	Lr Nawata	Lt.	P^4	23.7	21.4	23.9	23.5	32.6	1	5	4	2	6.5	5.2
LT 25436	Lr Nawata	Lt.	P^4	25.3	21.6	22.3	20.9	46.9	—	—	—	—	8.4	—
LT 25439	Lr Nawata	Lt.	P^4	23.2	19.8	23.9	21.2	31.4	3	4	4	1	7.5	3.6
LT 163	Ur Nawata	Lt.	P^4	26.2	23	—	—	49.2	—	—	—	—	—	—
LT 25469	Ur Nawata	Lt.	P^4	24.9	22.1	20.8	18.5	36.5	8	6	5	3	7.5	4.7
LT 25473	Ur Nawata	Rt.	P^4	20.2	18.6	21.3	20.9	21.6	2	5	2	2	6.5	4
LT 25486	Ur Nawata	Lt.	P^4	27.3	23.8	24.5	—	48.6	—	—	—	—	9.2	—
LT 25942	Ur Nawata	Lt.	P^4	23.5	21.4	20	22.3	45.7	1	4	2	1	6.9	3
LT 22868	Lr Nawata	Rt.	M^1	23.9	—	21.9	—	56.9	—	—	—	—	9.7	—
LT 25436	Lr Nawata	Lt.	M^1	26.1	20.1	—	—	49.7	0	3	2	0	—	—
LT 163	Ur Nawata	Lt.	M^1	24.2	19.8	—	—	46.7	1	5	1	1	—	—
LT 23999	Ur Nawata	Lt.	M^1	23.6	—	21.8	—	56.7	1	5	3	1	4.9	3.1
LT 25477	Ur Nawata	Rt.	M^1	19.7	18.6	21.4	21.1	15	1	6	3	1	7.8	3.8
LT 25486	Ur Nawata	Rt.	M^1	—	—	—	—	—	—	—	—	—	—	—
LT 25480	Lr Nawata	Rt.	M^2	—	—	21.5	22.2	52.5	1	4	1	—	6.7	4.1
LT 23687	Ur Nawata	Rt.	M^2	21.8	20	21.7	22	51.6	1	4	3	1	6.3	3.5
LT 25468	Ur Nawata	Rt.	M^2	22.6	21.8	20.6	22	46.4	—	—	—	—	8.1	—
LT 140	Lr Nawata	Rt.	M^3	19.6	20.3	15.9	18	41.5	1	4	3	0	6.6	2.8
LT 25439	Lr Nawata	Rt.	M^3	19.3	19.5	18.9	19.5	31	1	4	3	1	7.9	3.4
LT 25449	Lr Nawata	Rt.	M^3	22.2	24.7	20.8	22.8	33.9	1	—	0	3	6.9	3.5
LT 25461	Lr Nawata	Lt.	M^3	21.5	18.5	19.8	18.5	38.3	4	7	3	1	6.6	4.1
LT 25482	Ur Nawata	Lt.	M^3	21.8	21.6	16.1	19	41.4	1	—	—	1	6.4	2.9
LT 25485	Ur Nawata	Rt.	M^3	20.7	19.4	15.6	18.3	37.8	1	1	1	2	8.3	3

M1 = occlusal length.
M2 = length at 10 mm above roots.
M3 = occlusal width at mesostyle/protocone.
M4 = width at 10 mm above roots.
M5 = crown height along mesostyle.
M6 = number of plications on anterior face of prefossette.
M7 = number of plications on posterior face prefossette.
M8 = number plications on anterior face postfossette.
M9 = number plications on posterior face postfossette.
M10 = protocone length.
M11 = protocone width.
Source: Measurements from Bernor et al. (1997).

TABLE 9.12B Lower Teeth Measurements of *Eurygnathohippus feibeli*

Acc. No.	Member	Side	Specimen	M1	M2	M3	M4	M5	M6	M7	M8	M9	M10	M11	M12
LT 25477	Ur Nawata	Lt.	dP_3	24.3	21.8	14.1	7.6	7.7	9.8	12.2	9	7.4	12.7	10.5	8
LT 23995	Lr Nawata	Lt.	P_2	26.7	24.8	10.8	8.5	11	11.7	11.6	9	11.6	30.4	—	—
LT 25476	Ur Nawata	Lt.	P_2	23.4	22.8	9.8	6.7	9.6	14.5	11.5	10.1	11.1	17.7	—	—
LT 25469	Ur Nawata	Rt.	P_4	23.5	20.9	—	8.1	12.1	—	13.1	—	12.3	23.2	—	—
LT 392	Ur Nawata	Lt.	P_4	23.4	21.7	12.3	6.7	9.6	—	12.4	9.3	9	36.5	37	—
LT 25476	Ur Nawata	Rt.	M_1	21.2	19.3	12.1	5.6	7.2	14.9	12.6	10.3	10	24.9	28.1	—
LT 25477	Ur Nawata	Rt.	M_1	24.1	20	10.5	—	—	9.7	13.2	—	—	51.3	47.7	—
LT 25463	Lr Nawata	Rt.	M_3	23.2	25.8	9.5	7.2	7.7	10.3	11.3	8.5	7.4	30.3	33.4	—
LT 25476	Ur Nawata	Lt.	M_3	23.1	23.6	11	6.3	8.7	11.8	10.2	9	8.5	33.4	31.2	—
LT 25476	Ur Nawata	Rt.	M_3	22.9	25	11.6	6.8	9	11.8	10	9.1	7.3	32.8	33.4	—
LT 25479	Ur Nawata	Lt.	M_3	23.1	—	10.1	—	—	—	—	5.7	5.1	49.3	—	—

M1 = occlusal length.
M2 = length 10 mm above roots.
M3 = length of metaconid-metastylid.
M4 = length of the prefossettid.
M5 = length of the postfossettid.
M6 = width of tooth across plane of ectoflexid/linguaflexid.
M7 = width 10 mm above root.
M8 = width across plane of metaconid and enamel band labial to protoconid.
M9 = width across plane of metastylid and enamel band labial to hypoconid.
M10 = crown height on mesial face of tooth.
M11 = length ectostylid.
M12 = width ectostylid.
Source: Measurements from Bernor et al. (1997:41).

Table 9.12c Postcranial Measurements of *Eurygnathohippus feibeli*

Acc. No.	Member	Side	Specimen	M1	M2	M3	M4	M5	M6	M7[a]							
139	Ur Nawata	Lt.	Radius	—	—	—	—	58.7	31.8	60.7							
				M1	**M2**	**M3**	**M4**	**M5**	**M6**	**M7[b]**							
25938	Ur Nawata	Lt.	Astrag	50.0	51.2	22.9	48.5	38.3	27.1	41.0							
25453	Lr Nawata	Lt.	Astrag	50.3	49.5	24.4	46.1	37.0	25.3	37.1							
				M1	**M2**	**M3**	**M4**	**M5**	**M6**	**M7**	**M8**	**M9**	**M10**	**M11**	**M12**	**M13**	**M14[c]**
25450	Lr Nawata	Lt.	Mc III	—	—	—	—	—	—	—	—	—	38.8	34.6	28.8	23.0	24.8
139A	Ur Nawata	Lt.	Mc III	230.8	221.8	25.8	22.4	38.6	25.7	34.6	11.0	5.8	34.3	34.0	28.8	24.8	26.8
26582	Ur Nawata	Rt.	Mc III	—	—	—	—	—	—	—	—	—	32.4	33.5	26.9	23.0	24.1
				M1	**M2**	**M3**	**M4**	**M5**	**M6**	**M7**	**M8[c]**						
25451	Lr Nawata	Lt.	Mt III	—	—	29.5	26.7	42.0	34.3	41.6	8.5						
				M1	**M2**	**M3**	**M4**	**M5**	**M6**	**M7**	**M8**	**M9**	**M10**	**M11**	**M12**	**M13[d]**	
139B	Ur Nawata	Lt.	a1ph3	64.6	60.1	24.3	37.2	28.4	29.3	30.7	19.2	30.8	44.5	45.8	13.3	13.7	
25472	Ur Nawata	Lt.	a1ph3	65.0	59.1	24.4	36.3	28.8	30.4	30.7	18.8	31.4	48.0	46.8	15.6	13.3	
				M1	**M2**	**M3**	**M4**	**M5**	**M6[e]**								
139C	Ur Nawata	Lt.	a2ph3	40.5	31.0	28.2	35.7	24.5	32.5								
5472	Ur Nawata	Lt.	a2ph3	39.7	30.5	28.9	36.3	24.9	34.2								

[a] M1 = maximum length; M2 = medial length; M3 = minimum breadth; M4 = depth diaphysis at minimum breadth; M5 = proximal articular breadth; M6 = proximal articular depth; M7 = maximal proximal breadth.

[b] M1 = maximum length; M2 = maximum diameter of the medial condyle; M3 = breadth trochlea at the apex of each condyle; M4 = maximum breadth; M5 = distal articular breadth; M6 = distal articular depth; M7 = maximum medial depth.

[c] M1 = maximum length; M2 = medial length; M3 = minimum breadth; M4 = depth diaphysis at M3; M5 = proximal articular breadth; M6 = proximal articular depth; M7 = maximum diameter third carpal (tarsal) articular facet; M8 = diameter fourth carpal (tarsal) facet; M9 = diameter second carpal facet; M10 = maximum distal supraarticular breadth; M11 = maximum distal articular breadth; M12 = maximum distal depth of keel; M13 = minimum depth distal lateral condule; M14 = maximum depth distal medial conduyle.

[d] M1 = maximum length; M2 = anterior length; M3 = minimum breadth; M4 = proximal breadth; M5 = proximal depth; M6 = distal breadth at the tuberosities; M7 = distal articular breadth; M8 = distal articular depth; M9 = minimum length of the trigonum phalangis; M10 = medial supratuberosital length; M11 = lateral supratuberosital length; M12 = medial infratuberosital length; M13 = lateral infratuberosital length.

[e] M1 = maximum length; M2 = anterior length; M3 = minimum breadth; M4 = proximal maximum breadth; M5 = proximal maximum depth; M6 = distal articular maximum breadth.

Source: Parameters from Eisenmann et al. (1988).

TABLE 9.13 Tooth Measurements of *Eurygnathohippus* sp. indet. (Large)

Acc. No.	Member	Side	Specimen	M1	M2	M3	M4	M5	M6	M7	M8	M9	M10	M11[a]	
LT 26165	Kaiyumung	Rt.	dP^3	30.2	29	24.3	26.1	17.9	—	—	—	—	6.7	4.1	
LT 26164	Kaiyumung	Rt.	P^2	37	—	25.4	24	47	4	2	1	1	8.5	4.2	
LT 24052	Kaiyumung	Lt.	P^3	24.6	22.9	24.8	23.5	38.7	—	4	3	3	8.7	3.5	
LT 164	Kaiyumung	Lt.	P^4	28	—	—	—	34.7	2	4	3	2	—	—	
LT 23998	Kaiyumung	Rt.	P^4	28.8	22.1	26.5	24.7	66.3	4	5	2	3	9.2	4.6	
LT 25949	Kaiyumung	Rt.	M^3	24.1	22.5	21.7	22.7	46.1	4	5	1	1	6.3	4.6	
LT 25949	Kaiyumung	Lt.	M^3	23.4	22.3	—	—	43.8	4	3	3	3	—	—	
				M1	**M2**	**M3**	**M4**	**M5**	**M6**	**M7**	**M8**	**M9**	**M10**	**M11**	**M12[b]**
LT 25948	Kaiyumung	Lt.	P_4	27.2	24.8	17.4	9.9	14.1	14.3	15.2	14.3	13.8	54.5	53.3	54.3
LT 24947	Kaiyumung	Lt.	M_1	29	—	13.5	8.7	10	15	—	13.1	12.7	61	—	5.1

[a] M1 = occlusal length; M2 = length at 10 mm above roots; M3 = occlusal width at mesostyle/protocone; M4 = width at 10 mm above roots; M5 = crown height along mesostyle; M6 = number of plications on anterior face of prefossette; M7 = number of plications on posterior face prefossette; M8 = number plications on anterior face postfossette; M9 = number plications on posterior face postfossette; M10 = protocone length; M11 = protocone width (parameters from Bernor et al. 1997).

[b] M1 = occlusal length; M2 = length 10 mm above roots; M3 = length of metaconid-metastylid; M4 = length of the prefossettid; M5 = length of the postfossettid; M6 = width of tooth across plane of ectoflexid/linguaflexid; M7 = width 10 mm above root; M8 = width across plane of metaconid and enamel band labial to protoconid; M9 = width across plane of metastylid and enamel band labial to hypoconid; M10 = crown height on mesial face of tooth; M11 = length ectostylid; M12 = width ectostylid (parameters from Bernor et al. 1997:41).

TABLE 9.14 Tooth Measurements of *Eurygnathohippus* sp. indet. (Small)

Acc. No.	Member	Side	Specimen	M1	M2	M3	M4	M5	M6	M7	M8	M9	M10	M11[a]
LT 141	Apak	Rt.	P^3	—	—	25.1	24.2	47.5	—	4.0	2.0	1.0	—	—
LT 141	Apak	Lt.	P^4	22.1	20.3	21.7	22.6	52.8	1.0	4.0	1.0	1.0	7.0	3.6
LT 23966	Apak	Rt.	P^4	23.7	20.2	24.1	25.4	34.1	3.0	6.0	4.0	1.0	7.1	4.0
LT 25944	Apak	Lt.	P^4	24.9	21.7	23.4	24.3	58.3	0.0	2.0	0.0	0.0	8.3	3.8
LT 141	Apak	Lt.	M^1	—	22.0	—	—	53.3	—	2.0	1.0	1.0	—	—
LT 146	Apak	Rt.	M^1	21.7	21.3	18.6	22.3	46.8	1.0	4.0	2.0	1.0	7.3	3.1
LT 23689	Kaiyumung	Lt.	P^4	24.0	—	25.2	—	46.4	—	7.0	4.0	3.0	8.0	—
LT 23689	Kaiyumung	Lt.	M^1	23.2	—	24.3	—	48.0	5.0	8.0	7.0	3.0	6.9	3.8
LT 23689	Kaiyumung	Lt.	M^2	22.3	—	21.7	—	47.0	5.0	5.0	6.0	3.0	6.9	3.4
LT 23689	Kaiyumung	Lt.	M^3	20.7	—	19.2	—	48.0	4.0	3.0	3.0	2.0	8.4	3.3

				M1	M2	M3	M4	M5	M6	M7	M8	M9	M10	M11[b]
LT 23996	Apak	Rt.	M_1	22.4	20.1	12.1	7.2	10.1	11.6	10.7	10.3	10.3	34.4	37.0
LT 24000	Apak	Rt.	M_1	24.8	23.0	15.8	8.3	9.3	12.9	11.9	—	10.5	52.2	—
LT 25946	Apak	Lt.	M_2	23.9	22.5	12.7	6.4	10.3	9.1	11.1	10.3	10.1	33.1	—

[a] M1 = occlusal length; M2 = length at 10 mm above roots; M3 = occlusal width at mesostyle/protocone; M4 = width at 10 mm above roots; M5 = crown height along mesostyle; M6 = number of plications on anterior face of prefossette; M7 = number of plications on posterior face prefossette; M8 = number plications on anterior face postfossette; M9 = number plications on posterior face postfossette; M10 = protocone length; M11 = protocone width (parameters from Bernor et al. 1997).

[b] M1 = occlusal length; M2 = length 10 mm above roots; M3 = length of metaconid-metastylid; M4 = length of the prefossettid; M5 = length of the postfossettid; M6 = width of tooth across plane of ectoflexid/linguaflexid; M7 = width 10 mm above root; M8 = width across plane of metaconid and enamel band labial to protoconid; M9 = width across plane of metastylid and enamel band labial to hypoconid; M10 = crown height on mesial face of tooth; M11 = length ectostylid; M12 = width ectostylid (parameters from Bernor et al. 1997:41).

TABLE 9.15 Measurements of *Hippotherium* cf. *H. primigenium*

Acc. No.	Member	Side	Specimen	M1	M2	M3	M4	M5								
LT 25447	Lr Nawata	Lt.	M^3	24.5	20.3	—	—	51.6								
				M1	**M2**	**M3**	**M4**	**M5**	**M6**	**M7**	**M8**	**M9**				
LT 160	Lr Nawata	Rt.	Tibia	—	—	—	—	—	—	70.0	43.5	—				
LT 25448	Lr Nawata	Rt.	Tibia	—	—	—	—	—	102.0	—	—	56.1				
LT 26621	Ur Nawata	Lt.	Tibia	355.0	335.4	41.5	33.1	88.4	77.7	64.0	44.8	47.0				
				M1	**M2**	**M3**	**M4**	**M5**	**M6**	**M7**						
LT 25941	Ur Nawata	Rt.	Astragalus	52.1	52.4	27.0	51.6	39.5	28.3	41.8						
				M1	**M2**	**M3**	**M4**	**M5**	**M6**	**M7**	**M8**	**M9**	**M10**	**M11**	**M12**	**M13**
LT 25465	Lr Nawata	Lt.	a1ph3	60.9	54.7	32.0	43.6	33.3	34.3	35.3	19.4	27.0	46.5	39.5	14.6	14.9

Note: Abbreviations as in tables 9.8B and 9.8D.

10

HIPPOPOTAMIDAE AND SUIDAE

10.1

Fossil Hippopotamidae from Lothagam

Eleanor M. Weston

Hippos are the most frequently preserved mammals in the Lothagam assemblage, constituting 27 percent of the mammalian fauna. Several different species of hippo occur at Lothagam, all of which can be assigned to the genus *Hexaprotodon*. Much of the material can be attributed to *Hexaprotodon harvardi*. The new specimens throw light on the intraspecific variability, sexual dimorphism, and ontogeny of this extinct hippopotamus. There is a trend toward body size increase and extreme cursoriality in hippos, which has been interpreted as part of an evolving lineage of *Hexaprotodon harvardi*. A small proportion of the material, recovered from the lower member of the Nawata Formation, belongs to a new species of small- to medium-sized hippopotamus, *Hexaprotodon lothagamensis*, which possessed a relatively long, narrow muzzle. Several specimens described indicate that a third, extremely large, species of hippo also existed at Lothagam. Hippopotamid material from the Apak Member of the Nachukui Formation compares with *Hexaprotodon protamphibius* known from elsewhere in the Turkana Basin. Lothagam provides the most complete record to date of the morphology and diversity of early hippopotamids. During the course of evolution, some hippos underwent a transition from cursorial browsers to semiaquatic grazers, and *Hexaprotodon harvardi* appears to mark the beginning of this switch in diet and lifestyle.

Hippos are the most frequently preserved mammals in the Lothagam assemblage. The sample is comprised of more than 300 specimens that constitute 14 percent of the fossil vertebrate fauna and 27 percent of the mammalian fauna. Several different species of hippo are present in the Lothagam assemblage. Most of the material is from the Nawata Formation and ranges in age from 7.91 to 5.23 Ma (Powers 1980; Leakey et al. 1996; McDougall and Feibel 1999). A small portion of the fossil sample, 18 specimens of known provenance, is from the overlying Apak Member of the Nachukui Formation, estimated to range in age from 5.23 to 4.22 Ma (McDougall and Feibel 1999).

The dominant species at Lothagam is *Hexaprotodon harvardi* (Coryndon 1977), which is present in both the lower and upper members of the Nawata Formation. *Hexaprotodon harvardi* has been documented at a number of other Late Miocene and Early Pliocene localities in East Africa (Harrison 1997; Kalb et al. 1982a, 1982b). However, complete crania and associated postcranial elements are known only from Lothagam. The addition of over 100 new specimens of accurate provenance not only vastly improves our knowledge of this species but also permits correlation with the fragmentary sequences of hippopotamid remains known from other localities, such as those in the Tugen Hills in the Kenyan Rift Valley (Hill 1995; Coryndon 1978). The holotype of *Hexaprotodon harvardi* is a large cranium that was collected from the upper part of the Nawata sequence, whereas much of the new material has been collected from the lower part of the Nawata sequence; some of it being at least 7.4 Ma in age. The morphological variation that exists between early and late *Hexaprotodon harvardi* is described here for the first time. Not only does this species increase slightly in overall size through the succession, but also there is evidence to suggest that it was sexually dimorphic.

A second and contemporaneous species, *Hexaprotodon lothagamensis* (Weston 2000) which is distinctly smaller than *Hexaprotodon harvardi*, occurs in the Lower Nawata. This new species has previously been referred to as the Lothagam pygmy (Coryndon 1977;

Figure 10.1 Restoration of *Hexaprotodon harvardi* by Mauricio Antón. Shoulder height = 140 cm.

Leakey et al. 1996). The holotype, an almost complete mandible, was recovered from just above the lower markers and can be dated between 6.72 and 7.44 Ma (McDougall and Feibel 1999). This is an exciting discovery as not only does it permit the reevaluation of fragmentary hippopotamid specimens known from elsewhere in the Turkana Basin but also this species compares closely to Late Miocene hippos recovered from Spain and the Arabian Peninsula (Weston 2000).

Several specimens, including a large juvenile mandibular symphysis, cannot be assigned to either of the above species, indicating the existence of a third species at Lothagam. Finally, some specimens from the Apak Member can be attributed to a smaller, less gracile hippopotamus, here referred to as *Hexaprotodon* cf. *Hex. protamphibius*.

Systematic Description

Family Hippopotamidae Gray, 1821

Three genera within the family Hippopotamidae have gained universal acceptance. *Kenyapotamus* includes the first known representatives of the family, which have been described from teeth from Middle and Late Miocene sites in Kenya and Tunisia (Pickford 1983, 1990). *Hexaprotodon* and *Hippopotamus* encompass the subsequent members of the family, which in the past have populated much of the Old World, reaching as far east as Java and as far north as England. The latter genera both have living examples. *Hexaprotodon liberiensis* (Morton 1844), transferred by Coryndon (1977) from the genus *Choeropsis* Leidy, 1853, is a pygmy hippopotamus confined today to forest areas of West Africa. *Hippopotamus amphibius* Linnaeus, 1758, is a large amphibious sub-Saharan hippopotamus. In the light of new discoveries, it has become ever more difficult to assign species to separate genera based on their current diagnoses. This problem is discussed in the literature, the implication being that *Hippopotamus* and *Hexaprotodon* are congeneric (Hooijer 1950; Harris 1991). To a large extent, the two genera have been defined in terms of terrestrial and aquatic grades, implying that *Hippopotamus* and *Hexaprotodon* could have multiple origins. The phylogeny of the Hippopotamidae has received very little attention, and the evolutionary relationships are unclear. The new evidence from Lothagam reveals a new species of hippopotamus that is potentially the least derived *Hexaprotodon* ever to be described. The Lothagam material also expands our knowledge of the sexually dimorphic *Hexaprotodon harvardi,* a species that was prevalent during the Late Miocene in East Africa.

Genus *Hexaprotodon* Falconer and Cautley, 1836

Diagnosis

Muzzle short relative to the braincase. Lacrimal small and generally separated from the nasal by anterior extensions of the frontal. Upper canines with a deep posterior groove; lower canines with finely striated enamel. Incisor number variable. Limbs relatively long and gracile.

Falconer and Cautley (1836) first created *Hexaprotodon* as a subgenus of *Hippopotamus* when they described extinct hippopotamid material with six incisors in the lower jaw from the Siwalik Hills of India. The Siwalik fossils were later interpreted as a separate genus, and *Hexaprotodon* was given generic status (Matthew 1929; Colbert 1935). Subsequently, the extant pygmy hippo, previously assigned to its own genus *Choeropsis* Leidy, 1853, was reassigned to *Hexaprotodon,* the latter genus having nomenclatorial priority (Coryndon 1977). More recently, a number of African fossil species have been transferred from *Hippopotamus* to *Hexaprotodon* (Coryndon 1977; Gèze 1985; Harris 1991). In spite of the growing number of *Hexaprotodon* hippopotamids described, the generic characteristics assigned to this taxon are not shared by all the allocated species.

Coryndon (1977) gave priority to the arrangement of the facial bones, showing that the lacrimal in *Hexaprotodon* is small and separated from the nasal bone by an anterior extension of the frontal. Unfortunately, the position of the facial sutures, even on well-preserved fossil crania, can be difficult to distinguish. Other generic characteristics include unelevated orbits, positioned laterally, midway along the length of the cranium; upper canines with deep posterior grooves; lower canines coated with finely striated enamel; brachyodont molars and a gracile skeleton. These characters are the least conjectural. The generic significance of the following characteristics is disputable: the fusion of the premaxilla; upper incisors arranged in a shallow arc; reduced posterior expansion of the nasal; tip to tip occlusion of the incisors; and large premolars relative to the molars. There is a trend toward a smaller relative size of the premolars within *Hexaprotodon,* which invalidates the use of premolar size as a generic character.

The skeleton of *Hexaprotodon* is described as gracile and better adapted to a more mobile and less amphibious lifestyle (Coryndon 1977; Gèze 1985; Harris 1991). Gèze (1985) proposed that the external metapodials were more reduced in length relative to the internal metapodials in *Hexaprotodon.* This generalization is not valid, however; the observation is only strictly true for the two extant species, and the extreme reduction of the lateral digits of the foot in *Hexaprotodon liberiensis* appears to be a distinct specialization. In addition, in both genera, the reduction of the lateral relative to the medial digits tends to be more pronounced in the pes than in the manus. Another complicating factor is sexual dimorphism. *Hippopotamus amphibius* females exhibit less differentiation in manus digit length than the males (table 10.1).

Hexaprotodon harvardi Coryndon, 1977

(Figures 10.1–10.28; tables 10.1–10.9)

Diagnosis

A *Hexaprotodon* is a little smaller and more gracile than *Hippopotamus amphibius.* Orbits on the side of the skull; large pustulate premolars; double rooted P^1; hexaprotodont condition; upper and lower incisors arranged in a shallow arc; mandible with a long shallow symphysis and procumbent incisors.

Holotype

KNM-LT 4 (57–67K), almost complete adult cranium, from the Upper Nawata, Lothagam; housed in the National Museums of Kenya (figure 10.2, top).

Paratypes

KNM-LT 33 (246–67K), partial mandible, upper member of the Nawata Formation; KNM-LT 44 (279–67K), lower left canine, from the lower member of the Nawata Formation.

Locality

Lothagam, southwest of Lake Turkana, Kenya.

Lothagam Material

- *Lower Nawata:* 6, juvenile Lt. mandible symphysis (roots I_1, I_3, /C, dP_2 and dP_3); 9, Rt. mandible fragment (P_4–M_3); 11, Rt. mandible fragment (M_3); 18, adult Rt. maxilla (P^2–M^3) and Lt. max (P^3–M^3); 38, Lt. upper C/ and Rt. lower /I; 44, Lt. lower /C; 52, juvenile Rt. max fragment (erupting C/ and P^1, dP^3); 54, Lt. mandible fragment (M_3); 108, juvenile mandible (Lt. /C, dP_3–M_2 and Rt. /C–M_2); 120, juvenile Lt. dP^3; 398, Rt. dP_3; 402, Lt. P^1, Rt. P^3, M^2 and M^3, Lt. and Rt. M_2, Rt. M_3; 403, Rt. M^2; 8570, Lt. radioulna; 8626, Lt. P_2 or P_3; 8743, Rt. mandible

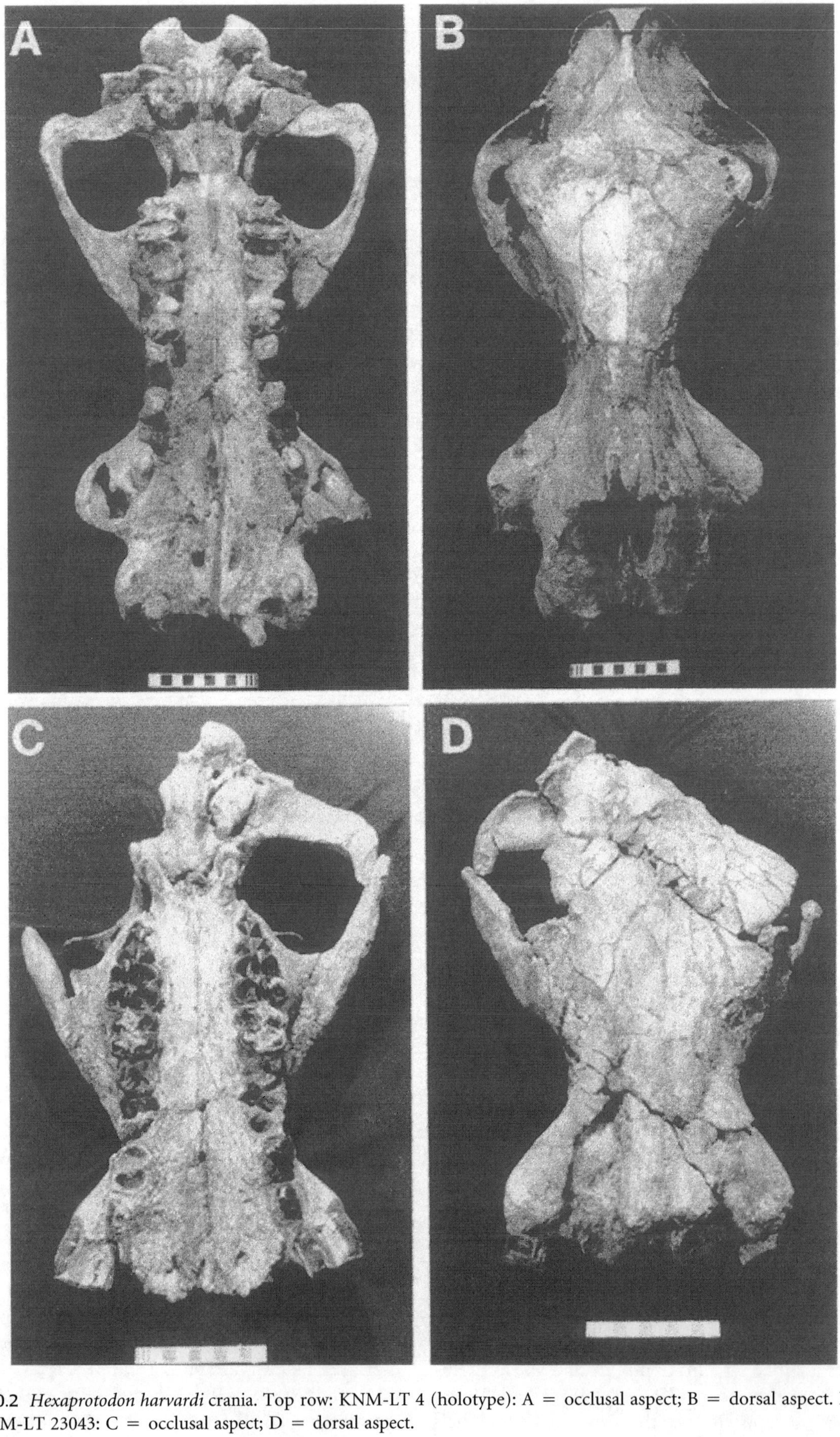

Figure 10.2 *Hexaprotodon harvardi* crania. Top row: KNM-LT 4 (holotype): A = occlusal aspect; B = dorsal aspect. Bottom row: KNM-LT 23043: C = occlusal aspect; D = dorsal aspect.

fragment (M_2–M_3); 22839, dP_4; 22862, Rt. M^3; 22863, Lt. M^3; 22870, Lt. dP_3; 22876, Rt. P_2, Rt. mandible (M_1–M_2) and Lt. M_3; 22964, juvenile mandible (Rt. and Lt. dP_2–dP_4, erupting M_1); 23038, proximal Mt V; 23097, Lt. lower /C; 23262, Lt. mandible fragment (M_3); 23263, Lt. M_2; 23264, Lt. P_4; 23265, Rt. dP_4; 23266, Rt. astragalus; 23268, Rt. upper C/; 23269, Rt. P^3; 23270, Lt. I^2; 23272, 3 phalanges, Rt. Mt III, Rt. Mc V, calcaneum and radioulna; 23285, juvenile Lt. and Rt. I^1, I^2; 23286, Rt. mandible fragment (M_3); 23830, Rt. and Lt. P^2; 23831, Lt. P^3; 23841, Rt. P_1; 23846, juvenile Lt. mandible (P_1–P_2, erupting P_3, broken dP_4 and M_1); 23847, Lt. P_2, Lt. mandible fragment (M_3), Rt. lower /C fragment, Rt. P_1 fragment, Rt. P_2 and Rt. M_1–M_2; 23849, adult max Lt. and Rt. (M^1–M^3); 23850, juvenile Rt. dP^4; 23864, Rt. premaxillary fragment (I^2), Rt. P^3, Lt. P^4, broken Rt. lower /C and P_1; 23865, juvenile Rt. upper incisor and premolar fragment; 23866, Rt. max (P^2–M^3); 23867, Rt. P_4 and Rt. mandible fragment (M_1–M_2); 23868, Lt. I^1 fragment, Rt. I^1, Rt. I^2 and Lt. P^1; 23876, juvenile Lt. dP^4; 23885, Lt. P^4; 23886, Lt. Mt III, Lt. Mt V, Lt. cuboid and Lt. internal cuneiform; 23891, semi-complete skeleton (no cranium); 23895, juvenile Lt. and Rt. upper incisors; 23896, adult Lt. mandible (P_3–M_3), Rt. mandible (P_4–M_3) and Lt. lower / C; 23897, A = Rt. I^2 B = Lt. I^1; 23899, lower incisor; 23901, Rt. cuboid; 23903, Rt. Mt V; 26190, Lt. dP_2; 26194, Lt. P_1; 26210, juvenile Rt. max (C/, P^1, dP^2–dP^4), Lt. M_2 and symphysis; 26225, Lt. P^4, Rt. M^2–M^3 and 2 malformed upper incisors; 26226, juvenile Rt. maxilla (C/, P^1, dP^2–dP^4, broken M^1 and unerupted M^2); 26227, adult Rt. mandible (broken M_1–M_3) and Rt. I_1; 26234, juvenile Lt. dP^4; 26236, adult posterior cranium (Lt. and Rt. M^2–M^3 and Lt. M^1); 26239, Lt. astragalus; 26243, Rt. astragalus; 26244, Rt. astragalus; 26245, Rt. astragalus; 26247, Rt. M^1 and broken M^2; 26248, Lt. astragalus; 26251, Rt. I^1 or I^2; 26252, Lt. astragalus; 26253, juvenile skeletal material, Lt. and Rt. (M^1, dP^2) and Rt. (P^2 and P^3); 26254, Lt. astragalus; 26255, Lt. P^4, M^1 and Rt. M^2; 26303, 2 distal radioulnae and proximal ulna fragment; 26304, Lt. and Rt. P_4, Rt. I_1, lower molar and teeth fragments; 26313, Rt. tibia and Rt. femur; 26315, Lt. radioulna; 26316, Lt. and Rt. astragali, Rt. navicular, Rt. external cuneiform and phalanx; 26415, Lt. mandible (post P_4–M_3) and Rt. P_2 or P_3; 26416, juvenile Lt. mandible (erupting M_2); 26583, juvenile Lt. mandible fragment (post M_2–M_3); 28764, maxillary fragments (Rt. P^2–P^4, M^1, M^2 and Lt. P^1, P^2, P^3) and upper C/ fragment.

- *Upper Nawata:* 1, juvenile Rt. mandible (dP_2–dP_4); 4, holotype adult cranium (Rt. I^1, I^3–M^3 and Lt. I^2, P^1–M^3); 5, juvenile Lt. mandible (dP_4); 33, adult Rt. mandible (P_3–M_3) and Lt. mandible (P_2–P_4) and Rt. upper C/; 34, Rt. radioulna; 48, juvenile semicomplete skeleton (no cranium); 134, adult cranium with broken teeth; 409, Rt. (lower /C, P_3, P^4) and Lt. (P^2, anterior P^3, P^4); 22873, juvenile Rt. maxilla fragment (post dP^3 and dP^4); 22957, juvenile mandible with damaged teeth; 22988, Lt. P^1; 23041, adult cranium lacking premaxilla (Rt. P^1–M^3 and Lt. P^1, P^3–M^3); 23043, subadult cranium (Rt. C/, P^1-erupting M^3 and Lt. C/–P^1, P^4-erupting M^3); 23105, subadult mandibular symphysis (Rt. I_1–/C and Lt. I_1–I_2, /C); 23127, adult Rt. mandible (P_2–M_3), Lt. mandible (P_3–M_3), Rt. lower /C and lower /I; 23602, Rt. mandible fragment (M_3), Lt. P_4, Lt. and Rt. M_3 and tooth fragments; 23832, juvenile Rt. dP_4; 23843, semicomplete skeleton (no cranium); 23844, juvenile Lt. mandible (erupting I_{1-3}, dI_{2-3}, dP_4 and erupting M_1); 23855, Lt. (cuboid, magnum, pisiform Mc II, Mt II and Mt V); 23856, Lt. upper incisor; 23859, Rt. lower teeth (I_1, I_2 or I_3, M_2 and M_3); 23861, Lt. astragalus; 23862, Lt. P_1; 23872, juvenile Rt. dP_4; 23878, Rt. astragalus; 23881, Rt. upper incisor; 23882, lower incisor; 23883, Rt. Mt V and two phalanges; 23887, Lt. mandible fragment (M_3); 23888, adult mandible (Rt. I_2–I_3, P_2–P_4 and Lt. I_1–I_3, P_2–P_4); 23898, Lt. proximal Mc III and V, Rt. scaphoid and Rt. internal cuneiform; 23902, adult max (Lt. M^{1-3} and Rt. P^4–M^3) and mandible (Lt. /C, /I, P_2–M_2 and Rt. P_3–P_4); 23905, Lt. Mc III; 23906, Rt. tibia; 23908, Lt. (P_2, P_4, M_1, M_2 and M_3); 24147, subadult cranium (Lt. and Rt. C/, P^4–M^2 and Lt. P^3); 26188, Rt. P^4; 26207, Rt. lower incisor; 26208, Lt. scaphoid; 26209, lower incisor; 26211, Rt. lower incisor; 26213, juvenile Rt. maxilla (erupting C/, P^1, broken dP^2, dP^3–dP^4) and M^1; 26214, Rt. P_2 or P_3; 26215, Lt. M_2; 26221, Lt. dP_4 and broken M_1; 26224, external phalanx; 26228, Lt. and Rt. hind limbs; 26229, posterior cranium; 26230, Lt. (P_3, P_4); 26232, Lt. and Rt. semilunar; 26238, Lt. Mc III; 26240, mandible fragments (Lt. and Rt. M_3); 26241, mandible fragments (P_2, Lt. and Rt. P_3–M_2); 26275, Lt. cuneiform; 26276, Lt. tibia; 26297, Lt. astragalus; 26314, Rt. humerus; 26585, Rt. scapula; 28654, Rt. scaphoid; 28655, Lt. astragalus; 28656, Lt. scaphoid; 28731, adult Lt. mandible (P_4–M_3).
- *Nawata Formation, level indet.:* 8542, Lt. Mt III; 8543, Rt. Mc II; 8547, Lt. cuneiform and phalanx; 8549, Lt. Mt IV; 23252, Lt. P^4.
- *Upper Nawata or Apak Member:* 397, juvenile back of cranium, occipital and basioccipital; 23877, adult Lt. mandible (P_4–M_3).
- *Apak Member:* 26, juvenile Rt. mandible fragment (roots I_1 and I_3, erupting I_2); 27, Lt. lower /C; 23616, Lt. upper C/; 23833, Lt. M_3; 23834, juvenile Lt. mandible fragment (dP_3–M_1); 23838, adult Rt. max (M^2–M^3), Lt. max (M^1–M^3) and Rt. mandible fragment (M_2–M_3); 23842, skull and teeth fragments (Lt.

M^1, M^2); 23848, juvenile skull fragments and premaxillae (Rt. I^3); 23904, Rt. proximal ulna; 26189, Lt. cuneiform; 26193, juvenile Lt. upper incisor; 26212, Rt. astragalus; 28713, adult Lt. max (M^1–M^3); 28718, Lt. max (P^1, P^3–M^3).

- *Horizon indet.:* 50, Lt. tibia; 56, partial cranium (Rt. and Lt. C/–M^3); 60, Lt. upper C/; 62, Rt. (P_4, M_3); 66, Rt. (M^2, M^3); 67, Rt. (P_4, M_2); 69, Lt. M_2 and Rt. M_3; 70, Rt. max (P^2–M^3) and Rt. zygomatic arch; 71, Lt. mandible (M_2–M_3), Rt. lower /C and Rt. P_4; 78, Lt. and Rt. P^4, Rt. P^2, Lt. P^2 fragment, M^3 fragment and dP^4 fragment; 102, juvenile cranium (Lt. and Rt. /C, P_1, dP_2–dP_4, M_1 and M_2 in crypt); 8610, partial post cranium; 8683, juvenile Rt. premaxilla (I^1–I^3).

The holotype (LT 4) remains the most complete cranium to be recovered from Lothagam, but several new cranial specimens improve our understanding of the sexual, ontogenetic, and intraspecific variation attributable to *Hexaprotodon harvardi.* The paratype specimens that form part of the earlier Lothagam collections lack accurate provenance, but more complete mandibular specimens have been recovered, together with much additional cranial and postcranial material.

Cranium

In addition to the holotype, there are eight partial skulls, two of which, LT 23043 and LT 24147, are subadult with M^3 and P^4 erupting. The dimensions are recorded in table 10.2. The existence of sexual dimorphism in *Hexaprotodon harvardi* is discussed later in this contribution. Essentially, the muzzle is less constricted in one sex relative to the other (figure 10.3 and table 10.3). Figure 10.2 compares the fossil cranium (LT 23043) with the holotype (LT 4). Although cranium LT 23043 has suffered a degree of compaction, the muzzle is still broader in the younger skull than in the holotype. LT 23041, a partial adult cranium, also possesses a broad muzzle (figures 10.3 and 10.4). Variation in zygomatic breadth is also evident in the fossil crania. Figure 10.5 illustrates how the relationship between braincase length and zygomatic width in *Hexaprotodon harvardi* compares closely with that of *Hexaprotodon liberiensis.* This variation can be attributed to sexual dimorphism in living pygmy hippos.

In most respects, other cranial material differs little from the holotype. Detail of the facial bones is only preserved in LT 23043, with the lacrimal and orbit situated on the side of the skull, separate from the nasal. The crushed dorsal surface of this specimen obscures detail of the muzzle shape that has vertical rather than rounded flanks in LT 4. The posterior cranium is long and rounded, and this feature is accentuated in LT 26236. LT 26236 exhibits further variation in skull proportions; the occipital is shorter and the condyles are small relative to the breadth of the mastoid and zygoma; the orbits are proportionately large for the size of the skull (figures 10.6 and 10.7). The specimen plots just beneath the other adult crania on the ontogenetic trajectory (figures 10.5 and 10.8). A slight degree of dwarfism (ontogenetic scaling) can be observed in the living *Hexaprotodon.* The crania of the Nigerian subspecies *Hexaprotodon liberiensis heslopi* (Corbet 1969) scale in a comparable fashion. In figure 10.6 it can also be observed that the roof of the orbit in cranium LT 4 is raised slightly above the roof of the skull, in contrast to the orbits of crania LT 23043 and LT 26236. In addition, the alignment of the zygomatic arch is horizontal in LT 26236 and LT 23043, as opposed to the oblique orientation seen in the holotype. Other characteristics of the holotype—such as the low position of the external au-

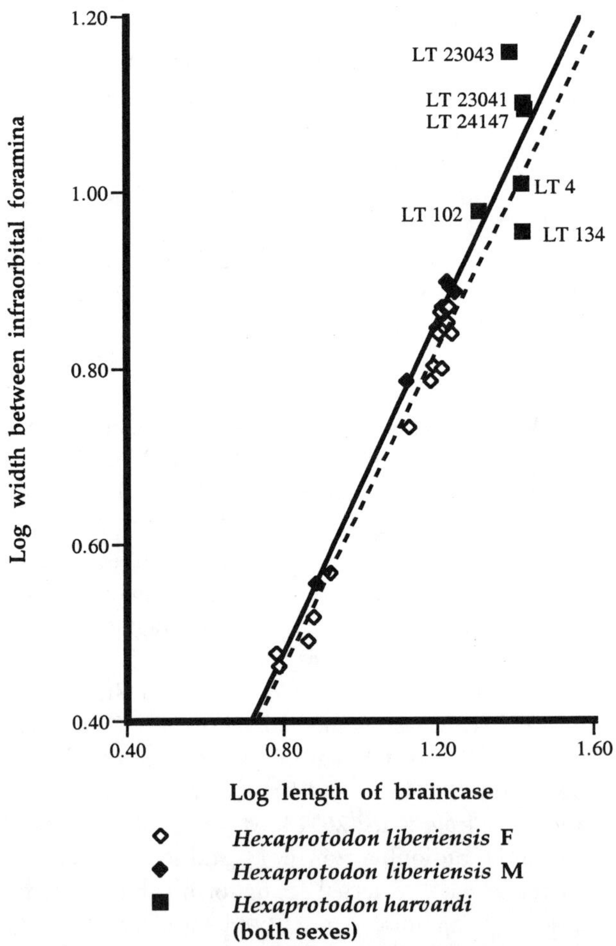

Figure 10.3 Plot of the width between the infraorbital foramina (rostral constriction) and the length of the braincase in fossil hippo crania to show their position relative to the male and female ontogenetic trajectories of *Hexaprotodon liberiensis.* F = female, M = male. Although the sample of *Hexaprotodon liberiensis* is small, due to limited specimens of known sex, the male and female intercepts are significantly different (see table 10.3).

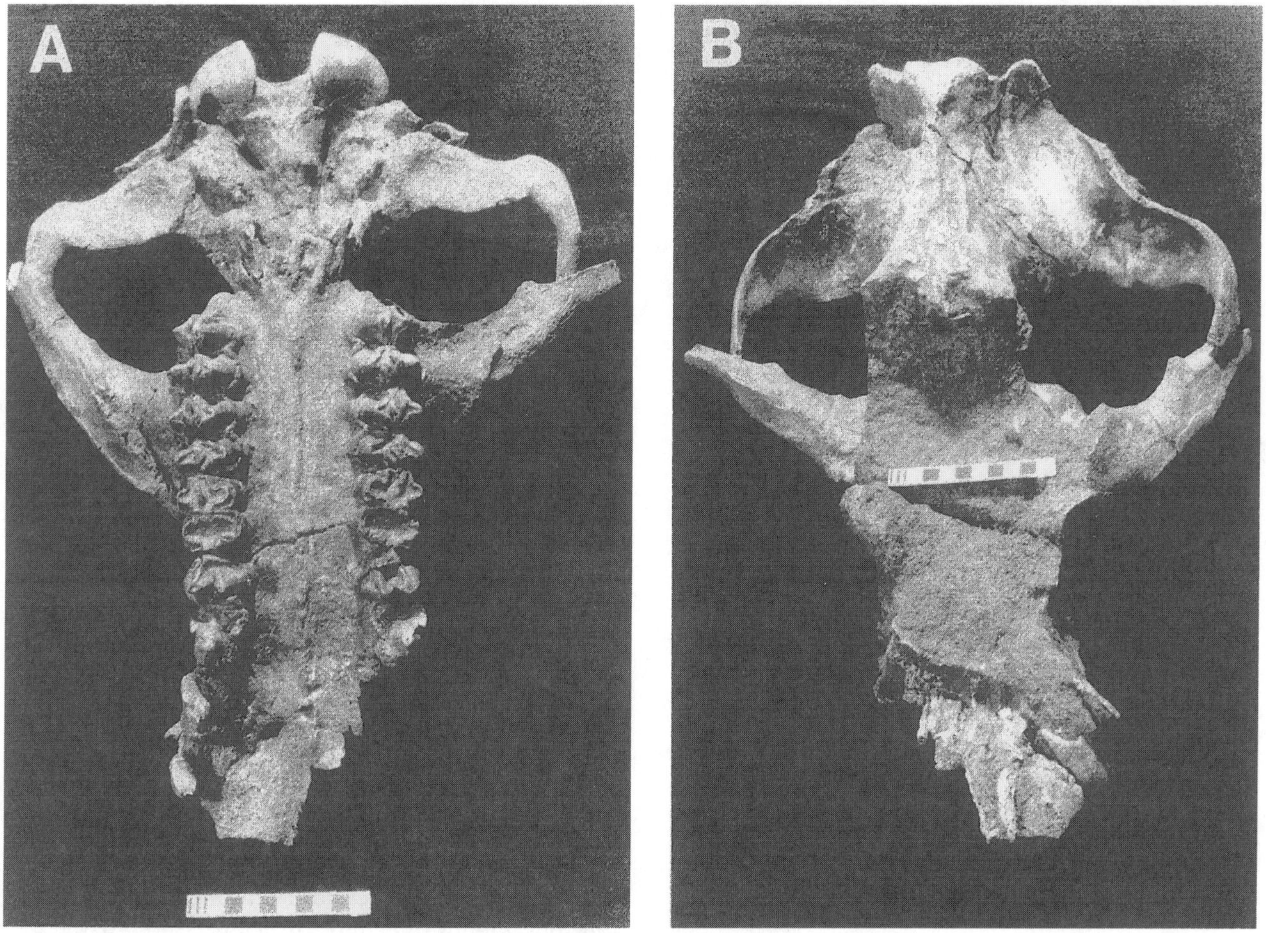

Figure 10.4 *Hexaprotodon harvardi* partial cranium, KNM-LT 23041: A = occlusal aspect; B = dorsal aspect.

ditory canal, slender zygomatic arches, sagittal crest only slightly elevated above the plane of the snout, and the relatively flat glenoid fossa—are shared by the other specimens where preserved.

The premaxillary is rarely preserved; in the holotype it is medially fused and downturned anteriorly, with peglike subequal incisors arranged in a shallow arc. The only additional premaxilla specimens are juvenile (LT 8683). Examples of isolated upper incisors have been recovered that differ from the holotype. The terminal wear facets in LT 4 that indicate tip to tip occlusion are not evident on all the upper incisors, the type of occlusion being subject to variation.

Upper incisors

A considerable amount of variation of upper incisor morphology exists at Lothagam, and the assessment of variation is complicated by the determination of I^1, I^2, and I^3. The holotype possesses simple peglike incisors with apical wear facets that range in shape from circular in I^3 to oval in I^1. I^3 appears to be slightly smaller than I^1 and I^2. Specimen LT 23897 consists of a right I^2 and left I^1 (figure 10.9). This fact has been concluded by comparing a premaxilla fragment with I^2 *in situ* (LT 23864). Both incisors are curved, with a distinct strip of enamel confined to the buccal surface. I^1 has a pronounced mesiolingual ridge. I^2 lacks the pronounced ridge, which results in an elliptical cross section and a tooth with a greater mesiodistal breadth. The curvature of the incisors is occasionally mesially directed as opposed to lingual (LT 23874). The profile and angle of the occlusal facets vary, indicating a degree of lateral occlusion. These teeth, all from the Lower Nawata, differ from the holotype. Only two isolated, upper incisors have been recovered from the Upper Nawata, and both of them are curved cylindrical teeth similar to the I^3 of the holotype, with the enamel lacking. This may represent a sampling artefact rather than a genuine distinction between the two horizons. In consequence, it remains unclear whether the two types of incisor can be attributed to intraspecific variation in *Hexaprotodon harvardi.*

Upper canines

The shape of the upper canine changes during the course of ontogeny, but a deep posterior groove is evi-

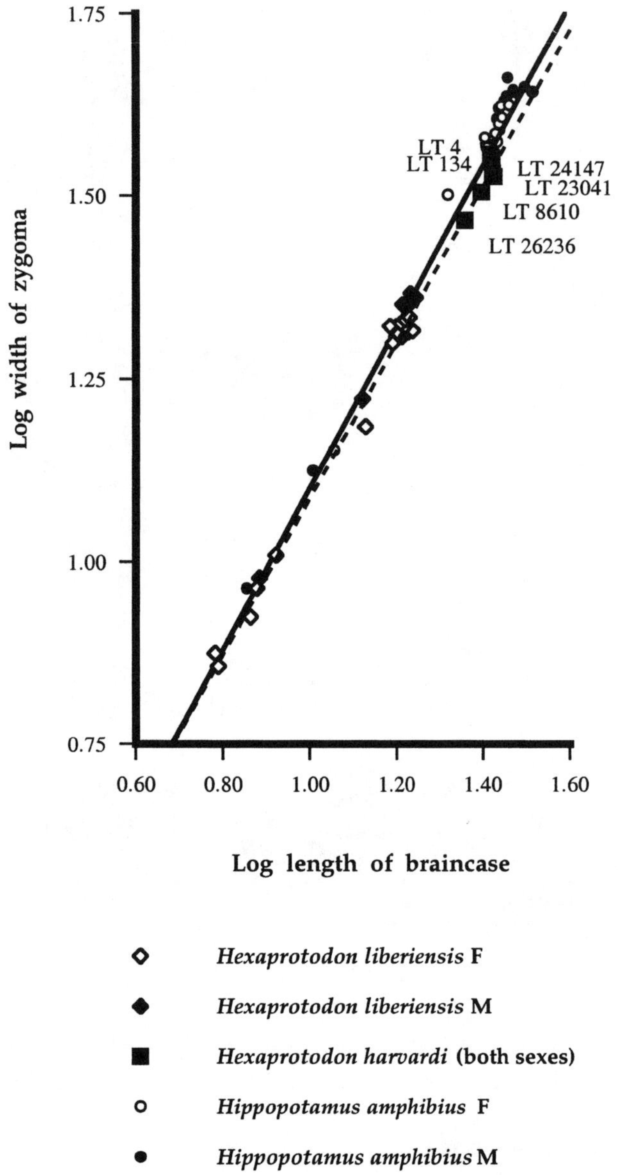

Figure 10.5 Plot of the width of the zygoma versus the length of the braincase in fossil hippo crania to show their position relative to the male and female ontogenetic trajectories of *Hexaprotodon liberiensis.* F = female, M = male. Although the sample of *Hexaprotodon liberiensis* is small, due to limited specimens of known sex, the male and female intercepts are significantly different (see table 10.3). In *Hippopotamus amphibius,* the males and females share the same ontogenetic trajectory (slope not illustrated; see table 10.3). *Hexaprotodon harvardi* compares more closely with *Hexaprotodon liberiensis,* the fossil specimens plotting on both growth trajectories. This allometric relationship indicates that KNM-LT 4 and 134 are males.

dent at all growth stages. A lateral groove, in the form of a sharp incision toward the anterior of the tooth, is present at eruption and gradually disappears as the canine continues to grow (figure 10.10). This feature is retained in LT 23043, an example of a subadult skull. The retention of this juvenile character in almost fully grown individuals gives the false impression of canine dimorphism. This observation is corroborated by the existence of the same trait in *Hexaprotodon liberiensis.* The upper canines of *Hippopotamus amphibius* lack a lateral groove. The finely striated enamel coating the fossil canines tends to be lost on the medial side of the tooth and on the lateral margin of the posterior groove.

Upper premolars

The quantitative characters recorded for the fossil upper premolars are listed in table 10.4. The P^1 is a double-rooted single-cusped tooth. The roots are commonly fused buccally and occasionally completely (LT 28764). LT 23868 is the only example with unfused roots. A characteristic triangle of pustulate enamel coats the anterior face of the crown, and the tooth broadens mesially. The cusp is curved lingually with a sharp distobuccal ridge. The cingulum is entire or lost buccally. LT 23148 has an additional short buccal ridge restricted to the apex of the cusp.

The P^2 and P^3 are similar, with one main cusp curved lingually toward the apex. The cingulum is well developed, especially lingually. There are two main ridges of pustulate enamel, running mesiolingually and distobuccally. A third subordinate ridge may form a loop inside the talon or, in the case of P^3, may terminate in a large cuspule. The development of the talon is variable, but it tends to be more developed in P^3 than in P^2. The expanded talon of P^3 in LT 23043 is supported by a third root. LT 23041 and LT 56 bear a large triangular cuspule on the distobuccal ridge.

The morphology and size of P^4 are extremely variable. The tooth is generally bicuspid, but occasionally a smaller cusp is present in a distal or mesial position. The buccal cusp is commonly quadrifoliate and larger than the trifoliate or globular lingual cusp as in LT 18 (figure 10.11). Occasionally, the cusps are equal in size (LT 28718), but the elevation of the buccal cusp is greater (figure 10.11). The cingulum is well developed, and it thins buccally. The tooth is usually broader buccolingually but can be rounded (LT 8636) and is often obliquely oriented to the long axis of the cheek tooth row. There is no correlation between size and stratigraphic horizon. (Note, however, that material from the Apak is not represented.) The P^4 in the two living species of hippo is very variable: both species exhibit unicuspid, bicuspid, and multicuspid forms (Weston 1998). There is no example of a unicuspid P^4 at Lothagam. Figure 10.12 illustrates some of the variation in P4 morphology.

Upper molars

The upper molars are similar in shape and brachyodont. Dimensions are listed in table 10.5. The distally tapering

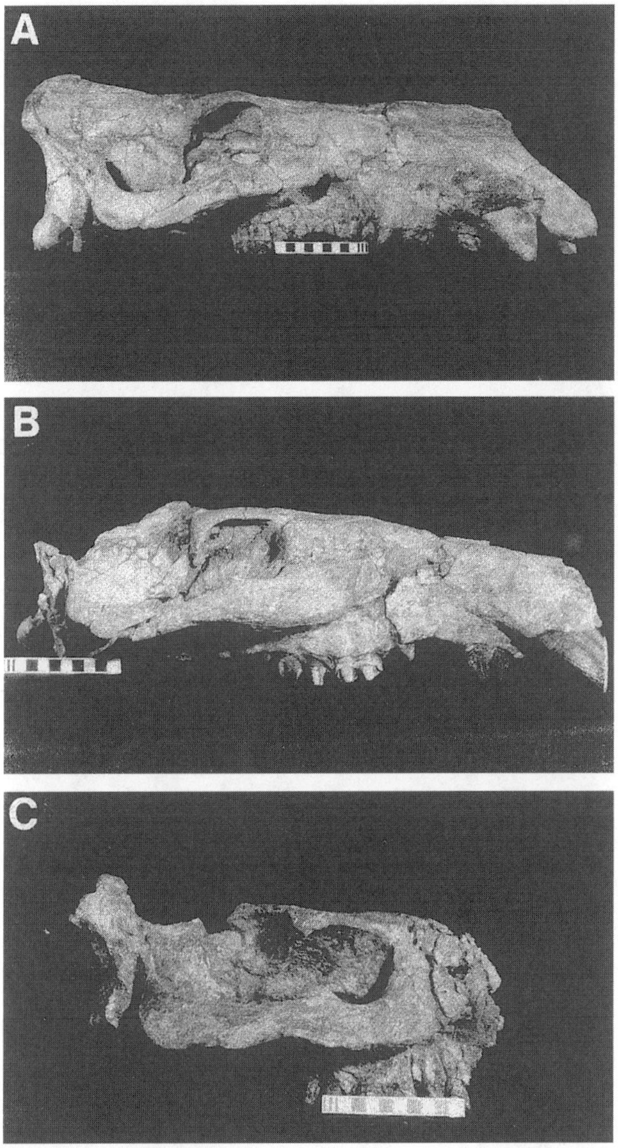

Figure 10.6 *Hexaprotodon harvardi* crania, right lateral aspect: A = KNM-LT 4 (holotype); B = KNM-LT 23043; C = KNM-LT 26236.

M^3 can be distinguished from M^1 and M^2, but all the molars have trifoliate cusps that narrow toward their apex. The cusp lobes are less pinched in appearance than are those of *Hippopotamus amphibius*, resulting in a more triangular wear facet. Occasionally, single cusps are tear-shaped and resemble the entoconid of the lower molars. This can be observed in both metaconules and paracones. The molars all have a strong pustulate cingulum, which may invaginate between the lingual cusps.

Upper deciduous dentition

A semideciduous upper dentition is preserved in the juvenile cranium LT 102 including P^1, dP^2–dP^4, and M^1 (figure 10.13). No example of a dP^1 has been recovered. The dP^2 has a single main cusp that is curved lingually toward its apex, with a sharp mesial ridge and a distobuccal ridge terminating in a cuspule; the cingulum is entire. There is an example of a dP^2 (LT 26226), which has a pustulate ridge that encircles the mesial crown of the cusp. This ridge is separate and parallel to the cingulum. The dP^3 tapers mesially, with one main central cusp and three smaller cusps, two positioned distally and one mesially. The cingulum is deep and well developed. A low ridge extends from the main cusp and joins the cingulum distolingually in LT 52, but this ridge is absent in LT 120. The dP^4 is molariform and low crowned, and it tapers mesiobuccally (figure 10.13). The buccal cusps tend to be slightly smaller, quadrifoliate, and more elevated than the lingual trifoliate cusps. The paracone has a W-shaped mesial border. The metacone can be enclosed by an invagination of the deep cingulum; it is occasionally quadrifoliate. The morphology of this tooth resembles that of *Hexaprotodon liberiensis* more closely than that of *Hippopotamus amphibius.*

Mandible

The shape of the mandible appears to be one of the most diagnostic differences between hippopotamid species. The symphysis, in particular, is very informative. There are four complete symphyses, representative of different stages of ontogeny, that can be assigned to *Hexaprotodon harvardi.* Mandible dimensions are recorded in table 10.6. These specimens are all more complete than paratype A described by Coryndon (1977). LT 23888 is a semicomplete adult mandible with the angle only partially preserved on the left ramus and the ascending rami completely absent (figure 10.14C). The symphysis is long, buttressed posteriorly, and thinning abruptly to form a convex shelf for the six procumbent, subequal incisors. The lower canines are slightly divergent, and the symphysis broadens anteriorly. The posterior border of the symphysis is V-shaped. The angle is preserved in LT 28731, where it protrudes inferiorly. The mandibular condyles are flatter and less elongated lateromedially than in *Hippopotamus amphibius.* LT 23105 is a well-preserved symphysis that lacks only the left I_3; however, the wear on the incisors is not as advanced as LT 23888, and the I_2 tapers toward the apex, which indicates that this is a particularly large subadult specimen (figure 10.15). In addition, the symphysis is particularly shallow for its length (table 10.6).

There are two reasonably complete juvenile mandibles (figure 10.14A–B). The youngest, LT 22964, resembles a 3-year-old extant hippopotamus, based on the degree of tooth eruption. The symphysis is shallow, and there is no lateral displacement of the canines; the anterior region is damaged. LT 108 is a slightly older individual, comparable with an 8- to 10-year-old extant specimen, and it bears a longer, deeper symphysis with laterally displaced canines. LT 23105 has a shallower symphysis than LT 108 has, suggesting an anomaly not linked with ontogeny. It is plausible that one of the mandibles does not belong to *Hexaprotodon harvardi.*

Lower incisors

I_1–I_3 are all preserved *in situ* in LT 23888. I_1 is slightly larger than I_2 and I_3, but the difference is almost negligible. The teeth are straight and cylindrical, with oblique wear facets that face dorsolaterally (figure 10.14C). Isolated examples of unworn incisors have a diagnostic asymmetrical enamel cap, with a tapering tail of enamel that extends toward the root. A trace of this cap is still visible on the incisors of specimen LT 23105, implying immaturity. The isolated lower incisors at Lothagam do vary. The teeth are commonly more mesiodistally compressed, with a greater buccolingual diameter. A few examples are curved, with longitudinal lateral grooves (e.g., LT 26207). The wear facets can be highly irregular.

Lower canines

The lower canines are mesiodistally compressed, and the cross-section of the tooth is diagnostic (figure 10.16). The tooth tapers lingually and has an expanded square buccal base. A gentle depression runs up the mesial tooth margin. The tooth is coated with finely crenulated enamel. The lower canines of LT 23105 appear to lack the mesial depression, and they possess a flattened mesial face.

Lower premolars

The single-cusped P_1 curves distally toward its apex, bearing two ridges that run mesiolingually and distolingually. The distal ridge can be serrated, and the cingulum can be pustulate. The majority of the specimens have a single root; only one specimen (LT 23879) possesses a buccally fused double root. The canine of the latter specimen compares more closely with that of *Hexaprotodon lothagamensis* (Weston 2000).

Dimensions recorded for the lower premolars are listed in table 10.7. The P_2 and P_3 are similar. One main cusp may bear a multiple set of pustulate ridges, and the mesiolingual ridge is common to all teeth. LT 22876

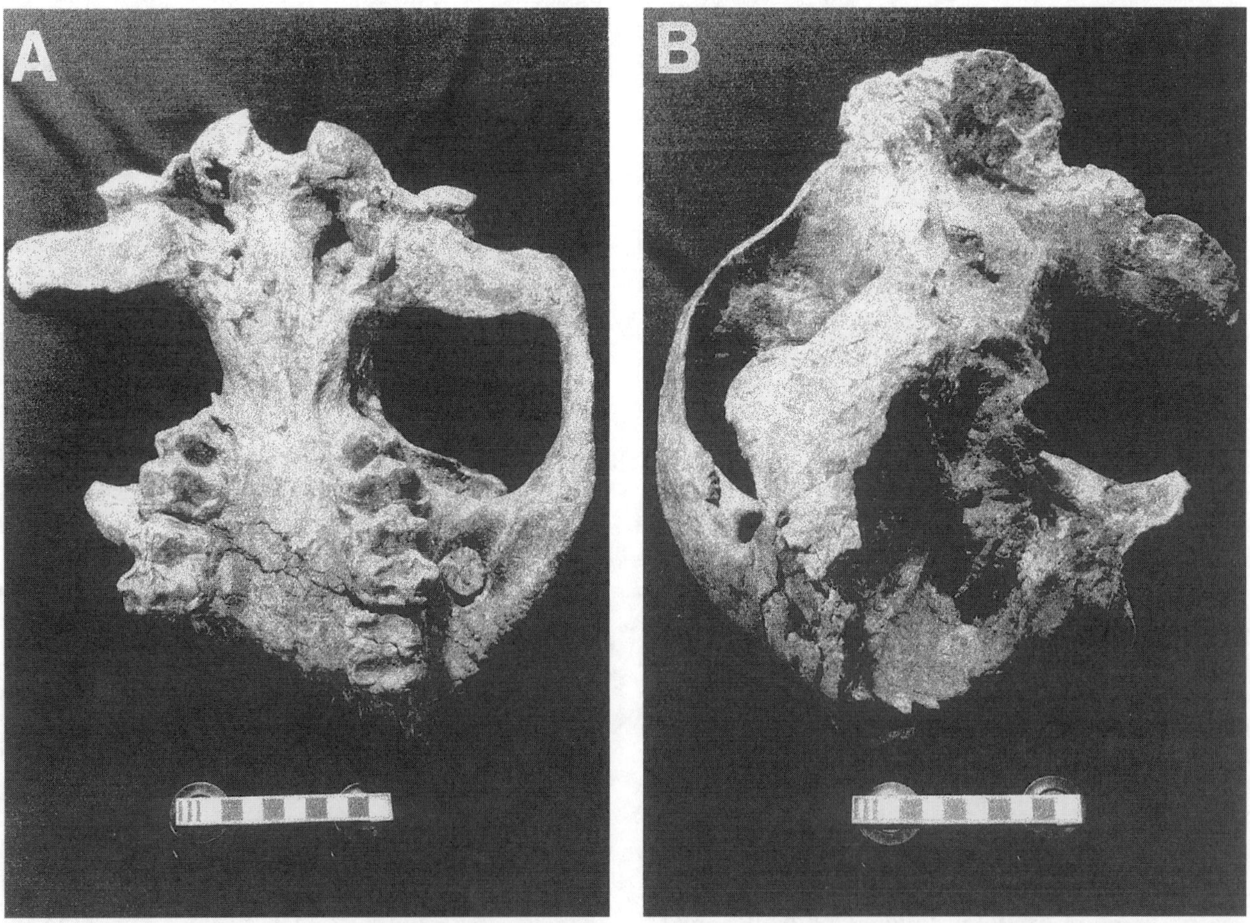

Figure 10.7 *Hexaprotodon harvardi* posterior cranium, KNM-LT 26236: A = occlusal aspect; B = dorsal aspect.

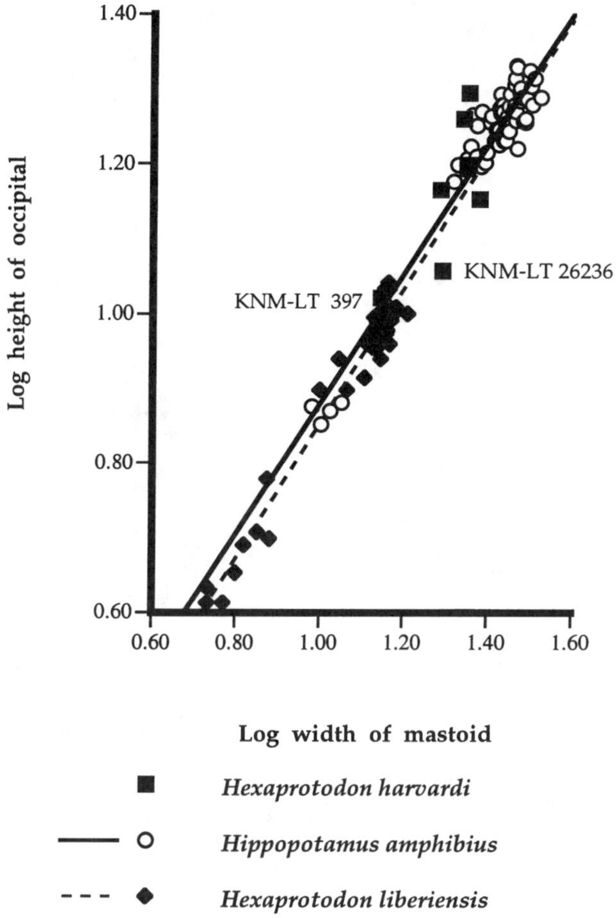

Figure 10.8 Comparison of the relationship between the height of the occipital and the width of the mastoid in fossil and modern hippopotamid species. The fossil cranium KNM-LT 26236 plots below the adult and subadult crania of *Hexaprotodon harvardi.* The proportional differences observed in KNM-LT 26236 appear to be the result of ontogenetic scaling. It is assumed that the fossils share a similar growth trajectory to that of the living species. KNM-LT 397 is a juvenile specimen. The slope and intercept values are recorded in table 10.3.

has a distinct distal face to the main cusp, which is bordered by two ridges that thus produce a prominent wedge-shaped depression (figure 10.17A–B). A less elevated lingual accessory cusp can often augment this morphology (figure 10.17C–D). Occasionally, a third ridge is present distobuccally (LT 23896). The cingulum is continuous. The P_3 may have a more developed distal cingular shelf and broaden posteriorly, relative to P_2. The distal enamel at the base of the main cusp can be extremely pustulate.

The P_4 is morphologically variable. It is commonly a bicuspid tooth, with a large buccal cusp and a smaller, less elevated lingual cusp that is positioned more distally. The main cusp has a sharp mesiolingual ridge and a distal ridge, and one or two cuspules occur close to its base. Another cuspule may be positioned at the base of the lingual cusp (figure 10.18). The cingulum is deep

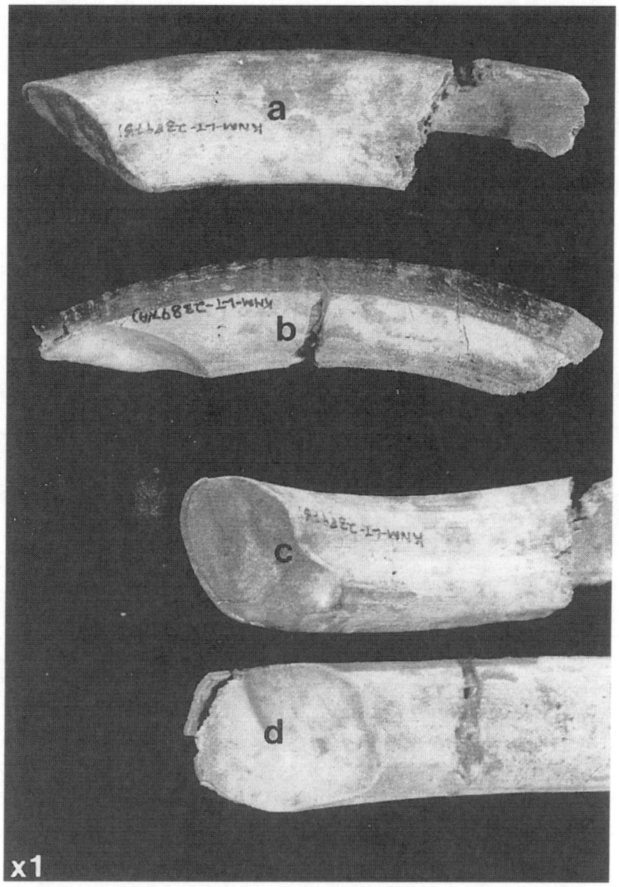

Figure 10.9 *Hexaprotodon harvardi* upper incisors, KNM-LT 23897: a = left I^1, distal aspect; b = Right I^2, mesial aspect; c = left I^1, lingual aspect; d = right I^2, lingual aspect. Note the pronounced mesiolingual ridge on the I^1 (see a and c).

and forms a thickened distal shelf; the tooth often broadens distally. LT 71 is an example of a single-cusped tooth with two distal ridges. This specimen may represent another species, but comparable intraspecific variation is common to both extant species.

Lower molars

Lower molar dimensions are listed in table 10.8. M_1 and M_2 are very similar in morphology, and the mesiodistal length always exceeds the breadth. M_1 is frequently ex-

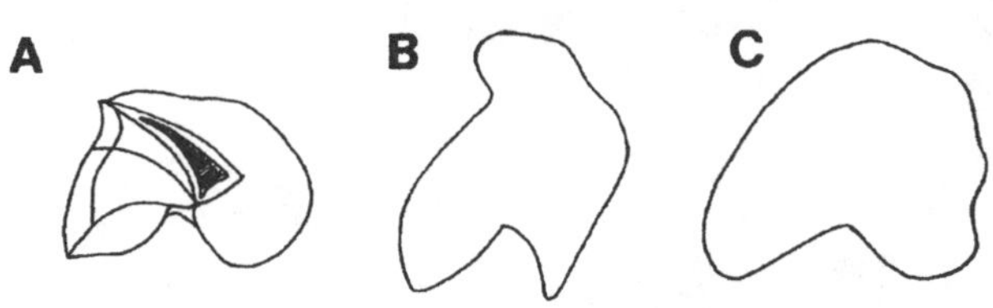

Figure 10.10 Diagrammatic representation of the growth stages of a right upper canine (occlusal aspect) of *Hexaprotodon harvardi:* A = juvenile; B = subadult; C = adult. The lateral groove gradually disappears during the course of ontogeny.

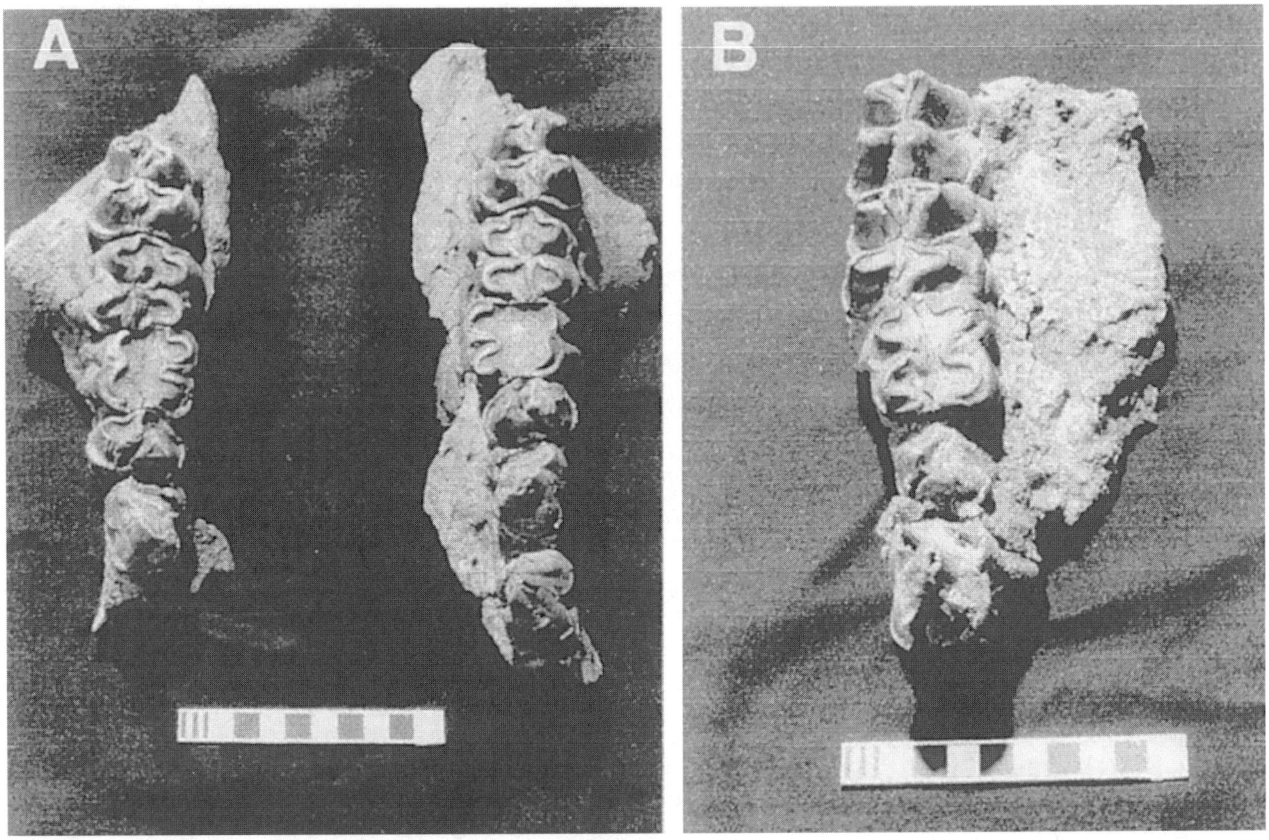

Figure 10.11 *Hexaprotodon harvardi* maxilla fragments illustrating P^4 morphology: A = KNM-LT 18, occlusal aspect of left and right cheek tooth rows; B = KNM-LT 28718, occlusal aspect of left cheek tooth row.

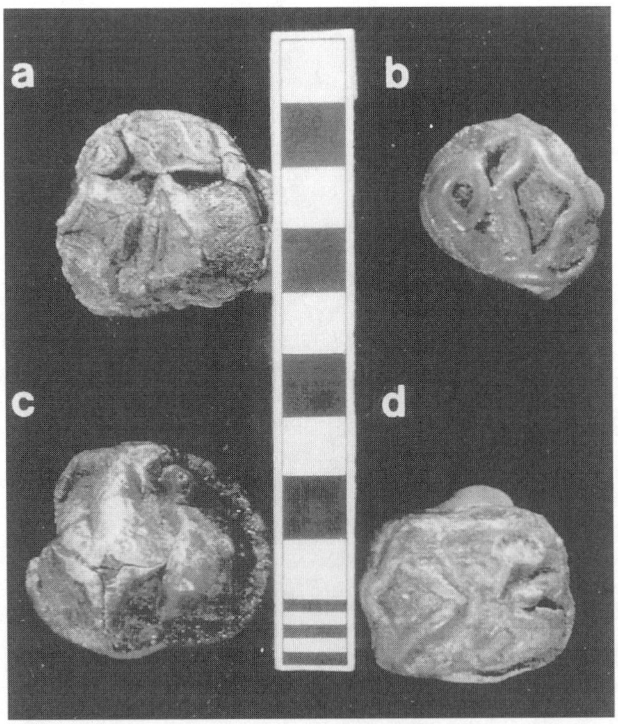

Figure 10.12 *Hexaprotodon harvardi* P^4s from Lothagam, illustrating the range of morphological variation. These teeth may belong to different species. a = KNM-LT 78; b = KNM-LT 8636; c = KNM-LT 23252; d = KNM-LT 23254.

tremely worn. The teeth are low crowned, with four cusps that taper toward their apex; all teeth are trifoliate in appearance, with the exception of the tear-shaped entoconid. The M_3 is differentiated from the other molars by the addition of a large hypoconulid (figure 10.19). The cingulum forms prominent shelves mesially and distally. Molar variation is evident. The cingulum may invaginate to form a ridge separating the mesial and distal cusps, or stylids may be present.

Lower deciduous dentition

The dP_2 consists of a single elongated cusp, with a sharp mesiolingual ridge that may bear a tubercle (LT 26190). Two distal ridges diverge from the cusp apex and border the pustulate talonid. The cingulum is raised buccally. The dP_3 has four cusps, two positioned distally and one mesially from a central main cusp. The main cusp possesses a centrally positioned distal ridge and a distolingual ridge with a circular tubercle at its base. A deep cingulum surrounds the tooth. The dP_3 is narrower than the dP^3, and the central cusp differs from that of the latter (figure 10.20). The dP_4 is a molariform mesially tapering tooth with three, serially arranged pairs of triangular cusps. The cingulum is more developed mesially and distally. The dP_4 is present in the juvenile mandible LT 22964 (figure 10.14A). The deciduous teeth compare more closely with those of *Hexaprotodon liberiensis* than with *Hippopotamus amphibius;* the latter possesses teeth with a relatively greater distal breadth, more convoluted cusps, and a smaller dP_2.

Postcranium

The long bones and metapodials of *Hexaprotodon harvardi* are slender and gracile. This is well illustrated by allometric plots of combined fossil and living skeletal samples (figure 10.21). The fossil metapodial dimensions are listed in table 10.9. Sexual dimorphism occurs in *Hippopotamus amphibius:* the male long bones and metapodials attain a greater length relative to their width. It is plausible that sexual variation of this kind accounts for some of the size variation in the fossil sample. In contrast, there is no marked difference in male versus female body size in *Hexaprotodon liberiensis* (Weston 1998).

Certain podial bones of *Hexaprotodon harvardi* have qualitative distinctions from those of *Hippopotamus amphibius,* but with the exception of the astragalus, calcaneum, lunar, and cuboid, all the carpals and tarsals are similar in size to those of *Hippopotamus amphibius.* The degree of curvature of the articulation facets does appear to vary, but this is very difficult to quantify. The intraspecific variation of articular facet shape and number is extensive in the two living species; hence, only features peculiar to either genus or to the fossils are discussed here.

Three semicomplete skeletons have been recovered from Lothagam: LT 23843 and LT 26228 from the Upper Nawata (the latter specimen is of the hind limbs only) and LT 23891 from the Lower Nawata. LT 23891 is a smaller adult individual, and smaller size is a characteristic common to postcranial specimens from the older horizon. The skeletons are all assignable to *Hexaprotodon harvardi,* in spite of the considerable intraspecific size variation (figure 10.22). However, a distinct smaller species is still discernible from this level. No cranial material was associated with the skeletal elements. The bones of the manus and pes are tightly packed together; the metapodials are arranged in an arch with the bone shafts largely in contact. The distal articular facets of the metapodials extend round into the shaft, forming a depression on the extensor surface (figure 10.23). The joint would have permitted considerable flexion, exceeding that of both living species. The prominence of the keel on the distal facets varies among metapodials. The collateral ligament scars on the phalanges are well developed, and the ligaments limit lateral movement. The development of such scars is apparent and variable in the living species. Adaptation to cursoriality has been studied extensively in recently extinct Mediterranean hippos (Houtekamer and Sondaar 1979; Spaan 1996; Caloi and Palombo 1996): a characteristic increase in sagittal movement associated with a restriction of lateral movement or deviation of the limbs is expected. On this basis, *Hexaprotodon harvardi* can be regarded as highly cursorial with an unguligrade posture.

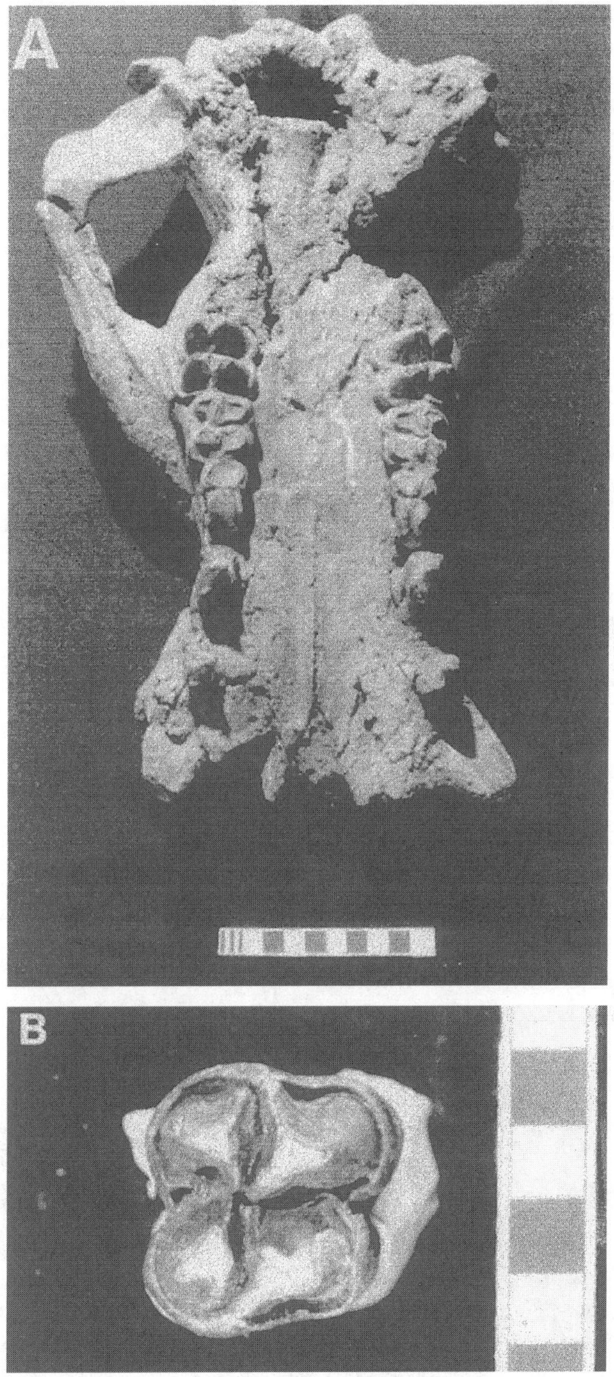

Figure 10.13 *Hexaprotodon harvardi* upper deciduous dentition: A = juvenile cranium KNM-LT 102, occlusal aspect; B = left dP^4, KNM-LT 26234, occlusal aspect.

The astragali of the two living species differ (Weston 1998). In *Hippopotamus amphibius,* the astragalus has a well-developed proximal stop facet, which forms the flexor margin of the medial trochlea. This process limits the degree of flexion of the tarsus to between 60 and 70°. The astragali of *Hexaprotodon liberiensis* completely lack stop facets, and this character is not subject to intraspecific variation. The potential flexion of the tarsus is greatly increased, exceeding 90°. A large sample of isolated astragali has been recovered from Lothagam; the range of sizes exceeds the expected ontogenetic range, in comparison with the sizes of *Hippopotamus amphibius* (figure 10.24). The astragali indicate the existence of more than one species, and they provide evidence for a trend toward increased size of hippos from the Nawata sediments (figure 10.24). The astragali are slender relative to their length, and they compare more closely with those of *Hexaprotodon liberiensis.* However, both types of morphology occur, and the presence or absence of stop facets is unrelated to size (figure 10.25). In the modern species, this character is not ontogenetically variable. LT 26228 is an example of a hind limb in which the degree of flexion of the tarsus exceeds 90°, whereas LT 26243 is a large astragalus with a well-developed stop facet that is analogous to that of *Hippopotamus amphibius.* If this character is of functional significance, it implies that two types of locomotory behavior coexisted at Lothagam, and the large cursorial hippo is assigned to *Hexaprotodon harvardi.*

The cuboid of *Hexaprotodon harvardi* has proportionately longer flexor arms than does *Hippopotamus amphibius* (figure 10.26). LT 23901 possesses a projected fifth metatarsal facet, which creates a socket for the articulation of the fourth metatarsal, and the distal articular face of the cuboid is stepped in appearance (figure 10.22, top). The fourth metatarsal would have been locked into position, thus greatly increasing the stability of the joint. In contrast, LT 26228 has a groove between the two facets, which deepens toward its base, and in this aspect it is more similar to the living species, although the groove is more pronounced in the fossils. The calcaneum facet has an S-shaped profile compared to the gentle convexity of the modern hippo cuboids, generating more stability during flexion.

The lunar of *Hexaprotodon liberiensis* possesses an extensor projection that tapers to a point proximally. In *Hippopotamus amphibius,* the projection is broad and square. The morphology of the lunar can be distinguished easily in the two living species. Figure 10.27 depicts two lunars: LT 23891 on the left, from the Lower Nawata, is similar to *Hexaprotodon liberiensis,* whereas LT 26232 on the right, from the Upper Nawata, tapers to a lesser degree and resembles an intermediate form; the shape of the extensor projection falls between that of the two living species.

There appears to be an association between the increase in size of *Hexaprotodon harvardi* and increased cursoriality. This contrasts directly with today's scenario, where the larger *Hippopotamus amphibius* is less mobile and more adapted to an amphibious lifestyle. The postcranial evidence from the Nawata Formation indicates that morphology, as well as size, varies in the fossil sample. In consequence, although the majority of the postcranial fossils have been referred to here as *Hexaprotodon harvardi,* based on the dominance of cranial

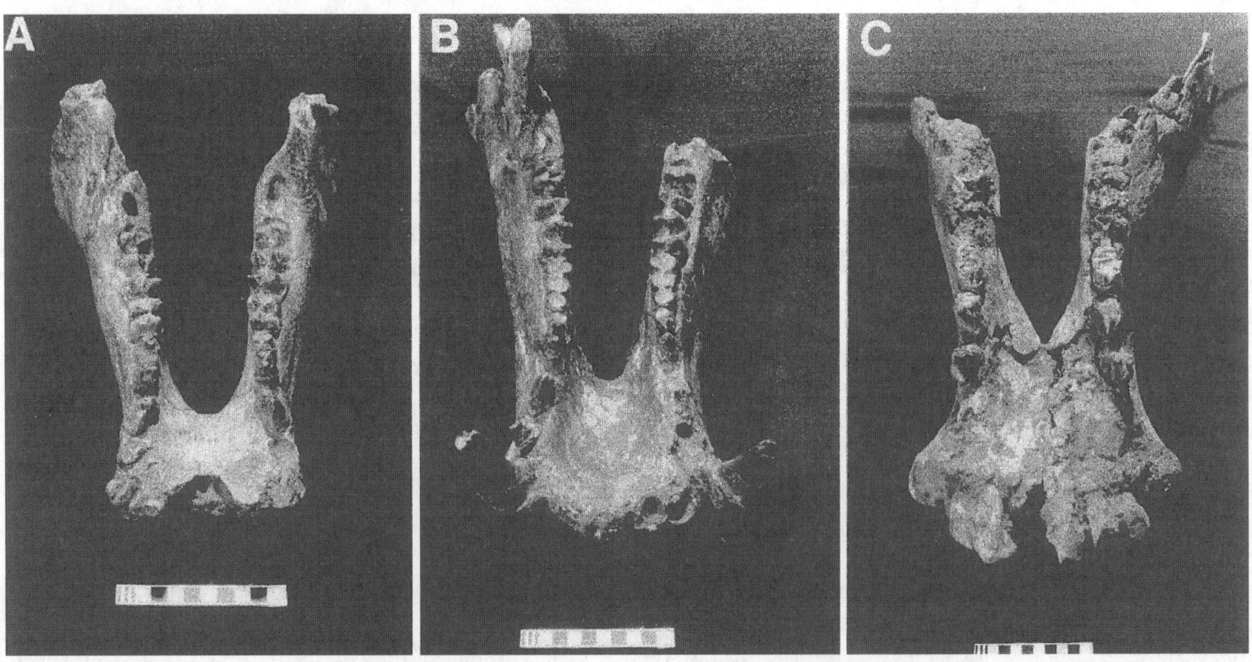

Figure 10.14 *Hexaprotodon harvardi* mandibles, occlusal aspect: A = KNM-LT 22964, young juvenile; B = KNM-LT 108, juvenile; C = KNM-LT 23888, adult.

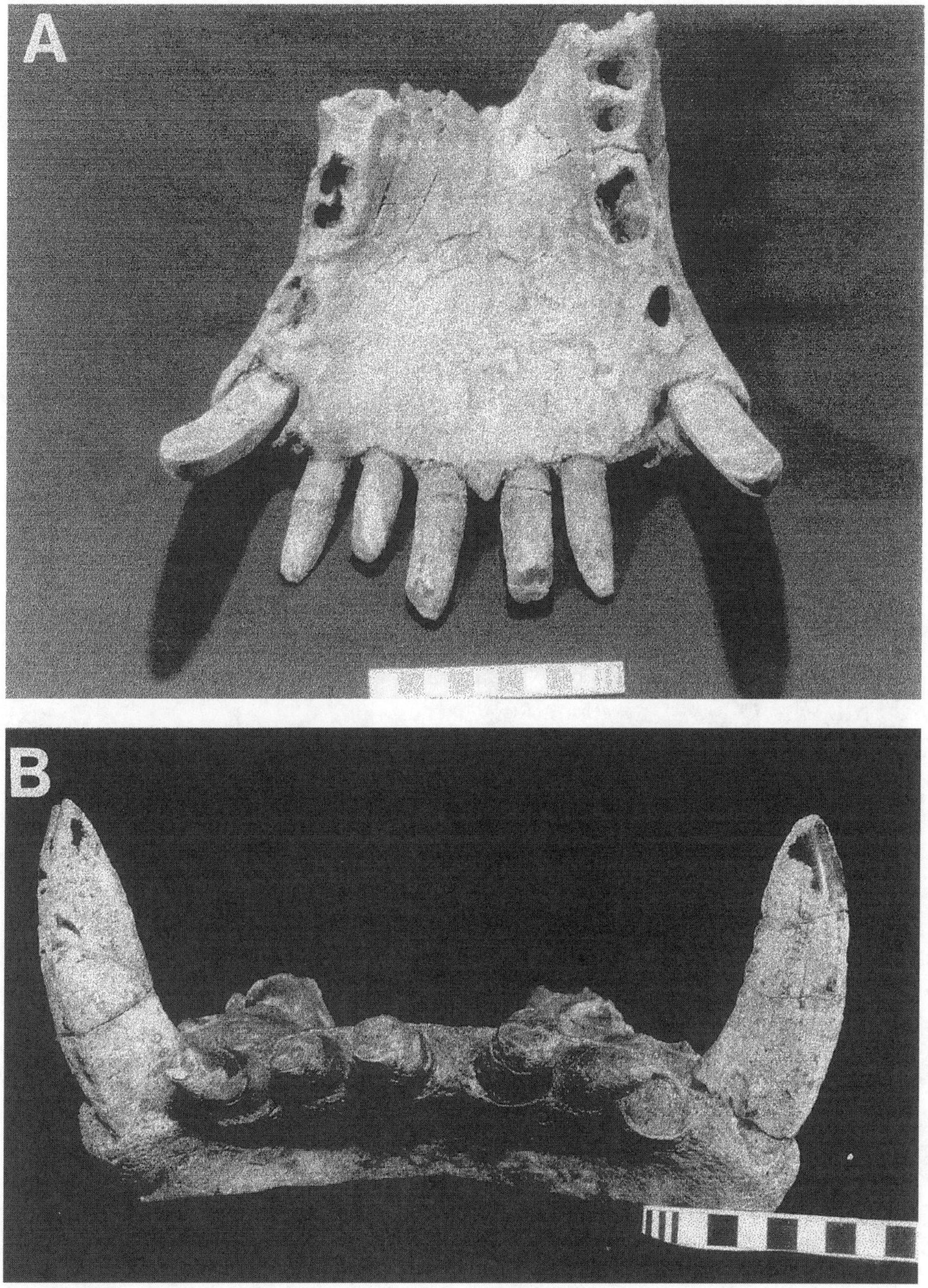

Figure 10.15 *Hexaprotodon harvardi* mandibular symphysis, KNM-LT 23105: A = occlusal aspect; B = anterior aspect.

evidence from this species, the postcranial material is essentially indeterminate. The functional significance of the postcranial material can only be fully interpreted when associated skeletons are recovered. In summary, there is evidence at Lothagam for one common species of hippo, which was highly variable and probably modified its lifestyle, possibly in response to changes in environmental conditions. However, some postcranial material from Lothagam does not fit this interpretation, and this corroborates the cranial evidence that other species also existed.

Hexaprotodon lothagamensis Weston, 2000
(Figures 10.28–10.31; tables 10.6–10.9)

Holotype

KNM-LT 23839, mandible with right P_4–M_3 and left P_4–M_2, coronoid and condyle (figure 10.29). The holotype was found below the Middle Markers (dated at 6.72 Ma) and just above the Lower Markers (dated at 7.44 Ma) of the lower member of the Nawata Formation (McDougall and Feibel 1999).

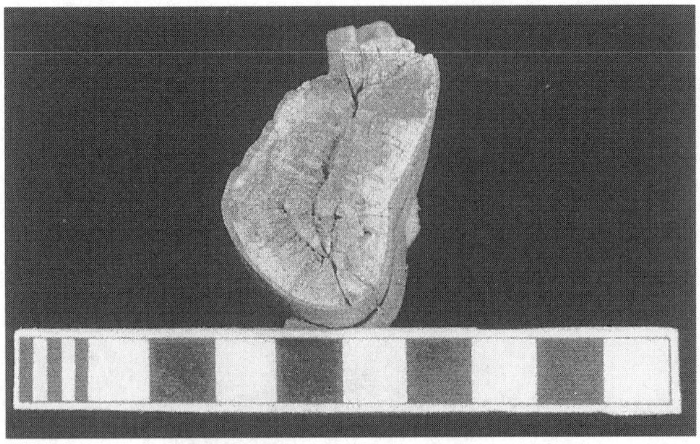

Figure 10.16 *Hexaprotodon harvardi,* KNM-LT 23864, transverse section of a right lower canine.

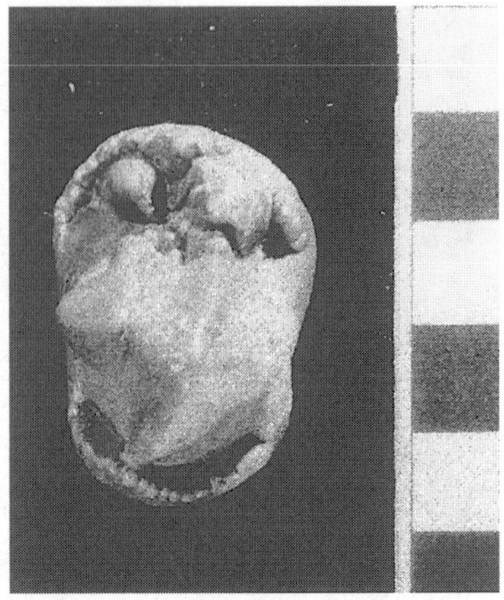

Figure 10.18 *Hexaprotodon harvardi* left P_4, KNM-LT 23908, occlusal aspect.

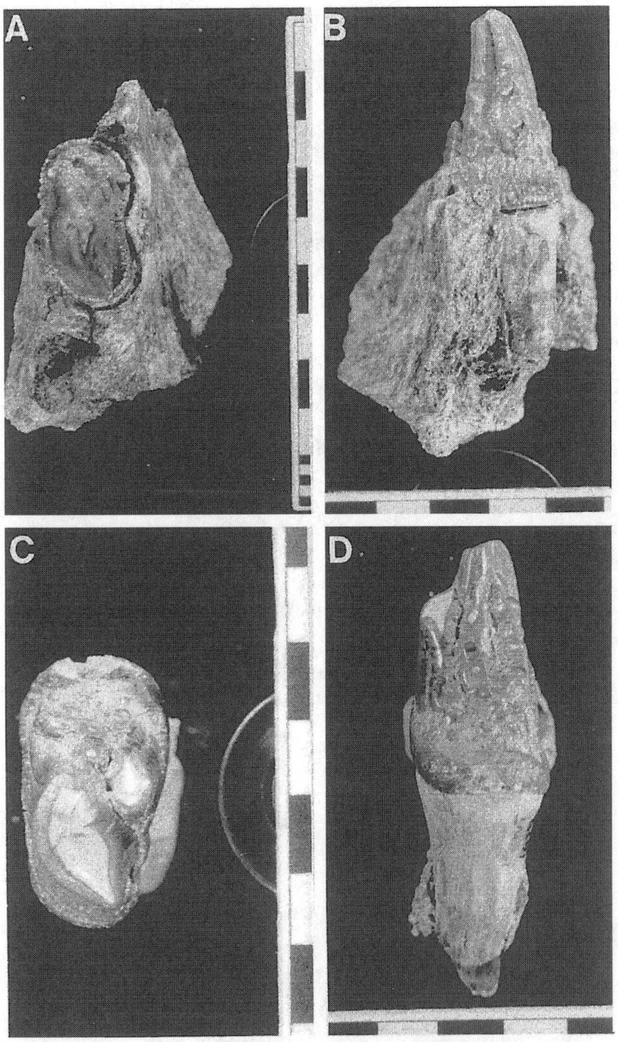

Figure 10.17 *Hexaprotodon harvardi* lower premolars: A = right P_2, KNM-LT 22876, occlusal aspect; B = right P_2, KNM-LT 22876, distal aspect; C = right P_3, KNM-LT 23864, occlusal aspect; D = right P_3, KNM-LT 23864, distal aspect.

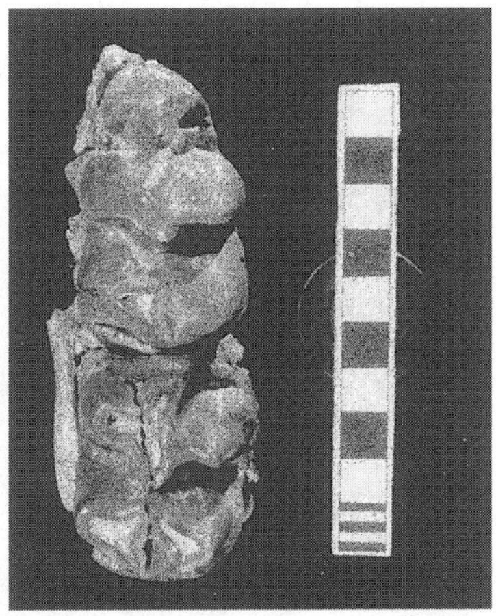

Figure 10.19 *Hexaprotodon harvardi* right M_2 and M_3, KNM-LT 23838, occlusal aspect.

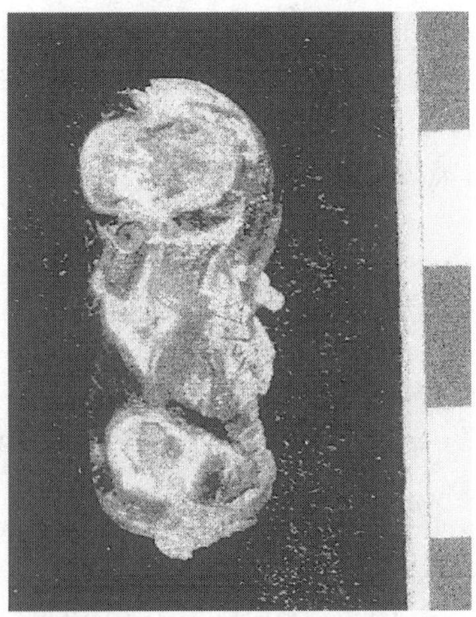

Figure 10.20 *Hexaprotodon harvardi* left dP_3, KNM-LT 22870, occlusal aspect.

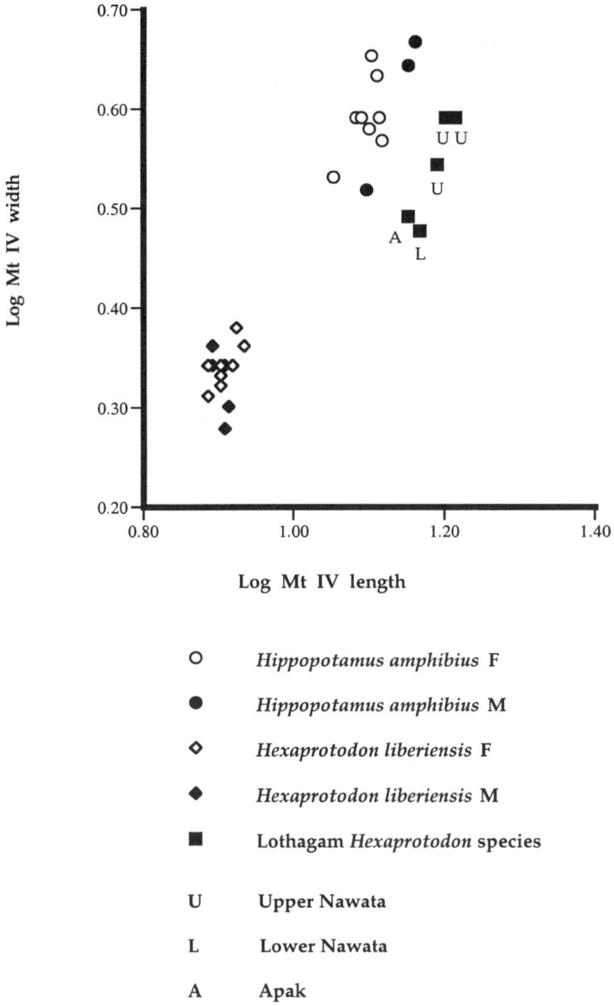

Figure 10.21 Allometric plot comparing the width and the length of fourth metatarsals in fossil and extant hippopotamids. Sexual dimorphism in *Hippopotamus amphibius* is also illustrated. The metatarsals of *Hexaprotodon harvardi* are long and slender relative to those of the two living species. F = female, M = male.

Lothagam Material

- *Lower Nawata:* holotype (see above); 22864, Lt. M_3; 23106, posterior Lt. ramus (M_3); 23879 juvenile Rt. /C fragment, Rt. P_1, Lt /I, and Lt. dP_3; 26235, Lt. Mc V; 26242, Rt. femur.
- *Upper Nawata:* 23871, Rt. mandible fragment (P_4–M_2).

Hexaprotodon lothagamensis constitutes a morphologically distinct hippo, much smaller in size than *Hexaprotodon harvardi,* but still larger than the living pygmy hippopotamus *Hexaprotodon liberiensis.* The holotype, a mandible, is the most complete specimen yet to be recovered. The right ramus is almost entire, with only the ascending ramus and angle absent (figure 10.29). The anterior symphysis is damaged, but the roots of three incisors (I_1–I_3) are clearly visible and appear to be undifferentiated in size. The procumbent incisors are aligned horizontally across the front of the jaw. The mandibular symphysis is half the breadth of *Hexaprotodon harvardi,* and the lateral displacement of the canines is almost negligible. The symphysis is extremely

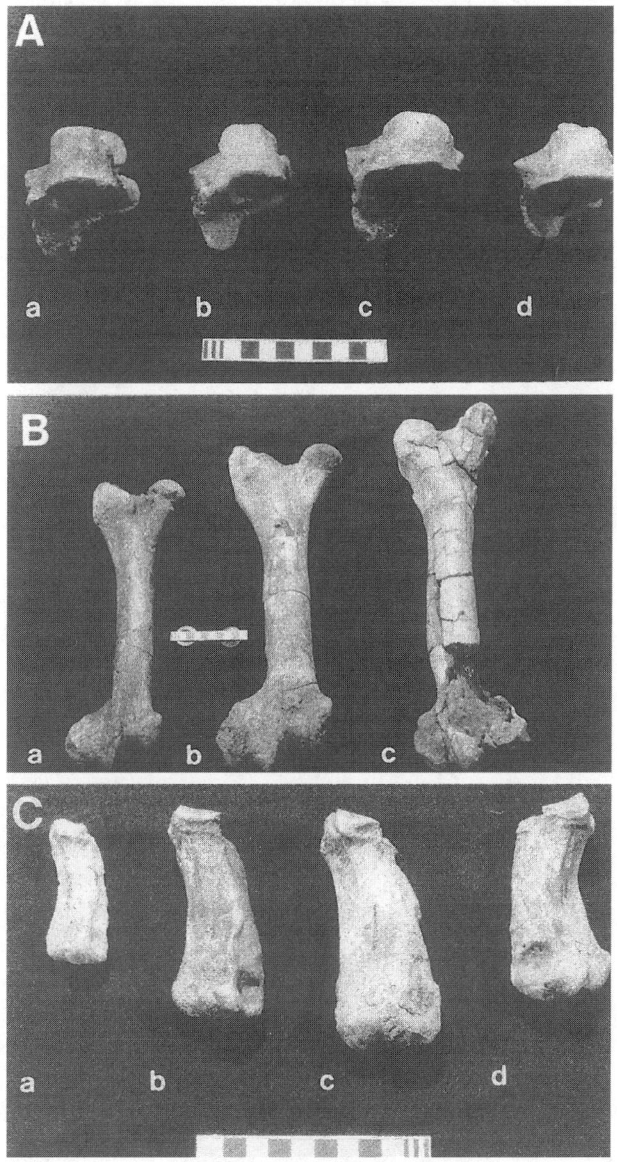

Figure 10.22 Comparison of hippopotamid postcranial elements from Lothagam. Specimens listed from left to right. A = extensor aspect of right cuboids: KNM-LT 23891 (Lower Nawata), 23901 (Lower Nawata), 26228 (Upper Nawata), 26219 (Apak). Note the stepped appearance of the distal articulation facets in 23901. B = extensor aspect of three right femora KNM-LT 26242: *Hexaprotodon lothagamensis,* 26313 *Hexaprotodon harvardi* (Lower Nawata), 26228 *Hexaprotodon harvardi* (Upper Nawata). C = flexor aspect of fifth metacarpals; KNM-LT 26235 *Hexaprotodon lothagamensis,* KNM-LT 23891 *Hexaprotodon harvardi* (Lower Nawata), LT 23843 *Hexaprotodon harvardi* (Upper Nawata), LT 26220 *Hexaprotodon harvardi* (Apak).

long and shallow, the anterior thinning being gradual and not abrupt. The canine is laterally compressed and tapers lingually, with a flattened mesial border. P_1–P_2 are absent, but the sockets indicate that the teeth were relatively large. The P_4 has two main cusps of equal height, but the lingual cusp is moderately smaller. Both cusps have mesial and distal ridges, and the distal ridges terminate in prominent cuspules. The cingulum is pronounced mesially and distally. The molars are low crowned and similar to those of *Hexaprotodon harvardi*, with the exception of size. The partially preserved angle in LT 23839 and LT 23106 protrudes inferiorly.

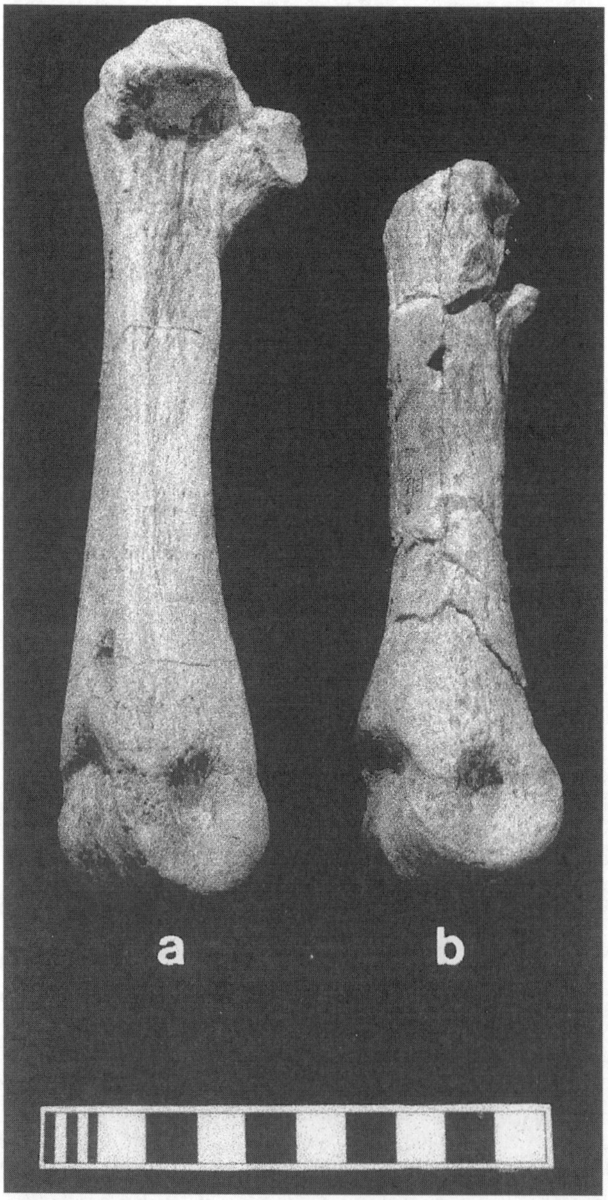

Figure 10.23 *Hexaprotodon harvardi* metapodials: a = KNM-LT 26238 left Mc III, lateral aspect; b = KNM-LT 26228 right Mt IV, medial aspect. There is a pronounced depression on the extensor surface of both metapodials formed where the distal articular facet curls into the shaft.

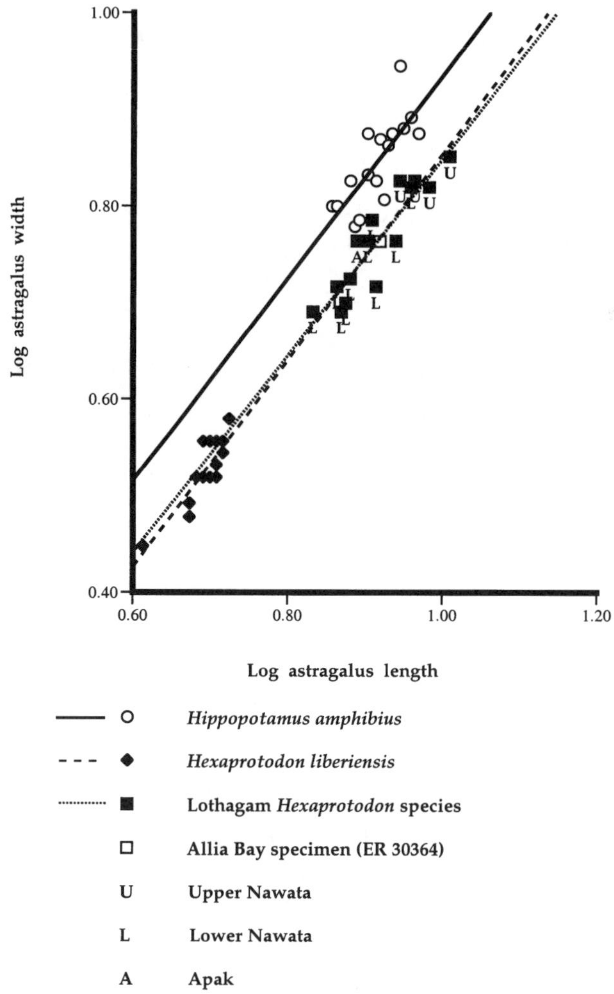

Figure 10.24 Comparison of growth allometries, illustrating the relationship between the width and the length of the astragalus. *Hexaprotodon liberiensis* and the Lothagam hippopotamids have similar ontogenetic trajectories (see table 10.3). There appears to be evidence for a trend toward increased size in the astragali from the Nawata Formation.

Hippopotamus imagunculus (Hopwood 1926; Cooke and Coryndon 1970), transferred to *Hexaprotodon* (Gèze 1985; Harris 1991), is a small hippopotamus described from the Western Rift that compares closely in size with *Hexaprotodon lothagamensis.* Until recently, it was known only from a limited sample of fragmentary specimens, and comparison was problematic. According to Faure (1994), further material assignable to *Hexaprotodon imagunculus* has been recovered from sediments ranging in age from 5 to 1.8 Ma, and this material can be distinguished readily from that of *Hexaprotodon harvardi.* The initial documentation of the new Ugandan material includes no cranial or mandibular specimens, and so the identification appears to be based on size and tooth morphology alone. The only symphysis described so far is from the Kazinga Channel, and this was initially referred to as *Hippopotamus* sp. (Cooke and Coryndon 1970) (table 10.10). Erdbrink and

Krommenhoek (1975) subsequently named it *Hexaprotodon imagunculus,* which they believed to be synonymous with *Hippopotamus (Hexaprotodon) hipponensis* (Gaudry 1876), from the Villefranchian of Algeria. This symphysis, though hexaprotodont, is clearly different from that of the Lothagam species, as the length of the symphysis is markedly shorter relative to its breadth. The allometric relationship between the width and the length of hippopotamid symphyses is illustrated in figure 10.30. *Hexaprotodon lothagamensis* appears to be an ontogenetically scaled-down version of *Hexaprotodon harvardi;* the mandible compares closely with LT 22964, a very young individual. In contrast, the Kazinga mandible falls on a separate ontogenetic trajectory shared by later hexaprotodonts and the two living species (figure 10.30). To augment the mandible shape character, there are other characters that differentiate *Hexaprotodon lothagamensis* from *Hexaprotodon imagunculus.* The lower canine is not bean-shaped, and the lingual cusp of P_4 is further reduced in size and elevation in *Hexaprotodon imagunculus.* The relative size of the premolars to molars and the incisor differentiation cannot be adequately compared. The molars are similar in size, however.

Other material should probably be assigned to this species, but tooth size alone is an unsatisfactory means of diagnosis. In addition, P_4 morphology may be subject to variation, and there are several specimens that deviate only slightly from the P_4 described here—for example, LT 23867 and LT 23864 (figure 10.31). However, there is one specimen (LT 23871), a mandibular fragment with P_4–M_2, which possesses a P_4 similar to that of *Hexaprotodon lothagamensis* (figure 10.32). This specimen is unusual as the molars are relatively large and compare more closely to those of *Hexaprotodon harvardi* (Weston 2000). Furthermore, the specimen was recovered from the upper boundary of the Nawata Formation, and it is at least a million years younger than the material assigned to *Hexaprotodon lothagamensis.*

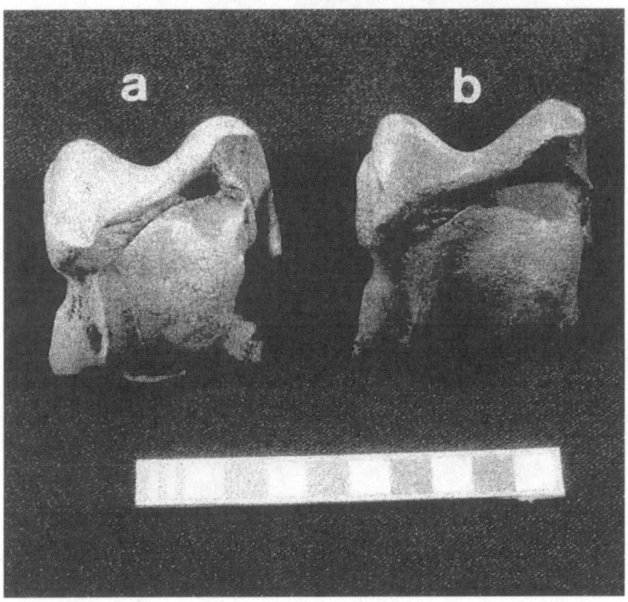

Figure 10.25 Astragali of fossil hippopotamids from the Lower Nawata, proximal aspect: a = KNM-LT 26243, possessing a stop facet; b = KNM-LT 23266, lacking a stop facet.

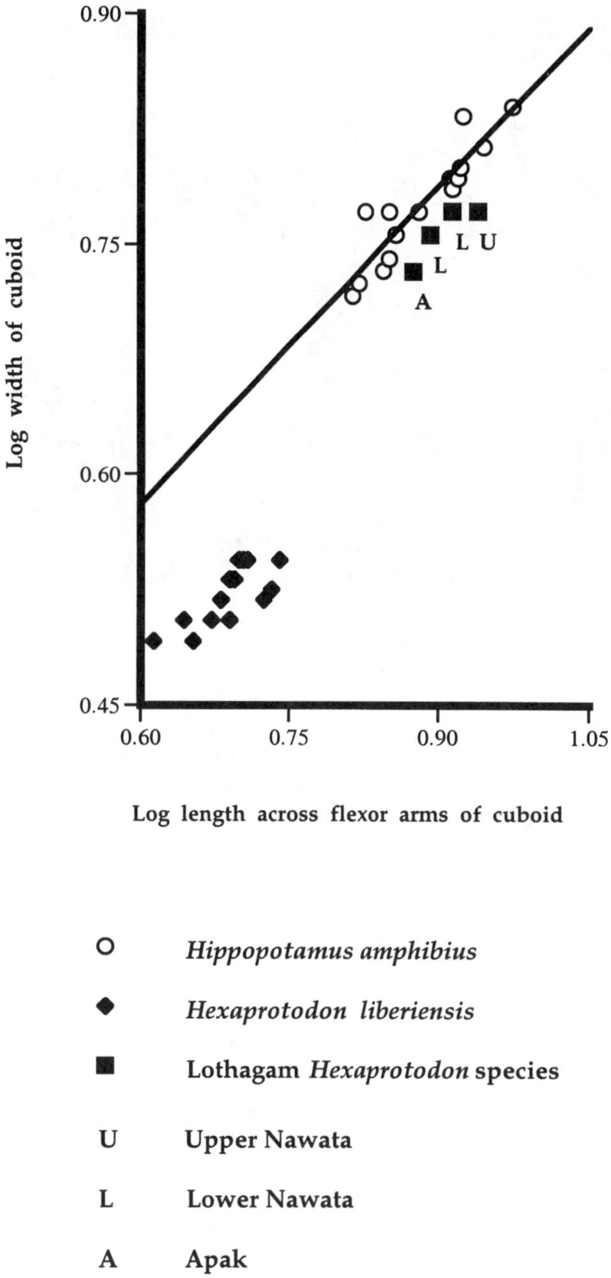

Figure 10.26 Allometric plot of width and length of fossil and extant hippopotamid cuboids. A = Apak, L = Lower Nawata, U = Upper Nawata. The fossil *Hexaprotodon* cuboids possess flexor arms that are more extended than those of *Hippopotamus amphibius.* A regression line has not been fitted to the *Hexaprotodon liberiensis* sample as the correlation coefficient is low (see table 10.3).

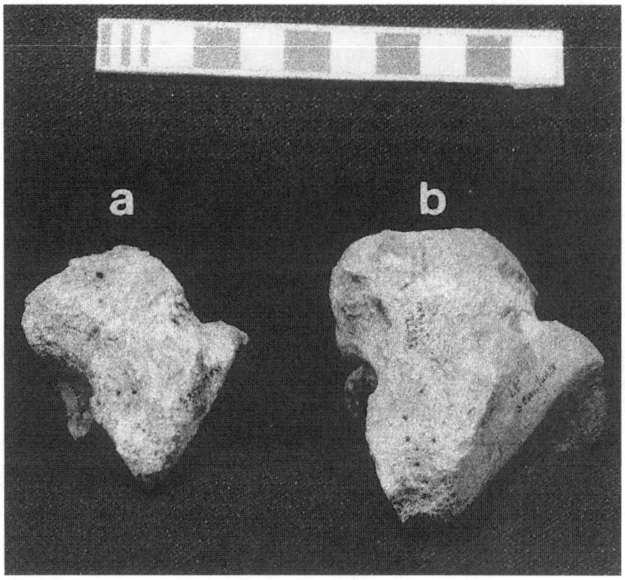

Figure 10.27 Lunar bones of fossil hippopotamids, extensor aspect: a = KNM-LT 23891 (Lower Nawata); b = KNM-LT 26232 (Upper Nawata).

LT 26235 is an exceptionally small fifth metacarpal; its diminutive size clearly distinguishes it from the other fossil material (figure 10.22C). The fifth metacarpal is smaller than those of *Hexaprotodon imagunculus* (Faure 1994). LT 26242 is a small gracile right femur (figure 10.22B). The length of the femur indicates that *Hexaprotodon lothagamensis* had relatively long limbs compared to the size of the lower jaw.

Several other species of hippopotamus described also possess a narrow muzzle (Colbert 1943:409; Lacomba et al. 1986; Gentry 1999). *Hexaprotodon lothagamensis* has the closest affinities with *Hexaprotodon* aff. *Hex. sahabiensis* (Gentry 1999), a species known from the Arabian Peninsula, and *Hexaprotodon crusafonti* (Aguirre 1963), a species known from Spain. The Arabian *Hexaprotodon* is larger than *Hexaprotodon lothagamensis,* and the symphysis is deeper and more trough-shaped. The six subequal incisors are arranged in a concave arc. The incisors are obliquely orientated as opposed to being procumbent and aligned across the front of the jaw (Weston 2000). The material assigned to *Hexaprotodon crusafonti* from Spain is not very complete. However, Lacomba et al. (1986) have described a mandibular specimen that has part of the symphysis still preserved. Although this symphysis possesses only four incisors in the lower jaw, which are a little larger than those of *Hexaprotodon lothagamensis,*

Figure 10.28 Restorations of *Hexaprotodon lothagamensis* (left) and *Hexaprotodon harvardi* (right) by Mauricio Antón.

the mandibular proportions of the two species are similar (Weston 2000). The dental morphology is also similar, except that the P_4 does not appear to have a metaconid (Lacomba et al. 1986: plate 1). A more detailed comparison of these Late Miocene *Hexaprotodon* species is given in Weston (2000). It is noteworthy that *Hexaprotodon lothagamensis* compares more closely with narrow-muzzled species of hippo from Europe and the Arabian Peninsula than to any other sub-Saharan hippopotamid.

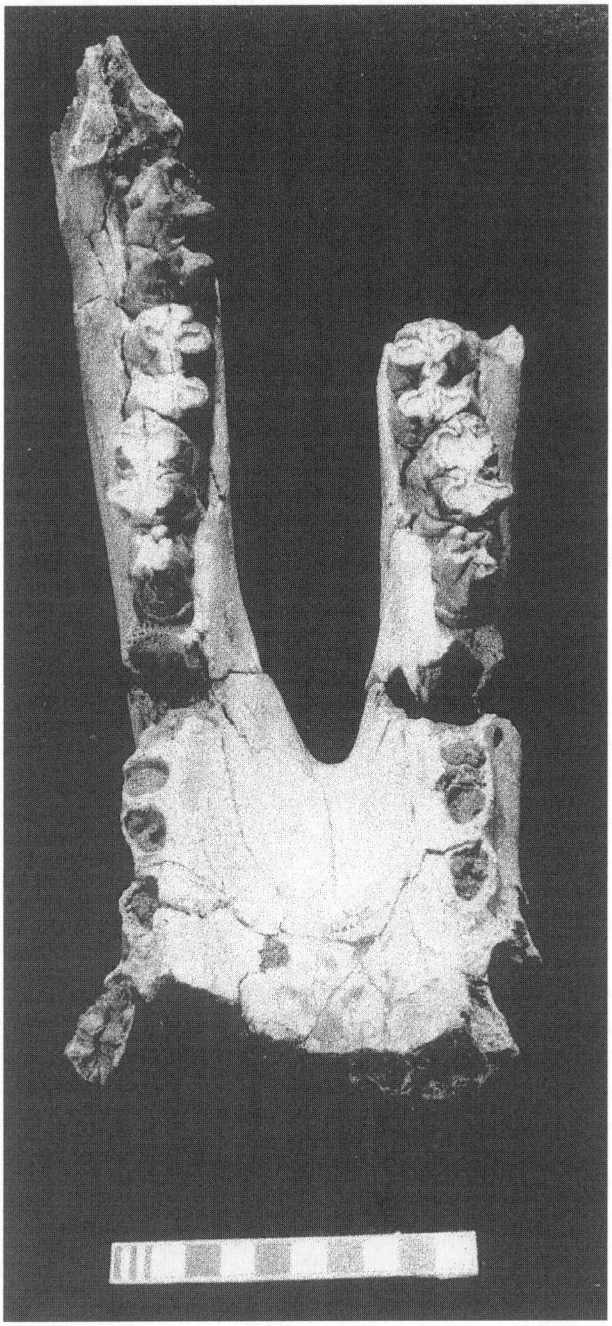

Figure 10.29 *Hexaprotodon lothagamensis* holotype, KNM-LT 23839, occlusal aspect.

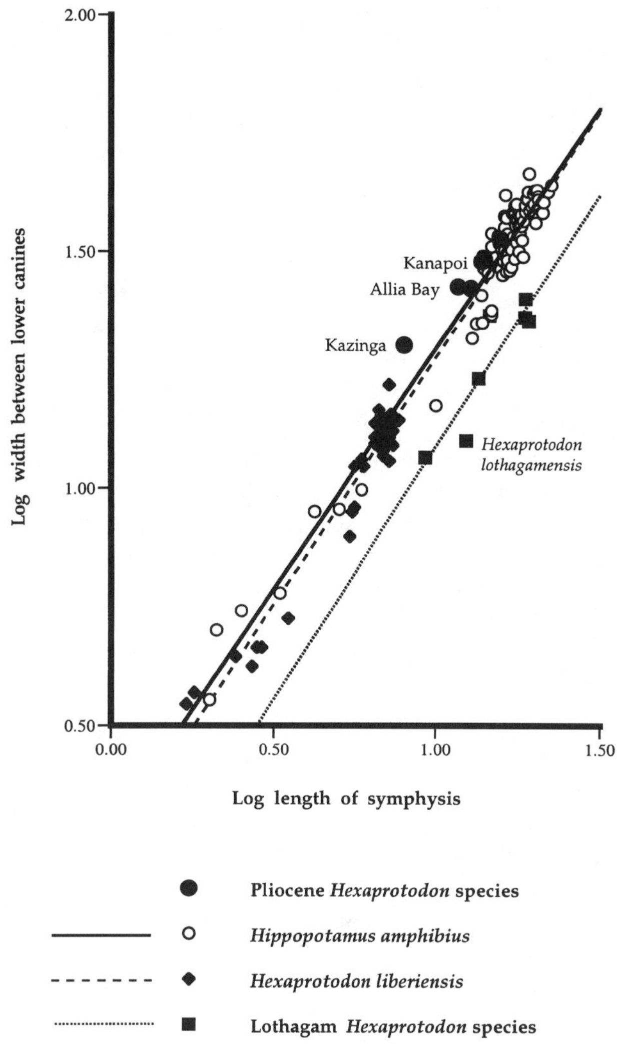

Figure 10.30 Allometric plot of the width between the lower canines and the length of the symphysis in extant and fossil hippopotamids. The Lothagam hippopotamids plot on a separate ontogenetic trajectory from that of the living hippopotamid species. Other Pliocene *Hexaprotodon* species have been included for comparison (see table 10.10). The slope and intercept values are recorded in table 10.3.

Hexaprotodon sp. indet.

(Figures 10.33–10.35; tables 10.4–10.8, 10.10)

Lothagam Material

- *Lower Nawata:* 317, juvenile Lt. mandible and symphysis (I_1, C, dP_3–dP_4, alveoli; M_1), Lt. premaxilla (I^2) and cranial fragments; 8585, Lt. (P^1, P^2, P^4) and Rt. (P^2, P_1, P_3); 23148, Lt. ramus fragment (dP_4–M_1), Lt. (dP^2, dP^4) and upper canine tip; 23278, Lt. P_4; 23874, Rt. (I^1, M^2, M^3, I_1).
- *Upper Nawata:* 51, Rt. ramus fragment (dP_4–M_1); 105, Lt. maxilla (P^2–M^3), Rt. maxilla (post P^2–P^3,

M^1–M^2) and Rt. upper canine; 26222, Lt. (P_1, P_2, P_3, P_4, M_1) and Rt. (P_2, P_3, P_4, M_2, M_3).

- *Nawata Formation level indet.:* 23254, Lt. P^4.
- *Horizon indet.:* 64, Lt. P_2; 79, juvenile mandibular symphysis (Rt. I_1–P_1, dP_2–dP_4, $M_1$1 and Lt. I_1–I_2); 8735, Rt. ramus fragment (P_4–M_3).

There is a significant amount of fossil evidence at Lothagam for a much larger hippopotamus than *Hexaprotodon harvardi.* Unfortunately, the variation in morphology, and the lack of associated material, makes it impossible to assign these specimens to a single species. LT 79 is a juvenile mandibular symphysis, with a size similar to that of an adult *Hexaprotodon harvardi* jaw (figure 10.33). The morphology is similar to LT 23105, and both specimens share a long, shallow symphysis, although the rami of LT 79 appear to be broader and more robust. In contrast, LT 108 is a juvenile mandible of equivalent age, and LT 23888 is an adult; both of them have deeper symphyses and more lightly built rami (figure 10.14). The robustness of LT 79 could well be linked with sexual dimorphism, and the complexity of ontogenetic variation within species is still poorly known. However, it is probable that the intraspecific variation previously attributed to *Hexaprotodon harvardi* includes at least one other species. The following specimens are exceptionally large or morphologically distinct.

LT 26222 is a collection of well-preserved associated teeth that are distinct in both size and morphology. The single-rooted P_1 possesses a pronounced distobuccal ridge (figure 10.34). P_2 possesses three pustulate, distal ridges, and the specimen compares with LT 64. P_3 is similar, with a triangular cuspule replacing the distolingual ridge. The P_4 is an extremely large, single-cusped tooth with one mesial and two distal ridges. The cingulum is deep and continuous, and it has a huge expansion of the distal shelf (figure 10.34). The buccal face of the cusp has a triangular band of secondary cingulum, and this characteristic is shared with LT 23278, another unicuspid P_4.

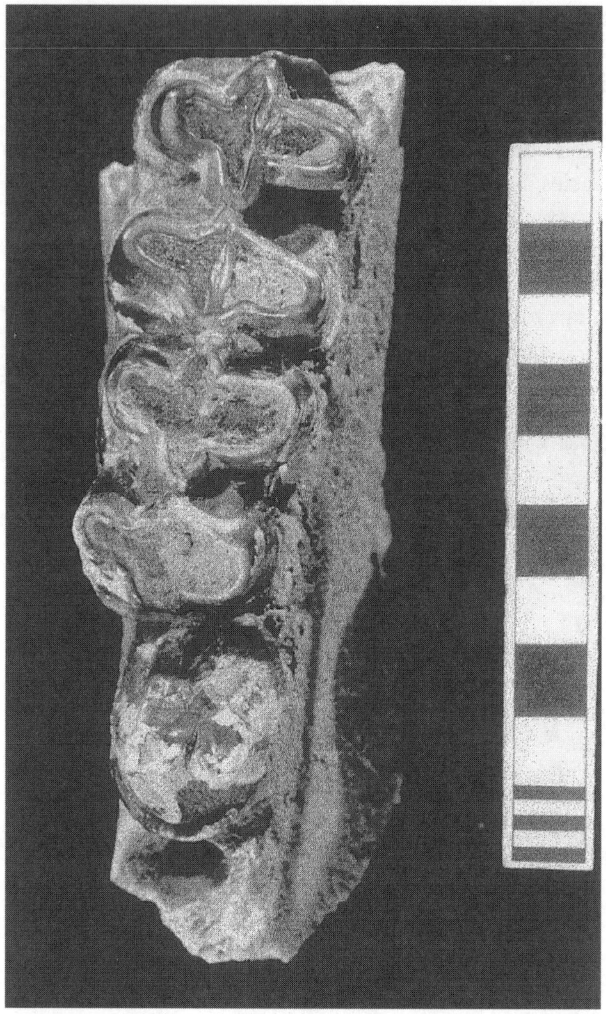

Figure 10.32 *Hexaprotodon* aff. *Hex. lothagamensis* right mandibular fragment, KNM-LT 23871, occlusal aspect.

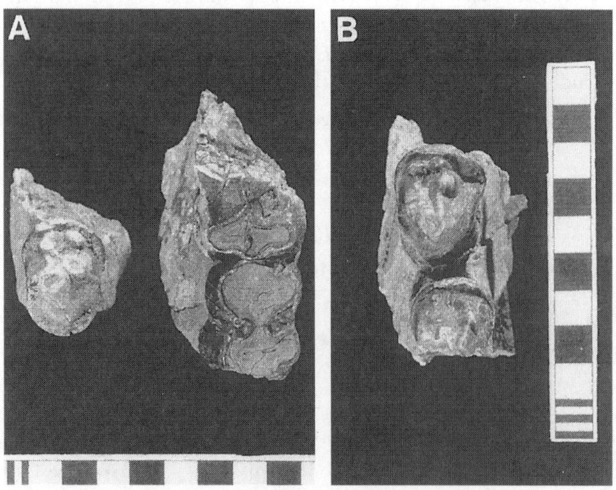

Figure 10.31 Examples of P_4 morphology that differ only slightly from the specimens assigned to *Hexaprotodon lothagamensis:* A = KNM-LT 23867, right P_4, M_1 and anterior M_2, occlusal aspect; B = KNM-LT 23864, left posterior P_{3-4}, occlusal aspect. Note that the P_4 main cusps are not equally elevated and are not aligned buccolingually.

LT 79, 51, 23148, and 317 are specimens of a large, subadult, robust hippopotamus. A left premaxillary fragment is associated with LT 317, which contains a huge, obliquely worn upper incisor; the occlusion is analogous to that of an upper canine. LT 23254 is a morphologically unique P^4 that possesses two lingual cusps separated by a ridge (figure 10.12). LT 8585 and LT 105 are both examples of associated large teeth.

Figure 10.33 *Hexaprotodon* sp. indet. juvenile mandibular symphysis, KNM-LT 79, occlusal aspect (Rt. I_1–P_1, dP_2–dP_4, M_1 and Lt. I_1–I_2).

Hexaprotodon protamphibius Arambourg, 1944 *Hexaprotodon* cf. *Hex. protamphibius*

(Figures 10.35–10.37; tables 10.9, 10.10)

Lothagam Material

- *Apak Member:* 23628, Lt. Mt IV; 26219, Rt. Mt V, Rt. and Lt. cuboid and proximal metapodial; 26220, Rt. Mc IV, Rt. Mc V, 2 phalanges and skeletal fragments; 26231, Lt. femur (distal end missing); 26348, juvenile symphysis fragments (broken /C and erupting P_2).

The sample of cranial material from the Apak Member is largely juvenile and poorly preserved. There are eight Apak hippopotamid specimens from the early collections, but the exact derivation and identity of this material remains uncertain. The new material, of known provenance, includes a fragmentary immature symphysis, LT 26348, with the P_2 partially erupted. The incisor root cavities indicate that the teeth were differentiated in size, and I_1 was larger than I_2 and I_3 (figure 10.35, top). The lower canine has a flattened mesial border. A symphysial specimen, LT 23854, from the Upper Nawata, possesses a reduced I_2 only, which erupts later than the adjacent incisors (figure 10.35, bottom). Differentiation of size and the eruption sequence of lower incisors are unique to these specimens. LT 23848 is a weathered premaxillary, similar to LT 4, with a peglike I^3.

There are seven postcranial specimens from the Apak Member, all of which differ from material assigned to *Hexaprotodon harvardi.* The metapodials are short and robust, and the bones are broader and less gracile, although they are similar in length to the Lower Nawata material (figure 10.22, bottom; table 10.9). A partial femur, with the distal end absent, is comparable with the *Hexaprotodon lothagamensis* femur LT 26242, with the exception of the broadened shaft. LT 26212 is the only astragalus known from the Apak, and its proximal surface is damaged. However, this specimen is relatively short and broad. A small, isolated cuneiform and a cuboid were also recovered from the Apak.

The cranial material is not complete enough to permit certain identification, but the abrupt appearance of

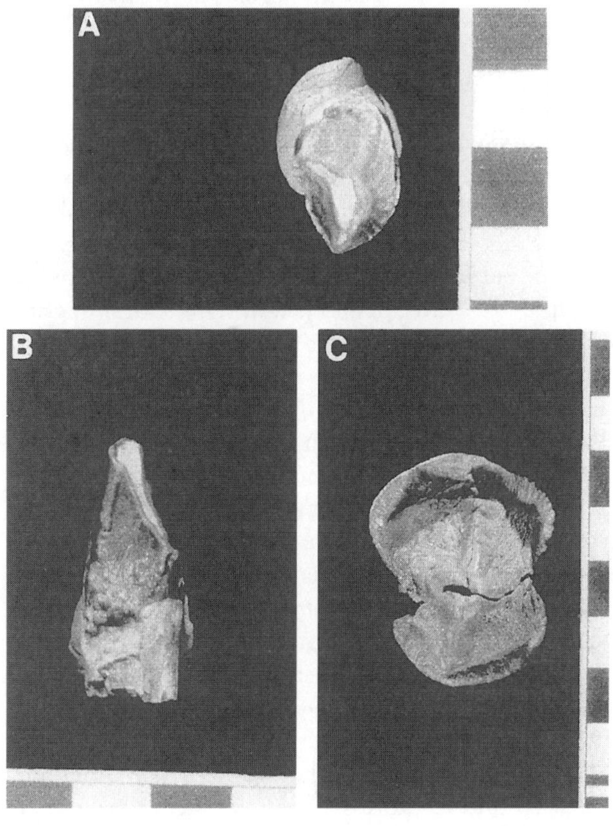

Figure 10.34 *Hexaprotodon* sp. indet. teeth, KNM-LT 26222: A = left P_1, occlusal aspect; B = left P_1, distal aspect; C = left P_4, occlusal aspect.

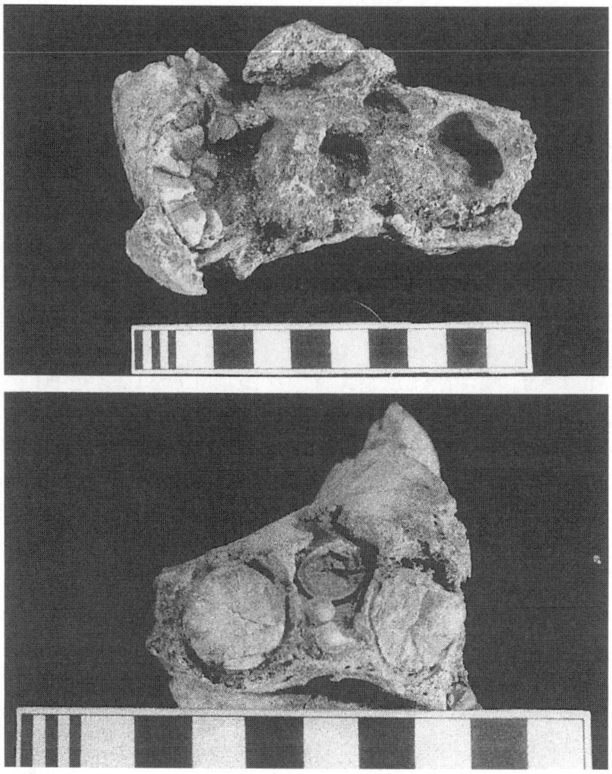

Figure 10.35 *Hexaprotodon* mandibular symphysis fragments: top = *Hexaprotodon* cf. *Hex. protamphibius,* KNM-LT 26348, left mandibular symphysis fragment, posterior aspect (note that the I_2 and I_3 sockets are smaller than the I_1 socket); bottom = *Hexaprotodon* sp. indet., KNM-LT 23854, right mandibular symphysis fragment, anterior aspect.

a shorter stockier hippo, together with the evidence for the initiation of incisor differentiation, indicates that this species is not *Hexaprotodon harvardi.* Incisor differentiation in fossil hippos recovered from other Pliocene sites (East and West Turkana sites, Allia Bay and Kanapoi, respectively) is linked with the shortening and broadening of the mandible (figure 10.30). The earliest described *Hexaprotodon* from the Turkana Basin, other than *Hexaprotodon harvardi,* is *Hexaprotodon protamphibius* (Arambourg 1944b). Harris (1991) regarded both hexaprotodont and tetraprotodont morphotypes of this medium-sized hippopotamus as being part of a single evolving *Hexaprotodon protamphibius* lineage, the early hexaprotodont material representing part of a basinal population. *Hexaprotodon protamphibius* has a wide geographical spread: it is known from the Omo Shungura sequence, Ethiopia, and the Koobi Fora and Nachukui Formation, Kenya (Coryndon and Coppens 1973; Gèze 1985; Harris et al. 1988a, 1988b; Harris 1991). Considerable morphological change differentiates early and late forms. Gèze (1985) considered early *Hexaprotodon protamphibius* to warrant placement in a separate subspecies, *Hexaprotodon protamphibius turkanensis.* Until recently, only a few fragmentary hippopotamus specimens were known from the lower part of the Koobi Fora sequence. A complete skeleton (ER 3661) has now been recovered from Allia Bay (table 10.10 and figure 10.36). The symphysial dimensions of the Allia Bay and Kanapoi hippopotamus compare more closely with early forms of *Hexaprotodon protamphibius* (figure 10.30). Other characters that distinguish this material from *Hexaprotodon harvardi* include a larger I_1 than I_2 and I_3, the arrangement of the incisors in a straight line across the front of the symphysis as opposed to a gentle arc (figures 10.36 and 10.15B), a distinctive lower canine with a bioconvex anteromedial border, the tooth tapering buccally, and a reduction in the size of the premolars relative to the molars. The Kanapoi mandible is not as derived as the Allia Bay specimen: the canines are not as laterally sited, and the

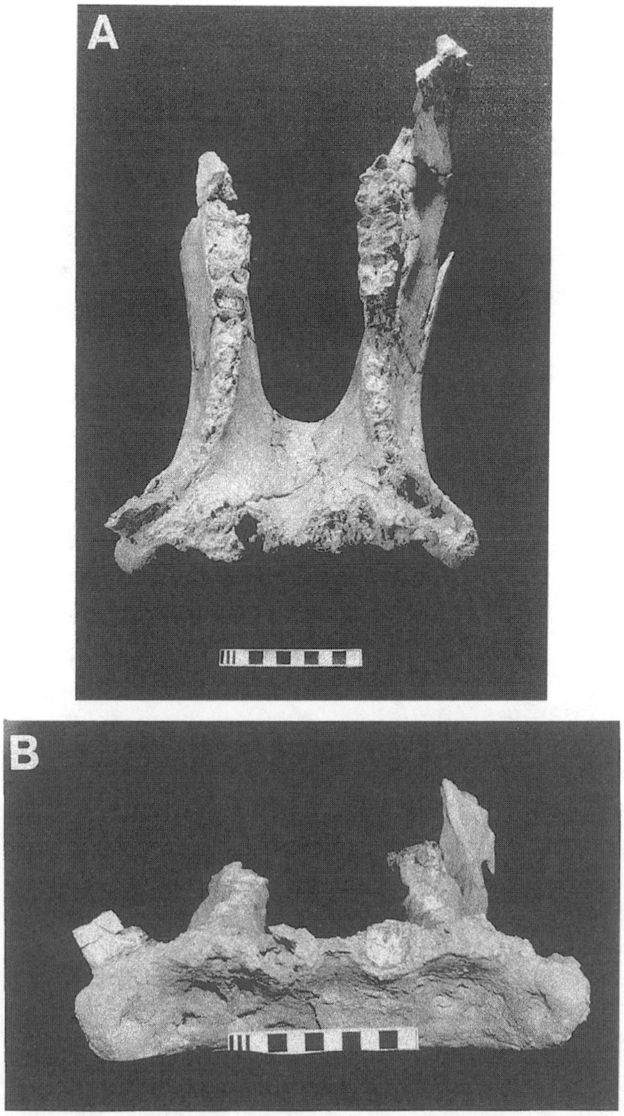

Figure 10.36 *Hexaprotodon* cf. *Hex. protamphibius* mandible, KNM-ER 3661 from Allia Bay: A = occlusal aspect; B = anterior aspect. The specimen is listed in table 10.10.

premolars are not as reduced in size relative to the molars (figures 10.36 and 10.37). A very slight degree of orbit elevation can be observed in the crania from both sites, evident from the strong depression present in the maxilla/lacrimal region of the facial skull, adjacent to the orbits. The muzzle possesses rounded, as opposed to vertical, flanks.

The range and variation of morphology in *Hexaprotodon protamphibius* is already extensive, and there is a need to fully describe the Allia Bay skeleton and assign it to a new species. The Apak material of known provenance at Lothagam does compare with material from both Kanapoi and Allia Bay, and, as a preliminary measure, prior to the full description of the Allia Bay hippo, it is referred to as *Hexaprotodon* cf. *Hex. protamphibius.*

Discussion

Geographic Distribution of Late Miocene and Early Pliocene *Hexaprotodon* Species

Hexaprotodon harvardi has been documented from other East African localities. A *Hexaprotodon* species from the Middle Awash Valley in Ethiopia had been provisionally identified as *Hexaprotodon* cf. *Hex. harvardi,* but the material has not yet been described (Kalb et al. 1982a, 1982b). The Middle Awash Valley sediments from the Adu-Asa Formation correlate with the Nawata Formation at Lothagam (Kalb 1993; Kalb and Mebrate 1993). More recently, *Hexaprotodon harvardi* has been identified from the Manonga Valley, Tanzania (Harrison 1997). These Neogene sediments have yielded a sizeable collection of postcranial and dental material. The hippopotamid material is described as being morphologically and metrically indistinguishable from that of Lothagam and the Lukeino Formation, Baringo, Kenya (Harrison 1997). The Lukeino material is roughly contemporary with the Upper Nawata at Lothagam, but similar specimens have also been recovered from the older Mpesida beds (Hill et al. 1985; Hill 1995), which are roughly contemporary with the Lower Nawata. In contrast to the findings at Lothagam, the older material at Baringo indicates the presence of a slightly larger hippo (Harrison 1997). The precise identification of the fragmentary, though abundant, hippopotamid specimens from the Baringo Basin has been

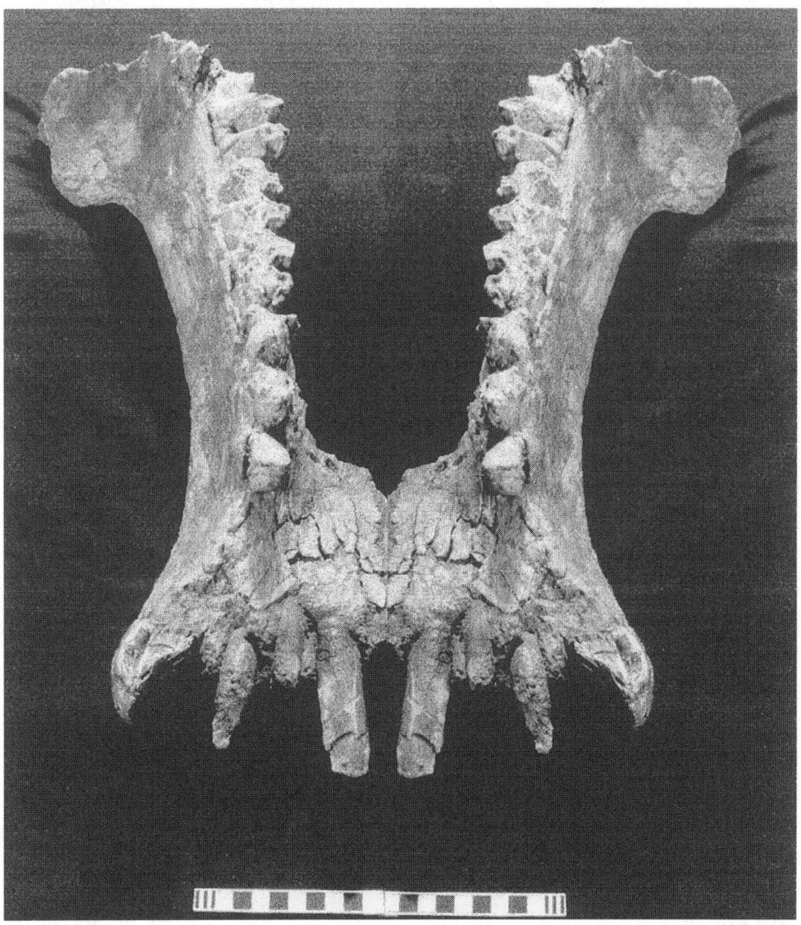

Figure 10.37 *Hexaprotodon* cf. *Hex. protamphibius* from Kanapoi, KNM-KP 1. A reconstruction of the mandible achieved by duplicating in reverse KP 1 (the right ramus) and joining the two images along the symphysis just to the right of I_1.

problematic (Coryndon 1978). Isolated teeth tend not to be taxonomically distinctive, resulting in a long period of apparent conservatism in the Late Miocene. A similar anomaly arises with the Late Miocene and Early Pliocene sites from the Western Rift (Uganda and Zaire) and Algeria (Arambourg 1944a; Cooke and Coryndon 1970; Pavlakis 1990; Faure 1994). These areas both yield teeth of a smaller *Hexaprotodon* species, the taxonomic affinities of which are difficult to ascertain without cranial specimens.

Gaziry (1987) described a new species, *Hexaprotodon sahabiensis* from Sahabi, a Libyan site presumed to be of Late Miocene and Early Pliocene age (Geraads 1987). The material described by Gaziry has affinities with *Hexaprotodon harvardi.* The worn P_4 of *Hexaprotodon sahabiensis* is almost identical to LT 23908, an unworn P_4 from Lothagam (figure 10.18). The P^4 is also similar to those found at Lothagam. It should be noted that Gaziry confused the buccal and lingual cusps in his description of the latter tooth. In addition, a mesiolingual ridge is evident on the upper incisor figured in Gaziry's paper, and this ridge is characteristic of the upper incisors that were recovered from the Lower Nawata (figure 10.9). Unfortunately, an unequivocal comparison of upper incisor dimensions is not possible, as Gaziry confused mesiodistal crown length with apicobasal length (Harrison 1997). However, the upper incisors of *Hexaprotodon sahabiensis* still appear to be larger than those from Lothagam. In contrast, the postcanine teeth of *Hexaprotodon sahabiensis* are relatively small, and they fall just within the range of size variation seen in *Hexaprotodon harvardi* (see Weston 2000:table 1). The depth of the ramus is also particularly shallow (Gaziry 1987). Without more complete mandibular or cranial evidence, it remains uncertain as to whether *Hexaprotodon sahabiensis* is taxonomically distinct from *Hexaprotodon harvardi.* Pertinent to this matter is the recent discovery of a Late Miocene *Hexaprotodon* species from Abu Dhabi on the Arabian Peninsula (Hill et al. 1990; Whybrow et al. 1990; Gentry 1999). In this species, although the tooth morphology can be regarded as similar to that of the latter hexaprotodonts, the mandibular symphysis is exceptionally narrow and more like that of *Hexaprotodon lothagamensis* (Weston 2000).

The Late Miocene marks the first appearance of *Hexaprotodon* in Eurasia. *Hexaprotodon sivalensis* (Falconer and Cautley 1836), from the Siwalik Hills of India (Colbert 1935; Hooijer 1950), is similar in size to *Hexaprotodon harvardi* but less gracile and more aquatically specialized. *Hexaprotodon iravaticus* (Falconer and Cautley 1847) is a smaller, poorly known species, from the upper Irrawaddy beds in Burma. This hexaprotodont species can be distinguished by its narrow muzzle. *Hexaprotodon* species are also known from Late Miocene and Early Pliocene sites in Spain, Italy, and France. Aguirre (1963) described a new species, *Hippopotamus (Hexaprotodon) crusafonti,* from the Granada Basin, Spain. The fragmentary remains of this medium-sized hippopotamus have been recovered from a number of localities, including La Mosson in France (Faure and Méon 1984). *Hexaprotodon crusafonti* is the earliest known tetraprotodont species, and it was mainly recovered from the Spanish Turolian (Lacomba et al. 1986). A second species, *Hexaprotodon primaevus* (Crusafont et al. 1963), was described from the Teruel Basin, Spain, but it is probably synonymous with *Hexaprotodon crusafonti.* Finally, a small hexaprotodont species has been recovered from Casino in Italy and also from Sicily. The former material has been referred to as *Hexaprotodon pantanellii* (Joleaud 1920) and the latter as *Hippopotamus siculus* (Hooijer 1946). These small hexaprotodont hippopotamids may be synonymous with African species, but the material is not sufficiently diagnostic to warrant such identification.

Sexual dimorphism

The recognition of sexually dimorphic traits within the family Hippopotamidae is a prerequisite of taxonomic determination. Size dimorphism between males and females has been well documented in the living *Hippopotamus amphibius* (Laws 1968; Kingdon 1979). This type of dimorphism is the result of the males growing for longer than the females (figure 10.38, top). The sexes can be seen to plot at different points along a common growth trajectory (figure 10.38, bottom). The male hippopotamus is proportionately larger than the female, with the exception of molar size. It was documented by Laws (1968) that there was no sex difference in weight of the molars. The undifferentiated molar size could merely result from tooth scaling during allometric size adjustments (Gould 1975; Shea and Gomez 1988). The latter explanation also implies that size dimorphism in *Hippopotamus amphibius* is highly derived. Sexual dimorphism in *Hexaprotodon liberiensis,* the living pygmy hippopotamus, is indicated by a distinction in cranial shape, as opposed to size. There is an increase in the breadth relative to length of the skull (figure 10.39). In addition, the muzzle in males is less constricted than in the females. Conversely, orbit size is larger in females. The latter characters can be heterochronically distinguished: the larger orbit size in females is generated by a different rate of growth rather than by growth duration (figure 10.40). Heterochronic distinctions may have some bearing on the plasticity of dimorphic characters.

Hexaprotodon harvardi crania are dimorphic in shape but not size. The posterior skull region is very similar to that of *Hexaprotodon liberiensis.* The length of the braincase is essentially similar in males and fe-

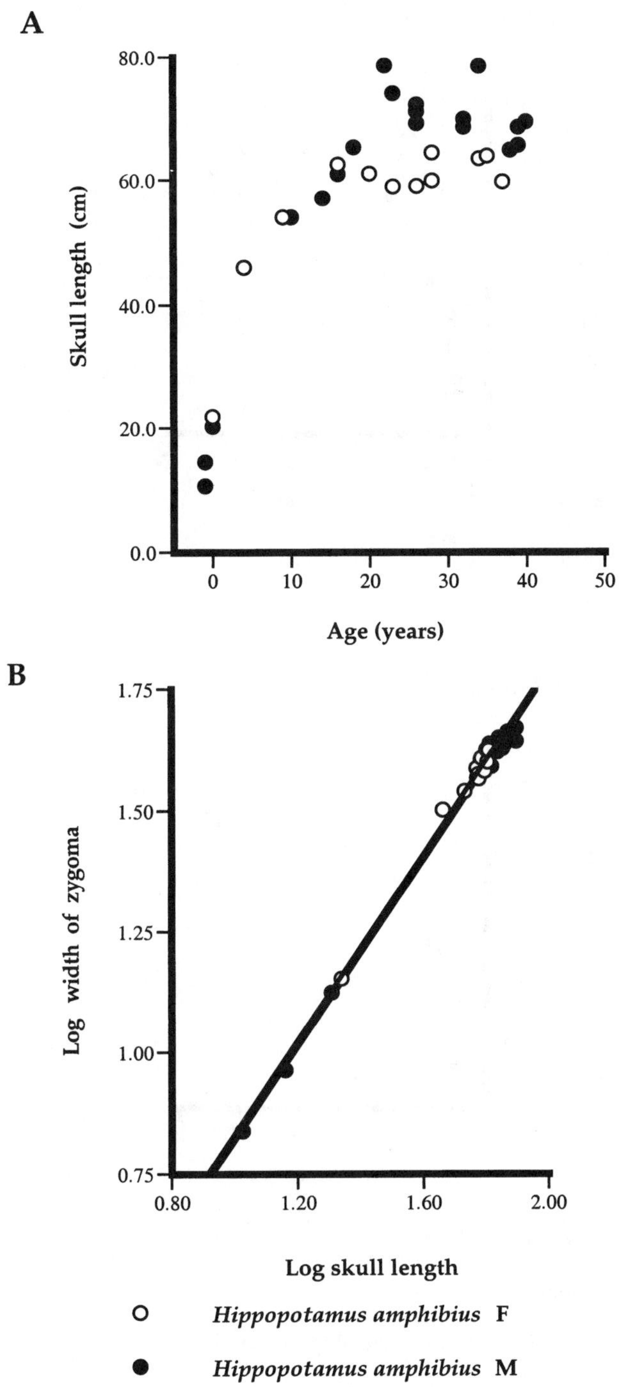

Figure 10.38 Comparison of male and female growth trajectories in *Hippopotamus amphibius,* illustrating sexual dimorphism as a result of growth that continues in males beyond that of the females: A = age plot; B = allometric plot. F = female, M = male. The slope and intercept values are recorded in table 10.3.

males (figure 10.5); this appears to be a distinction between *Hippopotamus* and *Hexaprotodon.* The muzzle of *Hexaprotodon harvardi* is more constricted in one sex than in the other. If the allometric adjustments due to differential size of the living and fossil species are taken into consideration, the dimorphism of the muzzle is more pronounced in *Hexaprotodon harvardi* (figure 10.3). However, sexual dimorphism of the orbit in *Hexaprotodon harvardi* is more difficult to evaluate, as the orbit size does not compare with that of *Hexaprotodon liberiensis* (figure 10.41A). The fossil specimens appear to plot on the *Hippopotamus amphibius* growth trajectory (figure 10.41B). Stuenes (1989) documented dimorphism in *Hippopotamus lemerlei* Grandidier, 1868, a subfossil species from Madagascar. In this species, the morph with the narrower muzzle was interpreted as a female. In contrast, varying degrees of muzzle constriction specific to each sex is not evident in *Hippopotamus amphibius.* It appears that constriction of the muzzle within the family Hippopotamidae is a particularly plastic trait, and this suggests that sexual dimorphism could be ecophenotypically linked. This is strongly supported by the Madagascan example, which is essentially an isolated population of *Hippopotamus amphibius* that has been subjected to different ecological conditions. It is clear that the existence of sexual dimorphism has considerable bearing on the interpretation of East African fossil hippos.

Evolutionary Relationships

The coexistence of two hippo species of different sizes is documented at several fossil localities and is evident today in West Africa. The available evidence indicates

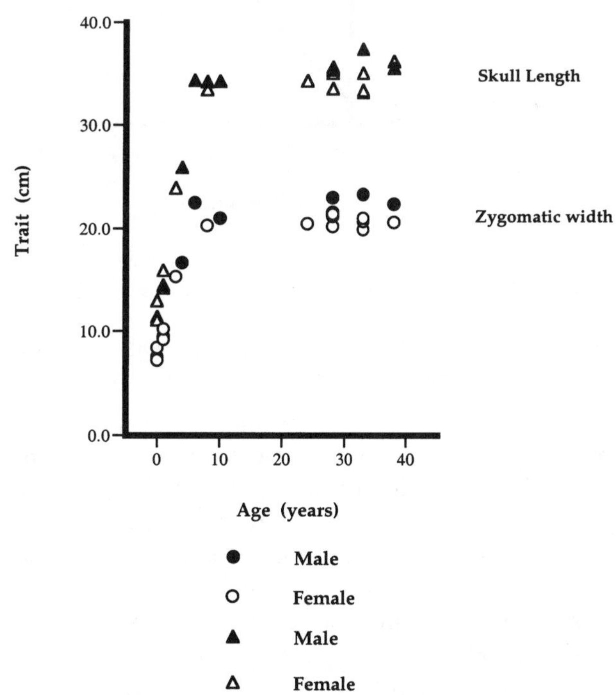

Figure 10.39 Comparison of male and female *Hexaprotodon liberiensis* growth trajectories, illustrating the differences in cranial width but not in cranial length between males and females.

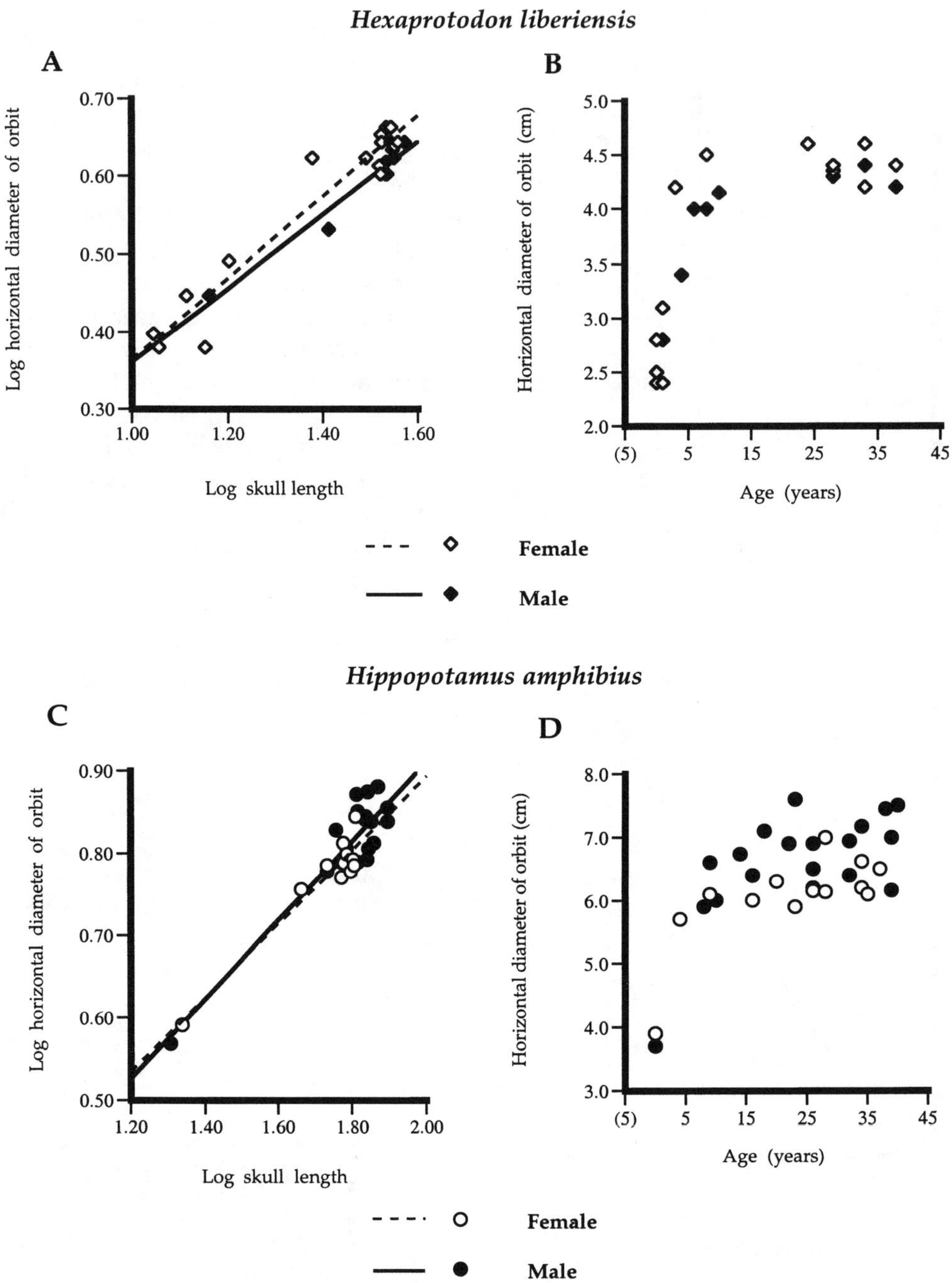

Figure 10.40 Comparison of allometric and age plots to illustrate sexual dimorphism of the hippopotamid orbit: top (A and B) = *Hexaprotodon liberiensis;* bottom (C and D) = *Hippopotamus amphibius.*

that the two living species represent separate evolutionary lineages diverging before the Pliocene. Coryndon (1977) postulated that *Hexaprotodon liberiensis* was the most primitive of any known hippopotamid. However, the discovery at Lothagam of a new species of narrow-muzzled hippopotamus, which appears to be less derived than *Hexaprotodon liberiensis,* has changed this perspective. *Hexaprotodon liberiensis* appears to be a scaled-down version of a *Hexaprotodon* species subsequent to or coeval with *Hexaprotodon harvardi.* The cranium of the living pygmy is very similar to that of early *Hexaprotodon harvardi,* and it is only the shortened symphysis that would have to be considered independently derived (Weston 2000).

In East Africa, *Hexaprotodon harvardi* appears to form the base of a *Hexaprotodon* lineage; it was succeeded by *Hexaprotodon protamphibius* and possibly culminated in *Hexaprotodon karumensis* Coryndon 1977 (Harris 1991). Lothagam provides a glimpse of the morphology and diversity of early hippopotamids. A large proportion of the material from the Nawata Formation can be interpreted to be part of an evolving *Hexaprotodon harvardi* lineage that is comparable with that of *Hexaprotodon protamphibius* in the Turkana Basin (Harris 1991). A trend toward size increase and extreme cursoriality seems to have occurred, and this trend was possibly linked with a change in diet, as indicated by the smaller upper incisors. Preliminary studies of the carbon isotopic composition in hippopotamid tooth enamel from Lothagam indicated a mixed diet (Leakey et al. 1996), but these results may have included enamel from different species.

Hexaprotodon lothagamensis is the first fossil hippopotamus with a long, narrow symphysis (i.e., the anterior jaw has not expanded laterally) to be described from Africa. However, narrow-muzzled species have been documented from outside Africa, notably *Hexaprotodon crusafonti* from Spain (Lacomba et al. 1986) and *Hexaprotodon* aff. *Hex. sahabiensis* from Abu Dhabi on the Arabian Peninsula (Gentry 1999). The holotype of *Hexaprotodon lothagamensis* is at least 7 million years old and probably predates both the Spanish and the Arabian material. The mandible described by Lacomba et al. (1986) from La Portera, Spain, is estimated to be of Late Miocene age, corresponding to the MN13 zone of the European Mammal Neogene (Mein 1975). Hippos appear to have spread from Africa into eastern Spain, probably via the Iberian Peninsula, for the first time during this period (Moyà-Solà and Agustí 1989). The Abu Dhabi material is estimated to be Late Miocene in age (Whybrow and Hill 1999). However, direct correlation of the Baynunah Formation with the Saudi Arabian Miocene is not possible, and more precise dates have not been obtained (Whybrow 1989; Whybrow et al. 1990; Whybrow and Hill 1999). The Abu Dhabi *Hexaprotodon* has affinities with both *Hexaprotodon sahabiensis* from Libya and *Hexaprotodon lothagamensis.* The earliest certain record of *Hexaprotodon* in Asia is 5.5 Ma (Barry 1995). It appears that both broad and narrow-muzzled hippos spread into Eurasia during the Late Miocene. The earliest evidence of *Hexaprotodon* is from the Tugen Hills, Baringo District, Kenya. Fragmentary, isolated teeth have been recovered from the top of the Ngorora Formation that probably dates between 8 and 10 Ma (Coryndon 1978; Hill 1995). Together with the remains of *Kenyapotamus,* known from earlier Miocene sites in Africa (Pickford 1983, 1990), this material supports the hypothesis that the hippopotamus arose in Africa.

The abundant remains of *Hexaprotodon harvardi* at Lothagam suggest that this species lived in close proximity to water; thus it favored a riparian environment. Conversely, the paucity of *Hexaprotodon lothagamensis* remains suggests that this species was less dependent on

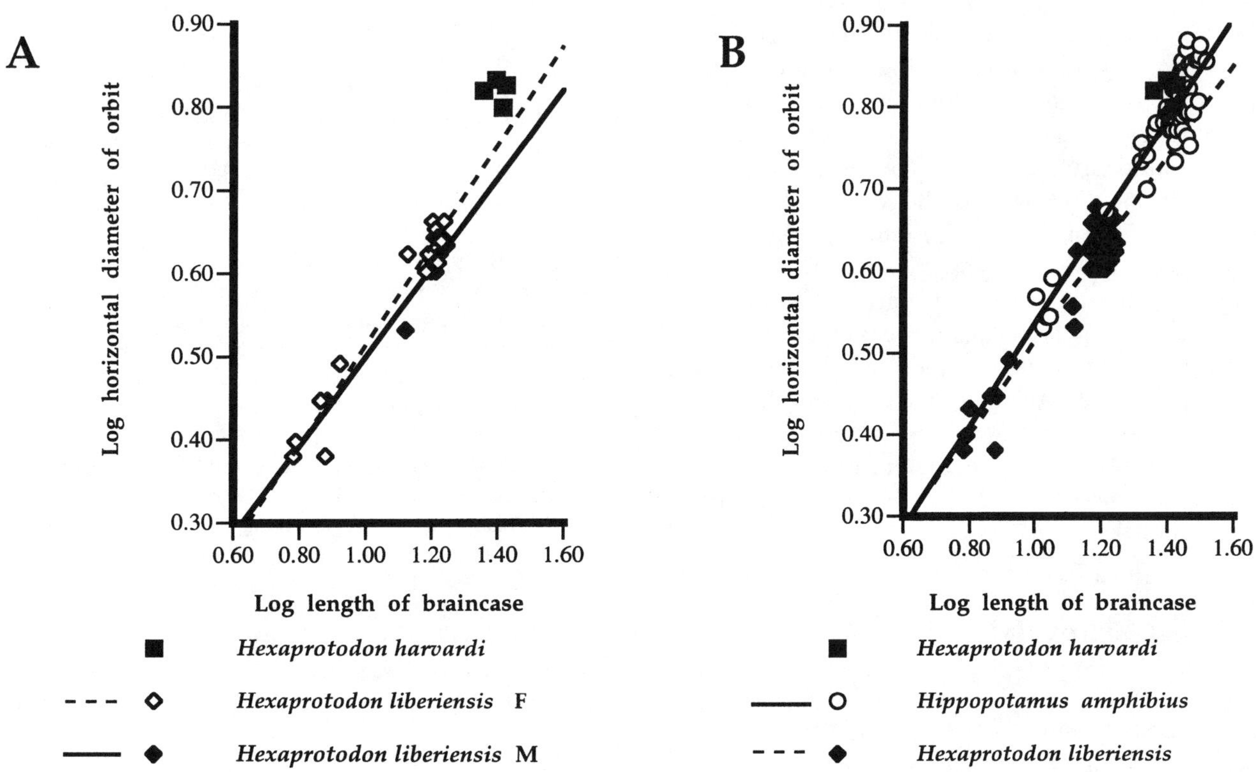

Figure 10.41 Comparison of hippopotamid growth allometries, illustrating the relationship between the horizontal diameter of the orbit and the length of the braincase. The slope and intercept values are recorded in table 10.3. A = comparison of *Hexaprotodon liberiensis* male and female growth trajectories; the *Hexaprotodon harvardi* specimens do not fall on either trajectory. B = comparison of *Hexaprotodon liberiensis* and *Hippopotamus amphibius* growth trajectories; the *Hexaprotodon harvardi* specimens compare more closely to the *Hippopotamus amphibius* growth trajectory.

water. Furthermore a narrow muzzle, large premolars, and shallow ramus are all characteristics that have been correlated with browsing as opposed to grazing in ungulates (Janis and Ehrhardt 1988; Janis 1995). Thus, within the family Hippopotamidae it appears that there was not only a transition from browsing to grazing but also a shift from terrestrial to progressively more aquatic habitats. *Hexaprotodon harvardi* appears to mark the beginning of this transition. However, narrow-muzzled species had quite a wide geographic spread and have been documented from the Late Miocene, Pliocene, and Recent (Weston 2000). The extant *Hexaprotodon liberiensis,* which has a relatively narrow (though short) symphysis, is a forest browser feeding on leaves, roots, and fallen fruit (Eltringham 1993). In addition, rapid dwarfing on islands led to the evolution of secondarily derived cursorial hippos that had readapted to a terrestrial environment (Houtekamer and Sondaar 1979; Stuenes 1989; Caloi and Palombo 1996; Spaan 1996).

Coryndon (1977) suggested that the two genera within the Hippopotamidae evolved independently. It is more probable to consider *Hippopotamus* species as derived members of a *Hexaprotodon* lineage. The implication here would be that the taxa are congeneric. The evidence is beginning to suggest that *Hexaprotodon* split into two lineages, which then independently gave rise to broad and narrow-muzzled hippos. The discovery of a new species at Lothagam, which has affinities with hippos outside Africa, radically changes our interpretation of the Late Miocene *Hexaprotodon* radiation. The evolutionary picture that Lothagam reveals about hippos is one of transition. The common hippo (*Hexaprotodon harvardi*) is a forerunner to the specialist semiaquatic grazing hippos common in the Pliocene and Pleistocene. The anterior muzzle of *Hexaprotodon harvardi* has begun to expand laterally, and the roof of the orbit is very slightly raised. This animal probably had a mixed diet. An additional testimony to the evolution of this remarkable dietary and habitat switch that occurred within the Hippopotamidae are the two modern survivors—one a dweller of forests and the other of rivers and lakes.

Acknowledgments

I thank Meave Leakey for her continual support, advice, and permission to work on the Lothagam specimens in the National Museums of Kenya. I also thank the following people for their help: Ken Joysey, Adrian Friday, Alan Gentry, William Anyonge, Mary Muungu, Frederick Kyalo Manthi, David Reese, Adrian Lister, Andrew Kitchener, Francis Renoult, Linda Gordon, Nancy Todd, Bill Stanley, and Bryn Mader. This research was supported by the Boise Fund, the Cambridge Philosophical Society, and a BBSRC studentship.

References Cited

Aguirre, E. 1963. *Hippopotamus crusafonti* n. sp. del Plioceno inferior de Arenas del Rey (Granada). *Notas y Comunicaciones del Instituto Geológico y Minero de España* 69:215–230.

Arambourg, C. 1944a. Au sujet de l'*Hippopotamus hipponensis* Gaudry. *Bulletin de la Société Géologique de France* 14: 147–152.

Arambourg, C. 1944b. Les hippopotames fossiles d'Afrique. *Comptes Rendus de l'Académie des Sciences* (Paris) 218: 602–604.

Barry, J. C. 1995. Faunal turnover and diversity in the terrestrial Neogene of Pakistan. In E. S. Vrba, G. H. Denton, T. C. Partridge, and L. H. Burckle, eds., *Paleoclimate and Evolution, with Emphasis on Human Origins,* pp. 115–134. New Haven: Yale University Press.

Caloi, L., and M. R. Palombo. 1996. Functional aspects and ecological implications in hippopotami and cervids of Crete. In D. S. Reese, ed., *Pleistocene and Holocene Fauna of Crete and Its First Settlers,* pp. 125–151. Monographs in World Archaeology 28. Madison, Wis.: Prehistory Press.

Colbert, E. H. 1935. Siwalik mammals in the American Museum of Natural History. *Transactions of the American Philosophical Society* 26:1–401.

Colbert, E. H. 1943. Pleistocene vertebrates collected in Burma by the American Southeast Asiatic Expedition. *Transactions of the American Philosophical Society* 32:395–429.

Cooke, H. B. S., and S. C. Coryndon. 1970. Pleistocene mammals from the Kaiso Formation and other related deposits in Uganda. In L. S. B. Leakey and R. J. G. Savage, eds., *Fossil Vertebrates of Africa,* vol. 2, pp. 107–224. London: Academic Press.

Corbet, G. B. 1969. The taxonomic status of the pygmy hippopotamus, *Choeropsis liberiensis,* from the Niger Delta. *Journal of Zoology* (London) 158:387–394.

Coryndon, S. C. 1977. The taxonomy and nomenclature of the Hippopotamidae (Mammalia, Artiodactyla) and a description of two new fossil species. *Proceedings of the Koninklijke Nederlandse Akademie van Wetenschappen,* ser. B, 80:61–88.

Coryndon, S. C. 1978. Fossil Hippopotamidae from the Baringo Basin and relationships within the Gregory Rift, Kenya. In W. W. Bishop, ed., *Geological Background to Fossil Man,* pp. 279–292. Edinburgh: Scottish Academic Press.

Coryndon, S. C., and Y. Coppens. 1973. Preliminary report on Hippopotamidae (Mammalia, Artiodactyla) from the Plio/Pleistocene of the lower Omo Basin, Ethiopia. In L. S. B. Leakey, R. J. G. Savage, and S. C. Coryndon, eds., *Fossil Vertebrates of Africa,* vol. 3, pp. 139–157. London: Academic Press.

Crusafont, M., R. Adrover, and J. M. Golpe. 1963. Hallazgo en España del mas primitivo hipopotamo fosil conocido hasta la actualidad. *Notas y Comunicaciones del Instituto Geológico y Minero de España* 70:71–79.

Eltringham, S. K. 1993. The pygmy hippopotamus. In W. L. R. Oliver, ed., *Status Survey and Conservation Action Plan: Pigs, Peccaries, and Hippos,* pp. 55–60. Gland, Switzerland: International Union for Conservation of Nature and Natural Resources.

Erdbrink, D., and W. Krommenhoek. 1975. Contribution to the knowledge of the fossil Hippopotamidae from the Kazinga Channel area (Uganda). *Säugetierkundliche Mitteilungen* 23:258–294.

Falconer, H., and P. T. Cautley. 1836. Note on the fossil hippopotamus of the Siválik Hills. *Asiatic Researches* 19:39–53.

Falconer, H., and P. T. Cautley. 1847. *Fauna Antiqua Sivalensis: Being the Fossil Zoology of the Sewalik Hills, in the North of India.* London: Smith, Elder.

Faure, M. 1994. Les Hippopotamidae (Mammalia, Artiodactyla) du rift occidental (bassin du Lac Albert, Ouganda): Etude préliminaire. In B. Senut and M. Pickford, eds., *Geology and Palaeobiology of the Albertine Rift Valley, Uganda-Zaire.* Vol. 2. *Palaeobiology/Paléobiologie,* pp. 321–337. Occasional Publication No. 29. Orléans: Centre International pour la Formation et les Echanges Géologiques.

Faure, M., and H. Méon. 1984. *L'Hippopotamus crusafonti* de La Mosson (près Montpellier): Première reconnaissance d'un hippopotame Néogène en France. *Comptes Rendus de l'Académie des Sciences* (Paris) 298:93–98.

Gaziry, A. W. 1987. *Hexaprotodon sahabiensis* (Artiodactyla, Mammalia): A new hippopotamus from Libya. In N. T. Boaz, A. El-Arnauti, A. W. Gaziry, J. de Heinzelin, and D. D. Boaz, eds., *Neogene Paleontology and Geology of Sahabi,* pp. 303–315. New York: Liss.

Gaudry, A. 1876. Sur un hippopotame à six incisives inférieures trouvé fossile en Algérie. *Comptes Rendus de l'Académie des Sciences* (Paris) 83:90–92.

Gentry, A. W. 1999. A fossil hippopotamus from the Emirate of Abu Dhabi, United Arab Emirates. In P. J. Whybrow and A. Hill, eds., *Fossil Vertebrates of Arabia,* pp. 271–289. New Haven: Yale University Press.

Geraads, D. 1987. Dating the northern African cercopithecid fossil record. *Human Evolution* 2:19–27.

Gèze, R. 1985. Répartition paléoécologique et relations phylogénétiques des Hippopotamidae (Mammalia, Artiodactyla) du Néogène d'Afrique orientale. In Y. Coppens, ed., *L'Environment des Hominides au Plio-Pléistocène,* pp. 81–100. Paris: Masson.

Gould, S. J. 1975. On the scaling of tooth size in mammals. *American Zoologist* 15:351–362.

Harris, J. M. 1991. Family Hippopotamidae. In J. M. Harris, ed., *Koobi Fora Research Project.* Vol. 3. *The Fossil Ungulates: Geology, Fossil Artiodactyls, and Palaeoenvironments,* pp. 31–85. Oxford: Clarendon Press.

Harris, J. M., F. H. Brown, and M. G. Leakey. 1988a. Stratigraphy and paleontology of Pliocene and Pleistocene localities west of Lake Turkana, Kenya. *Contributions in Science* 399:1–128.

Harris, J. M., F. H. Brown, M. G. Leakey, A. C. Walker, and R. E. Leakey. 1988b. Pliocene and Pleistocene hominid-bearing sites from west of Lake Turkana, Kenya. *Science* 239:27–33.

Harrison, T. 1997. The anatomy, paleobiology and phylogenetic relationships of the Hippopotamidae (Mammalia, Artiodactyla) from the Manonga Valley, Tanzania. In T. Harrison, ed., *Neogene Paleontology of the Manonga Valley, Tanzania: A Window into the Evolutionary History of East Africa,* pp. 137–190. New York: Plenum Press.

Hill, A. 1995. Faunal and environmental change in the Neogene of East Africa: Evidence from the Tugen Hills sequence, Baringo District, Kenya. In E. S. Vrba, G. H. Denton, T. C. Partridge, and L. H. Burckle, eds., *Paleoclimate and Evolution, with Emphasis on Human Origins,* pp. 178–193. New Haven: Yale University Press.

Hill, A., R. Drake, L. Tauxe, M. Monaghan, J. C. Barry, A. K. Behrensmeyer, G. Curtis, B. F. Jacobs, L. Jacobs, N. Johnson, and D. Pilbeam. 1985. Neogene palaeontology and geochronology of the Baringo Basin, Kenya. *Journal of Human Evolution* 14:759–773.

Hill, A., P. Whybrow, and W. Yasin al-Tikriti. 1990. Late Miocene primate fauna from the Arabian Peninsula: Abu Dhabi, United Arab Emirates. *American Journal of Physical Anthropology* 81:240–241.

Hooijer, D. A. 1946. Notes on some Pontian mammals from Sicily figured by Seguenza. *Archives Néerlandaises de Zoologie* 7:301–333.

Hooijer, D. A. 1950. The fossil Hippopotamidae of Asia with notes on the recent species. *Zoologische Verhandelingen* 8:1–124.

Hopwood, A. T. 1926. The geology and palaeontology of the Kaiso Bone Beds: Fossil mammalia. *Occasional Papers of the Geological Survey of the Uganda Protectorate* 2:13–36.

Houtekamer, J. L., and P. Y. Sondaar. 1979. Osteology of the forelimb of the Pleistocene dwarf hippopotamus from Cyprus with special reference to phylogeny and function. *Proceedings of the Koninklijke Nederlandse Akademie van Wetenschappen,* ser. B, 82:411–448.

Janis, C. M. 1995. Correlations between craniodental morphology and feeding behavior in ungulates: Reciprocal illumination between living and fossil taxa. In J. J. Thomason, ed., *Functional Morphology in Vertebrate Paleontology,* pp. 76–98. Cambridge: Cambridge University Press.

Janis, C. M., and D. Ehrhardt. 1988. Correlation of relative muzzle width and relative incisor width with dietary preference in ungulates. *Zoological Journal of the Linnean Society* 92:267–284.

Joleaud, L. 1920. Contribution a l'étude des hippopotames fossiles. *Bulletin de la Société Géologique de France* 20:13–26.

Kalb, J. E. 1993. Refined stratigraphy of the hominid-bearing Awash Group, Middle Awash Valley, Afar Depression, Ethiopia. *Newsletters on Stratigraphy* 29:21–62.

Kalb, J. E., and A. Mebrate. 1993. Fossil elephantoids from the hominoid-bearing Awash Group, Middle Awash Valley, Afar Depression, Ethiopia. *Transactions of the American Philosophical Society* 83:1–114.

Kalb, J. E., C. J. Jolly, A. Mebrate, S. Tebedge, C. Smart, E. B. Oswald, D. Cramer, P. Whitehead, C. B. Wood, G. C. Conroy, T. Adefris, L. Sperling, and B. Kana. 1982a. Fossil mammals and artefacts from the Middle Awash Valley, Ethiopia. *Nature* 298:25–29.

Kalb, J. E., C. J. Jolly, S. Tebedge, A. Mebrate, C. Smart, E. B. Oswald, P. F. Whitehead, C. B. Wood, T. Adefris, and V. Rawn-Schatzinger. 1982b. Vertebrate faunas from the Awash Group, Middle Awash Valley, Afar, Ethiopia. *Journal of Vertebrate Paleontology* 2:237–258.

Kingdon, J. 1979. Hippopotamids. In *East African Mammals: An Atlas of Evolution in Africa.* Vol. 3, pt. B. *Large Mammals,* pp. 250–277. London: Academic Press.

Lacomba, J. I., J. Morales, F. Robles, C. Santisteban, and M. T. Alberdi. 1986. Sedimentologia y paleontologia del yacimiento finimioceno de La Portera (Valencia). *Estudios Geo-*

lógicos del Instituto de Investigaciones Geológicas Lucas Mallada 42:167–180.

Laws, R. M. 1968. Dentition and aging of the hippopotamus. *East African Wildlife Journal* 8:19–52.

Leakey, M. G., C. S. Feibel, R. L. Bernor, J. M. Harris, T. E. Cerling, K. M. Stewart, G. W. Storrs, A. Walker, L. Werdelin, and A. J. Winkler. 1996. Lothagam: A record of faunal change in the Late Miocene of East Africa. *Journal of Vertebrate Paleontology* 16:556–570.

Matthew, W. D. 1929. Critical observations upon Siwalik mammals. *Bulletin of the American Museum of Natural History* 56:437–560.

McDougall, I., and C. S. Feibel. 1999. Numerical age control for the Miocene-Pliocene succession at Lothagam, a hominoid-bearing sequence in the northern Kenya Rift. *Journal of the Geological Society* (London) 156:731–745.

Mein, P. 1975. Résultats du groupe de travail des vertébrés. In J. Senes, ed., *Report on Activity of the RCMNS Working Groups (1971–1975), Bratislava: Vertebrata*, pp. 78–81. Bratislava: International Union of Geological Sciences, Commission on Stratigraphy, Subcommission on Neogene Stratigraphy.

Morton, S. G. 1844. On a supposed new species of *Hippopotamus. Proceedings of the National Academy of Sciences* (Philadelphia) 2:14–27.

Moyà-Solà, S., and J. Agustí. 1989. Bioevents and mammal successions in the Spanish Miocene. In E. H. Lindsay, V. Fahlbush, and P. Mein, eds., *European Neogene Mammal Chronology*, pp. 357–373. New York: Plenum Press.

Pavlakis, P. P. 1990. Plio-Pleistocene Hippopotamidae from the Upper Semliki. In N. T. Boaz, ed., *Evolution of Environments and Hominidae in the African Western Rift Valley*, pp. 203–223. Memoir No. 1. Martinsville: Virginia Museum of Natural History.

Pickford, M. 1983. On the origins of Hippopotamidae together with descriptions of two new species, a new genus and a new sub-family from the Miocene of Kenya. *Geobios* 16:193–217.

Pickford, M. 1990. Découverte de *Kenyapotamus* en Tunisie. *Annales de Paléontologie* 76:277–283.

Powers, D. W. 1980. Geology of Mio-Pliocene sediments of the lower Kerio River Valley, Kenya. Ph.D. diss., Princeton University.

Shea, B. T., and A. M. Gomez. 1988. Tooth scaling and evolutionary dwarfism: An investigation of allometry in human pygmies. *American Journal of Physical Anthropology* 77:117–132.

Spaan, A. 1996. *Hippopotamus creutzburgi:* The case of the Cretan hippopotamus. In D. S. Reese, ed., *Pleistocene and Holocene Fauna of Crete and Its First Settlers*, pp. 99–110. Monographs in World Archaeology 28. Madison, Wis.: Prehistory Press.

Stuenes, S. 1989. Taxonomy, habits and relationships of the subfossil Madagascan hippopotami, *Hippopotamus lemerlei* and *H. madagascariensis. Journal of Vertebrate Paleontology* 9:241–268.

Weston, E. M. 1998. A biometrical analysis of evolutionary change within the Hippopotamidae. Ph.D. diss., University of Cambridge.

Weston, E. M. 2000. A new species of hippopotamus *Hexaprotodon lothagamensis* (Mammalia: Hippopotamidae) from the late Miocene of Kenya. *Journal of Vertebrate Paleontology* 20:177–185.

Whybrow, P. J. 1989. New stratotype; the Baynunah Formation (Late Miocene). United Arab Emirates: Lithology and palaeontology. *Newsletters on Stratigraphy* 21:1–9.

Whybrow, P. J., and A. Hill, eds. 1999. *Fossil Vertebrates of Arabia.* New Haven: Yale University Press.

Whybrow, P. J., A. Hill, W. Yasin al-Tikriti, and E. A. Hailwood. 1990. Late Miocene primate fauna, flora and initial palaeomagnetic data from the Emirate of Abu Dhabi, United Arab Emirates. *Journal of Human Evolution* 19:583–588.

TABLE 10.1 Comparison of Relative Reduction of the External Metapodials in Fossil and Extant Hippopotamids

Species	Relative Reduction of External Metapodials[a]	
	Manus	Pes
Hippopotamus amphibius male (n = 4)	1.28	1.4
Hippopotamus amphibius female (n = 8)	1.19	1.4
Hexaprotodon liberiensis male and female (n = 16)	1.4	1.56
Hexaprotodon harvardi (n = 2)	1.33	1.4
Hexaprotodon cf. *Hex. protamphibius* ER 3661 (Allia Bay)	1.3	1.4

[a] Sum of internal metapodial lengths/sum of external metapodial lengths.

TABLE 10.2 Cranial Dimensions (in cm) of *Hexaprotodon harvardi*

Specimen	KNM-LT 4	KNM-LT 23043	KNM-LT 24147	KNM-LT 102	KNM-LT 8610	KNM-LT 56	KNM-LT 23041	KNM-LT 134	KNM-LT[a] 26236
Level	Upper Nawata	Upper Nawata	Upper Nawata	Horizon indet.	Horizon indet.	Horizon indet.	Upper Nawata	Upper Nawata	Lower Nawata
Age	Adult	Sub-adult	Sub-adult	Juvenile	Adult	Adult	Adult	Adult	Adult
Skull length	64.8	—	—	—	—	—	—	—	—
Length of sagittal crest	10.17	—	—	8.1	14.4	—	11.8	—	—
Width of palate at P^2	10.2	9.8	8.6	—	—	10.5	9.2	8.2	—
Width between I^2s	18.2	—	—	—	—	—	—	—	—
Width between upper canines	27.2	22.7	20.3	16.1	—	24.1	—	—	—
Width of zygoma	36.1	28.8[b]	33.6	—	32*	—	34	35.6	29.3
Width between infraorbital foramina	10.2	14.4	12.4	9.5	—	10.8	12.6	—	—
Minimum width between orbits	20.7	16.9	14.6	—	—	—	—	—	—
Maximum vertical diameter of orbit	6.7	—	—	—	6.2	—	—	—	5.5
Maximum horizontal diameter of orbit	6.3	—	6.7	—	6.8	—	—	—	6.6
Length of braincase	26.1	24.4	26.6	20.3	24.9	—	26.4	26.3	22.8
Height of occipital	19.7	—	15.8	—	18.2	—	15.6	14.2	11.4
Width of mastoid	22.6	—	22.6	20	22*	—	22.4	24	19.6
Width of occipital condyles	14.1	10.4[b]	13.6	10.9	12.4	—	13	12.2	10.3

[a] The adult cranium KNM-LT 26236 is ontogenetically scaled relative to the other adult crania.

[b] The dimension is an estimate. The measured distance of either the left or right side of the cranium has been doubled.

Table 10.3 Results of Regression and ANCOVA Analyses Comparing Allometries Between Sexes and Species

Comparison	Species	Sex	No. of Specimens	Slope (LS)	Major Axis/ Slope Value	Intercept (LS)	Correlation Coefficient (r)	ANCOVA	
								Slope	Intercept
Wd zygoma /BCL	*Hex. liberiensis*	Female	16	1.060	1.0659	0.028	0.995	NS	S*
		Male	8	1.097	1.1003	0.005	0.997	—	—
Wd between infraorbital foramina/BCL	*Hex. liberiensis*	Female	16	0.899	0.9065	–0.257	0.991	NS	S**
		Male	8	0.943	0.9463	–0.276	0.996	—	—
Horizontal diameter orbit/BCL	*Hex. liberiensis*	Female	16	0.600	0.6078	–0.087	0.976	NS	S*
		Male	8	0.536	0.5437	–0.037	0.969	—	—
Horizontal diameter orbit/SL	*Hex. liberiensis*	Female	16	0.538	0.5310	–0.176	0.971	NS	S*
		Male	8	0.470	0.4742	–0.109	0.978	—	—
Wd zygoma/BCL	*H. amphibius*	Female	10	1.155	1.1632	–0.055	0.993	NS	NS
		Male	15	0.990	1.1019	0.163	0.985	—	—
Horizontal diameter orbit/SL	*H. amphibius*	Female	12	0.449	0.4568	–0.005	0.955	NS	NS
		Male	17	0.485	0.5030	–0.056	0.917	—	—
Wd zygoma/SL	*H. amphibius*	Female	12	0.966	0.9734	–0.133	0.992	NS	NS
		Male	17	0.968	0.9700	–0.149	0.997	—	—
Wd between Lr Cs/length symphysis	*H. amphibius*	—	89	1.013	1.0369	0.281	0.978	NS	—
	Hex. liberiensis	—	45	1.039	1.0708	0.235	0.972	—	—
Wd between Lr Cs/length symphysis	*Hex. liberiensis*	—	45	1.039	1.0708	0.235	0.972	NS	—
	Lothagam spp.[a]	—	7	1.064	1.1916	0.023	0.908	—	—

continued

TABLE 10.3 Results of Regression and ANCOVA Analyses Comparing Allometries Between Sexes and Species (Continued)

Comparison	Species	Sex	No. of Specimens	Slope (LS)	Major Axis/ Slope Value	Intercept (LS)	Correlation Coefficient (r)	ANCOVA	
								Slope	Intercept
Height occipital/wd mastoid	*H. amphibius*	—	66	0.868	0.9010	0.011	0.959	NS	S*
	Hex. liberiensis	—	49	0.900	0.9169	–0.048	0.980	—	—
Horizontal diameter orbit/BCL	*H. amphibius*	—	72	0.625	0.6534	–0.090	0.929	NS	S**
	Hex. liberiensis	—	45	0.564	0.5799	–0.051	0.948	—	—
Wd astragalus/lth astragalus	*H. amphibius*	—	16	1.052	1.3330	–0.116	0.756	NS	S**
	Hex. liberiensis	—	19	1.073	1.3889	–0.218	0.819	—	—
Wd astragalus/lth astragalus	*Hex. liberiensis*	—	19	1.073	1.3889	–0.218	0.819	NS	S**
	Lothagam spp.[a]	—	16	1.023	1.1900	–0.173	0.910	—	—
Wd cuboid/lth cuboid	*H. amphibius*	—	15	0.690	0.7355	0.166	0.915	S**	—
	Hex. liberiensis	—	17	0.415	0.2719	0.234	0.732	—	—

[a] Includes all the fossil material from Lothagam that is attributable to more than one species.

LS = least squares regression.

NS = no significant difference.

S = significant difference.

BCL = braincase length.

Wd = width.

Lr C = lower canine.

Lth = length.

SL = skull length.

* $P \leq 0.05$.

** $P \leq 0.01$.

TABLE 10.4 Upper Premolar Dimensions (in cm) of Lothagam Hippopotamids

Specimen	Species	Premolar	MD[a]	BL[b]	Level
KNM-LT 23041	*Hex. harvardi*	P^2	3.90	3.10	Lr Nawata
KNM-LT 18	*Hex. harvardi*	P^2	3.50	3.10	Lr Nawata
KNM-LT 23866	*Hex. harvardi*	P^2	3.80	3.30	Lr Nawata
KNM-LT 28764	*Hex. harvardi*	P^2	4.10	3.10	Lr Nawata
KNM-LT 4	*Hex. harvardi*	P^2	4.10	3.10	Up Nawata
KNM-LT 23043	*Hex. harvardi*	P^2	3.90	2.70	Up Nawata
KNM-LT 56	*Hex. harvardi*	P^2	3.90	2.70	Horizon indet.
KNM-LT 409	*Hex. harvardi*	P^2	3.80	2.60	Up Nawata
KNM-LT 105	*Hex.* sp. indet.	P^2	4.50	3.10	Up Nawata
KNM-LT 23041	*Hex. harvardi*	P^3	4.10	3.20	Up Nawata
KNM-LT 18	*Hex. harvardi*	P^3	3.80	3.10	Lr Nawata
KNM-LT 23866	*Hex. harvardi*	P^3	3.60	3.40	Lr Nawata
KNM-LT 28764	*Hex. harvardi*	P^3	4.40	3.30	Lr Nawata
KNM-LT 70	*Hex. harvardi*	P^3	3.80	2.90	Horizon indet.
KNM-LT 4	*Hex. harvardi*	P^3	3.50	3.00	Up Nawata
KNM-LT 23043	*Hex. harvardi*	P^3	3.96	3.70	Up Nawata
KNM-LT 56	*Hex. harvardi*	P^3	4.20	2.70	Up Nawata
KNM-LT 24147	*Hex. harvardi*	P^3	4.00	3.20	Up Nawata
KNM-LT 28718	*Hex. harvardi*	P^3	3.90	3.20	Apak
KNM-LT 105	*Hex.* sp. indet.	P^3	4.50	3.30	Up Nawata
KNM-LT 23041	*Hex. harvardi*	P^4	3.10	4.20	Up Nawata
KNM-LT 18	*Hex. harvardi*	P^4	2.80	3.80	Lr Nawata
KNM-LT 23866	*Hex. harvardi*	P^4	3.15	3.70	Lr Nawata
KNM-LT 28764	*Hex. harvardi*	P^4	3.20	3.75	Lr Nawata
KNM-LT 70	*Hex. harvardi*	P^4	2.60	3.37	Horizon indet.
KNM-LT 26225	*Hex. harvardi*	P^4	2.40	3.00	Lr Nawata
KNM-LT 4	*Hex. harvardi*	P^4	3.20	4.10	Up Nawata
KNM-LT 56	*Hex. harvardi*	P^4	3.10	3.10	Horizon indet.
KNM-LT 24147	*Hex. harvardi*	P^4	2.80	3.70	Up Nawata
KNM-LT 409	*Hex. harvardi*	P^4	3.00	3.70	Up Nawata
KNM-LT 28718	*Hex. harvardi*	P^4	2.75	2.83	Apak
KNM-LT 105	*Hex.* sp. indet.	P^4	3.40	4.60	Up Nawata

[a] Maximum mesiodistal length.

[b] Maximum buccolingual width.

TABLE 10.5 Upper Molar Dimensions (in cm) of Lothagam Hippopotamids

Specimen	Species	Molar	MD[a]	BL[b]	Level
KNM-LT 23041	*Hex. harvardi*	M^1	4.55	3.90	Up Nawata
KNM-LT 18	*Hex. harvardi*	M^1	3.60	4.10	Lr Nawata
KNM-LT 23849	*Hex. harvardi*	M^1	3.50	3.80	Lr Nawata
KNM-LT 28764	*Hex. harvardi*	M^1	4.60	4.50	Lr Nawata
KNM-LT 56	*Hex. harvardi*	M^1	4.00	4.00	Horizon indet.
KNM-LT 24147	*Hex. harvardi*	M^1	4.60	4.10	Up Nawata
KNM-LT 23043	*Hex. harvardi*	M^1	4.37	4.20	Up Nawata
KNM-LT 28718	*Hex. harvardi*	M^1	3.70	3.80	Apak
KNM-LT 23838	*Hex. harvardi*	M^1	3.90	3.50	Apak
KNM-LT 102	*Hex. harvardi*	M^1	4.45	4.20	Horizon indet.
KNM-LT 105	*Hex.* sp. indet.	M^1	4.10	4.60	Up Nawata
KNM-LT 23041	*Hex. harvardi*	M^2	4.85	—	Up Nawata
KNM-LT 18	*Hex. harvardi*	M^2	4.10	—	Lr Nawata
KNM-LT 23849	*Hex. harvardi*	M^2	4.30	—	Lr Nawata
KNM-LT 28764	*Hex. harvardi*	M^2	5.00	—	Lr Nawata
KNM-LT 23866	*Hex. harvardi*	M^2	4.20	—	Lr Nawata
KNM-LT 70	*Hex. harvardi*	M^2	3.80	—	Horizon indet.
KNM-LT 26225	*Hex. harvardi*	M^2	3.70	—	Lr Nawata
KNM-LT 4	*Hex. harvardi*	M^2	3.90	—	Up Nawata
KNM-LT 23043	*Hex. harvardi*	M^2	4.60	—	Up Nawata
KNM-LT 23902	*Hex. harvardi*	M^2	4.70	—	Up Nawata
KNM-LT 56	*Hex. harvardi*	M^2	4.30	—	Horizon indet.
KNM-LT 24147	*Hex. harvardi*	M^2	4.70	—	Up Nawata
KNM-LT 28718	*Hex. harvardi*	M^2	4.50	—	Apak
KNM-LT 23838	*Hex. harvardi*	M^2	4.30	—	Apak
KNM-LT 105	*Hex.* sp. indet.	M^2	5.00	—	Up Nawata
KNM-LT 23041	*Hex. harvardi*	M^3	5.00	—	Up Nawata
KNM-LT 18	*Hex. harvardi*	M^3	4.70	—	Lr Nawata
KNM-LT 23849	*Hex. harvardi*	M^3	4.70	—	Lr Nawata
KNM-LT 23866	*Hex. harvardi*	M^3	4.85	—	Lr Nawata
KNM-LT 70	*Hex. harvardi*	M^3	4.30	—	Horizon indet.
KNM-LT 26225	*Hex. harvardi*	M^3	4.60	—	Lr Nawata
KNM-LT 4	*Hex. harvardi*	M^3	4.60	—	Up Nawata
KNM-LT 23902	*Hex. harvardi*	M^3	4.40	—	Up Nawata
KNM-LT 56	*Hex. harvardi*	M^3	4.40	—	Up Nawata
KNM-LT 28718	*Hex. harvardi*	M^3	4.40	—	Apak
KNM-LT 105	*Hex.* sp. indet.	M^3	4.84	—	Up Nawata

[a] Maximum mesiodistal length.

[b] Maximum buccolingual width.

TABLE 10.6 Mandible Dimensions (in cm) of Lothagam Hippopotamids

Specimen	KNM-LT 23888	KNM-LT 23105	KNM-LT 23877	KNM-LT 23127	KNM-LT 28731	KNM-LT 23896
Level	Upper Nawata	Upper Nawata	Upper Nawata/Apak	Upper Nawata	Upper Nawata	Lower Nawata
Species	*Hex. harvardi*	*Hex. harvardi?*	*Hex. harvardi*	*Hex. harvardi*	*Hex. harvardi*	*Hex. harvardi*
Age	Adult	Sub-adult	Adult	Adult	Adult	Adult
Maximum length of mandible	52.1	—	—	—	—	—
Height of ramus at M_1	10.4	—	11.8	11.7	9.4	8.9
Length of symphysis	18.8	18.7	—	—	—	—
Maximum depth of symphysis	12.8	6.2	—	—	—	—
Width between lower canines	25.1	22.9	—	—	—	—
Maximum diameter I_1	2.6	2.1	—	—	—	—
Maximum diameter I_2	2.3	1.8	—	—	—	—
Maximum diameter I_3	2.2	1.8	—	—	—	—

Specimen	KNM-LT 22964	KNM-LT 23846	KNM-LT 108	KNM-LT 33	KNM-LT 79	KNM-LT 23839
Level	Lower Nawata	Lower Nawata	Lower Nawata	Upper Nawata	Horizon indet.	Lower Nawata
Species	*Hex. harvardi*	*Hex. harvardi*	*Hex. harvardi*	*Hex. harvardi*	*Hex.* sp. indet.	*Hex. lothagamensis*
Age	Juvenile[a]	Juvenile	Juvenile[b]	Adult	Juvenile[c]	Adult
Maximum length of mandible	29.6	—	—	—	—	—
Height of ramus at M_1	5.9	8.5	8.6	9.6	—	6.6
Length of symphysis	9.3	—	13.5	14.6	19.2	12.4
Maximum depth of symphysis	—	8.8	7.9	9.6	5.8	4.7
Width between lower canines	11.6	—	17	23.1	22.5	12.6
Maximum diameter I_1	—	—	—	—	2.1	—
Maximum diameter I_2	—	—	—	—	2	—
Maximum diameter I_3	—	—	—	—	1.6	—

[a] 1 = Equivalent to a 3-year-old *H. amphibius.*
[b] 2 = Equivalent to an 8–10-year-old *H. amphibius.*
[c] 3 = Equivalent to an 8–10-year-old *H. amphibius.*

TABLE 10.7 Lower Premolar Dimensions (in cm) of Lothagam Hippopotamids

Specimen	Species	Premolar	MD[a]	BL[b]	Level
KNM-LT 23847	*Hex. harvardi*	P_2	3.40	2.10	Lr Nawata
KNM-LT 23888	*Hex. harvardi*	P_2	4.00	2.25	Up Nawata
KNM-LT 23127	*Hex. harvardi*	P_2	4.10	2.25	Up Nawata
KNM-LT 33	*Hex. harvardi*	P_2	3.70	2.40	Up Nawata
KNM-LT 26222	*Hex.* sp. indet.	P_2	4.10	2.70	Up Nawata
KNM-LT 23896	*Hex. harvardi*	P_3	3.90	2.40	Lr Nawata
KNM-LT 23902	*Hex. harvardi*	P_3	3.50	2.30	Up Nawata
KNM-LT 23888	*Hex. harvardi*	P_3	4.65	2.55	Up Nawata
KNM-LT 23127	*Hex. harvardi*	P_3	3.90	2.30	Up Nawata
KNM-LT 33	*Hex. harvardi*	P_3	4.00	2.60	Up Nawata
KNM-LT 26222	*Hex.* sp. indet.	P_3	4.40	2.80	Up Nawata
KNM-LT 23896	*Hex. harvardi*	P_4	3.50	2.60	Lr Nawata
KNM-LT 71	*Hex. harvardi*	P_4	3.75	2.50	Horizon indet.
KNM-LT 23902	*Hex. harvardi*	P_4	4.15	2.70	Up Nawata
KNM-LT 23877	*Hex. harvardi*	P_4	4.10	2.85	Up Nawata/Apak
KNM-LT 23888	*Hex. harvardi*	P_4	4.00	2.30	Up Nawata
KNM-LT 23127	*Hex. harvardi*	P_4	4.00	2.80	Up Nawata
KNM-LT 28731	*Hex. harvardi*	P_4	3.80	2.50	Up Nawata
KNM-LT 33	*Hex. harvardi*	P_4	4.10	3.10	Up Nawata
KNM-LT 26222	*Hex.* sp. indet.	P_4	4.40	3.55	Up Nawata
KNM-LT 8735	*Hex.* sp. indet.	P_4	3.90	2.70	Horizon indet.
KNM-LT 23839	*Hex. lothagamensis*	P_4	2.90	2.00	Lr Nawata

[a] Maximum mesiodistal length.

[b] Maximum buccolingual length.

TABLE 10.8 Lower Molar Dimensions (in cm) of Lothagam Hippopotamids

Specimen	Species	Molar	MD[a]	BL[b]	Level
KNM-LT 23896	*Hex. harvardi*	M_1	4.00	3.10	Lr Nawata
KNM-LT 23847	*Hex. harvardi*	M_1	3.55	2.90	Lr Nawata
KNM-LT 23846	*Hex. harvardi*	M_1	3.70	2.80	Lr Nawata
KNM-LT 23877	*Hex. harvardi*	M_1	3.70	3.60	Up Nawata/Apak
KNM-LT 23127	*Hex. harvardi*	M_1	3.90	2.90	Up Nawata
KNM-LT 28731	*Hex. harvardi*	M_1	4.10	3.50	Up Nawata
KNM-LT 33	*Hex. harvardi*	M_1	3.60	3.20	Up Nawata
KNM-LT 108	*Hex. harvardi*	M_1	4.00	2.75	Lr Nawata
KNM-LT 79	*Hex.* sp. indet.	M_1	5.00	3.50	Horizon indet.
KNM-LT 23839	*Hex. lothagamensis*	M_1	3.60	2.40	Lr Nawata
KNM-LT 23896	*Hex. harvardi*	M_2	4.30	—	Lr Nawata
KNM-LT 23847	*Hex. harvardi*	M_2	4.00	—	Lr Nawata
KNM-LT 8743	*Hex. harvardi*	M_2	4.80	—	Lr Nawata
KNM-LT 23902	*Hex. harvardi*	M_2	4.90	—	Up Nawata
KNM-LT 23877	*Hex. harvardi*	M_2	4.94	—	Up Nawata/Apak
KNM-LT 23127	*Hex. harvardi*	M_2	4.85	—	Up Nawata
KNM-LT 28731	*Hex. harvardi*	M_2	4.60	—	Up Nawata
KNM-LT 23838	*Hex. harvardi*	M_2	5.00	—	Apak
KNM-LT 33	*Hex. harvardi*	M_2	4.70	—	Up Nawata
KNM-LT 8735	*Hex.* sp. indet.	M_2	4.95	—	Horizon indet.
KNM-LT 23839	*Hex. lothagamensis*	M_2	4.00	—	Lr Nawata
KNM-LT 23896	*Hex. harvardi*	M_3	6.30	—	Lr Nawata
KNM-LT 23847	*Hex. harvardi*	M_3	6.03	—	Lr Nawata
KNM-LT 8743	*Hex. harvardi*	M_3	5.90	—	Lr Nawata
KNM-LT 71	*Hex. harvardi*	M_3	6.40	—	Horizon indet.
KNM-LT 23877	*Hex. harvardi*	M_3	6.40	—	Up Nawata/Apak
KNM-LT 23127	*Hex. harvardi*	M_3	6.30	—	Up Nawata
KNM-LT 28731	*Hex. harvardi*	M_3	6.30	—	Up Nawata
KNM-LT 23838	*Hex. harvardi*	M_3	6.25	—	Apak
KNM-LT 33	*Hex. harvardi*	M_3	6.25	—	Up Nawata
KNM-LT 8735	*Hex.* sp. indet.	M_3	7.10	—	Horizon indet.
KNM-LT 26222	*Hex.* sp. indet.	M_3	7.20	—	Up Nawata
KNM-LT 23839	*Hex. lothagamensis*	M_3	5.05	—	Lr Nawata

[a] Maximum mesiodistal length.

[b] Maximum buccolingual width.

TABLE 10.9 Metapodial Dimensions (in cm) of Lothagam Hippopotamids

Specimen	Species	Metapodial	LTH[a]	WD S[b]	WD D[c]	Level
KNM-LT 23891	*Hex. harvardi*	Mc II	13.40	2.70	3.60	Lr Nawata
KNM-LT 23843	*Hex. harvardi*	Mc II	13.50	3.00	3.90	Up Nawata
KNM-LT 23855	*Hex. harvardi*	Mc II	12.40	3.10	3.90	Up Nawata
KNM-LT 8543	*Hex. harvardi*	Mc II	13.00	2.90	3.20	Nawata; level indet.
KNM-LT 23843	*Hex. harvardi*	Mc III	17.70	3.20	4.20	Up Nawata
KNM-LT 23905	*Hex. harvardi*	Mc III	17.90	3.20	4.00	Up Nawata
KNM-LT 26238	*Hex. harvardi*	Mc III	18.50	3.20	4.50	Up Nawata
KNM-LT 23891	*Hex. harvardi*	Mc IV	15.40	3.20	—	Lr Nawata
KNM-LT 23843	*Hex. harvardi*	Mc IV	16.20	3.60	4.30	Up Nawata
KNM-LT 26220	*Hex.* cf. *protamphibius*	Mc IV	14.30	3.50	3.50	Apak
KNM-LT 23891	*Hex. harvardi*	Mc V	10.90	2.80	3.20	Lr Nawata
KNM-LT 23272	*Hex. harvardi*	Mc V	10.10	2.60	3.10	Lr Nawata
KNM-LT 23843	*Hex. harvardi*	Mc V	12.50	3.50	3.70	Up Nawata
KNM-LT 26220	*Hex.* cf. *protamphibius*	Mc V	10.30	2.70	3.60	Apak
KNM-LT 26235	*Hex. lothagamensis*	Mc V	7.60	2.10	2.30	Lr Nawata
KNM-LT 23891	*Hex. harvardi*	Mt II	10.60	1.80	2.90	Lr Nawata
KNM-LT 26228	*Hex. harvardi*	Mt II	11.10	2.70	3.40	Up Nawata
KNM-LT 23886	*Hex. harvardi*	Mt III	13.40	2.90	3.60	Lr Nawata
KNM-LT 23272	*Hex. harvardi*	Mt III	15.10	3.70	4.30	Lr Nawata
KNM-LT 26228	*Hex. harvardi*	Mt III	15.40	3.40	3.90	Up Nawata
KNM-LT 23843	*Hex. harvardi*	Mt III	16.60	3.40	4.40	Up Nawata
KNM-LT 8542	*Hex. harvardi*	Mt III	13.90	3.30	3.80	Nawata; level indet.
KNM-LT 23891	*Hex. harvardi*	Mt IV	14.70	3.00	3.90	Lr Nawata
KNM-LT 26228	*Hex. harvardi*	Mt IV	15.50	3.50	4.30	Up Nawata
KNM-LT 23843	*Hex. harvardi*	Mt IV	16.40	3.90	4.50	Up Nawata
KNM-LT 8549	*Hex. harvardi*	Mt IV	15.90	3.90	3.80	Nawata; level indet.
KNM-LT 23628	*Hex.* cf. *protamphibius*	Mt IV	14.20	3.10	3.90	Apak
KNM-LT 23903	*Hex. harvardi*	Mt V	9.50	1.80	2.80	Lr Nawata
KNM-LT 23886	*Hex. harvardi*	Mt V	9.80	2.00	2.90	Lr Nawata
KNM-LT 26228	*Hex. harvardi*	Mt V	10.90	2.30	3.20	Up Nawata
KNM-LT 23883	*Hex. harvardi*	Mt V	11.90	2.30	3.30	Up Nawata
KNM-LT 23855	*Hex. harvardi*	Mt V	10.10	2.30	3.40	Up Nawata
KNM-LT 26219	*Hex.* cf. *protamphibius*	Mt V	9.30	1.80	2.70	Apak

[a] Maximum length.

[b] Minimum width of shaft in lateral-medial section.

[c] Maximum distal width in lateral-medial section.

TABLE 10.10 Pliocene *Hexaprotodon* Specimens from Lothagam (Kaiyumung Member) and Other Localities That Have Been Included in the Analysis

Accession No.	Species	Location	Age	Material
KNM-LT 27703	*Hex.* sp. indet.	Lothagam (Kaiyumung)	Adult	Complete mandible (lt and rt I_1–M_3)
KNM-LT 28716	*Hex. protamphibius*	Lothagam (North Section)[d]	Juvenile	Mandible (lt I_1–C, M_2), lt P_1, rt P_2 or P_3 and lt P_4
KNM-WT 16588[a]	*Hex. protamphibius*	West Turkana (Lomekwi)	Adult	Complete cranium (rt I^1–I^2, lt and rt M^2–M^3)
KNM-ER 2738[b]	*Hex. protamphibius*	Koobi Fora Region	Adult	Partial mandible with complete symphysis
KNM-ER 3661	*Hex.* cf. *protamphibius*	Allia Bay (Koobi Fora)	Adult	Complete skeleton (partial dentition)
KNM-KP 8529	*Hex.* cf. *protamphibius*	Kanapoi	Adult	Complete cranium with dentition
KNM-KP 1	*Hex.* cf. *protamphibius*	Kanapoi	Adult	Complete rt ramus with dentition
UM P6202[c]	*Hex. imagunculus*	Kazinga (Uganda)	—	Mandibular symphysis (lt and rt roots I_1–C)

[a] Harris et al. (1988a).

[b] Harris (1991).

[c] Cooke and Coryndon (1970).

[d] Northern fault block (stratigraphy uncertain).

ER = East Turkana (≡ East Rudolf).

KP = Kanapoi.

UM = Ugandan Museum.

10.2

Lothagam Suidae

John M. Harris and Meave G. Leakey

The Lothagam suids are diverse. The Nawata Formation has yielded a large kubanochoerine, a medium-sized suid referable to *Potamochoerus,* and a small tayassuid-like form assigned to *Cainochoerus* cf. *C. africanus,* as well as the more frequently preserved *Nyanzachoerus syrticus tulotus* and *Nyanzachoerus devauxi.* In the Upper Nawata, *N. syrticus tulotus* displays progressive evolutionary changes that involve premolar reduction and increasing complexity of the third molar. The Apak Member yields the different and more derived tetraconodontine *Nyanzachoerus australis. Notochoerus euilus* abounds in the Kaiyumung Member, together with the less common *Nyanzachoerus pattersoni.* Fossilized teeth of bush pigs and warthogs found as surface specimens over much of the Lothagam locality represent lag fossils from erosion of the Galana Boi Beds. Lothagam documents the exploitation of riparian habitats by tetraconodontine suids that had migrated from Asia to dominate the latest Miocene and Pliocene of Africa. The tetraconodontines were eventually supplanted by Suinae that migrated from Asia at a time that postdates the main part of the Lothagam succession. Earlier tetraconodontines reported from the Namurungule Formation (Nakaya et al. 1984) include teeth indistinguishable from *N. devauxi* and *N. syrticus. Cainochoerus* was previously known only from Langebaanweg.

As of the early 1970s, some 23 different genera (77 species) of suids had been described from about 50 African Plio-Pleistocene localities. Subsequent revision by Cooke (1978), White and Harris (1977), and Harris and White (1979) reduced this number to more manageable proportions. Harris and White recognized 16 species within seven genera inclusive of the extant sub-Saharan suids. Although that analysis erred on the conservative side, and additional species now appear valid, it did identify the relationships of the major suid genera and provided a starting point for subsequent analysis. The evolution of the African Suidae was reasonably well documented from about 3.5 million years ago through to the end of the Early Pleistocene. The older part of the story was much less well understood—in part because of the relatively few samples then available and in part because of their uncertain temporal provenance. It is thus fortunate that subsequent fieldwork has recovered additional material from the Late Miocene and Early Pliocene sediments that crop out at Lothagam and has helped clarify the temporal distribution of species that characterize this interval of time in eastern Africa.

Harris and White (1979) recognized three species of *Nyanzachoerus:* the primitive *N. tulotus,* the more progressive *N. kanamensis,* and the most progressive *N. jaegeri* that was interpreted as the parent stock of the genus *Notochoerus.* Harris and White suggested that these three East African forms may be synonymous with representatives of previously described species from North Africa but deferred further taxonomic revision because they had not had access to relevant North African material. Subsequent work by Cooke (1987) confirmed some of Harris and White's predictions in this respect. Other work by Pickford (1989) provided new interpretation of the origin of the nyanzachoeres and recognized additional species elsewhere in eastern Africa. Most recently, van der Made (1999) reviewed biometrical trends in the Tetraconodontinae in a study that involved both Eurasian and African material. He came to a somewhat different conclusion as to the interrelationships within the subfamily and transferred many of the African species formerly assigned to the genus *Nyanzachoerus* to the genus *Sivachoerus* Pilgrim, 1926.

The latest Miocene of East Africa is represented by a different assemblage of suids than that typical of the earlier portions of this epoch. Genera of the subfamilies Hyotheriinae, Listriodontinae, and Sanitheriinae were replaced by tetraconodontine immigrants from Asia. *Nyanzachoerus* gave rise to the more hypsodont *Notochoerus.* These two genera are characteristic of the latest Miocene and Early Pliocene of sub-Saharan Africa, although, in the older part of the Lothagam succession, one or more representatives of mid Miocene assemblages persist and another Asian immigrant is a rare element of the Lower Nawata assemblage. Earlier tetraconodontines have been reported from the Namurungule Formation (Nakaya et al. 1984), where the small collection includes a lower third molar (KNM-SH 14758) that is indistinguishable from those referred to *Nyanzachoerus devauxi* at Lothagam. Larger upper and lower molars from the Namurungule Formation are very similar to those of *Nyanzachoerus syrticus.* The Middle Pliocene saw a second wave of immigration from Eurasia that founded the *Potamochoerus, Kolpochoerus,* and *Metridiochoerus* lineages, but these postdate the Lothagam sequence.

Definitive treatment of the Suidae from Lothagam was undertaken by Cooke and Ewer (1972), who recognized three species that they named *Nyanzachoerus tulotus, N. pattersoni,* and *N. plicatus.* Nearly simultaneous work on suids recovered from Hamada Damous, Tunisia, by Coppens (1971) led to the acceptance of *N. plicatus* as a junior synonym of *N. jaegeri.* Harris and White (1979) argued that *N. pattersoni* was a junior synonym of *N. kanamensis,* and this interpretation was accepted by Cooke (1987) in his revision of the fossil Suidae from Sahabi, Libya. Cooke (1987) also presented a case for accepting *N. tulotus* as a junior synonym of *N. syrticus.* More recent investigation by van der Made (1999) separated the Late Miocene African suids into two lineages represented by the tribes Tetraconodontini and Nyanzachoerini; he also resurrected some of the species previously placed into synonymy, and changed some of the generic assignations.

Interpretation of the roles and relationships of the Lothagam suid species has been hampered by uncertainty of their provenance within the Lothagam sequence, whose geologic age was not determined with precision until recently (McDougall and Feibel 1999). The material available to Cooke and Ewer was recovered by Harvard University expeditions to Kanapoi and Ekora in 1966 and to Lothagam in 1967 and 1968. The provenance of many of the specimens from the 1967 expedition can be reconstructed with the aid of copies of field notes subsequently deposited at the National Museum of Kenya in Nairobi. Documentation of the 1968 collection, however, is limited to a list (without localities) of specimens recovered during that year. Additional material collected in 1972 and 1973 in conjunction with an unpublished study of the geology of the Lothagam region by Dennis Powers (1980) did not become available until after Cooke and Ewer had completed their monograph. Where possible, the provenance of the 1972 and 1973 specimens has been reconstructed by Craig Feibel (Rutgers University) from Powers's unpublished field records.

The sample available to Cooke and Ewer has been considerably augmented as a result of further collecting from the Lothagam region between 1989 and 1993. Many new specimens have been retrieved, and the stratigraphy of the locality is now much better understood as a result of recent field investigations by Feibel. New material has permitted the recognition of several additional species and has helped clarify the temporal distribution of material collected in the 1960s. Here we document this new information and its bearing on our understanding of the evolution and distribution of the tetraconodont suids of Africa.

Systematic Description

Family Suidae Gray, 1821
Subfamily Listriodontinae Gervais, 1859
Tribe Kubanochoerini Gabunia, 1958
Kubanochoerus Gabunia, 1955

Diagnosis

Large bunodont Listriodontinae, some with cranial appendages (after van der Made 1996).

cf. *Kubanochoerus* sp.

(Figure 10.43; table 10.11)

Lothagam Material

- *Lower Nawata:* 318, mandible with Rt. P_3, roots Rt. M_2, Lt. M_3 and postcranial elements.

A partial suid mandible, LT 318, characterized by very large premolars and a more robust mandibular body than *N. syrticus,* was recovered in 1967 from low in the Lower Nawata. It has no anterior cheek teeth but the premolar roots show that the premolars were much bigger than those of *N. syrticus.* The incomplete third molar is heavily worn but is of comparable length to the equivalent teeth in *N. syrticus.* It probably represents the same species as two unpublished mandibles from the white sediments at Nakoret (KNM-NO 28806 and

28811), both of which have *Kubanochoerus*-like molars but huge premolars. The Nakoret specimens cannot be assigned to any previously described species of *Kubanochoerus* because of the extraordinarily large premolars. The Lothagam specimen could conceivably also represent the genus *Sivachoerus* but the premolars, again, are more massive than those of any previously named African species. It is smaller than Asian specimens of *Sivachoerus prior* but larger than those of *S. sindiense.*

A partial left forelimb of the same individual includes an olecranon fossa of a large ulna, unciform, magnum, metacarpal V (length 7.54 mm; distal anteroposterior 21.86 mm; distal transverse 18.18 mm), and elements of the two middle digits of the forefoot, including distal metacarpals, proximal phalanges, middle phalanges, and one distal phalanx. These bones are all relatively large and robustly built.

Subfamily Tetraconodontinae Lydekker, 1876

The tetraconodontines are suids with enlarged premolars; the lower fourth premolar is unicuspid, and the metacone of the upper fourth premolar is lacking or only feebly developed. van der Made (1999) recognized three tribes in this subfamily: the Tetraconodontini and the Nyanzachoerini occurred in Asia and Africa, but the Parachleuastochoerini was restricted to Europe. According to van de Made (1999), the Tetraconodontini was represented by two genera: *Tetraconodon* (Falconer 1868) from Asia and *Sivachoerus* (Pilgrim 1926) from Asia and Africa. *Sivachoerus* species were described from North African localities earlier this century (Leonardi 1952; Tobein 1936) but most of the recently collected Miocene and Pliocene tetraconodontines have been attributed to *Nyanzachoerus.* However, as van der Made (1999) has pointed out, when Leakey (1958) created the genus *Nyanzachoerus* he specified (1) that *Sivachoerus* has a three-rooted P_3 but that of *Nyanzachoerus* has two roots, and (2) the premolars of *Sivachoerus* are larger than those of *Nyanzachoerus.* On the basis of these and other reasons, van der Made (1999) transferred many of Miocene and Pliocene African species from *Nyanzachoerus* to *Sivachoerus,* recognizing an Asian lineage (*Sivachoerus sindiensis* > *S. indicus* > *S. prior*) that provided migrants to Africa on two separate occasions (van der Made 1999:figure 19). The Tribe Nyanzachoerini was created by van der Made (1999) to incorporate species of the genera *Conohyus* (Pilgrim 1926), *Nyanzachoerus* (Leakey 1958), *Notochoerus* (Broom 1925), and *Lophochoerus* (Pilgrim 1926) that form a sister lineage to the *Sivachoerus* species. Although recognizing that van der Made's interpretations are plausible, we are concerned that the basis for many of his interpretations rest on teeth rather than crania, and we therefore retain the genus *Nyanzachoerus* for the African tetraconodontines, pending the discovery of more complete Asian specimens.

Nyanzachoerus Leakey, 1958

Diagnosis

A tetraconodontine genus with teeth similar to *Potamochoerus* in basic structure but tending to be more hypsodont and with main cusps of molars distinctly more columnar. Third and fourth premolars relatively larger than in *Potamochoerus.* Upper canines oval to flattened oval in transverse section. Lower canines verrucose with thin, weakly grooved enamel on two lateral faces. Strong sexual dimorphism exhibited in size of canines and massiveness of skull; hollow bony protuberances or bosses on zygomatic arches present in male, weak or absent in female. Corpus of mandible heavy and contrasting markedly with unusually thin bone forming the angle (after Cooke and Ewer 1972; Harris and White 1979).

Nyanzachoerus syrticus (Leonardi, 1952)

Diagnosis

Sivachoerus of small size with relatively small and simple third molars. M_3 length about 36–50 mm. Ratio of length M_3 to M_1 ranges from 190 to 230 (after van der Made 1999:207).

Nyanzachoerus syrticus was named for a small to mid-sized tetraconodontine from Sahabi (Leonardi 1952; Kotsakis and Inigo 1980). Cooke and Ewer (1972) recognized a small tetraconodontine from Lothagam that they assigned to *Nyanzachoerus tulotus.* Cooke (1987) later placed the Sahabi species into the genus *Nyanzachoerus,* and Leakey et al. (1996) intimated that *N. syrticus* and *N. tulotus* were conspecific. Subsequently, van der Made (1999) treated the Sahabi and Lothagam samples as two subspecies: *S. syrticus syrticus* and *S. syrticus tulotus.* Recently Bishop and Hill (1999) described specimens from the Baynunah Formation, Abu Dhabi, attributed to *N. syrticus* and *Nyanzachoerus* aff. *N. syrticus.*

Nyanzachoerus syrticus tulotus (Cooke and Ewer, 1972)

(Figures 10.42–10.45; tables 10.12–10.15)

Diagnosis

About the size of the extant *Hylochoerus*. First molars present, third and fourth premolars large, P^4 about as broad as M^3. Third molars relatively low-crowned and composed of only two pairs of lateral cusps, together with a small talon(id). Combined length of third and fourth premolars more than 50 percent of the molar series length. Preorbital ridges strongly marked in male; maxillary flanges, naso-premaxillary thickenings, and zygomatic knobs sexually dimorphic (after Harris and White 1979). M_3 length between 36 and 50 mm (van der Made 1999).

Lothagam Material

- *Lower Nawata:* 107, male Lt. mandible (P_4 and M_3, broken P_3 and M_2); 284, edentulous female mandible (/C–M_2); 291, female Rt. mandible (P_2–partial M_2); 292, Rt. mandible (P_{3-4}); 295, female mandible (Lt. and Rt. P_3–M_3, Rt. P_2), Lt. I_1; 301, male mandible (worn Lt. and Rt. M_3, worn Lt. P_{3-4}, /C); 302, male Rt. mandible (P_3–M_3), Lt. mandible (P_2–M_1); 304, Lt. maxilla fragment (P^4–M^1); 305, male Rt. mandible (/C, P_4–M_3); 306, Rt. P_4; 316, [*tulotus* type] almost complete cranium (Rt. and Lt. P^2–M^3, Lt. P^1); 385, mandible (Lt. I_3, /C, P_3–M_3, Rt. /C–M_2); 386, Rt. maxilla fragment (M^2), Lt. M_2, M_3, premolar fragments; 387, Lt. mandible (P_4–M_2); 390, Rt. lower M_1; 554, Rt. I^1; 22877, Rt. mandible (M_1, erupting M_2); 22878, Rt. mandible (dP_4); 22880, Rt. P_3, Rt. P_4; 22882, Lt. M_2; 22953, female mandible (Lt. P_2, M_{1-3}, Rt. P_4, roots Lt. and Rt. P_1, P_3; Rt. mandible fragment, broken M_2, M_3, condyle; 22980, Lt. maxilla (P^2–M^1); 22985, Rt. M^1; 22986, mandible fragments (Rt. M_3, Lt. P_4); 22992, Lt. M^2; 22993, Lt. M^3; 22996, unworn Rt. /C, Lt. mandible (broken I_1, P_2, P_4–M_1, roots I^{2-3}, /C, P_3); 23001, cranial fragments, Lt. maxilla (P^4, roots M^{1-2}) occipital condyle; 23002, female mandible (roots Rt. and Lt. I_{1-2}, Rt. /C–M_3); 23004, I_1; 23006, female Lt. mandible, partial M_2, roots P_2–M_1); 23007, weathered Rt. mandible (P_4–M_3), Lt. condyle; 23009, Lt. mandible (M_1); 23010, Lt. mandible frag-

Figure 10.42 Restoration of male and female *Nyanzachoerus syrticus tulotus* by Mauricio Antón.

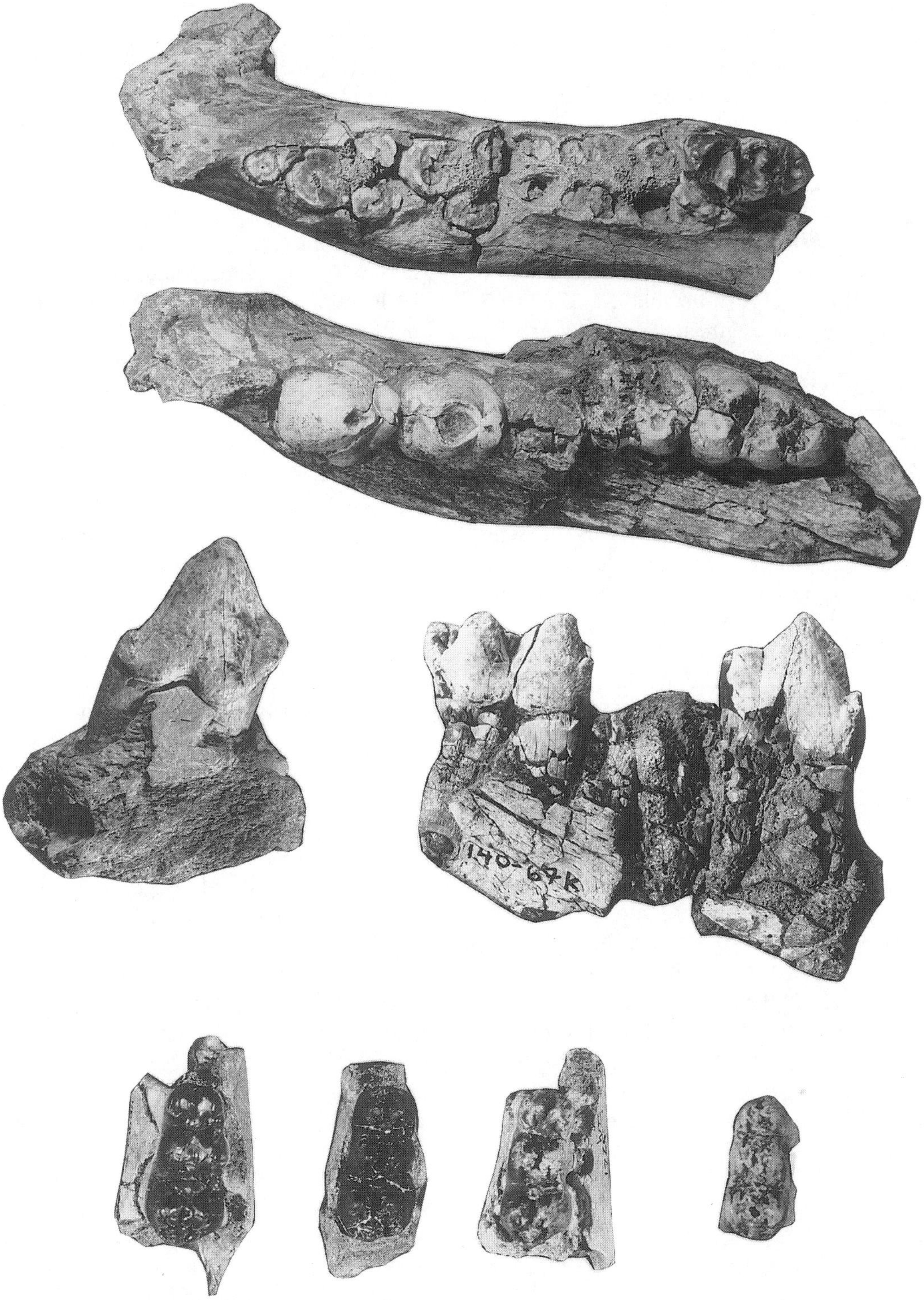

Figure 10.43 *Kubanochoerus, Potamochoerus,* and *Nyanzachoerus* dental specimens. Top = KNM-LT 318, cf. *Kubanochoerus* sp., left mandible (occlusal view), Lower Nawata; half natural size. Second row = KNM-WT 28806, cf. *Kubanochoerus* sp. left mandible with P_3–M_3 (occlusal view), Nakoret; half natural size. Third row, left = KNM-LT 23711, *Potamochoerus* sp., Lt. P_3 (lateral view), Lower Nawata; natural size. Third row, right = KNM-LT 294, *Potamochoerus* sp., right mandible with P_4 and M_2 (lateral view), Lower Nawata; natural size. Bottom, left = KNM-LT 23027, *Nyanzachoerus syrticus tulotus,* Rt. dP_4 (occlusal view), Lower Nawata; natural size. Bottom, left center = KNM-LT 22878, *Nyanzachoerus syrticus tulotus,* Rt. dP_4 (occlusal view), Lower Nawata; natural size. Bottom, right center = KNM-LT 23802, *Nyanzachoerus syrticus tulotus,* Rt. dP_4 (occlusal view), Lower Nawata; natural size. Bottom, right = KNM-LT 26118, *Nyanzachoerus devauxi,* Rt. dP_4 (occlusal view), Lower Nawata; natural size.

ment (M_1, P_4 in crypt); 23011, Lt. M^3, M^2 fragment, P^4 in crypt; 23012, Rt. M^3; 23016, Lt. I_1 and a tooth fragment; 23017, skull fragments and palate, edentulous Rt. maxilla (roots M^{2-3}), Lt. and Rt. M^3, Lt. P_3, Lt. P_4, Lt. M_1, Lt. M^2, P_1; 23018, Lt. and Rt. maxilla (P^3–M^2), Rt. M^3; 23021, Rt. mandible (M_{1-3}), Lt. M_3 (talonid missing in Rt. M_3); 23024, Rt. broken M_3; 23027, Rt. dP_4; 23030, Lt. I_1; 23054, Lt. mandible (roots I_2–P_2); 23119, Rt. M^2, I^2, partial M^2, tooth fragments; 23126, maxilla and basicranium (Lt. and Rt. dP^4, M^1–M^2, M^3 in crypt); 23584, mandible, worn Lt. and Rt. M_{1-2}, Rt. P_{3-4}, broken Lt. and Rt. M_3; 23590, mandible and maxilla fragments (Rt. I^3, Lt. and Rt. P^2–M^1, Lt. M^3, broken /P, Lt. and Rt. P_4; 23658, Lt. M_1; 23744, Rt. M^3, /I; 23750, Rt. mandible fragment, Rt. M_2; 23751, Rt. maxilla (P^4–M^2) and fragmentary M^3; 23752, Rt. mandible (P_3–M_3) and Lt. mandible (P_2–M_3), Lt. I_1, Lt. and Rt. I_2, /C; 23754, Lt. maxilla fragment (P^4); 23755, posterior portion Rt. M_3; 23757, Lt. mandible fragment (M_3); 23759, Rt. mandible (P_3–M_3); 23760, mandible symphysis (Lt. and Rt. I_1, Rt. I_2, roots Lt. I_2), root Lt. I_3; 23762, Lt. weathered mandible fragment (M_3), M_2; 23763, Lt. I^1, Lt. P_2, ?M_1; 23764, Lt. M^2; 23771, female cranium; 23801, Rt. I^3; 23802, Rt. dP_4; 23804, Rt. I_1; 26103, Rt. P_4 and M_3 talonid; 26104, Lt. M^2; 26105, dP_2; 26107, LdI_2; 26111, Rt. M^3 talonid; 26113, Rt. I_1, Rt. I_2, LI_3; 26114, mandible fragment (M_3 fragment); 26115, Rt. mandible (M_{2-3}); 26122, Lt. M_3, Lt. P_4; 26128, Lt. M^2, P^2; 26130, Lt. mandible (P_{3-4}); 26595, Lt. maxilla (P^{3-4}); 26598, Lt. maxilla (P^4–M^2), edentulous mandible symphysis; 28753, Lt. P^4; 28776, Lt. dP^4; 28782, Lt. and Rt. I^1, Lt. I^2; 28793, Lt. M^1.

- *Upper Nawata:* 286, Rt. M^3; 287, male mandible (Rt. I_2, /C–P_4, Lt. /C, P_{2-3}); 288, male Rt. mandible (Lt. and Rt. I_1, Rt. /C, Rt. broken P_{3-4}, Lt. /C, worn M_1, M_3; 293, Rt. maxilla (M^3 and broken M^2); 296, Rt. mandible (M_2); 300, male Rt. mandible (P_3–M_3); 307, Rt. M^3; 299, Lt. P_3; 319, unerupted Rt. M^3; 388, Rt. mandible (M_3); 10278, Lt. maxilla fragment (P^4–M^2); 22958, Lt. maxilla (P^2–M^3); 22982, Rt. M_3 talonid; 22989, Lt./dC; 23123, mandible (Rt. dP_1, dP_{3-4} and M_1, Lt. dP_{3-4}, canines in crypt); 23591, Lt. P_3; 23731, Rt. M^3; 23737, Rt. I_1; 23738, Rt. I^2, upper M fragment; 23739, Rt. mandible fragments (M_2); 23740, Lt. I^2; 23743, Lt. and Rt. I^1, Rt. M^{2-3}, Lt. and Rt. I_3; 23745, Rt. mandible fragment (M_1, broken M_{2-3}); 23747, maxilla fragments (Lt. P^2, Lt. and Rt. P^3, Rt. P^4, Lt. M^3); 23748, Lt. I_1; 23758, Rt. P_3, Rt. M_3 broken anteriorly; 23774, Lt. mandible (M_{2-3}); 23805, Rt. P_3; 24084, Rt. I_1, Rt. I_3, M^2 or M^3 fragment; 26077, Rt. maxilla fragment. I^2 P^4, M^1; 26083, mandible fragments and Lt. M_1; 26085, I^3; 26087, I^2; 26088, Lt. maxilla, P^4, M^{1-2}; 26089, maxilla (erupting P^2–M^2); 26091, Rt. M^3; 26093, upper incisor; 26095, Lt. mandible (roots P_2–M_3); 26108, Lt. I^3; 26109, mandible fragment (P_2–M_3) and Rt. M^1; 26117, Rt. dP_4; 26119, P^3; 26121, Rt. M^3; 26135, Rt. P_3; 26138, Rt. maxilla (dP^3–M^2); 26592, Rt. P_2; 26620, Lt. I^3; 28565, Rt. dP_3; 28779, Lt. M^1.
- *Upper Nawata or Apak Member:* 118, unerupted M_3 and M fragment.
- *Apak Member:* 309, Rt. mandible (M_3); 26110, Rt. M^3.
- *Horizon indet.:* 290, Rt. maxilla (P^{3-4}), Lt. worn M^3, Lt. P_4; 7709, Rt. upper I and tooth fragments; 10275, Lt. P^4; 10276, maxilla (Rt. M^3); 10277, Rt. M^3; 23003, male Rt. and Lt. maxilla (Lt. P^{2-4}, Rt. I^1–P^2 alveoli); 23005, Lt. P^4, Lt. P^3, Lt. M^2, M fragment; 23589, P^3, P^4, broken M^3, unerupted.

The morphology of the dentition and (holotype) male cranium has been thoroughly described in previous studies and need not be repeated here. Dentally this species is larger than but morphologically similar to *Nyanzachoerus devauxi* (= *Conohyus giganteus* of van der Made 1999) (figure 10.44). Additional material recovered by recent collecting expeditions at Lothagam has confirmed the virtual restriction of this species to the Nawata Formation, only a single specimen having been retrieved from the superjacent Apak Member. New material confirms the sexual dimorphism of the canines and documents the morphology of the female skull—the type and only previous cranium of this species from Lothagam being male. The female cranium, LT 23771, is shorter, narrower and, conceivably, less tall than that of the male, although the holotype cranium is incomplete posteriorly. The female canines are smaller, and the canine flanges on the cranium are less well developed. There are less prominent rugosities on the nasal platform of the female cranium, and the preorbital ridges are more gracile and less upwardly protuberant. The zygoma is less laterally protrusive and bears a smaller and less prominent zygomatic knob. The cranial vault of the female is narrower, and its dorsal surface is less concave. The female cranium is more complete posteriorly than that of the holotype male. The temporal ridges converge and nearly meet in the posterior part of the cranial vault, but sharply diverge laterally just in front of the nuchal crest. This portion of the cranial vault is much narrower than in any of the extant African suids, and it is possible that this is an attribute that characterizes the genus.

Also evident from the new dental material is the size increase in third molar specimens from the upper member of the Nawata Formation (figure 10.45). This size increase seems largely the result of addition of extra cusps to the talon(id) of the third molar. These extra cusps contribute to a morphological pattern that is intermediate between (1) the holotype and "typical"

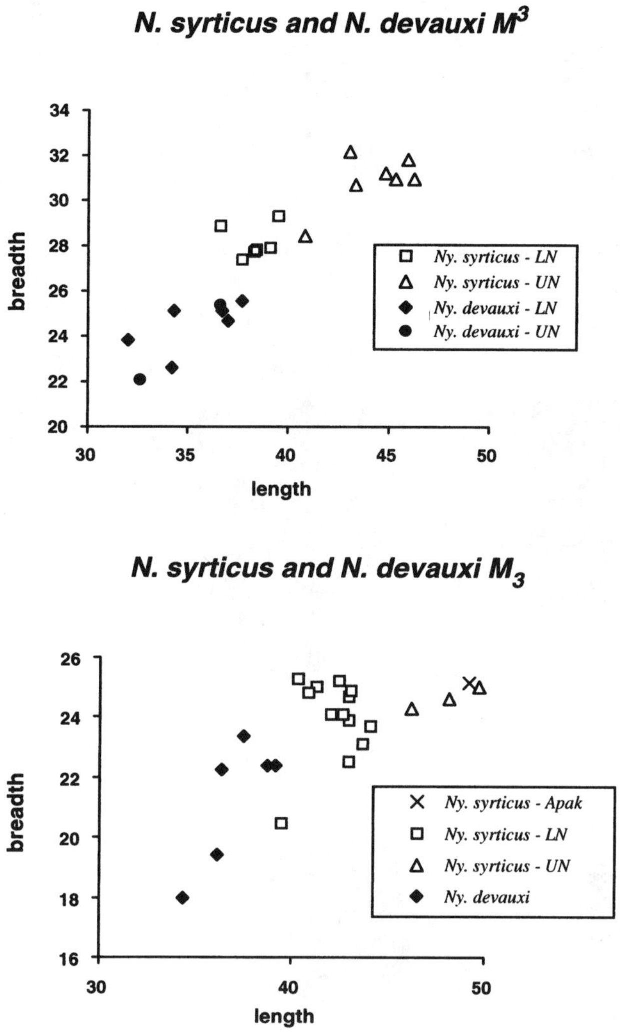

Figure 10.44 Plot of mesiodistal length against labiolingual breadth for *Nyanzachoerus syrticus tulotus* and *N. devauxi* third molars from Lothagam. LN = Lower Nawata; UN = Upper Nawata.

members of the species from the Lower Nawata and (2) the arrangement of talon(id) pillars in the more progressive *N. pattersoni* and *N. australis* from the Nachukui Formation. Although using different taxonomic names, Harris and White (1979) postulated that *N. syrticus tulotus* evolved into *N. pattersoni,* and the recovery of such "advanced" representatives of *N. syrticus tulotus* appears to confirm their interpretation. It is interesting that specimens of *N. syrticus tulotus* from Lukeino also display the progressive morphology seen in specimens from the Upper Nawata at Lothagam. Indeed, we interpret the nyanzachoere material from Lukeino to represent both the advanced form of *N. syrticus tulotus* found at Lothagam and *N. pattersoni* rather than *N. syrticus tulotus sensu strictu.* KNM-LU 515 (Rt. M^3) and LU 180 (mandible with Rt. and Lt. P_3–M_3) are allocated to *N. pattersoni.* One specimen, variously numbered KNM-LU 1, 2, 3, 4, 5, 509, and 980, but evidently comprising a single individual, is interpreted as the advanced form of *N. syrticus tulotus.* In contrast to the interpretations made by Cooke and Ewer (1972) based on the stratigraphic information then available, we no longer recognize the presence of *N. pattersoni* from the earlier portion of the Lothagam sequence. The specimens from the early part of the sequence that were assigned to "*N. pattersoni*" by Cooke and Ewer—LT 112, 285, 294, 304—constitute partial dentitions that lack the third molars and whose teeth fall within the size ranges of the more complete *N. syrticus tulotus* specimens (although 294 is now assigned to a species of *Potamochoerus*).

Nyanzachoerus cf. *N. syrticus*

Lothagam Material

- *Lower Nawata:* 26133, distal Lt. Mc III.
- *Upper Nawata:* 436, associated postcranial elements (Lt. humerus, Lt. proximal ulna and radius, partial Lt. and Rt. innominate and femur, Rt. tibia, Rt. distal fibula, Rt. astragalus, Rt. calcaneum, Rt. navicular and Rt. Mt IV, partial Lt. tibia, Lt. distal fibula, Lt. astragalus, Lt. calcaneum, Lt. cuboid, Lt. Mt III and IV, Rt. semilunar); 26090, Lt. navicular; 26097, distal Lt. Mc IV; 26125, Rt. Mt IV; 26192, Lt. Mc III.

Several isolated postcranial elements from the lower and upper members of the Nawata Formation that are evidently too large to belong to *N. devauxi* are here referred to *Nyanzachoerus* cf. *N. syrticus.* LT 26090 is a left navicular which is close in size, but of different proportions, to that of an associated skeleton LT 436. Other specimens include LT 26133, a distal left Mc III, and LT 26192, a complete but weathered left Mc III; LT 26097, a distal left Mc IV and right Mt III; and LT 26125, a complete right Mt IV. McCrossin (1987) divided postcranial elements from Sahabi into two size categories that he attributed to *Nyanzachoerus* cf. *N. devauxi* (the smaller) and *Nyanzachoerus* cf. *N. syrticus* (the larger). One of the Lothagam specimens, LT 26192, a very weathered left third metacarpal, is comparable to, but smaller in all dimensions than, that of partial skeleton P4A from Sahabi (McCrossin 1987:table 9). McCrossin attributed P4A, which includes several elements of a rather elongated cursorial forelimb, to *Nyanzachoerus* cf. *N. syrticus.*

A partial unpublished skeleton (LT 436) collected by the Harvard expedition represents a large suid that is here assigned to *Nyanzachoerus* cf. *N. syrticus.* Unfortunately, there is no associated cranial material. The provenance of this specimen was given by Cooke and Ewer as Lower Nawata, but the field catalogue and Kay

Figure 10.45 *Nyanzachoerus devauxi* and *Nyanzachoerus syrticus tulotus* third molars; all natural size. Top row: left = KNM-LT 26101, *N. devauxi*, Rt. M^3, Lower Nawata; center = KNM-LT 22970, *N. devauxi* Rt. M^3, Upper Nawata; right = KNM-LT 26094, *N. devauxi*, Rt M^3, Upper Nawata. Second row: left = KNM-LT 23018, *N. syrticus tulotus*, Rt. M^3, Lower Nawata; center = KNM-LT 26091, *N. syrticus tulotus*, Rt. M^3, Upper Nawata; right = KNM-LT 26110, *N. syrticus tulotus*, Rt. M^3, Apak. Third row: left = KNM-LT 23019, *N. devauxi*, Rt. M_3. Lower Nawata; center = KNM-LT 26075, *N. devauxi*, Lt. M_3, Lower Nawata; right = KNM-LT 282, *N. devauxi*, Lt. M_3, Lower Nawata. Bottom row: left = KNM-LT 26122, *N. syrticus tulotus*, Lt. M_3, Lower Nawata; center = KNM-LT 386, *N. syrticus tulotus*, Rt. M_3, Lower Nawata; right = KNM-LT 118, *N. syrticus tulotus*, Rt. M_3, Upper Nawata.

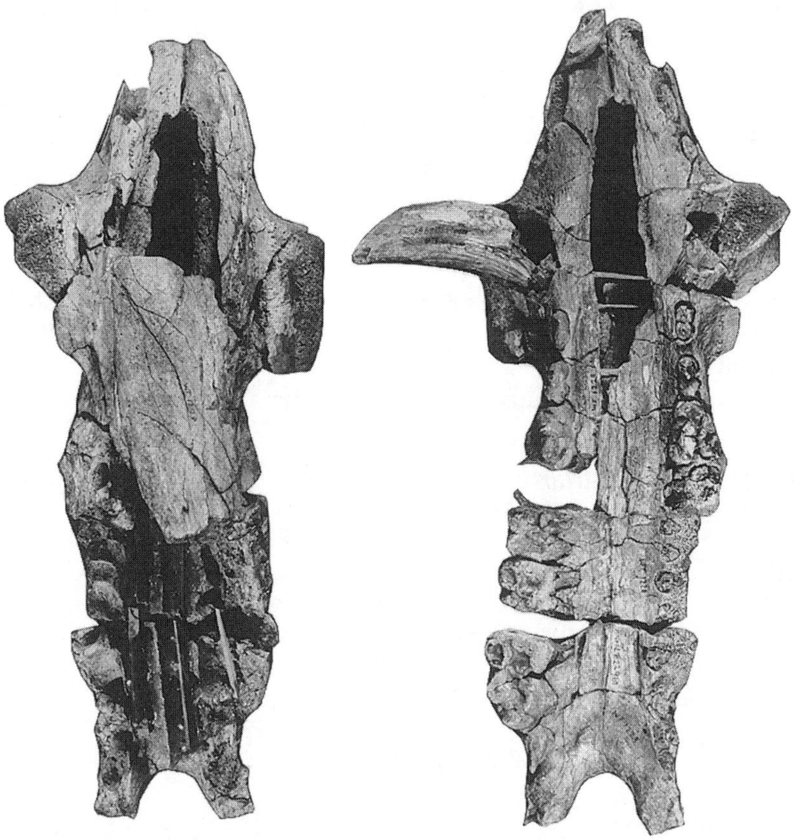

Figure 10.46 *Nyanzachoerus devauxi* partial cranium, KNM-LT 22967, Lower Nawata, half natural size: left = dorsal view; right = ventral view.

Behrensmeyer's chart indicate that it is almost certainly from the Upper Nawata. The three bones that form the elbow joint and the bones of both right and left ankles are preserved. The articular surfaces at these joints show deep grooves and elevated ridges, which indicate joint stability with movement restricted to the parasagittal plane that is typical of more cursorial, open country species. This is consistent with Bishop's (1994) conclusion from her measurements of the nearly complete humerus and third metatarsal. Her analysis of these elements was consistent with open and intermediate habitat preferences.

Comparison of the more complete elements of LT 436 with the larger Sahabi postcranial indicates several differences. McCrossin (1987) assigns the Sahabi sample to *Nyanzachoerus* cf. *N. syrticus.* Measurements show that the Lothagam humerus is longer than the Sahabi specimen 18aP4A but the mediolateral breadth of the proximal epiphysis is smaller. The minimum shaft breadth is only slightly larger as is the distal mediolateral breadth, but the trochlear breadth is marginally smaller (McCrossin 1987:table 2). Similarly, the Lothagam radius has a similar proximal breadth to Sahabi 31P4A but is significantly smaller distally. And the astragalus, when compared with three astragali from Sahabi, is also smaller. The indication from these comparisons is that the Lothagam skeleton has relatively long slender limbs with relatively smaller foot bones than the species from Sahabi. If this is the case, then LT 436 may represent a different species than *Nyanzachoerus* cf. *N. syrticus.*

Nyanzachoerus australis (Cooke and Hendey, 1992)

Diagnosis

Large *Nyanzachoerus* with relatively large third molars. M_1 length about 23 mm; M_3 length about 52–72 mm (after van der Made 1999).

Cooke and Hendey (1992) defined this South African form as a subspecies of *Nyanzachoerus kanamensis* differing from *N. kanamensis kanamensis* by the consistent retention of P^1 in a slightly longer diastema, the possession of relatively larger and more robust cheek teeth, and a distinct low hump (which may be rugose) on the nasals in front of the canine flanges in males. However, van der Made (1999:208, figure 19) interpreted the Langebaanweg material as species of *Sivachoerus* that represented a later immigration from Asia than the *S. syrticus* lineage and which was totally unrelated to *Nyanzachoerus kanamensis.* van der Made recognized two subspecies, *S. australis australis* and *S. australis megadens,* corresponding to the samples from the Quartzose Sand Member and Pelletal Phosphorite Members, respectively, of the Varswater Formation. Teeth that are indistinguishable from the South African representatives occur at Lothagam and at localities of comparable age in Ethiopia (J. Haile-Selassie personal communication), and we follow van de Made (1999) in recognizing *N. australis* as a species in its own right.

Nyanzachoerus cf. *N. australis*
(Figure 10.48; tables 10.16, 10.17)

Lothagam Material

- *Upper Nawata:* 285, Lt. mandible fragment (P_4–M_3), Rt. M^2; 313, Rt. maxilla (P^3–M^3; 23627, Rt. M^3 fragment; 23746, Rt. mandible fragment (Rt. M_3 talonid).
- *Apak Member:* 308, Rt. P_3, P_4 and fragments M_2; 310, broken Lt. P^3–M^1; 23742, Lt. mandible (P_4–M_3); 23749, Lt. mandible fragment (M_{1-3}) and mandible fragments; 23803, Lt. ?Mc V; 26076, Lt. M_3; 26084, Rt. M^3 talon; 26092, Lt. M^3 fragment; 26098, mandible fragment (Rt. P_{3-4}); 26099, Rt. P_3, Lt. I^1; 26134, Rt. M^2; 26191, Rt. Mc III fragment; 26591, Lt. M^2 fragment; 28715, Lt. M_2; 28720, M_3 talonid.

Teeth assigned to this taxon are generally larger and, in particular, broader than those assigned to *N. pattersoni* (figure 10.48). Two postcranial elements from the Apak Member, LT 26191, a right Mc III lacking the unfused distal epiphysis, and LT 23803, a complete ?Mc V, are also attributed here.

Nyanzachoerus pattersoni (Cooke and Ewer, 1972)
(Figures 10.47, 10.48; table 10.18)

Diagnosis

Sexually dimorphic nyanzachoere larger than *N. syrticus.* First premolars vestigial or absent. Compared to *N. syrticus* subspecies, third and fourth premolars robust but relatively smaller, and third molars elongate and with more complex talon(id) (after Harris and White 1979). M_3 length between 48 and 60 mm (van der Made 1999).

Lothagam Material

- *Kaiyumung Member:* 23768, Rt. mandible (P_3, M_{1-3}); 26078, Lt. M^3; 26137, P_3, P^4, M, M_2, M_3 broken, isolated teeth.

This species was originally described as *Nyanzachoerus pattersoni* by Cooke and Ewer (1972), but Harris and

Figure 10.47 Restoration of *Nyanzachoerus pattersoni* by Mauricio Antón.

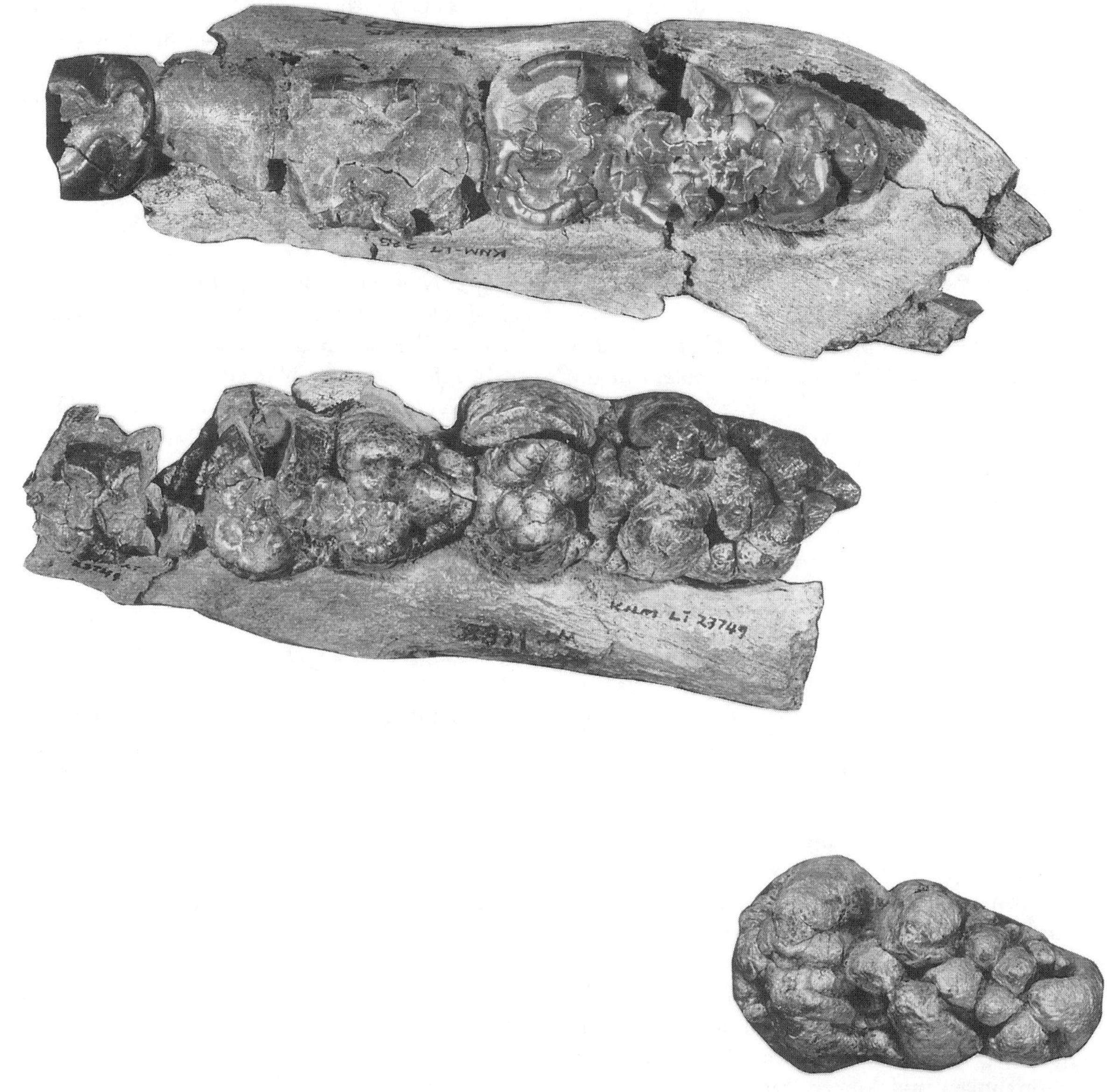

Figure 10.48 *Nyanzachoerus* cf. *N. australis* and *N. pattersoni* teeth; all natural size. Top = KNM-LT 23749, *Nyanzachoerus* cf. *N. australis,* left mandible, occlusal view, Apak Member. Second row = KNM-LT 285, *Nyanzachoerus* cf. *N. australis,* Lt. mandible, occlusal view; Upper Nawata Member. Third row = KNM-LT 26078, *N. pattersoni* Lt. M_3, occlusal view; Kaiyumung Member. Bottom = KNM-LT 313, *N. australis,* left maxilla, occlusal view; Upper Nawata Member.

White (1979) considered it conspecific with *Nyanzachoerus kanamensis,* a viewpoint that has been adopted by most subsequent workers. However, in his recent revision of the Tetraconodontinae, van de Made (1999) considered "*N. pattersoni*" to belong to the tetraconodontine genus *Sivachoerus,* whereas Leakey's (1958) *N. kanamensis* was regarded as a member of the Nyanzachoerini. We here follow van de Made's (1999) premise that the Lothagam and younger Kenyan specimens are specifically different from the narrow-toothed *N. kanamensis* from the Western Rift.

In the holotype female of *N. pattersoni,* the upper third molar resembles a large version of a "typical" *N. syrticus* with a talon comprised of a single main pillar plus subsidiary cusps. The lower third molars in the associated mandible, however, have a characteristic "*pattersoni*" talonid composed of a single pair of pillars plus a terminal pillar. In the paratype male of *N. pattersoni,* moreover, the upper third molar has a more complex talon that bears a pair of lateral pillars. Other specimens attributed to *N. pattersoni* by Cooke and Ewer (1972) have similarly complex talons (e.g., KP 223).

Nyanzachoerus pattersoni makes its initial Lothagam appearance in the Kaiyumung Member. Specimens here attributed to *Nyanzachoerus* cf. *N. australis* from the Apak have similar morphology to *N. pattersoni* but, although showing a great deal of variability in size, have larger and broader premolars and molars. In this they bear a strong resemblance to material from Langebaanweg for which Cooke and Hendey (1992) erected the separate subspecies *N. kanamensis australis* (= *Sivachoerus australis australis* of van der Made 1999). The few specimens of *N. pattersoni* that have been recovered from the Kaiyumung Member are similar in size to the "typical" *N. pattersoni* specimens, but, in keeping with their younger age, the Kaiyumung specimens have smaller premolars and longer third molars that are characterized by more complex talon(id) morphology.

Nyanzachoerus devauxi (Arambourg, 1968)

(Figures 10.43–10.46; tables 10.19–10.21)

This species was first described by Arambourg (1968) as *Propotamochoerus devauxi* from Bou Hanifia, Algeria. More recent workers including Cooke (1987), Pickford (1990), Hill et al. (1990), and Leakey et al. (1996) have assigned this species to *Nyanzachoerus.* Recently, van der Made (1999) retained the species in the Nyanzachoerini but interpreted the name as a junior synonym of *Conohyus giganteus* (Falconer and Cautley 1847). We here interpret it as a small species of *Nyanzachoerus* rather than a large species of *Conohyus.*

Diagnosis

A species of *Nyanzachoerus* that differs from *N. syrticus* by virtue of its smaller size, proportionately longer premaxilla, and (usually) simpler talon(id) morphology in the third molars. Differs from *N. waylandi* (Cooke and Coryndon 1970) by its larger premolars but smaller molars.

Lothagam Material

- *Lower Nawata:* 14, Lt. M^3; 110, squashed maxilla (Rt. P^1–broken M^3, Lt. P^3–M^3); 282, unerupted Lt. M_3; 283, broken M^3; 303, Lt. mandible (P_4, M_3); 314, Lt. M^3; 22885, Rt. mandible (P_4, M_1) partial M_2 and M_3; 22967, cranial fragments (Lt. and Rt. I^1, Rt. P^{1-4}, M^{1-2}, Lt. P^{2-3}), Lt. and Rt. C/, Lt. M^2, worn M^3; 23015, Rt. P_4; 23019, female mandible (Lt. P_4–M_2, Rt. P_4) broken Rt. P_3, Rt. and Lt. M_3, I_1, Rt. and Lt. /C, Lt. I_2; 23020, maxilla (Lt. and Rt. P^{3-4}, Rt. M^2) and Lt. and Rt. M^3; 23022, Rt. mandible (P_4–M_3, roots P_3); 23023, Lt. mandible (P_4–partial M_3), partial P_3, mandible fragments; 23025, Rt. M_2; 23029, broken Rt. M_3; 23161, Lt. I^3; 23753, Lt. M_3; 23957, Lt. M_2; 26075, Lt. M_3; 26101, Rt. M^3, I^3 and tooth fragments; 26102, Rt. maxilla fragment (P^3–P^4); 26106, mandible fragment (dP_4); 26112, juvenile mandible dP_2, partial dP_4 fragment; 26116, Lt. M_3 talonid (small); 26118, Rt. dP_4; 26126, Rt. M^3; 26127, Lt. M_3, teeth fragments; 26132, Lt. P^3.
- *Upper Nawata:* 22970, subadult cranial and postcranial fragments, Rt. I^1, Lt. and Rt. M^3, Lt. and Rt. I_1, Rt. M_3 and broken Lt. M_3; 26094, Rt. M^3; 26096, Rt. M^3, Lt. M^3 fragments; 26131, Rt. juvenile maxilla (dP^{2-3}); 26139, dP^3; 28708, Lt. dP^4; 28750, Lt. M^3.

This species was founded on a partial mandible from Bou Hanifia in Algeria by Arambourg (1968). Arambourg did not provide a specific diagnosis for this species, noting that it was similar to *Nyanzachoerus kanamensis* but smaller, and with the P_4 longer than the M_1. Pickford (1986) reported that the holotype specimen could not be located in the Museum National d'Histoire Naturelle, Paris, although a year later Cooke (1987) confirmed that he had seen it. The type mandible is said to be characterized by a P_4 in which the protoconid and metaconid are incompletely fused, but this is a variable character that is seen also in material identified as *N. syrticus* from Lukeino (Cooke 1987:265). The smallest nyanzachoere specimens from Sahabi were assigned to *Nyanzachoerus* cf. *N. devauxi* but are slightly smaller than the holotype (Cooke 1987). The species has also been recorded from Abu Dhabi (Hill et al. 1990; Bishop and Hill 1999). At Lothagam, *N. devauxi* is restricted to the Nawata Formation.

Cooke and Ewer did not recognize *N. devauxi* at Lothagam but assigned one tooth of this species (LT 314) to "*Sus* sp. or *Potamochoerus* sp." (Cooke and Ewer 1972:225). Other *N. devauxi* specimens from Lothagam comprise isolated teeth, mandible and maxilla fragments, and a partial skeleton. The teeth are almost identical in morphology to others formerly attributed to *S. syrticus tulotus* but seem distinctly smaller. The main difference in the teeth of these two species occurs in the talon(id) of the third molars. In the M^3 of *N. devauxi* the talon comprises an intermediate pillar (between and behind the metacone and hypocone), a single large posteromedial cusp, and two or three posterolateral cusplets. Identical morphology may be observed in the smaller M^3s of *S. syrticus,* although in the holotype of "*N. tulotus*" (LT 316) there is an intermediary cusplet between the hypocone and the posteromedial cusp.

A partial *N. devauxi* male cranium, LT 22967, differs from the male cranium of *S. syrticus,* LT 316, by its smaller size and relatively longer premaxilla—the length from the anterior premolar alveolus to the anterior canine alveolus is almost identical in the two specimens. However, when placed with their anterior extremities parallel, the anterior edge of the *N. devauxi* M^3 falls alongside the anterior edge of the *S. syrticus* M^1, and the U-shaped palatal notch of the *N. devauxi* palate lies level with the M^3 of *S. syrticus.* The anterior portions of the zygomatic arches of the *N. devauxi* cranium are swollen to form zygomatic knobs that are similar to but less laterally protuberant than those of the male *S. syrticus* cranium. On the basis of third molar size, palates LT 110 and LT 23020 are also male, whereas that of LT 22970 is female.

The M_3 of *N. devauxi* differs from that of the paratype mandible of "*N. tulotus*" by its talonid simplicity. The intermediate cusp is bordered by a tiny posterolateral cusplet, a large posterolateral cusp, and a smaller posteromedial cusp. In the paratype mandible of "*N. tulotus*" there is more than one posterolateral cusplet and posteromedial cusp.

Partial cranium LT 22970 was found associated with vertebrae, a partial scapula, humerus, ulna, radius, innominate, tibia, and cuboid. Comparisons were made with the same elements from an adult male modern bush pig (OM 3338) of similar size. In general, the limb bones seem of similar size to those of the bush pig, but perhaps the Lothagam specimen was more lightly built—though this is difficult to assess because none of the long bones is complete enough to estimate its length, and the unfused epiphyses indicate the specimen is immature. However, both the proximal articular surface of the radius and the distal articular surface of the tibia show more pronounced articular ridges and grooves, which indicates that these were strong joints with movement restricted to the parasagittal plane.

Preserved vertebrae include the atlas, axis, four other cervical vertebrae, and six lumbar vertebrae (the bush pig has only five). There are no major differences between the preserved vertebrae and those of the bush pig but the axis and atlas of the Lothagam specimen appear to be slightly taller dorsoventrally and narrower transversely and the lumbar vertebrae appear to be slightly longer and generally larger.

The glenoid fossa, supraglenoid tubercle, neck, and part of the spine of the right scapula (F) were preserved. The glenoid fossa is slightly more pear shaped than the rather circular glenoid fossa of the bush pig, and the Lothagam specimen's neck is long and narrow. The distal humerus differs from that of the bush pig by the narrower olecranon fossa (perforated by a large foramen 10.3 mm transverse and 8.6 mm proximodistal) and by articulating more tightly with the ulna, as indicated by the well-developed trochlea and capitulum grooves and trochlea crest. The right distal humerus (A) has carnivore tooth marks on the lateral face. The olecranon of the right ulna (K) also shows considerable evidence of carnivore chewing on the partially fused olecranon process. The anconeal process and the most proximal part of the semilunar notch are similar to that of the bush pig, but the entire proximal end is narrower transversely. The humeral articular surface of the proximal right radius (I) has a distinct sagittal ridge that separates the deep capitular fovea and also indicates a tight elbow articulation. As in the bush pig, the ulna articular facet is distinct and the area of attachment of the interosseous ligament is rugose. The right distal radius (G) has an incompletely fused epiphysis.

Partial left (C) and right (D) innominates preserve the complete acetabulum and partial ilium body; the wings are missing. The iliopubic eminence is marked and pointed, but the cross section of the break of the cranial branch of the pubis indicates that this was slender. The proximal right tibia (E) has a partially fused epiphysis and is broken close to midshaft. The distal portion of the right tibia also appears to be broken close to mid-shaft, but, judging by the dimensions of the broken surfaces, it is likely that several centimeters of the shaft are missing at the break; consequently, the complete tibia would have been longer than that of the modern bush pig. The left (H) and right (G) distal tibiae have more pronounced articular grooves and a sharp intermediate ridge. Only one tarsal, the right cuboid, is preserved; compared to that of the bush pig, it is relatively narrow mediolaterally relative to its dorsoplantar width and the astragalar facet is correspondingly narrow.

Unfortunately, very few of the Lothagam postcranial remains can be compared with equivalent elements attributed to *Nyanzachoerus* cf. *N. devauxi* from Sahabi (McCrossin 1987). The Lothagam distal humerus is larger than that from Sahabi (4P29A); maximum breadth is 43.35 mm compared to 34.5 mm for Sahabi, and the trochlea breadth is 32.6 mm compared to 23.3 mm for Sahabi. In contrast, the Lothagam distal tibia (29.7 mm mediolateral by 27.6 mm anteroposterior) is smaller than the small tibia from Sahabi (5P85B), which measures 34.7 mm in both dimensions. These differences could be due to sexual related size differences. The atlas measurements, however, indicate that there may be some difference in the morphology between these two suids although different methods of taking the measurements could contribute to some of the major differences, such as the dorsoventral thickness (McCrossin 1987:table 18). In general, the sizes of the comparable bones are roughly similar.

Bishop (1994) analyzed the morphological characters of the tibia and the humerus of LT 22970 in an attempt to ascertain the habitat preferences for the species; both bones grouped with mixed habitat species, from which she inferred that this species preferred habitats intermediate between open and closed.

Cooke and Coryndon (1970) referred teeth from several Ugandan localities to *"Sus" waylandi*. Pickford (1989) described additional Ugandan material that he reassigned to *Nyanzachoerus waylandi*, interpreting *N. waylandi* as a primitive member of the genus with premolar morphology approaching that of *Conohyus*. The molars of *N. waylandi* are larger than those of *N. devauxi* from Bou Hanifia in Algeria (Arambourg 1968), Sahabi in Libya (Cooke 1987), and the Beglia Formation of Tunisia (Pickford 1990). The premolars of *N. waylandi*, however, are 25–40 percent smaller than those of *N. devauxi*, whereas the M_3 of *N. waylandi* is proportionately longer. Thus, according to criteria employed by Cooke (1978) and Harris and White (1979), *N. waylandi* appears to be a more progressive nyanzachoere than *N. devauxi*. However, van der Made (1999) regarded *N. waylandi* as a subspecies of *N. kanamensis*.

It is also interesting to note that the premolars of *N. waylandi*, currently known only from the Western Rift, are proportionately narrower than most other tetraconodontine specimens reported from elsewhere in Africa. The premolars of the holotype of *N. kanamensis* are also narrower than specimens from the Turkana Basin that were originally described as *N. pattersoni* (Cooke and Ewer 1972). van der Made (1999) views this as one of the features that distinguish the Tetraconodontini from the Nyanzachoerini.

Nyanzachoerus jaegeri Coppens, 1971
(Table 10.22)

Diagnosis

A large progressive tetraconodontine possessing three pairs of premolars of which the third and fourth are proportionately smaller than *N. syrticus* or *N. pattersoni*. The third molar is longer and taller and has more pillars than those of either of the latter taxa. There is a tendency for the molar enamel to be folded. The cranium and mandible are larger and more elongate than in other *Nyanzachoerus* species; strong sexual dimorphism evident with the zygomatic swellings more localized but more protuberant in males (after Harris and White 1979).

Lothagam Material

- *Apak Member:* 311, Lt. M^3 broken mesially; 23741, Lt. mandible (P_3–broken M_3), Lt. maxilla (P^4–M^3).
- *Horizon indet.:* 10274, Lt. M^3 talon.

Two specimens from the Apak Member and one of unknown provenance may be referred to this species. The upper third molars may be distinguished from those of *N. australis* by the complex nature of the talons and the larger number of accessory cuspules. New and more complete material from Kanapoi and elsewhere indicates that this species does not belong to either *Nyanzachoerus* or *Sivachoerus*, but detailed revision of this species will be deferred until the Kanapoi biota is redescribed (Harris and Leakey in preparation).

Notochoerus Broom, 1925

Diagnosis

Tetraconodontines with small premolars, elongate molars, and very long and hypsodont third molars. The M_3 is the longest of all tetraconodontines (van der Made 1999).

Notochoerus euilus (Hopwood, 1926)
(Table 10.23)

Diagnosis

A species of *Notochoerus* with a deep and narrow cranium bearing sexually dimorphic zygomatic knobs. M^3 with two strong lingual talon pillars. M_3 with typically three pairs of talonid pillars plus single midline terminal pillar or series of pillars. Individual major lateral pillars of molars moderately tall, widely separated from adjacent lateral pillars, and tapering sharply from their bases (after Harris and White 1979).

Lothagam Material

- *Apak Member:* 23775, Lt. /C.
- *Kaiyumung Member:* 289, Lt. mandible fragments (M_{2-3}); 297, Lt. M_3 fragment; 298, Lt. M^3; 23679, Lt.

M_3 fragment; 23723, mandible and tooth fragments; 23765, Lt. mandible (M_2); 23766, partial M_3; 23767, posterior portion M_3; 23769, partial M^3; 23776, skull; 23777, tooth fragments; 23778, partial M^3 and proximal radius; 24050, Lt. M^3 talon; 26074, upper molar; 26079, very weathered lower P; 26080, lower incisor; 26081, M_3; 26124, M_3 and tooth fragments; 26136, mandible (P_4–M_3); 26596, talonid M_3; 37861, complete but badly weathered mandible.
- *Horizon indet.:* 574, Rt. mandible, broken M_3.

Notochoerus euilus is represented by about a score of teeth and tooth fragments from the Kaiyumung Member, making it the commoner suid from this interval. A left lower canine, LT 23775, from the upper part of the Apak Member also appears to represent this genus.

Subfamily Suinae Gray, 1821
Tribe Potamochoerini Gray, 1873
Potamochoerus Gray, 1854
Potamochoerus porcus (Linn.), 1758

Diagnosis

Small, sexually dimorphic suid similar in size and morphology to *Sus* but with more brachyodont and less trenchant premolars, simpler third molars, facial tuberosities, and more diverging zygomatic region.

Lothagam Material

- *Galana Boi Beds:* 26120, P^4; 23826, maxilla fragment (P^4–M^1); 23761, maxilla fragment (dP^{2-3}).

Three specimens of bush pig were collected from the surface of the upper member of the Nawata Formation. However, they and a number of fossilized wart hog specimens almost certainly represent lag from surficial deposits of the Galana Boi Beds. An immature maxilla fragment collected from the Galana Boi Beds can also be assigned to this taxon.

Potamochoerus sp.
(Figure 10.43; table 10.24)

Lothagam Material

- *Lower Nawata:* 294, Rt. Mandible (P_4–M_2); 23711, Lt. P_3.
- *Upper Nawata:* 28789, Rt. M_1.

One partial mandible, LT 294 from the Lower Nawata, cannot readily be assigned to *Nyanzachoerus.* The P_3 is narrow and tall, in striking contrast to the large and more bulbous *Nyanzachoerus* premolars. The M_2 has an intermediate cusp between the protoconid and hypoconid; when worn, this occlusal pattern forms an oblique ridge that has no counterpart in any *Sivachoerus* or *Nyanzachoerus* species. The specimen was collected in 1967 so has no exact provenance but Cooke and Ewer (1972) indicate that it comes from low in the sequence. The specimen is rather larger than the extant bush pigs or than the *Potamochoerus* specimens from Laetoli and low in the Koobi Fora and Nachukui Formations. In many respects it is similar to the cast of *Potamochoerus* sp. mandible (Y 6530, KNM-ZP 7292) from Yale Locality 551 near Jalapur, Pakistan, and whose age is believed to be between 5 and 6 Ma (J. C. Barry personal communication). A lower right first molar, LT 28789 from the Upper Nawata, displays the same crown morphology. A left lower third premolar, LT 23711, is more slender than those of *N. syrticus* or *N. kanamensis.* Both specimens appear closer to LT 294 than other possible alternatives.

Tribe Phacochoerini Gray, 1868
Phacochoerus Cuvier, 1817
Phacochoerus aethiopicus (Pallas, 1767)

Diagnosis

A sexually dimorphic species of Suidae of small to moderate size. Cranium with broad zygomatic arches lacking distinct knobs, elevated orbits, and short cranial region. Upper incisors reduced to one pair. Upper canines lacking enamel except at tips. Premolars reduced and commonly shed in adults. Molars hypsodont, formed of closely packed columnar elements and well cemented; lateral pillars flattened externally, elongate and oval to subtriangular in shape.

Lothagam Material

- *Galana Boi Beds:* 26123, M_3 talonid, /C and tooth fragments; 26141, M; 26142, tooth; 26143, M_3 talon; 26144, partial M_3; 26145, M_3 fragment; 26146, molar fragments; 26147, molar fragment; 26148, tooth fragment; 26149, M, C; 26150, tooth; 26151, tooth fragment; 26152, two M^3s; 26153, molar; 26154, molar; 26155, molar; 26156, mandible (M_3); 26157, M_3; 26159, M fragment; 26160, tooth fragment; 26161, tooth fragment.

Some 20 fully fossilized specimens of teeth or partial teeth representing the extant warthog species were collected from the surface of the Lothagam exposures. Due to their fully mineralized condition, some consideration was given to the possibility that they were derived from the Nawata and Nachukui Formations. However, they clearly belong to the extant species and it seems much more likely that, like the teeth of the extant bush pig species, they represent lag from now totally eroded exposures of the Galana Boi Beds.

Subfamily indet.
Cainochoerus Pickford, 1988
Cainochoerus africanus (Hendey, 1976)

Diagnosis

A small suid (5–9 kg) with proximal fusion of metatarsals III and IV. Dental formula 3:1:4:3/2:1:4:3 with no postcanine diastema; enamel smooth; I^1 much larger than I^2 and I^3; I_1 and I_2 almost equal in size; lingual surface of incisors with very thin or no enamel; C/ short, nearly straight, and pointed downward; /C short, curved, and pointed upward and slightly outward; P^{1-3} double rooted, single cusped, and relatively high-crowned; P^4 triple rooted with paired principal cusps flanked posteriorly by a prominent cingulum; P_4 double rooted with paired principal cusps flanked posteriorly by a smaller accessory cusp; M^{1-3} quadricuspid, each tooth with a transverse enamel ridge anteriorly and a smaller oblique one posteriorly; M_{1-3} quadricuspid, each tooth with a centrally situated posterior accessory cusp (talonid). Mandibular symphysis low and relatively compressed. Median metacarpals are 25 percent shorter than median metatarsals (after Hendey 1976 and Pickford 1988).

The smallest suid from the Lothagam assemblages is represented by an incomplete cranium and partial skeleton and closely resembles material from Langebaanweg that Hendey (1976) initially identified as a fossil peccary. Hendey assigned the Langebaanweg hypodigm to *?Pecarichoerus africanus* and interpreted it as having close affinities with *P. orientalis* from the Chinji deposits of the Siwalik Series in Pakistan. Unfortunately, at the time that Hendey described his new material, comparison was constrained to the upper third molars because these were the only relatively unworn teeth in the *P. orientalis* hypodigm. Hendey noted that the oblique ridges that link the principal cusps of the upper molars of *P. orientalis* are smaller or absent in *P. africanus.* He also pointed out that the upper canine was short and vertically orientated as in peccaries, and that proximal metatarsals III and IV were fused together.

Subsequently, Pickford (1988) reinterpreted the Langebaanweg material as a dwarf suid rather than a tayassuid, and he noted some dental peculiarities (interpreting the incisors to resemble those of rodents and the molars those of monkeys). He created the new genus *Cainochoerus* for this material.

The specimen from Lothagam appears closely related to the Langebaanweg sample and is similar in size and morphology. The dentition is clearly similar, although the oblique ridges that link the principal cusps are distinct, as Hendey described for *P. orientalis.* Unfortunately, the Lothagam material does not include lower teeth, upper incisors, posterior premolars, or proximal metatarsals, thus precluding detailed comparison with the diagnostic features of the Langebaanweg specimens. We identify the dwarf suid from Lothagam as *Cainochoerus* cf. *C. africanus* pending the recovery of further and more complete material.

Although dismissing tayassuid affinity for *Cainochoerus,* Pickford (1986) recognized several Tayassuidae representatives from the Middle Miocene of Kenya. Two small bunodont specimens from Fort Ternan that Pickford assigned to the subfamily Doliochoerinae are a little larger than the Lothagam and Langebaanweg specimens but are not morphologically similar. Pickford referred four other larger and lophodont specimens from the Maboko and Ngorora Formations to the Tribe Schizochoerini.

Cainochoerus cf. *C. africanus*
(Figures 10.49, 10.50; table 10.25)

Lothagam Material

- *Upper Nawata:* 26430: A, calvaria; B, Lt. temporal with partial petrous, occiput and condyle; C, Lt. maxilla fragment (M^{1-3}); D, Rt. maxilla fragment (C/, P^2, alveolus P^1); E, Rt. temporal fragment with petrous; F, petrous fragment; G, cranial fragments (8); H, axis vertebra; I, lumbar vertebra (LU?2); J, lumbar vertebra (LU?1); K, thoracic vertebra (TH13); L, centrum (?C7 or Th1); M, lumbar vertebra centrum (LU?5); N–Q, thoracic vertebrae centra; R, lumbar vertebra centrum (LU?4); S, thoracic vertebra centrum (TH14); T, lumbar vertebra centrum (LU?3); U, vertebral spine fragments (6); V, Lt. proximal scapula; W, Rt. proximal humerus; X, Rt. distal humerus; Y, Rt. proximal ulna; Z, Lt. radius; AA, Rt. proximal femur; AB, Rt. distal femur; AC, Lt. distal femur; AD, Rt. distal tibia; AE, Lt. partial cuboid; AF, Lt. innominate fragment; AG, Rt. astragalus; AH, Rt. MC III; AI, two distal metapodials; AJ, proximal phalanx; AK, terminal phalanx; AL, humerus head fragment; AM, two proximal ribs; AN, rib fragments (12); AO, Lt. proximal ulna; AP, Rt. proximal humerus shaft.

Figure 10.49 Restoration of *Cainochoerus* sp. from the Upper Nawata by Mauricio Antón.

The cranium lacks much of the left frontal and right temporal, although a fragment of right temporal is preserved with the mastoid. There is a very poor join between the left temporal (B) and the rest of the calvaria. Much of the cranial vault is preserved, but both zygomatics, most of the maxilla and palate, and the basicranium are lost. The occiput is relatively complete, although the posterior margin of the foramen magnum is missing. The cranium is lightly built with only slightly developed crests and ridges for muscle attachments. Many of the cranial sutures are visible. In contrast to modern suids in which the postorbital part of the cranium is relatively short, the minimum distance from a line drawn between the postorbital processes and the occipital protuberance (43 mm) is almost equal to the cranial width across the postorbital processes (44.8 mm). Also contrasting with modern suids, the braincase is inflated ventral to the temporal lines, and there is a relatively large area for the attachment of the temporal muscles. In lateral profile, the dorsal surface along the frontals at the sagittal suture is rather flat with very slight convexity posteriorly at the parietals toward the occipital protuberance. Viewed dorsally, the left and right coronal sutures form a deep "V" that is directed posteriorly, whereas the lambdoidal sutures form a more open "V" that points anteriorly. The temporal lines are clear, and, although they do not meet, they converge from a point just dorsal to the postoccipital process to a point 15 mm from the occipital protuberance. Here they continue in parallel 10.4 mm apart toward the lambdoid crest, where they again diverge briefly. Also in dorsal view, the posterior border is notched at the occipital protuberance due to the posterior extension of the lambdoid crest either side of the midline. More anteriorly, the right supraorbital foramen is visible dorsal to the anterior portion of the orbit, and a groove runs anteriorly from the foramen converging toward, but not meeting, a corresponding groove that is visible on the left frontal. The superior orbital margin is long and below the level of the braincase. The internal surface of the braincase is crisscrossed by many ridges. Viewed posteriorly, the occiput is relatively high (height occipital protuberance above foramen magnum is 24 mm) and with a roughly rectangular profile. Distinct rugose areas form the dorsolateral margins of the lambdoidal crest, and both sides of the midline mark strong muscle attachments for the superficial neck muscles. These give the dorsal portion of the oc-

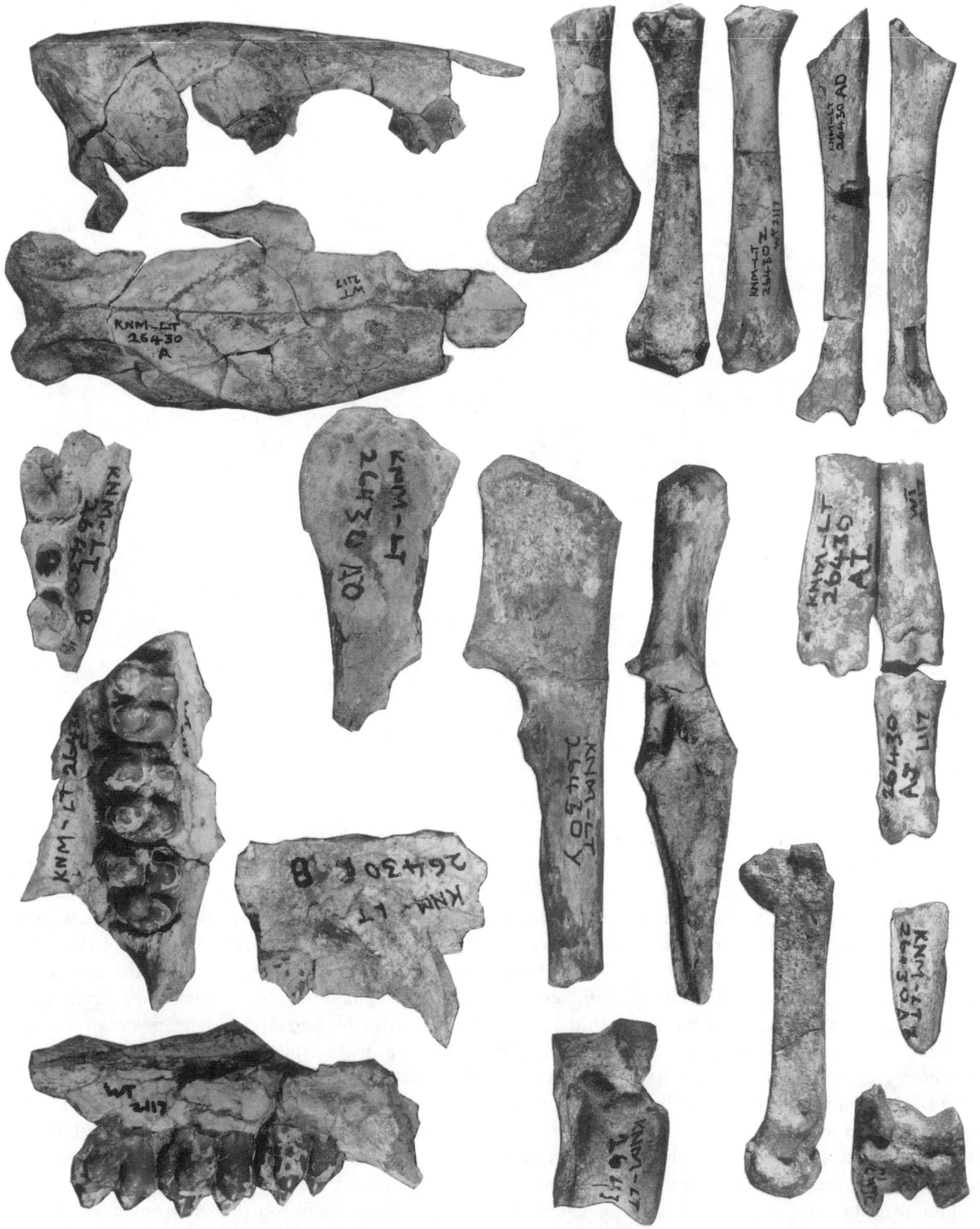

Figure 10.50 *Cainochoerus* sp. from the Upper Nawata, KNM-LT 26430. Top: calvaria (26430A), dorsal and right lateral views; distal right femur (26430AB), lateral view; left radius (26430Z), posterior and anterior views; right distal tibia with shaft (26430AD), medial and lateral views. Other specimens (twice natural size). Middle row from left: right maxilla fragment with C/, P^1 alveolus and P^2 (26430D) and right maxilla fragment with M^{1-3} (26430C), occlusal views; proximal left ulna (26430AO) and proximal right ulna (26430Y), medial views; proximal right ulna (26430Y), anterior view; two distal metapodials (26430AI), and proximal phalanx (26430AJ), all dorsal views; and terminal phalanx (26430AK), medial or lateral view. Bottom row from left: right maxilla fragment with C/, P^1 alveolus, and P^2 (26430D), lateral view; right maxilla fragment with M^{1-3} (26430C), medial view; right astragalus (26430AG), dorsal view; right metacarpal III (26430AH), lateral view; left partial navicular (26430W), medial view.

ciput a scooped appearance. There is no median ridge, but lateral ridges demarcate the lateral margins.

The fragment of left temporal (B) comprises some frontal with the postorbital process and some occipital including the left condyle. This cannot be confidently joined to the rest of the cranium because of the poor contacts at the frontal and foramen magnum. The fragment also includes the partial left petrous and base of the zygomatic, but the mandibular articular fossa is missing. An additional fragment of petrous (E) is poorly preserved.

A fragment of right maxilla (D) preserves the canine root and broken crown, the alveolus for P^1 and the P^2. The canine root can be seen through a break in the maxilla and is open, suggesting that the canine was still erupting. Although the canine crown is incomplete, the remaining portion suggests it was similar to that of the canine of the Langebaanweg specimens and to those of peccaries, being short, only slightly curved, and oriented almost vertically in relation to the occlusal row (Hendey 1976). The P^2 appears worn but is within the crypt, suggesting there must be some postmortem displacement of this tooth. A second maxilla fragment (C) preserves the lightly worn right M^{1-3}, the partial alveolus for P^4, and a small sliver of palate. The molars have four high, pointed cusps, each pair joined by a transverse crest. A median longitudinal ridge links the distinct crestlike anterior margin and the anterior transverse crest, and the anterior transverse and posterior transverse crests. Posteriorly there is a small hypoconule.

Several vertebrae are preserved; all are incomplete. The axis lacks the spinous process, and the transverse processes are broken off. The partial transverse foramen remains on the left side, whereas on the right side it is complete but matrix filled. The right and left cranial articular surfaces are preserved. The ventral surface preserves two distinct ridge-like processes that converge at the midline 5.5 mm from the markedly concave distal face of the centrum; in the bush pig these processes are fused into a single flattened rugosity. The maximum body dimension is 16.13 mm anteroposteriorly from the tip of the dens to the distal articular surface and 19.02 mm by 5.2 mm, respectively, transversely and dorsoventrally across the caudal articular surface.

A centrum (L), probably C7, lacks all processes. The circular caudal costal facets are clearly visible, but the absence of cranial costal facets indicates that this should be C7. However, the remaining portion of the transverse foramen is oriented dorsoventrally as in the extant bush pig's first thoracic rather than anteroposteriorly as in the cervical vertebrae. This dorsoventrally flattened centrum is 9.9 mm by 5.44 mm cranially and 9.1 mm by ~5.45 mm caudally.

Six thoracic vertebrae are preserved, all showing incomplete fusion of the vertebral discs. The most complete, Th13 (K), lacks much of the spinous process, and the craniad-directed accessory processes are broken superiorly. Only the caudad-directed mammillary processes are complete, and they protrude as two small prongs with laterally and superiorly directed articular facets similar to those of the extant bush pig Th13. The small transverse foramen is preserved on both sides. The centrum is transversely flattened; the caudal costal demifacets are small and indistinct, in contrast to the large and rather posteriorly oriented cranial costal demifacets. A wide but distinct ridge runs anteroposteriorly along the inferior face and is punctuated by two small foramina. The centrum measures 15.52 mm anteroposteriorly, 10.1 by 5.3 mm transversely by dorsoventrally across the cranial articular face and 11.0 by 5.87 mm across the caudal articular face. A second, even longer (15.7 mm) thoracic vertebra (S) preserves only the centrum. The cranial face is 9.76 by 5.8 mm and the caudal 10.65 by 6.36 mm. This is likely to be Th14, as there is a small cranial costal demifacet but no caudal demifacet. The centrum is less transversely compressed than Th13, and a sharp crest, also punctuated by two distinct foramina, runs anteroposteriorly along the ventral face. Four other thoracic vertebrae (N, O, P, and Q) are represented by centra. One (Q), which may be Th5, also preserves the accessory process with a transverse costal facet. All these thoracic vertebrae show slightly concave cranial and distinctly concave caudal demifacets, and inferiorly they all have distinct midline ridges. From the longest to the shortest (P, Q, O, N) measurements in millimeters of maximum anterior posterior length, transverse width by midline superior-inferior width for cranial, and lastly caudal centrum faces, respectively, are P: 12.3, 7.6 by 5.0, 8.2 by 5.2; Q: 11.8, 7.5 by 4.5, 8.0 by 5.2; O: 11.2, 7.6 by 5.2, 8.3 by 5.3; N: 10.5, 7.56 by 5.46, 7.32 by 5.1.

All five lumbar vertebrae were recovered. The centra have heart-shaped cranial faces and more flattened caudal faces; distinct midline keels along their inferior surfaces become more pronounced caudally. The most complete of the lumbar vertebrae (J) lacks both transverse processes, both caudal articular processes and part of the right cranial articular process, as well as a triangular chip from the front of the dorsal spinous process. Its low anteroposteriorly broad dorsal spine indicates this may be L1; its centrum measures 15.93 mm anteroposteriorly, the cranial articular face is 9.98 by 6.2 mm and the posterior face is 10.6 by 6.0 mm. A second, similarly sized and shaped vertebra (I), possibly L2, has a complete vertebral foramen and, except for part of the left transverse process, lacks all the processes. The centrum is 16.83 mm long, the cranial face is 10.16 by 7.3 mm, and the caudal face is 10.7 by 6.7 mm. Of the three other lumbar vertebrae, only one (T), possibly L3, preserves both vertebral discs. It measures 17.6 mm long, the cranial face is <10.5 by 6.9 mm, and the caudal face is 11.6 by 6.3 mm. Another (R), probably L4, is lacking the unfused caudal disc and is 16.7 mm long, and the cranial face is 11.24 by 6.73 mm. The shortest of the lumbar vertebrae (M) preserves the base of a slender transverse process, indicating it may be L5. The centrum lacks both unfused cranial and caudal vertebral discs and is 12.47 mm long. Several fragments of vertebral canal and dorsal and transverse processes (U) cannot be fitted to the preserved centra and more complete vertebrae.

The forelimb is represented by several elements. Only a small left scapula fragment (V) with the glenoid fossa, coracoid process, and neck of the left scapula was recovered. The coracoid process is more pronounced than that of the bush pig, and it extends cranially as a distinct peg rather than as a flattened tubercle. Fragments of proximal and distal right humerus remain. The proximal shaft (AO) is preserved, and a fragment of the articular surface of the head (AL) with adjoining sliver of lesser tuberosity. Little can be said of these small pieces: the shaft is mediolaterally flattened, the deltoid tuberosity lacks a well-defined pectoral ridge, and the posterior face is flattened due to the short but pronounced ridge marking the posterior edge of the teres attachment. The missing proximal epiphysis was unfused. The distal right humerus (X), which includes 22 mm of shaft, is almost complete, and only part of the lateral edge of the trochlear is abraded. Relative to the bush pig, the lateral epicondyle is very lightly developed, as is the supracondylar ridge. The anterior mesiolateral width of the distal articular surface (~11.7 mm) is almost equal to the maximum anterior posterior width (~12.6 mm) of the distal epiphysis, in contrast to the bush pig where this ratio is higher. The epiphysis and diaphysis are completely fused.

The complete, almost undamaged left radius (Z) is preserved. It is 54.8 mm long. The relative proportions of the respective proximal articular facets are similar to those of the bush pig. There is a marked increase in the width and thickness of the diaphysis distally. Distally, the articular facet for the scaphoid is preserved, but that for the lunar is an oval-shaped concavity, lacking the posterior portion typical of the bush pig. A rugose elongated area extends proximally and marks the area of contact with the ulna. The diaphysis is robust and has a strong medial ridge and a clearly defined lateral edge that extends proximally from the distal end for more than half the diaphysis, thus demarcating the relatively flat posterior face. The diaphysis has a less clearly defined anteroposterior curvature than is typical of the bush pig.

The proximal right ulna fragment (Y) lacks the unfused surface of the olecranon process, but this is present on a fragment of left ulna (AP), so that the anteroposterior length of the olecranon process from the anconeal process (20.9 mm) can be estimated. The maximum transverse width was only 7 mm. The proximal ulna has a rather straight shaft that lacks the marked anteroposterior curvature typical of the bush pig.

The complete right Mc III (AH) is preserved. It is 30.5 mm long, 5.74 mm mediolateral by 7.88 mm anteroposterior proximally, and the distal epiphysis is 5.5 mm mediolateral by 6.64 mm anteroposterior. The slender shaft (4.37 mm mediolateral by 4.48 mm anteroposterior) has a flattened lateral surface which in life, although not fused, must have been closely approximated to the medial surface of Mc IV. Relative to the bush pig in which the transverse measurements are greater than the anteroposterior measurements, this specimen has smaller transverse measurements than anteroposterior. Two distal epiphyses with the distal halves of the shafts (AI) are preserved. One, the left Mc III, measures 19 mm, and the other, probably the left Mc IV, is 19.2 mm long. The flattened lateral faces fit closely together, indicating they are both from the same appendage. The proximal epiphyses measure 5.5 by ~6.6 mm and 5.88 by 6.65 mm, respectively. A single proximal phalanx (AJ) and a single distal phalanx (AK) are preserved: the former is 15.5 mm long and the latter is 12.93 mm. Both are more slender than those of the bush pig.

The left ilium fragment (AF) is 43 mm long and preserves part of the auricular surface, a small part of the iliac crest, and part of the greater ischiatic notch. At the break across the body of the ilium, the cross section is oval and measures 9.7 by 5.8 mm. A marked ridge runs from the iliac crest toward the ischium.

The left (AC) and right (AB) distal and the right (AA) proximal femora are preserved. The proximal piece is 50.6 mm long, and the longest distal piece (left) is 44 mm. The femur appears to have been rather lightly built. Differences with the bush pig are notable, largely in the distal end where the patella articular surface extends proximally on the anterior face to a point above the most proximal extent of the condyles. On the posterior face proximal to the medial condyle a distinct ridge demarcates a groove. Proximal to the lateral surface, the shaft is rugose. The cross sectional area of the broken shaft of the proximal piece (maximum dimension 9.1 mm by minimum dimension 7.5 mm) is distinctly smaller than that of the broken shaft of the largest distal fragment (10.3 by 8.8 mm).

The tibia is represented by much of right shaft and distal articular surface (AD). The length of this piece is 71 mm. Only a small part of the tibial crest remains, and the anterior portion of the shaft adjacent to the distal articular surface is lost. The maximum dimensions of the distal epiphysis are 11.1 mm transverse and

10.6 mm anteroposterior. The tibia is longer and more slender than that of the bush pig.

The right astragalus (AG) is complete and matches that described and illustrated from Langebaanweg. The left cuboid (AE) is broken laterally and lacks most of the calcaneum articular facet proximally and much of the Mt IV facet distally. The astragalar articular facet is narrow and discontinuous, in contrast to that of the bush pig, which is wide and continuous.

Remarks

The three upper molars from Lothagam are smaller and labiolingually narrower relative to length than any of those recovered from Langebaanweg. The distal labiolingual width is less than the mesial in contrast to the Langebaanweg upper molars, which are more square in outline, and the M^3 is markedly constricted distally. Unfortunately, comparisons of the relative size of the postcrania cannot be assessed because no measurements have been published for the Langebaanweg material. However, the morphology of those elements that are known from both sites appears to be very similar, and it is likely that the Lothagam material represents a population of the Langebaanweg species but with a smaller average body size. Pickford (1988) estimated the average body size for the Langebaanweg *Cainochoerus* to be 7 ± 2 kg.

Pickford (1988) noted the unusual dentition and compared the cheek teeth with those of monkeys, concluding that *Cainochoerus* probably had a diet like that of monkeys of fruit, bark, and leaves. The dental morphology and dental wear makes rooting behavior improbable.

The postcranial skeleton shows many unusual features for a suid and indicates that there was some convergence with peccaries and with small bovids. These features include the lightly built skeleton with relatively long and slender limbs, the fusion of the metatarsals, and the tight joints with deep grooves and sharp ridges, which indicate that there was limited movement in the sagittal plane.

Cainochoerus was a small cursorial suid with a diet distinctly different from that of other contemporary suids. The convergence with peccaries in the postcranial skeleton may well indicate a similar lifestyle.

Discussion

The distribution of suids in the Lothagam succession documents the replacement of typical early and mid Miocene suid tribes by tetraconodontines that migrated to Africa from Eurasia to exploit the expanding grasslands characteristic of the Late Miocene. *Nyanzachoerus syrticus* and *N. devauxi* were the dominant suids in the Nawata Formation, but the local assemblages also included a kubanochoere relict that survived from earlier in the Miocene, as well as *Potamochoerus* and *Cainochoerus*—rare precursors of the suines that would initially coexist with and eventually replace the tetraconodontines in the African Plio-Pleistocene. The smaller *N. devauxi* shows no morphological change through the Nawata sequence, but the upper third molars of *N. syrticus* became larger and more complex in the Upper Nawata and closely resemble those in specimens recovered from Lukeino. In the Apak Member of the Nachukui Formation, the dominant suid is *Nyanzachoerus* cf. *N. australis*. This species is also represented by a single specimen from the Upper Nawata and was interpreted by van der Made (1999) to represent a later immigration from Eurasia. *Nyanzachoerus* cf. *N. australis* is replaced in the Kaiyumung Member by *N. pattersoni*, but the dominant suid in this member is *Notochoerus euilus*.

In terms of correlation, *Nyanzachoerus syrticus* and *Nyanzachoerus devauxi* also occur at Sahabi (Cooke 1987; McCrossin 1987), but the Lothagam representatives of *N. syrticus* are smaller (and hence less progressive) than those from Sahabi. Specimens similar to the two common Lower Nawata species have been recovered from the Namurungule Formation in the Samburu Hills. The more progressive examples of *S. syrticus* from the Upper Nawata appear very similar to specimens recovered from Lukeino. The type locality of *Nyanzachoerus australis* is the Quartzose Sand Member of the Varswater Formation at Langebaanweg; specimens that are similar to but a little smaller than the South African species derive from the Upper Nawata and Apak Members. At present, *Cainochoerus* is known only from Lothagam and Langebaanweg. If van der Made (1999) correctly construed *Nyanzachoerus kanamensis* to be restricted to the Western Rift, the majority of specimens formerly assigned to this taxon must now be assigned to *Nyanzachoerus pattersoni*, which thereby becomes a widely ranging and temporally long-lived species. At Lothagam, *N. pattersoni* occurs only in the Kaiyumung Member, but its precursor, *N. australis*, is the dominant suid in the Apak Member. Bishop (1997) assigned all the suids from the Manonga Valley of Tanzania (Harrison and Baker 1997; Harrison 1997) to *N. kanamensis* because the material was insufficient to determine diagnostic cranial characters, and the dental metrics did not differentiate between *N. australis* and *N. kanamensis*. *Nyanzachoerus jaegeri* is uncommon in the Lothagam sequence (only three specimens were recovered), but it is common at the nearby early Pliocene Kanapoi locality, and its descendent, *Notochoerus euilus*, is the common suid from the Kaiyumung Member. The changes in the suid component of the mammalian assemblages encountered between the Upper Nawata and Apak Members, and between the Apak and Kaiyumung Members, indicate that there were depositional hiatuses at the stratigraphic boundaries.

The increase in size and complexity of third molars in different suid lineages as noted by Cooke (1978), Harris and White (1979), and others would have enhanced the mastication of dietary resources. Low-crowned bunodont suids like *Nyanzachoerus* would be expected to have had a less abrasive diet than hypsodont suids such as *Notochoerus*. It is interesting, therefore, that tooth enamel samples of *Nyanzachoerus syrticus* from the Upper Nawata yield more positive $\delta^{13}C$ values than the simpler-crowned Lower Nawata samples, indicating that the specimens with the more complex third molars exploited a more abrasive (mixed C_3/C_4) diet. *Nyanzachoerus* specimens from the Apak Member evidently had a wholly grazing diet, despite the low-crowned nature of their cheek teeth. It is interesting, and not entirely unexpected, that the more hypsodont mammals such as equids and proboscideans, and also the hippos, exploited C_4 grass resources somewhat earlier than the suids did.

Acknowledgments

We thank the government of Kenya and the museum trustees of the National Museums of Kenya for permission to study the Lothagam suid material. We also thank the curatorial and preparation staff of the palaeontology division of the National Museum of Kenya, Nairobi, for making the material available for study. John Barry kindly provided information about casts of Siwalik suids in the collections of the National Museums of Kenya.

References Cited

Arambourg, C. 1968. Un suidé fossile nouveau du Miocène supérieure de l'Afrique du Nord. *Bulletin de la Société Géologique de France,* 7th ser., 10:110–115.

Bishop, L. C. 1994. Pigs and the ancestors: Hominids, suids, and the environment during the Plio-Pleistocene of East Africa. Ph.D. diss., Yale University.

Bishop, L. C. 1997. Fossil suids from the Manonga Valley. In T. Harrison, ed., *Neogene Paleontology of the Manonga Valley, Tanzania: A Window into the Evolutionary History of East Africa,* pp. 191–217. New York: Plenum Press.

Bishop, L. C., and A. Hill. 1999. Fossil Suidae from the Baynunah Formation, Emirate of Abu Dhabi, United Arab Emirates. In P. J. Whybrow and A. Hill, eds., *Fossil Vertebrates of Arabia,* pp. 254–270. New Haven: Yale University Press.

Broom, R. 1925. On evidence of a giant pig from the Late Tertiaries of South Africa. *Records of the Albany Museum* 3:307–308.

Cooke, H. B. S. 1978. Suid evolution and correlation of African hominid localities: An alternative taxonomy. *Science* 201:460–463.

Cooke, H. B. S. 1987. Fossil Suidae from Sahabi, Libya. In N. T. Boaz, A. El-Arnauti, A. W. Gaziry, J. de Heinzelin, and D. D. Boaz, eds., *Neogene Paleontology and Geology of Sahabi,* pp. 255–266. New York: Liss.

Cooke, H. B. S., and S. C. Coryndon. 1970. Pleistocene mammals from the Kaiso Formation and other related deposits in Uganda. In L. S. B. Leakey and R. J. G. Savage, eds., *Fossil Vertebrates of Africa,* vol. 2, pp. 107–224. London: Academic Press.

Cooke, H. B. S., and R. F. Ewer. 1972. Fossil Suidae from Kanapoi and Lothagam, Kenya. *Bulletin of the Museum of Comparative Zoology* 143:149–295.

Cooke, H. B. S., and Q. B. Hendey. 1992. *Nyanzachoerus* (Mammalia: Suidae: Tetraconodontinae) from Langebaanweg, South Africa. *Durban Museum Novitates* 17:1–20.

Coppens, Y. 1971. Une nouvelle espèce de Suidé du Villafranchien du Tunisie, *Nyanzachoerus jaegeri* nov. sp. *Comptes Rendus Hebdomedaires des Séances, Académie des Sciences* (Paris) 272:3264–3267.

Falconer, H. 1868. Description of a fragment of a jaw of an unknown extinct pachydermous animal, from the Valley of the Murkunda: *Tetraconodon magnum* vel *Choerithium.* In C. Murchison, ed., *Palaeontological Memoirs and Notes of the Late Hugh Falconer, A.M., M.D.* Vol. 1. *Fauna Antiqua Sivalensis,* pp. 149–156. London: Hardwicke.

Falconer, H., and P. T. Cautley. 1847. *Fauna Antiqua Sivalensis: Being the Fossil Zoology of the Sewalik Hills, in the North of India.* London: Smith, Elder.

Harris, J. M., and T. D. White. 1979. Evolution of the Plio-Pleistocene African Suidae. *Transactions of the American Philosophical Society* 69:1–128.

Harrison, T. H., ed. 1997. *Neogene Paleontology of the Manonga Valley, Tanzania: A Window into the Evolutionary History of East Africa.* New York: Plenum Press.

Harrison, T., and E. Baker. 1997. Paleontology and biochronology of fossil localities in the Manonga Valley, Tanzania. In T. Harrison, ed., *Neogene Paleontology of the Manonga Valley, Tanzania: A Window into the Evolutionary History of East Africa,* pp. 362–393. New York: Plenum Press.

Hendey, Q. B. 1976. Fossil peccary from the Pliocene of South Africa. *Science* 192:787–789.

Hill, A., P. Whybrow, and W. Yasin al-Tiktiti, 1990. Late Miocene fauna from the Arabian Peninsula: Abu Dhabi, United Arab Emirates. *American Journal of Physical Anthropology* 81:240–241.

Kotsakis, T., and S. Inigo. 1980. Osservazione sui *Nyanzachoerus* (Suidae, Artiodactyla) del Terziario superiore di Sahabi (Cyrenaica-Libia). *Bolletino del Servizio Geologico d'Italia* C:391–407.

Leakey, L. S. B. 1958. *Some East African Fossil Suidae.* Fossil Mammals of Africa No. 14. London: British Museum (Natural History).

Leakey, M. G., C. S. Feibel, R. L. Bernor, T. E. Cerling, J. M. Harris, K. M. Stewart, G. W. Storrs, A. Walker, L. Werdelin, and A. Walker. 1996. Lothagam: A record of faunal change in the Late Miocene of East Africa. *Journal of Vertebrate Paleontology* 16:556–570.

Leonardi, P. 1952. Resti fossili di *Sivachoerus* del Giacimento di

Sahabi in Cyrenaica (Africa settentrionale). Nitici preliminari. *Rendiconti Accademia Nazionale dei Lincei,* 8th ser., 13:166–169.

McCrossin, M. L. 1987. Postcranial remains of fossil Suidae from the Sahabi Formation, Libya. In N. T. Boaz, A. El-Arnauti, A. W. Gaziry, J. de Heinzelin, and D. D. Boaz, eds., *Neogene Paleontology and Geology of Sahabi,* pp. 267–286. New York: Liss.

McDougall, I., and C. S. Feibel 1999. Numerical age control for the Miocene-Pliocene succession at Lothagam, a hominoid-bearing sequence in the northern Kenya Rift. *Journal of the Geological Society* (London) 156:731–745.

Nakaya, H., M. Pickford, Y. Nakano, and H. Nishida. 1984. The Late Miocene large mammalian fauna from the Namurungule Formation, Samburu Hills, northern Kenya. *African Study Monographs,* Supplementary issue 2:87–131.

Pickford, M. 1986. *A Revision of the Miocene Suidae and Tayassuidae (Artiodactyla, Mammalia) of Africa.* Special Paper No. 7. London: Tertiary Research Group.

Pickford, M. 1988. Un étrange suidé nain du Néogène supérieure de Langebaanweg (Afrique du Sud). *Annales de Paléontologie (Vert.-Invert.)* 74:229–250.

Pickford, M. 1989. New specimens of *Nyanzachoerus waylandi* (Mammalia, Suidae, Tetraconodontinae) from the type area, Nyaburogo (Upper Miocene), Lake Albert Rift, Uganda. *Geobios* 5:641–651.

Pickford, M. 1990. Revision des suidés de la Formation de Beglia (Tunisie). *Annales de Paléontology* 76, fasc. 2:133–141.

Pilgrim, G. E. 1926. The fossil Suidae of India. *Memoirs of the Geological Survey of India,* n.s., 8:1–104.

Powers, W. D. 1980. Geology of Mio-Pliocene sediments of the lower Kerio River Valley, Kenya. Ph.D. diss., Princeton University.

Tobein, H. 1936. Mitteilungen über Wirbeltierreste aus dem MittelPliocän des Natrontales (Aegypten). 7 Artiodactyla. A Bunodontia. *Zeitschrift für die deutsche geologische Gesellschaft* (Berlin) 88:42–53.

van der Made, J. 1996. *Listriodontinae (Suidae, Mammalia): Their Evolution, Systematics, and Distribution in Time and Space.* Contributions to Tertiary and Quaternary Geology 33. Leiden: Werkgroep.

van der Made, J. 1999. Biometrical trends in the Tetraconodontinae, a subfamily of pigs. *Transactions of the Royal Society of Edinburgh: Earth Sciences* 89:199–225.

White, T. D., and J. M. Harris. 1977. Suid evolution and correlation of African hominid sites. *Science* 198:13–21.

Table Abbreviations

ant = anterior
* = approximate measurement
ap = anteroposterior length
bl = buccolingual
I = horizon indeterminate
L = Lower Nawata
Lr = lower
Lt. = left
md = mesiodistal
post = posterior
Rt. = right
tal = talon(id)
tr = transverse width
trig = trigon(id)
U = Upper Nawata
Ur = upper

TABLE 10.11 Tooth Measurements (in mm) of cf. *Kubanochoerus* sp.

	P_2 ap	P_2 tr	P_3 ap	P_3 tr	P_4 ap	P_4 tr
KNM-LT 318	—	—	26.5+	18.45	—	—
KNM-NO 28806	23.2	11.2	31.53	31.37	26.6	30.5

	M_1 ap	M_1 ant tr	M_1 post tr	M_2 ap	M_2 ant tr	M_2 post tr
KNM-LT 318	—	—	—	24.8	20.1	24.5
KNM-NO 28806	24.1	24.9	25.5	26.5	—	29.0

	M_3 ap	M_3 ant tr	M_3 post tr
KNM-LT 318	40.9	—	21.4
KNM-NO 28806	42.8	31.2	24.1

TABLE 10.12 Cranial Measurements (in mm) of *Nyanzachoerus syrticus tulotus*

Measurement	LT 316	LT 23771
Length of premaxilla-foramen magnum	410	375
Length of premaxilla, palatal notch	319	277
Length of premaxilla, nuchal crest	(434+)	467
Width of supracanine flange	143	70
Minimum width of orbits	126	104
Maximum width of zygomatic arch	315	242
Width of postorbital processes	171	139
Width of glenoid articular surfaces	243	200
Estimated width of occipital condyles	70	57
Width of palate at M^3	90	80
Width of palate at P^3	74	73

TABLE 10.13 Upper Teeth Measurements (in mm) of *Nyanzachoerus syrticus tulotus*

KNM-LT		I¹ md	I¹ bl	I² md	I² bl	I³ md	I³ bl
295	L	—	—	20.7	9.7	—	—
23119	L	—	—	20.1	9.5	—	—
23590	L	—	—	—	—	14.2	7.8
23763	L	17.3	9.6	—	—	—	—
23801	L	—	—	15.9	8.5	—	—
28782 Rt.	L	16.9	9.9	—	—	—	—
28782 Lt.	L	16.9	10.6	17.3	9.5	—	—
23738	U	—	—	19.0	9.2	—	—
23743	U	20.7	13.2	16.8	9.3	—	—
26085	U	—	—	—	—	12.9	9.2
26087	U	—	—	21.1	8.8	—	—
26108	U	—	—	—	—	15.1	9.7
7709	I	—	—	18.7	10.7	—	—
	Mean L	18.8	10.4	17.8	9.1	14.2	7.8
	Mean U	21.0	11	18.2	9.7	14.0	9.5

KNM-LT		P¹ ap	P¹ tr	P² ap	P² tr	P³ ap	P³ tr	P⁴ ap	P⁴ tr
316	L	11.9	6.4	15.6	10.3	23.0	23.5	21.0	28.0
22980	L	—	—	15.0	9.0	25.5	25.5	19.7	27.5
23001	L	—	—	—	—	—	—	17.6	22.4
23018	L	—	—	—	—	24.9	23.6	19.0	25.5
23590	L	—	—	13.4	7.1	22.6	21.7	—	24.7
23771	L	—	—	—	—	23.7	22.9	18.6	24.3
26598	L	—	—	—	—	—	—	21.5	27.7
28753	L	—	—	—	—	—	—	22.1	27.0
22958	U	—	—	15.5	8.8	22.8	20.9	17.0	25.2
26088	U	—	—	—	—	—	—	19.1	—
290	I	—	—	—	—	24.0	23.1	19.9	24.4
10275	I	—	—	—	—	—	—	18.6	22.0
23005	I	—	—	—	—	—	—	19.0	22.0
23589	I	—	—	—	—	22.8	24.4	19.6	25.1
	Mean L	11.9	6.4	14.6	8.8	23.9	23.44	19.9	—
	Mean U	—	—	15.5	—	22.8	20.9	18.0	—

KNM-LT		M¹ ap	M¹ ant tr	M¹ post tr	M² ap	M² ant tr	M² post tr
316	L	22.0	21.3	21.4	27.0	28.7	27.0
386	L	—	—	—	26.8	22.4	21.7
22985	L	19.7	20.4	19.7	—	—	—
22992	L	—	—	—	26.4	28.0	25.6
23011	L	—	—	—	—	26.8	—
23018	L	18.2	18.8	18.0	25.2	25.0	24.0
23119	L	—	—	—	25.6	27.5	26.7

continued

TABLE 10.13 Upper Teeth Measurements (in mm) of *Nyanzachoerus syrticus tulotus* (Continued)

KNM-LT		M^1 ap	M^1 ant tr	M^1 post tr	M^2 ap	M^2 ant tr	M^2 post tr
23126	L	20.8	20.8	19.6	28.0	25.6	24.3
23764	L	—	—	—	24.7	24.4	23.5
23771	L	20.9	—	—	25.9	26.0	25.9
26104	L	—	—	—	25.4	21.6	22.3
26128	L	—	—	—	25	24.8	23.2
28793	L	25.0	21.4	21.5	—	—	—
293	U	—	—	—	—	—	28.1
22958	U	21.0	19.0	18.6	25.6	24.0	23.0
23743	U	—	—	—	30.6	29.1	27.5
26088	U	22.0	—	20.6	30.0	28.3	—
26138	U	22.1	23.1	22.8	—	—	27.1
28779	U	25.1	19.4	20.2	—	—	—
23005	I	—	—	—	23.6	26.1	24.0
	Mean U	21.1	20.5	20.0	26.1	25.9	24.4
	Mean L	22.55	20.5	20.55	28.7	27.13	26.4

KNM-LT		M^3 ap	M^3 ant tr	M^3 post tr	M^3 trig	M^3 talon
316	L	36.6	28.9	23.7	28.7	17.4
22993	L	39.5	29.3	25.7	27.8	18.6
23011	L	42.1	—	23.6	32.2	18.8
23012	L	38.8	—	22.7	28.1	17.6
23017	L	—	—	25.8	31.0	—
23018	L	37.7	27.4	23.1	27.5	18.6
23590	L	38.4	27.8	23.5	29.0	19.6
23744	L	42.0	—	25.0	31.0	19.0
23771	L	39.1	27.9	22.6	29.5	17.5
286	U	45.9	31.8	26.5	34.0	23.4
293	U	43.3	30.7	27.7	31.6	20.0
307	U	45.3	31.0	25.8	34.8	20.0
319	U	44.8	—	27.4	32.2	21.9
22958	U	40.8	28.4	23.0	28.5	20.0
23743	U	43.0	32.2	28.5	30.7	22.4
26091	U	44.8	31.2	26.5	33.3	19.0
26121	U	46.2	31.0	—	31.3	21.0
	Mean L	39.3	28.3	24.0	29.4	18.4
	Mean U	44.3	30.9	26.5	32.1	21.0
26110	A	45+	29+	25+	—	—
290	I	—	26.5	24.5	—	—
10276	I	41.3	29.7	25.6	32.0	19.5
10277	I	40.0	29.0	23.6	29.8	20.6

TABLE 10.14 Lower Teeth Measurements (in mm) of *Nyanzachoerus syrticus tulotus*

KNM-LT		I_1 ap	I_1 tr	I_2 ap	I_2 tr	/C ap	/C tr
107	L	—	—	—	—	16.6	21.6
305	L	—	—	—	—	21.3	18.4
23016	L	10.6	14.5	—	—	—	—
23030	L	8.9	13.4	—	—	—	—
23752	L	10.1	14	10.4	18.1	8.4	10.8
23760	L	9.3	—	10.7	17	—	—
26113	L	9.48	—	10.1	9.47	—	—
288	U	—	—	—	—	22.5	21.2
23737	U	8.7	12.2+	—	—	—	—
23748	U	10.1	12.6+	—	—	—	—
28779	U	—	—	10.7	—	23	20.5
290	I	—	—	—	—	21.6	15
	Mean L	9.7	13.9	10.4	14.9	15.43	16.93
	Mean U	9.4	12.4+	10.7		22.75	20.85

KNM-LT		P_1 ap	P_1 tr	P_2 ap	P_2 tr	P_3 ap	P_3 tr	P_4 ap	P_4 tr
107	L	—	—	—	—	—	—	28.1	25+
291	L	—	—	14	8	27.5	21.8	23.3	22.4
295	L	—	—	15	7.7	24	18	21.7	19.5
302	L	—	—	15.4	8.4	28.6	23.6	23.4	23.4
305	L	—	—	—	—	—	—	24.8	22.2
306	L	—	—	—	—	—	—	25.5	22.3
385	L	12.6	5.8	16.7	7.6	28.9	23.3	22	23.4
386	L	—	—	—	—	—	—	24	—
387	L	—	—	—	—	—	—	24.3	21.9
22953	L	—	—	16.4	8.8	—	—	25.1	23.8
22986	L	—	—	—	—	—	—	23	21.3
22996	L	—	—	15.6	7.9	—	—	24.1	20.5
23007	L	—	—	—	—	—	—	23.5	24.8
23017	L	—	—	—	—	27.5	24.5	25.7	—
23584	L	—	—	—	—	26.3	21.1	24.9	—
23752	L	—	—	14.2	8.1	29.2	23.4	24	22.4
23759	L	—	—	—	—	25.6	19.3	24.4	20.6
23763	L	14	6.8	—	—	—	—	—	—
26103	L	—	—	—	—	24.4	22.2	—	—
26130	L			—	—	30.0	23+	25	21+
287	U	10.2	6.1	14.4	8.6	26.4	—	—	—
299	U	—	—	—	—	29.5	24.8	—	—
23591	U	—	—	—	—	26.2	20.8	—	—
23805	U	—	—	—	—	28.2	20.2	—	—
23758	U	—	—	—	—	28.9	23.5	—	—
26109	U	—	—	19.2	8.9	—	—	23.6	23.6

continued

TABLE 10.14 Lower Teeth Measurements (in mm) of *Nyanzachoerus syrticus tulotus* (Continued)

KNM-LT		P_1 ap	P_1 tr	P_2 ap	P_2 tr	P_3 ap	P_3 tr	P_4 ap	P_4 tr
309	A	—	—	—	—	22.4	19.6	—	—
	Mean L	13.3	6.3	15.3	8.1	27.5	21.9	24.3	22.2
	Mean U	10.2	6.1	16.8	8.7	27.84	22.3	23.6	23.6

KNM-LT		M_1 ap	M_1 ant tr	M_1 post tr	M_2 ap	M_2 ant tr	M_2 post tr
107	L	—	—	—	27.5	—	—
291	L	20.2	17.3	—	—	22.4	—
295	L	20.2	15.7	16	24.3	21	20
302	L	21.3	17.6	—	26.8	24	22.8
305	L	20.8	17	17.5	28.5	—	21.2
385	L	19	16.5	15.8	26.3	21.6	20.3
386	L	—	—	—	26.8	21	21.5
387	L	—	—	—	26.3	20.8	20.2
390	L	21.1	—	—	—	—	—
22882	L	—	—	—	26.7	20.6	19.3
22953	L	20.5	—	—	24.7	23.1	21.9
23007	L	—	—	—	23.5	20	21
23010	L	20	17	16.8	—	—	—
23017	L	—	—	—	24.7	23.1	21.2
23021	L	—	—	—	24.6	23.6	21.9
23658	L	19.5	16.3	15.3	—	—	—
23750	L	—	—	—	25.4	20	20.6
23752	L	19.8	16.9	17	25.7	22.6	22.2
23759	L	—	—	—	24	19.5	—
23763	L	20	15.6	15	—	—	—
26115	L	—	—	—	26.9	—	—
296	U	—	—	—	25.6+	19.6+	20.4+
300	U	17.8	16.3	—	25	21.5	—
23123	U	25.3	19.77	18.77	25	19.2	17
23739	U	—	—	—	—	28.4	23.4
23745	U	19.7	17.4	—	—	—	—
23774	U	—	—	—	—	26.4	21.7
26083	U	23.4	16.2	14.5	—	—	—
	Mean L	20.22	16.7	16.2	25.8	22.35	21.0
	Mean U	21.55	17.42	16.6	25	23.9	20.7

KNM-LT		M_3 ap	M_3 ant tr	M_3 post tr	M_3 tal	M_3 trig
107	L	44.1	23.7	20.5	32.7	17
295	L	42	24.1	21.1	32.6	18.7
302	L	42.9	24.67	21.8	29.7	17.7
305	L	—	24.9	—	—	—
385	L	43.6	23.1	20.4	29.5	19

TABLE 10.14 Lower Teeth Measurements (in mm) of *Nyanzachoerus syrticus tulotus* (Continued)

KNM-LT		M_3 ap	M_3 ant tr	M_3 post tr	M_3 tal	M_3 trig
386	L	42.9	23.9	21.9	29.7	20.1
22953	L	42.4	25.2	21.7	28.9	17.6
22986	L	41.2	25	19.3	28.5	16.8
23017	L	40.8	24.8	21.5	27	17.7
23021	L	43	24.9	21.7	30	18.5
23024	L	39.1	—	22.5	29	18.6
23752	L	40.28	25.3	22	27.6	17.6
23757	L	42.9	22.5	20.8	32	16.1
23759	L	39.5	20.5	18.5	27.2	16.5
26122	L	42.6	24.1	22.5	32.8	17
288	U	>43	22.9	21.7	32.5	17.5
300	U	—	24.7	—	—	18.1
388	U	49.7	25	23.7	37	19.4
22982	U	—	—	22.7	33.7	—
23774	U	46.1	24.3	—	—	—
26109	U	48.1	24.6	23	32.6	20.3
309	A	49.2	25.17	22.7	35.4	19.2
118	I	45.9	24.9	24.4	33.8	19
	Mean L	41.95	24.0	21.16	29.8	17.8
	Mean U	48.0	24.3	22.8	33.95	18.8

TABLE 10.15 Deciduous Teeth Measurements (in mm) of *Nyanzachoerus syrticus tulotus*

KNM-LT	dP^2 ap	dP^2 tr	dP^3 ap	dP^3 tr	dP^4 ap	dP^4 ant tr	dP^4 post tr
28776	—	—	—	—	17.9	15.23	14.73
26138	—	—	15.25	13.97	17.25	17.86	17.57

KNM-LT	dP_2 ap	dP_2 tr	dP_3 ap	dP_3 tr	dP_4 ap	dP_4 ant tr	dp_4 med tr	dP_4 post tr
22878	—	—	—	—	23.56	9.12	9.56	11.62
23027	—	—	—	—	22.07	8.47	10.73	12.4
23802	—	—	—	—	22.6	8.52	10.18	11.16
26105	12.37	7.09	—	—	—	—	—	—
26117	—	—	—	—	—	—	12.62	13.1
28565	—	—	12.87	—	—	—	—	—
23123L	—	—	12.66	6.64	22.75	11.12	12.5	12.8
23123R	13.53	6.91	—	—	22.16	11.42	—	14.11
Mean L	12.37	7.09	—	—	22.74	8.70	10.156	11.73
Mean U	13.53	6.91	12.66	6.64	22.45	11.27	12.56	13.22

TABLE 10.16 Upper Teeth Measurements (in mm) of *Nyanzachoerus* cf. *N. australis*

KNM-LT	I^1 ap	I^1 tr	P^3 ap	P^3 tr	P^4 ap	P^4 tr
26099	18.1	10.7	—	—	—	—
310	—	—	28.68	28.73	25.05	27.34*
313	—	—	27.92	26.05	23.64	28.87

KNM-LT	M^1 ap	M^1 ant tr	M^1 post tr	M^2 ap	M^2 ant tr	M^2 post tr
310	25.53	—	—	—	—	—
313	25.67	—	24.3	32.05	30.34	30.22
26134	—	—	—	39.85	—	26.56
26591	—	—	—	—	27.8	—

KNM-LT	M^3 ap	M^3 ant tr	M^3 post tr
313	49.98	32.54	28.82

TABLE 10.17 Lower Teeth Measurements (in mm) of *Nyanzachoerus* cf. *N. australis*

KNM-LT	P_1 ap	P_1 tr	P_2 ap	P_2 tr	P_3 ap	P_3 tr	P_4 ap	P_4 tr
285	—	—	—	—	—	—	—	20.98
308	—	—	—	—	30.57	19.6+	25.47	—
26098	—	—	—	—	29.1	21.9	26.5	22.9
26099	—	—	—	—	24.6	—	—	—

KNM-LT	M_1 ap	M_1 ant tr	M_1 post tr	M_2 ap	M_2 ant tr	M_2 post tr
285	—	—	—	—	22.16	22.5*
308	—	—	—	34.02	—	—
23749	—	—	—	30.67	22.27	22.96
23691	—	—	—	27.54	—	—
28715	—	—	—	—	—	21.7

KNM-LT	M_3 ap	M_3 ant tr	M_3 post tr
23749	57.49	28.52	27.38
285	56.93	25.58	23.06
23746	—	—	26.26
26076	—	—	23.2

TABLE 10.18 Teeth Measurements (in mm) of *Nyanzachoerus pattersoni*

KNM-LT	M^3 ap	M^3 ant tr	M^3 post tr
26078	53	32.14	28.17

KNM-LT	P_1 ap	P_1 tr	P_2 ap	P_2 tr	P_3 ap	P_3 tr	P_4 ap	P_4 tr
23768	—	—	—	—	22.63	16.52	—	—
26137 Rt.	—	—	—	—	24.5	23.54	23.42	20.58
26137 Lt.	—	—	—	—	25.4	23.74	—	—

KNM-LT	M_1 ap	M_1 ant tr	M_1 post tr	M_2 ap	M_2 ant tr	M_2 post tr
23768	—	—	—	30.87	19.56	20.46
26137 Rt.	—	—	—	29.96	22.72	24.23

KNM-LT	M_3 ap	M_3 ant tr	M_3 post tr
23768	60.72	23.22	24.32

TABLE 10.19 Upper Teeth Measurements (in mm) of *Nyanzachoerus devauxi*

KNM-LT		I¹ ap	I¹ tr	I² ap	I² tr	I³ ap	I³ tr
22967	L	17.0	11.3	—	—	—	—
23029	L	—	—	—	—	15.4	11.0
23161	L	—	—	—	—	7.6	7.9
22970	U	16.5	—	—	—	—	—

KNM-LT		P¹ ap	P¹ tr	P² ap	P² tr	P³ ap	P³ tr	P⁴ ap	P⁴ tr
110	L	10.7	5.6	14.0	6.6	19.8	19.1	16.7	23.0
22967	L	11.9	5.3	15+	8.6	22.0	—	—	—
23020	L	—	—	—	—	23.2	19.3	18.0	24.0
26102	L	—	—	—	—	23.6	17.7	17.4	20.7
26132	L	—	—	—	—	22.2	18.0	—	—

KNM-LT		M¹ ap	M¹ ant tr	M¹ post tr	M² ap	M² ant tr	M² post tr
110	L	18.0	17.9	17.9	25.0	23.5	21.6
22967	L	—	—	—	23.0	23.5	—
23020	L	—	—	—	25.0	24.5	22.0

KNM-LT		M³ ap	M³ ant tr	M³ post tr	M³ trig	M³ tal
110	L	37.0	24.7	22.1	26.3	19.3
283	L	39.6	—	—	—	—
314	L	>33.8	—	>19.4	24.9	11.8
22967	L	34.3	25.1	20.0	24.4	17.4
23020	L	36.7	25.1	21.3	28.0	17.7
26101	L	32	23.8	19.7	23.3	12.3
26126	L	37.7	25.6	21.5	28.7	17.5
28750	L	34.2	22.6	19.6	—	—
26094	U	36.6	25.4	20.2	26.7	16.9
26096	U	39.5	—	—	—	—
22970	U	32.6	22.1	18.8	26.6	9.2
	Mean L	36.5	24.6	20.6	27.0	17.5
	Mean U	38.1	25.4	20.2	26.7	16.9

TABLE 10.20 Lower Teeth Measurements (in mm) of *Nyanzachoerus devauxi*

KNM-LT		P_3 ap	P_3 tr	P_4 ap	P_4 tr
303	L	—	—	22.5	21.5
23015	L	22.6	20.3	—	—
23019	L	21.2	17.83	—	—
23022	L	22	17.5	20.4	19

KNM-LT		M_1 ap	M_1 ant tr	M_1 post tr	M_2 ap	M_2 ant tr	M_2 post tr
22885	L	20.5	14.8	14.2	—	—	—
23019	L	—	—	—	22.9	18.4	17.3
23022	L	18.9	16.5	15.2	24	20.5	18.8
23023	L	16.7	—	—	23.9	20.5	—
23025	L	—	—	—	23.2	18.3	18.4
23957	L	—	—	—	25.7	19.1	18.4

KNM-LT		M_3 ap	M_3 ant tr	M_3 post tr	M_3 trig	M_3 tal
282	L	38.7	22.4	20.7	27.6	16.7
303	L	37.5	23.4	19.5	27.8	16.1
22885	L	—	—	—	—	15.35
22970	U	36.1	19.4	18.3	24.5	15.5
23019	L	35.94	20.0	17.8	25	11.7
23022	L	36.4	22.3	19.5	25.7	16.7
23029	L	—	—	18.9	28.4	17.4
23753	L	34.3	19.0	16.9	25.7	12.8
26075	L	39.2	22.4	20.6	28.1	16.7
26127	L	38	—	20.1	—	—

TABLE 10.21 Deciduous Teeth Measurements (in mm) of *Nyanzachoerus devauxi*

KNM-LT	dP^2 ap	dP^2 tr	dP^3 ap	dP^3 tr	dP^4 ap	dP^4 ant tr	dP^4 post tr
26131	12.55	6.86	14.04	11.14	—	—	—
26139	—	—	14.2	9.91	—	—	—
28708	—	—	—	—	16.6	12.45	—

KNM-LT	dP_2 ap	dP_2 tr	dP_4 ap	dP_4 ant tr	dP_4 med tr	dP_4 post tr
26112	12.19	5.82	—	—	—	8.8
26118	—	—	21.54	7.78	8.75	9.92
26106	—	—	18.28	7.41	9.33	—

TABLE 10.22 Tooth Measurements (in mm) of *Nyanzachoerus jaegeri*

KNM-LT	M^3 ap	M^3 ant tr	M^3 post tr
311	—	—	33.99
23741	64.69	36.71	33.12

KNM-LT	P_1 ap	P_1 tr	P_2 ap	P_2 tr	P_3 ap	P_3 tr	P_4 ap	P_4 tr
23741 Lt.	—	—	—	—	26.45	20.42	20.46	18.53
23741 Rt.	—	—	—	—	—	—	22.67	20.47

TABLE 10.23 Tooth Measurements (in mm) of *Notochoerus euilus*

KNM-LT	M^3 ap	M^3 ant tr
298	—	32.69

KNM-LT	I_1 ap	I_1 tr
26080	10.69	12.0

KNM-LT	P_3 ap	P_3 tr	P_4 ap	P_4 tr
26136	—	—	18.9	13.9
26079	17.3*	11.6*	—	—

KNM-LT	M_1 ap	M_1 ant tr	M_1 post tr	M_2 ap	M_2 ant tr	M_2 post tr
23765	—	—	—	34.26	15.35	18.66
289	—	—	—	34.75	19.9	23.43

KNM-LT	M_3 ap	M_3 ant tr	M_3 post tr
574	82*	22.8	—
23723	80.46	23.43*	25.62
23766	—	26.1	23.2
26136	71.4	26.96	25.85
26124	88.48	22.88	25.11

TABLE 10.24 Tooth Measurements (in mm) of *Potamochoerus* sp.

KNM-LT	P_1 ap	P_1 tr	P_2 ap	P_2 tr	P_3 ap	P_3 tr	P_4 ap	P_4 tr
294	—	—	—	—	—	—	22.57	17.0
23711	—	—	—	—	23.5	18.2	—	—

KNM-LT	M_1 ap	M_1 ant tr	M_1 post tr
294	24.25	20.19	19.41
28789	21.95	16.63	16.85

TABLE 10.25 Tooth Measurements (in mm) of *Cainochoerus* sp.

KNM-LT	C/ md	C/ bl
26430	5.65	4.5

KNM-LT	M^1 ap	M^1 ant tr	M^1 post tr	M^2 ap	M^2 ant tr	M^2 post tr
26430	6.64	6.1	5.84	7.37	6.95	6.3

KNM-LT	M^3 ap	M^3 ant tr	M^3 post tr
26430	6.6	6.76	5.08

11

RUMINANTIA

11.1

Lothagam Giraffids

John M. Harris

Giraffids, as at most African Neogene localities, are only rare constituents of the Lothagam biota. Most of the recovered remains constitute partial dentitions or isolated postcranial elements. Two species of *Palaeotragus* appear to be represented in the Nawata Formation; the more progressive *Giraffa stillei* and *Sivatherium* occur in the younger strata.

Giraffoids are rare elements of middle Miocene faunas in East Africa and are best represented by *Climacoceras africanus* at Maboko, *Climacoceras gentryi* at Fort Ternan, and *Paleotragus primaevus* from Fort Ternan and Ngorora (Hamilton 1978). Churcher (1970) also described *Samotherium africanum* from Fort Ternan on the basis of ossicones and cervical vertebrae, but Hamilton (1978) considered the assigned material to be insufficient either to establish the identity of Churcher's species or to validate its purported generic identity. Recent fieldwork at Maboko, under the direction of Brenda Benefit and Monte McCrossin, has yielded a wealth of new *Climacoceras africanus* material. Dentally this material is very similar to that of *C. gentryi* from Fort Ternan, and the close relationship of the two species is further indicated by the similarity of the ossicones (those of *C. gentryi* being more derived). However, the appendicular skeletal elements of the Maboko species are bovid-like in their proportions, whereas those from Fort Ternan are giraffe-like; this may indicate rapid evolution in the *Climacoceras* lineage, but more likely indicates that Fort Ternan giraffoid postcranials formerly attributed to *Climacoceras* probably belong to a different genus.

The Late Miocene witnessed the transformation from more primitive giraffoids to more advanced giraffids in the East African assemblages. *Paleotragus germaini*, first described from Late Miocene localities in North Africa (Arambourg 1959), has been reported from Lothagam (Churcher 1979). The four-ossiconed *Giraffokeryx* was reported from Nakali by Aguirre and Leakey (1974) and from Lothagam (Churcher 1978:529), but the latter record has not been substantiated from the available Lothagam collections. Moreover, Hamilton (1978) considered the Nakali *Giraffokeryx* specimens to be misidentified *Paleotragus primaevus* material.

Pliocene and Pleistocene African biotas are characterized by the presence of *Sivatherium* and *Giraffa* species. A sivathere with short ossicones, *Sivatherium hendeyi*, was described from Langebaanweg (Harris 1976), but at most other localities only *Sivatherium maurusium* has been recognized. *Sivatherium maurusium* has very variable ossicone morphology (Harris 1974). Pliocene representatives of *Sivatherium* have elongate metapodials and were browsers, but Pleistocene forms have shorter metapodials and were grazers (Harris and Cerling 1998). Large and small *Giraffa* species, though not necessarily the same ones, were usually present at most Plio-Pleistocene localities in East Africa.

Systematic Description

Family Giraffidae Gray, 1821
Subfamily Giraffinae Gray, 1821
Tribe Palaeotragini Pilgrim, 1911
Palaeotragus Gaudrey, 1861
Palaeotragus germaini Arambourg, 1959

(Figure 11.1; tables 11.1–11.3)

Lothagam Material

- *Lower Nawata:* 92, cranial fragment; 414, Lt. M^3; 26256, Lt. unciform; 26262, Rt. M^3; 26302, distal Lt. tibia, metatarsal fragments.

Figure 11.1 Restoration of *Paleotragus* sp. by Mauricio Antón. Shoulder height estimated at just over 2 meters.

- *Upper Nawata:* 22865, distal Mc; 26259, mandible fragment (M_{2-3}); 26264, Rt. external cuneiform; 26265, proximal Lt. scapula, phalanx.

Paleotragus germaini is the largest *Paleotragus* species with long supraorbital ossicones, moderately elongated neck, and long limbs with the forelimb only slightly longer than the hind (Churcher 1978). A right upper first molar of *Paleotragus germaini* (LT 414) was collected by the 1967 expedition and described by Churcher (1979). An identical left upper third molar (LT 26262) has been recovered by the recent expeditions. Both specimens come from the Lower Nawata. A right mandibular fragment with incomplete second and third molars (LT 26259) appears to belong to the same taxon but derives from the Upper Nawata. A cranial fragment (LT 92), comprising the rear part of the cranial vault and the upper half of the occiput, can also be assigned to *P. germaini*—representing an individual intermediate in size between an okapi and an extant giraffe. The cranial vault is concave anteroposteriorly but convex mediolaterally. Bordered laterally by strong supratemporal ridges, the vault narrows from the parietal-frontal suture to the very thick and rugose nuchal crest that is about 14.2 cm wide. The occiput is triangular in shape, tapering downward to the strong bosses that surmount the foramen magnum. It was originally identified in the field as a rhino skull fragment, but the nuchal crest is too robust, the cranial vault too narrow, and the occiput too shallow to belong to a rhinocerotid, and the specimen was not listed by Hooijer and Patterson in their (1972) account of the Lothagam rhinos.

Paleotragus sp.

(Figures 11.2, 11.3; tables 11.1, 11.2, 11.4)

Lothagam Material

- *Lower Nawata:* 23150, maxilla (Rt. P^2–M^3, Lt. P_2); 23639, Rt. mandible (dP_4–M_2), Lt. P_3, lower partial M_1 and M_2; 23852, Rt. astragalus; 25451, Rt. astragulus; 26258, juvenile Lt. metatarsal; 26263, Lt. unciform; 26267, distal Lt. tibia, Lt. metatarsal, Lt. astragulus, Lt. naviculocuboid.

- *Upper Nawata:* 197, distal Lt. ossicone; 29066, Lt. mandible fragments (M_1? fragment); 26290, radioulna; 28659, Lt. unciform.

A partial palate (LT 23150) has teeth that are appreciably smaller than the specimens assigned to *P. germaini* and are of similar size and morphology to the extant *Okapia johnsoni* (figure 11.2). It is likely, however, that the specimen represents a smaller species of *Paleotragus* rather than an okapi. Mandibular fragments with deciduous premolars and molar germs (LT 23639) are comparable in size. A partial right mandibular ramus with associated enamel fragments (LT 29066) may also belong to this taxon.

LT 197 (figure 11.3) was at first identified as a boselaphine horn and does share mediolateral compression, flat lateral surface, and faint anterior and posterior keels with Lothagam boselaphine specimens. However, it seems shorter and more tightly curved backward and outward, and lacks any indication of a pedicel. This specimen is more likely to represent a paleotragine ossicone and is here assigned to the smaller species because of its small size. A similarly shaped ossicone fragment from the Baynunah Formation of Abu Dhabi was assigned to ?*Bramatherium* sp. (Gentry 1999).

Isolated giraffid limb bones from the Nawata Formation represent both species. The largest specimens, a left unciform (LT 26256) and a large incomplete tibia and metatarsal (LT 26302), from the Lower Nawata probably represent *P. germaini.* A slender proximal scapula, LT 26265A, is of comparable size to scapulae of a female *Sivatherium* from Koobi Fora (Harris 1991), as is an associated proximal phalanx from the pes, but these specimens are here assigned to *P. germaini.* An elongate but slender radioulna (LT 26290) from the Upper Nawata and other small postcranial elements from the Lower Nawata are assigned to the smaller, unnamed species.

Tribe Giraffini Gray, 1821
Giraffa Brunnich, 1771
Giraffa stillei (Dietrich, 1942)
(Figure 11.4; Tables 11.1, 11.2, 11.5, 11.6)

Lothagam Material

- *Apak Member:* 23669, Rt. mandible (P_4–M_1); 26261, proximal Rt. Mc; 26268, distal metapodial epiphysis fragment; 26269, proximal scapula, radius, fragments calcaneum and metapodials; 26270, proximal Lt. Mc and patella; 26271, cervical vertebra fragment; 26272, Lt. astragulus; 28646, Lt. astragalus; 28709, Rt. M^3.
- *Kaiyumung Member:* 24024, Lt. and Rt. ossicones; 26257, Lt. P_3; 23894, ossicone.

Two ossicones from the Kaiyumung Member are comparable in size and morphology to those of *Giraffa stillei* from later deposits elsewhere in the Turkana Basin (Harris 1991). LT 24024 is a right ossicone that was inserted above the rear of the orbit and slightly backwardly inclined (figure 11.4). It is anteroposteriorly compressed and tapers upward from its base but has a well-developed terminal knob. The ossicone is ornamented with deep grooves and ridges, but there is no

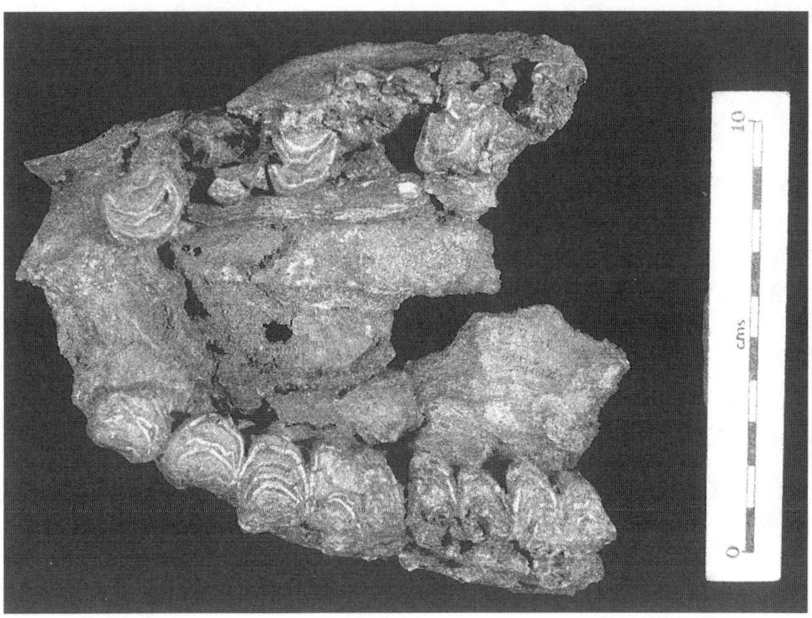

Figure 11.2 *Paleotragus* sp. maxilla from the Lower Nawata, KNM-LT 23150, occlusal view.

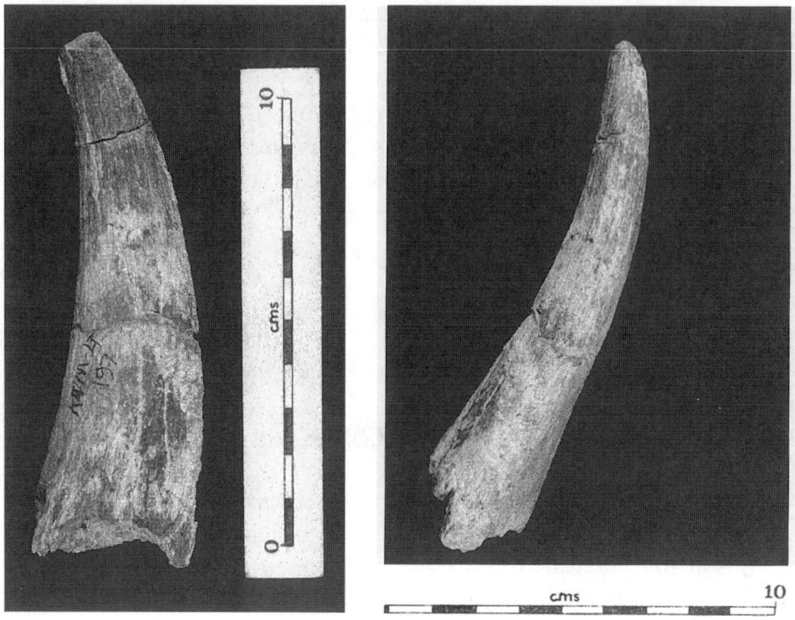

Figure 11.3 *Paleotragus* sp. left ossicone from the Upper Nawata, KNM-LT 197: top = medial view; bottom = anterior view.

indication of any secondary bone apposition. Associated with this specimen was the tip of the matching left ossicone. LT 23894 represented a slightly larger individual and was triangular in transverse section with a flattened posterior surface. Its anterior (carinate) ridge is rugose with secondary bone apposition. It is likely that 24024 was female but 23894 was male. Two dental specimens, a left P_3 (LT 26257) and a right mandible fragment with P_4–M_1 (LT 23669), represent *Giraffa stillei* or a giraffine of similar size. A number of postcranials might also belong to this species (table 11.6). LT 26270 is a metacarpal that is too long to be sivatheriine and is of comparable size to large specimens of *G. pygmaea* from Koobi Fora. Incomplete metacarpals LT 26261 and 26269 are similarly elongate, though slightly smaller. A left astragalus 26272 also represents a small giraffine.

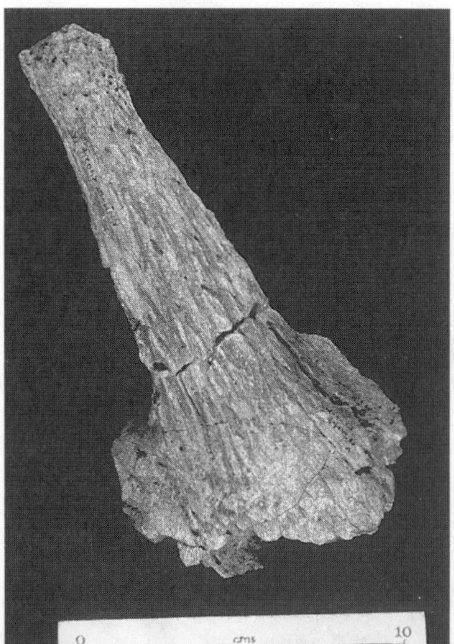

Figure 11.4 *Giraffa stillei*, left ossicone, from the Kaiyumung Member, KNM-LT 24024A

Subfamily Sivatheriinae Bonaparte, 1850
Sivatherium Falconer and Cautley, 1832
cf. *Sivatherium* sp.
(Table 11.7)

Lothagam Material

- *Lower Nawata:* 26266, distal metapodial epiphysis.
- *Horizon indet.:* 23892, Lt. naviculocuboid; 23893, Lt. astragulus.

Two very large tarsals collected by the Patterson expeditions, a left astragalus LT 23893 and a left naviculocuboid LT 23892, could conceivably represent *Sivatherium.* The specimens have no documented provenance, but their preservation is very similar to material recovered from the Apak Member by field personnel from the National Museums of Kenya. There is also a distal epiphysis of a metapodial (LT 26266) from the Lower Nawata that seems much too large for *Paleotragus.* To date, however, no cranial or dental specimens have been recovered to confirm the presence of *Sivatherium* in the Lothagam succession.

Discussion

Giraffid materials from Lothagam are tantalizing. They suffice to document that the paleotragines characteristic of the Nawata Formation were replaced by more progressive giraffids in the lower part of the Nachukui Formation but do not provide much information about the represented taxa. The replacement of the Nawata Formation paleotragines by the Nachukui Formation giraffines and sivatheriines is unlikely to directly reflect changing habitats, as all pre-Pleistocene giraffoids appear to have been browsers and to have obtained much of their water from their food (Cerling et al. 1997; Harris and Cerling 1998). More likely, the taxonomic change reflects the immigration of progressive giraffid species from Asia. Of the two sites frequently compared to Lothagam on the basis of age, Langebaanweg has yielded much *Sivatherium* and also some *Giraffa* material (Harris 1976), whereas Sahabi giraffoids comprise an unidentified species of *Samotherium* (Harris 1987). This is consistent with other correlative evidence that suggests Sahabi is temporally equivalent to the lower part of the Nawata Formation, whereas Langebaanweg equates with the Upper Nawata or later horizons.

Acknowledgments

I thank the government of Kenya and the museum trustees of the National Museums for permission to examine the Lothagam giraffid material. I thank also my friends and colleagues of the paleontology division curatorial and preparation staff at the National Museum of Kenya, Nairobi, for their assistance in making this material available for study. Stephen Mutaba took the photographs in Nairobi, and Dick Meier and Pete Mueller provided technical assistance in Los Angeles. I am indebted also to Nikos Solounias and Meave Leakey for their assistance during the preparation of the manuscript.

References Cited

Aguirre, E., and P. Leakey. 1974. Nakali: Nueva fauna de Hipparion del Rift Valley de Kenya. *Estudios Geológicos* 30:219–227.

Arambourg, C. 1959. *Vertébrés continentaux du Miocène supérieur de l'Afrique du nord.* Paléontologie Mémoires No. 4. Algiers: Carte Géologique d'Algérie.

Cerling, T. E., J. M. Harris, B. J. MacFadden, M. G. Leakey, J. Quade, V. Eisenmann, and J. R. Ehleringer. 1997. Pattern and significance of global ecologic change in the late Neogene. *Nature* 389:153–158.

Churcher, C. S. 1970. Two new upper Miocene giraffids from Fort Ternan, Kenya, East Africa: *Palaeotragus primaevus* n. sp. and *Samotherium africanum* n. sp. In L. S. B. Leakey and R. J. G. Savage, eds., *Fossil Vertebrates of Africa,* vol. 2, pp. 1–109. London: Academic Press.

Churcher, C. S. 1978. Giraffidae. In V. J. Maglio and H. B. S. Cooke, eds., *Evolution of African Mammals,* pp. 509–535. Cambridge, Mass.: Harvard University Press.

Churcher, C. S. 1979. The large paleotragine giraffid (*Palaeotragus germaini*) from the Late Miocene of Lothagam Hill. *Breviora* 453:1–8.

Gentry, A. W. 1999. Fossil pecorans from the Baynunah Formation, Emirate of Abu Dhabi, United Arab Emirates. In P. J. Whybrow and A. Hill, eds., *Fossil Vertebrates of Arabia,* pp. 290–318. New Haven: Yale University Press.

Hamilton, W. R. 1978. Fossil giraffes from the Miocene of Africa and a revision of the phylogeny of the Giraffoidea. *Philosophical Transactions of the Royal Society,* ser. B, 283:165–229.

Harris, J. M. 1974. Orientation and variability in the ossicones of African Sivatheriinae (Mammalia; Giraffidae). *Annals of the South African Museum* 65:189–198.

Harris, J. M. 1976. Pliocene Giraffoidea (Mammalia, Artiodactyla) from the Cape Province. *Annals of the South African Museum* 69:325–353.

Harris, J. M. 1987. Fossil Giraffidae from Sahabi, Libya. In N. T. Boaz, A. El-Arnauti, A. W. Gaziry, J. de Heinzelin, and D. D. Boaz, eds., *Neogene Paleontology and Geology of Sahabi,* pp. 317–322. New York: Liss.

Harris, J. M. 1991. Family Giraffidae. In J. M. Harris, ed., *Koobi Fora Research Project.* Vol. 3. *The Fossil Ungulates: Geology, Fossil Artiodactyls, and Palaeoenvironments,* pp. 93–138. Oxford: Clarendon Press.

Harris, J. M., and T. E. Cerling. 1998. Isotopic changes in the diet of giraffids. *Journal of Vertebrate Paleontology,* Supplement 18:49A.

Hooijer, D. A., and B. Patterson. 1972. Rhinoceroses from the Pliocene of northwestern Kenya. *Bulletin of the Museum of Comparative Zoology* 144:1–26.

Table Abbreviations

* = approximate
ap = anteroposterior
Dist = distal
Dt = depth
Dv = dorsoventral
ext = external
Lat = lateral
Lr = lower
lt = length
Lt = left
Med = medial
met = width posterior pair cusps
N/c tr = naviculocuboid trochlea
prot = width anterior pair cusps
Prox = proximal
Rt = right
Tib tr = tibial trochlea
tr = transverse
Ur = upper

TABLE 11.1 Upper Dentition Measurements (in mm) of Lothagam Giraffids

		LT 23150 Rt *Paleotragus* sp.	LT 23150 Lt *Paleotragus* sp.	LT 414 *P. germaini*	LT 26262 *P. germaini*	LT 287709 *G. stillei*
P^2	ap	16.91	16.78	—	—	—
	tr	17.04	16.25	—	—	—
P^3	ap	15.74	—	—	—	—
	tr	18.68	—	—	—	—
P^4	ap	17.53	—	—	—	—
	tr	22.28	—	—	—	—
M^1	ap	22.5*	—	—	—	—
	prot	24.09	—	—	—	—
	met	24.01	—	—	—	—
M^2	ap	24*	—	—	—	—
	prot	—	—	—	—	—
	met	23*	—	—	—	—
M^3	ap	20	—	30.67	30.34	27.77
	prot	23.6	—	29.6	32.8	28.29
	met	21.41	—	26.69	28.58	25.79

TABLE 11.2 Lower Dentition Measurements (in mm) of Lothagam Giraffids

		LT 29066 Ur Nawata *Paleotragus* sp.	LT 23639 Rt Lr Nawata *Paleotragus* sp.	LT 23639 Lt Lr Nawata *Paleotragus* sp.	LT 26259 Ur Nawata *P. germaini*	LT 23669 Apak *G. stillei*	LT 26257 Kaiyumung *G. stillei*
P_2	ap	—	—	—	—	—	—
	tr	—	—	—	—	—	—
P_3	ap	—	—	—	—	—	26.38
	tr	—	—	—	—	—	19.06
P_4	ap	—	—	—	—	23.16	—
	tr	—	—	—	—	19.93	—
M_1	ap	—	23.83	25.2	—	25.37	—
	prot	16.1	16.86	—	—	18.84	—
	met	—	16.23	—	—	17.95	—
M_2	ap	—	—	25.9	28.45	—	—
	prot	—	17.43	16.7	—	—	—
	met	—	—	17.42	20.85	—	—
M_3	ap	—	—	—	36.86*	—	—
	prot	—	—	—	20.36	—	—
	met	—	—	—	19.43	—	—
dP_3	ap	—	—	19.82	—	—	—
	tr	—	—	10.38	—	—	—
dP_4	ap	—	24.53	—	—	—	—
	prot	—	7.73	—	—	—	—
	met	—	13.18	—	—	—	—
	trit	—	13.09	—	—	—	—

TABLE 11.3 Postcranial Measurements (in mm) of *Paleotragus germaini*

Lower Nawata		Prox ap	Prox tr	Dist ap	Dist tr
LT 26302	Metatarsal	77.42	75.85	55.11	78.58
LT 26302	Tibia	—	—	68.38	96.51

		Max lt	Max tr	Dv
LT 26256	Unciform	88.81	71.32	45.41

Upper Nawata		lt	Prox ap	Prox tr	Dist ap	Dist tr
LT 22865	Metacarpal	—	—	—	50.74	80.19
LT 26265	Prox phalanx	108.2	53.36	43.4+	29.67	39.82

		Max lt	Max tr	Dv
LT 26264	Ext cuneiform	86.96	37.81	18.9
LT 28659	Unciform	73.06	66.65	40.64

TABLE 11.4 Measurements (in mm) of *Paleotragus* sp.

Lower Nawata		Lat lt	Med lt	Tib tr	N/c tr	Dt
LT 23852	Astragalus	88.92	81.42	57.19	58.55	53.04
LT 25451	Astragalus	—	74.72	—	55.79	48.56
LT 26267	Astragalus	91.5	85.3	62.14	57.53	54.12

		lt	Prox ap	Prox tr	Dist ap	Dist tr
LT 26258	Metatarsal	—	62.23	62.21	—	—
LT 26267	Metatarsal	—	60.55	62.64+	46.39	67.64
LT 26267	Tibia	—	—	—	60.98	78.43

		Max lt	Max tr	Dv
LT 26263	Unciform	64.32	62.56	36.45
LT 26267	Naviculocuboid	83.15	78.91	41.77

Upper Nawata		Prox ap	Prox tr
LT 197	Lt ossicone	27.17	27.13

		lt	Prox ap	Prox tr	Dist ap	Dist tr
LT 26290	Radius	760*	54.1	96.13	68.11	89.94
LT 26290	Ulna	—	—	44.99	—	—

Table 11.5 Ossicone Measurements (in mm) of *Giraffa stillei*

	LT 24024 Rt	LT 24024 Lt	LT 23894
Length	152.0	—	—
Prox ap	74.0	—	—
Prox tr	71.0	—	—
Dist ap	28.23	27.6	34.98
Dist tr	30.17	30.52	27.85

Table 11.6 Postcranial Measurements (in mm) of *Giraffa* sp.

		lt	Prox ap	Prox tr	Dist ap	Dist tr
LT 26261	Metacarpal	—	53.76	77.73	—	—
LT 26268	Metapodial	—	—	—	53.47+	—
LT 26270	Metacarpal	—	60.87	89.19	—	—
		Lat lt	**Med lt**	**Tib tr**	**N/c tr**	**Dt**
LT 28646	Astragalus	81.14	73.95	54.4	51.44	43.11
LT 26272	Astragalus	98.22	89.81	71.33	69.71	64.19

Table 11.7 Postcranial Measurements (in mm) of *Sivatherium* sp.

		Lat lt	Med lt	Tib tr	N/c tr	Dt
LT 23893	Astragalus	114.2	104	76.12	73.52	67.75

		Max lt	Max tr	Dv
LT 23892	Naviculocuboid	97.06	96.36	57.47

		Dist ap	Dist tr
LT 26266	Metapodial	60.7	111.55

11.2

Bovidae from the Lothagam Succession

John M. Harris

The Family Bovidae is represented in the Lothagam succession by 17 species belonging to eight tribes that are common in sub-Saharan Africa today (Tragelaphini, Bovini, Reduncini, Hippotragini, Alcelaphini, Aepycerotini, Antilopini, and Neotragini), and by one tribe (Boselaphini) that is now restricted to Asia. Impalas are the most abundant antelopes in all parts of the sequence. Boselaphines are common in the Lower Nawata but are less abundant in the Upper Nawata and are only rare elements of the Apak Member assemblage. Reduncines, alcelaphines, and hippotragines are more abundant in the Upper Nawata than at earlier or later levels. Tragelaphines and bovines are present in the Nawata Formation but are more abundant in the Apak Member. Habitat preferences of extant representatives of these tribes suggest a mixture of woodland and grassland biomes and this is supported by preliminary results from isotopic analysis of bovid tooth enamel. Impalas and tragelaphines were browsing on C_3 vegetation while boselaphines, hippotragines, and reduncines exploited C_4 grasses.

The Bovidae are sparsely represented at early Miocene localities in East and North Africa (Hamilton 1973; Gentry 1978) but bovids were the most numerous terrestrial mammals at the Mid-Miocene site of Fort Ternan (Gentry 1970) and they dominated younger vertebrate fossil assemblages from eastern Africa. Bovids occur in both Eurasia and Africa during the Early Miocene, and thereafter migrations between these continents occurred repeatedly (Vrba 1985, 1995), but Gentry (1990) argues for Africa as the origin of this family. Because the sole apomorphic character distinguishing the Bovidae is the presence of horn cores (Janis and Scott 1987), it would be difficult to substantiate a hornless bovid ancestor with any degree of certainty.

The Antilopini and Caprini, both of Eurasian origin, are first represented in Africa about 14 million years ago; the endemic Cephalophini and Neotragini make their appearance shortly thereafter. Toward the end of the Miocene, Ovibovini and Bovini migrated into Africa from Eurasia while the endemic Tragelaphini, Hippotragini, Alcelaphini, and Aepycerotini are documented for the first time. The Reduncini, whose continent of origin is uncertain, also appear in the Late Miocene, whereas the Boselaphini become extinct in Africa near the Mio-Pliocene boundary (Vrba 1985).

The bovid tribes may be considered in terms of two groups—the Böodontia and the Aegodontia—that depict major features of bovid evolution. Gentry (1978) considered the Tragelaphini, Bovini, Cephalophini, Reduncini, and Hippotragini to be böodonts whereas the Alcelaphini, Aepycerotini, Antilopini, Neotragini, Ovibovini, and Caprini are aegodonts. Böodonts have lower and wider skulls, cranial vaults that are more parallel to the axis of the face, horn cores that are more frequently keeled, less frequent internal sinuses in the frontals, less hypsodont teeth, more rugose enamel, more persistent basal pillars on the teeth, a slower rate of fusion between the pillars of the molars in ontogeny, longer premolar rows, shallower mandibles, and less cursorial limb bones. Aegodonts have converse versions of these characters. The differences between these two groups are best seen in Late Miocene to Early Pliocene representatives and it seems likely that the aegodonts evolved from early bovids with predominantly böodont characters (Gentry 1978:564) in order to exploit newly available, more open habitats.

Bovids are the most numerous terrestrial mammals represented in the Lothagam succession but the majority of specimens comprise partial dentitions and isolated horn cores. They seemingly represent some of the first representatives of the tribes characteristic of Africa today—Tragelaphini, Bovini, Hippotragini, Reduncini, Alcelaphini, Aepycerotini, Antilopini, and Neotragini—as well as some of the last African representatives of the Boselaphini, a tribe that is now restricted to Asia. In his preliminary assessment of the Lothagam Bovidae, Smart (1976) recognized 13 different taxa but did not assign any of them to species. The recent collections by the National Museums of Kenya expeditions have more than doubled the number of specimens and increased the bovid diversity. Although many of the represented species appear to be new, most of them are also incompletely represented and I have avoided introducing a plethora of new names that might require revision when more complete material is available and/or when the early history of African bovid tribes is better understood.

Abbreviations

ap = anteroposterior measurement
Lt. = left
Rt. = right
tr = transverse measurement

Systematic Description

Family Bovidae Gray, 1821
Tribe Tragelaphini Jerdon, 1874
Tragelaphus Blainville, 1816

Diagnosis

Medium to large tragelaphines with spiraled horn cores inserted close together and having an anterior keel and sometimes a stronger posterolateral one; small- to medium-sized supraorbital pits, which are frequently long and narrow; occipital surface tending to have a flat top edge and straight sides (Gentry 1985).

Type species

Tragelaphus scriptus Pallas, 1766.

The genus *Tragelaphus* is diverse, as it includes the kudus, the nyalas, the sitatunga, and the more ubiquitous bushbuck. Greater and lesser kudus occur today in the northern half of the Turkana Basin.

Tragelaphus kyaloae Harris, 1991
(Figure 11.5A; table 11.8)

Diagnosis

A medium-sized tragelaphine with horn cores inserted close together, at a low inclination, that diverge rapidly from their base but converge distally, and that spiral 180° anticlockwise in the right horn core. Proximally, there is a strong posterolateral keel and a fainter anterolateral keel; distally, these become anterolateral and posteromedial, respectively. The shape of the horn cores is reminiscent of the extant sitatunga (*T. spekei*); they are less helically coiled than in *T. strepsiceros, T. imberbis,* or *T. gaudreyi* but converge closer distally than in *T. nakuae* or the kudu-like species. The cranial vault bears a faint but distinct transverse bar immediately in front of the nuchal crest. The paroccipital processes are short but stout, and the posterior tuberosities of the basioccipital are much wider than the anterior ones.

Lothagam Material

- *Upper Nawata:* 25979, proximal Rt. horn core.
- *Apak Member:* 191, distal Lt. horn core fragment; 23617, Lt. horn core; 23618, distal Lt. horn core; 23629, distal Rt. horn core fragment; 23672, Rt. horn core fragment; 25958, Rt. horn core fragments; 26202, proximal Rt. horn core.
- *Kaiyumung Member:* 23102, horn core fragments; 23722, horn core; 26183, distal horn core fragments.
- *Horizon indet.:* 467, proximal Rt. horn core and distal Lt. horn core.

Although it is the commonest bovid from the early Pliocene site of Kanapoi, *Tragelaphus kyaloae* is represented at Lothagam by a single specimen (LT 25979) from the Upper Nawata and about a dozen incomplete horn cores from the Apak and Kaiyumung Members. Horn cores of *T. kyaloae* are of almost identical shape but are about twice as large as those of *Tragelaphus* cf. *T. spekei* (LU 852) from Lukeino (Thomas 1980). In addition to their larger size, the *T. kyaloae* horn cores are more compressed anteroposteriorly in their proximal portion, and they have a less prominent keel arising from their anterolateral corner (that of *T.* cf. *T. spekei* is more medially located) and their tips approach closer together distally. The supraorbital foramen is also located closer to the base of the horn core in *T. kyaloae.*

The horn cores of *T. kyaloae* are similar in cross section to those of *Tragelaphus nakuae* but have more pronounced torsion. Whereas it is entirely possible that these two species were closely related, *T. kyaloae* may also have been distantly related to *Tragelaphus gaudreyi*—the common kudu-like tragelaphine from the lower portion of the Shungura Formation and which has yet to be recognized elsewhere in the Lake Turkana Basin. In *Tragelaphus* aff. *T. nakuae* recorded from Hadar (Gentry 1981), the horn cores are more strongly spiraled and the braincase is longer and lacks the transverse excavation behind the orbits that is characteristic

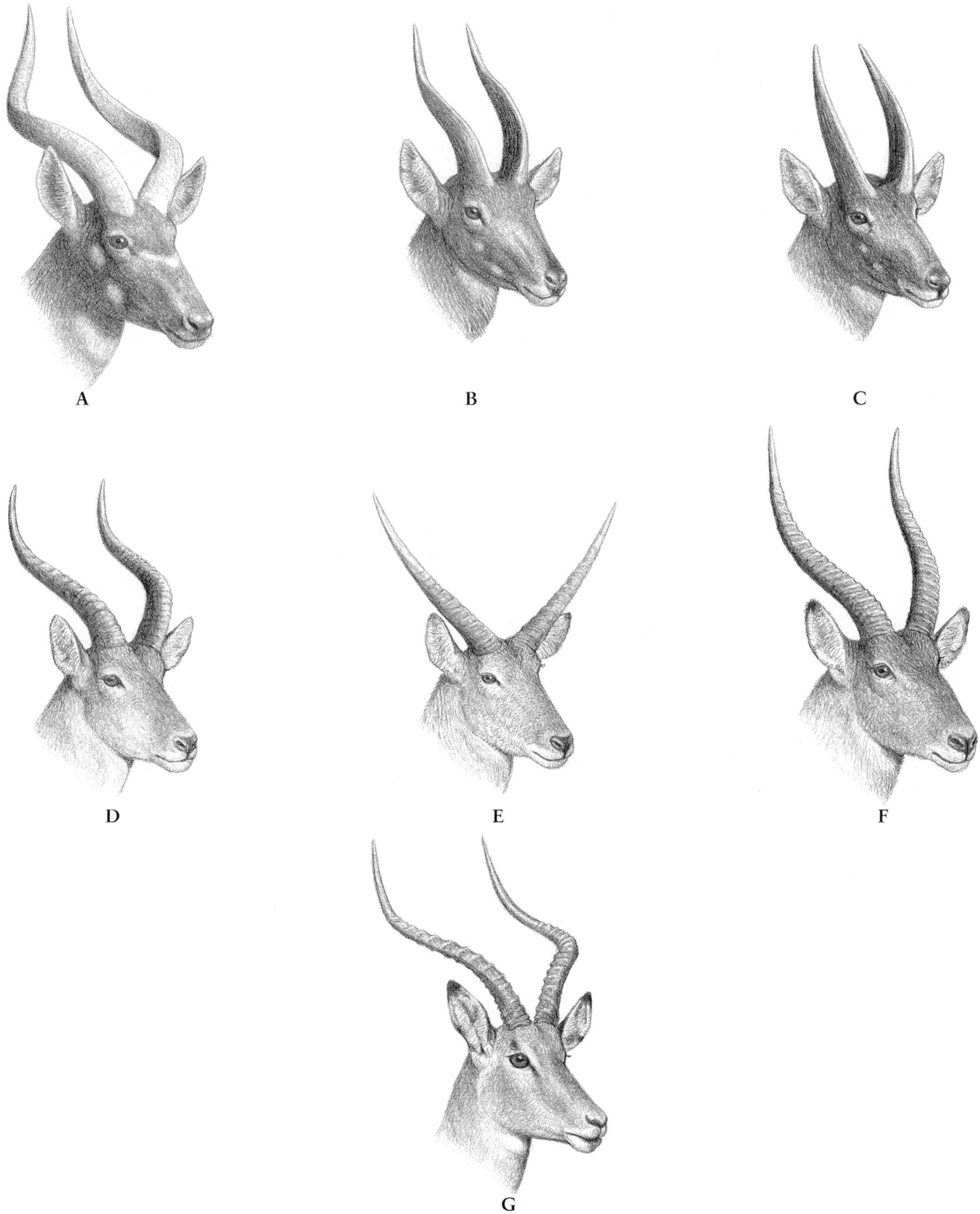

Figure 11.5 Restorations of Lothagam Bovidae by Mauricio Antón: A = *Tragelaphus kyaloae*; B = *Tragoportax* aff. *T. cyrenaicus*; C = *Tragoportax* sp. B; D = *Kobus presigmoidalis* sp. nov., E = *Kobus laticornis* sp. nov.; F = *Menelikia leakeyi*; G = *Aepyceros premelampus* sp. nov.

of *T. nakuae.* This transverse excavation is present in *T. kyaloae,* although it is fainter than in *T. nakuae.* Gentry (1997) noted the similarity of *T. kyaloae* horn cores to those of *T. nakuae* and pointed out that *T. kyaloae* horn cores are more compressed anteroposteriorly than those of tragelaphines from Lukeino or Langebaanweg.

Specimen L 40759 from Langebaanweg (Gentry 1980:221, figure 2) is of similar size to some of the Kanapoi horn cores. Gentry (1980:221–222) mentions the similarity of this specimen to others from the Mursi Formation, and it seems possible that the Langebaanweg specimen could be closely related to *T. kyaloae.*

Tragelaphus nakuae Arambourg, 1941

Diagnosis

A large species of *Tragelaphus* with a low and wide skull; horn cores strongly compressed anteroposteriorly, with a marked posterolateral keel and a weaker anterior one; horn cores inserted above the back of the orbits, inserted widely apart and at a low inclination in side view, and diverging strongly from the base. Dorsal part of the orbital rims projecting quite strongly, frontals raised between the horn core bases, probably a large and shallow preorbital fossa, supraorbital pits large, braincase not angled on the face, strong temporal ridges on the braincase roof, braincase short and not widening posteriorly, a transverse ridge across the back of the brain case roof, occipital surface in one backwardly facing and smoothly surfaced plane and tending to have a rounded top edge, mastoids narrow, basioccipital long and rather narrow with a central longitudinal groove, and anterior tuberosities very large. Small basal pillars on the lower molars decreasing in size from M_1 backward, and still smaller ones on the upper molars, upper molars with strong styles and some flattening of the lateral walls between the styles, paraconid/metaconid fusion on P_4 variable, an upstanding dorsal flange of the mandible just behind the incisors (Gentry 1985).

Lothagam Material

- *Horizon indet.:* 26576, cranial and horn core fragments.

Horn core fragments of a single individual were recovered from a faulted area at the north end of the Lothagam exposures where the sediments are believed to be somewhat younger than those of the Nawata Formation. These horn core fragments are much larger and are less tightly spiraled than those of *T. kyaloae.* The base of the right horn core measures 53.26 mm (ap) by 73.49 mm (tr). *Tragelaphus nakuae* is one of the most common bovids in the lower portions of the Nachukui, Koobi Fora, Shungura, and Mursi Formations, and the Lothagam individual may be of comparable age.

Tragelaphus scriptus (Pallas, 1766)
Tragelaphus cf. *T. scriptus*

Lothagam Material

- *Kaiyumung Member:* 24057, calvaria, horn core fragments.

A very eroded dorsal portion of a braincase from the Kaiyumung Member is similar in size and morphology to that of an extant male bushbuck except that the occiput seems to be less vertically inclined. The horn core pedicels are eroded and it is not possible to determine how the associated horn core was angled above and behind the orbit. The tightly spiraled horn core seems a little smaller than that of extant bushbucks but is similar in size to a small tragelaphine from the Shungura Formation provisionally identified by Gentry (1985) as *T. pricei* but thought to be directly ancestral to the extant bushbuck. The right horn core fragment is about 208 mm long, with proximal measurements of 26.7 mm (ap) by 29.4 mm (tr).

Tragelaphini gen. and sp. indet.
(Tables 11.9, 11.10)

Lothagam Material

- *Lower Nawata:* 182, Rt. mandible (P_2–M_3); 223, Lt. mandible fragment (M_3); 471, Rt. mandible fragment (M_3); 13020, Lt. M_2; 26022, Rt. M_3.
- *Upper Nawata:* 25964, Lt. M^2; 28777, Rt. mandible (M_{1-2}).
- *Apak Member:* 23613, Rt. maxilla (P_3–M_3); 23652, Lt. maxilla (P_4–M_3); 23718, Lt. mandible (M_{2-3}); 25955, Rt. M_3; 25991, Lt. mandible fragment (M_{1-2}); 26065, Rt. maxilla (M^{2-3}).

A number of isolated dentitions have been recovered from the Nawata Formation and Apak Member but are unassociated with cranial material and cannot be identified below tribal rank.

Tribe Bovini Gray, 1821

Bovines are not very prolific members of most sub-Saharan fossil assemblages. *Ugandax* and *Simatherium* have been reported from Late Miocene localities, and *Simatherium, Pelorovis,* and *Syncerus* occur in the Plio-Pleistocene. Vrba (1987) recognized a *Simatherium-Pelorovis* lineage that became extinct in the Early Holocene and a *Ugandax-Syncerus* lineage that today is represented by the Cape buffalo. The Lothagam bovine material is neither abundant nor well preserved but seems to be closest to *Simatherium.*

Simatherium Dietrich, 1941

Diagnosis

Extinct moderate- to large-sized African Bovini with short to moderately long horn cores, rather massive for the size of the skull. Horn cores slightly compressed mediolaterally or without compression, sometimes with an anterior keel, of irregular or rounded rather than neatly triangular cross section, inserted just behind the orbits, inserted widely apart, with moderate to strong divergence, gently curved backward in side view, and without torsion. Horn cores sometimes with deep longitudinal grooves. Frontals and horn pedicels with quite extensive, irregularly shaped sinuses, braincase short, braincase roof sloping a little downward posteriorly, temporal ridges present, a rugose raised area at the back of the braincase roof where temporal ridges converge toward the top of the occipital surface, occipital broad and low with horizontal top edge and some development of hollows dorsally on either side of the median vertical ridge. Nuchal crests strong, moderate to large mastoid. Basioccipital wide posteriorly and of a triangular shape, with a short central longitudinal valley between the posterior tuberosities, with a central longitudinal ridge in the area just behind the anterior tuberosities, with small localized anterior tuberosities and sometimes with low longitudinal ridges behind them (Gentry 1987).

Simatherium kohllarseni Dietrich, 1941

Diagnosis

A species of *Simatherium* with horn cores lacking keels, inserted fairly uprightly and widely or even very widely apart, so strongly divergent at the base as to emerge almost transversely from the sides of the skull, curved backward in side view only near the base if at all, with an irregular rather than a smooth surface and with longitudinal grooves anteroventrally. Temporal ridges present, weak development of a median vertical ridge on the occipital, and large mastoids. Frontals may be arched between the horn cores, and if the horn cores are sufficiently widely inserted there may be a rudimentary temporal fossa at the sides of the braincase (Gentry 1987).

Simatherium aff. *S. kohllarseni*

Lothagam Material

- *Apak Member:* 23597, proximal Rt. horn core.
- *Kaiyumung Member:* 23674, Rt. horn core, proximal Lt. horn core and cranial fragments; 23683, same specimen 23674; 23724, proximal Rt. horn core.

Two bovine specimens from the Kaiyumung Member and one from the Apak seem to be more closely related to *S. kohllarseni* than to other previously described contemporary bovine species. The most complete specimen is LT 23724, a proximal right horn core from Kaiyumung. This was backwardly inserted above the orbit. Massive at its base, the horn core has a flattened lateral surface and a slight anterior keel. The horn cores evidently diverged outward and are angled backward on the cranium. A second specimen, LT 23674 from Kaiyumung, is a poorly preserved but virtually complete right horn core. It is difficult to orientate accurately because of its poor preservation, but if orientated as LT 23724, the horn core curves round in an arc so that it diverges laterally at the base but points backward at its tip. LT 23683, a proximal left horn core, is probably the same individual as LT 23674. If this were the case, the posterolateral edge of the proximal portion of the horn core would have been swollen into a bulbous mass. A proximal right horn core, LT 23597 from Apak, is similar in shape to the Kaiyumung horn cores but larger. It is triangular in transverse section, with anterior keel and flattened posterior surface, and curved outward and backward from its pedicel; at its base it measures 61.76 mm anteroposteriorly and 72.59 mm transversely. It was initially mistaken for a specimen of *Tragelaphus nakuae* because of its triangular transverse section and prominent keel. The Lothagam bovine horn cores bear a superficial resemblance to specimens of *Simatherium kohllarseni* from the lower part of the Koobi Fora Formation (Harris 1991) but are not dissimilar in shape to those of *S. demissum* from Langebaanweg (Gentry 1980). Unlike these two species, the Lothagam bovine horn cores are wider transversely than anteroposteriorly but are closest in shape to the Koobi Fora specimens.

Vrba (1987) suggested that the *Simatherium-Pelorovis* lineage adapted to vegetationally more open environments earlier than the *Ugandax-Syncerus* lineage, in part because of the association of *Ugandax* and *Syncerus* with *Aepyceros* (Vrba 1987:42). Bovines are rare in the Nawata Formation (where impalas make up more than 40% of the bovid component of the biota) but are associated with impalas in the Apak and Kaiyumung Members.

Bovini gen. and sp. indet.
(Table 11.11)

A number of isolated dentitions have been recovered from both members of the Nawata Formation and from the Nachukui Formation, but these dentitions are unassociated with cranial material and cannot be identified below tribal rank.

Lothagam Material

- *Lower Nawata:* 458, Lt. M_2; 476, Lt. M^1 and Rt. M_2.
- *Upper Nawata:* 475, Rt. maxilla (dP^3–M^3) and Rt. M^2; 519, Lt. M_3; 489, Lt. M_1 fragment.
- *Apak Member:* 23608, Lt. maxilla (M^{2-3}); 23619, Lt. mandible (M_{2-3}); 25993, Lt. M3 fragment; 30648, Lt. M/, Lt. mandible (M_{2-3}); 24002, Rt. mandible (P_2–M_3).
- *Kaiyumung Member:* 505, Lt. M_1; 13006, Rt. mandible fragment (M_{2-3}); 26046, Lt. mandible (M_{2-3}).
- *Horizon indet.:* 192, Rt. mandible fragment (M_{1-2}); 480, Rt. M^1.

In general, the Lothagam bovine teeth are larger than those of *Simatherium kohllarseni* and overlap in size with those of *Pelorovis turkanensis.* They are also larger than representatives of *Ugandax* aff. *U. gautieri* from Lukeino (LU 490; Thomas 1980). The internal pillars of the upper molars are anteroposteriorly expanded, whereas those of the Lukeino specimen are simple and cylindrical in transverse section. The ribs on the external surface of the tooth are much more prominent in the Lothagam specimens than in that from Lukeino, and the inner lobes of the Lothagam molars are more pinched.

The teeth of *Simatherium demissum* from Langebaanweg were described as very similar to those of *Ugandax* aff. *gautieri* from Lukeino (Thomas 1980), but larger (Gentry 1980). The Lothagam bovine teeth are a little larger than those of the Langebaanweg bovine.

Tribe Boselaphini Knottnerus-Meyer, 1907

As discussed by Gentry (1999), boselaphines occur in the Middle Miocene but become abundant in the Late Miocene. They represent two generic or suprageneric groups. The Middle Miocene *Austroportax* from Europe, and allied forms like *Pachyportax latidens* and *Selenoportax vexcillarius* from Asia, are related to modern boselaphines and the bovine *Bubalus,* whereas the second group, represented by *Miotragoceras* and *Tragoportax,* appears in the Middle Miocene and is well represented in the Turolian, but becomes extinct at the end of the Miocene. In Kenya, boselaphines are absent from Maboko but present at Fort Ternan (Gentry 1970) and are rare elements in the Ngorora succession (Thomas 1981). The Lothagam specimens are among the last representatives from sub-Saharan Africa.

Solounias (1981) noted that the horn cores of *Miotragoceras* and *Mesembriportax* are different from those of other bovids and no more similar to those of other boselaphines than to any other tribe. According to Solounias, what made *Miotragoceras* and *Mesembriportax* horn cores unique was the change from a flat to a round cross section approximately two thirds of the distance above the base of the horn core. Solounias recognized an anterior demarcation at the point where the anterior keel of the proximal portion abuts against the cylindrical shape of the distal horn core. *Miotragocerus* was said to be similar to *Tragoportax* in general horn shape, but *Tragoportax* lacks the discontinuous anterior additions to the horn growth area (Gentry 1999; Solounias 1981). Moyà-Solà (1983) included *Miotragocerus cyrenaicus* (first described from Sahabi; Thomas 1980) into the genus *Tragoportax,* and, in his discussion of Miocene bovids from Abu Dhabi, Gentry (1999) followed suit with *Mesembriportax acrae* (originally described from Langebaanweg; Gentry 1974).

As discussed by Bouvrain (1988), Gentry (1999), and others, the horn cores of Late Miocene boselaphine species are notoriously variable. Because of this, and pending more detailed comparison with Asian and circum-Mediterranean representatives, no firm taxonomic assignment has yet been made for the Lothagam material.

Tragoportax Pilgrim, 1937

Revised Generic Diagnosis

Medium- to large-sized boselaphines; horn cores rather short, inserted fairly uprightly and widely apart, slightly to greatly divergent basally but less so distally; strongly compressed mediolaterally with a posterior keel and a slightly stronger helical anterior keel in their lower part, the anterior keel usually being stepped at its top and the more distal part of the horn core being of small circular cross section. Frontals extensively hollowed internally, with their dorsal surface being raised above the level of the (nonprojecting) orbital rim. Braincase slightly angled on the facial axis and with a rugose dorsal surface between the temporal ridges. Small supraorbital pits.

Tragoportax aff. *T. cyrenaicus* (Thomas, 1979)

(Figures 11.5B, 11.6; table 11.12)

Lothagam Material

- *Lower Nawata:* 213, proximal Rt. horn core fragment; 498, proximal Rt. horn core fragment; 22995, cranial and Rt. horn core fragments; 23149, frontlet with Rt. and Lt. horn cores; 26025, proximal Rt. horn core; 26042, proximal Lt. horn core; 26043, Lt. horn core and cranial fragments.
- *Upper Nawata:* 23980, proximal Rt. horn core.
- *Apak Member:* 23662, proximal Rt. horn core; 24021, Rt. proximal horn core.

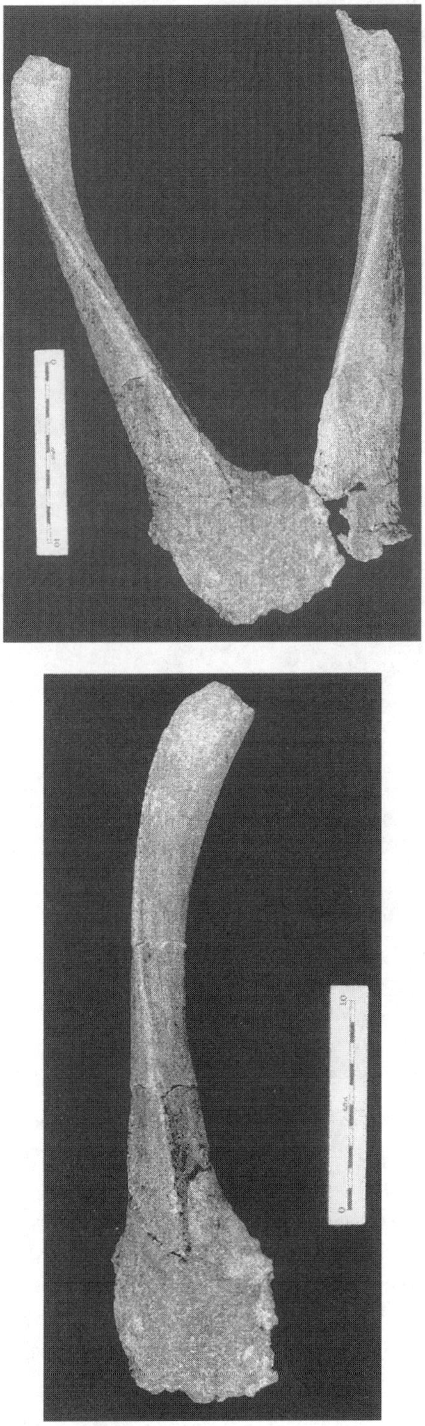

Figure 11.6 *Tragoportax* aff. *T. cyrenaicus*, frontlet with right and left horn cores, KNM-LT 23149: top = anterior view; bottom = right horn core, medial view.

Boselaphine species from the Lothagam sequence fall into two different size groups. Most large specimens are from the Lower Nawata, but two individuals (LT 23662, 24021) are from the Apak Member. However, there is no doubt that the large Apak specimens are conspecific with the Lower Nawata species.

The most complete example of the large species is the frontlet LT 23149. Its large horn cores arise above the orbit and are strongly mediolaterally compressed with flattened lateral surfaces and anterior and posterior keels. The anterior keel is the more strongly developed and is characterized by a rugose flange in its proximal portion. The torsion is clockwise upward in the right horn core and is greater than in the smaller species. There is a large basal sinus, above which the horn core tapers gently upward. The distal portion is less compressed. From the right horn core of LT 23149 it is possible to interpret the terminal portion as lacking keels and conical in shape (circular in transverse section). However, no tip is preserved in any of the specimens. Other specimens assigned to this taxon are all smaller and less complete. Cranial fragments of LT 23149 and 26043 indicate that the anterodorsal part of the cranial vault—that is, behind the horn cores and between the supratemporal lines—is concave and strongly sculpted, and this feature is also seen in the smaller species. The Lothagam horn cores are comparable in size to, but less laterally divergent than, specimens of *T. cyrenaicus* from Sahabi (Thomas 1979; Lehmann and Thomas 1987) and Abu Dhabi (Gentry 1999) or of *T. acrae* from Langebaanweg (Gentry 1974, 1980). Vrba considers the Lothagam material as probably conspecific with, but less advanced than, *T. cyrenaicus* from Sahabi (Vrba 1995:421).

Tragoportax sp. A

(Figure 11.7; table 11.13)

Lothagam Material

- *Lower Nawata:* 23621, proximal Rt. and Lt. horn cores and mandible fragment; 23820, proximal Rt. horn core and skull fragments; 24214, Rt. calvaria, horn core and mandible; 24293, proximal Rt. horn core.

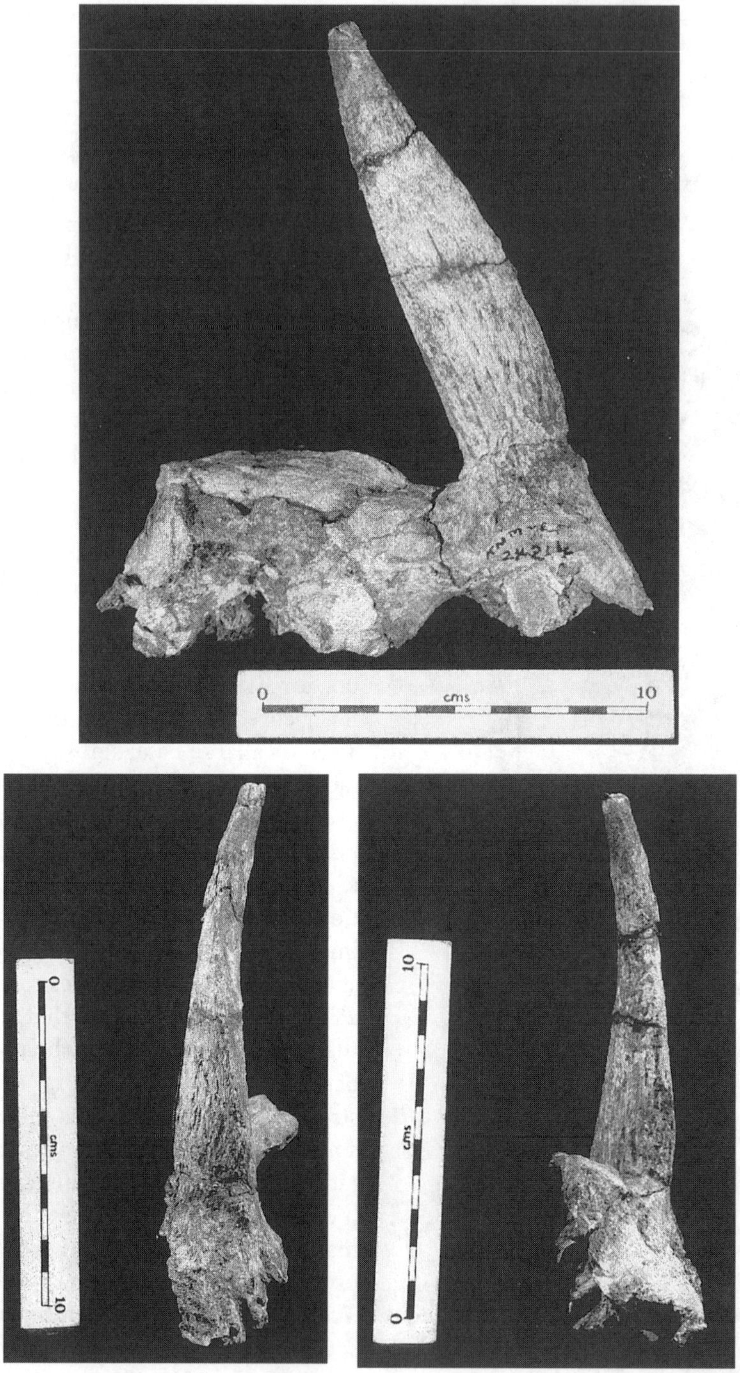

Figure 11.7 *Tragoportax* sp. A, partial calvaria with right horn, KNM-LT 24214, Lower Nawata: top = right lateral view; bottom left = anterior view; bottom right = posterior view.

- *Upper Nawata:* 214, proximal Rt. horn core; 23656, Lt. horn core.
- *Horizon indet.:* 22978, proximal Lt. horn core.

This small species is mostly from the Lower Nawata, but there are two Upper Nawata specimens (LT 214, 23656) and one from a younger horizon of undetermined age (LT 22978). The most complete specimen is LT 24214, a calvaria with right horn core from the Lower Nawata. LT 24214 has a mediolaterally compressed horn core with a flat lateral surface, a strong anterior keel, and a prominent posterior keel. The keels are less prominent at the apical extremity, which is cylindrical in shape. The horn core is uprightly inserted above the orbit. It tapers gradually upward in its proximal two-thirds, but the upward taper is more rapid in the distal third. There is slight clockwise torsion upward in the right horn core. The horn core extends slightly outward at the base but is more vertically orientated in its distal portion. Strong temporal lines extend posteromedially from behind the horn core bases, enclosing a strongly rugose triangular area between and behind the horn cores on the dorsal surface of the cranial vault.

Tragoportax sp. B
(Figures 11.5C, 11.8; table 11.14)

Lothagam Material

- *Lower Nawata:* 195, frontlet and horn cores.
- *Upper Nawata:* 196, frontlet and horn cores; 26197, Rt. horn core

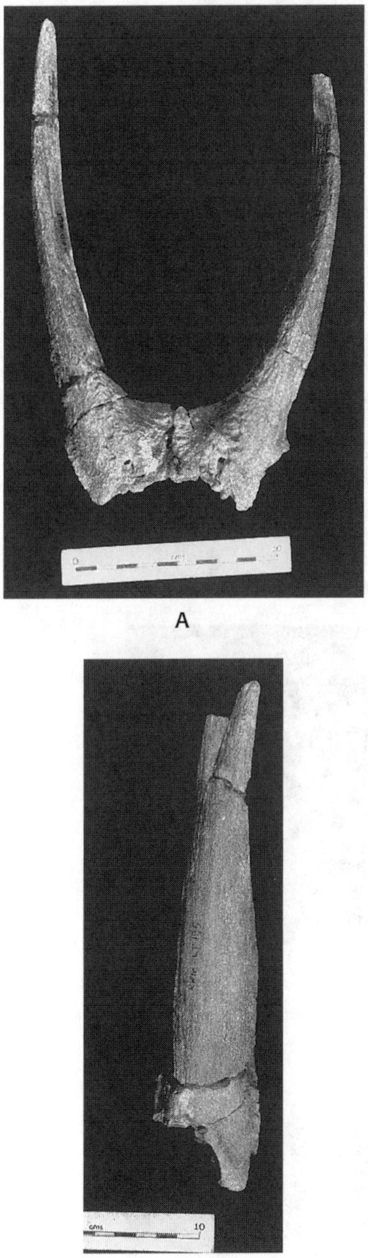

Figure 11.8 *Tragoportax* sp. B, frontlet with right and left horn cores, KNM-LT 195, Lower Nawata: A = anterior view; B = left lateral view.

A second small boselaphine species is represented by specimens LT 195, 196, and 26197. In LT 196, the curvature and orientation of the horn cores are similar, but the posterior keel is not well developed and a flattened posterior surface makes the horn core triangular in transverse section. In LT 195 and 26197, the horn cores are longer and do not taper so abruptly in their distal third. Bouvrain (1988) and others have pointed out that at least some *Tragoportax* species have differently shaped horn cores in males and females. However, pending the recovery of additional more complete specimens from Lothagam and elsewhere, the two Lothagam variants of small boselaphine are treated as different taxa.

Boselaphini gen. and sp. indet.
(Tables 11.15, 11.16)

A number of isolated dentitions have been recovered from the Nawata Formation but are unassociated with cranial material and cannot be identified below tribal rank.

Lothagam Material

- *Lower Nawata:* 193, Rt. mandible fragment (M_{1-3}); 202, Lt. mandible fragment (P_3–M_1); 203, Rt. mandible (P_3–M_3); 208, Lt. M_3; 209, Lt. mandible (M_{1-2}); 220, Lt. M^{1-3}, Rt. M^{1-2}; 221, Rt. M^{2-3}; 531, Lt. mandible fragment (P_4); 532, Rt. M_1, Lt. M_1, Rt. M_3 fragment; 13011, Lt. mandible fragment (P_4–M_1); 13014, Rt. M^3; 13015, Lt. M^2; 13016, Lt. M^2; 22875, Rt. maxilla (M_{1-3}), Lt. M_3; 23147, Rt. mandible (P_4–M_3); 23574, Lt. M^{1-2}, Rt. M^2; 23694, Lt. mandible (P_3–M_3); 23716, Lt. M^3; 24019, Lt. P_4 fragment; 26005, Lt. M/; 26021, Lt. maxilla (M^{2-3}); 26038, Lt. maxilla (P^3–M^1); 26055, Rt. M_3; 26070, Rt. M/; 38429, Lt. M_3.
- *Upper Nawata:* 204, Rt. mandible (P_2–M_3); 211, Lt. M_3 fragments (2 individuals); 13012, Lt. maxilla (P^4–M^3); 26572, Rt. M_1, Rt. M^2; 28574, Lt. and Rt. M/s; 194, Lt. mandible fragment (M_{2-3}).
- *Horizon indet.:* 201, Rt. mandible fragment (M_{2-3}).

Tribe Reduncini Lydekker and Blaine, 1914

Reduncines are medium- to large-sized antelopes represented today by two genera: *Redunca* (the reedbucks) and *Kobus* (waterbucks, lechwes, kob, and puku). A third genus (*Menelikia*) was a common constituent of Plio-Pleistocene assemblages from the Turkana Basin. Reduncines are represented at Lothagam by three species, two of them new.

Kobus Smith, 1840

Diagnosis

Larger-sized reduncines; horn cores usually long, their bases sometimes curving backward, instead of being concave anteriorly, usually with a flat lateral surface but with no tendency toward a flattened posteromedial surface; frontals sometimes with a small system of internal sinuses (Gentry 1985).

Type species

Kobus ellipsiprymnus (Ogilby 1833).

Remarks

Waterbucks were present in the southern part of the Turkana Basin at the beginning of the present century but no longer occur there today—in stark contrast to

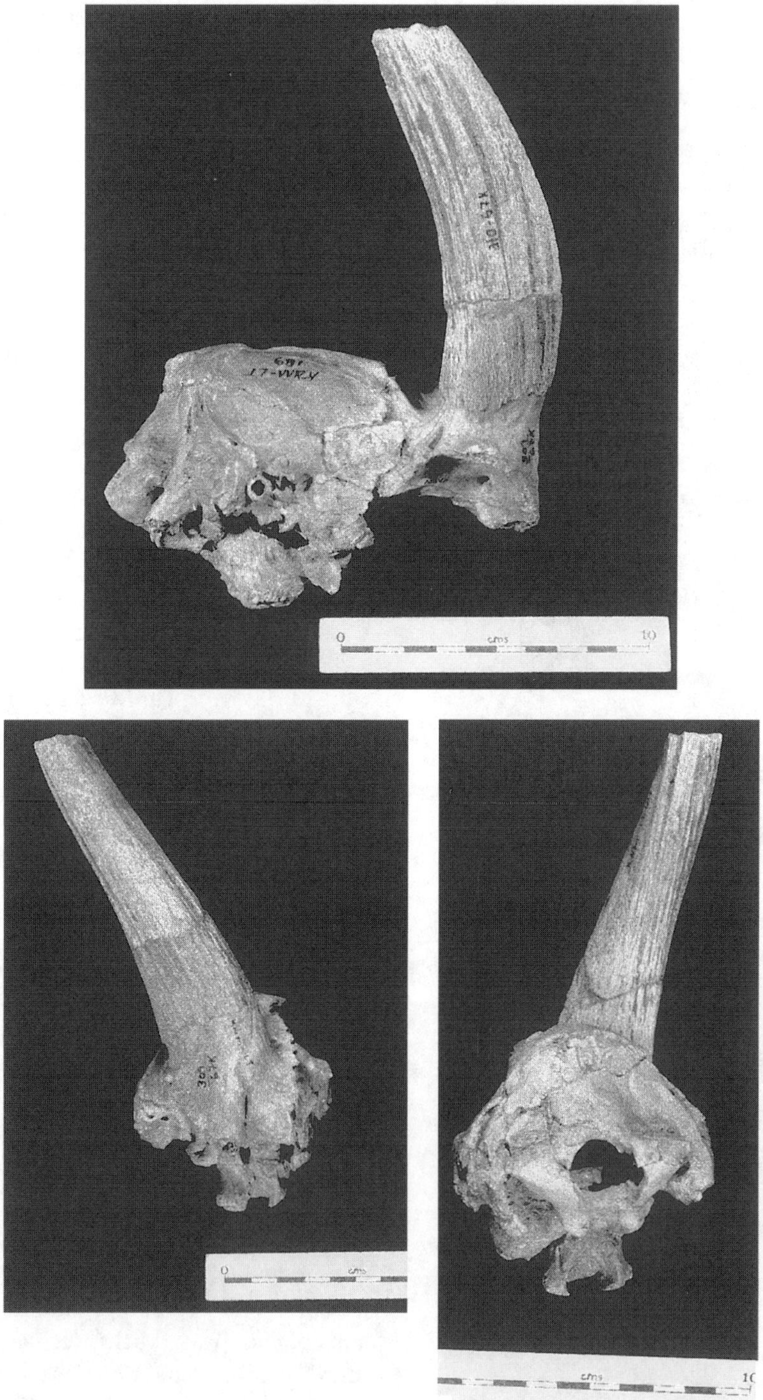

Figure 11.9 *Kobus presigmoidalis* sp. nov., calvaria with right and left horn cores, KNM-LT 189, Upper Nawata: top = right lateral view; bottom left = anterior view; bottom right = posterior view.

the fossil assemblages of the region in which the reduncines, and *Kobus* species in particular, featured prominently.

Kobus presigmoidalis sp. nov.

(Figures 11.5D, 11.9; tables 11.17, 11.18)

Diagnosis

Small reduncine with medium-length horn cores that are inserted uprightly and close together on moderately long pedicels above the orbit; face steeply angled on the braincase; supraorbital fossae small, level with the median edge of the horn core and located in shallow depressions. Horn core is compressed mediolaterally with a flattened lateral surface; tapers upward from the base with sigmoid curvature in lateral view; slightly lyrate in anterior view; ornamented with very strong longitudinal ridges and furrows that highlight the slight clockwise torsion (upward in right horn core); postcornual depression short but deep. Braincase fairly long and steep-sided, widening posteriorly; temporal ridges are flat; nuchal crest well defined; mastoid entirely lateral in its exposure; occiput shallow and wide, median ridge only faintly developed; paroccipital processes stout; auditory bullae large and well inflated.

Holotype

KNM-LT 189, calvaria with right and left horn cores, collected from the upper member of the Nawata Formation in 1967 by the Harvard University Expedition.

Etymology

The name reflects the morphological similarity to *K. sigmoidalis,* which was abundant in the Turkana Basin during the Pliocene and Early Pleistocene.

Lothagam Material

- *Lower Nawata:* 486, proximal Lt. horn core; 497, proximal Lt. horn core fragment; 13009, proximal Lt. horn core; 26575, Rt. and Lt. horn cores.
- *Upper Nawata:* 189, holotype; 212, frontlet and proximal horn cores; 224, proximal Rt. horn core; 233, proximal Lt. and distal Rt. horn cores; 466, proximal Lt. horn core; 483, proximal Rt. horn core; 496, proximal Lt. horn core; 23661, proximal Lt. horn core; 23671, proximal Lt. horn core; 23713, proximal Rt. horn core; 25951, proximal Rt. horn core; 25984, proximal Lt. horn core; 25997, proximal Lt. horn core.
- *Apak Member:* 25970, Rt. horn core; 25973, proximal Rt. horn core; 28710, proximal Rt. horn core; 30650, proximal Lt. horn core.
- *Kaiyumung:* 23726, proximal Lt. horn core.
- *Horizon indet.:* 23615, proximal Lt. horn core.

This species is represented by a calvaria and a score of incomplete horn cores. The holotype calvaria (LT 189) was missing the proximal portion of its left horn core when collected in 1967 but the missing piece was retrieved in 1992, so the provenance of this specimen is now secure. Most of the incomplete horn cores assigned to this species match well the morphology of the holotype, tapering upward from their base, curving backward in their proximal portion but recurving vertically near their tips. In anterior view they diverge strongly from their base but distally become more vertical. Most are ornamented with very strong longitudinal ridges and furrows. Two specimens differ slightly in morphology but are tentatively assigned to this taxon: LT 13009 is more mediolaterally compressed than the majority of *K. presigmoidalis* specimens, whereas in LT 26575 the outward divergence of the horn cores is less marked. Horn cores of this species from the Apak Member have stronger sigmoid curvature and have transverse ribbing on the anterior surfaces of the horn cores (e.g., LT 25973, 25970, 28170, 23726).

This species appears closely comparable with early specimens of *K. sigmoidalis* from the lower part of the Koobi Fora Formation such as KNM-ER 4552 from the Lokochot Member. The differences are that the Koobi Fora specimens are larger, the frontal bone is less thick at the intrafrontal suture, and the pedicels are a little longer.

The Lothagam specimens also compare quite closely to *Kobus* aff. *porrecticornis* (LU 941) from Lukeino (Thomas 1980). They differ in that the pedicel is shorter in LU 941, the horn core curves backward in a single plane whereas that of the Lothagam specimens is more lyrate (curving outward in its proximal portion and recurving medially toward its tip), and the backward curvature is slightly more pronounced in the Lothagam material.

The Lothagam specimens differ from *Kobus* aff. *porrecticornis* from Mpesida (Thomas 1980). The latter specimens share the outward and backward curvature from the base but are smaller, less elongate, and proportionately wider.

Kobus subdolus from Langebaanweg (Gentry 1980) has horn cores that are less mediolaterally compressed but are more backwardly inserted above the orbits. Gentry (1980) recognized *K. subdolus* from Sahabi, although Lehmann and Thomas (1987) attributed the same specimens to *Redunca* aff. *R. darti* after the South African species described by Wells and Cooke (1956).

Kobus laticornis sp. nov.

(Figures 11.5E, 11.10; tables 11.19, 11.20)

Diagnosis

Small reedbuck-sized reduncine, face angled on braincase at 120°, supraorbital fossae small, located in front of medial edges of horn cores and set in depressions. Horn cores long, inserted above orbits and fairly widely apart, arising from short pedicels and extending outward and upward from their base; anteroposteriorly compressed, tapering gently upward with slight clockwise torsion (in right horn core); ornamented with longitudinal ridges and furrows with strong groove in middle of anterior surface; lateral edge of horn core produced into slight keel; postcornual groove small but deep. Braincase low but wide, becoming progressively wider posteriorly; temporal ridges fairly prominent, converging posteriorly until level with the external auditory meati then diverging laterally to fuse with the strongly developed nuchal crest; mastoid is exposed laterally; occiput shallow but wide, with only a faint median ridge; paroccipital processes short and stout.

Holotype

KNM-LT 180, calvaria with proximal horn cores, collected from the upper member of the Nawata Formation in 1967 by the Harvard University Expedition.

Etymology

The name reflects the lateral divergence of the horn cores of this small *Kobus* species.

Lothagam Material

- *Upper Nawata:* 180, holotype; 199, proximal Rt. and Lt. horn cores; 200, frontlet Rt. horn core, proximal Lt. horn core; 227, frontlet and base Lt. horn core; 485, proximal Rt. and Lt. horn cores; 14129, proximal Rt. horn core; 14130, proximal Lt. horn core; 22969, frontlet with Rt. and Lt. horn core fragments; 23631, proximal Lt. horn core; 23704, Rt. horn core.
- *Upper Nawata or Apak:* 198, frontlet and proximal horn cores.

In addition to the holotype, *K. laticornis* is represented by several frontlets and horn core fragments. All except one specimen come from the Upper Nawata, but there is little doubt that the latter specimen belongs to this taxon because of the strongly divergent nature of its horn cores.

LT 180 is the most complete specimen. It is closer in size to a reedbuck than to a kob or waterbuck, and the facial region is angled on the braincase at an angle comparable to that of a reedbuck—that is, steeper than a waterbuck but flatter than a kob. The horn cores were long; none of the specimens is complete, but the longest specimen attains a length of 20 cm to the point where the tip is broken off. The widespread laterally divergent nature of the horn cores is strongly reminiscent of *Kobus oricornis* from the earlier portions of the Omo and Koobi Fora sequences. However, *K. laticornis* is smaller than *K. oricornis.* The horn cores of *K. laticornis* are more widely inserted, more laterally divergent, have more pronounced torsion, lack transverse ribbing, and seem to be more compressed than those of *K. oricornis.* The insertion of the horn cores on the cranium is almost identical to that of *Kobus ancystrocera,* and the braincase is very similar to, though smaller than, that of ER 1595 from the KBS Member of the Koobi Fora Formation. However the horn cores of *K. laticornis* are laterally orientated, whereas those of *K. ancystrocera* extend backward and outward.

K. laticornis may conceivably be related to *Kobus* sp. A from Hadar (Gentry 1981:910), which is also about the size of a kob and in which the horn cores are long and compressed anterolaterally to posteromedially and lack keels or transverse ridges. The horn cores of both species diverge strongly, but that of the Hadar form increases toward the tip, whereas the divergence of the Lothagam specimens appears uniform. The horn cores have little or no backward curvature, are inserted fairly uprightly, and may even curve forward from their base to their tip. The temporal lines on the cranial vault are prominent and approach closely together. The occipital surface is rather flat and faces wholly backward. The mastoids are wide but lie within the occipital surface. The braincase is wider than tall. Both Lothagam and Hadar specimens seem close to *K. oricornis* but represent different species.

Menelikia Arambourg, 1941

Menelikia was originally identified by Arambourg (1941) as a hippotragine; later it was placed into a subfamily of its own (Arambourg 1947). However, its reduncine affinities appear certain (Gentry 1976, 1985). The relationship of this genus to other reduncine genera is less clear, although Gentry (1985) remarked on some similarities to the extant *Kobus megaceros.*

Diagnosis

An extinct genus of medium-sized reduncines in which the horn cores are without keels, with transverse ridges, without a flattened lateral surface, inserted obliquely

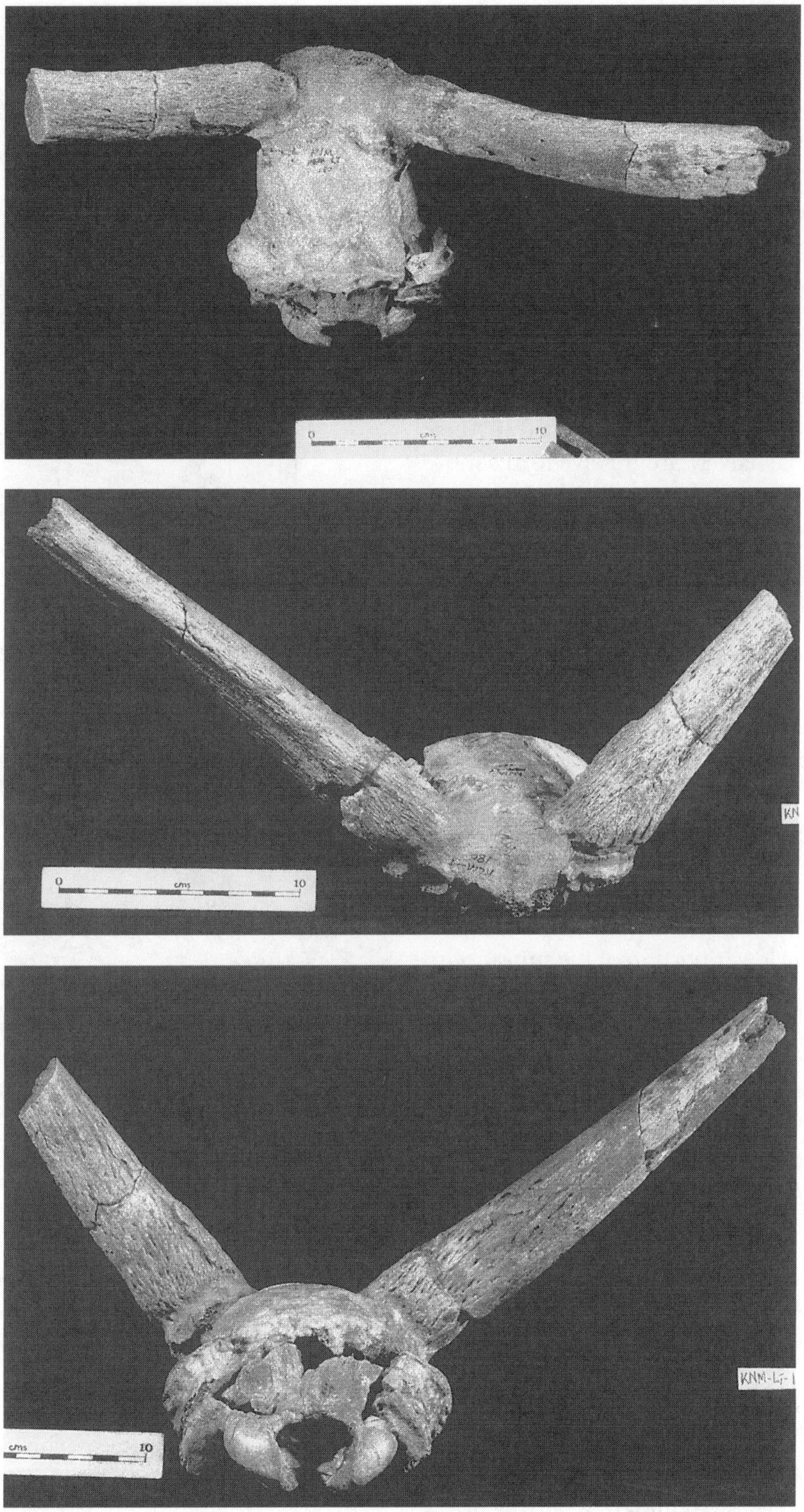

Figure 11.10 *Kobus laticornis* sp. nov., calvaria with right and left horn cores, KNM-LT 180, Upper Nawata: top = dorsal view; center = anterior view; bottom = posterior view.

and close together, backwardly curved, strongly divergent above the basal part, torsion clockwise (sometimes weakly so) in the right horn core from the base up, and often with strong longitudinal grooving. There is either a very small or no postcornual fossa. Frontals with extensive internal sinuses below and anteromedially to the horn insertions, and the level of the frontals between the horn bases usually high relative to the orbital rims. Supraorbital pits small, or small to moderate in size (Gentry 1985).

Type species

Menelikia lyrocera Arambourg, 1941

Menelikia leakeyi Harris, 1991
(Figures 11.5F, 11.11; table 11.21)

Diagnosis

A species of *Menelikia* with long horn cores that exhibit clockwise torsion from the base upward in the right horn core; early forms mediolaterally compressed with faint anterior and posterior keels and with transverse ridges in their proximal halves; progressive forms more mediolaterally compressed with a smaller degree of torsion, with stronger longitudinal grooves and ridges proximally (especially on the anterior surface), and with transverse ridges only in the distal half of the horn core. Horn cores of earlier examples inserted fairly uprightly above the orbits and fairly close together; those of more progressive examples inserted more obliquely and farther apart. Frontal sinuses do not extend into the base of the pedicel. Braincase angles quite steeply on the face, becoming wider posteriorly, with a gently rounded posterior edge and with prominent temporal ridges approaching closely posteriorly. Mastoids included on the occipital surface, in which the median ridge is abbreviated and the nuchal fossae are reduced to shallow indentations because the lower portion of the supraoccipital forms a backwardly projecting bulge above the foramen magnum. Basioccipital fairly short, anterior tuberosities large and closely sited, auditory bullae of moderate size and well inflated (Harris 1991).

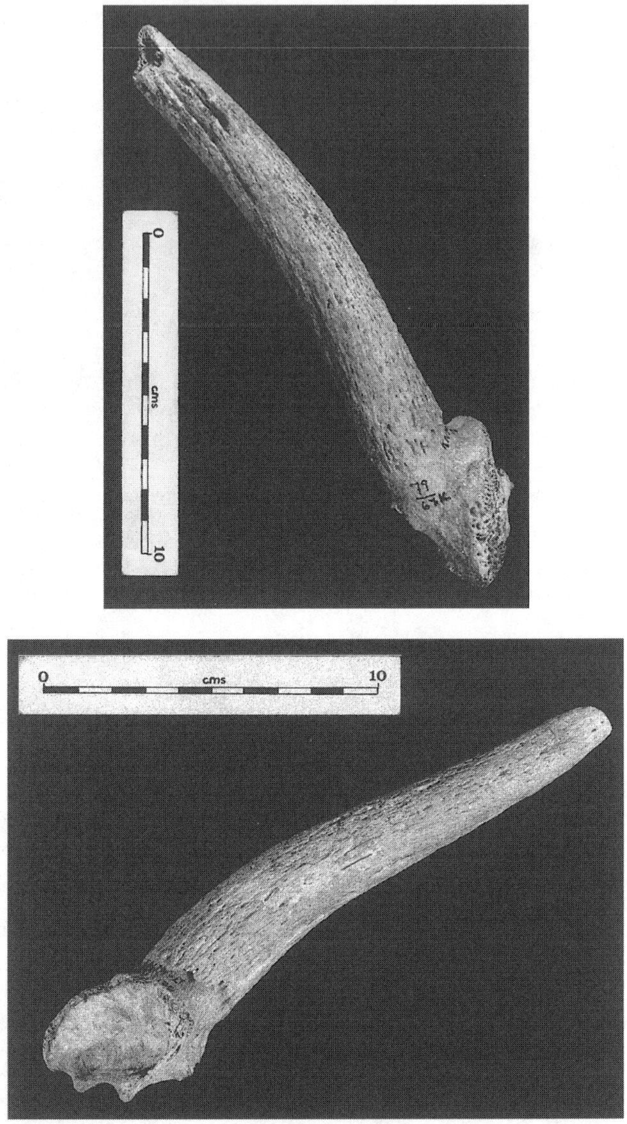

Figure 11.11 *Menelikia leakeyi,* right horn core, KNM-LT 178, Lower Nawata: top = posterior view; bottom = medial view.

Lothagam Material

- *Lower Nawata:* 178, Rt. horn core; 487, proximal Lt. horn core; 490, proximal Rt. horn core; 26026, proximal Rt. horn core.
- *Upper Nawata:* 25978, proximal Lt. horn core and M/ fragment; 23715, proximal Lt. horn core; 25966, proximal Rt. horn core.

The most complete specimen from Lothagam, LT 178 from the Lower Nawata, is seemingly identical to the material reported from the earlier portion of the Koobi Fora sequence. The horn core is fairly uprightly inserted above the orbit and is anteroposteriorly compressed at its base. With anticlockwise torsion upward in the right horn core, the horn core tapers gently as it spirals upward and outward. The *Menelikia* specimens may be distinguished from those of *K. laticornis* in that the bases of the horn cores are located very close together, by the more rapid tapering, by the much more pronounced torsion, and by the less laterally divergent orientation of the horn core axis. They differ from specimens of *M. leakeyi* from the Koobi Fora Formation in that the horn cores are smaller, shorter, and with lesser torsion.

The other Lothagam specimens attributed to this taxon are less complete and/or less well preserved. One of them—LT 25978—appears to display faint horizontal ribbing on its anterior surface. As this is not a distinctive feature of *Menelikia,* this horn core fragment may have been wrongly identified.

Reduncini gen. and sp. indet.
(Table 11.22)

A number of isolated dentitions have been recovered from the Nawata Formation and Apak Member but are unassociated with cranial material and cannot be identified below tribal rank.

Lothagam Material

- *Lower Nawata:* 207, Lt. mandible fragment (P_4–M_1); 534, Rt. mandible fragment (P_4); 13007, Rt. mandible fragment (P_3–M_1); 23579, Lt. mandible (P_3–M_3), Rt. mandible (P_4–M_3); 23624, Rt. mandible (M_{1-3}), M_1 broken; 23657, Lt. P_3; 26052, Rt. M_3; 26069, Lt. M_3; 28747, Lt. mandible (M_{1-2}).
- *Upper Nawata:* 205, Lt. mandible fragment (M_{2-3}); 23655, Lt. mandible (M_{2-3}); 23693, Lt. mandible (P_3–M_1); 26007, Rt. mandible (P_{2-4}); 28648, Lt. P_3.
- *Apak Member:* 23714, Lt. mandible (P_4–M_1); 25994, Rt. mandible (M_3).
- *Kaiyumung Member:* 26068, Lt. mandible (P_4–M_1).
- *Horizon indet.:* 181, Lt. mandible (P_3–M_2); 20671, Lt. and Rt. mandible fragments (M_{2-3}).

Tribe Hippotragini Retzius and Lovén, 1845

Hippotragini are characterized by horn cores that are long, of large basal area relative to skull size, and with moderately to well developed basal sinuses in their horn core pedicels. The horn cores are unkeeled, little divergent, and generally without transverse ridges. The postcornual fossae are shallow or absent, and there are bulbous, lingually extending metaconids on the P_4s (Vrba and Gatesy 1994).

The oldest recognized hippotragines include *Praedamalis howelli* from the Awash region of Ethiopia (Vrba and Gatesy 1994) and *P. deturi* from Hadar in Ethiopia (Gentry 1981) and Laetoli in Tanzania (Gentry 1987). Both described species of *Praedamalis,* however, have long, mediolaterally compressed horn cores with hardly any backward curvature (Gentry 1987; Vrba and Gatesy 1994). However, a complete left hippotragine horn core from Sahabi that was described by Gentry (1987) and illustrated by Lehmann and Thomas (1987) has strong backward curvature and was interpreted as a possible early representative of the genus *Hippotragus.* Vrba and Gatesy (1994) interpret this specimen as part of the clade that includes the extinct *Hippotragus gigas* and Vrba (1995:421) interprets the Sahabi specimen as of Late Pliocene age. Specimens from Lothagam appear to represent both *Praedamalis* and *Hippotragus.*

Praedamalis Dietrich, 1950

Diagnosis

Hippotragine having long horn cores with some mediolateral compression, no flattening of the medial or lateral surface, no keels or transverse ridges, and hardly any backward curvature. Horn cores are inserted rather upright above the back of the orbits, not very wide apart, and little divergent. Greatest horn core width lies slightly behind the anteroposterior midpoint. The base of the horn core proper is not well demarcated from the top of the pedicel, the frontals between the horn cores are about the same level as the dorsal part of the orbital rims, there is a shallow postcornual fossa, the pedicel and part of the frontal are hollowed internally, there is a tiny rudiment or vestige of a temporal ridge behind the horn cores, and a concavity in the surface of the frontals posteromedially to the pedicel. *Praedamalis* has exceptionally large sinuses in the frontals and pedicel at the base of the horn core and partly extending into the horn core. It has small skull size relative to other hippotragines, horn cores with an angulation of the maximum basal diameter relative to the metopic suture that is intermediate between *Oryx* and all described *Hippotragus* species, and widely separated supraorbital foramina as in modern oryxes and addaxes (after Gentry 1987 and Vrba and Gatesy 1994).

Praedamalis? sp.

(Figures 11.12, 11.13; table 11.23)

Lothagam Material

- *Upper Nawata:* 23709, proximal Lt. horn core; 25968, proximal Rt. horn core.
- *Kaiyumung Member:* 23681, proximal Lt. horn core; 23720, proximal Lt. horn core; 24022, distal horn core fragment; 24061, Rt. horn core.

Several specimens from the Upper Nawata and Kaiyumung Member represent a medium-sized bovid that had long horn cores inserted close together and arising from above the orbit. These horn cores are strongly compressed mediolaterally, and the long axis of the horn core base runs anteromedially-posterolaterally. They taper gently above their bases, curving slightly backward distally and diverging laterally. The identification of these horn cores is still uncertain. They could represent an extinct alcelaphine genus (e.g., *Damalacra*) or, alternatively, a hippotragine with flattened but laterally diverging horn cores. The horn cores differ from other Lothagam alcelaphine horn cores by their close insertion and from *Hippotragus* by their strong mediolateral compression. The horn cores are smaller than those of *Praedamalis deturi* from Laetoli and Hadar (Gentry 1987:387–388). They are of similar size to those of *Praedamalis howelli* from the Middle Awash of Ethiopia (Vrba and Gatesy 1994), but Lothagam specimens do not demonstrate

Figure 11.12 *Praedamalis*? sp. proximal left horn core, KNM-LT 23709, Upper Nawata: top = anterior view; bottom = medial view.

the huge basal sinuses characteristic of the Ethiopian species.

Hippotragus Sundevall, 1846

Diagnosis

Horn cores mediolaterally compressed, strongly curved backward, inserted uprightly above the orbits and closer together than in *Oryx;* ethmoidal fissures blocked by bone internally; nasals more domed than in *Oryx;* mastoid facing partly laterally as well as backward; longitudinal ridges behind the anterior tuberosities of the basioccipital are stronger than in *Oryx;* lower molars with stronger goat folds than in *Oryx* (Gentry and Gentry 1978).

Hippotragus sp.

(Figure 11.14; table 11.24)

Lothagam Material

- *Lower Nawata:* 26037, proximal Rt. horn core; 26599, proximal Rt. horn core.
- *Upper Nawata:* 13010, proximal Rt. horn core; 23131, skull and Rt. horn core fragment.
- *Apak Member:* 23598, partial calvaria and Rt. horn core.

A few specimens from the Lothagam sequence represent a medium-sized hippotragine characterized by its face strongly angled on the braincase and with backward curving, laterally flattened horn cores inserted close together. The stout base of the horn core arises above the orbit at a low angle. It tapers gently upward from its base, diverging slightly outward but curving strongly backward. A large basal sinus is evident in LT 23131. Specimens assigned to this species appear slightly larger than those of *Praedamalis howelli* and of comparable size to *P. deturi* and to ?*Hippotragus* sp. from Sahabi. Comparison of some specimens, particularly 13010 and 23598, with the cranium of *Hippotragus gigas* from Koobi Fora confirms a hippotragine identification on the basis of narrow width between the horn cores, strong angulation of the face on the brain case, horn core shape, and calvaria anatomy.

Hippotragini gen. and sp. indet.

(Table 11.25)

Isolated dentitions that are unassociated with cranial material and cannot be identified below tribal rank have been recovered throughout the Lothagam succession.

Lothagam Material

- *Lower Nawata:* 562, Lt. mandible fragment (M_{1-2}); 25432, Lt. M_3 fragment; 25452, Lt. M^3; 25960, Lt. M_3.
- *Upper Nawata:* 507, Lt. M^2; 23130, Rt. M^2.
- *Kaiyumung Member:* 23691, Rt. M^2; 26186, Lt. mandible (M_{2-3}).
- *Horizon indet.:* 450, Lt. M_2.

The molars assigned to this tribe are superficially similar to those of alcelaphines from which they differ by (1)

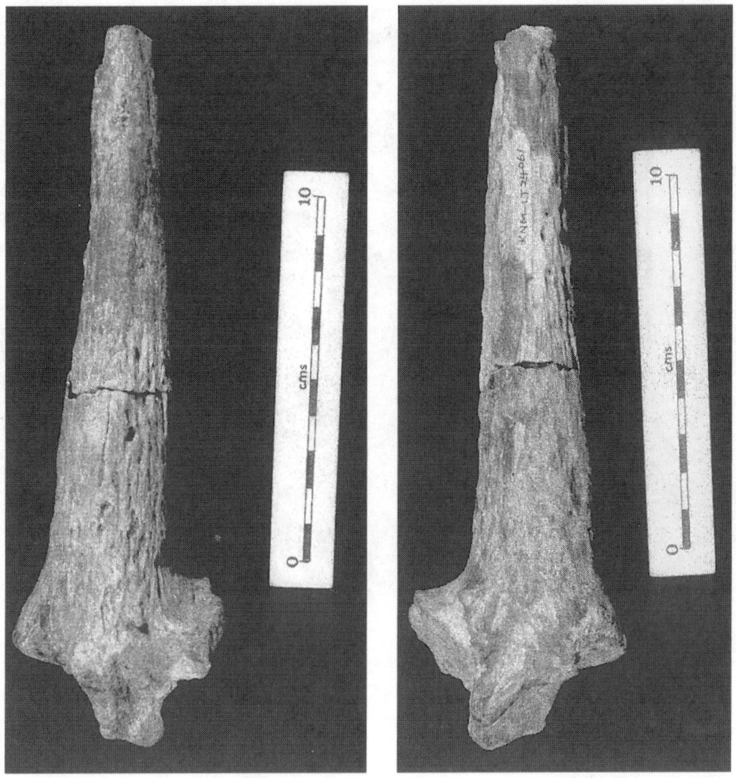

Figure 11.13 *Praedamalis*? sp., right horn core, KNM-LT 24061, Kaiyumung Member: left = anterior view; right = posterior view.

"pinching" of the external lobes of the lower molars (and internal lobes of upper molars), and (2) presence (variable) of a small internal style (external stylid).

Tribe Alcelaphini Rochebrune, 1883

The cladistic analysis of fossil and living Alcelaphini by Vrba (1997) suggested that two alcelaphine subtribes diverged, either during or before the Mio-Pliocene transition. *Damalacra neanica* was the first representative of the Alcelaphina, which consisted of two ancient subclades—the sister group of *D. neanica* and *Beatragus* (known since 5.0–4.5 Ma) and the large clade comprising *Damalops, Numidocapra, Alcelaphus, Rabaticeras, Megalotragus, Oreonagor,* and *Connochaetes* that first made an appearance at about 4.4 Ma. The Damalascina originated with *Damalacra acalla* and includes the *Damaliscus* and *Parmularius* lineages.

Damalacra Gentry, 1980

Diagnosis

Moderate-sized alcelaphines, a little smaller than the living *Alcelaphus buselaphus* or *Damaliscus lunatus.* Skull rather narrow as in these species and not wide as in *Connochaetes.* Horn cores moderately long and without keels or transverse ridges. Horn cores inserted fairly uprightly and close together. Shallow elongated postcornual fossa. Frontals set at high level between the horn core bases in comparison with the dorsal parts of the orbital rims, orbital rims project quite strongly, little or no central indentation of the parietofrontal suture, temporal lines on cranial roof do not approach closely posteriorly, braincase sides parallel, small postorbital pits set closely together. Occipital surface is wide and low and has a median vertical ridge, the mastoid has a large exposed area and is entirely contained in the occipital surface, the basioccipital is only slightly narrowed anteriorly, if at all, and has anterior tuberosities of moderate size and a central longitudinal groove, the basisphenoid rises fairly steeply in front of the basioccipital, and the auditory bullae are moderate to large sized.

Hypsodont cheek teeth with not very rugose enamel, small basal pillars on M_1s and dP_4s and occasionally on upper molars, central cavities of upper molars not very complicated in outline, upper molars with rather strong styles but poor development of ribs between them; medial lobes of upper molars less well rounded than in Pleistocene or Recent alcelaphines, medial walls of lower molars with less pronounced outbowings and with more prominent metastyles than in later alcela-

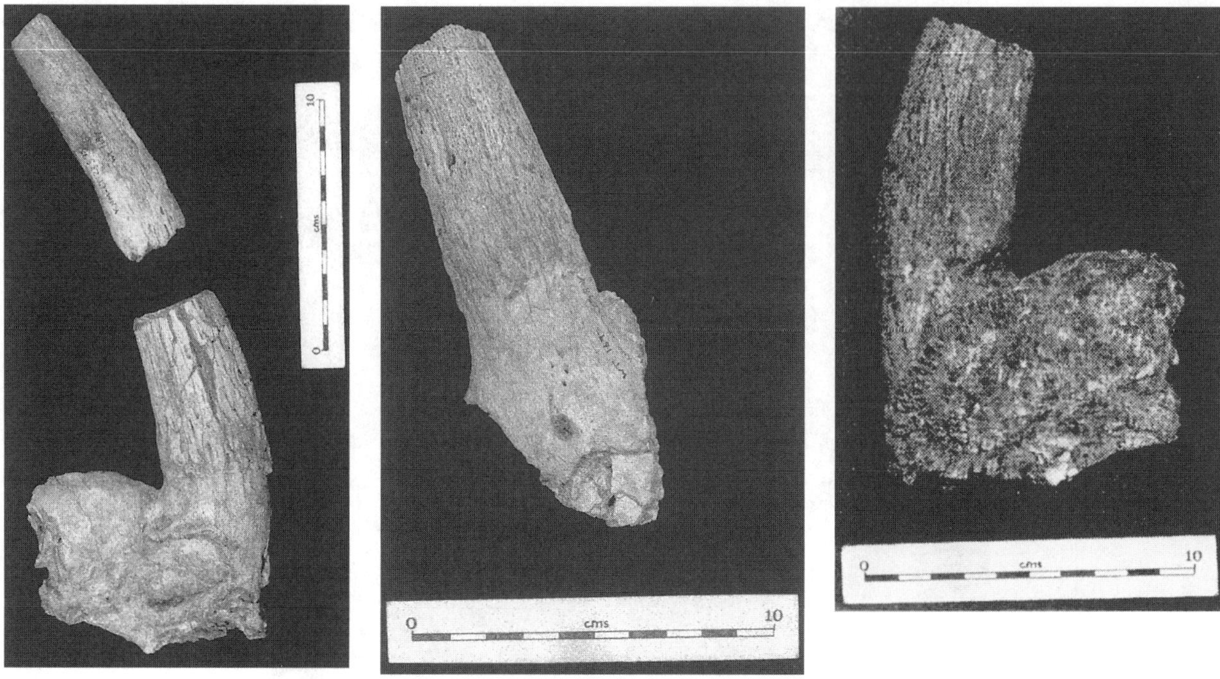

Figure 11.14 *Hippotragus* sp., partial calvaria and partial right horn core, KNM-LT 23598, Apak Member: left = right lateral view; center = anterior view; right = medial view.

phines, lower molars without goat folds, central cavities of lower molars with hardly any transverse constrictions centrally, P_2s reduced and often absent in life. P_4s with posterior part of tooth (behind level of metaconid) less reduced than in later alcelaphines, generally with transverse orientation of the valley between the entoconid and entostylid, and with paraconid and metaconid growing toward one another but not usually fusing (Gentry 1980).

Damalacra sp. A

(Figures 11.15, 11.16; table 11.26)

Lothagam Material

- *Lower Nawata:* 183, Lt. horn core; 238, Rt. horn core; 26030, proximal Rt. horn core; 26203, proximal Rt. horn core.
- *Upper Nawata:* 231, proximal Rt. horn core; 463, proximal Rt. horn core; 482, proximal Lt. horn core; 504, Rt. horn core fragment; 506, Lt. horn core; 514, proximal Rt. horn core; 23146, calvaria with proximal Rt. and Lt. horn cores; 23736, proximal Lt. horn core; 25977, proximal Rt. horn core; 25986, Lt. horn core; 26000, proximal Rt. horn core; 26014, proximal Lt. horn core; 26015, proximal Rt. and Lt. horn cores; 26016, horn core fragment; 26017, proximal Rt. horn core; 26018, proximal Lt. horn core; 26019, proximal Rt. horn core; 26032, proximal Rt. horn core; 26034, Lt. horn core; 26035, proximal Rt. horn core; 28790, proximal Lt. horn core; 38430, proximal Rt. horn core; 38432, proximal Rt. horn core.
- *Upper Nawata or Apak:* 493, Lt. horn core.
- *Apak Member:* 23649, proximal Rt. horn core; 23729, Rt. horn core; 26029, proximal Rt. horn core; 25975, Rt. horn core.
- *Kaiyumung Member:* 24049, proximal Rt. horn core.
- *Horizon indet.:* 179, Lt. horn core; 237, Rt. horn core; 225, proximal Rt. horn core.

As best seen from LT 237, the small and relatively short horn cores of this bovid are located close together and are inserted on a short pedicel above the orbit. The supraorbital foramen is small and located directly in front of the horn core base. From their base, the horn cores rise upward, backward, and outward, then converge slightly in their distal portion. In lateral view there is slight sigmoid curvature, the horn cores curving upward and backward from the base but recurving vertically at the tip. The horn cores are stout in their proximal portion, with slight mediolateral compression, and taper gradually toward their tips. The horn cores are stouter and less lyrate than those of *Damaliscus dorcas* and are smaller than, but closely approach in shape, those of *D. korrigum.* The cranium into which they inserted would have been about the size of *D. dorcas* but must have had appreciably shorter pedicels.

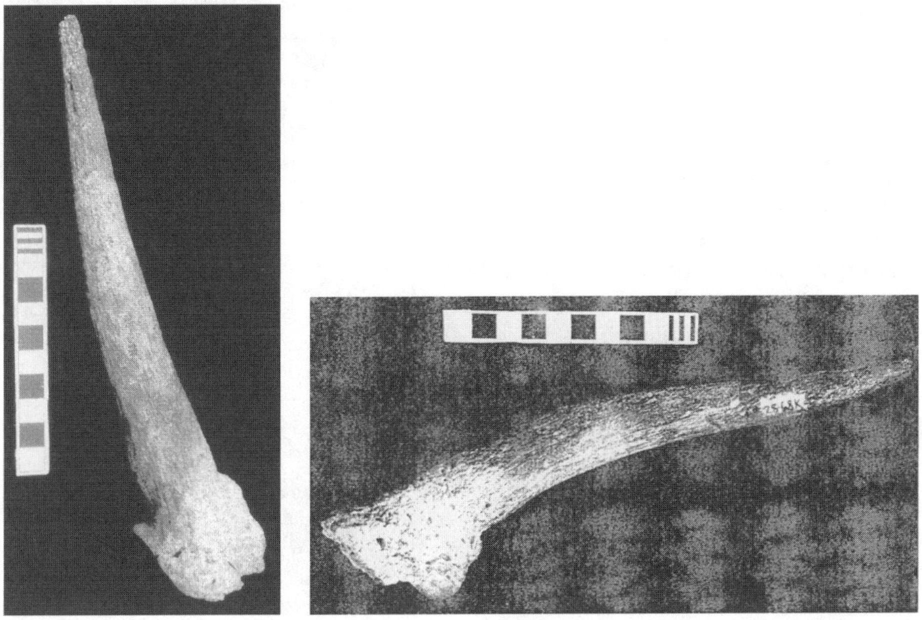

Figure 11.15 *Damalacra* sp. A, right horn core, KNM-LT 237, horizon indet.: left = anterior view; right = medial view.

This species is known only from horn cores, although presumably it is represented also by some of the Alcelaphini teeth. The horn cores display a fair amount of variation in size and stoutness, but all seem attributable to the same taxon. Most come from the Upper Nawata though there are a few from the lower part of the formation. There are also four specimens from the Apak Member that seem to be slightly smaller than the Nawata material.

If this material is indeed correctly interpreted as Alcelaphini, then it is one of the earliest representatives of this tribe. Vrba (1995) considered "aff. *Damaliscus*" of Smart (1976) to be hippotragine rather than alcelaphine.

Damalacra sp. B

(Figures 11.17, 11.18; table 11.27)

Lothagam Material

- *Lower Nawata:* 456, proximal Lt. horn core.
- *Apak Member:* 23647, Rt. horn core; 25971, proximal Lt. horn core.

Three horn cores (LT 456, 23647, 25971) are strongly reminiscent of alcelaphine specimens reported from the Koobi Fora Formation (Harris 1991). They all have large basal sinuses and were evidently inserted on moderately long pedicels. Unfortunately, the point of their insertion vis-à-vis the orbit cannot be determined. The horn core base is massive and swollen posteromedially but tapers rapidly in its proximal portion; it is almost D-shaped in transverse section because of the flattened lateral surface. In lateral view, the base of the horn core extends vertically from the pedicel but curves backward in its distal portion. When viewed from the front, there is a slight lyrate curvature as the horn core curves slightly outward above its base but recurves medially in its distal portion. In appearance the horn cores seem to be closest to ?*Damaliscus* sp. nov. from Koobi Fora (Harris 1991), but their bases seem somewhat less swollen than *Parmularius eppsi.* The two specimens of known provenance (LT 23647, 25971) come from the Apak Member; LT 23647 is the more complete.

Alcelaphini gen. and sp. indet.

(Table 11.28)

Isolated dentitions that are unassociated with cranial material and therefore cannot be identified below tribal rank have been recovered throughout the Lothagam succession.

Lothagam Material

- *Lower Nawata:* 235, Rt. mandible fragment (M_{1-2}); 530, Rt. mandible fragment (M_{1-2}).
- *Upper Nawata:* 222, Rt. M^3; 459, Lt. M_1.
- *Apak Member:* 26060, Lt. M^{2-3}; 26571, Lt. M^3; 23614, Rt. mandible (M_2).
- *Kaiyumung Member:* 217, Lt. mandible (P_2–M_3), Rt. M_3, Rt. M_2; 469, Rt. M_1.
- *Horizon indet.:* 381, Rt. M_2.

Tribe Aepycerotini Gray, 1872
Aepyceros Sundevall, 1847

Diagnosis

Small- or medium-sized antelopes. Horn cores only in males, not very compressed, with a flattened lateral surface and a posterolateral keel, with transverse ridges, inserted fairly uprightly, close together and above the back of the orbits, curving backward in side view, strongly divergent in their lower parts but with a change of course in their center, and with their distal parts parallel or slightly reapproaching. Postcornual fossae deep, frontals hollowed internally, and higher between the horn core bases than the dorsal rim of the orbits, complicated mid-frontals and parietofrontal sutures on the skull top, braincase roof bent downward posteriorly, side walls of brain case more or less parallel, temporal lines approaching fairly closely posteriorly, small supraorbital pits situated widely apart on a flat or slightly convex surface of the frontals, preorbital fossa very reduced or absent, ethmoidal fissure narrow or absent, a long narrow foramen between maxilla and premaxilla only known otherwise in *Neotragus moschatus* and *N. batesi* among the Bovidae; premaxilla with a long contact with the nasal; mastoids large, occipital surface facing partly laterally on each side as well as backward, and auditory bullae well inflated. Basioccipital with quite small anterior tuberosities, with slight to moderate longitudinal ridges behind them, and with a wide shallow central longitudinal groove. Hypsodont teeth without basal pillars on the molars, strong styles and poor ribs on the lateral walls of the upper molars, lower molars without anterior transverse flanges (goat folds). M_3 with a large rear (third) lobe. P_4 with paraconid/metaconid fusion. The tuber scapulae is far from the lateral edge of the glenoid facet in ventral view. On the proximal end of the radius, the lateral tubercle is small and

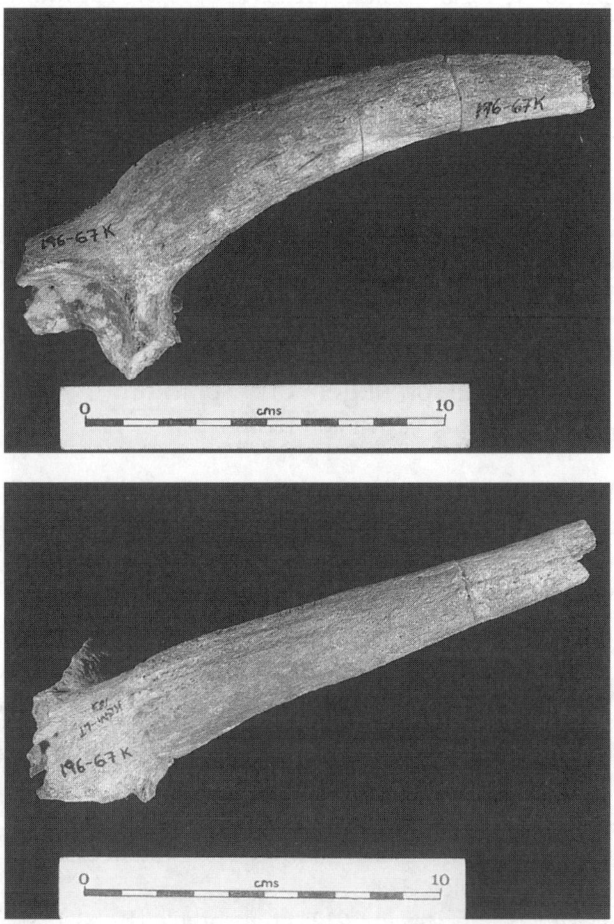

Figure 11.16 *Damalacra* sp. A, left horn core, KNM-LT 183, Lower Nawata: top = lateral view; bottom = anterior view.

Figure 11.17 *Damalacra* sp. B, proximal left horn core, KNM-LT 456, Lower Nawata: top = lateral view; bottom = anterior view.

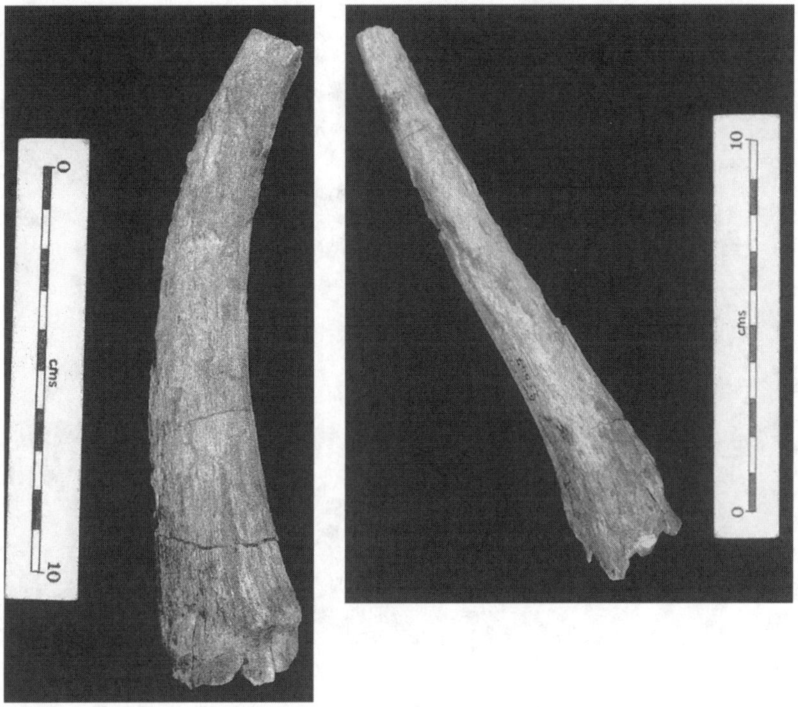

Figure 11.18 *Damalacra* sp. B, right horn core, KNM-LT 23647, Apak Member: left = medial view; right = anterior view.

situated at a lower level than the articular surface, the lateral articular facet is anteroposteriorly long, and noticeably concave, and there is a medial rim on the medial facet (Gentry 1985).

Type species

Aepyceros melampus (Lichtenstein, 1812).

Aepyceros premelampus sp. nov.
(Figures 11.5G, 11.19, 11.20; tables 11.29–11.31)

Diagnosis

Anatomically similar to that of the living *A. melampus,* but the cranial vault is proportionately longer, wider, and less steeply angled. The auditory bullae are more inflated, less mediolaterally compressed, and more transversely directed. The horn cores of the males are similar to or larger in size, arise from shorter pedicels, and are located immediately above the rear of the orbit, rather than behind it, as in the living species. As in *A. melampus,* the horn cores are mediolaterally compressed at their base, with a slight keel at the posterolateral edge. The lyrate shape and torsion of the proximal portions of the horn cores are similar to those of the living species, but the distal tips converge instead of extending parallel to each other. The supraorbital fossae are larger, set in a depressed area of the frontals, and separated by an elevated intrafrontal suture.

Holotype

KNM-LT 184, calvaria with horn cores, from the upper member of the Nawata Formation, collected in 1967 by the Harvard University expedition.

Paratype

KNM-LT 23153, partial cranium with horn cores and tooth roots, from the lower member of the Nawata Formation, collected in 1990 by the National Museums of Kenya expedition.

Etymology

The name reflects the presumed relationship of this species to the living impala.

Lothagam Material

- *Lower Nawata:* 185, Lt. horn core and proximal Rt. horn core; 210, frontlet and proximal horn cores; 218, Lt. horn core; 219, Rt. horn core; 228, Lt. M_1; 232, Rt. M_3; 457, proximal Rt. horn core; 465, proximal Rt. horn core; 468, proximal Lt. horn core; 472,

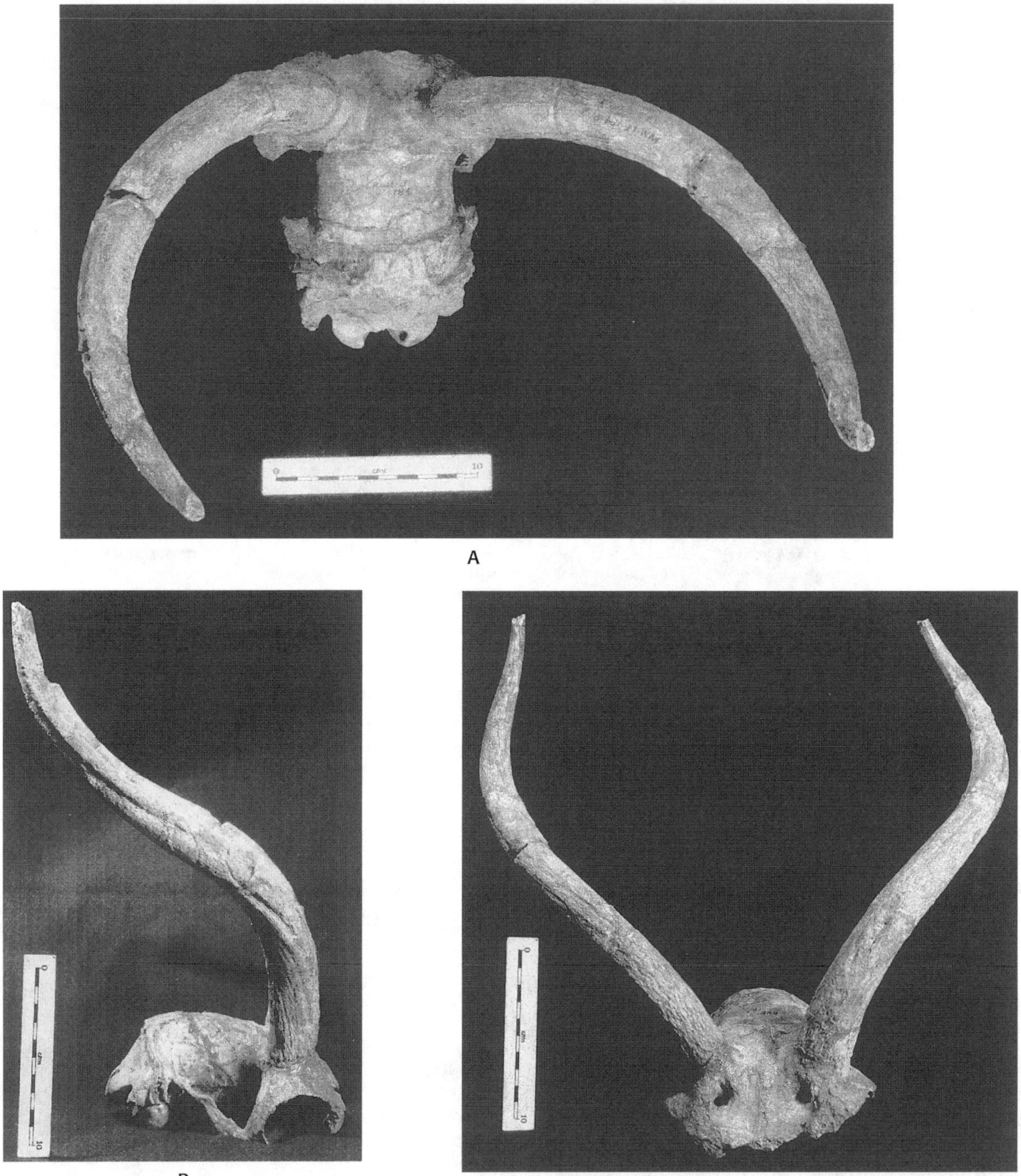

Figure 11.19 *Aepyceros premelampus* sp. nov. holotype, frontlet with right and left horn cores, KNM-LT 184, Upper Nawata: A = dorsal view; B = right lateral view; C = anterior view.

proximal Rt. horn core; 477, Rt. horn core fragment; 481, Rt. horn core fragments; 492, Lt. horn core; 495, proximal Lt. horn core; 502, proximal Rt. horn core; 515, proximal Lt. horn core; 22965, skull, horn core fragments, proximal Rt. horn core; 22977, horn core; 22983, proximal Lt. horn core; 22991, Lt. horn core; 22997, skull and Rt. horn core fragments; 23096, proximal Rt. horn core, cranial fragments; 23128, proximal Rt. horn core; 23132, proximal Rt. horn core; 23133, proximal Lt. horn core and horn core fragments; 23134, Lt. horn core; 23135, Rt. and Lt. horn core fragments; 23153, paratype (see above); 23573, Rt. horn core fragments; 23575, Rt. horn core; 23576, proximal Rt. horn core; 23578, proximal Lt. horn core; 23580, Rt. and Lt. horn cores; 23581, Rt. and Lt. horn core fragments; 23582, Rt. and Lt. horn

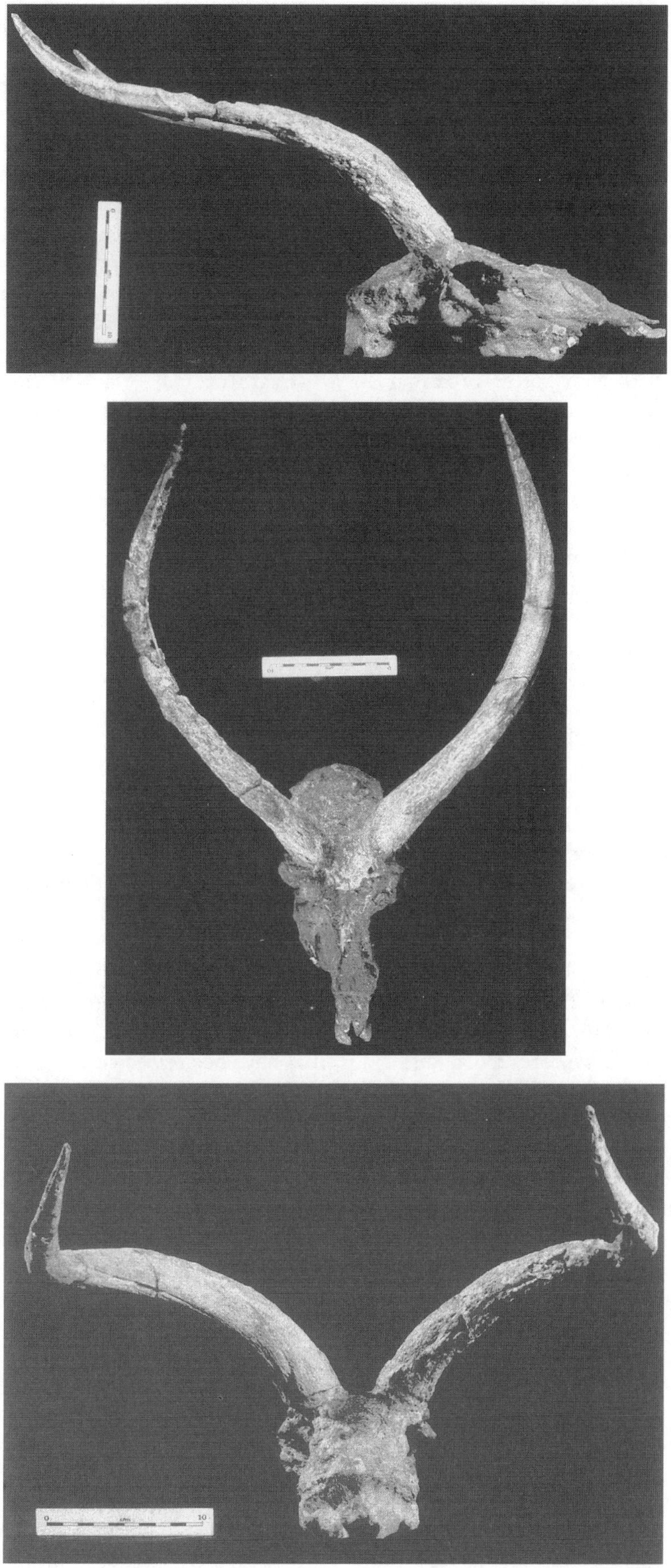

Figure 11.20 *Aepyceros premelampus* sp. nov. paratype, partial cranium with horn cores, KNM-LT 23153, Lower Nawata: top = right lateral view; center = dorsal view; bottom = posterior view.

cores; 23600, Rt. horn core fragments; 23607, proximal Lt. horn core; 23609, Rt. horn core; 23645, Lt. horn core; 23654, proximal Lt. horn core; 23701, proximal Rt. horn core; 23705, Rt. horn core; 23706, proximal Rt. and Lt. horn cores; 23710, proximal Rt. horn core; 23712, Rt. horn core; 23728, proximal Rt. horn core; 25963, Lt. horn core fragments; 26001, Rt. horn core fragment; 26002, distal Rt. horn core fragment; 26008, proximal Rt. horn core; 26009, proximal Rt. horn core; 26010, proximal Lt. horn core; 26011, Rt. horn core; 26012, proximal Lt. horn core and Lt. M_3; 26023, horn core fragments and Mc; 26024, proximal Lt. horn core; 26028, proximal Rt. horn core; 26031, proximal Lt. horn core; 26047, proximal Rt. horn core; 26051, proximal Lt. horn core; 26054, proximal Lt. horn core; 26056, proximal Rt. horn core; 26058, proximal Lt. horn core; 26062, proximal Lt. horn core; 26064, proximal Lt. horn core fragment; 26198, proximal Rt. and Lt. horn core; 26200, Rt. horn core fragment; 26201, Rt. horn core fragment; 26570, Rt. calvaria and horn core; 28729, maxilla (M^{2-3}); 38428, proximal Lt. horn core.

- *Upper Nawata:* 174, Lt. mandible (M_{1-3}); 184, holotype (see above); 187, Lt. horn core and proximal Rt. horn core; 188, calvaria and proximal Lt. horn core; 206, proximal Lt. horn core; 473, Lt. horn core; 510, proximal Rt. horn core; 13018, Rt. mandible (M_1); 13019, Rt. M2; 22990, Rt. mandible (P_2–M_1), Lt. mandible (M_1); 23098, proximal Lt. horn core; 23100, proximal Rt. horn core; 23129, proximal Rt. horn core; 23601, Rt. mandible (M_{1-3}); 23603, proximal Lt. horn core; 23604, Lt. and Rt. horn core fragments; 23605, Lt. and Rt. distal horn core fragments; 23610, Rt. mandible (M_3); 23611, Lt. M_3; 23620, proximal Lt. horn core; 23622, proximal Rt. horn core; 23625, proximal Rt. horn core; 23635, distal Lt. horn core; 23640, proximal Lt. horn core; 23641, proximal Lt. horn core; 23651, Lt. horn core; 23663, distal Lt. horn core fragment; 23670, proximal Rt. horn core; 23673, frontlet with Rt. and Lt. horn cores; 23698, skull and horn core fragments, Lt. horn core fragment; 23700, proximal Rt. horn core; 23702, Rt. and Lt. horn cores; 23721, skull, horn core, proximal humerus fragments, proximal Rt. horn core; 23727, proximal Lt. horn core; 24034, Lt. mandible (broken M_3); 25953, proximal Lt. horn core; 25967, proximal Rt. horn core; 25969, Rt. mandible (M_{1-2}); 25976, proximal Rt. horn core; 25980, proximal Lt. horn core; 25982, proximal Rt. horn core; 25985, proximal horn core, Lt. maxilla fragment (M^{2-3}); 25987, Rt. M_2; 25988, proximal Rt. horn core; 25995, proximal Lt. horn core; 25996, Rt. M_3; 25999, Lt. mandible fragment (M_2); 26027, proximal Rt. horn core; 26033, proximal Lt. horn core; 26036, proximal Rt. horn core; 26039, proximal Lt. horn core; 26057, proximal Lt. horn core; 26061, Lt. mandible (P_3); 26199, Lt. mandible (M_{2-3}); 28658, Lt. horn core fragment; 28752, calvaria and horn cores; 28754, proximal Rt. horn core; 30624, distal horn core.
- *Apak Member:* 173, Lt. mandible (M_{2-3}); 470, proximal Rt. horn core; 13017, Lt. mandible (M_{2-3}); 23630, Lt. mandible (P_{2-3}), and dP_4; 23642, Rt. horn core; 23643, proximal Rt. horn core; 38434, distal Rt. horn core; 23650, proximal Rt. horn core; 23664, Lt. mandible (P_2–M_3); 23668, Rt. mandible (M_3); 23697, Rt. mandible (M_3); 25952, Rt. mandible (dP_3–M_1); 25956, proximal Lt. horn core; 25983, Rt. maxilla fragment (M^{1-3}); 25989, proximal Lt. horn core; 25992, proximal Lt. horn core; 26067, proximal Lt. horn core; 28717, Rt. maxilla (P^4–M^2); 28719, Rt. mandible (P_3, dP_4, M_1); 28734, Lt. mandible (P_2–M_1).
- *Kaiyumung Member:* 226, proximal Rt. horn core; 236, proximal Lt. horn core; 462, proximal Lt. horn core; 491, proximal Lt. horn core; 23648, juvenile Rt. horn core fragment; 23703, Lt. horn core fragment; 24047, Rt. mandible (P_{3-4}); 25961, Lt. horn core; 25962, upper molars.
- *Horizon indet.:* 186, frontlet and horn cores; 190, partial calvaria and horn core fragments; 215, proximal Rt. horn core; 216, proximal and distal Rt. horn core; 234, Rt. and Lt. horn cores; 533, Rt. mandible fragment (M_{1-3}); 22979, Lt. horn core.

This is by far the most common bovid found at Lothagam, forming nearly half of the total bovid assemblage from the lower and upper members of the Nawata Formation and at least a fifth of the bovids from the Apak and Kaiyumung Members. The majority of specimens are partial horn cores, but there are two partial crania that have been selected as type material.

The most obvious differences from extant impalas—the size and orientation of the braincase and the size and shape of the horn cores—are listed in the diagnosis. Compared to the extant impala, the brain case of *A. premelampus* is relatively flat below the horn cores and makes an angle of about 120° with the medial/sagittal plane of the occiput; in the extant impala, the occiput makes an angle of 130° with the rear of the braincase, which makes an angle of 150° with that part of the braincase behind the horns.

Gentry (1985) described an extinct species of impala (*Aepyceros shungurae*) from the lower part of the Shungura succession, and this species was subsequently documented from the lower parts of the Koobi Fora and Nachukui Formations (Harris 1991; Harris et al. 1988). *Aepyceros shungurae* has horn cores that differ from the extant impala by being shorter, more slender, and less divergent in their lower part and with a less abrupt lyration. Similar differences separate *A. shungurae* from *A.*

premelampus, but the two extinct species share a shorter face, a longer braincase, lower frontals, and converging distal horn cores. The horn cores of *A. premelampus* decrease in mean size through the Lothagam succession, but the closeness of the relationship between it and *A. shungurae* remains a matter for conjecture.

Aepyceros sp. from Lukeino (LU 345; Thomas 1980) has horn cores that are comparable in size and curvature to some of those from Lothagam (e.g., LT 215). Harris (1991) noted that the Lothagam impala may be conspecific with specimens from the lower part of the Nachukui Formation cropping out at West Turkana (Harris et al. 1988:100) and with those from the Mursi Formation that Gentry (1985:183) surmised were conspecific with impala horn cores from the Karmosit Beds in the Lake Baringo Basin. The Lothagam impala material is among the oldest known from Africa, although Gentry (1978, 1990) suggested that specimens from Fort Ternan previously identified as *Gazella* may instead represent an early impala. Both Vrba (1975) and Thomas (1984) suggested that the impalas and alcelaphines diverged from the Caprini. Kaufulu et al. (1981) note the southernmost occurrence of impalas in the Chiwondo Beds sequence of Malawi.

Tribe Antilopini Gray, 1821
Gazella Blainville, 1816

Diagnosis

Horn cores subcircular or elliptical in cross section with some mediolateral compression, the lateral surface often flatter than the medial, fairly uprightly inserted with backward curvature in side view, generally more obliquely set in females than in males of the same species, slightly divergent in anterior view, without keels or torsion; frontals without or almost without internal sinuses, and the area between the horn core bases hardly raised above the level of the orbital rims; moderately large triangular supraorbital pits at the base of the horn core pedicels; ethmoidal fissure present; moderate to large preorbital fossae; premaxillae generally contacting the sides of the nasals, which have shortened with evolution; occipital low with each half facing partly laterally as well as backward; moderate to large auditory bullae; living species with hypsodont teeth but less hypsodont in fossil species; upper molars with moderately prominent styles and little development of the ribs between them; lower molars without goat folds; M_3s often with the rearmost (third) lobe enlarged; exception is the East Asian genus *Procapra,* P_4s are without paraconid-metaconid fusion to form a complete medial wall at the front of the tooth (Gentry and Gentry 1978).

Gazella sp. indet.
(Tables 11.32, 11.33)

Lothagam Material

- *Lower Nawata:* 13008, horn core fragments.
- *Upper Nawata:* 23599, Lt. mandible (P_3–M_3), naviculocuboid, tibia fragment; 23666, distal Rt.? horn core.
- *Apak Member:* 23653, Lt. horn core; 25972, proximal Rt. horn core; 38435, distal Lt. horn core.

Gazelles are represented by a handful of horn cores, a mandible and some postcranial elements. Two distinct horn core shapes are represented. Those specimens with pronounced mediolateral compression (LT 13008, 38435) are strongly reminiscent of *Gazella praethomsoni;* those with less marked compression (LT 23653, 25972) resemble *Gazella janenschi.* None of the specimens is complete, and the sample is too small to permit more precise identification. The horn cores are smaller than those of *Gazella* sp. recorded from Langebaanweg by Gentry (1980).

Raphiceras H. Smith, 1827

Raphiceras campestris (Thunberg, 1811), the steenbok (or steinbuck), is distributed discontinuously in eastern and southern Africa where it is found in open bushed plains or light woodland; 12 races have been described but are not distinct (Spinage 1986). Sharpe's grysbok, *R. sharpei* Thomas, 1897, inhabits the region separating the two steenbok populations and occurs in stony, hilly country (Spinage 1986). The Cape grysbok, *R. melanotis* (Thunberg, 1811), is restricted to the extreme south of the continent where it inhabits scrub-covered flats (Spinage 1986). The inclusion of the steenboks and grysboks in the same genus is more a matter of convention and convenience than an accurate reflection of close affinities (Kingdon 1982).

Diagnosis

Moderate-sized to large neotragines. Horn cores short to moderately long with little mediolateral compression, inserted widely apart above the back of the orbits, parallel to one another, and with a slightly concave front edge in profile. Postcornual fossae present. Supraorbital pits wide apart, back of braincase roof not very strongly turned down, temporal lines wide posteriorly on cranial roof, preorbital fossa moderately sized to large, premaxillae wide and rising to contact nasals, auditory bullae inflated, median indentation at back of palate level with or forward of lateral ones, palatal ridges on maxilla anterior to tooth row approach one another closely. Upper molars with quite small styles, central cavities of lower molars disappear early in wear, medial walls of lower molars fairly flat and metastylids not strong, M_3s with moderate to large back lobes. Metaconid of P_4 passes transversely then backward, front of lateral wall of P_4 bends round into transverse plane; P_2 not greatly reduced (Gentry 1980).

Type species

Raphiceras campestris (Thunberg, 1811).

Raphiceras sp.
(Table 11.34)

Lothagam Material

- *Lower Nawata:* 503, Rt. mandible fragment (P_4–M_1).
- *Apak Member:* 38433, Lt. mandible (M_3).
- *Kaiyumung Member:* 14125, Lt. mandible fragments (M_{1-2}); 26048, Rt. mandible (P_{1-4}).

Two neotragine mandible fragments (LT 503 and 14125) are larger than those of *Madoqua* species and also of *Raphiceras campestris* but of similar morphology. Both are larger than MP 129 from Mpesida, identified as *Gazella* aff. *vanhoepeni* or ?*Raphiceras* sp. by Thomas (1980) but smaller than *Raphiceras paralius* from Langebaanweg (Gentry 1980).

Madoqua Ogilby, 1837

Diagnosis

Small neotragines. Horn cores short, compressed anterolaterally to posteromedially, with some flattening of the posteromedial surface, with a tendency to keels, inserted widely apart above the back of the orbits, parallel to one another and not very upright in side view. Postcornual fossa small and shallow. Supraorbital pits small and wide apart. Temporal lines wide apart posteriorly on the cranial roof. Nasals shortened, ethmoidal fissure triangularly shaped, preorbital fossa wide and deep, premaxilla often contacting lachrymal as well as nasal or reduced and perhaps contacting neither. Infraorbital foramen above P^3, medial indentation at back of palate passing forward of lateral ones, mastoids small to moderate sized, auditory bullae large and inflated, basioccipital triangular. M^3s reduced posteriorly, central cavities of lower molars disappear early in wear, lower molars with fairly flat medial walls and weak metastylids, lateral lobes of lower molars not elongated transversely, M_3s with back lobes reduced or absent. Premolar row long but P_2 small relative to P_3 and P_4. P^2 and P^3 long relative to P^4. The metaconid of P_4 passes transversely and then backward. Diastema short and jaw angle projecting posteriorly (Gentry 1987).

Madoqua sp.
(Table 11.35)

Lothagam Material

- *Lower Nawata:* 28733, Rt. M^3, Rt. M/.
- *Upper Nawata:* 177, Rt. mandible fragment (M_{1-2}); 22981, Lt. mandible (P_3–M_2); 23612, Lt. mandible (P_3, M_1).
- *Apak Member:* 26066, Rt. mandible (M_{1-2}).
- *Horizon indet.:* 176, Lt. mandible fragment (P_4–M_3).

Half a dozen incomplete dentitions represent an indeterminate species of dikdik.

Discussion

Discounting the single specimen of *Tragelaphus nakuae,* whose age is uncertain, some 17 species of bovids representing nine tribes have been recovered from the Lothagam succession. Impalas are by far the most abundant, but they decline in numbers progressively through the succession from nearly 50 percent of the bovid assemblage in the Lower Nawata to about 30 percent in the Apak and Kaiyumung Members. Boselaphines decline more precipitously, comprising about 25 percent of the Lower Nawata bovid assemblage but only 8 and 5 percent in the Upper Nawata and Apak Member, respectively. Although present throughout the sequence, the reduncines, hippotragines, and alcelaphines reach their maximum abundance in the Upper Nawata. Tragelaphines and bovines were present in the Nawata Formation but are much more abundant in the Apak and Kaiyumung Members. If we draw on the habitat preferences of their nearest living relatives, it seems likely that there was an admixture of forest and open grassland throughout the sequence. The representation of the individual tribes would be consistent with the Upper Nawata being more open than the Lower Nawata or Apak Member habitats, and with the Apak Member environment being somewhat drier than that of the Lower Nawata.

The mixture of wooded and grassland habitats appears supported from isotopic analysis of Lothagam bovid enamel. Initially, analyses were undertaken on fragments of bovid teeth that were too incomplete to identify to tribal level (Cerling, Harris, and Leakey this volume:chapter 12.2), but a few samples from identified specimens were processed subsequently. The results indicate that bovids exploited C_4 grasses throughout the sequence. Teeth identified as reduncine (Lower and Upper Nawata, Kaiyumung), alcelaphine (Upper Nawata), and boselaphine (Upper Nawata) gave results indicative of a nearly pure C_4 diet, whereas two impala teeth (Upper Nawata, Kaiyumung) indicate a mainly C_3 diet. A tragelaphine from the Apak Member fed on C_3 browse, but a tooth identified as tragelaphine from the Upper Nawata was from a C_4 grazer.

Acknowledgments

I thank the government of Kenya and the museum trustees of the National Museums of Kenya for permission to study the Lothagam bovid material. I thank also the curatorial and preparation staff of the palaeontology division of the National Museum of Kenya, Nairobi, for making the material available for study. Discussions with Meave Leakey, Nikos Solounias, and Elisabeth Vrba on various aspects of the manuscript are gratefully acknowledged. The photographs were taken in Nairobi by Stephen Mutaba; technical assistance in Los Angeles was provided by Dick Meier and Pete Mueller.

References Cited

Arambourg, C. 1941. Antelopes nouvelles du Pléistocène ancien de l'Omo (Abysinnie). *Bulletin du Muséum National d'Histoire Naturelle* (Paris) 13:339–347.

Arambourg, C. 1947. Contribution à l'étude géologique et paléontologique du bassin de lac Rudolphe et de la Basse Vallée de l'Omo. In *Mission scientifique de l'Omo (1932–1933).* Vol. 1, fasc. 3. *Paléontologie,* pp. 231–562. Paris: Muséum National d'Histoire Naturelle.

Bouvrain, G. 1988. Les *Tragoportax* (Bovidae, Mammalia) des gisements du Miocène supérieur de Ditiko (Macedoine, Grèce). *Annales de Paléontologie* 74:43–63.

Gentry, A. W. 1970. The Bovidae of the Fort Ternan fossil fauna. In L. S. B. Leakey and R. J. G. Savage, eds., *Fossil Vertebrates of Africa,* vol. 2, pp. 243–323. London: Academic Press.

Gentry, A. W. 1974. A new genus and species of Pliocene boselaphine (Bovidae, Mammalia) from South Africa. *Annals of the South African Museum* 65:145–188.

Gentry, A. W. 1976. Bovidae of the Omo Group deposits. In Y. Coppens, F. C. Howell, G. L. Isaac, and R. E. Leakey, eds., *Earliest Man and Environments in the Lake Rudolf Basin: Stratigraphy, Paleoecology, and Evolution,* pp. 177–192. Chicago: University of Chicago Press.

Gentry, A. W. 1978. Bovidae. In V. J. Maglio and H. B. S. Cooke, eds., *Evolution of African Mammals,* pp. 540–572. Cambridge, Mass.: Harvard University Press.

Gentry, A. W. 1980. Fossil Bovidae from Langebaanweg, South Africa. *Annals of the South African Museum* 79:213–337.

Gentry, A. W. 1981. Notes on Bovidae (Mammalia) from the Hadar Formation and from Amado and Geraru, Ethiopia. *Kirtlandia* no. 33:1–33.

Gentry, A. W. 1985. The Bovidae of the Omo Group deposits, Ethiopia. In Y. Coppens and F. C. Howell, eds., *Les faunes Plio-Pléistocènes de la Basse Vallée de l'Omo (Ethiopie).* Vol. 1. *Perissodactyles, Artiodactyles (Bovidae),* pp. 119–191. Paris: Centre National de la Recherche Scientifique.

Gentry, A. W. 1987. Fossil Bovidae from Laetoli. In M. D. Leakey and J. M. Harris, eds., *Laetoli: A Pliocene Site in Northern Tanzania,* pp. 378–408. Oxford: Clarendon Press.

Gentry, A. W. 1990. Evolution and dispersal of African Bovidae. In G. A. Bubenik and A. B. Bubenik, eds., *Horns, Pronghorns, and Antlers: Evolution, Morphology, Physiology, and Social Significance,* pp. 195–227. New York: Springer-Verlag.

Gentry, A. W. 1997. Fossil ruminants (Mammalia) from the Manonga Valley, Tanzania. In T. Harrison, ed., *Neogene Paleontology of the Manonga Valley, Tanzania: A Window into the Evolutionary History of East Africa,* pp. 107–135. New York: Plenum Press.

Gentry, A. W. 1999. Fossil pecorans from the Baynunah Formation, Emirate of Abu Dhabi, United Arab Emirates. In P. J. Whybrow and A. Hill, eds., *Fossil Vertebrates of Arabia,* pp. 290–316. New Haven: Yale University Press.

Gentry, A. W., and A. Gentry. 1978. Fossil Bovidae of Olduvai Gorge, Tanzania. *Bulletin of the British Museum (Natural History)* 29:289–446.

Hamilton, W. R. 1973. The Lower Miocene ruminants of Gebel Zelten, Libya. *Bulletin of the British Museum (Natural History), Geology* 21:76–150.

Harris, J. M. 1991. Family Bovidae. In J. M. Harris, ed., *Koobi Fora Research Project.* Vol. 3. *The Fossil Ungulates: Geology, Fossil Artiodactyls, and Palaeoenvironments,* pp. 139–320. Oxford: Clarendon Press.

Harris, J. M., F. H. Brown, and M. G. Leakey. 1988. Stratigraphy and paleontology of Pliocene and Pleistocene localities west of Lake Turkana, Kenya. *Contributions in Science* 399:1–128.

Janis, C. M., and K. M. Scott. 1987. The interrelationships of higher ruminant families with special emphasis on the Cervoidea. *American Museum Novitates* 2893:1–85.

Kaufulu, Z., E. S. Vrba, and T. D. White. 1981. Age of the Chiwondo Beds, northern Malawi. *Annals of the Transvaal Museum* 33:1–8.

Kingdon, J. 1982. *East African Mammals.* Vol. 3, pts. C, D. *Bovids,* pp. 1–393, 394–746. London: Academic Press.

Lehmann, U., and H. Thomas. 1987. Fossil Bovidae (Mammalia) from the Mio-Pliocene of Sahabi. In N. T. Boaz, A. El-Arnauti, A. W. Gaziry, J. de Heinzelin, and D. D. Boaz, eds., *Neogene Paleontology and Geology of Sahabi,* pp. 323–335. New York: Liss.

Moyà-Solà, S. 1983. *Los Boselaphini (Bovidae, Mammalia) del Neogeno de la península Iberica.* Publicaciones de Geológia 18. Barcelona: Universidad Autonoma de Barcelona.

Smart, C. 1976. The Lothagam 1 fauna: Its phylogenetic, ecological and biogeographic significance. In Y. Coppens, F. C. Howell, G. L. Isaac, and R. E. Leakey, eds., *Earliest Man and*

Environments in the Lake Rudolf Basin: Stratigraphy, Paleoecology, and Evolution, pp. 361–369. Chicago: University of Chicago Press.

Solounias, N. 1981. *The Turolian Fauna from the Island of Samos, Greece.* Contributions to Vertebrate Evolution 6. Basel: Karger.

Spinage, C. A. 1986. *The Natural History of Antelopes.* New York: Facts on File.

Thomas, H. 1980. Les bovidés du Miocène supérieur des couches de Mpesida et de la formation de Lukeino (district de Baringo, Kenya). In R. E. F. Leakey and B. A. Ogot, eds., *Proceedings of the Eighth Pan-African Congress of Prehistory and Quaternary Studies, Nairobi 1977,* pp. 82–91. Nairobi: National Museums of Kenya.

Thomas, H. 1981. Les bovidés Miocènes de la formation de Ngorora du bassin de Baringo (Rift Valley, Kenya). *Proceedings of the Koninjlijke Nederlandsche Akademie van Wettenschappen,* ser. B, 84:335–410.

Thomas, H. 1984. Les origines africaines des Bovidae (Artiodactyla, Mammalia) Miocènes des lignites de Grosseto (Toscane, Italie). *Bulletin du Muséum National d'Histoire Naturelle* (Paris), 4th ser., 6:81–101.

Vrba, E. S. 1975. Some evidence of chronology and paleoecology of Sterkfontein, Swartkrans and Kromdraai from the fossil Bovidae. *Nature* 254:301–304.

Vrba, E. S. 1985. African Bovidae: Evolutionary events since the Miocene. *South African Journal of Science* 81:263–266.

Vrba, E. S. 1987. A revision of the Bovini (Bovidae) and a preliminary revised checklist of Bovidae from Makapansgat. *Palaeontologia Africana* 26:33–46.

Vrba, E. S. 1995. The fossil record of African antelopes (Mammalia, Bovidae) in relation to human evolution and climate. In E. S. Vrba, G. H. Denton, T. C. Partridge, and L. H. Burkle, eds., *Paleoclimate and Evolution, with Emphasis on Human Origins,* pp. 385–424. New Haven: Yale University Press.

Vrba, E. S. 1997. New fossils of Alcelaphini and Caprinae (Bovidae, Mammalia) from Awash, Ethiopia, and phylogenetic analysis of Alcelaphini. *Palaeontologia Africana* 34:127–198.

Vrba, E. S., and J. Gatesy. 1994. New antelope fossils from Awash (Ethiopia), and phylogenetic analysis of Hippotragini (Bovidae, Mammalia). *Palaeontologia Africana* 31:55–72.

Wells, L. H., and H. B. S. Cooke. 1956. Fossil Bovidae from the Limeworks Quarry, Makapansgat, Potgeitersrus. *Palaeontologia Africana* 4:1–55.

Table Abbreviations

+ = specimen incomplete
ant = anterior
ap = anteroposterior
* = approximate
cont. = continued
h/c = horn core
Kaiy = Kaiyumung
L = left (for tooth row)
Lr = lower
lt = length
Lt. = left
max = maximum
min = minimum
Naw = Nawata
occ = occipital
post = posterior
premax = premaxilla
prox = proximal
R = right (for tooth row)
Rt. = right
tr = transverse
Ur = upper

TABLE 11.8 Horn Core Measurements (in mm) of *Tragelaphus kyaloae*

	Rt. H/c		Lt. H/c	
	Prox ap	**Prox tr**	**Prox ap**	**Prox tr**
LT 467	41.15	58.01	—	—
LT 23102	42.04	46.89	41.61	—
LT 23617	—	—	36.7	48.28
LT 23722	32.4	46.32	—	—
LT 25979	34.7	45.03	—	—
LT 26202	34.87	44.67	—	—

TABLE 11.9 Upper Dentition Measurements (in mm) of Tragelaphini gen. and sp. indet.

	LT 23652 L	**LT 26065 R**	**LT 25964 L**	**LT 23613**
P^3 ap	—	—	—	9.17
tr	—	—	—	9.26
P^4 ap	11.1	—	—	10.02
tr	13.55	—	—	12.56
M^1 ap	16.18	—	—	15.72
tr	16.41	—	—	15.92
M^2 ap	19.57	15.42	24.35	16.51
tr	19.33	19.58	21.9	17.41
M^3 ap	19.55	14.15	—	20.63
tr	18.09	—	—	17.57

TABLE 11.10 Lower Dentition Measurements (in mm) of Tragelaphini gen. and sp. indet.

	LT 223 L	LT 471 R	LT 13020 L	LT 23718 L	LT 25955 R	LT 25991 L
M_1 ap	—	—	—	—	—	15.48
tr	—	—	—	—	—	10.24
M_2 ap	—	—	19.43	18.46	—	20.34
tr	—	—	10.84	12.34	—	10.58
M_3 ap	24.74	26.25	—	27.3	27.36	—
tr	11.14	10.09	—	13.02	9.55	—

	LT 182 R	LT 26022 R	LT 28777 R
P_2 ap	10.66	—	—
tr	5.59	—	—
P_3 ap	13.78	—	—
tr	7.74	—	—
P_4 ap	14.24	—	—
tr	8.24	—	—
M_1 ap	16.62	—	18.02
tr	10.58	—	12.38
M_2 ap	19.78	—	18.92
tr	11.77	—	12.73
M_3 ap	23.55+	25.77	—

TABLE 11.11 Dentition Measurements (in mm) of Bovini gen. and sp. indet.

	Lr Naw LT 476 L	Lr Naw LT 480 R	Apak LT 475 L	Apak LT 475 R	Apak LT 23608 L	Apak LT 30648 L	Kaiy LT 505 L
M^1 ap	—	—	—	26.47	—	—	—
tr	—	—	—	19.87	—	—	—
M^2 ap	26.52	28.45	29.61	29.91	27.89	29.97	30.54
tr	24.36	26.83	25.85	20.31	23.95	31.53	27.18
M^3 ap	—	—	—	—	27.99	—	—
tr	—	—	—	—	18.26	—	—

	LT 192 R	Lr Naw LT 458 L	Lr Naw LT 476 R	Ur Naw LT 519 L	Apak LT 23619 L	Kaiy LT 13006 R
M_1 ap	22.46	—	27.36	—	—	—
tr	15.84	—	17.52	—	—	—
M_2 ap	26.63	31.27	—	—	26.18	31.83
tr	14.73	21.59	—	—	17.81	16.77
M_3 ap	—	—	—	37.07	37.6	43.37
tr	—	—	—	18.04	16.47	15.8

	Apak LT 24002 R	Apak LT 25993 R	Kaiy LT 26046 L
P_3 ap	21.9	—	—
tr	13.0	—	—
P_4 ap	23.6	—	—
tr	15.0	—	—
M_1 ap	24.1	—	—
tr	16.8	—	—
M_2 ap	28.2	26.77	—
tr	17.79	16.79	16.52
M_3 ap	36.62	—	—
tr	15.96	—	16.43

TABLE 11.12 Horn Core Measurements (in mm) of *Tragoportax* aff. *T. cyrenaicus*

	Rt. H/c		Lt. H/c	
	Prox ap	Prox tr	Prox ap	Prox tr
LT 213	—	34.76	—	—
LT 498	—	31.26	—	—
LT 22995	—	—	47.88	28.11
LT 23149	74.2	42.79	67.12	37.1
LT 23662	46.49+	34.35	—	—
LT 23980	—	29.3	—	—
LT 24021	57.31	35.33	—	—
LT 26025	—	24.33	—	—
LT 26042	—	—	47.09	28.84
LT 26043	—	—	50.69	32.59

TABLE 11.13 Horn Core Measurements (in mm) of *Tragoportax* sp. A

	Rt. H/c			Lt. H/c		
	Prox ap	Prox tr	Lt	Prox ap	Prox tr	Lt
LT 214	44.14	27.18	—	—	—	—
LT 22978	—	—	—	30.97	17.68	86.98
LT 23656	—	—	—	38.65	23.69	132.37
LT 23820	38.61	23.62	—	—	—	—
LT 23621	40.84	24.15	—	37.09	23.68	—
LT 24214	36.36	19.23	126.23	—	—	—
LT 24293	38.77	24	—	—	—	—

TABLE 11.14 Horn Core Measurements (in mm) of *Tragoportax* sp. B

	Rt. H/c			Lt. H/c		
	Prox ap	Prox tr	Lt	Prox ap	Prox tr	Lt
LT 195	36.41	26.34	—	37.73	24.92	—
LT 196	34.4	—	—	35.17	23.72	—
LT 26197	38.25	24.7	—	—	—	—

TABLE 11.15 Upper Dentition Measurements (in mm) of Boselaphini gen. and sp. indet.

	LT 220 L	LT 220 R	LT221 R	LT 13012 L	LT 13014 R	LT 13015 L	LT 26005 L
P^4 ap	—	—	—	9.96	—	—	—
tr	—	—	—	11.45	—	—	—
M^1 ap	—	—	—	14.46	—	—	—
tr	—	—	—	16.42	—	—	—
M^2 ap	17.64	17.36	17.59	18.2	—	—	—
tr	17.24	17.55	17.3	15.44	—	—	—
M^3 ap	19.3	19.46	19.83	19.89	22.78	21.43	17.88
tr	15.89	17.6	18.24	16.03	17.99	11.83	—

	LT 13016 L	LT 22875 R	LT 23068 L	LT 23574 L	LT 23574 R	LT 23716
P^3 ap	—	—	9.51	—	—	—
tr	—	—	8.13	—	—	—
P^4 ap	—	—	9.42	—	—	—
tr	—	—	8.8	—	—	—
M^1 ap	—	—	14.32	13.19	—	—
tr	—	15.4	11.79+	11.14	—	—
M^2 ap	—	18.54	—	17.97	17.36	—
tr	—	15.98	—	14.27	16.52	—
M^3 ap	18.16	21.14	—	—	—	20.75
tr	17.97	19.15	—	—	—	20.17

	LT 26021	LT 26055 R	LT 26070 R	LT 26572 R	LT 28574 L
M^2 ap	—	18.4	21.99	19.49	20.1
tr	19.46	17.47	22.02	18.73	16.22

	LT 28574 R
M^3 ap	20.34
tr	16.26

TABLE 11.16 Lower Dentition Measurements (in mm) of Boselaphini gen. and sp. indet.

	LT 193 R	LT 194 L	LT 201 R	LT 202	LT 203 R
P_3 ap	—	—	—	9.75	10.63
tr	—	—	—	7.25	5.48
P_4 ap	—	—	—	11.47	11.44
tr	—	—	—	8.17	6.21
M_1 ap	11.76	—	—	—	13.73
tr	9.27	—	—	11.39	9.53
M_2 ap	15.24	16.67	14.47	—	17.46
tr	11.22	12.23	10.64	—	10.43
M_3 ap	22.92	26.32	25.51	—	24.29
tr	10.09	12.43	10.88	—	10.8

	LT 204 R	LT 209 L	LT 211 R	LT 523 L	LT 531 L	LT 532 R	LT 13011 L
P_2 ap	7.94	—	—	—	—	—	—
tr	5.09	—	—	—	—	—	—
P_3 ap	9.78	—	—	—	—	—	—
tr	6.67	—	—	—	—	—	—
P_4 ap	11.66	—	—	—	11.34	—	11.73
tr	7.55	—	—	—	6.39	—	6.98
M_1 ap	11.69	15.59	—	13.41	—	—	—
tr	9.73	7.86	—	9.55	—	—	9.21
M_2 ap	15.88	—	—	—	—	15.34	—
tr	10.72	—	—	—	—	11.33	—
M_3 ap	24.06	—	—	—	—	—	—
tr	11.44	—	9.86	—	—	—	—

	LT 22875 L	LT 23147 R	LT 23657 R	LT 23694 L	LT 24019 L	LT 26572 R	LT 38429 L
P_3 ap	—	—	—	10.22	—	—	—
tr	—	—	—	4.98	—	—	—
P_4 ap	—	10.75	—	11.34	12.61	—	—
tr	—	7.08	—	7.45	6.47	—	—
M_1 ap	—	13.8	—	14.71	—	12.58	—
tr	—	9.57	—	9.65	—	10.26	—
M_2 ap	—	14.95	—	19.32	—	—	—
tr	—	11.31	—	11.49	—	—	—
M_3 ap	26.39	25.88	21.71	26.88	—	—	23.36
tr	12.31	10.58	8.81	12.36	—	—	9.6

TABLE 11.17 Cranial Measurements (in mm) of *Kobus presigmoidalis* sp. nov.

	LT 189	LT 212	KNM-ER 4552
Intercornual width	(18)	(23)	(15)
Max width h/cs	(82)	(86)	(99)
Length h/c to nuchal crest	79	—	—
Mastoid width	80	—	99
Height basioccipital to nuchal crest	54	—	58
Width occ condyles	56	—	(59)
Width post tuberosities	28	—	34
Width ant tuberosities	30	—	32

TABLE 11.18 Horn Core Measurements (in mm) of *Kobus presigmoidalis* sp. nov.

	Rt. H/c		Lt. H/c	
	Prox ap	Prox tr	Prox ap	Prox tr
LT 189	40.93	29.77	41.1	29.34
LT 212	—	—	40.97	30.52
LT 224	36.57	29.75	—	—
LT 233	—	—	37.5	29.58
LT 466	—	—	37.33	28.68
LT 483	32.28	22.68	—	—
LT 486	—	—	37.24	29.04
LT 496	—	—	35.1	24.84
LT 497	30.71	—	—	—
LT 13009	39.86	28.08	—	—
LT 22884	35.24	27.2	—	—
LT 23615	—	—	32.28	22.02
LT 23661	—	—	39.56	28.7
LT 23671	—	—	34.49	25.73
LT 25951	40.51	32.23	—	—
LT 25984	—	—	37.09	26.48
LT 25997	—	—	40.65	26.88
LT 30650	—	—	39.09	30.62

TABLE 11.19 Cranial Measurements (in mm) of *Kobus laticornis* sp. nov.

	LT 119	LT 180	LT 198	LT 200	LT 227
Intercornual width	(44)	52	30	(47)	(150)
External width h/cs	113	116	96	116	(120)
Length h/c to nuchal crest	—	66	—	—	—
Mastoid width	(102)	112	—	—	—
Height basioccipital to nuchal crest	57	59	—	—	—
Width occ condyles	62	66	—	—	—
Width post tuberosities	37	41	—	39	—
Width ant tuberosities	—	25	—	21	—

TABLE 11.20 Horn Core Measurements (in mm) of *Kobus laticornis* sp. nov.

	Rt. H/c		Lt. H/c	
	Prox ap	Prox tr	Prox ap	Prox tr
LT 180	37.9+	31.88	42.89	34.66
LT 198	43.94	33.55	43.13	35.2
LT 199	38.9	29.15	39.48	29.99
LT 200	38.81	32.92	42.34	31.1
LT 227	—	—	44.42	28.69
LT 485	38.02	27.2	39.39	30.05
LT 14129	41.68	32.34	—	—
LT 14130	—	—	40.77	30.23
LT 22969	42.21	27.91	44.39	32.13
LT 23631	—	—	41.6	34.87
LT 23704	27.81	21.0	—	—

TABLE 11.21 Horn Core Measurements (in mm) of *Menelikia leakeyi*

	Rt. H/c			Lt. H/c	
	Prox ap	Prox tr	Lt	Prox ap	Prox tr
LT 00178	31.52	26.66	150.14+	—	—
LT 00487	—	—	—	34.06	26.95
LT 00490	30.96	23.9+	—	—	—
LT 23715	—	—	—	31.74	23.14+
LT 23726	—	—	—	35.44	29.13
LT 25966	34.72	25.95	—	—	—
LT 25978	—	—	—	36.96	28.65
LT 26026	37.01	30.82	—	—	—

TABLE 11.22 Dentition Measurements (in mm) of Reduncini gen. and sp. indet.

	LT 181	LT 205 L	LT 207 L	LT 534 R	LT 13007 L
P_3 ap	13.33	—	—	—	9.63
tr	7.52	—	—	—	6.37
P_4 ap	17.07	—	12.13	13.0	12.13
tr	9.19	—	8.22	7.24	7.72
M_1 ap	17.18	—	12.36	—	—
tr	11.19	—	9.26	—	—
M_2 ap	17.72	17.68	—	—	—
tr	12.08	11.57	—	—	—
M_3 ap	—	26.29	—	—	—
tr	—	11.08	—	—	—

	LT 20671 R	LT 20671 L	LT 23579 L	LT 23579 R	LT 23624 R	LT 23655 L
P_4 ap	—	—	—	12.91	—	—
tr	—	—	—	8.15	—	—
M_1 ap	—	—	—	—	13.4	—
tr	—	—	—	—	10.42	—
M_2 ap	19.16	18.27	19.44	18.21	18.39	14.85
tr	11.87	12.03	11.85	12.11	13.3	10.94
M_3 ap	27.91	28.12	28.52	27.08	27.8	23.28
tr	11.52	11.67	12.38	12.14	11.43	11.44

	LT 23657 L	LT 23714 L	LT 25994 R	LT 26007 R	LT 26052 R	LT 26068 L
P_2 ap	—	—	—	5.71	—	—
tr	—	—	—	3.86	—	—
P_3 ap	11.09	—	—	9.32	—	—
tr	7.07	—	—	6.02	—	—
P_4 ap	—	11.51	—	11.04	—	16.62
tr	—	8.08	—	7.48	—	8.62
M_1 ap	—	—	—	—	—	15.37
tr	—	9.06	—	—	—	11.72
M_3 ap	—	—	20.84	—	27.77	—
tr	—	—	9.44	—	13	—

	LT 26069 L	LT 28747	LT 28648
P_3 ap	—	—	22.61
tr	—	—	13.78
M^1 ap	—	13.9	—
tr	—	9.07	—
M_2 ap	—	17.49	—
tr	—	10.22	—
M_3 ap	27.04	—	—
tr	12.54	—	—

TABLE 11.23 Horn Core Measurements (in mm) of *Praedamalis?* sp.

	Rt. H/c		Lt. H/c	
	Prox ap	**Prox tr**	**Prox ap**	**Prox tr**
LT 23681	36.81	—	—	—
LT 23709	—	—	43.43	29.71
LT 23720	—	—	—	35.12
LT 24022	—	—	—	20.02
LT 24061	41.21	32.16	—	—
LT 25968	37.05	28.63	—	—

TABLE 11.24 Horn Core Measurements (in mm) of *Hippotragus* sp.

	Rt. H/c		Lt. H/c	
	Prox ap	**Prox tr**	**Prox ap**	**Prox tr**
LT 13010	39.58	30.07	—	—
LT 23131	44.81	33.42	—	—
LT 23598	45.26	36.57	—	—
LT 26599	—	—	48.32	37.51
LT 26037	—	25.9	—	—

TABLE 11.25 Dentition Measurements (in mm) of Hippotragini gen. and sp. indet.

	LT 507 L	**LT 23130 L**	**LT 23691 R**	**LT 25452 L**
M^2 ap	21.3	19.6	20.37	—
tr	15.34	15.69	—	—
M^3 ap	—	—	—	27.1
tr	—	—	—	15.93

	LT 450 R	**LT 562 L**	**LT 25432 L**	**LT 26186**	**LT 25960 R**
M_1 ap	15.31	13.05	—	—	—
tr	9.9	9.59	—	—	—
M_2 ap	—	21.3	—	20.89	—
tr	—	10.77	—	9.15	—
M_3 ap	—	—	—	25.65	25.13
tr	—	11.47	11.02	8.56	10.07

TABLE 11.26 Horn Core Measurements (in mm) of *Damalacra* sp. A

	Rt. H/c			Lt. H/c		
	Prox ap	Prox tr	Lt	Prox ap	Prox tr	Lt
LT 00179	—	—	—	30.87	23.91+	173+
LT 00183	—	—	—	34.11	26.29	—
LT 00225	33.16	25.99	—	—	—	—
LT 00231	36.34	28.59	—	—	—	—
LT 00237	35.08	30.43	206+	—	—	—
LT 00238	30.72	23.86	180+	—	—	—
LT 00463	30.31	24.8	—	—	—	—
LT 00482	—	—	—	34.14	28.47	—
LT 00493	—	—	—	37.5	27.07	214+
LT 00504	38.58+	26.77+	—	—	—	—
LT 00506	—	—	—	37.45	27.8	206+
LT 00514	34.44	26.46	—	—	—	—
LT 23146	31.95	23.55	—	—	25.03	—
LT 23649	—	—	—	—	—	—
LT 23729	32.23	29.51	—	—	—	—
LT 23736	34.74	26.14	—	—	—	—
LT 24049	—	—	—	30.65	29.56	—
LT 25963	—	—	—	—	—	—
LT 25975	33.18	26.97	—	—	—	—
LT 25977	38.99	29.92+	—	—	—	—
LT 25986	—	—	—	37.85	29.62	—
LT 38432	31.95	25.97	—	—	—	—
LT 26000	34.47	29.23	—	—	—	—
LT 26002	—	—	—	—	—	—
LT 26014	—	—	—	37.36	30.15	—
LT 26015	33.35	27.86	—	33.57	27.92	—
LT 26016	33.67	29.91	—	31.99	26.92	—
LT 26017	30.94	23.62	—	—	—	—
LT 26018	—	—	—	32.08	26.75	—
LT 26019	—	—	—	—	—	—
LT 38430	33.74	26.08	—	—	—	—
LT 26029	34.39	28.64	—	—	—	—
LT 26030	31.99	24.65	—	—	—	—
LT 26032	35.85	27.19	—	—	—	—
LT 26034	—	—	—	31.35	23.87	—
LT 26035	35.89	26.86	—	—	—	—
LT 26203	31.56	26.62	—	—	—	—
LT 28790	—	—	—	34.91	29.24	—

TABLE 11.27 Horn Core Measurements (in mm) of *Damalacra* sp. B

	Lt. H/c	
	Prox ap	Prox tr
LT 00456	39.66	30.96
LT 23647	34.02	28.02
LT 25971	40.11	33.44

TABLE 11.28 Dentition Measurements (in mm) of *Alcelaphini* gen. and sp. indet.

	LT 222 R	LT 26060 L	LT 26571 L
M^2 ap	—	17.7	—
tr	—	15.56+	—
M^3 ap	20.57	18.18	21.74
tr	15.96	15.09	14.46+

	LT 217 L	LT 217 R	LT 381 R	LT 459 L	LT 469 R
P_3 ap	7.2	—	—	—	—
tr	4.45	—	—	—	—
P_4 ap	11.72	—	—	—	—
tr	6.54	—	—	—	—
M_1 ap	13.35	16.73	18.74	—	—
tr	8.46	9.29	8.75	—	—
M_2 ap	18.23	—	—	20.47	22.46
tr	9.6	—	—	10.97	11.14
M_3 ap	25.1	24.59	—	—	—
tr	8.31	8.49	—	—	—

	LT 530 R	LT 235 R	LT 23614 R
M_1 ap	19.81	14.42	18.03
tr	11.16	8.47	7.9
M_2 ap	—	—	—
tr	—	8.81	—

TABLE 11.29 Cranial Measurements (in mm) of *Aepyceros premelampus*

	LT 184	LT 190	LT 23153	OM 688	OM 1894	OM 145 (F)
Max width orbits	125	—	104+	114	109	100
Max width base h/cs	100	105+	96.4	90	83	—
Width between h/cs	39	49	25.4	23	20	—
Length post h/c, nuchal crest	76	72	67.5	56	60	—
Mastoid width	90	110*	82	80	80	78
Height basioccipital, nuchal crest	52	—	53	52	52	52
Width occ condyles	54	—	54	52	50	51
Width post tuberosity	34	32	32	37	33	29
Width ant tuberosity	21	25	21	19	19	18
Length auditory bulla	33	—	31	35	32	35
Width auditory bulla	22	—	21	18	18	16
Min width cranial vault	73	78	64	74	67	71
Length premax-occ condyle	—	—	251	—	255	251
Length premax-nuchal crest	—	—	250	—	260	256

TABLE 11.30 Horn Core Measurements (in mm) of *Aepyceros premelampus*

	Rt. H/c			Lt. H/c		
	Prox ap	Prox tr	Lt	Prox ap	Prox tr	Lt
LT 184	35.59	28.28	—	36.52	29.49	—
LT 185	41.53	36.94	—	39.1	35.41	—
LT 186	31.14	27.35	—	31.33	27.98	—
LT 187	25.3	24.02	—	29.61	28.13	—
LT 188	—	—	—	39.45	30.02	—
LT 190	—	—	—	—	—	—
LT 206	—	—	—	33.92	29.27	—
LT 210	40.3	29.74	—	36.22	28.25	—
LT 215	39.7	30.86	—	—	—	—
LT 216	37.13	34.81	—	—	—	—
LT 218	—	—	—	41.17	37.95	—
LT 219	39.06	34.08	—	—	—	—
LT 226	30.1	27.96	—	—	—	—
LT 234	38.78	30.6	—	39.95	34.9	—
LT 236	—	—	—	38.91	36.48	—
LT 457	32	27.23	—	—	—	—
LT 462	—	—	—	36.25	32.47	—
LT 465	41.6	33.78	—	—	—	—
LT 468	—	—	—	32.75	25.99	—
LT 470	27.03	25.83	—	—	—	—
LT 472	31.76	29.57	—	—	—	—
LT 473	—	—	—	35.1	32.14	—
LT 477	—	—	—	31.59	26.65	—
LT 481	35.66	27.47	—	—	—	—
LT 491	—	—	—	32.89	32.35	—
LT 492	—	—	—	31.15	25.85	—
LT 495	—	—	—	37.06	30.94	—
LT 502	—	—	—	26.66	24.04	—
LT 510	33.19	26.78	—	—	—	—
LT 515	—	—	—	36.12	33.55	—
LT 22965	—	—	—	33.64	29.63	—
LT 22977	40.85	31.55	—	—	—	—
LT 22979	—	—	—	35.19	28.65	—
LT 22983	—	—	—	35.32	31.55	—
LT 22991	28.43	23.84	—	—	—	—
LT 22997	—	—	—	—	—	—
LT 23096	39.26	32.8	—	—	—	—
LT 23098	—	—	—	40.75	36.01	—
LT 23100	32.81	29.94	—	—	—	—
LT 23128	36.71	34.77	—	—	—	—
LT 23129	—	—	—	31.5	27.03	—

Table 11.30 Horn Core Measurements (in mm) of *Aepyceros premelampus* (Continued)

	Rt. H/c			Lt. H/c		
	Prox ap	**Prox tr**	**Lt**	**Prox ap**	**Prox tr**	**Lt**
LT 23132	—	—	—	39.34	29.39	—
LT 23133	—	—	—	—	—	—
LT 23134	—	—	—	34.77	24.59+	—
LT 23135	—	—	—	36.77	31.14	—
LT 23153	39.29	34.58	—	36.6	42.62	—
LT 23573	34.65	28.3	—	—	—	—
LT 23575	32.39	31.11	—	—	—	—
LT 23576	35.83	30.11	—	—	—	—
LT 23578A	—	—	—	—	30.53	—
LT 23578B	—	—	—	35.7	29.7	—
LT 23580	36.15	33.23	—	37.47	33.97	—
LT 23581	32.96	27.73	—	32.22	28.83	—
LT 23582	37.63	33.23	—	37.87	31.38	—
LT 23600	38.21+	31.86+	—	—	—	—
LT 23603	—	—	—	33.75	28.23	—
LT 23604	28.88	25.59	—	25.85	—	—
LT 23605	—	—	—	—	—	—
LT 23607	—	—	—	37.08	30.01	—
LT 23609	34.19	30.48	—	—	—	—
LT 23620	—	—	—	39.78	32.1	—
LT 23622	—	—	—	—	—	—
LT 23625	28.73	25.96	—	—	—	—
LT 23635	—	—	—	27.57	22.94	—
LT 23640	—	—	—	29.45	24.97	—
LT 23641	—	—	—	—	27.43	—
LT 23642	31.22	28.35	—	—	—	—
LT 23643	28.61	25.61	—	—	—	—
LT 23645	—	—	—	26.08	23.61	—
LT 23648	29.84	23.32	—	—	—	—
LT 38434	34.78	27.88	—	—	—	—
LT 23650	34.88	25.35	—	—	—	—
LT 23651	—	—	—	43.47	39.32	—
LT 23654	—	—	—	35.51	31.2	—
LT 23663	—	—	—	30.8	25.05	—
LT 23670	31.78	29.62	—	—	—	—
LT 23673	41.4	34.5	—	41.4	37.37	—
LT 23698	—	—	—	35.46	29.88	—
LT 23700	36.32	29.93	—	—	—	—
LT 23701	38.07	31.99	—	—	—	—
LT 23702	40.29	33.97	—	39.76	32.53	—
LT 23703	—	—	—	—	—	—

continued

TABLE 11.30 Horn Core Measurements (in mm) of *Aepyceros premelampus* (Continued)

	Rt. H/c			Lt. H/c		
	Prox ap	Prox tr	Lt	Prox ap	Prox tr	Lt
LT 23705	30.55	34.81	—	—	—	—
LT 23706	38.63	35.63	—	36.73	35.11	—
LT 23710	36.94	30.81	—	—	—	—
LT 23712	32	26.76	—	—	—	—
LT 23721	38.94	31.18	—	—	—	—
LT 23727	—	—	—	35.23	29.42	—
LT 23728	36.62	29.27	—	—	—	—
LT 25953	—	—	—	30.17	28.13	—
LT 25956	30.49	27.4	—	—	—	—
LT 25961	26.88	23.59	—	—	—	—
LT 25967	32	23.76	—	—	—	—
LT 25976	35.22	29.35	—	—	—	—
LT 25980	—	—	—	30.7	24.92	—
LT 25982	26.22	24.36	—	—	—	—
LT 25988	27.66	26.22	—	—	—	—
LT 25989	—	—	—	31.31	31.21	—
LT 25992	—	—	—	30.17	27.12	—
LT 25995	—	—	—	32.37	27.06	—
LT 26001	29.21	25.91	—	—	—	—
LT 26008	—	—	—	41.28	31.82	—
LT 26009	32.8	27.26	—	—	—	—
LT 26010	—	—	—	37.33	34.23	—
LT 26011	38.42	32.33	—	—	—	—
LT 26012	—	—	—	30.25	24.4	—
LT 26023	37.93	33.07	—	—	—	—
LT 26024	—	—	—	30.37	23.7	—
LT 26027	35.56	31.38	—	—	—	—
LT 26028	35.81	30.64	—	—	—	—
LT 26031	37.86	34.18	—	—	—	—
LT 26033	—	—	—	32.02	29.05	—
LT 26036	34.6	28.06	—	—	—	—
LT 26039	—	—	—	38.86	33.25	—
LT 26047	40.2	35.11	—	—	—	—
LT 26051	—	—	—	34.87	27.73	—
LT 26054	—	—	—	36.07	30.42	—
LT 26056	30	23.82	—	—	—	—
LT 26057	—	—	—	42.64	36.38	—
LT 26058	40.7	—	—	—	—	—
LT 26062	—	—	—	38.41	29.97	—
LT 26064	39.25	34.09	—	—	—	—
LT 26067	32.48	29.25	—	—	—	—

TABLE 11.30 Horn Core Measurements (in mm) of *Aepyceros premelampus* (Continued)

	Rt. H/c			Lt. H/c		
	Prox ap	**Prox tr**	**Lt**	**Prox ap**	**Prox tr**	**Lt**
LT 26198	35.28	31.87	—	33.49	30.07	—
LT 26200	35.77	29.25	—	—	—	—
LT 26201	—	—	—	40.7	37.79	—
LT 26570	38.11	34.69	—	—	—	—
LT 28754	28.11	22.57	—	—	—	—
LT 28752	26.77	26.65	—	24.34	27.38	—

TABLE 11.31 Dentition Measurements (in mm) of *Aepyceros premelampus*

	LT 25883 R	LT 25885 L	LT 25962 L	LT 28717	LT 28729
P^4 ap	—	—	—	10.65	—
tr	—	—	—	11.47	—
M^1 ap	10.21	—	15.85	—	—
tr	12.62	—	10.83	—	—
M^2 ap	15.59	12.99	—	14.59	—
tr	13.05	13.85	—	14.33	—
M^3 ap	15.88	18.19	19.92	—	17.64
tr	10.95	12.73	12.33	—	14.63

	LT 173 L	LT 174 L	LT 228 L	LT 232 R	LT 533 R	LT 28719	LT 24047
P_3 ap	—	—	—	—	—	8.74	7.62
tr	—	—	—	—	—	4.49	5.35
P_4 ap	—	—	—	—	—	—	8.63
tr	—	—	—	—	—	—	5.77
M_1 ap	—	12.97	13.02	—	—	13.96	—
tr	—	7.54	8.37	—	—	6.93	—
M_2 ap	14.49	16.12	—	—	13.28	—	—
tr	8.33	8.28	—	—	8.7	—	—
M_3 ap	20.74	22.82	—	21.11	22.39	—	—
tr	7.19	8.02	—	8.62	9.49	—	—

	LT 13017 L	LT 13018 R	LT 13019 R	LT 22920 L	LT 22920 R	LT 23061 R	LT 28734
P_2 ap	—	—	—	—	6.61	—	5.12
tr	—	—	—	—	3.0	—	3.4
P_3 ap	—	—	—	—	9.29	—	7.77
tr	—	—	—	—	5.12	—	5.28
P_4 ap	—	—	—	13.5	12.9	—	10.66
tr	—	—	—	6.14	5.72	—	6.67
M_1 ap	—	13.33	—	12.36	12.59	11.43	12.51
tr	—	8.01	—	6.67	7.3	8.53	8.05
M_2 ap	14.49	—	15.23	—	—	14.95	—
tr	8.71	—	9.28	—	—	8.87	—
M_3 ap	23.21	—	—	—	—	19.93	—
tr	8.33	—	—	—	—	7.95	—

	23610 R	23611	23664 R	23688 R	24034 L	23132A R	23630 R
P_3 ap	—	—	—	—	—	—	7.83
tr	—	—	—	—	—	—	5.53
P_4 ap	—	—	—	—	—	—	9.84
tr	—	—	—	—	—	—	6.74

TABLE 11.31 Dentition Measurements (in mm) of *Aepyceros premelampus* (Continued)

	23610 R	23611	23664 R	23688 R	24034 L	23132A R	23630 R
M_1 ap	—	—	—	—	—	—	—
tr	—	—	—	—	—	—	—
M_2 ap	—	—	—	—	—	14.28	—
tr	—	—	—	—	—	9.05	—
M_3 ap	22.2	21.77	20.69	24.33	23.31	23.21	—
tr	8.58	7.9	8.18	8.87	8.65	9.18	—

	25969 R	25987 R	25996 R	26061 L	26199 L	25999 L
P_3 ap	—	—	—	9.13	—	9.04
tr	—	—	—	5.12	—	5.78
P_4 ap	—	—	—	—	—	—
tr	—	—	—	—	—	—
M_1 ap	13.1	—	—	—	—	—
tr	7.75	—	—	—	—	—
M_2 ap	14.81	17.06	—	—	13.48	16.01
tr	7.58	8.79	—	—	8.22	8.67
M_3 ap	—	—	—	20.83	—	22.24
tr	—	—	—	8.92	—	7.99

Table 11.32 Horn Core Measurements (in mm) of *Gazella* sp.

	Rt. H/c		Lt. H/c	
	Prox ap	**Tr**	**Prox ap**	**Tr**
LT 13008	—	18.06	—	—
LT 23653	—	—	19.64	18.27
LT 23666	22.92	16.98	—	—
LT 25972	18.99	16.91	—	—

Table 11.33 Dentition Measurements (in mm) of *Gazella* sp.

	LT 23599
P_3 ap	8.86
tr	5.17
P_4 ap	9.87
tr	6.36
M_1 ap	12.18
tr	8.39
M_2 ap	14.81
tr	9.06
M_3 ap	21.89
tr	8.48

Table 11.34 Dentition Measurements (in mm) of *Raphiceros* sp.

	LT 503 R	**LT 14125 L**	**LT 38433 L**	**LT 26048 L**
P_2 ap	—	—	—	4.29
tr	—	—	—	2.6
P_3 ap	—	—	—	8.88
tr	—	—	—	3.66
P_4 ap	—	—	—	—
tr	—	—	—	—
M_1 ap	8.76	10.18	—	—
tr	—	6.35	—	—
M_2 ap	11.85	11.68	—	—
tr	6.57	7.62	—	—
M_3 ap	—	—	15.35	—
tr	—	—	5.27	—

TABLE 11.35 Dentition Measurements (in mm) of *Madoqua* sp.

	LT 28733 R
M^3 ap	7.54
tr	6.27

	LT 176 L	LT 177 R	LT 22981 L	LT 23612 L	LT 26066 R
P_3 ap	—	—	5.32	6.19	—
tr	—	—	2.83	2.92	—
P_4 ap	6.65	—	6.04	6.67	—
tr	3.45	—	3.77	4.26	—
M_1 ap	6.17	8.36	6.19	—	6.7
tr	3.83	4.42	4.51	—	3.69
M_2 ap	6.31	8.72	7.35	—	7.2
tr	3.85	4.61	—	—	3.72
M_3 ap	8.4	12.7	—	—	—
tr	3.74	4.84	—	—	—

12

ISOTOPES

12.1

Stable Isotope Ecology of Northern Kenya, with Emphasis on the Turkana Basin

Thure E. Cerling, John M. Harris, Meave G. Leakey, and Nina Mudida

The Lake Turkana Basin of northern Kenya supports a variety of wildlife in a dwarf shrubland ecosystem. Local meteoric water in the Turkana Basin probably averages about –3 permil (parts per thousand) for $\delta^{18}O$. The evaporated waters of Lake Turkana range from about +5 to +7 permil in $\delta^{18}O$ over the annual cycle. Local biomass includes significant C_3 and C_4 biomass, in subequal quantities, which contrasts with the wooded grasslands or grasslands of the modern Serengeti. The $\delta^{13}C$ of hypergrazers is more negative (+1 to +2 permil) than that of grazers in more mesic regions (+2 to +4 permil) indicating a subtle difference in their respective diets and may reflect the difference between arid adapted grasses (NAD, PCK, "*Aristida*-NADP") sampled by the Turkana Basin herbivores and the mesic grasses ("classical-NADP") found farther south. Hippos from the Turkana Basin have a higher fraction of C_4 biomass than other parts of East Africa studied thus far. Isotopically they are distinguishable from hypergrazers. Dikdik and giraffes are hyperbrowsers based on their $\delta^{13}C$ values, whereas eland, Grant's gazelle, kudu, and black rhino are browsers. The $\delta^{18}O$ of nonobligate drinkers is from about +4 to +14 permil and is enriched relative to some obligate drinkers. Hippos that drink the Lake Turkana water have $\delta^{18}O$ values about +2 to 4 permil. Comparison of $\delta^{13}C$ and $\delta^{18}O$ of fossil mammals and of fossil eggshells will prove useful for paleoecology and paleophysiology studies.

The stable isotopic composition of pedogenic carbonate and biogenic minerals is related to climate and ecology, and to dietary preferences and metabolic strategies, respectively. The $\delta^{18}O$ and $\delta^{13}C$ component of bioapatites has proven to be an important indicator of water sources and diet (Longinelli 1984; Lee-Thorp and van der Merwe 1987; Koch 1998). The $\delta^{18}O$ of apatite is measured either as the phosphate (e.g., Longinelli 1984; Crowson et al. 1991) or as the CO_3 component as released by reaction with H_3PO_4 (Lee-Thorp and van der Merwe 1987). On a global scale, the bioapatite $\delta^{18}O$ of a species is controlled by the isotopic composition of meteoric water (Longinelli 1984; Bryant et al. 1994), while differences between species at one locale is controlled by their different physiological strategies for water use (Bryant and Froelich 1995; Kohn 1996). The carbon isotopic composition of tooth enamel is determined by diet. In East Africa, grasses and dicots have very different $\delta^{13}C$ values, and thus grazers and browsers have correspondingly different $\delta^{13}C$ values.

Pedogenic carbonate records climatic information because the $\delta^{18}O$ of pedogenic carbonate is closely related to the isotopic composition of local meteoric water and to local evaporation effects (Cerling 1984; Quade et al. 1989; Cerling and Quade 1993; Cerling and Wang 1996). The carbon isotopic composition of pedogenic carbonate is primarily determined by the local ecology (Cerling 1984; Cerling et al. 1989; Cerling and Quade 1993). The controlling factors for the carbon isotopic composition of pedogenic carbonate are the fractions of biomass using the C_3 and C_4 photosynthetic pathways in the local ecosystem; these have average $\delta^{13}C$ values of about –26 and –12 permil, respectively. CAM plants have intermediate isotope values but are minor parts of most ecosystems and will be ignored in the following discussion. However, it should be noted in passing that, under the climatic conditions that favor C_4 plants, CAM plants have $\delta^{13}C$ values similar to those of C_4 plants and, under those climatic conditions that favor C_3 plants, CAM plants have $\delta^{13}C$ values similar to those of C_3 plants. In this contribution we review the isotope systematics of terrestrial ecology and present data on the meteoric waters and modern soils and animals from the Turkana Basin. This information provides a basis for interpreting the results of analysis of fossil tooth enamel from Lothagam that are described in the next contribution.

Methods

Soil carbonate, water, plant, and animal samples were collected from various Kenyan localities. Northern Kenya samples include those from the Koobi Fora region east of Lake Turkana, from the Lothagam and Kanapoi regions west of Lake Turkana, from the Samburu Game Reserve, and from Lewa Downs and the Mpala Research Center in the Laikipia region. We report the isotopic measurements on these samples and provide measurements on a few samples from other parts of East Africa.

The stable isotopic composition of waters, carbonates, and apatite were carried out using standard methods and are reported using the standard (permil) isotopic notation

$$\delta^{13}C\ (^{18}O) = (R_{sample}/R_{standard} - 1)\ 1000$$

where R_{sample} and $R_{standard}$ are the $^{13}C/^{12}C$ ($^{18}O/^{16}O$) ratios in the sample and standard, respectively. For carbon, the standard is PDB (= Pee Dee Belemnite). For oxygen, two different standards are used: waters are reported relative to the SMOW (= Standard Mean Ocean Water) standard, and carbonates and biogenic phosphates are reported relative to the PDB standard. Biogenic apatites were treated, following the procedure of Lee-Thorp and van der Merwe (1987), with an oxidizing agent (NaOCl or H_2O_2) and a weak acid (0.1 M acetic acid) treatment to remove organic matter or calcite, respectively. Biogenic apatites were reacted under vacuum for 36 hours at 25°C with 100 percent H_3PO_4, and their values are reported using the same fractionation factor as calcite for this temperature. After reaction, the cryogenically purified CO_2 was reacted with hot silver to remove the trace contaminant SO_2. Samples were analyzed on a Finnegan mass spectrometer.

Water, carbonates, and organic matter were extracted and purified using standard methods and analyzed on a mass spectrometer. Precision of analysis is about 0.1 to 0.2 permil.

The apparent isotope enrichment factor ε^* between diet and enamel is 14.1 permil where

$$\varepsilon^* = (R_{enamel}/R_{diet} - 1) * 1000$$

or

$$\varepsilon^* = [(1000 + \delta_{enamel})/(1000 + \delta_{diet}) - 1] * 1000$$

and the "*" implies that reversible isotopic equilibrium is not assumed (see Cerling and Harris 1999). It is significant to note that ε^* differs slightly from $\Delta_{enamel-diet}$ (= $\delta_{enamel}-\delta_{diet}$; see discussion in Cerling and Harris 1999).

Climatology and Ecology

The Turkana Basin is dominated by Lake Turkana, an alkaline lake with no surface outlet (Yuretich and Cerling 1983; Cerling 1986, 1996). The climate is very hot with a mean annual temperature of about 29°C, and there is little seasonal change in temperature. Annual precipitation is about 180 mm per year (East African Meteorological Department 1975). The lake is marginally potable with a TDS (Total Dissolved Solids) of about 3,000 ppm and a pH averaging about 9.1 (Yuretich and Cerling 1983; Cerling 1996). Lake Turkana is used as a source of drinking water by large mammals that live near its shores, including Burchell's zebra, carnivores, and, of course, the aquatic hippopotamus. Crocodiles also inhabit the lake. Large mammals that are rarely seen near the lake shore include greater and lesser kudu, giraffe, dikdik, oryx, and Grevy's zebra. Grant's gazelle, which are found both close to and far from the lake, are rarely seen actually drinking the lake water. Tiang are found near the lake and are often seen drinking from Lake Turkana. Tiang often prefer alkaline soils dominated by *Sporobolus spicatus* (Kingdon 1987), which is very common along the shores of Lake Turkana. The only observed black rhinoceros at Koobi Fora during the past three decades was never seen near the lake shore. Ostriches also are widespread in the region, but rarely drink any water. Other mammals are dependent on water holes; these include baboons and humans with their domestic animals.

The modern ecosystem is classified as dwarf shrub grassland, according to the terminology of Pratt et al. (1966)—an open environment in the semiarid regions of East Africa with a significant cover of grasses. In the context of this discussion, this dwarf shrub grassland is not dissimilar to the bushed grassland of Pratt et al. (1966). In addition, this ecosystem contrasts with the East African savanna in which grasses are the dominant component of the biomass, although they may include woody species that do not form a continuous cover (Sarmiento 1984).

Three other sites were compared with the Turkana Basin, two in northern Kenya and one near Nairobi. The Samburu Game Reserve and lower Laikipia region (Lewa Downs and Mpala Ranch) are significantly cooler and moister than the Turkana region (table 12.1). They have a somewhat similar isotopic signature in that they are dominated by NAD-me and PEP-ck grasses, whereas the Nairobi region is dominated by NADP-me grasses (the distinction between these C_4 subtypes is discussed later in this contribution).

$\delta^{18}O$ of Waters in the Turkana Basin

Meteoric waters in the Lake Turkana region are significantly more depleted in $\delta^{18}O$ than is Lake Turkana itself. Direct rainfall, collected at the beginning of the 1977 long rains, had $\delta^{18}O$ values that range from about –2 to +1 permil (table 12.2). The rain showers were of short duration and were not sufficient to infiltrate the soils. Small, ephemeral water holes (all collected at the beginning of the short rains) had $\delta^{18}O$ values similar to those of rain collected at the same time—that is, from –1 to +2 permil. A better estimate of the infiltration signal, which most likely represents the average $\delta^{18}O$ of local meteoric water, is obtained from water holes in the region. Table 12.2 shows that the $\delta^{18}O$ of water from permanent wells, water holes, and springs in the region (Derati, Huran Hura, Kubi Algi, Loiyengalani) varies from about –2 to –5.5 permil. This probably closely approximates the annual $\delta^{18}O$ of meteoric water in the region, which we estimate to be about –3 permil. As shown in figure 12.1, the perennial springs and the Omo River fall near the global meteoric water line (MWL), while ephemeral springs and Lake Turkana are more enriched in $\delta^{18}O$ compared to MWL, indicating that these waters have undergone evaporation.

For almost two years during the early 1980s, we collected Lake Turkana water monthly at the Koobi Fora spit. The $\delta^{18}O$ of the lake ranged from about +5 to +7 permil relative to SMOW and was weakly correlated with changes in salinity during the year (Cerling 1996); the lowest values occurred during the rise associated with the annual Omo River flood. The Omo River, which originates in the Ethiopian Highlands, provides the most important water source for the lake (Hopson 1982; Yuretich and Cerling 1983). It has a $\delta^{18}O$ of about 0 to –1 permil (table 12.2). Lake Turkana is considerably more evaporated than is either local meteoric water or the inflowing river waters. Craig (1961) noted that East African lake waters were evolved from meteoric waters, that they were significantly enriched in both deuterium and ^{18}O compared to meteoric waters, and

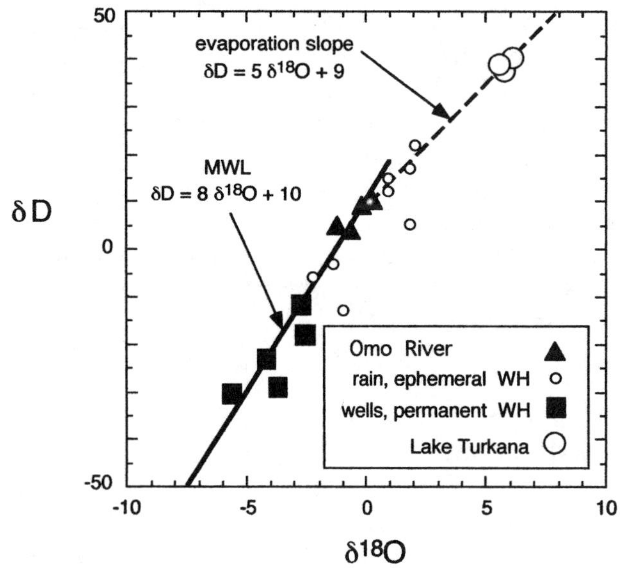

Figure 12.1 δD and $\delta^{18}O$ content of waters in the Turkana Basin. The solid line represents the global Meteoric Water Line (MWL) of $\delta D = 8\delta^{18}O + 10$ (Craig 1961), and the dashed line represents the evaporation trend, assuming Omo River water as the primary source of Lake Turkana water.

that δD and $\delta^{18}O$ from East African lakes differed significantly from the global Meteoric Water Line. In particular, highly evaporated waters are more enriched in ^{18}O than is their parent water, and they generally have a slope ranging between 4 and 6. The data from the Omo River and from Lake Turkana give the relationship $\delta D = 5\delta^{18}O + 9$.

Leaves represent highly evaporated waters, although suberized stem waters are not highly fractionated. Unsuberized stem water is evaporated, however. The $\delta^{18}O$ of leaf water can be 15 to 25 permil enriched relative to its source water. Based on the $\delta^{18}O$ values for groundwater and precipitation, the $\delta^{18}O$ of leaf water in the Turkana Basin is likely to be about +15 to +20 permil.

Soil water is closely related to local meteoric water, but in xeric environments shallow (<20 cm) soil water can also be enriched by many parts per thousand compared to the source water. In general, the longterm $\delta^{18}O$ value of soil water is highest near the soil-air interface in desert regions.

$\delta^{18}O$ in Soil Carbonate

The $\delta^{18}O$ of soil is preserved in pedogenic carbonates and that of lake water in lacustrine carbonates. Both types of carbonate have been used to investigate the paleoecology and sedimentology of the younger Koobi Fora sediments (Cerling et al. 1989). The $\delta^{18}O$ composition of pedogenic carbonate reflects the $\delta^{18}O$ of local meteoric waters (Cerling and Quade 1993; Cerling and Wang 1996); $\delta^{18}O$ values of the pedogenic carbonates range from about +2 to +6 permil (table 12.3). Based on the estimated soil temperature, these values are compatible with soil water being about +8 permil (Cerling and Quade 1993), indicating that significant evaporation of water occurred in the upper parts of these soils. Allison et al. (1987) documented that the water in the upper few tens of cms in arid soils can be significantly evaporated, whereas deeper waters are much less evaporated.

Stable Oxygen Isotopes of Extant Turkana Basin Animals

The $\delta^{18}O$ of body water is preserved in biogenic tooth enamel in mammals and in biogenic carbonate in eggshells. Other apatite phases (e.g., bone, dentine) are not suitable for isotopic analysis because they are more susceptible to diagenesis than is enamel (Wang and Cerling 1994). Therefore, the $\delta^{18}O$ gives information about the source of animal water, whether it is derived from unevolved meteoric water (e.g., water holes or rivers), from evolved meteoric waters (e.g., evaporated lakes), from highly fractionated leaf water, or from metabolic water (i.e., that derived from oxidation of carbohydrates in food). Metabolic water is buffered by the isotopic composition of the oxygen in the atmosphere and by the fact that for every oxygen in this water there are two oxygens in the associated CO_2 (simplified: $CH_2O + O_2 = CO_2 + H_2O$).

Kohn (1996) and Kohn et al. (1996) have modeled $\delta^{18}O$ in mammals by using an energy budget, but unfortunately, that model does not adequately predict observed $\delta^{18}O$ in modern mammals and needs revision. Although mammals from Lake Turkana were used to validate the model, Kohn (1996) used a $\delta^{18}O$ value of +6 permil for meteoric water in the Turkana Basin, which is not correct (table 12.2). Substitution of the correct value for meteoric water into his model gives $\delta^{18}O$ values for apatite that are 5 permil or more depleted in ^{18}O than were the observed values. In this contribution we will not discuss our $\delta^{18}O$ values using the Kohn model, even though, after revision, the model may prove applicable to arid ecosystems.

The $\delta^{18}O$ of tooth enamel and eggshell is related to the sources of water available to animals—that is, drinking water, plant water present in leaves or vegetation, and metabolic water created by oxidation of carbohydrates or fats. Drinking water sources in the Turkana Basin have variable $\delta^{18}O$ values; unevolved meteoric waters have $\delta^{18}O$ values between about 0 and –5 permil, but the highly evolved (due to evaporation) Lake Turkana water has a $\delta^{18}O$ value between 5 and 7 permil (table 12.2). Leaf water is generally about 15 to 20 permil enriched relative to local soil water. In the extremely hot, dry, and windy Turkana Basin the maximum fractionation is probably realized, and it is likely that leaf water is about +15 to +20 permil. Metabolic water ($CH_2O + O_2 = CO_2 + H_2O$) is buffered by the isotopic composition of the atmosphere (+23.5 permil) and by respired carbon dioxide that has been more enriched in $\delta^{18}O$ than that in associated CO_2 (Bryant and Froelich 1995; Kohn 1996). Cellulose oxygen is about 27 permil more enriched than leaf water (i.e., +42 to +47 permil). Metabolic water of mammals in the region can be estimated by assuming that the intake of oxygen during metabolism is +18 permil (Epstein and Zeiri 1988), that the O bound in food is +45 permil, and that the fractionation factor between CO_2 and H_2O is 1.038 at ca. 37°C. The buffering by atmospheric oxygen and the partitioning between H_2O and CO_2 results in metabolic water in the Lake Turkana Basin mammals being about +2 permil—which is more positive than unmodified meteoric water but significantly more negative than Lake Turkana water. Therefore, the $\delta^{18}O$ of mammals in the Turkana region will be related to the following water sources:

- meteoric –3 permil
- metabolic +2 permil
- Lake Turkana +6 permil
- leaf water +20 permil

In general, metabolic water makes up a small fraction of the total body water of large mammals but becomes important in smaller mammals. Because leaf water is more highly enriched than all other water sources in the tropics, obligate drinkers should have a $\delta^{18}O$ value more negative than mammals that rely on leaves for much of their water intake.

Giraffe, ostrich, gerenuk, kudu, Grevy's zebra, and dikdik require little or no drinking water, whereas the obligate drinkers—rhino, warthog, hippo, tiang, baboon, and Burchell's zebra—require water daily. Grant's gazelle have been observed to drink at the lake and at water holes although they do not appear to require daily access to water. In the Turkana region, the tiang are always restricted to areas associated with water and may even enter the lake in order to graze; it is possible that the leaf water in *Sporobolis spicatus* is "super-evaporated" if its source is the previously evaporated lake waters. Burchell's zebra occasionally drink from Lake Turkana. Warthogs and baboons do not appear to drink the lake water, which is alkaline and slightly brackish (ca. 3000 ppm TDS), but they do drink from water holes in the surrounding countryside.

Table 12.4 gives $\delta^{18}O$ values for tooth enamel for modern mammals in northern Kenya, and figure 12.2 shows some distinction between these groups. Obligate

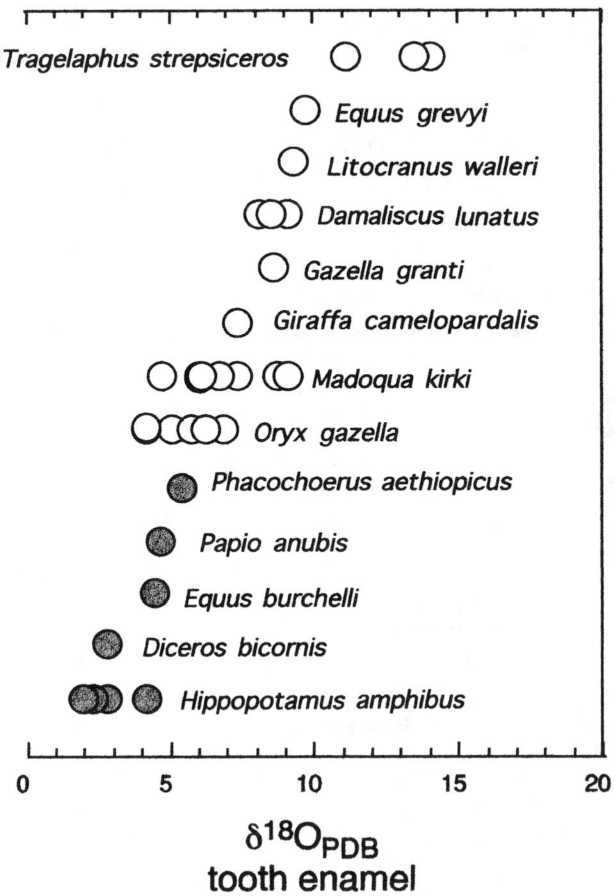

Figure 12.2 $\delta^{18}O$ values for tooth enamel from Turkana Basin mammals. Shaded symbols are obligate drinkers, and open symbols are those mammals thought to be largely independent of drinking water.

drinkers have $\delta^{18}O$ values between 2 and 5 permil. Hippos have $\delta^{18}O$ tooth enamel values between +2 and +4 permil; such high values are because of their reliance on the highly evaporated waters of Lake Turkana. At Amboseli Park, in contrast, the $\delta^{18}O$ of hippos is about –4 permil (Bocherens et al. 1996). Warthog, baboon, and zebra have higher $\delta^{18}O$ values than hippos, in the lower range of the nonobligate drinkers that have $\delta^{18}O$ values from +4 to +12 permil. Tiang have very enriched $\delta^{18}O$ values compared to the other obligate drinkers. Lake Turkana is highly evaporated; if it is the water source of their favored graze, *Sporobilis spicatis*, this would explain the very elevated $\delta^{18}O$ values documented for the tiang. It is interesting that tiang $\delta^{18}O$ is several permil more positive than that of hippos from this lake. A possible explanation is that the hippos also exploit grasses that grow at some distance (>100 meters) from the lake shore and thus from grasses that derive their water from soils with a meteoric source (i.e., about –3 permil).

The nonobligate drinkers from the Turkana region (giraffe, gerenuk, Grant's gazelle, dikdik, kudu, tiang, and Grevy's zebra) have elevated $\delta^{18}O$ values, from about 4 to +14 permil (table 12.4). The greater kudu has the highest $\delta^{18}O$ values measured so far. Interestingly, a single kudu from Isiolo has a much lower $\delta^{18}O$ value of +4.5 permil. Mammal species sampled from the Laikipia and Samburu regions are generally lower in $\delta^{18}O$ by several permil compared to the Turkana individuals (table 12.4).

$\delta^{13}C$ of Northern Kenya Plants

The $\delta^{13}C$ of individual plants is determined by their photosynthetic pathway and is modified by local conditions. The groupings recognized by botanists and stable isotope ecologists are the C_3 photosynthetic pathway, C_4 pathway, CAM photosynthesis, and C_3–C_4 intermediates. The C_3 and C_4 pathways are so named because the first carbon compound produced has either 3 or 4 carbon atoms. CAM (crassulacean acid metabolism) is an alternative photosynthetic pathway found in some succulents. C_3–C_4 intermediates can exhibit both pathways in different parts of the leaf or in different leaves; however, such plants are uncommon.

The C_3 photosynthetic pathway is the common pathway of dicotyledons. Most trees and shrubs use this pathway, the products of which have $\delta^{13}C$ values that range from about –22 to –36 permil (Bender 1971; Deines 1980). It is important to note, however, that a few dicots use the C_4 pathway (see the discussion that follows). Variations within C_3 plants are notable because they offer opportunity for reconstruction of the habits and habitats of fossils, but oversimplification of the system leads to erroneous conclusions about ancient ecologies. The average $\delta^{13}C$ of C_3 plants is about –27 permil. Some modelers have used this value in the reconstruction of ancient ecosystems, but the observed range is from about –22 permil to –36 permil. Enriched ^{13}C values invariably represent regions of high heat and water (or salt) stress (Ehleringer et al. 1986; Ehleringer and Cooper 1988). In contrast, very negative $\delta^{13}C$ values are associated with low light conditions under closed canopies. Thus, van der Merwe and Medina (1989) reported very negative $\delta^{13}C$ values on the forest floor below the Amazon rain forest canopy. Table 12.5 and figure 12.3 show that the average $\delta^{13}C$ of C_3 dicots in northern Kenya is about –27.2 ± 1.5 permil (n = 46). Plants collected after a prolonged drought were about 1 permil more enriched in ^{13}C than were those collected after the following rainy season. Thus the average $\delta^{13}C$ values of the January 1997 collection from the Turkana region was –26.9 ± 1.7 permil (n = 17; Koobi Fora) whereas the average $\delta^{13}C$ value for July 1997 was –27.7 ± 0.9 permil (n = 16; Kanapoi). Figure 12.4 shows that difference for individual plant species from the

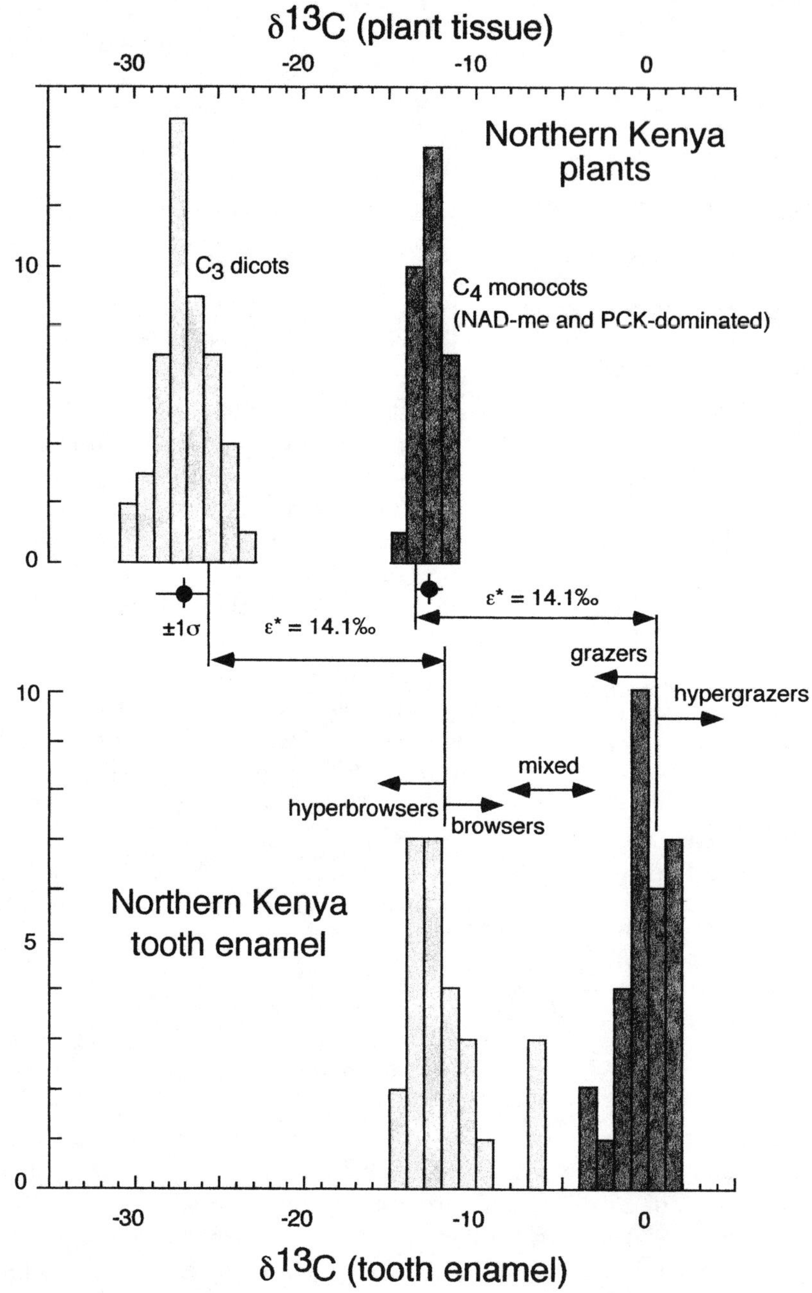

Figure 12.3 $\delta^{13}C$ of C_3 dicots and C_4 monocots from northern Kenya, and $\delta^{13}C$ values of tooth enamel from modern mammals in northern Kenya. The ε value for isotope enrichment between diet and tooth enamel is 14.1 permil (Cerling and Harris 1999).

Turkana Basin for these two periods. Thus, dry season $\delta^{13}C$ values for C_3 plants are significantly different from those sampled during moist seasons. C_3 grasses in East Africa include bamboo (*Arundinaria*), rice (*Oryza*), *Leersia*, papyrus (*Phragmites*), and some alpine grasses.

The C_4 pathway is prevalent in monocotyledons, especially grasses and sedges from regions with warm to hot growing seasons. However, there are some important C_4 dicots in Africa, including some species of *Euphorbia, Blepharis,* and *Tribulus.*

Three different C_4 subtypes are recognized from biochemical pathways (Kanai and Edwards 1999) and seven C_4 subtypes based on anatomy (Chapman 1996)—four of which are abundant in Africa. The three biochemical groups are the NAD-me (hereafter NAD), the PEP-ck (hereafter PCK), and the NADP-me (hereafter NADP) pathways. Aspartate is the first biochemical product in NAD and PCK, whereas malate is the first biochemical product in NADP. The NADP subpathway has two anatomical types: the "classical-NADP" and "*Aristida*-NADP" (Chapman 1996). The NAD and PCK subpathways also have characteristic anatomy. Classical-NADP grasses are common in more humid regions, whereas NAD, PCK, and *Aristida*-

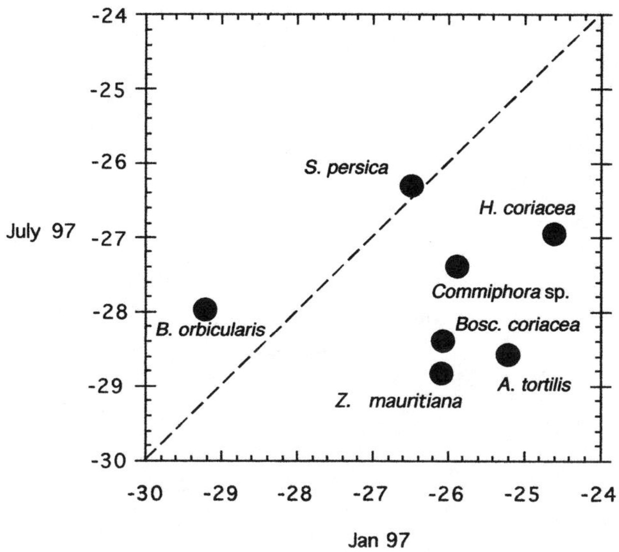

Figure 12.4 $\delta^{13}C$ for plant species from the Turkana Basin collected at the end of a prolonged drought (January, 1997) compared to those collected at the end of a rainy season (July, 1997). *A.* = *Acacia*; *B.* = *Balanites*; *Bosc.* = *Boscia*; *H.* = *Hyphaene*; *S.* = *Salvadora*; *Z.* = *Zizyphus*.

NADP grasses are more common in arid regions of both the Old World and the New World (Cowling 1983; Ellis et al. 1980; Hattersley and Roksandic 1983; Hattersley 1992; Schultze et al. 1996; Vogel et al. 1978, 1986). C_4 plants have a narrow range of $\delta^{13}C$ values, from about –14 to –11 permil; the classical-NADP subtype averages about –11.5 permil whereas the NAD, PCK, and *Aristida*-NADP subtypes average about –13 permil (Hattersley 1982). East African examples of NADP-me grasses include the genera *Andropogon, Digitaria, Themeda, Hyparrhenia, Paspalum, Pennisetum,* and *Setaria* (Ibrahim and Kabuye 1987; Sage et al. 1999). Examples of the NAD grasses that are more adapted to arid conditions include *Cynodon* and *Eragrostis.* The genera *Brachiaria, Chloris, Dactyloctenium,* and *Rhynchelytrum* use the PCK subpathway, whereas *Aristida* uses the *Aristida*-NADP subpathway. *Sporobolus* and *Harpachne* use either the PCK or NAD subpathway. Different species of the genus *Panicum* use the classical-NADP, NAD, PCK, or C_3 pathways (Sage et al. 1999).

Grasses from northern Kenya are predominantly C_4 species that use the NAD and PCK subpathways. Some classical-NADP grasses are found in riparian settings; however, NADP grasses other than *Aristida* are rare in the extremely hot and arid Turkana region. While most NADP grasses are found in mesic environments, *Aristida* generally occurs in more arid regions (Hattersley 1992). *Themeda* and *Pennesetum* (classical-NADP subpathway) are found in riparian environments in the Lewa and Samburu region, but Carr (1976) does not report them in the Omo Basin. Nor have they been reported in the Koobi Fora region or along the lower Turkwel or Kerio Rivers on the west side of Lake Turkana. Classical-NADP grasses found in riparian settings in northern Kenya have an average $\delta^{13}C$ value of about -12.0 ± 0.6 permil, whereas the more arid C_4 grasses (*Aristida*-NADP, NAD, and PCK) average -13.0 ± 0.7 permil (table 12.5).

The few dicots that use the C_4 pathway are generally found in arid or saline environments, or both. Northern Kenya C_4 dicots include *Blepharis, Sueda, Tribulus,* and some species of *Euphorbia.* A few plants are classified as C_3–C_4 intermediates; *Cleome,* a perennial forb, is an example of a C_3–C_4 intermediate.

Thus, in modern tropical Africa there is marked isotopic distinction between the dicots (which generally make up browse) and the monocots (which are generally grasses and sedges) (figure 12.5). In northern Kenya most grasses use the NAD and PCK C_4 subpathway and have an average $\delta^{13}C$ value of about –13 permil; this value can be compared to the dicot average of –27 permil.

Stable isotope studies of paleoecosystems have shown that, prior to about 8 million years ago, few C_4 plants were present in significant abundance anywhere (Cerling et al. 1997). It is also well accepted that the C_4 pathway is an adaptation to low atmospheric CO_2 (Ehleringer et al. 1991). Based on quantum yield, the "transition point" at which C_3 dicots adopt C_4 photosynthetic mechanisms is about 30°C at atmospheric CO_2 levels about 500 ppmV, or about 20°C under the conditions prevailing during the Holocene (ca. 280 ppmV) (Cerling et al. 1997; Ehleringer et al. 1997).

Mixing Model for C_3–C_4 Biomass Estimates

A mixing model for C_3 and C_4 biomass in ecosystems is shown in figure 12.5. For the modern Turkana region, we assume the C_4 end member to be -12.8 ± 0.8 permil (all C_4 plants from table 12.2) and the C_3 end member to be -27.2 ± 1.5 permil (all C_3 plants from table 12.2). The isotope enrichment factor for tooth enamel is 14.1 permil, as taken from Cerling and Harris (1999). We make the distinction based on isotopic analysis between hypergrazers, grazers, mixed feeders, browsers, and hyperbrowsers. Hypergrazers have no detectable C_3 component in their enamel: northern Kenya samples are represented by tooth enamel $\delta^{13}C$ values greater than +0.3 permil, which reflects a diet of xeric-adapted *Aristida*-NADP, NAD, and PCK C_4 grasses, along with some classical-NADP grasses from riparian habitats. Hyperbrowsers have no detectable C_4 component and are represented by tooth enamel $\delta^{13}C$ values more negative than –12 permil. The isotopic space between +0.3 permil and –12 permil is split among grazers (0.3 to –2.8 permil), mixed C_3–C_4 feeders (–2.8 permil to

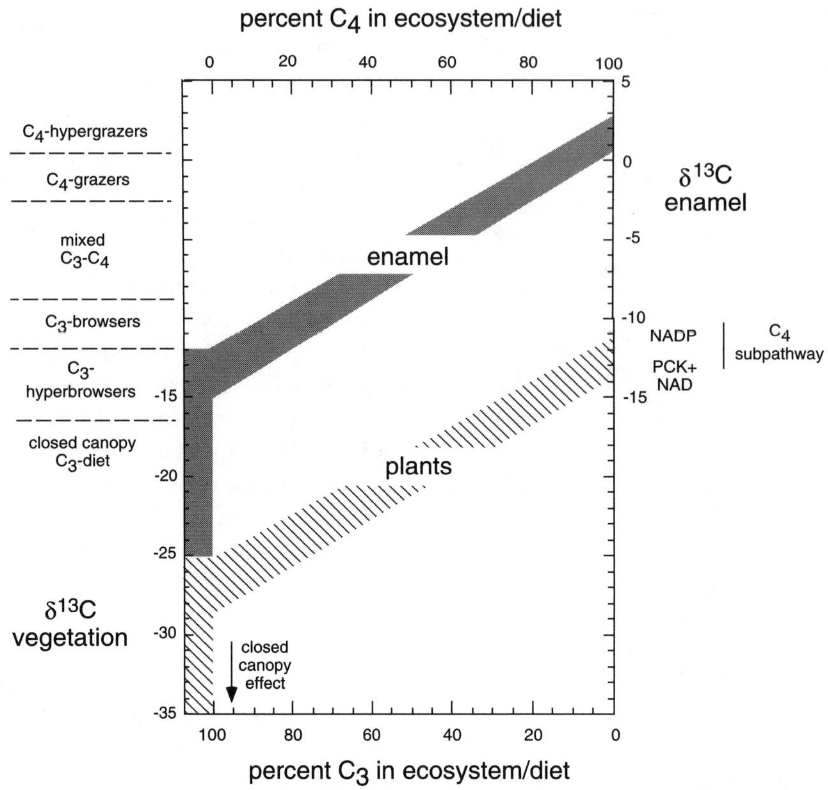

Figure 12.5 Mixing lines of C_3 and C_4 biomass for terrestrial ecosystems. The C_3 endmember has a large range of $\delta^{13}C$ values. The high end of the C_3 $\delta^{13}C$ values mixes with the C_4 endmember and represents mixing in xeric habitats. The lowest $\delta^{13}C$ values are for closed canopy situations and should not be mixed with the C_4 endmember because C_4 plants do not exist in closed canopy situations. Thus, the lower $\delta^{13}C$ values used for mixing with C_4 biomass are taken to be about –30 permil and represent more mesic habitats.

–8.9 permil), and browsers (–8.9 to –12 permil). However, as shown in figure 12.6, there are very few mixed feeding bovids in East Africa.

This mixing model shows that individual $\delta^{13}C$ values can be linked to dietary, behavioral, and ecological interpretations. The enrichment from diet to biogenic carbonate, such as in eggshells, is similar to that for biogenic apatite. However, the atmosphere has become 1.5 permil more negative in the last 150 years due to the burning of fossil fuel. Therefore, the mixing model shown in figures 12.5 and 12.6 should be considered in light of the estimated ambient $\delta^{13}C$ of the atmosphere for fossils for any given time period.

Stable Carbon Isotopes of Extant Animals from Other Ecosystems

The stable isotopic composition of tooth enamel from extant mammals in the Turkana Basin (table 12.4) provides a basis for interpretation of comparable results from fossil mammals. We analyzed a number of modern mammals from the Turkana Basin for $\delta^{13}C$ and $\delta^{18}O$, and thus we augment table 12.4 by analyses of several individuals from other parts of Kenya. Included are data from the Samburu Reserve and Archer's Post near Isiolo, and from Lewa and Mpala in the Laikipia region. In these areas the grasses are dominated by NAD and PCK grasses, although some classical-NADP grasses (*Themeda, Pennisetum*) grow in riparian environments.

The $\delta^{13}C$ of tooth enamel clearly represents the continuum from C_4 hypergrazers, C_4 grazers, mixed C_3–C_4 feeders, C_3 browsers, to C_3 hyperbrowsers (figure 12.6). East African hypergrazers (with a diet of ca. 100 percent grass) include cape buffalo, tiang, and Jackson's hartebeest; analysis produces $\delta^{13}C$ values that average 1.6 ± 0.3 permil (n = 7; table 12.4) which is a result of their NAD, PCK, and *Aristida*-NADP diet ($\delta^{13}C$ = –13.0 ± 0.7 permil). Of these bovids, only the tiang is found today at Lake Turkana, although buffalo were present in the early 1900s. The contemporaneous hypergrazer *Connochaetes taurinus* from a classical-NADP dominated ecosystem ($\delta^{13}C$ = –11.7 ± 0.7; Cerling and Harris 1999) near Nairobi has significantly more positive $\delta^{13}C$ values for tooth enamel; the average is 2.6 ± 0.6 permil (n = 9; table 12.6).

Grazers (oryx, hippo, and zebra) have less C_4 biomass in their diets than do hypergrazers. Lamprey (1963) noted that the diet of Burchell's zebra in Tan-

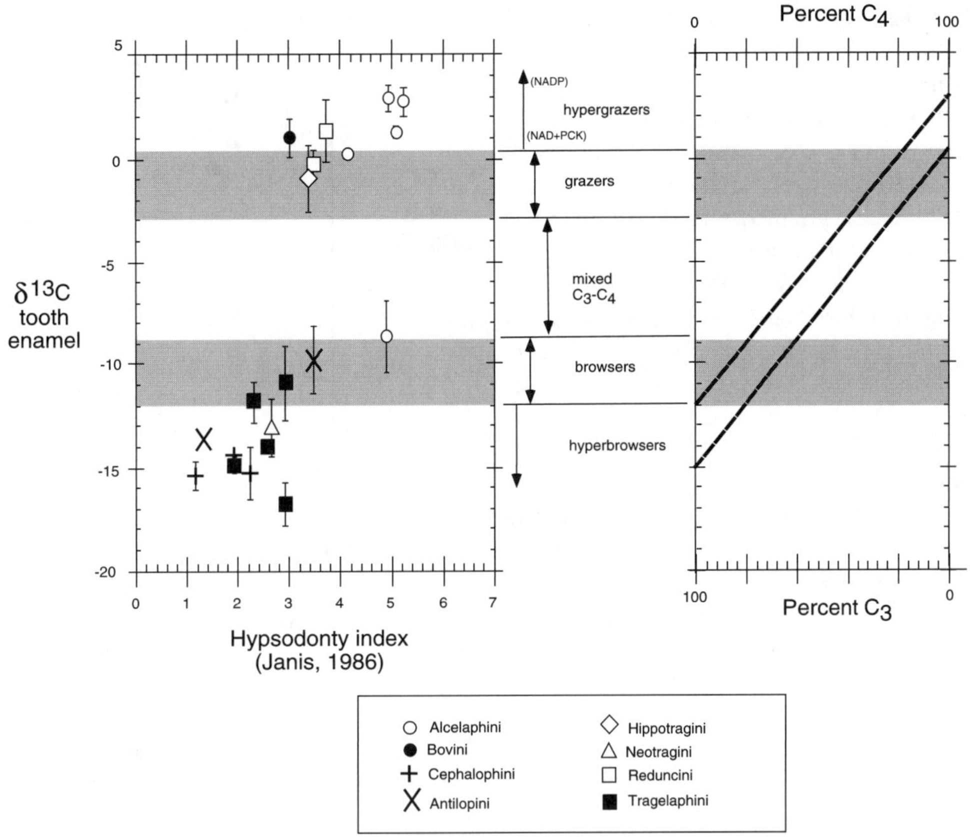

Figure 12.6 $\delta^{13}C$ of modern East African bovids from Kenya, Uganda, and eastern Congo compared with the hypsodonty index of Janis (1986). Isotope mixing lines are shown for reference. This figure shows that very few African bovids have a truly mixed diet—that is, one that makes up between about 25/75 and 75/25 C_3/C_4 mix. It is important to note that this figure is for the modern atmospheric $\delta^{13}C$ value of –8 permil; other $\delta^{13}C$ values of the atmosphere at other times (i.e., the pre–Industrial Revolution atmosphere was –6.5 permil) would result in an offset equivalent to the difference in the $\delta^{13}C$ value of the atmosphere.

zania consisted of about 93 percent grass, and this is consistent with the stable isotope values reported here. However, others (Gwynn and Bell 1968; McNaughton and Georgiadis 1986) have suggested that the zebra is a hypergrazer, although current isotopic evidence does not support that contention (but see discussion in Cerling and Harris 1999). The range of $\delta^{13}C$ values for Burchell's zebra is from +0.5 to –3.0 permil and that for a single Grevy's zebra is 0.7 permil (table 12.4), which indicates that grass made up from about 70 to 90 percent of the diets of the sampled individuals. Extant oryx from the east side of Lake Turkana have $\delta^{13}C$ values from 0.4 to –3.3 permil; this also indicating a significant fraction of browse in their diets. A single oryx from near Nanyuki, where there is more abundant grass than in the Turkana Basin, has a $\delta^{13}C$ value of +2 permil (unpublished data), and this is consistent with a higher proportion of classical-NADP grass in the diet than Turkana representatives. Although the sample size is limited, this preliminary data from the Turkana Basin indicates that grazers (except hippo: see the following discussion) have somewhat more negative $\delta^{13}C$ values than the same mammals in savanna grassland vegetation. This is compatible with the isotopic difference between the arid adapted grasses (*Aristida*-NADP, PCK, and NAD) and mesic grasses (classical-NADP). The $\delta^{13}C$ values between 0 and –3 permil for grazers and between 1 and 2 permil for hypergrazers reflect the presence of dwarf shrubland dominated by arid-adapted grasses. It is important to note that changes in the isotopic composition of the atmosphere would change the absolute values given here. For example, the pre-Industrial Revolution atmosphere was 1.5 permil, which is more enriched in ^{13}C than is the current (1999) atmospheric value of –8 permil.

Hippos (*Hippopotamus amphibius*) deserve separate consideration because of their aquatic habitat. The diet of extant hippos from the Turkana Basin has a very large grass component, with $\delta^{13}C$ values ranging from –0.5 to –2 permil. That hippos from other parts of East Africa (Thika, Queen Elizabeth Park, and Baringo: unpublished data) have more negative $\delta^{13}C$ values than those from Lake Turkana may reflect a diet with a significant component of browse or swamp grasses (some tropical swamp grasses—including *Leersia, Phragmites,* and *Oryza*—use the C_3 pathway).

Browsers from northern Kenya have $\delta^{13}C$ values that are much more negative than grazers. For example, elephants from Samburu Reserve (table 12.4; see also Cerling et al. 1999) have $\delta^{13}C$ values from about –10 to –11 permil, consistent with a diet composed of about 85 percent C_3 plants. Similarly, black rhino, greater kudu, eland, and Grant's gazelle are browsers according to our preliminary isotopic results (tables 12.4 and 12.6), and this finding is in agreement with previous field observation (McNaughton and Georgiadis 1986; Kingdon 1979, 1982a, 1982b). Dikdik and giraffe are hyperbrowsers. Based on the $\delta^{13}C$ value of their enamel, baboons have very little grass in their diet.

Domestic sheep and goats from the Turkana Basin are mixed feeders, and their flocks are present in the basin year round; the domestic cattle are grazers (table 12.4) and travel extensively in search of graze. Ostriches from the west side of Lake Turkana have a mixed diet that varies from about 50 to 100 percent of that of C_3 plants ($\delta^{13}C$ values from –4 to –9 permil). It will be interesting to compare ostriches from the east side to the west side of Lake Turkana because the vegetation of the west side has less grass due to over-grazing by domestic stock.

Although not represented in the extant or fossil assemblages of the Lake Turkana Basin, closed canopy dwellers such as bongo, okapi, and giant forest hog have very negative $\delta^{13}C$ values, from about –15 to –22 permil (unpublished data and table 12.6). In fossil samples, such very negative $\delta^{13}C$ values (<–16 permil) may be a distinguishing feature that represents closed canopy dwellers or habitats resembling closed canopy conditions.

Figure 12.7 shows the relationship between $\delta^{18}O$ and $\delta^{13}C$ for many of the large mammals in the Turkana Basin.

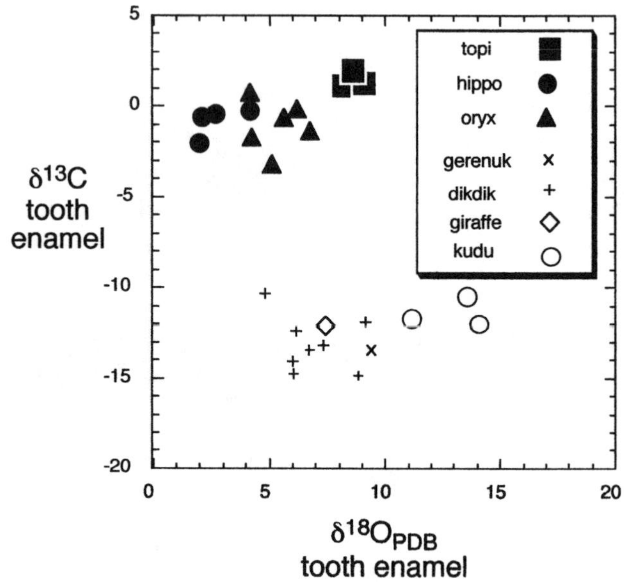

Figure 12.7 $\delta^{18}O$ and $\delta^{13}C$ tooth enamel from Recent mammals from the Koobi Fora region of Lake Turkana.

$\delta^{13}C$ in Soils

The enrichment in ^{13}C from organic matter to pedogenic carbonate varies from 14 to 17 permil, although greater enrichment is found in xeric ecosystems (Cerling 1984; Cerling et al. 1989; Quade et al. 1989; Cerling and Quade 1993). Accordingly, the C_3 end member for pedogenic carbonates in mesic and xeric regions is represented by $\delta^{13}C$ compositions of –12 and –9 permil respectively, whereas the end member for C_4 ecosystems ranges from +2 to +4 permil (Cerling and Quade 1993).

Koobi Fora is located in a semiarid region where dwarf shrubland dominates the ecosystem. Pedogenic carbonate from modern soils in the vicinity of Koobi Fora have $\delta^{13}C$ values from about –1 to –4 permil (table 12.3), which indicates a mixed C_3/C_4 biomass that is possibly skewed toward the C_4 end member. This is reasonable for the modern Turkana ecosystem which, on the east side of the lake, has only in the last two decades been subjected to intense grazing pressure that tends to reduce the fraction of grasses. On the east side of Lake Turkana, dominant food plants include C_4 grasses (e.g., *Aristida* spp., *Sporobolus* sp., *Dactyloctenium* sp.) and some other C_4 plants (e.g., *Euphorbia cuneata*), as well as C_3 dicots (*Acacia reficiens, A. nubica, A. mellifera*). The relatively undisturbed ecosystem by the east side of the lake has a much higher fraction of C_3 biomass than does the Serengeti grasslands of southern Kenya and northern Tanzania. We would predict that the biomass of the heavily grazed west side of the lake would have an even greater C_3 component.

Summary

The Lake Turkana Basin in northern Kenya is a very harsh environment but supports a variety of wildlife in spite of high mean annual temperature (29°C) and low annual rainfall (180 mm/year). The ecosystem is classified as a dwarf shrubland, similar in many respects to semidesert ecosystems in much of East Africa. Common mammals include tiang, oryx, Burchell's and Grevy's zebra, lion, leopard, cheetah, hippo, Grant's gazelle, dikdik, and spotted and striped hyena. Less common are the kudu, giraffe, gerenuk, black rhino, and, until the 1920s, elephant and buffalo.

Local meteoric waters range from about –30 to +15 permil in δD, and –5 to +1 permil in $\delta^{18}O$. Average local meteoric water in the Turkana Basin probably averages about –3 permil for $\delta^{18}O$. The evaporated waters of Lake Turkana range from about +5 to +7 in $\delta^{18}O$ over the annual cycle. Local biomass includes significant C_3 and C_4 biomass, in subequal quantities; this biomass contrasts with the wooded grasslands or grasslands of the Serengeti.

The $\delta^{13}C$ of hypergrazers in the region is more negative (+1 to +2 permil) than that of grazers in more mesic regions (+2 to +4 permil), indicating a subtle difference in their respective diets. This is probably due to the difference between arid adapted grasses (NAD, PCK, *Aristida*-NADP) and mesic grasses (classical-NADP). Alcelaphines (e.g., wildebeest, topi, tiang, hartebeest) and buffaloes are hypergrazers. In contrast, zebra, oryx, and warthogs are grazers that have a small but consistent fraction (ca. 10 to 15%) of C_3 biomass in their diets, perhaps in the form of forbs and herbs. Hippos from the Turkana Basin have a higher fraction of C_4 biomass than other parts of East Africa studied thus far. Isotopically they are distinguishable from hypergrazers. Based on their $\delta^{13}C$ values, dikdik and giraffes are hyperbrowsers whereas eland, Grant's gazelle, kudu, and the black rhino are browsers. Examples of closed canopy dwellers, which we have analyzed from other regions of East Africa and which exhibit very negative $\delta^{13}C$ values (–16 to –22 permil), are not found in the Turkana Basin today.

The $\delta^{18}O$ of nonobligate drinkers is from about +4 to +14 permil (relative to PDB for the carbonate fraction of apatite) and is more enriched than that of some obligate drinkers. Hippos that drink the Lake Turkana water have $\delta^{18}O$ values about +2 to +4 permil.

Comparison of $\delta^{13}C$ and $\delta^{18}O$ of fossil mammals and of fossil eggshells will be a useful indicator of paleoecology and paleophysiology.

Acknowledgments

We thank Craig Feibel, Nick Georgiadis, Louise Leakey, Katja Viehl, Jonathan Wynn, and the Kenya Wildlife Service for assistance in collecting some of the samples in this study; R. E. Leakey and R. Campbell for discussions of Lake Turkana wildlife; Alan Rigby and Ben Passey for sample preparation; F. H. Brown and T. P. Young for collection and identification of some of the plants in this study; and J. R. Ehleringer for access to the SIRFIR facility. This work was supported by the U.S. National Science Foundation.

References Cited

Allison, G. B., K. C. Colin, A. Filly, and J. C. Fontes. 1987. Measurement of isotopic equilibrium between water, water vapour and soil CO_2 in arid zone soils. *Journal of Hydrology* 95:131–141.

Bender, M. M. 1971. Variations in the $^{13}C/^{12}C$ ratios of plants in relation to the pathway of photosynthetic carbon dioxide fixation. *Phytochemistry* 10:1239–1245.

Bocherens, H., P. L. Koch, A. Marrioti, D. Geraads, and J. J. Jaeger. 1996. Isotope biogeochemistry of mammalian enamel from African Pleistocene hominid sites. *Palaios* 11:306–318.

Bryant, J. D., and P. N. Froelich. 1995. A model of oxygen isotope fractionation in body water of large mammals. *Geochimica et Cosmochimica Acta* 59:4523–4537.

Bryant, J. D., B. Luz, and P. N. Froelich. 1994. Oxygen isotopic composition of fossil horse tooth phosphate as a record of continental palaeoclimate. *Palaeogeography, Palaeoclimatology, Palaeoecology* 107:303–316.

Carr, M. J. 1976. Plant ecological variation and pattern in the lower Omo Basin. In Y. Coppens, F. C. Howell, G. L. Isaac, and R. E. Leakey, eds., *Earliest Man and Environments in the Lake Rudolf Basin: Stratigraphy, Paleoecology, and Evolution,* pp. 432–467. Chicago: University of Chicago Press.

Cerling, T. E. 1984. The stable isotopic composition of modern soil carbonate and its relationship to climate. *Earth and Planetary Science Letters* 71:229–240.

Cerling, T. E. 1986. A mass balance approach to basin sedimentation: Constraints on the recent history of the Turkana Basin. *Palaeogeography, Palaeoclimatology, Palaeoecology* 54:63–86.

Cerling, T. E. 1996. Pore water chemistry of an alkaline lake. In T. C. Johnson and E. Odada, eds., *The Limnology, Climatology, and Paleoclimatology of East African Lakes,* pp. 225–240. Cambridge: Cambridge University Press.

Cerling, T. E., and J. M. Harris. 1999. Carbon isotope fractionation between diet and bioapatite in ungulate mammals and implications for ecological and paleoecological studies. *Oecologia* 120:247–363.

Cerling, T. E., and J. Quade. 1993. Stable carbon and oxygen isotopes in soil carbonates. *American Geophysical Union Monograph* 78:241–248.

Cerling, T. E., and Y. Wang. 1996. Stable carbon and oxygen isotopes in soil CO_2 and soil carbonate: Theory, practice, and application to some prairie soils of the upper midwestern North America. In T. W. Boutton and S.-I. Yamasaki, eds., *Mass Spectrometry of Soils,* pp. 113–131. New York: Dekker.

Cerling, T. E., J. M. Harris, B. J. MacFadden, M. G. Leakey, J. Quade, V. Eisenmann, and J. R. Ehleringer. 1997. Global vegetation change through the Miocene/Pliocene boundary. *Nature* 389:153–158.

Cerling, T. E., J. Quade, Y. Wang, and J. R. Bowman. 1989. Carbon isotopes in soils and palaeosols as ecology and palaeoecology indicators. *Nature* 341:138–139.

Chapman, G. R. 1996. *The Biology of Grasses.* Wallingford, Australia: CAB International.

Cowling, R. M. 1983. The occurrence of C_3 and C_4 grasses in fynabos and allied shrublands in the South Eastern Cape, South Africa. *Oecologia* 58:121–127.

Craig, H. 1961. Isotopic variations in meteoric waters. *Science* 133:1702–1703.

Crowson, R. A., W. J. Showers, E. K. Wright, and T. C. Hoering. 1991. Preparation of phosphate samples for oxygen isotope analysis. *Analytical Chemistry* 63:2397–2400.

Deines, P. 1980. The isotopic composition of reduced organic carbon. In P. Fritz and J. Ch. Fontes, eds., *Handbook of En-*

vironmental Isotope Geochemistry. Vol. 1. *The Terrestrial Environment,* pp. 329–406. Amsterdam: Elsevier.

East African Meteorological Department. 1975. *Climatological Statistics for East Africa.* Part 1. *Kenya.* Nairobi: East African Meteorological Department.

Ehleringer, J. R., and T. A. Cooper. 1988. Correlations between carbon isotope ratio and microhabitat in desert plants. *Oecologia* 76:562–566.

Ehleringer, J. R., T. E. Cerling, and B. Helliker. 1997. C_4 photosynthesis, atmospheric CO_2, and climate. *Oecologia* 112: 285–299.

Ehleringer, J. R., C. B. Field, Z. F. Lin, and C. Y. Kuo. 1986. Leaf carbon isotope and mineral composition in subtropical plants along an irradiance cline. *Oecologia* 70:520–526.

Ehleringer, J. R., R. F. Sage, L. B. Flanagan, and R. W. Pearcy. 1991. Climate change and the evolution of C_4 photosynthesis. *Trends in Ecology and Evolution* 6:95–99.

Ellis, R. P., J. E. Vogel, and A. Fuls. 1980. Photosynthetic pathways and the geographical distribution of grasses in South West Africa/Namibia. *South African Journal of Science* 76: 307–314.

Epstein, S., and L. Zeiri. 1988. Oxygen and carbon isotopic composition of gases respired by humans. *Proceedings of the National Academy of Sciences* 85:1727–1731.

Gwynn, M. D., and R. H. V. Bell. 1968. Selection of vegetation components by grazing ungulates in the Serengeti National Park. *Nature* 220:390–393.

Hattersley, P. W. 1982. ^{13}C values of C_4 types in grasses. *Australian Journal of Plant Physiology* 9:139–154.

Hattersley, P. W. 1992. C_4 photosynthetic pathway variation in grasses (Poaceae): Its significance for arid and semi-arid lands. In G. W. Chapman, ed., *Desertified Grasslands: Their Biology and Management,* pp. 181–212. London: Linnean Society.

Hattersley, P. W., and Z. Roksandic. 1983. ^{13}C values of C_3 and C_4 species of Australian *Neurachne* and its allies (Poaceae). *Australian Journal of Botany* 31:317–321.

Hopson, A. J., ed. 1982. *Lake Turkana: A Report on the Findings of the Lake Turkana Project, 1972–1975.* London: Overseas Development Administration.

Ibrahim, K. M., and C. H. S. Kabuye. 1987. *An Illustrated Manuel of Kenya Grasses.* Rome: FAO.

Kanai, R., and G. F. Edwards. 1999. The biochemistry of C_4 photosynthesis. In R. F. Sage and R. K. Monson, eds., *C_4 Plant Biology,* pp. 49–87. San Diego: Academic Press.

Kingdon, J. 1979. *East African Mammals: An Atlas of Evolution in Africa.* Vol. 3, pt. B. *Large Mammals.* London: Academic Press.

Kingdon, J. 1982a. *East African Mammals: An Atlas of Evolution in Africa.* Vol. 3, pt. C. *Bovids.* London: Academic Press.

Kingdon, J. 1982b. *East African Mammals: An Atlas of Evolution in Africa.* Vol. 3, pt. D. *Bovids.* London: Academic Press.

Koch, P. J. 1998. Isotopic reconstruction of continental environments. *Annual Reviews of Earth and Planetary Science* 26:573–613.

Kohn, M. J. 1996. Predicting animal ^{18}O; accounting for diet and physiological adaptation. *Geochimica et Cosmochimica Acta* 60:4811–4829.

Kohn, M. J., M. J. Schoeninger, and J. W. Valley. 1996. Herbivore tooth oxygen isotope compositions: Effects of diet and physiology. *Geochimica et Cosmochimica Acta* 60: 3889–3896.

Lamprey, H. F. 1963. Ecological separation of large mammal species in the Tarangire Game Reserve, Tanganyika. *East African Wildlife Journal* 1:63–93.

Lee-Thorp, J. A., and N. J. van der Merwe. 1987. Carbon isotope analysis of fossil bone apatite. *South African Journal of Science* 83:712–715.

Longenelli, A. 1984. Oxygen isotopes in mammal bone phosphate: A new tool for paleohydrological and paleoclimatological research? *Geochimica et Cosmochimica Acta* 48: 385–390.

McNaughton, S. J., and N. J. Georgiadis. 1986. Ecology of African grazing and browsing mammals. *Annual Review of Systematics and Ecology* 17:39–65.

Pratt, D. J., Greenway, P. J., and M. D. Gwynne. 1966. A classification of East African rangeland with an appendix on terminology. *Journal of Applied Ecology* 3:369–382.

Quade, J., T. E. Cerling, and J. R. Bowman. 1989. Systematic variations in the carbon and oxygen isotopic composition of pedogenic carbonate along elevation transects in the southern Great Basin. *Geological Society of America Bulletin* 101:464–475.

Sage, R. F., M. Li, and R. K. Monson. 1999. The taxonomic distribution of C_4 photosynthesis. In R. F. Sage and R. K. Monson, eds., *C_4 Plant Biology,* pp. 551–596. San Diego: Academic Press.

Sarmiento, G. 1984. *The Ecology of Neotropical Savannas.* Cambridge, Mass.: Harvard University Press.

Schultze, E. D., R. P. Ellis, W. Schulze, P. Trimborn, and H. Ziegler. 1996. Diversity, metabolic types, and $\delta^{13}C$ carbon isotope ratios in the grass flora of Namibia in relation to growth form, precipitation, and habitat conditions. *Oecologia* 106:352–369.

van der Merwe, N. J., and E. Medina. 1989. Photosynthesis and $^{13}C/^{12}C$ ratios in Amazonian rain forests. *Geochimica et Cosmochimica Acta* 53:1091–1094.

Vogel, J. C., A. Fuls, and A. Danin. 1986. Geographical and environmental distribution of C_3 and C_4 grasses in the Sinai, Negev, and Judean deserts. *Oecologia* 70:258–265.

Vogel, J. C., A. Fuls, and R. P. Ellis. 1978. The geographical distribution of Kranz grasses in South Africa. *South African Journal of Science* 74:209–215.

Wang, Y., and T. E. Cerling. 1994. A model of fossil tooth enamel and bone diagenesis: Implications for stable isotope studies and paleoenvironment reconstruction. *Palaeogeography, Palaeoclimatology, Palaeoecology* 107:281–289.

Yuretich, R. F., and T. E. Cerling. 1983. Hydrogeochemistry of Lake Turkana, Kenya: Mass balance and mineral reactions in an alkaline lake. *Geochimica et Cosmochimica Acta* 47: 1099–1109.

TABLE 12.1 Climatological Data for Kenyan Sites Discussed

	Latitude	Longitude	Altitude	MAT	MAP	Source[b]
NAD-me and PEP-ck dominated						
Lodwar, Turkana	3.1	33.6	506	29.2	178	1
Samburu[a]	0.6	37.5	1016	24.0	600	2
Isiolo	0.4	37.6	1104	23.5	648	1
Mpala[a]	0.3	36.8	1700	18.5	500	2
Lewa[a]	0.2	37.5	1770	17.9	700	2
NADP-me dominated						
Nanyuki	0.0	37.1	1947	16.1	759	1
Nairobi (Wilson)	–1.3	36.8	1683	18.8	909	1

[a] Data for Samburu Game Reserve, Mpala, and Lewa Downs were estimated using comparative climatological data from nearby sites.

[b] 1 = East African Meteorological Department 1975; 2 = estimated from data in (1).

MAT = mean annual temperature.

MAP = mean annual precipitation.

TABLE 12.2 δD and $\delta^{18}O$ of Waters in the Turkana Basin

	δD	$\delta^{18}O$	Date
Precipitation			
Koobi Fora	5	1.9	Oct 75
Koobi Fora	15	1.0	Apr 77
Karari	17	1.9	Apr 77
Karari	–3	–1.4	Apr 77
Perennial water holes and pools			
Kubi Algi WH	–6	–2.2	Apr 77
Simba WH	12	1.0	Apr 77
Koobi Fora WH	22	2.1	Apr 77
Laga Bura Hasuma	–13	–1.0	Apr 77
Permanent springs, wells, and water holes			
Loiengalani spring	–23	–4.1	Aug 75
Derati well	–11	–2.9	Apr 77
Derati well	–29	–3.7	Aug 80
Derati well	—	–3.3	Aug 81
Burgi WH	–17	–2.7	Aug 80
Huran Hura WH	–31	–5.5	Apr 77
North Horr WH	—	–3.3	Aug 81
Flowing rivers			
Omo River	5	–1.2	Sep 79
Omo River	4	–0.7	Sep 79
Omo River[a]	9	–0.1	Feb 76
Turkwel River	10	0.4	Sep 79
Turkwel River	—	0.9	Aug 82
Lake Turkana			
Koobi Fora	37	5.8	Aug 75
Koobi Fora	40	6.1	Mar 77
Koobi Fora	39	5.6	Aug 80
Koobi Fora	—	5.5	Dec 80
Koobi Fora	—	6.4	Oct 81
Koobi Fora	—	5.8	Jun 82

[a] From Craig (1977).

TABLE 12.3 $\delta^{13}C$ and $\delta^{18}O$ of Modern Soils in the Turkana Basin

Sample	$\delta^{13}C$	$\delta^{18}O$
5103A	–2.3	3.7
5104	–3.8	2.7
5105	–1.7	4.8
5106	–1.8	5.2
5116	–3.3	2.5

Source: Data from Cerling et al. (1986).

TABLE 12.4 $\delta^{13}C$ and $\delta^{18}O$ of Enamel from Modern Mammals and Eggshell from Ostrich from Northern Kenya

Sample	$\delta^{13}C$	$\delta^{18}O$	Region	Common Name	Scientific Name	Tooth
Hyperbrowsers						
SU 3	−14.8	8.8	Turkana	Dik-dik	*Madoqua kirki*	M^3
SU 4	−10.3	4.8	Turkana	Dik-dik	*Madoqua kirki*	M^2
SU 2	−12.4	6.2	Turkana	Dik-dik	*Madoqua kirki*	M^2
SU96:10	−13.2	7.3	Turkana	Dik-dik	*Madoqua kirki*	M^3
SU96:11	−14.0	6.0	Turkana	Dik-dik	*Madoqua kirki*	M^3
SU96:17	−13.5	6.7	Turkana	Dik-dik	*Madoqua kirki*	M_3
SU96:18	−14.8	6.0	Turkana	Dik-dik	*Madoqua kirki*	M_3
SU96:21	−11.9	9.1	Turkana	Dik-dik	*Madoqua kirki*	M^3
K89.3	−12.2	7.4	Turkana	Giraffe	*Giraffa camelopardalis*	M_3
K97-390	−12.9	3.0	Lewa	Giraffe	*Giraffa camelopardalis*	M^3
SU96:12	−13.5	9.4	Turkana	Gerenuk	*Litocranius walleri*	M^3
Browsers						
K89.1	−10.2	2.3	Turkana	Black rhino	*Diceros bicornis*	M_3
96 JW 13	−12.8	8.6	Turkana	Gazelle	*Gazella grantii*	M^3
SU96:9	−9.5	7.3	Turkana	Porcupine	*Hystrix cristata*	M^3
K97-404E	−11.2	1.1	Samburu	Elephant	*Loxodonta africana*	P^4
K97-404E	−11.1	1.0	Samburu	Elephant	*Loxodonta africana*	P^4
K97-404E	−11.0	3.5	Samburu	Elephant	*Loxodonta africana*	M^1
K97-404E	−10.0	0.7	Samburu	Elephant	*Loxodonta africana*	P^4
SU96:5	−12.5	4.8	Turkana	Baboon	*Papio anubis*	M^3
K97-392	−12.3	2.3	Lewa	Eland	*Taurotragus oryx*	M^3
SU96:14	−12.0	14.1	Turkana	Greater kudu	*Tragelaphus strepsiceros*	M^3
SU96:23	−10.5	13.6	Turkana	Greater kudu	*Tragelaphus strepsiceros*	M_3
SU96:7	−11.7	11.1	Turkana	Greater kudu	*Tragelaphus strepsiceros*	M_3
OM 7764	−12.9	4.5	Isiolo	Greater kudu	*Tragelaphus strepsiceros*	M^2
Mixed: Ostriches						
75-neck	−7.0	4.9	Turkana	Ostrich	*Struthio camelus*	Eggshell
K95-209	−5.9	7.0	Turkana	Ostrich	*Struthio camelus*	Eggshell
KP 1	−5.0	9.4	Turkana	Ostrich	*Struthio camelus*	Eggshell
Loth-170	−8.6	5.9	Turkana	Ostrich	*Struthio camelus*	Eggshell
Mixed: Impala						
K97-383	−7.0	3.2	Lewa	Impala	*Aepyceros melampus*	M^3
K97-397	−7.6	3.2	Lewa	Impala	*Aepyceros melampus*	M^3
K97-404I	−7.4	2.0	Samburu	Impala	*Aepyceros melampus*	P^4
Grazers						
K89.6	−0.3	4.6	Turkana	Burchell's zebra	*Equus burchelli*	M^3
K97-233	−1.9	3.3	Mpala	Burchell's zebra	*Equus burchelli*	dP^4
K97-234	−0.6	4.0	Mpala	Burchell's zebra	*Equus burchelli*	P^4
K97-382	−0.1	5.3	Lewa	Burchell's zebra	*Equus burchelli*	M^2
K97-382	0.5	1.5	Lewa	Burchell's zebra	*Equus burchelli*	M^3
K97-384	−0.7	2.0	Lewa	Burchell's zebra	*Equus burchelli*	M^2
K97-387	−0.1	3.9	Lewa	Burchell's zebra	*Equus burchelli*	M^3

continued

TABLE 12.4 $\delta^{13}C$ and $\delta^{18}O$ of Enamel from Modern Mammals and Eggshell from Ostrich from Northern Kenya (Continued)

Sample	$\delta^{13}C$	$\delta^{18}O$	Region	Common Name	Scientific Name	Tooth
Grazers						
K97-388	0.2	3.6	Lewa	Burchell's zebra	*Equus burchelli*	M^3
20134	0.7	9.9	Turkana	Grevy's zebra	*Equus grevyi*	P^2
OM 7126	–0.1	1.6	Archer's Post	Grevy's zebra	*Equus grevyi*	M^2
OM 7127	–3.0	1.6	Archer's Post	Grevy's zebra	*Equus grevyi*	P^4
ET-161	–2.1	2.0	Turkana	Hippo	*Hippopotamus amphibius*	M^2
ET-162	–0.6	2.2	Turkana	Hippo	*Hippopotamus amphibius*	M_3
K89.2	–0.5	2.7	Turkana	Hippo	*Hippopotamus amphibius*	M^3
OM-6102B	–0.2	4.2	Turkana	Hippo	*Hippopotamus amphibius*	M_3
K97-393	0.1	1.2	Lewa	Waterbuck	*Kobus ellipsiprymnus*	M^2
K97-393	0.7	3.2	Lewa	Waterbuck	*Kobus ellipsiprymnus*	M^3
SU5	–0.3	6.1	Turkana	Oryx	*Oryx beisa*	M^2
SU 6	–3.3	5.1	Turkana	Oryx	*Oryx beisa*	M^2
SU-5	–1.8	4.3	Turkana	Oryx	*Oryx beisa*	M^3
SU96:15	–1.3	6.8	Turkana	Oryx	*Oryx beisa*	M_3
K89.9	–0.7	5.8	Turkana	Oryx	*Oryx beisa*	M^3
OM 1530	0.4	4.1	Turkana	Oryx	*Oryx beisa*	M^2
K89.5	–1.7	5.4	Turkana	Warthog	*Phacochoerus ethiopicus*	M^2
Hypergrazers						
K97-232	1.7	5.4	Mpala	Hartebeest	*Alcelaphus buselaphus jacksoni*	M^2
K97-120	1.2	8.8	Turkana	Topi	*Damaliscus lunatus*	M^1
K97-120	1.7	8.6	Turkana	Topi	*Damaliscus lunatus*	M^2
K97-121	1.1	0.5	Turkana	Topi	*Damaliscus lunatus*	M^2
K97-389	1.7	–2.6	Lewa	Buffalo	*Syncerus caffer*	P^2
K97-389	1.9	–1.8	Lewa	Buffalo	*Syncerus caffer*	P^3
K97-235	1.9	2.8	Mpala	Buffalo	*Syncerus caffer*	P^2
Carnivores						
K95-211	–7.1	1.9	Turkana	Jackel	*Canis mesomelis*	—
K98-347	–7.6	2.5	Mpala	Lion	*Panthera leo*	M^2
Domestic						
SU 7	0.4	7.8	Turkana	Cow	*Bos taurus*	M^2
SU96:20	2.4	2.9	Turkana	Cow	*Bos taurus*	M^3
96 JW 15	–14.4	6.1	Turkana	Camel	*Camelus dromedarius*	P^3
K89.4	–9.5	4.4	Turkana	Camel	*Camelus dromedarius*	M^2
96 JW 7	–13.1	8.7	Turkana	Goat	*Capra hircus*	M^2
K89.7	–2.7	0.6	Turkana	Sheep	*Ovis aries*	M_2
K95-205	–5.2	3.4	Turkana	Sheep or goat	*Ovis/Capra*	M_3
K95-207	–11.2	6.4	Turkana	Sheep or goat	*Ovis/Capra*	M_3
SU96:1	–4.2	1.9	Turkana	Sheep or goat	*Ovis/Capra*	M_2
SU96:13	–8.6	4.3	Turkana	Sheep or goat	*Ovis/Capra*	M_3
SU96:19	–4.1	3.6	Turkana	Sheep or goat	*Ovis/Capra*	M_3
SU96:6	–10.2	4.2	Turkana	Sheep or goat	*Ovis/Capra*	M^1

TABLE 12.5 $\delta^{13}C$ and $\delta^{15}N$ Values from Plants Collected in Northern Kenya, Including Koobi Fora, Lewa Downs, Mpala Ranch, and the Samburu Reserve

Family	Species	Region	Date	$\delta^{13}C$	$\delta^{15}N$	C/N
C_3 dicot[a]						
	Dicot	Koobi Fora	Jan 97	–25.9	9.2	7.5
	Dicot	Kanapoi	Jul 97	–27.8	5.6	11.1
	Dicot	Kanapoi	Jul 97	–25.6	5.2	16.6
Mimosaceae	*Acacia drepanolobium*	Mpala	Jul 97	–25.0	2.7	20.0
Mimosaceae	*Acacia elatior*	Koobi Fora	Jan 97	–26.2	—	—
Mimosaceae	*Acacia etbaica*	Mpala	Jul 97	–27.9	8.4	15.3
Mimosaceae	*Acacia horrida*	Samburu	Jul 97	–23.3	7.3	15.4
Mimosaceae	*Acacia nunica*	Kanapoi	Jul 97	–28.7	3.3	8.6
Mimosaceae	*Acacia reficiens*	Kanapoi	Jul 97	–26.7	3.8	17.3
Mimosaceae	*Acacia sennegal*	Kanapoi	Jul 97	–27.4	4.6	13.9
Mimosaceae	*Acacia seyel* var. *fistula*	Mpala	Jul 97	–28.4	6.4	10.9
Mimosaceae	*Acacia tortilis*	Koobi Fora	Jan 97	–25.8	6.2	24.7
Mimosaceae	*Acacia tortilis*	Koobi Fora	Jan 97	–24.6	7.0	21.5
Mimosaceae	*Acacia tortilis*	Samburu	Jul 97	–27.4	2.5	12.5
Mimosaceae	*Acacia tortilis*	Kanapoi	Jul 97	–29.2	2.8	17.3
Mimosaceae	*Acacia tortilis*	Kanapoi	Jul 97	–27.9	3.5	14.3
Balinitaceae	*Balanites orbicularis*	Koobi Fora	Jan 97	–29.2	—	—
Balinitaceae	*Balanites orbicularis*	Kanapoi	Jul 97	–28.0	3.4	16.2
Acanthanaceae	*Barleria acanthoides*	Koobi Fora	Jan 97	–27.0	—	—
Capparaceae	*Boscia coriacae*	Kanapoi	Jul 97	–28.4	5.9	13.1
Capparaceae	*Boscia coriacea*	Koobi Fora	Jan 97	–26.1	—	—
Capparaceae	*Cadaba rotundifolia*	Koobi Fora	Jan 97	–30.5	7.7	11.1
Combretaceae	*Combretum denhardtliorum*	Koobi Fora	Jan 97	–25.9	—	—
Burseraceae	*Commiphora* sp.	Samburu	Jul 97	–27.8	8.3	30.4
Burseraceae	*Commiphora schimperi*	Kanapoi	Jul 97	–27.4	5.4	17.2
Rubiaceae	*Conostyomium keniense*	Kanapoi	Jul 97	–27.9	12.6	9.0
Boraginaceae	*Cordia gharaf*	Koobi Fora	Aug 84	–24.9	—	—
Boraginaceae	*Cordia sinensis*	Koobi Fora	Jan 97	–26.4	5.3	18.2
Cucurbitaceae	*Cucumis* sp.	Kanapoi	Jul 97	–27.5	3.2	7.1
Salvadoraceae	*Dobera glabra*	Kanapoi	Aug 95	–25.4	—	—
Salvadoraceae	*Dobera glabra*	Kanapoi	Jul 97	–27.6	5.8	15.1
Asclepiadaceae	*Dregea abyssinica*	Kanapoi	Jul 97	–28.1	5.0	12.2
Euphorbiaceae	*Euphorbia cuneata*	Kanapoi	Jul 97	–27.4	6.7	11.6
Tiliaceae	*Grewia tembensis*	Koobi Fora	Jan 97	–27.6	6.1	17.5
Bignoniaceae	*Kigelia africana*	Samburu	Jul 97	–28.9	10.7	21.0
Euphorbiaceae	*Lawsonia inermis*	Koobi Fora	Jan 97	–27.1	—	—
Capparaceae	*Maerua angolensis*	Koobi Fora	Jan 97	–25.7	—	—
Malvaceae	*Pavonia* sp.	Koobi Fora	Aug 84	–25.6	—	—
Salvadoraceae	*Salvadora persica*	Koobi Fora	Jan 97	–26.5	—	—

continued

TABLE 12.5 $\delta^{13}C$ and $\delta^{15}N$ Values from Plants Collected in Northern Kenya, Including Koobi Fora, Lewa Downs, Mpala Ranch, and the Samburu Reserve (Continued)

Family	Species	Region	Date	$\delta^{13}C$	$\delta^{15}N$	C/N
C_3 dicot[a]						
Salvadoraceae	*Salvadora persica*	Kanapoi	Jul 97	–26.3	5.2	14.7
Salvadoraceae	*Salvadora*	Samburu	Jul 97	–27.9	7.3	12.7
Salvadoraceae	*Salvadora*	Samburu	Jul 97	–29.2	9.5	20.9
Amaranthaceae	*Sericocomopsis pallida*	Koobi Fora	Jan 97	–30.2	8.2	17.6
Pedaliaceae	*Sesamum atatum*	Kanapoi	Jul 97	–28.1	5.4	7.4
Rhamnaceae	*Zizyphus mauritania*	Kanapoi	Jul 97	–28.8	1.4	25.4
Rhamnaceae	*Zizyphus mauritania*	Koobi Fora	Jan 97	–26.1	5.1	17.5
C_3 monocot						
Palmae	*Hyphaene coriacea*	Koobi Fora	Jan 97	–24.6	5.6	77.8
Palmae	*Hyphaene coriacea*	Kanapoi	Jul 97	–27.0	0.8	29.5
CAM						
Liliaceae	*Aloe* sp.	Kanapoi	Jul 97	–12.7	3.8	68.5
C_4 dicot						
Compositae	*Blepheris linarfolia*	Kanapoi	Jul 97	–12.3	4.5	14.5
Zygophylleceae	*Tribulus cistoides*	Kanapoi	Jul 97	–12.4	1.1	8.0
C_3–C_4 intermediate						
Capparaceae	*Cleome allamanni*	Kanapoi	Jul 97	–18.0	3.6	9.3
Capparaceae	*Cleome* sp.	Koobi Fora	Aug 84	–16.8	—	—
C_4 monocot						
NADP-me[b]						
Gramineae	*Anthephora pubescens*	Turkana	Jul 98	–13.0	4.9	41.2
Gramineae	*Cenchrus* cf. *ciliarus*	Samburu	Jul 97	–12.8	10.5	17.2
Gramineae	*Cenchrus* cf. *ciliarus*	Samburu	Jul 97	–12.2	10.7	27.1
Gramineae	*Cenchrus ciliaris*	Mpala	Jul 97	–11.7	6.2	38.9
Gramineae	*Cenchrus pennisetiformia*	Turkana	Jul 98	–12.3	5.6	37.0
Gramineae	*Digetaria maitlandi*	Turkana	Jul 98	–12.9	4.7	19.5
Gramineae	*Digitaria* cf. *mombasana*	Mpala	Jul 97	–11.6	6.4	36.8
Gramineae	*Pennisetum schacelatum*	Samburu	Jul 97	–11.7	10.2	43.7
Gramineae	*Pennisetum schacelatum*	Samburu	Jul 97	–11.3	10.8	33.5
Gramineae	*Themeda triandra*	Lewa	Jul 97	–12.6	3.0	134.4
NADP-me[c]						
Gramineae	*Aristida kelleri*	Samburu	Jul 97	–14.4	6.2	26.8
Gramineae	*Aristida kelleri*	Samburu	Jul 97	–13.6	6.4	21.9
Gramineae	*Aristida adoensis*	Lewa	Jul 97	–11.8	5.1	61.4
Gramineae	*Aristida mutabilis*	Turkana	Jul 98	–14.1	6.6	21.6
Gramineae	*Aristida* sp.	Turkana	Jul 98	–12.9	0.8	42.3
Gramineae	*Aristida* sp. cf. *keniensis*	Mpala	Jul 97	–13.0	2.8	51.5
NAD-me[d]						
Gramineae	*Eragrositis racemosa*	Samburu	Jul 97	–13.2	5.1	58.0
Gramineae	*Eragrostis schweinfurthii*	Lewa	Jul 97	–13.3	3.8	94.4

TABLE 12.5 $\delta^{13}C$ and $\delta^{15}N$ Values from Plants Collected in Northern Kenya, Including Koobi Fora, Lewa Downs, Mpala Ranch, and the Samburu Reserve (Continued)

Family	Species	Region	Date	$\delta^{13}C$	$\delta^{15}N$	C/N
C_4 monocot						
NAD-me[d]						
Gramineae	*Eragrostis superba*	Mpala	Jul 97	–13.4	2.8	49.0
Gramineae	*Eragrostis superba*	Mpala	Jul 97	–12.9	3.7	77.0
Gramineae	*Harpachne schimperi*	Mpala	Jul 97	–13.0	4.2	56.6
Gramineae	*Harpachne schimperi*	Mpala	Jul 97	–12.6	5.2	68.6
Gramineae	*Sporobolus ioclades*	Samburu	Jul 97	–12.3	5.1	65.8
Gramineae	*Sporobolus kentrophyllus*	Turkana	Jul 98	–13.7	9.3	32.0
Gramineae	*Sporobolus spikatis*	Koobi Fora	Jan 97	–13.8	4.9	33.2
Gramineae	*Sporobolus spikatis*	Koobi Fora	Jan 97	–14.0	—	—
Gramineae	*Sporobolus spikatis*	Turkana	Jul 98	–12.5	4.8	41.5
PEP-ck[e]						
Gramineae	*Chloris roxberghiana*	Mpala	Jul 97	–13.2	3.8	28.1
Gramineae	*Chloris roxberghiana*	Mpala	Jul 97	–12.7	4.2	27.2
Gramineae	*Chloris virgata*	Mpala	Jul 97	–14.0	12.9	12.9
Gramineae	*Chloris virgata*	Mpala	Jul 97	–12.6	14.4	31.6
Gramineae	*Chloris virgata*	Mpala	Jul 97	–11.3	5.4	59.4
Gramineae	*Chloris virgata*	Mpala	Jul 97	–12.8	13.4	46.4
Gramineae	*Chloris virgata*	Turkana	Jul 98	–13.6	6.9	24.6
Gramineae	*Dactyloctenium aegyptum*	Mpala	Jul 97	–13.2	6.0	38.6
Gramineae	*Dactyloctenium aegyptum*	Mpala	Jul 97	–12.6	5.3	58.7
Gramineae	*Dactyloctenium aristatum*	Turkana	Jul 98	–12.6	5.9	22.6
Gramineae	*Eriochloa meyerana*	Turkana	Jul 98	–12.7	3.1	21.7
Gramineae	*Microchloa kunthii*	Mpala	Jul 97	–12.7	6.1	35.0
Gramineae	*Microchloa kunthii*	Mpala	Jul 97	–12.0	4.9	57.9
Gramineae	*Rhynchelytrum repens*	Mpala	Jul 97	–12.4	3.6	48.6
Gramineae	*Tragus berteronianus*	Mpala	Jul 97	–13.7	3.6	44.3

[a] (All. $\delta^{13}C$ = –27.2 ± 1.5 permil; n = 46); (Jan. 1997; $\delta^{13}C$ = –26.9 ± 1.7 permil; n = 16); (July 1997; $\delta^{13}C$ = –27.6 ± 1.3 permil; n = 27).

[b] (*Aristida*). ($\delta^{13}C$ = –12.2 ± 0.6 permil; n = 10).

[c] (*Aristida*). ($\delta^{13}C$ = –13.3 ± 0.9 permil; n = 6).

[d] ($\delta^{13}C$ = –13.2 ± 0.6 permil; n = 11).

[e] ($\delta^{13}C$ = –12.8 ± 0.7 permil; n = 15).

TABLE 12.6 $\delta^{13}C$ and $\delta^{18}O$ of Enamel from Some Other Modern Mammals from East Africa

	$\delta^{13}C$	$\delta^{18}O$	Region	Country	Year of death (approx)
Hypergrazers					
White-bearded wildebeest: *Connochaetes taurinus*					
K97-283	1.9	0.3	Nairobi	Kenya	1997
K97-283	1.9	–0.1	Nairobi	Kenya	1997
K97-305	1.9	3.1	Nairobi	Kenya	1997
K97-305	2.4	–0.2	Nairobi	Kenya	1997
K97-305	2.2	2.1	Nairobi	Kenya	1997
K97-310	3.2	0.6	Nairobi	Kenya	1997
K97-310	3.0	2.3	Nairobi	Kenya	1997
K97-311	3.1	1.6	Nairobi	Kenya	1997
K97-311	3.5	0.0	Nairobi	Kenya	1997
Grazers					
White rhino: *Ceratotherium simum*					
OM 7528	–2.1	0.3	Laikipia	Kenya	1988
GNP-wrh	0.6	1.3	Garamba	Zaire	1996
OM reg#655	1.1	4.0	W. Madi	Uganda	1950
OM 2186	1.4	4.6	—	Sudan	1951
Burchell's zebra: *Equus burchelli*					
OM 2380	0.3	–0.4	Kitengela	Kenya	1969
OM 2391	2.1	3.2	Kitengela	Kenya	1969
OM 2394	1.6	2.4	Kitengela	Kenya	1969
OM 2398	1.9	–0.1	Kitengela	Kenya	1969
OM 2399	1.7	1.3	Kitengela	Kenya	1969
Browsers					
Grant's gazelle: *Gazella granti*					
OM 1254	–12.2	2.3	Nairobi N.P.	Kenya	1969
OM 1264	–10.7	0.1	Nairobi N.P.	Kenya	1969
OM 1298	–8.5	0.4	Nairobi N.P.	Kenya	1969
OM 1309	–10.9	2.7	Nairobi N.P.	Kenya	1969
OM 1313	–7.6	–0.3	Nairobi N.P.	Kenya	1969
OM 1316	–10.9	2.8	Nairobi N.P.	Kenya	1969
Elephant: *Loxodonta africana*					
MGL-18	–13.0	–0.2	Meru	Kenya	1993
MGL-19	–13.0	0.7	Meru	Kenya	1993
MGL-20	–11.9	–0.4	Meru	Kenya	1993
MGL-21	–12.9	0.2	Meru	Kenya	1993
MGL-22	–12.1	–0.4	Meru	Kenya	1993
Black rhino: *Diceros bicornis*					
ODW 4	–11.5	1.6	Chyulu Hills	Kenya	1995
OM 2195	–12.7	1.2	Ngureman	Kenya	1969
OM 5482	–12.6	1.6	East Tsavo	Kenya	1976

TABLE 12.6 $\delta^{13}C$ and $\delta^{18}O$ of Enamel from Some Other Modern Mammals from East Africa (Continued)

	$\delta^{13}C$	$\delta^{18}O$	Region	Country	Year of death (approx)
Browsers					
Black rhino: *Diceros bicornis*					
OM 2180	−12.2	0.9	Makindu	Kenya	1956
OM 2742	−10.2	2.4	Ngureman	Kenya	1971
K89.1	−10.2	2.3	Koobi Fora	Kenya	1982
Eland: *Taurotragus oryx*					
ODW 7	−13.2	3.4	Chyulu Hills	Kenya	1995
OM 5219	−10.7	1.2	Athi plains	Kenya	1974
OM 5299	−10.2	3.4	Nairobi N.P	Kenya	1974
OM 5213	−7.5	4.9	Athi plains	Kenya	1974
K97-392	−12.3	2.3	Lewa	Kenya	1997
Hyperbrowsers					
Giraffes: *Giraffa camelopardalis*					
OM 2214	−14.9	0.5	Kitengela	Kenya	1970
OM 2278	−14.3	3.4	Athi River	Kenya	1968
ODW 6	−12.1	4.2	Chyulu Hills	Kenya	1995
94Olor253	−12.9	11.0	Olorgesailie	Kenya	1994
K89.3	−12.2	7.4	Koobi Fora	Kenya	1974
K97-201	−12.0	4.6	Laikipia	Kenya	1997
Forest hog: *Hylochoerus meinerzhageni*					
OM 2142	−18.8	−1.5	Kapenguria	Kenya	1965
OM 2143	−16.5	−0.6	Aberdares	Kenya	1939
QEP-Luigi	−14.5	−0.4	Queen Eliz. Park	Uganda	1998
Bongo: *Boocercus eurycerus*					
OM 1610	−14.6	−0.2	Aberdares?	Kenya	Unknown
OM 1628	−15.1	−2.6	Aberdares?	Kenya	Unknown
Okapi: *Okapia johnsoni*					
OM 2671	−20.0	−0.2	Unknown	Congo	Unknown
JAH-A-P3	−22.3	−0.4	Epulu	Congo	1992
JAH-A-P4	−20.7	−0.9	Epulu	Congo	1992
JAH-A-P2	−20.3	−0.9	Epulu	Congo	1992

12.2

Isotope Paleoecology of the Nawata and Nachukui Formations at Lothagam, Turkana Basin, Kenya

Thure E. Cerling, John M. Harris, and Meave G. Leakey

The Lothagam sequence is the oldest continuous section in East Africa that postdates the global ^{13}C event marking the expansion of C_4 plants. Its isotope ecology was investigated from pedogenic carbonate and fossil tooth enamel. Proboscideans record a change from a C_3-dominated diet at Nakali and in the Samburu Hills to a C_4 diet throughout the Lothagam sequence. Lothagam equids had a predominantly C_4 diet, but the earliest samples had a mixed C_3/C_4 diet. Suids were slower to exploit a C_4 diet than were equids or proboscideans. Rhinocerotids had developed browsing and grazing specializations by Apak Member times. The Lothagam hippopotamids generally had a mixed C_3/C_4 diet. In general, the diets of the fossil grazers were similar to those of grazers in the basin today, and it is likely that the Lothagam ecosystem was represented by a dwarf shrubland. The oxygen isotope evidence indicates that local meteoric water was enriched in ^{18}O by several permil compared to modern waters. The consistently depleted $\delta^{18}O$ values for hippopotamids suggests that they were riverine rather than lake dwellers. The most likely explanation for isotopic depletion of meteoric waters is a combination of slightly higher rainfall, slightly lower evaporation, and more rainfall derived from the west.

The stable carbon and oxygen isotope values of paleosols, fossil tooth enamel, and fossil eggshell provide important information about the paleoecology and paleophysiology of mammals and birds (Cerling 1984; Cerling et al. 1988, 1989, 1993, 1997; Lee-Thorp and van der Merwe 1987; Bryant et al. 1994). The $\delta^{13}C$ fraction in soils and paleosols is an indicator of the relative abundance of C_4 biomass in the ecosystem, whereas that in teeth and eggshell reflects the proportion of dietary C_4 plant matter. The $\delta^{18}O$ fraction is related to the isotopic composition of soil and body water and is a useful way to study paleoclimatology and paleophysiology, respectively.

Between them, the paleosols and fossils of the Lothagam sequence in the Turkana Basin of Kenya provide a nearly continuous record of ecology and climate from the Late Miocene into the Pliocene. The geology and dating of the Lothagam sequence is presented elsewhere (McDougall and Feibel 1999; Feibel this volume:chapter 2.1) and is not reviewed here. However, the samples are placed in the stratigraphic context according to section 107 of Powers (1980).

Methods

The stable isotopic composition of carbonates from paleosols and from eggshell, and of biogenic apatite from fossil mammals, were determined using standard methods (Cerling et al. this volume:chapter 12.1) and are reported using the standard permil (parts per thousand) isotopic notation

$$\delta^{13}C\ (^{18}O) = (R_{sample}/R_{standard} - 1) * 1000$$

where R_{sample} and $R_{standard}$ are the $^{13}C/^{12}C$ ($^{18}O/^{16}O$) ratios in the sample and standard, respectively. For carbon, the standard is PDB (= Pee Dee Belemnite). For oxygen, two different standards are used: waters are reported relative to the SMOW (Standard Mean Ocean

Water) standard, and carbonates and biogenic phosphates are reported relative to the PDB standard. Analytic procedures are listed in the preceding contribution (Cerling et al. this volume:chapter 12.1).

Isotope Ecology of the Lothagam Sequence

The study of paleoecology using stable isotopes is relatively new, and the rules are still being established. A great advantage of undertaking this study in the Turkana Basin is that there is a reasonably diverse extant fauna, and a large selection of fossil material is available for study. The preceding contribution (Cerling et al. this volume:chapter 12.1) represents the first attempt at an isotope ecology study of the Turkana Basin. A few other studies of modern isotope ecosystems have been published (e.g., Koch et al. 1991), but these are relatively incomplete and rarely do they include plants, animals, soils, and waters. However, important progress is being made (Koch 1998) and it is likely that within the next decade stable isotope analysis will become a routine part of paleoecological interpretation.

Diagenetic alteration of phosphate minerals is a potentially serious problem that affects isotopic analyses of bone and dentine samples (Kolodny et al. 1996), but it is thought to be much less of a factor for enamel samples (Quade et al. 1992; Wang and Cerling 1994). Wang and Cerling (1994) showed that oxygen isotope exchange is more seriously affected than carbon isotope exchange during diagenesis of biogenic apatites but that tooth enamel appears to be unaffected by diagenesis. Cerling and Sharp (1996) came to a similar conclusion based on the correlation between the $\delta^{18}O$ (CO_3) component and whole enamel $\delta^{18}O$. The Lothagam sequence provides another important test of the effect of diagenetic alteration. Those samples collected specifically for this project have detailed provenance. Hippo and equid teeth collected from the same stratigraphic bed should have undergone identical diagenetic histories. However, as shown by figure 12.8, equid teeth are always more enriched in ^{18}O than are those of hippos collected from the same bed, and this reflects the results obtained for their extant representatives from the Turkana Basin (see table 12.4 in the preceding contribution). The ^{18}O enrichment is expected because equids are likely to supplement more of their body fluids from the vegetation that provides their diet, and which is enriched in ^{18}O due to evaporation, than are hippos that spend much of their time in water. Lake Turkana waters are enriched in ^{18}O by about 6 permil compared to local meteoric waters because Lake Turkana is a closed basin lake that is undergoing evaporation. The teeth of hippos living in the lake today have an ^{18}O value of ca. +2 permil, whereas the results for zebra living on the lake shores are ca. +4 to +5 permil. However, most of the fossil equids from Lothagam are 4 to 7 permil more enriched in ^{18}O than are the contemporaneous hippos, probably because the Lothagam hippos were living in a river, or in a lake with an outlet, and hence had access to relatively unevaporated water.

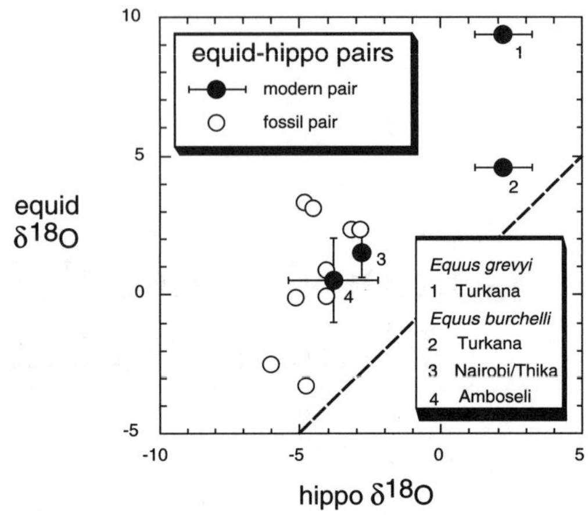

Figure 12.8 Plot of $\delta^{18}O$ content of fossil and recent hippo and equid teeth. Fossil hippopotamids and equids provide an important test for diagenesis of fossil tooth enamel. Fossil samples (open symbols) were collected from the same sedimentary bed in the Lothagam sequence and have undergone identical diagenetic histories. Modern coexisting samples from Lake Turkana are shown for reference (closed symbols). Hippopotamids always have lower $\delta^{18}O$ than equids have, and this depletion indicates that diagenesis has not homogenized the stable isotope distribution of the fossil samples. Diagenesis has a greater effect on $\delta^{18}O$ than on $\delta^{13}C$ (Wang and Cerling 1994).

Lothagam Paleosols

The Lothagam sequence has some well-preserved paleosols. These are recognized in mudstones by bioturbation features that destroy the original sedimentary fabric. Ped structure, with oriented clays, is preserved and is accompanied by large- and small-scale slickensided surfaces. The mudstone matrix is leached of primary carbonate for the paleosols selected for study. Bedded mudstones and sandstones at Lothagam are sometimes calcareous, although not always. Carbonate nodules are present in many of the paleosols and are found beginning about 30 to 60 cm below the top of the bioturbated zone. Nodules of similar appearance are sometimes found at the base of channel sandstones as reworked clasts. Paleosol carbonates selected for study were from clearly recognizable paleosols based on the presence of the features described in the preceding contribution. The paleosol carbonates had two forms: nodules that ranged in size from about 3 to 15 mm diameter, and soil rhizoliths.

There is a weak correlation between $\delta^{13}C$ and $\delta^{18}O$ in paleosol carbonates from the Lower Nawata (r^2 =

.45), a poor correlation in the Upper Nawata (r^2 = .25), and no correlation in the Apak Member (r^2 = .03). Cerling and Quade (1993) discussed the relationship between the $\delta^{13}C$ of paleosol carbonate and the fraction of C_4 biomass in the local ecosystem. Basically, the $\delta^{13}C$ in paleosol carbonate is enriched by 14 to 17 permil compared to organic matter. Characteristic $\delta^{13}C$ values for soil organic carbon for pure C_3 ecosystems are about –22 permil for arid ecosystems, about –26 permil for "normal" C_3 ecosystems, and about –33 permil for closed canopy ecosystems (see discussion in Cerling and Quade 1993). However, pedogenic carbonate rarely forms in dense closed canopy settings. Therefore, taking into account the 14–17 permil enrichment, we construe $\delta^{13}C$ values for pedogenic carbonates formed under C_3 ecosystems to range between –12 and –8 permil. C_4 plants have a more restricted range of $\delta^{13}C$ values, and we construe the $\delta^{13}C$ range of pedogenic carbonates formed under pure C_4 ecosystems to vary from +1 to +4 permil.

The $\delta^{13}C$ of pedogenic carbonate in the Lothagam sequence (table 12.7) provides evidence of habitats that include mixed C_3/C_4 ecosystems and C_3-dominated ecosystems. For comparison, modern pedogenic carbonate in the Koobi Fora region (dwarf shrubland) provides $\delta^{13}C$ values of about –2 to –4 permil, whereas Olduvai Gorge (grassland) yields positive $\delta^{13}C$ values.

The sequence from the Lower Nawata through the Apak Member includes both C_3-dominated (>75% C_3) paleosols (ca. –7 to –9 permil) and mixed C_3/C_4 paleosols with a significant C_4 fraction (–5 to –2 permil). Mixed ecosystems, such as wooded grasslands and grassy woodlands where significant grass was present, were common. Pure C_3 ecosystems, such as riparian forests, were also represented. No paleosol from anywhere in the sequence indicates pure grassland conditions. While this does not mean that no grasslands were present, it implies that long-lived grasslands were not abundant in the region. The average $\delta^{13}C$ values of –5.3, –4.4, and –5.6 permil for the Lower Nawata, Upper Nawata, and Apak Member, respectively, suggest that more than 50 percent of the total biomass was C_3 plants.

Previous studies in the Koobi Fora region of the Turkana Basin (Cerling et al. 1988) show similar, rather ^{13}C-depleted paleosols until about 1.8 Ma when C_4 plants became more abundant. However, at no time in the last 7.5 million years is there isotopic evidence from paleosols of C_4 grasslands (where $\delta^{13}C > 0$ permil). Grassland in the modern Koobi Fora ecosystem is mainly restricted to a 100 m swathe along the edge of the lake.

The $\delta^{18}O$ values of pedogenic carbonate are also instructive (table 12.7). The Lower Nawata and Upper Nawata have average $\delta^{18}O$ values of –3.6 and –3.4 permil, and the Apak Member has an average $\delta^{18}O$ value of –0.7 permil. In contrast, modern pedogenic carbonates in the Turkana Basin average +3.6 permil. The difference between the Mio-Pliocene and modern values is striking and implies significant differences in the isotopic composition of meteoric water.

Most (10 of 14) of the Lower Nawata pedogenic carbonates have $\delta^{18}O$ values between –3 and –6 permil. This is compatible with local meteoric water having $\delta^{18}O$ values of about –5 to –8 permil relative to SMOW (Cerling and Quade 1993). By comparison, local meteoric water in the region today is highly evaporated—about –3 to 0 permil. This preliminary paleosol evidence suggests that, compared to modern local water, for much of the Lower Nawata the isotopic content of meteoric water was several permil depleted in ^{18}O. This may be explained in several ways. A change in the intertropical convergence pattern may have brought more moisture from West Africa. Alternatively, higher rainfall with lower evaporation rates would result in a depletion of ^{18}O in local soil waters.

The $\delta^{18}O$ fraction of Upper Nawata pedogenic carbonates ranges from –1.6 to –5.6 permil and is significantly depleted in ^{18}O compared to modern soils. On the other hand, the Apak Member paleosols, where the $\delta^{18}O$ is significantly enriched compared to Nawata Formation paleosols but less so than in modern soils, may indicate a drier and possibly hotter regime than that represented by the underlying Nawata members.

Cerling et al. (1988) noted a significant change in the $\delta^{18}O$ values of paleosol carbonate in the Koobi Fora Formation at about 1.8 Ma. The values measured throughout most of the Nawata Formation and Apak Member are similar to the values measured below the KBS Tuff. Unfortunately, there is a major disconformity in the Koobi Fora sequence between about 1.9 and 2.4 Ma, and thus the timing of this major change in the isotopic composition of meteoric water in East Africa may not be more precisely bracketed.

In summary, the stable carbon isotope evidence from paleosols at Lothagam indicate a mosaic ecosystem with stands of pure C_3 vegetation interspersed with mixed C_3/C_4 floras. There is no paleosol evidence for pure C_4 grasslands, but our samples represent only a small part of the entire sequence and do not preclude the presence of local or short-lived grasslands. Oxygen isotopes indicate a different climate regime from today, with less evaporation and, probably, higher rainfall. It is possible that different wind patterns brought more moisture from the west than is encountered today.

$\delta^{13}C$ Analysis of Lothagam Mammals

As discussed by Cerling et al. (this volume:chapter 12.1), the $\delta^{13}C$ composition of mammalian enamel is an indicator of diet. Animals with a pure C_3 diet may have $\delta^{13}C$ values between about –21 and –8 permil. This enormous range reflects the total range of $\delta^{13}C$ in C_3 plants under different environmental conditions such as closed canopy conditions (about –16 to –22 permil), vegetation without significant water stress (about –11 to –15 permil), and vegetation impacted by water stress or containing a small fraction of C_4 biomass (about –11 to –8 permil) (Ehleringer et al. 1986; Ehleringer and Cooper 1988).

Mammals with a pure C_4 diet have $\delta^{13}C$ values between about +3 and +4 permil, as exemplified by extant wildebeest from Kitengela in Kenya (Cerling and Harris unpublished data). Sited in the Athi plains near Nairobi, the Kitengela region has abundant grasses, most of which use the NADP-me subpathway for C_4 photosynthesis. Other grazers from this locality, such as Burchell's zebra, *Equus burchelli*, have $\delta^{13}C$ values between about +1 and +3 permil, which indicates they have a small but important fraction of C_3 biomass in their diet. In contrast, *Equus burchelli* from regions such as the dwarf shrubland of the Koobi Fora region of Kenya, where grasses are less abundant and more grasses use the slightly ^{13}C-depleted PEP-ck or NAD-me C_4-subpathway, have $\delta^{13}C$ values that are slightly more negative (ca. –2 to 0 permil).

Mixed C_3/C_4 diets, with $\delta^{13}C$ values between –7 to –3 permil, are not common in modern mammals of East Africa (Cerling et al. this volume:chapter 12.1; Cerling and Harris unpublished data).

Equids

Equids first appear in East Africa at about 10.5 Ma, when hipparionines became widespread through the Old World. We analyzed samples of equids from Nakali and from the Samburu Hills, in addition to those from Lothagam. The Samburu Hills sequence overlaps in part with the lowest part of the Lower Nawata. The sediments at Nakali are older than either the Samburu Hills or the Lothagam successions. Results of the isotopic analysis of fossil equids are given in table 12.8.

The Nakali equids, the earliest from East Africa, have $\delta^{13}C$ values that indicate a pure C_3 diet ranging from –9.2 to –11.5 permil (table 12.8). This corroborates the work both of Morgan et al. (1994), who found that equids from the upper part of the Ngorora Formation in the Baringo Basin had a C_3 diet, and of Cerling et al. (1997), who showed a transition from C_3 to C_4 diets in equids in North America, Africa, and Asia in the latest Miocene. Transition to a C_4 diet took place at a time represented within the Samburu Hills succession and by the lowest part of the Lower Nawata. A mixed C_3/C_4 diet, with $\delta^{13}C$ values between –3 and –5 permil is found in about half (three of seven) of the samples from the Samburu Hills and in the lower 82 meters of the Lower Nawata (three of six samples; table 12.8). All samples from later in the Lower Nawata or younger horizons (n = 14; table 12.8) have $\delta^{13}C$ values more positive than –2 permil. The Late Miocene dietary transition for equids from a pure C_3 diet to one containing more than 75 percent C_4 is also observed in Pakistan and in southern North America (n = 38). In both Pakistan and North America, a few samples from the transition period have $\delta^{13}C$ values that indicate a mixed C_3/C_4 diet. The hypsodonty of Late Miocene equid teeth suggests that these animals were already grazers and feeding from C_3 grasses (but see MacFadden et al. 1999). The rapid nature of the transition, and the near synchroneity of the transition in widespread regions, indicates a rapid and nearly worldwide expansion of C_4 vegetation (mainly grasses) in the latest Miocene. Intermediate dietary values from this time interval may indicate a period of transition in the grasslands where a mixture of C_3 and C_4 grasses coexisted in the low elevation tropics for a few hundred thousand years.

After about 7 million years ago, virtually all the lowland grasses in Africa were C_4 grasses and East African equids had a diet that was dominated (>75%) by C_4 biomass. The upper part of the Lower Nawata, the Upper Nawata, and the Apak Member are characterized by a C_4-dominated diet for the equids, with values ranging from –1.8 to –0.2 permil. Isotopic values derived from the upper part of this range are similar to those of extant zebras from Koobi Fora, but are more negative than for extant zebras from the Athi Plains grassland region in Kenya (+0.5 to +2.5 permil, n = 15). This difference can be explained in one of two ways. Either the equids from the Lothagam sequence had a higher fraction of C_3 biomass in their diet than extant zebras from grasslands, or the local C_4 grasses were depleted in ^{13}C because they were dominantly NAD-me or PEP-ck C_4 grasses (see discussion in Cerling et al. this volume:chapter 12.1).

The stable carbon isotope results from fossil equids retrieved from Lothagam are compatible with a dwarf shrubland or bushland ecosystem and do not indicate extensive grasslands. Bernor and Harris (this volume:chapter 9.2) interpret *Eurygnathohippus turkanense* as adapted to closed woodland and not capable of sustained swift locomotion, whereas *E. feibeli* was interpreted as adapted to more open habitat and for swift locomotion. Isotopic analysis of the limited number of equid samples identified to species indicates that both equids were grazers, and the diets of both displayed an increase in C_4 fraction from Lower to Upper Nawata (i.e., reflecting the trend seen in suids). Despite behavioral differences inferred from limb proportions and cranial anatomy, the two equid species had similar diets.

Hippopotamids

Fossil hippopotamids are abundant throughout the Lothagam succession. Nearly all the fossils recovered to date have been assigned to the species *Hexaprotodon harvardi*; smaller and larger species are present but rare (Weston this volume:chapter 10.1). Isolated teeth collected from the Lothagam sequence for isotopic analysis almost certainly belong to *H. harvardi.* One sample from the Namurungule Formation in the Samburu Hills was also analyzed.

The two oldest samples, one from the lowest part of the Lower Nawata (+9 m) and one from the Namurungule Formation in the Samburu Hills, indicate C_3-dominated diets with $\delta^{13}C$ values of –7.8 and –9.1 permil, respectively. These results could represent a diet of water-stressed C_3 plants or, more likely, a C_3-dominated diet with a small component of C4 grasses and sedges.

Hippopotamids sampled from 62 m above the base of the Lower Nawata had a significant C_4 component in their diet (–5 permil; ca. 20 to 60% C_4; figure 12.8); by 76 m above the base of the Lower Nawata some individuals have a C_4-dominated diet (1.7 permil; ca 50 to 80% C_4; figure 12.8). Extant hippopotamuses from Lake Turkana (n = 3, table 12.9) have $\delta^{13}C$ values that range from –2.1 to –0.6 permil, which represents a diet dominated by C_4 grasses and sedges; at the other extreme, extant hippos from the Baringo region (n = 2; table 12.9) have more negative $\delta^{13}C$ values, which are indicative of a virtually pure C_3 diet. The most negative sample from Baringo, NL-1, was from an infant that died of starvation during a drought and may not be representative of the average population. The values measured for the Lake Baringo specimens are intriguing because the Baringo hippos are observed to come out at night to feed on the short grasses surrounding the lake; however, extensive bush has recently encroached on Lake Baringo so that grasses are less abundant than along the shores of Lake Turkana. Moreover, aquatic grasses and sedges can use the C_3 pathway.

From the middle of the Lower Nawata through the Apak Member the dietary preferences of hippopotamids range from virtually pure C_3 (–9 permil) to almost pure C_4 (+1 permil; table 12.9). Some East African wetland grasses and sedges (such as *Oriza* and *Phragmites*) use the C_3 pathway, and the high C_3 fraction in the diet of hippos could thus be due to ingestion of C_3 grasses or sedges from dambo wetlands or semi-aquatic C_3 grasses growing near shore in lakes or swamps, or they may be due to exploitation of C_3 browse. However, Kingdon (1979) reports that modern hippos only rarely eat semi-aquatic grasses.

Rhinocerotids

The two extant East African rhinos have very different diets, *Ceratotherium simum* being a grazer whereas *Diceros bicornis* is a browser (with $\delta^{13}C$ values of about +1 and –12 permil, respectively; table 12.10). The fossil record from Lothagam is somewhat difficult to interpret, in part because we have analyzed only a few rhinocerotids from Lothagam. Rhinocerotid samples from the earlier Nakali formation indicate a C_3-dominated diet.

Equids from the lowest part of the Lower Nawata had a significant C_4 component in their diet, so C_4 biomass was available at this time. However, samples of both *Brachypotherium lewisi* and *Ceratotherium praecox* from the Lower Nawata indicate a C_3 dietary preference (table 12.10). *Ceratotherium* specimens from the Upper Nawata (LT 23772) and Kaiyumung Member (LT 26283) were C_4 grazers. The sole *Brachypotherium* tooth from the Apak Member (LT 100) indicated a diet of mixed C_3 and C_4 vegetation. Of the unidentified rhino enamel fragments, those from the Lower Nawata were all C_3 browsers; one sample from the Apak Member was from a C_3 browser (perhaps *Diceros*?), but five others were C_4 grazers (perhaps *Ceratotherium*?).

Proboscideans

Fossil proboscideans from Lothagam include deinotheres, gomphotheres, and elephantids (Tassy this volume:chapter 8.1, Harris this volume:chapter 8.2). Deinotheres had a C_3 diet throughout their known history, which can be taken as further evidence for isotopic fidelity during diagenesis. In contrast, gomphotheres and elephantids have a very different dietary history.

Nakaya et al. (1984) report only *Tetralophodon* from the Namurungule Formation of the Samburu Hills, but specimens identified as *Stegotetrabelodon* are also present in the National Museums of Kenya collections. Both have a C_3-dominated diet, although relatively positive $\delta^{13}C$ values from –7.3 to –9.5 suggest a preference for more xeric vegetation (i.e., C_3 plants shifted to more positive $\delta^{13}C$ values) or for a small component of C_4 biomass in the diet (table 12.11). Fossil equids from the Samburu Hills had a C_4-dominated diet, so C_4 plants would have been available to the gomphotheres and elephantids but did not form a significant part of their diet.

Anancus kenyensis is the more common gomphothere from the Lothagam sequence where it is present, though not abundant, in the Lower Nawata, Upper Nawata, and Apak members. The elephantids *Stegotetrabelodon orbus* and *Primelephas gomphotheroides* are present in the Lower Nawata and are joined in the Upper Nawata by *Elephas nawataensis. Stegotetrabelodon* may persist into the Apak Member, but the common elephantids from that unit are *Elephas* cf. *E. ekorensis* and *Loxodonta* cf. *L. exoptata* (Tassy this volume:chapter 8.1).

Because of the fragmentary nature of the analyzed samples, many can only be identified as elephantoids. Three specimens from the Lower Nawata, including one identified as *Stegotetrabelodon*, had a significant component of C_3 plants in their diet (–3.9, –5.5, and –6.2 permil), but most proboscideans sampled from this unit were evidently feeding mainly from C_4 vegetation (range –2.3 to +0.01 permil; table 12.11). In later portions of the sequence, both elephantids and gomphotheres fed almost exclusively from C_4 plants. It is interesting that Lothagam documents the timing of the acquisition of a C_4 grazing habitus by East African proboscideans and that both gomphotheres and elephantids became grazers. This is in marked contrast to the extant species of *Elephas* and *Loxodonta*, both of which are dedicated browsers (Harris and Cerling 1996; Cerling et al. 1999).

Forty-four extant individuals of *Loxodonta africana* from Kenya have C_3-dominated diets with $\delta^{13}C$ values that range from –7 to –13 permil (where only two of these individuals have $\delta^{13}C$ values more positive than –9 permil; unpublished data). These results agree with previous measurements made on collagen and feces from African elephants (Tieszen et al. 1989; van der Merwe et al. 1990; Vogel et al. 1990; Koch et al. 1995). The forty-four individuals sampled from Kenya are savanna elephants, which belies the common belief that savanna elephants are grazers and forest elephants are browsers. The extinct loxodonts, presumably including the ancestors of the extant African elephant, were grazers, but the extant elephant is undoubtedly a browser by preference. Where, and when, did the dietary change take place—if indeed the presumed ancestry is correct? It has been speculated that the modern African elephant was derived from a forest elephant (see discussion in Kingdon 1979). The results from our study of fossil and modern representatives indicate that the evolutionary history of the African elephant still has some surprises.

Suids

The warthog (*Phacochoerus aethiopicus*), the only extant suid from the Turkana Basin, is a grazer with $\delta^{13}C$ values of about 0 permil. Warthogs from elsewhere in East Africa have similar values, but the bush pig and giant forest hog have $\delta^{13}C$ values compatible with their browsing diet (Cerling and Harris unpublished data). Fossil suids are represented thoughout the Lothagam sequence. *Nyanzachoerus syrticus* and *N. devauxi* are abundant in the Lower Nawata but were supplanted by more evolved tetraconodontines in the Apak. Given that –8 permil is the most positive composition for a pure C_3 biomass from semiarid to arid environments (Cerling and Harris unpublished data), the $\delta^{13}C$ content of suid specimens from the Lower Nawata ranges from –8.6 to –6.1 permil, and this indicates a diet dominated by C_3 plants but with some C_4 component (table 12.12).

Nyanzachoerus syrticus from the Upper Nawata has more positive $\delta^{13}C$ values than do the Lower Nawata samples, and evidently this suid exploited a mixed C_3/C_4 diet. Undoubted C_4–grazing suids do not occur until the Apak Member, where $\delta^{13}C$ values greater than –2 permil are found, although they average about –3.7 permil. None of the fossil suids from Lothagam have $\delta^{13}C$ values as positive as the Late Pleistocene *Phacochoerus aethiopicus* from Lothagam (ca. + 0.8 permil; table 12.12).

Bovids

Extant East African bovids exploit a very varied diet. Some alcelaphines (e.g., wildebeest) have the most positive $\delta^{13}C$ values yet reported for modern mammals (up to +3.9 permil; Cerling and Harris unpublished data), while other bovids have C_3-dominated diets (e.g., to –13 permil for some eland; Cerling and Harris unpublished data). Fossil bovids are more problematic than the other mammalian groups because only relatively complete specimens can be identified to species. Unfortunately, many analyzed bovid samples from Lothagam could not be identified to tribe. However, it is interesting to note that the bovid with the higher $\delta^{13}C$ value in the Lower Nawata was identified as an alcelaphine (+0.2 permil; table 12.13).

Ostriches

The enrichment from diet to ostrich eggshell is about 16 permil (Johnson et al. 1998). The $\delta^{13}C$ of ostrich eggshell from the Nawata Formation indicates a higher component of C_3 biomass in the diet than is found in modern ostriches from the region (table 12.14).

Dietary preferences

The Lothagam sequence is a very important locality for mammalian studies because it is one of the first African sites where C_4 biomass contributes an important component of the diet. The Samburu Hills succession overlaps in age with the oldest Lothagam sediments; both sites contain one of the earliest African mammalian assemblages to exploit C_4 plants. Within the Lothagam sequence, some interesting trends are worth noting:

1. Equids, with their previously evolved hypsodont teeth, quickly made the transition from a C_3-dominated diet (e.g., Nakali equids) to a C_4-dominated diet (e.g., equids from the Namurungule Formation of the Samburu Hills and from the Lower Nawata). It is evident that C_3 grasses were abundant before the expansion of C_4 biomass, but that scenario

has no modern analog because in the tropics today C_3 grasses are restricted to high elevations, wetlands, or the understory below closed canopies.

2. Gomphotheres and elephantids of the Lothagam sequence had C_4-dominated diets, in contrast to the modern African elephant, *Loxodonta africana*, which has a C_3-dominated diet. Deinotheres, however, maintained a C_3-diet throughout their known history. The consistent pure C_3 results from the deinothere samples provide support for isotopic fidelity during diagenesis.
3. Hippopotamids evidently exploited a diversity of plant resources and different individuals had C_3-dominated, mixed C_3/C_4, or C_4-dominated diets. This suggests hippos fed in a variety of habitats, but especially those characterized by the presence of C_3 browse or C_3 grasses. Extant hippos feed both on C_4 and C_3 grasses in wetlands and sometimes take C_3 browse. The depleted $\delta^{13}C$ values observed in the Lothagam assemblage seem to suggest that C_4 grasses were present, but C_3 grasses and sedges predominated in many of the habitats represented in the Nawata and Nachukui Formations.
4. Fossil suids were slow to adapt to the C_4 dietary resource. In spite of the presence of abundant C_4 biomass (i.e., see the discussion of the equids), suids were mixed feeders through the Nawata Formation and into the Apak Member where suids with a C_4-dominated diet (>–2 permil) first occur. Further study of the morphological changes and paleodiet may establish evidence of physical evolutionary changes in response to changing resources.

$\delta^{18}O$ Analysis of Lothagam Mammals

The $\delta^{18}O$ of tooth enamel provides important information about the sources of water available to mammals. Briefly, $\delta^{18}O$ values for meteoric water and for metabolic water is rather similar in East Africa, and leaf water is 15 to 20 permil enriched relative to local meteoric water (Cerling unpublished data). Closed basin waters of East Africa, such as lakes or even evaporating water holes, are typically 5–10 permil enriched in ^{18}O compared to their source meteoric water. Modern meteoric waters in the Turkana Basin are about 0 to –3 permil, and the modern Lake Turkana is between +5 and +7 permil (Cerling et al. this volume:chapter 12.1).

Equids

The $\delta^{18}O$ values for fossil equids in the Lothagam sequence average about –0.7 permil relative to PDB in the Lower Nawata, with slightly more positive values of about +1 permil in the Upper Nawata and Apak Member. In contrast, a single modern sample has an $\delta^{18}O$ value of about +4.6 permil (table 12.4). This implies that the isotopic composition of the body fluid of the fossil equids was in general about 5 permil more depleted in ^{18}O than in modern zebras. This depletion can be attributed to one of two factors: (1) the isotopic composition of local meteoric water may have been several permil more depleted in ^{18}O compared to the present, and (2) there may have been no isotopically enriched evaporated lake waters available. The pedogenic carbonate results indicate that local meteoric waters were probably depleted by a few more permil than are modern meteoric waters.

Hippopotamids

The $\delta^{18}O$ values for hippos are especially instructive. Lower Nawata hippos have relatively uniform $\delta^{18}O$ values, ranging from –2.8 to –7.2 permil and averaging –4.8 ± 1.0 permil (table 12.9). These Lower Nawata hippo values are more depleted in $\delta^{18}O$ than those of modern Turkana hippos by about 7 permil. Because hippos live in water for much of the day, their body fluids more closely approximate their habitat than other mammals. These values indicate that Lower Nawata hippos were probably riverine hippos because East African lakes, even nonsaline ones, are enriched in $\delta^{18}O$.

Upper Nawata and Apak Member hippos are, in general, also more enriched in $\delta^{18}O$ than are extant hippos from Lake Turkana. The few enriched $\delta^{18}O$ values, from samples collected from about +126 to +137 meters and from about 310 to 315 meters (table 12.9), may indicate that evaporated waters were present for brief intervals during deposition of the Nawata Formation and the Apak Member.

Rhinocerotids

Fossil rhinos from the Nawata Formation and Apak Member range between –5 and +2 permil. The sole extant rhino from Koobi Fora had an $\delta^{18}O$ value of +2.3 permil (table 12.10). The depleted $\delta^{18}O$ values for the Nawata and Apak are comparable, with meteoric water being several permil more depleted in ^{18}O compared to the modern meteoric waters.

Proboscideans

Proboscideans from the Nawata Formation and the Apak Member range from –6 to +1 permil (table 12.11) and average –1.8 ± 1.8 permil (n = 29). The last known elephant from the Turkana Basin south of the Omo delta was killed in the 1920s and hence is not available for analysis. Proboscideans from other parts of Kenya average –0.1 ± 1.3 permil (n = 44); because the Turkana Basin is one of the driest parts of Kenya, it is expected that the values for elephants in the

Turkana Basin would be more positive than –0.1 permil. Once again, these values suggest that the isotopic composition of meteoric waters during the Nawata and Apak intervals was more negative than it is today.

Suids

The $\delta^{18}O$ values for suids during the Lower Nawata, Upper Nawata, and Apak Member average –2.8 ± 1.6 permil (n = 6), –2.2 ± 1.1 permil (n = 3), and –1.3 ± 1.7 permil (n = 7), respectively (table 12.12). These are considerably more negative than the value of +5.4 permil measured for a single extant warthog from Koobi Fora. It is not known if the extant Koobi Fora suid frequented the lake. The consistent ^{18}O depletion in fossil suids compared to the modern suid again implies that the fossil meteoric waters were more depleted than are modern waters.

Bovids

The $\delta^{18}O$ values vary in both the extant and fossil bovid samples. Two extant kudus from Koobi Fora have $\delta^{18}O$ values of about +12 permil, while extant oryx and dikdik have $\delta^{18}O$ values about +5 permil (table 12.13). Kudu, dikdik, and oryx do not drink from the modern lake and represent values typical for animals that are not obligate drinkers. The range for fossil bovids from the Nawata and Apak sequences is from about –3 to +3 permil. This is probably due to meteoric water of the late Miocene and early Pliocene being depleted in ^{18}O compared to modern waters.

Ostriches

Eggshell from two ostriches from the Upper Nawata had $\delta^{18}O$ values of about +3 permil, compared to +7.3 ± 1.6 permil for modern ostriches (table 12.14). Ostriches do not drink Lake Turkana water and so, once again, the fossil evidence suggests that local meteoric waters were more depleted in ^{18}O relative to modern local meteoric waters.

Summary

We examined the isotope ecology of the Nawata Formation and Apak Member at Lothagam using pedo-

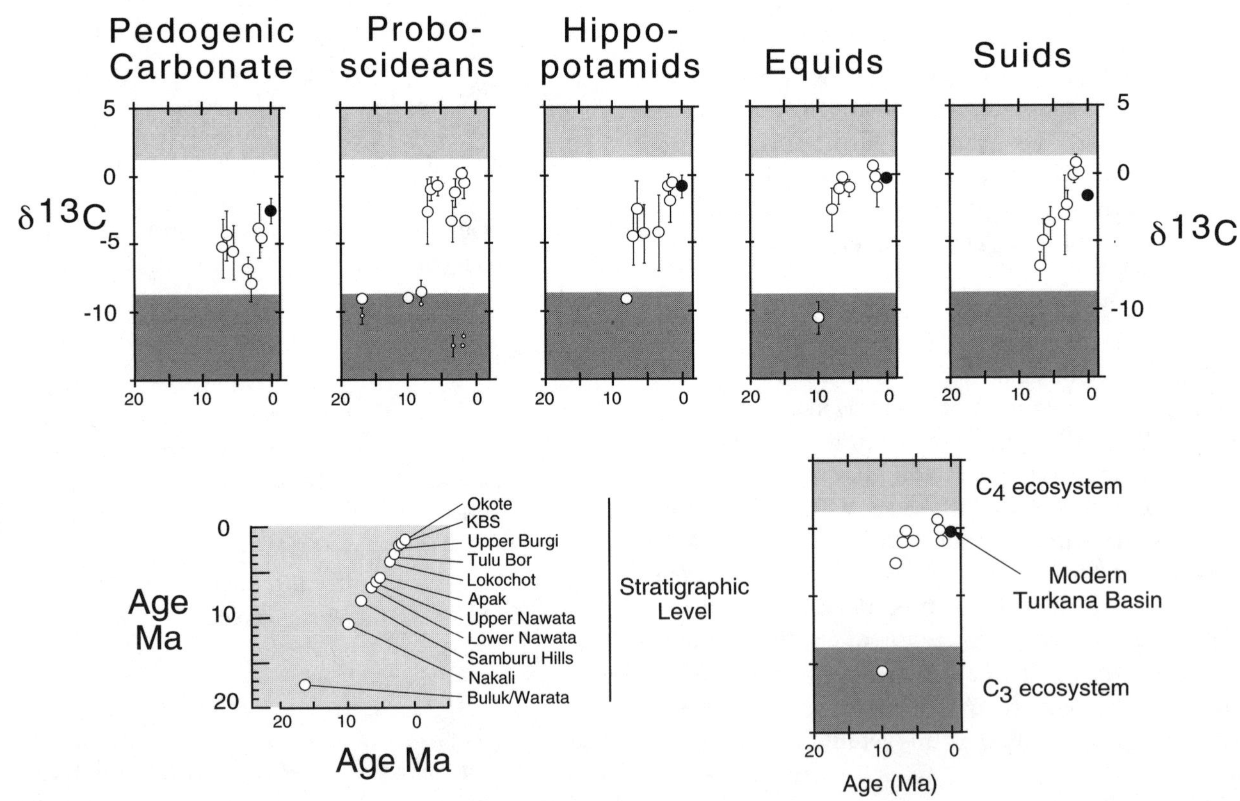

Figure 12.9 Plot of $\delta^{13}C$ of pedogenic and tooth enamel carbonate versus stratigraphic age in the greater Turkana Basin. Plotted carbonate values represent the averages for each stratigraphic interval, and the vertical bars represent one standard deviation for that interval. Pedogenic carbonate samples demonstrate a mixed C_3/C_4 flora throughout the succession, from Lower Nawata through upper Koobi Fora times. The oldest proboscideans had C_3-dominated diets; elephantoids (large open circles) acquire a significant C_4 component in their diet in Lower Nawata time, while deinotheres (small open circles) retain a pure C_3 diet. Hippos have mixed C_3/C_4 diets beginning in Lower Nawata time. Equids acquire a C_4 diet by the earliest portion of the Nawata sequence. Suids demonstrate a more gradual incorporation of a C_4 component in their diet. The black dots in the top graphs represent examples from the modern Lake Turkana Basin.

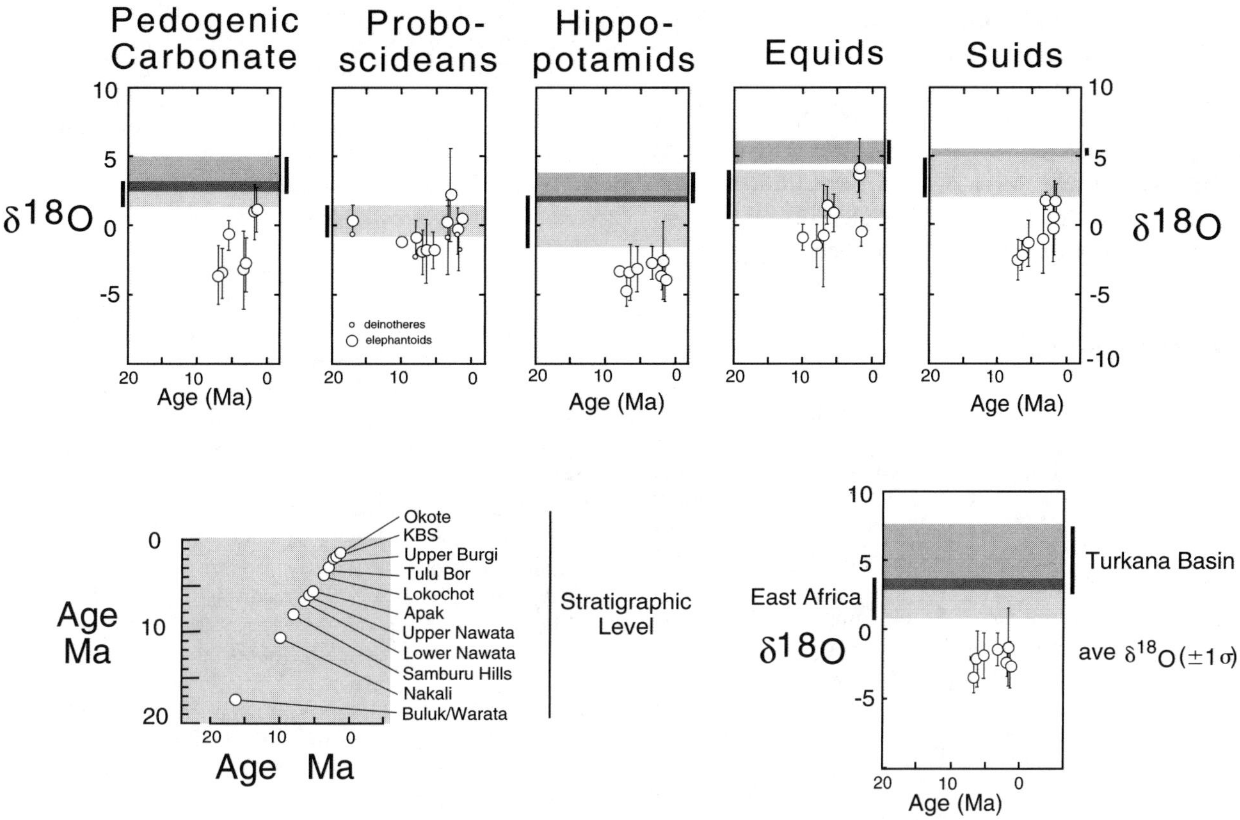

Figure 12.10 Plot of $\delta^{18}O$ of pedogenic and tooth enamel carbonates versus stratigraphic age in the greater Turkana Basin. Plotted carbonate values represent the averages for each stratigraphic interval, and the vertical bars represent one standard deviation for that interval. Pedogenic carbonate values have significantly lower $\delta^{18}O$ values than do modern carbonates from the Turkana Basin or from other parts of Kenya. This may indicate that modern meteoric water in the region has a significantly different source on average than did the meteoric waters of the Neogene. Similarly, $\delta^{18}O$ values for Neogene herbivore tooth enamel is also more depleted in $\delta^{18}O$ than in the modern representatives of these major mammalian groups.

genic carbonate and fossil tooth enamel (figures 12.9 and 12.10). The Lothagam sequence is particularly important because it is the oldest continuous section in East Africa that postdates the global ^{13}C event that marked the expansion of C_4 plants.

Equids at Lothagam all have a predominantly C_4 diet, but several samples from the lowest part of the section have $\delta^{13}C$ values that indicate a mixed C_3/C_4 diet. In contrast, equids from Nakali have a C_3-dominated diet, and equids from the Samburu Hills have $\delta^{13}C$ values similar to that of the lowest part of the Lower Nawata and which indicate mixed C_3/C_4 diets and C_4-dominated diets.

The hippopotamids from the Lothagam succession generally have a mixed C_3/C_4 diet. This probably indicates that C_3 semiaquatic grasses and sedges were available to them, unlike in the modern setting of Lake Turkana where *Hippopotamus amphibius* feeds almost exclusively on C_4 plants.

Proboscideans also record a change from a C_3-dominated diet at Nakali and in the Samburu Hills but a C_4 diet throughout the Lothagam sediments. Extant *Loxodonta*, however, is primarily a browser, and so these results suggest either that *Loxodonta* recently changed its diet back to C_3 plants or, as Beden (1987) intimated, the lineage that started at Lothagam and culminated in *Loxodonta exoptata* is not ancestral to the extant species.

Suids were slower than equids or proboscideans to exploit a C_4 diet, and the change is accompanied by corresponding changes in molar morphology. As is evident from the isotopic record, rhinocerotids had developed browsing and grazing specializations by Apak times.

In general, the diets of the fossil grazers are similar to that of grazers in the basin today. The $\delta^{13}C$ values of the extant grazers reflects the higher proportion of NAD-me and PEP-ck grasses (as opposed to the more mesic NADP-me grasses) in the local ecosystem, which is typical for modern scrub bushland in northern Kenya. Therefore, it is likely that the paleoecosystem did not include NADP-me grasslands but was more likely a dwarf shrubland similar to the modern ecosystem.

The oxygen isotope evidence, both from paleosols and from the tooth enamel of many different mammalian groups, indicates that the isotopic composition of local meteoric water was enriched in ^{18}O by several permil compared to modern waters. The consistently depleted $\delta^{18}O$ values for hippopotamids suggest that these animals were probably riverine rather than lake dwellers. Isotopic depletion of meteoric waters can be explained in three ways or a combination thereof: (1) the climate was significantly cooler; (2) rainfall was higher; and (3) more storms approached from the west.

References Cited

Beden, M. 1987. Fossil Elephantidae from Laetoli. In M. D. Leakey and J. M. Harris, eds., *Laetoli: A Pliocene Site in Northern Tanzania*, pp. 258–294. Oxford: Clarendon Press.

Bryant, J. D., B. Luz, and P. N. Froelich. 1994. Oxygen isotopic composition of fossil horse tooth phosphate as a record of continental paleoclimate. *Palaeogeography, Palaeoclimatology, Palaeoecology* 107:303–316.

Cerling, T. E. 1984. The stable isotopic composition of modern soil carbonate and its relationship to climate. *Earth and Planetary Science Letters* 71:229–240.

Cerling, T. E., and J. Quade. 1993. Stable carbon and oxygen isotopes in soil carbonates. In P. K. Swart, K. C. Lohmann, J. A. McKenzie, and S. M. Savin, eds., *Climate Change in Continental Isotopic Records*, pp. 217–231. Geophysical Monograph No. 78. Washington, D.C.: American Geophysical Union.

Cerling, T. E., and Z. D. Sharp. 1996. Stable carbon and oxygen isotope analysis of fossil tooth enamel using laser ablation. *Palaeogeography, Palaeoclimatology, Palaeoecology* 125:173–186.

Cerling, T. E., J. R. Bowman, and J. R. O'Neil. 1988. An isotopic study of a fluvial-lacustrine sequence: The Plio-Pleistocene Koobi Fora sequence, East Africa. *Palaeogeography, Palaeoclimatology, Palaeoecology* 63:335–356.

Cerling, T. E., J. M. Harris, and M. G. Leakey. 1999. Browsing and grazing in elephants: The isotope record of modern and fossil proboscideans. *Oecologia* 120:364–374.

Cerling, T. E., J. M. Harris, B. J. MacFadden, M. G. Leakey, J. Quade, V. Eisenmann, and J. R. Ehleringer. 1997. Pattern and significance of global ecologic change in the Late Neogene. *Nature* 389:153–158.

Cerling, T. E., J. Quade, Y. Wang, and J. R. Bowman. 1989. Carbon isotopes in soils and paleosols as ecologic and paleoecologic indicators. *Nature* 341:138–139.

Cerling, T. E., Y. Wang, and J. Quade. 1993. Expansion of C_4 ecosystems as an indicator of global ecological change in the late Miocene. *Nature* 361:344–345.

Ehleringer, J. R., and T. A. Cooper. 1988. Correlations between carbon isotope ratio and microhabitat in desert plants. *Oecologia* 76:562–566.

Ehleringer, J. R., C. B. Field, C. F. Lin, and C. Kuo. 1986. Leaf carbon isotope and mineral composition in subtropical plants along an irradiance cline. *Oecologia* 70:520–526.

Harris, J. M., and T. E. Cerling. 1996. Isotopic changes in the diet of African proboscideans. *Journal of Vertebrate Paleontology*, Supplement 16:40A.

Johnson, B. J., M. L. Fogel, and G. H. Miller. 1998. Stable isotopes in modern ostrich eggshell: A calibration for paleoenvironmental applications in semi-arid regions of southern Africa. *Geochimica et Cosmochimica Acta* 62:2451–2461.

Kingdon, J. 1979. *East African Mammals: An Atlas of Evolution in Africa*. Vol. 3, pt. B. *Large Mammals*. London: Academic Press.

Koch, P. L. 1998. Isotopic reconstruction of past continental environments. *Annual Reviews of Earth and Planetary Science* 26:573–613.

Koch, P. L., A. K. Behrensmeyer, and M. L. Fogel. 1991. The isotopic ecology of plants and animals in Amboseli National Park, Kenya. In *Annual Report of the Director*, pp. 163–171. Washington, D.C.: Geophysical Laboratory.

Koch, P. L., J. Heisinger, C. Moss, R. W. Carlson, M. L. Fogel, and A. K. Behrensmeyer. 1995. Isotopic tracking of change in diet and habitat use in African elephants. *Science* 267:1340–1343.

Kolodny, Y., B. Luz, M. Sander, and W. A. Clemens. 1996. Dinosaur bones: Fossils or pseudomorphs? The pitfalls of physiology reconstruction from apatitic fossils. *Palaeogeography, Palaeoclimatology, Palaeoecology* 126:161–171.

Lee-Thorp, J., and N. J. van der Merwe. 1987. Carbon isotope analysis of fossil bone apatite. *South African Journal of Science* 83:712–715.

MacFadden, B. J., N. Solounias, and T. E. Cerling. 1999. Ancient diets, ecology, and extinction of 5-million-year-old horses from Florida. *Science* 283:824–827.

McDougall, I., and C. S. Feibel. 1999. Numerical age control for the Miocene-Pliocene succession at Lothagam, a hominoid-bearing sequence in the northern Kenya Rift. *Journal of the Geological Society* (London) 156:731–745.

Morgan, M. E., J. D. Kingston, and B. D. Marino. 1994. Carbon isotopic evidence for the emergence of C_4 plants in the Neogene from Pakistan and Kenya. *Nature* 367:162–165.

Nakaya, H., M. Pickford, Y. Nakano, and H. Ishida. 1984. The Late Miocene large mammalian fauna from the Namurungule Formation, Samburu Hills, northern Kenya. *African Study Monographs*, Supplementary issue 2:87–131.

Powers, D. W. 1980. Geology of Mio-Pliocene sediments of the lower Kerio River Valley, Kenya. Ph.D. diss., Princeton University.

Quade, J., T. E. Cerling, M. M. Morgan, D. R. Pilbeam, J. Barry, A. R. Chivas, J. A. Lee-Thorp, and N. J. van der Merwe. 1992. A 16-million-year record of paleodiet using carbon and oxygen isotopes in fossil teeth from Pakistan. *Chemical Geology (Isotope Geoscience)* 94:183–192.

Tieszen, L. L., T. W. Boutton, W. K. Ottichilo, D. E. Nelson, and D. H. Brandt. 1989. An assessment of long-term food habits of Tsavo elephants based on stable carbon and nitrogen isotope ratios of bone collagen. *African Journal of Ecology* 27:219–226.

van der Merwe, N. J., J. A. Lee-Thorp, J. F. Thackeray, A. Hall-Martin, F. J. Kruger, H. Coetzee, R. H. V. Bell, and M. Lindeque. 1990. Source-area determination of elephant ivory by isotopic methods. *Nature* 346:744–746.

Vogel, J. C., B. Eglington, and J. M. Aurel. 1990. Isotope fingerprints in elephant bone and ivory. *Nature* 346:747–749.

Wang, Y., and T. E. Cerling. 1994. A model of fossil tooth enamel and bone diagenesis: Implications for stable isotope studies and paleoenvironment reconstruction. *Palaeogeography, Palaeoclimatology, Palaeoecology* 107:281–289.

TABLE 12.7 Isotopic Analyses of Pedogenic and Other Carbonates

$\delta^{13}C$	$\delta^{18}O$	Percent $CaCO_3$	Sample	Strat Level (m)	Material
Apak Member					
–9.4	–4.8	75.5	K.91-4815	330	Channel lag (nodule)
–5.5	–0.2	56.4	LOTH-32	315	Paleosol nodule
–4.2	0.6	57.9	LOTH-33	315	Channel lag (nodule)
–8.4	–1.4	74.9	LOTH-20	296	Paleosol nodule
–7.2	0.3	62.1	LOTH-20	296	Paleosol nodule
–3.0	–2.2	48.9	K.91-4797	290	Rhizolith
–5.4	–1.2	79.6	LOTH-17	274	Paleosol nodule
–6.2	–1.3	—	Average	—	—
±2.3	±1.9	—	±1σ	—	—
Upper Nawata Member					
–4.4	–5.6	59.7	K.91-4788	215	Rhizolith
–1.3	–1.6	58.5	K.91-4784	185	Paleosol nodule
–5.7	–2.2	67.0	K.91-4781	130	Paleosol nodule
–5.9	–5.1	66.9	LOTH-15	126	Paleosol nodule
–4.5	–2.8	78.2	LOTH-29	126	Paleosol nodule
–4.4	–3.4	—	Average	—	—
±1.8	±1.8	—	±1σ	—	—
Lower Nawata Member					
–7.4	–6.2	65.7	LOTH-23	122	Rhizolith
–3.2	–4.4	61.1	LOTH-24	122	Rhizolith
–7.8	–4.2	65.6	K.91-4711	120	Paleosol nodule
–5.7	–6.1	59.6	K.94-4729	120	Rhizolith
–6.1	–3.1	76.6	LOTH-27	109	Channel lag (nodule)
–3.4	0.9	73.1	K.91-4810	100	Paleosol nodule
–2.2	–1.8	60.2	K.91-4723	95	Paleosol nodule
–3.8	–2.7	71.0	K.91-4919	95	Paleosol nodule
–3.5	–0.1	58.4	LOTH-10	86	Paleosol nodule
–9.0	–5.2	81.5	K.91-4702	80	Rhizolith
–6.7	–5.2	42.8	K.91-4754	71	Rhizolith
–3.4	–3.1	72.5	K.91-4753	66	Paleosol nodule
–7.7	–5.2	68.3	K.93-5048	53	Paleosol nodule
–4.2	–3.9	81.5	LOTH-4	14	Paleosol nodule
–5.3	–3.6	—	Average	—	—
±2.2	±2.1	—	±1σ	—	—
Matrix or diagenetic carbonate					
–3.0	–8.6	26.2	LOTH-16	215	Diagenetic cement
–4.9	–1.7	62.8	K.91-4718	150	Carbonate concretion
–5.6	–1.8	54.2	K.91-4705	95	Undercarbonate
–1.4	–0.3	50.4	K.91-4766	85	Carbonate concretion
–2.6	–0.1	78.2	K.91-4767	85	Carbonate concretion
–12.1	–5.9	93.4	LOTH-8	80	Marrow mold
–4.8	–1.9	68.3	LOTH-25	55	Nodules
–8.6	–7.5	6.9	LOTH-2	12	Matrix

TABLE 12.8 $\delta^{13}C$ and $\delta^{18}O$ Values for Tooth Enamel from Fossil Equids from Lothagam, the Namarungule Formation in the Samburu Hills, and Nakali

$\delta^{13}C$	$\delta^{18}O$	Sample	Strat Level (m)	Species
Modern				
−0.3	4.6	K89.6	—	*Equus burchelli*
Koobi Fora (Okote)				
−2.1	−1.2	K92-4832	—	*Hipparion*
Upper Apak				
−1.0	2.3	LOTH-101	323	*Eurygnathohippus* sp.
Lower Apak				
−1.8	−0.1	LOTH-110	302	*Eurygnathohippus* sp.
−1.3	2.5	LOTH-97	266	*Eurygnathohippus* sp.
−0.5	0.0	LOTH-120	255	*Eurygnathohippus* sp.
−0.2	−0.2	LOTH-57	247	*Eurygnathohippus* sp.
−1.0	0.6	Average	—	*Eurygnathohippus* sp.
±0.7	±1.3	±1σ	—	*Eurygnathohippus* sp.
Upper Nawata				
−0.3	0.0	LOTH-50	151	*Eurygnathohippus* sp.
−0.1	0.9	LOTH-54	151	*Eurygnathohippus* sp.
−0.3	1.8	LOTH-48	149	*Eurygnathohippus* sp.
−0.5	3.2	LOTH-47	146	*Eurygnathohippus* sp.
−0.3	1.5	Average	—	—
±0.2	±1.4	±1σ	—	—
Lower Nawata				
−0.1	3.7	LOTH-91	108	*Eurygnathohippus* sp.
−0.8	−6.0	LOTH-94	107	*Eurygnathohippus* sp.
−1.1	−3.2	LOTH-40	105	*Eurygnathohippus* sp.
−1.4	−3.3	LOTH-77	99	*Eurygnathohippus* sp.
−0.7	3.3	LOTH-124	94	*Eurygnathohippus* sp.
−0.6	−2.5	LOTH-82	84	*Eurygnathohippus* sp.
−3.0	−4.3	LOTH-83	82	*Eurygnathohippus* sp.
−0.6	1.3	LOTH-123	72	*Eurygnathohippus* sp.
0.4	4.5	LOTH-127	57	*Eurygnathohippus* sp.
−2.9	−1.1	LOTH-128	57	*Eurygnathohippus* sp.
−0.2	0.7	LOTH-130	48	*Eurygnathohippus* sp.
−3.8	−1.1	LOTH-133	37	*Eurygnathohippus* sp.
−0.9	−0.7	Average	—	—
±1.0	±3.9	±1σ	—	—
Samburu Hills, Namarungule				
−1.3	−1.7	861009-1.1	—	*Eurygnathohippus* sp.
−3.0	−2.7	SH 82-334	—	*Eurygnathohippus* sp.
−1.1	−1.5	SH 82-348	—	*Eurygnathohippus* sp.
−2.6	1.9	SH-12243	—	*Hippotherium primigenium*
−4.2	−1.5	SH-12248	—	*Hippotherium primigenium*

TABLE 12.8 $\delta^{13}C$ and $\delta^{18}O$ Values for Tooth Enamel from Fossil Equids from Lothagam, the Namarungule Formation in the Samburu Hills, and Nakali (Continued)

$\delta^{13}C$	$\delta^{18}O$	Sample	Strat Level (m)	Species
Samburu Hills, Namarungule				
−5.1	−2.0	SH-15656	—	*Eurygnathohippus* sp.
−1.0	−2.8	SHS-23-84	—	*Eurygnathohippus* sp.
−2.6	−1.2	Average	—	—
±1.6	±1.8	±1σ	—	—
Nakali				
−10.0	−1.6	NA-152	—	*Hipparion africanum*
−11.5	−1.2	NA-240	—	*Hippotherium primigenium*
−9.2	0.2	NA-241	—	*Hippotherium primigenium*
−10.5	−0.9	Average	—	—
±1.2	±0.9	±1σ	—	—

TABLE 12.9 $\delta^{13}C$ and $\delta^{18}O$ Values for Tooth Enamel from Fossil Hippopotamids from Lothagam, from the Namurungule Formation in the Samburu Hills, and of Extant African Hippopotamids

$\delta^{13}C$	$\delta^{18}O$	Sample	Strat Level (m)	Species
Extant				
−13.7	0.0	NL1	Baringo, Kenya	*Hippopotamus amphibius*
−9.7	−2.0	OM2205	Baringo, Kenya	*Hippopotamus amphibius*
−5.8	−1.9	GNP hippo	Garamba, Zaire	*Hippopotamus amphibius*
−2.9	−2.8	OM 2054	Thika, Kenya	*Hippopotamus amphibius*
Modern Lake Turkana				
−2.1	2.0	ET-161	Koobi Fora, Kenya	*Hippopotamus amphibius*
−0.6	2.2	ET-162	Koobi Fora, Kenya	*Hippopotamus amphibius*
−0.5	2.7	TEC.K89.2	Koobi Fora, Kenya	*Hippopotamus amphibius*
−1.1	2.3	Average	—	—
±0.9	±0.4	±1σ	—	—
Apak Member				
−3.3	−4.5	LOTH-102	323	*Hexaprotodon* sp. indet.
−3.6	−3.2	LOTH-102	323	*Hexaprotodon* sp. indet.
−3.4	−2.9	LOTH-102	323	*Hexaprotodon* sp. indet.
−7.0	0.2	LOTH-99	315	*Hexaprotodon* sp. indet.
−4.6	−1.4	LOTH-104	310	*Hexaprotodon* sp. indet.
−0.9	−5.2	LOTH-109	305	*Hexaprotodon* sp. indet.
−8.2	−4.0	LOTH-112	300	*Hexaprotodon* sp. indet.
−4.4	−3.8	LOTH-62	251	*Hexaprotodon* sp. indet.
−3.3	−3.7	LOTH-63	251	*Hexaprotodon* sp. indet.
−4.3	−3.2	Average	—	—
±2.2	±1.7	±1σ	—	—
Upper Nawata				
−5.2	−4.2	LOTH-56	243	*Hexaprotodon* sp. indet.
−2.6	−4.3	LOTH-68	240	*Hexaprotodon* sp. indet.
−1.6	−4.6	LOTH-168	235	*Hexaprotodon* sp. indet.
−5.1	−6.2	LOTH-172	235	*Hexaprotodon* sp. indet.
−3.2	−2.8	LOTH-173	235	*Hexaprotodon* sp. indet.
−1.0	−6.8	LOTH-174	235	*Hexaprotodon* sp. indet.
−1.0	−3.4	LOTH-175	235	*Hexaprotodon* sp. indet.
−3.2	−2.9	LOTH-65	233	*Hexaprotodon* sp. indet.
−2.8	−5.0	LOTH-55	178	*Hexaprotodon* sp. indet.
−2.6	−2.5	LOTH-49	151	*Hexaprotodon* sp. indet.
−4.7	−2.4	LOTH-49	151	*Hexaprotodon* sp. indet.
−3.1	−4.1	LOTH-51	151	*Hexaprotodon* sp. indet.
0.9	−1.2	LOTH-165	147	*Hexaprotodon* sp. indet.
0.6	−4.6	LOTH-46	142	*Hexaprotodon* sp. indet.
−2.9	−4.4	LOTH-160	138	*Hexaprotodon* sp. indet.
−0.3	0.7	LOTH-157	137	*Hexaprotodon* sp. indet.
−0.4	−3.5	LOTH-156	137	*Hexaprotodon* sp. indet.

TABLE 12.9 $\delta^{13}C$ and $\delta^{18}O$ Values for Tooth Enamel from Fossil Hippopotamids from Lothagam, from the Namurungule Formation in the Samburu Hills, and of Extant African Hippopotamids (Continued)

$\delta^{13}C$	$\delta^{18}O$	Sample	Strat Level (m)	Species
Upper Nawata				
0.8	1.6	LOTH-153	128	*Hexaprotodon* sp. indet.
–3.7	–0.5	LOTH-151	126	*Hexaprotodon* sp. indet.
–3.1	–4.6	LOTH-150	125	*Hexaprotodon* sp. indet.
–7.6	–4.4	LOTH-42	125	*Hexaprotodon* sp. indet.
–2.4	–4.1	LOTH-43	125	*Hexaprotodon* sp. indet.
–4.3	–3.9	LOTH-44	125	*Hexaprotodon* sp. indet.
–2.5	–3.4	Average	—	—
±2.1	±2.0	$\pm 1\sigma$	—	—
Lower Nawata				
–4.1	–4.7	LOTH-138	—	*Hexaprotodon* sp. indet.
–9.1	–2.8	LOTH-92	121	*Hexaprotodon* sp. indet.
–5.4	–7.2	LOTH-93	121	*Hexaprotodon* sp. indet.
–5.8	–4.2	LOTH-93	121	*Hexaprotodon* sp. indet.
–5.7	–4.0	LOTH-93	121	*Hexaprotodon* sp. indet.
–0.7	–4.1	LOTH-73	105	*Hexaprotodon* sp. indet.
–5.1	–5.0	LOTH-95	102	*Hexaprotodon* sp. indet.
–2.1	–4.7	LOTH-69	101	*Hexaprotodon* sp. indet.
–3.3	–6.8	LOTH-69	101	*Hexaprotodon* sp. indet.
–2.8	–4.6	LOTH-71	101	*Hexaprotodon* sp. indet.
–7.2	–3.7	LOTH-78	101	*Hexaprotodon* sp. indet.
–5.5	–4.8	LOTH-74	99	*Hexaprotodon* sp. indet.
–5.4	–4.8	LOTH-126	97	*Hexaprotodon* sp. indet.
–4.0	–6.1	LOTH-81	85	*Hexaprotodon* sp. indet.
–3.8	–4.3	LOTH-85	79	*Hexaprotodon* sp. indet.
–3.0	–4.3	LOTH-86	79	*Hexaprotodon* sp. indet.
–3.2	–4.3	LOTH-86	79	*Hexaprotodon* sp. indet.
–1.7	–4.6	LOTH-87	76	*Hexaprotodon* sp. indet.
–5.0	–5.9	LOTH-122	62	*Hexaprotodon* sp. indet.
–7.8	–4.5	LOTH-135 (2699)	9	*Hexaprotodon* sp. indet.
–4.5	–4.8	Average	—	—
±2.1	±1.0	$\pm 1\sigma$	—	—
Namurungule Formation				
–9.1	–3.3	861009-1.2	Namurungule	Hippopotamidae gen. indet

TABLE 12.10 $\delta^{13}C$ and $\delta^{18}O$ Values for Tooth Enamel from Fossil Rhinocerotids from Lothagam, from the Namurungule Formation in the Samburu Hills, and of Some Extant East African Rhinocerotids

$\delta^{13}C$	$\delta^{18}O$	Sample	Strat Level (m)	Species
Modern				
–11.5	1.6	ODW 4	Extant	*Diceros bicornis*
–12.7	1.2	OM 2195	Extant	*Diceros bicornis*
–12.6	1.6	OM 5482	Extant	*Diceros bicornis*
–12.2	0.9	OM 2180	Extant	*Diceros bicornis*
–10.2	2.4	OM 2742	Extant	*Diceros bicornis*
–10.2	2.3	TEC.K89.1	Extant	*Diceros bicornis*
–2.1	0.3	OM 7528	Extant	*Ceratotherium simum*
0.6	1.3	GNP-wrh	Extant	*Ceratotherium simum*
1.1	4.0	OM reg#655	Extant	*Ceratotherium simum*
1.4	4.6	OM 2186	Extant	*Ceratotherium simum*
Kaiyumung Member				
0.8	–0.2	LT-26283	—	*Ceratotherium praecox*
Apak Member				
–2.1	2.3	LOTH-115	293	Rhinocerotidae gen. indet.
–2.0	0.6	LOTH-116	293	Rhinocerotidae gen. indet.
–2.2	–1.2	LOTH-117	293	—
–2.4	1.0	LOTH-117	293	Rhinocerotidae gen. indet.
–2.3	1.1	LOTH-117	293	Rhinocerotidae gen. indet.
–5.5	2.0	LT-90	—	*Brachypotherium lewisi*
–11.2	–1.0	LOTH-58	252	Rhinocerotidae gen. indet.
Upper Nawata				
–10.9	1.3	LT-95	—	*Brachypotherium lewisi*
–7.1	–2.8	LOTH-121	109	Rhinocerotidae gen. indet.
–8.9	–3.8	LOTH-70	101	Rhinocerotidae gen. indet.
Lower Nawata				
–11.6	–4.7	LT-86	—	*Brachypotherium lewisi*
–9.8	0.6	LT-89	—	*Ceratotherium praecox*
Unknown				
–7.7	–5.8	LT-100	—	*Brachypotherium lewisi*
–4.0	0.1	LT-23772	—	*Ceratotherium praecox*

TABLE 12.11 $\delta^{13}C$ and $\delta^{18}O$ Values for Tooth Enamel from Fossil Gomphotheres and Elephantids from Lothagam, from the Namurungule Formation in the Samburu Hills, and from the Koobi Fora Formation; and $\delta^{13}C$ and $\delta^{18}O$ Values for Tooth Enamel from Fossil Deinotheres from Northern Kenya

$\delta^{13}C$	$\delta^{18}O$	Sample	Strat Level (m)	Species
Koobi Fora Formation				
–3.4	0.5	ER 767	Okote	*Elephas recki recki*
–2.8	–3.9	ER 927	KBS	*Elephas recki ileretensis*
0.4	–3.2	ER 1302	KBS	*Elephas recki ileretensis*
0.5	–1.5	ER 4910	Upper Burgi	*Loxodonta adaurora kararae*
0.3	–1.0	ER 5871	Upper Burgi	*Elephas recki shungurensis*
–3.1	–1.2	ER 4939	Area 204	*Loxodonta exoptata*
0.2	2.7	ER 4106	TB Mb	*Elephas recki brumpti*
–2.1	2.5	ER 3201	Lokochot Mb	*Loxodonta adaurora adaurora*
–2.1	–0.1	ER 3196	Area 251	*Loxodonta adaurora adaurora*
Kanapoi				
–0.1	–1.1	KP 30442	400	*Anancus kenyensis*
Apak Member				
–0.4	–1.3	LOTH-118	293	Elephantidae gen. indet.
–0.2	–1.9	LOTH-61	251	Elephantidae gen. indet.
–0.8	–1.2	LT 26323	—	Elephantidae *incertae sedis* B
–0.8	–2.7	LT 26337	—	*Stegotetrabelodon orbus*
–0.9	–2.2	LT 26337	—	*Stegotetrabelodon orbus*
–1.1	–2.4	WT-2632	—	*Stegotetrabelodon*
0.2	0.9	LT 361	—	*Anancus kenyensis*
–2.1	–3.4	LT 28567	—	*Anancus kenyensis*
Upper Nawata				
0.2	–2.1	LOTH-60	249	—
–1.1	0.7	LOTH-64	240	*Stegotetrabelodon* or *Primelephas*
–2.0	0.3	LOTH-64.a	240	*Stegotetrabelodon* or *Primelephas*
–2.0	0.3	LOTH-64.b	240	*Stegotetrabelodon* or *Primelephas*
–1.3	0.5	LOTH-64C	240	*Stegotetrabelodon* or *Primelephas*
–0.2	–4.6	LOTH-171	235	Elephantidae gen. indet.
–0.9	–3.8	LOTH-176	232	Elephantidae gen. indet.
0.3	–0.3	LOTH-66	212	*Stegotetrabelodon*
–0.3	–4.1	LOTH-162	140	Elephantidae gen. indet.
–2.1	–0.4	LOTH-159	138	Elephantidae gen. indet.
–1.5	–0.9	LOTH-158	137	Elephantidae gen. indet.
–0.1	–6.1	LOTH-154	132	Elephantidae gen. indet.
–1.6	–3.6	LT 23783	—	*Elephas nawatensis* sp. nov.
Lower Nawata				
–5.5	–2.9	LOTH-137	—	Elephantidae gen. indet.
–6.2	–2.3	LT 26332	—	*Stegotetrabelodon* or *Primelephas*
–1.0	0.5	LT 26336	—	*Stegotetrabelodon* sp.
–1.0	0.3	LT 26336	—	*Stegotetrabelodon* sp.

continued

TABLE 12.11 $\delta^{13}C$ and $\delta^{18}O$ Values for Tooth Enamel from Fossil Gomphotheres and Elephantids from Lothagam, from the Namurungule Formation in the Samburu Hills, and from the Koobi Fora Formation; and $\delta^{13}C$ and $\delta^{18}O$ Values for Tooth Enamel from Fossil Deinotheres from Northern Kenya (Continued)

$\delta^{13}C$	**$\delta^{18}O$**	**Sample**	**Strat Level (m)**	**Species**
Lower Nawata				
–1.8	–4.1	LOTH-72	101	Gomphothere
–2.3	–2.7	LOTH-75	97	Gomphothere
–3.9	–2.3	LOTH-84	79	Gomphothere
0.7	–1.9	LOTH-90	78	Gomphothere
Namurungule Formation				
–9.5	–0.7	BGJ 506-84	—	Gomphothere
–9.0	–0.9	BGK 558-84	—	Gomphothere
–9.0	1.1	BGS 449-84	—	Gomphothere
–8.1	–2.2	SH 12308	—	*Tetralophodon*
–7.3	–1.8	SH 12381	—	*Stegotetrabelodon*
Nakali				
–9.0	–1.2	NA-260	—	*Anancus*
–8.0	0.3	NA-6	—	Gomphothere
Deinotheres: northern Kenya				
–12.5	–0.7	ER 4294	Upper Burgi	*Deinotherium bozasi*
–12.3	–2.4	WT 14987	Sub Tulu Bor	*Deinotherium bozasi*
–13.7	2.4	ER 3198	Lokochot	*Deinotherium bozasi*
–12.3	0.2	ER 2885	Lokochot	*Deinotherium bozasi*
–12.6	–6.4	261-1D	Lokochot	*Deinotherium bozasi*
–8.8	2.6	WT 3617	Pliocene	*Deinotherium bozasi*
–9.5	–2.2	SH 12306	Samburu Hills	*Deinotherium* sp.
–10.4	0.5	Buluk 5204	Buluk	*Prodeinotherium hobleyi*
–9.7	–0.7	Buluk 5212	Buluk	*Prodeinotherium hobleyi*
–10.9	1.4	WS 63	Buluk	*Prodeinotherium hobleyi*
–9.5	–2.2	SH 12306	Samburu Hills	*Deinotherium* sp.
–10.4	0.5	Buluk 5204	Buluk	*Prodeinotherium hobleyi*
–9.7	–0.7	Buluk 5212	Buluk	*Prodeinotherium hobleyi*
–10.9	1.4	WS 63	Buluk	*Prodeinotherium hobleyi*

TABLE 12.12 $\delta^{13}C$ and $\delta^{18}O$ Values for Tooth Enamel from Fossil Suids from Lothagam, from the Koobi Fora and Nachukui Formations, from Late Pleistocene Suids in Northern Kenya, and from a Modern Suid from the East Side of Lake Turkana

$\delta^{13}C$	$\delta^{18}O$	Sample	Strat Level (m)	Species
Modern, Koobi Fora				
–1.7	5.4	K89.5	—	*Phacochoerus ethiopicus*
Holocene				
0.9	0.3	LOTH-52	—	*Phacochoerus*
0.7	0.2	LOTH-53	—	*Phacochoerus*
Koobi Fora				
–2.06	–0.56	ER 3177	—	*Nyanzachoerus pattersoni*
Kanapoi				
–6.5	1.2	KP 205	—	*Nyanzachoerus jaegeri*
–2.0	1.1	KP 241	—	*Nyanzachoerus jaegeri*
–5.5	–2.2	KP 265	—	*Nyanzachoerus jaegeri*
Nachukui				
–2.6	–1.6	LT 26092	—	*Nyanzachoerus jaegeri*
–1.8	–1.2	LT 308	—	*Nyanzachoerus jaegeri*
Apak				
–2.5	–2.6	LOTH-113	310	Suidae gen. indet.
–4.2	–3.1	LOTH-114	310	Suidae gen. indet.
–4.6	–2.8	LOTH-105	305	Suidae gen. indet.
–4.8	0.8	LOTH-105	305	Suidae gen. indet.
–4.7	0.8	LOTH-105	305	Suidae gen. indet.
–2.9	–1.7	LOTH-107	300	Suidae gen. indet.
–1.9	–0.6	LOTH-98	—	*Nyanzachoerus pattersoni*
–3.7	–1.3	Average	—	—
1.2	1.7	$\pm 1\sigma$	—	—
Upper Nawata				
–5.8	–3.3	LOTH-67	237	*Nyanzachoerus syrticus*
–3.1	–2.0	LT 23743	—	*Nyanzachoerus syrticus*
–5.9	–1.2	LT 7709	—	*Nyanzachoerus syrticus*
–4.9	–2.2	Average	—	—
1.6	1.1	$\pm 1\sigma$	—	—
Lower Nawata				
–6.7	–3.0	LOTH-96	101	Suidae gen. indet.
–6.1	–1.1	LOTH-80	93	Suidae gen. indet.
–7.2	–2.8	LOTH-129	49	*Nyanzachoerus syrticus*
–6.1	–5.4	LOTH-131	48	Suidae gen. indet.
–7.7	–1.7	LOTH-132	45	*Nyanzachoerus syrticus*
–8.6	–2.0	LOTH-134	17	Suidae gen. indet.
–7.2	–2.8	Average	—	—
1.4	1.6	$\pm 1\sigma$	—	—
Unknown				
–5.6	–1.3	LT 22967	—	*Nyanzachoerus devauxi*

TABLE 12.13 $\delta^{13}C$ and $\delta^{18}O$ Values for Tooth Enamel from Fossil Bovids from Lothagam

$\delta^{13}C$	$\delta^{18}O$	Sample	Strat Level (m)	Species
Kaiyumung Member				
–2.1	–2.0	LT 26040	400	Bovini/Reduncini
–6.6	–0.2	LT 28736	400	*Aepyceros*
Upper Apak Member				
–0.4	2.0	LOTH-100	325	Bovidae gen. and sp. indet.
–2.8	1.7	LOTH-103	310	Bovidae gen. and sp. indet.
Lower Apak Member				
–2.1	3.4	LOTH-106	305	Bovidae gen. and sp. indet.
–1.7	0.4	LOTH-106	305	Bovidae gen. and sp. indet.
–3.2	0.0	LOTH-111	305	Bovidae gen. and sp. indet.
–0.9	–3.1	LOTH-119	293	Bovidae gen. and sp. indet.
–5.8	–0.7	LOTH-59	252	Bovidae gen. and sp. indet.
Upper Nawata				
–7.2	–0.5	LT 23613	—	Tragelaphini
–1.8	1.8	LT 23693	—	Reduncini
0.2	1.1	LT 517	—	Alcelaphini
–9.0	–0.6	LOTH-152	128	*Aepyceros*
–2.9	3.5	LOTH-41	126	Bovidae gen. and sp. indet.
–4.4	–0.6	LOTH-125	96	Bovidae gen. and sp. indet.
Unknown				
–1.1	2.9	LT 13013	Non loc.	Tragelaphini

TABLE 12.14 $\delta^{13}C$ and $\delta^{18}O$ Values for Ostrich Eggshell from Lothagam

$\delta^{13}C$	$\delta^{18}O$	Sample	Strat Level (m)	Species
Modern				
–8.6	5.9	LOTH-170	Extant	*Struthio*
–7.0	4.9	75-Struthio	Extant	*Struthio*
–6.0	7.5	LOTH-164	Extant	*Struthio*
–5.9	7.4	LOTH-166	Extant	*Struthio*
–5.9	7.0	KA-95-209-KP	Extant	*Struthio*
–5.0	9.4	KP 1	Extant	*Struthio*
–4.3	8.7	LOTH-167	Extant	*Struthio*
–6.1	7.3	Average	—	—
1.4	1.6	$\pm 1\sigma$	—	—
Kanapoi				
–8.7	10.9	WT 3597	400	*Struthio*
Upper Nawata				
–10.3	2.7	LOTH-163	142	*Struthio*
–12.5	3.6	LOTH-161	138	*Struthio*

13

LOTHAGAM: ITS SIGNIFICANCE AND CONTRIBUTIONS

Meave G. Leakey and John M. Harris

Lothagam, located to the southwest of Lake Turkana in northern Kenya, is an uplifted fault block composed of a gently westward-dipping sequence of volcanic and sedimentary rocks. These rocks have been subdivided into four major lithostratigraphic units, which document discrete stages in the large-scale tectonic and climatic evolution of the region and record a sequence of river systems and lakes (Feibel this volume:chapter 2.1). The Middle to Late Miocene Nabwal Arangan beds comprise conglomerates and lavas derived from a nearby high-relief volcanic source; to date, this basal unit has yielded only fossil wood. The main fossiliferous sequence begins with the superjacent Nawata Formation, which was deposited by a major fluvial system and with variations in fluvial facies that reflect changes in subsidence rate and water budget during the Late Miocene. The Early Pliocene Apak Member of the Nachukui Formation records a different source terrain and fluvial style and may represent the ancestral Kerio River system. The upper portion of the Apak Member and the subsequent Muruongori Member strata are lacustrine in character and represent the southern portion of the extensive Lonyumun Lake that filled much of the Turkana Basin during the Early Pliocene. Subsequent flu-

Figure 13.1 Restoration of the Nawata Formation habitat by Mauricio Antón.

vial deposits of the Kaiyumung Member appear to represent yet another fluvial system, possibly the ancestral Turkwel River, during the Mid- to Late Pliocene. Early Pleistocene strata attributed to the Kalochoro and Kaitio Members of the Nachukui Formation are primarily lacustrine in character at Lothagam, and they reflect conditions within the Late Pliocene Lorenyang Lake (Feibel et al. 1991:334). The Miocene and Pliocene strata exposed at Lothagam are overlain by the Galana Boi Formation, which was deposited during a Holocene highstand of Lake Turkana. Although of more limited lateral extent than the Plio-Pleistocene exposures in the Lake Turkana Basin, the age and the length of the Lothagam sequence contribute to an important record of environmental change in this portion of the African rift system.

The paleosol record documents a range of well to poorly drained alluvial settings (Wynn this volume: chapter 2.2). Ancient vertisols testify to regular annual or semiannual dry seasons throughout the Mio-Pliocene portion of the succession. Vegetation throughout the interval appears to have been a mosaic of floodplain savannas dissected by gallery woodland. It is interesting that evidence from changes in floodplain paleosol types documents a period of increased aridity between about 6.7 and 5 Ma. Extended intervals of depositional stasis are indicated at about 6.5 and 5.2 Ma by two very well developed luvisols.

The lower part of the sequence—lavas and coarse volcaniclastic sediments of the Nabwal Arangan beds—was deposited mainly between 14 Ma and 12 Ma (Middle Miocene), although the uppermost basalt has a K-Ar age of 9.1 Ma (McDougall and Feibel 1999). The overlying fluvial sediments of the lower Nawata Formation have yielded ages for five Late Miocene tuffaceous horizons that range from 7.4 $\pm$ 0.1 to 6.5 $\pm$ 0.1 Ma. A tuffaceous horizon in the superjacent Apak Member of the Nachukui Formation yields an age of 4.22 $\pm$ 0.03 Ma; ^{40}Ar-^{39}Ar age spectra on the overlying Lothagam Basalt indicate an age of 4.20 $\pm$ 0.03 Ma for its eruption. Much of the rich faunal assemblage from the Nawata Formation derives from the tightly dated lower intervals.

Aquatic elements of the Lothagam biota include crabs, fish, turtles, crocodiles, waterfowl, and hippos. Remains of crabs attributable to the Family Potamonautidae have been recovered from the Nawata Formation and the Apak Member (Martin and Trautwein this volume:chapter 3.1). The fish assemblages display considerable change throughout the sequence from the Lower Nawata to the Kaiyumung Member and document fluctuations in the aquatic regimes (Stewart this volume:chapter 3.2). The Nawata Formation fish assemblage is represented by archaic genera of small-sized fish. In the Apak Member, the archaic elements became scarce or were lost and modern genera predominate. Both Nawata and Apak assemblages are river-adapted. Fish recovered from the succeeding Muruongori and Kaiyumung Members comprise a predominantly lake fauna with several new taxa, including the first appearance in the basin of freshwater puffers. Three species new to the Turkana Basin—*Sindacharax deserti, Semlikiichthys rhachirhinchus,* and *Tetraodon* sp.—are also known from Mio-Pleistocene Egyptian localities and/or Western Rift deposits in Zaire and Uganda and must reflect a newly opened hydrological connection and exchange of faunas with those regions. Characids show considerable evolutionary change, with new species recognized from the Nawata Formation and from the Apak and Kaiyumung Members. The near-absence of tilapiine cichlids throughout the Lothagam succession may signify a later immigration from Asia than was previously thought.

Fossil turtles from the Nawata Formation and Apak Member include at least six different species and represent the Pelomedusidae, Testudinidae, and Trionychidae (Wood this volume:chapter 4.1). Remains of the newly recognized pelomedusid *Turkanemys pattersoni* outnumber those of all the other identifiable Lothagam chelonians combined. The pelomedusid *Kenyemys williamsi* and the trionychid *Cycloderma debroinae* are both restricted to Lothagam. Two other trionychid turtles and a giant tortoise have also been recovered. The Lothagam turtles represent a mixture of extinct and modern forms, including the earliest known occurrence of the living species *Cycloderma frenatum* and the last known continental African representative (*Turkanemys pattersoni*) of a lineage that survives today only as the Madagascan species *Erymnochelys madagascariensis.*

Four genera and five species of crocodilians document a high degree of ecological niche partitioning of the Late Miocene–Early Pliocene aquatic paleoenvironment (Storrs this volume:chapter 4.2). The extant *Crocodylus niloticus* and *C. cataphractus* persist from that time, although *C. cataphractus* has a very limited modern distribution and only the Nile crocodile, *C. niloticus,* occurs in the Turkana region today. Both species are first known from the Lower Nawata. Until the Late Neogene, the dominant crocodilian species at Lothagam and elsewhere in eastern Africa was the giant, brevirostrine *Rimasuchus lloydi.* A new species of *Eogavialis* from Lothagam is distinct from the earliest known gavialids (from the Fayum Basin Paleogene of Egypt) but is more primitive than *Gavialis* of the Indian subcontinent. The Lothagam *Eogavialis* specimens (and others newly recognized from the Early Miocene of Loperot, in the southwest Turkana Basin) are some of the few records of undoubted gavialids from the Miocene and to date the only ones from East Africa. The distinctive piscivorous longirostrine *Euthecodon* was present throughout the Lothagam sequence.

More than 30 bird specimens were retrieved by the recent National Museums of Kenya expeditions (Harris and Leakey this volume:chapter 4.3). The majority of bones recovered are those of waterfowl, although two, perhaps three, different ratite species are represented by shell fragments.

At least 15 species of rodents and lagomorphs are reported from Lothagam (Winkler this volume:chapter 5) (table 13.1). These include the earliest African record of the Hystricidae (Old World porcupines) and one of the earliest African records of the Leporidae (rabbits and hares). The extinct leporid *Alilepus* was previously recorded only from Eurasia and North America. The Lower Nawata assemblage includes the giant squirrel *Kubwaxerus.* The cane rats *Paraphiomys chororensis* and *Paraulacodus* cf. *P. johanesi* from the Lower Nawata were previously known only from Chorora, Ethiopia. Younger strata yielded the derived extant cane rat, *Thryonomys.* The gerbil *Abudhabia* is reported for the first time from sub-Saharan Africa. An unnamed new genus and species of murid from the Lower Nawata has affinity with *Myocricetodon magnus* from northern Africa. Murinae from Lothagam include the common extinct East African genus (*Saidomys*) and a new *Karnimata* species (*K. jacobsi*). *Karnimata* is poorly known from Africa but is better known from southern Asia.

Lothagam has provided the largest and most diverse collection of Cercopithecidae known from the Late Miocene of Africa (Leakey et al. this volume:chapter 6.1). Most specimens derive from the Nawata Formation in which papionins constitute 74 percent of the cercopithecid assemblage; three species of colobine and indeterminate species make up the remainder. Cercopithecids are rare in the Apak Member (only three specimens) but are abundant in the Kaiyumung Member where *Theropithecus brumpti* predominates. The postcrania indicate that early representatives of both colobines and cercopithecines were semiterrestrial. There are few postcranial characters that distinguish early representatives of these two subfamilies, and it seems likely that African colobines did not become fully arboreal until after the end of the Miocene. Studies of the molar microwear show that the Lothagam colobines were eating foods similar to those eaten by extant representatives. However, the lack of large pits on the molars of the Lothagam cercopithecines indicates that they, unlike extant species, were not ingesting hard objects. Differences in morphological features of the occlusal surface and the disparity in the size of the anterior dentition reflect subfamilial dietary differences. The colobines, with small anterior teeth and well-developed transverse molar lophs, ate seeds and some leaves, whereas the cercopithecines, with their large anterior teeth and less elevated molar cusps, probably ate mostly fruits. Primitive features of the cranium and deciduous dentition shared with *Victoriapithecus* but not with the Cercopithecinae do not support separation of Victoriapithecinae at the family level.

Only seven hominoid specimens have been recovered from Lothagam, and only three of these are from the Late Miocene deposits (Leakey and Walker this volume:chapter 6.2). The three older specimens represent populations from close to the time of the divergence between the human lineage and that leading to the chimpanzee and bonobo. Two isolated teeth from the Upper Nawata could represent either an early hominin or the ancestral morphotype of both lineages. The mandible from the lower Apak Member, KNM-LT 329, is older than 4.2 Ma but probably younger than 5.0 Ma. It resembles *Australopithecus anamensis* mandibles but, without comparative material from earlier populations, an attribution more secure than "Hominoidea indeterminate" cannot be justified. Four isolated teeth from the Kaiyumung Member are closest to specimens from Laetoli and from Hadar and are attributed to *Australopithecus* cf. *A. afarensis.*

At least 21 carnivoran taxa are represented at Lothagam: 15 from the Lower Nawata, 9 from the Upper Nawata, four from the Apak Member, and three from the Kaiyumung Member (Werdelin this volume:chapter 7). Six families are represented: Amphicyonidae (two species), Mustelidae (four species), Viverridae (five species), Hyaenidae (four species), Felidae (five species), and Canidae (one species). New mustelids include a giant-sized form with hypercarnivorous adaptations, a possible ancestor of the living honey badger *Mellivora capensis,* and a new species of the enhydrine genus *Vishnuonyx.* Hyaenids include a new species of *Ictitherium* with extreme cursorial adaptations represented by a complete skeleton. A new genus and species of machairodont felid is known from a partial skeleton. The Lothagam biota includes a number of first and last appearances of carnivoran taxa, including the youngest known record of Amphicyonidae and the first appearances of modern Mellivorinae, *Viverra, Genetta,* the *Hyaena* lineage, and *Dinofelis.* Several other taxa have a pivotal phylogenetic position with regard to later members of their group, making Lothagam a key site for interpretation of the evolution of the African Plio-Pleistocene carnivorans. The Lothagam carnivoran assemblage probably represents several biogeographic dispersals from Eurasia and the Indian subcontinent rather than *in situ* evolution from the Middle Miocene carnivorans of Africa.

Ten elephantoid taxa were present in the Lothagam succession (Tassy this volume:chapter 8.1); six can be identified at least tentatively at the species level. *Anancus kenyensis, Stegotetrabelodon orbus,* and *Primelephas gomphotheroides* were present in the Lower Nawata, together with an elephantid, previously described as *Stegotetrabelodon orbus* or *Primelephas gomphotheroides,* that may also persist in the Upper Nawata. Associated in the Upper Nawata are *Anancus kenyensis* and a second (trilophodont) gomphothere, *Stegotetrabelodon orbus, Elephas nawatensis* sp. nov., and an unidentified elephantid (species A). From the Apak Member have been recovered *Anancus kenyensis, Stegotetrabelodon orbus, Elephas* cf. *E. ekorensis, Loxodonta* ?aff. *L. exoptata,* and an unidentified elephantid, *Elephas* species B. The new early *Elephas* species from the Upper Nawata, *E. nawatensis,* demonstrates that the differentiation of the subfamily Elephantinae took place during the Late Miocene. Deinotheres were rare components of the Lothagam biota (Harris this volume:chapter 8.2), but deinothere enamel is distinctive and *Deinotherium bozasi* is documented from all four terrestrial members of the succession.

Two species of aardvark occur in the Lothagam sequence (Milledge this volume:chapter 8.3) as rare components of the faunal assemblage. The smaller *Leptorycteropus guilielmi,* represented by a partial skeleton and a femur fragment from the Lower Nawata, has a more generalized morphology than the extant aardvark but shows primitive fossorial adaptations. The larger *Orycteropus* sp. occurs in the Lower and Upper Nawata; it is about one-fifth smaller than the extant *O. afer* and displays some postcranial differences, but it undoubtedly belongs to the genus *Orycteropus.*

Three rhino species are represented in the Lothagam succession (Harris and Leakey this volume:chapter 9.1). *Brachypotherium lewisi* is the common rhino from the Nawata Formation, but *Ceratotherium praecox* is present in both the Lower and Upper Nawata and *Diceros bicornis* occurs in the Upper Nawata. Isotopic analysis of rhino tooth enamel suggests that *B. lewisi* was primarily a browser and that *C. praecox* had started exploiting C_4 grasses in the Upper Nawata.

Three hipparion species are present in the Lower Nawata: the rare *Hippotherium* cf. *H. primigenium*; the large *Eurygnathohippus turkanense*; and a new small species, *Eurygnathohippus feibeli,* that had been hitherto erroneously attributed to *E. sitifense* (Bernor and Harris this volume:chapter 9.2). The *Hippotherium* specimens constitute the last representatives of a lineage that entered Africa during the early Late Miocene. *Eurygnathohippus turkanense* is evidently closely related to *E. perimense* from the Siwaliks and represents a lineage that entered Africa during the middle Late Miocene; it is not closely related to the smaller *E. feibeli.* The two *Eurygnathohippus* species persist in the Upper Nawata, and very similar hipparions have been recovered from the Apak and Kaiyumung Members. Bernor and Harris (this volume:chapter 9.2) interpret *E. turkanense* as being adapted to closed woodland settings, whereas the smaller and more lightly built *E. feibeli* exploited more open habitats and was capable of sustained cursorial locomotion.

Lothagam provides the most complete record to date of the morphology and diversity of early hippopotamids (Weston this volume:chapter 10.1). Hippos are the most frequently preserved mammals in the Lothagam assemblage, and they constitute 27 percent of the mammalian fauna. Much of the material can be attributed to *Hexaprotodon harvardi,* which had begun the transition from cursorial browser to semiaquatic grazer. A few specimens from the Lower Nawata represent the small, narrow-muzzled *Hexaprotodon lothagamensis.* A third, extremely large, species was a rare component of the Nawata biota. Hippopotamid material from the Apak Member compares with *Hexaprotodon protamphibius* from elsewhere in the Turkana Basin.

The Lothagam suids are diverse (Harris and Leakey this volume:chapter 9.2). The Nawata Formation has yielded what appears to be a large kubanochoerine, a medium-sized suid referable to *Potamochoerus,* and a small tayassuid-like form assigned to *Cainochoerus* cf. *C. africanus,* as well as the more frequently preserved *Nyanzachoerus syrticus* and *Nyanzachoerus devauxi.* In the Upper Nawata, *N. syrticus* displays progressive evolutionary changes that involve premolar reduction and increasing complexity of the third molar. The Apak Member yields a different and more derived tetraconodontine, *Nyanzachoerus pattersoni,* whereas *Notochoerus euilus* abounds in the Kaiyumung Member. Fossilized teeth of bush pigs and warthogs found as surface specimens over much of the Lothagam locality represent lag fossils that were left behind after erosion of the Late Pleistocene and Holocene lacustrine deposits (Galana Boi beds). Lothagam documents the exploitation of riverine habitats by tetraconodontine suids that had migrated from Asia to dominate the latest Miocene and Pliocene of Africa. The presence of *N. australis* in the Upper Nawata may represent a second migratory wave.

As at most African Neogene localities, giraffids are only rare constituents of the Lothagam biota (Harris this volume:chapter 11.1). Most of the recovered remains constitute partial dentitions or isolated postcranial elements. Two species of *Palaeotragus* appear to be represented in the Nawata Formation; the more progressive *Giraffa stillei* and *Sivatherium* occur in the younger strata.

Bovids are the most abundant terrestrial mammals in the Lothagam succession, where they are represented by 17 species belonging to eight tribes that are common in sub-Saharan Africa today (Tragelaphini, Bovini, Reduncini, Hippotragini, Alcelaphini, Aepycerotini, Antilopini, and Neotragini), and by one tribe (Boselaphini) that is now restricted to Asia (Harris this volume:chapter 11.2). Impalas are the most numerous antelopes in all parts of the sequence. Boselaphines are common in the Lower Nawata, but they are less abundant in the Upper Nawata and are only rare elements of the Apak Member assemblage. Reduncines, alcelaphines, and hippotragines are more abundant in the Upper Nawata than at earlier or later levels. Tragelaphines and bovines are present in the Nawata Formation but are more abundant in the Apak Member.

The isotopic signals provided by soils, water, and mammalian herbivores of the present day Turkana Basin provide standards from which to contrast the Lothagam succession (Cerling et al. this volume:chapter 12.1). The dwarf shrubland ecosystem includes subequal quantities of C_3 and C_4 plants, which provide a notable contrast to the forests and wooded grasslands occurring in the highlands and plains to the south. The $\delta^{13}C$ fraction from dental enamel of grazing mammals of the Turkana Basin is more negative than that of grazers from more mesic regions and may reflect the contrast between the arid-adapted (NAD, PCK, "*Aristida*-NADP") grasses present in the Turkana Basin versus the mesic ("classical-NADP") grasses available farther south. Local meteoric water in the Turkana Basin averages about –3 permil for $\delta^{18}O$. The evaporated waters of the current Lake Turkana range from about +5 to +7 in $\delta^{18}O$ over the annual cycle.

The isotope ecology of the Lothagam succession was investigated from pedogenic carbonate and fossil tooth enamel (Cerling et al. this volume:chapter 12.2). Proboscideans record a change from a C_3-dominated diet at Nakali and in the Samburu Hills to a C_4 diet throughout the Lothagam sequence but the hippopotamids generally had a mixed C_3/C_4 diet thoughout the succession. Lothagam equids had a predominantly C_4 diet, but the earliest samples had a mixed C_3/C_4 diet. Suids were slower to exploit a C_4 diet than equids or proboscideans. Rhinocerotids had developed browsing and grazing specializations by Apak times. The isotopic signals of the fossil grazers were similar to those of grazers in the basin today, suggesting comparable diets, and it is likely that the Lothagam ecosystem was represented by a dwarf shrubland. The oxygen isotope evidence indicates that local meteoric water was enriched in ^{18}O by several parts permil compared to modern waters. The most likely explanation for isotopic depletion of meteoric waters is a combination of slightly higher rainfall, slightly lower evaporation, and more rainfall derived from the west. Consistently depleted $\delta^{18}O$ values for hippopotamids suggests that they were river rather than lake dwellers.

Paleoenvironmental Setting

Evidence from studies of the geology, reported by Feibel (this volume:chapter 2.1) and Wynn (this volume:chapter 2.2), indicate fluvial deposition throughout the Nawata Formation. The accumulation of the Lower Nawata was relatively rapid from a broad but shallow meandering and perennial river with extensive low gradient floodplains. Rainfall was relatively high (estimated to be over 1000 mm per year) but evidently seasonal, with a pronounced dry season for at least 4 months of each year (Wynn this volume:chapter 2.2). Below the Middle Markers (6.72 ± 0.06 Ma) the landscape would have included semideciduous thorn tree and tree savanna vegetation with gallery woodland. Above the Middle Markers, conditions became more arid. The floodplain vegetation was replaced by dry thorn bush savanna that was dissected by gallery woodlands. Annual rainfall during the accumulation of the Upper Nawata is estimated at less than 1000 mm per year. Decrease in the number of *Etheria* reefs indicates reduced river flow in the lower part of the Upper Nawata, but in the uppermost part of this member there was a return to wetter conditions that persisted into the Apak Member. The wetter conditions prevailing above the Purple Marker would have again supported wooded savanna.

The geology of the Apak Member is, however, more complex. The absence of *Etheria* reefs indicates that the river, which Feibel suggests may have been related to the modern Kerio drainage, perhaps had a more seasonal flow. There is evidence of a disconformity at about the time of the Purple Marker that may represent a considerable loss of time. The Lothagam paleosols indicate extended periods of depositional stasis at about 6.5 Ma and 5.2 Ma, and a drying episode between about 6.7 Ma and 5.0 Ma that is coincident with the Messinian salinity crisis (Zhang and Scott 1996). Recent evidence (Clauzon et al. 1996; Denton 1999) suggests that the drying of the Mediterranean was due more to local tectonic activity than to substantial climate changes, although there is evidence for a warm interval with fluctuating climate extending from 5 to 3 Ma (Denton 1999). Sediments from the top of the Apak Member mark the beginning of the transgression associated with expansion of the Lonyumun lake. These lacustrine sediments continue throughout the Muruongori Member, from which only aquatic species have been recovered. The Kaiyumung sediments are again fluviatile but may have been laid down by yet a third river system, the ancient Turkwel. It is interesting that the fish from the Kaiyumung Member suggest a lacustrine rather than fluviatile setting (Stewart this volume:chapter 3.2). Relatively few fossils have been recovered from this member (figure 13.2).

Evidence from oxygen isotope analyses of paleosols and of tooth enamel of many fossil mammalian taxa reported by Cerling et al. (this volume:chapter 12.2) complement the geological evidence and indicate a

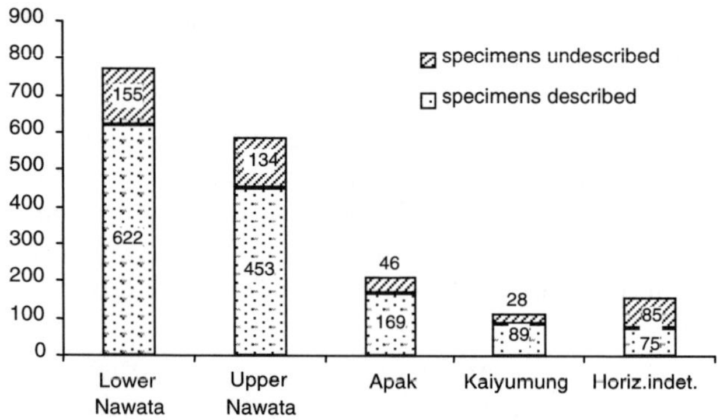

Figure 13.2 Number of specimens recovered from each member, showing the proportion of those described in this volume and those not yet identified.

mosaic ecosystem. Mixed habitats, such as wooded grasslands and grassy woodlands with a significant grass component, were common. Pure C_3 ecosystems, such as riparian forests, were also represented. There is no pedogenic carbonate evidence for pure C_4 grasslands, even though some mammals were feeding primarily from C_4 grasses; if C_4 grasslands were present, they must have been short-lived. Apak Member paleosols were significantly enriched in $\delta^{18}O$ compared to the Nawata Formation paleosols, indicating a drier and possibly hotter regime than in the subjacent strata. The pedogenic carbonates throughout the Lothagam succession are significantly more depleted in $\delta^{18}O$ than in modern soils. Comparably depleted values of $\delta^{18}O$ are typical of paleosols throughout the Koobi Fora Formation until about 1.8 Ma when there was a major change in isotopic composition of meteoric water (Cerling et al. 1988). Significant $\delta^{18}O$ depletion of local meteoric waters in the Lothagam succession (compared to modern waters) is also evident in all the mammalian tooth enamel sampled and in the ostrich egg shell. Cerling et al. (this volume:chapter 12.2) suggest that this can be explained by a combination of factors, including slightly higher rainfall, slightly lower evaporation, and more rainfall derived from the west. For example, it could be due to a change in the intertropical convergence pattern, with different wind patterns bringing more moisture from the west than is encountered today.

Complementary studies of the carbon isotopes show that the diets of the fossil grazers were similar to those of grazers in the basin today, and $\delta^{13}C$ values reflect the higher proportion of NAD-me and PEP-ck grasses as opposed to more mesic NADP-me grasses. This is consistent with the suggestion that the paleoecosystem was probably a dwarf shrubland similar to the modern ecosystem rather than an NADP-me grassland.

Lothagam is important because it represents the oldest continuous section in East Africa that postdates the global ^{13}C event when C_4 plants, most notably grasses, became a significant component of the biota. Although fossilized C_4 plants are known from about 13 Ma, they did not provide a significant component of the terrestrial biota until about 8 Ma, an event that Cerling et al. (1997) attributed to a global decline in atmospheric CO_2. This event was clearly of enormous significance in Neogene mammalian evolution. The subsequent spread of C_4 vegetation opened many new niches and heralded the appearance of an unprecedented diversity of herbivores. The Lothagam fauna documents exploitation of newly emergent grass-dominated habitats at the beginning of what Cerling at al. (1997) call "the C_4 world." It is tempting to speculate that increased reliance on the newly evolved grasslands, in a habitat subjected to marked seasonality, would have resulted in the evolution of migratory behavior. Today, a number of grazing herbivores, including wildebeest and the Uganda kob, depend on regular migrations to sustain sufficient graze. During the long dry spells, the grass is in short supply, so it is necessary for the herds to follow the rain to ensure a plentiful and nutritious food supply. The earliest record of migratory behavior is preserved in the prints and trails discovered at Laetoli, Tanzania (Leakey and Harris 1987). Here, a great diversity of mammal tracks in the Footprint Tuff of the Upper Laetolil beds documents migration into or through the Laetoli area at the onset of the rainy season (Hay 1987).

Paleoecological Implications of the Lothagam Fauna

The exceptional preservation of the large number of fossils recovered from Lothagam, especially those of vertebrates, provides the opportunity to reconstruct an unusually detailed picture of the faunal assemblages through time. However, sample size (figure 13.2) is an important factor in the assessment of faunal diversity because, for small samples, the number of species recognized in a fossil assemblage is directly related to the number of specimens recovered (figure 13.3). The assemblages of large mammals from the Nawata Formation are probably reasonably representative, whereas the smaller samples from the Kaiyumung and the Apak Members may not be.

Aquatic species are present throughout the Lothagam sequence, reflecting the prevailing mode of accumulation of the sediments. Large shoals of *Etheria* occur in the Lower Nawata and upper part of the Upper Nawata (Feibel this volume:chapter 2.1), testifying to the year-round flow of the Nawata river complex; their absence in the lower part of the Upper Nawata and from the Apak and Kaiyumung Members is attributed to a more seasonal flow regime in these units. Crabs are present in the Nawata Formation and persist into the

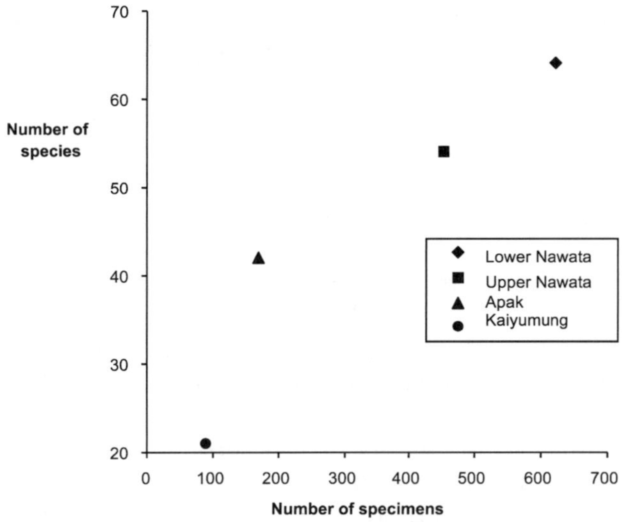

Figure 13.3 Number of specimens recovered from each member compared to the number of species represented.

Apak Member (Martin and Trautwein this volume:chapter 3.1); the presence of crabs is consistent with a well-oxygenated river system. Fish are abundant throughout the Lothagam sequence but there are fewer fish-bearing localities in the Upper Nawata than in the Lower Nawata (Stewart this volume:chapter 3.2). Fish from the Nawata Formation are small, archaic, and characteristic of shallow swampy water with considerable vegetation for spawning and for feeding. *Lates* requires well-oxygenated, well-mixed waters, and *Polypterus* is intolerant of even slightly saline water. The presence of *Hydrocynus*—and, to a lesser extent, large *Lates*—signifies the presence of open waters. The Nawata Formation fish assemblage suggests a large, slow-moving river with numerous well-vegetated back swamps and bays that were well-oxygenated and not brackish. The floodplains of this river system frequently supported shallow ponds, with a characteristic ostracod–*Lanistes*–*Pila* community (Feibel this volume:chapter 2.1).

In the Apak Member, the fish are also fluviatile in nature, but extant genera predominate. *Lates, Sindacharax,* and *Hydrocynus* became more common, suggesting a faster-flowing river with fewer vegetated backwaters than in Nawata times. Fish assemblages from the Muruongori and Kaiyumung Members represent a great diversity of habitats and trophic groups and are consistent with the presence of a large lake.

Aquatic reptiles are common throughout the succession. At least two crocodilian species—a piscivore and a more generalized carnivore—were represented in each member in contrast to the single species in Lake Turkana today (Storrs this volume:chapter 4.2). Chelonians too are common and diverse, with at least six different species representing three families. The exceptional abundance of *Turkanemys pattersoni* may result from annual congregations of this species at communal nesting beaches (Wood this volume:chapter 4.1).

Hippos make up between 18 and 20 percent of the mammalian fauna in the Nawata Formation, where *Hexaprotodon harvardi* is the dominant species. That hippos decrease in proportional representation in the Apak Member, and are only known from a few unidentifiable tooth fragments in the Kaiyumung (figures 13.4 and 13.5), may be attributable to the change in river systems from the broad, shallow, meandering perennial river of the Nawata Formation to the more seasonal predecessors of the Kerio and Turkwel Rivers in the Apak and Kaiyumung Members, respectively. Not unexpectedly, Lutrinae are rare—the sole representative being *Vishnuonyx angololensis* sp. nov. from the Lower Nawata (Werdelin this volume:chapter 7).

Bovidae are the commonest terrestrial mammals throughout the Lothagam succession (figure 13.4), although the tribal representation differs in each member (figure 13.6). Bovids form an exceptionally diverse family; many of the tribes display habitat preferences that make them useful indicators of paleoenvironment. Extant impalas are characteristic of forest edge ecotones, and the Aepycerotini, represented by *Aepyceros premelampus,* is the dominant tribe throughout the succession. In the Lower Nawata, the boselaphines are the second most common tribe, followed by the reduncines. The

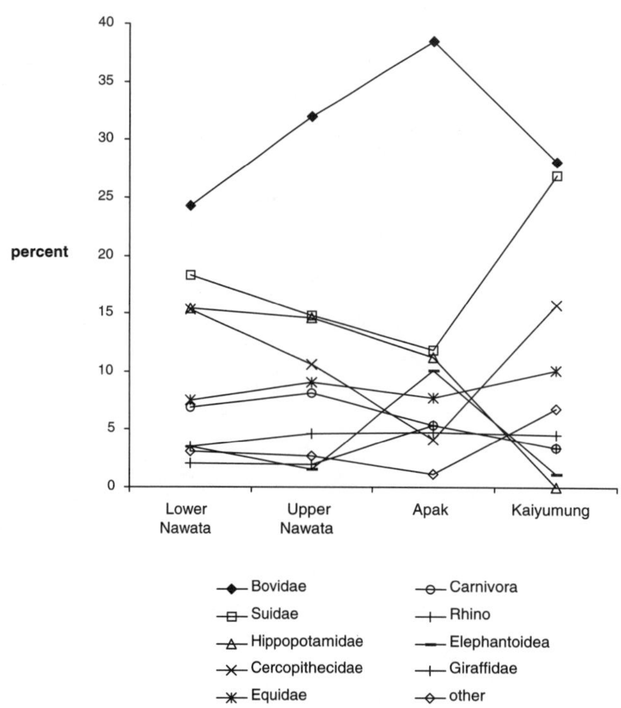

Figure 13.4 Percentage representation of taxa recognized in each member, using specimens published in this volume and unpublished specimens that can be attributed to higher taxonomic groups.

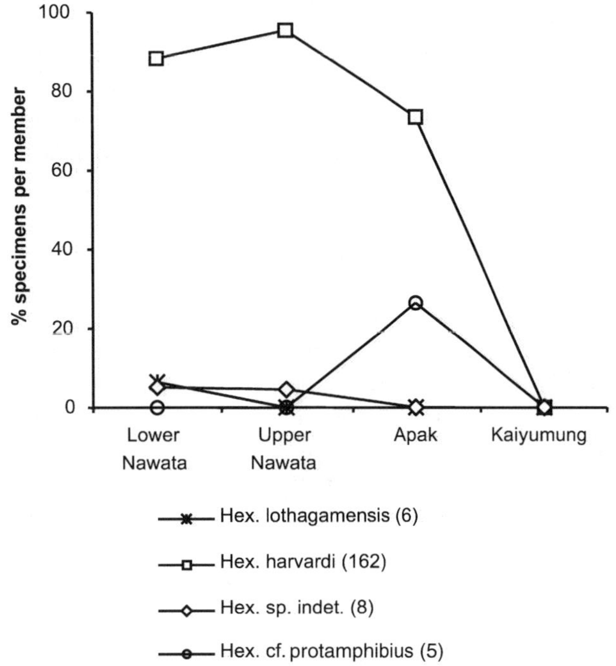

Figure 13.5 Distribution of Hippopotamidae in each member. Total number of specimens for each taxon is included in parentheses.

group of boselaphines to which the Lothagam boselaphines are related became extinct by the Early Pliocene but their presence would be consistent with forested conditions. Extant reduncines are water-frequenting grazers. In the Upper Nawata, the proportion of aepycerotines decreases and the boselaphines decline drastically, whereas alcelaphines and reduncines increase in number; these changes are consistent with more open conditions. The Apak Member sees a sharp increase in the proportion of tragelaphines and bovines but a corresponding decrease in alcelaphines and reduncines, which signals a return to more closed habitats. Aepycerotines, alcelaphines, and bovines are the dominant tribes in the Kaiyumung, suggesting a change to more open and drier conditions (Harris this volume:chapter 11.2).

The extant representatives of the second most common family, the Suidae, are dependent on closed vegetation for shelter, and Bishop (1994) has suggested that this lifestyle extends back to the Pliocene and Late Miocene. In the Lothagam sequence, there is a high turnover of species (figure 13.7). In the Lower Nawata, *Nyanzachoerus syrticus* is the dominant species and the smaller *Ny. devauxi* is the second most common suid. In the more open habitats of the Upper Nawata, *Ny. devauxi* is less abundant while increases in the length and complexity of the third molars of *Ny. syrticus* indicate a trend toward greater reliance on grazing that is supported by the $\delta^{13}C$ values of tooth enamel (Cerling et al. this volume:chapter 12.2). In the Apak Member there is a dramatic change in suid taxa, with the primarily grazing *Ny. australis* replacing *Ny. syrticus* and making up 85 percent of the suid assemblage. This is the earliest evidence for undoubted grazing suids in the succession. In the Kaiyumung there is again a marked

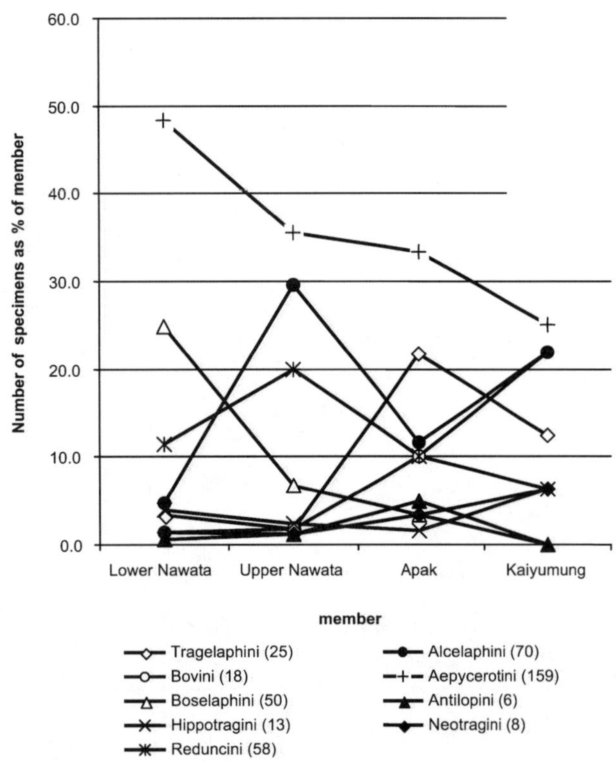

Figure 13.6 Distribution of Bovidae in each member. Total number of specimens for each taxon is included in parentheses.

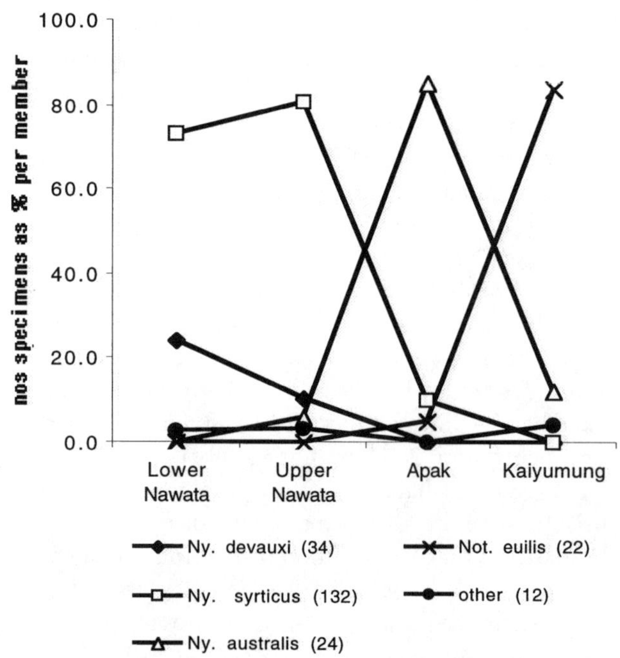

Figure 13.7 Distribution of Suidae in each member. Total number of specimens for each taxon is included in parentheses.

change in suids, with the more specialized grazing *Notochoerus euilus* replacing *Nyanzachoerus* as the dominant suid (Harris and Leakey this volume:chapter 10.2)

The third most common family of terrestrial mammals, the Cercopithecidae is dominated in the Nawata Formation by *Parapapio lothagamensis,* the only papionin (figure 13.8). Colobines are less numerous but more diverse, being represented by three species. In the Upper Nawata the relative proportion of *P. lothagamensis* decreases (64%), and colobines become more common (28%). This trend continues into the Apak Member where the numbers of colobines exceed those of cercopithecines, although the cercopithecid sample in this member is very small. In the Kaiyumung Member, *Theropithecus* cf. *T. brumpti* is the dominant monkey. The paleoecological indications from these early colobines and cercopithecines is less clear than for species of the modern subfamilies. Extant African colobines are normally indicative of forest, but Lothagam colobines had similar postcranial adaptations to the cercopithecines, indicating a terrestrial or semiterrestrial locomotion (Leakey et al. this volume:chapter 6.1).

The distribution of Equidae may well reflect changing habitats (figure 13.9). Bernor and Harris (this volume:chapter 9.2) interpret the larger, more heavily built *Eurygnathohippus turkanense* (the common equid in the Lower Nawata) as adapted to closed woodland habitats and incapable of sustained cursorial locomotion. The more lightly built, more open country–adapted *E. feibeli* replaces *E. turkanense* as the common equid in the Upper Nawata. In the Apak and Kaiyumung Members, large and small species of *Eurygnathohippus* persist, with the larger the most dominant, particularly in the Apak Member. In spite of the morphological differences that indicate different locomotory patterns and corresponding habitat preferences, the Lothagam equids all appear to have had similar diets. The carbon isotope analyses

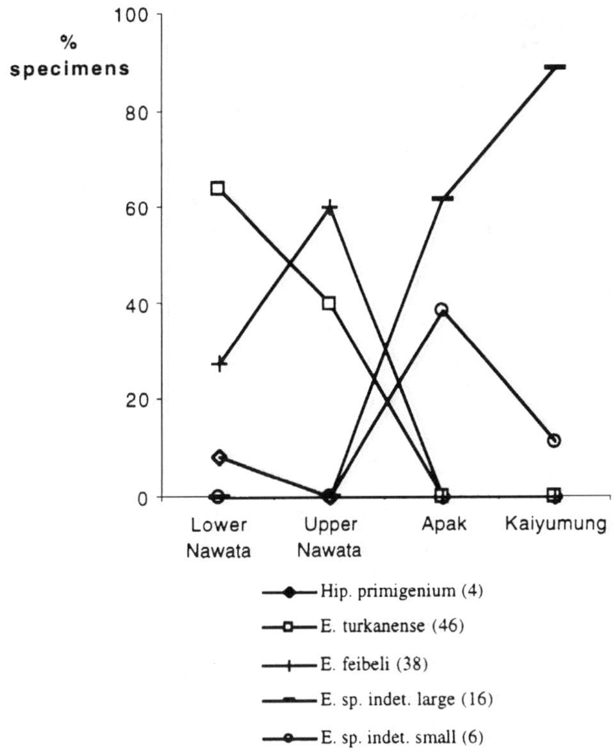

Figure 13.9 Distribution of Equidae in each member. Total number of specimens for each taxon is included in parentheses.

of the equids indicate an increasing C_4 fraction from the Lower to Upper Nawata. In the lowest levels of the Lower Nawata, these analyses document the change from a C_3-dominated diet to a C_4-dominated diet. A mixed C_3/C_4 diet with values between –3 and –5 permil is found in half of the samples from the lowest 82 meters of section. Comparable values indicating a mixed diet are rarely found in modern African herbivores and never in modern African equids. Similar intermediate values are also found in other mammals at this time and, as Cerling et al. suggest in this volume (chapter 12), they may indicate a period of transition where a mixture of C_3 and C_4 grasses coexisted for a few hundred thousand years in the low elevation tropics. Later in the Nawata Formation and all younger horizons, the equids all have more positive values, indicating a diet of more than 70 percent C_4.

Proboscideans are relatively infrequent throughout the succession but include a remarkable diversity of taxa (Tassy this volume:chapter 8.1), suggesting niche partitioning of food resources. *Stegotetrabelodon* is the dominant elephantid in the Lower Nawata, but in the Upper Nawata and Apak there is no clearly dominant species (figure 13.10). Surprisingly, *Deinotherium bozasi,* a committed browser, appears to be the dominant species in the Kaiyumung, but this is probably a collecting anomaly as this taxon is largely represented by tooth fragments collected as voucher specimens.

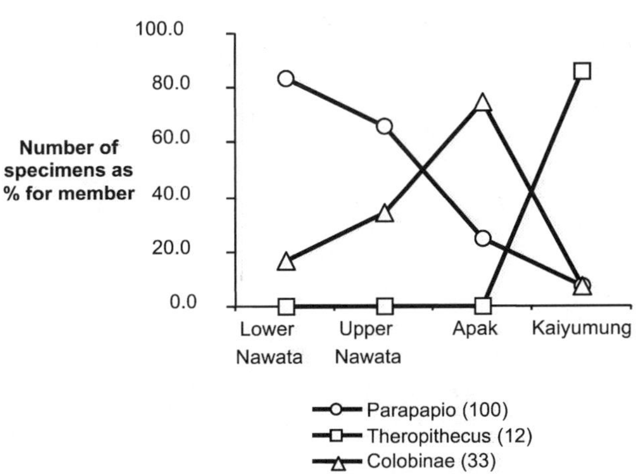

Figure 13.8 Distribution of Cercopithecidae in each member. Total number of specimens for each taxon is included in parentheses.

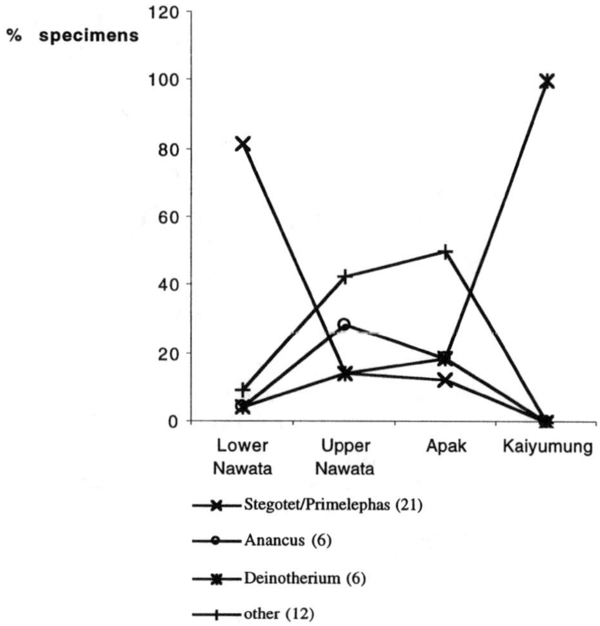

Figure 13.10 Distribution of Elephantoidea in each member. Total number of specimens for each taxon is included in paranthesis.

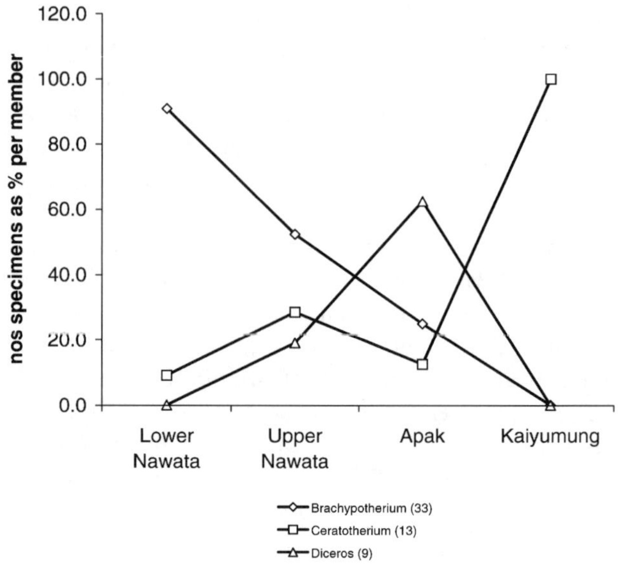

Figure 13.11 Distribution of Rhinocerotidae in each member. Total number of specimens for each taxon is included in parentheses.

Isotopic analyses show that deinotheres had an exclusively C_3 diet throughout their history. $\delta^{13}C$ analyses of gomphothere and elephantid teeth record three specimens from the Lower Nawata with mixed C_3/C_4 diets (–3.9, –5.5, and –6.2 permil) representing transitional diets, but most elephantids and gomphotheres were feeding mainly on C_4 vegetation; elephantids from higher in the sequence were almost exclusively C_4 feeders, a condition that continued throughout the Pliocene and Pleistocene (Cerling at al. this volume:chapter 12.2)

Rhinos are also relatively rare. The archaic Miocene genus, *Brachypotherium,* is dominant in the Lower Nawata, but later in the succession it is replaced by the extant genera *Ceratotherium* and *Diceros* (figure 13.11). The youngest *Brachypotherium* is recorded from the Apak. The extant genera become more abundant in the Upper Nawata. The browsing *Diceros* is the common rhino in the Apak, whereas only the grazing *Ceratotherium* has been recovered from the Kaiyumung (Harris and Leakey this volume:chapter 9.1). Isotopic analyses show both *Brachypotherium lewisi* and *Ceratotherium praecox* were browsing in the Lower Nawata, whereas *C. praecox* from the Upper Nawata and Kaiyumung was exploiting C_4 graze. A single *Brachypotherium* tooth analyzed from the Apak showed a mixed diet of C_3 and C_4 vegetation (Cerling et al. this volume:chapter 12.2).

Giraffids are also relatively uncommon but they, too, demonstrate a replacement of archaic taxa by more progressive forms (figure 13.12). *Palaeotragus* is represented by two species in the Nawata Formation, but the majority of giraffid specimens from the Apak Member are attributed to *Giraffa stillei.* The single specimen of *P. germaini* shows this species persisted into the lower part of the Apak Member (Harris this volume:chapter 11.1). Giraffids were anatomically adapted for browsing from tall vegetation; isotopic analyses of Miocene and Pliocene representatives confirm a pure C_3 diet (Harris and Cerling 1998) and indicate that giraffids obtained most of the water they required from the vegetation they ate.

The carnivores are diverse (figure 13.13), but they are necessarily less well represented than their herbiv-

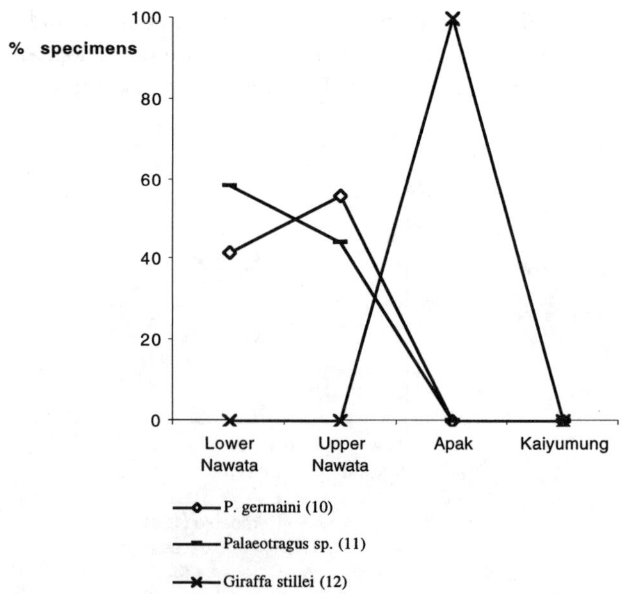

Figure 13.12 Distribution of Giraffidae in each member. Total number of specimens for each taxon is included in parentheses.

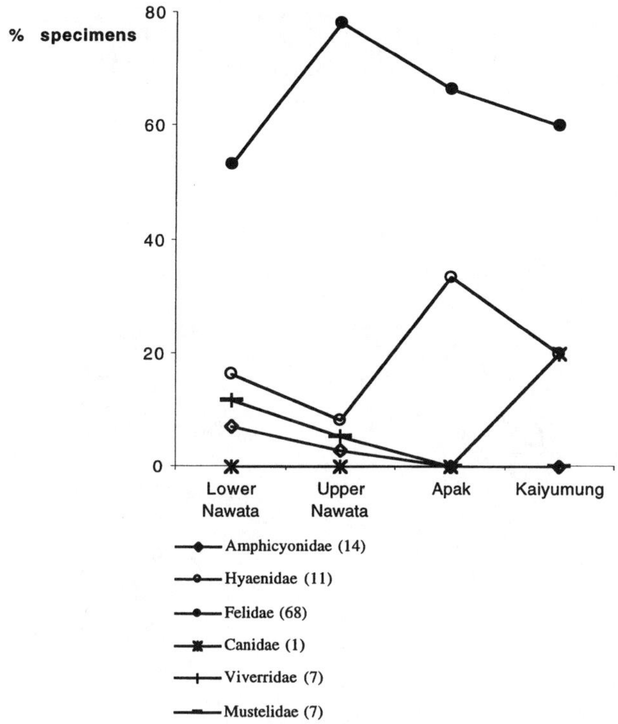

Figure 13.13 Distribution of Carnivora in each member. Total number of specimens for each taxon is included in parentheses.

orous prey. Werdelin (this volume:chapter 7) suggests that the presence of amphicyonids is an indication that the environment included large enough expanses of forest to accommodate large to very large forest-dwelling carnivores. At the same time, the postcranial skeleton of *Ictitherium ebu* shows adaptations to extreme cursoriality and that of *Ekorus ekakoran* indicates a degree of cursoriality not seen in other mustelids. The occurrence of *Ictitherium* and *Ekorus* in the Lower Nawata demonstrates the presence of open environments, although not necessarily adjacent to the Lothagam river. Among the smaller carnivores, the viverrids indicate the presence of arboreal or semiarboreal habitats and the small felids suggest the presence of closed woodland or possibly forest.

The rodents and hares are known from only 43 identifiable specimens, but they represent four families (13 genera and 15 species) of rodents and two genera of lagomorphs (Winkler this volume:chapter 5). The most diversity is seen in the Lower Nawata with eight species, but the sample size is low and is unlikely to accurately reflect the original biota. Winkler (this volume:chapter 5) urges caution in interpreting paleoenvironments based on such a small sample but feels that the presence of the giant squirrel *Kubwaxerus* in the Lower Nawata is a good indication of a closed forest habitat. *Thryonomys* today inhabits moist savannas, and Winkler interprets its presence in the Upper Nawata and Kaiyumung to suggest more open habitats.

The sequential assemblages of terrestrial mammals are consistent with the paleoenvironmental settings as interpreted from the geological studies (figure 13.4). Species from the Lower Nawata are consistent with a relatively lush closed woodland and wet swampy habitat, whereas in the Upper Nawata an increase in more open country species reflects increased aridity at this time (figure 13.1). In the Apak Member there is a dramatic change, with many of the archaic species found in the Nawata Formation being replaced by more modern taxa. The Apak assemblage is indicative of a woodland habitat but with abundant nearby grassland (figure 13.14). Several of the herbivores show more advanced grazing specializations than their earlier relatives. Even more specialized grazing adaptations are apparent in the Kaiyumung herbivores, and this suggests a more open habitat with increased grasslands and bushland (figure 13.15). It follows that the replacement of C_3 grasses by C_4 plants would have led to an evolutionary radiation and proliferation of C_4 grasses, thus opening many new niches for an array of specialized mammalian grazers; hence, the radiation of specialized grazing species that characterize the East African Pliocene and Pleistocene. The seasonality that is evident from the Lothagam paleosols indicates that droughts would have been a regular feature of East African habitats throughout the time of the Lothagam succession. This could have led to increased selection pressures when food and water were scarce and might account for the radiation of grazing species at this time.

In summary, the Bovidae, Hippopotamidae, Suidae, and Cercopithecidae are the most common elements of the fauna in the Lower Nawata (figure 13.4; table 13.1) and the proportional representation of these families is consistent with a well-vegetated habitat. In the Upper Nawata, an increase in bovids and equids, along with a decrease in suids and monkeys, can be attributed to the drier, more open conditions that are evident from the geological record. The Apak fossil sample is rather limited (figure 13.2) and may not be an accurate representation of the total fauna. Here the proportion of elephantids increases significantly, and giraffids and bovids also increase, whereas suids and equids decrease. The more open Apak environment with its seasonal river would suggest a more varied habitat. The Kaiyumung is the least fossiliferous member and thus possibly the least representative (figure 13.2). Here the number of suids and monkeys increase, but in each case there is one dominant rather specialized species, *Notochoerus euilus* and *Theropithecus brumpti*, respectively. Bovids decrease from their proportional representation in the Apak.

Figure 13.14 Restoration of the Apak Member habitat by Mauricio Antón.

Faunal Turnover

The Late Miocene and Early Pliocene were times of dramatic faunal change (Behrensmeyer et al. 1992), and this is well documented at Lothagam (tables 13.2 and 13.3). Contrasting first and last appearances from the four members of the Lothagam succession provides an indication of the tempo of this change (tables 13.2 and 13.3). The greatest turnover is seen between the Upper Nawata and the Apak Member. However, as noted ear-

Figure 13.15 Restoration of the Kaiyumung Member habitat by Mauricio Antón.

lier in this chapter, the smaller sample of fossils from the Apak Member than from the Upper Nawata may have resulted in fewer species being recognized from the Apak. Thus, some of the species last recorded from the Upper Nawata (34) may have persisted into the Apak but may not have been sampled. For the same reason, the Apak Member first appearance records (15) may be underestimated, with some of the less common taxa unrepresented; those that are recorded are likely to be an accurate reflection of change, however. A similar situation would have pertained between the Apak and the Kaiyumung where relatively few first appearances (9) are recorded and a large number of last appearances (24) are documented in the Apak. Behrensmeyer et al. (1997) and Bobe (1997) have undertaken analyses of rates of faunal turnover at later intervals of time (Shungura and Koobi Fora Formations) in the Turkana Basin.

The turnover between the Lower Nawata and the Upper Nawata is quite marked—16 species are first recorded from the Upper Nawata, whereas 19 Lower Nawata species do not continue in the Upper Nawata (tables 13.2 and 13.3). However, 11 of the 19 last appearance records are of small species (7 rodents, 1 hare, and 3 small viverrids), some of which may have actually been present in the Upper Nawata but were not sampled because of their small size. Eight species of large mammals found in the Lower Nawata do not persist into the Upper Nawata. Several of these—including Amphicyonidae sp. A, *Vishnuonyx angololensis*, *Hippotherium* cf. *H. primigenium*, and cf. *Kubanochoerus* sp.—are archaic taxa typical of the Miocene. The remainder are either very rare in the fossil record—for example, *Ekorus ekakoran*—or, like *Hexaprotodon lothagamensis* and *Primelephas gomphotheroides*, were unsuccessful competitors in the Late Miocene radiation of modern taxa. It is likely that the loss of some species in the Upper Nawata was due to environmental change.

Many first appearance records in the Upper Nawata comprise single specimens, four of which are of small or rare animals such as rodents and viverrids, but the 12 first appearance records of larger mammals are likely to reflect real changes in the biota (table 13.2). Two bovids—an alcelaphine and a reduncine, *Damalacra* sp. and *Kobus laticornis*, respectively—are probably indicative of more open habitats in contrast to the first records of the less common *Diceros bicornis*, *Nyanzachoerus* cf. *Ny. australis*, and *Tragelaphus kyaloae*. The significance of the first appearance of hominoids high in the Upper Nawata has yet to be determined.

Although sample size in the Apak Member is limited (figure 13.3), major faunal change is evident, with significant loss of larger mammalian species from the Lothagam biota at this time. Species that are commonly represented in the Nawata Formation but do not persist into the Apak include *Parapapio lothagamensis*, three species of colobine, the large machairodonts *Lokitunjailurus emergeritus* and cf. *Metailurus*, the two suids cf. *Potamochoeroides* sp. and *Nyanzachoerus devauxi*, the hippo *Hexaprotodon* sp. indet., the small giraffid *Palaeotragus* sp., and the small boselaphine *Tragoportax* species. The apparent disappearance of *Menelikia leakeyi* is of only local significance because this species occurs elsewhere in the basin later in the Pliocene. Similarly, hipparions *Eurygnathohippus turkanense* and *E. feibeli* are not represented by diagnostic material from the Apak Member, but closely related equids of comparable size do occur.

First appearances are also significant in the Apak Member, although several species first occurring in the Apak Member are more common in the superjacent Kaiyumung Member. Among the primates, two new colobines are recorded, together with a species of *Parapapio* that differs from *P. lothagamensis*. Other first appearances include the earliest recorded giraffid, *Giraffa stillei;* the earliest *Loxodonta* (aff. *L. exoptata*); the suids *Nyanzachoerus jaegeri* and *Notochoeus euilus*; and the bovine *Simatherium* aff. *S. kohllarsoni*. Species referred to taxa known better from later horizons include *Hexaprotodon* cf. *Hex. protamphibius* and *Gazella* cf. *G. janenschi*.

Some of the species last recorded from the Apak represent relicts from the Upper Nawata. *Stegotetrabelodon orbus*, *Brachypotherium lewisi*, *Nyanzachoerus syrticus*, and *Palaeotragus germaini* are only found in the lower horizons of the Apak Member. Others that persist higher in the Apak include *Nyanzachoerus* cf. *Ny. australis*, *Hexaprotodon harvardi*, *Tragoportax* cf. *T. cyrenaicus*, and *Ikelohyaena* cf. *I. abronia*. The absence of the herbivores *Anancus kenyensis*, *Diceros bicornis*, *Gazella* cf. *G. praethomsoni*, and *Madoqua* sp. from the Kaiyumung Member is conceivably due to environmental factors or small sample size, but all are known from later sediments elsewhere.

The Kaiyumung Member, too, records a number of first appearances in spite of the small sample size and rather fragmentary nature of the majority of the fossils. The seven species comprise an early, large-bodied colobine, *Theropithecus* cf. *T. brumpti*, *Australopithecus* cf. *A. afarensis*, a new species of *Dinofelis*, the oldest African canid, the cane rat *Thryonomys* cf. *T. gregorius*, and the bushbuck *Tragelaphus* cf. *T. scriptus*.

The Lothagam succession documents significant changes in the faunal assemblages through time, with turnover of species throughout the succession. However, the geological marker units do not necessarily coincide with faunal change. A relatively drastic change is apparent at the Upper Nawata–Apak Member boundary, and, in part, this change may be due to a hiatus in sedimentation just above the Purple Marker that, in turn, may be a manifestation of the climatic changes associated with the Messinian salinity crisis in the Mediterranean. In general, the faunal change in the Lothagam succession appears to have been relatively gradual, and species that are dominant in one member persist in lower frequency at subsequent horizons. Thus taxa typical of the Nawata Formation—such as the elephantid *Stegotetrabelodon orbus*, the rhinoceros *Brachypotherium lewisi*, the giraffid *Palaeotragus germaini*, and the suid *Nyanzachoerus syrticus*—all persist as rare elements of the Apak Member. Similarly some species that are rare elements at the top of one member become more dominant in the succeeding member. An example here is *Nyanzachoerus* cf. *Ny. australis*—rare in the Upper Nawata but the dominant suid of the Apak Member.

Comparison with Other Sites

The sequence of changes encountered in the Lothagam faunas reflect fluctuations in the local habitat. These changes were in turn manifestations of more widespread events that included the Messinian salinity crisis in the Mediterranean, the global radiation of C_4 grasses at approximately the same time, and faunal interchange between Africa and Eurasia. Comparisons with sites of similar age in Africa and Eurasia provide an indication of environmental conditions that prevailed at the transition between the Miocene and Pliocene epochs and furnish evidence of possible migration routes.

Meaningful comparisons with Lothagam can only be made with published sites that document a comparably rich and diverse fauna. Sites of broadly similar age in Africa with which Lothagam can be compared include the Lukeino Formation and the Mpesida Beds in the Baringo Basin of Kenya (Hill et al. 1985, 1986; Hill 1994, 1999), the Manonga Valley of Tanzania (Harrison 1997a), the Sinde Basin of Zaire (Yasui et al. 1992) and the adjacent Nkondo area in Uganda (Pickford et al. 1988, 1993), the Middle Awash region of Ethiopia (Renne et al. 1999), Langebaanweg in South Africa (Hendey 1970a, 1970b, 1974, 1981a, 1981b), Wadi el-Natrun in Egypt (Stromer 1907), Marceau in Algeria (Arambourg 1959), and Sahabi in Libya (Boaz et al. 1987). Some of these sites are still being investigated. Abu Dhabi, in the United Arab Emirates (Whybrow and Hill 1999), provides an important link with both the African and Eurasian faunas. Comparisons with the older sites of the Namurungule Formation in the Samburu Hills of Kenya (Nakaya et al. 1984; Nakaya 1990, 1993, 1994) estimated to be between 10.1 and 9.5 Ma (Sawada et al. 1998), and the richly fossiliferous eastern Mediterranean localities of Samos and Pikermi in Greece (Solounias 1981) and Maregheh in Iran (Bernor 1986; Bernor et al. 1996) are important for providing a deeper perspective on the Lothagam fauna.

African Sites

Not surprisingly, the Lothagam fauna shows closest affinities with sites in Africa of comparable age, most specifically those in East Africa.

Mpesida and Lukeino, Tugen Hills

The well-calibrated lengthy stratigraphic sequence in the Tugen Hills of the Baringo Basin, approximately 250 km to the south of Lothagam, includes the two Late Miocene sites of Mpesida and Lukeino. The Mpesida Beds mostly date to ~6.5 Ma, which is also approximately the age of the Marker Tuff, although site BPRP#133 has been dated at 6.7–7.2 Ma (Hill 1999). Fossil plants—including branches, roots, and leaves—suggest a lowland rainforest habitat for the majority of sites. The many genera in common with those from Lothagam include *Primelephas, Anancus, Stegotetrabelodon, Ceratotherium, Brachypotherium, Tragelaphus, Kobus, Madoqua, Gazella,* and/or *Raphiceras* (Hill 1999). The slightly younger Lukeino Formation, at 6.2–5.6 Ma, is of comparable age to the Upper Nawata and accordingly includes *Nyanzachoerus* and an increased number of modern elements (Hill et al. 1985; Hill et al. 1999; Bishop et al. 1999; Turner et al. 1999).

Manonga Valley

Although not well dated, the Wembere-Manonga Formation in the Manonga Valley of Tanzania includes the Ibole, Tinde, and Kilolele Members, which, on biostratigraphic evidence, are believed to extend from the Late Miocene (ca. 5.5 Ma) to the Early Pliocene (at 4.0 Ma) (Harrison 1997a). The sediments of the Wambere-Manonga Formation were deposited in a single transgression-regression cycle of a large shallow lake, approximately 10,000 km^2 in area, that filled the Manonga Basin (Harrison and Mbago 1997). Unfortunately, much of the Manonga material is too fragmentary for secure attribution to species, but most genera and some species are shared with Lothagam (figure 13.16).

The earliest sediments of the Ibole Member (5.5–5.0 Ma) are of broadly similar age to the uppermost Upper Nawata and lower Apak Member deposits at Lothagam. The fossiliferous uppermost portion of the member was subjected to cyclical drying and the fauna is dominated by proboscideans (30%), whereas suids (24%) and bovids (22%) were slightly less common (figure 13.16). Equids were rare and rhinos were absent (Harrison 1997b). The proboscideans include two known from the Apak Member at Lothagam, *Anancus kenyensis* and *Stegotetrabelodon* (Sanders 1997). At Lothagam proboscideans are more common in the Apak Member than at earlier horizons; *Primelephas,* which is present and not uncommon in the Ibole Member at Manonga, occurs only very rarely in the earlier horizons at Lothagam. The Ibole suid is interpreted to represent *Nyanzachoerus kanamensis,* although Bishop (1997) notes that the teeth are more robust than those of *Ny. kanamensis* elsewhere in East Africa and in this respect re-

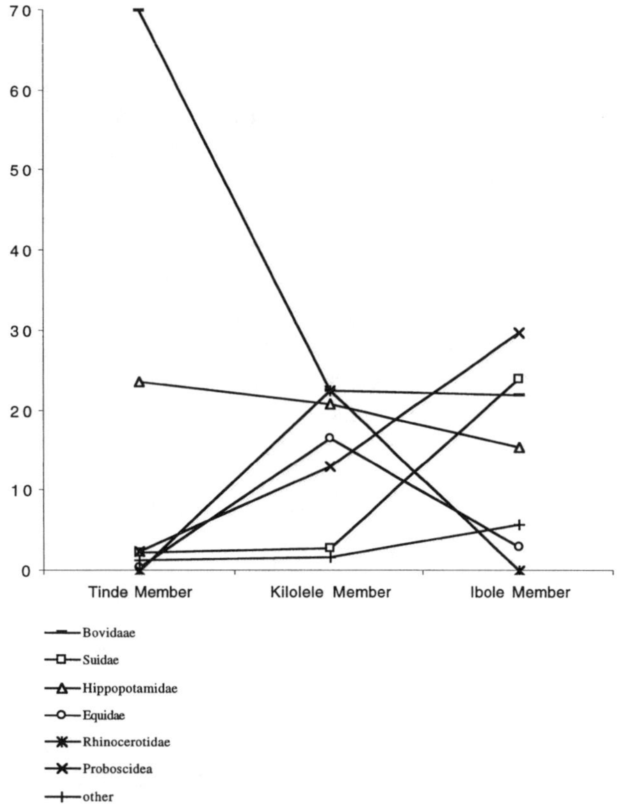

Figure 13.16 Distribution of taxa recognized in the Ibole, Tinde, and Kilolele Members of the Manonga Formation (data from Harrison 1997).

semble *Ny. australis.* At Lothagam some of the Apak Member suids are referred to *Nyanzachoerus* cf. *Ny. australis* because of their larger size and greater robusticity. The Manonga bovid assemblage differs from that at Lothagam because it is dominated by *Kobus* aff. *K. porrecticornis,* a species not recognized at Lothagam, and because hippotragines, reduncines, and tragelaphines are rare (Gentry 1997). Differences from the Lothagam bovid assemblages may be attributed to paleoecological differences associated with the lake margin setting.

The Tinde Member (5.0–4.5 Ma) correlates with the Apak Member at Lothagam; it accumulated at a time when the Manonga-Wambere lake was at its maximum extent with few emergent intervals represented in the sequence. The fauna shows very little diversity, and small mammals are absent. As in the Apak Member, the Tinde Member fauna was dominated by bovids (70%) and hippos (23.7%) (figure 13.16). The dominant bovid, making up 84 percent of the bovid assemblage, was the reduncine, *Kobus* aff. *K. subdolus* (Harrison 1997b). Although this bovid is not represented at Lothagam, other mammals including *Nyanzachoerus kanamensis* (= *N. pattersoni*), *Hexaprotodon harvardi,* and *Eurygnathohippus* are shared with the Lothagam biota.

The Kilolele Member (4.5–4.0 Ma) is of comparable age to the upper Apak Member sediments but several taxa are more progressive than those found in the Apak Member. The sediments were deposited in a prodeltaic or deltaic environment at a time when the lake basin was filling with sediment. Here alcelaphines make up 61 percent of the bovid assemblage, which suggests a drier and more open habitat than in the Tinde Member, although relatively densely vegetated bushland and woodland would have fringed the lake basin and lake margin habitats (Harrison 1997b). The fauna in this member is more advanced relative than that from earlier members at Manonga. *Eurygnathohippus hasumensis* is recognized in the Kilolele Member but not in the Apak Member, and the Apak Member elephantids—*Loxodonta* cf. *L. exoptata* and *Elephas ekorensis*—are less progressive than those from Kilolele.

Langebaanweg

The Late Miocene site of Langebaanweg is the only known South African locality with an abundant and diverse fauna that is of comparable age to Lothagam (Hendey 1974). The majority of fossils derive from the Quartzose Sand Member and Pelletal Phosphorite Member of the Varswater Formation (Hendey 1981b), which sample a coastal savanna environment crossed by a perennial river with seasonal differences in flow and with an estuary surrounded by low-lying sandy flats that flooded during the rainy season. The fresh water in the river would have attracted a concentration of land mammals in the long hot dry season, whereas periods of good rains supported large trees and relatively lush vegetation. The fauna is notable for the large number of ancestral white rhinos, *Ceratotherium praecox* (Hooijer 1972). The "E" Quarry of the Varswater Formation includes a number of species that also occur farther north in the upper part of the Upper Nawata and in the Apak Member at Lothagam.

Langebaanweg yields an extensive carnivore fauna (Hendey 1974) that seems a little younger than that of Lothagam. Phocids are absent from Lothagam for obvious ecological reasons, and canids are absent from the older part of the Lothagam sequence because they had not yet arrived in Africa from North America. The first record of a canid at Lothagam is from the Kaiyumung Member. Similarly, *Homotherium* is present at Langebaanweg but absent from Lothagam because of Lothagam's early age. Enhydrini are present at both localities, with *Vishnuonyx* at Lothagam and the more derived *Enhydriodon* at Langebaanweg. *Viverra leakeyi* is present at both sites, as is a species of *Genetta.* The Hyaenidae are similar, with common genera including *Hyaenictitherium, Hyaenictus,* and *Ikelohyaena.* However, Lothagam has the primitive *Ictitherium,* while Langebaanweg has the more derived *Chasmaporthestes.* Both faunas include a large machairodont, which may be the same species.

Proboscideans from Langebaanweg include *Anancus*, which occurs at Lothagam but persists at younger sites, and *Mammuthus subplanifrons*, which suggests that Langebaanweg may be younger than Lothagam. Of the horses, *Hippotherium* cf. *H. primigenium* has been recorded from the Gravel Member, which underlies the main fossiliferous portion of the Langebaanweg sequence. However, neither of the other two hipparion species (cf. *H. baardi* and cf. *H. namaquense*) is closely related to the Lothagam hipparions. The ancestral white rhino *Ceratotherium praecox* occurs at both localities. Only fragmentary hippo remains have been reported from Langebaanweg. The small *Cainochoerus* from the Pelletal Phosphorite Member is represented by a very similar form in the Upper Nawata. In contrast, the common suid from the Quartzose Sand Member is *Nyanzachoerus australis*, which does not appear in the Lothagam sequence until the Apak Member. Hendey (1981b) documents *Sivatherium hendeyi* and *Giraffa* sp. from the Quartzose Sand Member, and these two species are joined by *Paleotragus* cf. *P. germaini* in the Pelletal Phosphorite Member. At Lothagam, *Paleotragus* is restricted to the Nawata Formation and *Giraffa* does not appear until the Nachukui Formation. Gentry (1980) records from Langebaanweg an unnamed species of *Tragelaphus*, the large boselaphine *Tragoportax acrae*, the bovine *Simatherium demissum*, the reduncine *Kobus subdolus* and an unnamed *Kobus* species, the alcelaphines *Damalacra neanica* and *D. acalla*, the grysbok *Raphicerus paralius*, a gazelle, and an ovibovine. Congeneric tragelaphines, boselaphines, reduncines, alcelaphines, antilopines, and neotragines occur at Lothagam, but most of the species are different and the impalas, which predominate at Lothagam, are absent from Langebaanweg.

Sahabi

The North African site of Sahabi is not well dated but has been placed variously at MN13 (7.1–5.3 Ma) based on the carnivores (Howell 1987), post Messinian based on the elephantids (Gaziry 1987a), Middle Turolian and Late Pliocene based on microfauna (5.3 Ma) (Munthe 1987), and earliest Pliocene (5 Ma) based on the geology and limited palaeontological evidence (Bernor and Pavlakis 1987). Geraads (1989) suggested that the fauna is of mixed Late Miocene and Pliocene age on the basis of the bovids, a viewpoint that is shared by Vrba (1995).

The mammalian fauna of the Sahabi Formation represents a mixture of Eurasian, North African endemic, and pan-African–derived species (Boaz et al. 1987). Faunal cluster analysis links Sahabi with other Mio-Pliocene sites in Africa (Bernor and Pavlakis 1987). The site is exceptional in the diversity of habitats sampled, and the different members show differing facies and characteristic species adapted to that particular sedimentary environment (figure 13.17). Sirenians make up over 60 percent of the shallow marine deposits of the earliest Member T, whereas the fluvial deposits of the uppermost Member V are dominated by hippopotamids (56 percent of the mammals). The broad deltaic channel facies of the intervening deposits are dominated by bovids (Dechant Boaz 1987). The percentage of anthracotheres remains between 14 and 22 percent of the fauna in each member. With the exception of carnivores in Member U2, which make up almost 15 percent, the remaining families each make up less than 10 percent in all members.

Neither the sirenians nor the anthracotheres represented at Sahabi have been collected from Lothagam. Of the remaining taxa, many show Eurasiatic links but some also occur at Lothagam or have close affinities with Lothagam species. Sahabi Cercopithecidae include a form recognized as cf. *Macaca* that shows affinity with *Parapapio lothagamensis* and may well be synonymous. Two isolated teeth of an indeterminate colobine were also reported (Meikle 1987). Sahabi, like Lothagam, has a diverse assemblage of carnivores. Elements common to the two sites include *Hyaenictitherium*, one or more species of *Viverra*, and

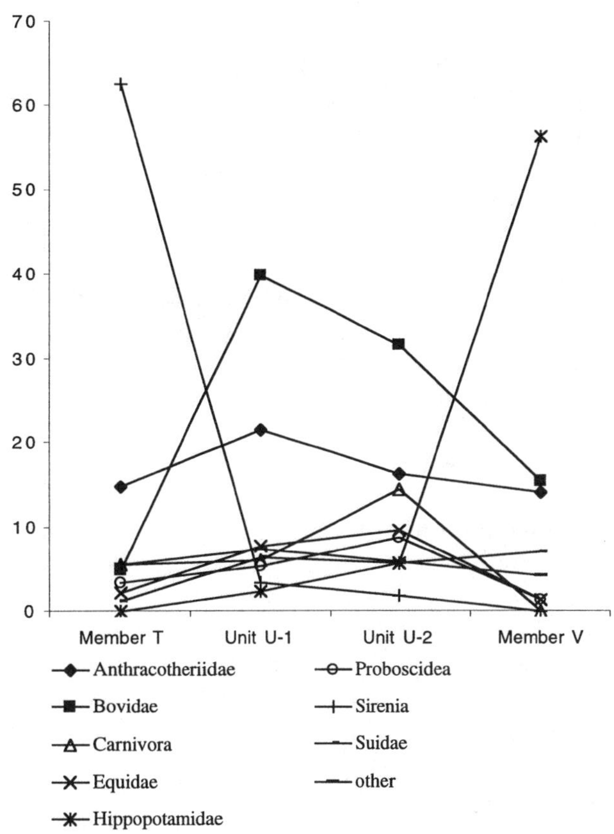

Figure 13.17 Distribution of taxa recognized in different units of the Sahabi Formation (data from Dechant Boaz 1987).

both large and small machairodonts. Differences include the presence of ursids at Sahabi versus amphicyonids at Lothagam and, importantly, the presence at Sahabi of several large, bone-cracking percrocutids and hyaenids that are absent from Lothagam. No true bone-cracking hyaenids are recorded in the Miocene of sub-Saharan Africa, and the percrocutids appear to have become extinct before the time interval of Lothagam. As at Lothagam, canids are absent (Howell 1987). Two proboscideans are reported from Sahabi: *Amebelodon cyrenaicus* and *Stegotetrabelodon lybicus* (Gaziry 1987a). The first is an advanced amebelodont that represents the latest occurrence of a shovel-tusker in the Old World; the second is restricted to North Africa and the Arabian Peninsula, but the genus is represented by a different species south of the Sahara. Tassy (this volume:chapter 8.1) notes that the single worn molar of a trilophodont gomphothere from the Apak could possibly be the same species as the Sahabi *Amebelodon.* As at Lothagam, two species of equid are present, but they show strikingly different postcranial proportions. The smaller Sahabi species is referred to as *Hipparion* cf. *sitifense*; Bernor and Harris (this volume:chapter 9.2) have renamed the Lothagam material that was formerly compared with that species. A single worn rhino premolar at Sahabi is referred to as *Diceros neumayri,* known from the Mediterranean region, but this attribution might be less secure than implied and the links could as well be with the African *Diceros.* Three species of *Nyanzachoerus* have been identified at Sahabi, but only *Nyanzachoerus syrticus* and *Ny. devauxi* are represented at Lothagam. A *Ny. kanamensis* skull was also identified at Sahabi by Cooke (1987). However, van de Made (1999) considered *Ny. kanamensis* to be restricted to the western Rift and thought that the morphologically similar *Ny. pattersoni* and *Ny. australis* were represented in the eastern Rift. Bishop (1997) followed Harris and White (1979) and Cooke (1978) in regarding *Ny. pattersoni* as a junior synonym of *Ny. kanamensis.* Harris and Leakey (this volume:chapter 10.2) report *Nyanzachoerus* cf. *Ny. australis* from the Apak Member and *Ny. pattersoni* from the Kaiyumung. The Sahabi hippopotamus *Hexaprotodon sahabiensis* (Gaziry 1987b) has closest affinities with *Hex. harvardi* from Lothagam (Weston this volume:chapter 10.1). The Sahabi giraffid is tentatively referred to *Samotherium,* but its similarities with *Palaeotragus* are noted. Eight species of bovid are recognized at Sahabi. The most abundant is the spiral-horned antelope *Prostrepsiceros,* which makes up about 50 percent of the specimens and is about the size of *Aepyceros melampus.* The impala, which dominates the Lothagam bovid assemblages, is absent. Bovids that are also found at Lothagam include cf. *Damalacra, Raphiceras, Gazella, Redunca,* and Boselaphini.

Arabian Peninsula Sites

Abu Dhabi, United Arab Emirates

Arabia lies at the junction of the classic Old World biogeographic divisions—the Ethiopian, Palaearctic, and Oriental regions. The Late Miocene faunal assemblage from the Baynunah Formation, Abu Dhabi, thus contributes important paleoenvironmental and systematic information relevant to faunal interchange between these adjoining regions. The Baynunah Formation consists of a sequence of predominantly fluviatile sediments that suggest a low-gradient river made up of numerous small shallow channels separated by low sandbanks. The entire braided river network was probably hundreds of meters wide, and the water flow was permanent. At that time, sea level was substantially lower than today and the marine coastline is believed to have been 300 km to the east of its present position (Hill and Whybrow 1999). Temperatures were warm and the climate was semiarid, with an annual rainfall of no more than 75 mm (Ditchfield 1999). Vegetation consisted of a mixture of grass, trees (including *Acacia*), and shrubs. The landscape was probably more open away from the river (Kingston 1999), but it would have included dense vegetation of reeds and shrubs on the banks of the channels. Analyses of stable carbon isotopes (Kingston 1999) in palaeosol carbonates, and analysis of the $\delta^{13}C$ in tooth enamel of 34 specimens of grazing and browsing herbivores, indicates that both C_3 and C_4 plants were available but there was a heavy reliance on C_4 grazing. Only a few specimens plotted within the range of obligate browsers. Those showing a mixed grazing and browsing strategy had a major C_4 component in the diet.

The more than 900 specimens indicate a rich and diverse fauna that is essentially African, and particularly North African, in character, although it includes elements with Asian affinities. No species have definitive links with Late Miocene European faunas, such as those from Greece and those known eastward through Turkey to northwestern Iran. The fauna thus appears to be part of a Late Miocene faunal belt that trends west and east (roughly between 15° N and 31° E), stretches across northern Africa and Arabia, and extends into parts of Asia, including Pakistan and India. This faunal belt was interpreted to suggest that animals could migrate readily eastward or westward, but movement in a northerly or southerly direction was hampered by geographical barriers such as deserts, mountains, and rivers (Hill and Whybrow 1999). This is a different geographic configuration from the North African and Sub-Paratethyan Provinces proposed by Bernor in 1983. The age of the Baynunah Formation is not securely dated, but it is believed to be between 8 and 6 Ma and probably closer to 6 Ma (Whybrow and Hill 1999).

Six species in the Baynunah Formation show affinities with those at Lothagam: *Abudhabia*, a large machairodont felid that may be congeneric with *Machairodus*, a cercopithecid that has lower canine proportions close to those of *Parapapio lothagamensis*, *Stegotetrabelodon*, the narrow-muzzled hippopotamus *Hexaprotodon* aff. *Hex. sahabiensis*, and the suid *Nyanzachoerus syrticus*. A possible phylogenetic relationship between *Nyanzachoerus* and *Sivachoerus* has been discussed frequently (Leakey 1958; Cooke and Ewer 1972; Cooke 1987; Pickford 1986, 1987) and more recently by van de Made (1999), who synonymized the two genera. The *Sivachoerus-Nyanzachoerus* clade is restricted to the Late Miocene of the Indian subcontinent, northern and northeastern and southern Africa, and Abu Dhabi. Bovids from the Baynunah Formation of Abu Dhabi include two boselaphines (*Tragoportax cyrenaicus* and *Pachyportax latidens*), two species of *Prostrepsiceros*, and a gazelle (Gentry 1999b). The Baynunah ruminant assemblage suggests an age of about 6 Ma (Gentry 1999b) and, although it has one species (*T. cyrenaicus*) in common with Lothagam, it is clearly closer in habitat to that from Sahabi and to assemblages from the Nagri and Dhok Pathan zones of the Siwaliks.

Asian Sites

The Siwalik series in the Potwar Plateau, Pakistan, records an exceptional, almost continuous Neogene record from 18 Ma to 2 Ma through a composite thickness of 5 km (Pilbeam et al. 1996). Permanent forests and woodlands with some interspersed grasslands were present before 9 Ma; thereafter, wooded grasslands became widespread on floodplains (Quade et al. 1989, 1992; Morgan et al. 1994). The Siwalik fauna is characterized by hipparionine equids, ruminant artiodactyls, hyaenids, archaic taxa such as creodonts and anthracotheres, and muroid, sciurid, and ctenodactyl rodents. Analysis of first and last appearance records indicates a trend toward reduction in species diversity between 13 and 8 Ma, together with an increase in ruminant and muroid body size (Flynn et al. 1990, 1995).

During the Early Neogene, the Indian subcontinent appears to have been at least partially isolated, and it is not until the Turolian that, for the first time, the Siwalik (Potwar Plateau) faunas resemble those of western Eurasia. In the Siwaliks, a faunal association that included archaic carnivores, rhinoceroses (*Brachypotherium*), proboscideans (*Deinotherium*), dormice, shrews, and hominoids (*Sivapithecus*), persisted until 8.3 Ma. Between 8.3 and 7.8 Ma, a major ecological change took place (Pilbeam et al. 1996) that led to the extinction of several bovid, tragulid, and cricetid species while *Sivapithecus* was replaced by cercopithecid monkeys. The lateness of this turnover in the Siwaliks (compared to that in western Europe) may reflect monsoon atmospheric dynamics in the Siwalik Hills that may be related to the uplift of the Tibetan Plateau and that could have maintained the subtropical forested conditions in the Siwaliks until 8.3 Ma. The Siwaliks faunal change also coincides with the shift in the $\delta^{13}C$ isotopic composition of palaeosol and dental carbonates; it thus indicates a change in dominance from C_3 to C_4 vegetation. Even more than at Lothagam, the subsequent evolutionary changes seen in the Siwaliks fauna were largely driven by the resultant spread of grasslands and more open habitat.

European Sites

Considerable faunal change is evident in Europe (Agustí et al. 1999a), but it follows a pattern different from that seen in Africa and the Siwaliks, where it is associated with the change from a C_3-dominated to a C_4-dominated vegetation. The changes in Europe are greatest during two time intervals, one at the Vallesian-Turolian boundary ~9.5–9.0 Ma (MN 9 to MN 11), and the other in the Turolian between ~6.5 and 5.0 Ma. Both are related to changes in vegetation and climate. European sites record the change from a forested condition in the mid-Vallesian, whereas Greek localities record changes associated with the Mediterranean salinity crisis. Neither change is directly related to the global C_3/C_4 event, since the Mediterranean-type climate does not favor C_4 vegetation. Hence any faunal turnover observed must be due to other influences. The transition from a woodland-forest fauna to an open country one occurred in Spain at 9.6 Ma, well before the 8–7 Ma age given for the global expansion of C_4 vegetation that is supported by isotopic evidence (Cerling et al. 1997).

Prior to the Vallesian, Europe was dominated by evergreen forest, but at the end of the Vallesian, drier and more arid conditions spread from the southwest, east, and southeast. Faunal change accompanied the change in prevailing habitat, but it is also evident in central Europe, which acted as a forest refuge (Franzen and Storch 1999). Notable faunal events at this time include the migration of hipparionine equids into Eurasia from the Americas by way of the Bering land bridge and thence almost immediately into Africa. This is believed to have taken place at 10.5 or 11.1 Ma, depending on which time scale is adopted (Jones 1999) and which region: 11.2 Ma in middle Europe (Rögle and Daxner-Höck 1996), 11.1 in Spain (Garcés et al. 1997), or 10.7 in the Siwaliks (Pilbeam et al. 1996). Later changes include the disappearance of the amphicyonids, the decline of the suids and cervoids, more than 50 percent replacement of carnivores, and the arrival of the hyaenids and percrocutids. Hominoids vanish completely from central Europe at the MN 10/11 boundary, although two lineages persist into the Turolian farther south (Andrews and Bernor 1999; Agustí et al. 1999b).

The Vallès-Penedès Basin in northeast Spain has a high density of mammalian fossil localities that represent most of the Miocene, and it documents these environmental changes in detail. The junction of the Early and Late Vallesian is estimated to be 9.7 Ma (Garcés et al. 1996; Agustí et al. 1997). In the Vallès-Penedès Basin, the replacement of the earlier tropical-subtropical warm forest by more open habitat characterizing the latest Miocene in western Eurasia is accompanied by a major faunal change termed by Agustí et al. (1999a) as the mid-Vallesian crisis (MVC). This saw the disappearance of most of the Middle Miocene elements that had adapted to the hitherto dominant warm and wet-subtropical conditions in western Europe, and it affected several rodents, carnivores, perissodactyls, suids, ceratomorphs, and primates. Carnivores that were particularly affected included amphicyonids, nimravids, and primitive hyaenids. European hominoids of the genus *Dryopithecus,* so abundantly represented in the Early Vallesian levels of the Vallès-Penedès Basin, also became extinct (Agustí et al. 1997; Moyà-Solà and Kohler 1993). The MVC particularly affected those species with forest affinities, and the latitudinal character of these sets of extinctions strongly suggests this event results from climate forcing. The MVC amounts to a deep overall restructuring of the Neogene mammalian faunas and coincides with the noticeable decay of the subtropical warm evergreen forest and the spread in western Europe of deciduous-dominated forests and dry-woodland biomass that characterize the Late Miocene all around western Eurasia (van de Burgh et al. 1993). The latest Vallesian saw the disappearance forever of the high mammal diversity characteristic of the Early Vallesian. Although there were no major extinction events, the loss of Middle Miocene taxa continued and was compensated for by the immigration and evolution of new Turolian taxa (Agustí et al. 1999b).

As drier, more arid conditions spread from the southwest, east, and southeast, it is likely that a decrease in annual temperature, a change to a seasonal more continental climate, and the disappearance of crucial plants led to the disappearance of many herbivores. Forests persisted in central Europe, where they became a refuge for forest-dwelling animals, but pollen evidence has shown that deciduous elements of the flora increased and replaced many of the evergreen elements (Kovar-Edar et al. 1996). Pollen records document a major change in vegetation between the Upper Serravallian and the Lower Tortonian (~12–11 Ma), and this change corresponded to a decrease in temperature (Suc et al. 1999). Prior to the change, the forests contained a great diversity of trees and were able to provide fruits all the year round. After the change, fruit production was reduced to a few months, which must have affected fruit-eating animals and may have led to the extinction of hominoid primates in western Europe at about 9 Ma (Andrews et al. 1996).

Global climatic events between 12 and 7 Ma include changes in the atmospheric and oceanic circulation and increasing latitudinal gradients of temperature, starting with the extensive Antarctic glaciation (Miller et al. 1991; Wright et al. 1992; Keller and Barron 1983) and the beginning of minor glacial processes in the Arctic regions (Zubhakov and Borzenkova 1990). Also of significant impact was the closure of the Tethys, by the late Middle to early Late Miocene, and the uplift of the Himalayas, Karakoram, and the Hindu Kush (Jones 1999). All contributed to a more seasonal continental climate, which resulted in significant vegetation change.

The closure of the Tethys opened migration routes for terrestrial mammals between Africa and Eurasia. Paleobotanical, palynological, and soil carbonate $\delta^{13}C$ analyses indicate that the dominant faunas of the Late Miocene "Pikermian Province" (France, Greece, Turkey, Bulgaria, Romania, Moldavia, Ukraine, Russia, and Iran) were of sclerophyllous evergreen woodland (similar to the modern Mediterranean) (Jones 1999).

The second major faunal change is seen in the latest Miocene through the Early Pliocene where it is associated with the Messinian salinity crisis. At the Miocene-Pliocene boundary, about 5.5 Ma, the Antarctic ice sheet may have exceeded its glacial maximum extent by as much as 50 percent (Shackleton and Kennet 1975; Denton 1985), resulting in an appreciable drop in sea level that, together with local tectonic activity, resulted in the Messinian salinity crisis, which expedited the isolation and eventual desiccation of the Mediterranean Sea (Hsu et al. 1977; Stein and Sarnthein 1983; Hoddell et al. 1986; Denton 1999). The circum-Mediterranean climate was mostly cool and dry at this time, and arid-adapted and semidesert elements came to dominate the fauna at the expense of the subtropical elements (Van Zinderen Bakker and Mercer 1986; Suc et al. 1999). In central Europe the climate was warm to cool. Migrations of mammals between Africa and Eurasia took place in the Late Turolian (MN 13) by way of a land bridge at the foot of the Iberian Peninsula, an island chain across the Mediterranean, or perhaps across the dried-up floor of the Mediterranean.

The rich Turolian sites in Greece, including Pikermi and Samos, record significant change in fauna between the Vallesian and the Turolian, and most of the Vallesian bovid species disappeared (de Bonis and Koufos 1999). Unfortunately, the provenance of the large collections from the nineteenth century and the first half of the twentieth century is not always secure, but ongoing studies to better document the faunas from these important Greek localities will provide a valuable comparison for African and Asian sites of this age.

Biogeographic Affinities

The Lothagam fauna shows affinities with fauna from almost all the sites mentioned in this chapter, but those with Europe are the least marked. Several families have exclusively African origins; these families include the Hominidae, Cercopithecidae, Elephantidae, and Hippopotamidae. Other families, such as the Equidae, invaded Africa from Eurasia on more than one occasion. The affinities of the Lothagam fauna with assemblages from sites in Africa and Eurasia are important in identifying the times and routes of such interchange.

The Leporidae and Muridae are first known in Africa from Late Miocene sites but they evidently originated in Eurasia. The leporid *Alilepus* has been recorded from just below the Marker Tuff at Lothagam (6.57–6.54 Ma), and a leporid is also recorded from the Mpesida Beds (~7–6.2 Ma) in the Tugen Hills. The earliest record of Hystricidae (African porcupines) is from low in the Lower Nawata at Lothagam; hystricids have also been recovered from the Late Miocene in Namibia. The Muridae is the most commonly represented family at Lothagam, and the Lothagam murids also show affinities with species in North Africa and Asia. *Abudhabia* is known from the Late Miocene of Abu Dhabi, United Arab Emirates, and Pakistan, and from the Early and Late Pliocene of India. It has also been suggested that *Protatera yardangi* from Sahabi is better referred to *Abudhabia.* The murine *Saidomys* recovered from the Upper Nawata is also widely distributed, being recorded from the Late Miocene to the Late Pliocene in North and East Africa and the Early Pliocene of Afghanistan (Winkler this volume:chapter 5). *Karnimata,* another widely distributed murine, is known from northern and southern Africa and Pakistan. *Karnimata* was probably an immigrant from Asia, where it is better known and is first recorded in the Late Miocene. The rodents appear to show affinities with Siwaliks and with Afghanistan.

The latest African sites where hominoids are well represented are the Middle Miocene localities of Fort Ternan (Andrews 1996) and Maboko (Benefit and McCrossin 1995). Their scarcity at Lothagam is thus not entirely unexpected. There appear to be similarities in the environments occupied by Middle Miocene apes in Africa and Europe. In Africa these apes inhabited seasonal woodlands and forests; their locomotory adaptations show they were partly terrestrial, and their large teeth with thick enamel indicate diets similar to those of some of the European apes (Andrews and Bernor 1999). Over much of Eurasia, hominoid primates enjoyed favorable environmental conditions in the late Early and Middle Miocene intervals until the replacement of evergreen subtropical forests by deciduous woodlands and, progressively into the Late Miocene, more seasonal warm, temperate woodlands with more open habitats. The hominoid distribution tracked these habitat changes closely during the 12–9 Ma interval, contracting their ranges from both west and east and finding temporary refuge in southeastern Europe, where favorable subtropical conditions persisted for a time even after they had disappeared elsewhere. Hominoids disappeared from Eurasia finally during MN 11, although they persisted until MN 12 in local insular habitats in Italy and the latest Miocene of China. The appearance of hominoids at Lothagam at the Upper Nawata–Apak Member boundary marks the beginning of a much more detailed subsequent fossil record.

The prevalence of *Parapapio lothagamensis* in the Nawata Formation shows the success of this early primitive baboon in the Late Miocene. The higher diversity of colobines at Lothagam is of interest because the earliest cercopithecids recorded in Europe are colobines, and, as in Africa, the cercopithecines are initially less diverse. The Cercopithecidae spread from Africa to Europe in the Turolian. The first unequivocal European cercopithecid occurrence is *Mesopithecus pentelicus* from Greece and southwestern Asia. Pikermi is the oldest certain *Mesopithecus*-bearing locality, ~8.3–8.2 Ma, although a single premolar tooth from the insecurely correlated site of Wissberg, ~10 Ma, is reported as the oldest from Europe; if the Wissberg correlation is confirmed, that specimen would compare with the oldest colobine from Africa, *Microcolobus tugenensis* (Benefit and Pickford 1986). *Mesopithecus pentelicus* is abundantly represented from MN 11–13, after which it was replaced by *Dolichopithecus.* A second species of *Mesopithecus, M. monspessulanus,* is known from the Pliocene of western and central Europe. *Macaca* is broadly represented geographically in the Pliocene after its first occurrence in the latest MN 13 (~5.3 Ma) site of Casablanca-M, in eastern Spain (Andrews et al. 1996). The affinities of the Lothagam monkeys with Eurasian species are unclear, but it does seem probable that the Sahabi cf. *Macaca* may be better attributed to *Parapapio.*

Although several of the Lothagam carnivores have African origins or at least long evolutionary histories in Africa, most have close relatives in the Eurasian Miocene (Werdelin this volume:chapter 7). This is true for Amphicyonidae sp. A, *Vishnuonyx, Viverra, Ictitherium, Hyaenictitherium, Hyaenictus, Lokitunjailurus,* and cf. *Metailurus*. The Lothagam records of these taxa suggest that there may have been several dispersal events, one from India in the early Late Miocene, bringing early *Vishnuonyx*, and two from Eurasia. The first dispersal from Eurasia would have included *Mellivora, Genetta,* and the *Ikelohyaena/Hyaena* lineages—taxa that have long evolutionary histories in Africa; the second dispersal would have brought the forebears of the other Lothagam taxa. The presence of amphicyonids in the Upper and Lower Nawata constitute one of the youngest records of this family in the Northern Hemisphere. Several mustelids of gigantic size are known from the Miocene of Eurasia and North America, but *Ekorus ekakoran* represents the earliest occurrence of a giant mustelid in Africa (Werdelin this volume:chapter 7).

The Lothagam proboscideans document the transition between the elephantoid-dominated assemblages characteristic of the Early and Middle Miocene and the elephantid-dominated assemblages of the African Plio-Pleistocene. Two gomphotheriids (*Anancus kenyensis* and an unnamed trilophodont gomphothere) occur in the Nawata Formation where the early elephantids *Primelaphus* and *Stegotetrabelodon orbus* predominate. A new early species of *Elephas, E. nawatensis,* is recognized from the Upper Nawata, supporting Maglio's (1973) contention that the elephantines differentiated during the Late Miocene. *Anancus* and *Stegotetrabelodon* persist in the Apak Member from which early representatives of other *Elephas* and *Loxodonta* species have been recovered.

Anancus and *Stegotetrabelodon* are present also at Lukeino (Bernor and Pavlakis 1987; Bishop et al. 1999), and *Stegotetrabelodon* occurs in the Mpesida Beds (Hill 1999) but the East African *Stegotetrabelodon* is a different species from that characteristic of North Africa and thus should properly be called *S. syrticus* (Pettrochi 1941; Tassy 1999) rather than *S. libycus* (e.g., Maglio 1973; Gaziry 1987a). Gaziry (1987a) considered *S. orbus* and *S. syrticus* to be conspecific but Tassy (1999) advocated retaining *S. orbus* as a small species that was more evolved than *S. syrticus* from Sahabi or Abu Dhabi.

The Lothagam hipparions are thought to be derived from two separate Eurasian hipparionine radiations into Africa. The earlier (between 10.2 and 10.8 Ma) saw the immigration of *"Hippotherium" primigenium,* the earliest African equid stock. The occurrence of *"Hippotherium" primigenium* in the Nawata Formation represents the last record of this relatively archaic equid. A subsequent migration, at approximately 8 Ma, brought the *Eurygnathohippus* lineage to East Africa (Bernor et al. 1989; Bernor and Lipscomb 1995; Bernor and Armour-Chelu 1999). *Eurygnathohippus turkanense* shares close phyletic affinities with the Indo-Pakistan Late Miocene species *Sivalhippus perimense* (*sensu* Bernor and Hussain 1985). Although neither *E. turkanense* nor *E. feibeli* has been recorded elsewhere, the genus *Eurygnathohippus* persisted after the Miocene and underwent a modest evolutionary radiation across Africa (Bernor and Armour-Chelu 1999). However, affinities between the *E. feibeli* and European Indo-Pakistan Late Miocene small horses do not appear to be close, and the similarities in postcranial morphology are most likely to be due to homoplasy (Bernor and Harris this volume:chapter 9.2).

Hooijer (1978) believed *Ceratotherium* evolved from a *Diceros* or *Paradiceros* stock, but the earliest occurrences of the extant African rhino genera are found in a number of Late Miocene circum-Mediterranean localities. The oldest *Ceratotherium* is reported at 10.6 Ma from Locality 12 in the Sinap Formation of Turkey (Lunkka et al. 1999), and *C. neumayri* is reported from the Miocene of Greece in MN 10 at Pentalophos 1 (de Bonis and Koufos 1999). *C. neumayri* is also known from Pikermi and Samos in Greece and Maregheh in Iran where its first occurrence is interpreted by Bernor et al. (1996) to be 8.3 Ma. The less derived *Diceros* is uncommon in Europe but is reported from the Beglia Formation in Tunisia (= 9.5 Ma; Woodburn et al. 1996). Hill (1999) claims the oldest sub-Saharan record of *Ceratotherium* is from the Mpesida Beds dated at about 6.5 Ma. The presence of *Diceros* and *Ceratotherium* in the Upper Nawata at Lothagam constitutes one of the earliest sub-Saharan records for both genera.

The Hippopotamidae are believed to originate in Africa, the earliest occurrence being *Kenyapotamus* from Ngorora in the Tugen Hills, and the earliest *Hexaprotodon* is reported from fragmentary material from the Ngorora Formation in the Tugen Hills (Hill 1995). The Lothagam collection represents the earliest large assemblage of fossil hippos. The broad-muzzled *Hexaprotodon harvardi* is probably derived from a primitive narrow-muzzled browsing ancestral morphotype prior to the Late Miocene, since both narrow-muzzled and broad-muzzled forms are found in the Lower Nawata at Lothagam. Both broad-muzzled and narrow-muzzled hippos appear to have spread out of Africa into Eurasia during the Late Miocene. A narrow-muzzled species, *Hex. sahabiensis,* is found in North Africa at Sahabi, Libya (Gaziry 1987b). A similar narrow-muzzled species, *Hexaprotodon* aff. *Hex. sahabiensis,* is found on the Arabian Peninsula at Abu Dhabi (Gentry 1999a). A narrow-muzzled hippo, *Hexaprotodon crusafonti,* is also reported from the MN 13 zone in Spain (Lacomba et al. 1986). The earliest certain record of *Hexaprotodon* in Asia is 5.7 Ma (Barry 1995).

The suids in the Lothagam succession are mostly tetraconodontines that migrated to Africa from Eurasia during the late Middle Miocene and replaced the suid tribes that had previously dominated the Early and Mid-Miocene. Teeth comparable to those of the two common Lower Nawata suids, *Nyanzachoerus syrticus* and *N. devauxi,* have been recovered from the Namurungule Formation in Kenya (~8–10 Ma). Several authors have drawn attention to the similarities between *Sivachoerus* and *Nyanzachoerus* (Leakey 1958; Cooke and Ewer 1972; Cooke 1987; Pickford 1986, 1987), and recently van de Made (1999) transferred many African species formerly assigned to *Nyanzachoerus* to the genus *Sivachoerus*. More derived tetraconodonts dominate the Lothagam suid assemblage later in the succession—*Nyanzachoerus australis* in the Apak Member and *Notochoerus euilus* in the Kaiyumung.

Tetraconodont suids of the *Sivachoerus-Nyanzachoerus* clade are restricted to the Late Miocene of the Indian subcontinent, northern and northeastern Africa, and the Arabian Peninsula. *Nyanzachoerus devauxi* and *N. syrticus* also occur at Sahabi in North Africa (Cooke 1987; McCrossin 1987), but the Lothagam representatives are smaller and less progressive, although specimens of *N. syrticus* from the Upper Nawata are more progressive than those from the Lower Nawata and are similar to those from Lukeino. *N. syrticus* is also reported from the Baynunah Formation in Abu Dhabi, and Bishop and Hill (1999) suggest that the Abu Dhabi *N. syrticus* material represented an immigrant from Africa.

With the notable exception of Langebaanweg, giraffids were rare at Late Miocene and Early Pliocene localities of sub-Saharan Africa. *Climacoceras, Samotherium,* and *Paleotragus* characterized the late Middle Miocene of Africa (Churcher 1970; Hamilton 1978), but only species of *Paleotragus* have been recovered from Lothagam. *Paleotragus* was present in both sub-Saharan and North Africa during the Late Miocene but was more common in the eastern Mediterranean. Gentry (1999b) reported a sivathere (?*Bramatherium* sp.) and a smaller giraffid (Giraffidae sp. indet.) from Abu Dhabi. Sivatheres were not recovered from Sahabi, where Harris (1987) recognized only *Samotherium* sp., but were abundant at Langebaanweg, where they are represented by *Sivatherium hendeyi* (Harris 1976) and were accompanied by the large *Giraffa* cf. *G. jumae* (Harris 1976) and smaller *Paleotragus* cf. *P. germaini* (Hendey 1981b). Gentry (1997) documented ?*Sivatherium* sp. and *Giraffa* sp. from the Manonga Valley of Tanzania. Sivatheres may have been of Eurasian origin, but the giraffines probably evolved in Africa. Gentry (1997) recognized *Giraffa* from Ngorora and validated Pickford's (1978) report of this genus from Lukeino. Giraffines are thus known from sites that are older than the horizons where they first appear at Lothagam. *Giraffa punjabicus* has been recognized from the Siwalik localities of Dhok Pathan (7.1–5.0 Ma) and Hasnot (7.0 Ma), but Gentry (1997) seemed to support Matthew's (1929) interpretation that the species represented an immigrant from Africa to Eurasia.

Bovids have been the commonest constituents of the terrestrial biota of sub-Saharan Africa since the late Middle Miocene. Antilopini, Boselaphini, and Caprini—all of Eurasian origin—are known from about 14 million years ago; the endemic Cephalophini and Neotragini appeared shortly thereafter. As discussed by Vrba (1985), Ovibovini and Bovini migrated into Africa from Eurasia toward the end of the Miocene, which is when the endemic Tragelaphini, Hippotragini, Alcelaphini, and Aepycerotini are documented for the first time. Reduncines, whose continent of origin is uncertain, also appear in the Late Miocene. Hitherto, some of the earliest records of modern African tribes come from Lukeino and Mpesida in the Tugen Hills of Kenya (Thomas 1980, 1984; Gentry 1990). At Lukeino (6.2–5.6 Ma; Hill 1999), the somewhat fragmentary bovid material includes tragelaphines, bovines, reduncines, and *Aepyceros*. At Mpesida (6.5–7.2 Ma; Hill 1999), documented bovid genera comprise *Tragelaphus, Kobus, Madoqua,* and *Gazella* or *Raphiceras*. Hill (1999) claims reduncine bovids from the top of the Ngorora sequence at about 9 Ma.

Bovids from the Baynunah Formation of Abu Dhabi include two boselaphines (*Tragoportax cyrenaicus* and *Pachyportax latidens*), two species of *Prostrepsiceros*, and a gazelle (Gentry 1999b). The Baynunah ruminant assemblage suggests an age of about 6 Ma (Gentry 1999b) but is clearly closer in habitat to that from Sahabi and to assemblages from the Nagri and Dhok Pathan zones of the Siwaliks than it is to Lothagam. At Sahabi, the spiral-horned antelope *Prostrepsiceros libycus*, a genus and species unrepresented at Lothagam, is by far the most abundant antelope (Lehman and Thomas 1987). Similarities of these two assemblages include a boselaphine (*Tragoportax cyrenaicus*), ?*Hippotragus* sp., the alcelaphine *Damalacra*, a gazelle (*Gazella* sp.), and *Raphiceras* sp., but the bovines and reduncines are different and Sahabi lacks *Aepyceros* (the most abundant species at Lothagam) and also tragelaphines.

Aepyceros is absent also from the Langebaanweg localities, as are hippotragines, but Lothagam and Langebaanweg share tragelaphines, boselaphines, bovines, reduncines, alcelaphines, gazelles, and steinbuck although there are no species in common. Ovibovines, uncommon elements of the Langebaanweg biota, have yet to be recovered from Lothagam. Only fragmentary material has been recovered from the Manonga Valley. Genera in common with Lothagam include *Tragelaphus, Kobus, Praedamalis, Damalacra,* and *Aepyceros* (Gentry 1997), but only the reduncines can be identified to species and neither of the Manonga taxa (*Kobus* aff. *K. porrecticornis* and *Kobus* aff. *K. subdolus*) has been recognized at Lothagam. The absence of boselaphines at the Manonga localities suggests that these are younger than those of Lothagam, Sahabi, and Langebaanweg, although Vrba (1995:414) documents boselaphines in the Late Pliocene of southern Africa.

Comparison with Late Miocene–Early Pliocene sites elsewhere in Africa and Eurasia emphasizes that Lothagam has its closest affinities with fauna from other African sites and is less close to fauna from Arabia and Asia. Even less affinity is seen with the European faunas except, perhaps, those from the eastern Mediterranean. These relationships support the suggestion of Whybrow and Hill (1999), based on the Abu Dhabi fauna, that animals could migrate out of Africa to Asia more easily in an east–west direction through Arabia and that geographical barriers prevented significant movement in a north–south direction. However, Solounias et al. (1999) believed that much of the modern African fauna migrated into Africa from more northerly latitudes and that these immigrants replaced endemic African species. They argue that species adapted for mixed feeding in the Pikermi biome of sclerophyllous evergreen woodlands were preadapted to move into the African grasslands as these began to open up at the end of the Miocene. Thus many of the Pliocene African savanna-dwelling mammals originated not from forest-dwelling African Miocene relatives but, rather, from the Pikermian biome that was located north of Africa. With the drying out of Africa, large mammals from the sclerophyllous woodland of more northern climates extended their range southward. It follows that later Miocene faunas such as Pikermi and Samos should be more closely related to modern African taxa than are Middle Miocene and Late Miocene African faunas, but further studies are needed to test this hypothesis.

Evolution of Early Hominins

Molecular studies suggest that the time of divergence of the hominin lineage from that of the chimpanzee and bonobo was between 5 and 6 Ma and close to 5.5 Ma (Kumer and Hedges 1998), which is close to the time that the earliest hominoids are found at Lothagam. Evidence of bipedality is generally a prerequisite for confirmation of hominin status but, unfortunately, neither the fragmentary Upper Nawata dental specimens nor the mandible from the lower Apak provides any bearing on locomotory function. Until recently, the earliest secure evidence for bipedality was the *Australopithecus anamensis* tibia from the Kanapoi Formation dated at 4.12 Ma (Leakey et al. 1995, 1998; Ward et al. 1999). *Ardipithecus ramidus* has been claimed to have hominin status based on evidence of the occipital, which points to a bipedal gait (White et al. 1994, 1995), and more recently additional evidence of this species has been reported from 5.8 Ma to 5.2 deposits in the Middle Awash (Haile-Selassie 2001). Publication of the now much more complete *A. ramidus* sample will be important in evaluating the bipedal capabilities of this early human ancestor.

The apparent absence of hominins from the Lower Nawata and lower levels of the Upper Nawata is significant. Much of the faunal change in the Lothagam sequence represents the appearance and adaptive radiation of essentially modern groups and can be attributed to the spread of more open habitats characterized by C_4 grasses. The emergence of the Homininae at this time would be consistent with the early representatives exploiting these newly available habitats. That hominins are rare or absent until the Early Pliocene points to a slower adaptive radiation into more open habitats than is seen in the majority of mammalian families, including the Cercopithecidae. However, at a time when many of the carnivores were large, fast, and ferocious, the relatively smaller canines of hominins afforded little means of defense, and the adoption of a bipedal gait is an unlikely adaptation for escaping fast predators. That no other primate made this change is perhaps salutory. The subsequent evolution of manual dexterity and increased encephalization, with all the inherent advantages, was only possible after our ancestors walked upright and the hands were freed from a locomotory function. What prompted the adoption of a bipedal gait is a key question for understanding the early evolution of our ancestors.

The scenario that the Lothagam evidence offers could be as follows: The skeletal morphology of the early australopithecines, including *A. afarensis* and *A. africanus,* shows a unique configuration of morphological traits that indicate their locomotor and postural adaptations were unlike those of other living forms (McHenry 1986; Susman et al. 1984). The body proportions too—of at least some early hominins—indicate relatively large forelimbs and short hindlimbs (McHenry 1998). The common ancestor of the common chimpanzee, bonobo, and humans probably inhabited a closed woodland or forest habitat and spent considerable time on the ground foraging, as do African apes today. Reliance on large trees for suitable refuge would not necessarily have favored the adoption of committed bipedality. But, as environments opened and new niches appeared with the spread of C_4 grasslands, selective pressure would have favored those individuals that moved fastest on the ground. Other selective pressures relating to energetics (Rodman and McHenry 1980) and thermoregulation (Wheeler 1991a, 1991b, 1993) would also have contributed to the evolution of an increasingly efficient bipedal mode of locomotion. Although there are differences in opinion as to the mode of australopithecine bipedality (Jungers 1992; Lovejoy 1975), it appears that the australopithecines retained an effective arboreal component in their locomotory repertoire (McHenry 1992; Susman et al. 1984). We also know that manual dexterity evolved subsequent to the adoption of bipedality, presumably only once the hand was freed from a locomotor function (Susman 1998). The hand bones of *A. anamensis* (Leakey et al. 1998) and *A. afarensis* (Marzke 1983, 1997; McHenry 1983) show the retention of a more primitive morphology than that known for *Homo habilis* in Lower Bed 1 (Napier 1962).

Late in the Upper Nawata, conditions became more seasonal, and a series of extreme climatic changes (Zhang and Scott 1996) is represented at Lothagam in the lower Apak Member by the apparent hiatus in sedimentation just above the Purple Marker (Feibel this volume:chapter 2.1). It is possible that (1) prior to the time interval represented by the top of the Upper Nawata, ancestral hominids were mainly restricted to relatively closed forests with tall trees, and that (2) the dwarf shrubland and dry thornbush savanna of the Upper Nawata lacked this crucial element of their habitat. It seems entirely possible that by the time of the first appearance of hominids at Lothagam some form of effective defense involving hand-wielded clubs and stones had been developed, and that their bipedal gait was sufficiently efficient to enable them to forage in more open country (although they remained rare in such habitats). Perhaps, too, the climatic oscillations led to times of extreme drought and high selective pressures that favored a species that was more flexible and could exploit a wider variety of habitats. Whatever the cause, we know that by 4.1 Ma the australopithecines were efficient bipeds that lived in more seasonal open country habitats. The number of hominin specimens from the Kaiyumung is consistent with their prevalence at other African sites after 4.1 Ma (White 1977, 1980; Johanson et al. 1982; Kimbel et al. 1985, 1994). Although definitive evidence is lacking at Lothagam, it is likely that the appearance of the hominin lineage was related to the environmental changes that resulted in the initial appearance at Lothagam of the many other modern mammalian groups.

Conclusions

A dramatic change took place in the African biota between 7 and 5 million years ago. Shrinkage of the equatorial forests coincided with expansion of the modern C_4 savanna grassland flora—expedited by an expansion of the polar ice caps or a reduction in atmospheric concentration of carbon dioxide, or perhaps a combination of these and other events. Expansion of the C_4 biomass was a worldwide phenomenon, and in Africa it resulted in the emergence of the faunal elements that would dominate the later Cenozoic—including hippos, giant pigs, grazing antelopes, true giraffes and elephants, and, of course, man.

Early humans are restricted in distribution to Africa, and their acquisition of an upright bipedal striding gait, the hallmark of humanity, appears to be at least circumstantially linked to the reduction of the equatorial forests and spread of grasslands on that continent. Only a few sites in Africa are representative of the time interval in which this ecological transition took place—among them the coastal site of Langebaanweg at the southern tip of Africa (Hendey 1981a, 1981b), the North African coastal site of Sahabi in Libya (Boaz et al. 1987), the Manonga Valley in Tanzania (Harrison 1997a), the Tugen Hills of the Baringo Basin in Kenya (Hill 1999), and Lothagam at the southwest edge of Lake Turkana in northern Kenya. Only Lothagam currently combines a lengthy stratigraphic sequence, diverse and evolving vertebrate assemblages, and the presence of early human remains.

The Lothagam biota documents a sequence of assemblages preserved in northern Kenya at a time when climatic conditions favored replacement of forest by more open habitats and when C_4 grasses were becoming a significant part of the flora. The Lothagam fauna reflects the changing nature of the habitats that prevailed in what is now the southwestern part of the Lake Turkana Basin—changes that combine both local and continent-wide environmental fluctuations with the establishment of new continental waterways and the immigration into Africa of Eurasian mammal stocks. The interval represented by the Lothagam succession saw the final vestiges of fauna characteristic of the East African Miocene being replaced by more open habitat assemblages that incorporated the ancestors of the extant East African biota. Sadly, the search for early human remains was less successful than was promised by the otherwise bountiful mammalian fossils. To date, the results serve mainly to confirm the rare presence in horizons a little older than 5 Ma of creatures that could be either early humans or the common ancestors of humans and apes. However, the search continues, both in the Lake Turkana Basin and in other areas of similar age in eastern Africa, for the elusive evidence of the first creatures that crossed the bipedality rubicon to found the lineage that led to ourselves.

References Cited

Agustí, J., L. Cabrera, M. Garcés, and M. Llenas. 1999a. Mammal turnover and global climate change in the Late Miocene terrestrial record of the Vallès-Penedès Basin (NE Spain). In J. Agustí, L. Rook, and P. Andrews, eds., *Hominoid Evolution and Climatic Change in Europe*. Vol. 1. *The Evolution of Neogene Terrestrial Ecosystems in Europe*, pp. 397–412. Cambridge: Cambridge University Press.

Agustí, J., L. Cabrera, M. Garcés, and J. M. Parés. 1997. The Vallesian mammal succession in the Vallès-Penedès Basin

(NE Spain): Paleomagnetic calibration and correlation with global events. *Palaeogeography, Palaeoclimatolology, Palaeoecology* 133:3–4.

Agustí, J., L. Rook, and P. Andrews, eds. 1999b. *Hominoid Evolution and Climatic Change in Europe.* Vol. 1. *The Evolution of Neogene Terrestrial Ecosystems in Europe.* Cambridge: Cambridge University Press.

Agustí, J., L. Rook, and P. Andrews. 1999c. Introduction. In J. Agustí, L. Rook, and P. Andrews, eds., *Hominoid Evolution and Climatic Change in Europe.* Vol. 1. *The Evolution of Neogene Terrestrial Ecosystems in Europe,* pp. 1–6. Cambridge: Cambridge University Press.

Andrews, P. 1996. Palaeoecology and hominoid palaeoenvironments. *Biological Review* 71:257–300.

Andrews, P., and R. Bernor. 1999. Vicariance biogeography and paleoecology of Eurasian Miocene hominoid primates. In J. Agustí, L. Rook, and P. Andrews, eds., *Hominoid Evolution and Climatic Change in Europe.* Vol. 1. *The Evolution of Neogene Terrestrial Ecosystems in Europe,* pp. 454–487. Cambridge: Cambridge University Press.

Andrews, P., T. Harrison, E. Delson, R. L. Bernor, and L. Martin. 1996. Distribution and biochronology of European and Southwest Asian Miocene catarrhines. In R. L. Bernor, V. Fahlbusch, and H.-W. Mittmann, eds., *The Evolution of Western Eurasian Neogene Mammal Faunas,* pp. 168–207. New York: Columbia University Press.

Arambourg, C. 1959. *Vertébrés continentaux du Miocène superieur de l'Afrique du nord.* Paléontologie Mémoires No. 4. Algiers: Carte Géologique d'Algérie.

Barry, J. C. 1995. Faunal turnover and diversity in the terrestrial Neogene of Pakistan. In E. S. Vrba, G. H. Denton, T. C. Partridge, and L. H. Burkle, eds., *Paleoclimate and Evolution, with Emphasis on Human Origins,* pp. 115–134. New Haven: Yale University Press.

Behrensmeyer, A. K., J. D. Damuth, W. A. DiMichele, R. Potts, H.-D. Sues, and S. L. Wing. 1992. *Terrestrial Ecosystems Through Time: Evolutionary Paleoecology of Terrestrial Plants and Animals.* Chicago: University of Chicago Press.

Behrensmeyer, A. K., N. E. Todd, R. Potts, and G. E. McBrinn. 1997. Late Pliocene faunal turnover in the Turkana Basin, Kenya and Ethiopia. *Science* 278:1589–1594.

Benefit, B. R., and M. L. McCrossin. 1995. Miocene hominoids and hominid origins. *Annual Review of Anthropology* 24:237–256.

Benefit, B. R., and M. Pickford. 1986. Miocene fossil cercopithecoids from Kenya. *American Journal of Physical Anthropology* 69:441–464.

Bernor, R. L. 1983. Geochronology and zoogeographic relationships of Miocene Hominoidea. In R. L. Ciochon and R. S. Corruccini, eds., *New Interpretations of Ape and Human Ancestry,* pp. 21–64. New York: Plenum Press.

Bernor, R. L. 1986. Mammalian biostratigraphy, geochronology, and zoogeographic relationships of the Late Miocene Maragheh fauna, Iran. *Journal of Vertebrate Paleontology* 6:76–95.

Bernor, R. L., and M. Armour-Chelu. 1999. Toward an evolutionary history of African hipparionine horses. In T. Brommage and F. Schrenk, eds., *African Biogeography and Human Evolution,* pp. 189–215. Oxford: Oxford University Press.

Bernor, R. L., and S. T. Hussain. 1985. An assessment of the systematic, phylogenetic and biogeographic relationships of Siwalik hipparionine horses. *Journal of Vertebrate Paleontology* 5:32–87.

Bernor, R. L., and D. Lipscomb. 1995. A consideration of Old World hipparionine horse phylogeny and global abiotic processes. In E. Vrba, G. H. Denton, T. C. Partridge, and L. H. Burkle, eds., *Palaeoclimate and Evolution, with Emphasis on Human Origins,* pp. 164–177. New Haven: Yale University Press.

Bernor, R. L., and P. P. Pavlakis. 1987. Zoogeographic relationships of the Sahabi large mammal fauna (Early Pliocene, Libya). In N. T. Boaz, A. El-Arnauti, A. W. Gaziry, J. de Heinzelin, and D. D. Boaz, eds., *Neogene Paleontology and Geology of Sahabi,* pp. 349–383. New York: Liss.

Bernor, R. L., N. Solounias, C. C. Swisher III, and J. A. Van Couvering. 1996. The correlation of three classical "Pikermian" mammal faunas—Maragheh, Samos, and Pikermi—with the European MN unit system. In R. L. Bernor, V. Fahlbusch, and H.-W. Mittmann, eds., *The Evolution of Western Eurasian Neogene Mammal Faunas,* pp. 137–154. New York: Columbia University Press.

Bernor, R. L., H. Tobein, and M. O. Woodburne. 1989. Patterns of Old World hipparionine evolutionary diversification and biogeographic extension. In E. H. Lindsay, V. Fahlbusch, and P. Andrews, eds., *Topics on European Mammalian Chronology,* pp. 263–319. New York: Plenum Press.

Bishop, L. C. 1994. Pigs and their ancestors: Hominids, suids and environments during the Plio-Pleistocene of East Africa. Ph.D. diss., Yale University.

Bishop, L. C. 1997. Fossil suids from the Manonga Valley, Tanzania. In T. Harrison, ed., *Neogene Paleontology of the Manonga Valley, Tanzania: A Window into the Evolutionary History of East Africa,* pp. 191–217. New York: Plenum Press.

Bishop, L., and A. Hill. 1999. Fossil Suidae from the Baynunah Formation, Emirate of Abu Dhabi, United Arab Emirates. In P. J. Whybrow and A. Hill, eds., *Fossil Vertebrates of Arabia,* pp. 254–270. New Haven: Yale University Press.

Bishop, L. C., A. Hill, and J. Kingston. 1999. Palaeoecology of Suidae from the Tugen Hills, Baringo, Kenya. In P. Andrews and P. Banham, eds., *Late Cenozoic Environments and Hominid Evolution: A Tribute to Bill Bishop,* pp. 99–111. London: Geological Society.

Boaz, N. T., A. El-Arnauti, A. W. Gaziry, J. de Heinzelin, and D. D. Boaz, eds. 1987. *Neogene Paleontology and Geology of Sahabi.* New York: Liss.

Bobe, R. 1997. Hominid environments in the Pliocene: An analysis of fossil mammals from the Omo Valley, Ethiopia. Ph.D. diss., University of Washington.

Cerling, T. E., J. R. Bowman, and J. R. O'Neil. 1988. An isotopic study of a fluvial-lacustrine sequence: The Plio-Pleistocene Koobi Fora sequence, East Africa. *Palaeogeography, Palaeoclimatology, Palaeoecology* 63:335–356.

Cerling, T. E., J. M. Harris, B. J. MacFadden, M. G. Leakey, J. Quade, V. Eisenmann, and J. R. Ehleringer. 1997. Pattern and significance of global ecologic change in the Late Neogene. *Nature* 389:153–158.

Churcher, C. S. 1970. Two new upper Miocene giraffids from Fort Ternan, Kenya, East Africa: *Palaeotragus primaevus* n. sp. and *Samotherium africanum* n. sp. In L. S. B. Leakey and R. J. G. Savage, eds., *Fossil Vertebrates of Africa,* vol. 2, pp. 1–109. London: Academic Press.

Clauzon, G., J.-P. Suc, F. Gautier, A. Berger, and M.-F. Loutre.

1996. Alternate interpretation of the Messinian salinity crisis: Controversy resolved? *Geology* 24:363–366.

Cooke, H. B. S. 1978. Suid evolution and correlation of hominid localities: An alternative taxonomy. *Science* 201:460–463.

Cooke, H. B. S. 1987. Fossil Suidae from Sahabi, Libya. In N. T. Boaz, A. El-Arnauti, A. W. Gaziry, J. de Heinzelin, and D. D. Boaz, eds., *Neogene Paleontology and Geology of Sahabi,* pp. 255–266. New York: Liss.

Cooke, H. B. S., and R. F. Ewer. 1972. Fossil Suidae from Kanapoi and Lothagam, northwestern Kenya. *Bulletin of the Museum of Comparative Zoology* 43:149–296.

de Bonis, L., and G. D. Koufos. 1999. The Miocene large mammal succession in Greece. In J. Agustí, L. Rook, and P. Andrews, eds., *Hominoid Evolution and Climatic Change in Europe.* Vol. 1. *The Evolution of Neogene Terrestrial Ecosystems in Europe,* pp. 205–237. Cambridge: Cambridge University Press.

Dechant Boaz, D. 1987. Taphonomy and palaeoecology at the Miocene site of Sahabi, Libya. In Boaz, N. T., A. El-Arnauti, A. W. Gaziry, J. de Heinzelin, and D. D. Boaz, eds., *Neogene Paleontology and Geology of Sahabi,* pp. 337–348. New York: Liss.

Denton, G. H. 1985. Did the Antarctic ice sheet influence late Cainozoic climate and evolution in the Southern Hemisphere? *South African Journal of Science* 81:224–229.

Denton, G. H. 1999. Cenozoic climatic change. In T. G. Bromage and F. Schrenk, eds., *African Biogeography, Climatic Change, and Human Evolution,* pp. 94–114. Oxford: Oxford University Press.

Ditchfield, P. W. 1999. Diagenesis of the Baynunah, Shuwaihat, and Upper Dam Formations sediments exposed in the western region, Emirate of Abu Dhabi, United Arab Emirates. In P. J. Whybrow and A. Hill, eds., *Fossil Vertebrates of Arabia,* pp. 61–74. New Haven: Yale University Press.

Feibel, C. S., J. M. Harris, and F. H. Brown. 1999. Palaeoenvironmental context for the Late Neogene of the Turkana Basin. In J. M. Harris, ed., *Koobi Fora Research Project.* Vol. 3. *The Fossil Ungulates: Geology, Fossil Artiodactyls, and Palaeoenvironments,* pp. 321–370. Oxford: Clarendon Press.

Flynn, L. J., J. C. Barry, M. E. Morgan, D. Pilbeam, L. L. Jacobs, and E. H. Lindsay. 1995. Neogene Siwalik mammalian lineages: Species longevities, rates of change and modes of speciation. *Palaeogeography, Palaeoclimatology, Palaeoecology* 115:249–264.

Flynn, L. J., D. Pilbeam, L. L. Jacobs, J. C. Barry, A. K. Behrensmeyer, and J. W. Kappleman. 1990. The Siwaliks of Pakistan: Time and faunas in a Miocene terrestrial setting. *Journal of Geology* 98:589–604.

Franzen, J. L., and G. Storch. 1999. Late Miocene mammals from central Europe. In J. Agustí, L. Rook, and P. Andrews, eds., *Hominoid Evolution and Climatic Change in Europe.* Vol. 1. *The Evolution of Neogene Terrestrial Ecosystems in Europe,* pp. 165–190. Cambridge: Cambridge University Press.

Garcés, M., J. Agustí, L. Cabrera, and J. M. Parés. 1996. Magnetostratigraphy of the Vallesian (Late Miocene) in the Vallès-Penedès Basin (northeastern Spain). *Earth and Planetary Science Letters* 142:381–196.

Garcés, M., L. Cabrera, J. Agustí, and J. M. Parés. 1997. Old World first appearance datum of "*Hipparion*" horses: Late Miocene large mammal dispersal and global events. *Geology* 25:19–22.

Gaziry, A. W. 1987a. Remains of Proboscidea from the Early Pliocene of Sahabi, Libya. In N. T. Boaz, A. El-Arnauti, A. W. Gaziry, J. de Heinzelin, and D. D. Boaz, eds., *Neogene Paleontology and Geology of Sahabi,* pp. 183–203. New York: Liss.

Gaziry, A. W. 1987b. *Hexaprotodon sahabiensis* (Artiodactyla, Mammalia): A new hippopotamus from Libya. In N. T. Boaz, A. El-Arnauti, A. W. Gaziry, J. de Heinzelin, and D. D. Boaz, eds., *Neogene Paleontology and Geology of Sahabi,* pp. 305–315. New York: Liss.

Gentry, A. 1980. Fossil Bovidae from Langebaanweg, South Africa. *Annals of the South African Museum* 79:213–337.

Gentry, A. 1990. Evolution and dispersal of African Bovidae. In G. A. Bubenik and A. B. Bubenik, eds., *Horns, Pronghorns, and Antlers: Evolution, Morphology, Physiology, and Social Significance,* pp. 195–229. New York: Springer-Verlag.

Gentry, A. 1997. Fossil ruminants (Mammalia) from the Manonga Valley, Tanzania. In T. Harrison, ed., *Neogene Paleontology of the Manonga Valley, Tanzania: A Window into the Evolutionary History of East Africa,* pp. 107–135. New York: Plenum Press.

Gentry, A. W. 1999a. A fossil hippopotamus from the Emirate of Abu Dhabi, United Arab Emirates. In P. J. Whybrow and A. Hill, eds., *Fossil Vertebrates of Arabia,* pp. 271–289. New Haven: Yale University Press.

Gentry, A. W. 1999b. Fossil pecorans from the Baynunah Formations, Emirate of Abu Dhabi, United Arab Emirates. In P. J. Whybrow and A. Hill, eds., *Fossil Vertebrates of Arabia,* pp. 290–316. New Haven: Yale University Press.

Geraads, D. 1989. Phylogenetic analysis of the Tribe Bovini (Mammalia, Artiodactyla). *Zoological Journal of the Linnean Society* 104:193–207.

Haile-Selassie, Y. 2001. Late Miocene hominids from the Middle Awash, Ethiopia. *Nature* 412:178–181.

Hamilton, W. R. 1978. Fossil giraffes from the Miocene of Africa and a revision of the phylogeny of the Giraffoidea. *Philosophical Transactions of the Royal Society of London,* ser. B, 283:165–229.

Harris, J. M. 1976. Pliocene Giraffoidea (Mammalia, Artiodactyla) from the Cape Province. *Annals of the South African Museum* 65:189–198.

Harris, J. M. 1987. Fossil Giraffidae from Sahabi, Libya. In N. T. Boaz, A. El-Arnauti, A. W. Gaziry, J. de Heinzelin, and D. D. Boaz, eds., *Neogene Paleontology and Geology of Sahabi,* pp. 317–322. New York: Liss.

Harris, J. M., and T. E. Cerling. 1998. Isotopic changes in the diet of giraffids. *Journal of Vertebrate Paleontology,* Supplement 18:49A.

Harris, J. M., and T. D. White. 1979. Evolution of the Plio-Pleistocene African Suidae. *Transactions of the American Philosophical Society* 69:1–128.

Harrison, T. H., ed. 1997a. *Neogene Paleontology of the Manonga Valley, Tanzania: A Window into the Evolutionary History of East Africa.* New York: Plenum Press.

Harrison, T. H. 1997b. Paleoecology and taphonomy of fossil localities in the Manonga Valley, Tanzania. In T. Harrison, ed., *Neogene Paleontology of the Manonga Valley, Tanzania: A Window into the Evolutionary History of East Africa,* pp. 79–105. New York: Plenum Press.

Harrison, T. H., and M. L. Mbago. 1997. Introduction: Palaeontological and geological research in the Manonga Valley,

Tanzania. In T. Harrison, ed., *Neogene Paleontology of the Manonga Valley, Tanzania: A Window into the Evolutionary History of East Africa*, pp. 1–32. New York: Plenum Press.

Hay, R. L. 1987. Geology of the Laetoli area. In M. D. Leakey and J. M. Harris, eds., *Laetoli: A Pliocene Site in Northern Tanzania*, pp. 23–47. Oxford: Clarendon Press.

Hendey, Q. B. 1970a. The age of the fossiliferous deposits at Langebaanweg, Cape Province. *Annals of the South African Museum* 56:119–131.

Hendey, Q. B. 1970b. A review of the geology and paleontology of the Plio/Pleistocene deposits at Langebaanweg, Cape Province. *Annals of the South African Museum* 56:75–117.

Hendey, Q. B. 1974. The late Cenozoic Carnivora of the South-Western Cape Province. *Annals of the South African Museum* 63:1–369.

Hendey, Q. B. 1981a. Geological succession at Langebaanweg, Cape Province, and global events of the Late Tertiary. *South African Journal of Science* 77:33–38.

Hendey, Q. B. 1981b. Paleoecology of the Late Tertiary fossil occurrences in "E" Quarry, Langebaanweg, South Africa, and a reinterpretation of their geological context. *Annals of the South African Museum* 84:1–104.

Hill, A. 1994. Late Miocene and Early Pliocene hominids from Africa. In R. S. Corrucini and R. L. Ciochon, eds., *Integrative Paths to the Past*, pp. 123–145. Englewood Cliffs, N.J.: Prentice Hall.

Hill, A. 1995. Faunal and environmental change in the Neogene of East Africa: Evidence from the Tugen Hills sequence, Baringo District, Kenya. In E. S. Vrba, G. H. Denton, T. C. Partridge, and L. H. Burkle, eds., *Paleoclimate and Evolution, with Emphasis on Human Origins*, pp. 178–193. New Haven: Yale University Press.

Hill, A. 1999. The Baringo Basin, Kenya: From Bill Bishop to BPRP. In P. Andrews and P. Banham, eds., *Late Cenozoic Environments and Hominid Evolution: A Tribute to Bill Bishop*, pp. 85–97. London: Geological Society.

Hill, A., and P. J. Whybrow. 1999. Summary and overviews of the Baynunah fauna, Emirate of Abu Dhabi, and its context. In P. J. Whybrow and A. Hill, eds., *Fossil Vertebrates of Arabia*, pp. 7–14. New Haven: Yale University Press.

Hill, A., G. Curtis, and R. Drake. 1986. Sedimentary stratigraphy of the Tugen Hills, Baringo, Kenya. In L. E. Frosdick, R. W. Renault, I. Ried, and J. J. Tiercelin, eds., *Sedimentation in the African Rifts*, pp. 285–295. Geological Society Special Publication No. 25. Oxford: Blackwell.

Hill, A., R. Drake, L. Tauxe, M. Monaghan, J. C. Barry, A. K. Behrensmeyer, G. Curtis, B. F. Jacobs, N. Johnson, and D. Pilbeam. 1985. Neogene palaeontology and geochronology of the Baringo Basin, Kenya. *Journal of Human Evolution* 14:749–773.

Hoddell, D. A., K. M. Elmstrom, and J. Kennett. 1986. Latest benthic ^{18}O changes, global ice volume, sea level and the "Messinian salinity crisis." *Nature* 320:411–414.

Hooijer, D. A. 1972. A Late Pliocene rhinocerotid from Langebaanweg, Cape Province. *Annals of the South African Museum* 59:21–34.

Hooijer, D. A. 1978. Rhinocerotidae. In V. J. Maglio and H. B. S. Cooke, eds., *Evolution of African Mammals*, pp. 371–378. Cambridge, Mass.: Harvard University Press.

Howell, F. C. 1987. Preliminary observations on Carnivora from the Sahabi Formation. In N. T. Boaz, A. El-Arnauti, A. W. Gaziry, J. de Heinzelin, and D. D. Boaz, eds., *Neogene Paleontology and Geology of Sahabi*, pp. 153–181. New York: Liss.

Hsu, K. J., L. Montadert, D. Bernoulli, M. B. Cita, A. Erickson, R. E. Garrison, R. B. Kidd, F. Milieres, C. Mueller, and R. Wright. 1977. History of the Messinian salinity crisis. *Nature* 267:399–403.

Johanson, D. C., M. Taieb, and Y. Coppens. 1982. Pliocene hominids from the Hadar Formation, Ethiopia (1973–1977): Stratigraphic, chronologic, and paleoenvironmental contexts, with notes on hominid morphology and systematics. *American Journal of Physical Anthropology* 57:373–402.

Jones, R. W. 1999. Marine invertebrate (chiefly foraminiferal) evidence for the palaeogeography of the Oligocene-Miocene of western Eurasia, and consequences for terrestrial vertebrate migration. In J. Agustí, L. Rook, and P. Andrews, eds., *Hominoid Evolution and Climatic Change in Europe*. Vol. 1. *The Evolution of Neogene Terrestrial Ecosystems in Europe*, pp. 274–308. Cambridge: Cambridge University Press.

Jungers, W. L. 1992. Lucy's limbs: Skeletal allometry and locomotion in *Australopithecus afarensis*. *Nature* 297:676–678.

Keller, G., and J. A. Barron. 1983. Paleocenographic implications of Miocene deep-sea hiatuses. *Geological Society of America Bulletin* 94:590–613.

Kimbel, W. H., D. C. Johanson, and Y. Rak. 1994. The first skull and other new discoveries of *Australopithecus afarensis* at Hadar, Ethiopia. *Nature* 368:449–451.

Kimbel, W. H., T. D. White, and D. C. Johanson. 1985. Craniodental morphology of the hominids from Hadar and Laetoli: Evidence of "*Paranthropus*" and *Homo* in the mid-Pliocene of eastern Africa? In E. Delson, ed., *Ancestors: The Hard Evidence*, pp. 120–137. New York: Liss.

Kingston, J. D. 1999. Isotopes and environments of the Baynunah Formation, Emirate of Abu Dhabi, United Arab Emirates. In P. J. Whybrow and A. Hill, eds., *Fossil Vertebrates of Arabia*, pp. 354–372. New Haven: Yale University Press.

Kovar-Edar, J., Z. Kvacek, E. Zastawniak, R. Givulescu, L. Hably, D. Mihajlovic, J. Teslenko, and H. Walther. 1996. Floristic trends in the vegetation of the Paratethys and surrounding areas during Neogene time. In R. L. Bernor, V. Fahlbusch, and H.-W. Mittmann, eds., *The Evolution of Western Eurasian Neogene Mammal Faunas*, pp. 395–413. New York: Columbia University Press.

Kumer, S., and S. B. Hedges. 1998. A molecular timescale for vertebrate evolution. *Nature* 392:917–920.

Lacomba, J. I., J. Morales, F. Robles, C. Santisteban, and M. T. Alberdi. 1986. Sedimentologia y paleontologia del yacimiento finimioceno de La Portera (Valencia). *Estudios Geológicos. Instituto de Investigaciones Geológicas Lucas Mallada* 42:167–180.

Leakey, L. S. B. 1958. *Some East African Pleistocene Suidae*. Fossil Mammals of Africa No. 14. London: British Museum (Natural History).

Leakey, M. D., and J. M. Harris, eds. 1987. *Laetoli: A Pliocene Site in Northern Tanzania*. Oxford: Clarendon Press.

Leakey, M. G., C. S. Feibel, I. McDougall, and A. Walker. 1995. New four-million-year-old hominid species from Kanapoi and Allia Bay, Kenya. *Nature* 376:565–571.

Leakey, M. G., C. S. Feibel, I. McDougall, C. Ward, and A. Walker. 1998. New specimens and confirmation of an early age for *Australopithecus anamensis*. *Nature* 393:62–66.

Lehmann, U., and H. Thomas. 1987. Fossil Bovidae from the

Mio-Pliocene of Sahabi, Libya. In P. J. Whybrow and A. Hill, eds., *Fossil Vertebrates of Arabia,* pp. 323–335. New Haven: Yale University Press.

Lovejoy, C. O. 1975. Biochemical perspectives on the lower limb of early hominids. In R. H. Tuttle, ed., *Primate Functional Morphology and Evolution,* pp. 291–326. The Hague: Mouton.

Lunkka, J. P., M. Fortelius, J. Kappleman, and S. Sen. 1999. Chronology and mammal faunas of the Miocene Sinap Formation, Turkey. In J. Agustí, L. Rook, and P. Andrews, eds., *Hominoid Evolution and Climatic Change in Europe.* Vol. 1. *The Evolution of Neogene Terrestrial Ecosystems in Europe,* pp. 238–264. Cambridge: Cambridge University Press.

Maglio, V. J. 1973. Origin and evolution of the Elephantidae. *Transactions of the American Philosophical Society,* n.s., 63:1–149.

Marzke, M. W. 1983. Joint function and grips of the *Australopithecus afarensis* hand, with special reference to the region of the capitate. *Journal of Human Evolution* 12:197–211.

Marzke, M. W. 1997. Precision grips, hand morphology and tools. *American Journal of Physical Anthropology* 102:91–110.

Matthew, W. D. 1929. Critical observations upon Siwalik mammals. *Bulletin of the American Museum of Natural History* 56:437–560.

McCrossin, M. L. 1987. Postcranial remains of fossil Suidae from the Sahabi Formation, Libya. In N. T. Boaz, A. El-Arnauti, A. W. Gaziry, J. de Heinzelin, and D. D. Boaz, eds., *Neogene Paleontology and Geology of Sahabi,* pp. 267–286. New York: Liss.

McDougall, I., and C. S. Feibel. 1999. Numerical age control for the Miocene-Pliocene succession at Lothagam, a hominoid-bearing sequence in the northern Kenya Rift. *Journal of the Geological Society* (London) 156:731–745.

McHenry, H. M. 1983. The capitate of *Australopithecus afarensis* and *A. africanus. American Journal of Physical Anthropology* 62:187–198.

McHenry, H. M. 1986. The first bipeds: A comparison of the *A. afarensis* and *A. africanus* postcranium and implications for the evolution of bipedalism. *Journal of Human Evolution* 15:177–191.

McHenry, H. M. 1992. Body size and proportions in early hominids. *American Journal of Physical Anthropology* 87:407–431.

McHenry, H. M. 1998. Body proportions in *Australopithecus afarensis* and *A. africanus* and the origin of the genus *Homo. Journal of Human Evolution* 35:1–22.

Meikle, W. E. 1987. Fossil Cercopithecidae from the Sahabi Formation. In N. T. Boaz, A. El-Arnauti, A. W. Gaziry, J. de Heinzelin, and D. D. Boaz, eds., *Neogene Paleontology and Geology of Sahabi,* pp. 119–127. New York: Liss.

Miller, K. G., J. D. Wright, and R. G. Fairbanks. 1991. Unlocking the icehouse: Oligocene-Miocene oxygen isotopes, eustasry and margin erosion. *Journal of Geophysical Research* 96:6829–6848.

Morgan, M. E., J. D. Kingston, and B. D. Marino. 1994. Carbon isotopic evidence for the emergence of C_4 plants in the Neogene of Pakistan and Kenya. *Nature* 267:162–165.

Moyà-Solà, S., and M. Köhler. 1993. Recent discoveries of *Dryopithecus* shed new light on evolution of great apes. *Nature* 365:543–545.

Munthe, J. 1987. Small-mammal fossils from the Pliocene Sahabi Formation of Libya. In N. T. Boaz, A. El-Arnauti, A. W. Gaziry, J. de Heinzelin, and D. D. Boaz, eds., *Neogene Paleontology and Geology of Sahabi,* pp. 135–144. New York: Liss.

Nakaya, H. 1990. Late Miocene mammalian fauna of Samburu Hills, northern Kenya. Ph.D. diss., Kagawa University.

Nakaya, H. 1993. Les faunes de mammifères du Miocène supérieur de Samburu Hills, Kenya, Afrique de l'est et l'environnement des pré-hominidés. *L'Anthropologie* 97:9–16.

Nakaya, H. 1994. Faunal change of Late Miocene Africa and Eurasia: Mammalian fauna from the Namurungule Formation, Samburu Hills, northern Kenya. *African Study Monographs,* Supplement 20:1–112.

Nakaya, H., M. Pickford, Y. Nakano, and H. Ishida. 1984. The Late Miocene large mammalian fauna from the Namurungule Formation, Samburu Hills, northern Kenya. *African Study Monographs,* Supplement 2:87–131.

Napier, J. R. 1962. Fossil hand bones from Olduvai Gorge. *Nature* 196:409–411.

Pettrochi, C. 1941. Il giacimento fossilifero di Sahabi. *Bollettino della Società Geologia Italiana* 60:107–114.

Pickford, M. 1978. Stratigraphy and mammalian palaeontology of the Late Miocene Lukeino Formation, Kenya. In W. W. Bishop, ed., *Geological Background to Fossil Man,* pp. 263–278. Edinburgh: Scottish Academic Press.

Pickford, M. 1986. *A Revision of the Miocene Suidae and Tayassuidae (Artiodactyla, Mammalia) of Africa.* Tertiary Research Special Paper No. 7. Leiden: Brill.

Pickford, M. 1987. Miocene Suidae from Ad Dabtiyah, eastern Saudi Arabia. *Bulletin of the British Museum (Natural History), Geology* 41:441–446.

Pickford, M., B. Senut, and D. Hadoto, eds. 1993. *Geology and Palaeobiology of the Albertine Rift Valley, Uganda-Zaire.* Vol. 1. *Geology.* Occasional Publication No. 24. Orléans: Centre Internationale pour la Formation et les Echanges Géologiques.

Pickford, M., B. Senut, I. Ssemmanda, D. Elepu, and P. Obwona. 1988. Premiers résultats de la mission de l'Uganda Palaeontology Expedition à Nkondo (Pliocène du bassin du lac Albert, Ouganda). *Comptes Rendus de l'Académie des Sciences* (Paris), 3rd ser., 306:315–320.

Pilbeam, D., M. Morgan, J. C. Barry, and L. Flynn. 1996. European MN units and the Siwalik faunal sequence of Pakistan. In R. L. Bernor, V. Fahlbusch, and H.-W. Mittmann, eds., *The Evolution of Western Eurasian Neogene Mammal Faunas,* pp. 96–105. New York: Columbia University Press.

Quade, J., T. E. Cerling, J. C. Barry, M. E. Morgan, D. R. Pilbeam, A. R. Chivas, J. A. Lee-Thorp, and N. J. van der Merwe. 1992. A 16-million-year record of paleodiet using carbon and oxygen isotopes in fossil teeth from Pakistan. *Chemical Geology (Isotope Geoscience)* 94:183–192.

Quade, J., T. E. Cerling, and J. R. Bowman. 1989. Development of Asian monsoon revealed by marked ecological shift during the latest Miocene in northern Pakistan. *Nature* 342:163–166.

Renne, P. R., G. WoldeGabriel, W. K. Hart, G. Heiken, and T. D. White. 1999. Chronostratigraphy of the Miocene-Pliocene Sagantole Formation, Middle Awash Valley, Afar Rift, Ethiopia. *Geological Society of America Bulletin* 111:869–885.

Rodman, P. S., and H. M. McHenry. 1980. Bioenergetics and the origin of hominid bipedalism. *American Journal of Physical Anthropology* 52:103–106.

Rögle, F., and G. Daxner-Höck. 1996. Late Miocene Paratethys correlations. In R. L. Bernor, V. Fahlbusch, and H.-W. Mittmann, eds., *The Evolution of Western Eurasian Neogene Mammal Faunas,* pp. 47–55. New York: Columbia University Press.

Sawada, Y., M. Pickford, T. Itaya, T. Makinouchi, M. Tateishi, K. Kabeto, S. Ishida, and H. Ishida. 1998. K-Ar ages of Miocene Hominoidea (*Kenyapithecus* and *Samburupithecus*) from Samburu Hills, northern Kenya. *Comptes Rendus de l'Académie des Sciences* (Paris) 326:445–451.

Sanders, W. J. 1997. Fossil proboscideans from the Wembere-Manonga Formation, Manonga Valley, Tanzania. In T. Harrison, ed., *Neogene Paleontology of the Manonga Valley, Tanzania: A Window into the Evolutionary History of East Africa,* pp. 265–310. New York: Plenum Press.

Senut, B., M. Pickford, D. Gommery, P. Mein, C. Cheboi, and Y. Coppens. 2001. First hominid from the Miocene (Lukeino Formation, Kenya). *Comptes Rendus de l'Académie des Sciences* (Paris) 332:137–144.

Shackleton, N. J., and J. P. Kennett. 1975. Late Cenozoic oxygen and carbon isotopic changes at DSDP Site 284: Implications for glacial history of the Northern Hemisphere and Antarctica. In J. P. Kennett and R. E. Houtz, eds., *Initial Reports of the Deep Sea Drilling Project,* pp. 801–807. Washington, D.C.: Government Printing Office.

Solounias, N. 1981. The Turolian fauna from the island of Samos, Greece. *Contributions to Vertebrate Evolution* 6:1–231.

Solounias, N., J. M. Plavcan, J. Quade, and L. Witmer. 1999. The palaeoecology of the Pikermian biome and the savanna myth. In J. Agustí, L. Rook, and P. Andrews, eds., *Hominoid Evolution and Climatic Change in Europe.* Vol. 1. *The Evolution of Neogene Terrestrial Ecosystems in Europe,* pp. 436–453. Cambridge: Cambridge University Press.

Stein, R., and M. Sarnthein. 1983. Late Neogene events of atmospheric and oceanic circulation offshore northwest Africa: High resolution record from deep-sea sediments. In J. A. Coetzee and E. M. van Zinderen Bakker, eds., *Palaeoecology of Africa,* pp. 9–36. Rotterdam: Balkema.

Stromer, E. 1907. Fossile Wirbeltier-Rests aus der Uadi Faregh and Uadi Natrun in Ägypten. *Abhandllungen herausgegeben von der Senckenbergischen naturforschenden Gesellschaft* 29:99–132.

Suc, J.-P., S. Fauquette, M. Bessedik, A. Bertini, Z. Zheng, G. Clauzon, D. Suballyova, F. Diniz, P. Quezel, N. Feddi, M. Clet, E. Bessais, N. B. Taoufiq, H. Meon, and N. Combourieu-Nebout. 1999. Neogene vegetation changes in west European and west circum-Mediterranean areas. In J. Agustí, L. Rook, and P. Andrews, eds., *Hominoid Evolution and Climatic Change in Europe.* Vol. 1. *The Evolution of Neogene Terrestrial Ecosystems in Europe,* pp. 378–389. Cambridge: Cambridge University Press.

Susman, R. L. 1998. Hand function and tool behavior in early hominids. *Journal of Human Evolution* 35:23–46.

Susman, R. L., J. T. Stern, and W. L. Jungers. 1984. Arboreality and bipedality in the Hadar hominids. *Folia Primatologia* 43:113–156.

Tassy, P. 1999. Miocene elephantids (Mammalia) from the Emirate of Abu Dhabi, United Arab Emirates: Palaeobiogeographic implications. In P. J. Whybrow and A. Hill, eds., *Fossil Vertebrates of Arabia,* pp. 209–233. New Haven: Yale University Press.

Thomas, H. 1980. Les bovidés du Miocène supérieur des couches de Mpesida et de la formation de Lukeino (district de Baringo, Kenya). In R. E. F. Leakey and B. A. Ogot, eds., *Proceedings of the Eighth Pan-African Congress of Prehistory and Quaternary Studies, Nairobi 1977,* pp. 82–91. Nairobi: National Museums of Kenya.

Thomas, H. 1984. Les Bovidae du Miocène du sous-continent Indien, de la peninsule Arabique, et de l'Afrique. *Palaeogeography, Palaeoclimatology, Palaeoecology* 45:251–299.

Turner, A., L. C. Bishop, C. Denys, and J. K. McKee. 1999. Appendix A: A locality-based listing of African Plio-Pleistocene mammals. In T. G. Bromage and F. Schrenk, eds., *African Biogeography, Climatic Change, and Human Evolution,* pp. 369–399. Oxford: Oxford University Press.

van der Burgh, J., H. Vesscher, D. L. Dilcher, and W. M. Kürschner. 1993. Palaeoatmospheric signatures in Neogene fossil leaves. *Science* 260:1788–1790.

van der Made, J. 1999. Biometrical trends in the Tetraconodontinae, a subfamily of pigs. *Transactions of the Royal Society of Edinburgh: Earth Sciences* 89:199–225.

Van Zinderen Bakker, E. M., and J. H. Mercer. 1986. Major Late Cenozoic climatic events and palaeoenvironmental changes in Africa viewed in a worldwide context. *Palaeogeography, Palaeoclimatology, Palaeoecology* 56:217–235.

Vrba, E. S. 1995. The fossil record of African antelopes (Mammalia, Bovidae) in relation to human evolution and climate. In E. S. Vrba, G. H. Denton, T. C. Partridge, and L. H. Burkle, eds., *Paleoclimate and Evolution, with Emphasis on Human Origins,* pp. 385–424. New Haven: Yale University Press.

Ward, C., M. Leakey, and A. Walker. 1999. The new hominid species *Australopithecus anamensis. Evolutionary Anthropology* 7:191–226.

Wheeler, P. E. 1991a. The influence of bipedalism on the energy and water budgets of early hominids. *Journal of Human Evolution* 21:117–136.

Wheeler, P. E. 1991b. The thermoregulatory advantages of hominid bipedalism in open equatorial environments: The contribution of increased convective heat loss and cutaneous evaporative cooling. *Journal of Human Evolution* 21: 107–115.

Wheeler, P. E. 1993. The influence of stature and body form on hominid energy and water budgets: A comparison of *Australopithecus* and early *Homo* physiques. *Journal of Human Evolution* 24:13–28.

White, T. D. 1977. New fossil hominids from Laetoli, Tanzania. *American Journal of Physical Anthropology* 46:197–230.

White, T. D. 1980. Additional fossil hominids from Laetoli, Tanzania: 1976–1979 specimens. *American Journal of Physical Anthropology* 53:487–504.

White, T. D., G. Suwa, and B. Asfaw. 1994. *Australopithecus ramidus,* a new species of early hominid from Aramis, Ethiopia. *Nature* 371:306–312.

White, T. D., G. Suwa, and B. Asfaw. 1995. Corrigendum. *Nature* 375:88.

Whybrow, P. J., and A. Hill, eds. 1999. *Fossil Vertebrates of Arabia.* New Haven: Yale University Press.

Woodburn, M. O., R. L. Bernor, and C. C. Swisher III. 1996. An appraisal of the stratigraphic and phylogenetic bases for the "*Hipparion*" Datum in the Old World. In R. L. Bernor, V. Fahlbusch, and H.-W. Mittmann, eds., *The Evolution of the Western Eurasian Neogene Mammal Faunas,* pp. 124–136. New York: Columbia University Press.

Wright, J. D., K. G. Miller, and R. G. Fairbanks. 1992. Early and

Middle Miocene stable isotopes; implications for deep water circulation and climate. *Paleoceanography* 7:357–389.

Yasui, K., Y. Kunimatsu, N. Kuga, B. Bajope, and H. Ishida. 1992. Fossil mammals from the Neogene strata in the Sinda Basin, eastern Zaire. *African Study Monographs,* Supplementary issue 17:87–107.

Zhang, J., and D. B. Scott. 1996. Messinian deep-water turbidites and glacioeustatic sea-level changes in the North Atlantic: Linkage to the Mediterranean salinity crisis. *Paleoceanography* 11:277–297.

Zubhakov, V. A., and I. I. Borzenkova, eds. 1990. *Global Palaeoclimates of the Late Cenozoic.* Amsterdam: Elsevier.

TABLE 13.1 Distribution of Mammalian Taxa Through the Lothagam Sequences

		Lower Nawata	Upper Nawata	Apak Member	Kaiyumung Member	Horizon indet.
Leporidae	*Alilepus* sp.	2	—	—	—	—
	Serengetilagus praecapensis	—	—	1	—	—
Hystricidae	*Hystrix* sp. (small)	1	—	—	—	—
	Hystrix sp. (large)	—	—	—	1	—
Sciuridae	*Kubwaxerus pattersoni*	6	—	—	—	1
Muridae	*Abudhabia* sp.	—	1	—	—	—
	Karnimata jacobsi	4	—	—	—	—
	Saidomys sp.	—	1	—	—	—
Thryonomidae	*Paraphiomys chororensis*	1	—	—	—	—
	Paraulacodus cf. *P. johanesi*	1	—	—	—	—
	Thryonomys cf. *T. gregorianus*	—	—	—	1	—
	Thryonomys sp. (small)	—	6	—	—	—
Muridae subfamily *incertae sedis*	Gen. and sp. unknown	1	—	—	—	—
Hominoidae	Hominoidea indet.	—	2	1	—	—
	Australopithecus cf. *A. afarensis*	—	—	—	4	—
Cercopithecidae	*Parapapio lothagamensis*	76	31	—	—	2
	cf. *Parapapio* sp. indet.	—	—	4	1	2
	Theropithecus cf. *T. brumpti*	—	—	—	12	—
	Cercopithecoides kerioensis	—	—	?1	—	—
	Colobinae sp. A	4	3	—	—	1
	Colobinae sp. B	8	8	—	—	—
	Colobinae sp. C	—	—	2	—	—
	Colobinae gen. and sp. indet. (small)	4	4	—	—	—
	Colobinae gen. and sp. indet. (large)	—	—	—	1	—
	Cercopithecidae gen. and sp. indet.	3	2	1	—	—
Amphicyonidae	Amphicyonidae sp. A (large)	3	—	—	—	—
	Amphicyonidae sp. B (small)	—	1	—	—	—
Mustelidae	*Ekorus ekakoran*	3	—	—	—	—
	Erokomellivora lothagamensis	—	1	—	—	—
	Mellivorinae gen. and sp. indet.	1	1	—	—	—
	Vishnuonyx angololensis	1	—	—	—	—
Viverridae	*Viverra* cf. *V. leakeyi*	1	—	—	—	—
	Viverrinae gen. and sp. indet.	1	—	—	—	—
	cf. *Genetta* sp. A	2	—	—	—	—
	cf. *Genetta* sp. B	—	1	—	—	—
	Viverridae gen. and sp. indet. (large)	1	1	—	—	—

TABLE 13.1 Distribution of Mammalian Taxa Through the Lothagam Sequences (Continued)

		Lower Nawata	Upper Nawata	Apak Member	Kaiyumung Member	Horizon indet.
Hyaenidae	*Ictitherium ebu*	2	—	1	—	—
	Hyaenictitherium cf. *H. parvum*	1	2	—	—	—
	cf. *Hyaenictus* sp.	3	1	1	—	—
	Ikelohyaena cf. *I. abronia*	1	—	1	—	—
Felidae	*Lokotunjailurus emageritus*	18	15	—	—	2
	Dinofelis sp. A	4	8	6	—	1
	Dinofelis sp. B	—	—	—	1	—
	cf. *Metailurus* sp.	1	6	—	—	—
	Leptailurus/Caracal sp.	—	—	—	1	—
Canidae	cf. *Canis* sp.	—	—	—	1	—
Gomphotheriidae	*Anancus kenyensis*	1	2	3	—	2
	Gomphotheriidae gen. and sp. indet.	—	—	1	—	—
Elephantidae	*Stegotetrabelodon orbus*	8	1	2	—	5
	Primelephas gomphotheroides	3	—	—	—	1
	Primelephas or *S. orbus*	7	—	—	—	—
	Elephas nawatensis	—	2	—	—	—
	Elephas cf. *E. ekorensis*	—	—	2	—	—
	Loxodonta sp. indet. (?aff. *L. exoptata*)	—	—	3	—	—
	Elephantidae gen. and sp. indet.	2	—	1	—	1
	Elephantidae gen. and sp. *incertae sedis* A	—	1	1	—	—
	Elephantidae gen. and sp. *incertae sedis* B	—	—	1	—	—
Deinotheriidae	*Deinotherium bozasi*	1	1	3	1	—
Tubulidentata	*Leptorycteropus guilielmi*	2	—	—	—	—
	Orycteropus sp. indet.	1	2	—	—	—
Rhinocerotidae	*Brachypotherium lewisi*	20	11	2	—	1
	Ceratotherium praecox	2	6	1	4	—
	Diceros bicornis	—	4	5	—	—
Equidae	*Hippotherium* cf. *H. primigenium*	4	—	—	—	—
	Eurygnathohippus turkanense	30	16	—	—	—
	Eurygnathohippus feibeli	13	24	—	—	—
	Eurygnathohippus sp. indet. (large)	—	—	8	8	1
	Eurygnathohippus sp. indet. (small)	—	—	5	1	—

continued

TABLE 13.1 Distribution of Mammalian Taxa Through the Lothagam Sequences (Continued)

		Lower Nawata	Upper Nawata	Apak Member	Kaiyumung Member	Horizon indet.
Suidae	cf. *Kubanochoerus* sp.	1	—	—	—	—
	Nyanzachoerus syrticus	82	49	2	—	9
	Nyanzachoerus cf. *Ny. syrticus*	1	5	—	—	—
	Nyanzachoerus cf. *Ny. australis*	—	4	15	—	—
	Nyanzachoerus pattersoni	—	—	—	3	—
	Nyanzachoerus devauxi	28	7	—	—	—
	"Nyanzachoerus" jaegeri	—	—	2	—	1
	Notochoerus euilus	—	—	1	21	1
	cf. *Potamochoerus* sp.	2	1	—	—	—
	Cainochoerus cf. *C. africanus*	—	1	—	—	—
Hippopotamidae	*Hexaprotodon lothagamensis*	6	—	—	—	—
	Hexaprotodon harvardi	85	63	14	—	15
	Hexaprotodon sp. indet.	5	3	—	—	3
	Hexaprotodon cf. *Hex. protamphibius*	—	—	5	—	—
Giraffidae	cf. *Sivatherium* sp.	1	—	—	—	2
	Palaeotragus germaini	5	5	—	—	—
	Palaeotragus sp.	7	4	—	—	—
	Giraffa stillei	—	—	9	3	—
Bovidae	*Tragelaphus kyaloae*	—	1	7	3	1
	Tragelaphus nakuae	—	—	—	—	1
	Tragelaphus cf. *T. scriptus*	—	—	—	1	—
	Tragelaphini gen. and sp. indet.	5	2	6	—	—
	Simatherium aff. *S. kohllarsoni*	—	—	1	4	—
	Bovini gen. and sp. indet.	2	3	5	3	2
	Tragoportax cf. *T. cyrenaicus*	7	1	2	—	—
	Tragoportax sp. A	4	2	—	—	1
	Tragoportax sp. B	1	2	—	—	—
	Boselaphini gen. and sp. indet.	25	6	—	—	1
	Kobus presigmoidalis	4	13	4	1	1
	Kobus laticornis	—	10	—	—	1
	Menelikia leakeyi	4	3	—	—	—
	Reduncini gen. and sp. indet.	9	5	2	1	2
	Praedamalis ?sp.	2	4	—	—	—
	Hippotragus sp.	2	2	1	—	—
	Hippotragini gen. and sp. indet.	4	2	2	—	1
	Damalacra sp. A	4	23	4	1	4
	Damalacra sp. B	1	—	2	—	—
	Alcelaphini gen. and sp. indet.	2	2	3	2	1
	Aepyceros premelampus	72	58	20	9	7
	Gazella sp. indet.	1	2	3	—	—
	Raphiceros sp.	1	1	2	—	—
	Madoqua sp.	1	3	1	—	1
Total no. specimens		622	452	169	89	75

TABLE 13.2 First Appearance Records for the Mio-Pliocene Members of the Lothagam Succession

Upper Nawata	Apak Member	Kaiyumung Member
Thryonomys sp. (small)	*Serengetilagus praecapensis*	*Hystrix* sp. (large)
Abudhabia sp.	cf. *Parapapio* sp.	*Thryonomys* cf. *T. gregorianus*
Saidomys sp.	*Cercopithecoides kerioensis*	*Theropithecus* cf. *T. brumpti*
Hominoidea indet.	Colobinae sp. C	Colobinae gen. and sp. indet. (large)
Amphicyonidae sp. B	Gomphotheriidae gen. and sp. indet.	*Australopithecus* cf. *A. afarensis*
Erokomellivora lothagamensis	*Elephas* cf. *E. ekorensis*	*Dinofelis* sp. B
cf. *Genetta* sp. B	*Loxodonta* sp. indet. (aff. *L. exoptata*)	*Leptailurus/Caracal* sp.
Elephas nawatensis	Elephantidae gen. and sp. *incertae sedis* B	cf. *Canis* sp.
Elephantidae gen. and sp. *incertae sedis* A	*Eurygnathohippus* sp. indet. (large)	*Nyanzachoerus pattersoni*
Diceros bicornis	*Eurygnathohippus* sp. indet. (small)	*Tragelaphus* cf. *T. scriptus*
Nyanzachoerus cf. *N. australis*	"*Nyanzachoerus*" *jaegeri*	
Cainochoerus cf. *C. africanus*	*Notochoerus euilus*	
Tragelaphus kyaloae	*Hexaprotodon* cf. *H. protamphibius*	
Kobus laticornis	*Giraffa stillei*	
	Simatherium aff. *S. kohllarsoni*	

TABLE 13.3 Last Appearance Records for the Mio-Pliocene Members of the Lothagam Succession

Lower Nawata	Upper Nawata	Apak Member
Alilepus sp.	*Thryonomys* sp. (small)	*Serengetilagus praecapensis*
Hystrix sp. (small)	*Abudhabia* sp.	*Cercopithecoides kerioensis*
Kubwaxerus pattersoni	*Saidomys* sp.	Colobinae sp. C
Paraphiomys chororensis	*Parapapio lothagamensis*	Hominoidea indet.
Paraulacodus cf. *P. johanesi*	Colobinae sp. A	*Ictitherium ebu*
Karnimata jacobsi	Colobinae sp. B	cf. *Hyaenictis* sp.
Muridae subfamily *incertae sedis*	Colobinae gen. and sp. indet. (small)	*Ikelohyaena* cf. *I. abronia*
Amphicyonidae sp. A	Amphicyonidae sp. B	*Dinofelis* sp. A
Ekorus ekakoran	*Erokomellivora lothagamensis*	*Anancus kenyensis*
Vishnuonyx angololensis	Mellivorinae gen. and sp. indet.	Gomphotheriidae gen. and sp. indet.
Viverra leakeyi	Viverridae gen. and sp. indet. (large)	*Stegotetrabelodon orbus*
Viverrinae gen. and sp. indet.	cf. *Genetta* sp. B	*Elephantidae* gen. and sp. indet.
cf. *Genetta* sp. A	*Hyaenictitherium* cf. *H. parvum*	*Elephas* cf. *E. ekorensis*
Primelephas gomphotheroides	*Lokotunjailurus emageritus*	*Loxodonta* sp. (aff. *L. exoptata*)
Leptorycteropus guilielmi	cf. *Metailurus* sp.	*Elephantidae* gen. and sp. *incertae sedis* A
Hippotherium cf. *H. primigenium*	*Elephas nawatensis*	*Elephantidae* gen. and sp. *incertae sedis* B
Hexaprotodon lothagamensis	*Orycteropus* sp.	*Brachypotherium lewisi*
cf. *Kubanochoerus* sp.	*Eurygnathohippus turkanense*	*Diceros bicornis*
cf. *Sivatherium* sp.	*Eurygnathohippus feibeli*	*Hexaprotodon harvardi*
	Hexaprotodon sp.	*Hexaprotodon* cf. *H. protamphibius*
	Nyanzachoerus cf. *N. syrticus*	*Nyanzachoerus syrticus*
	Nyanzachoerus devauxi	*Nyanzachoerus* cf. *N. australis*
	cf. *Potamochoerus* sp.	*"Nyanzachoerus" jaegeri*
	Cainochoerus cf. *C. africanus*	*Tragoportax* cf. *T. cyrenaicus*
	Palaeotragus germaini	*Hippotragus* sp.
	Palaeotragus sp.	*Damalacra* sp. B
	Tragoportax sp. A	*Gazella* sp.
	Tragoportax sp. B	*Raphiceros* sp.
	Kobus laticornis	*Madoqua* sp.
	Menelikia leakeyi	
	Praedamalis ?sp.	

Appendix

NOTES ON THE RECONSTRUCTIONS OF FOSSIL VERTEBRATES FROM LOTHAGAM

Mauricio Antón

Figure 1.4. Paw of *Lokotunjailurus emageritus.* The remarkable preservation of the manus in the Lothagam machairodont leaves no doubt as to the relative proportions of the claws. The huge dewclaw was absolutely larger than that of a modern lion of greater body size, whereas the claws of digits II to V were actually smaller than the corresponding elements in a modern leopard, which is, of course, much smaller than *Lokotunjailurus.*

To show more clearly the size differences between the claws of different digits, I have drawn the claws protracted, although during normal locomotion on the ground they would be retracted, as in modern felids. The shape and size of the claw sheaths are broadly determined by those of the ungual, or distal, phalanges. The position of the foot pads is determined by the skeletal elements of the manus: the phalangeal pads in digits II to IV are placed under the articulation between phalanges 2 and 3 of each digit, while the main pad is placed under the articulations between the metacarpals and the proximal phalanges. The carpal pad is slightly distal to the pisiform. The morphology of the main pad is constant among modern felids, and fossil footprints attributable to early felids from European sites of Early Miocene age suggest that this morphology was already well established by that time.

Figure 4.14a. Brevirrostrine crocodiles. These reconstructed heads are based on well-preserved skull material from Lothagam, complemented when necessary by material from other sites. In the case of *Rimasuchus lloydi,* cranial material from Koobi Fora was used to reconstruct parts of the skull that were less well preserved in the Lothagam specimens. The external features (scales, skin texture) in Lothagam *Crocodylus niloticus* and *C. cataphractus* can be safely reconstructed because there are extant representatives of both species. For the reconstruction of *Rimasuchus,* external features were modeled from extant *C. niloticus.*

Figure 4.14b. Longirostrine crocodiles. Skull morphology in *Euthecodon* is mostly based on Lothagam cranial material, but skulls from Koobi Fora have also been used for this reconstruction. *Eogavialis* is based on the holotype skull from Lothagam, and part of the posterior skull had to be restored. External features in *Euthecodon* are based on modern Crocodylidae, while those of *Eogavialis* follow those of *Gavialis,* the only surviving gavialid genus.

Figure 5.1. *Kubwaxerus pattersoni.* This restoration is based primarily on the holotype partial skeleton from Lothagam. Additional material from Lothagam, including hindlimb bones, helps provide a complete image of the animal's body proportions. External features such as coat pattern, ears, and vibrissae are based on living members of the tribe Protoxerini (African giant and sun squirrels), to which *Kubwaxerus* is closely related. *Kubwaxerus* is shown walking on the ground to reflect terrestrial adaptations of its postcranial skeleton that suggest that this animal, although largely arboreal, would have spent a significant amount of time foraging on the ground, perhaps searching for fallen nuts in the forest floor.

Figure 6.1. *Parapapio lothagamensis.* Although the cranial material from Lothagam provides a fairly clear picture of the head and face of this cercopithecid, the available postcranials are fragmentary and mostly unassociated, which makes it difficult to reconstruct body proportions. Consequently, the relative proportions in this restoration are based on modern species of papionines, mostly of the genus *Macaca.* Video footage of various species of wild macaques and my own observations of captive specimens show that, like baboons (and unlike vervet monkeys), their forefeet are digitigrade while moving on the ground, although the angle of the metacarpals with the horizontal appears to be lower than in baboons. *Parapapio* is thus shown with digitigrade hands as well.

Figure 6.15. Hominoidea indet. The patches of light that pass through the foliage and glint off the animal's teeth emphasize that the Lothagam hominoid is known mostly from its dentition. Consequently, this restoration is basically a hypothetical re-creation of how a hominoid close to the hominid–great ape dichotomy might look. What can be seen of the face is a combination of primitive facial features: prognathous muzzle, flattened nose, and prominent brow ridges. Although the eyes of modern chimpanzees may occasionally show a white or very light area surrounding the iris, I have deliberately avoided giving the Lothagam hominoid such a humanizing feature.

Figure 7.1. *Ekorus ekakeran.* The holotype skeleton from Lothagam is remarkably complete and allows very precise restoration of body proportions. Compared to the modern honey badger (*Mellivora capensis*), which is the largest extant terrestrial mustelid from Africa, the differences are many and striking in addition to *E. ekakoran*'s much larger size. Most relevant for the restoration of the fossil animal are differences in the morphology of the appendicular skeleton, including the shape of the scapula, the elbow articulation, the shape of the radius, the orientation and shape of the ilium, the morphology of the proximal femur, and the morphology and relative size of the phalanges. In all these features, *Ekorus* differs from *Mellivora* and the other badgers that are robust carnivores with an ambulatory locomotion and developed fossorial abilities. *Ekorus* more closely resembles cursorial and subcursorial carnivores like the cats and the hyenas. As a result, the articulated skeleton of *Ekorus* is restored to be more upright and gracile than that of any badger, even more than that of the wolverine *Gulo.*

In line with the relatively gracile build of *Ekorus,* I have shown the animal in a dynamic, trotting gait. Even the robust extant honey badger moves along with a steady trot during its nightly forages, as shown by video footage of wild specimens. It remains to be determined if *Ekorus* was digitigrade. The morphology of the feet suggests that, even if *Ekorus* was digitigrade, the metapodials would normally be at a low angle to the ground, so I have shown the "landing" hind foot (the left one) almost horizontal.

The muscle insertion areas in the forelimbs of *Ekorus* indicate that many of the muscles involved in fossorial activities (the teres major, along with the extensors and rotators of the forearm) were much less developed than in badgers, although in general the animal would have been more muscular than any dog or hyena, as well as more muscular than most cats.

For the coat pattern, I have used for reference the living species of terrestrial, predatory mustelids.

Figure 7.20. *Ictitherium ebu.* The holotype skeleton is complete enough to permit a confident restoration of body proportions. The elongation of the limb bones is striking, greater even than in the gracile living hyaenids—the striped hyena (*Hyaena hyaena*) and the aardwolf (*Proteles cristatus*). As in living hyaenids, the neck was very elongated, although it was not nearly as heavily muscled as in the larger extant species. A single lumbar vertebra indicates that, also as in living hyaenids, the back was considerably shortened. The hindlimbs were as long relative to the forelimbs as in the extant aardwolf, which implies that the back would be less sloping than in the extant spotted, brown, and striped hyenas. These proportions suggest that, as living hyaenids, *Ictitherium ebu* would normally use lateral sequence gaits for slow and medium-speed locomotion. The trot would normally not be employed because each forelimb would tend to get in the way of the advancing hindlimb of the same side in this diagonal-sequence gait. Accordingly, I have restored *Ictitherium* employing the pace, a gait often used by modern hyaenids.

The coat pattern in this restoration is based on that of modern hyaenids, especially *Hyaena* and *Proteles,* which probably retain a pattern that is primitive for the whole family. Some features incorporated in this restoration—such as the presence of stripes, especially in the legs, and the presence of long, erectile hairs in the upper neck and along the back—are present in all recent hyaenids except *Crocuta* (which does have a much less developed mane). Other features—such as a dark muzzle, a rather unpatterned face, and large ears lacking a bursa—are common to all living hyaenids, and it seems thus reasonable to infer their presence in *Ictitherium.*

In contrast to this reconstruction of *Ictitherium ebu,* the skeletal restoration of *Ictitherium viverrinum* from Pikermi figured by Gaudry in the 1860s, and later reproduced in several textbooks, shows an animal with a primitive, civet-like vertebral column, with moderately long neck and long lumbars. In fact, no vertebral elements were available to Gaudry, who based his reconstruction solely on cranial and appendicular material; although his guess was a very reasonable one given the information available at the time, it now seems likely that the vertebral proportions of *I. viverrinum* were more hyena-like and less civet-like than shown in that restoration, which has been the "official" image of *Ictitherium* for generations of paleontologists.

Figure 7.22. *Lokotunjailurus emageritus.* As was the case with *Ekorus ekakeran* and *Ictitherium ebu,* the holotype skeleton of the Lothagam machairodont is remarkably complete and permits an accurate reconstruction of body proportions. *Lokotunjailurus* was a rather cursorial felid, with gracile limb bones, more elongated metapodials, and a relatively smaller head than in a modern lion. In contrast to those of later machairodonts such as *Homotherium*, the lumbar vertebrae were not particularly shortened. All these features contribute to make this an especially gracile, elegant cat. The enormous dewclaw would probably be a visible feature of the living animal in spite of being partly covered with flesh and fur (figure 1.4). The upper canines, although long and flattened, would have scarcely protruded beyond the upper lip in the living animal, at least when it had a relaxed face, as shown in this restoration.

The reconstructed coat pattern is based on that of several species of modern felids, which is probably similar to the primitive pattern for the whole family Felidae. Of course, attributing a primitive pattern to an extinct species is just a conservative, probabilistic choice, and *Lokotunjailurus* might well have developed a more derived design.

Figure 8.1. *Stegotetrabelodon orbus.* Very little is known at present of the postcranial skeleton of *Stegotetrabelodon,* so the body proportions shown in this restoration are mostly conjectural and broadly intermediate between the long-bodied, short-limbed gomphotheres and the taller living elephants. The space between the two rami of the mandible was probably occupied by a well-developed tongue, and it is very likely that the oral cavity was delimited by a well-developed lower lip.

Figure 8.11. *Deinotherium bozasi.* As in the case of *Stegotetrabelodon,* the postcranial skeleton of *Deinotherium bozasi* is very poorly known, so for this restoration the body proportions of the better known European species *Deinotherium giganteum* were used for reference. The anatomy of the more primitive deinothere genus *Prodeinotherium* from Africa and Europe was also studied, but it is important to remember that this was a smaller and more primitive animal, so we should be cautious when transferring some of its features to the more evolved *D. bozasi.*

The presence of a developed lower lip is suggested by the morphology of the mandible and, as in the case of *Deinotherium,* it would have helped to close the mouth and keep the food within the oral cavity during mastication. In modern elephants the lower lip appears as a continuation of the pointed shape of the toothless mandibular symphysis.

Differences in the nasal region of the skull indicate that the trunk of deinotheres was not structurally identical to that of modern elephants, and it seems likely that the muscular control of trunk movements was considerably less sophisticated. In this restoration the trunk is shown as relatively shorter than in modern elephants, but still long enough to have allowed the animal to drink from shallow water holes while standing in the typical elephantine posture indicated by the graviportal anatomy of deinothere forelimbs.

Figure 8.12. *Leptorycteropus guilielmi.* The holotype skeleton of this species is rather fragmentary, but enough is preserved to show that the animal was not only much smaller but also considerably more lightly built than the modern aardvark. The back had the arched shape typical of the extant species, but the limbs were relatively longer and less muscular. The length of the toothrow indicates that the muzzle would have been relatively shorter. The gait shown in this restoration is the same observed in modern aardvarks, a single-foot walk in diagonal sequence, but the more gracile proportions of the animal provide for a more dynamic picture.

Figure 9.1. *Brachypotherium lewisi.* The head of *B. lewisi* as shown in this illustration is based on the skull restoration offered by Hooijer and Patterson (1972), which was in turn based on two skulls, each of them crushed in different and somehow complementary ways. The resulting restoration showed a skull that differs from typical European brachypotheres in having a straighter and less concave, dorsal profile.

As the postcranial skeleton of *B. lewisi* is poorly known, body proportions in this restoration are mostly based on European brachypotheres, especially *Brachypotherium (Diaceratherium) aurelianense,* which is probably close to the ancestry of *B. lewisi.* The more complete remains of an earlier brachypothere, *B. heinzelini,* from Rusinga Island, indicate that Early Miocene African brachypotheres had already developed the short-limbed anatomy characteristic of their European relatives.

Figure 9.6. *Eurygnathohippus turkanense.* The larger Lothagam hipparionine was a robustly built animal, as indicated by the preserved limb bones. Parts of the anatomy not preserved in the Lothagam species were restored following the well-known, and more primitive, European taxon *Hippotherium primigenium.*

The presence of at least some degree of striping, especially in the legs and shoulders, is possibly primitive for the Equidae and is thus incorporated in this restoration.

Figure 9.8. *Eurygnathohippus feibeli.* The preserved limb bones of this hipparionine (including a complete tibia and metacarpal besides more fragmentary material) show that it was not only smaller but much more gracile than *E. turkanense.*

Figure 10.1. *Hexaprotodon harvardi.* The limb proportions of this species can be confidently restored based on the large collection of limb bones from Lothagam, which includes many complete but isolated bones and a complete associated hind limb. These show that this animal, although unmistakably hippo-like, would have been considerably more gracile than an extant hippopotamus. Another clear difference would be the outline of the head in side view, where the orbits would not be nearly as protuberant as in the extant species.

Figure 10.28. *Hexaprotodon harvardi* (*right*) and *H. lothagamensis.* These two hippo species from Lothagam are drawn here to scale, in frontal view and with heads lowered, because it is the dorsal aspect of the head that most clearly shows the differences between them. The remains of *H. lothagamensis* are unfortunately much more fragmentary, and the anterior part of the skull is not known, but its shape can be restored with some confidence based on the preserved anterior part of the mandible. The anterior mandible of *H. lothagamensis* is comparatively very narrow, so it follows that the maxilla would have been correspondingly narrow. Besides, the lower canines are not flaring, and since the upper canines would have occluded directly with the lower ones, the position of the latter is also an indication of the maximum width of the muzzle. The resulting picture shows *H. lothagamensis* as not only distinctly smaller, but also much less hippo-like than *H. harvardi,* looking even more primitive than the extant pigmy hippo.

There are also some postcranial remains from Lothagam that, although not directly associated with cranial remains of *H. lothagamensis,* are of the right size to have belonged to this species, and so they too have been used as a reference for this reconstruction.

Figure 10.42. *Nyanzachoerus syrticus tulotus,* male (*left*) and female. The beautiful and well-preserved cranial remains from Lothagam allow an accurate reconstruction of the head of these enormous pigs. All the knobs and protuberances of the head are based on the osteological evidence, and although a little extra thickness has been added to reflect the presence of soft tissue, this has been a conservative estimate and it is possible that at least some of these protuberances would have been even more spectacular in the living animal.

The body proportions in this restoration are based on sets of associated limb bones from Lothagam and Sahabi that by their large size are likely to represent *N. syrticus.* Both samples point to a huge animal, larger than the extant giant forest hog (*Hylochoerus*), and with proportionally longer metapodials, suggesting a slightly more cursorial adaptation.

Figure 10.47. *Nyanzachoerus pattersoni.* Lothagam fossils assigned to this species are rather fragmentary, so this restoration is based on material from the Kenyan sites of Kanapoi and the Baringo Basin. Beautiful cranial material from these sites clearly shows the differences with *N. syrticus,* especially the enormously developed zygomatic protuberances, the more moderate ornamentations in the muzzle, and the somewhat larger tusks, all of which would be readily appreciable in the living animal. Postcranial elements from Kanapoi suggest a body plan similar to that of the earlier species.

Figure 10.49. *Cainochoerus* sp. Most parts of the body of this tiny suid are represented in the Lothagam sample, but unfortunately most of the elements are fragmentary. The sample of *Cainochoerus africanus* from the South African site of Langebaanweg includes many postcranial elements as well and, combining both sources, it is possible to get a reasonable picture of the cainochoeres as living animals. The Lothagam specimens were even smaller than the South African ones, so that the reconstructed height at shoulder is only 26 cm! The shape of the limb bone articulations shows that the locomotion was more cursorial than in typical pigs. This, coupled with the gracile build of the skeleton, would surely have made the living animal look remarkably similar to a modern peccary even if there was no direct relationship.

Figure 11.1. *Paleotragus* sp. This restoration is based mainly on the remarkable sample of *Paleotragus germaini* from the Algerian site of Oued el Hammam, which includes ossicones, dental series, many complete long bones, and several vertebral elements. The resulting picture is that of a large giraffid with a reconstructed shoulder height a little over 2 meters, with considerably elongated neck and limbs that give the animal an appearance intermediate between an okapi and a giraffe. Tooth row length is absolutely longer than in modern giraffes, indicating a rather elongate head. Skull morphology is otherwise poorly known in this species, so it was restored following the better represented *Paleotragus microdon* and the closely related genus *Samotherium,* both from China.

Figure 11.5. Lothagam Bovidae. Complete bovid skulls are very rare, and most of these head restorations are based on frontlets or calvariae. This meant that skull shape had to be restored in each case from related forms, either fossil or extant.

The tragelaphine *Tragelaphus kyaloae* is based on a pair of horn cores from West Turkana, and the skull morphology follows that of closely related extant species like the sitatunga (*T. spekei*). The pattern of face markings is rather constant among modern tragelaphines, including white spots on the sides and a V-shaped marking in front of the eyes, so a similar pattern is inferred for the fossil species. The texture of the horn sheaths is also borrowed from modern tragelaphines.

The two species of *Tragoportax* are restored on the basis of frontlets from Lothagam, with skull morphology following that of the better known European species of the genus, such as *T. gaudryi* from the Turolian of Spain. The pattern of face markings follows that of the extant boselaphine *Boselaphus tragocamelus*, the Indian nilgai, which, it may be noted, is similar to that of the tragelaphines except for the absence of the V-shaped frontal marking.

The horn shape of the two species of *Kobus* is based on calvariae from Lothagam, while skull morphology follows that of modern species like *Kobus kob*. The restoration of the other reduncine, *Menelekia leakeyi*, is based on horn cores from Lothagam and Koobi Fora. The convex dorsal profile of the nasals is borrowed from a closely related species, *Menelekia lyrocera*, whose skull is well known from other localities in the northern Turkana Basin. All three reduncine species are reconstructed with annulated horn sheaths, and this condition is observed in all extant members of this tribe.

The small impala, *Aepyceros premelampus*, is restored based on the holotype of the species, a beautifully preserved frontlet with horn cores. Skull morphology, horn sheath texture, and face markings are based on the extant impala (*Aepycerus melampus*).

Figure 13.1. Nawata Formation habitat. This restoration attempts to show the likely transition of habitats from river and riverine woodland on the left side of the scene, to edaphic grassland and more open woodland on the right. On the right side background, beyond the grassland, another thicket of woodland corresponds to a distant meander of the river or to an oxbow lake or abandoned meander. The environments are broadly similar to those found today around large African rivers like the Luangwa in Zambia and the Chobe in Botswana. The scene is set in the dry season, when ungulates like the impalas (*Aepyceros premelampus*) and large hipparions (*Eurygnathohippus turkanense*) shown here are attracted to the permanent water of the river, and thus they become more prone to predation by the large, carnivorous crocodile *Rimasuchus lloydi*. In contrast, adult *Stegotetrabelodon orbus* would have little to fear from crocodiles and would wade and drink at leisure. Suids such as *Nyanzachoerus syrticus* (*right, mid-distance*) would frequent the transitional habitats, and the boselaphine bovids of the species *Tragoportax cyrenaicus* (*right, background*) would alternately graze in the floodplains and browse in the woodlands, depending on the season. Papionine monkeys of the genus *Parapapio* (*center, mid-distance*) would probably forage both on the ground and in the trees, seeking protection from terrestrial predators in the high branches.

Figure 13.14. Apak Member habitat. This scene is set in a nearly dry riverbed, where the larger mammals would gather to drink from the shrinking water holes. Bovids like *Tragelaphus kyaloae* (*right, foreground*) and the impalas *Aepyceros premelampus* (*mid-distance*), would normally dwell in the gallery forest and thus be among the frequent visitors to these water holes. Water-dependent proboscideans would be attracted to the sand river; we see here the deinotheres, *Deinotherium bozasi* (*left, mid-distance*), and the long-tusked gomphotheres of the genus *Anancus* (*right, mid-distance*).

Figure 13.15. Kaiyumung Member habitat. Shown in the foreground is the transition zone between dry woodlands and floodplain grasslands, while in the background denser vegetation indicates the vicinity of the river. A troop of large baboons of the species *Theropithecus brumpti* appears in the foreground, while two giant pigs of the species *Notochoerus eulius* approach from the right. In life, notochoeres would be readily distinguished from members of the earlier genus *Nyanzachoerus* by their huge tusks and less prominent facial protuberances, but, like the nyanzachoeres, notochoeres were huge pigs that stood more than 1 meter at the shoulder. A white rhino of the extinct species *Ceratotherium praecox* grazes in the left mid-distance.

References Cited

Gaudry, A. 1862–1867. *Animaux fossiles et géologie de l'Attique.* Paris: Savy.

Hooijer, D. A., and B. Patterson. 1972. Rhinoceroses from the Pliocene of northwestern Kenya. *Bulletin of the Museum of Comparative Zoology* 144:1–26.

CONTRIBUTORS

Raymond L. Bernor
Department of Anatomy
College of Medicine
Howard University
520 W Street N.W.
Washington, D.C. 20059

Thure E. Cerling
Department of Geology and Geophysics
University of Utah
Salt Lake City, Utah 84112

Craig S. Feibel
Anthropology Department
Rutgers University
131 George Street
New Brunswick, N.J. 08901

John M. Harris (Editor)
George C. Page Museum
5801 Wilshire Boulevard
Los Angeles, Calif. 90036

Meave G. Leakey (Editor)
Division of Paleontology
National Museums of Kenya
P.O. Box 40658
Nairobi, Kenya

Joel W. Martin
Research and Collections Branch
Natural History Museum of Los Angeles County
900 Exposition Boulevard
Los Angeles, Calif. 90007

Ian McDougall
Research School of Earth Sciences
The Australian National University
Canberra, ACT 0200
Australia

Simon A. H. Milledge
The Cider House
Holmer, Hereford HR4 9RH
England

Nina Mudida
Osteology Department
National Museums of Kenya
P.O. Box 40658
Nairobi, Kenya

Kathlyn M. Stewart
Canadian Museum of Nature
P.O. Box 3443, Station D
Ottawa, Ontario K1P 6P4
Canada

Glenn W. Storrs
Cincinnati Museum of Natural History
Geier Collections and Research Center
1720 Gilbert Avenue
Cincinnati, Ohio 45202

Pascal Tassy
Laboratoire de Paléontologie
Muséum National d'Histoire Naturelle
8 rue de Buffon
75005 Paris
France

Mark F. Teaford
Department of Cell Biology and Anatomy
Johns Hopkins University
School of Medicine
725 North Wolfe Street
Baltimore, Md. 21205

Sandra Trautwein
Department of Biology
University of California at Los Angeles
405 Hilgard Avenue
Los Angeles, Calif. 90024

Alan C. Walker
Departments of Anthropology and Biology
Pennsylvania State University
409 Carpenter Building
University Park, Pa. 16802

Carol V. Ward
Departments of Anthropology and Pathology
and of Anatomical Sciences
University of Missouri
Columbia, Mo. 65211

Lars Werdelin
Swedish Museum of Natural History
Department of Palaeozoology
Box 50007
S-10405 Stockholm
Sweden

Eleanor M. Weston
Department of Zoology
University of Cambridge
Downing Street
Cambridge CB2 3EJ
England

Alisa J. Winkler
Shuler Museum of Paleontology
Department of Geological Sciences
Southern Methodist University
Dallas, Tex. 75275

Roger C. Wood
Stockton State College
Jim Leeds Road
Pomona, N.J. 08240

Jonathan G. Wynn
Department of Geological Sciences
University of Oregon
Eugene, Ore. 97403

Index